W9-ART-948

Test Yourself

Take a chapter quiz on the *Inquiry into Life* ARIS site. Each quiz is specially constructed to test your comprehension of key concepts. Feedback on your responses helps you gauge your mastery of the material.

Access to Premium Learning Materials

The *Inquiry into Life* ARIS site is your portal to exclusive study tools such as McGraw-Hill's animations, Virtual Labs, and ScienCentral videos.

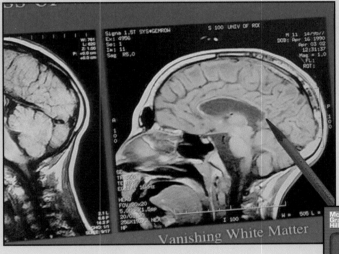

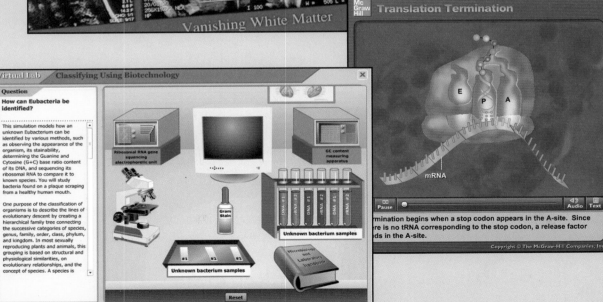

Twelfth Edition

Inquiry into *Life*

Sylvia S. Mader

With significant contributions by

Jeffrey A. Isaacson
Nebraska Wesleyan University

Kimberly G. Lyle-Ippolito
Anderson University

Andrew T. Storfer
Washington State University

McGraw Hill **Higher Education**

Boston   Burr Ridge, IL   Dubuque, IA   New York   San Francisco   St. Louis
Bangkok   Bogotá   Caracas   Kuala Lumpur   Lisbon   London   Madrid   Mexico City
Milan   Montreal   New Delhi   Santiago   Seoul   Singapore   Sydney   Taipei   Toronto

The McGraw-Hill Companies

Higher Education

INQUIRY INTO LIFE, TWELFTH EDITION

 This book is printed on recycled, acid-free paper containing 10% postconsumer waste.

4 5 6 7 8 9 0 QPD/QPD 0 9

ISBN 978–0–07–298675–4
MHID 0–07–298675–1

Publisher: *Janice Roerig-Blong*
Sponsoring Editor: *Thomas C. Lyon*
Developmental Editor: *Rose M. Koos*
Marketing Manager: *Tamara Maury*
Senior Project Manager: *Jayne Klein*
Lead Production Supervisor: *Sandy Ludovissy*
Senior Media Project Manager: *Jodi K. Banowetz*
Senior Media Producer: *Eric A. Weber*
Designer: *Laurie B. Janssen*
Cover/Interior Designer: *Christopher Reese*
Senior Photo Research Coordinator: *Lori Hancock*
Photo Research: *Evelyn Jo Hebert*
Supplement Producer: *Melissa M. Leick*
Compositor: *Electronic Publishing Services Inc., NYC*
Art Studio: *Electronic Publishing Services Inc., NYC*
Typeface: 10/12 *Palatino*
Printer: *Quebecor World Dubuque, IA*

(USE) Cover Image: © *Getty Images, Father and Son Examining a Starfish by Jim Arbogast*

The credits section for this book begins on page C-1 and is considered an extension of the copyright page.

Library of Congress Cataloging-in-Publication Data
Mader, Sylvia S.
  Inquiry into life / Sylvia Mader. -- Twelfth ed.
    p. cm.
  Includes index.
  ISBN 978-0-07-298675-4 --- ISBN 0-07-298675-1 (hard copy : alk. paper)
  1. Biology. I. Title.
QH308.2.M363 2008
570--dc22
                                            2006034783

About the Author

From left to right: Drs. Isaacson, Mader, Storfer, and Lyle-Ippolito

DR. SYLVIA S. MADER has written numerous biology textbooks for WCB/McGraw-Hill in addition to her most famous book, *Inquiry into Life.* Her 30-year relationship with the company has yielded such titles as *Biology, Human Biology, Understanding Anatomy and Physiology,* and *Essentials of Biology.*

A brilliant and prolific writer, Dr. Mader was a respected and well-loved biology instructor before she began her writing career. She developed some of her well-known teaching and learning techniques while helping science-shy students appreciate and learn biology at Lowell University and Massachusetts Bay Community College.

About the Contributors

JEFFREY A. ISAACSON, D.V.M., PH.D., is an Associate Professor of Biology at Nebraska Wesleyan University. His main teaching responsibilities include Microbiology, Immunology, and Cell Biology, and he also supervises senior research projects, and conducts research on the immunological effects of bovine leukemia virus. Dr. Isaacson received a B.S. in Biology from Nebraska Wesleyan University, a D.V.M from Kansas State College of Veterinary Medicine, and a Ph.D. in Immunobiology from Iowa State University. He worked as a small-animal veterinarian for several years, and also completed a post-doctoral fellowship in the Department of Immunology at the Mayo Clinic. From 1998 to 2000, Dr. Isaacson worked on vaccine development as a Senior Scientist at Schering Plough Animal Health in Elkhorn, Nebraska. Dr. Isaacson revised chapters 11-21, and 28.

KIMBERLY G. LYLE-IPPOLITO, PH.D., is a Professor of Biology at Anderson University, Anderson, Indiana where she teaches Genetics, Molecular Biology, Biochemistry, Microbiology, Immunology, and Virology. She has twice received the Outstanding Teacher of the Year (Nicholson) Award at Anderson University. Dr. Lyle-Ippolito received a B.S. in Biology at Wright State University, and a M.S. in Genetics and a Ph.D. from The Ohio State University. During various positions at Case Western University, the Cleveland Clinic Foundation, and the Dayton (Ohio) Veterans Administration Hospital, she has been involved in vaccine research and oversight of a clinical virology laboratory. Dr. Lyle-Ippolito revised chapters 2-8, 22, 23, 25, and 26, and assisted with chapters 9, 10, and 29.

ANDREW T. STORFER, PH.D., is an Assistant Professor of Biology at Washington State University, and he teaches Introductory Biology to a class of more than 1,000 non-major students, Biology of Amphibians and Reptiles, and a graduate course in Conservation Biology. Dr. Storfer received a B.S. in Biology from State University of New York at Binghamton and a Ph.D. in Biology from the University of Kentucky, specializing in ecology and evolution. His current research program is at the interface of two "grand challenges in environmental sciences" for the 21st century as recognized by the National Research Council: 1) understanding the causes and consequences of the Earth's diminishing biodiversity; and 2) understanding the ecology and evolution of infectious diseases. Dr. Storfer revised chapters 1, 24, 27, and 30-36.

Brief Contents

Contents

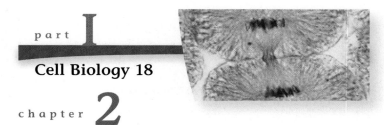

part I

Cell Biology 18

part II

Plant Biology 130

part **V**

Continuance
of the Species 416

part **VI**

Evolution
and Diversity 548

part **VII**

Behavior and Ecology 674

Readings

ix

Preface

MY WRITING CAREER began as I worked with my teaching colleagues to develop a course that reached out to science-shy students. We all believed in the human approach, and regardless of that week's topic, we sought human applications that would make the material more relevant to the student. After teaching for several years, I began to develop a methodology that enabled most students to learn biology, and from those teaching experiences, I developed the first edition of *Inquiry into Life*.

The goal of *Inquiry into Life* has always been, and continues to be, to blend the classic with the new, and develop a book that is both time-tested and current. The levels of biological organization serve as a guide, and the text begins with chemistry and ends with ecology. Students need to know not only about themselves but also about the other organisms that share this planet with us. The systems chapters in *Inquiry into Life* pertain to humans, and the anatomy and physiology of other animals is discussed in the Animal Kingdom chapters. I firmly believe that students must develop an understanding and appreciation of how the biosphere works in order to make decisions that will help the biosphere endure.

Biology has changed rapidly since *Inquiry into Life* was first published in 1976. New findings, ideas, and concepts have emerged over the years and created an excitement that I have always worked hard to convey. To continue with this tradition, I enlisted the help of three major contributors for the twelfth edition of *Inquiry into Life*. Each contributor has a different area of expertise, and shares a genuine enthusiasm for teaching biology to non-major students.

What Sets This Book Apart

Inquiry into Life covers the entire field of general biology to provide the fundamental principles. What sets the book apart is the application of those principles to human concerns.

As with previous editions, the central theme of *Inquiry into Life* is understanding the workings of the human body and how humans fit into the world of living things. This is accomplished through

- the continued use of short stories to begin each chapter, many of which are stories that apply chapter material to real-life situations,
- analogies to help students relate the concepts to something familiar, and
- the Health, Ecology, Science, and Bioethical Focus readings that present current and applicable information for discussion. These focus readings cover topics such as DNA Fingerprinting and the Criminal Justice System, Bioterrorism, and Emerging Viral Pathogens, and are presented with the core content to encourage the application of the concepts to current happenings in our world.

Overview of Changes
to *Inquiry into Life*, Twelfth Edition

- **Extensively revised art program.** The content and clarity that have been a hallmark of Dr. Mader's art have been retained and improved through the use of vibrant colors and added dimension. Each figure has been updated for content clarification and visual appeal.

- **Content revised throughout by contributing experts in the field.** Changes were made throughout all chapters to clarify and update information. See pages xi and xii for a listing of the significant changes.

- **Collation of human disease coverage.** As requested by adopters, coverage of diseases and disorders has been collated and placed at the end of each systems chapter. These sections will allow an instructor to focus on disorders after students have learned about the normal anatomy and physiology of a system.

- **Emphasis on Critical-Thinking and Relevance.** Questions that encourage students to apply what they have learned are now provided with every boxed reading and at the end of each chapter. Vignettes and boxed readings increase the relevancy of the material, including topics such as Mate Choice and Smelly T-Shirts, Medicinal Leeches: Medicine Meets *Fear Factor*, Bioterrorism, and DNA Fingerprinting.

New to this edition, a "Disease and Disorder" section has been added to the end of each human system chapter. These sections will give instructors an opportunity to incorporate diseases into a review of the chapter to enhance student interest and learning.

What's New to This Edition

Illustrations

The illustrations in this edition have been greatly enhanced. Colors are brilliant and figures are designed to be visually eye-catching. Each figure is new or significantly revised to convey the concepts with greater clarity and effectiveness. Review the Guided Tour on pages xiv and xv for an overview of the art program in this edition.

Content Changes and Updates

Each chapter has been revised by the contributors in accordance with their areas of expertise, and with the guidance of Dr. Sylvia Mader. Many of the changes are in direct response to reviewer feedback and the following gives a summary of those changes.

Chapter 1 includes a new chapter-opening vignette that captures three major themes of the book: 1) an emphasis on humans; 2) the commonality of living things; and, 3) environmental concerns. The section explaining adaptation has been revised for clarity, so students get a sense that evolution is not "directional"—that is, the environmental changes to which organisms respond are random. In addition, the controlled experiment section has been revised with better artwork and a clearer description of the experiment.

Chapter 2 The sections entitled "Basic Chemistry" and "Water and Living Things" have been rewritten and reorganized to simplify the information and provide more familiar examples to students.

Chapter 3 has been reorganized. The discussion of prokaryotic cells now precedes that of eukaryotic cells. This gives students a clearer, evolutionary prospective of cell structure and function.

Chapter 4 has been consolidated and simplified. The section "Permeability of the Plasma Membrane" has been rewritten to include diffusion and osmosis, transport by carrier proteins, and also endocytosis and exocytosis.

Chapter 5 Prokaryotic binary fission has been added to the chapter and clearer differentiation is made between cytokinesis in plant and animal cells. The section on oogenesis has been revised for clarity.

Chapter 6 The concepts of enzymes and energy have been integrated into the overarching theme of metabolism. Some of the more difficult concepts have been explained in the text and correspondingly illustrated with new artwork. Familiar analogies are used whenever possible.

Chapter 7 was revised to simplify concepts and to make the level of detail appropriate for non-majors. The very difficult topic of redox reactions was deleted in order to provide more examples and analogies to reinforce the overall aspect of metabolism.

Chapter 8 emphasizes the value of plants and photosynthesis to students. In response to reviewers' comments, the section on light harvesting pigments has been expanded and the section on C3, C4, and CAM plants was rewritten to increase relevancy.

Chapter 10 contains updated data on transgenic crops and the concept of totipotency. The hormone section has been significantly improved for clarity and accuracy.

Chapter 13 has been updated and clarified to reflect recent terminology changes in immunology. For example, sections have been renamed and divided into manageable segments. The discussion of B and T cells and their relationship to self antigens has been improved.

Chapter 14 Material related to disorders of the digestive tract has been moved into a new section at the end of the chapter. This section is divided into:
- Disorders of the Digestive Tract, including stomach ulcers, diarrhea and constipation, and colon cancer.
- Disorders of the Accessory Organs, including pancreatitis, hepatitis, and cirrhosis.

Chapter 15 The section on "Disorders of the Respiratory System" has been reorganized into two sections by anatomical region:
- Disorders of the Upper Respiratory Tract, which covers very common conditions such as the common cold, pharyngitis, tonsillitis, laryngitis, sinusitis, and otitis media (middle ear infections).
- Disorders of the Lower Respiratory Tract, including bronchitis, asthma, pneumonia, emphysema, cystic fibrosis, and lung cancer.

Chapter 16 The text explaining the function of the distal convoluted tubule has been rewritten to improve clarity. A new section, "Disorders of the Urinary System," has been added, which includes:
- Disorders of the Kidneys, such as pyelonephritis, kidney stones.
- Disorders of the Bladder and Urethra, including bladder infections, bladder stones. A new image of a bladder stone, as well as a newly developed artificial bladder, has also been added.

Chapter 17 The section on Drug Abuse has been updated and rewritten, including a new discussion of "club drugs" such as Ecstasy and Rohypnol. A new section entitled "Disorders of the Nervous System" has been written, which contains the following three sections:
- Disorders of the Brain, including such common conditions as Alzheimer disease, Parkinson disease, MS, stroke, and meningitis, plus prion diseases.
- Disorders of the Spinal Cord, such as trauma and ALS.
- Disorders of the Peripheral nerves, such as Guillian-Barre syndrome and myasthenia gravis.

Chapter 18 Information on taste receptors has been updated, to include recently discovered "umami" receptors. New examples have been added for relevancy, such as hearing loss from overuse of iPods.

Chapter 19 Based on reviewer comments, the description of the role of actin and myosin in muscle contraction has been clarified.

Chapter 20 The section on chemical signals and hormone action has been moved to the beginning of the chapter so students will have a better idea of how hormones work before learning about specific endocrine organs.

Chapter 21 The descriptions of the male and female reproductive anatomy and physiology have been clarified. A new section on "Disorders of the Reproductive System" has been added and covers:

- Sexually Transmitted Diseases, which includes information on the global AIDS pandemic; antiretroviral therapy, or HAART; and an updated discussion of human papillomaviruses.
- Disorders of the Male Reproductive System, including new discussions on erectile dysfunction (ED), the prostate, and testicular cancer.
- Disorders of the Female Reproductive System, including new discussions on endometriosis, ovarian cancer, and problems with menstruation.

Chapter 23 includes a more thorough explanation of probability to help students understand the formation of gametes. Terminology was clarified in order to reduce student confusion.

Chapter 24 The discussion and artwork explaining the genetic code and translation have been revised for clarity, as students often struggle with these concepts. In addition, the section on "DNA Fingerprinting" has been updated to include modern technology.

Chapter 25 has been extensively revised to simplify the concepts and to make the level of detail more consistent and appropriate for non-majors. The chapter begins with an introduction to cellular differentiation and then describes gene expression in prokaryotes. The terminology in the section on gene expression in eukaryotes has been simplified.

Chapter 26 The section covering genomics has been updated to keep pace with this rapidly changing field.

Chapter 27 has been reorganized and extensively revised for clarity, precision, and flow. Definitions were made more precise and appropriate for the non-majors level.

Chapter 28 has been reorganized so viruses, viroids, and prions are discussed last, since these organisms are acellular, and thus atypical. In addition, it is likely that viruses evolved from cells that already existed, so this chronology is more consistent with that line of thinking.

Chapter 29 Dates significant to evolution have been updated, life cycle explanations have been clarified, and the section on the ferns has been streamlined.

Chapter 31 The section on human evolution has been revised to include new artwork and current information. For example, the recent discovery of *Homo florensis* fossils is discussed.

Chapter 32 was reorganized to include the discussion of the human society under "Sociobiology" for a more logical presentation. The section on human mate choice was rewritten to point out that many ideas about human mate choice are controversial. The section on altruism now stresses the term "kin selection," an idea of importance in ecology and social behavior.

Chapter 33 The section on "Ecological Succession" was moved to the end of the chapter, because this chapter focuses on population ecology and succession is a ecological community concept. The reorganization creates a smooth transition into the next chapter on ecosystems. The section on "Demographic Transition" was expanded because this topic is an important concept for understanding human population sizes and urbanization, a central theme of the chapter.

Chapter 35 The discussion of solar radiation effects has been revised for clarity. The Ecology Focus reading has been revised to include the most current tissue techniques pertinent to forensic genetics.

Chapter 36 includes additional examples to increase the relevancy of the chapter. For example, discussions were added about insecticide resistance and meat consumption, hurricane Katrina, and hybrid cars. Details about the difference between "exotic" and "invasive" species were made clearer, and the section on disease was modified, including a timely discussion of Avian flu.

Guided Tour. Students and instructors can become acquainted with the key features of this book by browsing through the Guided Tour starting on the next page. These pages constitute a visual exposition of the book's pedagogy and art program.

Acknowledgments

It would be impossible for a single individual to publish a book of the quality of *Inquiry into Life*. I want to thank the contributors who joined me in revising this edition. Jeff Isaacson, Kimberly Lyle-Ippolito, and Andrew Storfer met with me at the onset of the project to discuss my overall vision for *Inquiry into Life,* and together we formulated a plan for developing the twelfth edition. Their unique and specialized backgrounds, in addition to their current experiences in the classroom, have been extremely helpful in shaping this edition. I also want to extend a special thank you to Dr. Donna Becker, Northern Michigan University, for her expertise in revising chapters 9, 10, and 29.

It has been a pleasure to work with my associates at McGraw-Hill publishing: Tom Lyon, my editor; Tamara Maury, my marketing manager; Rose Koos, my developmental editor; and Jayne Klein, my production manager. We supported one another and urged each other on to reach a higher level of accomplishment than would have otherwise been possible.

The design of the book was managed by Rick Noel, and the illustration program was carried to completion by the artists at Electronic Publishing Services. Evelyn Jo Hebert and Lori Hancock did a superb job of securing quality photographs.

Textbook writing is very time consuming, and I have always appreciated the continued patience and encouragement of my family. By now my children are grown, but they never fail to keep up with how I am progressing with my work. My husband, Arthur Cohen, himself a biology teacher, gives me considerable support each day and discusses with me the details of my work. His input is invaluable to me.

Inquiry into Life is the product of my own efforts and those of the many reviewers that give me guidance on particular parts of the book. I am extremely grateful to these reviewers whose comments were so valuable as we worked on this edition.

Michelle Alexander
MidSouth Community College

Lawrence W. Anderson
Westminster College

Nahel W. Awadallah
Sampson Community College

Adébiyi Banjoko
Chandler Gilbert Community College

Harry Bass
Virginia Union University

Robert E. Bast
Cleveland State University

Lisa L. Boggs
Southwestern Oklahoma State University

Susan M. Bornstein-Forst
Marian College

Steven D. Brown
Redlands Community College

Barbara Y. Bugg
Northwest Mississippi Community College

David Burns
Carl Sandburg College

Beth Campbell
Itawamba Community College

John S. Campbell
Northwest College

John R. Capeheart
University of Houston

Ann M. Chiesa
Castleton State College

William H. Coleman
University of Hartford

Susan J. Cook
Indiana University, South Bend

Anthony P. Cooper
Albany State University

Larry D. Corpus
Dodge City Community College

Michael Crandell
Carl Sandburg College

David G. Dalbow
Redlands Community College

Gregg Dieringer
Northwest Missouri State University

Darren Divine
Community College of Southern Nevada

Claudia B. Douglass
Central Michigan University

JodyLee Estrada Duek
Pima Community College

Renata H. Dusenbury
Saint Augustine's College

Gladstone Dzata
Alabama State University

Michael J. Farabee
Estrella Mountain Community College

Eugene J. Fenster
Longview Community College

Stephen Fleckenstein
Sullivan County Community College

Sylvia Fromherz
University of Northern Colorado

Anne Galbraith
University of Wisconsin, La Crosse

Patrick Galliart
North Iowa Area Community College

Sandi B. Garnder
Triton College

Larry E. Gibson
Georgia Military College

Amy Goode
Illinois Central College

Marcella W. Hackney
Baton Rouge Community College

Reba Harrell
Hinds Community College

Bobbie Jo Hill
Coastal Bend College

Juliana G. Hinton
McNeese State University

Lauren D. Howard
Norwich University

Sara Huang
Los Angeles Valley College

Kenneth J. Hudiburg
Fort Scott Community College

Diann Jordan
Alabama State University

Daniel Kainer
Montgomery College

Erica Kipp
Pace University

Scott M. Layton
Cowley County Community College

Thomas A. Lonergan
University of New Orleans

Gregory J. Long
Olivet Nazarene University

Xiao Li Ma
Coastal Bend College

Paul Madtes Jr.
Mount Vernon Nazarene University

Blase Maffia
University of Miami

Luis Malaret
Community College of Rhode Island

June F. Manderfield
Community College of Allegheny County

Tonya McKinley
Concord University, Athens WV

Diane L. Melroy
University of North Carolina, Wilmington

Mark A. Melton
Saint Augustine's College

Jennifer Metzler
Ball State University

Jeanne Mitchell
Truman State University

Brenda Moore
Truman State University

Carla S. Murray
Carl Sandburg College

Eric R. Myers
South Suburban College

Jonas E. Okeagu
Fayetteville State University

Tammy D. Oliver
Eastfield College

Ariela Olivos Vader
Messiah College

Sandy Pace
Rappahannock Community College

Jack Pennington
Forest Park College

Karen Persky
College of DuPage

Carolyn J. Peters
Spoon River College

Shyamal Premaratne
Virginia Union University

Jerry Purcell
San Antonio College

Kirsten Raines
San Jacinto College, South Campus

Ramona Gail Rice
Georgia Military College

Kayla Rihani
Northwestern Illinois University

Daryl Ritter
Okaloosa-Walton College

Tatiana Roth
Coppin State University

Mark Sandheinrich
University of Wisconsin, La Crosse

Yolanda K. Sankey
Pellissippi State Technical Community College

A.K.M. Shahjahan
Baton Rouge Community College

Phillip Simpson
Shepherd University

Linda Smith-Staton
Pellissippi State Technical Community College

Stephanie R. Songer
North Georgia College and State University

Julian Stark
College of New Rochelle

Rafael Torres
San Antonio College

Martin A. Vaughan
Indiana University Purdue University, Indianapolis

Keti Venovski
Lake Sumter Community College

Tracy Ware
Salem State College

Gary Wassmer
Bloomsburg University of Pennsylvania

Jamie D. Welling
South Suburban College

Michael Wenzel
California State University, Sacramento

George Williams, Jr.
Southern University, Baton Rouge

Susan M. Willis
University of Cincinnati Raymond Walters College

James A. Wise
Hampton University

Guided Tour

Vivid and Captivating Illustrations Contribute to Learning

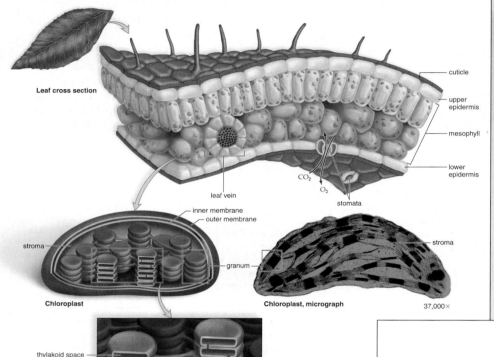

Leaf cross section

cuticle

upper epidermis

mesophyll

CO_2

O_2

lower epidermis

leaf vein

stomata

inner membrane
outer membrane

stroma

stroma

granum

Chloroplast

Chloroplast, micrograph

37,000×

thylakoid space
thylakoid membrane

Grana

channel between thylakoids

Multi-Level Perspective

Illustrations depicting complex structures show macroscopic and microscopic views to help students connect the two levels.

Combination Art

Drawings of structures are paired with micrographs to give students the best of both perspectives: the realism of photos and the explanatory clarity of line drawings.

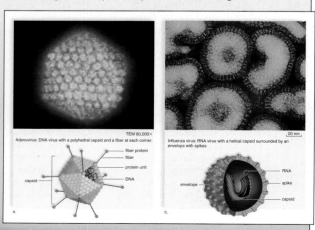

TEM 80,000×

Adenovirus: DNA virus with a polyhedral capsid and a fiber at each corner.

Influenza virus: RNA virus with a helical capsid surrounded by an envelope with spikes.

20 nm

fiber protein
fiber
protein unit
capsid
DNA

a.

RNA
envelope
spike
capsid

b.

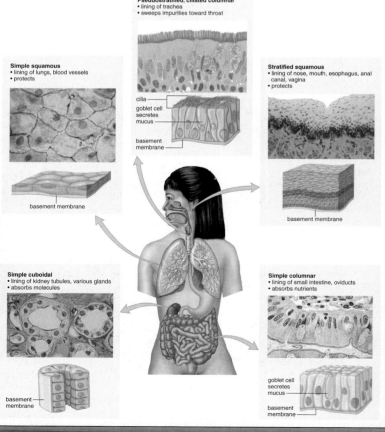

Pseudostratified, ciliated columnar
• lining of trachea
• sweeps impurities toward throat

Simple squamous
• lining of lungs, blood vessels
• protects

Stratified squamous
• lining of nose, mouth, esophagus, anal canal, vagina
• protects

cilia
goblet cell secretes mucus
basement membrane

basement membrane

basement membrane

Simple cuboidal
• lining of kidney tubules, various glands
• absorbs molecules

Simple columnar
• lining of small intestine, oviducts
• absorbs nutrients

basement membrane

goblet cell secretes mucus

basement membrane

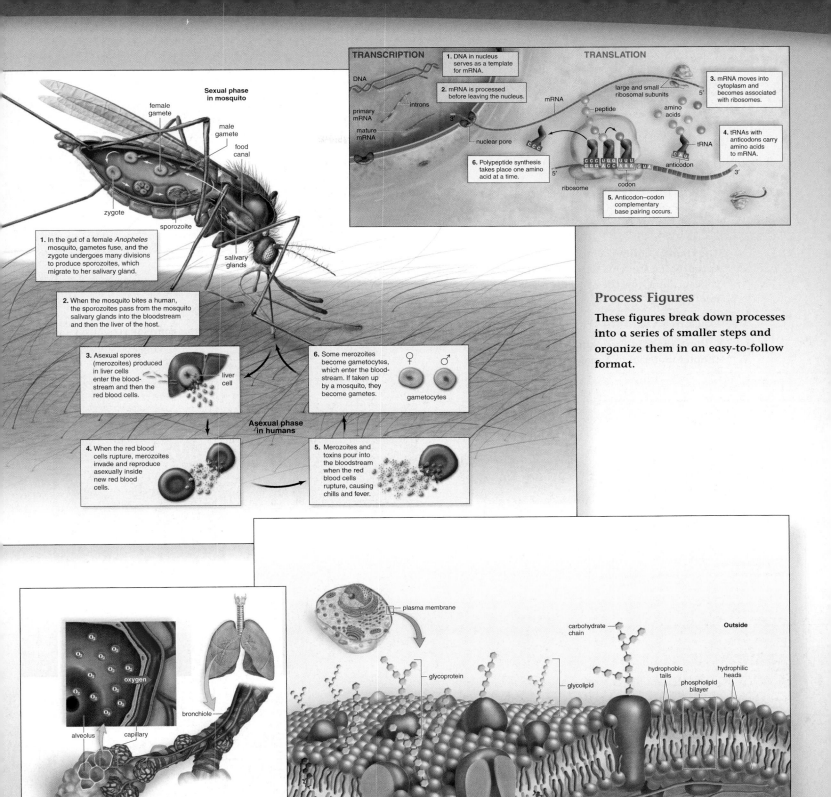

Sexual phase in mosquito

female gamete

male gamete

food canal

zygote

sporozoite

salivary glands

1. In the gut of a female *Anopheles* mosquito, gametes fuse, and the zygote undergoes many divisions to produce sporozoites, which migrate to her salivary gland.

2. When the mosquito bites a human, the sporozoites pass from the mosquito salivary glands into the bloodstream and then the liver of the host.

3. Asexual spores (merozoites) produced in liver cells enter the bloodstream and then the red blood cells.

liver cell

6. Some merozoites become gametocytes, which enter the bloodstream. If taken up by a mosquito, they become gametes.

♀ ♂

gametocytes

Asexual phase in humans

4. When the red blood cells rupture, merozoites invade and reproduce asexually inside new red blood cells.

5. Merozoites and toxins pour into the bloodstream when the red blood cells rupture, causing chills and fever.

TRANSCRIPTION

DNA

primary mRNA

introns

mature mRNA

nuclear pore

1. DNA in nucleus serves as a template for mRNA.

2. mRNA is processed before leaving the nucleus.

TRANSLATION

mRNA

large and small ribosomal subunits

peptide

amino acids

tRNA

anticodon

codon

ribosome

3. mRNA moves into cytoplasm and becomes associated with ribosomes.

4. tRNAs with anticodons carry amino acids to mRNA.

5. Anticodon–codon complementary base pairing occurs.

6. Polypeptide synthesis takes place one amino acid at a time.

CCC UGG UUU
GGG ACC AAA GUA

Process Figures

These figures break down processes into a series of smaller steps and organize them in an easy-to-follow format.

Icons

Icons help orient the student.

oxygen

alveolus

capillary

bronchiole

plasma membrane

carbohydrate chain

Outside

glycoprotein

glycolipid

hydrophobic tails

phospholipid bilayer

hydrophilic heads

filaments of cytoskeleton

Inside

peripheral protein

integral protein

cholesterol

The Learning System: Time-Proven Pedagogical Features Facilitate Your Understanding of Biology

Chapter Concepts

The chapter outline contains questions, encouraging students to concentrate on concepts and how the concepts relate to one another.

Opening Vignette

A short, thought-provoking vignette applies chapter material to a real life situation.

Lance Armstrong overcame cancer to win the Tour de France bicycle race seven times. Prior to being diagnosed with testicular cancer that had spread to his lungs and brain, Lance was one of the world's most highly trained athletes. He had conditioned his body to withstand hours of grueling physical activity. His heart was 30% larger than normal, and his resting heart rate was 32–34 beats per minute, compared to the average 60–100 beats per minute. For any athlete to function at such a high level, the cells, tissues, organs, and organ systems of the body must work together in an extremely coordinated fashion. The increased demand for oxygen by muscle tissues must be balanced by the amount of heat generated by muscles and the amount of fluid lost through sweating. Once cancer struck, Lance had to rely on his immune system to help him fight the cancer cells, and his liver and kidneys to help his normal cells survive the toxic effects of chemotherapy.

Although athletes' incredible feats are a clear demonstration of the human body's ability to deal with extreme circumstances, even as you read this, your body is performing tasks that are, in many ways, just as amazing. Your body temperature is maintained in a relatively narrow range, just as a thermostat-controlled heating and air conditioning system controls the temperature in a house. Your blood pressure, heart rate, breathing rate, and urinary output are all being monitored and maintained for you without any conscious effort. Beyond this, the numbers and types of cells in your blood as well as

chapter **11**

Human Organization

Chapter Concepts

11.1 Types of Tissues
- What are the four major tissue types in the human body? 198
- Where are epithelial tissues found, and what are their functions? 198–200
- What tissue type binds and supports other body tissues? What are some examples of this type of tissue? 200–202
- Which tissue type is responsible for body movements? What are some examples of this type of tissue? 202–203
- Which tissue type coordinates the activities of the other tissue types in the body? 204

11.2 Body Cavities and Body Membranes
- What are the two major body cavities? What smaller cavities are in each of these? 205
- What are the four types of body membranes? 205

11.3 Organ Systems
- In general, what is the function of each of the body systems? 206–207

11.4 Integumentary System
- What are three specific functions of the skin? 208
- What are three accessory organs of the skin? 209

11.5 Homeostasis
- What is homeostasis, and why is it important to body function? How does the body maintain homeostasis? 209–212
- What is disease, in relation to homeostasis? What are some general types of disease? 212

---

124 Part One Cell Biology

Once NADH has delivered electrons to the electron transport chain, the NAD⁺ that results can pick up more hydrogen atoms. In like manner, once FADH₂ gives up electrons to the chain, the FAD that results can pick up more hydrogen atoms. The recycling of coenzymes and ADP increases cellular efficiency since it does away with the necessity to synthesize NAD⁺ and FAD everytime.

Many poisons act on the electron transport chain. The Bioethical Focus essay discusses one such poison, cyanide.

The preparatory reaction and citric acid cycle take place in the matrix of the mitochondria. The electron transport chain is located in the cristae of mitochondria. As electrons pass down the electron transport chain, energy is released and captured for the production of ATP.

Organization of Cristae

The electron transport chain is located within the cristae of the mitochondria. The cristae increase the internal surface area of a mitochondrion, thereby increasing the area devoted to ATP formation. Figure 7.9 is a simplified overview of the electron transport chain. Figure 7.10 shows how the components of the electron transport chain are arranged in the cristae.

We have been stressing that the carriers of the electron transport chain accept electrons, which they pass from one to the other. What happens to the hydrogen ions from NADH + H⁺? The complexes in the cristae use the energy released by electrons as they move down the electron transport chain to pump H⁺ from the mitochondrial matrix into the space between the outer and inner membrane of a mitochondrion. This space is called the **intermembrane space.** The pumping of H⁺ into the intermembrane space establishes an unequal distribution of H⁺ ions; there are many H⁺ in the intermembrane space and few in the matrix of a mitochondrion.

The cristae also contain an *ATP synthase complex.* The H⁺ ions flow through an ATP synthase complex from the intermembrane space into the matrix. The flow of three H⁺ through an ATP synthase complex brings about a change in shape, which causes the enzyme ATP synthase to synthesize ATP from ADP + ℗. Mitochondria produce ATP by **chemiosmosis,** so called because ATP production is tied to an electrochemical gradient, namely the unequal distribution of H⁺ across the cristae. Once formed, ATP molecules pass into the cytoplasm.

Chemiosmosis is similar to using water behind a dam to generate electricity. The pumping of H⁺ out of the matrix into the intermembrane space is like pumping water behind the dam. The floodgates of the dam are like the ATP synthase. When the floodgates are open, the water rushes through, generating electricity. In the same way, H⁺ rushing through the ATP synthase is used to produce ATP.

Chapter Seven Cellular Respiration 125

Figure 7.11 Accounting of energy yield per glucose molecule breakdown.
Substrate-level ATP synthesis during glycolysis and the citric acid cycle accounts for four ATP. Chemiosmosis accounts for 32 or 34 ATP, and the grand total of ATP is therefore 36 or 38 ATP. Cells differ as to how the electrons are delivered outside the mitochondria. If these two electrons are delivered by a shuttle mechanism to the start of the electron transport chain, six ATP result; otherwise, four ATP result.

Energy Yield from Cellular Respiration

Figure 7.11 calculates the ATP yield for the complete breakdown of glucose to CO₂ and H₂O. Per glucose molecule, there is a net gain of two ATP from glycolysis, which takes place in the cytoplasm. The citric acid cycle, which occurs in the matrix of mitochondria, accounts for two ATP per glucose molecule. This means that a total of four ATP are formed by substrate-level ATP synthesis outside the electron transport chain.

The most ATP is produced by the electron transport chain and chemiosmosis. Per glucose molecule, ten NADH + H⁺ and two FADH₂ take electrons to the electron transport chain. For each NADH + H⁺ formed *inside* the mitochondria by the citric acid cycle, three ATP result, but for each FADH₂, only two ATP are produced. Figure 7.9 shows the reason for this difference: FADH₂ delivers its electrons to the transport chain lower in the chain than does NADH + H⁺, and therefore these electrons cannot account for as much ATP production.

What about the ATP yield of NADH + H⁺ generated *outside* the mitochondria by the glycolytic pathway? NADH + H⁺ cannot cross mitochondrial membranes, but a "shuttle" mechanism allows its electrons to be delivered to the electron transport chain inside the mitochondria. The shuttle consists of an organic molecule, which can cross the

outer membrane, accept the electrons, and in most, but not all cells, deliver them to a FAD molecule in the inner membrane. If FAD is used, only two ATP result because the electrons have not entered at the start of the electron transport chain.

Efficiency of Cellular Respiration

It is interesting to calculate how much of the energy in a glucose molecule eventually becomes available to the cell. The difference in energy content between the reactants (glucose and O₂) and the products (CO₂ and H₂O) is 686 kcal. An ATP phosphate bond has an energy content of 7.3 kcal, and 36 of these are usually produced during glucose breakdown; 36 phosphates are equivalent to a total of 263 kcal. Therefore, 263/686, or 39%, of the available energy is usually transferred from glucose to ATP. The rest of the energy dissipates in the form of heat.

The preparatory reaction produces the molecule that enters the citric acid cycle, which in turn produces ATP, NADH + H⁺, and FADH₂. NADH + H⁺ and FADH₂ pass electrons to the electron transport chain, which generates ATP via chemiosmosis.

Internal Summary Statements

A summary statement appears at the end of each major section of the chapter to help students focus on the key concepts.

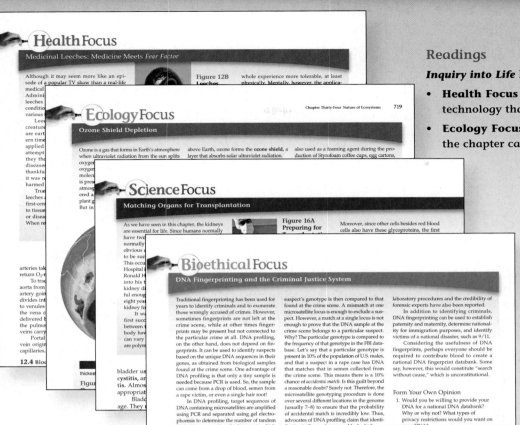

Readings

Inquiry into Life has four types of boxed readings.

- **Health Focus** readings review procedures and technology that can contribute to our well-being.
- **Ecology Focus** readings show how the concepts of the chapter can be applied to ecological concerns.
- **Science Focus** readings describe how experimentation and observations have contributed to our knowledge about the living world.
- **Bioethical Focus** readings describe modern situations that call for value judgments and challenge students to develop a point of view.

Chapter Summary

The summary is organized according to the major sections in the chapter and helps students review the important concepts and topics.

Quiz Questions

Challenging quiz questions close each chapter. In addition, a page-referenced *Understanding the Terms* section reinforces the scientific vocabulary used in the chapter, and critical-thinking questions encourage students to apply what they have learned.

Teaching and Learning Supplements

For Instructors

Laboratory Manual

ISBN (13) 978-0-07-298682-2
ISBN (10) 0-07-298682-4
The *Inquiry into Life Laboratory Manual*, Twelfth Edition, was written by Dr. Sylvia Mader and revised by Terry Damron and Eric Rabitoy of Citrus College. With few exceptions, each chapter in the text has an accompanying laboratory exercise in the manual. Every laboratory has been written to help students learn the fundamental concepts of biology and the specific content of the chapter to which the lab relates, as well as gain a better understanding of the scientific method.

McGraw-Hill Presentation Center

Build instructional materials where-ever, when-ever, and how-ever you want! ARIS Presentation Center is an online digital library containing assets such asphotos, artwork, PowerPoints, animations, and other media types that can be used to create customized lectures, visually enhanced tests and quizzes, compelling course websites, or attractive printed support materials.

Nothing could be easier! Accessed from the instructor side of your textbook's ARIS website, Presentation Center's dynamic search engine allows you to explore by discipline, course, textbook chapter, asset type, or keyword. Simply browse, select, and download the files you need to build engaging course materials. All assets are copyright McGraw-Hill Higher Education but can be used by instructors for classroom purposes.

McGraw-Hill's ARIS—Assessment, Review, and Instruction System

McGraw-Hill's ARIS is a complete, online electronic homework and course management system, designed for greater ease of use than any other system available. Created specifically for *Inquiry into Life*, Twelfth Edition, instructors can create and share course materials and assignments with colleagues with a few clicks of the mouse. For instructors, personal response system questions, all PowerPoint lectures, and assignable content are directly tied to text-specific materials in *Inquiry into Life*. Instructors can also edit questions, import their own content, and create announcements and due dates for assignments.

ARIS has automatic grading and reporting of easy-to-assign homework, quizzing, and testing. All student activity within McGraw-Hill's ARIS is automatically recorded and available to the instructor through a fully integrated grade book that can be downloaded to Excel.

For students, there are multiple-choice quizzes, animations with quizzing, videos with quizzing, and even more materials that may be used for self-study or in combination with assigned materials.

Go to aris.mhhe.com to learn more and register!

Instructor's Testing and Resource CD-ROM

ISBN (13) 978-0-07-298679-2
ISBN (10) 0-07-298679-4
This cross-platform CD features a computerized test bank that uses testing software to quickly create customized exams. The user-friendly program allows instructors to search for questions by topic, format, or difficulty level, edit existing questions or add new ones; and scramble questions for multiple versions of the same test. An Instructor's Manual is also included on the CD-ROM.

Transparencies

ISBN (13) 978-0-07-298677-8
ISBN (10) 0-07-298677-8
This set of overhead transparencies includes every piece of line art in the textbook plus every table. The images are printed with better visibility and contrast than ever before, and labels are large and bold for clear projection.

ScienCentral Videos

McGraw-Hill is pleased to announce an exciting new partnership with ScienCentral, Inc. to provide brief biology news videos for use in lecture or for student study and assessment purposes. A complete set of ScienCentral videos are located within this text's ARIS course management system, and each video includes a learning objective and quiz questions These active learning tools enhance a biology course by engaging students in real life issues and applications such as developing new cancer treatments and understanding how methamphetamine damages the brain. ScienCentral, Inc., funded in part by grants from the National Science Foundation, produces science and technology content for television, video and the Web.

McGraw-Hill: Biology Digitized Video Clips

ISBN (13) 978-0-07-312155-0
ISBN (10) 0-07-312155-X

McGraw-Hill is pleased to offer adopting instructors a new presentation tool--digitized biology video clips on DVD! Licensed from some of the highest-quality science video producers in the World, these brief segments range from about five seconds to just under three minutes in length and cover all areas of general biology from cells to ecosystems. Engaging and informative, McGraw-Hill's digitized biology videos will help capture students' interest while illustrating key biological concepts and processes such as mitosis, how cilia and flagella work, and how some plants have evolved into carnivores.

Photo Atlas for General Biology

ISBN (13) 978-0-07-284610-2
ISBN (10) 0-07-284610-0

This atlas was developed to support our numerous general biology titles. It can be used as a supplement for a general biology lecture or laboratory course.

PageOut

McGraw-Hill's exclusive tool for creating your own website for your introductory biology course. It requires no knowledge of coding and is hosted by McGraw-Hill.

Course Management Systems

ARIS content compatible with online course management systems like WebCT and Blackboard makes putting together your course website easy. Contact your local McGraw-Hill sales representative for details.

For Students

ARIS (Assessment Review and Instruction System)

Explore this dynamic website for a variety of study tools.

- **Self-quizzes** test your understanding of key concepts.
- **Flash cards** ease learning of new vocabulary.
- **Animations** and **Videos** bring key genetic concepts to life and are followed by a **quiz** to test your understanding.
- **Essential Study Partner** contains hundreds of animations, learning activities, and quizzes designed to help students grasp complex concepts.

Go to aris.mhhe.com to learn more or go directly to this book's ARIS site at www.mhhe.com/maderinquiry12.

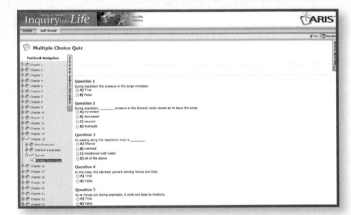

Student Study Guide

ISBN (13) 978-0-07-298680-8
ISBN (10)0-07-298680-8

Dr. Sylvia Mader has written the Student Study Guide that accompanies *Inquiry into Life*, and Kimberly Lyle-Ippolito has revised it for the twelfth edition, thereby ensuring close coordination with the text. Each text chapter has a corresponding study guide chapter that includes a chapter review, learning objectives and study questions for each section of the chapter, and a chapter test. Answers to the study guide and the chapter tests are provided to give students immediate feedback.

How to Study Science

ISBN (13) 978-0-07-234693-0
ISBN (10) 0-07-234693-0

This workbook offers students helpful suggestions for meeting the considerable challenges of a science course. It gives practical advice on such topics as how to take notes, how to get the most out of laboratories, and how to overcome science anxiety.

The diversity of life on Earth is astonishing. *Nearly 2 million species have been described, and scientists estimate that as many as 5–30 million species may exist. Yet all living things—from bacteria to trees to dogs to mosquitoes to birds to humans— share certain common characteristics. Why? Living things are similar because all forms of life can be traced back to an ancient common ancestor, the very first living thing.*

Because of this common ancestry, humans and plants, although quite different in appearance, share a common genetic code and basic life processes. Their many obvious differences, on the other hand, have been shaped by millions of years of evolution. It is evolutionary forces that have resulted in the vast diversity of life on Earth.

Currently, Earth's biodiversity is being threatened by human activities. Some biologists estimate that as many as half the world's species may become extinct in the next 100 years. Why should we care? Living things depend on one another in complex ways, and thus the loss of one species can lead to the loss of other species that depend on it. Since humans use various plants and animals for food and medicines, the loss of 50% of the world's species would most likely include many species that we depend on. Do we want to let this happen? The future of the planet is in our hands . . . .

The Study of Life

Figure 1.1 Living things on planet Earth.
If aliens ever visit our corner of the universe, they will be amazed at the diversity of life on our planet. They will also note that it is the only known place where oxygen is present in the atmosphere—and in huge quantities. This is one of the criteria for the origin of life and its diversity as we know it.

1.1   The Characteristics of Life

Life. Except for the most desolate and forbidding regions of the polar ice caps, planet Earth is teeming with life. Without life, our planet would be nothing but a barren rock hurtling through space. The variety of life on Earth is staggering, and human beings are a part of it. So are butterflies, trees, birds, mushrooms, and giraffes (Fig. 1.1). The variety of living things ranges in size from bacteria, much too small to be seen by the naked eye, all the way up to giant sequoia trees that can reach heights of 100 meters or more.

The diversity of life seems overwhelming, and yet all living things have certain characteristics in common. Taken together, these characteristics give us insight into the nature of life and help us distinguish living things from nonliving things. Living things generally (1) are organized, (2) acquire materials and energy, (3) reproduce, (4) respond to stimuli, (5) are homeostatic, (6) grow and develop, and (7) have the capacity to adapt to their environment. Let us consider each of these characteristics in order.

Living Things Are Organized

Living things can be organized in a hierarchy of levels (Fig. 1.2). In trees, humans, and all other organisms, atoms join together to form molecules, such as DNA molecules that occur within cells. A **cell** is the smallest unit of life, and some organisms are single-celled. In multicellular organisms, a cell is the smallest structural and functional unit. For example, a human nerve cell is responsible for conducting electrical impulses to other nerve cells. A **tissue** is a group of similar cells that perform a particular function. Nervous tissue is composed of millions of nerve cells that transmit signals to all parts of the body. Several tissues join together to form an **organ.** The main organs in the nervous system are the brain, the spinal cord, and nerves. Organs then work together to form an **organ system.** The brain sends messages to the spinal cord, which in turn sends them to body parts through spinal nerves. Complex **organisms** such as trees and humans are a collection of organ systems. The human body contains a digestive system, a cardiovascular system, and several other systems in addition to the nervous system.

Living Things Acquire Materials and Energy

Living things cannot maintain their organization or carry on life's other activities without an outside source of materials and energy. Plants, such as trees, use carbon dioxide, water, and solar energy to make their own food. Human beings and other animals acquire materials and energy when they eat food.

Food provides nutrient molecules, which cells use as building blocks or for energy. **Energy** is the capacity to do work, and it takes work to maintain the organization of a cell as well as an organism. Cells use energy from nutrient molecules to carry out everyday activities and functions. Some

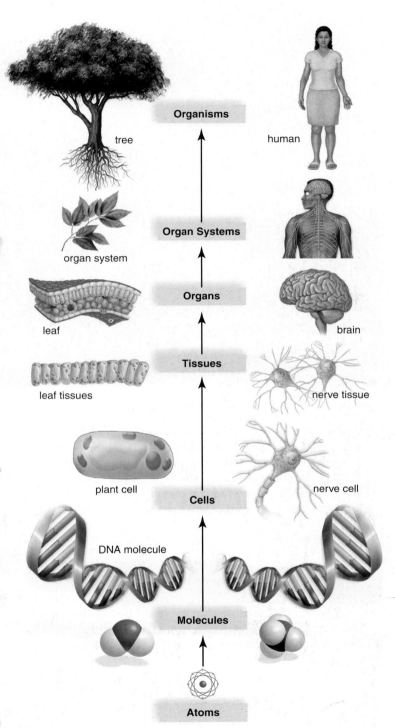

Figure 1.2 Levels of biological organization.
At the lowest level of organization, atoms join to form the molecules (including DNA) that are found in cells. Cells, the smallest units of life, differ in size, shape, and function. Tissues are groups of cells that serve the same function. Organs, which are composed of different tissues, make up the organ systems of complex organisms. Each organ and organ system performs functions that keep the organism alive and well.

Figure 1.3 Living things acquire materials and energy, and they reproduce. A red-tailed hawk has captured a rabbit, which it is feeding to its young.

nutrient molecules are broken down completely to provide the necessary energy for cells to carry out synthetic reactions, such as building proteins. Above, a red-tailed hawk feeds its young to provide the energy they need (Fig. 1.3).

Most living things can convert energy into motion. Self-directed movement, as when we decide to rise from a chair, is even considered by some to be one of life's characteristics.

Living Things Reproduce

Life comes only from life. The presence of **genes** in the form of DNA molecules allows cells and organisms to **reproduce**—that is, to make more of themselves. DNA contains the hereditary information that directs the structure of the cell and its **metabolism,** all the chemical reactions in the cell. Before reproduction occurs, DNA is replicated so that exact copies of genes are produced.

Unicellular organisms reproduce asexually simply by dividing. The new cells have the same genes and structure as the single parent. Multicellular organisms usually reproduce sexually. Each parent, male and female, contributes roughly one-half the total number of genes to the offspring, which then does not resemble either parent exactly.

Living Things Respond to Stimuli

Living things respond to external stimuli, often by moving toward or away from a stimulus, such as the sight of food. Movement in animals, including humans, is dependent upon their nervous and musculoskeletal systems. Other living things use a variety of mechanisms in order to move. The leaves of plants track the passage of the sun during the day, and when a houseplant is placed near a window, hormones help its stem bend to face the sun.

The movement of an organism, whether self-directed or in response to a stimulus, constitutes a large part of its **behavior.** Behavior is largely directed toward minimizing injury, acquiring food, and reproducing.

Living Things Are Homeostatic

Homeostasis means "staying the same." Actually, the internal environment of an organism stays *relatively* constant; for example, human body temperature fluctuates slightly throughout the day. Also, the body's ability to maintain a normal internal temperature is somewhat dependent on the external temperature—we will die if the external temperature becomes too hot or cold.

All organ systems contribute to homeostasis. The digestive system provides nutrient molecules; the cardiovascular system transports them about the body; and the urinary system rids blood of metabolic wastes. The nervous and endocrine systems coordinate the activities of the other systems. One of the major purposes of this text is to show how all the systems of the human body help maintain homeostasis.

Living Things Grow and Develop

Growth, recognized by an increase in the size of an organism and often in the number of cells, is a part of development. In

Figure 1.4 Living things grow and develop.
A small acorn gives rise to a very large oak tree by the process of growth and development.

humans, **development** includes all the changes that take place between conception and death. First, the fertilized egg develops into a newborn, and then a human goes through the stages of childhood, adolescence, adulthood, and aging. Development also includes the repair that takes place following an injury.

All organisms undergo development. Figure 1.4 illustrates that an acorn progresses to a seedling before it becomes an adult oak tree.

Living Things Have the Capacity to Adapt

Throughout the nearly 4 billion years that life has been on Earth, the environment has constantly been changing. For example, glaciers that once covered much of the world's surface 10,000–15,000 years ago have since receded, and many areas that were once covered by ice are now habitable. On a smaller scale, a hurricane or fire could drastically change the landscape in an area in a short period of time.

As the environment changes, some individuals of a **species** (a group of organisms that can successfully interbreed and produce fertile offspring) may possess certain features that make them better suited to the new environment; we call such features **adaptations.** For example, consider the hawk in Figure 1.3, which catches and eats rabbits. A hawk, like other birds, can fly because it has hollow bones, which can be considered an adaptation. Similarly, its

strong feet can take the shock of a landing after a hunting dive, and its sharp claws can grab and hold onto the prey.

Individuals of a species that are better adapted to their environment tend to live longer and produce more offspring than other individuals. This differential reproductive success, called **natural selection,** results in changes in the characteristics of a population (all the members of a species within a particular area) through time. That is, adaptations that result in higher reproductive success will increase in frequency in a population across generations. This change in the frequency of traits in populations and species is called **evolution.**

Evolution explains both the unity and diversity of life. As stated at the beginning of this chapter, all organisms share the same basic characteristics of life because we all share a common ancestor—the first cell or cells—that arose nearly 4 billion years ago. During the past 4 billion years, the Earth's environment has changed drastically, and the diversity of life has been shaped by the evolutionary responses of organisms to these changes.

Living things share certain characteristics. Evolution explains this unity of life and also its diversity.

1.2 The Classification of Living Things

Since life is so diverse, it is helpful to have a classification system to group organisms according to their similarities.

Systematics is the discipline of identifying and classifying organisms according to specific criteria. Each type of organism is placed in a **species, genus, family, order, class, phylum, kingdom,** and finally **domain.**

DOMAIN ARCHAEA

Methanosarcina mazei

a. Archaea are capable of living in extreme environments.

1.6 µm

DOMAIN BACTERIA

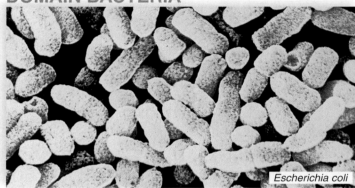

Escherichia coli

b. Bacteria are found nearly everywhere.

1.5 µm

DOMAIN EUKARYA

Kingdom	Organization	Type of Nutrition	Representative Organisms				
Protista	Complex single cell, some multicellular	Absorb, photosynthesize, or ingest food	paramecium	euglenoid	slime mold	dinoflagellate	Protozoans, algae, water molds, and slime molds
Fungi	Some unicellular, most multicellular filamentous forms with specialized complex cells	Absorb food	black bread mold	yeast	mushroom	bracket fungus	Molds, yeasts, and mushrooms
Plantae	Multicellular form with specialized complex cells	Photosynthesize food	moss	fern	pine tree	nonwoody flowering plant	Mosses, ferns, nonwoody and woody flowering plants
Animalia	Multicellular form with specialized complex cells	Ingest food	sea star	earthworm	finch	raccoon	Invertebrates, fishes, reptiles, amphibians, birds, and mammals

c. Eukaryotes are divided into four kingdoms.

Figure 1.5 Pictorial representation of the three domains of life. Archaea (**a**) and bacteria (**b**) are both prokaryotes but are so biochemically different that they are not believed to be closely related. **c.** Eukaryotes are biochemically similar but structurally dissimilar. Therefore, they have been categorized into four kingdoms. Many protists are unicellular, but the other three kingdoms are characterized by multicellular forms.

Domains

Domains are the largest classification category. Biochemical and genetic evidence tells us that there are three domains: **Archaea, Bacteria,** and **Eukarya** (Fig. 1.5). Both domain Archaea and domain Bacteria contain unicellular **prokaryotes,** which lack the membrane-bounded nucleus found in the cells of **eukaryotes** in domain Eukarya. The genes of eukaryotes are found in the nucleus, which thereby controls the cell.

Prokaryotes are structurally simple (Fig. 1.5*a, b*) but metabolically complex. The *archaea* live in aquatic environments that lack oxygen or are too salty, too hot, or too acidic for most other organisms. Perhaps these environments are similar to those of the primitive Earth, and archaea represent the first cells that evolved. *Bacteria* are found almost anywhere—in the water, soil, and atmosphere, as well as on our skin and in our mouths and digestive tracts. Although some bacteria cause diseases, others are useful, both environmentally and commercially. For example, bacteria can be used to develop new medicines, to clean up oil spills, or to help purify water in sewage treatment plants.

Kingdoms

Systematists are in the process of deciding how to categorize archaea and bacteria into kingdoms. The *eukaryotes* are currently classified into four kingdoms with which you may be familiar (Fig. 1.5*c*). **Protists** (kingdom Protista) range from unicellular to a few multicellular organisms. Some can make their own food (photosynthesizers), while others must ingest their food. Among the **fungi** (Kingdom Fungi) are the familiar molds and mushrooms that, along with bacteria, help decompose dead organisms. **Plants** (kingdom Plantae) are well known as multicellular photosynthesizers, while **animals** (kingdom Animalia) are multicellular and ingest their food.

Other Categories

The other classification categories are phylum, class, order family, genus, and species. Each classification category is more specific than the one preceding it. For example, the species within one genus share very similar characteristics, while those within the same kingdom share only general characteristics. Modern humans are the only living species in the genus *Homo,* but many different types of animals are in the animal kingdom (Table 1.1). To take another example, all species in the genus *Pisum* (pea plants) look pretty much the same, while the species in the plant kingdom can be quite different, as is evident when we compare grasses to trees.

Systematics makes sense out of the bewildering variety of life on Earth because organisms are classified according to their presumed evolutionary relationships. Those organisms placed in the same genus are the most closely related, and those placed in separate domains are the most distantly related. Therefore, all eukaryotes are more closely related to one another than they are to bacteria or archaea. Similarly, all animals are more closely related to one another than they are to plants. As more is learned about evolutionary

TABLE 1.1	Classification of Humans
Classification Category	**Characteristics**
Domain Eukarya	Cells with nuclei
Kingdom Animalia	Multicellular, motile, ingestion of food
Phylum Chordata	Dorsal supporting rod and nerve cord
Class Mammalia	Hair, mammary glands
Order Primates	Adapted to climb trees
Family Hominidae	Adapted to walk erect
Genus *Homo*	Large brain, tool use
Species *Homo sapiens**	Body proportions of modern humans

* *To specify an organism, you must use the full binomial name, such as Homo sapiens.*

relationships between species, systematic relationships change. Systematists are even now making observations and performing experiments that will one day bring about changes in the classification system adopted by this text.

Scientific Names

Within the broad field of systematics, **taxonomy** is the assignment of a binomial, or two-part name, to each species. For example, the scientific name for human beings is *Homo sapiens,* and for the garden pea, *Pisum sativum.* The first word is the genus to which the species belongs, and the second word is the species name. (Note that both words are in italics, but only the genus is capitalized.) The genus name can be used alone to refer to a group of related species. Also, a genus can be abbreviated to a single letter if used with the species name (e.g., *P. sativum*).

Scientific names are in a common language—Latin—and biologists use them universally to avoid confusion. Common names, by contrast, tend to overlap and often are in the language of a particular country.

Systematics places species into classification categories according to their evolutionary relationships. The categories are species, genus, family, order, class, phylum, kingdom, and domain (the most inclusive). Taxonomy assigns names to species.

1.3 The Organization of the Biosphere

The organization of life extends beyond the individual to the **biosphere,** the zone of air, land, and water at the surface of Earth where living organisms are found. Individual organisms belong to a **population,** all the members of a species within a particular area. All the different populations in the same area make up a **community,** in which organisms interact among themselves and with the physical environment (soil, atmosphere, etc.), thereby forming an **ecosystem.**

Figure 1.6 depicts a grassland inhabited by populations of rabbits, mice, snakes, hawks, and various types of plants. These populations exchange gases with and give off heat to the atmosphere. They also take in water from and give off water to the physical environment. In addition, the populations interact with each other by forming food chains in which one population feeds on another. For example, mice feed on plants and seeds, snakes feed on mice, and hawks feed on rabbits and snakes. The interactions among the various food chains make up a food web.

Ecosystems are characterized by chemical cycling and energy flow, both of which begin when photosynthetic plants, algae, and some bacteria take in solar energy and inorganic nutrients to produce food in the form of organic nutrients. The gray arrows in Figure 1.6 represent chemical cycling—chemicals move from one species to another in a food chain, until with death and decomposition, inorganic nutrients are returned to living plants once again. The yellow to red arrows represent energy flow. Energy flows from the sun through plants and other members of the food chain as one species feeds on another. With each transfer, some energy is lost as heat. Eventually, all the energy

taken in by photosynthesizers has dissipated into the atmosphere. Because energy flows and does not cycle, ecosystems could not stay in existence without a constant input of solar energy and the ability of photosynthesizers to absorb it.

Climate largely determines where different ecosystems are found around the globe. For example, deserts are found in areas of almost no rain, grasslands require a minimum amount of rain, and forests require much rain. The two most biologically diverse ecosystems—tropical rain forests and coral reefs—occur where solar energy is most abundant. Coral reefs, which are found just offshore from the continents and islands of the Southern and Northern Hemispheres, are built up from the skeletons of sea animals called corals. Inside the tissues of corals are tiny, one-celled protists that carry on photosynthesis and provide food to their hosts. Reefs provide a habitat for many other animals, including jellyfish, sponges, snails, crabs, lobsters, sea turtles, moray eels, and some of the world's most colorful fishes (Fig. 1.7).

The Human Species

The human species tends to modify existing ecosystems for its own purposes. Humans clear forests or grasslands in order to grow crops; later, they build houses on what was once farmland; and finally, they convert small towns into cities. As coasts are developed, humans send sediments, sewage, and other pollutants into the sea.

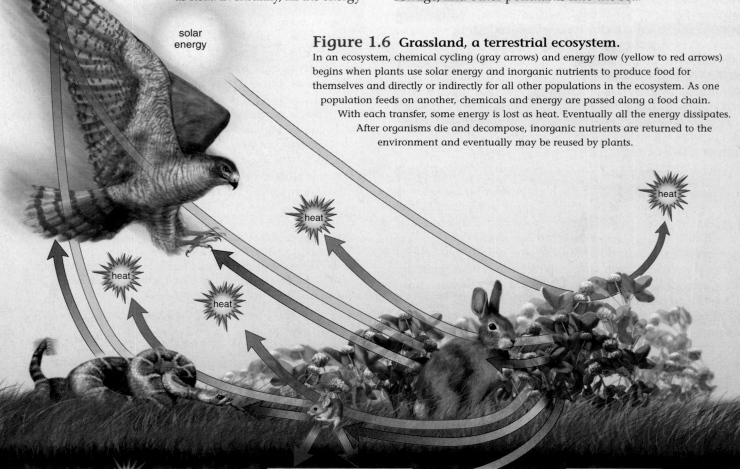

solar energy

Figure 1.6 Grassland, a terrestrial ecosystem.
In an ecosystem, chemical cycling (gray arrows) and energy flow (yellow to red arrows) begins when plants use solar energy and inorganic nutrients to produce food for themselves and directly or indirectly for all other populations in the ecosystem. As one population feeds on another, chemicals and energy are passed along a food chain. With each transfer, some energy is lost as heat. Eventually all the energy dissipates. After organisms die and decompose, inorganic nutrients are returned to the environment and eventually may be reused by plants.

heat

WASTE MATERIAL, DEATH, AND DECOMPOSITION

Chemical cycling
Energy flow

Like tropical rain forests, coral reefs are severely threatened as the human population increases in size. Some reefs are 50 million years old, and yet in just a few decades, human activities have destroyed 10% of all coral reefs and seriously degraded another 30%. At this rate, nearly three-quarters could be destroyed within 50 years. Similar levels of destruction have occurred in tropical rain forests.

It has long been clear that human beings depend on healthy ecosystems for food, medicines, and various raw materials. We are only now beginning to realize that we depend on them even more for the services they provide. Just as chemical cycling occurs within ecosystems, so do ecosystems keep chemicals cycling throughout the entire biosphere. The workings of ecosystems ensure that the environmental conditions of the biosphere are suitable for the continued existence of humans. And many biologists believe that ecosystems cannot function properly unless they remain biologically diverse.

Biodiversity

Biodiversity encompasses the total number of species, the variability of their genes, and the ecosystems in which they live. The present biodiversity of our planet has been estimated at as high as 5–30 million species, and so far, fewer than 2 million have been identified and named. Extinction is the death of a species or larger taxonomic group. It is estimated that we are presently losing as many as 400 species per day due to human activities. For example, several species of fishes have all but disappeared from the coral reefs of Indonesia and along the African coast because of overfishing. Many biologists are alarmed about the present rate of extinction and believe it may eventually rival the rates of the five mass extinctions that have occurred during our planet's history. The dinosaurs became extinct during the last mass extinction, 65 million years ago.

Most biologists agree that the primary bioethical issue of our time is conservation of biodiversity. Just as a native fisherman who assists in overfishing a reef is doing away with his own food source, so we as a society are contributing to the destruction of our home, the biosphere. If instead we adopt an ethic that conserves the biosphere, we are helping to ensure the continued existence of our species.

Living things belong to ecosystems where populations interact among themselves in communities and with the physical environment. Conservation of biodiversity is of primary importance because species and ecosystems perform services that ensure our continued existence.

Figure 1.7 Coral reef, a marine ecosystem.

Coral reefs, a type of ecosystem found in tropical seas; contain many diverse forms of life, a few of which are shown here. Coral reefs are now threatened because of certain human activities. Saving biodiversity is a modern-day challenge of great proportions.

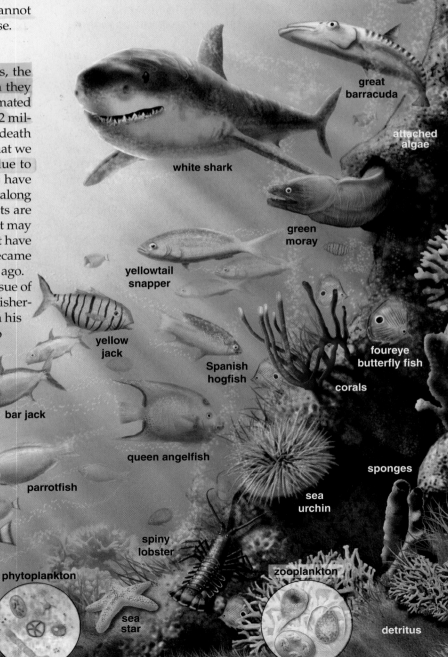

great barracuda

attached algae

white shark

green moray

yellowtail snapper

foureye butterfly fish

yellow jack

Spanish hogfish

corals

bar jack

queen angelfish

sponges

parrotfish

sea urchin

Bermuda chub

spiny lobster

zooplankton

surgeonfish

phytoplankton

sea star

detritus

yellowtail damselfish

sea grass

1.4   The Process of Science

Biology is the scientific study of life. Biologists can be found almost anywhere studying life-forms; often, they perform experiments, either in the field (outside the laboratory) or in the laboratory.

Biologists—and all scientists—use various investigative methods depending on the topic under study. Still, in general, scientists test hypotheses using the scientific method based on previous observations (Fig. 1.8).

Observation

Scientists ultimately are curious about nature and how the world works. They believe that natural events, or **phenomena,** can be understood more fully by observing and studying them. Scientists use all their senses to make **observations**. We can observe with our noses that dinner is almost ready; observe with our fingertips that a surface is smooth and cold; and observe with our ears that a piano needs tuning. Scientists also extend the ability of their senses by using instruments; for example, the microscope enables us to see objects too small to view with the naked eye. Finally, scientists may expand their understanding even further by taking advantage of the knowledge and experiences of other scientists. For instance, they may look up past studies in scientific journals or in the library, or they may attend scientific meetings and hear reports from others who are researching similar topics.

Chance alone can help a scientist get an idea. The most famous case pertains to penicillin. When examining a petri dish, Alexander Fleming observed an area around a mold that was free of bacteria. Upon investigating, Fleming found that the mold produced an antibacterial substance he called penicillin. This observation led Fleming to think that perhaps penicillin would be useful in humans.

Hypothesis

After making observations and gathering knowledge about a phenomenon, a scientist uses inductive reasoning. **Inductive reasoning** occurs whenever a person uses creative thinking to combine isolated facts into a cohesive whole. In this case, the scientist comes up with a **hypothesis,** a tentative explanation for the natural event. The scientist presents the hypothesis as a falsifiable statement.

All of a scientist's past experiences, no matter what they might be, most likely influence the formation of a hypothesis. But a scientist only considers hypotheses that can be tested. Moral and religious beliefs, while very important to our lives, differ between cultures and through time, and are not testable scientifically.

Experiment/Further Observations

Testing a hypothesis involves either conducting an experiment or making further observations. To determine how to test a hypothesis, a scientist uses deductive reasoning. **Deductive reasoning** involves "if, then" logic. For example, a scientist might reason, *if* organisms are composed of cells,

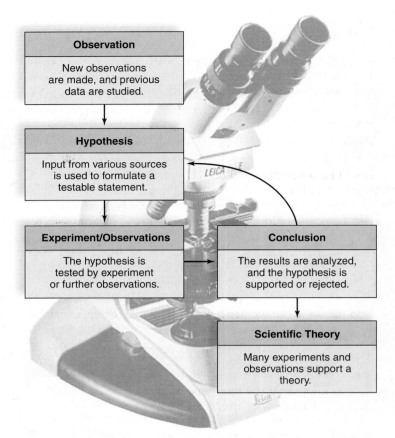

Figure 1.8 Flow diagram for the scientific method.
On the basis of new and/or previous observations, a scientist formulates a hypothesis. The scientist then tests the hypothesis by making further observations and/or performing experiments. The resulting new data either support or do not support the hypothesis. The return arrow indicates that a scientist often chooses to retest the same hypothesis or to test an alternative hypothesis. Conclusions from many different but related experiments may lead to the development of a scientific theory. For example, studies of development, anatomy, and fossil remains all support the theory of evolution.

then microscopic examination of any part of an organism should reveal cells. We can also say that the scientist has made a **prediction** that the hypothesis can be supported by doing microscopic studies. Making a prediction helps a scientist know what to do next.

Further observations may be used to test the hypothesis. Many times, however, scientists perform experiments. The manner in which a scientist intends to conduct an experiment is called the **experimental design.** A good experimental design ensures that scientists are testing what they want to test and that their results will be meaningful. It is always best for an experiment to include a control group. A control group, or simply the **control,** goes through all the steps of an experiment but lacks the factor, or is not exposed to the factor, being tested. For example, in a drug trial the treatment group receives the actual drug being tested while the control group receives a placebo (sugar pill).

In their experiments, scientists often use a **model,** a representation of an actual subject. For example, scientists can use computer modeling to suggest future ways human activities may affect global climate change, or they may use mice as a model, instead of humans, when doing cancer research. When a medicine is effective in mice, researchers still need to test it in humans before making recommendations or using it in widespread treatment. Thus, results obtained using models are considered a hypothesis until the experiment can be performed using the actual subject.

Data

The results of an experiment are referred to as the **data.** Data should be observable and objective, rather than subjective or based on opinion. Mathematical data are often displayed in the form of a graph or table. Many studies rely on statistical data. Let's say an investigator wants to know if eating onions can prevent women from getting osteoporosis (weak bones). The scientist conducts a survey asking women about their onion-eating habits and then correlates these data with the condition of their bones. Other scientists critiquing this study would want to know: How many women were surveyed? How old were the women? What were their exercise habits? What proportion of the diet consisted of onions? And what criteria were used to determine the condition of their bones? Should the investigators conclude that eating onions does protect a woman from osteoporosis, other scientists would want to know that the relationship did not occur simply by chance. The larger the sample size (in this case, the number of women studied) and the larger the effect (in this case, strength of bones in women eating more onions compared to those eating fewer onions), the less likely that the result is due to chance alone. Statistical tests are used to evaluate how likely it is that the results of an experiment are due to chance. And in the end, statistical data of this sort would only be suggestive until we know of some ingredient in onions that has a direct biochemical or physiological effect on bone strength. Therefore, you can see that scientists are skeptics who always pressure one another to keep investigating a particular topic.

Conclusion

Scientists must analyze the data in order to reach a **conclusion** as to whether the hypothesis is supported or not. Because science progresses, the conclusion of one experiment can lead to the hypothesis for another experiment, as represented by the return arrow in Figure 1.8. Sometimes, results that do not support one hypothesis can help a scientist formulate another hypothesis to be tested. Scientists report their findings in scientific journals, subject to critical review by their peers, so that their methodology and data are available to other scientists. Experiments and observations must be repeatable—that is, the reporting scientist and any scientist who repeats the experiment must get the same results, or else the data are suspect. This scientific process is continuous, and in biology scientists are discovering new things all the time.

Scientific Theory

The ultimate goal of science is to understand the natural world in terms of **scientific theories,** concepts that join together well-supported and related hypotheses. In ordinary speech, the word *theory* refers to a speculative idea. But in science, a theory is supported by a broad range of observations, experiments, and data and is thus similar to a law, such as the theory of gravity.

Some of the basic theories of biology are as follows:

Theory	Concept
Cell	All organisms are composed of cells, and new cells only come from preexisting cells.
Homeostasis	The internal environment of an organism stays relatively constant—within a range that is protective of life.
Gene	Organisms contain coded information that dictates their form, function, and behavior.
Ecosystem	Organisms are members of populations, which interact with each other and with the physical environment within a particular locale.
Evolution	A change in the frequency of traits that affect reproductive success in a population or species across generations.

The theory of evolution is the unifying concept of biology because it pertains to many different aspects of living things. For example, the theory of evolution enables scientists to understand the history of life; the diversity of living things; and the anatomy, physiology, and development of organisms—even their behavior.

The theory of evolution has been a very fruitful theory scientifically, meaning that it has helped scientists generate new hypotheses. Because this theory has been supported by so many observations and experiments for over 100 years, some biologists refer to the **principle** of evolution, a term considered appropriate for theories that are generally accepted by an overwhelming number of scientists. Other biologists prefer to use the term **law** instead of principle. For instance, in a subsequent chapter concerning energy relationships, we will examine the laws of thermodynamics.

Scientists carry out experiments in which they test hypotheses based on their observations of nature. If the conclusions of many related experiments are similar, scientists formulate the general, core ideas as a theory.

A Controlled Study

Most investigators do controlled studies in which some groups receive a treatment (experimental), while other groups receive no treatment (control). To ensure that the outcome is due to the **experimental variable** (independent variable), all other conditions between the groups in the experiment are identical. The result is called the **response variable** (or dependent variable) because it is due to the experimental treatment.

Experimental Variable (Independent Variable)	**Response Variable** (Dependent Variable)
Factor of the experiment being tested	Result or change that occurs due to the experimental variable

In the study we are about to discuss, researchers performed an experiment in which nitrogen fertilizer is the experimental variable and crop yield is the response variable. Nitrogen fertilizer in the short run has long been known to enhance yield and increase food supplies. However, excessive nitrogen fertilizer application can cause pollution by adding toxic levels of nitrates to water supplies. Also, applying nitrogen fertilizer year after year may alter soil properties to the point that crop yields may decrease instead of increase. Then the only solution is to let the land remain unplanted for several years until the soil recovers naturally.

An alternative to nitrogen fertilizers is the use of legumes, plants such as peas and beans, that increase soil nitrogen. Legumes provide a home for bacteria that convert atmospheric nitrogen to a form usable by plants. The bacteria live in nodules on the roots (Fig. 1.9). The products of photosynthesis move from the leaves to the root nodules; in turn, the nodules supply the plant with nitrogen compounds the plant can use to make proteins.

There are numerous legume crops that can be rotated (planted every other season) with any number of cereal crops. The nitrogen added to the soil by the legume crop is a natural fertilizer that increases cereal crop yield. The particular rotation used depends on the location, climate, and market demand.

The Experiment

The pigeon pea plant is a legume with a high rate of atmospheric nitrogen conversion. This plant is widely grown as a food crop in India, Kenya, Uganda, Pakistan, and other subtropical countries. Researchers formulated the hypothesis that rotating pigeon pea and winter wheat crops would be a reasonable alternative to using nitrogen fertilizer to increase the yield of winter wheat.

HYPOTHESIS: A pigeon pea/winter wheat rotation will cause winter wheat production to increase as well as or better than the use of nitrogen fertilizer.
PREDICTION: Wheat biomass following the growth of pigeon peas will surpass wheat biomass following nitrogen fertilizer treatment.

In this study, the investigators decided on the following experimental design (Fig. 1.10a):

CONTROL POTS

- Winter wheat was planted in pots of soil that received no fertilization treatment (i.e., no nitrogen fertilizer and no preplanting of pigeon peas)

TEST POTS

- Winter wheat was grown in clay pots in soil treated with nitrogen fertilizer equivalent to 45 kilograms (kg)/hectare (ha).
- Winter wheat was grown in clay pots in soil treated with nitrogen fertilizer equivalent to 90 kg/ha.
- Pigeon pea plants were grown in clay pots in the summer. The pigeon pea plants were then tilled into the soil, and winter wheat was planted in the same pots.

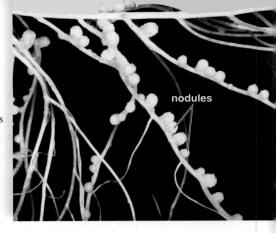

Figure 1.9
Root nodules.
Bacteria that live in nodules on the roots of legumes, such as pea plants, convert nitrogen in the soil to a form that plants can use to make proteins and other nitrogen-containing molecules.

nodules

To ensure a controlled experiment, the conditions for the control pots and the test pots were identical with the exception of the treatments noted; the plants were exposed to the same environmental conditions and watered equally. During the following spring, the wheat plants were dried and weighed to determine total wheat production (biomass) in each of the pots.

The Results   After the first year, wheat biomass was higher in certain test pots than in the control pots (Fig. 1.10b). Specifically, test pots with 45 kg/ha of nitrogen fertilizer (orange) had only slightly more wheat biomass production than the control pots, but test pots that received 90 kg/ha treatment (green) had nearly twice the biomass production of the control pots. Thus, application of nitrogen fertilizer had the potential to increase wheat biomass. To the surprise of investigators, wheat production following summer planting of pigeon peas did not result in as high a biomass production as the control pots.

CONCLUSION: The hypothesis is not supported. Wheat biomass following the growth of pigeon peas is not as high as that obtained with nitrogen fertilizer treatments.

Continuing the Experiment

The researchers decided to continue the experiment using the same design and the same pots as before, to see if the buildup of residual soil nitrogen from pigeon peas would eventually increase wheat biomass. They formed a new hypothesis.

HYPOTHESIS: A sustained pigeon pea/winter wheat rotation will eventually cause an increase in winter wheat production.
PREDICTION: Wheat biomass following two years of pigeon pea/winter wheat rotation will surpass wheat biomass following nitrogen fertilizer treatment.

The Results   After two years, the yield following 90 kg/ha nitrogen treatment (*green*) was not as much as it was the first year (Fig. 1.10b). As predicted, wheat biomass following summer planting of pigeon peas (*brown*) was the highest of all treatments, suggesting that buildup of residual nitrogen from pigeon peas had the potential to provide fertilization for winter wheat growth.

CONCLUSION: The hypothesis was supported. At the end of two years, the yield of winter wheat following a

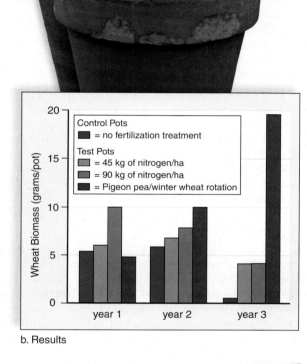

a. Control pot and three types of test pots

Control pot no fertilization treatment

Test pot 90 kg of nitrogen/ha

Test pot Pigeon pea/winter wheat rotation

Test pot 45 kg of nitrogen/ha

b. Results

Figure 1.10 Pigeon pea/winter wheat rotation study.

a. The experiment involves control pots that receive no fertilization and three types of test pots: test pots that received 45 kg/ha of nitrogen; test pots that received 90 kg/ha of nitrogen; and test pots in which pigeon peas rotated with winter wheat. **b.** The graph compares wheat biomass for each of three years. Wheat biomass in test pots that received the most nitrogen fertilizer declined, while wheat biomass in test pots with pigeon pea/winter wheat rotation increased dramatically.

pigeon pea/winter wheat rotation was better than for the other types of pots.

The researchers continued their experiment for still another year. After three years, winter wheat biomass production had decreased in the control pots and in the pots treated with nitrogen fertilizer. Pots treated with nitrogen fertilizer still had increased wheat biomass production compared with the control pots, but not nearly as much as pots following summer planting of pigeon peas. Compared to the first year, wheat biomass increased almost fourfold in pots having a pigeon pea/winter wheat rotation (Fig. 1.10*b*). The researchers suggested that the soil had been improved by the organic matter, as well as by the addition of nitrogen from the pigeon peas, and published their results in a scientific journal.[1]

Ecological Importance of This Study

This study showed that the use of a legume, namely pigeon peas, to improve the soil produced a far better yield than the use of a nitrogen fertilizer over the long haul. Legumes are plants that house bacteria capable of converting atmospheric nitrogen into a form a plant can use. When the pigeon pea plants were turned over into the soil, the winter wheat plants could make use of this nitrogen.

Rotation of crops, as was done in this study, is an important part of organic farming. Sometimes there are adverse economic results when farmers switch from chemical-intensive to organic farming practices. But, because the benefit of making the transition may not be realized for several years, some researchers advocate that the process be gradual. This study suggests, however, that organic farming will eventually be beneficial both for farmers and the environment!

> A controlled study usually has a control group in addition to a number of test groups. When conducting the study, investigators formulate a hypothesis, carry out an experiment, and collect data. The data indicate whether the hypothesis is supported. Supported hypotheses become a part of the knowledge of biology, while refuted hypotheses require revision and new experiments.

[1] Bidlack, J. E., Rao, S. C., and Demezas, D. H. 2001. Nodulation, nitrogenase activity, and dry weight of chickpea and pigeon pea cultivars using different Bradyrhizobium strains. Journal of Plant Nutrition 24:549–60.

The Pros and Cons of DDT

The widespread use of DDT began in 1945 to control agricultural pests and mosquitoes. DDT is extremely toxic to insects—both those that benefit crops and those that destroy them. DDT is inexpensive, kills most pests, and provides lasting protection because it breaks down slowly. DDT is believed to have helped save millions of human lives. The military used it during World War II to control body lice and thus typhus fever, and the World Health Organization (WHO) used it to control mosquitoes and consequently the malaria they carry.

In 1962, Rachel Carson wrote the book *Silent Spring,* which opened our eyes to the fact that long-term pesticide use in the United States could be harmful. Subsequent studies showed that DDT (and other pesticides) can become concentrated in the food chain. DDT is known to have led to the decline of birds of prey, such as the bald eagle, because it causes thinning of eggshells. Adults then inadvertently crush the eggs when attempting to incubate them, killing the embryos before they hatch.

Because of public concerns about DDT's effects on the environment and possible long-term effects on human health, the U.S. Environmental Protection Agency banned DDT in 1972. Recently, the United Nations passed the international Persistent Organic Pollutants treaty to cut back on DDT in the environment by permitting its use only under strict regulations.

Still, DDT is the most cost-effective chemical we have for preventing malaria because it kills the mosquitoes that carry the disease. According to WHO, malaria causes 300 million acute illnesses and over 1 million deaths each year in the tropics, subtropics, and portions of the Middle East. Researchers state in the *Journal of Infectious Diseases,* "Today, DDT is still needed for malaria control. If the pressure to abandon this effective insecticide continues, millions of additional malaria cases world-wide [will result]." So, should we continue to cut back on the use of DDT because of long-term environmental effects on animals and people? Or should we consider allowing its use to help control malaria and prevent some of the many deaths from that disease that occur each year?

Form Your Own Opinion

1. Present regulations ban the manufacture and use of DDT in many, but not all, countries. Should it be banned everywhere? Why or why not?
2. Under what circumstances might the use of DDT be acceptable? Explain.
3. Currently in the United States, the insecticide malathion is widely used to control mosquitoes that may carry West Nile virus. Although malathion breaks down more quickly than DDT, some evidence suggests it can be harmful to amphibians and fish. What data are needed before making decisions on continued widespread use of malathion?

1.5 Science and Social Responsibility

Many scientists are engaged in fields of study that sometimes seem remote from our everyday lives. Other scientists are interested in using the findings of past and present scientists to produce a product or develop a technique that does affect our lives. The application of scientific knowledge for a practical purpose is called technology. For example, virology, the study of viruses and molecular chemistry, led to the discovery of new drugs that extend the life span of people who have HIV/AIDS. And cell biology research, which attempts to find the causes of cancer, has allowed physicians to develop various cancer treatments.

Most technologies have benefits but also drawbacks. Research has led to modern agricultural practices that are helping to feed the burgeoning world population. However, the use of nitrogen fertilizers leads to water pollution, and the use of pesticides, as you may know, kills not only pests but also other types of organisms.

Who should decide how, and even whether, a technology should be put to use? Making value judgments is not a part of science. Ethical and moral decisions must be made by all people. Therefore, the responsibility for how to use the fruits of science must reside with people from all walks of life, not with scientists alone. Scientists should provide the public with as much information as possible, but all citizens, including scientists, should decide how technologies will be used.

Presently, we need to decide whether we want to stop producing bioengineered organisms that may be harmful to the environment. Also, through gene therapy, we are developing the ability to cure diseases and to alter the genes of our offspring. Perhaps one day we might even be able to clone ourselves. Should we do these things? So far, as a society, we continue to believe in the sacredness of human life, and therefore we have passed laws against doing research with fetal tissues or using fetal tissues to cure human ills. Even if the procedure is perfected, we may also continue to rule against human cloning.

The Bioethical Focus at the end of this chapter presents the complexities of banning the use of DDT. Each of us must wrestle with this and the other bioethical issues discussed in this text and hopefully make decisions that are beneficial to society.

Technology has many benefits, but sometimes its products and techniques also have drawbacks.

Summarizing the Concepts

1.1 The Characteristics of Life

Evolution accounts for both the diversity and the unity of life we see about us. All organisms share the following characteristics of life:

- Living things are organized. The levels of biological organization extend as follows: atoms and molecules → cells → tissues → organs → organ systems → organisms → populations → communities → ecosystems. In an ecosystem, populations interact with one another and with the physical environment.
- Living things take materials and energy from the environment; they need an outside source of nutrients.
- Living things reproduce; they produce offspring that resemble themselves.
- Living things respond to stimuli; they react to internal and external events.
- Living things are homeostatic; internally they stay just about the same despite changes in the external environment.
- Living things grow and develop; during their lives, they change—most multicellular organisms undergo various stages from fertilization to death.
- Living things have the capacity to adapt to a changing environment.

1.2 The Classification of Living Things

Living things are classified according to their evolutionary relationships into these ever-more-inclusive categories: species, genus, family, order, class, phylum, kingdom, and domain. Species in different domains are only distantly related; species in the same genus are very closely related.

Each type of organism has a scientific name in Latin that consists of the genus and species name. Both the genus and the species name are italicized; only the genus is capitalized, as in *Homo sapiens.*

1.3 The Organization of the Biosphere

Living things generally belong to a population, which is defined as all the members of a single species that occur in a particular area. Populations interact with each other within a community and with the physical environment, forming an ecosystem. Ecosystems are characterized by chemical cycling and energy flow, which begin when a photosynthesizer becomes food (organic nutrients) for an animal. Food chains tell who eats whom in an ecosystem. As one population feeds on another, the energy dissipates, but the nutrients do not. Eventually, inorganic nutrient molecules are decomposed and return to photosynthesizers, which use them and solar energy to produce more food.

Human activities have totally altered many ecosystems and are putting stress on most of the others. Coral reefs and tropical rain forests, for example, are quickly disappearing. We now realize that the health of the biosphere is essential to the future continuance of the human species.

1.4 The Process of Science

When studying the natural world, scientists use the scientific method. Observations, along with previous data, are used to formulate a hypothesis. New observations and/or experiments are carried out in order to test the hypothesis. A good experimental design includes a control group. The experimental and observational results are analyzed, and the scientist comes to a conclusion as to whether the results support the hypothesis or prove it false. Science is always open to change; therefore, hypotheses can be supported and modified based on new experiments.

Several related conclusions resulting from similar scientific experiments may allow scientists to arrive at a theory, such as the cell theory, the gene theory, or the theory of evolution. The theory of evolution is a unifying theory of biology.

1.5 Science and Social Responsibility

Science does not consider moral or ethical questions; those are the responsibility of all people. It is up to all of us to decide how the various technologies that grow out of basic science should be used or regulated.

Testing Yourself

Choose the best answer for each question.

1. Which sequence represents the correct order of increasing complexity in living systems?
   a. cell, molecule, organ, tissue
   b. organ, tissue, cell, molecule
   c. molecule, cell, tissue, organ
   d. cell, organ, tissue, molecule

2. What is wrong with the scientific name for humans as written here: *Homo Sapiens?*
   a. *Homo* should not be capitalized.
   b. Neither *Homo* nor *Sapiens* should be capitalized.
   c. *Sapiens* should not be capitalized.
   d. Nothing is wrong.

3. Classification of organisms reflects
   a. similarities.
   b. evolutionary history.
   c. Neither a nor b is correct.
   d. Both a and b are correct.

4. A population is defined as
   a. the number of species in a given geographic area.
   b. all of the individuals of a particular species in a given area.
   c. a group of communities.
   d. all of the organisms in an ecosystem.

5. Information collected during scientific inquiry is called
   a. results.
   b. conclusions.
   c. observations.
   d. data.

6. The ultimate source of energy in any ecosystem is (are)
   a. the sun.
   b. green plants.
   c. fungi.
   d. animals.

7. Which are the most biologically diverse ecosystems?
   a. deserts and rain forests   c. deserts and coral reefs
   b. rain forests and coral reefs   d. None of these are correct.

   tropical rain forests and coral reefs.

8. Having hair and mammary glands places an organism in the
   a. family Hominidae.
   b. class Mammalia.
   c. genus Homo
   d. None of these are correct.

9. Humans belong to the domain
   a. Archaea.
   b. Bacteria.
   c. Eukarya.
   d. All of these are correct.

10. What is the unifying theory in biology that explains the relationships of all living things?
    a. ecology
    b. evolution
    c. biodiversity
    d. taxonomy

11. If a biologist studies biodiversity, she will be interested in
    a. extinction.
    b. species diversity.
    c. genetic differences.
    d. All of these are correct.

12. Making a prediction in science is an example of using
    a. deductive logic.
    b. inductive logic.
    c. Both a and b are correct.
    d. Neither a nor b is correct.

13. Which sequence exhibits an increasingly more-inclusive scheme of classification?
    a. kingdom, phylum, class, order
    b. phylum, class, order, family
    c. class, order, family, genus
    d. genus, family, order, class

14. The best explanation as to why "big, fierce carnivores (animals that eat other animals) are rare" would be the fact that
    a. energy flows and dissipates.
    b. energy is stored in living systems.
    c. nutrients are cycled.
    d. nutrients flow.

15. Which sequence represents the correct order of increasing complexity?
    a. biosphere, community, ecosystem, population
    b. population, ecosystem, biosphere, community
    c. community, biosphere, population, ecosystem
    d. population, community, ecosystem, biosphere

16. Features that make an organism suited to its way of life are called
    a. ecosystems.
    b. populations.
    c. adaptations.
    d. None of these are correct.

17. The use and application of knowledge gained through scientific discoveries is known as
    a. adaptation.
    b. technology.
    c. conclusion.
    d. prediction.

18. A scientist who studies _____ could be an ecologist.
    a. populations
    b. communities
    c. ecosystems
    d. All of these are correct.

For questions 19–23, match the statements with the correct terms.
    a. photosynthetic ability
    b. absorb food
    c. ingest food
    d. all populations within a given area
    e. all the interactions of all organisms with each other and with the environment in a given area

19. Animals
20. Plants
21. Ecosystem
22. Community
23. Fungi

24. Label the following diagram with the terms *energy flow* and *chemical cycling*.

25. What's the difference between a test group and a control group?
    a. There is no real difference because both are part of a study.
    b. The test group is testing something, but the control group is not testing anything.
    c. The test group is exposed to the factor being tested, but the control group is not exposed.
    d. The control group proves the results true, but the test group does not.
    e. All of these are correct.

26. In the controlled study of nitrogen fertilizers and legumes, (see pages 12–13)
    a. pigeon peas and winter wheat were treated similarly; for example, winter wheat was kept inside to stabilize the temperature.
    b. no pigeon peas were grown in the control pots, and no fertilizer was added.
    c. there were two test groups; therefore, there were two test pots.
    d. the investigators were surprised that the control group didn't show better results.
    e. All of these are correct.

27. The study described in question 26 contributes to our knowledge of
    a. the theory of evolution.
    b. how to do a controlled study.
    c. the benefits of pesticides
    d. how to use biological controls.

28. Which term and definition are mismatched?
    a. data—factual information
    b. hypothesis—the idea to be tested
    c. prediction—how to test the hypothesis
    d. conclusion—what the data tell us
    e. All of these are properly matched.

Understanding the Terms

adaptation 5

animal 7

Archaea 7

Bacteria 7

behavior 4

biodiversity 9

biology 10

biosphere 7

cell 3

class 6

community 7

conclusion 11

control 10

data 11

deductive reasoning 10

development 5

domain 6

ecosystem 8

energy 3

Eukarya 7

eukaryote 7

evolution 5

experimental design 10

experimental variable 11

family 6

fungi 7

gene 4

genus 6

homeostasis 4

hypothesis 10

inductive reasoning 10

kingdom 6

law 11

metabolism 4

model 10

natural selection 5

observation 10

order 6

organ 3

organism 3

organ system 3

phenomenon 10

phylum 6

plant 7

population 7

prediction 10

principle 11

prokaryote 7

protist 7

reproduce 4

response variable 11

scientific theory 11

species 5, 6

systematics 6

taxonomy 7

techology 14

tissue 3

Match the terms to these definitions:

a._____ Scientific theory that is generally accepted by an overwhelming number of scientists.

b._____ Statement that is capable of explaining present observations and will be tested by further experimentation and observations.

c._____ Capacity to do work and bring about change; occurs in a variety of forms.

d._____ Suitability of an organism for its environment, enabling it to survive and produce offspring.

e._____ Maintenance of the internal environment of an organism within narrow limits.

Thinking Critically

1. How can evolution explain both the unity and diversity of life?

2. Viruses are infectious particles. Viruses can reproduce, but only inside a living (host) cell. Outside a living cell, some viruses can be stored in the environment, while others cannot survive. Viruses do contain genetic material and can evolve, thereby evading destruction by the host's immune system. Should a virus be considered alive? Explain the criteria upon which you base your answer.

3. You are a scientist working at a pharmaceutical company and have developed a new cancer medication that has the potential for use in humans. Outline a series of experiments, including the use of a model, to test whether the cancer medication works.

4. One of the biggest bioethical issues we face in the twenty-first century is the loss of biodiversity, much of which is due to human activities, such as development and agriculture. These activities benefit human society by providing shelter, food, and other necessities of life, and yet they also threaten many of the world's species with extinction. Given the benefits of development to humans, why should we conserve biodiversity?

5. Technology can be considered a "double-edged sword." That is, many technological inventions, such as medicines, have provided great benefits to society. Other technological innovations, such as pesticides and fertilizers, can be considered both advances, because they increase crop yields, and drawbacks, because they can have adverse effects on the health of humans and wildlife. Using the Internet or other resources, come up with your own example of a technological invention that has both advantages and drawbacks. Discuss the pros and cons of this invention.

ARIS, the *Inquiry into Life* Website

ARIS, the website for *Inquiry into Life*, provides a wealth of information organized and integrated by chapter. You will find practice quizzes, interactive activities, labeling exercises, flashcards, and much more that will complement your learning and understanding of general biology.

www.aris.mhhe.com

Cell Biology

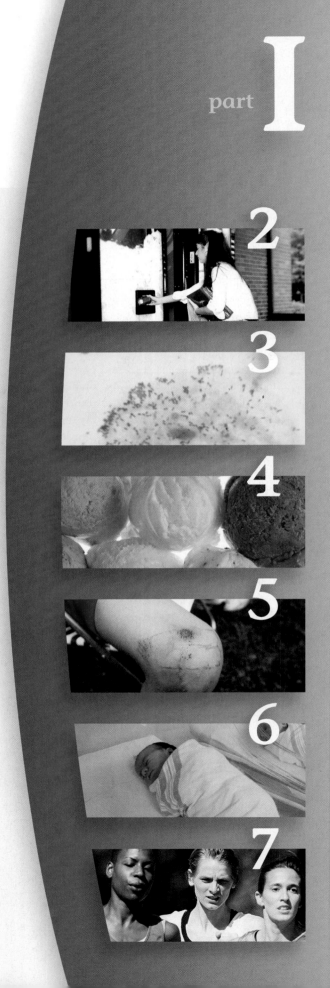

questions on page 42

chapter **2**

The Molecules of Cells

Chapter Concepts

Ashley bought her second can of diet cola. *These 8 A.M. classes were a killer. She probably should have gone to bed before 3 A.M., but she had been studying for a test. Most of her friends drank coffee, but Ashley preferred a nice, cold diet cola for her caffeine in the morning. As she gathered her books, she thought about also grabbing something to eat, but she really didn't have time, so she just ran to class. By the end of that class, Ashley's stomach was burning. The pit of her stomach felt like it was on fire. This was becoming a common feeling, so she reached for the antacids she kept in her purse and chewed them on her way to the next class.*

Colas are acidic, similar in strength to vinegar (although they taste a lot better!). By drinking large amounts of acid on an empty stomach, Ashley was setting herself up for stomach problems. Antacids are actually bases, designed to counter the effects of acids (ant-, against) by neutralizing them. Other foods we eat, including tomatoes and citrus fruits, are also very acidic. If consumed in moderation with other foods, these acidic foods should not cause stomach problems. Antacids are fine to use occasionally, after a spicy meal. However, if Ashley regularly has burning in her stomach, she should see her physician.

2.1 Basic Chemistry

Turn the page, throw a ball, pat your dog, rake leaves; everything that we touch—from the water we drink to the air we breathe—is composed of matter. **Matter** refers to anything that takes up space and has mass. Although matter has many diverse forms—anything from molten lava to kidney stones—it only exists in three distinct states: solid, liquid, and gas.

All matter, both nonliving and living, is composed of certain basic substances called **elements.** An element is a substance that cannot be broken down to simpler substances with different properties by ordinary chemical means. (A property is a physical or chemical characteristic, such as density, solubility, melting point, and reactivity.) It is quite remarkable that only 92 naturally occurring elements serve as the building blocks of all matter. Other elements have been "human-made" and are not biologically important.

Earth's crust as well as all organisms are composed of elements, but they differ as to which ones are predominant (Fig. 2.1). Only six elements—carbon, hydrogen, nitrogen, oxygen, phosphorus, and sulfur—are basic to life and make up about 95% of the body weight of organisms. The acronym **CHNOPS** helps us remember these six elements. The properties of these elements are essential to the uniqueness of cells and organisms. Other elements are also important to living things, including sodium, potassium, calcium, iron, and magnesium.

All living and nonliving things are matter composed of elements. Six elements (CHNOPS) in particular are basic to life.

Atomic Structure

In the early 1800s, the English scientist John Dalton championed the atomic theory, which says that elements consist of tiny particles called **atoms.** An atom is the smallest part of an element that displays the properties of the element. An element and its atoms share the same name. One or two letters create the atomic symbol, which stands for this name. For example, the symbol H means a hydrogen atom, the symbol Cl stands for chlorine, and the symbol Na (for *natrium* in Latin) is used for a sodium atom.

Physicists have identified a number of subatomic particles that make up atoms. The three best-known subatomic particles are positively charged **protons,** uncharged **neutrons,** and negatively charged **electrons.** Protons and neutrons are located within the nucleus of an atom, and electrons move about the nucleus. Figure 2.2 shows the arrangement of the subatomic particles in a helium atom, which has only two electrons. In Figure 2.2a, the stippling shows the probable location of electrons, and in Figure 2.2b, the circle represents an electron shell, the average location of electrons.

Figure 2.1 Elements that make up Earth's crust and its organisms.
Scarlet and red-blue-green macaws gather on a salt lick in South America. The graph inset shows that Earth's crust primarily contains the elements silicon (Si), aluminum (Al), and oxygen (O). Organisms primarily contain the elements oxygen, nitrogen (N), carbon (C), and hydrogen (H). Along with sulfur (S) and phosphorus (P), these elements make up biological molecules.

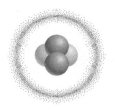
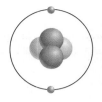
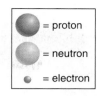

a.

b.

| = proton |
| = neutron |
| = electron |

Subatomic Particles			
Particle	Electric Charge	Atomic Mass	Location
Proton	+1	1	Nucleus
Neutron	0	1	Nucleus
Electron	−1	0	Electron shell

c.

Figure 2.2 Model of helium (He).

Atoms contain subatomic particles called protons, neutrons, and electrons. Protons and neutrons are within the nucleus, and electrons are outside the nucleus. **a.** The stippling shows the probable location of the electrons in the helium atom. **b.** A circle termed an electron shell sometimes represents the average location of an electron. **c.** The electric charge and the atomic mass units of the subatomic particles vary as shown.

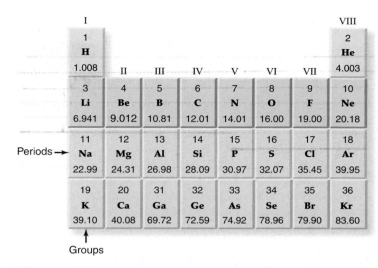

Periods →

Groups

Figure 2.3 A portion of the periodic table.

In the periodic table, the elements, and therefore atoms, are arranged in the order of their atomic numbers but placed in groups (vertical columns) and periods(horizontal rows). All the atoms in a particular group have certain chemical characteristics in common. These four periods contain the elements that are most important in biology; the complete periodic table is in Appendix D.

The concept of an atom has changed greatly since Dalton's day. If an atom could be drawn the size of a football field, the nucleus would be like a gumball in the center of the field, and the electrons would be tiny specks whirling about in the upper stands. Most of an atom is empty space. We should also realize that we can only indicate where the electrons are expected to be most of the time. In our analogy, the electrons might very well stray outside the stadium at times.

All atoms of an element have the same number of protons. This is called the **atomic number.** The number of protons housed in the nucleus makes each atom unique. The atomic number is often written as a subscript to the lower left of the atomic symbol.

Each atom has its own specific mass. The **atomic mass,** or mass number, of an atom is essentially the sum of its protons and neutrons. Protons and neutrons are assigned one atomic mass unit each. Electrons are so small that their mass is considered zero in most calculations (Fig. 2.2c). The term *atomic mass* is used, rather than *atomic weight,* because mass is constant while weight changes according to the gravitational force of a body. The gravitational force of Earth is greater than that of the moon; therefore, substances weigh less on the moon even though their mass has not changed. The atomic mass is often written as a superscript to the upper left of the atomic symbol. For example, the carbon atom can be noted in this way:

atomic mass———————12
atomic number———————6

$\mathbf{C}$ ————atomic symbol

The Periodic Table

Once chemists discovered a number of the elements, they began to realize that even though each element consists of a different atom, certain chemical and physical characteristics recur. The periodic table was constructed as a way to group the elements, and therefore atoms, according to these characteristics. Notice in Figure 2.3 that the periodic table is arranged according to increasing atomic number. The vertical columns in the table are groups; the horizontal rows are periods, which cause each atom to be in a particular group. For example, all the atoms in group VII react with one atom at a time, for reasons we will soon explore. The atoms in group VIII are called the noble gases because they are inert and rarely react with another atom. Notice that helium and krypton are noble gases.

In Figure 2.3 the atomic number is above the atomic symbol and the atomic mass is below the atomic symbol. The atomic number tells you the number of positively charged protons, and also the number of negatively charged electrons if the atom is electrically neutral. To determine the number of neutrons, subtract the number of protons from the atomic mass, and take the closest whole number.

Atoms have an atomic symbol, an atomic number, and an atomic mass. The subatomic particles (protons, neutrons, and electrons) determine the characteristics of atoms.

Isotopes

Isotopes are atoms of the same element that differ in their number of neutrons. Isotopes have the same number of protons, but they have different atomic masses. Because the number of protons gives an atom its identity, changing the number of neutrons affects the atomic mass but not the name of the atom. For example, the element carbon has three common isotopes:

$$^{12}_{6}\text{C} \qquad ^{13}_{6}\text{C} \qquad ^{14}_{6}\text{C*}$$

*radioactive

Carbon 12 has six neutrons, carbon 13 has seven neutrons, and carbon 14 has eight neutrons. Unlike the other two isotopes of carbon, carbon 14 is unstable; it changes over time into nitrogen 14, which is a stable isotope of the element nitrogen. As carbon 14 decays, it releases various types of energy in the form of rays and subatomic particles, and therefore it is a **radioactive isotope.** Today, biologists use radiation to date objects, create images, and trace the movement of substances.

Low Levels of Radiation

The chemical behavior of a radioactive isotope is essentially the same as that of the stable isotopes of an element. This means that you can put a small amount of radioactive isotope in a sample and it becomes a tracer by which to detect molecular changes.

The importance of chemistry to medicine is nowhere more evident than in the many medical uses of radioactive isotopes. Specific tracers are used in imaging the body's organs and tissues. For example, after a patient drinks a solution containing a minute amount of ^{131}I, it becomes concentrated in the thyroid—the only organ to take it up. A subsequent image of the thyroid indicates whether it is healthy in structure and function (Fig. 2.4a). Positron emission tomography (PET) is a way to determine the comparative activity of tissues. Radioactively labeled glucose, which emits a subatomic particle known as a positron, is injected into the body. The radiation given off is detected by sensors and analyzed by a computer. The result is a color image that shows which tissues took up glucose and are metabolically active (Fig. 2.4b). A PET scan of the brain can help diagnose a brain tumor, Alzheimer disease, epilepsy, or whether a stroke has occurred.

High Levels of Radiation

Radioactive substances in the environment can harm cells, damage DNA, and cause cancer. When Marie Curie was studying radiation, its harmful effects were not known, and she and many of her co-workers developed cancer. The

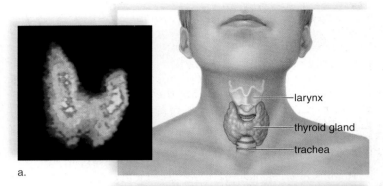

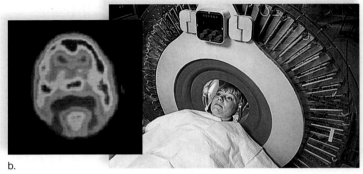

larynx

thyroid gland

trachea

a.

b.

Figure 2.4 Low levels of radiation.
a. The missing area in this thyroid scan (*upper left*) indicates the presence of a tumor that does not take up the radioactive iodine. **b.** A PET (positron emission tomography) scan reveals which portions of the brain are most active (yellow and red colors).

release of radioactive particles following a nuclear power plant accident can have far-reaching and long-lasting effects on human health. However, the harmful effects of radiation can also be put to good use (Fig. 2.5). Radiation from radio-

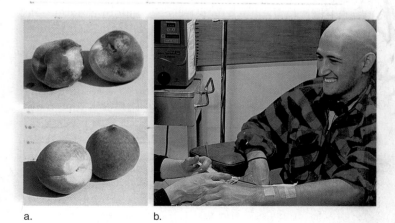

a.   b.

Figure 2.5 High levels of radiation.
a. Radiation kills bacteria and fungi. Irradiated peaches (*bottom*) spoil less quickly and can be kept for a longer length of time. **b.** Physicians use radiation therapy to kill cancer cells.

active isotopes has been used for many years to sterilize medical and dental products. Now it can be used to sterilize the U.S. mail and other packages to free them of possible pathogens, such as anthrax spores. The ability of radiation to kill cells is often applied to cancer cells. Targeted radio-isotopes can be introduced into the body so that the sub-atomic particles emitted destroy only cancer cells, with little risk to the rest of the body.

> Isotopes of an element have the same atomic number but differ in mass due to a different number of neutrons. Radioactive isotopes, which emit radiation, have the potential to do harm but also have many beneficial uses.

Electrons   (+ and −) = O charge.

In an electrically neutral atom, the positive charges of the protons in the nucleus are balanced by the negative charges of electrons moving about the nucleus. Various models in years past have attempted to illustrate the precise location of elec-trons. Figure 2.6 uses the Bohr model, which is named after the physicist Niels Bohr. The Bohr model is useful, but today's physicists tell us it is not possible to determine the precise location of any individual electron at any given moment.

In the diagrams in Figure 2.6, the energy levels (also termed electron shells) are drawn as concentric rings about the nucleus. For atoms up through number 20 (i.e.,

calcium), the first shell (closest to the nucleus) can contain two electrons; thereafter, each additional shell can contain eight electrons. For these atoms, each lower level is filled with electrons before the next higher level contains any electrons.

The sulfur atom, with an atomic number of 16, has two electrons in the first shell, eight electrons in the second shell, and six electrons in the third, or outer, shell. Revisit the periodic table (see Fig. 2.3), and note that sulfur is in the third period. In other words, the horizontal row tells you how many shells an atom has. Also note that sulfur is in group VI. The group tells you how many electrons an atom has in its outer shell.

If an atom has only one shell, the outer shell is com-plete when it has two electrons. Otherwise, atomic shells follow the octet rule, which states that the outer shell is most stable when it has eight electrons. As mentioned previously, atoms in group VIII of the periodic table are called the noble gases because they do not ordinarily react. Atoms with fewer than eight electrons in the outer shell react with other atoms in such a way that after the reaction, each has a stable outer shell. Atoms can give up, accept, or share electrons in order to have eight electrons in the outer shell.

> The number of electrons in the outer shell determines whether an atom reacts with other atoms.

Figure 2.6 Bohr models of atoms.
Electrons orbit the nucleus at particular energy levels (electron shells): the first shell contains up to two electrons, and each shell thereafter can contain up to eight electrons as long as we consider only atoms with an atomic number of 20 or below. Each shell is filled before electrons are placed in the next shell. Why does carbon have only two shells while phosphorus and sulfur have three shells?

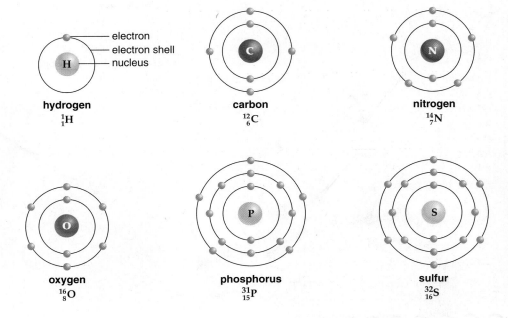

2.2   Molecules and Compounds

Atoms, except for noble gases, routinely bond with one another. A **molecule** is formed when two or more atoms bond together. For example, oxygen does not exist in nature as a single atom, O; instead, two oxygen atoms are joined to form a molecule of oxygen, O_2. When atoms of two or more different elements bond together, the product is called a **compound.** Water (H_2O) is a compound that contains atoms of hydrogen and oxygen. We can also speak of molecules of water because a molecule is the smallest part of a compound that still has the properties of that compound.

Electrons possess energy, and the bonds that exist between atoms also contain energy. Organisms are directly dependent on chemical-bond energy to maintain their organization. When a chemical reaction occurs, electrons shift in their relationship to one another, and energy may be given off or absorbed. This same energy is used to carry on our daily lives.

Ionic Bonding

Ions form when electrons are transferred from one atom to another. For example, sodium (Na), with only one electron in its third shell, tends to be an electron donor (Fig. 2.7a). Once it gives up this electron, the second shell, with eight electrons, becomes its outer shell. Chlorine (Cl), on the other hand, tends to be an electron acceptor. Its outer shell has seven electrons, so if it acquires only one more electron, it has a completed outer shell. When a sodium atom and a chlorine atom come together, an electron is transferred from the sodium atom to the chlorine atom. Now both atoms have eight electrons in their outer shells.

This electron transfer, however, causes a charge imbalance in each atom. The sodium atom has one more proton than it has electrons; therefore, it has a net charge of +1 (symbolized by Na^+). The chlorine atom has one more electron than it has protons; therefore, it has a net charge of –1 (symbolized by Cl^-). Such charged particles are called **ions.** Sodium (Na^+) and chloride (Cl^-) are not the only biologically important ions. Some, such as potassium (K^+), are formed by the transfer of a single electron to another atom; others, such as calcium (Ca^{2+}) and magnesium (Mg^{2+}), are formed by the transfer of two electrons.

Ionic compounds are held together by an attraction between negatively and positively charged ions called an **ionic bond.** When sodium reacts with chlorine, an ionic compound called sodium chloride (NaCl) results. Sodium chloride is a salt, commonly known as table salt because it is used to season our food (Fig. 2.7b). Salts can exist as a dry solid, but when salts are placed in water, they release ions as they dissolve. NaCl separates into Na^+ and Cl^-. Ionic compounds are most commonly found in this dissociated (ionized) form in biological systems because these systems are 70–90% water.

The transfer of electron(s) between atoms results in ions that are held together by an ionic bond, the attraction of negative and positive charges.

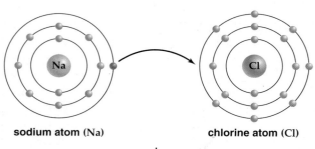

sodium atom (Na)   chlorine atom (Cl)

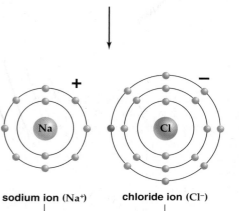

+   −

sodium ion (Na^+)   chloride ion (Cl^-)

sodium chloride (NaCl)

a.

Figure 2.7 Formation of sodium chloride (table salt).

a. During the formation of sodium chloride, an electron is transferred from the sodium atom to the chlorine atom. At the completion of the reaction, each atom has eight electrons in the outer shell, but each also carries a charge as shown. **b.** In a sodium chloride crystal, ionic bonding between Na^+ and Cl^- causes the atoms to assume a three-dimensional lattice in which each sodium ion is surrounded by six chloride ions, and each chloride ion is surrounded by six sodium ions. The result is crystals of salt as in table salt.

Na^+ Cl^-

b.

Covalent Bonding

A **covalent bond** results when two atoms share electrons in such a way that each atom has an octet of electrons in the outer shell. In a hydrogen atom, the outer shell is complete when it contains two electrons. If hydrogen is in the presence of a strong electron acceptor, it gives up its electron to become a hydrogen ion (H^+). But if this is not possible, hydrogen can share with another atom and thereby have a completed outer shell. For example, one hydrogen atom will share with another hydrogen atom. Their two orbitals overlap, and the electrons are shared between them (Fig. 2.8a). Because they share the electron pair, each atom has a completed outer shell.

A more common way to symbolize that atoms are sharing electrons is to draw a line between the two atoms, as in the structural formula H—H. In a molecular formula, the line is omitted and the molecule is simply written as H_2.

Sometimes, atoms share more than one pair of electrons to complete their octets. A double covalent bond occurs when two atoms share two pairs of electrons (Fig. 2.8b). To show that oxygen gas (O_2) contains a double bond, the molecule can be written as O=O.

It is also possible for atoms to form triple covalent bonds, as in nitrogen gas (N_2), which can be written as N≡N. Single covalent bonds between atoms are quite strong, but double and triple bonds are even stronger.

Shape of Molecules

Structural formulas make it seem as if molecules are one-dimensional, but actually molecules have a three-dimensional shape that often determines their biological function. Molecules consisting of only two atoms are always linear, but a molecule such as methane with five atoms (Fig. 2.8c) has a tetrahedral shape. Why? Because, as shown in the ball-and-stick model, each bond is pointing to the corners of a tetrahedron (Fig. 2.8d, left). The space-filling model comes closest to the actual shape of the molecule. In space-filling models, each type of atom is given a particular color—carbon is always black and hydrogen is always off-white (Fig. 2.8d, right).

The shapes of molecules are necessary to the structural and functional roles they play in living things. For example, hormones have specific shapes that allow them to be recognized by the cells in the body. Antibodies combine with disease-causing agents, like a key fits a lock, to keep us well. Similarly, homeostasis is maintained only when enzymes have the proper shape to carry out their particular reactions in cells.

> In a "covalently bonded molecule" atoms share electrons; the final shape of the molecule often determines the role it plays in cells and organisms.

Electron Model	Structural Formula	Molecular Formula
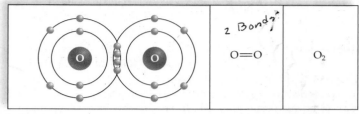 a. Hydrogen gas (one Bond)	H—H	H_2
b. Oxygen gas (2 Bonds)	O=O	O_2
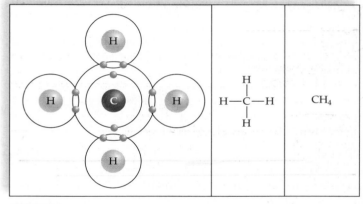 c. Methane	H—C—H (with H above and below)	CH_4

Ball-and-stick Model	Space-filling Model
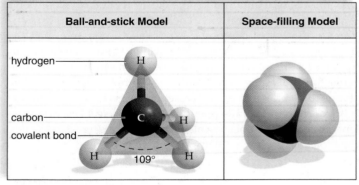 hydrogen — H; carbon — C; covalent bond; 109°	

d. Methane—continued

Figure 2.8 Covalently bonded molecules.

In a covalent bond, atoms share electrons, allowing each atom to have a completed outer shell. **a.** A molecule of hydrogen (H_2) contains two hydrogen atoms sharing a pair of electrons. This single covalent bond can be represented in any of these three ways. **b.** A molecule of oxygen (O_2) contains two oxygen atoms sharing two pairs of electrons. This results in a double covalent bond. **c.** A molecule of methane (CH_4) contains one carbon atom bonded to four hydrogen atoms. **d.** When carbon binds to four other atoms, as in methane, each bond actually points to one corner of a tetrahedron. Ball-and-stick models and space-filling models are three-dimensional representations of a molecule—in this case, methane.

Electron Model	Ball-and-stick Model	Space-filling Model
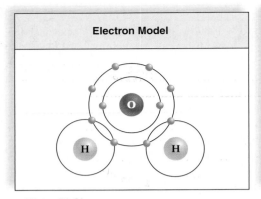	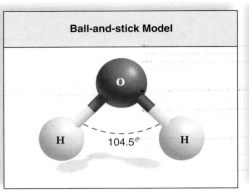	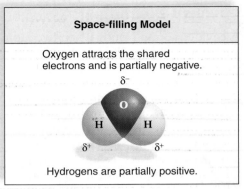

a. Water (H_2O)

Figure 2.9 Water molecule.

a. Three models for the structure of water. The electron model does not indicate the shape of the molecule. The ball-and-stick model shows that the two bonds in a water molecule are angled at 104.5°. The space-filling model also shows the V shape of a water molecule. **b.** Hydrogen bonding between water molecules. A hydrogen bond is the attraction of a slightly positive hydrogen to a slightly negative atom in the vicinity. Each water molecule can hydrogen-bond to four other molecules in this manner. When water is in its liquid state, some hydrogen bonds are forming and others are breaking at all times.

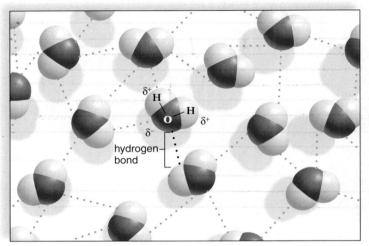

b. Hydrogen bonding between water molecules

Nonpolar and Polar Covalent Bonds

When the sharing of electrons between two atoms is fairly equal, the covalent bond is said to be a nonpolar covalent bond. All the molecules in Figure 2.8, including methane (CH_4), are nonpolar. In the case of water (H_2O), however, the sharing of electrons between oxygen and each hydrogen is not completely equal. The attraction of an atom for the electrons in a covalent bond is called its electronegativity. The larger oxygen atom, with the greater number of protons, is more electronegative than the hydrogen atom. The oxygen atom can attract the electron pair to a greater extent than each hydrogen atom can. In a water molecule, this causes the oxygen atom to assume a slightly negative charge (δ^-), and it causes the hydrogen atoms to assume slightly positive charges (δ^+). The unequal sharing of electrons in a covalent bond creates a polar covalent bond, and in the case of water, the molecule itself is a polar molecule (Fig. 2.9a).

> The water molecule is a polar molecule with an asymmetrical distribution of charge: one end of the molecule (the oxygen atom) carries a slightly negative charge, and the other ends of the molecule (the hydrogen atoms) carry slightly positive charges.

Hydrogen Bonding

Polarity within a water molecule causes the hydrogen atoms in one molecule to be attracted to the oxygen atoms in other water molecules (Fig. 2.9b). This attraction, although weaker than an ionic or covalent bond, is called a **hydrogen bond.** Because a hydrogen bond is easily broken, it is often represented by a dotted line. Hydrogen bonding is not unique to water. Many biological molecules have polar covalent bonds involving an electropositive hydrogen and usually an electronegative oxygen or nitrogen. In these instances, a hydrogen bond can occur within the same molecule or between different molecules.

Although a hydrogen bond is more easily broken than a covalent bond, many hydrogen bonds taken together are quite strong. Hydrogen bonds between cellular molecules help maintain their proper structure and function. For example, hydrogen bonds hold the two strands of DNA together. When DNA makes a copy of itself, each hydrogen bond easily breaks, allowing the DNA to unzip. On the other hand, the hydrogen bonds acting together add stability to the DNA molecule. As we shall see, many of the important properties of water are the result of hydrogen bonding.

> A hydrogen bond occurs between a slightly positive hydrogen atom of one molecule and a slightly negative atom of another molecule, or between atoms of the same molecule.

<table>
</table>

2.3	## Chemistry of Water

The first cell(s) evolved in water, and all living things are 70–90% water. Water is a polar molecule, and water molecules are hydrogen-bonded to one another (see Fig. 2.9b). Due to hydrogen bonding, water molecules cling together. Without hydrogen bonding between molecules, water would melt at –100°C and boil at –91°C, making most of the water on Earth steam, and life unlikely. But because of hydrogen bonding, water is a liquid at temperatures typically found on Earth's surface. It melts at 0°C and boils at 100°C. These and other unique properties of water make it essential to the existence of life.

Properties of Water

Water has a high heat capacity. A **calorie** is the amount of heat energy needed to raise the temperature of 1 gram (g) of water 1°C. In comparison, other covalently bonded liquids require input of only about half this amount of energy to rise in temperature 1°C. The many hydrogen bonds that link water molecules help water absorb heat without a great change in temperature.

Converting 1 g of the coldest liquid water to ice requires the loss of 80 calories of heat energy (Fig. 2.10a). Water holds onto its heat, and its temperature falls more slowly than that of other liquids. This property of water is important not only for aquatic organisms but also for all living things. Because the temperature of water rises and falls slowly, organisms are better able to maintain their normal internal temperatures and are protected from rapid temperature changes.

Water has a high heat of vaporization. Converting 1 g of the hottest water to a gas requires an input of 540 calories of heat energy. Water has a high heat of vaporization because hydrogen bonds must be broken before water boils and water molecules vaporize—that is, evaporate into the environment. Water's high heat of vaporization gives animals in a hot environment an efficient way to release excess body heat. When an animal sweats, or gets splashed, body heat is used to vaporize the water, thus cooling the animal (Fig. 2.10b).

Because of water's high heat capacity and high heat of vaporization, temperatures along coasts are moderate. During the summer, the ocean absorbs and stores solar heat, and during the winter, the ocean releases it slowly. In contrast, the interior regions of continents can experience severe changes in temperature.

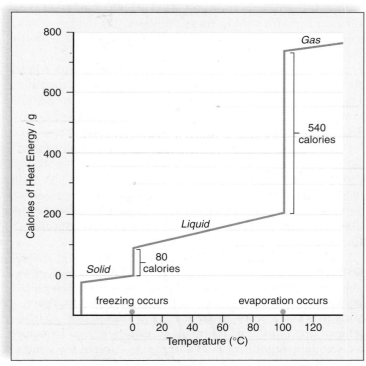

a. Calories lost when 1 g of liquid water freezes and calories required when 1 g of liquid water evaporates.

b. Bodies of organisms cool when their heat is used to evaporate water.

Figure 2.10 Temperature and water.

a. Water can be a solid, a liquid, or a gas at naturally occurring environmental temperatures. At room temperature and pressure, water is a liquid. When water freezes and becomes a solid (ice), it gives off heat (80 calories), and this heat can help keep the environmental temperature higher than expected. On the other hand, when water vaporizes, it takes up a large amount of heat (540 calories) as it changes from a liquid to a gas. **b.** This means that splashing water on the body will help keep body temperature within a normal range. Can you also see why water's properties help keep temperatures at the east and west coasts of the United States moderate in both winter and summer?

Water is a solvent. Due to its polarity, water facilitates chemical reactions, both outside and within living systems. It dissolves a great number of substances. A solution contains dissolved substances, which are then called **solutes.** When ionic salts—for example, sodium chloride (NaCl)—are put into water, the negative ends of the water molecules are attracted to the sodium ions, and the positive ends of the water molecules are attracted to the chloride ions. This causes the sodium ions and the chloride ions to separate, or dissociate, in water:

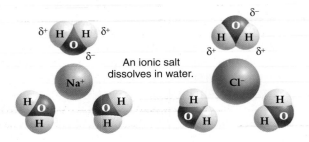

An ionic salt dissolves in water.

Water is also a solvent for larger molecules that contain ionized atoms or are polar molecules.

Those molecules that can attract water are said to be **hydrophilic.** When ions and molecules disperse in water, they move about and collide, allowing reactions to occur. Nonionized and nonpolar molecules, such as oil, that cannot attract water are said to be **hydrophobic.**

Water molecules are cohesive and adhesive. Cohesion is apparent because water flows freely, and yet water molecules do not separate from each other. They cling together because of hydrogen bonding. Water exhibits adhesion because its positive and negative poles allow it to adhere to polar surfaces. Cohesion and adhesion allow water to fill a tubular vessel. Therefore, water is an excellent transport system, both outside of and within living organisms. Unicellular organisms rely on external water to transport nutrient and waste molecules, but multicellular organisms often contain internal vessels through which water transports nutrients and wastes. For example, the liquid portion of our blood, which transports dissolved and suspended substances throughout the body, is 90% water.

Cohesion and adhesion also contribute to the transport of water in plants. The roots of plants are anchored in the soil, where they absorb water, but the leaves are uplifted and exposed to solar energy. How is it possible for water to rise to the top of even very tall trees? A plant contains a system of vessels that reaches from the roots to the leaves. Water evaporating from the leaves is immediately replaced with water molecules from the vessels. Because water molecules are cohesive, a tension is created that pulls a water column up from the roots. Adhesion of water to the walls of the vessels also helps prevent the water column from breaking apart.

Water has a high surface tension. The stronger the force between molecules in a liquid, the greater the surface tension. As with cohesion, hydrogen bonding causes water

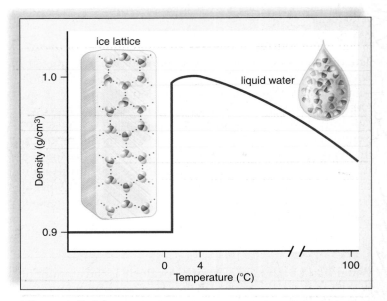

Figure 2.11 Density of water versus temperature.
Remarkably, water is more dense at 4°C than at 0°C. Most substances contract when they solidify. But water expands when it freezes because the water molecules in ice form a lattice in which the hydrogen bonds are farther apart than in liquid water.

to have a high surface tension. This property makes it possible for humans to skip rocks on water. The water strider, a common insect, can even walk on top of a pond without breaking the surface.

Frozen water (ice) is less dense than liquid water. As liquid water cools, the molecules come closer together. They are densest at 4°C, but they are still moving about, bumping into each other (Fig. 2.11). At temperatures below 4°C, including at 0°C when water is frozen, the water forms a regular crystal lattice that is rigid and more open since the water molecules are no longer moving about. For this reason water expands as it freezes, which is why cans of soda burst when placed in a freezer or why frost heaves make northern roads bumpy in the winter. It also means that ice is less dense than liquid water, and therefore ice floats on liquid water.

If ice did not float on water, it would sink, and ponds, lakes, and perhaps even the ocean would freeze solid, making life impossible in the water and also on land. Instead, bodies of water always freeze from the top down. When a body of water freezes on the surface, the ice acts as an insulator to prevent the water below it from freezing. This protects aquatic organisms so that they can survive the winter. As ice melts in the spring, it draws heat from the environment, helping to prevent a sudden change in temperature that might be harmful to life.

Water has unique properties that allow cellular activities to occur and make life on Earth possible.

Acids and Bases

When water ionizes, it releases an equal number of hydrogen ions (H^+) and hydroxide ions (OH^-):

$$H—O—H \rightleftharpoons H^+ + OH^-$$
water   hydrogen   hydroxide
ion   ion

Only a few water molecules at a time dissociate, and the actual number of H^+ and OH^- is very small (1×10^{-7} moles/liter).[1]

Acidic Solutions (High H^+ Concentrations)

Lemon juice, vinegar, tomatoes, and coffee are all acidic solutions. What do they have in common? **Acids** are substances that dissociate in water, releasing hydrogen ions (H^+).[2] For example, hydrochloric acid (HCl) is an important inorganic acid that dissociates in this manner:

$$HCl \rightarrow H^+ + Cl^-$$

Dissociation is almost complete; therefore, HCl is called a strong acid. If hydrochloric acid is added to a beaker of water, the number of hydrogen ions (H^+) increases greatly.

Basic Solutions (Low H^+ Concentration)

Baking soda and antacids are common basic solutions familiar to most people. **Bases** are substances that either take up hydrogen ions (H^+) or release hydroxide ions (OH^-). For example, sodium hydroxide (NaOH) is an important inorganic base that dissociates in this manner:

$$NaOH \rightarrow Na^+ + OH^-$$

Dissociation is almost complete; therefore, sodium hydroxide is called a strong base. If sodium hydroxide is added to a beaker of water, the number of hydroxide ions increases.

pH Scale

The **pH scale** is used to indicate the acidity or basicity (alkalinity) of solutions.[3] The pH scale (Fig. 2.12) ranges from 0 to 14. A pH of 7 represents a neutral state in which the hydrogen ion and hydroxide ion concentrations are equal. A pH below 7 is an acidic solution because the hydrogen ion concentration [H^+] is greater than the hydroxide concentration [OH^-]. A pH above 7 is basic because [OH^-] is greater than [H^+]. Further, as we move down the pH scale from pH 14 to pH 0, each unit has ten times the [H^+] of the previous

---

[1] In chemistry, a mole is defined as the amount of matter that contains as many objects (atoms, molecules, ions) as the number of atoms in exactly 12 g of ^{12}C.

[2] A hydrogen atom contains one electron and one proton. A hydrogen ion has only one proton, so it is often simply called a proton.

[3] pH is defined as the negative log of the hydrogen ion concentration [H^+]. A log is the power to which ten must be raised to produce a given number.

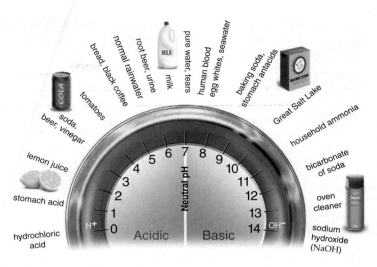

Figure 2.12 The pH scale.
The dial of this pH meter indicates that pH ranges from 0 to 14, with 0 the most acidic and 14 the most basic. pH 7 (neutral pH) has equal amounts of hydrogen ions (H^+) and hydroxide ions (OH^-). An acidic pH has more H^+ than OH^-, and a basic pH has more OH^- than H^+.

unit. As we move up the scale from 0 to 14, each unit has ten times the [OH^-] of the previous unit.

The pH scale was devised to eliminate the use of cumbersome numbers. For example, the possible hydrogen ion concentrations of a solution are on the left in the following listing, and the pH is on the right:

[H^+] (moles per liter)		pH
0.000001	$= 1 \times 10^{-6}$	6
0.0000001	$= 1 \times 10^{-7}$	7
0.00000001	$= 1 \times 10^{-8}$	8

To further illustrate the relationship between hydrogen ion concentration and pH, consider the following question. Which of the pH values listed indicates a higher hydrogen ion concentration [H^+] than pH 7, and therefore would be an acidic solution? A number with a smaller negative exponent indicates a greater quantity of hydrogen ions than one with a larger negative exponent. Therefore, pH 6 is an acidic solution.

The Ecology Focus on page 30 describes detrimental environmental consequences to nonliving and living things as rain and snow have become more acidic. In humans, pH needs to be maintained within a narrow range or there are health consequences.

Buffers and pH

A **buffer** is a chemical or a combination of chemicals that keeps pH within normal limits. Many commercial products, such as Bufferin®, shampoos, or deodorants, are buffered as an added incentive for us to buy them. Buffers resist pH changes because they can take up excess hydrogen ions (H^+) or hydroxide ions (OH^-).

Normally, rainwater has a pH of about 5.6 because the carbon dioxide in the air combines with water to produce a weak solution of carbonic acid. Rain falling in the northeastern United States and southeastern Canada now has a pH of between 5 and 4. To comprehend the increase in acidity this represents, we have to remember that a pH of 4 is ten times more acidic than a pH of 5.

Strong evidence indicates that this observed increase in rainwater acidity is a result of the burning of fossil fuels such as coal, oil, and gasoline derived from oil. When fossil fuels are burned, sulfur dioxide and nitrogen oxides are produced, and they combine with water vapor in the atmosphere to form sulfuric and nitric acids. These acids return to Earth dissolved in rain or snow, a process properly called wet deposition but more often called acid rain. During dry deposition, dry particles of sulfate and nitrate salts descend from the atmosphere.

Unfortunately, regulations that require the use of tall smokestacks to reduce local air pollution only cause pollutants to be carried farther from their place of origin. For example, acid deposition in southeastern Canada results from the burning of fossil fuels in factories and power plants in the midwestern United States. Acid deposition adversely affects lakes, particularly in areas where the soil is thin and lacks limestone (calcium carbonate, $CaCO_3$), a buffer to acid deposition. Acid deposition leaches aluminum from the soil, carries aluminum into the lakes, and converts mercury deposits in lake bottom sediments to soluble and toxic methyl mercury. Lakes not only become more acidic, but they also accumulate toxic substances. The increasing deterioration of thousands of lakes and rivers in southern Norway and Sweden during the past two decades has been attributed to acid deposition. Some lakes contain no fish, and others have decreasing numbers of fish. The same phenomenon has been observed in Canada and the United States (mostly in the Northeast and upper Midwest).

In forests, acid deposition weakens trees because it leaches away nutrients and releases aluminum (Fig. 2A*a*). By 1988, most spruce, fir, and other conifers atop North Carolina's Mt. Mitchell were dead from being bathed in ozone and acid fog for years. The soil was so acidic that new seedlings could not survive. Many countries in northern Europe have also reported woodland and forest damage, most likely due to acid deposition.

Lake and forest deterioration aren't the only effects of acid deposition. Reduction of agricultural yields, damage to marble and limestone monuments and buildings (Fig. 2A*b*), and even human illnesses have been reported.

Discussion Questions

1. What responsibility should the United States assume for damage caused by its pollution to Canada?
2. Describe the advantages and disadvantages of reducing our dependence on fossil fuels for energy.
3. Visit the U.S. Environmental Protection Agency web site (www.epa.gov) and determine what you can do individually to reduce acid rain.

a.   b.

Figure 2A Effects of acid deposition.
The burning of gasoline derived from oil, a fossil fuel, leads to acid deposition, which causes (*a*) trees to die and (*b*) statues and monuments to deteriorate.

In animals, the pH of body fluids is maintained within a narrow range, or else health suffers. The pH of our blood when we are healthy is always about 7.4—that is, just slightly basic (alkaline). If the blood pH drops to about 7, acidosis results. If the blood pH rises to about 7.8, alkalosis results. Both conditions can be life threatening. Normally, pH stability is possible because the body has built-in mechanisms to prevent pH changes. Buffers are the most important of these mechanisms. For example, carbonic acid (H_2CO_3) is a weak acid that minimally dissociates and then re-forms in the following manner:

$$\underset{\text{carbonic acid}}{H_2CO_3} \underset{\underset{\text{re-forms}}{\longleftarrow}}{\overset{\text{dissociates}}{\rightleftharpoons}} H^+ + \underset{\text{bicarbonate ion}}{HCO_3^-}$$

Blood always contains a combination of some carbonic acid and some bicarbonate ions. When hydrogen ions (H^+) are added to blood, the following reaction occurs:

$$H^+ + HCO_3^- \longrightarrow H_2CO_3$$

When hydroxide ions (OH^-) are added to blood, this reaction occurs:

$$OH^- + H_2CO_3 \longrightarrow HCO_3^- + H_2O$$

These reactions prevent any significant change in blood pH.

A pH value is the hydrogen ion concentration [H^+] of a solution. Buffers act to keep the pH within normal limits.

2.4   Organic Molecules

Inorganic molecules constitute nonliving matter, but even so, inorganic molecules such as salts (e.g., NaCl) and water play important roles in living things. The molecules of life, however, are organic molecules. **Organic molecules** always contain carbon (C) and hydrogen (H). The chemistry of carbon accounts for the formation of the very large variety of organic molecules found in living things. A carbon atom has four electrons in the outer shell. In order to achieve eight electrons in the outer shell, a carbon atom shares electrons covalently with as many as four other atoms, as in methane (CH_4). A carbon atom can share with another carbon atom, and in so doing, a long hydrocarbon chain can result:

$$
\begin{array}{c}
\;\;\;\;\;\text{H}\;\;\text{H}\;\;\text{H}\;\;\text{H}\;\;\text{H}\;\;\text{H}\;\;\text{H}\;\;\text{H} \\
\;\;\;\;\;|\;\;\;\;|\;\;\;\;|\;\;\;\;|\;\;\;\;|\;\;\;\;|\;\;\;\;|\;\;\;\;| \\
\text{H}-\text{C}-\text{C}-\text{C}-\text{C}-\text{C}-\text{C}-\text{C}-\text{C}-\text{H} \\
\;\;\;\;\;|\;\;\;\;|\;\;\;\;|\;\;\;\;|\;\;\;\;|\;\;\;\;|\;\;\;\;|\;\;\;\;| \\
\;\;\;\;\;\text{H}\;\;\text{H}\;\;\text{H}\;\;\text{H}\;\;\text{H}\;\;\text{H}\;\;\text{H}\;\;\text{H}
\end{array}
$$

As we shall see, the chain can turn back on itself to form a ring compound. So-called functional groups can be attached to carbon chains. A **functional group** is a particular cluster of atoms that always behaves in a certain way. One functional group of interest is the acidic (carboxyl) group —COOH because it can give up a hydrogen (H^+) and ionize to —COO⁻.

Many molecules of life are macromolecules—that is, they contain many molecules joined together. A **monomer** (*mono*, one) is a simple organic molecule that exists individually or can link with other monomers to form a **polymer** (*poly*, many). The polymers in cells form from monomers as follows:

Polymer	Monomer
carbohydrate (e.g., starch)	monosaccharide
protein	amino acid
nucleic acid	nucleotide

Aside from carbohydrates, proteins, and nucleic acids, the other organic molecules in cells are lipids. You are very familiar with carbohydrates, lipids, and proteins because certain foods are known to be rich in these molecules, as illustrated in Figure 2.13. The nucleic acid DNA makes up our genes, which are hereditary units that control our cells and the structure of our bodies.

Cells have a common way of joining monomers to build polymers. During a **dehydration reaction,** an —OH (hydroxyl group) and an —H (hydrogen atom), the equivalent of a water molecule, are removed as the reaction proceeds (Fig. 2.14*a*). To degrade polymers, the cell uses a **hydrolysis reaction,** in which the components of water are added (Fig. 2.14*b*).

Figure 2.13 Common foods.
Carbohydrates, such as bread, potatoes and pasta, are digested to sugars; lipids, such as oils and butter are digested to glycerol and fatty acids; and proteins, such as meat, are digested to amino acids. Cells use these subunit molecules to build their own macromolecules.

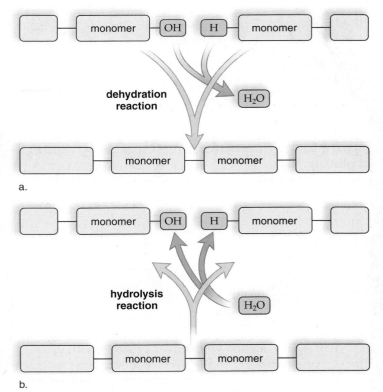

Figure 2.14 Synthesis and degradation of polymers.
a. In cells, synthesis often occurs when monomers bond during a dehydration reaction (removal of H_2O). **b.** Degradation occurs when the monomers in a polymer separate during a hydrolysis reaction (addition of H_2O).

2.5   Carbohydrates

Carbohydrates first and foremost function for quick fuel and short-term energy storage in all organisms, including humans. Carbohydrates play a structural role in woody plants, bacteria, and animals such as insects. In addition, carbohydrates on cell surfaces are involved in cell-to-cell recognition, as we learn in Chapter 3.

Carbohydrate molecules are characterized by the presence of the atomic grouping H—C—OH, in which the ratio of hydrogen atoms (H) to oxygen atoms (O) is approximately 2:1. Since this ratio is the same as the ratio in water, the name "hydrates of carbon" seems appropriate.

Simple Carbohydrates

If the number of carbon atoms in a molecule is low (from three to seven), then the carbohydrate is a simple sugar, or **monosaccharide.** The designation **pentose** means a 5-carbon sugar, and the designation **hexose** means a 6-carbon sugar. **Glucose,** a hexose, is blood sugar (Fig. 2.15); our bodies use glucose as an immediate source of energy. Other common hexoses are fructose, found in fruits, and galactose, a constituent of milk. These three hexoses (glucose, fructose, and galactose) all occur as ring structures with the molecular formula $C_6H_{12}O_6$. The exact shape of the ring differs, as does the arrangement of the hydrogen (—H) and hydroxyl (—OH) groups attached to the ring.

A **disaccharide** (*di,* two; *saccharide,* sugar) contains two monosaccharides that have joined during a dehydration reaction. Figure 2.16 shows how the disaccharide maltose forms when two glucose molecules bond together. Note the position of this bond; our hydrolytic digestive juices can break this bond, and the result is two glucose molecules. When glucose and fructose join, the disaccharide sucrose forms. Sucrose is another disaccharide of special interest because we use it at the table to sweeten our food. We acquire the sugar from plants such as sugarcane and sugar beets. You may also have heard of lactose, a disaccharide found in milk. Lactose is glucose combined with galactose. Some people are lactose intolerant because they cannot break down lactose. This leads to unpleasant gastrointestinal symptoms when they drink milk. Lactose-free milk is available in stores.

Polysaccharides

Long polymers such as starch, glycogen, and cellulose are **polysaccharides** that contain many glucose subunits.

Starch and Glycogen

Starch and **glycogen** are ready storage forms of glucose in plants and animals, respectively. Some of the polymers in starch are long chains of up to 4,000 glucose units. Starch has fewer side branches, or chains of glucose that branch off from the main chain, than does glycogen, as shown in Figures 2.17 and 2.18. Flour, which we use for baking and usually acquire by grinding wheat, is high in starch, and so are potatoes.

After we eat starchy foods such as potatoes, bread, and cake, glucose enters the bloodstream, and the liver stores glucose as glycogen. In between eating, the liver releases glucose so that the blood glucose concentration is always about 0.1%.

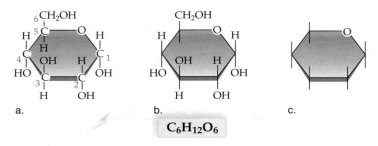

$C_6H_{12}O_6$

a.   b.   c.

Figure 2.15 Three ways to represent the structure of glucose.
$C_6H_{12}O_6$ is the molecular formula for glucose. The *far left* structure (**a**) shows the carbon atoms, but the middle structure (**b**) does not show the carbon atoms. The *far right* structure (**c**) is the simplest way to represent glucose.

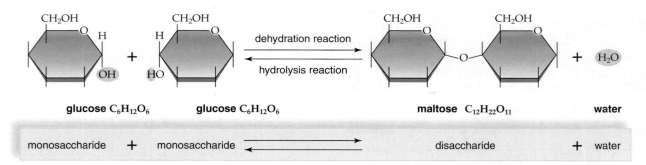

glucose $C_6H_{12}O_6$ glucose $C_6H_{12}O_6$ maltose $C_{12}H_{22}O_{11}$ water

monosaccharide   +   monosaccharide   disaccharide   +   water

Figure 2.16 Synthesis and degradation of maltose, a disaccharide.
Synthesis of maltose occurs following a dehydration reaction when a bond forms between two glucose molecules and water is removed. Degradation of maltose occurs following a hydrolysis reaction when this bond is broken by the addition of water.

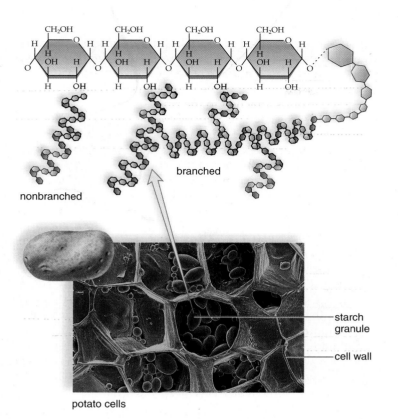

Figure 2.17 Starch structure and function.
Starch is composed of straight chains of glucose molecules. Some chains are branched, as indicated. Starch is the storage form of glucose in plants. The electron micrograph shows starch granules in potato cells. When we eat starch-containing foods such as corn, potatoes, bread, white rice, and pasta, much glucose enters our bloodstream.

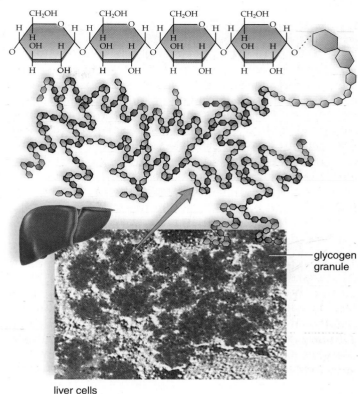

Figure 2.18 Glycogen structure and function.
Glycogen is a highly branched polymer of glucose molecules. Glycogen is the storage form of glucose in animals. The electron micrograph shows glycogen granules in liver cells. Muscle cells also store glycogen.

Cellulose

Some types of polysaccharides function as structural components of cells. The polysaccharide **cellulose** is found in plant cell walls, and this accounts, in part, for the strong nature of these walls.

In cellulose (Fig. 2.19), the glucose units are joined by a slightly different type of linkage than that in starch or glycogen. (Observe the alternating position of the oxygen atoms in the linked glucose units.) While this might seem to be a technicality, it is significant because we are unable to digest foods containing this type of linkage; therefore, cellulose largely passes through our digestive tract as fiber, or roughage. Medical scientists believe that fiber in the diet is necessary to good health, and some have suggested it may even help prevent colon cancer.

Chitin, which is found in the exoskeleton (shell) of crabs and related animals, is another structural polysaccharide. Scientists have discovered that chitin can be made into a thread and used as a suture material.

Cells typically use the monosaccharide glucose as an energy source. The polysaccharides starch and glycogen are storage compounds in plant and animal cells, respectively, and the polysaccharide cellulose is found in plant cell walls.

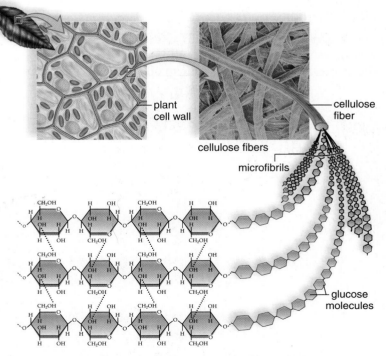

Figure 2.19 Cellulose structure and function.
In cellulose, the linkage between glucose molecules is slightly different from that in starch or glycogen. Plant cell walls contain cellulose, and the rigidity of these cell walls permits nonwoody plants to stand upright as long as they receive an adequate supply of water.

2.6   Lipids

Lipids contain more energy per gram than other biological molecules, and fats and oils function as energy storage molecules in organisms. Phospholipids form a membrane that separates the cell from its environment and forms its inner compartments as well. The steroids are a large class of lipids that includes, among others, the sex hormones.

Lipids are diverse in structure and function, but they have a common characteristic: they do not dissolve in water. This is because lipids are neutral—they lack polar groups. Therefore, lipids do not undergo electrical attraction to other groups.

Fats and Oils

The most familiar lipids are those found in fats and oils. **Fats,** which are usually of animal origin (e.g., lard and butter), are solid at room temperature. **Oils,** which are usually of plant origin (e.g., corn oil and soybean oil), are liquid at room temperature. Fat has several functions in the body: it is used for long-term energy storage, it insulates against heat loss, and it forms a protective cushion around major organs.

Fats and oils form when one glycerol molecule reacts with three fatty acid molecules (Fig. 2.20). A fat molecule is sometimes called a **triglyceride** because of its three-part structure, and the term *neutral fat* is sometimes used because the molecule is nonpolar.

Emulsification

Emulsifiers can cause fats to mix with water. They contain molecules with a nonpolar end and a polar end. The molecules position themselves about an oil droplet so that their nonpolar ends project inward and their polar ends project outward. Now the droplet disperses in water, which means that **emulsification** has occurred:

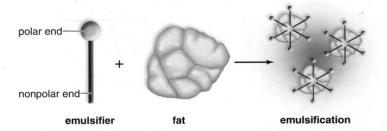

Emulsification takes place when dirty clothes are washed with soaps or detergents. It explains why some salad dressings are uniform in consistency (emulsified) while others separate into two layers. Also, prior to the digestion of fatty foods, fats are emulsified by bile. The gallbladder stores bile for use when a meal is eaten, and a person who has had the gallbladder removed may have trouble digesting fatty foods.

Saturated and Unsaturated Fatty Acids

A **fatty acid** is a hydrocarbon chain that ends with the acidic group —COOH (Fig. 2.20). Most of the fatty acids in cells contain 16 or 18 carbon atoms per molecule, although smaller ones with fewer carbons are also known.

Fatty acids are either saturated or unsaturated. **Saturated fatty acids** have no double covalent bonds between carbon atoms. The carbon chain is saturated, so to speak, with all the hydrogens it can hold. Saturated fatty acids account for the solid nature at room temperature of fats such as lard

Figure 2.20 Synthesis and degradation of a fat molecule.
Fatty acids can be saturated or unsaturated. Saturated fatty acids have no double bonds between carbon atoms, whereas unsaturated fatty acids have some double bonds (colored yellow) between carbon atoms. When a fat molecule (triglyceride) forms, three fatty acids combine with glycerol, and three water molecules are produced.

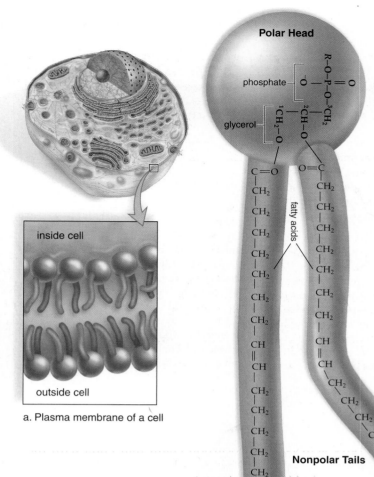

Polar Head

phosphate

glycerol

fatty acids

Nonpolar Tails

a. Plasma membrane of a cell

inside cell

outside cell

b. Phospholipid structure

Figure 2.21 Phospholipids form membranes.

a. Phospholipids arrange themselves as a bilayer in the plasma membrane that surrounds cells. **b.** Phospholipids are constructed like fats, except that in place of the third fatty acid, they have a polar phosphate group. The bilayer structure forms because the polar (hydrophilic) head is soluble in water, whereas the two nonpolar (hydrophobic) tails are not.

and butter. **Unsaturated fatty acids** have double bonds between carbon atoms wherever the number of hydrogens is less than two per carbon atom. Unsaturated fatty acids account for the liquid nature of vegetable oils at room temperature. Hydrogenation, the chemical addition of hydrogen to vegetable oils, converts them into a solid. This type of fat, called *trans fat,* is often found in processed foods.

Phospholipids

Phospholipids, as their name implies, contain a phosphate group. Essentially, they are constructed like fats, except that in place of the third fatty acid, there is a polar phos-

phate group or a grouping that contains both phosphate and nitrogen. These molecules are not electrically neutral, as are fats, because the phosphate and nitrogen-containing groups are ionized. They form the so-called polar (hydrophilic) head of the molecule, while the rest of the molecule becomes the nonpolar (hydrophobic) tails (Fig. 2.21).

Phospholipids illustrate that the chemistry of a molecule helps determine its function. Phospholipids are the primary components of cellular membranes; they spontaneously form a bilayer in which the hydrophilic heads face outward toward watery solutions and the tails form the hydrophobic interior. Plasma membranes separate extracellular from intracellular materials and thus are absolutely vital to the form and function of a cell.

Steroids

Steroids have a backbone of four fused carbon rings. Each one differs primarily by the arrangement of the atoms in the rings and the type of functional groups attached to them. Cholesterol is a steroid formed by the body that also enters the body as part of our diet. Cholesterol has several important functions. It is a component of an animal cell's plasma membrane and is the precursor of several other steroids, such as bile salts and the sex hormones testosterone and estrogen (Fig. 2.22).

Now we know that a diet high in saturated fats, trans fats, and cholesterol can cause fatty material to accumulate inside the lining of blood vessels thereby reducing blood flow. The Health Focus on page 36 discusses which sources of carbohydrates, fats, and proteins are recommended for inclusion in the diet.

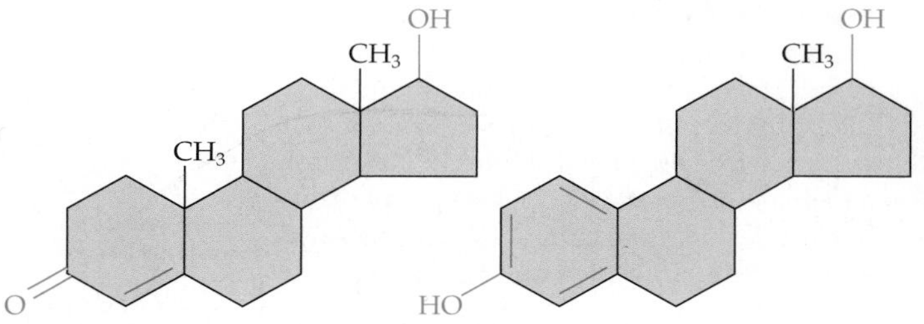

a. Testosterone

b. Estrogen

Figure 2.22 Steroids.

All steroids have four adjacent rings, but their attached groups differ. The effects of (**a**) testosterone and (**b**) estrogen on the body largely depend on the difference in the attached groups (shown in blue).

Lipids include fats and oils (for long-term energy storage), phospholipids, and steroids. Phospholipids, unlike other lipids, are soluble in water because they have a hydrophilic group. The steroid cholesterol performs several important functions in the body.

Health Focus

A Balanced Diet

Everyone agrees that we should eat a balanced diet, but just what is a balanced diet? The U.S. Department of Agriculture (USDA) released a new food pyramid in April of 2005 (Fig. 2B). The new pyramid contains vertical bands showing that all types of foods are needed for a balanced diet. Some bands are wider than others because you need more of some foods and less of others in your daily diet.

Progressing up the pyramid, the bands get smaller. The bands at the base represent food with little or no added fat or sugar. The narrower top area stands for foods with more added sugars and fats—which should be eaten more sparingly. For example, apple pie may contain fruit, but it also contains fat and sugar and should be eaten in smaller quantities than raw apples.

Carbohydrates

Carbohydrates (sugars and polysaccharides) are the quickest, most readily available source of energy for the body. Complex carbohydrates, such as those in whole-grain breads and cereals, are preferable to simple carbohydrates, such as candy and ice cream, because they contain dietary fiber (nondigestible plant material), plus vitamins and minerals. Insoluble fiber has a laxative effect, and soluble fiber combines with the cholesterol in food and prevents cholesterol from exiting the intestinal tract and entering the blood. However, researchers have found that the starch in potatoes and processed foods, such as white bread and white rice, leads to a high blood glucose level just as simple carbohydrates do. Researchers ask, Is this why many adults are now coming down with adult-onset diabetes?

Fats

We have known for many years that saturated fats in animal products contribute to the formation of deposits called plaque, which clog arteries and lead to high blood pressure and heart attacks.

Today, processed foods often contain trans fatty acids, which are especially capable of causing cardiovascular disease. So-called "trans fats" are oils that have been hydrogenated to solidify them.

Notice that the new food pyramid advocates an intake of certain liquid oils. These oils contain monounsaturated and polyunsaturated fatty acids, which researchers have found are protective against the development of cardiovascular disease.

Other Nutrients

Red meat is rich in protein, but it is usually also high in saturated fat; therefore, fish and chicken are preferred sources of protein. Also, a combination of rice and *legumes* (a group of plants that includes peas and beans) can provide all of the amino acids the body needs to build cellular proteins.

Nutritionists agree that eating fruits and vegetables is beneficial. At the very least, they provide us with the vitamins we need in our diet.

Discussion Questions

1. List the number and types of fruits and vegetables you eat in a typical day. Are you getting the recommended amounts of these important foods? Do you have variety in the types of fruits and vegetables you consume?
2. Why are some fats "good" for you and some "bad" for you? What is the difference in these fats?
3. Chemically, what is the difference between a "whole-grain" carbohydrate and a simple carbohydrate?

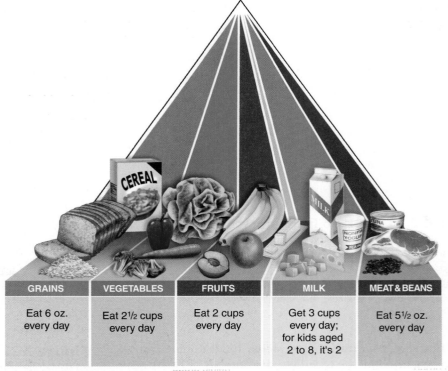

Source: U.S. Dept. of Agriculture

Figure 2B Food guide pyramid.
The United States Department of Agriculture (USDA) developed this pyramid as a guide to better health. The different widths of the food group bands suggest how much food a person should choose from each group. The six different colors illustrate that foods from all groups are needed each day for good health. The yellow band represents oils; the USDA recommends consuming 1–2 tablespoons every day. The wider base is supposed to encourage the selection of foods containing little or no solid fats or added sugars.

2.7 | Proteins

Proteins are polymers with amino acid monomers. An **amino acid** has a central carbon atom bonded to a hydrogen atom and three groups. The name of the molecule is appropriate because one of these groups is an amino group ($-NH_2$) and another is an acidic group ($-COOH$). The third group is called an R group, and amino acids differ from one another by their R group.

Name	Structural Formula	R Group
alanine (ala)	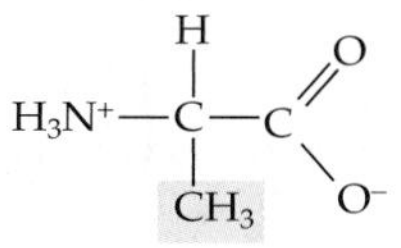	R group has a single carbon atom
valine (val)	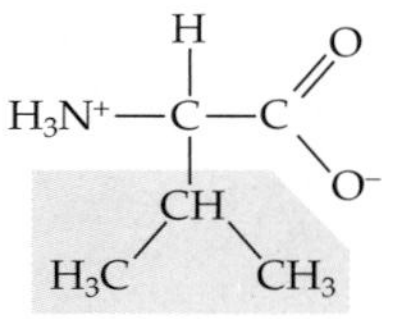	R group has a branched carbon chain
cysteine (cys)	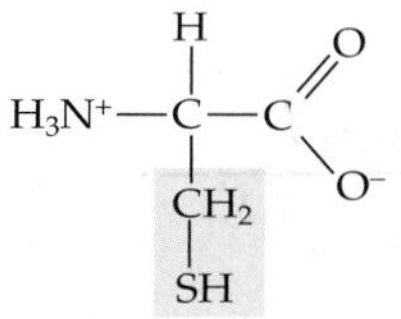	R group contains sulfur
phenylalanine (phe)	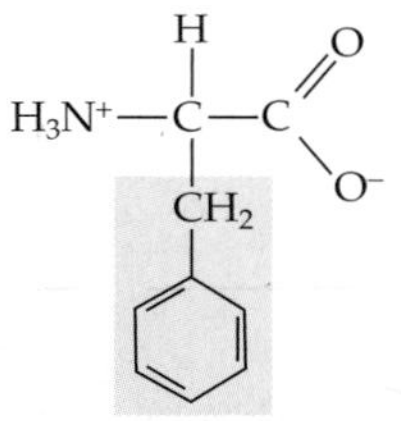	R group has a ring structure

Figure 2.23 Representative amino acids.
Amino acids differ from one another by their R group; the simplest R group is a single hydrogen atom (H). R groups (blue) containing carbon vary as shown.

The R group varies from having a single carbon to being a complicated ring structure (Fig. 2.23).

Proteins perform many functions. Proteins such as keratin, which makes up hair and nails, and collagen, which lends support to ligaments, tendons, and skin, are structural proteins. Many hormones, messengers that influence cellular metabolism, are also proteins. The proteins actin and myosin account for the movement of cells and the ability of our muscles to contract. Some proteins transport molecules in the blood; hemoglobin is a complex protein in our blood that transports oxygen. Antibodies in blood and other body fluids are proteins that combine with foreign substances, preventing them from destroying cells and upsetting homeostasis.

Proteins in the plasma membrane of cells have various functions: some form channels that allow substances to enter and exit cells; some are carriers that transport molecules into and out of the cell; and some are enzymes. Enzymes are necessary contributors to the chemical workings of the cell, and therefore of the body. **Enzymes** speed chemical reactions; they work so quickly that a reaction that normally takes several hours or days without an enzyme takes only a fraction of a second with an enzyme.

Peptides

Figure 2.24 shows that a synthesis reaction between two amino acids results in a dipeptide and a molecule of water. A **polypeptide** is a single chain of amino acids. The bond that joins any two amino acids is called a **peptide bond.** The atoms associated with a peptide bond—oxygen (O), carbon (C), nitrogen (N), and hydrogen (H)—share electrons in such a way that the oxygen has a partial negative charge (δ^-) and the hydrogen has a partial positive charge (δ^+):

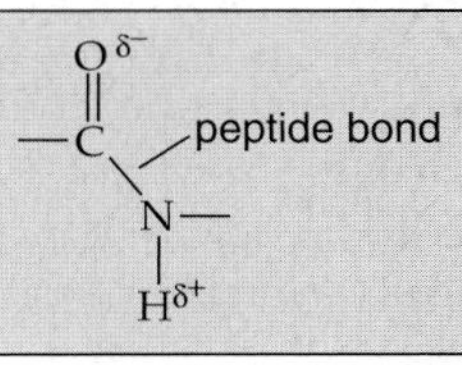

Therefore, the peptide bond is polar, and hydrogen bonding is possible between the C=O of one amino acid and the N—H of another amino acid in a polypeptide.

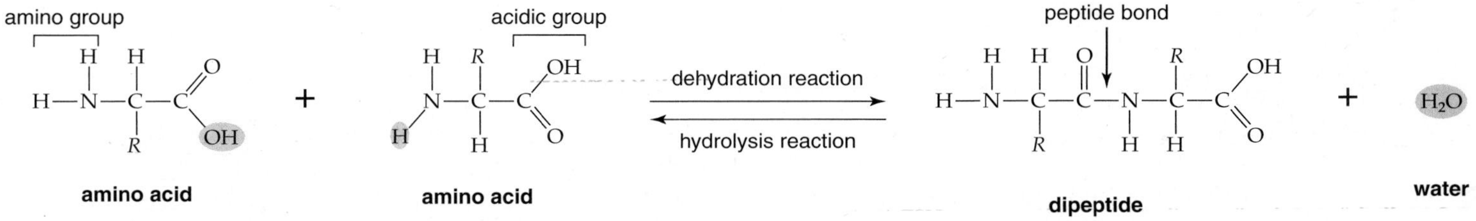

Figure 2.24 Synthesis and degradation of a dipeptide.
Following a dehydration reaction, a peptide bond joins two amino acids, and a water molecule is released. Following a hydrolysis reaction, the bond is broken with the addition of water.

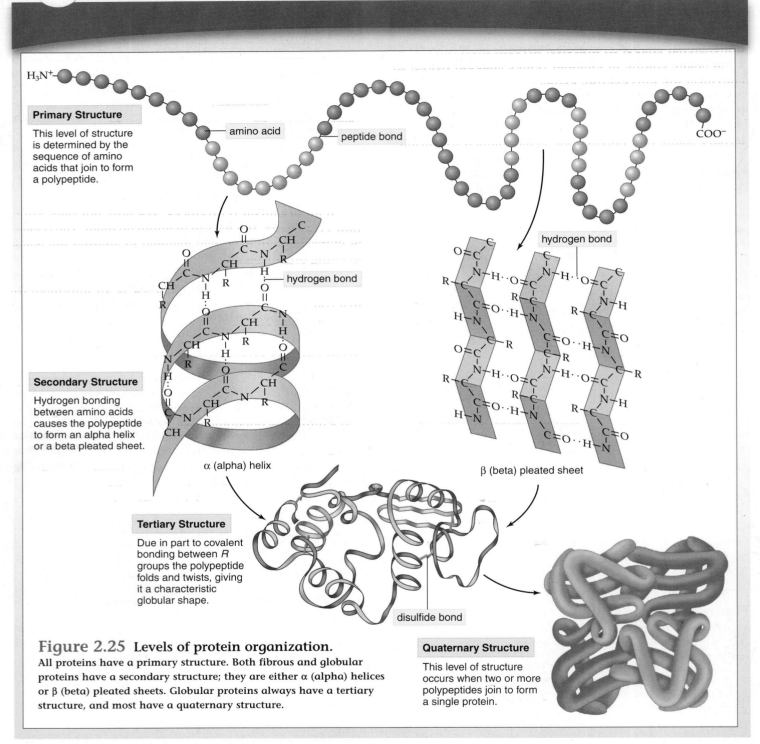

Primary Structure

This level of structure is determined by the sequence of amino acids that join to form a polypeptide.

amino acid

peptide bond

H₃N⁺

COO⁻

hydrogen bond

Secondary Structure

Hydrogen bonding between amino acids causes the polypeptide to form an alpha helix or a beta pleated sheet.

α (alpha) helix

β (beta) pleated sheet

Tertiary Structure

Due in part to covalent bonding between *R* groups the polypeptide folds and twists, giving it a characteristic globular shape.

disulfide bond

Quaternary Structure

This level of structure occurs when two or more polypeptides join to form a single protein.

Figure 2.25 Levels of protein organization.
All proteins have a primary structure. Both fibrous and globular proteins have a secondary structure; they are either α (alpha) helices or β (beta) pleated sheets. Globular proteins always have a tertiary structure, and most have a quaternary structure.

Levels of Protein Organization

The structure of a protein has at least three levels of organization and can have four levels (Fig. 2.25). The first level, called the *primary structure,* is the linear sequence of the amino acids joined by peptide bonds. Polypeptides can be quite different from one another. If a polysaccharide is like a necklace that contains a single type of "bead," namely

glucose, then polypeptides make use of 20 different possible types of "beads," namely amino acids. Each particular polypeptide has its own sequence of amino acids. Therefore, each polypeptide differs by the sequence of its *R* groups.

The *secondary structure* of a protein comes about when the polypeptide takes on a certain orientation in space. A coiling of the chain results in an α (alpha) helix, or a right-handed spiral, and a folding of the chain results in a β (beta) pleated sheet. Hydrogen bonding between peptide bonds holds the shape in place.

The *tertiary structure* of a protein is its final three-dimensional shape. In muscles, myosin molecules have a rod shape ending in globular (globe-shaped) heads. In enzymes, the polypeptide bends and twists in different ways. Invariably, the hydrophobic portions are packed mostly on the inside, and the hydrophilic portions are on the outside where they can make contact with water. The tertiary shape of a polypeptide is maintained by various types of bonding between the *R* groups; covalent, ionic, and hydrogen bonding all occur. One common form of covalent bonding between *R* groups is a disulfide (S—S) linkage between two cysteine amino acids.

Some proteins have only one polypeptide, and others have more than one polypeptide, each with its own primary, secondary, and tertiary structures. These separate polypeptides are arranged to give some proteins a fourth level of structure, termed the *quaternary structure*. Hemoglobin is a complex protein having a quaternary structure; most enzymes also have a quaternary structure. Thus, proteins can differ in many ways, such as in length, sequence, and structure. Each individual protein is chemically unique.

The final shape of a protein is very important to its function. As we will discuss in Chapter 6, for example, enzymes cannot function unless they have their usual shape. When proteins are exposed to extremes in heat and pH, they undergo an irreversible change in shape called **denaturation.** For example, we are all aware that adding acid to milk causes curdling and that heating causes egg white, which contains a protein called albumin, to coagulate. Denaturation occurs because the normal bonding between the *R* groups has been disturbed. Once a protein loses its normal shape, it is no longer able to perform its usual function. Researchers hypothesize that an alteration in protein organization is related to the development of Alzheimer disease and Creutzfeldt-Jakob disease (the human form of "mad cow disease").

A polypeptide is a chain of amino acids joined by peptide bonds. Proteins are important in the structure and the function of cells. Some proteins are enzymes, which speed chemical reactions. A protein can be composed of one or more polypeptides, each with levels of organization. The three-dimensional shape of a polypeptide/protein is important to its function in the body.

2.8 Nucleic Acids

The two types of nucleic acids are **DNA (deoxyribonucleic acid)** and **RNA (ribonucleic acid).** The discovery of the structure of DNA has had an enormous influence on biology and on society in general. DNA stores genetic informa-

TABLE 2.1	**DNA Structure Compared to RNA Structure**	
	DNA	**RNA**
Sugar	Deoxyribose	Ribose
Bases	Adenine, guanine, thymine, cytosine	Adenine, guanine, uracil, cytosine
Strands	Double stranded with base pairing	Single stranded
Helix	Yes	No

tion in the cell and in the organism. Further, it replicates and transmits this information when a cell reproduces and when an organism reproduces. We now not only know how genes work, but we can manipulate them. The science of biotechnology is largely devoted to altering the genes in living organisms.

DNA codes for the order in which amino acids are to be joined to form a protein. RNA is an intermediary that conveys DNA's instructions regarding the amino acid sequence of a protein.

Structure of DNA and RNA

Both DNA and RNA are polymers of nucleotides. Every **nucleotide** is a molecular complex of three types of subunit molecules—phosphate (phosphoric acid), a pentose sugar, and a nitrogen-containing base:

Nucleotide structure

The nucleotides in DNA contain the sugar deoxyribose and the nucleotides in RNA contain the sugar ribose; this difference accounts for their respective names (Table 2.1). There are four different types of bases in DNA: **adenine (A), thymine (T), guanine (G),** and **cytosine (C).** The base can have two rings (adenine or guanine) or one ring (thymine or cytosine). In RNA, the base **uracil (U)** replaces the base thymine. These structures are called bases because their presence raises the pH of a solution.

The nucleotides form a linear molecule called a strand, which has a backbone made up of phosphate-sugar-phosphate-sugar, with the bases projecting to one side of the backbone. The nucleotides and their bases occur in a definite order. After many years of work, researchers now know the sequence of all the bases in human DNA—the human genome. This breakthrough is expected to lead to improved genetic counseling, gene therapy, and medicines to treat the causes of many human illnesses.

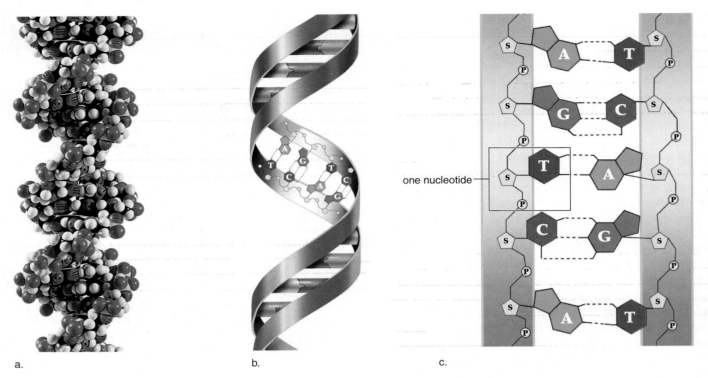

a.   b.   c.

Figure 2.26 Overview of DNA structure.

The structure of DNA is absolutely essential to its ability to replicate and to serve as the genetic material. **a.** Space-filling model of DNA's double helix. **b.** Complementary base pairing between strands. **c.** Ladder configuration. Notice that the uprights are composed of phosphate and sugar molecules and that the rungs are complementary nitrogen-containing paired bases. The heredity information stored by DNA is the sequence of its bases, which determines the primary structure of the cell's proteins.

DNA is double stranded, with the two strands twisted about each other in the form of a **double helix** (Fig. 26a, b). In DNA, the two strands are held together by hydrogen bonds between the bases. When unwound, DNA resembles a ladder. The uprights (sides) of the ladder are made entirely of phosphate and sugar molecules, and the rungs of the ladder are made only of complementary paired bases. Thymine (T) always pairs with adenine (A), and guanine (G) always pairs with cytosine (C). Complementary bases have shapes that fit together (Fig. 2.26c).

Complementary base pairing allows DNA to replicate in a way that ensures the sequence of bases will remain the same. This sequence of the DNA bases contains a code that specifies the sequence of amino acids in the proteins of the cell.

RNA is single stranded and is formed by complementary base pairing with one DNA strand. One type of RNA, mRNA or messenger RNA, carries the information from the DNA strand to the ribosome where it is translated into the sequence of amino acids specified by the DNA.

DNA has a structure like a twisted ladder: sugar and phosphate molecules make up the uprights of the ladder, and hydrogen-bonded bases make up the rungs. The sequence of bases contains the DNA code.

ATP (Adenosine Triphosphate)

In addition to being the monomers of nucleic acids, nucleotides have other metabolic functions in cells. When adenosine (adenine plus ribose) is modified by the addition of three phosphate groups instead of one, it becomes **ATP (adenosine triphosphate),** an energy carrier in cells. Glucose is broken down in a step-wise fashion so that the energy of glucose is converted to that of ATP molecules. ATP molecules serve as small "energy packets" suitable for supplying energy to a wide variety of a cell's chemical reactions.

ATP is a high-energy molecule because the last two phosphate bonds are unstable and easily broken. In cells, the terminal phosphate bond usually is hydrolyzed, leaving the molecule **ADP (adenosine diphosphate)** and a molecule of inorganic phosphate Ⓟ (Fig. 2.27). The cell uses the energy released by ATP breakdown to synthesize macromolecules such as carbohydrates and proteins. Muscle cells use the energy for muscle contraction, and nerve cells use it for the conduction of nerve impulses. After ATP breaks down, it is rebuilt by the addition of Ⓟ to ADP. Notice in Figure 2.27 that an input of energy is required to re-form ATP.

ATP is a high-energy molecule. ATP breaks down to ADP + Ⓟ, releasing energy, which is used for all the metabolic work done in a cell.

Bioethical Focus

Blue Gold

Environmentalists believe that the world is running out of clean drinking water. Over 97% of the world's water is salt water found in the oceans. Salt water is unsuitable for drinking without expensive desalination. Of the fresh water in the world, most is locked in frozen form in the polar ice caps and glaciers and therefore unavailable. This leaves only a small percentage in groundwater, lakes, and rivers that could be available for drinking, industry, and irrigation. However, some of that water is polluted and unsuitable.

Water has always been the most valuable commodity in the Middle East, even more valuable than oil. But as fresh water becomes limited and the world's population grows, the lack of sufficient clean water is becoming a worldwide problem.

The combination of increasing demand and dwindling supply has attracted global corporations who want to sell water. Water is being called the "blue gold" of the twenty-first century, and an issue has arisen regarding whether the water industry should be privatized. That is, could water rights be turned over to private companies to deliver clean water and treat wastewater at a profit, similar to the way oil and electricity are handled? Private companies have the resources to upgrade and modernize water delivery and treatment systems, thereby conserving more water. However, opponents of this plan claim that water is a basic human right required for life, not a need to be supplied by the private sector. In addition, a corporation can certainly own the pipelines and treatment facilities, but who owns the rights

to the water? For example, North America's largest underground aquifer, the Ogallala, covers 175,000 square miles under several states in the southern Great Plains. If water becomes a commodity, do we allow water to be taken away from people who cannot pay in order to give it to those who can?

Form Your Own Opinion

1. Do you agree that the water industry should be privatized? Why or why not?
2. Is access to clean water a "need" or a "right"? If it is a right, who pays for that right?
3. Since water is a shared resource, everyone believes they can use water, but few people feel responsible for conserving it. What can you do to conserve water?

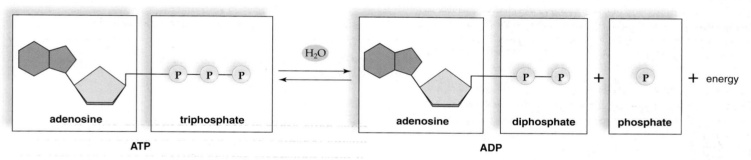

Figure 2.27 ATP reaction.
ATP, the universal energy currency of cells, is composed of adenosine and three phosphate groups (called a triphosphate). When cells require energy, ATP undergoes hydrolysis, producing ADP + Ⓟ, with the release of energy.

Summarizing the Concepts

2.1 Basic Chemistry

Both living and nonliving things are composed of matter consisting of elements. The acronym CHNOPS stands for the most significant elements found in living things: carbon, hydrogen, nitrogen, oxygen, phosphorus, and sulfur. Elements contain atoms, and atoms are composed of subatomic particles. Protons and neutrons in the nucleus determine the atomic mass of an atom. The atomic number indicates the number of protons and the number of electrons in electrically neutral atoms. Protons have positive charges, neutrons are uncharged, and electrons have negative charges. Isotopes are atoms of a single element that differ in their numbers of neutrons. Radioactive isotopes have many uses, including serving as tracers in biological experiments and medical procedures.

The number of electrons in the outer shell determines the reactivity of an atom. The first shell is complete when it is occupied by two electrons. In atoms up through calcium, number 20, every

shell beyond the first shell is complete with eight electrons. The octet rule states that atoms react with one another in order to have a completed outer shell.

2.2 Molecules and Compounds

Ions form when atoms lose or gain one or more electrons to achieve a completed outer shell. An ionic bond is an attraction between oppositely charged ions. A covalent bond is one or more shared pairs of electrons.

When carbon combines with four other atoms, the resulting molecule has a tetrahedral shape. The shape of a molecule is important to its biological role.

In polar covalent bonds, the sharing of electrons is not equal; one of the atoms exerts greater attraction for the shared electrons than the other, and a slight charge on each atom results. A hydrogen bond is a weak attraction between a slightly positive hydrogen atom and a slightly negative oxygen or nitrogen atom within the same or a different molecule. Hydrogen bonds help maintain the structure and function of cellular molecules.

2.3 Chemistry of Water

Water is a polar molecule. Its polarity allows hydrogen bonding to occur between water molecules. Water's polarity and hydrogen bonding account for its unique properties. These features allow living things to exist and carry on cellular activities.

A small fraction of water molecules dissociate to produce an equal number of hydrogen ions and hydroxide ions. Solutions with equal numbers of H^+ and OH^- are termed neutral. In acidic solutions, there are more hydrogen ions than hydroxide ions; these solutions have a pH less than 7. In basic solutions, there are more hydroxide ions than hydrogen ions; these solutions have a pH greater than 7. Cells are sensitive to pH changes. Biological systems often contain buffers that help keep the pH within a normal range.

2.4 Organic Molecules

The chemistry of carbon accounts for the chemistry of organic compounds. Carbohydrates, lipids, proteins, and nucleic acids are macromolecules with specific functions in cells.

Polymers are formed by the joining together of monomers. Long-chain carbohydrates, proteins, and nucleic acids are all polymers. For each bond formed during a dehydration reaction, a molecule of water is removed, and for each bond broken during a hydrolysis reaction, a molecule of water is added.

2.5 Carbohydrates

Glucose is the 6-carbon sugar most utilized by cells for "quick" energy. Like the rest of the macromolecules to be studied, a dehydration reaction joins two or more sugar monomers to form polysaccharides, and hydrolysis splits the bond. Plants store glucose as starch, and animals store glucose as glycogen. Humans cannot digest cellulose, which forms plant cell walls.

2.6 Lipids

Lipids are varied in structure and function. Fats and oils, which function in long-term energy storage, contain glycerol and three fatty acids. Fatty acids can be saturated or unsaturated. Plasma membranes in cells contain phospholipids that have a polarized end. Certain hormones are derived from cholesterol, a complex ring compound.

2.7 Proteins

Proteins have numerous functions in cells; some are enzymes that speed chemical reactions. The primary structure of a polypeptide is its own particular sequence of the possible 20 types of amino acids. The secondary structure is often an alpha (α) helix or a beta (β) pleated sheet. Tertiary structure occurs when a polypeptide bends and twists into a three-dimensional shape. A protein can contain several polypeptides, and this accounts for a possible quaternary structure.

2.8 Nucleic Acids

Nucleic acids are polymers of nucleotides. Each nucleotide has three components: a sugar, a base, and phosphate (phosphoric acid). DNA, which contains the sugar deoxyribose, is the genetic material that stores information for its own replication and for the sequence of amino acids in proteins. DNA, with the help of RNA, specifies protein synthesis.

ATP, with its unstable phosphate bonds, is the energy currency of cells. Hydrolysis of ATP to ADP + ℗ releases energy that the cell uses to do metabolic work.

Testing Yourself

Choose the best answer for each question.

1. Which of these ions is important for acid-base balance?
   a. K^+
   b. Na^+
   c. Cl^-
   d. H^+

2. Carbon can engage in
   a. only ionic bonds.
   b. only covalent bonds with hydrogen.
   c. four covalent bonds.
   d. only bonds with oxygen.

3. Which of these offers the most accurate information concerning the shape of a molecule?
   a. electron-dot formula
   b. structural formula
   c. molecular formula
   d. space-filling model

4. An example of a hydrogen bond would be
   a. the bond between a carbon atom and a hydrogen atom.
   b. the bond between two carbon atoms.
   c. the bond between sodium and chlorine.
   d. the bond between two water molecules.

5. When water dissociates, it releases
   a. equal amounts of H^+ and OH^-.
   b. more H^+ than OH^-.
   c. more OH^- than H^+.
   d. only H^+.

6. Carbon and oxygen atoms can be found in
   a. amino acids.
   b. glucose.
   c. starch.
   d. All of these are correct.

7. Plants store glucose as
   a. maltose.
   b. glycogen.
   c. starch.
   d. None of these are correct.

8. Covalent bonding between *R* groups in proteins is associated with the _____ structure.
   a. primary
   b. secondary
   c. tertiary
   d. None of these are correct.

9. The polysaccharide found in plant cell walls is
   a. glucose.
   b. starch.
   c. maltose.
   d. cellulose.

10. Ionic bonding involves
    a. loss of electrons.
    b. gain of electrons.
    c. attraction of opposite charges.
    d. All of these are correct.

11. Phosphates can be found in
    a. DNA.
    b. RNA.
    c. glucose.
    d. Both a and b are correct.

12. Which of these cannot be considered an organic molecule?
    a. table salt
    b. glucose
    c. DNA
    d. proteins

13. Most _____ are_____ .
    a. proteins, enzymes
    b. sugars, monosaccharides
    c. enzymes, proteins
    d. sugars, polysaccharides

14. Removal of the gallbladder may cause difficulty in digesting
    a. fats.
    b. proteins.
    c. sugars.
    d. water.

15. _____ is the precursor of_____ .
    a. Estrogen, cholesterol
    b. Cholesterol, glucose
    c. Testosterone, cholesterol
    d. Cholesterol, testosterone and estrogen

16. If the pH in the blood drops below the normal 7.4,
    a. H^+ may have been added, and nothing can be done to raise the pH.
    b. OH^- may have been added, and nothing can be done to raise the pH.
    c. OH^- may have been added, and only the addition of more OH^- can help get back to pH 7.4.
    d. H^+ may have been added, and buffers may be able to help get back to pH 7.4.

17. An example of a dehydration reaction would be
    a. building amino acids from proteins.
    b. building lipids from fatty acids.
    c. breaking down glucose into starch.
    d. breaking down starch into glucose.

18. Radioactive elements differ in their
    a. number of protons.
    b. atomic number.
    c. number of neutrons.
    d. number of electrons.

19. A 5-carbon sugar is associated with
    a. glucose.   c. DNA.
    b. protein.   d. lipids.

In questions 20–23, indicate whether the statement is true or false.

20. Methane is a polar molecule. _____

21. Sodium chloride contains covalent bonds. _____

22. H^+ affects pH. _____

23. Nucleic acids and carbohydrates both contain phosphate groups. _____

24. What type of bonding is represented by the dotted lines in the following diagram?

Hydrogen bonding between water molecules

25. Nucleotides
    a. contain sugar, a nitrogen-containing base, and a phosphate molecule.
    b. are the monomers for fats and polysaccharides.
    c. join together by covalent bonding between the bases.
    d. are found in DNA, RNA, and proteins.

26. The joining of two adjacent amino acids is called
    a. a peptide bond.   c. a covalent bond.
    b. a dehydration reaction. d. All of these are correct.

27. Which of the following pertains to an RNA nucleotide and not to a DNA nucleotide?
    a. contains the sugar ribose
    b. contains a nitrogen-containing base
    c. contains a phosphate molecule
    d. becomes bonded to other nucleotides by condensation

28. Saturated fatty acids and unsaturated fatty acids differ in
    a. the number of carbon-to-carbon bonds.
    b. their consistency at room temperature.
    c. the number of hydrogen atoms present.
    d. All of these are correct.

29. Which of the following is an example of a polysaccharide used for energy storage?
    a. cellulose   c. cholesterol
    b. glycogen   d. glucose

Understanding the Terms

acid 29
adenine (A) 39
ADP (adenosine
    diphosphate) 40
amino acid 37
atom 20
atomic mass 21
atomic number 21
ATP (adenosine
    triphosphate) 40
base 29
buffer 29
calorie 27
carbohydrate 32
cellulose 33
compound 24
covalent bond 25
cytosine (C) 39
dehydration reaction 31
denaturation 39
disaccharide 32
DNA (deoxyribonucleic
    acid) 39
double helix 40
electron 20
element 20
emulsification 34
enzyme 37
fat 34
fatty acid 34
functional group 31
glucose 32
glycogen 32
guanine (G) 39
hexose 32
hydrogen bond 26

hydrolysis reaction 31
hydrophilic 28
hydrophobic 28
inorganic molecule 31
ion 24
ionic bond 24
isotope 22
lipid 34
matter 20
molecule 24
monomer 31
monosaccharide 32
neutron 20
nucleotide 39
oil 34
organic molecule 31
pentose 32
peptide bond 37
phospholipid 35
pH scale 29
polymer 31
polypeptide 37
polysaccharide 32
protein 37
proton 20
radioactive isotope 22
RNA (ribonucleic acid) 39
saturated fatty acid 34
solute 28
starch 32
steroid 35
thymine (T) 39
triglyceride 34
unsaturated fatty acid 35
uracil (U) 39

Match the terms to these definitions:

a._____Polymer of many amino acids linked by peptide bonds.
b._____Organic catalyst that speeds up a reaction in cells due to its particular shape.
c._____Polysaccharide composed of glucose molecules; the chief constituent of a plant's cell wall.
d._____Subatomic particle that has the mass of one atomic mass unit, carries no charge, and is found in the nucleus of an atom.
e._____Solution in which pH is less than 7; a substance that contributes or liberates hydrogen ions in a solution.

Thinking Critically

1. Why can we use potatoes as a food source but not wood?
2. If the pH of a substance changes from 7 to 5, how many more or fewer hydrogen ions are present?
3. Since proteins are composed of the same limited number of amino acids, why are they all so different?
4. What feature of the structure of DNA makes it uniquely qualified to function as an information carrier?
5. Given the following ions, can you determine the number of electrons normally present in the outer shell of each atom?
   a. K^+
   b. Ca^{+2}
   c. F^{-1}
   d. N^{-3}

ARIS, the *Inquiry into Life* Website

ARIS, the website for *Inquiry into Life,* provides a wealth of information organized and integrated by chapter. You will find practice quizzes, interactive activities, labeling exercises, flashcards, and much more that will complement your learning and understanding of general biology.

www.aris.mhhe.com

Laurel looked through the **microscope,** *fascinated by the sight of her own cells. She had obtained the cells by gently scraping them from the inner lining of her mouth, smearing them onto a microscope slide, and then staining them with blue dye. Laurel noticed that each cell contained a darker blue spot. She remembered from biology lecture that this region is known as the nucleus, and it is especially important because it contains instructions in the form of DNA for all the cell's functions. As Laurel carefully sketched one of the cells, she realized that each cell had a distinct boundary that separated it from its surroundings. This, she determined, must be the plasma membrane.*

Laurel then observed numerous smaller blue dots all over her cheek cells. Adjusting her microscope to a higher magnification, she viewed the dots again, wondering if they, too, might be cells of some kind. Laurel checked with her lab instructor, who said that the little dots were indeed cells. In fact, they were various types of bacteria that live in everyone's mouth.

This chapter is an introduction to the study of cells. As we will see, all life on Earth consists of cells, which come in an astounding variety of shapes, sizes, and arrangements. Laurel's body is made up of trillions of cells, many with specialized functions, but all working together to keep her alive and healthy. The bacteria thriving in the environment of Laurel's mouth are tiny, single-celled organisms. They are smaller than any of Laurel's own cells, and yet they are also capable of carrying out a versatile range of life activities.

chapter 3

Cell Structure and Function

Chapter Concepts

3.1 The Cellular Level of Organization

The cell marks the boundary between the nonliving and the living. The molecules that serve as food for a cell and the macromolecules that make up a cell are not alive, and yet the cell is alive. The cell is the structural and functional unit of an organism, the smallest structure capable of performing all the functions necessary for life. Thus, the answer to what life is must lie within the cell, because the smallest living organisms are unicellular, while larger organisms are multicellular—that is, composed of many cells.

Cells can be classified as either prokaryotic or eukaryotic. Prokaryotic cells do not contain the membrane-enclosed structures found in eukaryotic cells. Therefore, eukaryotic cells are thought to have evolved from prokaryotic cells (see Section 3.4). Prokaryotic cells are exemplified by the bacteria.

The diversity of cells is illustrated by the many types in the human body, such as muscle cells and nerve cells. But despite a variety of forms and functions, human cells contain the same components. The basic components that are common to all eukaryotic cells, regardless of their specializations, are the subject of this chapter. Viewing these components requires a microscope. The Science Focus on page 53 introduces you to the microscopes most used today to study cells. Electron microscopy and biochemical analyses have revealed that eukaryotic cells actually contain tiny, specialized structures called **organelles.** Each organelle performs specific cellular functions.

Today, we are accustomed to thinking of living things as being constructed of cells. But the word cell didn't enter biology until the seventeenth century. Antonie van Leeuwenhoek of Holland is now famous for making his own microscopes and observing all sorts of tiny things that no one had seen before. Robert Hooke, an Englishman, confirmed Leeuwenhoek's observations and was the first to use the term "cell." The tiny chambers he observed in the honeycomb structure of cork reminded him of the rooms, or cells, in a monastery.

Over 150 years later—in the 1830s—the German microscopist Matthias Schleiden stated that plants are composed of cells; his counterpart, Theodor Schwann, stated that animals are also made up of living units called cells. This was quite a feat, because aside from their own exhausting examination of tissues, both had to take into consideration the studies of many other microscopists. Rudolf Virchow, another German microscopist, later came to the conclusion that cells don't suddenly appear; rather, they come from preexisting cells. Think about how we reproduce. The sperm fertilizes an egg and a human being develops from a resulting cell, called a zygote.

The **cell theory** states that all organisms are made up of basic living units called cells, and that all cells come only from previously existing cells. Today, the cell theory is a basic theory of biology.

The cell theory states the following:
- All organisms are composed of one or more cells.
- Cells are the basic living unit of structure and function in organisms.
- All cells come only from other cells.

Cell Size

Cells are quite small. A frog's egg, at about 1 millimeter (mm) in diameter, is large enough to be seen by the human eye. But most cells are far smaller than 1 mm; some are even as small as 1 micrometer (μm)—one thousandth of a millimeter. Cell inclusions and macromolecules are smaller than a micrometer and are measured in terms of nanometers (nm). Figure 3.1 outlines the visual range of the eye, light microscope, and electron microscope, and the discussion of microscopy in the Science Focus explains why the electron microscope allows us to see so much more detail than the light microscope does.

The fact that cells are so small is a great advantage for multicellular organisms. Nutrients, such as glucose and oxygen, enter a cell, and wastes, such as carbon dioxide, exit a cell at its surface; therefore, the amount of surface area affects the ability to get material into and out of the cell. A large cell requires more nutrients and produces more wastes than a small cell. But, as cells get larger in volume, the proportionate amount of surface area actually decreases. For example, for a cube-shaped cell, the volume increases by the cube of the sides (height × width × depth), while the surface area increases by the square of the sides and number of sides (height × width × 6). If a cell doubles in size, its surface area only increases fourfold while its volume increases eightfold. Therefore, small cells, not large cells, are likely to have an adequate surface area for exchanging nutrients and wastes. As Figure 3.2 demonstrates, cutting a large cube into smaller cubes provides a lot more surface area per volume.

Most actively metabolizing cells are small. The frog's egg is not actively metabolizing. But once the egg is fertilized and metabolic activity begins, the egg divides repeatedly without growth. These cell divisions restore the amount of surface area needed for adequate exchange of materials. Further, cells that specialize in absorption have modifications that greatly increase their surface-area-to-volume ratio. For example, the columnar epithelial cells along the surface of the intestinal wall have surface foldings called microvilli (sing., microvillus) that increase their surface area.

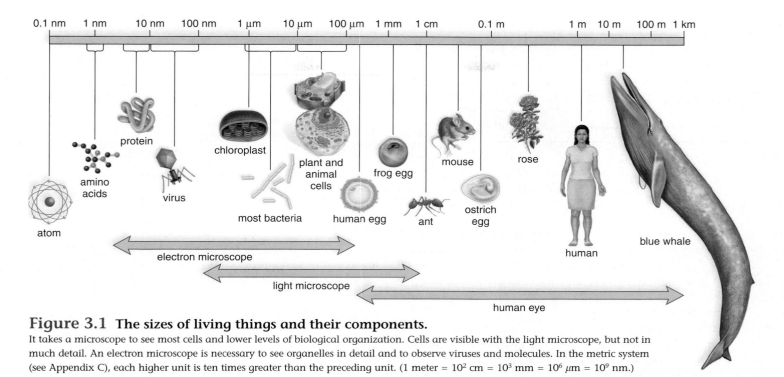

Figure 3.1 The sizes of living things and their components.

It takes a microscope to see most cells and lower levels of biological organization. Cells are visible with the light microscope, but not in much detail. An electron microscope is necessary to see organelles in detail and to observe viruses and molecules. In the metric system (see Appendix C), each higher unit is ten times greater than the preceding unit. (1 meter = 10^2 cm = 10^3 mm = 10^6 μm = 10^9 nm.)

Plasma Membrane and Cytoplasm

All cells are surrounded by a **plasma membrane** that consists of a phospholipid bilayer in which some protein molecules are embedded:

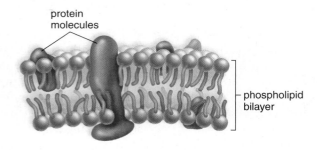

The plasma membrane is a living boundary that separates the living contents of the cell from the nonliving surrounding environment. Inside the cell is a semifluid medium called the **cytoplasm.** The cytoplasm is composed of water, salts, and dissolved organic molecules. The plasma membrane regulates the entrance and exit of molecules into and out of the cytoplasm.

A cell needs a surface area that can adequately exchange materials with the environment. Surface-area-to-volume considerations require that cells stay small. All cells contain a plasma membrane and cytoplasm.

| One 4-cm cube | Eight 2-cm cubes | Sixty-four 1-cm cubes |

Figure 3.2 Surface-area-to-volume relationships.

All three have the same volume, but the group on the right has four times the surface area.

3.2 Prokaryotic Cells

Cells can be classified by the presence or absence of a nucleus. **Prokaryotic cells** lack a membrane-bounded nucleus. The domains Archaea and Bacteria consist of prokaryotic cells. Prokaryotes generally exist as unicellular organisms (single cells) or as simple strings and clusters. Many people think of germs when they hear the word bacteria, but not all bacteria cause disease. In fact, most bacteria are beneficial and are essential for other living organisms' survival.

Figure 3.3 illustrates the main features of bacterial anatomy. The **cell wall,** located outside of the plasma membrane, contains peptidoglycan, a complex molecule that is unique to bacteria and composed of chains of disaccharides joined by peptide chains. The cell wall protects the bacteria. Some antibiotics, such as penicillin, interfere with the synthesis of

Ribosome:
site of protein synthesis

Nucleoid:
location of the bacterial
chromosome

Plasma membrane:
sheath around cytoplasm
that regulates entrance
and exit of molecules

Fimbriae:
hairlike bristles that
allow adhesion to
surfaces

Cell wall:
covering that supports,
shapes, and protects cell

Capsule:
gel-like coating outside
cell wall

Flagellum:
rotating filament present
in some bacteria that
pushes the cell forward

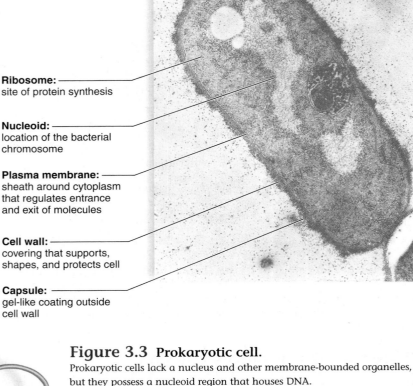

Figure 3.3 Prokaryotic cell.
Prokaryotic cells lack a nucleus and other membrane-bounded organelles,
but they possess a nucleoid region that houses DNA.

peptidoglycan. In some bacteria, the cell wall is further sur-rounded by a **capsule** and/or a gelatinous sheath called a **slime layer.** Some bacteria have long, very thin appendages called flagella (sing., **flagellum**), which are composed of subunits of the protein flagellin. The flagella rotate like propellers, allowing the bacterium to move rapidly in a fluid medium. Some bacteria also have **fimbriae,** which are short appendages that help them attach to an appropriate surface. The capsule and fimbriae often give pathogenic bacteria increased ability to cause disease.

Prokaryotes have a single chromosome (loop of DNA and associated proteins) located within a region of the cyto-plasm called the **nucleoid.** The nucleoid is not bounded by a membrane. Many prokaryotes also have small accessory rings of DNA called **plasmids.** The cytoplasm has thousands of **ribosomes** for the synthesis of proteins. The ribosomes of prokaryotic organisms are smaller and structurally differ-ent from those of eukaryotic cells, which makes ribosomes a good target for antibacterial drugs. In addition, the photosyn-thetic cyanobacteria have light-sensitive pigments, usually within the membranes of flattened disks called **thylakoids.**

Although prokaryotes are structurally simple, they are much more metabolically diverse than eukaryotes. Many of them can synthesize all their structural compo-nents from very simple, even inorganic molecules. Indeed, humans exploit the metabolic capability of bacteria by using them to produce a wide variety of chemicals and products. Prokaryotes have also adapted to living in almost every envi-ronment on Earth. In particular, archaea have been found liv-ing under conditions that would not support any other form of life—for example, in water at temperatures above boiling. Archaeal membranes have unique membrane-spanning lip-ids that help them survive in extremes of heat, pH, and salin-ity. **Table 3.1** compares the major structures of prokaryotes (archaea and bacteria) with those of eukaryotes.

TABLE 3.1	Comparison of Major Structural Features of Archaea, Bacteria, and Eukaryotes		
	Archaea	**Bacteria**	**Eukaryotes**
Cell wall	Usually present, no peptidoglycan	Usually present, with peptidoglycan	Sometimes present, no peptidoglycan
Plasma membrane	Yes	Yes	Yes
Nucleus	No	No	Yes
Membrane-bounded organelles	No	No	Yes
Ribosomes	Yes	Yes	Yes, larger than prokaryotic

3.3 Eukaryotic Cells

Eukaryotic cells are structurally very complex. The principal distinguishing feature of eukaryotic cells is the presence of a nucleus, which separates the chromosomes (DNA) from the cytoplasm of the cell. In addition, eukaryotic cells possess a variety of other organelles, many of which are surrounded by membranes. Animals, plants, fungi, and protists are all composed of eukaryotic cells.

Cell Walls

Some eukaryotic cells have a permeable but protective cell wall, in addition to a plasma membrane. Many plant cells have both a primary and a secondary cell wall. A main constituent of a primary cell wall is cellulose molecules. Cellulose molecules form fibrils that lie at right angles to one another for added strength. The secondary cell wall, if present, forms inside the primary cell wall. Such secondary cell walls contain lignin, a substance that makes them even stronger than primary cell walls. The cell walls of some fungi are composed of cellulose and chitin, the same type of polysaccharide found in the exoskeleton of insects. Algae, members of the kingdom Protista, contain cell walls composed of cellulose.

Organelles of Animal and Plant Cells

Originally the term organelle referred to only membranous structures, but we will use it to include any well-defined subcellular structure (Table 3.2). Just as all the assembly lines of a factory operate at the same time, so all the organelles of a cell function simultaneously. Raw materials enter a factory where different departments turn them into various products. In the same way, the cell takes in chemicals, and then the organelles process them.

Both animal cells (Fig. 3.4) and plant cells (Fig. 3.5) contain mitochondria, but only plant cells have chloroplasts. Only animal cells have centrioles. In the illustrations throughout this text, note that each of the organelles has an assigned color.

The Nucleus

The **nucleus,** which has a diameter of about 5 μm, is a prominent structure in the eukaryotic cell. The nucleus is of primary importance because it stores the genetic material, DNA, which governs the characteristics of the cell and its metabolic functioning. Every cell in an individual contains the same DNA, but in each cell type, certain genes are turned on and certain others are turned off. Activated DNA, with RNA acting as an intermediary, specifies the sequence of amino acids when a protein is synthesized. The proteins of a cell determine its structure and the functions it can perform.

TABLE 3.2	Eukaryotic Structures in Animal Cells and Plant Cells	
Structure	**Composition**	**Function**
Cell wall*	Contains cellulose fibrils	Support and protection
Plasma membrane	Phospholipid bilayer with embedded proteins	Defines cell boundary; regulates molecule passage into and out of cells
Nucleus	Nuclear envelope, nucleoplasm, chromatin, and nucleoli	Storage of genetic information; synthesis of DNA and RNA
Nucleolus	Concentrated area of chromatin, RNA, and proteins	Ribosomal subunit formation
Ribosome	Protein and RNA in two subunits	Protein synthesis
Endoplasmic reticulum (ER)	Membranous flattened channels and tubular canals	Synthesis and/or modification of proteins and other substances, and distribution by vesicle formation
Rough ER	Network of folded membranes studded with ribosomes	Folding, modification, and transport of proteins
Smooth ER	Having no ribosomes	Various; lipid synthesis in some cells
Golgi apparatus	Stack of membranous saccules	Processing, packaging, and distribution of proteins and lipids
Lysosome**	Membranous vesicle containing digestive enzymes	Intracellular digestion
Vacuole and vesicle	Membranous sacs	Storage of substances
Peroxisome	Membranous vesicle containing specific enzymes	Various metabolic tasks
Mitochondrion	Inner membrane (cristae) bounded by an outer membrane	Cellular respiration
Chloroplast*	Membranous grana bounded by two membranes	Photosynthesis
Cytoskeleton	Microtubules, intermediate filaments, actin filaments	Shape of cell and movement of its parts
Cilia and flagella	9 + 2 pattern of microtubules	Movement of cell
Centriole**	9 + 0 pattern of microtubules	Formation of basal bodies

Plant cells only.
**Animal cells only.*

Figure 3.4 Animal cell anatomy.

Micrograph of liver cell and drawing of a generalized animal cell. See Table 3.2 for a description of these structures, along with a listing of their functions.

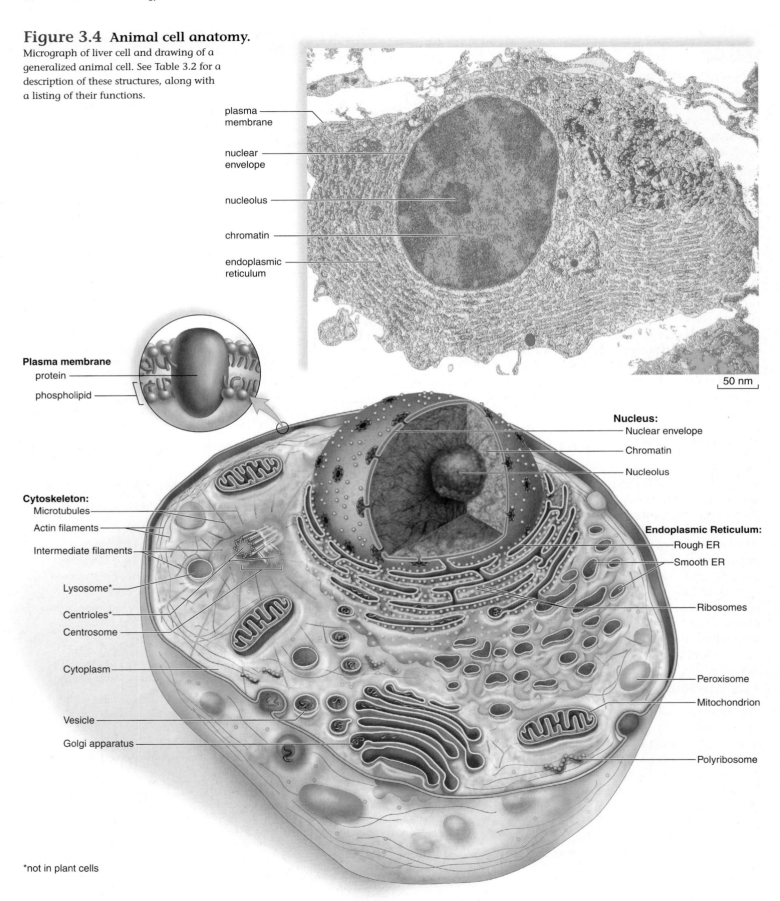

plasma membrane

nuclear envelope

nucleolus

chromatin

endoplasmic reticulum

50 nm

Plasma membrane

protein

phospholipid

Cytoskeleton:

Microtubules

Actin filaments

Intermediate filaments

Lysosome*

Centrioles*

Centrosome

Cytoplasm

Vesicle

Golgi apparatus

Nucleus:

Nuclear envelope

Chromatin

Nucleolus

Endoplasmic Reticulum:

Rough ER

Smooth ER

Ribosomes

Peroxisome

Mitochondrion

Polyribosome

*not in plant cells

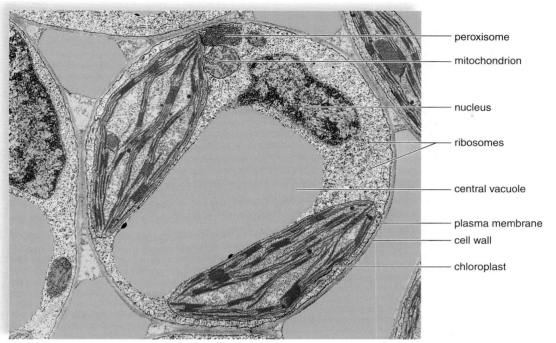

peroxisome
mitochondrion
nucleus
ribosomes
central vacuole
plasma membrane
cell wall
chloroplast

1 µm

Figure 3.5 Plant cell anatomy.

False-colored micrograph of a young plant cell and drawing of a generalized plant cell. See Table 3.2 for a description of these structures, along with a listing of their functions.

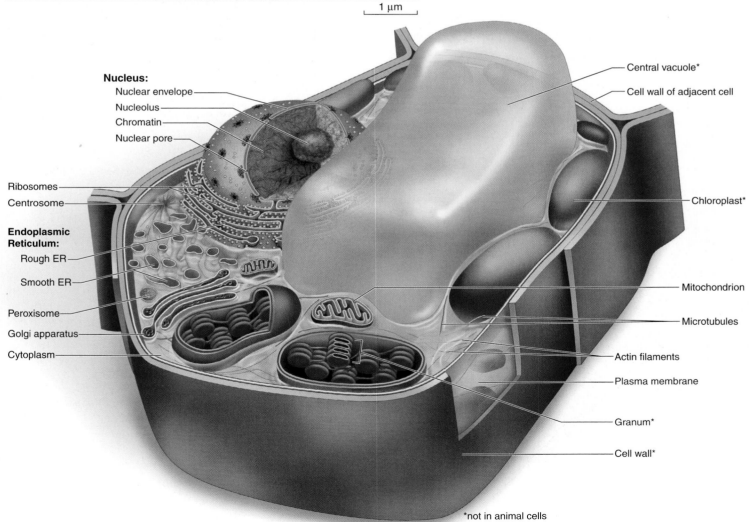

Nucleus:
Nuclear envelope
Nucleolus
Chromatin
Nuclear pore

Ribosomes
Centrosome

Endoplasmic Reticulum:
Rough ER
Smooth ER

Peroxisome
Golgi apparatus
Cytoplasm

Central vacuole*
Cell wall of adjacent cell
Chloroplast*
Mitochondrion
Microtubules
Actin filaments
Plasma membrane
Granum*
Cell wall*

*not in animal cells

When you look at the nucleus, even in an electron micrograph, you cannot see a DNA molecule. You can see **chromatin,** which consists of DNA and associated proteins (Fig. 3.6). Chromatin looks grainy, but actually it is a threadlike material that undergoes coiling to form rodlike structures, called **chromosomes** during the initial stages of cell division. Chromatin is immersed in a semifluid medium called the **nucleoplasm.** A difference in pH between the nucleoplasm and the cytoplasm suggests that the nucleoplasm has a different composition.

Most likely, too, when you look at an electron micrograph of a nucleus, you will see one or more regions that look darker than the rest of the chromatin. These are nucleoli (sing., **nucleolus**), where another type of RNA, called ribosomal RNA (rRNA), is produced and where rRNA joins with proteins to form the subunits of ribosomes described in the next section.

The nucleus is separated from the cytoplasm by a double membrane known as the **nuclear envelope,** which is continuous with the endoplasmic reticulum (also discussed next). The nuclear envelope has **nuclear pores** of sufficient size (100 nm) to permit proteins to pass into the nucleus and ribosomal subunits to pass out.

The structural features of the nucleus include:

Chromatin:	DNA and proteins
Nucleolus:	Chromatin and ribosomal subunits
Nuclear envelope:	Double membrane with pores

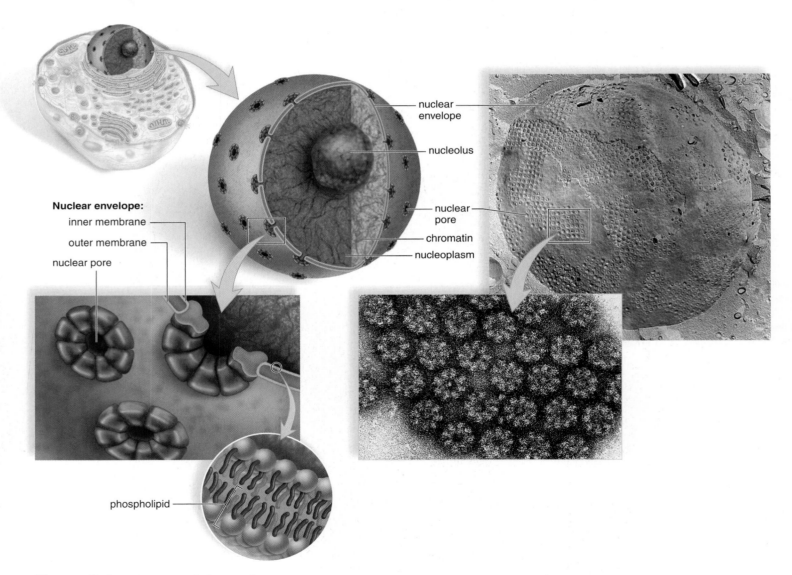

Figure 3.6 Anatomy of the nucleus.
The nucleus contains chromatin. The nucleolus is a region of chromatin where rRNA is produced, and ribosomal subunits are assembled. The nuclear envelope contains pores, as shown in the larger micrograph of a freeze-fractured nuclear envelope. Nuclear pores serve as passageways for substances to pass into and out of the nucleus.

ScienceFocus

Microscopy Today

Today, scientists use two types of microscopes: light microscopes and electron microscopes. Figure 3A depicts one type of light microscope and two types of electron microscopes, along with micrographs obtained with each one.

The most commonly used light microscope is the compound light microscope. "Compound" refers to the presence of more than one lens. In a compound light microscope, light rays pass through a specimen such as a blood smear, a tissue section, or even living cells. The light rays are brought to a focus by a set of glass lenses and the resulting image is then viewed. Specimens are often stained to provide viewing contrast.

In an electron microscope, electrons rather than light rays are brought to focus by a set of magnetic lenses, and the resulting image is projected onto a viewing screen or photographic film. Electron microscopes produce images of much higher magnification than may be obtained with even the best compound

light microscope. The ability to distinguish two points as separate points (resolution) is much greater with electron microscopes. The greater resolving power is due to the fact that electrons travel at a much shorter wavelength than do light rays. Live specimens, however, cannot be viewed and must be dried out. Two commonly used types of electron microscopes are transmission and scanning.

A transmission electron microscope (a TEM) may magnify an object's image over 100,000 times, compared to approximately 1,000 times with a compound light microscope. Also, a TEM has a much greater ability to make out detail. Much of our knowledge of the internal structures of cells comes from with a TEM.

The magnifying and resolving powers of a typical scanning electron microscope a (SEM) are not as great as with a TEM although they are still greater than those of a compound light microscope. A SEM provides a striking three-dimensional view of the surface of an object.

A picture obtained using a compound light microscope is sometimes called a photomicrograph, and a picture resulting from the use of an electron microscope is called a transmission or scanning electron micrograph, depending on the type of microscope used. Color may be added to electron micrographs, such as those in Figure 3A, using a computer.

Discussion Questions

1. What distortions or artifacts might be seen in electron microscopy when viewing a dried, nonliving specimen?
2. What happens to an image if the magnification is increased without increasing the resolution?
3. Why are electron micrographs always in black and white (or artificially colored)?

Figure 3A Diagram of microscopes with accompanying micrographs. The compound light microscope and the transmission electron microscope provide an internal view of an organism. The scanning electron microscope provides an external view of an organism.

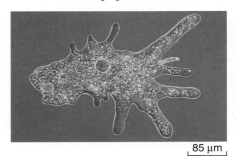

85 µm

amoeba, light micrograph, stained with dye

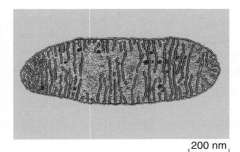

200 nm

mitochondrion, TEM, artificially colored

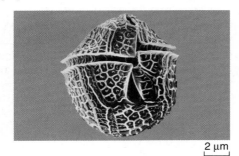

2 µm

dinoflagellate, SEM, artificially colored

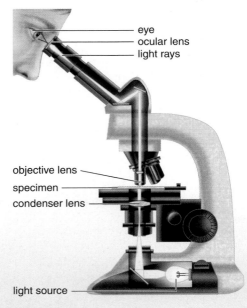

a. Compound light microscope

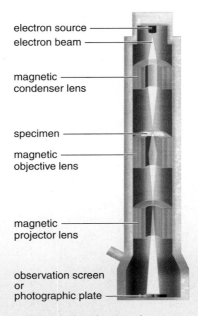

b. Transmission electron microscope

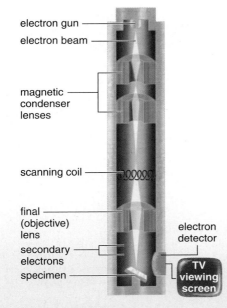

c. Scanning electron microscope

Ribosomes

Ribosomes are composed of two subunits, called "large" and "small" because of their sizes relative to each other. Each subunit is a complex of unique ribosomal RNA (rRNA) and protein molecules. Ribosomes can be found individually in the cytoplasm, as well as in groups called **polyribosomes** (several ribosomes associated simultaneously with a single mRNA molecule). Ribosomes can also be found attached to the endoplasmic reticulum, a membranous system of saccules and channels discussed in the next section. Proteins synthesized at ribosomes attached to the endoplasmic reticulum have a different destination from that of proteins synthesized at ribosomes free in the cytoplasm.

> Ribosomes are small organelles where protein synthesis occurs. Ribosomes occur in the cytoplasm, both singly and in groups (i.e., polyribosomes). Numerous ribosomes are also attached to the endoplasmic reticulum.

The Endomembrane System

The endomembrane system consists of the nuclear envelope, the endoplasmic reticulum, the Golgi apparatus, and several **vesicles** (tiny membranous sacs). This system compartmentalizes the cell so that particular enzymatic reactions are restricted to specific regions. Organelles that make up the endomembrane system are connected either directly or by transport vesicles.

The Endoplasmic Reticulum

The **endoplasmic reticulum (ER),** a complicated system of membranous channels and saccules (flattened vesicles), is physically continuous with the outer membrane of the nuclear envelope. Rough ER is studded with ribosomes on the side of the membrane that faces the cytoplasm (Fig. 3.7). Here, proteins are synthesized and enter the ER interior, where processing and modification begin.

Proper folding, processing, and transport of proteins are critical to the functioning of the cell. For example, in cystic fibrosis a mutated plasma membrane channel protein is retained in the endoplasmic reticulum because it is folded incorrectly. Without this protein in its correct location, the cell is unable to regulate the transport of the chloride ion, resulting in the various symptoms of the disease.

Smooth ER, which is continuous with rough ER, does not have attached ribosomes. Smooth ER synthesizes the phospholipids that occur in membranes and performs various other functions depending on the particular cell. In the testes, it produces testosterone, and in the liver, it helps detoxify drugs. Regardless of any specialized function, smooth ER also forms vesicles in which products are transported to the Golgi apparatus.

> ER is involved in protein synthesis (rough ER) and various other processes, such as lipid synthesis (smooth ER). Vesicles transport products from the ER to the Golgi apparatus.

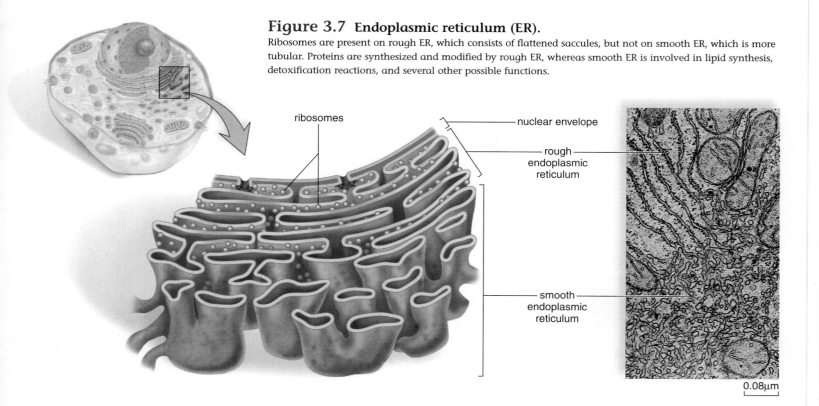

Figure 3.7 Endoplasmic reticulum (ER).
Ribosomes are present on rough ER, which consists of flattened saccules, but not on smooth ER, which is more tubular. Proteins are synthesized and modified by rough ER, whereas smooth ER is involved in lipid synthesis, detoxification reactions, and several other possible functions.

ribosomes

nuclear envelope

rough endoplasmic reticulum

smooth endoplasmic reticulum

0.08μm

The Golgi Apparatus

The **Golgi apparatus** is named for Camillo Golgi, who discovered its presence in cells in 1898. The Golgi apparatus consists of a stack of three to twenty slightly curved saccules whose appearance can be compared to a stack of pancakes (Fig. 3.8). The Golgi apparatus is referred to as the post office of the cell because it collects, sorts, packages, and distributes materials such as proteins and lipids. In animal cells, one side of the stack (the inner face) is directed toward the ER, and the other side of the stack (the outer face) is directed toward the plasma membrane. Vesicles can frequently be seen at the edges of the saccules.

The Golgi apparatus receives proteins and also lipid-filled vesicles that bud from the ER. These molecules then move through the Golgi from the inner face to the outer face. How this occurs is still being debated. According to the maturation saccule model, the vesicles fuse to form an inner face saccule, which matures as it gradually becomes a saccule at the outer face. According to the stationary saccule model, the molecules move through stable saccules from the inner face to the outer face by shuttle vesicles. It is likely that both models apply, depending on the organism and the type of cell.

During their passage through the Golgi apparatus, proteins and lipids can be modified before they are repackaged in secretory vesicles. Secretory vesicles proceed to the plasma membrane, where they discharge their contents. This action is termed **secretion**.

The Golgi apparatus is also involved in the formation of lysosomes, vesicles that contain proteins and remain within the cell. How does the Golgi apparatus direct traffic—in other words, what makes it direct the flow of proteins to different destinations? It now seems that proteins made at the rough ER have specific molecular tags that serve as "zip codes" to tell the Golgi apparatus whether they belong inside the cell in some membrane-bounded organelle or in a secretory vesicle.

Figure 3.8
Endomembrane system.

The organelles in the endomembrane system work together to carry out the functions noted.

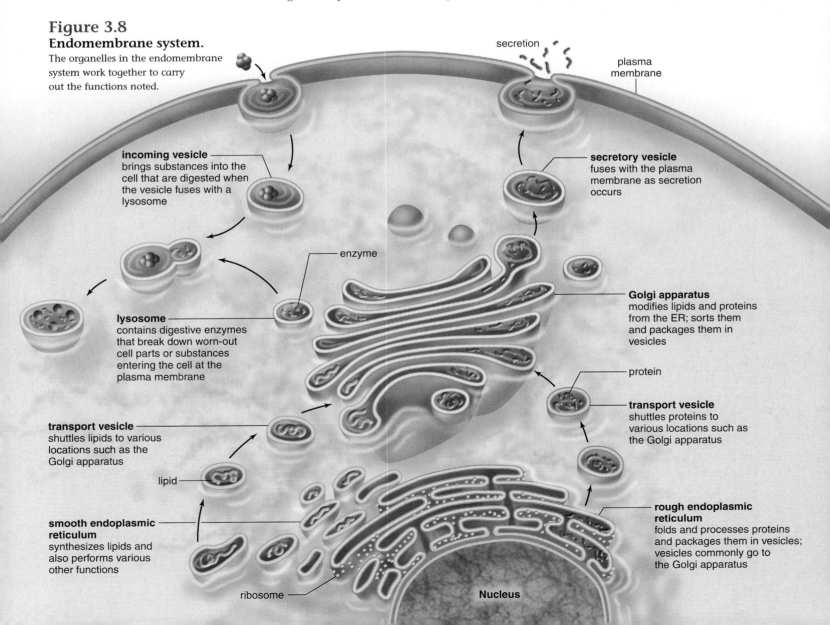

secretion

plasma membrane

incoming vesicle brings substances into the cell that are digested when the vesicle fuses with a lysosome

secretory vesicle fuses with the plasma membrane as secretion occurs

enzyme

Golgi apparatus modifies lipids and proteins from the ER; sorts them and packages them in vesicles

lysosome contains digestive enzymes that break down worn-out cell parts or substances entering the cell at the plasma membrane

protein

transport vesicle shuttles proteins to various locations such as the Golgi apparatus

transport vesicle shuttles lipids to various locations such as the Golgi apparatus

lipid

rough endoplasmic reticulum folds and processes proteins and packages them in vesicles; vesicles commonly go to the Golgi apparatus

smooth endoplasmic reticulum synthesizes lipids and also performs various other functions

ribosome

Nucleus

Lysosomes

Lysosomes are membrane-bounded vesicles produced by the Golgi apparatus. Lysosomes contain hydrolytic digestive enzymes.

Sometimes macromolecules are brought into a cell by vesicle formation at the plasma membrane (see Fig. 3.8). When a lysosome fuses with such a vesicle, its contents are digested by lysosomal enzymes into simpler subunits that then enter the cytoplasm. Some white blood cells defend the body by engulfing pathogens via vesicle formation. When lysosomes fuse with these vesicles, the bacteria are digested. Even parts of a cell are digested by its own lysosomes (called autodigestion). For example, the finger webbing found in the human embryo is later dissolved by lysosomes so that the fingers are separated.

Lysosomes contain many enzymes for digesting all sorts of molecules. Occasionally, a child inherits the inability to make a lysosomal enzyme, and therefore has a lysosomal storage disease. For example, in Tay Sachs disease, the cells that surround nerve cells cannot break down a particular lipid, which then accumulates inside lysosomes and affects the nervous system. At about six months, the infant can no longer see and, then, gradually loses hearing and even the ability to move. Death follows at about three years of age.

> The endomembrane system consists of the endoplasmic reticulum, Golgi apparatus, lysosomes, and transport vesicles.

Vacuoles

A **vacuole** is a large membranous sac. A vacuole is larger than a vesicle. Although animal cells have vacuoles, they are much more prominent in plant cells. Typically, plant cells have a large central vacuole so filled with a watery fluid that it gives added support to the cell (see Fig. 3.5).

Vacuoles store substances. Plant vacuoles contain not only water, sugars, and salts but also pigments and toxic molecules. The pigments are responsible for many of the red, blue, or purple colors of flowers and some leaves. The toxic substances help protect a plant from herbivorous animals. The vacuoles present in unicellular protozoans are quite specialized; they include contractile vacuoles for ridding the cell of excess water and digestive vacuoles for breaking down nutrients.

> Vacuoles are larger than vesicles. Plants typically have a large central vacuole for storage of various molecules.

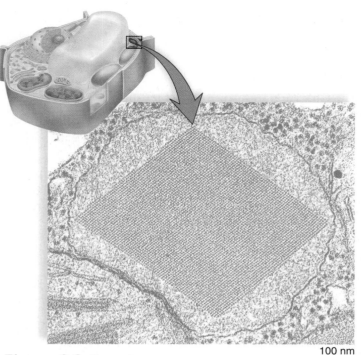

Figure 3.9 Peroxisomes.
Peroxisomes contain one or more enzymes that can oxidize various organic substances. Peroxisomes also contain the enzyme catalase, which breaks down the hydrogen peroxide (H_2O_2) that builds up after organic substances are oxidized.

100 nm

Peroxisomes

Peroxisomes, similar to lysosomes, are membrane-bounded vesicles that enclose enzymes (Fig. 3.9). However, the enzymes in peroxisomes are synthesized by cytoplasmic ribosomes and transported into a peroxisome by carrier proteins. Typically, peroxisomes contain enzymes whose action results in hydrogen peroxide (H_2O_2), a toxic molecule:

$$RH_2 + O_2 \longrightarrow R + H_2O_2$$

(R = remainder of molecule)

Hydrogen peroxide is immediately broken down to water and oxygen by another peroxisomal enzyme called catalase.

The enzymes present in a peroxisome depend on the function of the cell. Peroxisomes are especially prevalent in cells that are synthesizing and breaking down fats. In the liver, some peroxisomes break down fats and others produce bile salts from cholesterol. In the movie *Lorenzo's Oil*, Lorenzo's cells lacked a carrier protein to transport an enzyme into peroxisomes. As a result, long-chain fatty acids accumulated in his brain and he suffered from neurological damage.

Plant cells also have peroxisomes. In germinating seeds, they oxidize fatty acids into molecules that can be converted to sugars needed by the growing plant. In leaves, peroxisomes can carry out a reaction that is opposite to photosynthesis—the reaction uses up oxygen and releases carbon dioxide.

> Typically, the enzymes in peroxisomes break down molecules and as a result produce hydrogen peroxide molecules.

Energy-Related Organelles

Life is possible only because of a constant input of energy. Organisms use this energy for maintenance and growth. Chloroplasts and mitochondria are the two eukaryotic membranous organelles that specialize in converting energy to a form the cell can use. **Chloroplasts** use solar energy to synthesize carbohydrates, and carbohydrate-derived products are broken down in mitochondria (sing., **mitochondrion**) to produce ATP molecules, as shown in the following diagram:

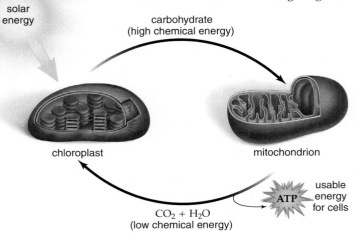

This diagram shows that chemicals recycle between chloroplasts and mitochondria, but energy flows from the sun

through these organelles to ATP. When cells use ATP as an energy source, energy dissipates as heat. Life could not exist without a constant input of solar energy.

Only plants, algae, and cyanobacteria are capable of carrying on **photosynthesis** in this manner:

solar energy + carbon dioxide + water ⟶ carbohydrate + oxygen

Plants and algae have chloroplasts (Fig. 3.10), while cyanobacteria carry on photosynthesis within independent thylakoids. Solar energy is the ultimate source of energy for cells because nearly all organisms, either directly or indirectly, use the carbohydrates produced by photosynthesizers as an energy source.

All organisms carry on **cellular respiration**, the process by which the chemical energy of carbohydrates is converted to that of ATP (adenosine triphosphate), the common energy carrier in cells. All organisms, except bacteria, complete the process of cellular respiration in mitochondria. Cellular respiration can be represented by this equation:

carbohydrate + oxygen ⟶ carbon dioxide + water + energy

Here, *energy* is in the form of ATP molecules. When a cell needs energy, ATP supplies it. The energy of ATP is used for all energy-requiring processes in cells.

Figure 3.10 Chloroplast structure.

Chloroplasts carry out photosynthesis. **a.** Electron micrograph of a longitudinal section of a chloroplast. **b.** Generalized drawing of a chloroplast in which the outer and inner membranes have been cut away to reveal the grana, each of which is a stack of membranous sacs called thylakoids. In some grana, but not all, it is obvious that thylakoid spaces are interconnected.

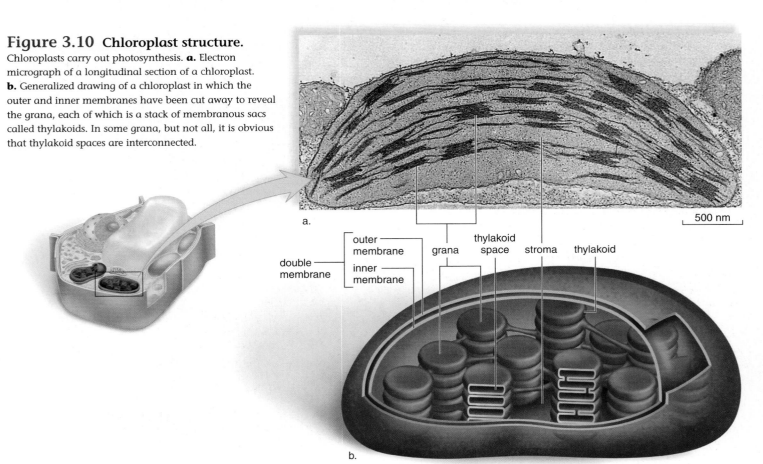

Chloroplasts

Plant and algal cells contain chloroplasts, the organelles that allow them to produce their own organic food. Chloroplasts are about 4–6 μm in diameter and 1–5 μm in length; they belong to a group of organelles known as plastids. Among the plastids are also the *amyloplasts*, common in roots, which store starch, and the *chromoplasts*, common in leaves, which contain red and orange pigments. A chloroplast is green, of course, because it contains the green pigment chlorophyll.

A chloroplast is bounded by two membranes that enclose a fluid-filled space called the **stroma** (see Fig. 3.10). A membrane system within the stroma is organized into interconnected flattened sacs called thylakoids. In certain regions, the thylakoids are stacked up in structures called grana (sing., **granum**). There can be hundreds of grana within a single chloroplast (see Fig. 3.10). Chlorophyll, which is located within the thylakoid membranes of grana, captures the solar energy needed to enable chloroplasts to produce carbohydrates. The stroma also contains DNA, ribosomes, and enzymes that synthesize carbohydrates from carbon dioxide and water.

Mitochondria

All eukaryotic cells, including those of plants and algae, contain mitochondria. This means that plant cells contain both chloroplasts and mitochondria. Mitochondria are usually 0.5–1.0 μm in diameter and 2–5 μm in length.

Mitochondria, like chloroplasts, are bounded by a double membrane (Fig. 3.11). In mitochondria, the inner fluid-filled space is called the **matrix.** The matrix contains DNA, ribosomes, and enzymes that break down carbohydrate products, releasing energy to be used for ATP production.

The inner membrane of a mitochondrion invaginates to form **cristae.** Cristae provide a much greater surface area to accommodate the protein complexes and other participants that produce ATP.

Mitochondria and chloroplasts are able to make some proteins, encoded by their own genes—but other proteins, encoded by nuclear genes, are imported from the cytoplasm.

In humans, all mitochondria come from the maternal line through the egg. The father's sperm does not contribute any mitochondria to the offspring. Mutations in the mitochondrial DNA usually affect high-energy-demand tissues such as the eye, brain, central nervous system, and muscles.

Chloroplasts and mitochondria are membranous organelles with structures that lend themselves to the energy transfers occurring within them.

Figure 3.11 Mitochondrion structure.
Mitochondria are involved in cellular respiration.
a. Electron micrograph of a longitudinal section of a mitochondrion. **b.** Generalized drawing in which the outer membrane and portions of the inner membrane have been cut away to reveal the cristae.

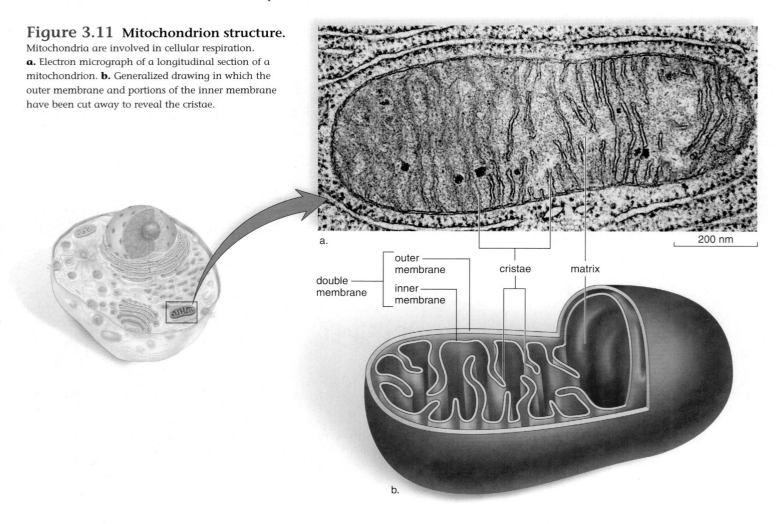

a.

200 nm

double membrane
outer membrane
inner membrane
cristae
matrix

b.

The Cytoskeleton

The protein components of the **cytoskeleton** interconnect and extend from the nucleus to the plasma membrane in eukaryotic cells. Prior to the 1970s, scientists believed that the cytoplasm was an unorganized mixture of organic molecules. Then, high-voltage electron microscopes, which can penetrate thicker specimens, showed instead that the cytoplasm is highly organized. The technique of immunofluorescence microscopy identified the makeup of the protein components within the cytoskeletal network (Fig. 3.12).

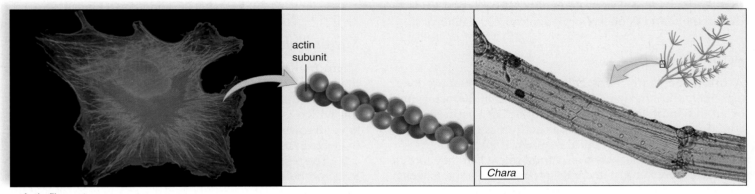

a. Actin filaments

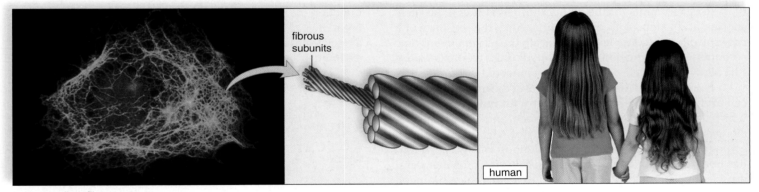

b. Intermediate filaments

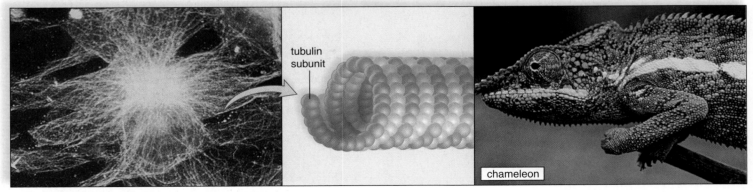

c. Microtubules

Figure 3.12 The cytoskeleton.

The cytoskeleton maintains the shape of the cell and allows its parts to move. Three types of protein components make up the cytoskeleton. They can be identified in cells by using a special fluorescent technique that detects only one type of component at a time. **a.** *Left to right:* Fibroblasts in animal tissue have been treated so that actin filaments can be microscopically detected; the drawing shows that actin filaments are composed of a twisted double chain of actin subunits. The giant cells of the green alga *Chara* rely on actin filaments to move organelles from one end of the cell to another. **b.** *Left to right:* Fibroblasts in an animal tissue have been treated so that intermediate filaments can be microscopically detected; the drawing shows that fibrous proteins account for the ropelike structure of intermediate filaments. Human hair is strengthened by the presence of intermediate filaments. **c.** *Left to right:* Fibroblasts in an animal tissue have been treated so that microtubules can be microscopically detected; the drawing shows that microtubules are hollow tubes composed of tubulin subunits. The skin cells of a chameleon rely on microtubules to move pigment granules around so that they can take on the color of their environment.

The cytoskeleton contains actin filaments, intermediate filaments, and microtubules, which maintain cell shape and allow the cell and its organelles to move. Therefore, the cytoskeleton is often compared to the bones and muscles of an animal. However, the cytoskeleton is dynamic, especially because its protein components can assemble and disassemble as appropriate. Apparently a number of different mechanisms regulate this process, including protein kinases that add phosphates to proteins. Phosphorylation leads to disassembly, and dephosphorylation causes assembly.

Actin Filaments

Actin filaments (formerly called microfilaments) are long, extremely thin, flexible fibers (about 7 nm in diameter) that occur in bundles or meshlike networks. Each actin filament contains two chains of globular actin monomers twisted about one another in a helical manner.

Actin filaments play a structural role when they form a dense, complex web just under the plasma membrane, to which they are anchored by special proteins. They are also seen in the microvilli that project from intestinal cells, and their presence most likely accounts for the ability of microvilli to alternately shorten and extend into the intestine. Also, the presence of a network of actin filaments lying beneath the plasma membrane accounts for the formation of pseudopods (false feet), extensions that allow certain cells to move in an amoeboid fashion.

How are actin filaments involved in the movement of the cell and its organelles? They interact with **motor molecules,** which are proteins that can attach, detach, and reattach farther along an actin filament. In the presence of ATP, the motor molecule myosin pulls actin filaments along in this way. Myosin has both a head and a tail. In muscle cells, the tails of several muscle myosin molecules are joined to form a thick filament. In nonmuscle cells, cytoplasmic myosin tails are bound to membranes, but the heads still interact with actin:

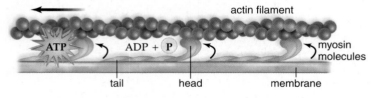

During animal cell division, the two new cells form when actin, in conjunction with myosin, pinches off the cells from one another.

Intermediate Filaments

Intermediate filaments (8-11 nm in diameter) are intermediate in size between actin filaments and microtubules and perform a structural role in the cell. They are a ropelike assembly of fibrous polypeptides, but the specific type varies according to the tissue. Some intermediate filaments support the nuclear envelope, whereas others support the plasma membrane and take part in the formation of cell-to-cell junctions. In skin cells, intermediate filaments, made of the protein keratin, give great mechanical strength. We now know that intermediate filaments are also highly dynamic and will disassemble when phosphate is added by a kinase.

Microtubules

Microtubules are small, hollow cylinders about 25 nm in diameter and 0.2–25 μm in length.

Microtubules are made of the globular protein tubulin, which is of two types, called α and β. Microtubules have 13 rows of tubulin dimers, surrounding what appears in electron micrographs to be an empty central core.

The regulation of microtubule assembly is controlled by a microtubule organizing center (MTOC). In most eukaryotic cells, the main MTOC is in the **centrosome,** which lies near the nucleus. Microtubules radiate from the centrosome, helping to maintain the shape of the cell and acting as tracks along which organelles can move. Whereas the motor molecule myosin is associated with actin filaments, the motor molecules kinesin and dynein are associated with microtubules:

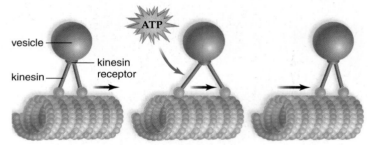

vesicle moves, not microtubule

Before a cell divides, microtubules disassemble and then reassemble into a structure called a spindle that distributes chromosomes in an orderly manner. At the end of cell division, the spindle disassembles, and microtubules reassemble once again into their former array.

The cytoskeleton is an internal skeleton composed of actin filaments, intermediate filaments, and microtubules that maintain the shape of the cell and assist movement of its parts.

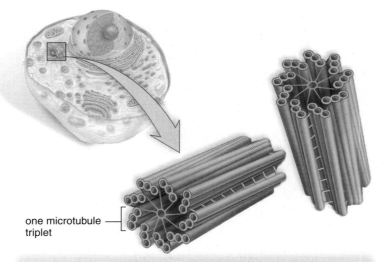

one microtubule — triplet

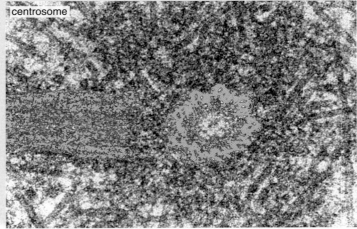

centrosome

one pair of centrioles

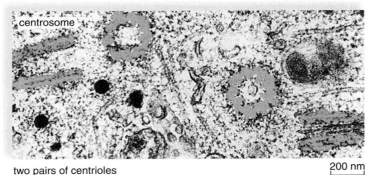

centrosome

two pairs of centrioles   200 nm

Figure 3.13 Centrioles.

In a nondividing animal cell, a single pair of centrioles lies in the centrosome located just outside the nucleus. Just before a cell divides, the centrioles replicate, producing two pairs of centrioles. During cell division, centrioles in their respective centrosomes separate so that each new cell has one centrosome containing one pair of centrioles.

Centrioles

Centrioles are short cylinders with a 9 + 0 pattern of microtubule triplets—that is, a ring having nine sets of triplets, with none in the middle (Fig. 3.13). In animal cells, a centrosome contains two centrioles lying at right angles to each other. The centrosome is the major microtubule organizing center for the cell, and centrioles may be involved in the process of microtubule assembly and disassembly.

Before an animal cell divides, the centrioles replicate such that the members of each pair are again at right angles to one another (Fig. 3.13). Then, each pair becomes part of a separate centrosome. During cell division, the centrosomes move apart and may function to organize the mitotic spindle. Plant cells have the equivalent of a centrosome, but it does not contain centrioles, suggesting that centrioles are not necessary to the assembly of cytoplasmic microtubules.

Centrioles, which are short cylinders with a 9 + 0 pattern of microtubule triplets, may be involved in microtubule formation.

Cilia and Flagella

Cilia (sing., **cilium**) and flagella are hairlike projections that can move either in an undulating fashion, like a whip, or stiffly, like an oar. Cells that have these organelles are capable of movement. For example, unicellular organisms called paramecia move by means of cilia, whereas sperm cells move by means of flagella. In the human body, the cells that line our upper respiratory tract have cilia that sweep debris trapped within mucus back up into the throat, where it can be swallowed or ejected. This action helps keep the lungs clean.

In eukaryotic cells, cilia are much shorter than flagella, but they have a similar construction. Both are membrane-bounded cylinders enclosing a matrix area. In the matrix are nine microtubule doublets arranged in a circle around two central microtubules. Therefore, they have a 9 + 2 pattern of microtubules. Cilia and flagella move when the microtubule doublets slide past one another (Fig. 3.14).

Cilia and flagella, which have a 9 + 2 pattern of microtubules, enable some cells to move.

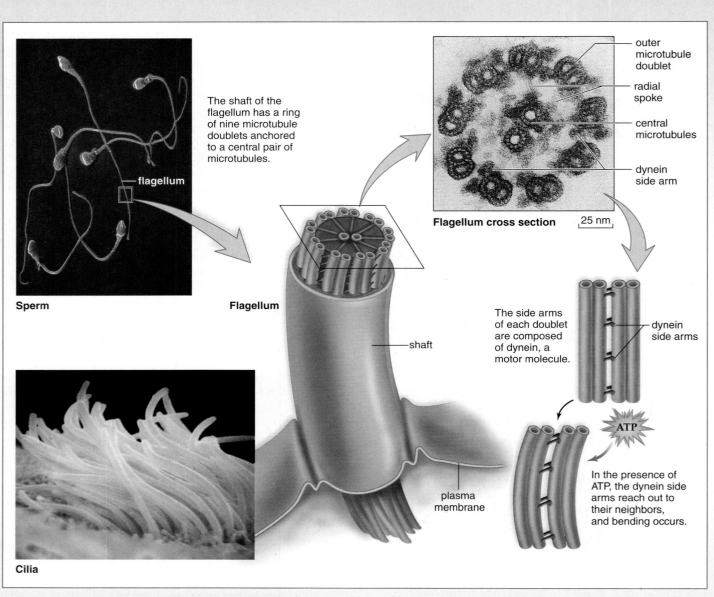

The shaft of the flagellum has a ring of nine microtubule doublets anchored to a central pair of microtubules.

outer microtubule doublet

radial spoke

central microtubules

dynein side arm

Flagellum cross section   25 nm

Sperm

flagellum

Flagellum

shaft

The side arms of each doublet are composed of dynein, a motor molecule.

dynein side arms

ATP

In the presence of ATP, the dynein side arms reach out to their neighbors, and bending occurs.

plasma membrane

Cilia

Figure 3.14 Structure of a flagellum or cilium.

The shaft of a flagellum (or cilium) contains microtubule doublets whose side arms are motor molecules that cause the projection to move. Sperm have flagella. Without the ability of sperm to move to the egg, human reproduction would not be possible. Cilia cover the surface of the cells of the respiratory system where they beat upward to remove foreign matter.

3.4 Origin and Evolution of the Eukaryotic Cell

The fossil record, which is based on the remains of ancient life, suggests that the first cells were prokaryotes. Therefore, scientists believe that eukaryotic cells evolved from prokaryotic cells. Both bacteria and archaea are candidates, but biochemical data suggest that eukaryotes are more closely related to the archaea. The eukaryotic cell probably evolved from a prokaryotic cell in stages. Invagination of the plasma membrane might explain the origin of the nuclear envelope and such organelles as the endoplasmic reticulum and the Golgi apparatus. Some believe that the other organelles could also have arisen in this manner.

Another, more interesting hypothesis of organelles has been put forth. Observations in the laboratory indicate that an amoeba infected with bacteria can become dependent upon them. Some investigators believe mitochondria and chloroplasts are derived from prokaryotes that were taken up by a much larger cell (Fig. 3.15). Perhaps mitochondria were originally aerobic heterotrophic bacteria, and chloroplasts were originally cyanobacteria. The eukaryotic host cell would have benefited from an ability to utilize oxygen or synthesize organic food when, by chance, the prokaryote was taken up and not destroyed. After the prokaryote entered the host cell, the two would have begun living together cooperatively. This proposal is known as the **endosymbiotic theory** (*endo-*, in; *symbiosis*, living together) some of the evidence supporting this hypothesis is as follows:

1. Mitochondria and chloroplasts are similar to bacteria in size and in structure.
2. Both organelles are bounded by a double membrane—the outer membrane may be derived from the engulfing vesicle, and the inner one may be derived from the plasma membrane of the original prokaryote.
3. Mitochondria and chloroplasts contain a limited amount of genetic material and divide by splitting. Their DNA (deoxyribonucleic acid) is a circular loop like that of prokaryotes.
4. Although most of the proteins within mitochondria and chloroplasts are now produced by the eukaryotic host, they do have their own ribosomes and they do produce some proteins. Their ribosomes resemble those of prokaryotes.
5. The RNA (ribonucleic acid) base sequence of the ribosomes in chloroplasts and mitochondria also suggests a prokaryotic origin of these organelles.

It is also just possible that the flagella of eukaryotes are derived from an elongated bacterium with a flagellum that became attached to a host cell. However, the flagella of eukaryotes are constructed differently than those of modern bacteria.

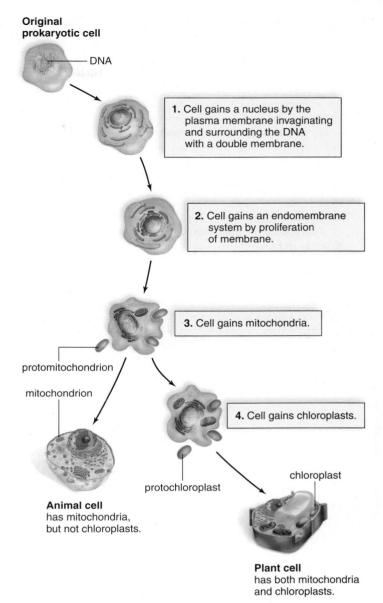

Figure 3.15 Origin of organelles.
Invagination of the plasma membrane could have created the nuclear envelope and an endomembrane system. The endosymbiotic theory suggests that mitochondria and chloroplasts were once independent prokaryotes that took up residence in a eukaryotic cell.

According to the endosymbiotic theory, heterotrophic bacteria became mitochondria, and cyanobacteria became chloroplasts after being taken up by precursors to modern-day eukaryotic cells.

Cloning refers to the production of identical copies, whether of a particular molecule, a type of cell, or an entire organism. Cloning of particular molecules through biotechnology has become fairly commonplace. Insulin, vaccines, and other pharmaceuticals are currently produced in this manner. However, two types of cloning associated with humans raise several difficult bioethical questions: therapeutic cloning and reproductive cloning.

Therapeutic cloning involves genetically manipulating human cells and then cloning those cells in culture. The intent is to eventually provide a source of "good" or "cured" cells for a diseased or injured person. For example, the ability to alter specific genes and selectively produce a given tissue type offers strong promise for the treatment of conditions such as Parkinson disease and Alzheimer disease. Scientists would like to use embryonic stem cells as the starting material for therapeutic cloning, and currently the best sources of such cells are early (approximately one week old) human embryos. Some of these embryos are a by-product of in vitro fertilization attempts, while others come from abortions.

Embryonic stem cells are unspecialized cells that can be cultivated in the laboratory and, in theory, coaxed with the addition of particular chemical signals to develop into any one of the many cell types in the human body. Currently, much of this work is in the developmental stage, and thus cloned cells are used primarily for research purposes. Those opposed to the use of embryonic stem cells have suggested that other possible sources of stem cells be pursued, including stored newborn umbilical cord cells and adult stem cells.

Reproductive cloning does not currently have much support in our society. In humans, this type of cloning would involve producing another human being with the exact genetic material of an existing person. This type of cloning has been successful with other animals, including some endangered species and household cats, but currently the United States and 19 European nations have a moratorium on the cloning of humans.

Form Your Own Opinion

1. Should scientists be allowed to use early human embryos as a source of stem cells?
2. With in vitro fertilization, extra embryos are produced that are frozen, and ultimately destroyed. Is there a problem with using these embryos for research if they would be destroyed anyway?
3. What could be the advantages of cloning animals, such as endangered species?
4. Can you think of any good reasons to lift the moratorium on human reproductive cloning?

Summarizing the Concepts

3.1 The Cellular Level of Organization
All organisms are composed of cells, the smallest units of living matter. Cells are capable of self-reproduction, and new cells come only from preexisting cells. Cells must remain small in order to have an adequate ratio of surface-area-to-volume for exchange of molecules with the environment. All cells have a plasma membrane consisting of a phospholipid bilayer with embedded proteins. The membrane regulates the movement of molecules into and out of the cell. The inside of the cell is filled with a fluid called cytoplasm.

3.2 Prokaryotic Cells
Prokaryotic cells do not have a nucleus, but they do have a nucleoid that is not bounded by a nuclear envelope. They also lack most of the other organelles that compartmentalize eukaryotic cells. Prokaryotic cells, as exemplified by the bacteria and archaea, are structurally less complex than eukaryotic cells but metabolically very diverse.

3.3 Eukaryotic Cells
The nucleus of eukaryotic cells, which include animal, plant, fungi, and protist cells, is bounded by a nuclear envelope containing pores. These pores serve as passageways between the cytoplasm and the nucleoplasm. Within the nucleus, the chromatin is a complex of DNA and protein. In dividing cells, the DNA is found in separate structures called chromosomes. The nucleolus is a special region of the chromatin where rRNA is produced and where proteins from the cytoplasm gather to form ribosomal subunits.

Ribosomes are organelles that function in protein synthesis. They can be bound to the endoplasmic reticulum (ER) or can exist within the cytoplasm singly or in groups called polyribosomes.

The endomembrane system includes the ER, the Golgi apparatus, the lysosomes, and other types of vesicles and vacuoles. The endomembrane system compartmentalizes the cell. The rough endoplasmic reticulum (RER) is covered with ribosomes and is involved in the folding, modification, and transport of proteins. The smooth ER has various metabolic functions depending on the cell type, but it also forms vesicles that carry products to the Golgi apparatus. The Golgi apparatus processes proteins and repackages them into lysosomes, which carry out intracellular digestion, or into vesicles for transport to the plasma membrane or other organelles. Vacuoles are large storage sacs, and vesicles are smaller ones. The large single plant cell vacuole not only stores substances but also lends support to the plant cell.

Peroxisomes contain enzymes that oxidize molecules by producing hydrogen peroxide, which is subsequently broken down.

Cells require a constant input of energy to maintain their structure. Chloroplasts capture the energy of the sun and carry on photosynthesis, which produces carbohydrates. Carbohydrate-derived products are broken down in mitochondria at the same time as ATP is produced. This is an oxygen-requiring process called cellular respiration.

The cytoskeleton contains actin filaments, intermediate filaments, and microtubules. These maintain cell shape and allow the cell and its organelles to move. Actin filaments, the thinnest filaments, interact with the motor molecule myosin in muscle cells to bring about contraction; in other cells, they pinch off daughter cells and have other dynamic functions. Intermediate filaments support the nuclear envelope and the plasma membrane. Microtubules radiate out from the centrosome and are present in centrioles, cilia, and flagella. In the cytoplasm they serve as tracks along which vesicles and other organelles move due to the action of specific motor molecules.

3.4 Origin and Evolution of the Eukaryotic Cell

The first cells were probably prokaryotic cells. Eukaryotic cells most likely arose from prokaryotic cells in stages. Biochemical data suggest that eukaryotic cells are closer evolutionarily to the archaea than to the bacteria. The nuclear envelope, most likely, evolved through invagination of the plasma membrane, but mitochondria and chloroplasts may have arisen through endosymbiotic events.

Testing Yourself

Choose the best answer for each question.

1. Proteins are produced
   a. in the cytoplasm or the ER.   c. in the Golgi apparatus.
   b. in the nucleus.   d. None of these are correct.

2. Which is associated with DNA?
   a. chromatin   c. nucleus
   b. chromosome   d. All of these are correct.

3. Cell walls are found in _____ and contain _____.
   a. animals, chitin   c. plants, chitin
   b. animals, cellulose   d. plants, cellulose

4. Which structure is characteristic of prokaryotic cells?
   a. nucleus   c. nucleoid
   b. mitochondria   d. All of these are correct.

5. Eukaryotic cells are associated with _____, and prokaryotic cells are associated with the _____.
   a. chloroplasts, nucleus
   b. chloroplasts, chloroplasts
   c. individual thylakoids, chloroplasts
   d. chloroplasts, individual thylakoids

For questions 6–9, match the statements to the terms in the key.

Key:

   a. chloroplasts   c. vacuoles
   b. lysosomes   d. nucleus

6. All eukaryotic cells

7. Photosynthetic site

8. Store water

9. Contain enzymes for digestion

10. The Golgi apparatus can be found in _____ cells.
    a. animal   c. bacterial
    b. plant   d. Both a and b are correct.

11. _____ are (is) produced by the smooth ER.
    a. Proteins   c. Lipids
    b. DNA   d. Ribosomes

12. Mitochondria and chloroplasts contain _____ and are able to synthesize _____.
    a. RNA, fatty acids   c. DNA, fatty acids
    b. DNA, proteins   d. cholesterol, fatty acids

13. Which of these are involved in the movement of structures inside a cell?
    a. actin   c. centrioles
    b. microtubules   d. All of these are correct.

14. Lysosomes function in _____
    a. protein synthesis.   d. lipid synthesis.
    b. processing and packaging.   e. All of these are correct.
    c. intracellular digestion.

15. Which of these could you see with a light microscope?
    a. atom   c. proteins
    b. amino acid   d. None of these are correct.

For questions 16–20, match the structure to the function in the key.

Key:

   a. movement of cell   d. ribosome formation
   b. transport of proteins   e. folds and processes proteins
   c. photosynthesis

16. Chloroplasts

17. Flagella

18. Golgi apparatus

19. Rough ER

20. Nucleolus

21. Which of these sequences depicts a hypothesized evolutionary scenario?
    a. cyanobacteria ⟶ mitochondria
    b. Golgi ⟶ mitochondria
    c. mitochondria ⟶ cyanobacteria
    d. cyanobacteria ⟶ chloroplast

22. Vacuoles are most prominent in
    a. animal cells.   c. Both a and b are correct.
    b. plant cells.   d. Neither a nor b is correct.

23. A "9 + 2" formation refers to
    a. cilia.   c. centrioles.
    b. flagella.   d. Both a and b are correct.

24. The products of photosynthesis are
    a. carbohydrates and oxygen.   c. oxygen and water.
    b. carbon dioxide and water.   d. carbohydrates and water.

25. Label these parts of the cell that are involved in protein synthesis and modification. Give a function for each structure.

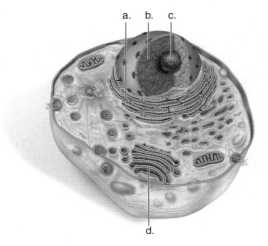

26. The cell theory states:
    a. Cells form as organelles and molecules become grouped together in an organized manner.
    b. The normal functioning of an organism does not depend on its individual cells.
    c. The cell is the basic unit of life.
    d. Only eukaryotic organisms are made of cells.
27. The small size of cells is best correlated with
    a. the fact that they are self-reproducing.
    b. their prokaryotic versus eukaryotic nature.
    c. an adequate surface area for exchange of materials.
    d. their vast versatility.
    e. All of these are correct.
28. Mitochondria
    a. are involved in cellular respiration.
    b. break down ATP to release energy for cells.
    c. contain grana and cristae.
    d. have a convoluted outer membrane.
    e. All of these are correct.
29. Which of these compounds is broken down during cellular respiration?
    a. carbon dioxide   d. oxygen
    b. water   e. Both c and d are correct
    c. carbohydrate
30. Which of the following is not one of the three components of the cytoskeleton?
    a. flagella   c. microtubules
    b. actin filaments   d. intermediate filaments
31. Which of the following structures would be found in both plant and animal cells?
    a. centrioles
    b. chloroplasts
    c. cell wall
    d. mitochondria
    e. All of these are found in both types of cells.

Understanding the Terms

capsule 48
cell theory 46
cell wall 47
cellular respiration 57
centriole 61
centrosome 60
chloroplast 57
chromatin 52
chromosome 52
cilium (pl., cilia) 61
cristae 58
cytoplasm 47
cytoskeleton 59
endoplasmic reticulum (ER) 54
endosymbiotic theory 63
eukaryotic cell 49

fimbriae 48
flagellum (pl., flagella) 48
Golgi apparatus 55
granum (pl., grana) 58
lysosome 56
matrix 58
mitochondrion 57
motor molecule 60
nuclear envelope 52
nuclear pore 52
nucleoid 48
nucleolus (pl., nucleoli) 52
nucleoplasm 52
nucleus 49
organelle 46
peroxisome 56
photosynthesis 57

plasma membrane 47
plasmid 48
polyribosome 54
prokaryotic cell 47
ribosome 48
secretion 55

slime layer 48
stroma 58
thylakoid 48
vacuole 56
vesicle 54

Match the terms to these definitions:

a._____Unstructured semifluid substance that fills the space inside organelles.
b._____Dark-staining, spherical body in the nucleus that produces ribosomal subunits.
c._____Internal framework of the cell, consisting of microtubules, actin filaments, and intermediate filaments.
d._____Organelle consisting of saccules and vesicles that processes, packages, and distributes molecules about or from the cell.
e._____System of membranous saccules and channels in the cytoplasm, often with attached ribosomes.

Thinking Critically

1. At the beginning of this chapter, you met Laurel, who was examining cells scraped from the inside of her mouth, placed on a slide, and stained with blue dye. What type of microscope was Laurel most likely using to look at her cells? What was the purpose of the dye?
2. In the 1958 movie *The Blob,* a giant, single-celled alien creeps and oozes around, attacking and devouring helpless humans. Why couldn't there be a real single-celled organism as large as the Blob?
3. A red blood cell circulating in your bloodstream has no nucleus. It had one, but it ejected it during the maturation process. Is the mature red blood cell a prokaryotic cell or a eukaryotic cell? What characteristics, apart from a nucleus, would help you distinguish between a prokaryotic cell and a eukaryotic cell?
4. Calculate the surface-area-to-volume ratio of a 1 mm cube and a 2 mm cube. Which has the smaller ratio?
5. Some of the proteins in the mitochondria are encoded by mitochondrial DNA, while others are encoded by nuclear DNA. If the mitochondria were once free-living prokaryotic cells, how would you explain this?

ARIS, the *Inquiry into Life* Website

ARIS, the website for *Inquiry into Life,* provides a wealth of information organized and integrated by chapter. You will find practice quizzes, interactive activities, labeling exercises, flashcards, and much more that will complement your learning and understanding of general biology.

www.aris.mhhe.com

Chad decided to skip dessert after lunch *even though most of his friends were helping themselves to the ice cream bar. Chad watches what he eats now because he has type 2, or adult-onset, diabetes. Diabetes is a condition in which the level of sugar (glucose) in the blood is too high. Long-term complications of diabetes include eye and kidney damage, heart problems, high blood pressure, and nerve damage.*

Insulin is the normal signal sent to cells to tell them to take up glucose from the blood. In type 1 diabetes, also called juvenile-onset diabetes, the body does not make enough insulin, so the cells do not take up glucose. Type 1 diabetics need to take insulin to maintain normal blood levels of glucose. Chad has normal amounts of insulin, so he does not need to take it. Having type 2 diabetes means that his body does not respond correctly to insulin, either because the receptors on his cells do not recognize the insulin or because the carrier proteins that allow glucose to enter the cells do not work correctly. Right now Chad can control his blood sugar levels with exercise and diet. He works out regularly and eats healthy foods, at the right time and in the right amounts. He knows that the complications of diabetes are not worth the momentary enjoyment ice cream might bring. In this chapter, you will learn how glucose and many other molecules get into and out of the cell through the plasma membrane.

chapter 4

Membrane Structure and Function

Chapter Concepts

4.1 Plasma Membrane Structure and Function

The plasma membrane separates the internal environment of the cell from the external environment. It regulates the entrance and exit of molecules into and out of the cell. In this way, it helps the cell and the organism maintain a steady internal environment. The plasma membrane is a phospholipid bilayer in which protein molecules are either partially or wholly embedded. The phospholipid bilayer has a fluid consistency, comparable to that of light oil. The proteins are scattered either just outside or within the membrane; therefore, they form a *mosaic* pattern. This description of the plasma membrane is called the **fluid-mosaic model** of membrane structure (Fig. 4.1).

The hydrophilic (water-loving) polar heads of the phospholipid molecules face the outside and inside of the cell where water is found, and the hydrophobic (water-fearing) nonpolar tails face each other (Fig. 4.1). Cholesterol is another lipid found in animal plasma membranes; related steroids are found in the plasma membranes of plants. Cholesterol stiffens and strengthens the membrane, thereby helping to regulate its fluidity.

The proteins in a membrane consist of peripheral proteins and integral proteins. Peripheral proteins are associated with only one side of the plasma membrane; peripheral proteins on the inside of the membrane are often held in place by cytoskeletal filaments. In contrast, integral proteins span the membrane, and can protrude from one or both sides. They are embedded in the membrane, but they can move laterally, changing their position in the membrane.

Both phospholipids and proteins can have attached carbohydrate (sugar) chains. If so, these molecules are called **glycolipids** and **glycoproteins,** respectively. Since the carbohydrate chains occur only on the outside surface and peripheral proteins occur asymmetrically on one surface or the other, the two halves of the membrane are not identical.

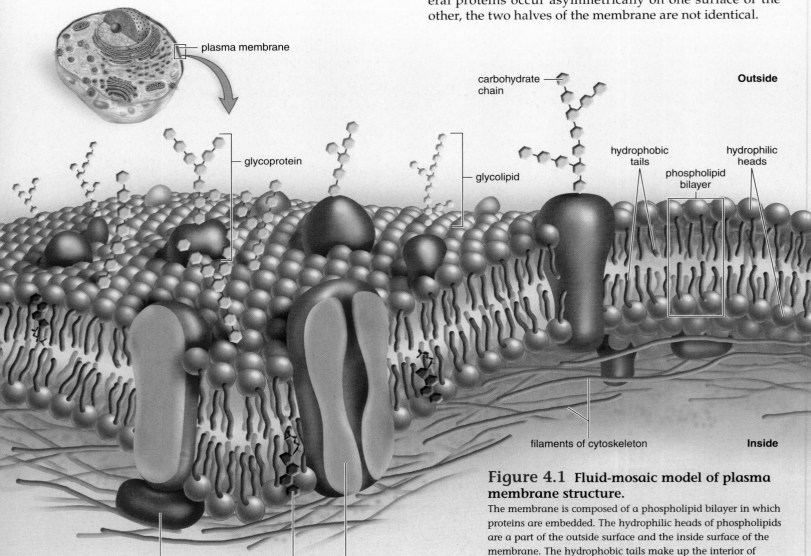

Figure 4.1 Fluid-mosaic model of plasma membrane structure.

The membrane is composed of a phospholipid bilayer in which proteins are embedded. The hydrophilic heads of phospholipids are a part of the outside surface and the inside surface of the membrane. The hydrophobic tails make up the interior of the membrane. Note the plasma membrane's asymmetry— carbohydrate chains are attached to the outside surface, and cytoskeleton filaments are attached to the inside surface.

Functions of the Proteins

The plasma membranes of various cells and the membranes of various organelles each have their own unique collections of proteins. For example, the plasma membrane of a red blood cell contains over 50 different types of proteins, and each has a specific function.

The integral proteins largely determine a membrane's specific functions. Integral proteins can be of the following types:

Channel proteins Channel proteins are involved in the passage of molecules through the membrane. They have a channel that allows a substance to simply move across the membrane (Fig. 4.2a). For example, a channel protein allows hydrogen ions to flow across the inner mitochondrial membrane. Without this movement of hydrogen ions, ATP would never be produced.

Carrier proteins Carrier proteins are also involved in the passage of molecules through the membrane. They combine with a substance and help it move across the membrane (Fig. 4.2b). For example, a carrier protein transports sodium and potassium ions across a nerve cell membrane. Without this carrier protein, nerve conduction would be impossible.

Cell recognition proteins Cell recognition proteins are glycoproteins (Fig. 4.2c). Among other functions, these proteins help the body recognize when it is being invaded by pathogens so that an immune

reaction can occur. Without this recognition, pathogens would be able to freely invade the body.

Receptor proteins Receptor proteins have a shape that allows a specific molecule to bind to it (Fig. 4.2d). The binding of this molecule causes the protein to change its shape and thereby bring about a cellular response. The coordination of the body's organs is totally dependent on such signal molecules. For example, the liver stores glucose after it is signaled to do so by insulin.

Enzymatic proteins Some plasma membrane proteins are enzymatic proteins that carry out metabolic reactions directly (Fig. 4.2e). Without the presence of enzymes, some of which are attached to the various membranes of the cell, a cell would never be able to perform the metabolic reactions necessary to its proper function.

The peripheral proteins often have a structural role in that they help stabilize and shape the plasma membrane.

> The plasma membrane consists of a phospholipid bilayer that has the consistency of light oil and accounts for the fluidity of the membrane. The integral proteins have specific functions. Some are involved in the passage of molecules through the membrane, others are receptors for signal molecules, and still others play an enzymatic role.

a.

Channel Protein
Allows a particular molecule or ion to cross the plasma membrane freely. Cystic fibrosis, an inherited disorder, is caused by a faulty chloride (Cl^-) channel; a thick mucus collects in airways and in pancreatic and liver ducts.

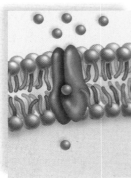

b.

Carrier Protein
Selectively interacts with a specific molecule or ion so that it can cross the plasma membrane. The inability of some persons to use energy for sodium-potassium (Na^+-K^+) transport has been suggested as the cause of their obesity.

c.

Cell Recognition Protein
The MHC (major histocompatibility complex) glycoproteins are different for each person, so organ transplants are difficult to achieve. Cells with foreign MHC glycoproteins are attacked by white blood cells responsible for immunity.

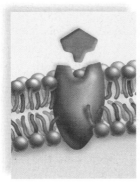

d.

Receptor Protein
Shaped in such a way that a specific molecule can bind to it. Some types of dwarfism result not because the body does not produce enough growth hormone, but because the plasma membrane growth hormone receptors are faulty and cannot interact with growth hormone.

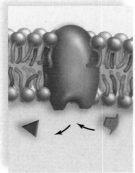

e.

Enzymatic Protein
Catalyzes a specific reaction. The membrane protein, adenylate cyclase, is involved in ATP metabolism. Cholera bacteria release a toxin that interferes with the proper functioning of adenylate cyclase, which eventually leads to severe diarrhea.

Figure 4.2 Membrane protein diversity.
These are some of the functions performed by integral proteins found in the plasma membrane.

4.2   The Permeability of the Plasma Membrane

The plasma membrane regulates the passage of molecules into and out of the cell. This function is critical because the life of the cell depends on maintenance of its normal composition. The plasma membrane can carry out this function because it is **differentially permeable,** meaning that certain substances can move across the membrane while others cannot.

Table 4.1 lists, and Figure 4.3 illustrates, which types of molecules can freely (i.e., passively) cross a membrane and which may require transport by a carrier protein and/or an expenditure of energy. In general, water and small, noncharged molecules, such as carbon dioxide, oxygen, glycerol, water, and alcohol, can freely cross the membrane. They are able to slip between the hydrophilic heads of the phospholipids and pass through the hydrophobic tails of the membrane. These molecules are said to follow their **concentration gradient** as they move from an area where their concentration is high to an area where their concentration is low.

Consider that a cell is always using oxygen when it carries on cellular respiration. Therefore, the concentration of oxygen is always lower inside a cell than outside a cell, and so oxygen has a tendency to enter a cell. Carbon dioxide, on the other hand, is produced when a cell carries on cellular respiration. Therefore, carbon dioxide is also following a concentration gradient when it moves from inside the cell to outside the cell.

Large molecules and some ions and charged molecules are unable to freely cross the membrane. They can cross the plasma membrane through channel proteins, with the assistance of carrier proteins, or in vesicles. A channel protein forms a pore through the membrane that allows molecules of a certain size and/or charge to pass. Carrier proteins are specific for the substances they transport across the plasma membrane—for example, sodium ions or glucose.

Vesicle formation is another way a molecule can exit a cell (called exocytosis) or enter a cell (called endocytosis).

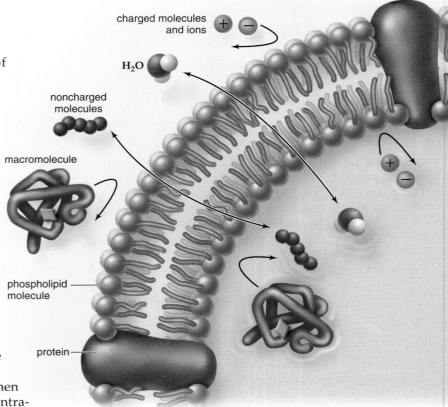

Figure 4.3 How molecules cross the plasma membrane. The curved arrows indicate substances that cannot diffuse across the plasma membrane, and the long back-and-forth arrows indicate substances that can diffuse across the plasma membrane.

This method of crossing a plasma membrane is reserved for macromolecules or even larger materials, such as a virus.

The plasma membrane is differentially permeable. Certain substances can freely pass through the membrane, and others cannot. Those that cannot freely cross the membrane may move through channel proteins or be transported across by carrier proteins or by vesicle formation.

TABLE 4.1	Passage of Molecules Into and Out of the Cell			
	Name	**Direction**	**Requirement**	**Examples**
Energy Not Required	Diffusion	Toward lower concentration	Concentration gradient	Lipid-soluble molecules, water, and gases
	Facilitated transport	Toward lower concentration	Channels or carrier and concentration gradient	Some sugars and some amino acids
Energy Required	Active transport	Toward higher concentration	Carrier plus energy	Sugars, amino acids, and ions
	Exocytosis	Toward outside	Vesicle fuses with plasma membrane	Macromolecules
	Endocytosis	Toward inside	Vesicle formation	Macromolecules

Figure 4.4 Process of diffusion.

Diffusion is spontaneous, and no chemical energy is required to bring it about. **a.** When a dye crystal is placed in water, it is concentrated in one area. **b.** The dye dissolves in the water, and there is a net movement of dye molecules from a higher to a lower concentration. There is also a net movement of water molecules from a higher to a lower concentration. **c.** Eventually, the water and the dye molecules are equally distributed throughout the container.

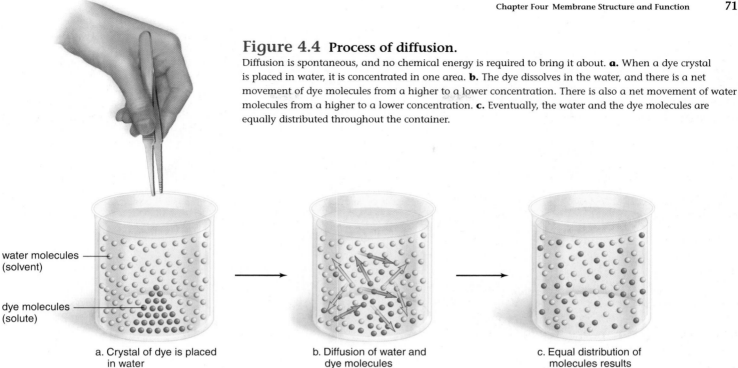

water molecules
(solvent)

dye molecules
(solute)

a. Crystal of dye is placed
   in water

b. Diffusion of water and
   dye molecules

c. Equal distribution of
   molecules results

Diffusion and Osmosis

Diffusion is the movement of molecules from a higher to a lower concentration—that is, down their concentration gradient—until equilibrium is achieved and they are distributed equally. Diffusion is a physical process that can be observed with any type of molecule. For example, when a crystal of dye is placed in water (Fig. 4.4), the dye and water molecules move in various directions, but their net movement, which is the sum of their motion, is toward the region of lower concentration. Eventually, the dye is dissolved in the water, resulting in equilibrium and a colored solution.

A solution contains both a solute, usually a solid, and a **solvent,** usually a liquid. In this case, the solute is the dye and the solvent is the water molecules. Once the solute and solvent are evenly distributed, their molecules continue to move about, but there is no net movement of either one in any direction.

The chemical and physical properties of the plasma membrane allow only a few types of molecules to enter and exit a cell simply by diffusion. Gases can diffuse through the lipid bilayer; this is the mechanism by which oxygen enters cells and carbon dioxide exits cells. Also, consider the movement of oxygen from the alveoli (air sacs) of the lungs to the blood in the lung capillaries (Fig. 4.5). After inhalation (breathing in), the concentration of oxygen in the alveoli is higher than that in the blood; therefore, oxygen diffuses into the blood.

Several factors influence the rate of diffusion. Among these factors are temperature, pressure, electrical currents, and molecular size. For example, as temperature increases, the rate of diffusion increases.

Molecules diffuse down their concentration gradients. A few types of small molecules can simply diffuse through the plasma membrane.

Figure 4.5 Gas exchange in lungs.

Oxygen (O_2) diffuses into the capillaries of the lungs because there is a higher concentration of oxygen in the alveoli (air sacs) than in the capillaries.

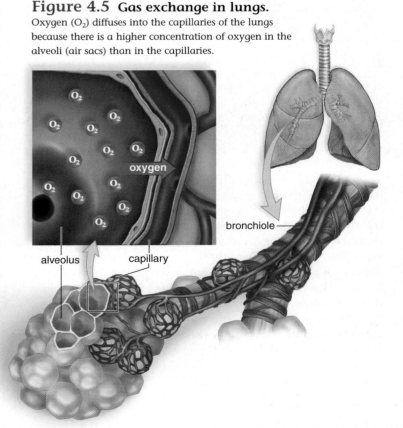

O_2

oxygen

alveolus   capillary

bronchiole

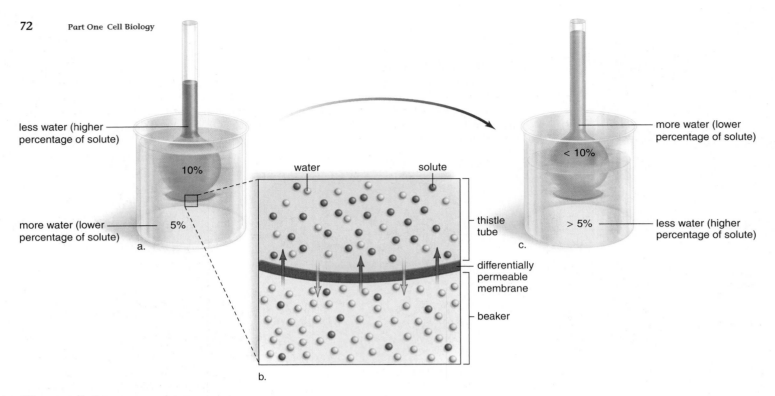

less water (higher percentage of solute)

10%

more water (lower percentage of solute)

5%

a.

water

solute

thistle tube

differentially permeable membrane

beaker

b.

more water (lower percentage of solute)

< 10%

less water (higher percentage of solute)

> 5%

c.

Figure 4.6 Osmosis demonstration.
a. A thistle tube, covered at the broad end by a differentially permeable membrane, contains a 10% solute solution. The beaker contains a 5% solute solution.
b. The solute (green circles) is unable to pass through the membrane, but the water (blue circles) passes through in both directions. There is a net movement of water toward the inside of the thistle tube, where the percentage of water molecules is lower. **c.** Due to the incoming water molecules, the level of the solution rises in the thistle tube.

Osmosis

The diffusion of water across a differentially permeable membrane due to concentration differences is called **osmosis**. To illustrate osmosis, a thistle tube containing a 10% solute solution[1] is covered at one end by a differentially permeable membrane and then placed in a beaker containing a 5% solute solution (Fig. 4.6). The beaker has a higher concentration of water molecules (lower percentage of solute), and the thistle tube has a lower concentration of water molecules (higher percentage of solute). Diffusion always occurs from higher to lower concentration. Therefore, a net movement of water takes place across the membrane from the beaker to the inside of the thistle tube.

The solute does not diffuse out of the thistle tube. Why not? Because the membrane is not permeable to the solute. As water enters and the solute does not exit, the level of the solution within the thistle tube rises (Fig. 4.6c). In the end, the concentration of solute in the thistle tube is less than 10%. Why? Because there is now less solute per unit volume of solution. And, the concentration of solute in the beaker is greater than 5% because there is now more solute per unit volume.

Water enters the thistle tube due to the osmotic pressure of the solution within the thistle tube. **Osmotic pressure** is the pressure that develops in a system due to osmosis.[2] In other words, the greater the possible osmotic pressure, the more likely it is that water will diffuse in that direction. Due to osmotic pressure, water is absorbed by the kidneys and taken up by capillaries in the tissues. Osmosis also occurs across the plasma membrane, as we shall now see (Fig. 4.7).

Isotonic Solution In the laboratory, cells are normally placed in **isotonic solutions**—that is, the solute concentration and the water concentration both inside and outside the cell are equal, and therefore there is no net gain or loss of water. The prefix *iso* means "the same as," and the term tonicity refers to the strength of the solution. A 0.9% solution of the salt sodium chloride (NaCl) is known to be isotonic to red blood cells. Therefore, intravenous solutions medically administered usually have this tonicity. Terrestrial animals can usually take in either water or salt as needed to maintain the tonicity of their internal environment. Many animals living in an estuary, such as oysters, blue crabs, and some fishes, are able to cope with changes in the salinity (salt concentrations) of their environment. Their kidneys, gills, and other structures help them do this.

Hypotonic Solution Solutions that cause cells to swell, or even to burst, due to an intake of water are said to be **hypotonic solutions**. The prefix *hypo* means "less than" and refers to a solution with a lower concentration of solute (higher concentration of water) than inside the cell. If a cell is placed in a hypotonic solution, water enters the cell; the net movement of water is from the outside to the inside of the cell.

[1]*Percent solutions are grams of solute per 100 ml of solvent. Therefore, a 10% solution is 10 g of sugar with water added to make 100 ml of solution.*

[2]*Osmotic pressure is measured by placing a solution in an osmometer and then immersing the osmometer in pure water. The pressure that develops is the osmotic pressure of a solution.*

Animal cells

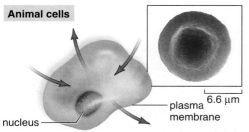 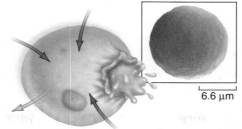 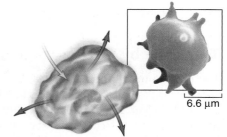

nucleus — plasma membrane

In an isotonic solution, there is no net movement of water.

In a hypotonic solution, water enters the cell, which may burst (lysis).

In a hypertonic solution, water leaves the cell, which shrivels (crenation).

Plant cells

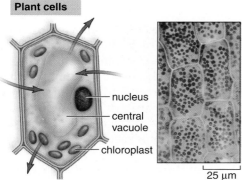 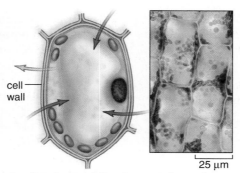 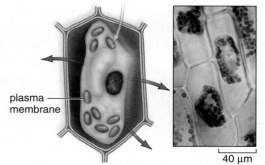

nucleus
central vacuole
chloroplast

cell wall

plasma membrane

In an isotonic solution, there is no net movement of water.

In a hypotonic solution, the central vacuole fills with water, turgor pressure develops, and chloroplasts are seen next to the cell wall.

In a hypertonic solution, the central vacuole loses water, the cytoplasm shrinks (plasmolysis), and chloroplasts are seen in the center of the cell.

Figure 4.7 Osmosis in animal and plant cells.
The arrows indicate the movement of water molecules. To determine the net movement of water, compare the number of arrows that are taking water molecules into the cell with the number that are taking water out of the cell. In an isotonic solution, a cell neither gains nor loses water; in a hypotonic solution, a cell gains water; and in a hypertonic solution, a cell loses water.

Any concentration of a salt solution lower than 0.9% is hypotonic to red blood cells. Animal cells placed in such a solution expand and sometimes burst due to the buildup of pressure. The term **cytolysis** is used to refer to disrupted cells; hemolysis, then, is disrupted red blood cells.

The swelling of a plant cell in a hypotonic solution creates **turgor pressure.** When a plant cell is placed in a hypotonic solution, the cytoplasm expands because the large central vacuole gains water and the plasma membrane pushes against the rigid cell wall. The plant cell does not burst because the cell wall does not give way. Turgor pressure in plant cells is extremely important to the maintenance of the plant's erect position. If you forget to water your plants, they wilt due to decreased turgor pressure.

Organisms that live in fresh water have to prevent their internal environment from becoming hypotonic. Many protozoans, such as paramecia, have contractile vacuoles that rid the body of excess water. Freshwater fishes have well-developed kidneys that excrete a large volume of dilute urine. Even so, they have to take in salts at their gills. Even though freshwater fishes are good osmoregulators, they would not be able to survive in either distilled water or a marine environment.

Hypertonic Solution Solutions that cause cells to shrink or shrivel due to loss of water are said to be **hypertonic solutions.** The prefix *hyper* means "more than" and refers to a solution with a higher percentage of solute (lower concentration of water) than the cell. If a cell is placed in a hypertonic solution, water leaves the cell; the net movement of water is from the inside to the outside of the cell.

Any concentration of a salt solution higher than 0.9% is hypertonic to red blood cells. If animal cells are placed in this solution, they shrink. The term **crenation** refers to the shriveling of a cell in a hypertonic solution. Meats are sometimes preserved by salting them. The salt kills any bacteria present because it makes the meat a hypertonic environment.

When a plant cell is placed in a hypertonic solution, the plasma membrane pulls away from the cell wall as the large central vacuole loses water. This is an example of **plasmolysis,** shrinking of the cytoplasm due to osmosis. The dead plants you may see along a salted roadside died because they were exposed to a hypertonic solution during the winter. Also, when salt water invades coastal marshes due to storms or human activities, coastal plants die. Without roots to hold the soil, it washes into the sea, thereby losing many acres of valuable wetlands.

Marine animals cope with their hypertonic environment in various ways that prevent them from losing water. Sharks increase or decrease the urea in their blood until their blood is isotonic with the environment. Marine fishes and other types of animals excrete salts across their gills. Have you ever seen a marine turtle cry? It is ridding its body of salt by means of glands near the eye.

In an isotonic solution, a cell neither gains nor loses water. In a hypotonic solution, a cell gains water. In a hypertonic solution, a cell loses water and the cytoplasm shrinks.

Transport by Carrier Proteins

The plasma membrane impedes the passage of all but a few substances. Yet, biologically useful molecules are able to enter and exit the cell at a rapid rate because of carrier proteins in the membrane. Carrier proteins are specific; each can combine with only a certain type of molecule or ion, which is then transported through the membrane. Scientists do not completely understand how carrier proteins function, but after a carrier combines with a molecule, the carrier is believed to undergo a change in shape that moves the molecule across the membrane. Carrier proteins are required for both facilitated transport and active transport (see Table 4.1).

Some of the proteins in the plasma membrane are carriers. They transport biologically useful molecules into and out of the cell.

Facilitated Transport

Facilitated transport explains the passage of such molecules as glucose and amino acids across the plasma membrane even though they are not lipid-soluble. The passage of glucose and amino acids is facilitated by their reversible combination with carrier proteins, which in some manner transport them through the plasma membrane. These carrier proteins are specific. For example, various sugar molecules of identical size might be present inside or outside the cell, but glucose can cross the membrane hundreds of times faster than the other sugars. As stated earlier, this is the reason the membrane can be called differentially permeable.

A model for facilitated transport (Fig. 4.8) shows that after a carrier has assisted the movement of a molecule to the other side of the membrane, it is free to assist the passage of other similar molecules. Neither diffusion nor facilitated transport requires an expenditure of energy (use of ATP) because the molecules are moving down their concentration gradient in the same direction they tend to move anyway.

Active Transport

During **active transport,** molecules or ions move through the plasma membrane, accumulating either inside or outside the cell. For example, iodine collects in the cells of the thyroid gland; glucose is completely absorbed from the gut by the cells lining the digestive tract; and sodium can be almost completely withdrawn from urine by cells lining the kidney tubules. In these instances, molecules have moved to the region of higher concentration, exactly opposite to the process of diffusion.

Both carrier proteins and an expenditure of energy are needed to transport molecules against their concentration gradient. In this case, chemical energy, usually in the form of ATP, is required for the carrier to combine with the substance to be transported. Therefore, it is not surprising that cells involved primarily in active transport, such as kidney cells, have a large number of mitochondria near membranes where active transport is occurring.

Proteins involved in active transport are often called pumps because, just as a water pump uses energy to move water against the force of gravity, proteins use energy to move a substance against its concentration gradient. One type of pump that is active in all animal cells, but is especially associ-

Figure 4.8 Facilitated transport.

During facilitated transport, a carrier protein speeds the rate at which the solute crosses the plasma membrane toward a lower concentration. Note that the carrier protein undergoes a change in shape as it moves a solute across the membrane.

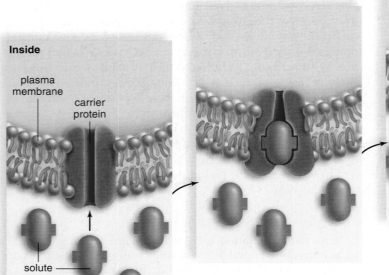
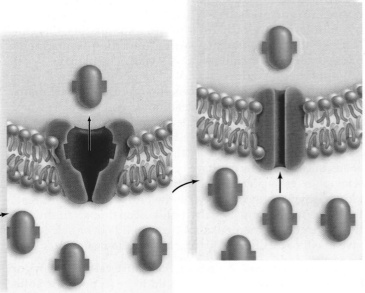

ated with nerve and muscle cells, moves sodium ions (Na⁺) to the outside of the cell and potassium ions (K⁺) to the inside of the cell. These two events are linked, and the carrier protein is called a **sodium-potassium pump.** A change in carrier shape after the attachment of a phosphate group, and again after its detachment, allows the carrier to combine alternately with sodium ions and potassium ions (Fig. 4.9). The phosphate group is donated by ATP when it is broken down enzymatically by the carrier. The sodium-potassium pump results in both a solute concentration gradient and an electrical gradient for these ions across the plasma membrane.

The passage of salt (NaCl) across a plasma membrane is of primary importance to most cells. The chloride ion (Cl⁻) usually crosses the plasma membrane because it is attracted by positively charged sodium ions (Na⁺). First sodium ions are pumped across a membrane, and then chloride ions simply diffuse through channels that allow their passage.

As noted in Figure 4.2a, the genetic disorder cystic fibrosis results from a faulty chloride channel. Ordinarily, after chloride ions have passed through the membrane, sodium ions (Na⁺) and water follow. In cystic fibrosis, Cl⁻ transport is reduced, and so is the flow of Na⁺ and water. Researchers

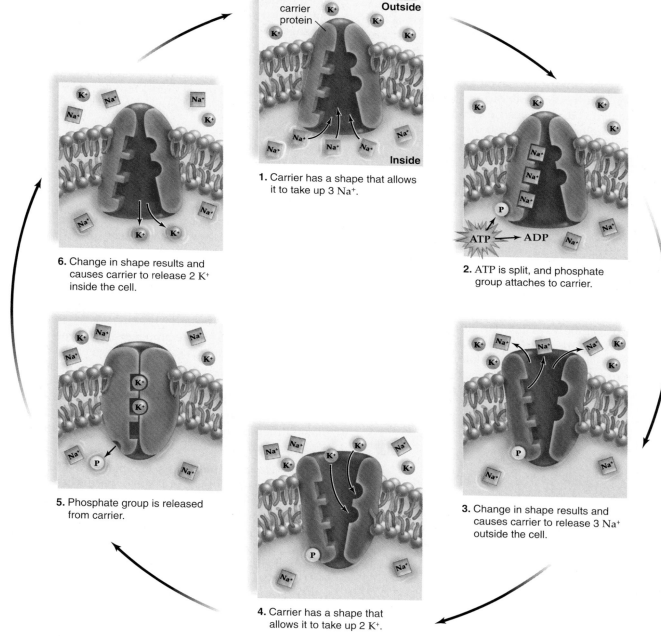

1. Carrier has a shape that allows it to take up 3 Na⁺.

2. ATP is split, and phosphate group attaches to carrier.

3. Change in shape results and causes carrier to release 3 Na⁺ outside the cell.

4. Carrier has a shape that allows it to take up 2 K⁺.

5. Phosphate group is released from carrier.

6. Change in shape results and causes carrier to release 2 K⁺ inside the cell.

Figure 4.9 The sodium-potassium pump.
The same carrier protein transports sodium ions (Na⁺) to the outside of the cell and potassium ions (K⁺) to the inside of the cell because it undergoes an ATP-dependent change in shape. Three sodium ions are carried outward for every two potassium ions carried inward; therefore, the inside of the cell is negatively charged compared to the outside.

believe that the lack of water causes the mucus in the bronchial tubes and pancreatic ducts to be abnormally thick, thus interfering with the function of the lungs and pancreas.

During facilitated transport, small molecules follow their concentration gradient. During active transport, small molecules and ions move against their concentration gradient.

Vesicle Formation

How do macromolecules such as polypeptides, polysaccharides, or polynucleotides enter and exit a cell? Because they are too large to be transported by carrier proteins, macromolecules are transported into and out of the cell by vesicle formation. Vesicle formation is called membrane-assisted transport because membrane is needed to form the vesicle. Vesicle formation requires an expenditure of cellular energy, but an added benefit is that the vesicle membrane keeps the contained macromolecules from mixing with molecules within the cytoplasm. Exocytosis is a way substances can exit a cell, and endocytosis is a way substances can enter a cell.

Exocytosis

During **exocytosis,** a vesicle fuses with the plasma membrane as secretion occurs (Fig. 4.10). Hormones, neurotransmitters, and digestive enzymes are secreted from cells in this manner. The Golgi apparatus often produces the vesicles that carry these cell products to the membrane. Notice that during exocytosis, the membrane of the vesicle becomes a part of the plasma membrane, which is thereby enlarged. For this reason, exocytosis can be a normal part of cell growth. The proteins released from the vesicle adhere to the cell surface or become incorporated in an extracellular matrix.

Cells of particular organs are specialized to produce and export molecules. For example, pancreatic cells produce

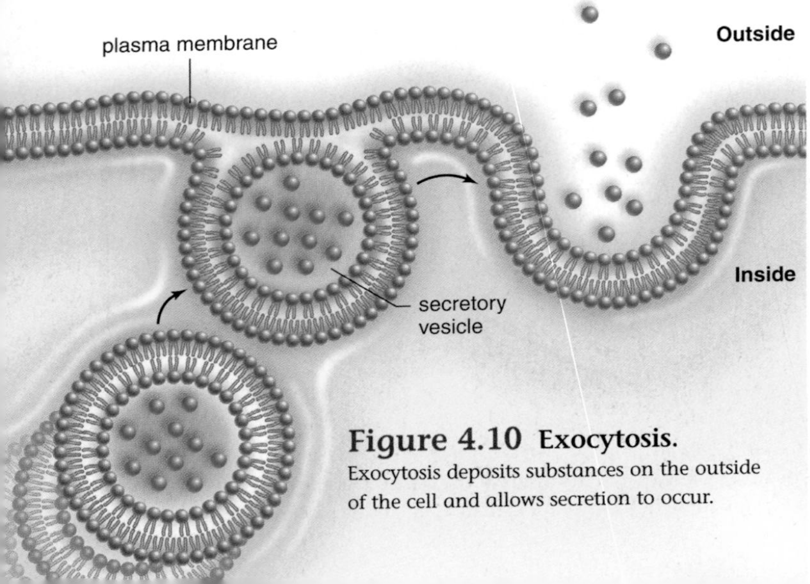

Figure 4.10 Exocytosis.
Exocytosis deposits substances on the outside of the cell and allows secretion to occur.

digestive enzymes or insulin, and anterior pituitary cells produce growth hormone, among other hormones. In these cells, secretory vesicles accumulate near the plasma membrane, and the vesicles release their contents only when the cell is stimulated by a signal received at the plasma membrane. A rise in blood sugar, for example, signals pancreatic cells to release the hormone insulin. This is called regulated secretion, because vesicles fuse with the plasma membrane only when it is appropriate to the needs of the body.

Endocytosis

During **endocytosis,** cells take in substances by vesicle formation. A portion of the plasma membrane invaginates to envelop the substance, and then the membrane pinches off to form an intracellular vesicle. Endocytosis occurs in one of three ways, as illustrated in Figure 4.11. Phagocytosis transports large substances, such as viruses, and pinocytosis transports small substances, such as macromolecules, into cells. Receptor-mediated endocytosis is a special form of pinocytosis.

Phagocytosis When the material taken in by endocytosis is large, such as a food particle or another cell, the process is called **phagocytosis.** Phagocytosis is common in unicellular organisms such as amoebas (Fig. 4.11a). It also occurs in humans. Certain types of human white blood cells are amoeboid—that is, they are mobile like an amoeba, and they are able to engulf debris such as worn-out red blood cells or viruses. When an endocytic vesicle fuses with a lysosome, digestion occurs. We will see that this process is a necessary and preliminary step toward the development of immunity to bacterial diseases.

Pinocytosis Pinocytosis occurs when vesicles form around a liquid or around very small particles (Fig. 4.11b). Blood cells, cells that line the kidney tubules or the intestinal wall, and plant root cells all use pinocytosis to ingest substances.

Whereas phagocytosis can be seen with the light microscope, an electron microscope is required to observe pinocytic vesicles, which are no larger than 0.1–0.2 µm. Still, pinocytosis involves a significant amount of the plasma membrane because it occurs continuously. The loss of plasma membrane due to pinocytosis is balanced by the occurrence of exocytosis, however.

Receptor-Mediated Endocytosis Receptor-mediated endocytosis is a form of pinocytosis that is quite specific because it uses a receptor protein shaped so that a specific molecule, such as a vitamin, peptide hormone, or lipoprotein, can bind to it (Fig. 4.11c). The receptors for these substances are found at one location in the plasma membrane. This location is called a coated pit because there is a layer of protein on the cytoplasmic side of the pit. Once formed, the vesicle becomes uncoated and may fuse with a lysosome. When empty, used vesicles fuse with the plasma membrane, the receptors return to their former location.

Receptor-mediated endocytosis is selective and much more efficient than ordinary pinocytosis. It is involved in uptake and also in the transfer and exchange of substances between cells. Such exchanges take place when substances move from maternal blood into fetal blood at the placenta, for example.

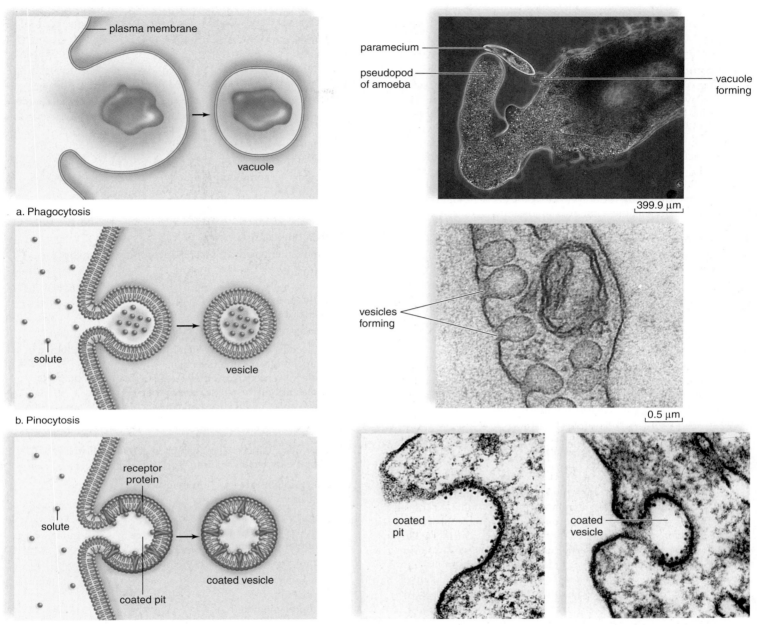

a. Phagocytosis

b. Pinocytosis

c. Receptor-mediated endocytosis

Figure 4.11 Three methods of endocytosis.
a. Phagocytosis occurs when the substance to be transported into the cell is large; amoebas ingest by phagocytosis. Digestion occurs when the resulting vacuole fuses with a lysosome. **b.** Pinocytosis occurs when a macromolecule such as a polypeptide is transported into the cell. The result is a vesicle (small vacuole). **c.** Receptor-mediated endocytosis is a form of pinocytosis. Molecules first bind to specific receptor proteins, which migrate to or are already in a coated pit. The vesicle that forms contains the molecules and their receptors.

The importance of receptor-mediated endocytosis is demonstrated by a genetic disorder called familial hyper-cholesterolemia. Cholesterol is transported in the blood by a complex of lipids and proteins called low-density lipoprotein (LDL). Ordinarily, body cells take up LDL when LDL receptors gather in a coated pit. But in some individuals, the LDL receptor is unable to properly bind to the coated pit, and the cells are unable to take up cholesterol. Instead, cholesterol accumulates in the walls of arterial blood vessels, leading to high blood pressure, occluded (blocked) arteries, and heart attacks.

Substances are secreted from a cell by exocytosis. Substances enter a cell by endocytosis. Receptor-mediated endocytosis allows cells to take up specific kinds of molecules, which are then released within the cell.

Summarizing the Concepts

4.1 Plasma Membrane Structure and Function

According to the fluid-mosaic model of the plasma membrane, a lipid bilayer is fluid and has the consistency of light oil. The hydrophilic heads of phospholipids form the inner and outer surfaces, and the hydrophobic tails form the interior. Proteins within the membrane are the mosaic portion. The integral proteins form channels, act as carriers to move substances across the membrane, function in cell recognition, act as receptors, and carry on enzymatic reactions.

Carbohydrate chains are attached to some of the lipids and proteins in the membrane; these are glycolipids and glycoproteins.

4.2 The Permeability of the Plasma Membrane

Some substances, such as gases and water, freely cross a plasma membrane, while others—particularly ions, charged molecules, and macromolecules—have to be assisted across. Passive ways of crossing a plasma membrane (diffusion and facilitated transport) do not require an expenditure of chemical energy (ATP). Active ways of crossing a plasma membrane (active transport and vesicle formation) do require an expenditure of chemical energy.

Lipid-soluble compounds, water, and gases simply diffuse across the plasma membrane from the area of higher concentration to the area of lower concentration.

The diffusion of water across a differentially permeable membrane is called osmosis. Water moves across the membrane into the area of lower water (higher solute) content. When cells are in an isotonic solution, they neither gain nor lose water; when they are in a hypotonic solution, they gain water; and when they are in a hypertonic solution, they lose water.

Some molecules are transported across the membrane by carrier proteins that span the membrane. During facilitated transport, a carrier protein assists the movement of a molecule down its concentration gradient. No energy is required. During active transport, a carrier protein acts as a pump that causes a substance to move against its concentration gradient. The sodium-potassium pump carries Na^+ to the outside of the cell and K^+ to the inside of the cell. Energy in the form of ATP molecules is required for active transport to occur.

Larger substances can exit and enter a membrane by exocytosis and endocytosis. Exocytosis involves secretion. Endocytosis includes phagocytosis and pinocytosis. Receptor-mediated endocytosis, a type of pinocytosis, makes use of receptor molecules in the plasma membrane and a coated pit, which pinches off to form a vesicle. After losing the coat, the vesicle can join with a lysosome, or after freeing its contents, the vesicle can fuse with the plasma membrane and thereby return the receptors to their former location.

Testing Yourself

Choose the best answer for each question.

1. Phagocytosis is common in
   a. amoebas.
   b. white blood cells.
   c. red blood cells.
   d. Both a and b are correct.

2. Energy is used for
   a. diffusion.
   b. osmosis.
   c. active transport.
   d. None of these are correct.

3. When mixed together, water is a _____, and dye is a _____.
   a. solute, solvent
   b. solvent, solvent
   c. solute, solute
   d. solvent, solute

4. Pinocytosis is common in
   a. blood cells.
   b. kidney tubules.
   c. plant root cells.
   d. All of these are correct.

5. Na^+ movement across a plasma membrane occurs by
   a. diffusion.
   b. endocytosis.
   c. active transport.
   d. None of these are correct.

6. The order of events during active transport is
   a. molecules move across membrane, ATP causes change in shape, molecule enters carrier.
   b. molecule enters carrier, ATP causes change in shape, molecules move across membrane.
   c. ATP causes change in shape, molecules move across membrane, molecule enters carrier.
   d. None of these are correct.

7. True or false: Glucose is transported across a plasma membrane by facilitated transport.

8. Carrier proteins are _____ in their action.
   a. specific
   b. not specific
   c. involved in diffusion
   d. None of these are correct.

9. Which of these are methods of endocytosis?
   a. phagocytosis
   b. pinocytosis
   c. receptor-mediated
   d. All of these are correct.

10. An isotonic solution has the/a _____ concentration of water as the cell and the/a _____ concentration of solute as the cell.
    a. same, same
    b. same, different
    c. different, same
    d. different, different

11. The mosaic part of the fluid-mosaic model of membrane structure refers to
    a. the phospholipids.
    b. the proteins.
    c. the glycolipids.
    d. the cholesterol.

12. Cholesterol is found in the _____ of _____ cells.
    a. cytoplasm, plant
    b. plasma membrane, animal
    c. plasma membrane, plant
    d. None of these are correct.

13. Glycolipids contain _____ and _____.
    a. proteins, lipids
    b. sugars, proteins
    c. carbohydrates, lipids
    d. cholesterol, proteins

14. The carbohydrate chains on lipids and proteins are found
    a. on the inside of the membrane.
    b. on the outside of the membrane.
    c. on both sides of the membrane.

15. A coated pit is associated with
    a. diffusion.
    b. osmosis.
    c. receptor-mediated endocytosis.
    d. pinocytosis.

16. Pumps used in active transport are made of
    a. sugars.
    b. proteins.
    c. lipids.
    d. cholesterol.

17. Exocytosis involves
    a. fusion with the plasma membrane.
    b. fusion with the nucleus.
    c. fusion with the mitochondria.
    d. fusion with a ribosome.

18. When mitochondria are found near the surface of kidney cells, the cells are probably
    a. using active transport.
    b. using facilitated transport.
    c. pumping a molecule across the membrane.
    d. allowing a molecule to diffuse into the cell.
    e. Both a and c are correct.

19. Accumulation of cholesterol can lead to
    a. high blood pressure.
    b. heart attacks.
    c. blocked arteries.
    d. All of these are correct.

20. Cystic fibrosis is caused by the malfunction of _____ channels.
    a. sodium ion
    b. chloride ion
    c. calcium ion
    d. potassium ion

21. Which of these passes through a plasma membrane by way of diffusion?
    a. CO_2
    b. oxygen
    c. water
    d. All of these are correct.

In questions 22–24, match the diagrams to the correct descriptions.

22. Receptor protein

23. Cell recognition protein

24. Enzymatic protein

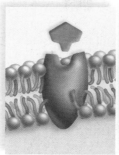

a.   b.   c.

25. A phospholipid molecule has a head and two tails. The tails are found
    a. at the surfaces of the membrane.
    b. in the interior of the membrane.
    c. spanning the membrane.
    d. where the environment is hydrophilic.
    e. Both a and b are correct.

26. Besides phospholipids, the other lipid molecule that is vital to an animal's plasma membrane is
    a. cholesterol.
    b. glycogen.
    c. triglycerides.
    d. glycerol.

27. Which of the following is not a function of proteins present in the plasma membrane?
    a. assisting the passage of materials into the cell
    b. interacting and recognizing other cells
    c. binding with specific hormones
    d. carrying out specific metabolic reactions
    e. producing lipid molecules

28. The saline (NaCl solution) used in hospitals as an IV solution (intravenous) would be
    a. hypotonic to red blood cells.
    b. isotonic to red blood cells.
    c. hypertonic to red blood cells.

29. When a cell is placed in a hypotonic solution,
    a. solute exits the cell to equalize the concentration on both sides of the plasma membrane.
    b. water exits the cell toward the area of lower solute concentration.
    c. water enters the cell toward the area of higher solute concentration.
    d. solute exits and water enters the cell.
    e. Both c and d are correct.

30. When a cell is placed in a hypertonic solution,
    a. solute exits the cell to equalize the concentration on both sides of the plasma membrane.
    b. water exits the cell toward the area of higher solute concentration.
    c. water enters the cell toward the area of higher solute concentration.
    d. solute exits and water enters the cell.
    e. Both a and c are correct.

31. After salt is put on roads during icy weather, why does the grass along the roadside often die?
    a. The melting ice lowers the temperature of the roadside, which kills the grass.
    b. The salt creates a hypertonic environment, which kills the grass.
    c. The salt beside the road attracts many animals that trample the grass.
    d. The excess water creates a hypotonic environment, which kills the grass.
    e. The salt mixes with road grime, which washes onto the roadside and kills the grass.

32. Plants wilt on a hot summer day because of a decrease in
    a. turgor pressure.
    b. evaporation.
    c. condensation.
    d. diffusion.

33. Pickles appear wrinkled and shriveled because they have been placed in a(n) _____ solution.
    a. hypertonic
    b. hypotonic
    c. isotonic
    d. basic

34. The sodium-potassium pump
    a. is active in nerve and muscle cells.
    b. concentrates sodium on the outside of the membrane.
    c. utilizes a carrier protein and energy.
    d. is present in the plasma membrane.
    e. All of these are correct.

35. The movement of water from an area of high to low water concentration would be called
    a. active transport.
    b. osmosis.
    c. facilitated transport.
    d. diffusion.
    e. Both b and c are correct.

Understanding the Terms

Match the terms to these definitions:

a._____ Movement of molecules from a region of higher concentration to a region of lower concentration.

b._____ Internal pressure that adds to the strength of a cell and builds up when water moves by osmosis into a plant cell.

c._____ Solution that contains the same concentration of solute and water as the cell.

d._____ Passive transfer of a substance into and out of a cell along a concentration gradient by a process that requires a carrier.

e._____ Process in which an intracellular vesicle fuses with the plasma membrane so that the vesicle's contents are released outside the cell.

Thinking Critically

1. When a signal molecule such as a growth hormone binds to a receptor protein in the plasma membrane, it stays on the outside of the cell. How might the inside of the cell know that the signal has bound?

2. Some antibiotics interfere with the formation of the bacterial cell wall, thus weakening the cell wall. How might this cause a bacterium to be killed?

3. How could a channel protein regulate what enters the cell?

4. An integral membrane protein must be composed of amino acids that interact with the phospholipids in order to span the membrane. What types of amino acids would be found in the different areas of the protein?

5. What would happen if you drank seawater?

ARIS, the *Inquiry into Life* Website

ARIS, the website for *Inquiry into Life,* provides a wealth of information organized and integrated by chapter. You will find practice quizzes, interactive activities, labeling exercises, flashcards, and much more that will complement your learning and understanding of general biology.

www.aris.mhhe.com

chapter 5

Cell Division

Jason removed the bandage from his knee. Tripping and falling down the steps had given him a nasty scrape and a bruise. His knee was still a little purple, but the skin was healing nicely. It was certainly healing better than his ego. He was still hoping that no one had seen him fall.

Skin is one of the tissues in the body that can regenerate itself. The outer layer of the skin, called the epidermis, is actually replacing itself all the time. The epidermis is divided into several layers. The cells in the surface layer are filled with keratin, a water-proofing substance, and are no longer living. Cells in the lower part of the epidermis are constantly dividing and pushing out the dead cells above them. These cells divide by mitosis, which we will learn about in this chapter.

Other tissues in the body, such as the blood, can also replace cells via mitosis. Blood cells of different types, including red blood cells and the various types of white blood cells, are constantly arising from stem cells via mitosis. Certain tissues, however, such as the brain and heart, do not replace themselves. When those cells are damaged, as by a stroke or heart attack, they cannot be replaced and the damage is permanent. Thankfully for Jason, skin heals very well and the only lasting part of his fall will be some slight embarrassment.

Chapter Concepts

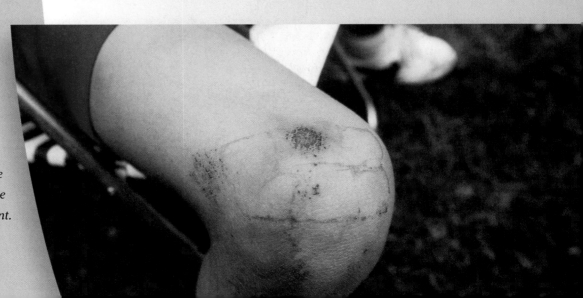

5.1   Cell Increase and Decrease

Cell division increases the number of **somatic cells** (body cells). Each of us started out as a single-celled fertilized egg, but now we are composed of trillions of cells. Each of those cells contains the same genetic material, except for sperm and eggs. Cell division continues to occur during your entire life. Even now, your body is producing thousands of new red blood cells, skin cells, and cells that line your respiratory and digestive tracts. Also, if you suffer a cut, cell division will repair the injury. Cell division consists of **mitosis,** which is division of the nucleus, and **cytokinesis,** which is division of the cytoplasm. This process results in two daughter cells that are genetically identical to the mother cell.

Apoptosis, programmed cell death, decreases the number of cells. Both cell division and apoptosis are normal parts of growth and development. Apoptosis occurs during development to remove unwanted tissue—for example, the tail of a tadpole as it matures into a frog. In humans, the fingers and toes of the embryo are initially webbed, but are normally freed from one another later in development as a result of apoptosis. Apoptosis also plays an important role in preventing cancer. An abnormal cell that becomes cancerous will often die via apoptosis, thus preventing a tumor from developing.

The Cell Cycle

Cell division is a part of the cell cycle. The **cell cycle** is an orderly set of stages that take place between the time a cell divides and the time the resulting cells also divide. Some cells become specialized and no longer enter the cell cycle.

The Stages of Interphase

As Figure 5.1*a* shows, most of the cell cycle is spent in **interphase.** This is the time when a cell carries on its usual functions, which are dependent on its location in the body. It also gets ready to divide: it grows larger, the number of organelles doubles, and the amount of DNA doubles. For mammalian cells, interphase lasts for about 20 hours, which is 90% of the cell cycle.

Interphase is divided into three stages: the G_1 stage occurs before DNA synthesis, the S stage includes DNA synthesis, and the G_2 stage occurs after DNA synthesis. Originally G stood for the "gaps" that occur in DNA synthesis during interphase. But now that we know growth occurs during these stages, the G can be thought of as standing for growth.

During the G_1 stage, a cell doubles its organelles (such as mitochondria and ribosomes), and it accumulates the materials needed for DNA synthesis.

During the S stage, DNA replication occurs. At the beginning of the S stage, each chromosome is composed of one DNA molecule, which is called a chromatid. At the end of this stage, each chromosome consists of two identical DNA molecules, and therefore is composed of two sister chromatids. Another way of expressing these events is to say that DNA replication has resulted in duplicated chromosomes.

During the G_2 stage, the cell synthesizes the proteins needed for cell division, such as the protein found in microtubules. The role of microtubules in cell division is described in section 5.2.

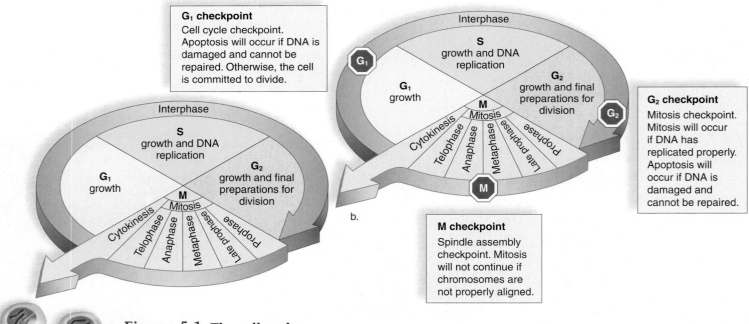

Figure 5.1 The cell cycle.
a. Cells go through a cycle that consists of four stages: G_1, S, G_2, and M. Each stage is characterized by a major activity. **b.** The cell cycle may stop at any of three checkpoints if necessary.

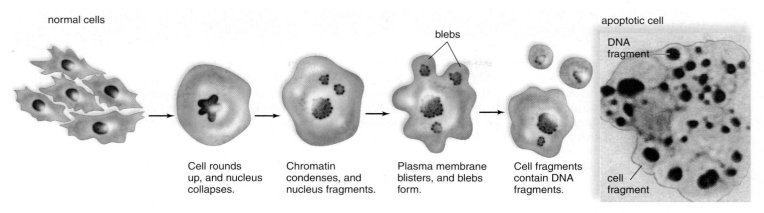

normal cells

blebs

apoptotic cell

DNA fragment

Cell rounds up, and nucleus collapses.

Chromatin condenses, and nucleus fragments.

Plasma membrane blisters, and blebs form.

Cell fragments contain DNA fragments.

cell fragment

Figure 5.2 Apoptosis.

Apoptosis is a sequence of events that results in a fragmented cell. The fragments are engulfed by white blood cells and neighboring tissue cells.

The amount of time the cell takes for interphase varies widely. Some cells, such as nerve and muscle cells, typically do not complete the cell cycle and are permanently arrested in G_1. Embryonic cells spend very little time in G_1 and complete the cell cycle in a few hours.

The Mitotic Stage

Following interphase, the cell enters the M (for mitotic) stage. This stage not only includes mitosis, but also cytokinesis, if it occurs. During mitosis, as we shall see, the sister chromatids of each chromosome separate, becoming daughter chromosomes that are distributed to two daughter nuclei. When cytokinesis is complete, two daughter cells are present. Mammalian cells usually require only about four hours to complete the mitotic stage.

Control of the Cell Cycle

The cell cycle is controlled by internal and external signals. These signals ensure that the stages follow one another in the normal sequence and that each stage is properly completed before the next stage begins. Growth factors are hormones that act as external signals received at the plasma membrane. When blood platelets release a growth factor, skin fibroblasts in the vicinity are stimulated to finish the cell cycle so an injury can be repaired.

The red stop signs in Figure 5.1b represent three checkpoints when the cell cycle possibly stops. Researchers have identified a signal called **cyclin** that increases and decreases as the cell cycle continues. Cyclin has to be present for the cell to proceed from the G_2 stage to the M stage and for the cell to proceed from the G_1 stage to the S stage.

DNA damage, such as by solar radiation or X ray, can stop the cell cycle at the G_1 checkpoint. In mammalian cells, the protein p53 stops the cycle at the G_1 checkpoint when DNA is damaged. First, p53 attempts to initiate DNA repair, but if that is not possible, it brings about apoptosis.

The cell cycle stops at the G_2 checkpoint if DNA has not finished replicating. This prevents the initiation of the M stage before completion of the S stage. Also, if DNA is damaged, stopping the cell cycle at this checkpoint allows time for the damage to be repaired so that it is not passed on to daughter cells. If repair is not possible, apoptosis occurs.

Another cell cycle checkpoint occurs during the mitotic (M) stage. The cycle stops if the chromosomes are not going to be distributed accurately to the daughter cells.

These checkpoints are critical for preventing cancer development. A damaged cell should not complete mitosis, but instead should undergo apoptosis. Tumors develop when the checkpoints malfunction. We now know that many forms of tumors contain cells that lack an active *p53* gene. In other words, when the *p53* gene is faulty and unable to promote apoptosis, cancer develops.

Apoptosis

During apoptosis, the cell progresses through a typical series of events that bring about its destruction (Fig. 5.2). The cell rounds up and loses contact with its neighbors. The nucleus fragments, and the plasma membrane develops blisters. Finally, the cell fragments, and its bits and pieces are engulfed by white blood cells and/or neighboring cells.

A remarkable finding of the past few years is that cells routinely harbor the enzymes, now called caspases, that bring about apoptosis. The enzymes are ordinarily held in check by inhibitors, but they can be unleashed either by internal or external signals. There are two sets of caspases. The first set, the "initiators," receive the signal to activate the second set, the "executioners," which then activate the enzymes that dismantle the cell. For example, executioners turn on enzymes that tear apart the cytoskeleton and enzymes that chop up DNA.

Cell division and apoptosis are two opposing processes that keep the number of healthy cells in balance.

Science Focus

What's in a Chromosome?

When early investigators decided that the genes are contained in the chromosomes, they had no idea of chromosome composition. By the mid-1900s, it was known that chromosomes are made up of both DNA and protein. Only in recent years, however, have investigators been able to produce models suggesting how chromosomes are organized.

A eukaryotic chromosome is more than 50% protein. Some of these proteins are concerned with DNA and RNA synthesis, but a large proportion, termed histones, seem to play primarily a structural role. The five primary types of histone molecules are designated H1, H2A, H2B, H3, and H4. Remarkably, the amino acid sequences of H3 and H4 vary little between organisms.

For example, the H4 of peas is only two amino acids different from the H4 of cattle. This similarity suggests that few mutations in the histone proteins have occurred during the course of evolution and that the histones therefore have important functions.

A human cell contains at least 2 meters of DNA. Yet all of this DNA is packed into a nucleus that is about 5 μm in diameter. The histones are responsible for packaging the DNA so that it can fit into such a small space. First the DNA double helix is wound at intervals around a core of eight histone molecules (two copies each of H2A, H2B, H3, and H4), giving the appearance of a string of beads (Fig. 5A*a,b*). Each bead is called a nucleosome, and the nucleosomes are said to be joined by "linker" DNA. This

string is coiled tightly into a fiber that has six nucleosomes per turn (Fig. 5A*c*). The H1 histone appears to mediate this coiling process. The fiber loops back and forth (Fig. 5A*d,e*) and can condense to produce a highly compacted form (Fig. 5A*f*) characteristic of metaphase chromosomes. During interphase, extended chromatin makes DNA available for RNA synthesis and subsequent protein synthesis.

Some regions of chromatin remain tightly coiled and compacted throughout the cell cycle and appear as dark-stained fibers known as heterochromatin (Fig. 5A*e*). Heterochromatin is considered inactive chromatin. Euchromatin (Fig. 5A*d*), the active chromatin, is condensed only during cell division.

Discussion Questions

1. Why would histones be conserved throughout evolution? In other words, why are the histones of peas and cattle very similar?
2. Why would you expect very little, if any, RNA synthesis to occur during mitosis?
3. The packing ratio (greatest length of DNA divided by its length when packed) of human DNA is approximately 7000. The shortest human chromosome is approximately 14,000 μm long. How long is this chromosome when it is packed?

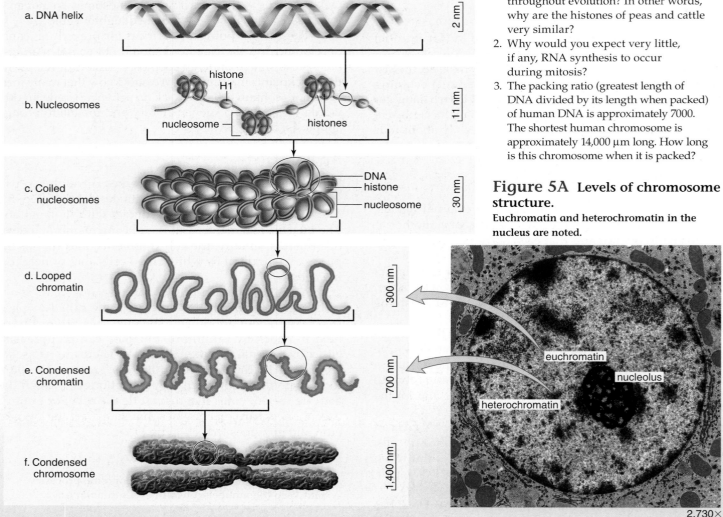

Figure 5A **Levels of chromosome structure.**
Euchromatin and heterochromatin in the nucleus are noted.

a. DNA helix — 2 nm

b. Nucleosomes — 11 nm

histone H1 — nucleosome — histones

c. Coiled nucleosomes — 30 nm

DNA — histone — nucleosome

d. Looped chromatin — 300 nm

e. Condensed chromatin — 700 nm

f. Condensed chromosome — 1,400 nm

euchromatin — nucleolus — heterochromatin

2,730×

5.2 Maintaining the Chromosome Number

When a eukaryotic cell is not undergoing division, the DNA (and associated proteins) within a nucleus is a tangled mass of thin threads called **chromatin.** At the time of cell division, chromatin coils, loops, and condenses to form highly compacted structures called **chromosomes.** The Science Focus on the opposite page describes the transition from chromatin to chromosomes in eukaryotic cells.

Each species has a characteristic chromosome number; for instance, human cells contain 46 chromosomes, corn has 20 chromosomes, and the crayfish has 200! This number is called the **diploid (2n) number** because it contains two (a pair) of each type of chromosome. Humans have 23 pairs of chromosomes. One member of each pair originates from the mother, and the other member originates from the father. In Figure 5.3, the blue chromosomes are inherited from one parent, and the red are inherited from the other.

Half the diploid number, called the **haploid (n) number** of chromosomes, contains only one of each kind of chromosome. In the life cycle of humans, only sperm and eggs have the haploid number of chromosomes.

Overview of Mitosis

Mitosis is nuclear division in which the chromosome number stays constant. A 2n nucleus divides to produce daughter nuclei that are also 2n (Fig. 5.3). Before nuclear division takes place, DNA replication occurs, duplicating the chromosomes. A **duplicated chromosome** is composed of two sister chromatids held together in a region called the **centromere. Sister chromatids** are genetically identical—they contain the same genes. At the completion of mitosis, each chromosome consists of a single chromatid:

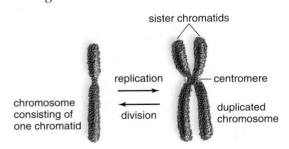

Figure 5.3 gives an overview of mitosis; for simplicity, only four chromosomes are depicted. (*In determining the number of chromosomes, it is only necessary to count the number of independent centromeres.*) During mitosis, the centromeres divide and then the sister chromatids separate, becoming **daughter chromosomes.** Therefore, each daughter nucleus gets a complete set of chromosomes and has the same number of chromosomes as the parental cell. This makes the daughter cells genetically identical to each other and to the parental cell.

Following mitosis, a 2n parental cell gives rise to two 2n daughter cells. It would be possible to diagram this as 2n ⟶ 2n. The cells of some organisms, such as algae and fungi, are

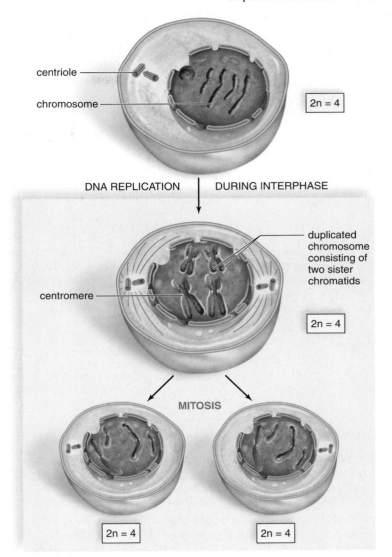

Figure 5.3 Mitosis overview.

Following DNA replication during interphase, each chromosome in the parental nucleus is duplicated and consists of two sister chromatids. During mitosis, the centromeres divide and the sister chromatids separate, becoming daughter chromosomes that move into the daughter nuclei. Therefore, daughter cells have the same number and kinds of chromosomes as the parental cell. (The blue chromosomes were inherited from one parent, and the red chromosomes were inherited from the other.)

haploid as adults. The n parental cell gives rise to two daughter cells that are also haploid. In these organisms, n ⟶ n.

As stated earlier, mitosis is the type of nuclear division that occurs when tissues grow or when repair occurs. Following fertilization, the zygote begins to divide mitotically, and mitosis continues during development and the life span of the individual.

After the chromosomes duplicate prior to mitosis, they consist of two sister chromatids. The sister chromatids separate during mitosis so that each of two daughter cells has the same number and kinds of chromosomes as the parental cell.

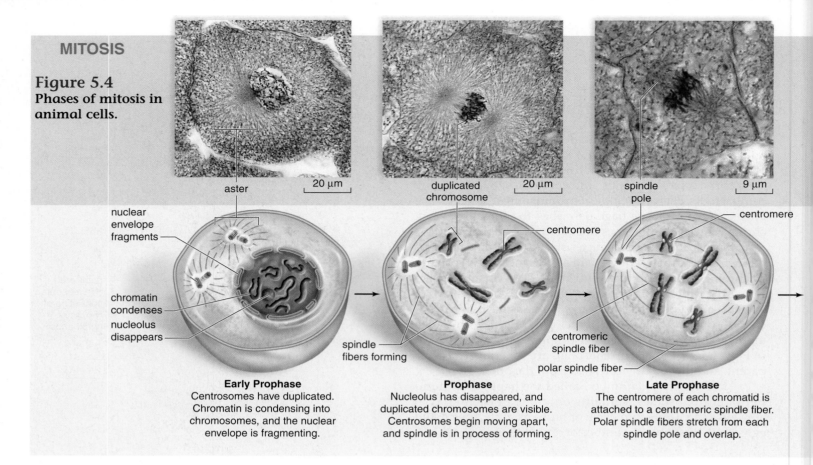

MITOSIS

Figure 5.4
Phases of mitosis in animal cells.

aster   20 µm

duplicated chromosome   20 µm

spindle pole   9 µm

nuclear envelope fragments

chromatin condenses

nucleolus disappears

spindle fibers forming

centromere

centromeric spindle fiber

polar spindle fiber

spindle pole

centromere

Early Prophase
Centrosomes have duplicated. Chromatin is condensing into chromosomes, and the nuclear envelope is fragmenting.

Prophase
Nucleolus has disappeared, and duplicated chromosomes are visible. Centrosomes begin moving apart, and spindle is in process of forming.

Late Prophase
The centromere of each chromatid is attached to a centromeric spindle fiber. Polar spindle fibers stretch from each spindle pole and overlap.

Mitosis in Detail

Mitosis is nuclear division that produces *two daughter nuclei, each with the same number and kinds of chromosomes as the parental nucleus.*

During mitosis, a **spindle** brings about an orderly distribution of chromosomes to the daughter cell nuclei. The spindle contains many fibers, each composed of a bundle of microtubules. Microtubules are able to disassemble and assemble. When microtubules assemble, protein molecules join together, and when they disassemble, protein molecules separate.

The centrosome, which is the main microtubule organizing center of the cell, divides during late interphase. It is believed that centrosomes are responsible for organizing the spindle. In animal cells each centrosome contains a pair of barrel-shaped organelles called **centrioles** and an **aster,** which is an array of short microtubules that radiate from the centrosome. The fact that plant cells lack centrioles suggests that centrioles are not required for spindle formation.

Mitosis in Animal Cells

Mitosis is a continuous process that is arbitrarily divided into four phases for convenience of description: prophase, metaphase, anaphase, and telophase (Fig. 5.4).

Prophase It is apparent during early **prophase** that cell division is about to occur. The centrosomes begin moving away from each other toward opposite ends of the nucleus. Spindle fibers appear between the separating centrosomes as the nuclear envelope begins to fragment, and the nucleolus begins to disappear.

The chromosomes are now visible. Each is duplicated and composed of sister chromatids held together at a centromere. The spindle begins forming during late prophase, and the chromosomes become attached to the spindle fibers. Their centromeres attach to fibers called centromeric (or kinetochore) fibers. As yet, the chromosomes have no particular orientation, moving first one way and then the other.

Metaphase By the time of **metaphase,** the fully formed spindle consists of poles, asters, and fibers. The **metaphase plate** is a plane perpendicular to the axis of the spindle and equidistant from the poles. The chromosomes attached to centromeric spindle fibers line up at the metaphase plate during metaphase. Polar spindle fibers reach beyond the metaphase plate and overlap.

Anaphase At the beginning of **anaphase,** the centromeres uniting the sister chromatids divide. Then the sister chromatids separate, becoming daughter chromosomes that move toward the opposite poles of the spindle. Daughter chromosomes have a centromere and a single chromatid.

What accounts for the movement of the daughter chromosomes? First, the centromeric spindle fibers shorten, pulling the daughter chromosomes toward the poles. Second, the polar spindle fibers push the poles apart as they lengthen and slide past one another.

Telophase During **telophase,** the spindle disappears, and nuclear envelope components reassemble around the daughter chromosomes. Each daughter nucleus contains the same number and kinds of chromosomes as the original parental cell. Remnants of the polar spindle fibers are still visible between the two nuclei.

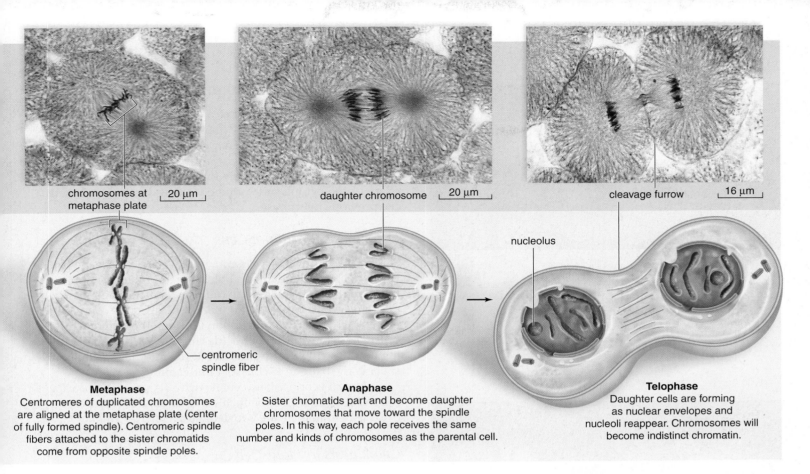

chromosomes at metaphase plate   20 μm

daughter chromosome   20 μm

cleavage furrow   16 μm

nucleolus

centromeric spindle fiber

Metaphase
Centromeres of duplicated chromosomes are aligned at the metaphase plate (center of fully formed spindle). Centromeric spindle fibers attached to the sister chromatids come from opposite spindle poles.

Anaphase
Sister chromatids part and become daughter chromosomes that move toward the spindle poles. In this way, each pole receives the same number and kinds of chromosomes as the parental cell.

Telophase
Daughter cells are forming as nuclear envelopes and nucleoli reappear. Chromosomes will become indistinct chromatin.

The chromosomes become more diffuse chromatin once again, and a nucleolus appears in each daughter nucleus. Cytokinesis is under way, and soon there will be two individual daughter cells, each with a nucleus that contains the diploid number of chromosomes.

Mitosis in Plant Cells

As with animal cells, mitosis in plant cells permits growth and repair. A particular plant tissue called meristematic tissue retains the ability to divide throughout the life of a plant. Meristematic tissue is found at the root tip and also at the shoot tip of stems. Lateral meristematic tissue accounts for the ability of trees to increase their girth each growing season.

Figure 5.5 illustrates mitosis in plant cells; exactly the same phases occur in plant cells as in animal cells. During early prophase, the chromatin condenses into scattered chromosomes, which are already duplicated. The nuclear envelope can no longer be seen, and the spindle forms. During late prophase, the chromosomes move about the cell and become attached to the spindle fibers. During metaphase, the chromosomes are at the metaphase plate of the spindle. During anaphase, the sister chromatids separate, becoming daughter chromosomes that start moving toward the poles.

MITOSIS

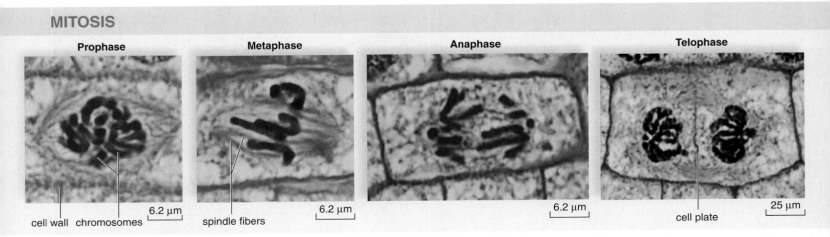

Prophase

Metaphase

Anaphase

Telophase

cell wall chromosomes   6.2 μm

spindle fibers   6.2 μm

6.2 μm

cell plate   25 μm

Figure 5.5 Phases of mitosis in plant cells.
Note the presence of the cell wall and the absence of centrioles and asters. Even so, plant cells have a spindle that brings about the movement of the chromosomes and distributes them so that the two daughter nuclei will each have the same number of chromosomes.

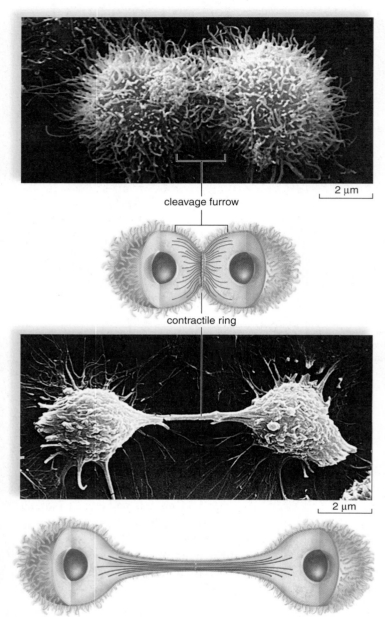

2 µm

cleavage furrow

contractile ring

2 µm

Figure 5.6 Cytokinesis in animal cells.
A single cell becomes two cells by a furrowing process. A contractile ring composed of actin filaments gradually gets smaller, and the cleavage furrow pinches the cell into two cells.

Copyright by R.G. Kessel and C.Y. Shih, Scanning Electron Microscopy in Biology: A Student's Atlas on Biological Organization, *Springer-Verlag, 1974.*

During telophase, the daughter chromosomes are at the poles, and then the daughter nuclei start forming. Although plant cells have a centrosome and spindle, there are no centrioles or asters during cell division.

Mitosis in plant and animal cells ensures that the daughter cells have the same number and kinds of chromosomes as the parental cell.

Cytokinesis in Animal and Plant Cells

Cytokinesis, or cytoplasmic cleavage, usually accompanies mitosis, but they are separate processes. Division of the cytoplasm begins in anaphase and continues in telophase but does not reach completion until just before the next interphase. By that time, the newly forming cells have received a share of the cytoplasmic organelles that duplicated during the previous interphase.

Cytokinesis in Animal Cells

As anaphase draws to a close in animal cells, a **cleavage furrow,** which is an indentation of the membrane between the two daughter nuclei, begins to form. The cleavage furrow deepens when a band of actin filaments, called the contractile ring, slowly forms a constriction between the two daughter cells. The action of the contractile ring can be likened to pulling a drawstring ever tighter about the middle of a balloon, causing the balloon to constrict in the middle.

A narrow bridge between the two cells can be seen during telophase, and then the contractile ring continues to separate the cytoplasm until there are two daughter cells (Fig. 5.6).

Cytokinesis in Plant Cells

Cytokinesis in plant cells occurs by a process different from that seen in animal cells (Fig. 5.7). The rigid cell wall that surrounds plant cells does not permit cytokinesis by furrowing. Instead, cytokinesis in plant cells involves building new cell walls between the daughter cells.

Cytokinesis is apparent when a small, flattened disk appears between the two daughter plant cells. Electron micrographs reveal that the disk is at right angles to a set of microtubules. The Golgi apparatus produces vesicles, which move along the microtubules to the region of the disk. As more vesicles arrive and fuse, a **cell plate** can be seen. The membrane of the vesicles completes the plasma membrane

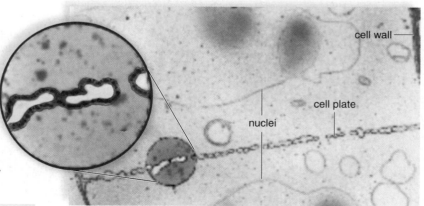

cell wall

cell plate

nuclei

Figure 5.7 Cytokinesis in plant cells.
During cytokinesis in a plant cell, a cell plate forms midway between two daughter nuclei and extends to the plasma membrane.

Copyright by R.G. Kessel and C.Y. Shih, Scanning Electron Microscopy in Biology: A Student's Atlas on Biological Organization, *Springer-Verlag, 1974.*

for both cells, and they release molecules that form the new plant cell walls. These cells walls are later strengthened by the addition of cellulose fibrils.

Cytokinesis in animal cells is accomplished by furrowing. Cytokinesis in plant cells involves the formation of a cell plate.

Cell Division in Prokaryotes

Prokaryotes, such as bacteria, have a single chromosome. Prokaryotes divide by a process called binary fission because division (fission) produces two (binary) daughter cells that are identical to the original parental cell (Fig. 5.8). Before division occurs, DNA replicates and the single chromosome duplicates. Thus, there are two chromosomes that separate as the cell elongates. When the cell is approximately twice its original length, the plasma membrane grows inward and a new cell wall forms, dividing the cell into two approximately equal portions.

Binary fission is a form of asexual reproduction. Asexual reproduction requires only a single parent, and the offspring are identical to the parent because they contain the same genes.

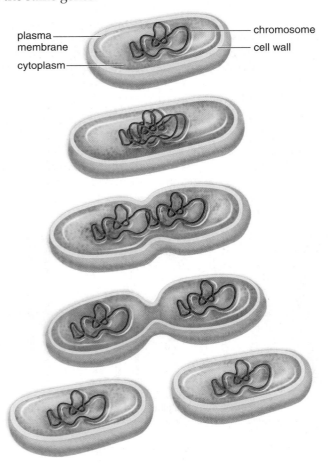

Figure 5.8 Binary fission.
First, DNA replicates. Then the cell lengthens, the two chromosomes separate, and the cell divides. The two resulting bacteria are identical.

Asexual reproduction in prokaryotes is accomplished by binary fission. Following DNA replication, the two resulting chromosomes separate as the cell elongates.

5.3 Reducing the Chromosome Number

Meiosis occurs in any life cycle that involves sexual reproduction. **Meiosis** reduces the chromosome number in such a way that the daughter nuclei receive only one of each kind of chromosome. The process of meiosis ensures that the next generation of individuals will have only the diploid number of chromosomes and a combination of traits different from that of either parent.

Overview of Meiosis

At the start of meiosis, the parental cell has the diploid number of chromosomes. In Figure 5.9, the diploid number is 4, which you can verify by counting the number of centromeres. Notice that meiosis requires two cell divisions and that the four daughter cells have the haploid number of chromosomes.

Recall that when a cell is 2n, or diploid, the chromosomes occur in pairs. For example, the 46 chromosomes of humans occur in 23 pairs. In Figure 5.9, there are two pairs of chromosomes. The members of a pair are called **homologous chromosomes** or **homologues.**

The two cell divisions of meiosis are called meiosis I and meiosis II. Prior to meiosis I, DNA replication has occurred, and the chromosomes are duplicated.

Meiosis I

During meiosis I, the homologous chromosomes come together and line up side by side. This so-called **synapsis** results in an association of four chromatids that stay in close proximity during the first two phases of meiosis I.

Because of synapsis, there are pairs of homologous chromosomes at the metaphase plate during meiosis I. Notice that only during meiosis I is it possible to observe paired chromosomes at the metaphase plate. When the members of these pairs separate, each daughter nucleus receives one member of each pair. Therefore, each daughter cell now has the haploid number of chromosomes, as you can verify by counting its centromeres.

Meiosis II and Fertilization

Replication of DNA does not occur between meiosis I and meiosis II because the chromosomes are still duplicated. During meiosis II, the centromeres divide and the sister chromatids separate, becoming daughter chromosomes that are distributed to daughter nuclei. In the end, each of four

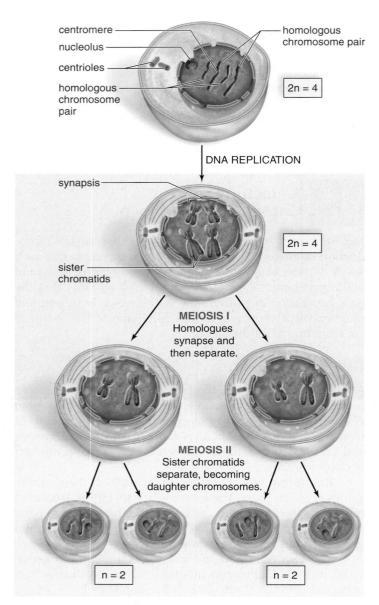

Figure 5.9 Overview of meiosis.
Following DNA replication, each chromosome is duplicated. During meiosis I, the homologous chromosomes pair during synapsis and then separate. During meiosis II, the centromeres divide and the sister chromatids separate, becoming daughter chromosomes that move into the daughter nuclei.

daughter cells has the haploid number of chromosomes, and each chromosome consists of one chromatid.

In some life cycles, such as that of humans (see Fig. 5.15), the daughter cells mature into **gametes** (sex cells—sperm and egg) that fuse during fertilization. **Fertilization** restores the diploid number of chromosomes in a cell that will develop into a new individual. If the gametes carried the diploid instead of the haploid number of chromosomes, the chromosome number would double with each fertilization.

> During meiosis I, homologous chromosomes pair and then separate. Each daughter nucleus receives one member from each pair of chromosomes. Following meiosis II, there are four haploid daughter cells, and each chromatid is now a chromosome.

Meiosis in Detail

Meiosis, which requires two nuclear divisions, *results in four daughter nuclei, each having one of each kind of chromosome and therefore half the number of chromosomes as the parental cell.*

Meiosis I helps ensure that **genetic recombination** of the parental genes occurs through two key events: crossing-over and independent assortment of chromosome pairs. These events are highly significant because the members of a homologous pair can carry slightly different instructions for the same genetic trait. For example, one homologue may carry instructions for brown eyes, while the corresponding homologue may carry instructions for blue eyes. Upon fertilization, the offspring will have different combinations of genes than either parent.

Prophase I

Meiosis I has four phases, called prophase I, metaphase I, anaphase I, and telophase I. In prophase I, synapsis occurs, and then the spindle appears while the nuclear envelope fragments and the nucleolus disappears. During synapsis, the homologous chromosomes come together and line up side by side. Now an exchange of genetic material may occur between the *nonsister* chromatids of the homologues. This exchange is called **crossing-over.** Crossing-over means that the chromatids held together by a centromere are no longer identical. As a result of crossing-over, the daughter cells receive chromosomes with recombined genetic material. In Figure 5.10, crossing-over is represented by an exchange of color. Because of crossing-over, first the nonsister chromatids, and then the chromosomes they give rise to, have a different combination of genes than the parental cell.

Metaphase I

During metaphase I, the homologues align at the metaphase plate. Depending on how they align, the maternal or paternal member of each pair may be oriented toward either pole.

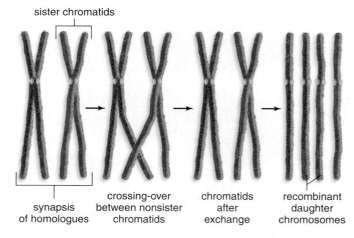

Figure 5.10 Synapsis and crossing-over.
During prophase I, *from left to right,* duplicated homologous chromosomes undergo synapsis when they line up with each other. During crossing-over, nonsister chromatids break and then rejoin. The two resulting daughter chromosomes will have a different combination of genes than they had before.

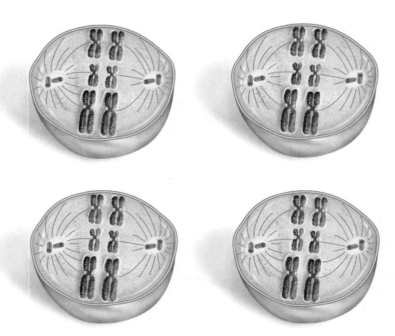

Figure 5.11 Independent assortment.

Four possible orientations of homologous chromosome pairs at the metaphase plate are shown for metaphase I. Each of these will result in daughter nuclei with a different combination of parental chromosomes (independent assortment). In a cell with three pairs of homologous chromosomes, there are 2^3 possible combinations of parental chromosomes in the daughter nuclei.

Independent assortment occurs when these pairs separate from each other, generating cells with different combinations of maternal and paternal chromosomes. Figure 5.11 shows four possible orientations for a cell that contains only three pairs of chromosomes (2n=6). Gametes from the first cell, but not the others, will have these three paternal or these three maternal chromosomes. The gametes from the other cells will have different combinations of maternal and paternal chromosomes.

Once all possible orientations are considered, the result will be 2^3, or eight, possible combinations of maternal and paternal chromosomes in the resulting gametes from this cell. In humans, where there are 23 pairs of chromosomes, the number of possible chromosomal combinations in the gametes is a staggering 2^{23}, or 8,388,608. And this does not even consider the genetic recombinations that are introduced due to crossing-over. Together, prophase I and metaphase I introduce the genetic recombination that helps ensure that no two offspring have the same combination of genes as the parents.

First Division: All Phases

The phases of meiosis I for an animal cell are diagrammed in Figure 5.12. Crossing-over is represented by an exchange of color. During metaphase I, homologous chromosome pairs align independently at the metaphase plate. During anaphase I, homologous chromosomes undergo **independent assortment,** and all possible combinations of chromosomes occur in the daughter nuclei. Each chromosome still consists of two chromatids.

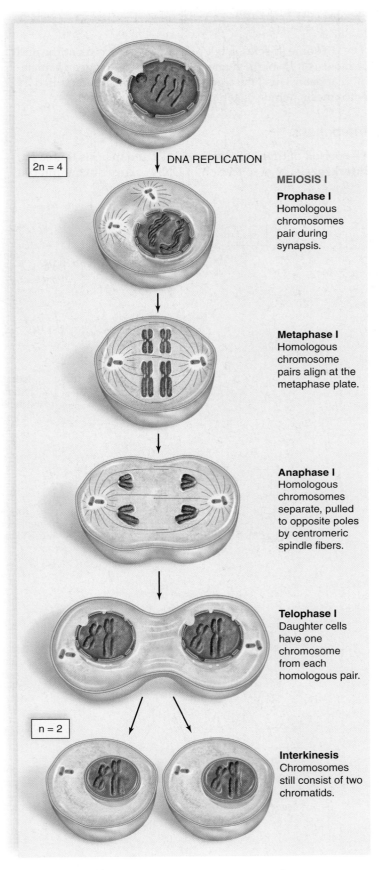

$2n = 4$

DNA REPLICATION

MEIOSIS I

Prophase I
Homologous chromosomes pair during synapsis.

Metaphase I
Homologous chromosome pairs align at the metaphase plate.

Anaphase I
Homologous chromosomes separate, pulled to opposite poles by centromeric spindle fibers.

Telophase I
Daughter cells have one chromosome from each homologous pair.

$n = 2$

Interkinesis
Chromosomes still consist of two chromatids.

Figure 5.12 Meiosis I in an animal cell.

The exchange of color between nonsister chromatids represents crossing-over. Notice that the daughter cells have only 2 centromeres compared to the parental cell's 4.

In some species, a telophase I phase occurs at the end of meiosis I. If so, the nuclear envelopes re-form, and nucleoli reappear. This phase may or may not be accompanied by cytokinesis, which is separation of the cytoplasm.

Interkinesis

The period of time between meiosis I and meiosis II is called **interkinesis.** No replication of DNA occurs during interki-

nesis. Why is this appropriate? Because the chromosomes are already duplicated.

Second Division

Phases of meiosis II for an animal cell are diagrammed in Figure 5.13. At the beginning of prophase II, a spindle appears while the nuclear envelope disassembles and the nucleolus disappears. Each duplicated chromosome attaches

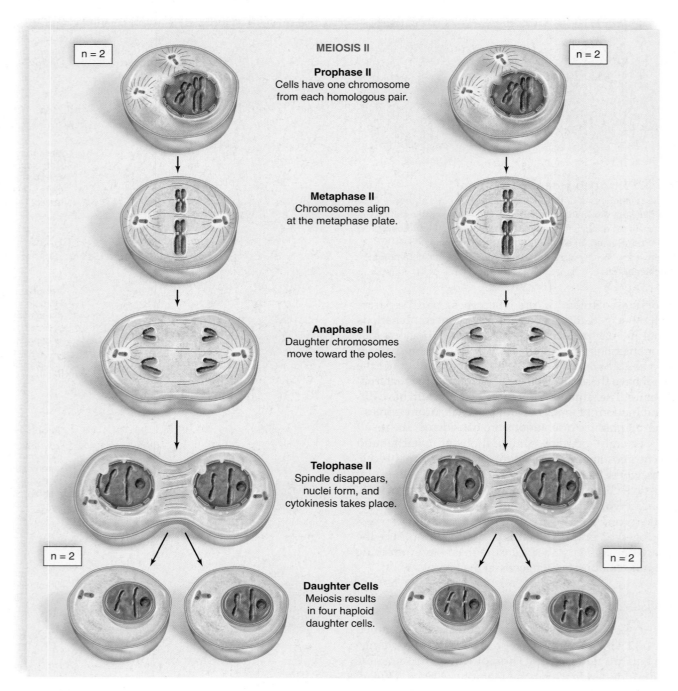

MEIOSIS II

n = 2

Prophase II
Cells have one chromosome from each homologous pair.

n = 2

Metaphase II
Chromosomes align at the metaphase plate.

Anaphase II
Daughter chromosomes move toward the poles.

Telophase II
Spindle disappears, nuclei form, and cytokinesis takes place.

n = 2

n = 2

Daughter Cells
Meiosis results in four haploid daughter cells.

Figure 5.13 Meiosis II in an animal cell.
During meiosis II, sister chromatids separate, becoming daughter chromosomes that are distributed to the daughter nuclei. Following meiosis II, there are four haploid daughter cells. Comparing the number of centromeres in each daughter cell with the number in the parental cell at the start of meiosis I (Fig. 5.12) verifies that each daughter cell is haploid.

to the spindle, and then they align at the metaphase plate during metaphase II. During anaphase II, sister chromatids separate, becoming daughter chromosomes that move into the daughter nuclei. In telophase II, the spindle disappears as nuclear envelopes re-form.

During the cytokinesis that can follow meiosis II, the plasma membrane furrows to produce two complete cells, each of which has the haploid number of chromosomes. Because each cell from meiosis I undergoes meiosis II, there are four daughter cells altogether.

Meiosis ensures that gametes for sexual reproduction contain a haploid number of chromosomes. After gametes fuse during fertilization, the diploid number is restored.

Nondisjunction

Nondisjunction is the failure of homologous chromosomes to separate during anaphase I, or the failure of sister chromatids to separate during anaphase II. This event results in gametes that have one more or one less than the correct number of chromosomes. Nondisjunctions cause various syndromes in humans (a syndrome is a combination of symptoms that always occur together). Down syndrome results when the egg, or occasionally the sperm, has two number 21 chromosomes instead of one. Turner syndrome results when either the egg or the sperm has no X chromosome. These conditions are discussed further in Chapter 26.

Genetic Recombination

As you've seen, meiosis produces haploid cells. One of the hallmarks of meiosis is the possibility of genetic recombination: the haploid cells produced are no longer identical to the diploid parent cell from which they came.

Recombination occurs in two ways. First, during prophase I, crossing-over between nonsister chromatids of the paired homologous chromosomes rearranges genes, with the result that the sister chromatids of each homologue may no longer be identical. Second the independent assortment of chromosomes during anaphase I means that the gametes produced by meiosis II may have different combinations of chromosomes than the parental cell.

Upon fertilization, the combining of chromosomes from genetically different gametes, even if the gametes are from a single individual (as in many plants), helps ensure that offspring are not identical to their parents. This genetic variability is the main advantage of sexual reproduction.

During sexual reproduction, meiosis leads to genetic recombination in two ways:
- **Crossing-over between nonsister chromatids exchanges genetic material.**
- **Independent assortment of chromosomes may give gametes new chromosome combinations.**

5.4 Comparison of Meiosis with Mitosis

Figure 5.14 compares meiosis to mitosis. Notice the following differences:

- DNA replication takes place only once prior to either meiosis or mitosis. However, meiosis requires two nuclear divisions, while mitosis requires only one nuclear division.
- Meiosis produces four daughter nuclei, and following cytokinesis, there are four daughter cells. Mitosis followed by cytokinesis results in two daughter cells.
- The four daughter cells following meiosis are haploid and have half the chromosome number as the parental cell. The daughter cells following mitosis have the same chromosome number as the parental cell.
- The daughter cells resulting from meiosis are not genetically identical to each other or to the parental cell. The daughter cells resulting from mitosis are genetically identical to each other and to the parental cell.

The specific differences between these nuclear divisions can be categorized according to when they occur and the processes involved.

Occurrence

Meiosis occurs only at certain times in the life cycle of sexually reproducing organisms. In humans, meiosis occurs only in the reproductive organs and produces the gametes. Mitosis occurs almost continously in all tissues during growth and repair.

Processes

To summarize the differences in process, Tables 5.1 and 5.2 separately compare meiosis I and meiosis II to mitosis.

TABLE 5.1	Comparison of Meiosis I with Mitosis
Meiosis I	**Mitosis**
Prophase I	*Prophase*
Pairing of homologous chromosomes	No pairing of chromosomes
Metaphase I	*Metaphase*
Homologous duplicated chromosomes at metaphase plate	Duplicated chromosomes at metaphase plate
Anaphase I	*Anaphase*
Homologous chromosomes separate.	Sister chromatids separate, becoming daughter chromosomes that move to the poles.
Telophase I	*Telophase/Cytokinesis*
Two haploid daughter cells	Two daughter cells, identical to the parental cell

Comparison of Meiosis I to Mitosis

Notice that these events distinguish meiosis I from mitosis:

- Homologous chromosomes pair and undergo crossing-over during prophase I of meiosis, but not during mitosis.

- Paired homologous chromosomes align at the metaphase plate during metaphase I in meiosis. These paired chromosomes have four chromatids altogether. Individual chromosomes align at the metaphase plate during metaphase in mitosis. They each have two chromatids.

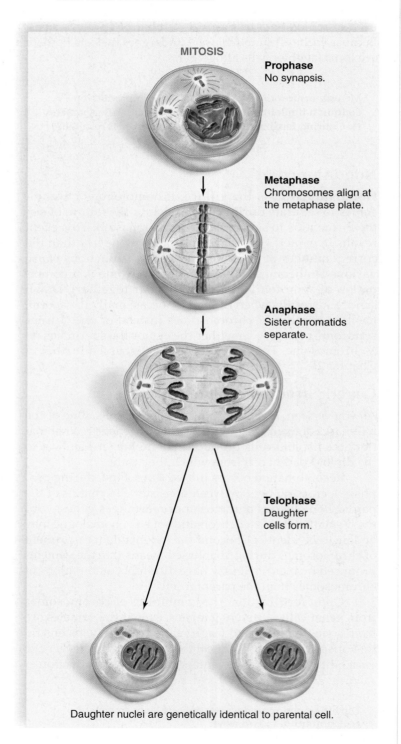

Figure 5.14 Meiosis compared to mitosis.
Compare metaphase I of meiosis to metaphase of mitosis. In meiosis metaphase I, homologues are paired at the metaphase plate, but in mitosis metaphase, all chromosomes align individually at the metaphase plate. Individual homologues separate during meiosis anaphase I, so that the daughter cells are haploid. (Blue chromosomes are from one parent and red from the other; exchange of color represents crossing-over.)

Bioethical Focus

Delaying Childbirth

More and more women are delaying childbirth until their 30s and 40s. The proportion of women over 30 in the United States having their first child has increased from 5% in 1975 to 22% in 1995. Why such an increase in older first-time mothers? Some women are focusing on their education and getting established in a career. Other women are marrying later or experiencing divorce and remarriage. Also, the availability of birth control methods has made it easier to delay childbirth.

However, delaying the start of a family is not without risks. For some women, increasing age results in reduced fertility, and others may not be able to get pregnant at all. Women in their 40s have one-and-a-half times the miscarriage rate of mothers in their 20s. Older women tend to experience more health complications, such as high blood pressure and diabetes, during pregnancy as well. But there are also risks to the baby, including low birth weight or chromosomal abnormalities.

The risk of a chromosomal abnormality is higher for the child of an older mother because all of a female's eggs are arrested in prophase I when she is a fetus. After 40 years, the chromosomes in those eggs become "sticky" and tend to exhibit increased nondisjunction, the failure to separate during meiosis. For example, the risk of having a child with Down syndrome, caused by three copies of chromosome 21, is approximately 1 in 100 for a 40-year-old mother compared to 1 in 1,250 for a 20-year-old. There is also evidence that the father's age may have a role in Down syndrome. Other abnormalities may increase with increasing paternal or maternal age.

Form Your Own Opinion

1. Do the advantages of delaying childbirth outweigh the disadvantages?
2. If delaying pregnancy means increased health and fertility problems, should insurance companies pay for the extra costs associated with such pregnancies?
3. Should doctors strongly discourage couples from trying to get pregnant after they reach a certain age?

TABLE 5.2	Comparison of Meiosis II with Mitosis
Meiosis II	**Mitosis**
Prophase II	*Prophase*
No pairing of chromosomes	No pairing of chromosomes
Metaphase II	*Metaphase*
Haploid number of duplicated chromosomes at metaphase plate	Duplicated chromosomes at metaphase plate
Anaphase II	*Anaphase*
Sister chromatids separate, becoming daughter chromosomes that move to the poles.	Sister chromatids separate, becoming daughter chromosomes that move to the poles.
Telophase II	*Telophase/Cytokinesis*
Four haploid daughter cells, not identical to parental cell	Two daughter cells, identical to the parental cell

- Homologous chromosomes (with centromeres intact) separate and move to opposite poles during anaphase I in meiosis. Centromeres split, and sister chromatids, now called daughter chromosomes, move to opposite poles during anaphase in mitosis.

Comparison of Meiosis II to Mitosis

The events of meiosis II are just like those of mitosis except that in meiosis II, the nuclei contain the haploid number of chromosomes.

> Meiosis is a specialized process that reduces the chromosome number and occurs during the production of gametes. Mitosis is a process that occurs during growth and repair of all tissues.

5.5 The Human Life Cycle

The human life cycle requires both meiosis and mitosis (Fig. 5.15). A haploid sperm (n) and a haploid egg (n) join at fertilization, and the resulting **zygote** has the full, or diploid (2n), number of chromosomes. During development of the fetus before birth, mitosis keeps the chromosome number constant in all the cells of the body. After birth, mitosis is involved in the continued growth of the child and repair of tissues at any time. As a result of mitosis, each somatic cell in the body has the same number of chromosomes.

Spermatogenesis and Oogenesis in Humans

In human males, meiosis is a part of **spermatogenesis,** which occurs in the testes and produces sperm. In human females, meiosis is a part of **oogenesis,** which occurs in the ovaries and produces eggs (Fig. 5.16).

Spermatogenesis

Following puberty, the time of life when the sex organs mature, spermatogenesis is continual in the testes of human males. As many as 300,000 sperm are produced per minute, or 400 million per day.

During spermatogenesis, primary spermatocytes, which are diploid, divide during the first meiotic division to form two **secondary spermatocytes,** which are haploid. Secondary spermatocytes divide during the second meiotic division to produce four **spermatids,** which are also haploid. What's the difference between the chromosomes in haploid secondary spermatocytes and those in haploid spermatids? The chromosomes in secondary spermatocytes are duplicated and consist of two chromatids, while those in spermatids consist of only one chromatid. Spermatids then mature into **sperm (spermatozoa).**

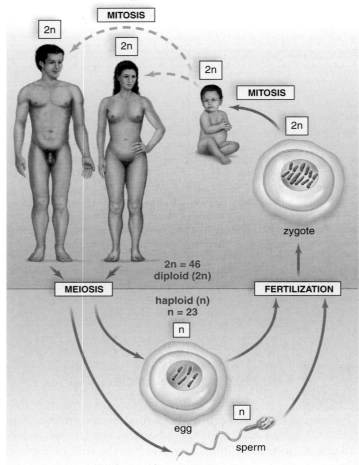

Figure 5.15   Life cycle of humans.
Meiosis in males is a part of sperm production, and meiosis in females is a part of egg production. When a haploid sperm fertilizes a haploid egg, the zygote is diploid. The zygote undergoes mitosis as it develops into a newborn child. Mitosis continues throughout life during growth and repair.

Oogenesis

Meiosis in human females begins in the fetus. In the ovaries of the fetus, all of the primary oocytes (diploid cells) begin meiosis but become arrested in prophase I. Following puberty and the initiation of the menstrual cycle, one primary oocyte begins to complete meiosis. It finishes the first meiotic division as two cells, each of which is haploid, although the chromosomes are still duplicated. One of these cells, termed the **secondary oocyte,** receives almost all of the cytoplasm. The other is the first **polar body,** a nonfunctioning cell. The polar body contains duplicated chromosomes and may or may not divide again; eventually it disintegrates. If the secondary oocyte is fertilized by a sperm, it completes the second meiotic division, in which it again divides unequally, forming an **egg** and a second polar body. The egg and sperm nuclei then fuse to form the 2n zygote. If the secondary oocyte is not fertilized by a sperm, it disintegrates and passes out of the body with the menstrual flow.

> **Sexual reproduction ensures that each generation has the same number of chromosomes and that each individual has a different genetic makeup from that of either parent.**

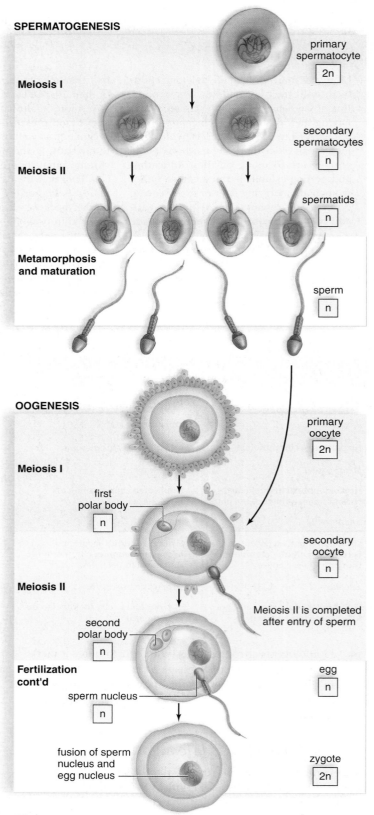

Figure 5.16   Spermatogenesis and oogenesis in mammals.
Spermatogenesis produces four viable sperm, whereas oogenesis produces one egg and at least two polar bodies. In humans, both the sperm and the egg have 23 chromosomes each; therefore, following fertilization, the zygote has 46 chromosomes.

Summarizing the Concepts

5.1 Cell Increase and Decrease
Cell division increases the number of cells in the body, and apoptosis reduces this number when appropriate. Cells go through a cell cycle that includes (1) interphase and (2) cell division, consisting of mitosis and cytokinesis. Interphase, in turn, includes G_1 (growth as certain organelles double), S (DNA synthesis), and G_2 (growth as the cell prepares to divide). Cell division occurs during the mitotic stage (M) when daughter cells receive a full complement of chromosomes.

Internal and external signals control the cell cycle, which can stop at any of three checkpoints: in G_1 prior to the S stage, in G_2 prior to the M stage, and near the end of mitosis. DNA damage is one reason the cell cycle stops; the p53 protein is active at the G_1 checkpoint, and if DNA is damaged and can't be repaired, this protein initiates apoptosis. During apoptosis, enzymes called caspases bring about destruction of the nucleus and the rest of the cell.

5.2 Maintaining the Chromosome Number
Each species has a characteristic number of chromosomes. The total number is the diploid number, and half this number is the haploid number. Among eukaryotes, cell division involves nuclear division and division of the cytoplasm (cytokinesis).

Replication of DNA precedes cell division. The duplicated chromosome is composed of two sister chromatids held together at a centromere. During mitosis, the centromeres divide, and daughter chromosomes go into each new nucleus.

Mitosis has the following phases: prophase, which includes early prophase, when chromosomes have no particular arrangement, and late prophase, when the chromosomes are attached to spindle fibers; metaphase, when the chromosomes are aligned at the metaphase plate; anaphase, when the chromatids separate, becoming daughter chromosomes that move toward the poles; and telophase, when new nuclear envelopes form around the daughter chromosomes and cytokinesis is well under way.

5.3 Reducing the Chromosome Number
Meiosis occurs in any life cycle that involves sexual reproduction. During meiosis I, homologues separate, and this leads to daughter cells with the haploid number of homologous chromosomes. Crossing-over and independent assortment of chromosomes during meiosis I ensure genetic recombination in daughter cells. During meiosis II, chromatids separate, becoming daughter chromosomes that are distributed to daughter nuclei. In some life cycles, the daughter cells become gametes, and upon fertilization, the offspring have the diploid number of chromosomes, the same as their parents.

Meiosis utilizes two nuclear divisions. During meiosis I, homologous chromosomes undergo synapsis, and crossing-over between nonsister chromatids occurs. When the homologous chromosomes separate during meiosis I, each daughter nucleus receives one member from each pair of chromosomes. Therefore, the daughter cells are haploid. Distribution of daughter chromosomes derived from sister chromatids during meiosis II then leads to a total of four new cells, each with the haploid number of chromosomes.

5.4 Comparison of Meiosis with Mitosis
Figure 5.14 contrasts the phases of mitosis with the phases of meiosis.

5.5 The Human Life Cycle
The human life cycle involves both mitosis and meiosis. Mitosis ensures that each somatic cell has the diploid number of chromosomes.

Meiosis is a part of spermatogenesis and oogenesis. Spermatogenesis in males produces four viable sperm, while oogenesis in females produces one egg and two polar bodies. Oogenesis does not go on to completion unless a sperm fertilizes the secondary oocyte.

Testing Yourself

Choose the best answer for each question.

1. Spindle fibers begin to appear during
   a. prophase.
   b. metaphase.
   c. anaphase.
   d. interphase.
2. If the haploid number is 7, what is the diploid number?
   a. 12
   b. 14
   c. 16
   d. 32
3. DNA synthesis occurs during
   a. interphase.
   b. prophase.
   c. metaphase.
   d. telophase.
4. Mitosis involves
   a. gametes.
   b. somatic cells.
   c. Both a and b are correct.
   d. Neither a nor b is correct.

In questions 5–8, match the part of the diagram of the cell cycle to the statements provided.

5. DNA replication
6. Organelles double
7. Cell prepares to divide
8. Cytokinesis occurs

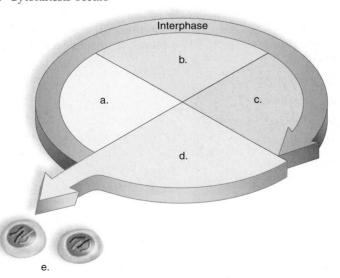

9. Which is not a part of the mitosis stages?

    a. anaphase
    b. telophase
    c. interphase
    d. metaphase

10. What is the checkpoint for the completion of DNA synthesis?

    a. G_1 c. G_2
    b. S   d. M

In questions 11–15, match the part of the diagram that follows to the appropriate statement.

11. Meiosis II
12. Primary spermatocyte
13. Sperm
14. Secondary spermatocyte
15. Spermatids

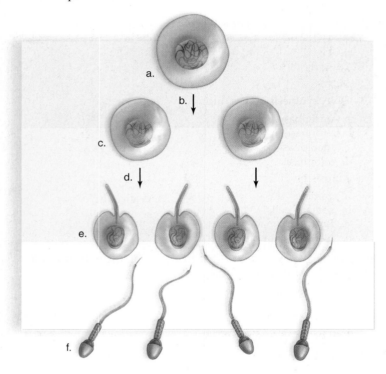

In questions 16–20, match the part of the diagram to the appropriate statement. More than one answer may be required.

16. Haploid cell

17. Diploid cell

18. Mitosis

19. Meiosis

20. Fertilization

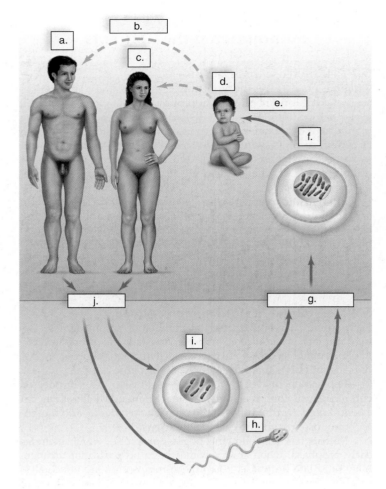

21. The end products of _____ are _____ cells.

    a. mitosis, diploid   c. meiosis, diploid
    b. meiosis, haploid   d. Both a and b are correct.

22. Programmed cell death is called

    a. mitosis.   c. cytokinesis.
    b. meiosis.   d. apoptosis.

23. Which helps to ensure that genetic diversity is maintained?

    a. independent assortment
    b. crossing-over
    c. genetic recombination
    d. All of these are correct.

24. Which cell contributes genetically to the zygote?

    a. secondary oocyte   c. second polar body
    b. first polar body   d. None of these are correct.

25. _____ ensures the same number of chromosomes from one _____ to the next.

    a. Meiosis, somatic cell   c. Mitosis, generation
    b. Meiosis, generation   d. None of these are correct.

26. In human beings, mitosis is necessary to
    a. growth and repair of tissues.
    b. formation of the gametes.
    c. maintaining the chromosome number in all body cells.
    d. the death of unnecessary cells.
    e. Both a and c are correct.

For questions 27–30, match the descriptions that follow to the terms in the key. Answers may be used more than once or not at all.

Key:
    a. centriole   c. chromosome
    b. microtubule   d. centromere

27. Point of attachment for sister chromatids

28. Found at a pole in the center of an aster

29. Coiled and condensed chromatin

30. Component of the spindle

31. If a parental cell has 14 chromosomes prior to mitosis, how many chromosomes will the daughter cells have?
    a. 28
    b. 14
    c. 7
    d. any number between 7 and 28

32. In which phase of mitosis are chromosomes moving toward the poles?
    a. prophase
    b. metaphase
    c. anaphase
    d. telophase
    e. Both b and c are correct.

33. If a parental cell has 12 chromosomes, the daughter cells following meiosis II will have
    a. 12 chromosomes.
    b. 24 chromosomes.
    c. 6 chromosomes.
    d. Any of these could be correct.

34. At the metaphase plate during metaphase I of meiosis, there are
    a. single chromosomes.
    b. unpaired duplicated chromosomes.
    c. homologous pairs.
    d. always 23 chromosomes.

35. Crossing-over occurs between
    a. sister chromatids of the same chromosomes.
    b. chromatids of nonhomologous chromosomes.
    c. nonsister chromatids of a homologous pair.
    d. two daughter nuclei.
    e. Both b and c are correct.

36. Plant cells
    a. do not have centrioles.
    b. have a cleavage furrow like animal cells do.
    c. form a cell plate by fusion of several vesicles.
    d. do not have the four phases of mitosis: prophase, metaphase, anaphase, and telophase.
    e. Both a and c are correct.

Understanding the Terms

anaphase 86
apoptosis 82
aster 86
cell cycle 82
cell plate 88
centriole 86
centromere 85
chromatin 85
chromosome 85
cleavage furrow 88
crossing-over 90
cyclin 83
cytokinesis 82
daughter chromosome 85
diploid (2n) number 85
duplicated chromosome 85
egg 96
fertilization 90
gamete 90
genetic recombination 90
haploid (n) number 85
homologous chromosome (homologue) 89
independent assortment 91

interkinesis 92
interphase 82
meiosis 89
metaphase 86
metaphase plate 86
mitosis 82
nondisjunction 93
oogenesis 95
polar body 96
prophase 86
secondary oocyte 96
secondary spermatocyte 95
sister chromatid 85
somatic cell 82
sperm (spermatozoa) 95
spermatid 95
spermatogenesis 95
spindle 86
synapsis 89
telophase 86
zygote 95

Match the terms to these definitions:

a. _____ Production of sperm in males by the process of meiosis and maturation.

b. _____ In oogenesis, a nonfunctional product; two to three meiotic products are of this type.

c. _____ In oogenesis, the functional product of meiosis I; becomes the egg.

d. _____ Pairing of homologous chromosomes during meiosis I.

e. _____ Division of the cytoplasm following mitosis and meiosis.

Thinking Critically

1. How many mitotic divisions would be necessary for one cell to produce an organism with over one million cells?
2. Why would a cell cycle checkpoint need to be nonfunctional in order for cancer to develop?
3. Which organism, one with 4 chromosomes or one with 20 chromosomes, would have a greater potential for genetic recombination? Justify your answer.
4. Compare and contrast sister chromatids and homologous chromosomes.
5. Nondisjunction increases as maternal age increases. From what you know about oogenesis, why would you expect this?

ARIS, the *Inquiry into Life* Website

ARIS, the website for *Inquiry into Life,* provides a wealth of information organized and integrated by chapter. You will find practice quizzes, interactive activities, labeling exercises, flashcards, and much more that will complement your learning and understanding of general biology.

www.aris.mhhe.com

Before leaving the hospital, *every newborn infant in the United States is screened for the disease PKU (full name, phenylketonuria). The test for PKU, developed in 1959, was the first newborn screening test implemented in the United States. PKU is a genetic disease that affects about one infant in 14,000. In these individuals, the enzyme responsible for breaking down the amino acid phenylalanine is missing or defective.*

Phenylalanine is a common component of proteins, and eating protein results in accumulation of phenylalanine, an amino acid, in the blood, and derived phenylketones in the urine.

Before 1959, infants born with PKU went untreated, and they developed nervous system damage, including mental retardation and convulsions. Now infants who test positive for PKU are placed on a special diet within the first 7–10 days of life. The diet does contain some phenylalanine, because this amino acid is required for growth to occur. All meat, fish, poultry, milk, cheese, and eggs must be avoided. Special formulations provide all the necessary amino acids for a child's growth and development. Strict adherence to the diet is required throughout childhood, and perhaps for life. The artificial sweetener aspartame, found in diet drinks and foods, contains phenylalanine. You may have seen the warning label for phenylketonurics on these products. There are thousands of enzymes in the body and each has a specific function. In this chapter we will discuss how enzymes work and their role in metabolism.

Metabolism: Energy and Enzymes

Chapter Concepts

6.1 Cells and the Flow of Energy

6.2 Metabolic Reactions and Energy Transformations

6.3 Metabolic Pathways and Enzymes

6.4 Oxidation-Reduction and the Flow of Energy

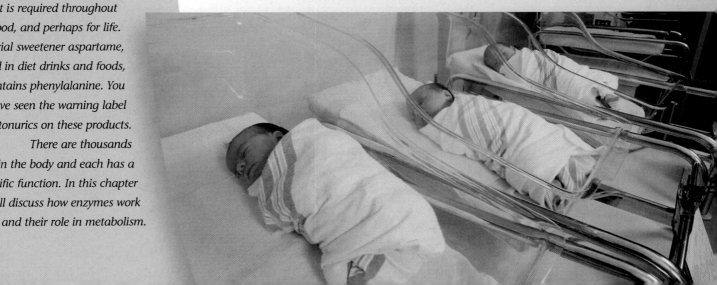

6.1   Cells and the Flow of Energy

Energy is the ability to do work or bring about a change. In order to maintain their organization and carry out metabolic activities, cells as well as organisms need a constant supply of energy. This energy allows living things to carry on the processes of life, including growth, development, locomotion, metabolism, and reproduction.

The majority of organisms get their energy from organic nutrients produced by photosynthesizers (algae, plants, and some bacteria). Photosynthesizers use solar energy to produce organic nutrients; therefore, life on Earth is ultimately dependent on solar energy.

Forms of Energy

Energy occurs in two forms: kinetic and potential. **Kinetic energy** is the energy of motion, as when a ball rolls down a hill or a moose walks through grass. **Potential energy** is stored energy—its capacity to accomplish work is not being used at the moment. The food we eat has potential energy because it can be converted into various types of kinetic energy. Food is specifically called **chemical energy** because it contains energy in the chemical bonds of organic molecules. When a moose walks, it has converted chemical energy into a type of kinetic energy called **mechanical energy** (Fig. 6.1).

Two Laws of Thermodynamics

Figure 6.1 illustrates the flow of energy in a terrestrial ecosystem. Plants capture only a small portion of solar energy, and much of it dissipates as heat. When plants photosynthesize and then make use of the food they produce, more heat results. Still, there is enough remaining to sustain a moose and the other organisms in an ecosystem. As living things metabolize nutrient molecules, all the captured solar energy eventually dissipates as heat. Therefore, energy flows and does not cycle. Two laws of thermodynamics explain why energy flows in ecosystems and in cells. These laws were formulated by early researchers who studied energy relationships and exchanges.

The first law of thermodynamics—the law of conservation of energy—states that energy cannot be created or destroyed, but it can be changed from one form to another.

When leaf cells photosynthesize, they use solar energy to form carbohydrate molecules from carbon dioxide and water. (Carbohydrates are energy-rich molecules, while carbon dioxide and water are energy-poor molecules.) Not all of the captured solar energy is used to form carbohydrates; some becomes heat:

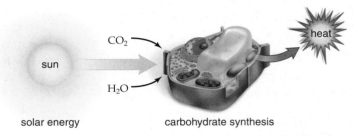

solar energy   carbohydrate synthesis

Obviously, plant cells do not create the energy they use to produce carbohydrate molecules; that energy comes from the sun. Is any energy destroyed? No, because heat is

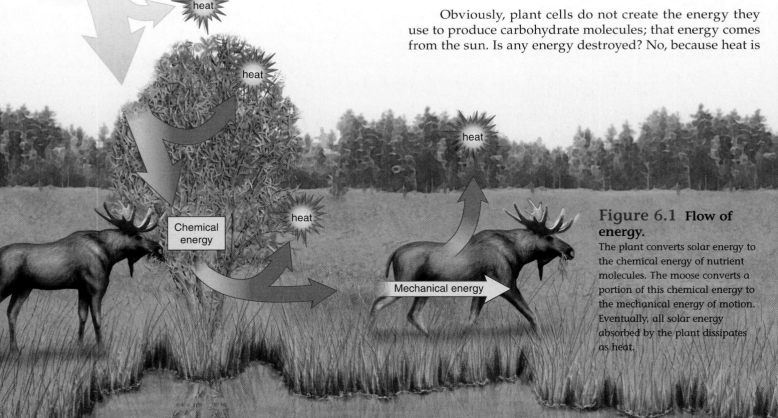

Figure 6.1 Flow of energy.
The plant converts solar energy to the chemical energy of nutrient molecules. The moose converts a portion of this chemical energy to the mechanical energy of motion. Eventually, all solar energy absorbed by the plant dissipates as heat.

also a form of energy. Similarly, a moose uses the energy derived from carbohydrates to power its muscles. And as its cells use this energy, none is destroyed, but some becomes heat, which dissipates into the environment:

carbohydrate   muscle contraction

The second law of thermodynamics, therefore, applies to living systems.

The second law of thermodynamics states that energy cannot be changed from one form to another without a loss of usable energy.

In our sample ecosystem, this law is upheld because some of the solar energy taken in by the plant and some of the chemical energy within the nutrient molecules taken in by the moose become heat. When heat dissipates into the environment, it is no longer usable—that is, it is not available to do work. With transformation upon transformation, eventually all usable forms of energy become heat that is lost to the environment. Heat that dissipates into the environment cannot be captured and converted to one of the other forms of energy.

As a result of the second law of thermodynamics, no process requiring a conversion of energy is ever 100% efficient. Much of the energy is lost in the form of heat. In automobiles, the gasoline engine is between 20% and 30% efficient in converting chemical energy into mechanical energy. The majority of energy is obviously lost as heat. Cells are capable of about 40% efficiency, with the remaining energy given off to their surroundings as heat.

Cells and Entropy

The second law of thermodynamics can be stated another way: Every energy transformation makes the universe less organized and more disordered. The term **entropy** is used to indicate the relative amount of disorganization. Since the processes that occur in cells are energy transformations, the second law means that every process that occurs in cells always does so in a way that increases the total entropy of the universe. Then, too, any one of these processes makes less energy available to do useful work in the future.

Figure 6.2 shows two processes that occur in cells: glucose breakdown and hydrogen ion diffusion. The second law of thermodynamics tells us that glucose tends to break apart into carbon dioxide and water. Why? Because glucose is more organized, and therefore less stable, than its breakdown products. Also, hydrogen ions on one side of a membrane tend to move to the other side unless they are prevented from doing so. Why? Because when they are distributed randomly, entropy has increased. As an analogy, you know from experience that a neat room is more organized but less stable than a messy room, which is dis-

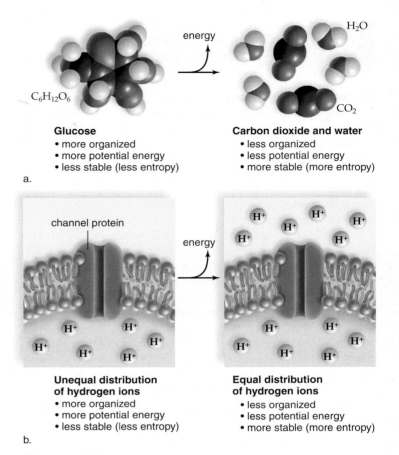

Glucose
- more organized
- more potential energy
- less stable (less entropy)

Carbon dioxide and water
- less organized
- less potential energy
- more stable (more entropy)

a.

channel protein

Unequal distribution of hydrogen ions
- more organized
- more potential energy
- less stable (less entropy)

Equal distribution of hydrogen ions
- less organized
- less potential energy
- more stable (more entropy)

b.

Figure 6.2 Cells and entropy.

The second law of thermodynamics tells us that (**a**) glucose, which is more organized, tends to break down to carbon dioxide and water, which are less organized. **b.** Similarly, hydrogen ions (H^+) on one side of a membrane tend to move to the other side so that the ions are randomly distributed. Both processes result in a loss of potential energy and an increase in entropy.

organized but more stable. How do you know a neat room is less stable than a messy room? Consider that a neat room always tends to become more messy.

On the other hand, you know that some cells can make glucose out of carbon dioxide and water, and all cells can actively move ions to one side of the membrane. How do they do it? These cellular processes obviously require an input of energy from an outside source. This energy ultimately comes from the sun. Living things depend on a constant supply of energy from the sun because the ultimate fate of all solar energy in the biosphere is to become randomized in the universe as heat. A living cell is a temporary repository of order purchased at the cost of a constant flow of energy.

Energy exists in several different forms. When energy transformations occur, energy is neither created nor destroyed. However, some usable energy is always lost. For this reason, living things are dependent on an outside source of energy that ultimately comes from the sun.

6.2   Metabolic Reactions and Energy Transformations

A chemical reaction involves the breaking and making of covalent bonds. Cellular **metabolism** is the sum of all the chemical reactions that occur in a cell—both the anabolic (building up) and catabolic (breaking down) reactions. **Reactants** are substances that participate in a reaction, while **products** are substances that form as a result of a reaction. For example, in the reaction A + B $\longrightarrow$ C + D, A and B are the reactants while C and D are the products. How do you know whether this reaction will go forward as written?

Using the concept of entropy, it is possible to state that a reaction will go forward if it increases the entropy of the universe. But in cell biology, we do not usually wish to consider the entire universe. We simply want to consider a particular reaction. In such instances, cell biologists use the concept of free energy instead of entropy. **Free energy** is the amount of energy available—that is, energy that is still "free" to do work—after a chemical reaction has occurred. Free energy is denoted by the symbol G after Josiah Gibbs, who first developed the concept. The change (Δ) in free energy (G) after a reaction occurs (ΔG) is calculated by subtracting the free energy content of the reactants from that of the products. A negative ΔG means that the products have less free energy than the reactants, and the reaction will go forward. In our reaction, if C and D have less free energy than A and B, the reaction will "go."

In **exergonic reactions,** ΔG is negative and energy is released; in **endergonic reactions,** ΔG is positive and the products have more free energy than the reactants. Endergonic reactions can only occur if there is an input of energy. In the body, many reactions, such as protein synthesis, nerve impulse conduction, or muscle contraction, are endergonic, and they are driven by the energy released by exergonic reactions. ATP is a carrier of energy between exergonic and endergonic reactions.

ATP: Energy for Cells

ATP (adenosine triphosphate) is the common energy currency of cells; when cells require energy, they "spend" ATP. A sedentary oak tree as well as a flying bat requires vast amounts of ATP. The more active the organism, the greater the demand for ATP. However, the amount on hand at any one moment is minimal because ATP is constantly being generated from **ADP (adenosine diphosphate)** and a molecule of inorganic phosphate P (Fig. 6.3). A cell is assured of a supply of ATP, because glucose breakdown during cellular respiration provides the energy for the buildup of ATP in mitochondria. Only 39% of the free energy of glucose is transformed to ATP; the rest is lost as heat.

There are many biological advantages to the use of ATP as an energy carrier in living systems. ATP is a common and universal energy currency because it can be used in many different types of reactions. Also, when ATP is converted to energy, ADP, and P, the amount of energy released is sufficient for a particular biological function, and little energy is wasted. In addition, ATP breakdown can be coupled to endergonic reactions in such a way that it minimizes energy loss.

Figure 6.3 The ATP cycle.

a. In cells, ATP carries energy between exergonic reactions and endergonic reactions. When a phosphate group is removed by hydrolysis, ATP releases the appropriate amount of energy for most metabolic reactions. **b.** In order to produce light, a firefly breaks down ATP.

adenosine triphosphate

P — P — P

Energy from exergonic reactions (e.g., cellular respiration)

ATP

Energy for endergonic reactions (e.g., protein synthesis, nerve impulse conduction, muscle contraction)

ADP + P

P — P + P

adenosine diphosphate   +   phosphate

a.

b.   2.25×

Structure of ATP

ATP is a nucleotide composed of the nitrogen-containing base adenine and the 5-carbon sugar ribose (together called adenosine) and three phosphate groups. ATP is called a "high-energy" compound because of the energy stored in the chemical bonds of the phosphates. Under cellular conditions, the amount of energy released when ATP is hydrolyzed to ADP + Ⓟ is about 7.3 kcal per mole.[1]

Coupled Reactions

In **coupled reactions,** the energy released by an exergonic reaction is used to drive an endergonic reaction. ATP breakdown is often coupled to cellular reactions that require an input of energy. Coupling, which requires that the exergonic reaction and the endergonic reaction be closely tied, can be symbolized like this:

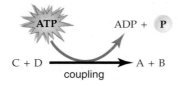

---

[1] A mole is the number of molecules present in the molecular weight of a substance (in grams).

Notice that the word *energy* does not appear following ATP breakdown. Why not? Because this energy was used to drive forward the coupled reaction. Figure 6.4 shows that ATP breakdown provides the energy necessary for muscular contraction to occur. The transfer of energy is not complete, and some energy is lost as heat.

Function of ATP

In living systems, three obvious uses of ATP are:

Chemical work ATP supplies the energy needed to synthesize macromolecules that make up the cell, and therefore the organism.

Transport work ATP supplies the energy needed to pump substances across the plasma membrane.

Mechanical work ATP supplies the energy needed to permit muscles to contract, cilia and flagella to beat, chromosomes to move, and so forth.

In most cases, ATP is the immediate source of energy for these processes.

> ATP is a carrier of energy in cells. It is the common energy currency because it supplies energy for many different types of reactions.

Figure 6.4 Coupled reactions.

Muscle contraction occurs only when it is coupled to ATP breakdown.

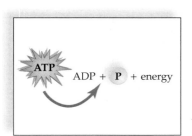

a. ATP breakdown is exergonic.

b. Muscle contraction is endergonic and cannot occur without an input of energy.

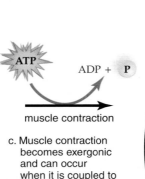

c. Muscle contraction becomes exergonic and can occur when it is coupled to ATP breakdown.

6.3   Metabolic Pathways and Enzymes

Reactions do not occur haphazardly in cells; they are usually part of a **metabolic pathway,** a series of linked reactions. Metabolic pathways begin with a particular reactant and terminate with an end product. While it is possible to write an overall equation for a pathway as if the beginning reactant went to the end product in one step, actually many specific steps occur in between. In the pathway, one reaction leads to the next reaction, which leads to the next reaction, and so forth in an organized, highly structured manner. This arrangement makes it possible for one pathway to lead to several others, because various pathways have several molecules in common. Also, metabolic energy is captured and utilized more easily if it is released in small increments rather than all at once.

A metabolic pathway can be represented by the following diagram:

$$A \xrightarrow{E_1} B \xrightarrow{E_2} C \xrightarrow{E_3} D \xrightarrow{E_4} E \xrightarrow{E_5} F \xrightarrow{E_6} G$$

In this diagram, the letters A − F are reactants, and the letters B − G are products in the various reactions. In other words, the products from the previous reaction become the reactants of the next reaction. The letters E_1 − E_6 are enzymes.

An **enzyme** is a protein (or sometimes an RNA molecule) that functions to speed a chemical reaction. An enzyme is a catalyst—it participates in the chemical reaction, but is not used up by the reaction. Note that the enzyme does not determine whether the reaction goes forward; that is determined by the free energy of the reaction. Enzymes simply increase the rate of the reaction.

The reactants in an enzymatic reaction are called the **substrates** for that enzyme. In the first reaction, A is the substrate for E_1, and B is the product. Now B becomes the substrate for E_2, and C is the product. This process continues until the final product (G) forms.

Any one of the molecules (A − G) in this linear pathway could also be a substrate for an enzyme in another pathway. A diagram showing all the possibilities would be highly branched.

Energy of Activation

Molecules frequently do not react with one another unless they are activated in some way. In the lab, for example, in the absence of an enzyme, activation is very often achieved by heating a reaction flask to increase the number of effective collisions between molecules. The energy that must be added to cause molecules to react with one another is called the **energy of activation (E_a).** Even though the reaction will go forward (ΔG is negative), the energy of activation must be overcome. The burning of firewood is a very exergonic reaction, but firewood in a pile does not spontaneously combust. The input of some energy, perhaps a lit match, is required to overcome the energy of activation.

Figure 6.5 shows E_a when an enzyme is not present compared to when an enzyme is present, illustrating that enzymes lower the amount of energy required for activation to occur. Nevertheless, the addition of the enzyme does not change the end result of the reaction. Notice that the energy of the products is less than the energy of the reactants. This results in a negative ΔG, so the reaction will go forward. But the reaction will not go at all unless the energy of activation is overcome. Without the enzyme, the reaction rate will be very slow. By lowering the energy of activation, the enzyme increases the rate of the reaction.

Sugar in your kitchen cupboard will break down to carbon dioxide and water because the energy of the products (carbon dioxide and water) is much less than the free energy of the reactant (sugar). However, the rate of this reaction is so slow that you never see it. If you eat the sugar, the enzymes in your digestive system greatly increase the speed at which the sugar is broken down. However, the end result (carbon dioxide and water) is still the same.

How Enzymes Function

The following equation, which is pictorially shown in Figure 6.6, is often used to indicate that an enzyme forms a complex with its substrate:

$$\underset{\text{substrate}}{S} + \underset{\text{enzyme}}{E} \longrightarrow \underset{\substack{\text{enzyme-substrate}\\\text{complex}}}{ES} \longrightarrow \underset{}{E} + \underset{\text{product}}{P}$$

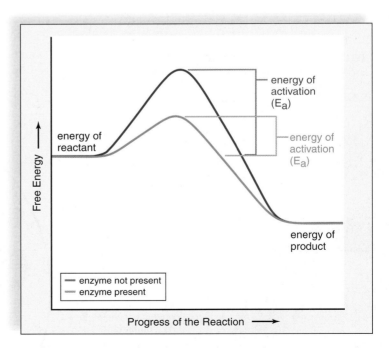

Figure 6.5   Energy of activation (E_a).
Enzymes speed the rate of reactions because they lower the amount of energy required for the reactants to react. Even reactions like this one, in which the energy of the product is less than the energy of the reactant (ΔG is negative), speed up when an enzyme is present.

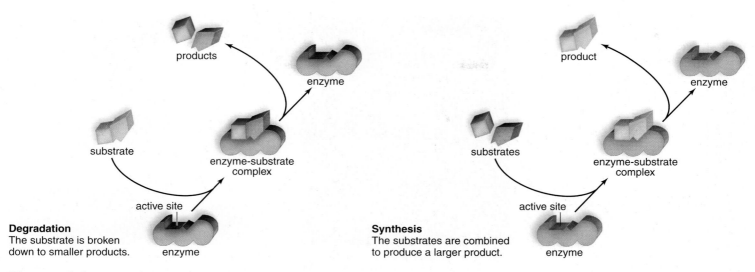

Figure 6.6 Enzymatic action.
An enzyme has an active site where the substrates and enzyme fit together in such a way that the substrates react. Following the reaction, the products are released, and the enzyme is free to act again. The enzymatic reaction can result in the degradation of a substrate into multiple products or the synthesis of a product from multiple substrates.

In most instances, only one small part of the enzyme, called the **active site,** complexes with the substrate(s). It is here that the enzyme and substrate fit together, seemingly like a key fits a lock; however, cell biologists now know that the active site undergoes a slight change in shape in order to accommodate the substrate(s). This is called the **induced fit model** because the enzyme is induced to undergo a slight alteration to achieve optimum fit (Fig. 6.7).

The change in shape of the active site facilitates the reaction that now occurs. After the reaction has been completed, the product(s) is released, and the active site returns to its original state, ready to bind to another substrate molecule. Only a small amount of enzyme is actually needed in a cell because enzymes are not used up by the reaction.

Some enzymes do more than simply complex with their substrate(s); they participate in the reaction. For example, trypsin digests protein by breaking peptide bonds. The active site of trypsin contains three amino acids with *R* groups that actually interact with members of the peptide bond—first to break the bond and then to introduce the components of water. This illustrates that the formation of the enzyme-substrate complex is very important in speeding up the reaction.

Sometimes it is possible for a particular reactant(s) to produce more than one type of product(s). The presence or absence of an enzyme determines which reaction takes place. If a substance can react to form more than one product, then the enzyme that is present and active determines which product is produced.

Every reaction in a cell requires that its specific enzyme be present. Because enzymes complex only with

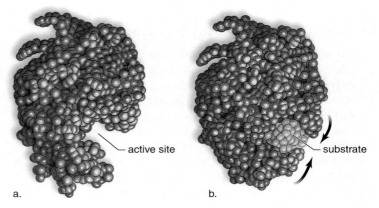

Figure 6.7 Induced fit model.
These computer-generated images show an enzyme called lysozyme that hydrolyzes its substrate, a polysaccharide that makes up bacterial cell walls. **a.** Shape of enzyme when no substrate is bound to it. **b.** After the substrate binds, the shape of the enzyme changes so that hydrolysis can better proceed.

their substrates, they are often named for their substrates with the suffix -*ase,* as in the following examples:

Substrate	Enzyme
Lipid	Lipase
Urea	Urease
Maltose	Maltase
Ribonucleic acid	Ribonuclease
Lactose	Lactase

Enzymes are protein molecules that speed chemical reactions by lowering the energy of activation. They do this by forming an enzyme-substrate complex.

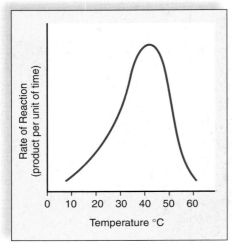

a. Rate of reaction as a function of temperature.

b. Body temperature of ectothermic animals often limits rates of reactions.

c. Body temperature of endothermic animals promotes rates of reactions.

Figure 6.8 **The effect of temperature on rate of reaction.**
a. Usually, the rate of an enzymatic reaction doubles with every 10°C rise in temperature. This enzymatic reaction is maximum at about 40°C; then it decreases until the reaction stops altogether, because the enzyme has become denatured. **b.** The body temperature of ectothermic animals, which take on the temperature of their environment, often limits rates of reactions. **c.** The body temperature of endothermic animals promotes rates of reaction.

Factors Affecting Enzymatic Speed

Enzymatic reactions proceed quite rapidly. Consider, for example, the breakdown of hydrogen peroxide (H_2O_2) as catalyzed by the enzyme catalase: $2 H_2O_2 \longrightarrow 2 H_2O + O_2$. The breakdown of hydrogen peroxide can occur 600,000 times a second when catalase is present. To achieve maximum product per unit time, there should be enough substrate to fill active sites most of the time. In addition to substrate concentration, the rate of an enzymatic reaction can also be affected by environmental factors (temperature and pH) or by cellular mechanisms, such as enzyme activation, enzyme inhibition, and cofactors.

Substrate Concentration

Generally, enzyme activity increases as substrate concentration increases because there are more collisions between substrate molecules and the enzyme. As more substrate molecules fill active sites, more product results per unit time. But when the enzyme's active sites are filled almost continuously with substrate, the enzyme's rate of activity cannot increase any more. Maximum rate has been reached.

Temperature and pH

As the temperature rises, enzyme activity increases (Fig. 6.8). This occurs because higher temperatures cause more effective collisions between enzyme and substrate. However, if the temperature rises beyond a certain point, enzyme activity eventually levels out and then declines rapidly because the enzyme is **denatured.** An enzyme's shape changes during denaturation, due to the loss of secondary and tertiary structure, and then it

can no longer bind its substrate(s) efficiently. As the temperature decreases, enzyme activity decreases (see Ecology Focus).

Each enzyme also has a preferred pH at which the rate of the reaction is highest. Figure 6.9 shows the preferred pH for the enzymes pepsin and trypsin. At this pH value, these enzymes have their normal configurations. The globular structure of an enzyme is dependent on interactions, such as hydrogen bonding, between R groups (see Fig. 2.23). A change in pH can alter the ionization of R groups and disrupt normal interactions, and under extreme pH conditions, denaturation eventually occurs. Again, the enzyme's shape is altered, and it is then unable to combine efficiently with its substrate.

Enzyme Activation

Not all enzymes are needed by the cell all the time. Genes can be turned on to increase the concentration of an enzyme in a cell or turned off to decrease the concentration. But enzymes can also be present in the cell in an inactive form. Activation of enzymes occurs in many different ways. Some enzymes are covalently modified by the addition or removal of phosphate groups. An enzyme called a kinase adds phosphates to proteins, as shown below, and an enzyme called a phosphatase removes them. In some proteins, adding phosphates activates them; in others, removing phosphates activates them. Enzymes can also be activated by cleaving or removing part of the protein, or by associating with another protein or cofactor.

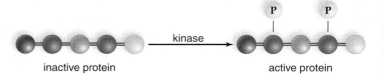

inactive protein　　　　　　　　active protein

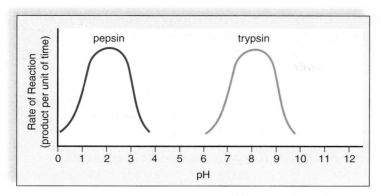

Figure 6.9 The effect of pH on rate of reaction.
The preferred pH for pepsin, an enzyme that acts in the stomach, is about 2, while the preferred pH for trypsin, an enzyme that acts in the small intestine, is about 8. The preferred pH of an enzyme maintains its shape so that it can bind with its substrates.

Enzyme Inhibition

Enzyme inhibition occurs when the substrate is unable to bind to the active site of an enzyme. The activity of almost every enzyme in a cell is regulated by feedback inhibition. In the simplest case, when there is plenty of product, it binds to the enzyme's active site, and then the substrate is unable to bind. As the product is used up, inhibition is reduced, and more product can be produced. In this way, the concentration of the product is always kept within a certain range.

Most metabolic pathways in cells are regulated by a more complicated type of feedback inhibition (Fig. 6.10). In these instances, the end product of an active pathway binds to a site other than the active site of the first enzyme. The binding changes the shape of the active site so that the substrate is unable to bind to the enzyme, and the pathway shuts down (inactive). Therefore, no more product is produced.

Poisons are often enzyme inhibitors. Cyanide is an inhibitor for an essential enzyme (cytochrome *c* oxidase) in all cells, which accounts for its lethal effect on humans. Penicillin is an *antimicrobial agent* that blocks the active site of an enzyme unique to bacteria. Therefore, penicillin is a poison for bacteria.

Enzyme Cofactors

Many enzymes require an inorganic ion or an organic, but nonprotein, helper to function properly. The inorganic ions are metals such as copper, zinc, or iron; these helpers are called **cofactors.** The organic, nonprotein molecules are called **coenzymes.** These cofactors assist the enzyme and may even accept or contribute atoms to the reactions.

Vitamins are often components of coenzymes. **Vitamins** are relatively small organic molecules that are required in trace amounts in our diet and in the diets of other animals for synthesis of coenzymes that affect health and physical fitness. The vitamin becomes a part of the

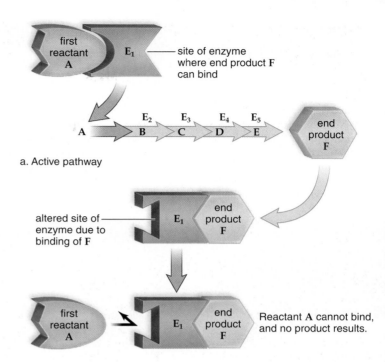

a. Active pathway

b. Inactive pathway

Reactant **A** cannot bind, and no product results.

Figure 6.10 Feedback inhibition.
a. In an active pathway, the first reactant (A) is able to bind to the active site of enzyme E_1. **b.** Feedback inhibition occurs when the end product (F) of the metabolic pathway binds to the first enzyme of the pathway—at a site other than the active site. This binding causes the active site to change its shape. Now reactant A is unable to bind to the enzyme's active site, and the whole pathway shuts down.

coenzyme's molecular structure. For example, the vitamin niacin is part of the coenzyme NAD, and riboflavin (B_2) is part of the coenzyme FAD.

A deficiency of any one of these vitamins results in a lack of the coenzyme and therefore a lack of certain enzymatic actions. In humans, this eventually results in vitamin-deficiency symptoms. For example, niacin deficiency results in a skin disease called pellagra, and riboflavin deficiency results in cracks at the corners of the mouth.

Various factors affect enzymatic speed, including substrate concentration, temperature, pH, enzyme activation, enzyme inhibition, or presence of necessary cofactors.

6.4 Oxidation-Reduction and the Flow of Energy

When oxygen (O) combines with a metal such as iron or magnesium (Mg), oxygen receives electrons and becomes an ion that is negatively charged; the metal loses electrons and becomes an ion that is positively charged. When magnesium oxide (MgO) forms, it is appropriate to say that

Ecology Focus

Why Do Leaves Change Color in the Fall?

During the summer, leaves are green because chloroplasts contain the green pigment chlorophyll. Chlorophyll is not a very stable pigment, and as it breaks down, metabolic energy (ATP) is needed for enzymes to rebuild it. As the hours of sunlight lessen in the fall, the energy needed to rebuild chlorophyll is not as plentiful. Further, enzymes are working at a reduced speed because of lower temperatures. Therefore, the amount of chlorophyll in leaves slowly disintegrates. When that happens, we begin to see the other pigments in the leaves (Fig. 6A).

The yellow pigment carotene is another pigment frequently found in leaves. Carotene, a helper to chlorophyll in absorbing solar energy, is more stable than chlorophyll. Therefore, it persists in leaves even when chlorophyll has disappeared. Now, the leaf appears yellow.

A class of pigments called anthocyanins accounts for the red and purple color of leaves in the fall. Before a leaf falls from the tree, an abscission, or separation, layer appears between the leaf and the stem. The development of this layer interferes with the flow of water into the leaf and the exit of sap from the leaf. Glucose produced by photosynthesis is in the sap, which is carried by tiny tubes. When glucose cannot exit the leaf, it begins to react with certain proteins, and the result is anthocyanins. The more acidic the sap, the brighter the red color produced.

Discussion Questions

1. Why do leaves appear green even when other pigments are present in the leaves?
2. What is the advantage of having several pigments present in leaves?
3. What happens to energy production in deciduous trees when they lose their leaves?

Figure 6A Fall foliage.
The beautiful colors we see in the fall are due to a slowdown in leaf metabolism. Lower temperatures and lack of metabolic energy (due to fewer hours of sunlight) mean that enzymes cannot function as they once did to make chlorophyll.

magnesium has been oxidized and has therefore lost electrons. On the other hand, oxygen has been reduced because it has gained negative charges (i.e., electrons).

Today, the terms oxidation and reduction are applied to many reactions, whether or not oxygen is involved. Very simply, **oxidation** is the loss of electrons, and **reduction** is the gain of electrons. In the ionic reaction $Na + Cl \rightarrow NaCl$, sodium has been oxidized (loss of electron), and chlorine has been reduced (gain of electron). Because oxidation and reduction go hand-in-hand, the entire reaction is called a **redox reaction.**

The terms oxidation and reduction also apply to covalent reactions in cells. In this case, however, oxidation is the loss of hydrogen atoms ($e^- + H^+$), and reduction is the gain of hydrogen atoms. A hydrogen atom contains one proton and one electron; therefore, when a molecule loses a hydrogen atom, it has lost an electron, and when a molecule gains a hydrogen atom, it has gained an electron. This form of oxidation-reduction is exemplified by the overall reactions for photosynthesis and cellular respiration, pathways that permit energy to flow from the sun through all living things.

Photosynthesis

The overall reaction for photosynthesis can be written like this:

$$\text{energy} + \underset{\substack{\text{carbon} \\ \text{dioxide}}}{6\,CO_2} + \underset{\text{water}}{6\,H_2O} \rightarrow \underset{\text{glucose}}{C_6H_{12}O_6} + \underset{\text{oxygen}}{6\,O_2}$$

This equation shows that during photosynthesis, hydrogen atoms are transferred from water to carbon dioxide, and glucose is formed. Therefore, water has been oxidized, and carbon dioxide has been reduced. It takes energy to form glucose, and this energy is supplied by solar energy. Chloroplasts are able to capture solar energy and convert it to the chemical energy of ATP molecules that are used along with hydrogen atoms to reduce carbon dioxide.

The reduction of carbon dioxide to form a mole of glucose stores 686 kcal in the chemical bonds of glucose. This is the energy that living things use to support themselves. But glucose must be broken down during cellular respiration for this energy to be made available. Only in this way is the energy stored in glucose made available to cells.

Cellular Respiration

The overall equation for cellular respiration is the opposite of the one we used to represent photosynthesis:

$$C_6H_{12}O_6 + 6\,O_2 \longrightarrow 6\,CO_2 + 6\,H_2O + energy$$

glucose   oxygen   carbon   water

           dioxide

In this reaction, glucose has lost hydrogen atoms (been oxidized), and oxygen has gained hydrogen atoms (been reduced). When oxygen gains electrons, it becomes water. The complete oxidation of a mole of glucose releases 686 kcal of energy, and this energy is used to synthesize ATP molecules.

If the energy within glucose and for that matter amino acids and fats, which can also respired, were released all at once, most of it would dissipate as heat. Instead, cells oxidize glucose, amino acids, and fatty acids step by step. Their energy is gradually converted to that of ATP molecules.

Organelles and the Flow of Energy

During photosynthesis, the chloroplasts in plants capture solar energy and use it to convert water and carbon dioxide into carbohydrates, which serve as food for all living things. Oxygen is one by-product of photosynthesis.

Mitochondria, present in both plants and animals, complete the breakdown of carbohydrates and use the released energy to build ATP molecules. Cellular respiration consumes oxygen and produces carbon dioxide and water, the very molecules taken up by chloroplasts.

It is actually the cycling of molecules between chloroplasts and mitochondria that allows a flow of energy from the sun through all living things (Fig. 6.11). As mentioned at the beginning of the chapter, this flow of energy occurs in cells as well as in ecosystems. In keeping with the laws of thermodynamics, energy dissipates as chemical transformations occur, and eventually the solar energy captured by plants is lost in the form of heat. Therefore, most living things are dependent upon an input of solar energy.

Human beings are involved in the cycling of molecules between plants and animals and in the flow of energy from the sun through ecosystems. We inhale oxygen and eat plants, or we eat other animals that have eaten plants. Oxygen and nutrient molecules enter our mitochondria, which produce ATP and release carbon dioxide and water, the molecules plants use to produce carbohydrates. Without a supply of energy-rich molecules produced by plants, we could not produce the ATP molecules needed to maintain our bodies.

The oxidation-reduction pathways of photosynthesis in chloroplasts and cellular respiration in mitochondria permit energy to flow from the sun through all living things.

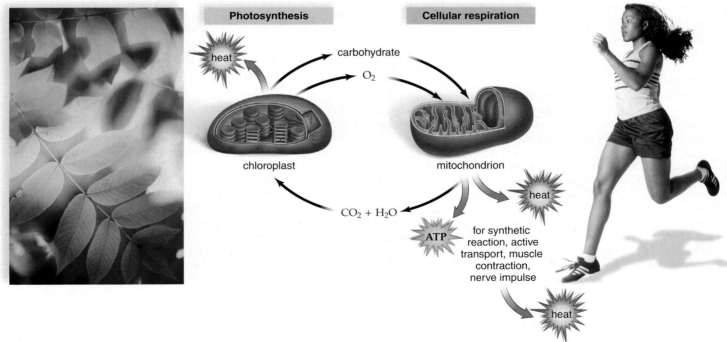

Figure 6.11 Relationship of chloroplasts to mitochondria.
Chloroplasts produce energy-rich carbohydrates. These carbohydrates are broken down in mitochondria, and the energy released is used for the buildup of ATP. Mitochondria can also respire molecules derived from amino acids and fats for the build-up of ATP. Usable energy is lost as heat due to the energy conversions of photosynthesis, cellular respiration, and the use of ATP.

You have been told to eat your fruits and vegetables, but have you ever been told to eat your enzymes? Enzyme therapy is a booming business. In enzyme therapy, patients swallow supplements containing enzymes purified from natural sources in order to aid the enzymes of their own bodies. Some of these are digestive enzymes isolated from plant and animal sources. Others are enzymes that are intended to fight inflammation, increase energy, or enhance antioxidant activity. And still others are touted to improve or even cure serious diseases. But opponents of enzyme therapy are becoming more and more vocal about the dangers of this type of alternative medicine.

Medical science has learned that taking certain digestive enzymes can be beneficial for people who have genetic defects. For example, people with cystic fibrosis take pancreatin, a mix of pancreatic enzymes, to assist their digestion. But for other people, opponents of enzyme therapy believe that purchasing supplemental enzymes is a waste of money. They say that

other enzymes would be broken down by the stomach acid and the digestive enzymes in the small intestine and would never be able to help patients. Opponents also caution that just because something is "natural" doesn't mean it cannot be harmful. Worse, they believe, is that some people fail to seek traditional medical attention for serious ailments because they believe these enzymes will heal them—claims that are unsubstantiated by medical research.

Proponents of enzyme therapy strongly believe that the goal of medicine should be to keep people healthy instead of only to treat disease. They believe everyone could benefit from extra enzymes, not just people who have defects. Aging and poor diets contribute to a need for extra enzymes. Cooking food destroys the natural enzymes in the food, so all of us are lacking in these enzymes. Enzyme therapy proponents claim that because enzyme therapies are labeled "alternative," funding is unavailable for the types of studies that need to be done. And enzyme therapists can point to many

satisfied customers who feel much better after taking enzyme supplements. Just because traditional medicine doesn't currently understand how enzyme therapy works doesn't mean that it won't move from alternative to traditional medical practice in a few years.

Form Your Own Opinion

1. Should manufacturers of enzyme supplements be required to complete controlled medical research studies before selling their product?
2. If taking enzyme supplements helps you feel better, should you be prevented from taking them just because they are not traditional medicine?
3. If someone chooses enzyme therapy for a serious medical problem and later dies, can/should the enzyme therapist and/or manufacturer be sued for treating this patient in a nontraditional manner? Keep in mind that traditional medicine does not have a 100% cure rate, either.

Summarizing the Concepts

6.1 Cells and the Flow of Energy
Two energy laws are basic to understanding energy-use patterns in cells and ecosystems. The first law states that energy cannot be created or destroyed but can only be changed from one form to another. The second law of thermodynamics states that one usable form of energy cannot be converted into another form without loss of usable energy. Therefore, every energy transformation makes the universe less organized. As a result of these laws, we know that the entropy of the universe is increasing and that only a constant input of energy maintains the organization of living things.

6.2 Metabolic Reactions and Energy Transformations
The term metabolism encompasses all the chemical reactions occurring in a cell. Considering individual reactions, only those that result in products that have less usable energy than the reactants go forward. Such reactions, called exergonic reactions, release energy. Endergonic reactions, which require an input of energy, occur in cells only because it is possible to couple an exergonic process with an endergonic process. For example, glucose breakdown is an exergonic metabolic pathway that drives the buildup of many ATP molecules. These ATP molecules then supply energy for cellular work. Therefore, ATP goes through a cycle of constantly being built up from, and then broken down to, ADP + Ⓟ.

6.3 Metabolic Pathways and Enzymes
A metabolic pathway is a series of reactions that proceed in an orderly, step-by-step manner. Each reaction requires a specific enzyme. Reaction rates increase when enzymes form a complex

with their substrates. Generally, enzyme activity increases as substrate concentration increases; once all active sites are filled, the maximum rate has been achieved.

Any environmental factor, such as temperature or pH, affects the shape of a protein, and therefore also affects the ability of an enzyme to do its job. Cellular mechanisms regulate enzyme quantity and activity. The activity of most metabolic pathways is regulated by feedback inhibition. Many enzymes have cofactors or coenzymes that help them carry out a reaction.

6.4 Oxidation-Reduction and the Flow of Energy
The overall equation for photosynthesis is the opposite of that for cellular respiration. Both processes involve oxidation-reduction reactions. Redox reactions are a major way in which energy is transformed in cells.

During photosynthesis, carbon dioxide is reduced to glucose, and water is oxidized. Glucose formation requires energy, and this energy comes from the sun. Chloroplasts capture solar energy and convert it to the chemical energy of ATP molecules, which are used along with hydrogen atoms to reduce carbon dioxide to glucose.

During cellular respiration, glucose is oxidized to carbon dioxide, and oxygen is reduced to water. This reaction releases energy, which is used to synthesize ATP molecules in all types of cells.

Molecules move from plants to animals, and energy flows through all living things. Photosynthesis is a metabolic pathway in chloroplasts that transforms solar energy to the chemical energy within carbohydrates, and cellular respiration is a metabolic pathway completed in mitochondria that transforms this energy into that of ATP molecules. Eventually, the energy within ATP molecules becomes heat. The world of living things depends on a constant input of solar energy.

Testing Yourself

Choose the best answer for each question.

1. ATP is used for
    a. chemical work.
    b. transport work.
    c. mechanical work.
    d. All of these are correct.

2. The _____ of energy would break the first law of thermodynamics.
    a. creation
    b. transformation
    c. Both a and b are correct.
    d. Neither a nor b is correct.

3. The fact that energy transformation is never 100% efficient is stated in the _____ law of thermodynamics.
    a. second
    b. first
    c. third
    d. None of these are correct.

4. ATP is a
    a. modified protein.
    b. modified amino acid.
    c. nucleotide.
    d. modified fat.

5. The amount of energy needed to get a chemical reaction started is known as the
    a. starter energy.
    b. activation energy.
    c. reaction energy.
    d. product energy.

In questions 6–9, match the statements concerning enzyme action to the appropriate graph.

6. Progress of reaction in the absence of an enzyme.

7. Enzymatic action with varying temperature.

8. Progress of reaction in the presence of an enzyme.

9. Enzymes with varying pH optima

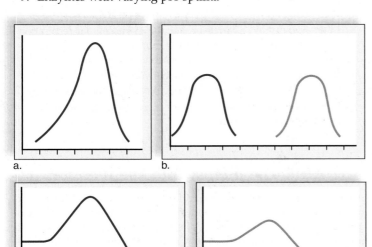

a.   b.

c.   d.

10. Enzymatic action is sensitive to changes in
    a. temperature.
    b. pH.
    c. substrate concentration.
    d. All of these are correct.

11. Vitamins can be components of
    a. coreactants.
    b. cosugars.
    c. coenzymes.
    d. None of these are correct.

12. An example of mechanical work would be
    a. muscle contraction.
    b. synthesis of macromolecules.
    c. photosynthesis.
    d. None of these are correct.

13. ATP is a "high-energy" molecule because it
    a. easily accepts electrons.
    b. contains high-energy bonds.
    c. can be broken down into ADP + ℗.
    d. Both b and c are correct.

14. The end products of photosynthesis are
    a. CO_2 and H_2O.
    b. O_2 and CO_2.
    c. O_2 and H_2O.
    d. O_2 and carbohydrates.

15. Which has more potential energy: glucose or CO_2 and H_2O?
    a. glucose
    b. CO_2 and H_2O
    c. Both are the same because reactants and products are equal.
    d. Neither is correct.

16. The vitamin niacin is part of the
    a. coenzyme NAD.
    b. metal Mg^{2+}.
    c. vitamin B_2.
    d. None of these are correct.

17. _____ is the loss of electrons, and _____ is the gain of electrons.
    a. Oxidation, reduction
    b. Reduction, oxidation
    c. Neither a nor b is correct.
    d. Need more information to answer.

18. Entropy is a term used to indicate the level of
    a. usable energy.
    b. disorganization.
    c. enzyme action.
    d. None of these are correct.

19. Based upon the reactions of glycolysis and cellular respiration, it can be inferred that energy
    a. is stored in bonds and can be released when bonds are broken.
    b. is stored in molecules but not in bonds.
    c. is stored in bonds but cannot be released because of the laws of thermodynamics.
    d. None of these are correct.

20. Muscle contraction involves converting _____ energy into _____ energy.
    a. chemical, potential
    b. mechanical, chemical
    c. mechanical, potential
    d. chemical, mechanical

21. Enzymes catalyze reactions by
    a. bringing the reactants together.
    b. lowering the activation energy.
    c. Both a and b are correct.
    d. Neither a nor b is correct.

For questions 22–25, match the statements with the molecule in the key. There may be more than one answer per statement.

Key:
    a. oxygen
    b. carbon dioxide
    c. glucose
    d. water

22. 6-carbon sugar

23. Product of photosynthesis

24. 1-carbon compound

25. Product of respiration

26. If the reaction A + B ⟶ C + D + energy occurs in a cell,
    a. the reaction is exergonic.
    b. an enzyme could still speed the reaction.
    c. ATP is not needed to make the reaction go.
    d. A and B are reactants; C and D are products.
    e. All of these are correct.

27. The energy of an ATP molecule is released when
    a. ATP becomes ADP + Ⓟ.
    b. ATP is hydrolyzed.
    c. ATP is coupled to an endergonic reaction.
    d. All of these are correct.

28. The active site of an enzyme is determined by its
    a. primary structure.
    b. secondary structure.
    c. quaternary structure.
    d. Ultimately, all of these determine the shape of the active site.

29. The active site of an enzyme
    a. is identical to that of any other enzyme.
    b. is the part of the enzyme where its substrate can fit.
    c. can be used over and over again.
    d. is not affected by environmental factors such as pH and temperature.
    e. Both b and c are correct.

30. If you want to increase the amount of product per unit time of an enzymatic reaction, increase
    a. the amount of substrate.   c. the temperature somewhat.
    b. the amount of enzyme.   d. All of these are correct.

31. O_2 is _____ compared to H_2O, which is _____.
    a. reduced, oxidized   c. oxidized, reduced
    b. neutral, a coenzyme   d. active, denatured

32. During photosynthesis, carbon dioxide
    a. is oxidized to oxygen.
    b. is reduced to glucose.
    c. gives up water to the environment.
    d. is a coenzyme of oxidation-reduction.
    e. All of these are correct.

33. Any energy transformation involves the loss of some energy as
    a. electricity.   c. light.
    b. heat.   d. motion.

34. In a metabolic pathway A ⟶ B ⟶ C ⟶ D ⟶ E,
    a. A, B, C, and D are substrates.
    b. B, C, D, and E are products.
    c. each reaction requires its own enzyme.
    d. All of these are correct.

35. The enzyme-substrate complex.
    a. indicates that an enzyme has denatured.
    b. accounts for why enzymes lower the energy of activation.
    c. is nonspecific.
    d. All of these are correct.

Understanding the Terms

active site 107
ADP (adenosine diphosphate) 104
ATP (adenosine triphosphate) 104
chemical energy 102
coenzyme 109
cofactor 109
coupled reactions 105
denatured 108
endergonic reaction 104
energy 102
energy of activation (E_a) 106
entropy 103
enzyme 106
enzyme inhibition 109
exergonic reaction 104
free energy 104
induced fit model 107
kinetic energy 102
mechanical energy 102
metabolic pathway 106
metabolism 104
oxidation 110
potential energy 102
product 104
reactant 104
redox reaction 110
reduction 110
substrate 106
vitamin 109

Match the terms to these definitions:

a._____ All of the chemical reactions that occur in a cell during growth and repair.

b._____ Nonprotein adjunct required by an enzyme in order to function. Many are metal ions; others are coenzymes.

c._____ Energy associated with motion.

d._____ Essential requirement in the diet needed in small amounts; often a part of coenzymes.

e._____ Loss of an enzyme's normal shape so that it no longer functions; caused by an extreme change in pH and temperature.

Thinking Critically

1. If cellular respiration and photosynthesis are the same chemical reaction, just going in different directions, can you reverse photosynthesis and get respiration in cells? Why or why not?

2. Why is it said that "energy does not cycle"?

3. In the body, glucose can be broken down via cellular respiration to provide energy, or it can be built into chains of glycogen and stored. What determines which reaction glucose undergoes?

4. Why don't the complex molecules of your body spontaneously break down to less-ordered forms in order to increase entropy?

5. What would be the advantage of having preformed but inactivated enzymes in a cell?

ARIS, the *Inquiry into Life* Website

ARIS, the website for *Inquiry into Life*, provides a wealth of information organized and integrated by chapter. You will find practice quizzes, interactive activities, labeling exercises, flashcards, and much more that will complement your learning and understanding of general biology.

www.aris.mhhe.com

chapter **7**

Cellular Respiration

Jessica was getting ready to run in her city's marathon, *a distance of 26.2 miles. She had been on a specific training regimen since last year's marathon and was hoping to better her time. This year, she had worked with a sports trainer who had recommended carbohydrate loading.*

Carbohydrate loading is a strategy practiced by endurance athletes to increase the amount of glycogen in their muscles. Glycogen is the stored form of glucose, which is most often used for energy. In this chapter we will learn how glucose and also amino acids and fats are broken down to generate the ATP that is used for muscle contraction. This process, called cellular respiration, occurs in the cytoplasm and mitochondria of the cell. The first source of glucose for energy production is the blood. When that source of glucose is exhausted, the body turns to stored glycogen and fat. For endurance athletes, who are exercising for longer than 90 minutes, excess glycogen in the muscle is a bonus.

Other aspects of Jessica's training involved aerobic versus anaerobic runs. In aerobic runs, athletes run shorter distances at slower rates. This enables muscle cells to completely break down glucose and fat to water and carbon dioxide, producing more ATP than otherwise. Anaerobic training, in contrast, requires running short sprints as fast as possible. Without oxygen (anaerobic), glucose cannot be broken down completely. It is changed into lactate, which is responsible for that muscle burn we sometimes feel after strenuous exercise. Both types of training should help Jessica finish the marathon with a faster time than last year.

Chapter Concepts

7.1 Metabolism
- What nutrients can be catabolized to drive ATP production? 116
- The pathways of what metabolic process provides molecules for anabolism? 116–117
- What is the overall reaction when glucose is used for cellular respiration? 116–117

7.2 Overview of Cellular Respiration
- What coenzymes participate in cellular respiration? What do they do? 118–119
- What are the four phases of cellular respiration? Which phases are aerobic? Anaerobic? 119

7.3 Outside the Mitochondria: Glycolysis
- What metabolic pathway partially breaks down glucose outside the mitochondria? 120–121
- How does this pathway contribute to ATP buildup from glucose breakdown? 120–121
- What happens to the end product of this pathway if oxygen is present? 120

7.4 Inside the Mitochondria
- How do the preparatory reaction and the citric acid cycle contribute to ATP buildup from glucose breakdown? 122–123
- How many molecules of CO_2 result from the preparatory reaction and from the citric acid cycle per glucose breakdown? 122–123
- How does the electron transport chain generate ATP molecules? 123–124
- How is water produced from the electron transport chain? 123
- How many ATP molecules are produced per each phase of cellular respiration? 125

7.5 Fermentation
- How is fermentation different from cellular respiration? 126

7.1   Metabolism

A significant part of cellular metabolism involves the breaking down and the building up of molecules. The term **catabolism** is sometimes used to refer to the breaking down of molecules and the term **anabolism** is sometimes used to refer to the building up (synthesis) of molecules. ATP is the energy currency these reactions use. How does the food we eat get broken down to provide us with ATP and basic building blocks for anabolic reactions? For example, how does the pizza you had for lunch provide you with the ATP needed to walk up the stairs to your next class? And how is the pizza changed into the amino acids you need to build muscles?

Catabolism

Our food contains three nutrients that our bodies use as energy sources: carbohydrates, proteins, and fats (Fig. 7.1). Catabolism involves the breaking down of these nutrients. Carbohydrates such as bread and potatoes are broken down in our digestive systems to glucose. Glucose is the molecule that directly undergoes cellular respiration to produce ATP, as we shall see in subsequent sections of this chapter. Complete breakdown of glucose requires the pathways shown in Figure 7.1 and defined on page 119.

Fats, such as oils and butter, are broken down in the digestive tract to glycerol and fatty acids. Glycerol can enter cellular respiration at glycolysis, while fatty acids enter respiration at acetyl CoA. Fats are a good source of energy because the oxidation of fatty acids produces considerably more ATP per carbon atom than does the oxidation of carbohydrates.

If we are on a diet that restricts carbohydrates, or we are fasting, our bodies will use amino acids as an energy source. When we are starving, the body will catabolize protein from the muscles attached to our bones and even the heart, as an energy source. The amino groups from the amino acids are removed in a process called **deamination.** The amino group becomes part of urea, the primary excretory product of humans. The remainder of the amino acid can enter cellular respiration at various places in the pathway, depending on the length and type of molecule.

Anabolism

Some of the building blocks for the synthesis of larger molecules come directly from our food, as when glucose is used to form glycogen or amino acids are used to form proteins. However, synthesis nearly always requires energy, and this energy is supplied by ATP built up by the pathways of cellular respiration shown in Figure 7.1.

Anabolism is also related to catabolism in another way. The substrates of glycolysis and the citric acid cycle can be the starting point for synthetic reactions. For example, a substrate from glycolysis can be converted to glycerol. Acetyl groups that ordinarily enter the citric acid cycle can instead be joined to form fatty acids. Fat synthesis follows when glycerol joins with three fatty acids. This explains why we gain weight from eating too much candy, ice cream, or cake.

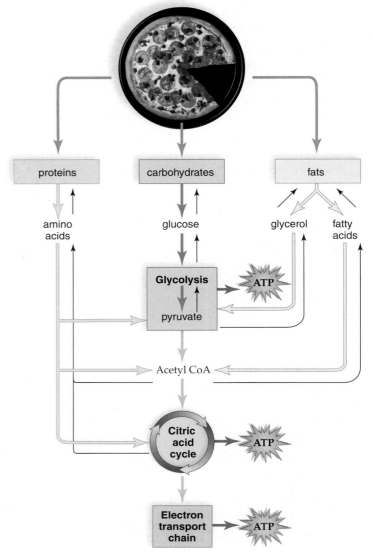

Figure 7.1 Metabolism.
Carbohydrates, fats, and proteins can be used as energy sources when their components are degraded. These pathways produce molecules that can also serve as the starting point for the synthesis of carbohydrates, fats, and proteins.

Some substrates of the citric acid cycle can become amino acids through the addition of an amino group. Plants are able to synthesize all of the amino acids they need in this way. Animals, however, lack some of the enzymes necessary to synthesize all amino acids. The amino acids they cannot synthesize must be supplied by the diet; these are called the essential amino acids. (The amino acids that can be synthesized are called nonessential.) It is quite possible for animals to suffer from protein deficiency if the diet does not contain adequate quantities of all the essential amino acids.

All the reactions involved in cellular respiration are a part of metabolism, and substrates can be used for catabolism or for anabolism.

Breathing, Eating, and Cellular Respiration

Cellular respiration is the step-wise release of energy from molecules such as glucose, accompanied by the use of this energy to synthesize ATP molecules. Cellular respiration is an **aerobic** process that requires oxygen (O_2) and gives off carbon dioxide (CO_2). It usually involves the complete breakdown of glucose to carbon dioxide and water, as illustrated in Figure 7.2.

The air we inhale when we breathe contains oxygen (O_2), and the food we digest after eating contains glucose (Fig. 7.3). These enter the bloodstream at the lungs and digestive tract, respectively. The bloodstream carries them through the body, and from there oxygen and glucose enter each and every cell. Glucose metabolism begins in the cytoplasm, but its breakdown products enter the mitochondria where the preparatory reaction, citric acid cycle, and electron transport chain are located. As oxidation of substrates occurs, carbon dioxide (CO_2) results. Oxygen that has entered the mitochondria accepts electrons and hydrogen ions at the end of the electron transport chain. The result is water molecules. The purpose of cellular respiration is to produce ATP molecules. When cellular respiration is finished, not only does the cell have a supply of ATP molecules, but water and carbon dioxide have also been produced.

Carbon dioxide diffuses out of the mitochondria and the cell to enter the bloodstream. The bloodstream takes the CO_2 to the lungs, where it is exhaled. The water can remain in the mitochondria or in the cell, or it can enter the bloodstream to be excreted by the kidneys as necessary.

Metabolism involves both catabolism and anabolism. Catabolism is the breaking down of molecules while anabolism involves the synthesis of molecules. Breathing and eating are required for metabolism.

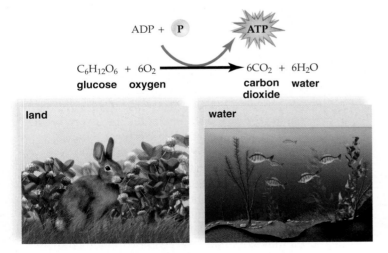

$$ADP + P \longrightarrow ATP$$

$$C_6H_{12}O_6 + 6O_2 \longrightarrow 6CO_2 + 6H_2O$$

glucose   oxygen   carbon   water
                 dioxide

land   water

Figure 7.2 Cellular respiration.
Almost all organisms, whether they reside on land or in the water, carry on cellular respiration, which is most often glucose breakdown coupled to ATP synthesis.

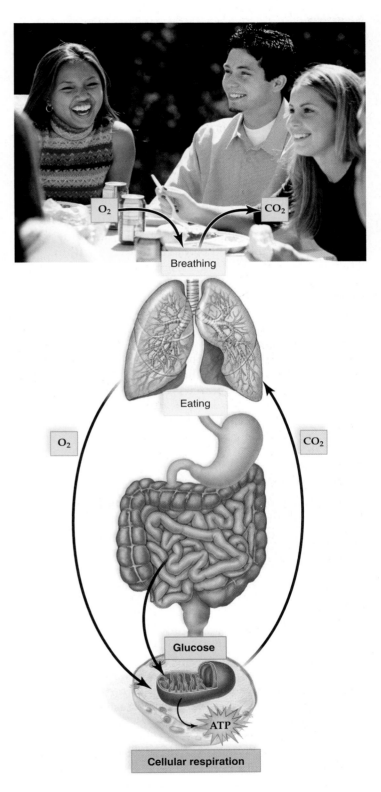

Figure 7.3 Relationship between breathing, eating, and cellular respiration.
The O_2 we inhale and the glucose we ingest with our food are carried by the bloodstream into our cells, where they enter mitochondria. Following cellular respiration, the ATP stays in the cell, but the CO_2 is carried to the lungs for exhalation.

7.2   Overview of Cellular Respiration

The metabolic pathways of cellular respiration allow the energy within a glucose molecule to be released in a step-wise fashion so that it can be coupled to ATP production. Cells would lose a tremendous amount of energy if glucose breakdown occurred all at once—much of that energy would become nonusable heat.

The overall equation for cellular respiration (see Fig. 7.2) shows the coupling of glucose breakdown to ATP buildup. As glucose is broken down, ATP is built up, and this is why the ATP reaction is drawn using a curved arrow above the glucose reaction arrow. The breakdown of one glucose molecule results in 36 or 38 ATP molecules. This represents about 40% of the potential energy within a glucose molecule; the rest of the energy dissipates. This conversion is more efficient than many others; for example, only about 25% of the energy within gasoline is converted to the motion of a car.

Why does a cell go to the bother of converting the energy within the bonds of glucose to energy within the bonds of ATP? ATP contains just about the amount of energy required for most cellular reactions. This energy is released by a relatively simple procedure: the removal of a single phosphate group. The cell uses the reverse reaction for building up ATP again.

NAD+ and FAD

The metabolic pathways that make up cellular respiration involve many individual reactions, each one catalyzed by its own enzyme. Certain of these enzymes are redox enzymes that utilize the coenzyme **NAD+ (nicotinamide adenine dinucleotide).** (You may recall from Chapter 6 that coenzymes are organic, nonprotein molecules that help an enzyme function properly.) When a substrate is oxidized and hydrogen atoms ($H^+ + e^-$) are removed, NADH + H+ results. The electrons received by NAD+ are high-energy electrons that are usually carried to an electron transport chain (Fig. 7.4)

FAD (flavin adenine dinucleotide) is another frequently used coenzyme of oxidation-reduction. When FAD accepts hydrogen atoms, $FADH_2$ results. NAD+ and FAD are analogous to electron shuttle buses. They pick up electrons (and their accompanying hydrogen nuclei) at specific enzymatic reactions in either the cytoplasm or the matrix of the mitochondria. They each can only hold two electrons and two hydrogen nuclei. They then carry these electrons to the electron transport chain in the cristae of the mitochondria, where they drop them off. The empty NAD+ or FAD is then free to go back and pick up more electrons.

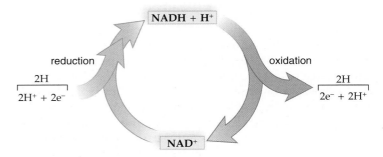

Figure 7.4 The NAD+ cycle.
The coenzyme NAD+ accepts two hydrogen atoms ($H^+ + e^-$), and NADH + H+ results. When NADH passes on electrons, usually to the electron transport chain, NAD+ results. Only a small amount of NAD+ need be present in a cell, because each NAD+ molecule is used over and over again.

NAD+ and FAD are two coenzymes of oxidation-reduction that are active during cellular respiration.

Phases of Cellular Respiration

The oxidation of glucose by removal of hydrogen atoms (cellular respiration) involves four phases (Fig. 7.5). The first phase, glycolysis, takes place outside the mitochondria and does not utilize oxygen. The other phases take place inside the mitochondria, where oxygen is the final acceptor of electrons.

- **Glycolysis** is the breakdown of glucose ($C_6H_{12}O_6$) to two molecules of pyruvate, a C_3 molecule. Oxidation by removal of hydrogen atoms ($H^+ + e^-$) provides enough energy for the immediate buildup of two ATP.
- During the **preparatory (prep) reaction,** pyruvate is oxidized to a C_2 acetyl group carried by CoA (coenzyme A), and CO_2 is removed. Since glycolysis ends with two molecules of pyruvate, the prep reaction occurs twice per glucose molecule.
- The **citric acid cycle** is a cyclical series of oxidation reactions that give off CO_2 and produce one ATP. The citric acid cycle used to be called the Krebs cycle in honor of the man who worked out most of the steps. The cycle is now named for citric acid (or citrate), the first molecule in the cycle. The citric acid cycle turns twice because two acetyl CoA molecules enter the cycle per glucose molecule. Altogether, the citric acid cycle accounts for two immediate ATP molecules per glucose molecule.
- The **electron transport chain** is a series of carriers that accept the electrons (but not the H+) removed from glucose and pass them along from one carrier to the next until they are finally received by O_2, which then combines with H+ and becomes water. As the electrons

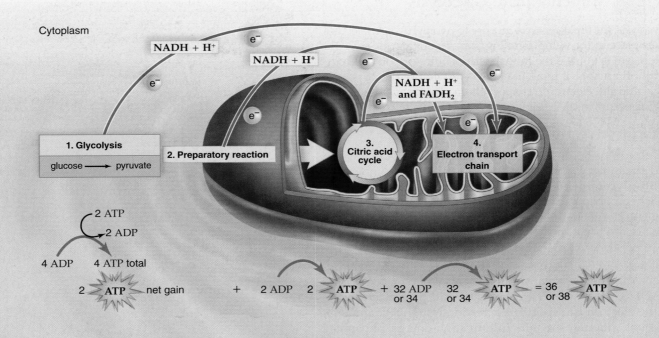

Figure 7.5 The four phases of complete glucose breakdown.

The complete breakdown of glucose consists of four phases. (1) Glycolysis in the cytoplasm produces pyruvate, which enters mitochondria if oxygen is available. (2,3) The preparatory reaction and the citric acid cycle that follow occur inside the mitochondria. (4) Also inside mitochondria, the electron transport chain receives the electrons that were removed from glucose breakdown products. The result of glucose breakdown is 36 or 38 ATP, depending on the particular cell.

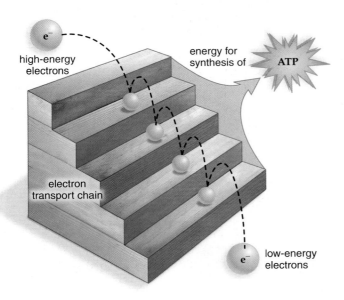

Figure 7.6 Electron transport chain.

High-energy electrons are delivered to the chain. As they pass from carrier to carrier, energy is released and used for ATP production.

pass from a higher-energy to a lower-energy state, energy is released and used for ATP synthesis. The electrons from one glucose molecule result in 32 or 34 ATP, depending on certain conditions. The electron transport chain is like stair steps. As electrons fall down each step, they lose energy (Fig. 7.6).

Pyruvate is a pivotal metabolite in cellular respiration. If oxygen is not available to the cell, **fermentation** occurs in the cytoplasm instead of continued aerobic cellular respiration. During fermentation, pyruvate is reduced to lactate or to carbon dioxide and alcohol, depending on the organism. As we shall see in section 7.5 fermentation results in a net gain of only two ATP per glucose molecule.

Cellular respiration involves the oxidation of glucose to carbon dioxide and water. As glucose breaks down, energy is made available for ATP synthesis. For every glucose molecule in cellular respiration, a total of 36 or 38 ATP molecules are produced (2 from glycolysis, 2 from the citric acid cycle, and 32 or 34 from the electron transport chain).

Glycolysis, the breakdown of glucose to two pyruvate molecules, takes place in the cytoplasm (Fig. 7.7a). In eukaryotic cells, the cytoplasm is outside the mitochondria. Since glycolysis occurs in all organisms, it most likely evolved before the citric acid cycle and the electron transport chain. This may be why glycolysis occurs in the cytoplasm and does not require oxygen.

Energy-Investment Steps

As glycolysis begins, two ATP are used to activate glucose, and the molecule that results splits into two C_3 molecules **(G3P, glyceraldehyde 3-phosphate),** each of which has an attached phosphate group. From this point on, each C_3 molecule undergoes the same series of reactions (Fig. 7.7b).

Energy-Harvesting Steps

Oxidation of G3P occurs by the removal of hydrogen atoms $(H^+ + e^-)$. The hydrogen atoms are picked up by NAD^+, and $NADH + H^+$ results. Later, NADH will pass electrons on to the electron transport chain. Oxidation of G3P and subsequent substrates results in four high-energy phosphate groups, which are used to synthesize four ATP. This is **substrate-level ATP synthesis,** in which an enzyme passes a high-energy phosphate to ADP, and ATP results:

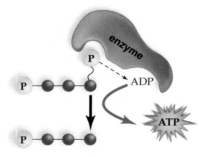

Subtracting the two ATP that were used to get started, glycolysis yields a net gain of two ATP (Fig. 7.7b).

When oxygen is available, the end product, pyruvate, enters the mitochondria, where it undergoes further breakdown. Altogether, the inputs and outputs of glycolysis are as follows:

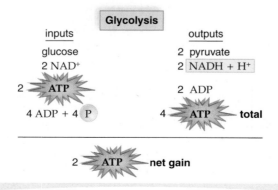

Glycolysis, which occurs in the cytoplasm, breaks down glucose to pyruvate resulting in a net gain of 2 ATP.

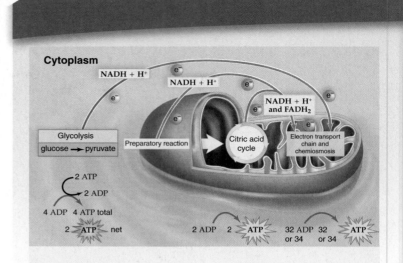

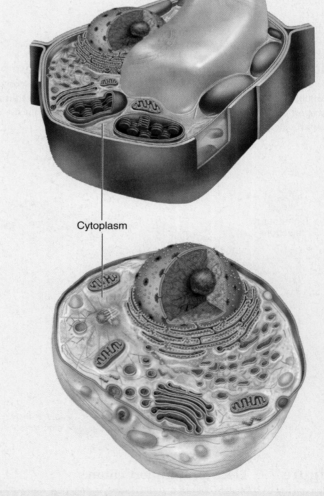

Cytoplasm

a.

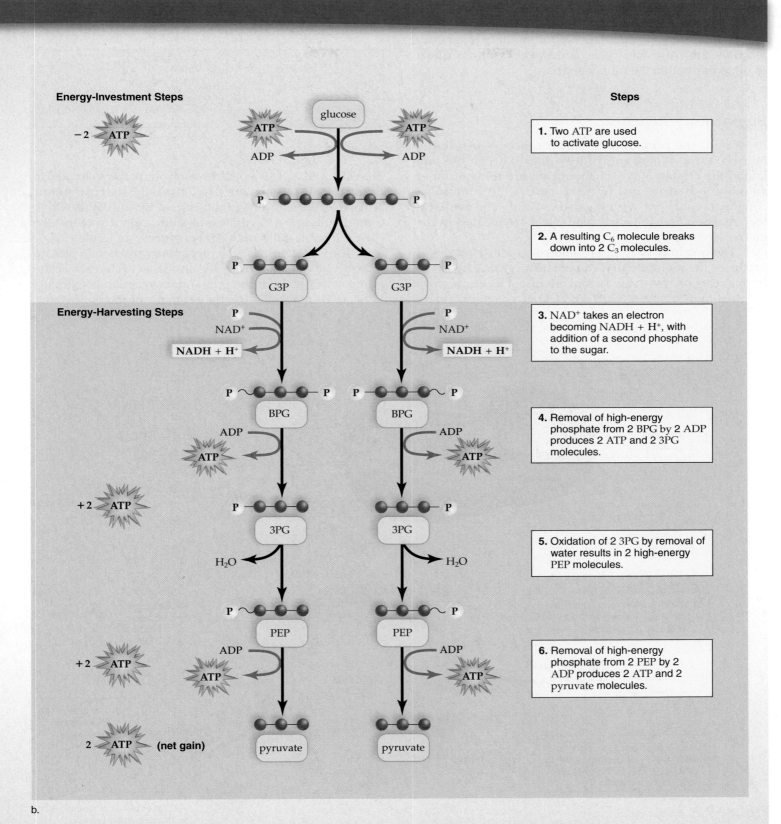

Energy-Investment Steps

− 2 ATP

glucose

ATP → ADP

ATP → ADP

P ●●●●●● P

1. Two ATP are used to activate glucose.

2. A resulting C_6 molecule breaks down into 2 C_3 molecules.

P ●●● G3P

●●● P G3P

Energy-Harvesting Steps

P
NAD⁺
NADH + H⁺

P
NAD⁺
NADH + H⁺

3. NAD⁺ takes an electron becoming NADH + H⁺, with addition of a second phosphate to the sugar.

P ~●●● P BPG

P ●●●~ P BPG

ADP → ATP

ADP → ATP

4. Removal of high-energy phosphate from 2 BPG by 2 ADP produces 2 ATP and 2 3PG molecules.

+ 2 ATP

P ●●● P 3PG

P ●●● P 3PG

H_2O

H_2O

5. Oxidation of 2 3PG by removal of water results in 2 high-energy PEP molecules.

P ~●●● PEP

●●●~ P PEP

ADP → ATP

ADP → ATP

+ 2 ATP

6. Removal of high-energy phosphate from 2 PEP by 2 ADP produces 2 ATP and 2 pyruvate molecules.

2 ATP **(net gain)**

●●● pyruvate

●●● pyruvate

b.

Figure 7.7 Glycolysis.

a. Glycolysis takes place in the cytoplasm of almost all cells. **b.** This pathway begins with glucose and ends with two pyruvate molecules. There is a gain of two NADH + H⁺ and a net gain of two ATP from glycolysis.

7.4   Inside the Mitochondria

The final reactions of cellular respiration, the preparatory reaction, the citric acid cycle, and the electron transport chain, occur within the mitochondria.

Preparatory Reaction

As stated, the preparatory reaction occurs inside the mitochondria. Where specifically does it occur? As you know, the **cristae** of a mitochondrion are folds of inner membrane that jut out into the **matrix,** an innermost compartment filled with a gel-like fluid. The preparatory reaction and the citric acid cycle are located in the matrix (see Fig. 7.5).

The preparatory (prep) reaction is so called because it produces the molecule that can enter the citric acid cycle. In this reaction, pyruvate is converted to a C_2 *acetyl group* attached to *coenzyme A (CoA),* and CO_2 is given off. This is

an oxidation reaction in which hydrogen atoms ($H^+ + e^-$) are removed from pyruvate by NAD^+ and $NADH + H^+$ results. This reaction occurs twice per glucose molecule:

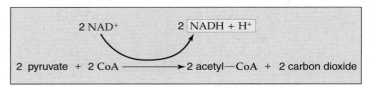

Citric Acid Cycle

The citric acid cycle is a cyclical metabolic pathway located in the matrix of mitochondria (Fig. 7.8). At the start of the citric acid cycle, the C_2 acetyl group carried by CoA joins with a C_4 molecule, and a C_6 citrate molecule results. (Note that the citric acid cycle will turn twice per glucose molecule.)

The CoA returns to the preparatory reaction to pick up more acetyl groups. During the citric acid cycle, each acetyl group received from the preparatory reaction is oxidized to two CO_2 molecules. As the reactions of the cycle occur,

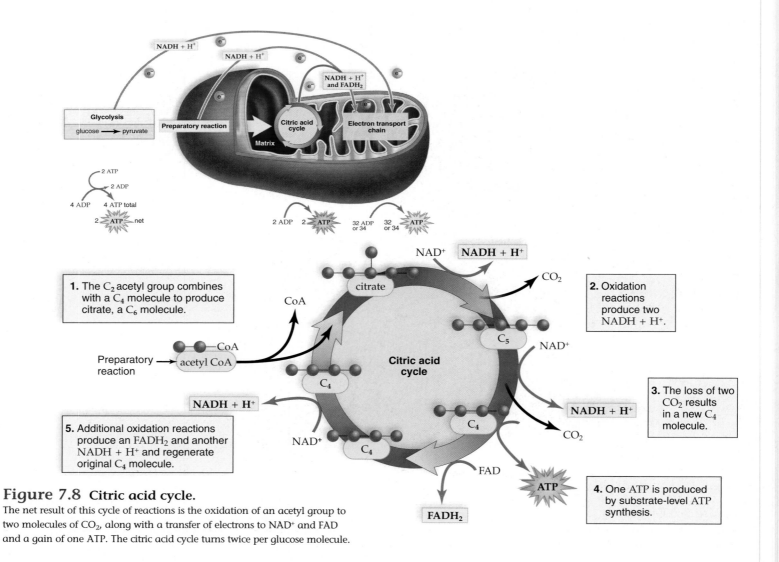

Figure 7.8 Citric acid cycle.
The net result of this cycle of reactions is the oxidation of an acetyl group to two molecules of CO_2, along with a transfer of electrons to NAD^+ and FAD and a gain of one ATP. The citric acid cycle turns twice per glucose molecule.

oxidation is carried out by the removal of hydrogen atoms ($H^+ + e^-$). In three instances, $NADH + H^+$ results, and in one instance, $FADH_2$ is formed.

Substrate-level ATP synthesis is also an important event of the citric acid cycle. In substrate-level ATP synthesis, an enzyme passes a high-energy phosphate to ADP, and ATP results.

Because the citric acid cycle turns twice for each original glucose molecule, the inputs and outputs of the citric acid cycle per glucose molecule are as follows:

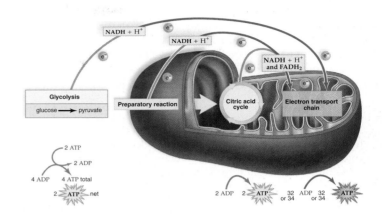

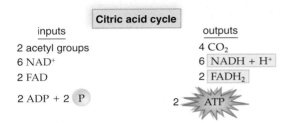

Citric acid cycle		
inputs		outputs
2 acetyl groups		4 CO_2
6 NAD^+		6 $NADH + H^+$
2 FAD		2 $FADH_2$
2 ADP + 2 ⓟ		2 ⭐ ATP

The six carbon atoms originally located in a glucose molecule have now become CO_2. The preparatory reaction produces two CO_2, and the citric acid cycle produces four CO_2 per glucose molecule.

Electron Transport Chain

The electron transport chain located in the cristae of the mitochondria is a series of carriers that pass electrons from one to the other. Some of the electron carriers of the system are called **cytochrome** molecules. Cytochromes are a class of iron-containing proteins important in redox reactions.

Two electrons per each NADH and $FADH_2$ enter the chain. Figure 7.9 is arranged to show that high-energy electrons enter the system and low-energy electrons leave the system. When NADH gives up its electrons, it becomes NAD^+; the next carrier gains the electrons and is reduced. This oxidation-reduction reaction starts the process, and each of the carriers, in turn, becomes reduced and then oxidized as the electrons move down the system.

As the electrons pass from one carrier to the next, energy that will be used to produce ATP molecules is captured and stored as a hydrogen ion gradient (see Fig. 7.10). Oxygen receives the energy-spent electrons from the last of the carriers. After receiving electrons, oxygen combines with hydrogen ions and forms water.

When NADH delivers electrons to the first carrier of the electron transport chain, enough energy is released by the time the electrons are received by O_2 to permit the production of three ATP molecules. When $FADH_2$ delivers electrons to the electron transport chain, only two ATP are produced.

Figure 7.9 The electron transport chain.

NADH and $FADH_2$ bring electrons to the electron transport chain. As the electrons move down the chain, energy is captured and used to form ATP. For every pair of electrons that enters by way of NADH, three ATP result. For every pair of electrons that enters by way of $FADH_2$, two ATP result. Oxygen, the final acceptor of the electrons, becomes a part of water.

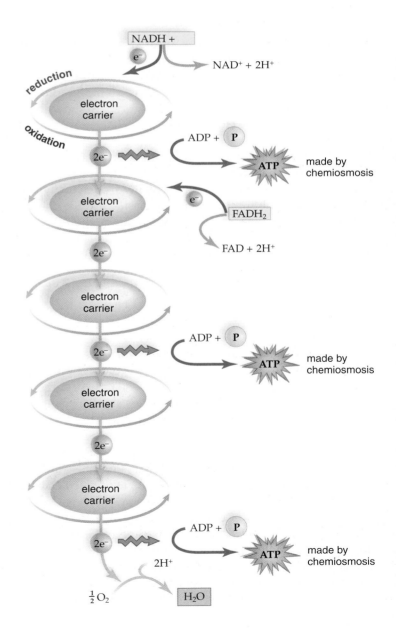

Once NADH has delivered electrons to the electron transport chain, the NAD$^+$ that results can pick up more hydrogen atoms. In like manner, once FADH$_2$ gives up electrons to the chain, the FAD that results can pick up more hydrogen atoms. The recycling of coenzymes and ADP increases cellular efficiency since it does away with the necessity to synthesize NAD$^+$ and FAD everytime.

Many poisons act on the electron transport chain. The Bioethical Focus essay discusses one such poison, cyanide.

> **The preparatory reaction and citric acid cycle take place in the matrix of the mitochondria. The electron transport chain is located in the cristae of mitochondria. As electrons pass down the electron transport chain, energy is released and captured for the production of ATP.**

Organization of Cristae

The electron transport chain is located within the cristae of the mitochondria. The cristae increase the internal surface area of a mitochondrion, thereby increasing the area devoted to ATP formation. Figure 7.9 is a simplified overview of the electron transport chain. Figure 7.10 shows how the components of the electron transport chain are arranged in the cristae.

We have been stressing that the carriers of the electron transport chain accept electrons, which they pass from one to the other. What happens to the hydrogen ions from NADH + H$^+$? The complexes in the cristae use the energy released by electrons as they move down the electron transport chain to pump H$^+$ from the mitochondrial matrix into the space between the outer and inner membrane of a mitochondrion. This space is called the **intermembrane space.** The pumping of H$^+$ into the intermembrane space establishes an unequal distribution of H$^+$ ions; there are many H$^+$ in the intermembrane space and few in the matrix of a mitochondrion.

The cristae also contain an *ATP synthase complex.* The H$^+$ ions flow through an ATP synthase complex from the intermembrane space into the matrix. The flow of three H$^+$ through an ATP synthase complex brings about a change in shape, which causes the enzyme ATP synthase to synthesize ATP from ADP + Ⓟ. Mitochondria produce ATP by **chemiosmosis,** so called because ATP production is tied to an electrochemical gradient, namely the unequal distribution of H$^+$ across the cristae. Once formed, ATP molecules pass into the cytoplasm.

Chemiosmosis is similar to using water behind a dam to generate electricity. The pumping of H$^+$ out of the matrix into the intermembrane space is like pumping water behind the dam. The floodgates of the dam are like the ATP synthase. When the floodgates are open, the water rushes through, generating electricity. In the same way, H$^+$ rushing through the ATP synthase is used to produce ATP.

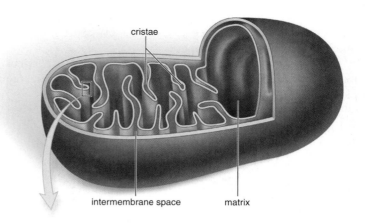

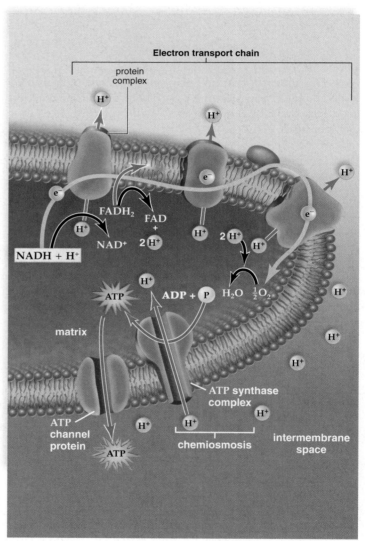

Figure 7.10 Organization and function of cristae in mitochondria.

The electron transport chain is located in the cristae. As electrons move from one protein complex to the other, hydrogen ions (H$^+$) are pumped from the mitochondrial matrix into the intermembrane space. As hydrogen ions flow down a concentration gradient from the intermembrane space into the matrix, ATP is synthesized by the enzyme ATP synthase. ATP leaves the matrix by way of a channel protein.

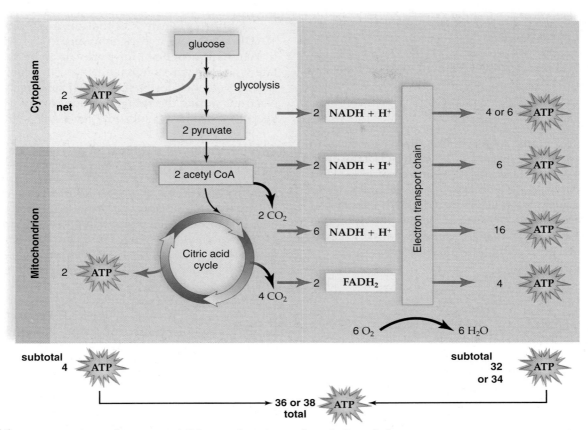

Figure 7.11 **Accounting of energy yield per glucose molecule breakdown.**
Substrate-level ATP synthesis during glycolysis and the citric acid cycle accounts for four ATP. Chemiosmosis accounts for 32 or 34 ATP, and the grand total of ATP is therefore 36 or 38 ATP. Cells differ as to how the electrons are delivered from the two NADH + H^+ generated outside the mitochondria. If these two electrons are delivered by a shuttle mechanism to the start of the electron transport chain, six ATP result; otherwise, four ATP result.

Energy Yield from Cellular Respiration

Figure 7.11 calculates the ATP yield for the complete breakdown of glucose to CO_2 and H_2O. Per glucose molecule, there is a net gain of two ATP from glycolysis, which takes place in the cytoplasm. The citric acid cycle, which occurs in the matrix of mitochondria, accounts for two ATP per glucose molecule. This means that a total of four ATP are formed by substrate-level ATP synthesis outside the electron transport chain.

The most ATP is produced by the electron transport chain and chemiosmosis. Per glucose molecule, ten NADH + H^+ and two $FADH_2$ take electrons to the electron transport chain. For each NADH + H^+ formed *inside* the mitochondria by the citric acid cycle, three ATP result, but for each $FADH_2$, only two ATP are produced. Figure 7.9 shows the reason for this difference: $FADH_2$ delivers its electrons to the transport chain lower in the chain than does NADH + H^+, and therefore these electrons cannot account for as much ATP production.

What about the ATP yield of NADH + H^+ generated *outside* the mitochondria by the glycolytic pathway? NADH + H^+ cannot cross mitochondrial membranes, but a "shuttle" mechanism allows its electrons to be delivered to the electron transport chain inside the mitochondria. The shuttle consists of an organic molecule, which can cross the outer membrane, accept the electrons, and in most, but not all cells, deliver them to a FAD molecule in the inner membrane. If FAD is used, only two ATP result because the electrons have not entered at the start of the electron transport chain.

Efficiency of Cellular Respiration

It is interesting to calculate how much of the energy in a glucose molecule eventually becomes available to the cell. The difference in energy content between the reactants (glucose and O_2) and the products (CO_2 and H_2O) is 686 kcal. An ATP phosphate bond has an energy content of 7.3 kcal, and 36 of these are usually produced during glucose breakdown; 36 phosphates are equivalent to a total of 263 kcal. Therefore, 263/686, or 39%, of the available energy is usually transferred from glucose to ATP. The rest of the energy dissipates in the form of heat.

The preparatory reaction produces the molecule that enters the citric acid cycle, which in turn produces ATP, NADH + H^+, and $FADH_2$. NADH + H^+ and $FADH_2$ pass electrons to the electron transport chain, which generates ATP via chemiosmosis.

7.5   Fermentation

Most organisms carry on cellular respiration when oxygen is available. When oxygen is not available, cells turn to fermentation (Fig. 7.12). In human cells, like other animal cells, the pyruvate formed by glycolysis accepts two hydrogen atoms and is reduced to lactate. Other types of organisms instead produce alcohol with the release of CO_2. Bacteria vary as to whether they produce an organic acid, such as lactate, or an alcohol and CO_2. Yeasts are good examples of organisms that generate ethyl alcohol and CO_2 as a result of fermentation. When yeast is used to leaven bread, the CO_2 produced makes bread rise. Yeast is also used to ferment wine; in that case, it is the ethyl alcohol that is desired. Eventually, yeasts are killed by the very alcohol they produce.

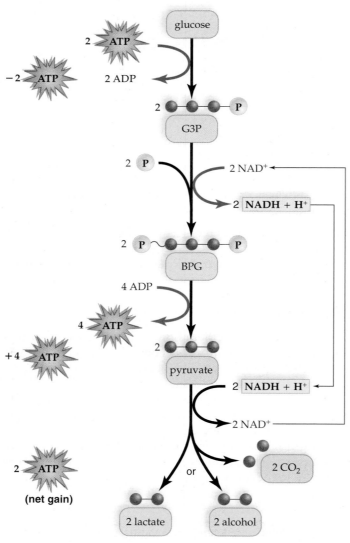

Figure 7.12 Fermentation.
Fermentation consists of glycolysis followed by a reduction of pyruvate by NADH + H⁺. The resulting NAD⁺ returns to the glycolytic pathway to pick up more hydrogen atoms.

Notice in Figure 7.12 that during fermentation, two NADH + H⁺ pass hydrogen atoms to pyruvate, reducing it. Why is it beneficial for pyruvate to be reduced to lactate when oxygen is not available? The reason is that this reaction regenerates NAD⁺, which can then pick up more electrons during the earlier reactions of glycolysis, and this keeps glycolysis and substrate-level ATP synthesis going.

Advantages and Disadvantages of Fermentation

Despite its low yield of only two ATP, fermentation is essential to humans. It can provide a rapid burst of ATP, and thus muscle cells more than other cells are apt to carry on fermentation. When our muscles are working vigorously over a short period of time, as when we run, fermentation is a way to produce ATP even though oxygen is temporarily in limited supply.

Lactate, however, is toxic to cells. At first, blood carries away all the lactate formed in muscles. But eventually lactate begins to build up, changing the pH and causing the muscles to "burn." When we stop running, our bodies are in **oxygen debt,** as signified by the fact that we continue to breathe very heavily for a time. Recovery is complete when the lactate is transported to the liver, where it is reconverted to pyruvate. Some of the pyruvate is broken down completely, and the rest is converted back to glucose.

Energy Yield of Fermentation

Fermentation produces only two ATP by substrate-level ATP synthesis. These two ATP represent only a small fraction of the potential energy stored in a glucose molecule. As noted earlier, complete glucose breakdown during cellular respiration results in at least 36 ATP. Therefore, following fermentation, most of the potential energy a cell can capture from the respiration of a glucose molecule is still waiting to be released.

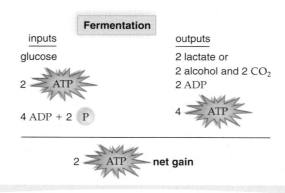

When oxygen is unavailable, some cells ferment glucose instead of carrying on cellular respiration. Fermentation only yields 2 ATP per molecule of glucose.

Illegal Cyanide Fishing

Cyanide is an electron transport chain poison. It binds to cytochrome *c,* one of the electron carriers, and short-circuits the production of ATP. Cyanide has quite a history as a tool of murder and suicide. Unfortunately, cyanide is now being used illegally to catch live fish.

The very lucrative live fish trade provides fish for both food and aquarium use. In Hong Kong alone, live fish for food is estimated to bring in close to $1 billion per year. The retail market for marine aquarium fish is estimated at $500 million worldwide. The majority of these fish are colorful coral reef fish that come from Southeast Asia, especially the Philippines and Indonesia. However, because of overexploitation of coral reef fishes, the fish yields are decreasing. This leads to illegal harvesting practices in order to keep up with demand.

One of the illegal practices used to harvest live fish is called cyanide fishing. In cyanide fishing, crushed cyanide tablets are dissolved in a squirt bottle of water, which is then sprayed toward the fish on top of coral

reefs. The fish are stunned and partially paralyzed, and they float to the surface, making them easy to catch. But cyanide is not just a poison for stunning fish; it also kills the coral animals and their mutualistic algae, and disrupts the marine ecosystem. As coral reefs are destroyed, illegal fishermen move to more pristine reefs in order to harvest fish, eventually destroying those reefs too. Many factors besides cyanide fishing affect the health of the coral reefs, but scientists estimate that the combination of such factors has killed over 25% of the world's coral reefs, and another 30% are in danger.

The live fish industry is very interested in identifying practices that will lead to a sustainable industry, one in which the fish yields can be stabilized, and fish can be caught without harm to the environment or the consumer. The Marine Aquarium Council (MAC) certifies collectors and suppliers of tropical aquarium fish and ensures that these fish were caught "without cyanide." Other nongovernmental organizations are teaching fishermen how to fish

with a net or a hook and line. Agencies such as the WWF (formerly World Wildlife Fund) have proposed international legislation to manage coral reefs and the fishing industry. Aquaculture involves growing fish in tanks or ponds for food or aquarium use. The demand for some fish, such as tilapia, salmon, and carp, is now being filled by fish farming as opposed to catching wild populations.

Form Your Own Opinion

1. Would you be willing to pay higher prices in order to purchase aquarium fish from dealers that have been certified by the MAC?
2. Are you willing to limit your fish consumption to those that are farmed and not caught off coral reefs? How could you determine where the fish came from?
3. Who should pay for the restoration of the damaged coral reefs? Would you be in favor of a tax on live fish to subsidize the restoration?

Summarizing the Concepts

7.1 Metabolism
Carbohydrates, proteins, and fats are broken down via various pathways. These pathways also provide molecules needed for the synthesis of various important substances. The same reactants are used during catabolism and anabolism. Breathing and eating provide the raw materials for cellular respiration and remove some of its products.

7.2 Overview of Cellular Respiration
During cellular respiration, glucose is oxidized to CO_2 and H_2O. This exergonic reaction drives ATP buildup, an endergonic reaction. Four phases are required: glycolysis, the preparatory reaction, the citric acid cycle, and the electron transport chain. Oxidation occurs by the removal of hydrogen atoms ($H^+ + e^-$) from substrate molecules.

7.3 Outside the Mitochondria: Glycolysis
Glycolysis, the breakdown of glucose to two pyruvates, is a series of enzymatic reactions that occur in the cytoplasm. Oxidation by NAD^+ releases enough energy immediately to give a net gain of two ATP by substrate-level ATP synthesis. Two NADH + H^+ are formed.

7.4 Inside the Mitochondria
Pyruvate from glycolysis enters a mitochondrion, where the preparatory reaction takes place. During this reaction, oxidation occurs as CO_2 is removed. NAD^+ is reduced, and CoA receives the

C_2 acetyl group that remains. Since the reaction must take place twice per glucose, two NADH + H^+ result.

The acetyl group enters the citric acid cycle, a series of reactions located in the mitochondrial matrix. Complete oxidation follows, as two CO_2, three NADH + H^+, and one $FADH_2$ are formed. The cycle also produces one ATP. The entire cycle must turn twice per glucose molecule.

The final stage of glucose breakdown involves the electron transport chain located in the cristae of the mitochondria. The electrons received from NADH + H^+ and $FADH_2$ are passed down a chain of electron carriers until they are finally received by O_2, which combines with H^+ to produce H_2O. As the electrons pass down the chain, ATP is produced.

The carriers of the electron transport chain are located in protein complexes on the cristae of the mitochondria. Each protein complex receives electrons and pumps H^+ into the intermembrane space, setting up an electrochemical gradient. When H^+ ions flow down this gradient through the ATP synthase complex, energy is released and used to form ATP molecules from ADP and ⑫. This is ATP synthesis by chemiosmosis.

To calculate the total number of ATP produced per glucose molecule, consider that for each NADH + H^+ formed inside the mitochondrion, three ATP are produced. In most eukaryotic cells, each NADH + H^+ formed in the cytoplasm results in only two ATP. This occurs because the carrier that shuttles the hydrogen atoms across the mitochondrial outer membrane usually passes them to a FAD. Each molecule of $FADH_2$ results in the formation of only two ATP because the electrons enter the electron transport chain at a lower energy level than

$NADH + H^+$. Of the 36 or 38 ATP formed by cellular respiration, four are produced outside the electron transport chain: two by glycolysis and two by the citric acid cycle. The rest are produced by the electron transport chain.

7.5 Fermentation

Fermentation involves glycolysis, followed by the reduction of pyruvate by $NADH + H^+$ to either lactate or alcohol and CO_2. The reduction process regenerates NAD^+ so that it can accept more electrons during glycolysis.

Although fermentation results in only two ATP, it still serves a purpose: in humans, it provides a quick burst of ATP energy for short-term, strenuous muscular activity. The accumulation of lactate puts the individual in oxygen debt because oxygen is needed to completely metabolize lactate to CO_2 and H_2O.

Testing Yourself

Choose the best answer for each question.

For questions 1–5, match the letters in the diagram to the appropriate statement.

1. Preparatory reaction

2. Electron transport chain

3. Glycolysis

4. Citric acid cycle

5. $NADH + H^+$ and $FADH_2$

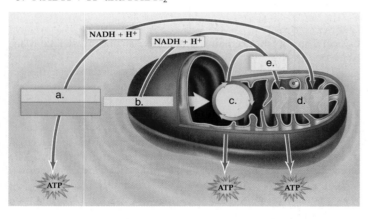

6. Which type of human cells carry on the most fermentation?
   a. fat   c. nerve
   b. muscle   d. bone

7. Which reaction of cellular respiration occurs in the cytoplasm?
   a. glycolysis
   b. citric acid cycle
   c. preparatory reaction
   d. electron transport chain

8. _____ reactions release energy and _____ molecules.
   a. Catabolic, build up   c. Anabolic, break down
   b. Catabolic, break down   d. None of these are correct.

9. When fats are used for energy production, glycerol is used in
   a. glycolysis.   c. the electron transport chain.
   b. the citric acid cycle.   d. none of these.

For questions 10–13, label each molecule as either a substrate or a product of respiration.

10. Glucose

11. Oxygen

12. CO_2

13. H_2O

14. Fermentation occurs in the absence of
    a. CO_2.   c. oxygen.
    b. H_2O.   d. sodium.

15. Cellular respiration cannot occur without
    a. sodium.   c. lactic acid.
    b. oxygen.   d. All of these are correct.

16. Mitochondria produce ATP by
    a. glycolysis.   c. chemiosmosis.
    b. fermentation.   d. All of these are correct.

17. An ATP synthase complex can be found in the
    a. cytoplasm.   c. Both a and b are correct.
    b. mitochondria.   d. Neither a nor b is correct.

18. FAD and NAD^+ are both
    a. adenine trinucleotides.   c. adenine dinucleotides.
    b. cytosine dinucleotides.   d. None of these are correct.

19. Which play a role in cellular respiration?
    a. oxidation and reduction   c. the citric acid cycle
    b. mitochondria   d. All of these are correct.

20. The correct order of these processes would be
    a. citric acid cycle, glycolysis, prep reaction.
    b. glycolysis, prep reaction, citric acid cycle.
    c. prep reaction, citric acid cycle, glycolysis.
    d. None of these are correct.

21. Anabolic reactions are _____ and _____ molecules.
    a. exergonic, break down
    b. endergonic, break down
    c. endergonic, build up
    d. None of these are correct.

For questions 22–25, match the labeled diagram to the appropriate statement.

22. Reduction

23. NAD^+

24. Oxidation

25. $NADH + H^+$

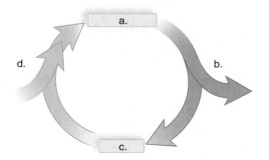

26. Which of the following is needed for glycolysis to occur?
    a. pyruvate
    b. glucose
    c. NAD⁺
    d. ATP
    e. All of these are needed except (a)

27. Which of the following is *not* a product, or end result, of the citric acid cycle?
    a. carbon dioxide
    b. pyruvate
    c. NADH + H⁺
    d. ATP
    e. FADH₂

28. How many ATP molecules are produced by the reduction and oxidation of one molecule of NAD⁺?
    a. 1
    b. 3
    c. 36
    d. 10

29. How many NADH + H⁺ molecules are produced during the complete breakdown of one molecule of glucose?
    a. 5
    b. 30
    c. 10
    d. 6

30. What is the name of the process that adds the third phosphate to an ADP molecule using the flow of hydrogen ions?
    a. substrate-level ATP synthesis
    b. fermentation
    c. reduction
    d. chemiosmosis

31. Which are possible products of fermentation?
    a. lactic acid
    b. alcohol
    c. CO₂
    d. All of these are correct.

32. The metabolic process that produces the most ATP molecules is
    a. glycolysis.
    b. the citric acid cycle.
    c. the electron transport chain.
    d. fermentation.

33. The greatest contributor of electrons to the electron transport chain is
    a. oxygen.
    b. glycolysis.
    c. the citric acid cycle.
    d. the prep reaction.
    e. fermentation.

34. Substrate-level ATP synthesis takes place during
    a. glycolysis and the citric acid cycle.
    b. the electron transport chain and the prep reaction.
    c. glycolysis and the electron transport chain.
    d. the citric acid cycle and the prep reaction.

35. Fatty acids are broken down to
    a. pyruvate molecules, which take electrons to the electron transport chain.
    b. acetyl groups, which enter the citric acid cycle.
    c. glycerol, which is found in fats.
    d. amino acids, which are excreted by ammonia.
    e. All of these are correct.

36. Which of the following is not true? Fermentation
    a. has a net gain of only two ATP.
    b. occurs in the cytoplasm.
    c. donates electrons to the electron transport chain.
    d. begins with glucose.
    e. is carried on by yeast.

Understanding the Terms

aerobic 117
anabolism 116
catabolism 116
cellular respiration 117
chemiosmosis 124
citric acid cycle 119
cristae 122
cytochrome 123
deamination 116
electron transport chain 119
FAD (flavin adenine dinucleotide) 119
fermentation 119
G3P (glyceraldehyde 3-phosphate) 120
glycolysis 119
intermembrane space 124
matrix 122
NAD⁺ (nicotinamide adenine dinucleotide) 118
oxygen debt 126
preparatory (prep) reaction 119
pyruvate 119
substrate-level ATP synthesis 120

Match the terms to these definitions:

a._____ Cycle of reactions in mitochondria that begins with citric acid; it produces CO₂, ATP, NADH + H⁺, and FADH₂.
b._____ Growing or metabolizing in the presence of oxygen.
c._____ Anaerobic breakdown of glucose that results in a gain of two ATP and end products such as alcohol and lactate.
d._____ End product of glycolysis.
e._____ Passage of electrons along a series of carrier molecules from a higher to lower energy level; the energy released is used for the synthesis of ATP.

Thinking Critically

1. Bacteria do not have mitochondria, and yet they contain an electron transport chain. Where is this located?
2. Rotenone is a broad-spectrum insecticide that inhibits the electron transport chain. Why might it be toxic to humans?
3. What would happen to the electron transport chain if the passage of electrons was uncoupled from the transport of H⁺ into the intermembrane space? Would O₂ still be required?
4. If the breakdown of one 18-carbon fatty acid chain produces 9 acetyl CoA molecules, how many ATP are produced from this single fatty acid chain if all the molecules complete the citric acid cycle and electron transport chain?
5. Organisms that ferment have a greatly increased rate of glycolysis compared to those that do not ferment. Why?

ARIS, the *Inquiry into Life* Website

ARIS, the website for *Inquiry into Life,* provides a wealth of information organized and integrated by chapter. You will find practice quizzes, interactive activities, labeling exercises, flashcards, and much more that will complement your learning and understanding of general biology.

www.aris.mhhe.com

Plant Biology

chapter 8

Photosynthesis

Tropical rain forests occur near the equator. *They can exist wherever temperatures are above 26°C and rainfall is heavy (100–200 cm annually) and regular. Brazil's Amazon region contains the largest rain forest in the world. Today, tropical rain forests only cover approximately 6% of Earth, where once they covered 14%. At the present deforestation rate, all of the rain forests will be gone by 2050. What do we lose if the tropical rain forests disappear?*

Half of all species on Earth live in a tropical rain forest. An incredible biodiversity of plants and animals, including insects, can be found there, many of which scientists are yet to discover. For example, a typical area equivalent to 4 square miles may shelter over 1,500 species of flowering plants, 750 species of trees, 400 species of birds, and 150 species of butterflies.

Tropical rain forests consume enormous amounts of CO_2 and thereby reduce the amount of atmospheric CO_2, which is responsible for global warming. Tropical rain forests have been called the "world's largest pharmacy." Many of the drugs we use today, such as quinine for malaria, were discovered in rain forests. The U.S. National Cancer Institute estimates that 70% of the plants known to be useful for treating cancer only grow in rain forests. Many consumer products, such as coffee, rubber, and cacao beans, have rain forest origins. We simply can't afford to lose the tropical rain forests.

Chapter Concepts

8.1   Overview of Photosynthesis

Photosynthesis converts solar energy into the chemical energy of a carbohydrate. Photosynthetic organisms, including plants, algae, and cyanobacteria, are called **autotrophs** because they produce their own food (Fig. 8.1). Each year, photosynthesizing organisms produce approximately 170 billion metric tons of carbohydrates that are available for other organisms. No wonder photosynthetic organisms are able to sustain themselves and all other living things on Earth.

With few exceptions, it is possible to trace any food chain back to plants and algae. In other words, producers, which have the ability to synthesize carbohydrates, feed not only themselves but also consumers, which must take in preformed organic molecules. Collectively, consumers are called **heterotrophs.** Both autotrophs and heterotrophs use organic molecules produced by photosynthesis as a source of building blocks for growth and repair and as a source of chemical energy for cellular work. The Bioethical Focus at the end of this chapter discusses the human population's increasing need for plants as a food source.

Pigments are the means by which these organisms capture solar energy, the "fuel" that makes photosynthesis (and ultimately life) possible. Many organisms that photosynthesize contain the pigment chlorophyll, which gives them a green color. But not all organisms that photosynthesize are green. Different pigments allow organisms to capture other wavelengths of the sun's light. The class of pigments called **carotenoids** contains red, orange, and yellow pigments. The class of pigments called phycobilins contains the red pigment phycoerythrin, which gives red algae their color, and the blue pigment phycocyanin, which gives cyanobacteria their bluish color.

Flowering Plants as Photosynthesizers

Although many different organisms can photosynthesize, we will limit our discussion to the green flowering plants. The green portions of plants, particularly the leaves, carry on photosynthesis. The leaf of a flowering plant contains mesophyll tissue in which cells are specialized for photosynthesis (Fig. 8.2). The raw materials for photosynthesis are water and carbon dioxide. The roots of a plant absorb water,

Figure 8.1
Photosynthetic organisms.

Photosynthetic organisms include **(a)** cyanobacteria such as *Oscillatoria*, which are a type of bacterium; **(b)** algae such as kelp, which typically live in water and can range in size from microscopic to macroscopic; and **(c)** plants such as the sequoia, which typically live on land.

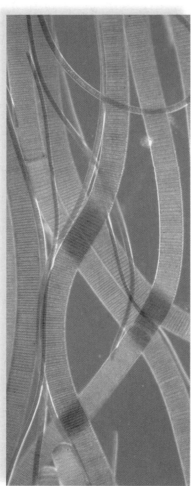

a. *Oscillatoria*   100×   b. Kelp   c. Sequoia

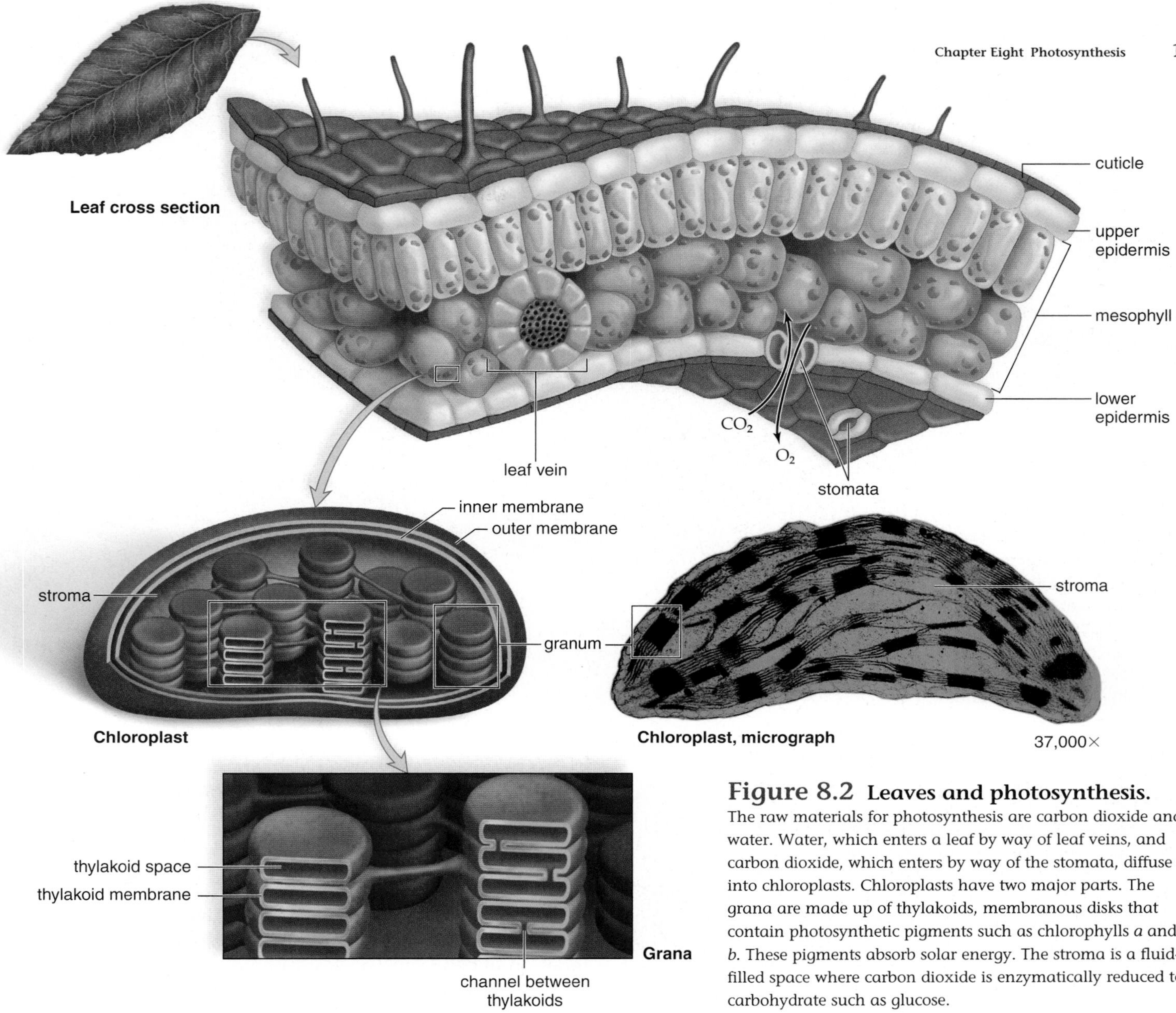

Leaf cross section

cuticle

upper epidermis

mesophyll

lower epidermis

CO_2

O_2

stomata

leaf vein

inner membrane

outer membrane

stroma

granum

Chloroplast

stroma

Chloroplast, micrograph   37,000×

thylakoid space

thylakoid membrane

Grana

channel between thylakoids

Figure 8.2 Leaves and photosynthesis.

The raw materials for photosynthesis are carbon dioxide and water. Water, which enters a leaf by way of leaf veins, and carbon dioxide, which enters by way of the stomata, diffuse into chloroplasts. Chloroplasts have two major parts. The grana are made up of thylakoids, membranous disks that contain photosynthetic pigments such as chlorophylls *a* and *b*. These pigments absorb solar energy. The stroma is a fluid-filled space where carbon dioxide is enzymatically reduced to a carbohydrate such as glucose.

which then moves in vascular tissue up the stem to a leaf by way of the leaf veins. Carbon dioxide in the air enters a leaf through small openings called **stomata** (sing., stoma). After entering a leaf, carbon dioxide and water diffuse into **chloroplasts,** the organelles that carry on photosynthesis.

A double membrane surrounds a chloroplast and its fluid-filled interior, which is called the **stroma.** A different membrane system within the stroma forms flattened sacs called **thylakoids.** In some places, thylakoids are stacked to form **grana** (sing., granum), so called because they looked like piles of seeds to early microscopists. Scientists believe the space of each thylakoid is connected to the space of every other thylakoid within a chloroplast, thereby forming a continuous inner compartment within chloroplasts called the thylakoid space.

Chlorophyll and other pigments that are part of a thylakoid membrane are capable of absorbing solar energy. This is the energy that drives photosynthesis. The stroma, filled with enzymes, is where carbon dioxide is first attached to an

organic compound and is then reduced to a carbohydrate. Therefore, it is proper to associate the absorption of solar energy with the thylakoid membranes making up the grana, and to associate the reduction of carbon dioxide to a carbohydrate with the stroma of a chloroplast.

Human beings, and indeed nearly all organisms, release carbon dioxide into the air. This is some of the same carbon dioxide that enters a leaf through the stomata and is converted to carbohydrate. Carbohydrate, in the form of glucose, is the chief energy source for most organisms.

Photosynthesis, which occurs in chloroplasts, is critically important because photosynthetic organisms use solar energy to produce carbohydrate, an organic nutrient. Almost all organisms depend either directly or indirectly on these organic nutrients to sustain themselves.

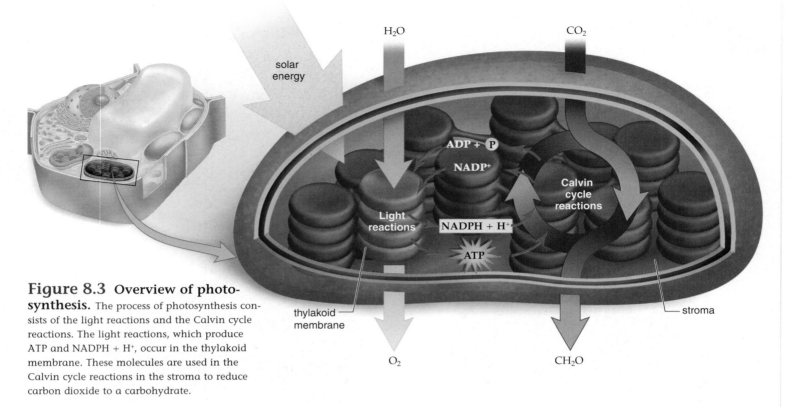

Figure 8.3 Overview of photosynthesis. The process of photosynthesis consists of the light reactions and the Calvin cycle reactions. The light reactions, which produce ATP and NADPH + H⁺, occur in the thylakoid membrane. These molecules are used in the Calvin cycle reactions in the stroma to reduce carbon dioxide to a carbohydrate.

Photosynthetic Reaction

For convenience, the overall equation for photosynthesis is sometimes simplified in this manner:

$$6\,CO_2 + 6\,H_2O \xrightarrow{\text{solar energy}} C_6H_{12}O_6 + 6\,O_2$$

This equation shows glucose and oxygen as the products of photosynthesis.

The oxygen given off by photosynthesis comes from water. This was proven experimentally by exposing plants first to CO_2 and then to H_2O that contained an isotope of oxygen called heavy oxygen (^{18}O). Only when heavy oxygen was a part of water did this isotope appear in O_2 given off by the plant. Therefore, O_2 released by chloroplasts comes from H_2O, not from CO_2.

This reaction also shows that during photosynthesis, CO_2 gains hydrogen atoms and becomes a carbohydrate, and that *photosynthesis is driven by solar energy.*

The overall chemical equation for photosynthesis explains some of the reasons that plants are so important to our lives. Plants are responsible for producing the oxygen we breathe. Plants are also responsible, either directly or indirectly, for the glucose we use for energy. The Ecology Focus later in this chapter describes some of the many specific ways that plants sustain and enhance our lives.

Two Sets of Reactions

An overall equation for photosynthesis tells us the beginning reactants and the end products of the pathway. But much goes on in between. The word photosynthesis suggests that the process requires two sets of reactions: *photo,* which means light, refers to the reactions that capture solar energy, and *synthesis* refers to the reactions that produce carbohydrate.

The two sets of reactions are called the **light reactions** and the **Calvin cycle reactions** (Fig. 8.3). We will see that the coenzyme **NADP⁺ (nicotinamide adenine dinucleotide phosphate)** carries hydrogen atoms from the light reactions to the Calvin cycle reactions. When NADP⁺ accepts hydrogen atoms, it becomes NADPH + H⁺. Also, ATP carries energy between the two sets of reactions.

Photosynthesis is an energy-requiring process that is divided into two sets of reactions. Solar energy is captured in thylakoid membranes and converted to a form that allows carbon dioxide to be reduced in the stroma.

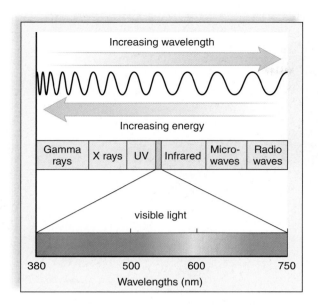

Figure 8.4 The electromagnetic spectrum.
The electromagnetic spectrum extends from the very short wavelengths of gamma rays through the very long wavelengths of radio waves. Visible light, which drives photosynthesis, is expanded to show its component colors. The components differ according to wavelength and energy content.

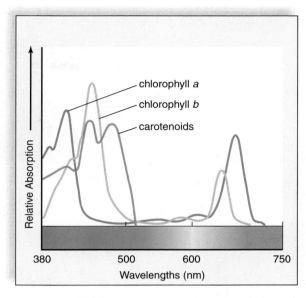

Figure 8.5 Photosynthetic pigments and photosynthesis. The photosynthetic pigments in chlorophylls *a* and *b* and the carotenoids absorb certain wavelengths within visible light. This is their absorption spectrum. Notice that they do not absorb green light; that's why the leaves of plants appear green to us. In other words, you observe the color that is *not* absorbed by the leaf.

8.2 Solar Energy Capture

During the light reactions, the pigments within the thylakoid membranes absorb solar energy. Solar energy (radiant energy from the sun) can be described in terms of its wavelength and its energy content. Figure 8.4 lists the different types of radiant energy, from the shortest wavelength, gamma rays, to the longest, radio waves. White, or *visible*, light is only a small portion of this spectrum.

Visible Light

Visible light itself contains various wavelengths of light. When it is passed through a prism, we see all the different colors that make up visible light. (Actually, of course, it is our brains that interpret these wavelengths as colors.) The colors in visible light range from violet (the shortest wavelength) to indigo, blue, green, yellow, orange, and red (the longest wavelength). The energy content is highest for violet light and lowest for red light.

Only about 42% of the solar radiation that hits Earth's atmosphere ever reaches the surface, and most of this radiation is within the visible-light range. Higher-energy wavelengths are screened out by the ozone layer in the atmosphere, and lower-energy wavelengths are screened out by water vapor and carbon dioxide (CO_2) before they reach Earth's surface. Both the organic molecules within organisms and certain life processes, such as vision and photosynthesis, are adapted to the solar radiation that is most prevalent in the environment.

The pigments found within most types of photosynthesizing cells, the chlorophylls *a* and *b* and the carotenoids, are capable of absorbing various portions of visible light. The absorption spectrum for these pigments is shown in Figure 8.5. Both chlorophyll *a* and chlorophyll *b* absorb violet, indigo, blue, and red light better than the light of other colors. Because green light is reflected and only minimally absorbed, leaves appear green to us. The yellow or orange carotenoids are able to absorb light in the violet-blue-green range. These pigments and others become noticeable in the fall when chlorophyll breaks down and the other pigments are uncovered (see the Ecology Focus in Chapter 6). Photosynthesis begins when the pigments within thylakoid membranes absorb solar energy.

Light Reactions

The light reactions that occur in the thylakoid membrane consist of two electron pathways called the noncyclic electron pathway and the cyclic electron pathway. Both electron pathways produce ATP, but only the noncyclic pathway also produces NADPH + H^+.

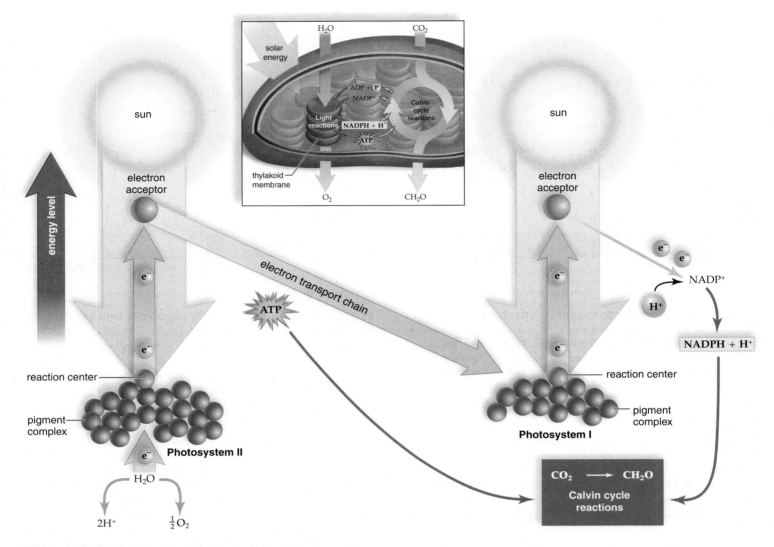

Figure 8.6 Noncyclic electron pathway: Electrons move from water to NADP⁺.

Energized electrons (replaced from water, which splits, releasing oxygen) leave photosystem II and pass down an electron transport chain, leading to the formation of ATP. Energized electrons (replaced by photosystem II) leave photosystem I and pass to NADP⁺, which then combines with H⁺, becoming NADPH + H⁺.

Noncyclic Electron Pathway

The **noncyclic electron pathway** is so named because the electron flow can be traced from water to a molecule of NADP⁺ (Fig. 8.6). This pathway uses two photosystems, called photosystem I (PS I) and photosystem II (PS II). The photosystems are named for the order in which they were discovered, not for the order in which they participate in the photosynthetic process. A **photosystem** consists of a pigment complex (molecules of chlorophyll *a*, chlorophyll *b*, and the carotenoids) and an electron acceptor within the thylakoid membrane. The pigment complex serves as an "antenna" for gathering solar energy.

The noncyclic pathway begins with photosystem II. The pigment complex absorbs solar energy, which is then passed from one pigment to the other until it is concentrated in a particular pair of chlorophyll *a* molecules called the *reaction center*. Electrons (e⁻) in the reaction center become so energized that they escape from the reaction center and move to a nearby electron acceptor.

Photosystem II would disintegrate without replacement electrons, and these are removed from water, which splits, releasing oxygen to the atmosphere. Many organisms, including plants and even ourselves, use this oxygen within their mitochondria. The hydrogen ions (H⁺) stay in the thylakoid space and contribute to the formation of a hydrogen ion gradient.

An electron acceptor sends energized electrons, received from the reaction center, down an electron transport chain, a series of carriers that pass electrons from

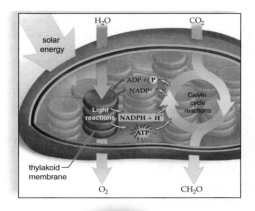

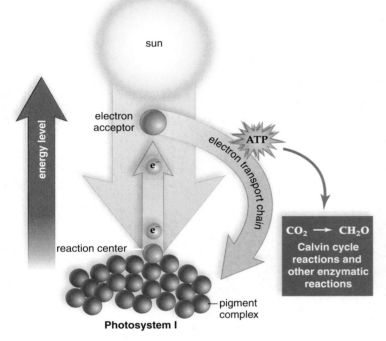

Figure 8.7 Cyclic electron pathway: Electrons leave and return to photosystem I. Energized electrons leave the photosystem I reaction center and are taken up by an electron acceptor, which passes them down an electron transport chain before they return to photosystem I. Only ATP production occurs as a result of this pathway.

one to the other. As the electrons pass from one carrier to the next, energy is captured and stored in the form of a hydrogen ion (H^+) gradient. When these hydrogen ions flow down their electrochemical gradient through **ATP synthase** complexes, ATP production occurs, as will be described shortly. Notice that this ATP will be used in the Calvin cycle reactions in the stroma to reduce carbon dioxide to a carbohydrate.

When the photosystem I pigment complex absorbs solar energy, energized electrons leave its reaction center and are captured by a different electron acceptor.

(Low-energy electrons from the electron transport chain adjacent to photosystem II replace those lost by photosystem I.) The electron acceptor in photosystem I passes its electrons to $NADP^+$ molecules. Each one accepts two electrons and two H^+ to become a reduced form of the molecule—that is, NADPH + H^+. This NADPH + H^+ will also be used by the Calvin cycle reactions in the stroma to reduce carbon dioxide to a carbohydrate.

> As a result of noncyclic electron flow in the thylakoid membrane, water is oxidized (split), yielding H^+, e^-, and O_2, which is released. Also, ATP is produced, and $NADP^+$ becomes NADPH + H^+. In the stroma, ATP and NADPH + H^+ reduce CO_2 to CH_2O, a carbohydrate.

Cyclic Electron Pathway

The **cyclic electron pathway** (Fig. 8.7) begins when the photosystem I pigment complex absorbs solar energy and it is passed from one pigment to the other until it is concentrated in a reaction center. As with photosystem II, electrons (e^-) become so energized that they escape from the reaction center and move to nearby electron acceptor molecules.

Energized electrons (e^-) taken up by an electron acceptor are sent down an electron transport chain. As the electrons pass from one carrier to the next, their energy becomes stored as a hydrogen (H^+) gradient. Again, the flow of hydrogen ions down their electrochemical gradient through ATP synthase complexes produces ATP.

This time, instead of the electrons moving on to $NADP^+$, they return to photosystem I; this is how photosystem I receives replacement electrons and why this electron pathway is called cyclic. It is also why the cyclic pathway produces ATP but does not produce NADPH + H^+.

In plants, the reactions of the Calvin cycle can use any extra ATP produced by the cyclic pathway because the Calvin cycle reaction requires more ATP molecules than NADPH + H^+ molecules. Also, other enzymatic reactions, aside from those involving photosynthesis, are occurring in the stroma and can use this ATP. It's possible that the cyclic flow of electrons is used alone when carbohydrate is not being produced. At this time, there would be no need for NADPH + H^+, which is produced only by the noncyclic electron pathway.

> The cyclic electron pathway, from photosystem I back to photosystem I, has only one effect: production of ATP.

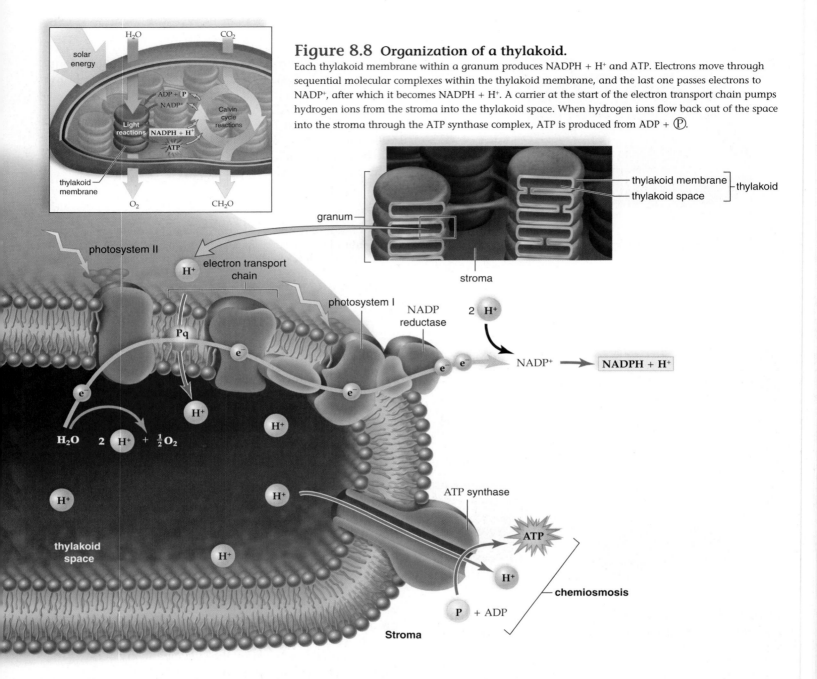

Figure 8.8 Organization of a thylakoid.
Each thylakoid membrane within a granum produces NADPH + H⁺ and ATP. Electrons move through sequential molecular complexes within the thylakoid membrane, and the last one passes electrons to NADP⁺, after which it becomes NADPH + H⁺. A carrier at the start of the electron transport chain pumps hydrogen ions from the stroma into the thylakoid space. When hydrogen ions flow back out of the space into the stroma through the ATP synthase complex, ATP is produced from ADP + Ⓟ.

The Organization of the Thylakoid Membrane

As we have discussed, the following molecular complexes are present in the thylakoid membrane (Fig. 8.8):

Photosystem II, which consists of a pigment complex and an electron acceptor molecule, receives electrons from water and it splits, releasing oxygen.

The electron transport chain carries electrons from photosystem II to photosystem I, and pumps H⁺ from the stroma into the thylakoid space.

Photosystem I, which also consists of a pigment complex and an electron acceptor molecule, is adjacent to NADP reductase, which reduces NADP⁺ to NADPH + H⁺.

The ATP synthase complex has a channel and a protruding ATP synthase, an enzyme that joins ADP + Ⓟ.

ATP Production

The thylakoid space acts as a reservoir for hydrogen ions (H⁺). First, each time oxygen is removed from water, two H⁺ remain in the thylakoid space. Second, as the electrons move from carrier to carrier along the electron transport chain, the electrons give up energy, which is used to pump H⁺ from the stroma into the thylakoid space. Therefore, there are more H⁺ in the thylakoid space than in the stroma. The flow of H⁺ from high to low concentration across the thylakoid membrane provides the energy that allows an ATP synthase enzyme to enzymatically produce ATP from ADP + Ⓟ. This method of producing ATP is called chemiosmosis because ATP production is tied to the establishment of an H⁺ gradient.

Chemiosmosis is the production of ATP due to an H⁺ gradient.

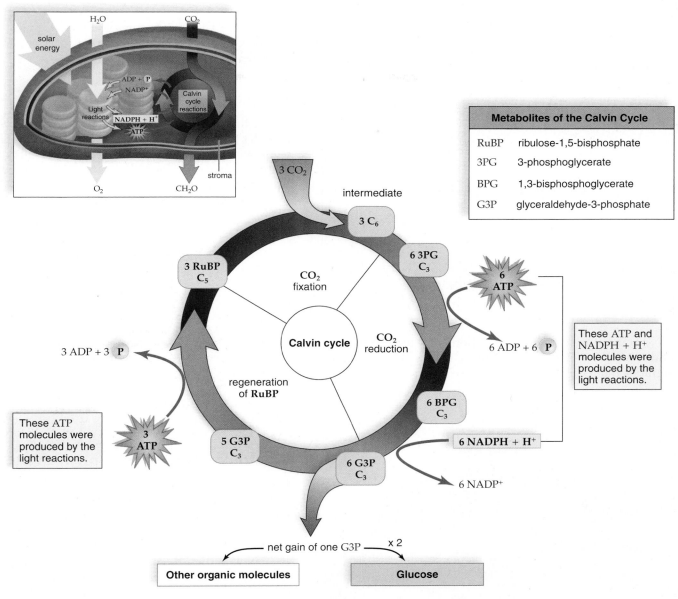

Metabolites of the Calvin Cycle	
RuBP	ribulose-1,5-bisphosphate
3PG	3-phosphoglycerate
BPG	1,3-bisphosphoglycerate
G3P	glyceraldehyde-3-phosphate

Figure 8.9 The Calvin cycle reactions.
The Calvin cycle is divided into three portions: CO_2 fixation, CO_2 reduction, and regeneration of RuBP. Because five G3P are needed to re-form three RuBP, it takes three turns of the cycle to have a net gain of one G3P. Two G3P molecules are needed to form glucose.

<table>
<tr><td>

8.3 | Calvin Cycle Reactions

The Calvin cycle reactions follow the light reactions. The Calvin cycle is a series of reactions that produce carbohydrate before returning to the starting point once more (Fig. 8.9). Therefore, this set of reactions is called a cycle. The cycle is named for Melvin Calvin, who, with colleagues, used the radioactive isotope ^{14}C as a tracer to discover the reactions making up the cycle.

This series of reactions uses carbon dioxide from the atmosphere to produce carbohydrate. How does carbon dioxide get into the atmosphere? Humans and most other organisms take in oxygen from the atmosphere and release carbon dioxide to the atmosphere. The Calvin cycle includes (1) carbon dioxide fixation, (2) carbon dioxide reduction, and (3) regeneration of **RuBP** (ribulose-1, 5-bisphosphate).

</td><td>

Fixation of Carbon Dioxide

Carbon dioxide (CO_2) fixation is the first step of the Calvin cycle. During this reaction, carbon dioxide from the atmosphere is attached to RuBP, a 5-carbon molecule. The result is one 6-carbon molecule, which splits into two 3-carbon molecules.

The enzyme that speeds this reaction, called RuBP carboxylase, is a protein that makes up about 20–50% of the protein content in chloroplasts. The reason for its abundance may be that it is unusually slow (it processes only a few molecules of substrate per second compared to thousands per second for a typical enzyme), and so there has to be a lot of it to keep the Calvin cycle going.

</td></tr>
</table>

Reduction of Carbon Dioxide

The first 3-carbon molecule in the Calvin cycle is called 3PG (3-phosphoglycerate). Each of two 3PG molecules undergoes reduction to G3P in two steps:

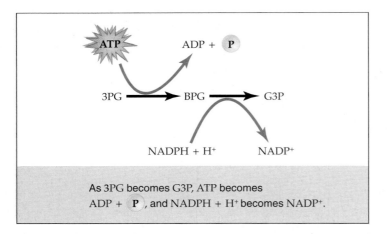

As 3PG becomes G3P, ATP becomes ADP + P, and NADPH + H⁺ becomes NADP⁺.

This is the sequence of reactions that uses some ATP and NADPH + H⁺ from the light reactions. This sequence signifies the reduction of carbon dioxide to a carbohydrate because $R—CO_2$ has become $R—CH_2O$. Energy and electrons are needed for this reduction reaction, and they are supplied by ATP and NADPH + H⁺.

Regeneration of RuBP

Notice that the Calvin cycle reactions in Figure 8.9 are multiplied by three because it takes three turns of the Calvin cycle to allow one G3P to exit. Why? Because, for every three turns of the Calvin cycle, five molecules of G3P are used to re-form three molecules of RuBP and the cycle continues. Notice that 5×3 (carbons in G3P) $= 3 \times 5$ (carbons in RuBP):

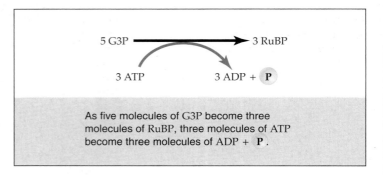

As five molecules of G3P become three molecules of RuBP, three molecules of ATP become three molecules of ADP + P.

This reaction also uses some of the ATP produced by the light reactions.

The Importance of the Calvin Cycle

G3P (glyceraldehyde-3-phosphate) is the product of the Calvin cycle that can be converted to all sorts of organic molecules. Compared to animal cells, algae and plants have enormous biochemical capabilities. They use G3P for the purposes described in Figure 8.10.

Notice that glucose phosphate is among the organic molecules that result from G3P metabolism. This is of interest to us because glucose is the molecule that plants and animals most often metabolize to produce the ATP molecules they require for their energy needs. Glucose is blood sugar in human beings.

Glucose phosphate can be combined with fructose (and the phosphate removed) to form sucrose, the molecule plants use to transport carbohydrates from one part of the plant to the other.

Glucose phosphate is also the starting point for the synthesis of starch and cellulose. Starch is the storage form of glucose. Some starch is stored in chloroplasts, but most starch is stored in roots. Cellulose is a structural component of plant cell walls and becomes fiber in our diet because we are unable to digest it.

A plant can use the hydrocarbon skeleton of G3P to form fatty acids and glycerol, which are combined in plant oils. We are all familiar with corn oil, sunflower oil, or olive oil used in cooking. Also, when nitrogen is added to the hydrocarbon skeleton derived from G3P, amino acids are formed.

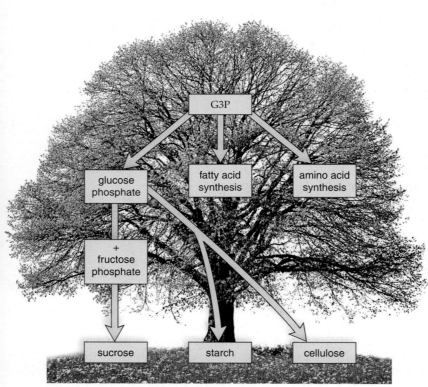

Figure 8.10   Fate of G3P.

G3P is the first reactant in a number of plant cell metabolic pathways. Two G3Ps are needed to form glucose phosphate; glucose is often considered the end product of photosynthesis. Sucrose is the transport sugar in plants; starch is the storage form of glucose; and cellulose is a major constituent of plant cell walls.

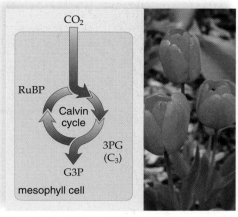

a. CO_2 fixation in a C_3 plant, tulip

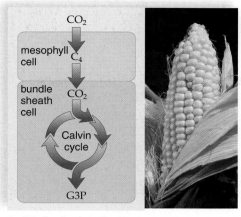

b. CO_2 fixation in a C_4 plant, corn

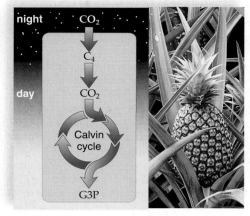

c. CO_2 fixation in a CAM plant, pineapple

Figure 8.11 Alternative pathways for photosynthesis.
Photosynthesis can be categorized according to how, where, and when CO_2 fixation occurs. **a.** In C_3 plants, CO_2 is taken up by the Calvin cycle directly in mesophyll cells. **b.** C_4 plants form a C_4 molecule in mesophyll cells prior to releasing CO_2 to the Calvin cycle in bundle sheath cells. **c.** CAM plants fix CO_2 at night, forming a C_4 molecule.

8.4 Alternative Pathways for Photosynthesis

Plants are able to live under all sorts of environmental conditions, and one reason is that various modes of photosynthesis have evolved (Fig. 8.11).

C_3 Pathway

When RuBP carboxylase binds to carbon dioxide, it produces carbohydrates in what has been called C_3 photosynthesis, because a 3-carbon molecule is detectable immediately following CO_2 fixation. **C_3 plants,** such as wheat, rice, and oats, contain mesophyll cells that are in parallel layers. The bundle sheath cells around the plant veins do not contain chloroplasts. However, the enzyme RuBP carboxylase can also bind with oxygen. When the enzyme binds oxygen, it undergoes a nonproductive, wasteful reaction called photorespiration because it uses oxygen and releases carbon dioxide. This decreases the overall efficiency of the enzyme. Researchers believe that photorespiration has increased during the age of Earth because of rising O_2 concentrations in the atmosphere, a result of the increase in photosynthetic organisms.

C_4 Pathway

An alternative pathway, called C_4 photosynthesis, evolved to bypass the photorespiration problem. C_4 plants have both the C_4 pathway and the C_3 pathway. In **C_4 plants,** such as sugarcane and corn, the mesophyll cells are arranged in concentric rings around the bundle sheath cells, which also contain chloroplasts.

This diagram shows how these arrangements differ in C_3 and C_4 plants:

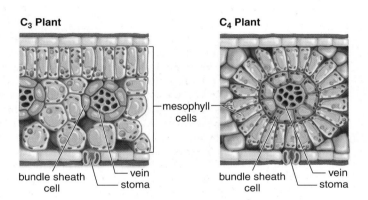

In the mesophyll cells, CO_2 is initially fixed into a 4-carbon molecule, hence the name. That C_4 molecule is transported to the bundle sheath cells where it is broken down to carbon dioxide and a 3-carbon molecule. The carbon dioxide enters the Calvin cycle, and the 3-carbon molecule is transported back to the mesophyll cells. The bundle sheath cells are deeper in the leaf and therefore do not contain as much oxygen as the mesophyll cells due to diffusion.

C_4 plants tend to grow where the weather is hot and dry. That is because when the stomata on the leaf surfaces close in order to conserve water due to the hot, dry environment, oxygen increases in the leaf air spaces due to photosynthesis. In C_3 plants, this oxygen competes with carbon dioxide for the active site of RuBP carboxylase. In C_4 plants, the carbon dioxide is delivered to the Calvin cycle, which is located in bundle sheath cells that are sheltered from the leaf air spaces.

Because of the need to transport precursors back and forth between where CO_2 is fixed and where the Calvin cycle operates, this C_4 pathway would appear to require extra energy. However, the prevention of water loss and the increased efficiency of photosynthesis apparently compensate for this.

Ecology Focus

So Why Is Photosynthesis Important to Me?

Life on this planet cannot exist without photosynthesis. Several futuristic movies depict humans living on a planet without the sun or plants (and thus without photosynthesis). This fictional scenario may make for great drama, but scientifically, life without sun or plants is not possible.

With a few exceptions, photosynthesizers (including plants) are found at the beginning of every food chain. One of the products of the photosynthetic reaction is the oxygen we breathe. As we know, life would not be possible without oxygen. The other product of the photosynthetic reaction is carbohydrate. Almost all organisms depend on this organic nutrient to sustain themselves. So, directly or indirectly, humans and all other heterotrophs owe their lives to photosynthesis.

We eat other animals that eat plants, or we eat the plants themselves. Plants provide us with a great deal of variety in our food. Fruits, vegetables, nuts, grains, herbs, and spices all contribute in wonderful ways to our diet, enhancing our enjoyment of foods through different tastes, textures, and colors.

Photosynthesis also provides the fuel we use. A small percentage of the world's fuel supply comes from burning wood, but a much larger percentage comes from coal, oil, and natural gas, which are fossil fuels. Fossil fuels were formed nearly 300 million years ago, primarily from the bodies of protists and photosynthetic plants. So when you drive a car, heat your home, or fly in an airplane, remember that photosynthesis has made these activities possible.

You would not be reading this book if not for photosynthesis—approximately one-third of harvested wood that is not used for fuel is made into paper. Lumber is also used for con-

Plants provide foods and other materials we need to sustain life; they also add beauty to our lives and provide a means of recreation and relaxation.

struction of our homes and other buildings. Plants provide other products we use daily, such as cotton for the clothes we wear, oil for producing plastics, and rubber for tires.

Photosynthesis contributes to our health—plants are the source of many of the medicines that fight disease and increase our longevity. And not only are plants beautiful to look at, they also provide exercise and relaxation through outdoor activities such as hiking and camping. Many fragrances that please our senses come from plants, as do natural dyes for wool and fabrics. These factors all contribute to our enjoyment and quality of life. As one bumper sticker reads, "Have you thanked a green plant today?"

Discussion Questions

1. Can you think of any other ways in which photosynthesis benefits you personally, besides the ways listed in this Ecology Focus?
2. Using outside resources, research one plant that is important to you. What does this plant produce? What does it add to your life?
3. Considering how important plants are, why do you think that students are not as interested in learning about plants as they are about animals?

CAM Pathway

Another alternative pathway, called the CAM pathway, has also evolved because of environmental pressures. CAM stands for crassulacean-acid metabolism; Crassulaceae is a family of flowering succulent plants that live in warm, arid regions of the world. We now know that CAM photosynthesis is prevalent among most succulent plants that grow in desert environments, including the cacti.

While the C_4 pathway separates components of photosynthesis by location, the CAM pathway separates them in time. CAM plants fix CO_2 into a 4-carbon molecule at night, when they can keep their stomata open without losing much water. The C_4 molecule is stored in large vacuoles in their mesophyll

cells until the next day. During the day, the C_4 molecule releases the CO_2 to the Calvin cycle within the same cell.

Plants are capable of fixing carbon by more than one pathway. These pathways appear to be the result of adaptation to different climates.

C_3 plants fix carbon dioxide into a three carbon molecule, but they sometimes use oxygen instead in a wasteful process called photorespiration. In C_4 plants, the Calvin cycle is located in bundle sheath cells where O_2 cannot accumulate when stomata close. CAM plants carry on carbon dioxide fixation at night and release CO_2 to the Calvin cycle during the day.

142

8.5 Photosynthesis versus Cellular Respiration

Both plant and animal cells carry on cellular respiration, but only plant cells photosynthesize. The organelle for cellular respiration is the mitochondrion while the organelle for photosynthesis is the chloroplast. Photosynthesis is the building up of glucose, while cellular respiration is the breaking down of glucose. Figure 8.12 compares the two processes.

The following overall equation for photosynthesis is the opposite of that for cellular respiration. The reaction in the forward direction represents photosynthesis, and the word *energy* stands for solar energy. The reaction in the opposite direction represents cellular respiration, and the word *energy* then stands for ATP:

$$\text{energy} + 6\,CO_2 + 6\,H_2O \underset{\text{cellular respiration}}{\overset{\text{photosynthesis}}{\rightleftarrows}} C_6H_{12}O_6 + 6\,O_2$$

Both photosynthesis and cellular respiration are metabolic pathways within cells, and therefore consist of a series of reactions that the overall equation does not indicate. Both pathways, which use an electron transport chain located in membranes, produce ATP by chemiosmosis. Both also use an electron carrier; photosynthesis uses $NADP^+$, and cellular respiration uses NAD^+.

Both pathways utilize the following reaction, but in opposite directions. For photosynthesis, read the reaction

from left to right; for cellular respiration, read the reaction from right to left:

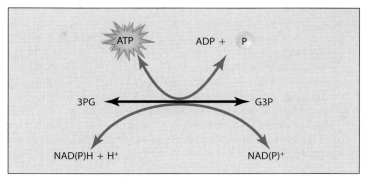

Both photosynthesis and cellular respiration occur in plant cells. In plants, both processes occur during the daylight hours, while only cellular respiration occurs at night. During daylight hours, the rate of photosynthesis exceeds the rate of cellular respiration, resulting in a net increase and storage of glucose. The stored glucose is used for cellular metabolism, which continues during the night. Also during daylight hours, the release of O_2 is greater than the release of CO_2, whereas at night the release of O_2 ceases and the release of CO_2 increases.

Photosynthesis and cellular respiration utilize the same overall chemical reaction, but in opposite directions.

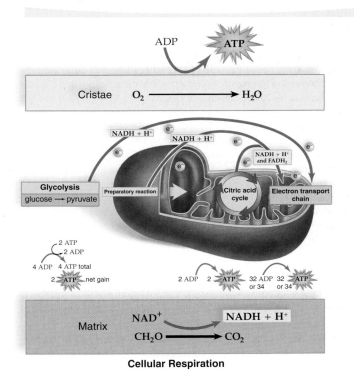

Photosynthesis   **Cellular Respiration**

Figure 8.12 Photosynthesis versus cellular respiration. Both photosynthesis and cellular respiration have an electron transport chain located within membranes, where ATP is produced. Both processes have enzyme-catalyzed reactions located within the fluid interior of respective organelles. In photosynthesis, hydrogen atoms are donated by NADPH + H⁺ when CO_2 is reduced in the stroma of a chloroplast. During cellular respiration, NADH + H⁺ forms when glucose is oxidized in the cytoplasm and glucose breakdown products are oxidized in the matrix of a mitochondrion.

Food for the Human Population

The world's population is constantly increasing. Will there be enough food for all its people by the middle of the twenty-first century? Over the past 40 years, the world's food supply has expanded faster than the population, due to the development of high-yielding plants and the increased use of irrigation, pesticides, and fertilizers. One of the most worrisome threats to food security, however, is an increasing degradation of agricultural land. Soil erosion in particular is robbing the land of its topsoil and reducing its productivity.

In 1961, 1.3 billion hectares of arable land (land capable of sustaining agriculture) were available globally. Ten percent of that required irrigation. In 1990, there were 1.4 billion hectares of arable land, with 17% requiring irrigation. In that time, the world population grew from 3 billion to 5.3 billion people. And the world population continues to rise having recently passed the 6-billion mark.

Most population growth will occur in the less-developed countries of Africa, Asia, and Latin America that are only now becoming industrialized. In these countries, the technical gains needed to prevent a disaster will be enormous. But even the United States will face an increased drain on its economic resources and increased pollution problems due to population growth.

Some people feel that technology will continue to make great strides for many years to come. For example, they say that the new genetically modified (GM) crops could play a major role in increasing production. GM crops have been genetically engineered by the insertion of a single gene that either reduces their need for pesticides or water, or increases their tolerance of saline soils. However, other people feel that technology's successes are self-defeating; for example, reliance on pesticides and fertilizers to increase agricultural yields has resulted in pollution of the air, water, and land. And they fear that GM foods may also pose risks to human health and the environment. These people are in favor of calling a halt to human population increase by all possible measures and as quickly as possible. Which of these approaches do you favor?

Form Your Own Opinion

1. The population could very well double from 6 billion to 12 billion in your lifetime. Do you approve of measures to limit population growth? Why or why not?
2. Do you believe instead that we should rely on technology, such as the use of GM crops, to increase our agricultural yields?

Summarizing the Concepts

8.1 Overview of Photosynthesis
Photosynthesis is absolutely essential for the continuance of life because it supplies the biosphere with food and energy. Chloroplasts carry on photosynthesis. During photosynthesis, solar energy is converted to chemical energy within carbohydrates. The overall equation for photosynthesis is $CO_2 + H_2O \rightarrow (CH_2O) + O_2$. A chloroplast contains two main portions: the stroma and membranous grana made up of thylakoid sacs. Chlorophyll and other pigments within the thylakoid membrane absorb solar energy, and enzymes in the fluid-filled stroma reduce CO_2.

Photosynthesis consists of two reactions: the light reactions and the Calvin cycle reactions. The Calvin cycle reactions, located in the stroma, use NADPH + H⁺ and ATP to reduce carbon dioxide. These molecules are produced by the light reactions located in the thylakoid membranes of the grana, after chlorophyll captures the energy of sunlight.

8.2 Solar Energy Capture
Photosynthesis uses solar energy in the visible-light range; chlorophylls a and b and the carotenoids largely absorb violet, indigo, blue, and red wavelengths and reflect green wavelengths. This causes leaves to appear green to us.

Photosynthesis begins when pigment complexes within photosystem I and photosystem II absorb radiant energy. In the noncyclic electron pathway, electrons are energized in photosystem II before they enter an electron transport chain. Electrons from H_2O replace those lost in photosystem II. The electrons energized in photosystem I pass to NADP⁺, which becomes NADPH + H⁺. Electrons from the electron transport chain replace those lost by photosystem I.

In the cyclic electron pathway of the light reactions, electrons energized by the sun leave photosystem I and enter an electron transport chain that produces ATP, and then the energy-spent electrons return to photosystem I.

8.3 Calvin Cycle Reactions
The ATP and NADPH + H⁺ made in thylakoid membranes pass into the stroma, where carbon dioxide is reduced during the Calvin cycle reactions. Specifically, ATP and NADPH + H⁺ from the light reactions are used to reduce 3PG to G3P. G3P is synthesized to various molecules, including carbohydrates such as glucose.

8.4 Alternative Pathways for Photosynthesis
C_3 plants fix carbon dioxide into a 3-carbon molecule. They also sometimes undergo a wasteful reaction called photorespiration. Plants utilizing C_4 photosynthesis have a different construction than plants using C_3 photosynthesis. C_4 plants fix CO_2 in mesophyll cells and then deliver the CO_2 to the Calvin cycle in bundle sheath cells. Now, O_2 cannot compete for the active site of RuBP carboxylase when stomata are closed due to hot and dry weather. This represents a partitioning of pathways in space: carbon dioxide fixation occurs in mesophyll cells, and the Calvin cycle occurs in bundle sheath cells. CAM plants fix carbon dioxide at night when their stomata remain open. This represents a partitioning of pathways in time: carbon dioxide fixation occurs at night, and the Calvin cycle occurs during the day.

8.5 Photosynthesis versus Cellular Respiration
Both photosynthesis and cellular respiration utilize an electron transport chain and ATP synthesis. However, photosynthesis reduces CO_2 to a carbohydrate. The oxidation of H_2O releases O_2. Cellular respiration oxidizes carbohydrate, and CO_2 is given off. Oxygen is reduced to water.

Testing Yourself

Choose the best answer for each question.

1. Photosynthetic activity can be found in
   a. plants.   c. bacteria.
   b. algae.   d. All of these are correct.

2. If carbohydrates are the intended product of photosynthesis, then _____ can be considered the "by-product" of the reaction.
   a. CO_2 c. water
   b. oxygen   d. nitrogen

3. Cellular respiration and photosynthesis both
   a. use oxygen.
   b. produce carbon dioxide.
   c. contain an electron transport chain.
   d. occur in the chloroplast.

4. The light reactions
   a. take place in the stroma.   c. Both a and b are correct.
   b. consist of the Calvin cycle. d. Neither a nor b is correct.

5. The cyclic electron pathway, but not the noncyclic pathway, generates only
   a. oxygen.   c. chlorophyll.
   b. ATP.   d. NADPH + H⁺.

6. In plants, G3P is used to form
   a. fatty acids.   c. amino acids.
   b. oils.   d. All of these are correct.

7. _____ produces ATP by chemiosmosis.
   a. Cellular respiration   c. Both a and b are correct.
   b. Photosynthesis   d. Neither a nor b is correct.

8. The different types of photosynthesis are dependent upon the timing and location of
   a. CO_2 fixation.   c. H_2O fixation.
   b. nitrogen fixation.   d. All of these are correct.

9. When leaves change color in the fall, _____ light is absorbed for photosynthesis.
   a. orange range   c. violet-blue-green range
   b. red range   d. None of these are correct.

10. CAM plants
    a. have evolved strategies   c. include cacti.
       for reducing water loss.   d. All of these are correct.
    b. are C_4 plants.

For questions 11–15, match each definition with a type of plant in the key.

Key:
   a. C_3 plants   c. C_4 plants
   b. CAM plants

11. Contain chloroplasts in bundle sheath cells

12. Mesophyll cells arranged in parallel

13. Open stomata at night

14. Corn

15. Can be inefficient

16. In the absence of sunlight, plants are not able to engage in the Calvin cycle due to a lack of
    a. ATP.   c. oxygen.
    b. NADPH + H⁺.   d. Both a and b are correct.

17. Disrupting the flow of H⁺ across the thylakoid membrane will also disrupt
    a. CO_2 production.   c. water production.
    b. ATP production.   d. None of these are correct.

18. In respiration, electrons are carried by _____, while in photosynthesis they are carried by _____.
    a. NADP⁺, NAD⁺
    b. ATP, ADP
    c. NAD⁺, NADP⁺
    d. ATP, FAD

19. Label a, b, and c in the following diagram of the cyclic electron pathway.

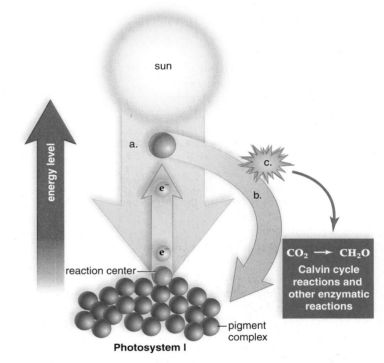

20. Which light rays are used for photosynthesis?
    a. light having long wavelengths
    b. light having short wavelengths
    c. violet, indigo, and blue light
    d. ultraviolet light
    e. Both b and c are correct.

21. Label a, b, c, d, and e in the following diagram of a chloroplast.

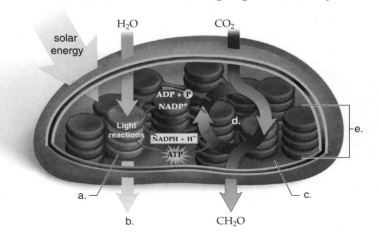

22. The absorption spectrum of chlorophyll *a* and chlorophyll *b*
    a. is not the same as that of the carotenoids.
    b. approximates the action spectrum of photosynthesis.
    c. explains why chlorophyll is a green pigment.
    d. shows that some colors of light are absorbed more than others.
    e. All of these are correct.

23. The oxygen given off by photosynthesis comes from
    a. H_2O.   c. glucose.
    b. CO_2.   d. RuBP.

24. The final acceptor of electrons during the noncyclic electron pathway is
    a. photosystem I.   d. ATP.
    b. photosystem II.   e. $NADP^+$.
    c. water.

25. A photosystem contains
    a. pigments, a reaction center, and an electron acceptor.
    b. ADP, Ⓟ, and hydrogen ions (H^+).
    c. protons, photons, and pigments.
    d. cytochromes only.
    e. Both b and c are correct.

26. Which of these should not be associated with an electron transport chain?
    a. chloroplasts
    b. protein complexes
    c. movement of H^+ into the thylakoid space
    d. formation of ATP
    e. the RuBP complex

27. The function of the light reactions is to
    a. obtain CO_2.
    b. make carbohydrate.
    c. convert light energy into usable forms of chemical energy.
    d. regenerate RuBP.

28. The NADPH + H^+ and ATP from the light reactions are used to
    a. split water.
    b. cause RuBP carboxylase to fix CO_2.
    c. re-form the photosystems.
    d. cause electrons to move along their pathways.
    e. convert 3PG to G3P.

29. Chemiosmosis depends on
    a. protein complexes in the thylakoid membrane.
    b. a difference in H^+ concentration between the thylakoid space and the stroma.
    c. ATP breaking down to ADP + Ⓟ.
    d. the action spectrum of chlorophyll.
    e. Both a and b are correct.

30. The Calvin cycle reactions
    a. produce carbohydrate.
    b. convert one form of chemical energy into a different form of chemical energy.
    c. regenerate more RuBP.
    d. use the products of the light reactions.
    e. All of these are correct.

31. Carbon dioxide fixation occurs when CO_2 combines with
    a. ATP.   c. BPG.
    b. NADPH + H^+.   d. RuBP.

Understanding the Terms

ATP synthase 137
autotroph 132
C_3 plant 141
C_4 plant 141
Calvin cycle reactions 134
carbon dioxide (CO_2) fixation 139
carotenoid 132
chlorophyll 133
chloroplast 133
cyclic electron pathway 137
grana (sing., granum) 133

heterotroph 132
light reactions 134
$NADP^+$ (nicotinamide adenine dinucleotide phosphate) 134
noncyclic electron pathway 136
photosynthesis 132
photosystem 136
RuBP 139
stomata (sing., stoma) 133
stroma 133
thylakoid 133

Match the terms to these definitions:

a._____ Energy-capturing portion of photosynthesis that takes place in thylakoid membranes of chloroplasts and cannot proceed without solar energy; it produces ATP and NADPH + H^+.

b._____ Flattened sac with a granum whose membrane contains chlorophyll and where the light reactions of photosynthesis occur.

c._____ Plant that directly uses the Calvin cycle; the first detected molecule during photosynthesis is 3PG, a 3-carbon molecule.

d._____ Green pigment that absorbs solar energy and is important in photosynthesis.

e._____ Photosynthetic unit where solar energy is absorbed and high-energy electrons are generated; contains a pigment complex and an electron acceptor.

Thinking Critically

1. What adaptations allow organisms that photosynthesize on land to increase their ability to absorb solar energy?

2. Based on what you know about pigments, why do people who live in very hot, sunny climates wear white?

3. Why does the electron transport chain in chloroplasts pump the H^+ into the thylakoid space instead of into the stroma?

4. Why do you think carrots contain so much carotene (one of the carotenoids) that they are bright orange?

5. Compare and contrast C_4 and CAM plants.

ARIS, the *Inquiry into Life* Website

ARIS, the website for *Inquiry into Life,* provides a wealth of information organized and integrated by chapter. You will find practice quizzes, interactive activities, labeling exercises, flashcards, and much more that will complement your learning and understanding of general biology.

www.aris.mhhe.com

chapter 9

Plant Organization and Function

We eat many parts of plants: *stems, leaves, fruits, and roots. Many vegetables, including carrots, are actually plant roots. With the introduction of "baby" carrots on grocery shelves in the 1980s, carrots have become one of the staples of the American diet. Although real baby carrots are sold as specialty items, the familiar bags in the grocery store are actually baby-cut carrots, full-sized carrots cut into two inch sections and abraded to round the ends and remove the outer layer.*

Carrots are thought to have originated in Afghanistan. The original carrots were purple on the outside, and tough and fibrous. They were not very popular as a food item; instead, they were grown as a medicine. The Greeks used carrot juice as a love potion. However, with modern breeding, carrots are less fibrous and much sweeter. Today's carrots can be orange, purple, red, yellow, or white.

Carrots are considered one of the world's healthiest foods. Carrots are an excellent source of antioxidants and the best vegetable source for carotenes, responsible for their orange color. Our liver converts beta-carotene into vitamin A. A cup of raw carrots contains the equivalent of 686% of the recommended daily allowance of vitamin A with a mere 52 calories. And yes, eating carrots can improve your eyesight because vitamin A is part of the pigment responsible for night vision.

9.1 Plant Organs

From cacti living in hot deserts to water lilies growing in a nearby pond, the flowering plants, or **angiosperms,** are extremely diverse. But despite their great diversity in size and shape, flowering plants share many common structural features. Most flowering plants possess a root system and a shoot system (Fig. 9.1). The **root system** simply consists of the roots, while the **shoot system** consists of the stem and leaves. An **organ** is a structure that contains different tissues and performs one or more specific functions. The roots, stems, and leaves are the vegetative organs common to plants. Flowers, seeds, and fruits are structures involved in reproduction.

Roots

The root system in the majority of plants is located underground. As a rule of thumb, the root system is at least equivalent in size and extent to the shoot system. An apple tree has a much larger root system than a corn plant, for example. A single corn plant may have roots as deep as 2.5 m and spread out over 1.5 m, while a mesquite tree that lives in the desert may have roots that penetrate to a depth of over 20 m.

The extensive root system of a plant anchors it in the soil and gives it support (Fig. 9.2a). The root system absorbs water and minerals from the soil for the entire plant. The cylindrical shape of a root allows it to penetrate the soil as it grows and to absorb water from all sides. The absorptive capacity of a root is also increased by its many branch (lateral) roots and root hairs located in a special zone near a root tip. Root hairs, which are projections from epidermal root-hair cells, are especially responsible for the absorption of water and minerals. Root hairs are so numerous that they increase the absorptive surface of a root tremendously. It has been estimated that a single rye plant has about 14 billion hair cells, and if placed end to end, the root hairs would stretch 10,626 km. Root-hair cells are constantly being replaced, so this same rye plant forms about 100 million new root-hair cells every day. A plant roughly pulled out of the soil will not fare well when transplanted; this is because small lateral roots and root hairs are torn off. Transplantation is more apt to be successful if you take a part of the surrounding soil along with the plant, leaving as much of the branch roots and the root hairs intact as possible.

Roots have still other functions. Roots produce hormones that stimulate the growth of stems and coordinate their size with the size of the root. It is most efficient for a plant to have root and stem sizes that are appropriate to each other. **Perennial** plants have vegetative structures that live year after year. Herbaceous perennials, which live in temperate areas and die back in autumn, store the products of photosynthesis in their roots.

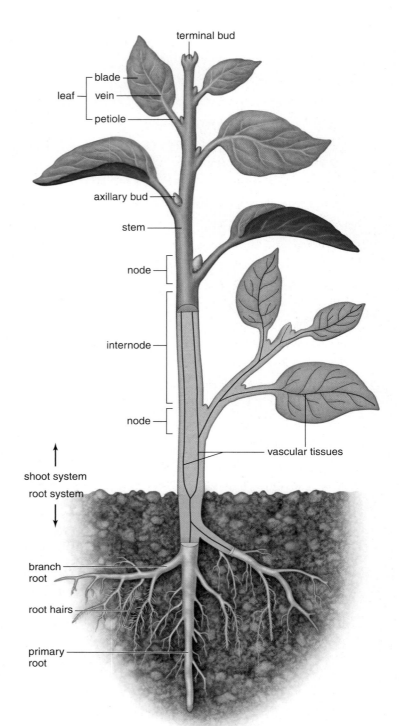

Figure 9.1 Organization of a plant body.

The body of a plant consists of a root system and a shoot system. The shoot system contains the stem and leaves, two types of plant vegetative organs. Axillary buds can develop into branches of stems or flowers, the reproductive structures of a plant. The root system is connected to the shoot system by vascular tissue (brown lines) that extends from the roots to the leaves.

Stems

The shoot system of a plant is composed of the stem, the branches, and the leaves. A **stem,** the main axis of a plant, terminates in tissue that allows the stem to elongate and produce leaves (Fig. 9.2b). If upright, as most are, stems

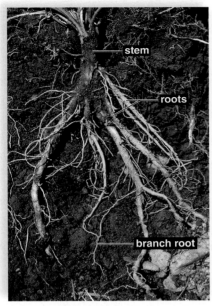

a. Root system, dandelion

b. Shoot system, bean seedling

c. Leaves, pumpkin seedling

Figure 9.2 Vegetative organs of several eudicots.
a. The root system anchors the plant and absorbs water and minerals. **b.** The shoot system consists of a stem and its branches, which support the leaves and transport water and organic nutrients. **c.** The leaves, which are often broad and flat, carry on photosynthesis.

support leaves in such a way that each leaf is exposed to as much sunlight as possible. A **node** occurs where leaves are attached to the stem, and an **internode** is the region between the nodes (see Fig. 9.1). The presence of nodes and internodes is used to identify a stem, even if it happens to be an underground stem. In some plants, the nodes of horizontal stems asexually produce new plants.

In addition to supporting the leaves, a stem has vascular tissue that transports water and minerals from the roots through the stem to the leaves and transports the products of photosynthesis, usually in the opposite direction. **Xylem** cells consist of nonliving cells that form a continuous pipeline for water and mineral transport, while **phloem** cells consist of living cells that join end to end for organic nutrient transport. A stem can sometimes expand in girth as well as length. As trees grow taller each year, they accumulate woody tissue that adds to the strength of their stems.

Stems may have functions other than transport. In some plants (e.g., cactus), the stem is the primary photosynthetic organ. In succulent plants, the stem is a water reservoir. Tubers are underground horizontal stems that store nutrients.

Leaves

Leaves are the major part of a plant that carries on photosynthesis, a process that requires water, carbon dioxide, and sunlight. Leaves receive water from the root system by way of the stem. Stems and leaves function together to transport water from the roots.

The size, shape, color, and texture of leaves are highly variable. These characteristics are fundamental in plant identification. The leaves of some aquatic duckweeds may be less than 1 mm in diameter, while some palms may have leaves that exceed 6 m in length. The shapes of leaves can vary from cactus spines to deeply lobed white oak leaves. Leaves can exhibit a variety of colors, from many shades of green to deep purple. The texture of leaves varies from smooth and waxy like a magnolia to coarse like a sycamore. Plants that lose their leaves every year are called **deciduous.** This is in contrast to most of the gymnosperm plants (conifers), which retain their leaves for usually two to seven years. These plants are called **evergreens.**

Plant leaves that are broad and flat have the maximum surface area for the absorption of carbon dioxide and the collection of solar energy needed for photosynthesis. Also, unlike stems, leaves are almost never woody. With few exceptions, their cells are living, and the bulk of a leaf contains tissue specialized to carry on photosynthesis.

The wide portion of a foliage leaf is called the **blade,** and the stalk that attaches the blade to the stem is the **petiole** (Fig. 9.2*c*). The upper acute angle between the petiole and stem is the leaf axil, where an **axillary bud** (or lateral bud) originates. This bud may become a branch or a flower. Not all leaves are foliage leaves. Some are specialized to protect buds, attach to objects (tendrils), store food (bulbs), or even capture insects.

A flowering plant has three vegetative organs. The root absorbs water and minerals, the stem supports and services leaves, and the leaf carries on photosynthesis.

9.2   Monocot versus Eudicot Plants

Flowering plants are divided into two groups, depending on the number of **cotyledons,** or seed leaves, in the embryonic plant (Fig. 9.3). Some plants have one cotyledon, and are known as monocotyledons, or **monocots.** Other embryos have two cotyledons, and are known as eudicotyledons, or **eudicots.**

The vascular (transport) tissue is organized differently in monocots and eudicots. In the monocot root, vascular tissue occurs in a ring. In the eudicot root, phloem, which transports organic nutrients, is located between the arms of xylem, which transports water and minerals, and has a star shape. In the monocot stem, the vascular bundles, which contain vascular tissue surrounded by a bundle sheath, are scattered. In a eudicot stem, the vascular bundles occur in a ring.

Leaf veins are vascular bundles within a leaf. Monocots exhibit parallel venation, and eudicots exhibit netted venation, which may be either pinnate or palmate. Pinnate venation means that major veins originate from points along the centrally placed main vein, and palmate venation means

that the major veins all originate at the point of attachment of the blade to the petiole:

Netted venation: pinnately veined   palmately veined

Adult monocots and eudicots have other structural differences, such as the number of flower parts and the number of apertures (thin areas in the wall) of pollen grains. The flower parts of monocots are arranged in multiples of three, and the flower parts of eudicots are arranged in multiples of four or five. Eudicot pollen grains usually have three apertures, and monocot pollen grains usually have one aperture.

Although the division between monocots and eudicots may seem of limited importance, it does in fact affect many aspects of their structure. The eudicots are the larger group and include some of our most familiar flowering plants—from dandelions to oak trees. The monocots include grasses, lilies, orchids, and palm trees, among others. Some of our most significant food sources are monocots, including rice, wheat, and corn.

Figure 9.3 Flowering plants are either monocots or eudicots. Five features are used to distinguish monocots from eudicots: the number of cotyledons in the seed; the arrangement of vascular tissue in roots, stems, and leaves; and the number of flower parts.

Flowering plants are divided into monocots and eudicots on the basis of structural differences.

	Seed	Root	Stem	Leaf	Flower
Monocots	One cotyledon in seed	Root xylem and phloem in a ring	Vascular bundles scattered in stem	Leaf veins form a parallel pattern	Flower parts in threes and multiples of three
Eudicots	Two cotyledons in seed	Root phloem between arms of xylem	Vascular bundles in a distinct ring	Leaf veins form a net pattern	Flower parts in fours or fives and their multiples

Plant Tissues

A plant has the ability to grow its entire life because it possesses meristematic (embryonic) tissue. Apical **meristems** are located at or near the tips of stems and roots, where they increase the length of these structures. This increase in length is called *primary growth.* In addition to apical meristems, monocots have a type of meristem called intercalary meristem, which allows them to regrow lost parts. Intercalary meristems occur between mature tissues, and they account for why grass can so readily regrow after being grazed by a cow or cut by a lawnmower.

Apical meristem continually produces three types of meristem—protoderm, ground meristem, and procambium. These develop into the three types of specialized primary tissues in the body of a plant:

1. Protoderm gives rise to **epidermal tissue,** which forms the outer protective covering of a plant.
2. Ground meristem produces **ground tissue,** which fills the interior of a plant.
3. Procambium produces **vascular tissue,** which transports water and nutrients in a plant and provides support.

Epidermal Tissue

The entire body of both nonwoody (herbaceous) and young woody plants is covered by a layer of **epidermis,** which contains closely packed epidermal cells. The walls of epidermal cells that are exposed to air are covered with a waxy **cuticle** to minimize water loss. The cuticle also protects against bacteria and other organisms that might cause disease.

In roots, certain epidermal cells have long, slender projections called **root hairs** (Fig. 9.4*a*). As mentioned, the hairs increase the surface area of the root for absorption of water and minerals; they also help anchor the plant firmly in place.

On stems, leaves, and reproductive organs, epidermal cells produce hairs called **trichomes** that have two important functions: protecting the plant from too much sun and conserving moisture. Sometimes trichomes, particularly glandular ones, help protect a plant from herbivores by producing a toxic substance. Under the slightest pressure, the stiff trichomes of the stinging nettle lose their tips, forming "hypodermic needles" that inject an intruder with a stinging secretion.

In leaves, the epidermis contains specialized cells called **guard cells.** Guard cells, which are epidermal cells with chloroplasts, surround microscopic pores called stomata (sing., stoma) (Fig. 9.4*b*). When the stomata are open, gas exchange and water loss occur.

In older woody plants, the epidermis of the stem is replaced by **periderm,** which is composed primarily of boxlike **cork** cells. At maturity, cork cells can be sloughed off, and new cork cells are made by a meristem called cork cambium (Fig. 9.4*c*). As the new cork cells mature, they increase slightly in volume, and their walls become encrusted with suberin, a lipid material, so that they are waterproof and chemically inert. These nonliving cells protect the plant and make it resistant to attack by fungi, bacteria, and animals. Some cork tissues are commercially used for bottle corks and other products.

The cork cambium overproduces cork in certain areas of the stem surface; this causes ridges and cracks to appear. These areas where the cork layer is thin and the cells are loosely packed are called **lenticels.** Lenticels are important in gas exchange between the interior of a stem and the air.

Epidermal tissue forms the outer protective covering of a herbaceous plant. It is modified in roots, stems, and leaves.

a. Root hairs

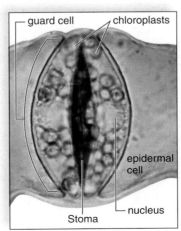

b. Stoma of leaf

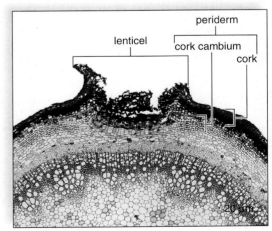

c. Cork of older stem

Figure 9.4 Modifications of epidermal tissue.
a. Root epidermis has root hairs to absorb water. **b.** Leaf epidermis contains stomata (sing., stoma) for gas exchange. **c.** Periderm includes cork and cork cambium. Lenticels in cork are important in gas exchange.

Figure 9.5
Ground tissue cells.

a. Parenchyma cells are the least specialized of the plant cells. **b.** The walls of collenchyma cells are much thicker than those of parenchyma cells. **c.** Sclerenchyma cells have very thick walls and are nonliving—their primary function is to give strong support.

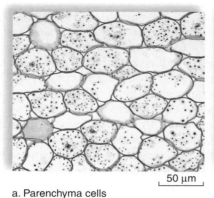
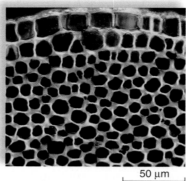

a. Parenchyma cells 50 μm

b. Collenchyma cells 50 μm

c. Sclerenchyma cells 50 μm

Ground Tissue

Ground tissue forms the bulk of a plant and contains parenchyma, collenchyma, and sclerenchyma cells (Fig. 9.5). **Parenchyma** cells are the most abundant and correspond best to the typical plant cell. These are the least specialized of the cell types and are found in all the organs of a plant. They may contain chloroplasts and carry on photosynthesis, or they may contain colorless plastids that store the products of photosynthesis. A juicy bite from an apple yields mostly storage parenchyma cells. Parenchyma cells line the connected air spaces of a water lily and other aquatic plants. Parenchyma cells can divide and give rise to more specialized cells, as when roots develop from stem cuttings placed in water.

Collenchyma cells are like parenchyma cells except they have thicker primary walls. The thickness is irregular and usually involves the corners of the cell. Collenchyma cells often form bundles just beneath the epidermis and give flexible support to immature regions of a plant body. The familiar strands in celery stalks (leaf petioles) are composed mostly of collenchyma cells.

Sclerenchyma cells have thick secondary cell walls (an additional cell wall layer produced between the primary wall and the plasma membrane) impregnated with **lignin,** which is a highly resistant organic substance that makes the walls tough and hard. If we compare a cell wall to reinforced concrete, cellulose fibrils would play the role of steel rods, and lignin would be analogous to the cement. Most sclerenchyma cells are nonliving; their pri-

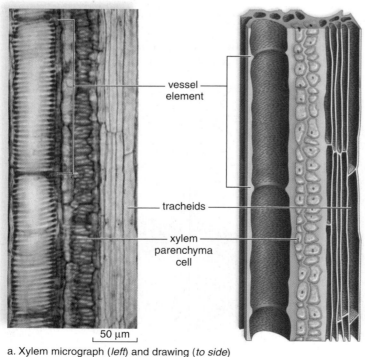

vessel element

tracheids

xylem parenchyma cell

50 μm

a. Xylem micrograph (*left*) and drawing (*to side*)

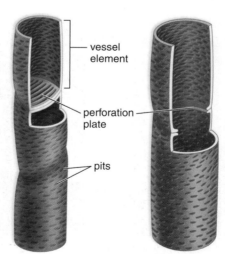

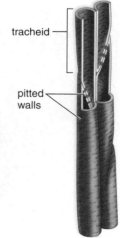

vessel element

perforation plate

pits

tracheid

pitted walls

b. Two types of vessels

c. Tracheids

Figure 9.6 Xylem structure.

a. Photomicrograph of xylem vascular tissue and drawing showing its general organization. **b.** Drawing of two types of vessels, both composed of vessel elements, but with differing perforation plates. **c.** Drawing of tracheids.

mary function is to support the mature regions of a plant. Two types of sclerenchyma cells are *fibers* and *sclereids.* Although fibers are occasionally found in ground tissue, most are in vascular tissue, which is discussed next. Fibers are long and slender and may be grouped in bundles that are sometimes commercially important. Hemp fibers can be used to make rope, and flax fibers can be woven into linen. Flax fibers, however, are not lignified, which is why linen is soft. Sclereids, which are shorter than fibers and more varied in shape, are found in seed coats and nut-shells. Sclereids, or "stone cells," are responsible for the gritty texture of pears. The hardness of nuts and peach pits is also due to sclereids.

Vascular Tissue

There are two types of vascular (transport) tissue. Xylem transports water and minerals from the roots to the leaves, and phloem transports sugar and other organic compounds, including hormones, usually from the leaves to the roots. Both xylem and phloem are considered complex tissues because they are composed of two or more kinds of cells. Xylem contains two types of conducting cells: tracheids and vessel elements (VE) (Fig. 9.6). Both types of conducting cells are hollow and nonliving, but the **vessel elements** are larger, have perforation plates in their end walls, and are arranged to form a continuous vessel for water and mineral transport. The elongated **tracheids,** with tapered ends, form a less obvious means of transport, but water can move

across the end walls and side walls because there are pits, or depressions, where the secondary wall does not form. In addition to vessel elements and tracheids, xylem contains parenchyma cells that store various substances. Vascular rays, which are flat ribbons or sheets of parenchyma cells located between rows of tracheids, conduct water and minerals across the width of a plant. Xylem also contains fibers that lend support.

The conducting cells of phloem are **sieve-tube members** arranged to form a continuous sieve tube (Fig. 9.7). Sieve-tube members contain cytoplasm but no nuclei. The term *sieve* refers to a cluster of pores in the end walls, collectively called a sieve plate. Each sieve-tube member has a **companion cell,** which does have a nucleus. The two are connected by numerous **plasmodesmata,** strands of cytoplasm extending from one sieve tube member to another, through the sieve plate. The nucleus of the companion cell controls and maintains the life of both cells. The companion cells are also believed to be involved in phloem's transport function.

It is important to realize that vascular tissue (xylem and phloem) extends from the root through stems to the leaves and vice versa (see Fig. 9.1). In the roots, the vascular tissue is located in the **vascular cylinder;** in the stem, it forms **vascular bundles;** and in the leaves, it is found in **leaf veins.**

The vascular tissues are xylem and phloem. Xylem transports water and minerals, and phloem transports organic nutrients.

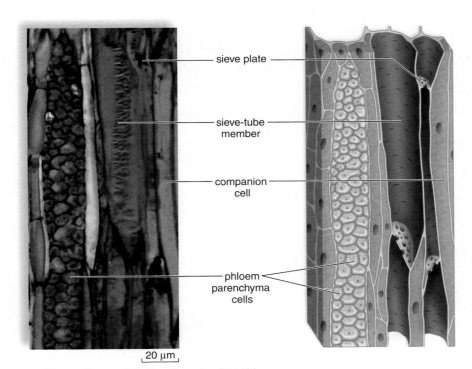

a. Phloem micrograph (*left*) and drawing (*to side*)

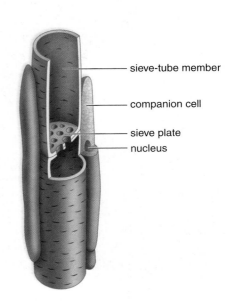

b. Sieve-tube member and companion cells

Figure 9.7 Phloem structure.
a. Photomicrograph of phloem vascular tissue and drawing showing its general organization. **b.** Drawing of sieve-tube member and companion cells.

9.4 Organization of Leaves

Leaves are the organs of photosynthesis in vascular flowering plants. As mentioned earlier, a leaf usually consists of a flattened blade and a petiole connecting the blade to the stem. The blade may be single or composed of several leaflets. Externally, it is possible to see the pattern of the leaf veins, which contain vascular tissue. Leaf veins have a net pattern in eudicot leaves and a parallel pattern in monocot leaves (see Fig. 9.3).

Figure 9.8 shows a cross section of a typical eudicot leaf of a temperate zone plant. At the top and bottom are layers of epidermal tissue that often bear trichomes, protective hairs sometimes modified as glands that secrete irritating substances. These features may prevent the leaf from being eaten by insects. The epidermis characteristically has an outer, waxy cuticle that helps keep the leaf from drying out. The cuticle also prevents gas exchange because it is not gas permeable. The epidermal layers contain stomata that allow gases to move into and out of the leaf. Water loss also occurs at stomata, but each stoma has two guard cells that

regulate its opening and closing, and stomata close when the weather is hot and dry.

The body of a leaf is composed of **mesophyll** tissue. Most eudicot leaves have two distinct regions: **palisade mesophyll,** containing elongated cells, and **spongy mesophyll,** containing irregular cells bounded by air spaces. The parenchyma cells of these layers have many chloroplasts and carry on most of the photosynthesis for the plant. The loosely packed arrangement of the cells in the spongy layer increases the amount of surface area for gas exchange.

Leaf Diversity

The blade of a leaf can be simple or compound (Fig. 9.9a). A simple leaf has a single blade in contrast to a compound leaf, which is divided in various ways into leaflets. For example, a magnolia tree has simple leaves, and a buckeye tree has compound leaves. In pinnately compound leaves, the leaflets occur in pairs, as in a black walnut tree, while in palmately compound leaves, all of the leaflets are attached to a single point, as in a buckeye tree. Plants such as the mimosa have

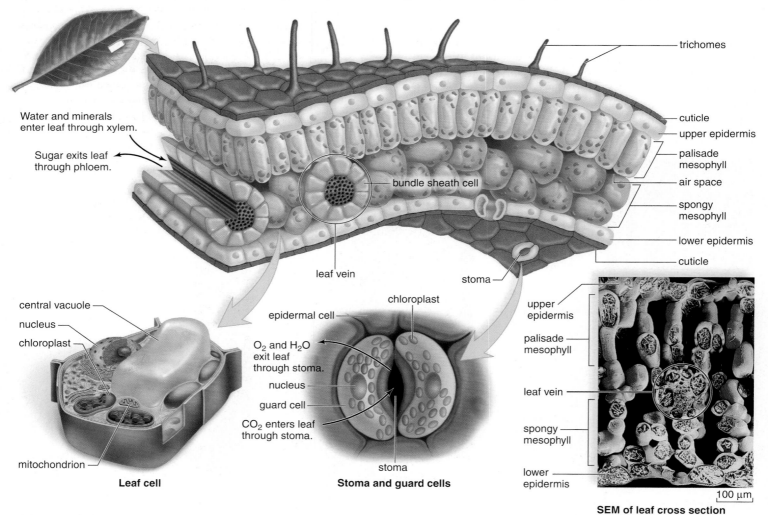

Figure 9.8 Leaf structure. Photosynthesis takes place in the mesophyll tissue of leaves. The leaf is enclosed by epidermal cells covered with a waxy layer, the cuticle. Trichomes (leaf hairs) are also protective. The veins contain xylem and phloem for the transport of water and solutes. A stoma is an opening in the epidermis that permits the exchange of gases.

Figure 9.9 Classification of leaves.

a. Leaves are simple or compound, being either pinnately compound or palmately compound. Note the one axillary bud per compound leaf. **b.** The leaf arrangement on a stem can be alternate, opposite, or whorled.

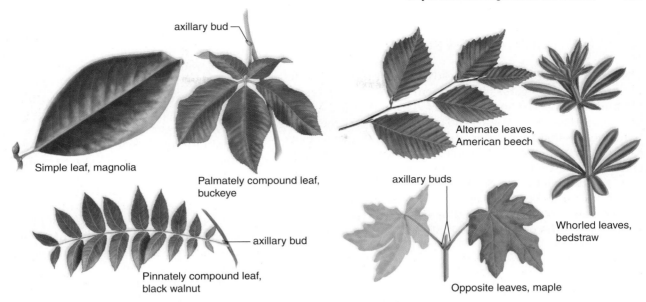

axillary bud

Simple leaf, magnolia

Palmately compound leaf, buckeye

Pinnately compound leaf, black walnut

axillary bud

Alternate leaves, American beech

axillary buds

Whorled leaves, bedstraw

Opposite leaves, maple

a. Simple versus compound leaves

b. Arrangement of leaves on stem

bipinnately compound leaves, with leaflets subdivided into even smaller leaflets.

Leaves can be arranged on a stem in three ways: alternate, opposite, or whorled (Fig. 9.9b). The American beech has alternate leaves; the maple has opposite leaves, being attached to the same node; and bedstraw has a whorled leaf arrangement, with several leaves originating from the same node.

Leaves are adapted to environmental conditions. Shade plants tend to have broad, wide leaves, and desert plants tend to have reduced leaves with sunken stomata. The leaves of a cactus are the spines attached to the succulent (fleshy) stem (Fig. 9.10a). Other succulents have leaves adapted to hold moisture.

An onion bulb is composed of leaves surrounding a short stem. In a head of cabbage, large leaves overlap one another. The petiole of a leaf can be thick and fleshy, as in celery and rhubarb. Leaves of climbing plants, such as those of peas and cucumbers, are modified into tendrils that can attach to nearby objects (Fig. 9.10b).

The leaves of a few plants are specialized for catching insects. A sundew has sticky trichomes that trap insects and others that secrete digestive enzymes. The Venus flytrap has hinged leaves that snap shut and interlock when an insect triggers sensitive trichomes that project from inside the leaves (Fig. 9.10c). The leaves of a pitcher plant resemble a pitcher and have downward-pointing hairs that lead insects into a pool of digestive enzymes secreted by trichomes. Insectivorous plants commonly grow in marshy regions, where the supply of soil nitrogen is severely limited. The digested insects provide the plants with a source of organic nitrogen.

Leaves carry out photosynthesis and are composed of epidermal, mesophyll, and vascular tissues. Leaves adapt to their environment, and their shape is extremely diverse.

stem   spine

tendril

hinged leaves

c. Venus flytrap

a. Cactus

b. Cucumber

Figure 9.10 Leaf diversity. a. The spines of a cactus are modified leaves that protect the fleshy stem from animal predation. **b.** The tendrils of a cucumber are modified leaves that attach the plant to a physical support. **c.** The modified leaves of the Venus flytrap serve as a trap for insect prey. When triggered by an insect, the leaf snaps shut. Once shut, the leaf secretes digestive juices that break down the soft parts of the insect's body.

9.5   Organization of Stems

The anatomy of a woody twig ready for next year's growth reviews for us the organization of a stem (Fig. 9.11). The **terminal bud** contains the shoot tip protected by bud scales, which are modified leaves. Leaf scars and bundle scars mark the location of leaves that have dropped. Dormant axillary buds that can give rise to branches or flowers are also found here. Each spring when growth resumes, bud scales fall off and leave a scar. You can tell the age of a stem by counting these groups of bud scale scars because there is one for each year's growth.

When growth resumes, the apical meristem at the shoot tip produces new cells that elongate and thereby increase the height of the stem (primary growth). The shoot **apical meristem** is protected within the terminal bud, where immature leaves called leaf primordia (sing., primordium) envelop it (Fig. 9.12). The leaf primordia mark the location of a node; the portion of stem between nodes is an internode. As a stem grows, the internodes increase in length.

In addition to leaf primordia, three specialized types of primary meristem develop from a shoot apical meristem (Fig. 9.12b). These primary meristems contribute to the length of a shoot. The protoderm, the outermost primary meristem, gives rise to epidermis. The ground meristem produces two tissues composed of parenchyma cells. The parenchyma tissue in the center of the stem is the pith, and the parenchyma tissue between the epidermis and the vascular tissue is the cortex.

The procambium, shown as an orange strand of tissue in Figure 9.12a, produces the first xylem cells, called primary xylem, and the first phloem cells, called primary phloem. Differentiation continues as certain cells become the first tracheids or vessel elements of the xylem within a vascular bundle. The first sieve-tube members of a vascular bundle do not have companion cells and are short-lived (some live only a day before being replaced). Mature vascular bundles contain fully differentiated xylem, phloem, and a lateral meristem called **vascular cambium.** Vascular cambium is discussed more fully later in this section.

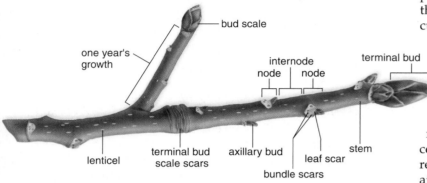

Figure 9.11 Woody twig. The major parts of a stem are illustrated by a woody twig collected in winter.

Figure 9.12 Shoot tip and primary meristems.
a. The shoot apical meristem within a terminal bud is surrounded by leaf primordia. **b.** The shoot apical meristem produces the primary meristems. Protoderm gives rise to epidermis; ground meristem gives rise to pith and cortex; and procambium gives rise to vascular tissue, including primary xylem, primary phloem, and vascular cambium.

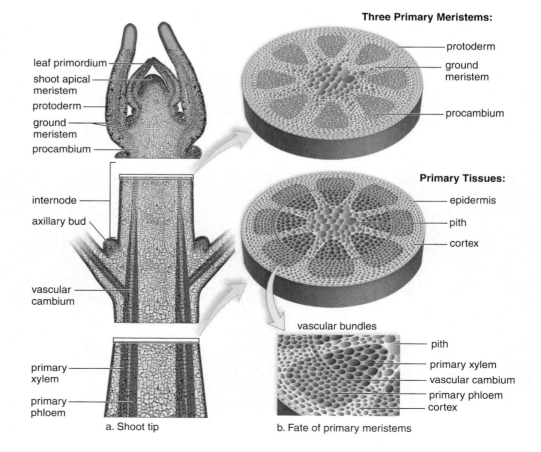

a. Shoot tip

b. Fate of primary meristems

Herbaceous Stems

Mature nonwoody stems, called **herbaceous stems,** exhibit only primary growth. The outermost tissue of herbaceous stems is the epidermis, which is covered by a waxy cuticle to prevent water loss. These stems have distinctive vascular bundles containing xylem and phloem. In each bundle, xylem is typically found toward the inside of the stem, and phloem is found toward the outside.

In a herbaceous eudicot stem, such as a sunflower, the vascular bundles are arranged in a distinct ring that separates the cortex from the central pith, which stores water and the products of photosynthesis (Fig. 9.13). The cortex is sometimes green and carries on photosynthesis. In a monocot stem such as corn, the vascular bundles are scattered throughout the stem, and often there is no well-defined cortex or well-defined pith (Fig. 9.14).

As a stem grows, the shoot apical meristem produces new leaves and primary meristems. The primary meristems produce the other tissues found in herbaceous stems.

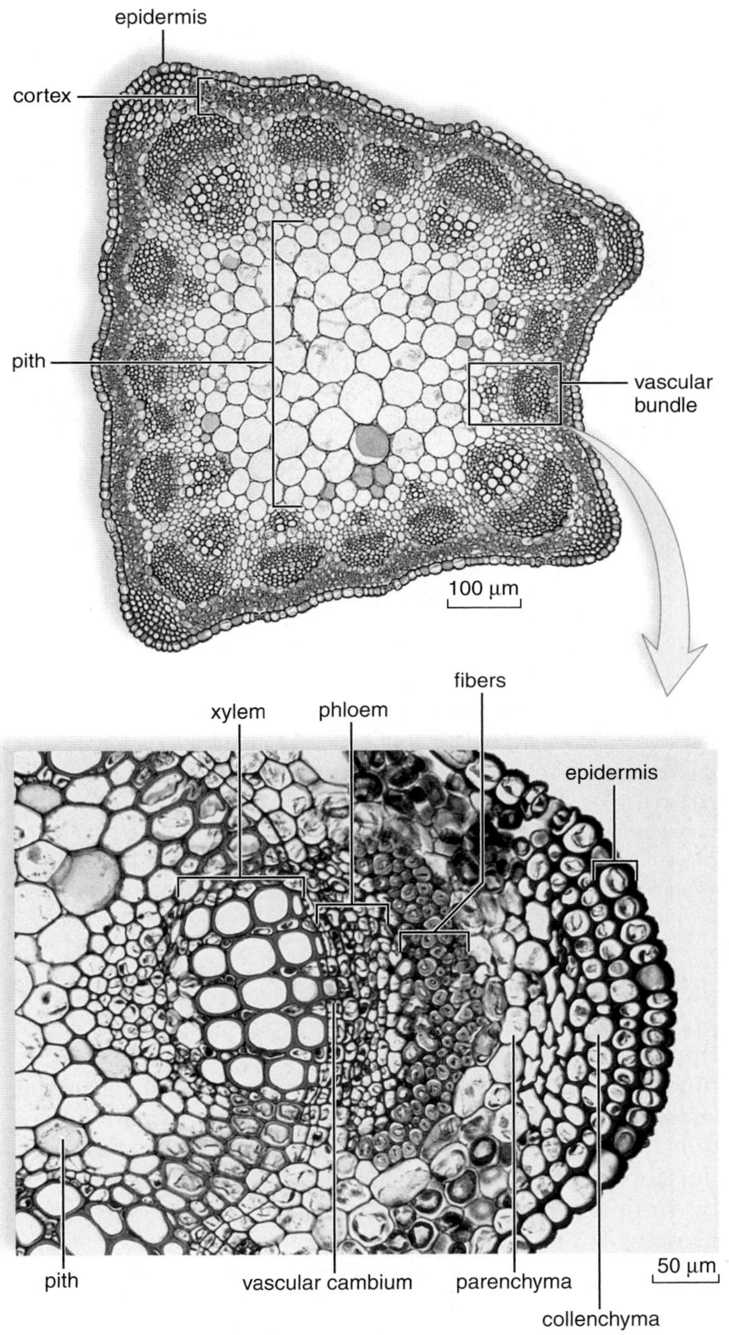

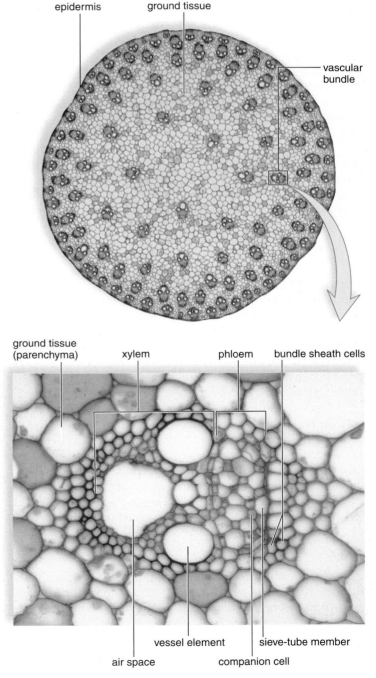

Figure 9.13 Herbaceous eudicot stem.

Figure 9.14 Monocot stem.

Woody Stems

A woody plant, such as an oak tree, has both primary and secondary tissues. Primary tissues are those new tissues formed each year from primary meristems right behind the shoot apical meristem. Secondary tissues develop during the first and subsequent years of growth from lateral meristems: vascular cambium and cork cambium. Primary growth occurs in all plants; *secondary growth,* which occurs only in conifers and woody eudicots, increases the girth of trunks, stems, branches, and roots.

Trees and shrubs undergo secondary growth due to activities of the vascular cambium (Fig. 9.15). In herbaceous plants, vascular cambium is present between the xylem and phloem of each vascular bundle. In woody plants, the vascular cambium develops to form a ring of meristem that divides parallel to the surface of the plant, and produces new xylem and phloem each year. Eventually, a woody eudicot stem has an entirely different organization from that of a herbaceous eudicot stem. A woody stem has no distinct vascular bundles and instead has three distinct areas: the bark, the wood, and the pith. Vascular cambium lies between the bark and the wood.

You will also notice in Figure 9.15c the xylem rays and phloem rays that are visible in the cross section of a woody stem. Rays consist of parenchyma cells that permit lateral conduction of nutrients from the pith to the cortex and some storage of food. A phloem ray is actually a continuation of a xylem ray. Some phloem rays are much broader than other phloem rays.

Bark

The **bark** of a tree contains both periderm (cork, cork cambium, and a single layer of cork cells filled with suberin called the phelloderm) and phloem. Although secondary phloem is produced each year by vascular cambium, phloem does not build up from season to season. The bark of a tree can be removed; however, doing so is very harmful because without phloem organic nutrients cannot be transported. Girdling, removing a ring of bark from around a tree, can be lethal to the tree.

Cork cambium develops beneath the epidermis. When cork cambium first begins to divide, it produces tissue that disrupts the epidermis and replaces it with cork cells. Cork cells are impregnated with suberin, a waxy layer that makes them waterproof but also causes them to die. This is protective because now the stem is less edible. But an impervious barrier means that gas exchange is impeded except at lenticels, which are pockets of loosely arranged cork cells not impregnated with suberin.

Wood

Wood is secondary xylem that builds up year after year, thereby increasing the girth of trees. In trees that have a growing season, vascular cambium is dormant during the winter. In the spring, when moisture is plentiful and leaves require much water for growth, the secondary xylem contains wide vessel elements with thin walls. In this so-called *spring wood,* wide vessels transport sufficient water to the growing leaves. Later in the season, moisture is scarce, and

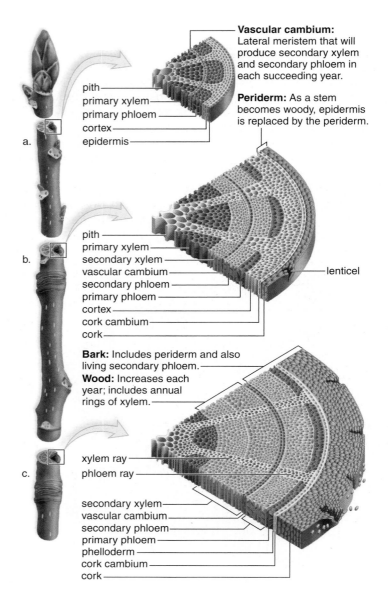

Figure 9.15 Secondary growth of stems.
a. Diagram showing a eudicot herbaceous stem just before secondary growth begins. **b.** Secondary growth has begun, and periderm has replaced the epidermis. Vascular cambium produces secondary xylem and secondary phloem each year. **c.** In a two-year-old stem, the primary phloem and cortex have disappeared, and only the secondary phloem (within the bark) produced by vascular cambium will be active that year. Secondary xylem builds up to become the annual rings of a woody stem.

the wood at this time, called *summer wood,* has a lower proportion of vessels (Fig. 9.16). Strength is required because the tree is growing larger and summer wood contains numerous thick-walled tracheids. At the end of the growing season, just before the cambium becomes dormant again, only heavy fibers with especially thick secondary walls may develop. When the trunk of a tree has spring wood followed by summer wood, the two together make up one year's growth, or an **annual ring.** You can tell the age of a tree by counting the annual rings (Fig. 9.17a). The outer annual rings, where transport occurs, are called sapwood.

In older trees, the inner annual rings, called the heartwood, no longer function in water transport. The cells become plugged with deposits such as resins, gums, and other substances that inhibit the growth of bacteria and fungi. Heartwood may help support a tree, although some trees stand erect and live for many years after the heartwood has rotted away. Figure 9.17*b* shows the layers of a woody stem in relation to one another.

The annual rings are important not only in telling the age of a tree, but also in serving as a historical record of tree growth. For example, if rainfall and other conditions were extremely favorable during a season, the annual ring may be wider than usual. If the tree was shaded on one side by another tree or building, the rings may be wider on the favorable side.

Woody Plants Is it advantageous to be woody? With adequate rainfall, woody plants can grow taller than herbaceous plants and increase in girth because they have adequate vascular tissue to support and service their leaves. However, it takes energy to produce secondary growth and prepare the body for winter if the plant lives in the temperate zone. Also, woody plants need more defense mechanisms because a long-lasting plant that stays in one spot is likely to be attacked by herbivores and parasites. Then, too, trees don't usually repro-

duce until they have grown for several seasons, by which time they may have succumbed to an accident or disease. In certain habitats, it is more advantageous for a plant to put most of its energy into simply reproducing rather than being woody.

> Woody plants grow in girth due to the presence of vascular cambium and cork cambium. Their bodies have three main parts: bark (which contains cork, cork cambium, phelloderm, and phloem), wood (which contains secondary xylem), and pith.

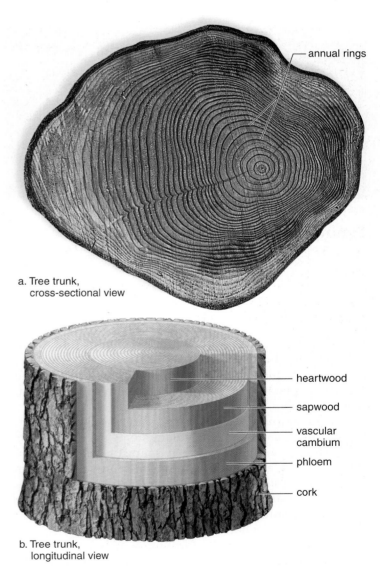

a. Tree trunk, cross-sectional view

b. Tree trunk, longitudinal view

Figure 9.17 Tree trunk.
a. A cross section of a 39-year-old tree trunk. The xylem within the darker heartwood is inactive; the xylem within the lighter sapwood is active. **b.** The relationship of bark, vascular cambium, and wood is retained in a mature stem. The pith has been buried by the growth of layer after layer of new secondary xylem.

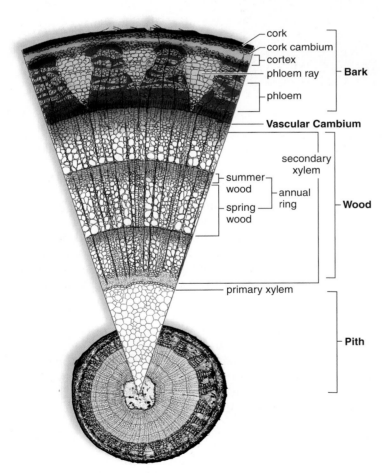

Figure 9.16 Three-year-old woody twig.
The buildup of secondary xylem in a woody stem results in annual rings, which tell the age of the stem. The rings can be distinguished because each one begins with spring wood (large vessel elements) and ends with summer wood (smaller and fewer vessel elements).

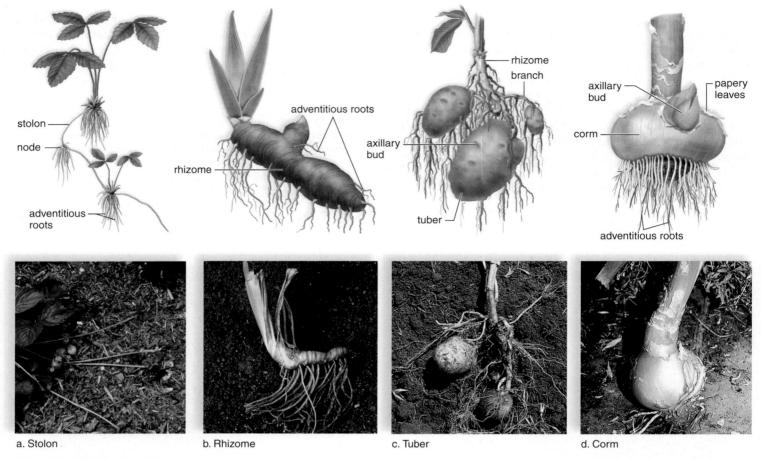

a. Stolon   b. Rhizome   c. Tuber   d. Corm

Figure 9.18 Stem diversity.

a. A strawberry plant has aboveground horizontal stems called stolons. Every other node produces a new shoot system. **b.** The underground horizontal stem of an iris is a fleshy rhizome. **c.** The underground stem of a potato plant has enlargements called tubers. We call the tubers potatoes. **d.** The corm of a gladiolus is a thick stem covered by papery leaves.

Stem Diversity

Stems exist in diverse forms, or modifications (Fig. 9.18). Aboveground horizontal stems, called **stolons** or runners, produce new plants where nodes touch the ground. The strawberry plant is a common example of this type of stem, which functions in vegetative reproduction.

Aboveground vertical stems can also be modified. For example, cacti have succulent stems specialized for water storage, and the tendrils of grape plants (which are stem branches) allow them to climb. Morning glory and its relatives have stems that twine around support structures. Such tendrils and twining shoots help plants expose their leaves to the sun.

Underground horizontal stems, called **rhizomes,** may be long and thin, as in sod-forming grasses, or thick and fleshy, as in irises. Rhizomes survive the winter and contribute to asexual reproduction because each node bears a bud. Some rhizomes have enlarged portions called tubers, which function in food storage. Potatoes are tubers; the eyes are buds that mark the nodes.

Corms are bulbous underground stems that lie dormant during the winter, just as rhizomes do. They have thin, papery leaves and a thick stem. They also produce new plants the next growing season. Gladiolus corms are referred to as bulbs by laypersons, but the botanist reserves the term *bulb* for a structure composed of thick modified leaves attached to a short vertical stem. An onion is a bulb.

Humans make use of stems in many ways. The stem of the sugarcane plant is a primary source of table sugar. The spice cinnamon and the drug quinine are derived from the bark of *Cinnamomum verum* and various *Cinchona* species, respectively. And the cellulose in plant cell walls is necessary for the production of paper, as discussed in the Ecology Focus, "Paper Comes from Plants."

Plants use modified stems for such functions as vegetative reproduction, climbing, survival, and food storage.
Modified stems aid adaptation to different environments.

EcologyFocus

Paper Comes from Plants

The word *paper* takes its origin from papyrus, the plant Egyptians used to make the first form of paper. The Egyptians manually placed thin sections cut from papyrus at right angles and pressed them together to make a sheet of writing material. From that beginning some 5,500 years ago the production of paper has grown to become a worldwide industry of major importance (Fig. 9A). The process is fairly simple. Plant material is ground up mechanically to form a pulp that contains "fibers," which biologists know come from vascular tissue. The fibers automatically form a sheet when they are screened from the pulp.

If wood is the source of the fibers, the pulp must be chemically treated to remove lignin. If only a small amount of lignin is removed, the paper is brown, as in paper bags. Acid processing removes more lignin and results in white paper, but acidic paper is not very durable. Alkaline processes have been developed to provide highly durable, acid-free paper. Paper is more durable when it is made from cotton or linen because the fibers from these plants are lignin-free.

Other major plants used to make paper include:

Eucalyptus trees In recent years, Brazil has devoted huge areas of the Amazon region to the growing of cloned eucalyptus seedlings, specially selected and engineered to be ready for harvest after about seven years.

Temperate hardwood trees Plantation cultivation in Canada provides birch, beech, chestnut, poplar, and particularly aspen wood for paper making. Tropical hardwoods, usually from Southeast Asia, are also used.

Softwood trees In the United States, several species of pine trees have been genetically improved to have a higher wood density and to be harvestable five years earlier than ordinary pines. Southern Africa, Chile, New Zealand, and Australia also devote thousands of acres to growing pines for paper pulp.

Bamboo Several Asian countries, especially India, provide vast quantities of bamboo pulp for the making of paper. Because bamboo,

Figure 9A Paper production.
Today, a revolving wire-screen belt is used to deliver a continuous wet sheet of paper to heavy rollers and heated cylinders, which remove most of the remaining moisture and press the paper flat.

which is actually a grass, is harvested without destroying the roots, and the growing cycle is favorable, this plant is expected to be a significant source of paper pulp despite high processing costs to remove impurities.

Flax and cotton rags Linen and cotton cloth from textile and garment mills are used to produce *rag paper,* which has flexibility and durability that are desirable in legal documents, high-grade bond paper, and high-grade stationery, as well as in paper currency.

Paper largely consists of the cellulose within plant cell walls. It seems reasonable to suppose, then, that paper could be made from synthetic polymers (e.g., rayon). Indeed, synthetic polymers produce a paper that has qualities superior to those of paper made from natural sources, but the cost thus far is prohibitive. Another consideration, however, is the ecological impact of making paper from trees. Plantations containing stands of uniform trees replace natural ecosystems, and when the trees are

clear-cut, the land is laid bare. Paper mill wastes, which include caustic chemicals, add significantly to the pollution of rivers and streams.

The use of paper for packaging and making all sorts of products has increased dramatically in the last century. Each person in the United States uses about 318 kg of paper products per year, compared to only 2.3 kg of paper per person in India. It is clear, then, that we should take the initiative in recycling paper. When newspaper, office paper, and photocopies are soaked in water, the fibers are released, and they can be used to make a new batch of paper. It's estimated that recycling the Sunday newspapers alone would save approximately 500,000 trees each week!

Discussion Questions

1. What are the advantages of engineering trees to produce less lignin?
2. Why is cellulose such a good substance for paper to be made of?
3. Why does paper deteriorate over time?

9.6   Organization of Roots

Figure 9.19*a*, a longitudinal section of a eudicot root, reveals zones where cells are in various stages of differentiation as primary growth occurs. The root apical meristem is in the region protected by the **root cap.** Root cap cells have to be replaced constantly because they are worn away by rough soil particles as the root grows. The primary meristems are located in the zone of cell division, which continuously provides cells to the zone of elongation above by mitosis. In the zone of elongation, the cells lengthen as they become specialized. The zone of maturation, which contains fully differentiated cells, is recognizable by the presence of root hairs on many of the epidermal cells.

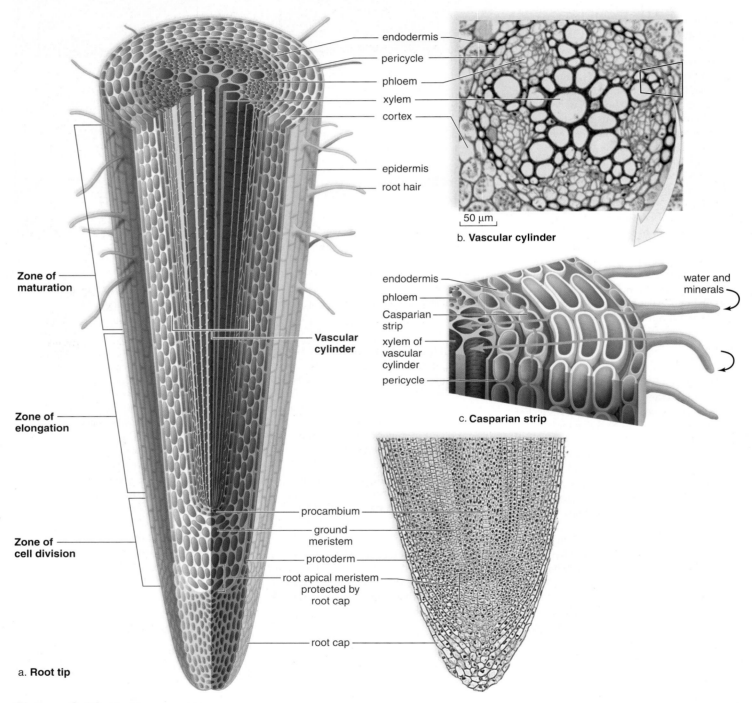

b. **Vascular cylinder**

c. **Casparian strip**

a. **Root tip**

Figure 9.19 Eudicot root tip.

a. The root tip is divided into three zones, best seen in a longitudinal section such as this. **b.** The vascular cylinder of a eudicot root contains the vascular tissue. Xylem is typically star-shaped, and phloem lies between the points of the star. **c.** Water and minerals either pass between cells, or they enter root cells directly. In any case, because of the Casparian strip, water and minerals must pass through the cytoplasm of endodermal cells in order to enter the xylem. In this way, endodermal cells regulate the passage of minerals into the vascular cylinder.

Tissues of a Eudicot Root

Figure 9.19*a, b* also shows a cross section of a root at the zone of maturation. These specialized tissues are identifiable:

Epidermis The epidermis, which forms the outer layer of the root, consists of only a single layer of cells. The majority of epidermal cells are thin-walled and rectangular, but in the zone of maturation, many epidermal cells have root hairs that project as far as 5–8 mm into the soil.

Cortex Moving inward, next to the epidermis, large, thin-walled parenchyma cells make up the **cortex** of the root. These irregularly shaped cells are loosely packed, so that water and minerals can move through the cortex without entering the cells. The cells contain starch granules, and the cortex functions in food storage.

Endodermis The **endodermis** is a single layer of rectangular cells that forms a boundary between the cortex and the inner vascular cylinder. The endodermal cells fit snugly together and are bordered on four sides by a layer of impermeable lignin and suberin known as the **Casparian strip** (Fig. 9.19*c*). This strip prevents the passage of water and mineral ions between adjacent cell walls. The two sides of each cell that contact the cortex and the vascular cylinder respectively remain permeable. Therefore, the only access to the vascular cylinder is through the endodermal cells themselves, as shown by the arrows in Figure 9.19*c*. This arrangement regulates the entrance of minerals into the vascular cylinder.

Vascular tissue The **pericycle,** the first layer of cells within the vascular cylinder, has retained its capacity to divide and can start the development of branch roots (Fig. 9.20). The main portion of the vascular cylinder contains xylem and phloem. The xylem appears star-shaped in eudicots because several arms of tissue radiate from a common center (see Fig. 9.19). The phloem is found in separate regions between the arms of the xylem.

Organization of Monocot Roots

Monocot roots have the same growth zones as eudicot roots, but they do not undergo secondary growth as many eudicot roots do. Also, the organization of their tissues is slightly different. The ground tissue of a monocot root's **pith,** which is centrally located, is surrounded by a vascular ring composed of alternating xylem and phloem bundles (Fig. 9.21). Monocot roots also have pericycle, endodermis, cortex, and epidermis.

The root system of a plant absorbs water and minerals. These nutrients cross the epidermis and cortex before entering the endodermis, the tissue that regulates their entrance into the vascular cylinder.

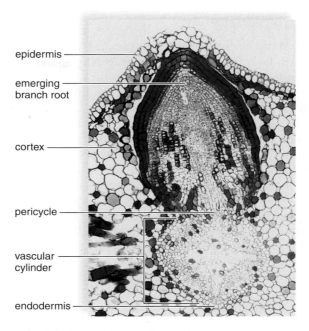

Figure 9.20 Branching of eudicot root.
This cross section of a willow shows the origination and growth of a branch root from the pericycle.

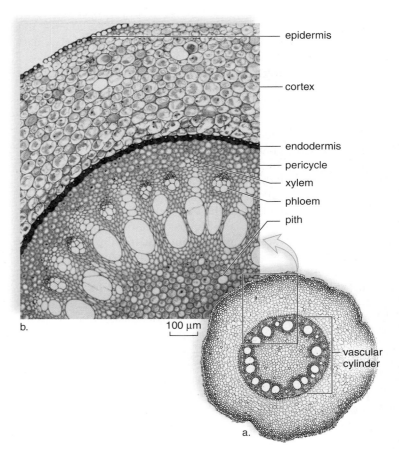

Figure 9.21 Monocot root.
a. This overall cross section of a monocot root shows that a vascular ring surrounds a central pith. **b.** The enlargement shows the exact placement of various tissues.

Root Diversity

Roots have various adaptations and associations to better perform their functions: anchorage, absorption of water and minerals, and storage of carbohydrates.

In some plants, notably eudicots, the first or **primary root** grows straight down and remains the dominant root of the plant. This so-called **taproot** is often fleshy and stores food (Fig. 9.22*a*). Carrots, beets, turnips, and radishes have taproots that we consume as vegetables. Sweet potato plants don't have taproots, but they do have roots that expand to store starch. We call these storage areas sweet potatoes.

Other plants, notably monocots, have no single, main root, but rather a large number of slender roots. These grow from the lower nodes of the stem when the first (primary) root dies. These slender roots make up a **fibrous root system** (Fig. 9.22*b*). Many grasses have fibrous root systems that strongly anchor the plants to the soil.

Root Specializations

When roots develop from organs of the shoot system instead of the root system, they are known as **adventitious roots.** Some adventitious roots emerge above the soil line, as in corn plants, where their main function is to help anchor the plant. If so, they are called prop roots (Fig. 9.22*c*). Other examples of adventitious roots are those found on horizontal stems or the rootlets at the nodes of climbing English ivy. As the vines climb, the rootlets attach the plant to any available vertical structure.

Black mangroves live in swampy water and have pneumatophores, root projections that rise above the water and acquire oxygen for cellular respiration (Fig. 9.22*d*).

Some plants, such as dodders and broomrapes, are parasitic on other plants. Their stems have rootlike projections called haustoria (sing., haustorium) that grow into the host plant and make contact with vascular tissue from which they extract water and nutrients (Fig. 9.22*e*). As discussed in section 9.7, **mycorrhizae** are associations between roots and fungi that can extract water and minerals from the soil better than roots that lack a fungus partner. This is a mutualistic relationship because the fungus receives sugars and amino acids from the plant, while the plant receives water and minerals via the fungus.

Peas, beans, and other legumes have **root nodules** where nitrogen-fixing bacteria live. Plants cannot extract nitrogen from the air, but the bacteria within the nodules can take up and reduce atmospheric nitrogen. This means that the plant is no longer dependent on a supply of nitrogen (i.e., nitrate or ammonium) from the soil, and indeed these plants are often planted just to bolster the nitrogen content of the soil.

Humans have learned ways to take advantage of some plant roots' absorption capacities to clean up pollution (see the Ecology Focus, "Plants Can Clean Up Toxic Messes").

> **Roots have various adaptations and associations to enhance their ability to anchor a plant, absorb water and minerals, and store the products of photosynthesis.**

a. Taproot

b. Fibrous root system

c. Prop roots, a type of adventitious root

d. Pneumatophores of black mangrove trees

dodder

haustorium   dodder

e. Dodder

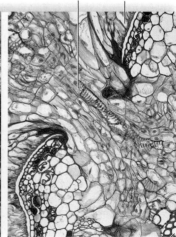
host's vascular tissue

Figure 9.22 Root diversity.

a. A taproot may have branch roots in addition to a main root. **b.** A fibrous root has many slender roots with no main root. **c.** Prop roots are specialized for support. **d.** The pneumatophores of a black mangrove tree allow it to acquire oxygen even though it lives in swampy water. **e.** *Left:* Dodder is a parasitic plant consisting mainly of orange-brown twining stems. (The green in the photograph is the host plant.) *Right:* Haustoria are rootlike projections of the stem that tap into the host's vascular system.

EcologyFocus

Plants Can Clean up Toxic Messes

Most trees planted along the edges of farms are intended to break the wind. But a mile-long stand of spindly poplars outside Amana, Iowa, serves a different purpose. It cleans up pollution.

The poplars act like vacuum cleaners, sucking up nitrate-laden runoff from a fertilized cornfield before this runoff reaches a nearby brook—and perhaps other waters. Nitrate runoff into the Mississippi River from midwestern farms, after all, is a major cause of the large "dead zone" of oxygen-depleted water that develops each summer in the Gulf of Mexico.

Before the trees were planted, the brook's nitrate levels were as much as ten times the amount considered safe. But then Louis Licht, a University of Iowa graduate student, had the idea that poplars, which absorb lots of water and tolerate pollutants, could help. In 1991, Licht tested his hunch by planting the trees along a field owned by a corporate farm. The brook's nitrate levels subsequently dropped more than 90%, and the trees have thrived, serving as a prime cleanup method known as **phytoremediation.**

This method uses plants—many of them common species such as poplar, mustard, and mulberry—that have an appetite for lead, uranium, and other pollutants. These plants' genetic makeups allow them to absorb and then store, degrade, or transform substances that kill or harm other plants and animals. "It's an elegantly simple solution," says Licht, who now runs Ecolotree, an Iowa City phytoremediation company.

The idea behind phytoremediation is not new; scientists have long recognized certain plants' abilities to absorb and tolerate toxic substances. But the idea of using these plants on contaminated sites has just gained support in the last decade. The plants clean up sites in two basic ways, depending on the substance involved. If it is an organic contaminant, such as spilled oil, the plants or microbes around their roots break down the substance. The remainders can either be absorbed by the plant or left in the soil or water. For an inorganic contaminant such as cadmium or zinc, the plants absorb the substance and trap it. The plants must then be harvested and disposed of, or processed to reclaim the trapped contaminant.

Different plants work on different contaminants. The mulberry bush, for instance, is effective on industrial sludge; some grasses attack petroleum wastes; and sun-

Figure 9B Canola plants. Scientist Gary Bañuelos recommended planting canola to pull selenium out of the soil.

flowers (together with soil additives) remove lead. Canola plants, meanwhile, are grown in California's San Joaquin Valley to soak up excess selenium in the soil, thus helping to prevent an environmental catastrophe like the one that occurred there in the 1980s.

Back then, irrigated farming caused naturally occurring selenium to rise to the soil surface. When excess water was pumped onto the fields, some selenium would flow off into drainage ditches, eventually ending up in Kesterson National Wildlife Refuge. The selenium in ponds at the refuge accumulated in plants and fish and subsequently deformed and killed waterfowl, says Gary Bañuelos, a plant scientist with the U.S. Department of Agriculture who helped remedy the problem. He recommended that farmers add selenium-accumulating canola plants to their crop rotations (Fig. 9B). As a result, selenium levels in runoff are being managed. Although the underlying problem of excessive selenium in soils has not been solved, Bañuelos says, "This is a tool to manage mobile selenium and prevent another unlikely selenium-induced disaster."

Phytoremediation has also helped clean up badly polluted sites, in some cases at a fraction of the usual cost. Edenspace Systems Corporation of Reston, Virginia, just concluded a phytoremediation demonstration at a Superfund site on an Army firing range in Aberdeen, Maryland. The company successfully used mustard plants to remove uranium from the firing range, at as little as 10% of the cost of traditional cleanup methods. Depending on the contaminant involved,

traditional cleanup costs can run as much as $1 million per acre, experts say.

Phytoremediation does have its limitations, however. One of them is its slow pace. Depending on the contaminant, it can take several growing seasons to clean a site—much longer than with conventional methods. "We normally look at phytoremediation as a target of one to three years to clean a site," notes Edenspace's Mike Blaylock. "People won't want to wait much longer than that."

Phytoremediation is also only effective at depths that plant roots can reach, making it useless against deep-lying contamination unless the contaminated soils are excavated. Phytoremediation will not work on lead and other metals unless chemicals are added to the soil. In addition, it is possible that animals may ingest pollutants by eating the leaves of plants in some projects.

Phytoremediation is a promising solution to pollution problems but, says the EPA's Walter W. Kovalick, "It's not a panacea. It's another arrow in the quiver. It takes more than one arrow to solve most problems."

Discussion Questions

1. In phytoremediation, which part of the plant takes up the toxins?
2. Why did the selenium runoff into the ponds eventually cause death to waterfowl but not to fish? What does this imply if humans were to eat the fish?
3. What are the major advantages to the use of phytoremediation? What are the major disadvantages?

9.7 Uptake and Transport of Nutrients

In order to produce a carbohydrate, a plant requires carbon dioxide from the air and water from the soil. One of the major functions of roots is to supply the plant with water and also minerals. Humans make use of a plant's ability to take minerals from the soil. We depend on them for our basic supply of such minerals as calcium, to build bones and teeth, and iron, to help carry oxygen to our cells. Minerals such as copper and zinc are cofactors for the functioning of enzymes.

As you know, water and minerals are transported through a plant in xylem, while the products of photosynthesis are transported in phloem.

Water Uptake and Transport

Osmosis of water and diffusion of minerals aid the entrance of water and minerals. Eventually, however, a plant uses active transport to bring minerals into root cells. Water and minerals enter a plant at the root, primarily through the root hairs. From there, water, along with minerals, moves across the tissues of a root until it enters xylem.

Water entering root cells creates a positive pressure called root pressure that tends to push xylem sap upward. Atmospheric pressure (101 kPa) can support a column of water to a maximum height of approximately 10.3 meters. However, since some trees can exceed 90 meters in height, other factors must be involved in causing water to move from the roots to the leaves. Figure 9.23 illustrates the accepted model for the transport of water and, therefore, minerals in a plant, called the **cohesion-tension model.**

Cohesion-Tension Model of Xylem Transport

Once water enters xylem, it must be transported upward to the top of a tree. This can be a daunting task.

Recall that the vascular tissue of xylem contains the hollow conducting cells called tracheids and vessel elements (see Fig. 9.6). The vessel elements are larger than the tracheids, and they are stacked one on top of the other to form a pipeline that stretches from the roots to the leaves. It is an open pipeline because the vessel elements have no end walls, just perforation plates, separating one from the other (Fig. 9.24a, b). The tracheids, which are elongated with tapered ends, form a less obvious means of transport. Water can move across the end and side walls of tracheids because of pits, or depressions, where the secondary wall does not form (Fig. 9.24c).

Explanation of the Model Unlike animals, which rely on a pumping heart to move blood through their vessels, plants utilize a passive, not active, means of transport to move water in xylem. The cohesion-tension model of xylem transport relies on the properties of water (see section 2.3). The term *cohesion* refers to the tendency of water

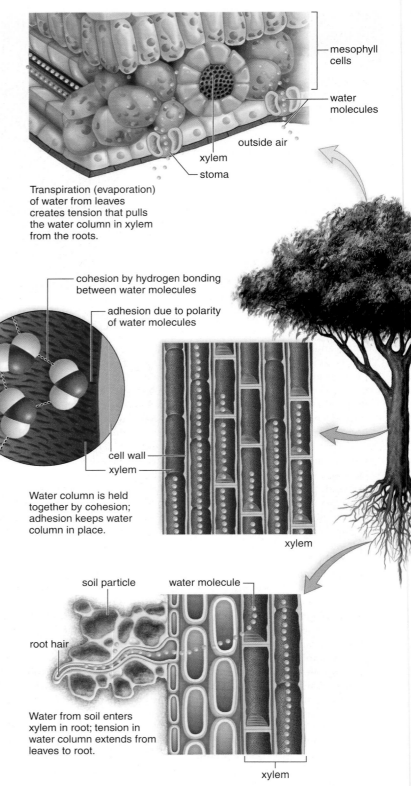

Transpiration (evaporation) of water from leaves creates tension that pulls the water column in xylem from the roots.

cohesion by hydrogen bonding between water molecules

adhesion due to polarity of water molecules

Water column is held together by cohesion; adhesion keeps water column in place.

Water from soil enters xylem in root; tension in water column extends from leaves to root.

Figure 9.23 Cohesion-tension model of xylem transport.

Tension created by evaporation (transpiration) at the leaves pulls water along the length of the xylem—from the root hairs to the leaves.

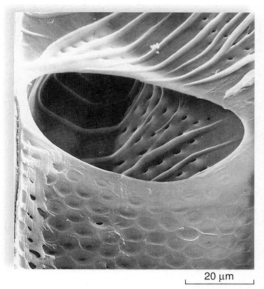

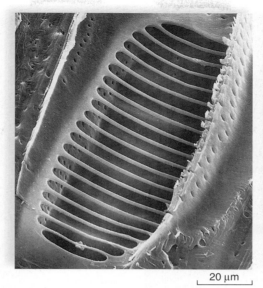

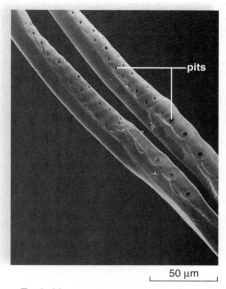

a. Perforation plate with a single, large opening   b. Perforation plate with a series of openings   c. Tracheids

Figure 9.24 Conducting cells of xylem.

a., b. Water can move from vessel element to vessel element through perforation plates. Vessel elements can also exchange water with tracheids through pits. **c.** Tracheids are long, hollow cells with tapered ends. Water can move into and out of tracheids through pits only.

molecules to cling together. Because of hydrogen bonding, water molecules interact with one another, forming a continuous water column in xylem from the leaves to the roots that is not easily broken. *Adhesion* refers to the ability of water, a polar molecule, to interact with the molecules making up the walls of the vessels in xylem. Adhesion gives the water column extra strength and prevents it from slipping back.

Why does the continuous water column move passively upward? Consider the structure of a leaf. When the sun rises, stomata open, and carbon dioxide enters a leaf. Within the leaf, the mesophyll cells—particularly the spongy layer—are exposed to the air, which can be quite dry. Water now evaporates from mesophyll cells. Evaporation of water from leaf cells is called **transpiration.** At least 90% of the water taken up by the roots is eventually lost by transpiration. This means that the total amount of water lost by a plant over a long period of time is surprisingly large. A single *Zea mays* (corn) plant loses somewhere between 135 and 200 liters of water through transpiration during a growing season.

The water molecules that evaporate are replaced by other water molecules from the leaf veins. In this way, transpiration exerts a driving force—that is, a *tension*—which draws the water column up in vessels from the roots to the leaves. As transpiration occurs, the water column is pulled upward, first within the leaf, then from the stem, and finally from the roots.

Tension can reach from the leaves to the root only if the water column is continuous. What happens if the water column within xylem is broken, as by cutting a stem? The water column "snaps back" in the xylem vessel, away from the site of breakage, making it more difficult for conduction to occur. This is why it is best to maximize water conduction by cutting flower stems under water. This effect has also allowed

investigators to measure the tension in stems. A device called the pressure bomb measures how much pressure it takes to push the xylem sap back to the cut surface of the stem.

There is an important consequence to the way water is transported in plants. When a plant is under water stress, the stomata close. Now the plant loses little water because the leaves are protected against water loss by the waxy cuticle of the upper and lower epidermis. When stomata are closed, however, carbon dioxide cannot enter the leaves, and plants are unable to photosynthesize. (CAM plants are a notable exception.) Photosynthesis, therefore, requires an abundant supply of water so that the stomata remain open and allow carbon dioxide to enter. Next, we discuss how the opening and closing of stomata is regulated.

> Water is transported from the roots to the leaves in xylem. If the water column remains intact, transpiration pulls water from the roots to the leaves.

Opening and Closing of Stomata

As we learned in section 9.3, the leaf epidermis contains stomata. Each stoma, a small pore in leaf epidermis, is bordered by guard cells. When water enters the guard cells and turgor pressure increases, the stoma opens; when water exits the guard cells and turgor pressure decreases, the stoma closes. Notice in Figure 9.25 that the guard cells are attached to each other at their ends and that the inner walls are thicker than the outer walls. When water enters, a guard cell's radial

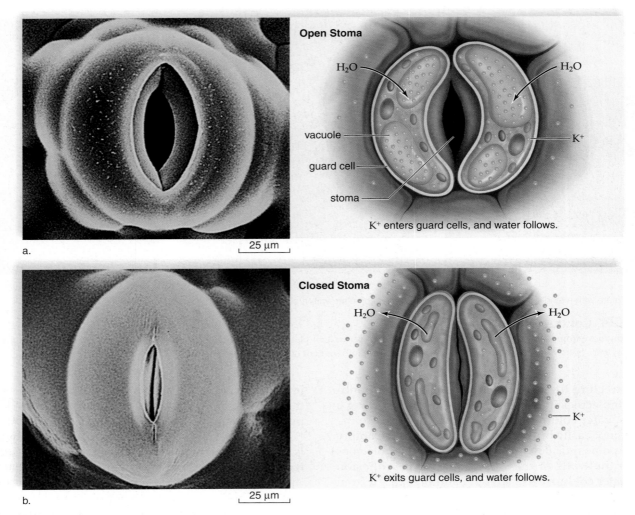

Figure 9.25   Opening and closing of stomata.
a. A stoma opens when turgor pressure increases in guard cells due to the entrance of K$^+$ followed by the entrance of water. **b.** A stoma closes when turgor pressure decreases due to the exit of K$^+$ followed by the exit of water.

expansion is restricted because of cellulose microfibrils in the walls, but lengthwise expansion of the outer walls is possible. When the outer walls expand lengthwise, they buckle out from the region of their attachment, and the stoma opens.

Since about 1968, it has been clear that potassium ions (K$^+$) accumulate within guard cells when stomata open. In other words, active transport of K$^+$ into guard cells causes water to follow by osmosis and stomata to open. Also interesting is the observation that hydrogen ions (H$^+$) accumulate outside guard cells as K$^+$ moves into them. A proton pump run by the hydrolysis of ATP transports H$^+$ to the outside of the cell. This establishes an electrochemical gradient that allows K$^+$ to enter by way of a channel protein.

What regulates the opening and closing of stomata? It appears that the blue-light component of sunlight is a signal that can cause stomata to open. Evidence suggests that a flavin pigment absorbs blue light, and then this pigment sets in motion the cytoplasmic response that leads to activation of the proton pump. Similarly, there could be a receptor in the plasma membrane of guard cells that brings about inactivation of the pump when car-

bon dioxide (CO$_2$) concentration rises, as might happen when photosynthesis ceases. Abscisic acid (ABA), which accumulates in cells of wilting leaves, can also cause stomata to close (see Chapter 10). Although photosynthesis cannot occur, water is conserved.

If plants are kept in the dark, stomata open and close about every 24 hours, just as if they were responding to the presence of sunlight in the daytime and the absence of sunlight at night. This means that some sort of internal biological clock must be keeping time. Circadian rhythms (behaviors that occur nearly every 24 hours) and biological clocks are areas of intense investigation at this time. Other factors that influence the opening and closing of stomata include temperature, humidity, and stress.

When stomata open, first K$^+$ and then water enter guard cells. Stomata open and close in response to environmental signals; the exact mechanism is being investigated.

a. Root nodule

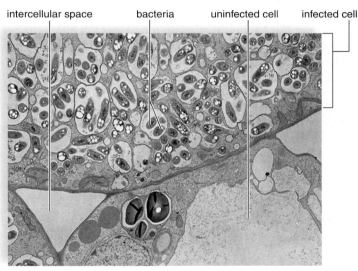

b. Cross section of nodule

Figure 9.26 Root nodules.

a. Nitrogen-fixing bacteria live in nodules on the roots of plants, particularly legumes. **b.** In infected nodule cells, bacteria fix atmospheric nitrogen and make reduced nitrogen available to a plant. The plant passes carbohydrates to the bacteria.

Mineral Uptake and Transport

Some plants have poorly developed roots or no roots at all because minerals and water are supplied by other mechanisms. **Epiphytes** are "air plants"; they do not grow in soil but on larger plants, which give them support, but no nutrients. Some epiphytes have roots that absorb moisture from the atmosphere, and many catch rain and minerals in special pockets at the base of their leaves. Parasitic plants, such as dodders, broomrapes, and pinedrops, send out root-like projections called haustoria that tap into the xylem and phloem of the host stem (see Fig. 9.22e). Carnivorous plants, such as the Venus flytrap and the sundew, obtain some nitrogen and minerals when their leaves capture and digest insects (see Fig. 9.10c).

Two symbiotic relationships assist roots in taking up mineral nutrients. For example, legumes (soybeans and alfalfa) have roots infected by nitrogen-fixing *Rhizobium* bacteria. These bacteria can fix atmospheric nitrogen (N_2) by breaking the $N \equiv N$ bond and reducing nitrogen to NH_4^+ for incorporation into organic compounds. The bacteria live in root nodules and are supplied with carbohydrates by the host plant (Fig. 9.26). The bacteria, in turn, furnish their host with nitrogen compounds.

The second type of symbiotic relationship, called a mycorrhizal association, involves fungi and almost all plant roots (Fig. 9.27). Only a small minority of plants do not have mycorrhizae, sometimes called fungus roots. Ectomycorrhizae form a mantle that is exterior to the root, and they grow between cell walls. Endomycorrhizae can penetrate cell walls. In any case, the fungus increases the surface area available for mineral and water uptake and breaks down organic matter, releasing nutrients the plant can use. In return, the root furnishes the fungus with sugars and amino acids. Plants are extremely dependent on mycorrhizae. Orchid seeds, which are quite small and contain limited nutrients, do not germinate until a mycorrhizal fungus has invaded their cells. Nonphotosynthetic plants, such as Indian pipe, use their mycorrhizae to extract nutrients from

Figure 9.27 Mycorrhizae.

Left: Experimental results show that plants grown with mycorrhizae (two plants on right) grow much larger than a plant (left) grown without mycorrhizae. *Right:* Ectomycorrhizae on red pine roots.

nearby trees. Plants without mycorrhizae are usually limited as to the environment in which they can grow.

> Mineral uptake follows the same pathway as water uptake; however, mineral uptake requires an expenditure of energy. Most plants are assisted in acquiring minerals by a symbiotic relationship with microorganisms and/or fungi.

Organic Nutrient Transport

Not only do plants transport water and minerals from the roots to the leaves, but they also transport organic nutrients to the parts of plants that need them. This includes young leaves that have not yet reached their full photosynthetic potential, flowers that are in the process of making seeds and fruits, and the roots, whose location in the soil prohibits them from carrying on photosynthesis.

a. An aphid feeding on a plant stem

Figure 9.28
Acquiring phloem sap.
Aphids are small insects that remove nutrients from phloem by means of a needlelike mouthpart called a stylet. **a.** Excess phloem sap appears as a droplet after passing through the aphid's body. **b.** Micrograph of a stylet in plant tissue. When an aphid is cut away from its stylet, phloem sap becomes available for collection and analysis.

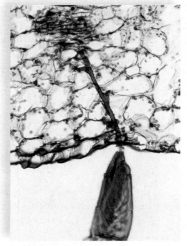

b. Aphid stylet in place

Role of Phloem

As long ago as 1679, Marcello Malpighi suggested that bark is involved in translocating sugars from leaves to roots. He observed the results of removing a strip of bark from around a tree, a procedure called **girdling.** If a tree is girdled below the level of the majority of its leaves, the bark swells just above the cut, and sugar accumulates in the swollen tissue. We know today that when a tree is girdled, the phloem is removed, but the xylem is left intact. Therefore, the results of girdling suggest that phloem is the tissue that transports sugars.

Radioactive tracer studies with carbon 14 (^{14}C) have confirmed that phloem transports organic nutrients. When ^{14}C-labeled carbon dioxide (CO_2) is supplied to mature leaves, radioactively labeled sugar is soon found moving down the stem into the roots. It's difficult to get samples of sap from phloem without injuring the phloem, but this problem is solved by using aphids, small insects that are phloem feeders. The aphid drives its stylet, a sharp mouthpart that functions like a hypodermic needle, between the epidermal cells, and sap enters its body from a sieve-tube member (Fig. 9.28). If the aphid is anesthetized using ether, its body can be carefully cut away, leaving the stylet. Phloem can then be collected and analyzed. The use of radioactive tracers and aphids has revealed that the movement of nutrients through phloem can be as fast as 60–100 cm per hour and possibly up to 300 cm per hour.

Pressure-Flow Model of Phloem Transport

The **pressure-flow model** is the current explanation for the movement of organic materials in phloem (Fig. 9.29). Consider the following experiment in which two bulbs are connected by a glass tube. The left-hand bulb contains solute at a higher concentration than the right-hand bulb. Each bulb is bounded by a differentially permeable membrane, and the entire apparatus is submerged in distilled water:

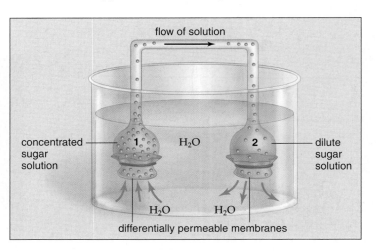

flow of solution

concentrated sugar solution

H_2O

dilute sugar solution

H_2O H_2O

differentially permeable membranes

Distilled water flows into the left-hand bulb to a greater extent because it has the higher solute concentration. The entrance of water creates a positive *pressure*, and

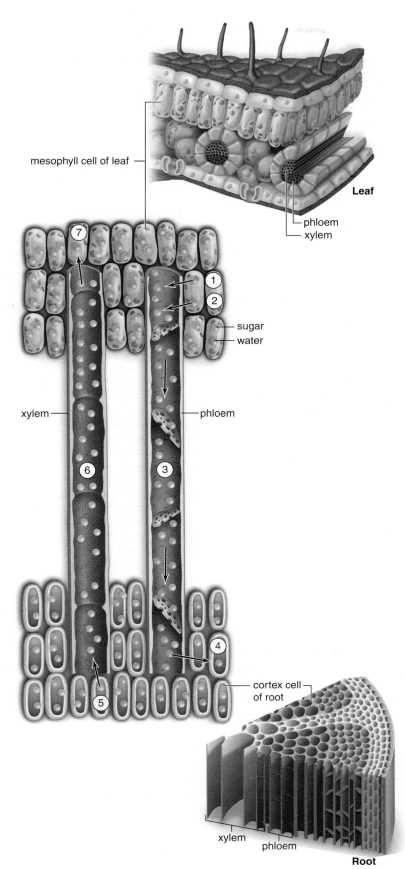

mesophyll cell of leaf

Leaf

phloem
xylem

xylem

phloem

sugar
water

cortex cell
of root

xylem   phloem

Root

Figure 9.29 Pressure-flow model of phloem transport.

At a source, ① sugar is actively transported into sieve tubes. ② Water follows by osmosis. ③ A positive pressure causes phloem contents to flow from the source to a sink. At a sink, ④ sugar is actively transported out of sieve tubes, and cells use it for cellular respiration. Water exits by osmosis. ⑤ Some water returns to the xylem, where it mixes with more water absorbed from the soil. ⑥ Xylem transports water to the mesophyll of the leaf. ⑦ Most of the water is transpired, but some is used for photosynthesis, and some reenters the phloem by osmosis.

water *flows* toward the second bulb. This flow not only drives water toward the second bulb, but it also provides enough force for water to move out through the membrane of the second bulb—even though the second bulb contains a higher concentration of solute than the distilled water.

In plants, sieve tubes are analogous to the glass tube that connects the two bulbs. Sieve tubes are composed of sieve-tube members, each of which has a companion cell. It is possible that the companion cells assist the sieve-tube members in some way. The sieve-tube members align end to end, and strands of plasmodesmata (cytoplasm) extend through sieve plates from one sieve-tube member to the other. Sieve tubes, therefore, form a continuous pathway for organic nutrient transport throughout a plant.

During the growing season, photosynthesizing leaves are producing sugar. Therefore, they are a **source** of sugar. This sugar is actively transported into phloem. Again, transport is dependent on an electrochemical gradient established by a proton pump, a form of active transport. Sugar is carried across the membrane in conjunction with hydrogen ions (H^+), which are moving down their concentration gradient. After sugar enters sieve tubes, water follows passively by osmosis. The buildup of water within sieve tubes creates the positive pressure that starts a flow of phloem contents. The roots (and other growth areas) are a **sink** for sugar, meaning that they are removing sugar and using it for cellular respiration. After sugar is actively transported out of sieve tubes, water exits phloem passively by osmosis (which reduces pressure at the sink) and is taken up by xylem, which transports water to leaves, where it is used for photosynthesis. Now, phloem contents continue to flow from the leaves (source) to the roots (sink).

The pressure-flow model of phloem transport can account for any direction of flow in sieve tubes if we consider that the direction of flow is always from source to sink. For example, recently formed leaves can be a sink, and they will receive sucrose until they begin to maximally photosynthesize.

The pressure-flow model of phloem transport suggests that phloem contents can move either up or down as appropriate for the plant at a particular time in its life cycle.

Summarizing the Concepts

9.1 Plant Organs
A flowering plant has three vegetative organs. Roots anchor a plant, absorb water and minerals, and store the products of photosynthesis. Stems support leaves, conduct materials to and from roots and leaves, and help store plant products. Leaves are specialized for gas exchange, and they carry on most of the photosynthesis in the plant.

9.2 Monocot versus Eudicot Plants
Flowering plants are divided into monocots and eudicots according to the number of cotyledons in the seed; the arrangement of vascular tissue in roots, stems, and leaves; and the number of flower parts.

9.3 Plant Tissues
Flowering plants have apical meristem plus three types of primary meristem. Protoderm produces epidermal tissue. In the roots, epidermal cells bear root hairs. In the leaves, the epidermis contains guard cells. In a woody stem, epidermis is replaced by periderm.

Ground meristem produces ground tissue. Ground tissue is composed of parenchyma cells, which are thin-walled and capable of photosynthesis when they contain chloroplasts. Collenchyma cells have thicker walls for flexible support. Sclerenchyma cells are hollow, nonliving support cells with secondary walls fortified by lignin.

Procambium produces vascular tissue. Vascular tissue consists of xylem and phloem. Xylem contains two types of conducting cells: vessel elements and tracheids. Vessel elements, which are larger and have perforation plates, form a continuous pipeline from the roots to the leaves. In elongated tracheids with tapered ends, water must move through pits in end walls and side walls. Xylem transports water and minerals. In phloem, sieve tubes are composed of sieve-tube members, each of which has companion cell. Phloem transports sugar and other organic compounds, including hormones.

9.4 Organization of Leaves
The bulk of a leaf is mesophyll tissue bordered by an upper and lower layer of epidermis. The epidermis is covered by a cuticle and may bear trichomes. Stomata tend to be in the lower layer. Vascular tissue is present within leaf veins. Leaves are diverse. The spines of a cactus are leaves. Other succulents have fleshy leaves. An onion is a bulb with fleshy leaves, and the tendrils of peas are leaves. The Venus flytrap has leaves that trap and digest insects.

9.5 Organization of Stems
The activity of the shoot apical meristem within a terminal bud accounts for the primary growth of a stem. A terminal bud contains internodes and leaf primordia at the nodes. When stems grow, the internodes lengthen.

In a cross section of a nonwoody eudicot stem, epidermis is the outermost layer of cells, followed by cortex tissue, vascular bundles in a ring, and an inner pith. Monocot stems have scattered vascular bundles, and the cortex and pith are not well defined.

Secondary growth of a woody stem is due to activities of the vascular cambium, which produces new xylem and phloem every year, and cork cambium, which produces new cork cells when needed. Cork, a part of the bark, replaces epidermis in woody plants. In a cross section of a woody stem, the bark is composed of all the tissues outside the vascular cambium: secondary phloem, cork cambium, phelloderm, and cork. Wood consists of secondary xylem, which builds up year after year and forms annual rings.

Stems are diverse. Modified stems include horizontal aboveground and underground stems, corms, and some tendrils.

9.6 Organization of Roots
A root tip has a zone of cell division (containing the primary meristems), a zone of elongation, and a zone of maturation.

A cross section of a herbaceous eudicot root reveals the epidermis, which protects; the cortex, which stores food; the endodermis, which regulates the movement of minerals; and the vascular cylinder, which is composed of vascular tissue. In the vascular cylinder of a eudicot, the xylem appears star-shaped, and the phloem is found in separate regions, between the arms of the xylem. In contrast, a monocot root has a ring of vascular tissue with alternating bundles of xylem and phloem surrounding the pith.

Roots are diversified. Taproots are specialized to store the products of photosynthesis; a fibrous root system covers a wider area. Prop roots are adventitious roots specialized to provide increased anchorage.

9.7 Uptake and Transport of Nutrients
Water transport in plants occurs within xylem. The cohesion-tension model of xylem transport states that transpiration (evaporation of water at stomata) creates tension, which pulls water upward in xylem. This method works only because water molecules are cohesive. Transpiration and carbon dioxide uptake occur when stomata are open. Stomata open when guard cells take up potassium (K^+) ions and water follows by osmosis. Stomata open because the entrance of water causes the guard cells to buckle out.

Plants often expend energy to concentrate minerals, but sometimes they are assisted in mineral uptake by specific adaptations—formation of root nodules or a mycorrhizal association.

Transport of organic nutrients in plants occurs within phloem. The pressure-flow model of phloem transport states that sugar is actively transported into phloem at a source, and water follows by osmosis. The resulting increase in pressure creates a flow, which moves water and sugar to a sink.

Testing Yourself

Choose the best answer for each question.

1. _____ forms the outer protective covering of a plant.
   a. Ground tissue   c. Epidermal tissue
   b. Root tissue   d. Vascular tissue

2. Which of these is an incorrect contrast between monocots (stated first) and eudicots (stated second)?
   a. one cotyledon—two cotyledons
   b. leaf veins parallel—net veined
   c. vascular bundles in a ring—vascular bundles scattered
   d. flower parts in threes—flower parts in fours or fives
   e. All of these are correct contrasts.

3. Vegetative organs ensure _____ survival, and reproductive organs ensure _____ survival.
   a. species, individual   c. species, species
   b. individual, species   d. individual, individual

4. Which of these types of cells is most likely to divide?

   a. parenchyma
   d. xylem
   b. meristem
   e. sclerenchyma
   c. epidermis

5. Primary tissues are associated with _____, and secondary tissues are associated with _____.

   a. apical meristem, apical meristem
   b. apical meristem, lateral meristem
   c. lateral meristem, lateral meristem
   d. lateral meristem, apical meristem

6. Which of these cells in a plant is apt to be nonliving?

   a. parenchyma
   d. epidermal
   b. collenchyma
   e. guard cells
   c. sclerenchyma

7. Root hairs are found in the zone of

   a. cell division.
   d. apical meristem.
   b. elongation.
   e. All of these are correct.
   c. maturation.

8. Label this root using these terms: endodermis, phloem, xylem, cortex, and epidermis.

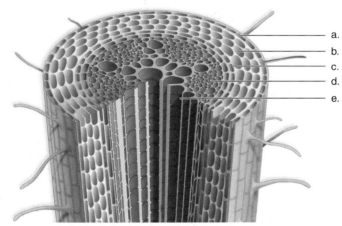

a.
b.
c.
d.
e.

9. Label this leaf using these terms: leaf vein, lower epidermis, palisade mesophyll, spongy mesophyll, and upper epidermis.

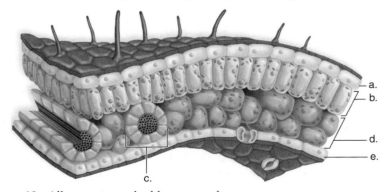

a.
b.
d.
e.
c.

10. All _____ trees shed leaves yearly.

   a. herbaceous
   c. perennial
   b. evergreen
   d. deciduous

11. Insect-eating plants are common in regions where _____ is in limited supply.

   a. water
   c. nitrogen
   b. CO_2
   d. oxygen

12. Stomata are open when

   a. guard cells are filled with water.
   b. potassium ions move into the guard cells.
   c. the hydrogen ion pump is working.
   d. All of these are correct.

13. Annual rings are the number of

   a. internodes in a stem.
   b. rings of vascular bundles in a monocot stem.
   c. layers of xylem in a stem.
   d. bark layers in a woody stem.
   e. Both b and c are correct.

14. Bark is made of which of the following?

   a. phloem
   d. cork cambium
   b. cork
   e. All of these are correct.
   c. cortex

15. What role do cohesion and adhesion play in xylem transport?

   a. Like transpiration, they create a tension.
   b. Like root pressure, they create a positive pressure.
   c. Like sugars, they cause water to enter xylem.
   d. They create a continuous water column in xylem.
   e. All of these are correct.

16. The sugar produced by mature leaves moves into sieve tubes by way of _____, while water follows by _____.

   a. osmosis, osmosis
   b. active transport, active transport
   c. osmosis, active transport
   d. active transport, osmosis

17. Which term(s) would apply to all plants?

   a. herbaceous
   c. meristem
   b. bark
   d. All of these are correct.

For questions 18–21, match the definitions with the structure in the key.

Key:

   a. stoma
   c. xylem
   b. node
   d. parenchyma cell

18. Major constituent of ground tissue

19. Region where leaves attach to stem

20. Opens when transpiration occurs

21. Contains tracheids

22. Label the following leaf diagrams.

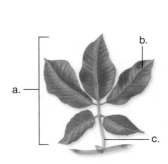

a.
b.
c.
d.

23. The Casparian strip is found
    a. between all epidermal cells.
    b. between xylem and phloem cells.
    c. on four sides of endodermal cells.
    d. within the secondary wall of parenchyma cells.
    e. in both endodermis and pericycle.
24. In a cross section of which of the following structures would you find scattered vascular bundles?
    a. monocot root
    b. monocot stem
    c. eudicot root
    d. eudicot stem
25. Secondary growth would result in which of the following?
    a. new leaves
    b. new shoots
    c. new root
    d. cork, bark, and wood
    e. All but d are correct.
26. Sclerenchyma, parenchyma, and collenchyma are cells found in what type of plant tissue?
    a. epidermal
    b. ground
    c. vascular
    d. meristem
    e. All but d are correct.

Understanding the Terms

adventitious root 164
angiosperms 148
annual ring 158
apical meristem 156
axillary bud 149
bark 158
blade 149
Casparian strip 163
cohesion-tension model 166
collenchyma 152
companion cell 153
cork 151
cork cambium 158
cortex 163
cotyledon 150
cuticle 151
deciduous 149
endodermis 163
epidermal tissue 151
epidermis 151
epiphyte 169
eudicot 150
evergreen 149
fibrous root system 164
girdling 170
ground tissue 151
guard cell 151
herbaceous stem 157
internode 149
leaf 149
leaf vein 153
lenticel 151
lignin 152
meristem 151
mesophyll 154
monocot 150
mycorrhizae 164
node 149
organ 148
palisade mesophyll 154
parenchyma 152
perennial 148
pericycle 163
periderm 151
petiole 149
phloem 149
phytoremediation 165
pith 163
plasmodesmata (sing., plasmodesma) 153
pressure-flow model 170

primary root 164
rhizome 160
root cap 162
root hair 151
root nodule 164
root system 148
sclerenchyma 152
shoot system 148
sieve-tube member 153
sink 171
source 171
spongy mesophyll 154
stem 148
stolon 160
taproot 164
terminal bud 156
tracheid 153
transpiration 167
trichome 151
vascular bundle 153
vascular cambium 156
vascular cylinder 153
vascular tissue 151
vessel element 153
wood 158
xylem 149

Match the terms to these definitions:

a._____Inner, thickest layer of a leaf consisting of palisade and spongy layers; the site of most photosynthesis.
b._____Outer covering of tree bark; made up of dead cells that may be sloughed off.
c._____Extension of a root epidermal cell that increases the surface area for the absorption of water and minerals.
d._____Vascular tissue that conducts organic solutes in plants; contains sieve-tube members and companion cells.
e._____The evaporation of water from leaves.

Thinking Critically

1. Why do you think collenchyma tissue is found toward the outside of the stem rather than in the center of the stem?
2. What is the advantage of having a bundle of xylem cells rather than just one big conduit for water transport?
3. Why are there more vessels in spring wood than in summer wood?
4. Think about a wet environment versus a dry environment. Compare and contrast how a plant would adapt to these very different conditions.
5. What are some strategies that reduce transpiration rates in a plant?
6. Can you think of a situation in which forming a relationship with mycorrhizal fungi might be detrimental for the plant?

ARIS, the *Inquiry into Life* Website

ARIS, the website for *Inquiry into Life,* provides a wealth of information organized and integrated by chapter. You will find practice quizzes, interactive activities, labeling exercises, flashcards, and much more that will complement your learning and understanding of general biology.

www.aris.mhhe.com

It is nice to be able to eat a slice of cold watermelon *on a very hot summer day—without bothering with the seeds. Seedless watermelons, once a novelty, have become a common convenience. But, how are seedless watermelons produced?*

Normally-seeded watermelons have 22 chromosomes—that is, they are diploid (2n). Genetic manipulation of a 2n plant can produce a plant with 44 chromosomes (4n). When a 2n plant (pollen parent) is crossed with a 4n plant (seed parent), the resulting fruit has 3n seeds. The plant that germinates from this 3n seed is sterile, and unable to develop any seeds— giving us the seedless watermelon!

3n seeds must be grown in a field with normal 2n plants in order for seedless watermelons to develop. The 2n plants produce pollen, and honeybees pollinate the flowers of the 3n plants. This pollination is necessary for the correct development of seedless watermelons on the 3n plants.

Although seedless watermelons do not produce viable seeds, they develop small white "seeds." These are actually soft, tasteless seedcoats that can be eaten with the watermelon. Before marketing, the 2n seeded watermelons must be sorted from the 3n seedless watermelons, so the two varieities should look different from each other.

Obviously, the seeds from a 3n seedless watermelon can't be saved to start next year's crop. Each new crop must be produced from new crosses of 2n and 4n plants, and the resulting seeds are expensive and germinate poorly. Much time and energy is needed, all to save us the inconvenience of spitting out watermelon seeds!

chapter 10

Plant Reproduction and Responses

Chapter Concepts

10.1 Sexual Reproduction in Flowering Plants

In contrast to animals, which have only one multicellular stage in their life cycle, all plants have two. Plants exhibit an **alternation of generations** life cycle in which two adult forms alternate in producing each other (Fig. 10.1). A diploid stage alternates with a haploid one. The diploid plant, called the **sporophyte,** produces haploid **spores** by meiosis. The spores divide by mitosis to become haploid **gametophytes** that produce gametes by mitosis. In flowering plants, the sporophyte is dominant, and it is the generation that bears flowers.

A **flower,** which is the reproductive structure of angiosperms, produces two types of spores: microspores and megaspores. A **microspore** develops into a male gametophyte; a **megaspore** develops into a female gametophyte. A microspore undergoes mitosis and becomes a pollen grain. In the meantime, the megaspore has undergone mitosis to become the female gametophyte, an embryo sac located within an ovule that is in turn within an ovary. At maturity,

a pollen grain contains nonflagellated sperm, which travel by way of a pollen tube to the embryo sac. After a sperm fertilizes an egg, the zygote becomes an embryo, still within an ovule. The ovule develops into a **seed,** which contains the embryo and stored food surrounded by a seed coat. The ovary becomes a **fruit,** which aids in dispersing the seeds. When a seed germinates, a new sporophyte emerges and develops into a mature organism.

The life cycle of flowering plants is adapted to a land existence. The microscopic gametophytes develop within the sporophyte and are thereby protected from desiccation. Pollen grains are not released until they develop a thick wall. No external water is needed to bring about fertilization; instead, the pollen tube provides a passage for a sperm. Following fertilization, the seed coat protects the embryo until conditions favor germination.

The flower is unique to angiosperms. Aside from producing the spores and protecting the gametophytes, flowers often attract pollinators. Flowers also produce the fruits that enclose the seeds. The evolution of the flower was a major factor leading to the success of angiosperms, which now number over 240,000 described species.

Flowers

A flower develops in response to environmental signals such as the length of the day (see section 10.4). In many plants, shoot apical meristem suddenly stops producing leaves and starts producing a flower enclosed within a bud. In other plants, axillary buds develop directly into flowers.

A typical flower has four whorls of modified leaves attached to a receptacle at the end of a flower stalk. The receptacle bearing a single flower is attached to a structure called a peduncle, while the structure that bears one of several flowers is called a pedicel. Figure 10.2 shows the following structures of a typical flower:

1. The **sepals,** which are the most leaflike of all the flower parts, are usually green, and they protect the bud as the flower develops within. Sepals can also be the same color as the flower petals.
2. An open flower next has a whorl of **petals,** whose color accounts for the attractiveness of many flowers. The size, shape, and color of petals are attractive to a specific pollinator. Wind-pollinated flowers may have no petals at all.
3. **Stamens** are the "male" portion of the flower. Each stamen has two parts: the **anther,** a saclike container, and the **filament,** a slender stalk. Pollen grains develop from the microspores produced in the anther.
4. At the very center of a flower is the **carpel** (also called the pistil), a vaselike structure that represents the "female" portion of the flower. A carpel usually has three parts: the **stigma,** an enlarged sticky knob; the **style,** a slender stalk; and the **ovary,** an enlarged base that encloses one or more ovules.

In monocots, flower parts occur in threes and multiples of three; in eudicots, flower parts are in fours or fives and multiples of four or five (Fig. 10.3).

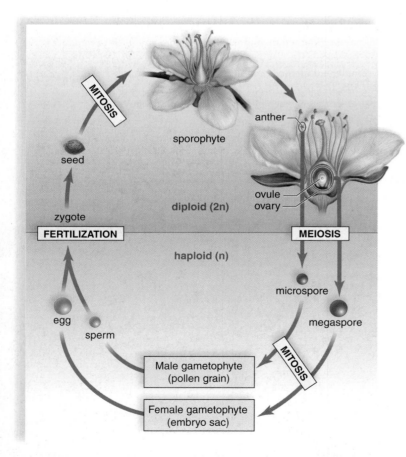

Figure 10.1 Alternation of generations in flowering plants. The sporophyte bears flowers. The flower produces microspores within anthers and megaspores within ovules by meiosis. A megaspore becomes a female gametophyte, which produces an egg within an embryo sac, and a microspore becomes a male gametophyte (pollen grain), which produces sperm. Fertilization results in a seed-enclosed zygote and stored food.

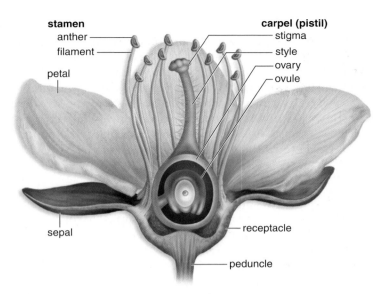

Figure 10.2 Anatomy of a flower. A complete flower has all flower parts: sepals, petals, stamens, and at least one carpel.

A flower can have a single carpel or multiple carpels. Sometimes several carpels are fused into a single structure, in which case this compound ovary has several chambers containing **ovules,** each of which will house a megaspore. For example, an orange develops from a compound ovary, and every section of the orange is a chamber:

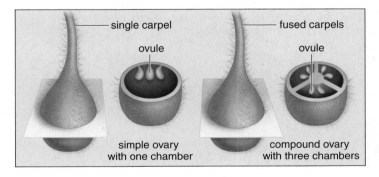

Not all flowers have sepals, petals, stamens, and carpels. Those that do, such as the flower in Figure 10.2, are said to be *complete,* and those that do not are said to be *incomplete.* Flowers that have both stamens and carpels are called perfect (bisexual) flowers; those with only stamens or only carpels are imperfect (unisexual) flowers. If staminate and carpellate (pistillate) flowers are on one plant, the plant is monoecious (one house) (Fig. 10.4). If staminate and carpellate flowers are on separate plants, the plant is dioecious (two houses).

Plants undergo alternation of generations. The flower is the reproductive structure of angiosperms, producing pollen in stamens and eggs in ovaries. Flowers produce fruits that enclose seeds. Flower structure is highly variable.

a. Daylily

b. Festive azalea

Figure 10.3 Monocot versus eudicot flowers.
a. Monocots, such as daylilies, have flower parts usually in threes. In particular, note the three petals and three sepals. **b.** Azaleas are eudicots. They have flower parts in fours or fives; note the five petals of this flower. (p = petal; s = sepal.)

a. Staminate flowers

b. Carpellate flowers

Figure 10.4 Corn plants are monoecious.
A corn plant has **(a)** clusters of staminate flowers and **(b)** clusters of carpellate flowers. Staminate flowers produce the pollen that is carried by wind to the carpellate flowers, where an ear of corn develops.

Life Cycle of Flowering Plants

In plants, the sporophyte produces haploid spores by meiosis. The haploid spores grow and develop into haploid gametophytes, which produce gametes by mitotic division.

Flowering plants are heterosporous, meaning that they produce microspores and megaspores. Microspores become mature male gametophytes (sperm-bearing pollen grains), and megaspores become mature female gametophytes (egg-bearing embryo sacs).

Development of Male Gametophyte

Microspores are produced in the anthers of flowers (Fig. 10.5). An anther has four pollen sacs, each containing many **microsporocytes** (microspore mother cells). A microsporocyte undergoes meiosis to produce four haploid microspores. In each, the haploid nucleus divides mitotically,

followed by unequal cytokinesis, and the result is two cells enclosed by a finely sculptured wall. This structure, called the **pollen grain,** is at first an immature **male gametophyte** that consists of a tube cell and a generative cell. The larger tube cell will eventually produce a *pollen tube.* The smaller generative cell divides mitotically either now or later to produce two sperm. Once these events take place, the pollen grain has become the mature male gametophyte.

Development of Female Gametophyte

The ovary contains one or more ovules. An ovule has a central mass of parenchyma cells almost completely covered by layers of tissue called integuments except where there is an opening, the micropyle. One parenchyma cell enlarges to become a **megasporocyte** (megaspore mother cell), which undergoes meiosis, producing four haploid megaspores (Fig. 10.5). Three of these megaspores are nonfunctional. In a typical pattern, the nucleus of the functional megaspore divides mitotically until there are eight nuclei in the **female gametophyte.** When cell walls form later, there are seven

Figure 10.5 Life cycle of flowering plants.

Development of gametophytes: Four microspores and four megaspores are produced in a pollen sac and an ovule, respectively. The microspores undergo mitosis to become male gametophytes (mature pollen grains). The megaspores also undergo mitosis to produce the female gametophyte (an embryo sac containing seven cells). *Development of sporophyte:* A pollen grain germinates on the stigma. One cell forms the pollen tube, and two sperm cells travel down the tube. During double fertilization, one sperm fertilizes the egg and the other fuses with the polar nuclei to form a triploid endosperm cell.

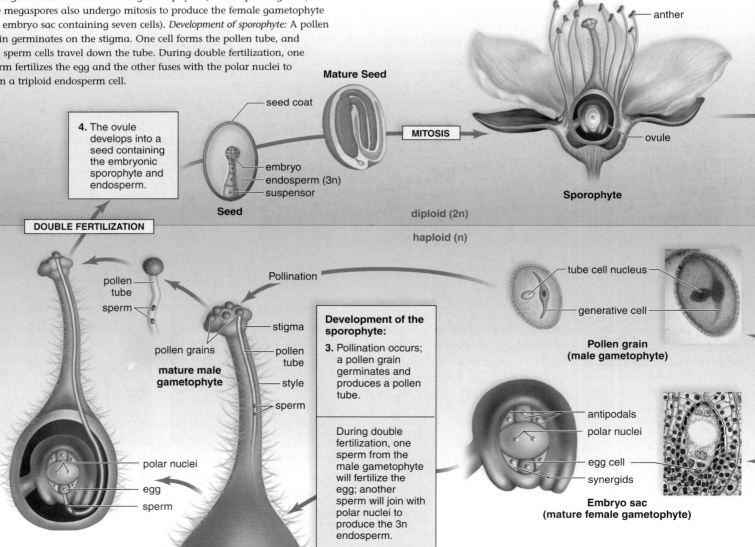

Mature Seed

seed coat

MITOSIS

4. The ovule develops into a seed containing the embryonic sporophyte and endosperm.

embryo
endosperm (3n)
suspensor

Seed

ovule

Sporophyte

anther

diploid (2n)

haploid (n)

DOUBLE FERTILIZATION

pollen tube
sperm

Pollination

tube cell nucleus

generative cell

**Pollen grain
(male gametophyte)**

stigma

pollen grains

pollen tube

**mature male
gametophyte**

style

sperm

**Development of the
sporophyte:**

3. Pollination occurs; a pollen grain germinates and produces a pollen tube.

During double fertilization, one sperm from the male gametophyte will fertilize the egg; another sperm will join with polar nuclei to produce the 3n endosperm.

polar nuclei

egg

sperm

antipodals
polar nuclei

egg cell
synergids

**Embryo sac
(mature female gametophyte)**

cells, one of which is binucleate. The female gametophyte, also called the **embryo sac,** consists of these seven cells:

1 egg cell;
2 synergid cells;
1 central cell containing two polar nuclei; and
3 antipodal cells

Development of Sporophyte

The walls separating the pollen sacs in the anther break down when the pollen grains are ready to be released (Fig. 10.6). **Pollination** is simply the transfer of pollen from an anther to the stigma of a carpel. Self-pollination occurs if the pollen is from the same plant, and cross-pollination occurs if the pollen is from a different plant of the same species.

Angiosperms often have adaptations to foster cross-pollination; for example, the carpels may mature only after the anthers have released their pollen. As discussed in the Ecology Focus on page 184, cross-pollination may also be brought about with the assistance of a particular pollinator.

b. 118 μm

a. c. 8 μm

Figure 10.6 Pollen grains.

a. Cocksfoot grass releasing pollen. **b.** Pollen grains of Canadian goldenrod. **c.** Pollen grains of pussy willow. The shape and pattern of pollen grain walls are quite distinctive, and experts can use them to identify the genus, and sometimes even the species, that produced a particular pollen grain. Pollen grains have strong walls resistant to chemical and mechanical damage; therefore, they frequently become fossils.

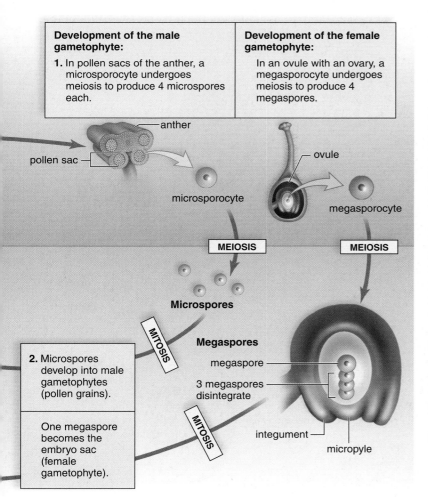

Development of the male gametophyte:	Development of the female gametophyte:
1. In pollen sacs of the anther, a microsporocyte undergoes meiosis to produce 4 microspores each.	In an ovule with an ovary, a megasporocyte undergoes meiosis to produce 4 megaspores.

anther

pollen sac

microsporocyte

ovule

megasporocyte

MEIOSIS

MEIOSIS

MITOSIS

Microspores

Megaspores

megaspore

3 megaspores disintegrate

MITOSIS

integument

micropyle

2. Microspores develop into male gametophytes (pollen grains).

One megaspore becomes the embryo sac (female gametophyte).

If a pollinator goes from flower to flower of only one type of plant, cross-pollination is more likely to occur in an efficient manner. Color, odor, and secretion of nectar are ways that pollinators are attracted to plants. Over time, certain pollinators have become adapted to reach the nectar of only one type of flower. In the process, pollen is inadvertently picked up and taken to another plant of the same type.

When a pollen grain lands on the stigma of the same species, it germinates, forming a pollen tube (see Fig. 10.5). The germinated pollen grain, containing a tube cell and two sperm, is the mature male gametophyte. As it grows, the pollen tube passes between the cells of the stigma and the style to reach the micropyle, a pore of the ovule. Now **double fertilization** occurs. One sperm nucleus unites with the egg nucleus, forming a 2n zygote, and the other sperm nucleus migrates and unites with the polar nuclei of the central cell, forming a 3n endosperm cell. The zygote divides mitotically to become the **embryo,** a young sporophyte, and the endosperm cell divides mitotically to become the endosperm. **Endosperm** is the tissue that will nourish the embryo and seedling as they develop.

Flowering plants are heterosporous. Microspores develop into sperm-bearing mature male gametophytes. A megaspore develops into an egg-bearing mature female gametophyte. During double fertilization, one sperm nucleus unites with the egg nucleus, producing a zygote, and the other sperm nucleus unites with the polar nuclei, forming a 3n endosperm cell.

10.2   Growth and Development

Within a seed, the zygote undergoes development to become an embryo.

Development of the Eudicot Embryo

The endosperm cell, shown in Figure 10.7a divides to produce the endosperm tissue, shown in Figure 10.7b. The zygote also divides, but asymmetrically. One of the resulting cells is small, with dense cytoplasm. This cell is destined to become the embryo, and it divides repeatedly in different planes, forming a ball of cells. The other, larger cell also divides repeatedly, but it forms an elongated structure called a suspensor, which has a basal cell. The suspensor pushes the embryo deep into the endosperm tissue; the suspensor then disintegrates as the seed matures.

During the globular stage, the embryo is a ball of cells (Fig. 10.7c). The root-shoot axis of the embryo is already established at this stage because the embryonic cells near the suspensor will become a root, while those at the other end will ultimately become a shoot.

The embryo has a heart shape when the **cotyledons,** or seed leaves, appear (Fig. 10.7d). As the embryo continues to enlarge and elongate, it takes on a torpedo shape (Fig. 10.7e). Now the root tip and shoot tip are distinguishable. The shoot apical meristem in the shoot tip is responsible for aboveground growth, and the root apical meristem in the root tip is responsible for underground growth. In a mature embryo, the epicotyl is the portion between the cotyledon(s) that contributes to shoot development (Fig. 10.7f). The hypocotyl is that portion below the cotyledon(s) that contributes to stem development; the radicle contributes to root development. Now, the embryo is ready to develop the two main parts of a plant the shoot system and the root system.

The cotyledons are quite noticeable in a eudicot embryo and may fold over. As the embryo develops, the wall of the ovule becomes the seed coat.

Monocot versus Eudicot Embryos

Whereas eudicots have two cotyledons, monocots have only one cotyledon. In monocots, the cotyledon rarely stores food; instead, it absorbs food molecules from the endosperm and passes them to the embryo. In eudicots, the cotyledons

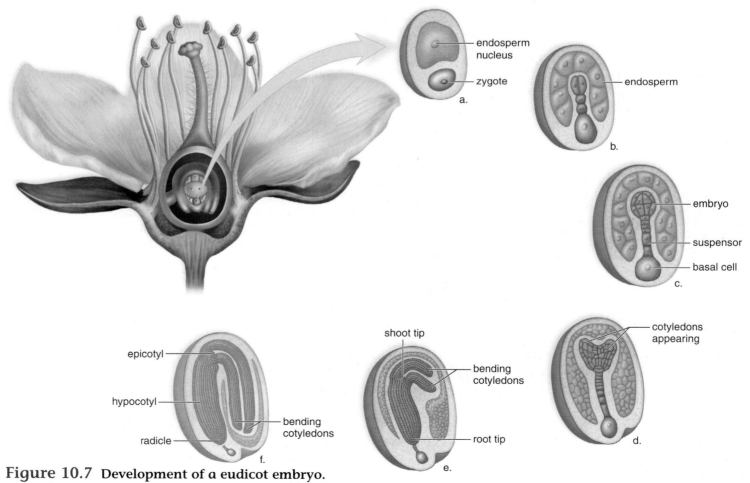

Figure 10.7 Development of a eudicot embryo.

a. The unicellular zygote lies beneath the endosperm nucleus. **b., c.** The endosperm is a mass of tissue surrounding the embryo. The embryo is located above the suspensor. **d.** The embryo becomes heart shaped as the cotyledons begin to appear. **e.** There is progressively less endosperm as the embryo differentiates and enlarges. As the cotyledons bend, the embryo takes on a torpedo shape. **f.** The embryo consists of the epicotyl, the hypocotyl, and the radicle.

usually store the nutrient molecules that the embryo uses. Therefore, the endosperm disappears because it has been taken up by the two cotyledons.

Figures 10.9 and 10.10, on pages 182–183, contrast the structure of a bean seed (eudicot) and a cornkernel (monocot).

The embryo, plus its stored food, is contained within a seed coat.

Fruit Types and Seed Dispersal

A fruit—derived from an ovary and sometimes other flower parts—protects and helps disperse offspring. As a fruit develops, the ovary wall thickens to become the pericarp (Fig. 10.8a). The pericarp can have as many as three layers: exocarp, mesocarp, and endocarp.

Simple fruits are derived from a simple ovary of a single carpel or from a compound ovary of several fused carpels. A pea pod is a fruit; at maturity, the pea pod breaks open on both sides of the pod to release the seeds. Peas and beans, you will recall, are legumes. A *legume* is a fruit that splits along two sides when mature.

Like legumes, cereal grains of wheat, rice, and corn are *dry fruits.* Sometimes, these grainlike fruits are mistaken for seeds because a dry pericarp adheres to the seed within. These dry fruits are indehiscent, meaning that they don't split open. Humans gather grains before they are released from the plant and then process them to acquire their nutrients.

In some simple fruits, mesocarp becomes fleshy. When ripe, *fleshy fruits* often attract animals and provide them with food (Fig. 10.8b). Peaches and cherries are examples of fleshy fruits that have a hard endocarp. This type of endocarp protects the seed so it can pass through the digestive system of an animal and remain unharmed. In a tomato, the entire pericarp is fleshy. If you cut open a tomato, you see several chambers because the flower's carpel is composed of several fused carpels.

Among fruits, apples are an example of an *accessory fruit* because the bulk of the fruit is not from the ovary, but from the receptacle. Only the core of an apple is derived from the ovary. If you cut an apple crosswise, it is obvious that an apple, like a tomato, came from a compound ovary with several chambers.

Compound fruits develop from several individual ovaries. For example, each little part of a raspberry or blackberry is derived from a separate ovary. Because the flower had many separate carpels, the resulting fruit is called an *aggregate fruit* (Fig. 10.8c). The strawberry is also an aggregate fruit, but each ovary becomes a one-seeded fruit called an achene. The flesh of a strawberry is from the receptacle. In contrast, a pineapple comes from many different carpels that belong to separate flowers. As the ovaries mature, they fuse to form a large, *multiple fruit* (Fig. 10.8d).

In flowering plants, the seed develops from the ovule, and the fruit develops from the ovary. Fruits aid dispersal of seeds.

seed covered by pericarp   wing

a.

b.   flesh is from receptacle   one fruit

one fruit

fruits from ovaries of one flower

c.

fruits from ovaries of many flowers

one fruit

d.

Figure 10.8 Structure and function of fruits.
a. Maple trees produce a winged fruit. The wings rotate in the wind and keep the fruit aloft. **b.** Strawberry plants produce an accessory fruit; each "seed" is actually a fruit on a fleshy, expanded receptacle. (A slug is snacking on this strawberry.) **c.** Like strawberries, raspberries are aggregate fruits. Each "berry" is derived from an ovary within the same flower. **d.** A pineapple is a multiple fruit derived from the ovaries of many flowers.

Dispersal of Seeds

For plants to be widely distributed, their seeds have to be dispersed—that is, distributed, preferably long distances, from the parent plant.

Plants have various means of ensuring that dispersal takes place. The hooks and spines of clover, bur, and cocklebur attach to the fur of animals and the clothing of humans. Birds and mammals sometimes eat fruits, including the seeds, which are then defecated (passed out of the digestive tract with the feces) some distance from the parent plant. Squirrels and other animals gather seeds and fruits, which they bury some distance away.

The fruit of the coconut palm, which can be dispersed by ocean currents, may land many hundreds of kilometers away from the parent plant. Some plants have fruits with trapped air or seeds with inflated sacs that help them float in water. Many seeds are dispersed by wind. Woolly hairs, plumes, and wings are all adaptations for this type of dispersal. The seeds of an orchid are so small and light that they need no special adaptation to carry them far away. The somewhat heavier dandelion fruit uses a tiny "parachute"

for dispersal. The winged fruit of a maple tree, which contains two seeds, has been known to travel up to 10 kilometers from its parent (see Fig. 10.8a). A touch-me-not plant has seed pods that swell as they mature. When the pods finally burst, the ripe seeds are hurled out.

Animals, water, and wind help plants disperse their seeds.

Germination of Seeds

Following dispersal, the seeds **germinate**—that is, they begin to grow so that a seedling appears. Some seeds do not germinate until they have been dormant for a period of time. For seeds, **dormancy** is the time during which no growth occurs, even though conditions may be favorable for growth. In the temperate zone, seeds often have to be

Figure 10.9 Common garden bean seed structure and germination.
The common garden bean is a eudicot. **a.** Seed structure. **b.** Germination and development of the seedling.

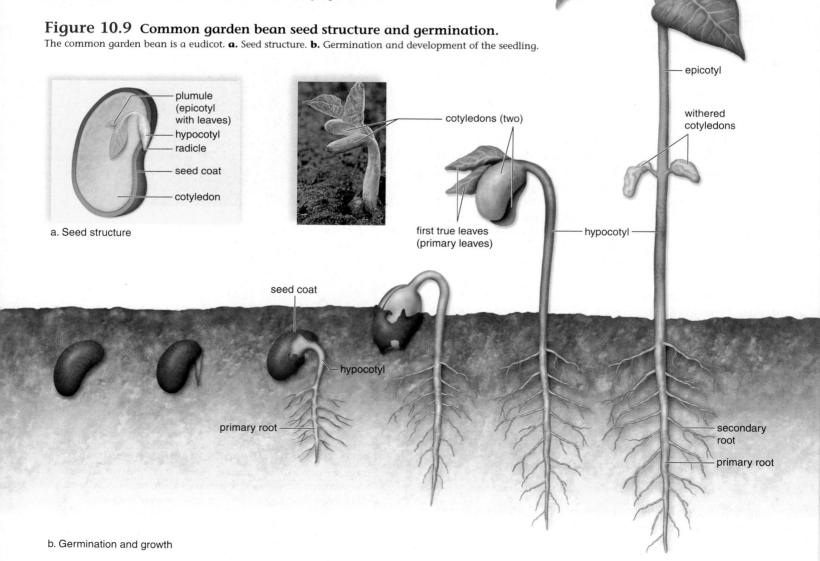

a. Seed structure

b. Germination and growth

exposed to a period of cold weather before dormancy is broken. In deserts, germination does not occur until there is adequate moisture. This requirement helps ensure that seeds do not germinate until the most favorable growing season has arrived. Germination takes place if there is sufficient water, warmth, and oxygen to sustain growth.

Germination requires regulation, and both inhibitors and stimulators are known to exist. Fleshy fruits (e.g., apples, oranges, and tomatoes) contain inhibitors so that germination does not occur until the seeds are removed and washed. In contrast, stimulators are present in the seeds of some temperate-zone woody plants. Mechanical action may also be required. Water, bacterial action, and even fire can act on the seed coat, allowing it to become permeable to water. The uptake of water causes the seed coat to burst.

Eudicot Versus Monocot Germination

As mentioned, the embryo of a eudicot has two seed leaves called cotyledons. The cotyledons, which have absorbed the endosperm, supply nutrients to the embryo and seedling

and eventually shrivel and disappear. If the two cotyledons of a bean seed are parted, you can see a rudimentary plant (Fig. 10.9a). The epicotyl bears young leaves and is called a **plumule.** As the eudicot seedling emerges from the soil, the shoot is hook shaped to protect the delicate plumule. The hypocotyl becomes part of the stem, and the radicle develops into the roots.

A corn plant is a monocot that contains a single cotyledon. In monocots, the endosperm is the food-storage tissue, and the cotyledon does not have a storage role. Corn kernels are actually fruits, and therefore, the outer covering is the pericarp (Fig. 10.10). The plumule and radicle are enclosed in protective sheaths called the coleoptile and the coleorhiza, respectively. The plumule and the radicle burst through these coverings when germination occurs.

Germination is a complex event regulated by many factors. The embryo breaks out of the seed coat and becomes a seedling with leaves, a stem, and roots.

Figure 10.10 Corn kernel structure and germination.
A corn plant is a monocot. **a.** Grain structure. **b.** Germination and development of the seedling.

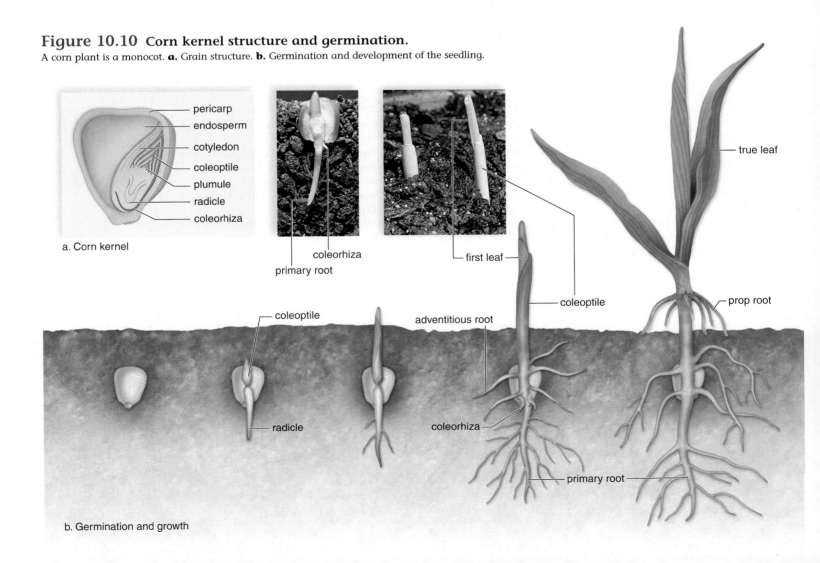

Bees, butterflies, and other pollinators play a critical role in our world. They make possible the fruits and vegetables, and yes, even the chocolate and coffee, we enjoy every day. Many agricultural crops are dependent upon insects to carry their pollen between flowers. Only then can these plants produce a fruit, which becomes food for humans and/or farm animals.

Plants and their pollinators have a symbiotic relationship. In return for pollinator services, the flower provides the pollinator with food, often a sweet liquid called nectar. The flower and the pollinator are suited to one another; floral coloring and odor are agreeable to the pollinator, and the structure of the flower fits its mouthparts. The pollinator is accustomed to foraging at the time of day its specific type of flower is open.

Important pollinators include a variety of insects as well as birds and mammals, particularly bats. The best-known pollinator is the honeybee (Fig. 10A*a*). Beekeepers maintain around three million colonies of the European honeybee in the United States. Wild honeybee colonies also occur. Drs. Roger Morse and Nicholas Calderone of Cornell University calculated that the honeybee is responsible for pollinating food crops valued at $14.6 billion a year. It is also estimated that about one-third of human nutrition is due to bee pollination.

Unlike humans, bees cannot see red, but they can use ultraviolet light to see. Bee-pollinated flowers often have ultraviolet shadings, called nectar guides, which highlight the reproductive portion of the flower. The nectar guides point to a narrow floral tube large enough for the bee's feeding apparatus, but too small for other insects to reach the nectar. Bee-pollinated flowers are sturdy and irregular in shape because they often have a landing platform, where the bee can alight. Here, the bee brushes up against the anther and stigma as it moves toward the floral tube. Pollen clings to its hairy body, and the bee then stores the pollen in pollen baskets on the third pair of its legs. Bees feed pollen to their larvae.

Butterflies have a weak sense of smell, so their flowers tend to be odorless. The flower has bright colors, even red, because butterflies, unlike bees, can see the color red. Unable to hover, butterflies need a place to land. Composite flowers (composed of a compact head of numerous individual flowers) are especially favored by butterflies because these flowers provide a flat landing platform. Each flower has a long, slender floral tube, accessible to the long, thin butterfly proboscis (Fig. 10A*b*).

a.

b.

Figure 10A Two types of pollinators. a. A bee-pollinated flower is a color other than red (bees can't see this color) and has a landing platform where the reproductive structures of the flower brush up against the bee's body. **b.** A butterfly-pollinated flower is often a composite, containing many individual flowers. The broad expanse provides room for the butterfly to land, after which it lowers its proboscis into each flower, in turn.

Today, there are some 240,000 described species of flowering plants and over 900,000 described species of insects. This diversity suggests that the success of angiosperms has contributed to the success of insects, and vice versa.

In recent years, however, the populations of bees and other pollinators have been declining worldwide. Consequently, some plants are endangered because they have lost their normal pollinator. Research suggests that the decline in pollinator populations has been caused by a variety of factors, including pollution, habitat loss, and emerging diseases. Although insecticides are not supposed to be applied to crops that are blooming, they frequently are. And while beekeepers can quickly move their beehives, native wild bees have no protection

whatsoever. Because of indiscriminate pesticide applications, bumblebee populations have seriously declined in cotton-growing areas.

Then, too, widespread aerial applications for mosquitoes, med-flies, grasshoppers, gypsy moths, and other insects leave no region for wild insect pollinators to reproduce and repopulate. Our present "chemlawn philosophy" decrees that the dandelions and clover favored by bees are weeds and, furthermore, that lawns should be treated with herbicides. Few farms, suburbs, and cities provide a habitat where bees and butterflies like to live! Migratory pollinators, such as monarch butterflies and some hummingbirds, are also threatened, because their nectar corridors no longer exist.

Two serious emerging diseases of honeybees are caused by mites. The tracheal mite, a stowaway from Europe, lives in the breathing tubes of adult honeybees and sucks their blood, causing adult bees to become disoriented and weak. The other mite, called the varoa mite, from Asia, is an external parasite that feeds on pupae, larvae, and adult stages, causing the death of entire colonies.

Today, professional farmers rent bee colonies and have them trucked long distance to ensure crop pollination. Small-scale farmers and backyard gardeners can only expect to see smaller yields and smaller, lower-quality fruits and vegetables as a result of the decline in pollinator populations. Keep in mind that pollinators cannot be exchanged on a one-for-one basis, because they are specific to certain plants. Therefore, each species of pollinator should be preserved.

Discussion Questions

1. If a flower is white and emits a strong odor, do you think its pollinator visits the flower during the day or night? Explain your answer.
2. Based on what you know of evolution, what type of insect do you hypothesize might have been the first to pollinate flowers?
3. As a citizen, what could you do to better ensure the survival of important pollinator species such as bees and butterflies?
4. You discover a new species of insect. It is rather large (weighs about the same as a small rabbit), has a hairy head, a very long, curved tongue, and cannot fly. It is more active during the day. What type of flower would it pollinate? Describe as fully as possible the flower's characteristics.

10.3   Asexual Reproduction

Because plants contain nondifferentiated meristem tissue, they routinely reproduce asexually by vegetative propagation. In asexual reproduction, there is only one parent, instead of two as in sexual reproduction. For example, complete strawberry plants will grow from the nodes of stolons (see Fig. 9.18a) and violets will grow from the nodes of rhizomes (see Fig. 9.18b). White potatoes are actually portions of underground stems, and each eye is a bud that will produce a new potato plant if it is planted with a portion of the swollen tuber. Sweet potatoes are modified roots; they can be propagated by planting sections of the root. You may have noticed that the roots of some fruit trees, such as cherry and apple trees, produce "suckers," small plants that can grow into new trees.

Today, seedless fruits are sometimes available in supermarkets. How can they be reproduced? Many types of plants are now propagated from stem cuttings. The discovery that the plant hormone auxin can cause roots to develop from a stem, as discussed in section 10.4, has expanded the list of plants that can be propagated from stem cuttings.

Propagation of Plants in Tissue Culture

Tissue culture is the growth of a *tissue* in an artificial liquid or solid *culture* medium. Tissue culture now allows botanists to breed a large number of plants from somatic tissue and to select any that have superior hereditary characteristics. These and any plants with desired genotypes can be propagated rapidly in tissue culture.

Plant cells are **totipotent,** which means that from most plant cells, an entire plant can be produced. The only exceptions are the plant cells that lose their nuclei (sieve-tube members) or that are dead at maturity (xylem, sclerenchyma, and cork cells). Commercial micropropagation methods now produce thousands, even millions, of identical seedlings in a small vessel. One favorite such method is meristem culture. If the correct proportions of hormones are added to a liquid medium, many new shoots will develop from a single shoot tip. When these are removed, more shoots form. Because the shoots are genetically identical, the adult plants that develop from them, called clonal plants, all have the same traits. Another advantage to meristem culture is that meristem, unlike other portions of a plant, is virus-free; therefore, the plants produced are also virus-free. (The presence of plant viruses weakens plants and makes them less productive.)

When mature plant cells, as opposed to meristem cells, are used to grow entire plants, enzymes are often used to digest the cell walls of a small piece of tissue, usually mesophyll tissue from a leaf, and the result is cells without walls, called **protoplasts** (Fig. 10.11a). The protoplasts regenerate a new cell wall (Fig. 10.11b) and begin to divide, forming aggregates of cells and then a callus (Fig. 10.11c, d). These clumps of cells can be manipulated to produce **somatic embryos** (asexually produced embryos) (Fig. 10.11e). Somatic embryos that are encapsulated in a protective hydrated gel (and sometimes called artificial seeds) can be shipped anywhere. It's possible to produce millions of somatic embryos at once in large tanks called bioreactors. This is done for certain vegetables, such as tomato, celery, and asparagus, and for ornamental plants, such as lilies, begonias, and African violets. A mature plant develops from each somatic embryo (Fig. 10.11f). Plants generated from the somatic embryos vary somewhat because of mutations that arise during the production process. These so-called somaclonal variations are another way to produce new plants with desirable traits.

Recall that pollen grains are produced in the anthers of a flower. Anther culture is a technique in which mature anthers

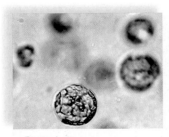

a. Protoplasts

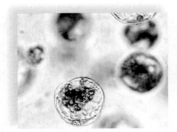

b. Cell wall regeneration

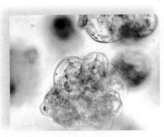

c. Aggregates of cells

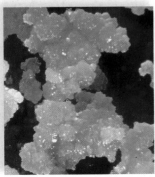

d. Callus (undifferentiated mass)

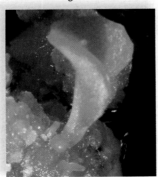

e. Somatic embryo

f. Plantlet

Figure 10.11 Tissue culture of plants.
a. When plant cell walls are removed by digestive enzyme action, the resulting cells are called protoplasts. **b.** Cell walls regenerate, and cell division begins. **c.** Cell division produces an aggregate of cells. **d.** An undifferentiated mass, called a callus, forms. **e.** Somatic cell embryos appear. **f.** The embryos develop into plantlets that can be transferred to soil for growth into adult plants.

are cultured in a medium containing vitamins and growth regulators. The haploid tube cells within the pollen grains divide, producing proembryos consisting of as many as 20 to 40 cells. Finally, the pollen grains rupture, releasing haploid embryos. The experimenter can now generate a haploid plant, or chemical agents can be added that encourage chromosomal doubling. After chromosomal doubling, the resulting plants are diploid, and the homologous chromosomes carry the same genes. Anther culture is a direct way to produce a line of plants whose homologous chromosomes have the same genes.

The culturing of plant tissues has led to a technique called cell suspension culture. Rapidly growing calluses are cut into small pieces and shaken in a liquid nutrient medium so that single cells or small clumps of cells break off and form a suspension. These cells will produce the same chemicals as the entire plant. For example, cell suspension cultures of *Cinchona ledgeriana* produce quinine, and those of *Digitalis lanata* produce digitoxin.

Genetic Engineering of Plants

Traditionally, **hybridization,** the crossing of different varieties of plants or even species, was used to produce plants with desirable traits. Hybridization, followed by vegetative propagation of the mature plants, generated a large number of identical plants with these traits. Today, it is possible to directly alter the genes of organisms and, in that way, produce new varieties with desirable traits.

Tissue Culture and Genetic Engineering

Protoplasts growing in a tissue culture medium can be genetically altered. A foreign gene isolated from any type of organism—plant, animal, or bacteria—is placed in the tissue culture medium. High-voltage electric pulses are then used to create pores in the plasma membrane so that the foreign gene enters the cells. In one such procedure, a gene for the production of the firefly enzyme luciferase was inserted into tobacco protoplasts, and the adult plants glowed when sprayed with the substrate luciferin.

Unfortunately, the regeneration of cereal grains from protoplasts has been difficult. As a result, other methods are used to introduce DNA into plant cells with intact cell walls. Today, it is possible to use a gene gun to bombard a callus (undifferentiated mass of cells) with DNA-coated microscopic metal particles. Then genetically altered somatic embryos develop into genetically altered adult plants.

Many plants, including corn and wheat varieties, have been genetically engineered by these methods. Such plants are called **transgenic plants** (or **genetically modified plants**) because they carry a foreign gene and have new and different traits. Figure 10.12 shows two types of transgenic plants. In another technique, foreign DNA is inserted into the plasmid of the bacterium *Agrobacterium,* which normally infects plant cells. The plasmid then contains *recombinant DNA* because it has genes from different sources, namely those of the plasmid and also the foreign genes of interest. When a bacterium with a recombinant plasmid infects a plant, the recombinant plasmid enters the cells of the plant and can incorporate into the plant's DNA.

Agricultural Plants with Improved Traits

Corn and cotton plants, in addition to soybean and potato plants, have been engineered to be resistant to either herbicides or insect pests. Some corn and cotton plants have been developed that are both insect and herbicide resistant. According to the not-for-profit group International Service for the Acquisition of Agri-biotech Applications (ISAAA), the global planted area of biotech crops has increased by more than 50-fold, from 4.2 million acres in six countries since 1996 to 222 million acres in 21 countries

a. Herbicide-resistant soybean plant   b. Nonresistant potato plant   c. Pest-resistant potato plant

Figure 10.12 Transgenic plants. a. These soybean plants have been given a gene that causes them to be resistant to a particular herbicide. **b.** This potato plant is not resistant to a pest and does poorly. **c.** This potato plant is resistant to a pest and grows beautifully.

Transgenic Crops of the Future	
Improved Agricultural Traits	
Herbicide resistant	Wheat, rice, sugar beets, canola
Salt tolerant	Cereals, rice, sugarcane, canola
Drought tolerant	Cereals, rice, sugarcane
Cold tolerant	Cereals, rice, sugarcane
Improved yield	Cereals, rice, corn, cotton
Modified wood pulp	Trees
Disease protected	Wheat, corn, potatoes
Improved Food-Quality Traits	
Fatty acid/oil content	Corn, soybeans
Protein/starch content	Cereals, potatoes, soybeans, rice, corn
Amino acid content	Corn, soybeans

a. Desirable traits

b. Salt-intolerant   Salt-tolerant

**Figure 10.13
Transgenic crops of
the future.**
a. Transgenic crops of the future
include those with improved
agricultural or food quality
traits. **b.** A salt-tolerant plant
has been engineered. The plant
to the left does poorly when
watered with a salty solution,
but the engineered plant to the
right is tolerant of the solution.
The development of salt-tolerant
crops would increase food
production in the future.

in 2005. This group predicts that the dramatic growth in commercialization of biotech crops seen during the first decade will be surpassed in the second decade.

If crops are resistant to a broad-spectrum herbicide and weeds are not, then the herbicide can be used to kill the weeds. When herbicide-resistant plants were planted, weeds were easily controlled, less tillage was needed, and soil erosion was minimized. However, wildlife and native plants declined because increased amounts of herbicides were used. Another concern is the potential transfer of herbicide-resistant genes from the modified crop plant to a related wild, "weedy" species through natural hybridization.

Crops with other improved agricultural and food-quality traits are desirable (Fig. 10.13). A salt-tolerant tomato will soon be field tested. First, scientists identified a gene coding for a channel protein that transports Na^+ into a vacuole, preventing it from interfering with plant metabolism. Then the scientists used the gene to engineer tomato plants that overproduce the channel protein. The modified plants thrived when watered with a salty solution. Irrigation, even with fresh water, inevitably leads to salinization of the soil, which reduces crop yields. Salt-tolerant crops would increase yield on such land. Salt- and also drought- and cold-tolerant cereals, rice, and sugarcane might help provide enough food for a growing world population.

Potato blight is the most serious potato disease in the world. About 150 years ago, it was responsible for the Irish potato famine that caused the death of millions of people. By placing a gene from a naturally blight-resistant wild potato into a farmed variety, researchers have now made potato plants that are no longer vulnerable to a range of blight strains.

Some progress has also been made in increasing the food quality of crops. Soybeans have been developed that mainly produce monounsaturated fatty acid, a change that may improve human health. These altered plants also produce acids that can be used as hardeners in paints and plastics. The necessary genes were derived from *Vernonia* and castor bean seeds and were transferred into the soybean DNA.

Other types of genetically engineered plants are also expected to increase productivity. Stomata might be altered to take in more carbon dioxide or lose less water. A team of Japanese scientists is working on introducing the C_4 photosynthetic cycle into rice. Unlike C_3 plants, C_4 plants do well in hot, dry weather (see Chapter 8). These modifications would require a more complete reengineering of plant cells than the single-gene transfers that have been done so far.

Some people have expressed health and environmental concerns regarding the growing of transgenic crops, as discussed in the Bioethical Focus at the end of this chapter.

Commercial Products

Single-gene transfers have allowed plants to produce various products, including human hormones, clotting factors, and antibodies. One type of antibody made by corn can deliver radioisotopes to tumor cells, and another made by soybeans may be developed to treat genital herpes.

The tobacco mosaic virus has been used as a vector to introduce a human gene into adult tobacco plants in the field. (Note that this technology bypasses the need for tissue culture completely.) Tens of grams of α-galactosidase, an enzyme needed for the treatment of a human lysosomal storage disease, were harvested per acre of tobacco plants. And it took only 30 days to get tobacco plants to produce antibodies to treat non-Hodgkin's lymphoma after being sprayed with a genetically engineered virus.

Because plants can reproduce asexually, scientists have developed tissue culture techniques to grow plants from somatic plant tissue. This has led to the development of genetically engineered plants that express new traits to improve their growth in adverse environments and to reduce disease levels, improve nutrient content, and serve as tools for the manufacture of important drugs.

10.4   Control of Growth and Responses

Plants respond to environmental stimuli such as light, gravity, and seasonal changes, usually by a change in their pattern of growth. Hormones are involved in these responses.

Plant Hormones

Plant hormones are small organic molecules produced by the plant that serve as chemical signals between cells and tissues. Currently, the five commonly recognized groups of plant hormones are auxins, gibberellins, cytokinins, abscisic acid, and ethylene. Other chemicals, some of which differ only slightly from the natural hormones, also affect the growth of plants. These and the naturally occurring hormones are sometimes grouped together and called plant growth regulators.

Plant hormones bring about a physiological response in target cells after binding to a specific receptor protein in the plasma membrane. Figure 10.14 shows how the hormone auxin brings about elongation of a plant cell, a necessary step toward differentiation and maturation.

Auxins

The most common naturally occurring **auxin** is indoleacetic acid (IAA). It is produced in shoot apical meristem and is found in young leaves, flowers, and fruits. Therefore, you would expect auxin to affect many aspects of plant growth and development.

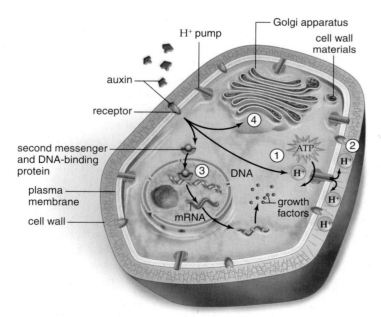

Figure 10.14 Auxin mode of action.

(**1**) The binding of auxin to a receptor stimulates an H⁺ pump, moving hydrogen ions into the cell wall. (**2**) The resulting acidity causes the cell wall to weaken, and water enters the cell. (**3**) At the same time, a second messenger stimulates production of growth factors, and (**4**) new cell wall materials are synthesized. The result of all these actions is that the cell elongates, causing the stem to elongate.

Effects of Auxin Auxin is present in the apical meristem of a plant shoot, where it causes the shoot to grow from the top, a phenomenon called **apical dominance.** Only when the terminal bud is removed, deliberately or accidentally, are the axillary buds able to grow, allowing the plant to branch (Fig. 10.15). Pruning the top (apical meristem) of a plant generally achieves a fuller look because of increased branching of the main body of the plant.

The application of a weak solution of auxin to a woody cutting causes adventitious roots to develop more quickly than they would otherwise. Auxin production by seeds also promotes the growth of fruit. As long as auxin is concentrated in leaves or fruits rather than in the stem, leaves and fruits do not fall off. Therefore, trees can be sprayed with auxin to keep mature fruit from falling to the ground.

How Auxins Work When a plant is exposed to unidirectional light, auxin moves to the shady side, where it binds to receptors and activates an ATP-driven pump that transports hydrogen ions (H⁺) out of the cell (see Fig. 10.14). The acid environment weakens cellulose fibrils, and activated enzymes further degrade the cell wall. Water now enters the cell, and the resulting increase in turgor pressure causes the cell to elongate and the stem to bend toward the light. This growth response to unidirectional light is called **phototropism.** Besides phototropism, auxins are also involved in **gravitropism,** in which roots curve downward and stems curve upward in response to gravity, as discussed later in this section.

Gibberellins

Gibberellins were discovered in 1926 while a Japanese scientist was investigating a fungal disease of rice plants called "foolish seedling disease." The plants elongated too quickly, causing the stem to weaken and the plant to collapse. The fungus infecting the plants produced an excess of a chemical called gibberellin, named after the fungus *Gibberella fujikuroi.* It wasn't until 1956 that gibberellic acid was isolated from a flowering plant rather than from a fungus. Sources of gibberellin in flowering plant parts are young leaves, roots, embryos, seeds, and fruits.

Gibberellins are growth-promoting hormones that bring about elongation of the resulting cells. We know of about 70 gibberellins, and they differ chemically only slightly. The most common of these is gibberellic acid, GA_3 (the subscript designation distinguishes it from other gibberellins).

Effects of Gibberellins When gibberellins are applied externally to plants, the most obvious effect is stem elongation (Fig. 10.16). Gibberellins can cause dwarf plants to grow, cabbage plants to become 2 m tall, and bush beans to become pole beans.

During dormancy a plant does not grow, even though conditions may be favorable for growth. The dormancy of seeds and buds can be broken by applying gibberellins, and research with barley seeds has shown how GA_3 influences the germination of seeds. Barley seeds have a large, starchy endosperm that must be broken down into sugars to provide

a. Plant with apical bud intact   b. Plant with apical bud removed

Figure 10.15 Apical dominance.
a. Auxin in the apical bud inhibits axillary bud development, and the plant exhibits apical dominance. **b.** When the apical bud is removed, lateral branches develop.

energy for the embryo to grow. It is hypothesized that after GA_3 attaches to a receptor in the plasma membrane, calcium ions (Ca^{2+}) combine with a protein. This complex is believed to activate the gene that codes for amylase. Amylase then acts on starch to release the sugars needed for the seed to germinate. Gibberellins have been used to promote barley seed germination for the beer brewing industry.

Cytokinins

Cytokinins were discovered as a result of attempts to grow plant tissue and organs in culture vessels in the 1940s. It was found that cell division occurred when coconut milk (a liquid endosperm) and yeast extract were added to the culture medium. Although the effective agent or agents could not be isolated, they were collectively called cytokinins because, as you may recall, cytokinesis means division of the cytoplasm. A naturally occurring cytokinin was not isolated until 1967. Because it came from the kernels of maize (*Zea*), it was called zeatin.

Effects of Cytokinins The cytokinins promote cell division. These substances, which are derivatives of adenine, one of the purine bases in DNA and RNA, have been isolated from actively dividing tissues of roots and also from seeds and fruits. A synthetic cytokinin, called kinetin, also promotes cell division.

Researchers have found that **senescence** (aging) of leaves can be prevented by applying cytokinins. When a plant organ, such as a leaf, loses its natural color, it is most likely undergoing senescence. During senescence, large molecules within the leaf are broken down and transported to other parts of the plant. Senescence does not always affect the entire plant at once; for example, as some plants grow taller, they naturally lose their lower leaves. Not only can

Figure 10.16 Effect of gibberellins.
The cabbage plants on the right were treated with gibberellins; the cabbage plants on the left were not treated. Gibberellins are often used to promote stem elongation in economically important plants, but the exact mode of action remains unclear.

cytokinins prevent the death of leaves, but they can also initiate leaf growth. Axillary buds begin to grow despite apical dominance when cytokinin is applied to them.

Researchers are well aware that the ratio of auxin to cytokinin and the acidity of the culture medium determine whether a plant tissue forms an undifferentiated mass, called a callus, or differentiates to form roots, vegetative shoots, leaves, or floral shoots. Researchers have reported that chemicals called oligosaccharins (chemical fragments released from the cell wall) are effective in directing differentiation. They hypothesize that reception of auxin and cytokinins, which leads to the activation of enzymes, releases these fragments from the cell wall.

Abscisic Acid

Abscisic acid (ABA) is produced by any "green tissue" with chloroplasts, monocot endosperm, and roots. Abscisic acid was once thought to function in **abscission**, the dropping of leaves, fruits, and flowers from a plant. But even though the external application of abscisic acid promotes abscission, researchers no longer believe this hormone functions naturally in this process. Instead, they think the hormone ethylene, discussed next, brings about abscission.

Effects of Abscisic Acid   Abscisic acid is sometimes called the stress hormone because it initiates and maintains seed and bud dormancy and brings about the closure of stomata when a plant is under water stress (Fig. 10.17). Dormancy has begun when a plant stops growing and prepares for adverse conditions (even though conditions at the time are favorable for growth). For example, researchers believe that abscisic acid moves from leaves to vegetative buds in the fall, and thereafter, these buds are converted to winter buds. A winter bud is covered by thick, hardened scales. A reduction in the level of abscisic acid and an increase in the level of gibberellins are believed to break seed and bud dormancy. Then seeds germinate, and buds send forth leaves.

Ultimately abscisic acid causes potassium ions (K^+) to leave guard cells. Thereafter, the guard cells lose water, and stomata close.

Ethylene

Ethylene is a gas that can move freely in the air. Like the other hormones studied, ethylene works with other hormones to bring about certain effects.

a. No abscission

b. Abscission

c. Ripening

Figure 10.18 **Functions of ethylene.**
a. Normally, there is no abscission when a holly twig is placed under a glass jar for a week. **b.** When an ethylene-producing ripe apple is also under the jar, abscission of the holly leaves occurs. **c.** Similarly, ethylene given off by this one ripe tomato will cause the others to ripen.

Effects of Ethylene   Ethylene is involved in abscission. Low levels of auxin and perhaps gibberellin in the leaf, compared to the stem, probably initiate abscission. But once the process of abscission has begun, ethylene stimulates certain enzymes, such as cellulase, which cause leaf, fruit, or flower drop (Fig. 10.18a,b). Cellulase hydrolyzes cellulose in plant cell walls.

In the early 1990s, it was common practice to prepare citrus fruits for market by placing them in a room with a kerosene stove. Only later did researchers realize that ethylene, an incomplete combustion product of kerosene, was ripening the fruit (Fig. 10.18c). Because it is a gas, ethylene can act from a distance. A barrel of ripening apples can induce the ripening of a bunch of bananas, even if they are in different containers. Ethylene is released at the site of a plant wound due to physical damage or infection (which is why one rotten apple spoils the whole bushel).

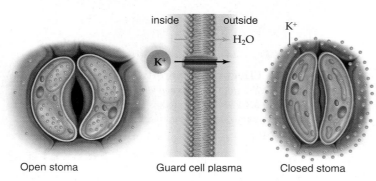

inside   outside
→ H_2O
K^+
K^+

Open stoma   Guard cell plasma   Closed stoma

Figure 10.17 **Abscisic acid control of stoma closing.**
K^+ is concentrated inside the guard cells, and the stoma is open (*left*). With the loss of K^+, water (H_2O) exits the guard cells, and the stoma closes (*right*).

Hormones help control the growth patterns of plants. Auxins, gibberellins, and cytokinins usually stimulate growth. Abscisic acid and ethylene usually inhibit growth.

a.

b.

Figure 10.19 Tropisms.

a. In positive phototropism, the stem of a plant curves toward the light. This response is due to the accumulation of auxin on the shady side of the stem. **b.** In negative gravitropism, the stem of a plant curves away from the direction of gravity 24 hours after the plant was placed on its side. This response is due to the accumulation of auxin on the lower side of the stem.

Plant Responses to Environmental Stimuli

Environmental signals determine the seasonality of growth, reproduction, and dormancy in plants. Plant responses are strongly influenced by such environmental stimuli as light, day length, gravity, and touch. The ability of a plant to respond to environmental signals fosters the survival of the plant and the species in a particular environment.

Plant responses to environmental signals can be rapid, as when stomata open in the presence of light, or they can

take some time, as when a plant flowers in season. Despite their variety, most plant responses to environmental signals are due to growth and sometimes differentiation, brought about at least in part by particular hormones.

Plant Tropisms

Plant growth toward or away from a directional stimulus is called a **tropism.** Tropisms are due to differential growth—one side of an organ elongates faster than the other, and the result is a curving toward or away from the stimulus. The following three well-known tropisms were each named for the stimulus that causes the response:

phototropism	growth in response to a light stimulus
gravitropism	growth in response to gravity
thigmotropism	growth in response to touch

Growth toward a stimulus is called a positive tropism, and growth away from a stimulus is called a negative tropism. For example, in positive phototropism, stems curve toward the light, and in negative gravitropism, stems curve away from the direction of gravity (Fig. 10.19). Roots, of course, exhibit positive gravitropism.

The role of auxin in the positive phototropism of stems has been studied for quite some time. Because blue light, in particular, causes phototropism to occur, researchers believe that a yellow pigment related to the vitamin riboflavin acts as a photoreceptor for light. Following reception, auxin migrates from the bright side to the shady side of a stem. The cells on that side elongate faster than those on the bright side, causing the stem to curve toward the light. Negative gravitropism of stems occurs because auxin moves to the lower part of a stem when a plant is placed on its side. Figure 10.14 explains how auxin brings about elongation.

Flowering

Flowering is a striking response in angiosperms to environmental seasonal changes. In some plants, flowering occurs according to the **photoperiod,** which is the ratio of the length of day to the length of night over a 24-hour period. Plants can be divided into three groups:

1. **Short-day plants** flower when the day length is shorter than a critical length. (Examples are cocklebur, poinsettia, and chrysanthemum.)
2. **Long-day plants** flower when the day length is longer than a critical length. (Examples are wheat, barley, clover, and spinach.)
3. **Day-neutral plants** do not depend on day length for flowering. (Examples are tomato and cucumber.)

Experiments have shown that the length of continuous darkness, not light, controls flowering. For example, the cocklebur does not flower if a suitable length of darkness

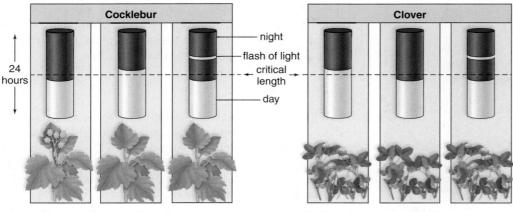

Figure 10.20
Photoperiodism and flowering.
a. Short-day plant. When the day is shorter than a critical length, this type of plant flowers. The plant does not flower when the day is longer than the critical length. It also does not flower if the longer-than-critical-length night is interrupted by a flash of light. **b.** Long-day plant. The plant flowers when the day is longer than a critical length. When the day is shorter than a critical length, this type of plant does not flower. However, it does flower if the slightly longer-than-critical-length night is interrupted by a flash of light.

is interrupted by a flash of light. In contrast, clover does flower when an unsuitable length of darkness is interrupted by a flash of light (Fig. 10.20). (Interrupting the light period with darkness has no effect on flowering.)

> **Plants respond to external stimuli to adapt and survive in their environment. Plants respond to light, gravity, and touch through a growth response called a tropism. To flower, short-day plants require a period of darkness longer than a critical length, and long-day plants require a period of darkness shorter than a critical length.**

Phytochrome and Plant Flowering

If flowering is dependent on day and night length, plants must have some way to detect these periods. This appears to be the role of **phytochrome,** a blue-green leaf pigment that alternately exists in two forms—P_r and P_{fr}:

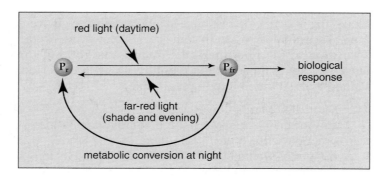

Direct sunlight contains more red light than far-red light; therefore, P_{fr} (far red) is apt to be present in plant leaves during the day. In the shade and at sunset, there is more far-red light than red light; therefore, P_{fr} is converted to P_r (red) as night approaches. There is also a slow metabolic replacement of P_{fr} by P_r during the night.

It is possible that phytochrome conversion is the first step in a signaling pathway that results in flowering. A flowering hormone has never been discovered.

Other Functions of Phytochrome

The $P_r \rightarrow P_{fr}$ conversion cycle has other functions. The presence of P_{fr} indicates to seeds of some plants that sunlight is present and conditions are favorable for germination. Such seeds must be only partly covered with soil when planted. Following germination, the presence of P_r indicates that stem elongation may be needed to reach sunlight. Seedlings that are grown in the dark etiolate—that is, the stem increases in length, and the leaves remain small (Fig. 10.21). Once the seedling is exposed to sunlight and P_r is converted to P_{fr}, the seedling begins to grow normally—the leaves expand, and the stem branches.

> **Phytochrome alternates between two forms—P_{fr} during the day and P_r during the night—and this conversion allows a plant to detect photoperiod changes.**

Figure 10.21
Phytochrome control of growth pattern.
a. If far-red light is prevalent, as it is in the shade, etiolation occurs.
b. If red light is prevalent, as it is in bright sunlight, normal growth occurs. These effects are due to phytochrome.

a. Etiolation   b. Normal growth

Bioethical Focus

Transgenic Crops

Transgenic plants (genetically modified plants) may allow crop yields to keep up with the ever-increasing worldwide demand for food. And some of these plants have the added benefit of requiring less fertilizer and/or pesticides, which are harmful to human health and the environment.

But some scientists believe that transgenic crops pose their own threat to the environment, and many activists believe that transgenic plants are themselves dangerous to our health. Studies have shown that wind-carried pollen can cause transgenic crops to hybridize with nearby weedy relatives. Although it has not happened yet, some fear that characteristics acquired in this way might cause weeds to become uncontrollable pests. Or perhaps a toxin produced by transgenic crops could pos-

sibly hurt other organisms in the field. Many researchers are conducting tests to see if this might occur. Also, although transgenic crops have not caused any illnesses in humans, some scientists concede the possibility that people could be allergic to a transgene's protein product. After unapproved genetically modified corn was detected in Taco Bell taco shells, a massive recall pulled about 2.8 million boxes of the product from grocery stores.

Already, transgenic plants must be approved by the Food and Drug Administration before they are considered safe for human consumption, and they must meet certain Environmental Protection Administration standards. Some people believe safety standards for transgenic crops should be further strengthened, while others fear stricter standards will

result in less food production. Another possibility is to retain the current standards but require clear labeling so buyers can choose whether to eat biotech foods.

Form Your Own Opinion

1. If plant pollen is transferred by an insect rather than by the wind, do you think there would still be a risk of these plants hybridizing with weedy relatives?
2. What are the advantages and disadvantages of genetically engineering a plant versus using artificial breeding methods to develop new traits?
3. If a food label states that the food has been genetically modified, do you think most people would still buy the product? Why or why not?

Summarizing the Concepts

10.1 Sexual Reproduction in Flowering Plants

Flowering plants exhibit an alternation of generations life cycle. Flowers borne by the sporophyte produce microspores and megaspores by meiosis. Microspores develop into a male gametophyte, and megaspores develop into the female gametophyte. The gametophytes produce gametes by mitotic cell division. Following fertilization, the sporophyte is enclosed within a seed covered by fruit.

The flowering plant life cycle is adapted to a land existence. The microscopic gametophytes are protected from desiccation by the sporophyte; the pollen grain has a protective wall; and fertilization does not require external water. The seed has a protective seed coat, and seeds do not germinate until conditions are favorable.

A typical flower has several parts. Sepals form an outer whorl; petals are the next whorl; and stamens form a whorl around the base of at least one carpel. The carpel, in the center of a flower, consists of a stigma, style, and ovary. The ovary contains ovules.

The anthers contain microsporocytes, each of which divides meiotically to produce four haploid microspores. Each of these divides mitotically to produce a two-celled pollen grain. One cell is the tube cell, and the other is the generative cell. The generative cell later divides mitotically to produce two sperm cells. The pollen grain is the male gametophyte. Pollination is simply the transfer of pollen from anther to stigma. After pollination, the pollen grain germinates, and as the pollen tube grows, the sperm cells travel to the embryo sac.

Each ovule contains a megasporocyte, which divides meiotically to produce four haploid megaspores, only one of which survives. This megaspore divides mitotically to produce the female gametophyte (embryo sac), which usually has seven cells.

Flowering plants undergo double fertilization. One sperm nucleus unites with the egg nucleus, forming a 2n zygote, and the other unites with the polar nuclei of the central cell, forming a 3n endosperm cell. The zygote becomes the sporophyte embryo, and the endosperm cell divides to become endosperm tissue.

10.2 Growth and Development

Prior to seed formation, the zygote undergoes growth and development to become an embryo. The seeds enclosed by a fruit contain the embryo (hypocotyl, epicotyl, plumule, radicle) and stored food (endosperm and/or cotyledons). Following dispersal, a seed germinates. The root appears below, and the shoot appears above the seed.

10.3 Asexual Reproduction

Many flowering plants reproduce asexually, as when buds located on stems (either aboveground or underground) give rise to entire plants, or when roots produce new shoots. The ability of plants to reproduce asexually means that they can be propagated in tissue culture. Micropropagation, the production of clonal plants utilizing tissue culture, is now a commercial venture. In cell suspension cultures, plant cells produce chemicals of medical importance. The practice of plant tissue culture facilitates genetic engineering of plants. One aim of genetic engineering is to produce plants that have improved agricultural or food-quality traits.

10.4 Control of Growth and Responses

Plant hormones are chemical signals produced by plants that are involved in growth and responses to environmental signals. Plant hormones bind to receptor proteins, and this leads to physiological changes within the cell. The five commonly recognized groups of plant hormones are auxins, gibberellins, cytokinins, abscisic acid, and ethylene. Auxins cause apical dominance, the growth of adventitious roots, and two types of tropisms—phototropism and gravitropism. Gibberellins promote stem elongation and break seed dormancy so that germination occurs. Cytokinins promote cell division, prevent senescence of leaves, and along with auxin, influence differentiation of plant tissues. Abscisic acid initiates and maintains seed and bud dormancy and brings about the closing of stomata when a plant is water stressed. Ethylene causes abscission of leaves, fruits, and flowers. It also causes some fruits to ripen.

Environmental signals play a significant role in plant growth and development. Tropisms are growth responses toward or away from unidirectional stimuli. When a plant is exposed to light, auxin moves laterally from the bright to the shady side of a stem. Thereafter, the cells on the shady side elongate, and the stem bends toward the light. Similarly, auxin is responsible for the negative gravitropism exhibited by stems. Stems grow upward opposite the direction of gravity.

Flowering is a striking response to seasonal changes. Short-day plants flower when the days are shorter (nights are longer) than a critical length, and long-day plants flower when the days are longer (nights are shorter) than a critical length. Some plants are day-neutral. Phytochrome, a plant pigment that responds to daylight, is believed to be part of a biological clock system that in some unknown way brings about flowering. Phytochrome has various other functions in plant cells, such as seed germination, leaf expansion, and stem branching.

Testing Yourself

Choose the best answer for each question.

1. Stigma is to carpel as anther is to
   a. sepal.
   b. stamen.
   c. ovary.
   d. style.

2. _____ always promotes cell division.
   a. Auxin
   b. Phytochrome
   c. Cytokinin
   d. None of these are correct.

3. The embryo of the plant is stored in the
   a. pollen.
   b. anther.
   c. microspore.
   d. seed.

4. Which is not a plant hormone?
   a. auxin
   b. cytokinin
   c. gibberellin
   d. All of these are plant hormones.

5. The term totipotent means
   a. that each plant cell can become an entire plant.
   b. hormones control all plant growth.
   c. all cells develop from the same tissue.
   d. None of these are correct.

For questions 6–10, match the definitions with the structure in the key.

Key:
   a. flower
   b. megaspore
   c. gibberellin
   d. style
   e. senescence

6. Develops into the female gametophyte

7. Growth hormone that promotes elongation

8. Aging in plants

9. Produces seeds enclosed by fruits

10. Slender stalk of the carpel

11. Nondifferentiated meristem allows for _____ in _____.
    a. elongation, meiosis
    b. sexual reproduction, vegetative propagation
    c. asexual reproduction, tissue culture
    d. asexual reproduction, meiosis

12. Double fertilization refers to the formation of a __ and a __.
    a. zygote, zygote
    b. zygote, pollen grain
    c. zygote, megaspore
    d. zygote, endosperm

13. Plant biotechnology can lead to
    a. increased crop production.
    b. disease-resistant plants.
    c. treatment of human disease.
    d. All of these are correct.

14. Which is the correct order of the following events: (1) megaspore becomes embryo sac, (2) embryo formed, (3) double fertilization, (4) meiosis?
    a. 1, 2, 3, 4
    b. 4, 1, 3, 2
    c. 4, 3, 2, 1
    d. 2, 3, 4, 1

15. The _____ forms leaves, and the _____ forms roots.
    a. radicle, radicle
    b. radicle, plumule
    c. plumule, radicle
    d. plumule, plumule

16. In an accessory fruit, such as an apple, the bulk of the fruit is from the
    a. ovary.
    b. style.
    c. pollen.
    d. None of these are correct.

17. Ethylene stimulates the action of _____ to produce _____.
    a. auxins, root formation
    b. gibberellins, flower formation
    c. cellulase, flower formation
    d. cellulase, flower dropping

18. In the absence of abscisic acid, plants may have difficulty
    a. forming winter buds.
    b. closing the stomata.
    c. Both a and b are correct.
    d. Neither a nor b is correct.

19. Phytochrome plays a role in
    a. flowering.
    b. stem growth.
    c. leaf growth.
    d. All of these are correct.

20. Label the following diagram of a flower.

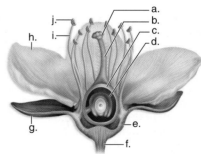

21. Label the following diagram of alternation of generations in flowering plants.

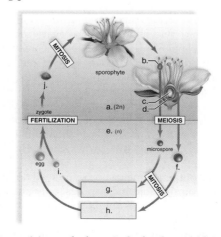

22. A plant requiring a dark period of at least 14 hours will
    a. flower if a 14-hour night is interrupted by a flash of light.
    b. not flower if a 14-hour night is interrupted by a flash of light.
    c. not flower if the days are 14 hours long.
    d. Both b and c are correct.

23. Short-day plants
    a. are the same as long-day plants.
    b. are apt to flower in the fall.
    c. do not have a critical photoperiod.
    d. will not flower if a short day is interrupted by bright light.
    e. All of these are correct.
24. Which of these is a correct statement?
    a. Both stems and roots show positive gravitropism.
    b. Both stems and roots show negative gravitropism.
    c. Only stems show positive gravitropism.
    d. Only roots show positive gravitropism.
25. Which of the following plant hormones causes plants to grow in an upright position?
    a. auxin   d. abscisic acid
    b. gibberellins   e. ethylene
    c. cytokinins
26. Which of the following plant hormones is responsible for a plant losing its leaves?
    a. auxin
    b. gibberellins
    c. cytokinins
    d. abscisic acid
    e. ethylene
27. The function of the flower is to _____, and the function of fruit is to _____.
    a. produce fruit; provide food for humans
    b. aid in seed dispersal; attract pollinators
    c. attract pollinators; assist in seed dispersal
    d. produce the ovule; produce the ovary
28. How is the megaspore in the plant life cycle similar to the microspore? Both
    a. have the diploid number of chromosomes.
    b. become an embryo sac.
    c. become a gametophyte that produces a gamete.
    d. are necessary for seed production.
    e. Both c and d are correct.

Understanding the Terms

abscisic acid (ABA) 190
abscission 190
alternation of generations 176
anther 176
apical dominance 188
auxin 188
carpel 176
cotyledon 180
cytokinin 189
day-neutral plant 191
dormancy 182
double fertilization 179
embryo 179
embryo sac 179
endosperm 179
ethylene 190
female gametophyte 178
filament 176
flower 176
fruit 176
gametophyte 176
genetically modified plant 186
germinate 182
gibberellin 188
gravitropism 188
hybridization 186
long-day plant 191
male gametophyte 178
megaspore 176
megasporocyte 178
microspore 176
microsporocyte 178
ovary 176
ovule 177
petal 176
photoperiod 191
phototropism 188
phytochrome 192
plant hormone 188
plumule 183
pollen grain 178
pollination 179
protoplast 185
seed 176
senescence 189
sepal 176
short-day plant 191
somatic embryo 185
spore 176
sporophyte 176
stamen 176
stigma 176
style 176
tissue culture 185
totipotent 185
transgenic plant 186
tropism 191

Match the terms to these definitions:

a._____ In plants, a growth response toward or away from a directional stimulus.
b._____ In the life cycle of a plant, the haploid generation that produces gametes.
c._____ In flowering plants, portion of the carpel where pollen grains adhere and germinate before fertilization can occur.
d._____ In seed plants, a small spore that develops into the sperm-producing male gametophyte; pollen grain.
e._____ Plant pigment that induces a photoperiodic response in plants.

Thinking Critically

1. Under what environmental conditions would it be advantageous for a plant to carry out asexual reproduction?
2. Why do you think plants have evolved to "recruit" usually just one type of pollinator rather than having many different types of pollinators?
3. You want to grow larger and sweeter grapes. What type of hormone would you spray on the grapevine and why?
4. You bought green bananas at the grocery store this morning. However, you want a ripe banana for breakfast tomorrow morning. What could you do to accomplish this?
5. Humans produce hormones. It is known that exposure to too much of some hormones can lead to cancer. Plants also produce hormones. Why is it not dangerous to eat plants if they produce hormones?
6. If someone offered you unlimited funding to genetically engineer a plant in order to benefit humankind, what idea would you pursue and why?

ARIS, the *Inquiry into Life* Website

ARIS, the website for *Inquiry into Life*, provides a wealth of information organized and integrated by chapter. You will find practice quizzes, interactive activities, labeling exercises, flashcards, and much more that will complement your learning and understanding of general biology.

www.aris.mhhe.com

Maintenance of the Human Body

part **III**

Lance Armstrong overcame cancer to win the Tour de France bicycle race seven times. Prior to being diagnosed with testicular cancer that had spread to his lungs and brain, Lance was one of the world's most highly trained athletes. He had conditioned his body to withstand hours of grueling physical activity. His heart was 30% larger than normal, and his resting heart rate was 32–34 beats per minute, compared to the average 60–100 beats per minute. For any athlete to function at such a high level, the cells, tissues, organs, and organ systems of the body must work together in an extremely coordinated fashion. The increased demand for oxygen by muscle tissues must be balanced by the amount of heat generated by muscles and the amount of fluid lost through sweating. Once cancer struck, Lance had to rely on his immune system to help him fight the cancer cells, and his liver and kidneys to help his normal cells survive the toxic effects of chemotherapy.

Although athletes' incredible feats are a clear demonstration of the human body's ability to deal with extreme circumstances, even as you read this, your body is performing tasks that are, in many ways, just as amazing. Your body temperature is maintained in a relatively narrow range, just as a thermostat-controlled heating and air conditioning system controls the temperature in a house. Your blood pressure, heart rate, breathing rate, and urinary output are all being monitored and maintained for you without any conscious effort. Beyond this, the numbers and types of cells in your blood as well as the concentrations of hundreds of enzymes, electrolytes, and other molecules are being regulated. As a popular T-shirt once read, "It might look like I'm doing nothing, but at the cellular level I'm really quite busy."

chapter **11**

Human Organization

11.1   Types of Tissues

Recall the biological levels of organization described in Chapter 1. Cells are composed of molecules; a tissue has similar types of cells; an organ contains several types of tissues; and several organs make up an organ system. In this chapter, we consider the tissue, organ, and organ system levels of organization.

A **tissue** is composed of similarly specialized cells that perform a common function in the body. The tissues of the human body can be categorized into four major types:

Epithelial tissue covers body surfaces and lines body cavities.
Connective tissue binds and supports body parts.
Muscular tissue moves the body and its parts.
Nervous tissue receives stimuli and conducts nerve impulses.

Cancers are classified according to the type of tissue from which they arise. **Carcinomas,** the most common type, are cancers of epithelial tissue; the Health Focus later in this chapter describes the different types of carcinomas. Sarcomas are cancers arising in muscle or connective tissue, especially bone or cartilage. Leukemias are cancers of the blood cells, and lymphomas are cancers of the lymphoid tissue. The chance of developing cancer in a particular tissue shows a positive correlation to the rate of cell division. Epithelial cells reproduce at a high rate, and 2,500,000 new blood cells appear each second. Therefore, carcinomas and leukemias are common types of cancers.

Epithelial Tissue

Epithelial tissue, also called epithelium, consists of tightly packed cells that form a continuous layer. Epithelial tissue covers surfaces and lines body cavities. Epithelial tissue has numerous functions in the body. Usually, it has a protective function, but it can also be modified to carry out secretion, absorption, excretion, and filtration.

On the external surface, epithelial tissue protects the body from injury, drying out, and possible invasion by microbes such as bacteria and viruses. On internal surfaces, modifications help epithelial tissue carry out both its protective and specific functions. Epithelial tissue secretes mucus along the digestive tract and sweeps up impurities from the lungs by means of **cilia** (sing., **cilium**). It efficiently absorbs molecules from kidney tubules and from the intestine because of minute cellular extensions called **microvilli** (sing., **microvillus**).

A **basement membrane** often joins an epithelium to underlying connective tissue. We now know that the basement membrane consists of glycoprotein secreted by epithelial cells and collagen fibers that belong to the connective tissue.

Epithelial tissue is classified according to the type of cell it is composed of (squamous, cuboidal, or columnar) and the number of layers in the tissue (simple or stratified) (Fig. 11.1). **Squamous epithelium** is composed of flattened cells and is found lining the lungs and blood vessels. **Cuboidal epithelium** contains cube-shaped cells and lines the kidney tubules. **Columnar epithelium** has cells resembling rectangular pillars or columns, with nuclei usually located near the bottom of each cell. This epithelium lines the digestive tract. Ciliated columnar epithelium lines the oviducts, where it propels the egg toward the uterus, or womb.

Based on the number of layers in the tissue, *simple* means the tissue has a single layer of cells, and *stratified* means the tissue has layers of cells piled one on top of another. The walls of the smallest blood vessels, called **capillaries,** are composed of simple squamous epithelium. The permeability of capillaries allows exchange of substances between the blood and tissue cells. Simple cuboidal epithelium lines kidney tubules and the cavities of many internal organs. Stratified squamous epithelium lines the nose, mouth, esophagus, anal canal, and vagina. As you will see, the outer layer of skin is also stratified squamous epithelium, but the cells have been reinforced by keratin, a protein that provides strength.

When an epithelium is *pseudostratified*, it appears to be layered, but true layers do not exist because each cell touches the basement membrane. The lining of the windpipe, or trachea, is pseudostratified ciliated columnar epithelium. A secreted covering of mucus traps foreign particles, and the upward motion of the cilia carries the mucus to the back of the throat, where it may either be swallowed or expelled. Smoking can cause a change in mucous secretion and inhibit ciliary action, and the result is a chronic inflammatory condition called bronchitis.

When an epithelium secretes a product, it is said to be glandular. A **gland** can be a single epithelial cell, as are the mucus-secreting goblet cells within the columnar epithelium lining the digestive tract, or a gland can contain many cells. Glands that secrete their product into ducts are called exocrine glands, and those that secrete their product directly into the bloodstream are called endocrine glands. The pancreas is both: an exocrine gland, because it secretes digestive juices into the small intestine via ducts, and an endocrine gland, because it secretes insulin into the bloodstream.

> Epithelial tissue is classified according to the shape of the cell as squamous, cuboidal, or columnar. These tightly packed protective cells can occur in one layer (simple epithelium) or in more than one layer (stratified epithelium). The cells lining a cavity can be ciliated or glandular.

Junctions Between Epithelial Cells

The cells of a tissue can function in a coordinated manner when the plasma membranes of adjoining cells interact. The junctions between cells help cells function as a tissue. A **tight junction** forms an impermeable barrier because adjacent plasma membrane proteins actually join, producing a

Visual Focus

Figure 11.1 Epithelial tissue.

Certain types of epithelial tissue—squamous, cuboidal, and columnar—are named for the shapes of their cells. They all have a protective function in addition to other specific functions.

Pseudostratified, ciliated columnar
• lining of trachea
• sweeps impurities toward throat

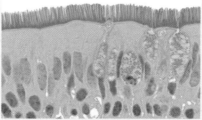

Simple squamous
• lining of lungs, blood vessels
• protects

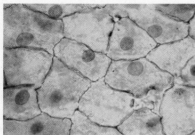

cilia
goblet cell secretes mucus
basement membrane

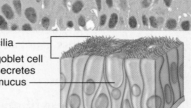

basement membrane

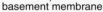

Stratified squamous
• lining of nose, mouth, esophagus, anal canal, vagina
• protects

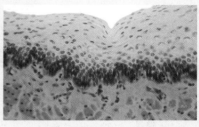

basement membrane

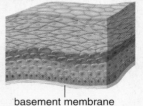

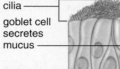

Simple cuboidal
• lining of kidney tubules, various glands
• absorbs molecules

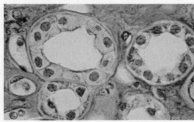

basement membrane

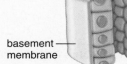

Simple columnar
• lining of small intestine, oviducts
• absorbs nutrients

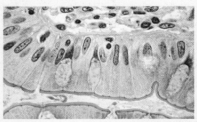

goblet cell secretes mucus
basement membrane

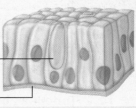

zipperlike fastening (Fig. 11.2*a*). In the intestine, the gastric juices stay out of the body, and in the kidneys, the urine stays within kidney tubules because epithelial cells are joined by tight junctions.

A **gap junction** forms when two adjacent plasma membrane channels join (Fig. 11.2*b*). This lends strength, but it also allows ions, sugars, and small molecules to pass between the two cells. Gap junctions in heart and smooth muscle ensure synchronized contraction. In an **adhesion junction** (also called a desmosome), the adjacent plasma

membranes do not touch but are held together by intercellular filaments firmly attached to cytoplasmic plaques (Fig. 11.2*c*). Adhesion junctions act like rivets or "spot welds" that enhance the strength of a tissue, such as the skin.

Connective Tissue

Connective tissue binds organs together, provides support and protection, fills spaces, produces blood cells, and stores fat. As a rule, connective tissue cells are widely separated by a **matrix,** consisting of a noncellular material that varies in consistency from solid to jellylike to fluid. A nonfluid matrix may have fibers of three possible types: White **collagen fibers** contain collagen, a protein that gives them flexibility and strength. **Reticular fibers** are very thin collagen fibers that are highly branched and form delicate supporting networks. Yellow **elastic fibers** contain elastin, a protein that is not as strong as collagen but more elastic.

Loose Fibrous and Dense Fibrous Tissues

Loose fibrous connective tissue supports epithelium and also many internal organs (Fig. 11.3*a*). Its presence in lungs, arteries, and the urinary bladder allows these organs to expand. It forms a protective covering enclosing many internal organs, such as muscles, blood vessels, and nerves.

Dense fibrous connective tissue contains many collagen fibers that are packed together (Fig. 11.3*b*). This type of tissue has more specific functions than does loose connective tissue. For example, dense fibrous connective tissue is found in **tendons,** which connect muscles to bones, and in **ligaments,** which connect bones to other bones at joints.

Both loose fibrous and dense fibrous connective tissues have cells called **fibroblasts** located some distance from one another and separated by a jellylike matrix containing white collagen fibers and yellow elastic fibers.

Adipose Tissue and Reticular Connective Tissue

In **adipose tissue** (Fig. 11.3*c*), the fibroblasts enlarge and store fat. The body uses this stored fat for energy, insulation, and organ protection. Adipose tissue is found beneath the skin, around the kidneys, and on the surface of the heart. **Reticular connective tissue** forms the supporting meshwork of lymphoid tissue in lymph nodes, the spleen, the thymus, and the bone marrow. All types of blood cells are produced in red bone marrow, but a certain type of lymphocyte (T lymphocyte) completes its development in the thymus. The lymph nodes are sites of lymphocyte responses (see Chaper 13).

Cartilage

In **cartilage,** the cells lie in small chambers called lacunae (sing., **lacuna**), separated by a matrix that is solid yet flexible. Unfortunately, because this tissue lacks a direct blood supply, it heals very slowly. There are three types of cartilage, distinguished by the type of fiber in the matrix.

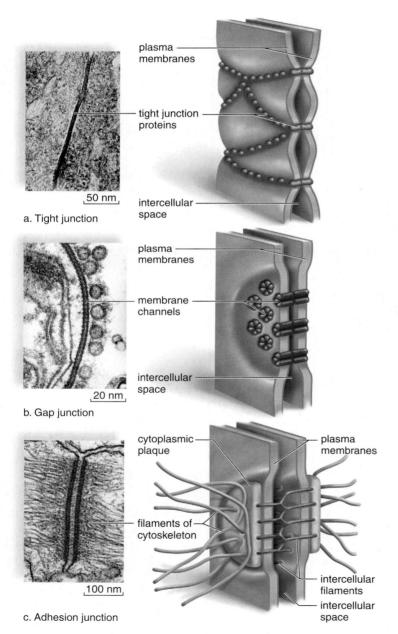

a. Tight junction

50 nm

plasma membranes

tight junction proteins

intercellular space

b. Gap junction

20 nm

plasma membranes

membrane channels

intercellular space

c. Adhesion junction

100 nm

cytoplasmic plaque

plasma membranes

filaments of cytoskeleton

intercellular filaments

intercellular space

Figure 11.2 Junctions between epithelial cells.
Epithelial tissue cells are held tightly together by (**a**) tight junctions that hold cells together; (**b**) gap junctions that allow materials to pass from cell to cell; and (**c**) adhesion junctions that increase tissue strength.

Hyaline cartilage (Fig. 11.3*d*), the most common type of cartilage, contains only very fine collagen fibers. The matrix has a white, translucent appearance. Hyaline cartilage is found in the nose and at the ends of the long bones and the ribs, and it forms rings in the walls of respiratory passages. The fetal skeleton is also made of this type of cartilage. Later, the cartilaginous fetal skeleton is replaced by bone.

Elastic cartilage has more elastic fibers than hyaline cartilage. For this reason, it is more flexible and is found, for example, in the framework of the outer ear.

Fibrocartilage has a matrix containing strong collagen fibers. Fibrocartilage is found in structures that withstand tension and pressure, such as the pads between the vertebrae in the backbone and the wedges in the knee joint.

Bone

Bone is the most rigid connective tissue. It consists of an extremely hard matrix of inorganic salts, notably calcium salts, deposited around protein fibers, especially collagen fibers. The inorganic salts give bone rigidity, and the protein fibers provide elasticity and strength, much as steel rods do in reinforced concrete.

Compact bone makes up the shaft of a long bone (Fig. 11.3*e*). It consists of cylindrical structural units called osteons (Haversian systems). The central canal of each osteon is surrounded by rings of hard matrix. Bone cells are located in spaces called lacunae between the rings of matrix. In the central canal, nerve fibers carry nerve impulses and blood vessels carry nutrients that allow bone to renew itself. Nutrients can reach all of the bone cells because they are connected by thin processes within canaliculi (minute canals) that also reach to the central canal. The shaft of long bones such as the femur (thigh bone) has a hollow chamber filled with bone marrow, a site where blood cells develop.

The ends of a long bone contain spongy bone, which has an entirely different structure. **Spongy bone** contains numerous bony bars and plates, separated by irregular spaces. Although lighter than compact bone, spongy bone is still designed for strength. Just as braces are used for support in buildings, the solid portions of spongy bone follow lines of stress. Spongy bone occurs at the ends of long bones, where it supports and protects the cells of red bone marrow. These cells are in the process of becoming mature blood cells.

Connective tissue differs according to the type of matrix. Some types of connective tissue, such as fibrous connective tissue, bind organs together. Others, such as cartilage and bone, support body parts.

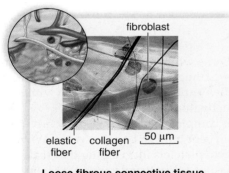

elastic   collagen   50 µm
fiber   fiber

Loose fibrous connective tissue
• has space between components.
• occurs beneath skin and most epithelial layers.
• functions in support and binds organs.

a.

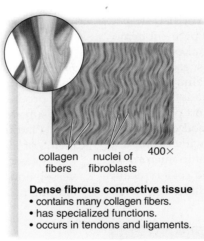

collagen   nuclei of   400×
fibers   fibroblasts

Dense fibrous connective tissue
• contains many collagen fibers.
• has specialized functions.
• occurs in tendons and ligaments.

b.

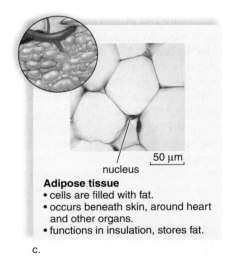

nucleus   50 µm

Adipose tissue
• cells are filled with fat.
• occurs beneath skin, around heart and other organs.
• functions in insulation, stores fat.

c.

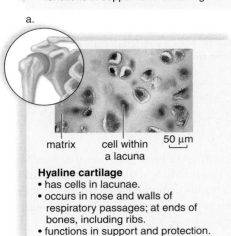

matrix   cell within   50 µm
          a lacuna

Hyaline cartilage
• has cells in lacunae.
• occurs in nose and walls of respiratory passages; at ends of bones, including ribs.
• functions in support and protection.

d.

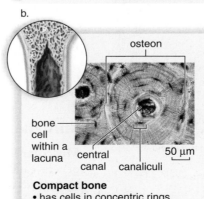

osteon

bone
cell
within a
lacuna   central   50 µm
          canal   canaliculi

Compact bone
• has cells in concentric rings.
• occurs in bones of skeleton.
• functions in support and protection.

e.

Figure 11.3 Connective tissue examples. a. In loose fibrous connective tissue, cells called fibroblasts are separated by a jellylike matrix, which contains both collagen and elastic fibers. **b.** Dense fibrous connective tissue contains tightly packed collagen fibers for added strength. **c.** Adipose tissue cells have nuclei pushed to one side because the cells are filled with fat. **d.** In hyaline cartilage, the flexible matrix has a white, translucent appearance. **e.** In compact bone, the hard matrix contains calcium salts. Concentric rings of cells in lacunae form an elongated cylinder called an osteon (Haversian system). An osteon has a central canal that contains blood vessels and nerve fibers.

Blood

Blood is unlike other types of connective tissue in that the matrix (i.e., plasma) is not made by the cells. In fact, some scientists do not classify blood as connective tissue, suggesting instead a separate category called vascular tissue.

The internal environment of the body consists of blood and the fluid between the body's cells. The systems of the body help keep the composition and chemistry of blood within normal limits, and blood, in turn, creates tissue fluid. Blood transports nutrients and oxygen to tissue fluid and removes carbon dioxide and other wastes. It helps distribute heat and also plays a role in fluid, ion, and pH balance. Various components of blood help protect us from disease, and blood's ability to clot prevents fluid loss.

If blood is transferred from a person's vein to a test tube and prevented from clotting, it separates into two layers (Fig. 11.4a). The upper, liquid layer, called **plasma,** represents about 55% of the volume of whole blood and contains a variety of inorganic and organic substances dissolved or suspended in water (Table 11.1). The lower layer consists of red blood cells (erythrocytes), white blood cells (leukocytes), and blood platelets (thrombocytes) (Fig. 11.4b). Collectively, these are called the formed elements, and they represent about 45% of the volume of whole blood. Formed elements are manufactured in the red bone marrow of the skull, ribs, vertebrae, and ends of the long bones.

The **red blood cells** are small, biconcave, disk-shaped cells without nuclei. The presence of the red pigment hemoglobin makes the cells red and, in turn, makes the blood red. Hemoglobin is composed of four units; each contains the protein globin and a complex, iron-containing structure called heme. The iron forms a loose association with oxygen, and in this way, red blood cells transport oxygen.

White blood cells differ from red blood cells in that they are usually larger, have a nucleus, and without staining would appear translucent. When smeared onto microscope slides and stained, the nuclei of white blood cells typically look bluish (Fig. 11.4b). White blood cells fight infection, pri-

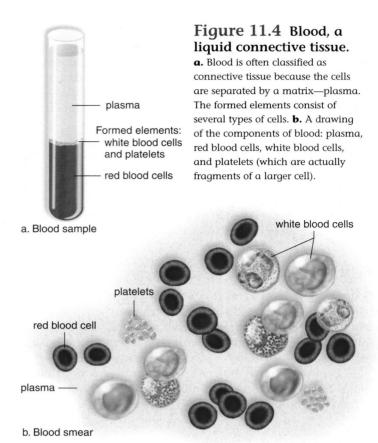

Figure 11.4 Blood, a liquid connective tissue.

a. Blood is often classified as connective tissue because the cells are separated by a matrix—plasma. The formed elements consist of several types of cells. **b.** A drawing of the components of blood: plasma, red blood cells, white blood cells, and platelets (which are actually fragments of a larger cell).

plasma

Formed elements:
white blood cells
and platelets

red blood cells

a. Blood sample

white blood cells

platelets

red blood cell

plasma

b. Blood smear

marily in two ways. Some white blood cells are phagocytic and engulf infectious pathogens, while other white blood cells produce antibodies, molecules that combine with foreign substances to inactivate them.

Platelets are not complete cells; rather, they are fragments of giant cells present only in bone marrow. When a blood vessel is damaged, platelets form a plug that seals the vessel, and injured tissues release molecules that help the clotting process.

> Blood is a connective tissue in which the matrix is plasma. Red blood cells carry oxygen to tissues; white blood cells fight infection. Platelets, which are fragments of a very large cell, are involved in clotting.

TABLE 11.1	Components of Blood Plasma
Water (90–92% of total)	**H_2O**
Solutes (8–10% of total):	
Inorganic ions (electrolytes)	Na^+, Ca^{2+}, K^+, Mg^{2+}, Cl^-, HCO_3^-, HPO_4^{2+}, SO_4^{2-}
Gases	O_2, CO_2
Plasma proteins	Albumin, globulins, fibrinogen, transport proteins
Organic nutrients	Glucose, lipids, phospholipids, amino acids, etc.
Nitrogen-containing waste products	Urea, ammonia, uric acid
Regulatory substances	Hormones, enzymes

Muscular Tissue

Muscular tissue (also called contractile tissue) is composed of cells called muscle fibers. Muscle fibers contain actin filaments and myosin filaments, whose interaction accounts for movement. The three types of muscle tissue are skeletal, smooth, and cardiac.

Skeletal muscle, also called voluntary muscle (Fig. 11.5*a*), is attached by tendons to the bones of the skeleton, and when it contracts, body parts move. Contraction of skeletal muscle is under voluntary control and occurs faster than in the other muscle types. Skeletal muscle fibers are cylindrical and quite long—sometimes they run the length of the muscle. They arise during development when several cells fuse, resulting in one fiber with multiple nuclei. The nuclei are located at the periphery of the cell, just inside the plasma membrane. The fibers have alternating light and dark bands that give them a **striated** appearance. These bands are due to the placement of actin filaments and myosin filaments in the cell.

Smooth muscle (also called visceral muscle) is so named because the cells lack striations. The spindle-shaped fibers, each with a single nucleus, form layers in which the thick middle portion of one cell is opposite the thin ends of adjacent cells. Consequently, the nuclei form an irregular pattern in the tissue (Fig. 11.5*b*). Smooth muscle is not under voluntary control, and therefore is said to be involuntary. Smooth muscle, found in the walls of viscera (intestine, stomach, and other internal organs) and blood vessels, contracts more slowly than skeletal muscle but can remain contracted for a longer time. When the smooth muscle of the intestine contracts, food moves along its lumen (central cavity). When the smooth muscle of the blood vessels contracts, blood vessels constrict, helping to raise blood pressure.

Cardiac muscle (Fig. 11.5*c*) is found only in the walls of the heart. Its contraction pumps blood and accounts for the heartbeat. Cardiac muscle combines features of both smooth muscle and skeletal muscle. Like skeletal muscle, it has striations, but the contraction of the heart is involuntary. Cardiac muscle cells also differ from skeletal muscle cells in that they have a single, centrally placed nucleus. The cells are branched and seemingly fused one with the other, and the heart appears to be composed of one large interconnecting mass of muscle cells. Actually, cardiac muscle cells are separate and individual, but they are bound end to end at **intercalated disks,** areas where folded plasma membranes between two cells contain adhesion junctions and gap junctions.

> All muscular tissue contains actin filaments and myosin filaments; these form a striated pattern in skeletal and cardiac muscle, but not in smooth muscle.

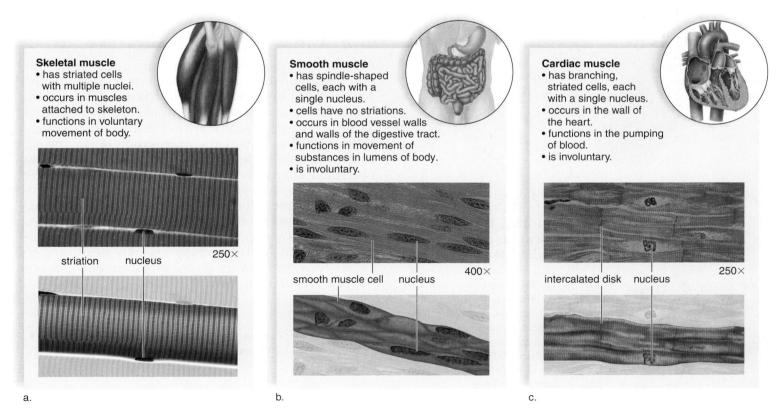

Skeletal muscle
- has striated cells with multiple nuclei.
- occurs in muscles attached to skeleton.
- functions in voluntary movement of body.

striation   nucleus   250×

Smooth muscle
- has spindle-shaped cells, each with a single nucleus.
- cells have no striations.
- occurs in blood vessel walls and walls of the digestive tract.
- functions in movement of substances in lumens of body.
- is involuntary.

smooth muscle cell   nucleus   400×

Cardiac muscle
- has branching, striated cells, each with a single nucleus.
- occurs in the wall of the heart.
- functions in the pumping of blood.
- is involuntary.

intercalated disk   nucleus   250×

a.   b.   c.

Figure 11.5 Muscular tissue.
a. Skeletal muscle is voluntary and striated. **b.** Smooth muscle is involuntary and nonstriated. **c.** Cardiac muscle is involuntary and striated. Cardiac muscle cells branch and fit together at intercalated disks.

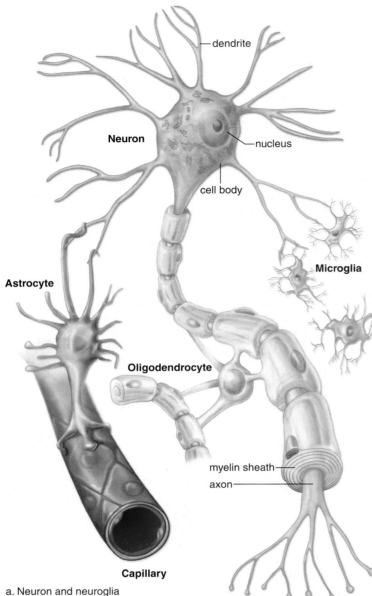

Neuron

dendrite

nucleus

cell body

Astrocyte

Microglia

Oligodendrocyte

myelin sheath

axon

Capillary

a. Neuron and neuroglia

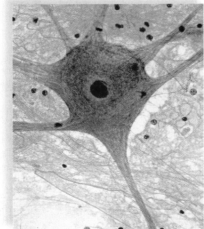

b. Micrograph of a neuron

200×

Figure 11.6
Neurons and neuroglia.
a. Neurons conduct nerve impulses. Neuroglia consist of cells that support and nourish neurons and have various functions. Microglia become mobile in response to inflammation and phagocytize debris. Astrocytes lie between neurons and a capillary; therefore, nutrients entering neurons from the blood must first pass through astrocytes. Oligodendrocytes form the myelin sheaths around fibers in the brain and spinal cord. **b.** A neuron cell body as seen with a light microscope.

Nervous Tissue

Nervous tissue, which contains nerve cells called neurons (Fig. 11.6*a*), is present in the brain and spinal cord. A **neuron** is a specialized cell that has three parts: a cell body, dendrites, and an axon (Fig. 11.6*a, b*). The cell body contains the major concentration of the cytoplasm and the nucleus of the neuron. A dendrite is a process that conducts signals toward the cell body. An axon is a process that typically conducts nerve impulses away from the cell body. Long axons are covered by myelin, a white fatty substance. The term *fiber*[1] is used here to refer to an axon along with its myelin sheath if it has one. Outside the brain and spinal cord, fibers bound by connective tissue form **nerves.**

The nervous system has just three functions: sensory input, integration of data, and motor output. Nerves conduct impulses from sensory receptors to the spinal cord and the brain where integration occurs. The phenomenon called sensation occurs only in the brain, however. Nerves also conduct nerve impulses away from the spinal cord and brain to the muscles and glands, causing them to contract and secrete, respectively. In this way, a coordinated response to the stimulus is achieved.

Neuroglia

In addition to neurons, nervous tissue contains neuroglia. **Neuroglia** are cells that outnumber neurons nine to one and take up more than half the volume of the brain. Although the primary function of neuroglia is to support and nourish neurons, research is currently being conducted to determine how much they directly contribute to brain function. The three types of neuroglia in the brain are microglia, astrocytes, and oligodendrocytes (Fig. 11.6*a*). Microglia, in addition to supporting neurons, engulf bacterial and cellular debris. Astrocytes provide nutrients to neurons and produce a hormone known as glia-derived growth factor, which has potential as a treatment for Parkinson disease and other diseases caused by neuron degeneration. Oligodendrocytes form myelin sheaths. Neuroglia don't have a long process, but even so, researchers are now beginning to gather evidence that they do communicate among themselves and with neurons!

Nerve cells, called neurons, have processes called axons and dendrites. In general, neuroglia support and nourish neurons.

---

[1] *In connective tissue, a fiber is a component of the matrix; in muscle tissue, a fiber is a muscle cell; in nervous tissue, a fiber is an axon and its myelin sheath.*

11.2 Body Cavities and Body Membranes

The human body is divided into two main cavities: the ventral cavity and the dorsal cavity (Fig. 11.7a). The ventral cavity, which is called a **coelom** during development, becomes divided into the thoracic, abdominal and pelvic cavities. The thoracic cavity contains the right and left lungs and the heart. The thoracic cavity is separated from the abdominal cavity by a horizontal muscle called the diaphragm. The stomach, liver, spleen, gallbladder, and most of the small and large intestines are in the upper portion of the abdominal cavity. The pelvic cavity contains the rectum, the urinary bladder, the internal reproductive organs, and the rest of the large intestine. Males have an external extension of the abdominal wall, called the scrotum, containing the testes.

The dorsal cavity also has two parts: the cranial cavity within the skull contains the brain; the vertebral canal, formed by the vertebrae, contains the spinal cord.

Body Membranes

Body membranes line cavities and the internal spaces of organs and tubes that open to the outside.

Mucous membranes line the tubes of the digestive, respiratory, urinary, and reproductive systems. They are composed of an epithelium overlying a loose fibrous connective tissue layer. The epithelium contains goblet cells that secrete mucus. This mucus ordinarily protects the body from invasion by bacteria and viruses; hence, more mucus is secreted and expelled when a person has a cold and has to blow her/his nose. In addition, mucus usually protects the walls of the stomach and small intestine from digestive juices; but this protection breaks down when a person develops an ulcer.

Serous membranes, which line the thoracic and abdominal cavities and cover the organs they contain, are also composed of epithelium and loose fibrous connective tissue. They secrete a watery fluid that keeps the membranes lubricated. Serous membranes support the internal organs and compartmentalize the large thoracic and abdominal cavities. This helps hinder the spread of any infection.

Serous membranes have specific names according to their location (Fig. 11.7b). The pleurae (sing., pleura) line the thoracic cavity and cover the lungs; the pericardium covers the heart; the peritoneum lines the abdominal cavity and covers its organs. A double layer of peritoneum, called mesentery, supports the abdominal organs and attaches them to the abdominal wall. Peritonitis is a potentially life-threatening infection of the peritoneum that may occur if an inflamed appendix bursts before it is removed, or if the digestive tract is perforated for any other reason.

Synovial membranes, composed only of loose connective tissue, line freely movable joint cavities. They secrete synovial fluid that lubricates the ends of the bones so that they can move freely in the joint cavity. In rheumatoid arthritis, the synovial membrane becomes inflamed and grows thicker, restricting movement.

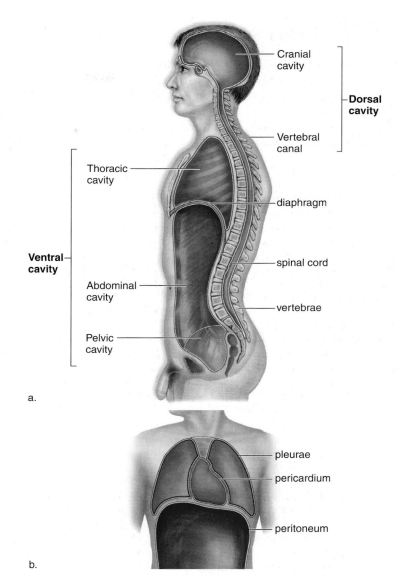

a.

b.

Figure 11.7 Mammalian body cavities.

a. Side view. The dorsal (toward the back) cavity contains the cranial cavity and the vertebral canal. The brain is in the cranial cavity, and the spinal cord is in the vertebral canal. In the ventral (toward the front) cavity, the diaphragm separates the thoracic cavity and the abdominal cavity. The heart and lungs are in the thoracic cavity, and most other internal organs are in the abdominal cavity. **b.** Types of serous membranes.

The **meninges** are membranes within the dorsal cavity. They are composed only of connective tissue and serve as a protective covering for the brain and spinal cord. Meningitis is an inflammation of the meninges.

Different types of membranes line body cavities and spaces. Mucous membranes secrete a protective layer of mucus, serous membranes are lubricated with fluid, synovial membranes line movable joints, and meninges protect the brain and spinal cord.

11.3 Organ Systems

Figure 11.8 illustrates the organ systems of the human body.

It should be emphasized that just as organs work together in an organ system, so do organ systems work together in the body. In a sense, it is arbitrary to assign a particular organ to one system when it also assists the functioning of many other systems.

Integumentary System

The **integumentary system** contains skin, which is made up of two main types of tissue: the epidermis is composed of stratified squamous epithelium, and the dermis is composed of fibrous connective tissue. The integumentary system also includes nails, located at the ends of the fingers and toes (collectively called the digits), hairs, muscles that move hairs, the oil and sweat glands, blood vessels, and nerves leading to sensory receptors. Besides having a protective function, skin also synthesizes vitamin D, collects sensory data, and helps regulate body temperature.

Digestive System

The **digestive system** consists of the mouth, esophagus, stomach, small intestine, and large intestine (colon), along with the following associated organs: teeth, tongue, salivary glands, liver, gallbladder, and pancreas. This system receives food and digests it into nutrient molecules, which can enter the body's cells. The nondigested remains are eventually eliminated.

Cardiovascular System

In the **cardiovascular system,** the heart pumps blood and sends it under pressure into the blood vessels. While blood is moving throughout the body, it distributes heat produced by the muscles. Blood transports nutrients and oxygen to the cells and removes their waste molecules, including carbon dioxide. Despite the movement of molecules into and out of the blood, it has a fairly constant volume and pH. The blood is also a route by which cells of the immune system can be distributed throughout the body.

Lymphatic and Immune Systems

The **lymphatic system** consists of lymphatic vessels (which transport lymph), lymph nodes, and other lymphatic (lymphoid) organs. This system protects the body from disease by purifying lymph and storing lymphocytes, the white blood cells that produce antibodies. Lymphatic vessels absorb fat from the digestive system and collect excess tissue fluid, which is returned to the cardiovascular system.

The **immune system** consists of all the cells in the body that protect us from disease, especially those caused by infectious agents. The lymphocytes, in particular, belong to this system.

Respiratory System

The **respiratory system** consists of the lungs and the tubes that take air to and from them. The respiratory system brings oxygen into the body for cellular respiration and removes carbon dioxide from the body at the lungs, restoring pH.

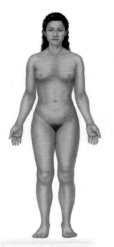
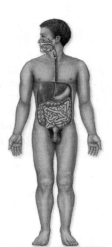
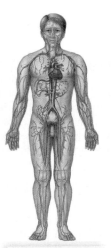
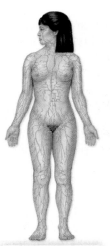
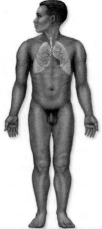
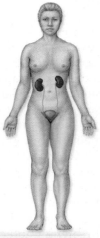

Integumentary system
- protects body.
- receives sensory input.
- helps control temperature.
- synthesizes vitamin D.

Digestive system
- ingests food.
- digests food.
- absorbs nutrients.
- eliminates waste.

Cardiovascular system
- transports blood, nutrients, gases, and wastes.
- helps control temperature, fluid, and pH balance.

Lymphatic and immune systems
- helps control fluid balance.
- absorbs fats.
- defend against infectious disease.

Respiratory system
- maintains breathing.
- exchanges gases at lungs and tissues.
- helps control pH balance.

Urinary system
- excretes metabolic wastes.
- helps control fluid balance.
- helps control pH balance.

Figure 11.8 Organ systems of the body.

Urinary System

The **urinary system** contains the kidneys, the urinary bladder, and the tubes that carry urine. This system rids the body of metabolic wastes, particularly nitrogenous (nitrogen-containing) wastes, and helps regulate the fluid balance and pH of the blood.

Skeletal System

The bones of the **skeletal system** protect body parts. For example, the skull forms a protective encasement for the brain, as does the rib cage for the heart and lungs. The skeleton also helps move the body because it serves as a place of attachment for the skeletal muscles.

In addition, the skeletal system stores minerals, notably calcium, and it produces blood cells within red bone marrow.

Muscular System

In the **muscular system,** skeletal muscle contraction maintains posture and accounts for the movement of the body and its parts. Cardiac muscle contraction results in the heart-beat. The walls of internal organs contract due to the presence of smooth muscle.

Nervous System

The **nervous system** consists of the brain, spinal cord, and associated nerves. The nerves conduct nerve impulses from sensory receptors to the brain and spinal cord where integration occurs. Nerves also conduct nerve impulses from the brain and spinal cord to the muscles and glands, allowing us to respond to both external and internal stimuli.

Endocrine System

The **endocrine system** consists of the hormonal glands, which secrete chemical messengers called hormones. Hormones have a wide range of effects, including regulating cellular metabolism, regulating fluid and pH balance, and helping us respond to stress. Both the nervous and endocrine systems coordinate and regulate the functioning of the body's other systems. The endocrine system also helps maintain the functioning of the male and female reproductive organs.

Reproductive System

The **reproductive system** has different organs in the male and female. The male reproductive system consists of the testes, other glands, and various ducts that conduct semen to and through the penis. The testes produce sex cells called sperm. The female reproductive system consists of the ovaries, oviducts, uterus, vagina, and external genitals. The ovaries produce sex cells called eggs, or oocytes. When a sperm fertilizes an oocyte, an offspring begins to develop.

The integumentary system protects the body and receives sensory input. The digestive, cardiovascular, lymphatic, respiratory, and urinary systems perform processing and transporting functions that maintain the normal conditions of the body. The immune system protects the body against infectious disease. The muscular and skeletal systems support the body and permit movement. The nervous system directs responses to outside stimuli. The endocrine system produces hormones that affect other body systems, and the reproductive system allows humans to make more of their own kind.

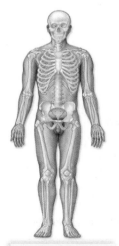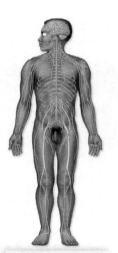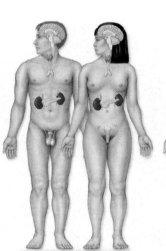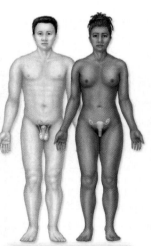

Skeletal system
- supports the body.
- protects body parts.
- helps move the body.
- stores minerals.
- produces blood cells.

Muscular system
- maintains posture.
- moves body and internal organs.
- produces heat.

Nervous system
- receives sensory input.
- integrates and stores input.
- initiates motor output.
- helps coordinate organ systems.

Endocrine system
- produces hormones.
- helps coordinate organ systems.
- responds to stress.
- helps regulate fluid and pH balance.
- helps regulate metabolism.

Reproductive system
- produces gametes.
- transports gametes.
- produces sex hormones.
- in females, nurtures and gives birth to offspring.

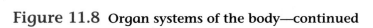

Figure 11.8 Organ systems of the body—continued

11.4   Integumentary System

The skin and its accessory organs (nails, hair, oil glands, and sweat glands) are collectively called the integumentary system. Skin covers the body, protecting underlying tissues from physical trauma, pathogen invasion, and water loss; it also helps regulate body temperature. Therefore, skin plays a significant role in homeostasis. The skin even synthesizes vitamin D with the aid of ultraviolet radiation. Vitamin D is required for proper bone growth. Skin also contains sensory receptors, which help us to be aware of our surroundings and to communicate through touch.

Regions of the Skin

The **skin** has two regions: the epidermis and the dermis (Fig. 11.9). A subcutaneous layer is present between the skin and any underlying structures, such as muscle or bone. The subcutaneous layer is not a part of the skin.

The **epidermis** is made up of stratified squamous epithelium. New cells derived from *basal cells* become flattened and hardened as they push to the surface. Hardening takes place because the cells produce keratin, a waterproof protein. Dandruff occurs when the rate of keratinization in the skin of the scalp is two or three times the normal rate. A thick layer of dead keratinized cells, arranged in spiral and concentric patterns, forms fingerprints and footprints. Specialized cells in the epidermis called **melanocytes** produce melanin, the main pigment responsible for skin color.

The **dermis** is a region of fibrous connective tissue beneath the epidermis. The dermis contains collagen and elastic fibers. The collagen fibers are flexible but offer great resistance to overstretching; they prevent the skin from being torn. The elastic fibers maintain normal skin tension but also stretch to allow movement of underlying muscles and joints. (The number of collagen and elastic fibers decreases with exposure to the sun, causing the skin to become less supple and more prone to wrinkling.) The dermis also contains blood vessels that nourish the skin. When blood rushes into these vessels, a person blushes, and when blood is minimal in them, the skin turns "blue."

Sensory receptors are specialized free nerve endings in the dermis that respond to external stimuli. There are

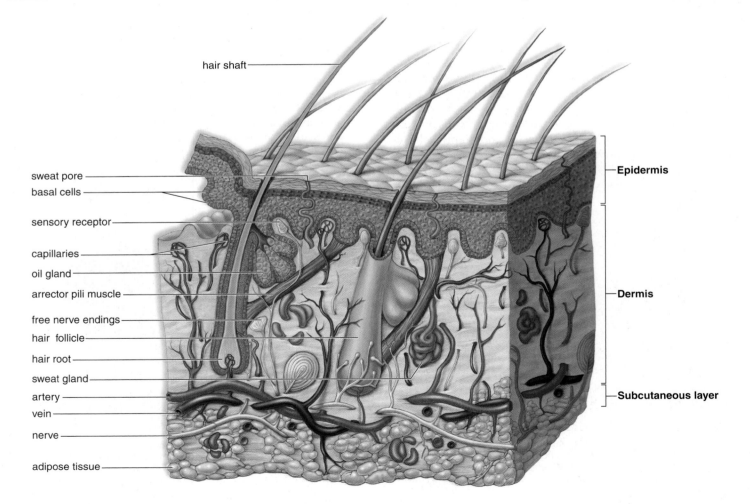

Figure 11.9 Human skin anatomy.
Skin consists of two regions, the epidermis and the dermis. A subcutaneous layer lies below the dermis.

sensory receptors for touch, pressure, pain, and temperature. The fingertips contain the most touch receptors, and these add to our ability to use our fingers for delicate tasks.

The **subcutaneous layer,** which lies below the dermis, is composed of loose connective tissue and adipose tissue, which stores fat. Fat is a stored source of energy in the body. Adipose tissue helps thermally insulate the body from either gaining heat from the outside or losing heat from the inside. A well-developed subcutaneous layer gives the body a rounded appearance and provides protective padding against external assaults. Excessive development of the subcutaneous layer accompanies obesity.

Skin has two regions: the epidermis and the dermis. A subcutaneous layer lies beneath the dermis.

Accessory Organs of the Skin

Nails, hair, and glands are structures of epidermal origin, even though some parts of hair and glands are largely found in the dermis.

Nails are a protective covering of the distal part of the digits. Nails can help pry things open or pick up small objects. They are also used for scratching oneself or others. Nails grow from special epithelial cells at the base of the nail in the portion called the nail root. These cells become keratinized as they grow out over the nail bed. The visible portion of the nail is called the nail body. The cuticle is a fold of skin that hides the nail root. The whitish color of the half-moon-shaped base, or lunula, results from the thick layer of cells in this area (Fig. 11.10).

Hair follicles are in the dermis and continue through the epidermis where the hair shaft extends beyond the skin. Contraction of the arrector pili muscles attached to hair follicles causes the hairs to "stand on end" and goose bumps to develop. Epidermal cells form the root of hair, and their division causes a hair to grow. The cells become keratinized and dead as they are pushed farther from the root.

Each hair follicle has one or more **oil glands,** also called sebaceous glands, which secrete sebum, an oily substance that lubricates the hair within the follicle and the skin itself. If the oil glands fail to discharge, the secretions collect and form "whiteheads" or "blackheads." The color of blackheads is due to oxidized sebum. Acne is an inflammation of the oil glands that most often occurs during adolescence. Hormonal changes during this time cause the oil glands to become more active.

Sweat glands, also called sudoriferous glands, are quite numerous and are present in all regions of the skin. A sweat gland is a coiled tubule within the dermis that straightens out near its opening, or pore. Some sweat glands open into hair follicles, but most open onto the surface of the skin. Sweat glands play a role in modifying body temperature. When body temperature starts to rise, sweat glands become active.

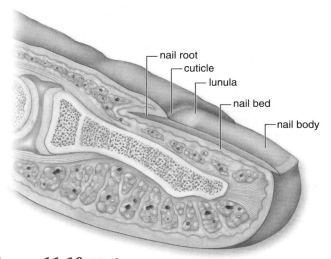

Figure 11.10 **Nail anatomy.**
Cells produced by the nail root become keratinized, forming the nail body.

Sweat absorbs body heat as it evaporates. Once body temperature lowers, sweat glands are no longer active.

The accessory organs of the skin—nails, hair, and glands—are of epidermal origin, even those portions located in the dermis.

11.5 Homeostasis

Homeostasis is the relative constancy of the body's internal environment. Because of homeostasis, even though external conditions may change dramatically, internal conditions stay within a narrow range. For example, regardless of how cold or hot it gets, the temperature of the body stays around 98.6°F (37°C). No matter how acidic your meal, the pH of your blood is usually about 7.4, and even if you eat a candy bar, the amount of sugar in your blood stays between 0.05% and 0.08%.

It is important to realize that internal conditions are not absolutely constant; they tend to fluctuate above and below a particular value. Therefore, the internal state of the body is often described as one of *dynamic* equilibrium. If internal conditions change to any great degree, illness results. This makes the study of homeostatic mechanisms medically important.

Negative Feedback

Negative feedback is the primary homeostatic mechanism that keeps a variable close to a particular value, or set point. A homeostatic mechanism has at least two components: a sensor and a control center (Fig. 11.11). The sensor detects a change in internal conditions; the control center then directs a response that brings conditions back to normal again. Now, the sensor is no longer activated. In other words, a **negative feedback** mechanism is present when the output of the system dampens the original stimulus.

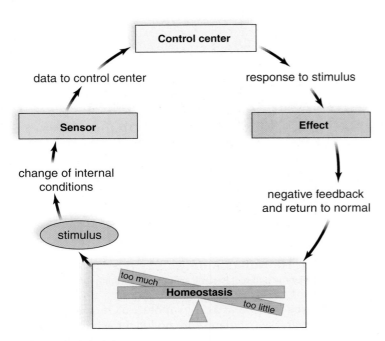

Figure 11.11 Negative feedback mechanism.
The sensor and control center of a feedback mechanism work together to keep a variable close to a particular value.

Let's take a simple example. When the pancreas detects that the blood glucose level is too high, it secretes insulin, a hormone that causes cells to take up glucose. Now the blood sugar level returns to normal, and the pancreas is no longer stimulated to secrete insulin.

Mechanical Example

A home heating system is often used to illustrate how a more complicated negative feedback mechanism works (Fig. 11.12).

You set the thermostat at, say, 68°F. This is the *set point*. The thermostat contains a thermometer, a sensor that detects when the room temperature is above or below the set point. The thermostat also contains a control center; it turns the furnace off when the room is too hot and turns it on when the room is too cold. When the furnace is off, the room cools, and when the furnace is on, the room warms. In other words, typical of negative feedback mechanisms, there is a fluctuation above and below normal:

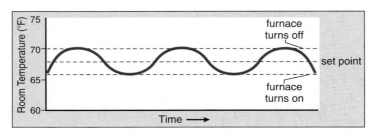

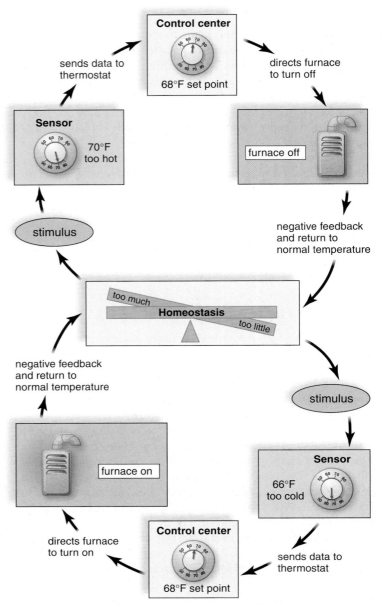

Figure 11.12 Complex negative feedback mechanism. When a room becomes too warm, negative feedback allows the temperature to return to normal. A contrary cycle, in which the furnace turns on and gives off heat, returns the room temperature to normal when the room becomes too cool.

Human Example: Regulation of Body Temperature

The sensor and control center for body temperature are located in a part of the brain called the hypothalamus. When the body temperature is above normal, the control center directs the blood vessels of the skin to dilate. This allows more blood to flow near the surface of the body, where heat can be lost to the environment. In addition, the nervous system activates the sweat glands, and the evaporation of sweat helps lower body temperature. Gradually, body temperature decreases to 98.6°F (37°C). When the body temperature

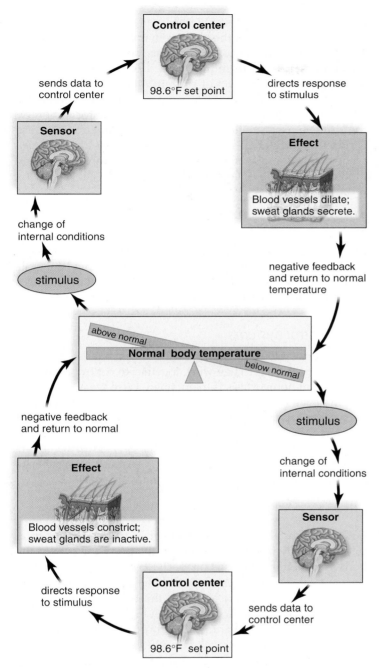

Figure 11.13 Regulation of body temperature.
Normal body temperature is maintained by a negative feedback system.

Notice that a negative feedback mechanism prevents change in the same direction; that is, body temperature does not get warmer and warmer because warmth brings about a change toward a lower body temperature. Also, body temperature does not get colder and colder because a body temperature below normal brings about a change toward a warmer body temperature.

> **Negative feedback is the primary homeostatic mechanism that keeps an internal condition close to a set value. A sensor detects that some condition is too high or too low, and a control center responds to correct that condition.**

Positive Feedback

Positive feedback is a mechanism that brings about an ever greater change in the same direction. A positive feedback mechanism can be harmful, as when a fever causes metabolic changes that push the fever still higher. Death occurs at a body temperature of 113°F (45°C) because cellular proteins denature at this temperature and metabolism stops.

Still, positive feedback loops such as those involved in blood clotting, the stomach's digestion of protein, and childbirth assist the body in completing a process that has a definite cutoff point.

Consider that when a woman is giving birth, the head of the baby begins to press against the cervix, stimulating sensory receptors there. When nerve impulses reach the brain, the brain causes the pituitary gland to secrete the hormone oxytocin. Oxytocin travels in the blood and causes the uterus to contract. As labor continues, the cervix is ever more stimulated, and uterine contractions become ever stronger until birth occurs.

> **In homeostatic systems of the body that use positive feedback, ever-increasing changes occur in the same direction. Such systems are normally only used when there is a definite cutoff or end point to a process, such as during childbirth.**

Homeostasis and Body Systems

The internal environment of the body consists of blood and tissue fluid. Tissue fluid, which bathes all the cells of the body, is refreshed when molecules such as oxygen and nutrients move into tissue fluid from the blood and when carbon dioxide and wastes move from tissue fluid into the blood (Fig. 11.14). The composition of tissue fluid remains constant only as long as blood composition remains constant. All systems of the body contribute toward maintaining homeostasis and, therefore, a relatively constant internal environment.

falls below normal, the control center (via nerve impulses) directs the blood vessels of the skin to constrict (Fig. 11.13). This conserves heat. If body temperature falls even lower, the control center sends nerve impulses to the skeletal muscles, and shivering occurs. Shivering generates heat, and gradually body temperature rises to 98.6°F. When the temperature rises to normal, the control center is inactivated.

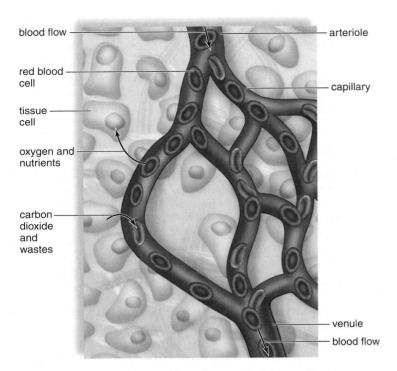

Figure 11.14 Regulation of tissue fluid composition.
Cells are surrounded by tissue fluid, which is continually refreshed because oxygen and nutrient molecules constantly exit, and carbon dioxide and waste molecules continually enter the bloodstream.

The Transport Systems

The cardiovascular system conducts blood to and away from capillaries, where exchange of gases and wastes occurs. The heart pumps the blood and thereby keeps it moving toward the capillaries. The formed elements also contribute to homeostasis. Red blood cells transport oxygen and participate in the transport of carbon dioxide. Platelets participate in the clotting process.

The lymphatic system is an accessory to the cardiovascular system. Lymphatic capillaries collect excess tissue fluid, and this is returned via lymphatic veins to the cardiovascular veins. Lymph nodes help purify lymph and are sites where the immune system responds to invading microorganisms. These responses are conducted by the white blood cells that are housed within lymph nodes.

The Maintenance Systems

The respiratory system adds oxygen to and removes carbon dioxide from the blood. It also plays a role in regulating blood pH because removal of CO_2 causes the pH to rise and helps prevent acidosis. The digestive system takes in and digests food, providing nutrient molecules that enter the blood to replace the nutrients that are constantly being used by the body cells. The liver, an organ that assists the digestive process by producing bile, also plays a significant role in regulating blood composition. Immediately after glucose enters the blood, any excess is removed by the liver and stored as glycogen. Later, the glycogen can be broken down

to replace the glucose used by the body cells; in this way, the glucose composition of the blood remains relatively constant. The liver also removes toxic chemicals, such as ingested alcohol and other drugs from the blood. The liver makes urea, a nitrogenous end product of protein metabolism. Urea and other metabolic waste molecules are excreted by the kidneys, which are a part of the urinary system. Urine formation by the kidneys is extremely critical to the body, not only because it rids the body of unwanted substances, but also because urine formation offers an opportunity to carefully regulate blood volume, salt balance, and pH.

The Support Systems

The integumentary, skeletal, and muscular systems protect the internal organs. In addition, the integumentary system produces vitamin D, while the skeletal system stores minerals and produces the blood cells. The muscular system produces the heat that maintains the internal temperature.

The Control Systems

The nervous system and the endocrine system regulate the other systems of the body. They work together to control body systems so that homeostasis is maintained. We have already seen that in negative feedback mechanisms, sensory receptors send nerve impulses to control centers in the brain, which then direct effectors to become active. Effectors can be muscles or glands. Muscles bring about an immediate change. Endocrine glands secrete hormones that bring about a slower, more lasting change that keeps the internal environment relatively stable.

Disease

Disease is an abnormality or upset in the body's normal processes. When homeostasis fails, the body (or part of the body) no longer functions properly. For example, different types of cancer can affect virtually any system in the body, and thus may hinder the ability of these systems to perform their normal homeostatic functions. A malignant cancer often becomes **systemic,** meaning that it affects the entire body or at least several organ systems. Other diseases, including certain types of infections or degenerative diseases such as arthritis or Alzheimer disease, are more **localized** to a specific part of the body. Diseases may also be classified on the basis of their severity and duration. **Acute diseases,** such as poison ivy dermatitis or influenza, tend to occur suddenly and generally last a short time, although some can be life threatening. **Chronic diseases,** such as multiple sclerosis or AIDS, tend to develop slowly and last a long time, even for the rest of a person's life, unless an effective cure is available.

Most organ systems of the body contribute to homeostasis. Four major functions of these systems are transport, maintenance, support, and control. Diseases can be thought of as abnormalities in the body's normal processes.

Health Focus

Skin Cancer on the Rise

In the nineteenth century, and earlier, it was fashionable for upper-class Caucasian women, who did not labor outdoors, to keep their skin fair by using parasols when they went out. But early in the twentieth century, some fair-skinned people began to prefer the golden-brown look and took up sunbathing as a way to achieve a tan.

A few hours of exposure to the sun cause the pain and redness characteristic of an inflammatory reaction due to damage to skin cells. Tanning occurs when melanin granules increase in keratinized cells at the surface of the skin as a way to prevent any further damage by ultraviolet (UV) rays. The sun gives off two types of UV rays: UV-A rays and UV-B rays. UV-A rays penetrate the skin deeply, affect connective tissue, and cause the skin to sag and wrinkle. UV-A rays are also believed to increase the effects of the UV-B rays, which are the cancer-causing rays. UV-B rays are more prevalent at midday.

Skin cancer is categorized as either nonmelanoma or melanoma. Nonmelanoma cancers are of two types: basal cell carcinoma and squamous cell carcinoma. Basal cell carcinoma, the most common type, begins when DNA damage induced by UV radiation causes epidermal basal cells to form a tumor, while at the same time suppressing the immune system's ability to detect the tumor. The signs of a basal cell tumor are varied. They include an open sore that will not heal; a recurring reddish patch; a smooth, circular growth with a raised edge; a shiny bump; or a pale mark (Fig. 11A*a*). In about 95% of patients, the tumor can be excised surgically, but recurrence is common.

Squamous cell carcinoma begins in the epidermis proper. Squamous cell carcinoma is five times less common than basal cell carcinoma, but if the tumor is not excised promptly, it is more likely to spread to nearby organs. The death rate from squamous cell carcinoma is about 1% of cases. The signs of a tumor are the same as for basal cell carcinoma, except that a squamous cell carcinoma may also show itself as a wart that bleeds and scabs (Fig. 11A*b*).

Melanoma is most likely to appear in fair-skinned persons, particularly if they suffered occasional severe sunburns as children. It affects pigmented cells and often has the appearance of an unusual mole. But unlike a mole that is circular and confined, melanoma moles often look like spilled ink spots (Fig. 11A*c*). A variety of shades can be seen in a single mole, and it can itch, hurt, or feel numb. The skin around the melanoma turns gray, white, or red.

The chance of melanoma increases with the number of moles a person has. Most moles appear before the age of 14, and their appearance is linked to sun exposure. Any moles that become malignant are removed surgically; if the cancer has spread, chemotherapy and a number of other treatments are also available.

Scientists have developed a UV index to determine how powerful the solar rays are in different U.S. cities. In general, the more southern the city, the higher the UV index and the greater the risk of skin cancer. Regardless of where you live, for every 10% decrease in the ozone layer, the risk of skin cancer rises 13–20%. To prevent the occurrence of skin cancer, take the following precautions:

- Use a broad-spectrum sunscreen, which protects you from both UV-A and UV-B radiation, with an SPF (sun protection factor) of at least 15. (This means, for example, that if you usually burn after a 20-minute exposure, it will take 15 times that long before you will burn.)
- Stay out of the sun altogether between the hours of 10 A.M. and 3 P.M. This will reduce your annual exposure by as much as 60%.
- Wear protective clothing. Choose fabrics with a tight weave, and wear a wide-brimmed hat.
- Wear sunglasses that have been treated to absorb both UV-A and UV-B radiation. Otherwise, sunglasses can expose your eyes to more damage than usual because pupils dilate in the shade.
- Avoid tanning machines. Although most tanning devices use high levels of only UV-A, these rays cause the deep layers of the skin to become more vulnerable to UV-B radiation when you are later exposed to the sun.

Discussion Questions

1. Is it hard for you to imagine that pale, fair skin was once considered the most desirable look for Caucasian women? Can you think of any other social customs from the 1800s that might seem very strange, or even offensive, today?
2. In addition to the increased risk of cancer, skin damage from sun exposure is known to make the skin age faster and become more wrinkled. Which is more important to you—having a "healthy-looking" tan now, or preserving your skin's health later in life? Why?
3. What are the three main types of skin cancer? Which do you think would be the worst to have? Why?

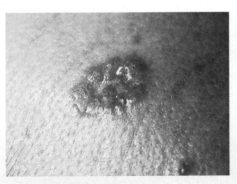

a. Basal cell carcinoma

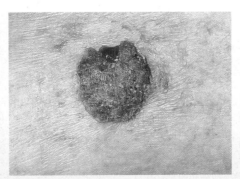

b. Squamous cell carcinoma

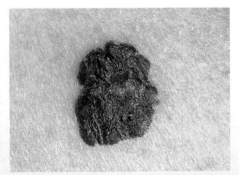

c. Melanoma

Figure 11A Skin cancer.
a. Basal cell carcinoma occurs when basal cells proliferate abnormally. **b.** Squamous cell carcinoma arises in epithelial cells derived from basal cells. **c.** Malignant melanoma results from a proliferation of pigmented cells.

After death, it is possible to give the "gift of life" to someone else. A single cadaver can provide over 25 organs and tissues for transplants. Transplantation of the kidney, heart, liver, pancreas, lung, and other organs is now possible due to two major breakthroughs. First, solutions have been developed that preserve donor organs for several hours, and second, effective immunosuppressive drugs prevent rejection of transplanted organs. These advances make donor organs usable for longer periods of time, and organs can be donated by unrelated individuals, living or dead.

Although these advances have made many more organs available for donation, there is still a limited availability of organs for transplantation. At any one time, at least 27,000 people in the United States are waiting for a donated organ. Keen competition for organs can lead to many bioethical inequities. When the governor of Pennsylvania received a heart and lungs within a relatively short period of time, it appeared that his social status might have played a role in his seemingly special treatment. In contrast, a heart transplant performed on a prison inmate in California in early 2002 stirred a debate about whether some individuals deserve access to organ transplantation at all.

Xenotransplantation—the transplanting of animal tissues into humans—sparked the public interest in 1984 when a 15-month-old baby with a defective heart lived for 20 days with a transplanted baboon's heart. Since then, surgeons have achieved limited success in transplanting a baboon liver and pig heart valves, as well as in using fetal pig nerve cells to treat Parkinson and Huntington diseases. This raises many ethical issues. Will these sorts of transplants be safe for humans or possibly spread new and dangerous diseases? And is it right to genetically alter animals to serve as a source of organs for humans? Such organs will most likely be for sale, and does this make the wealthy more likely to receive a transplant than those who cannot pay? Or, if these organs allow everyone who needs a transplant to receive one, might this have unforeseen consequences on our health-care system?

Form Your Own Opinion

1. Because the supply of organs currently limits the number of patients who can receive needed transplants, what types of factors should be considered in determining who gets the organs: age of the patient, severity of the illness, possession of health insurance/ability to pay, or reason the organ is needed? For example, should a person who has developed cirrhosis of the liver from excessive alcohol consumption get priority for a liver transplant?
2. If you were in desperate need of a liver transplant and the only available liver was a xenotransplant (i.e., from a pig that has been "humanized" so that the organ is less likely to be rejected), would you have any hesitation? Why or why not?
3. Are you signed up to be an organ donor? Why or why not?

Summarizing the Concepts

11.1 Types of Tissues

Human tissues are categorized into four groups. Epithelial tissue covers the body and lines its cavities. The different types of epithelial tissue (squamous, cuboidal, and columnar) can be simple, stratified, or pseudostratified and have cilia or microvilli. Epithelial cells sometimes form glands that secrete either into ducts or into the blood.

Connective tissues, in which cells are separated by a matrix, often bind body parts together. More solid connective tissues have both white and yellow fibers and reticular fibers and may also have fat (adipose) cells. Loose fibrous connective tissue supports epithelium and encloses organs. Dense fibrous connective tissue, such as that of tendons and ligaments, contains closely packed collagen fibers. Adipose tissue stores fat. Both cartilage and bone have cells within lacunae, but the matrix for cartilage is more flexible than that for bone, which contains calcium salts. Blood is a connective tissue in which the matrix is a liquid called plasma.

Muscular tissue is of three types. Both skeletal and cardiac muscle are striated; both cardiac and smooth muscle are involuntary. Skeletal muscle is found in muscles attached to bones, and smooth muscle is found in internal organs. Cardiac muscle makes up the heart.

Nervous tissue has one main type of conducting cell, the neuron, and several types of supporting neuroglia. Each neuron has a cell body, dendrites, and an axon. Axons are specialized to conduct nerve impulses.

11.2 Body Cavities and Body Membranes

The internal organs occur within cavities. The thoracic cavity contains the heart and lungs; the abdominal and pelvic cavities contain organs of the digestive, urinary, and reproductive systems, among others. Membranes line body cavities and the internal spaces of organs. Mucous membrane lines the tubes of the digestive system, while serous membrane lines the thoracic and abdominal cavities and covers the organs they contain. Synovial membranes line movable joint cavities. The meninges cover the brain and spinal cord.

11.3 Organ Systems

The digestive, cardiovascular, lymphatic, respiratory, and urinary systems perform processing and transport functions that maintain the normal conditions of the body. The immune system protects the body against infectious disease. The skeletal and muscular systems support the body and permit movement. The nervous system receives sensory input and directs muscles and glands to respond. The endocrine system produces hormones, some of which influence the functioning of the reproductive system. The integumentary system serves protective functions and also makes vitamin D, collects sensory data, and helps regulate body temperature.

11.4 Integumentary System

The integumentary system is composed of skin and the accessory organs (nails, hair, oil glands, and sweat glands). Skin has two regions. The epidermis contains basal cells that produce new epithelial cells, which become keratinized as they move toward the surface. The dermis, a largely fibrous connective tissue, contains epidermally derived glands and hair follicles, nerve endings, and blood vessels. Sensory receptors for touch, pressure, temperature, and pain are also in the dermis. A subcutaneous layer, made up of loose connective tissue containing adipose cells, lies beneath the skin.

11.5 Homeostasis

Homeostasis is the relative constancy of the internal environment. Negative feedback mechanisms keep the environment relatively stable. When a sensor detects a change above or below a set point, a control center brings about an effect that reverses the change and brings conditions back to normal again. In contrast, a positive feedback mechanism brings about rapid change in the same direction as the stimulus. Still, positive feedback mechanisms are useful under certain conditions, such as during childbirth.

The internal environment consists of blood and tissue fluid. All organ systems contribute to the constancy of tissue fluid and blood. Special contributions are made by the liver, which keeps the level of blood glucose constant, and the kidneys, which regulate the blood pH. The nervous and endocrine systems regulate the other systems.

Testing Yourself

Choose the best answer for each question.

1. A grouping of similar cells that perform a specific function is called a
   a. sarcoma.   c. tissue.
   b. membrane.   d. None of these are correct.

2. The smallest blood vessels are called
   a. veins.   c. arterioles.
   b. arteries.   d. capillaries.

3. _____ glands secrete directly into the _____.
   a. Exocrine, blood   c. Endocrine, urine
   b. Exocrine, arteries d. Endocrine, blood

4. Without a contribution from the connective tissue, the basement membrane would lack
   a. glycoprotein.   c. elastin.
   b. collagen fibers.   d. All of these are correct.

5. Tight junctions are associated with
   a. connective tissue. c. cartilage.
   b. adipose tissue.   d. epithelium.

6. The loss of _____ will prevent proper blood clotting.
   a. red blood cells   c. platelets
   b. white blood cells d. oxygen

7. Blood is a(n) _____ tissue because it has a _____.
   a. connective, gap junction   c. epithelial, gap junction
   b. muscle, matrix   d. connective, matrix

8. Skeletal muscle is
   a. striated.   c. multinucleated.
   b. under voluntary control.   d. All of these are correct.

9. A reduction in red blood cells would cause problems with
   a. fighting infection. c. blood clotting.
   b. carrying oxygen.   d. None of these are correct.

10. Which of these systems plays the biggest role in fluid balance?
    a. cardiovascular   c. digestive
    b. urinary   d. integumentary

11. A thoracic surgeon would be experienced in medicine pertaining to the
    a. heart.   c. bladder.
    b. lungs.   d. Both a and b are correct.

12. Which choice is true of both cardiac and skeletal muscle?
    a. striated   c. multinucleated cells
    b. single nucleus per cell   d. involuntary control

13. The skeletal system functions in
    a. blood cell production.   c. movement.
    b. mineral storage.   d. All of these are correct.

14. Which layer of the skin is involved in storing energy?
    a. epidermis   c. subcutaneous layer
    b. dermis   d. None of these are correct.

15. The correct order for homeostatic processing is
    a. sensory detection, control center, effect brings about change.
    b. control center, sensory detection, effect brings about change in environment.
    c. sensory detection, control center, effect causes no change in environment.
    d. None of these are correct.

16. Which allows rapid change in one direction and does not achieve stability?
    a. homeostasis   c. negative feedback
    b. positive feedback d. All of these are correct.

17. Which of these correctly describes a layer of the skin?
    a. The epidermis is simple squamous epithelium in which hair follicles develop and blood vessels expand when we are hot.
    b. The subcutaneous layer lies between the epidermis and the dermis. It contains adipose tissue, which keeps us warm.
    c. The dermis is a region of connective tissue that contains sensory receptors, nerve endings, and blood vessels.
    d. The skin has a special layer, still unnamed, in which there are all the accessory structures such as nails, hair, and various glands.

18. Without melanocytes, skin would
    a. be too thin.   c. lack color.
    b. lack nerves.   d. None of these are correct.

19. Label the following diagram of human skin.

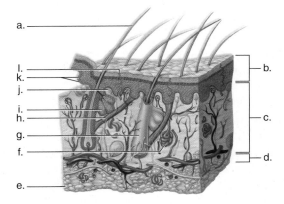

20. Label the following diagram of body cavities.

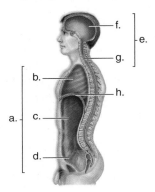

21. Which of the following is an example of negative feedback?
    a. Air conditioning goes off when room temperature lowers.
    b. Insulin decreases blood sugar levels after eating a meal.
    c. Heart rate increases when blood pressure drops.
    d. All of these are examples of negative feedback.
22. Which of the following is a function of skin?
    a. temperature regulation
    b. manufacture of vitamin D
    c. collection of sensory input
    d. protection from invading pathogens
    e. All of these are correct.

Understanding the Terms

acute disease 212	lacuna 200
adhesion junction 200	ligament 200
adipose tissue 200	localized 212
basement membrane 198	loose fibrous connective tissue 200
blood 202	
bone 201	lymphatic system 206
capillary 198	matrix 200
carcinoma 198	melanocyte 208
cardiac muscle 203	meninges 205
cardiovascular system 206	microvillus 198
cartilage 200	mucous membrane 205
chronic disease 212	muscular system 207
cilium 198	muscular tissue 202
coelom 205	negative feedback 209
collagen fiber 200	nerve 204
columnar epithelium 198	nervous system 207
compact bone 201	nervous tissue 204
connective tissue 200	neuroglia 204
cuboidal epithelium 198	neuron 204
dense fibrous connective tissue 200	oil gland 209
	plasma 202
dermis 208	platelet 202
digestive system 206	positive feedback 211
disease 212	red blood cell 202
elastic cartilage 201	reproductive system 207
elastic fiber 200	respiratory system 206
endocrine system 207	reticular connective tissue 200
epidermis 208	
epithelial tissue 198	reticular fiber 200
fibroblast 200	serous membrane 205
fibrocartilage 201	skeletal muscle 203
gap junction 200	skeletal system 207
gland 198	skin 208
hair follicle 209	smooth muscle 203
homeostasis 209	spongy bone 201
hyaline cartilage 201	squamous epithelium 198
immune system 206	striated 203
integumentary system 206	subcutaneous layer 208
intercalated disk 203	sweat gland 209

synovial membrane 205	tissue 198
systemic 212	urinary system 207
tendon 200	white blood cell 202
tight junction 198	

Match the terms to these definitions:

a._____ Fibrous connective tissue that joins bone to bone at a joint.
b._____ Outer region of the skin composed of stratified squamous epithelium.
c._____ Having bands, as in cardiac and skeletal muscle.
d._____ Self-regulatory mechanism that is activated by an imbalance and results in a fluctuation above and below a mean.
e._____ Porous bone found at the ends of long bones where blood cells are formed.

Thinking Critically

1. In what way(s) is blood like a tissue?
2. Think about what types of environmental conditions germs might favor. What are some features of human skin that make it a good protection against most infectious agents, such as bacteria and viruses?
3. Which of these homeostatic mechanisms in the body are examples of positive feedback, and which are examples of negative feedback? Why?
   a. The adrenal glands produce epinephrine in response to a hormone produced by the pituitary gland in times of stress; the pituitary gland senses the epinephrine in the blood and stops producing the hormone.
   b. As the bladder fills with urine, pressure sensors send messages to the brain with increasing frequency signaling that the bladder must be emptied. The more the bladder fills, the more messages are sent.
   c. When you drink an excess of water, specialized cells in your brain as well as stretch receptors in your heart detect the increase in blood volume. Both signals are transmitted to the kidneys, which increase the production of urine.
4. Homeostatic systems in the body can be categorized as transport systems, maintenance systems, support systems, and control systems. What organ systems (from section 11.3) could be placed into more than one of these categories?
5. What are one or two characteristics that all types of cancer have in common? Why do they often have such different manifestations, such as leukemia compared to a melanoma?

ARIS, the *Inquiry into Life* Website

ARIS, the website for *Inquiry into Life*, provides a wealth of information organized and integrated by chapter. You will find practice quizzes, interactive activities, labeling exercises, flashcards, and much more that will complement your learning and understanding of general biology.

www.aris.mhhe.com

As the nurse gently slipped a needle into a vein of her outstretched arm, *Beth thought about her friend who desperately needed this gift of life. Beth had the same blood type as her friend, who had been in an automobile accident. As the blood flowed into the donor bag, Beth realized the importance of her decision to help her friend. Blood, a vital fluid, carries oxygen from the lungs and nutrients from the intestines to all the body's cells. It also takes carbon dioxide to the lungs and wastes to the kidneys. Still other functions of blood include helping to fight infection, regulating body temperature, and coordinating the activity of body tissues. A severe loss of blood must be replaced by transfusion if life is to continue.*

Humans have a closed cardiovascular system: the blood never runs free but is conducted to and from the tissues by blood vessels. Only the capillaries have walls thin enough to allow exchange of molecules with the tissues. The heart is the organ that keeps the blood moving through the vessels to the capillaries. Beth knew that if the heart fails to pump the blood for even a few minutes, an individual's life is in danger. The body has many mechanisms for ensuring that blood remains in the vessels and under a pressure that will maintain its transport function.

chapter 12

Cardiovascular System

Chapter Concepts

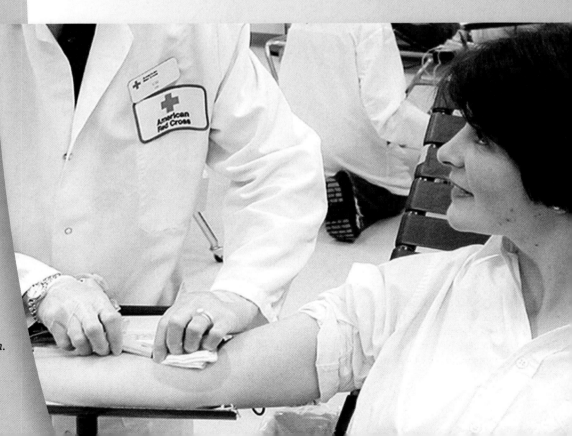

12.1 The Blood Vessels

The cardiovascular system has three types of blood vessels: the **arteries** (and arterioles), which carry blood away from the heart to the capillaries; the **capillaries,** which permit exchange of material with the tissues; and the **veins** (and venules), which return blood from the capillaries to the heart.

Like other tissues, the blood vessels require oxygen and nutrients, and therefore the larger ones have blood vessels in their own walls.

The Arteries

An arterial wall has three layers (Fig. 12.1a). The inner layer is a simple squamous epithelium called **endothelium** with a connective tissue basement membrane that contains elastic fibers. The middle layer is the thickest layer and consists of smooth muscle that can contract to regulate blood flow and blood pressure. The outer layer is fibrous connective tissue near the middle layer, but it becomes loose connective tissue at its periphery. The largest artery in the human body is the aorta. It is approximately 25 mm wide and carries O$_2$-rich blood from the heart to other parts of the body.

Smaller arteries branch into a number of arterioles. **Arterioles** are small arteries just visible to the naked eye, being under 0.5 mm in diameter. The inner layer of arterioles is endothelium, and the middle layer is composed of some elastic tissue but mostly of smooth muscle with fibers that encircle the arteriole. When these muscle fibers are contracted, the vessel has a smaller diameter (is constricted); when these muscle fibers are relaxed, the vessel has a larger diameter (is dilated). Whether arterioles are constricted or dilated affects blood pressure. The greater the number of vessels dilated, the lower the blood pressure.

Arteries and arterioles carry blood away from the heart. They have three layers: an outer, fibrous layer; a middle layer of smooth muscle; and an inner endothelium. Constriction and dilation of arterioles regulates blood flow and blood pressure.

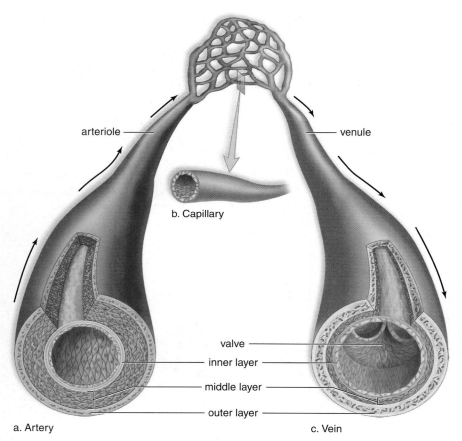
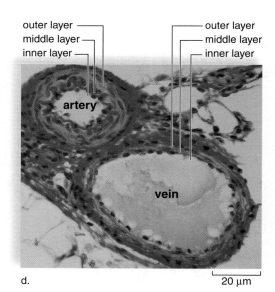

Figure 12.1 Blood vessels.
The walls of arteries and veins have three layers. The inner layer is composed largely of endothelium, with a basement membrane that has elastic fibers; the middle layer is smooth muscle tissue; the outer layer is connective tissue (largely collagen fibers). **a.** Arteries have a thicker wall than veins because they have a larger middle layer than veins. **b.** Capillary walls are one-cell-thick endothelium. **c.** Veins are larger in diameter than arteries, so collectively, veins have a larger holding capacity than arteries. **d.** Light micrograph of an artery and a vein.

The Capillaries

Capillaries join arterioles to venules (Fig. 12.1*b*). Capillaries are extremely narrow—about 8–10 µm wide—and have thin walls composed only of a single layer of endothelium with a basement membrane. Although each capillary is small, they form vast networks; their total surface area in a human is about 6,000 square meters. Capillary beds (networks of many capillaries) are present in nearly all regions of the body; consequently, a cut to almost any body tissue draws blood.

Capillaries are a very important part of the human cardiovascular system because an exchange of substances takes place across their thin walls. Oxygen and nutrients, such as glucose, diffuse out of a capillary into the tissue fluid that surrounds cells. Wastes, such as carbon dioxide, diffuse into the capillary. Some water also leaves a capillary; any excess is picked up by lymphatic vessels, as discussed in Section 12.4. The relative constancy of tissue fluid is absolutely dependent upon capillary exchange.

Because capillaries serve the cells, the heart and the other vessels of the cardiovascular system can be thought of as the means by which blood is conducted to and from the capillaries. Only certain capillary beds are completely

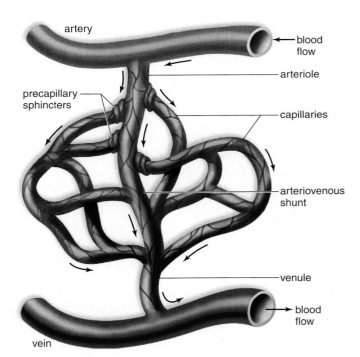

Figure 12.2 Anatomy of a capillary bed.
A capillary bed forms a maze of capillary vessels that lies between an arteriole and a venule. When precapillary sphincter muscles are relaxed, the capillary bed is open, and blood flows through the capillaries. When sphincter muscles are contracted, blood flows through a shunt that carries blood directly from an arteriole to a venule. As blood passes through a capillary in the tissues, it gives up its oxygen (O_2). Therefore, blood goes from being O_2-rich in the arteriole (red color) to being O_2-poor in the vein (blue color).

open at any given time. For example, after eating, the capillary beds that serve the digestive system are mostly open, and those that serve the muscles are mostly closed. Each capillary bed has an arteriovenous shunt that allows blood to go directly from the arteriole to the venule, bypassing the bed (Fig. 12.2). Contracted precapillary sphincter muscles prevent the blood from entering the capillary vessels.

A decreased flood flow to the brain after meals has also been offered as an explanation for "postprandial somnolence," or the sleepiness that many people feel after eating. However, recent evidence suggests that the blood supply to the brain is maintained under most physiological conditions, including ingestion of a heavy meal, and that hormones released by the digestive tract may instead be the culprit.

Capillaries join arterioles to venules and consist of a single layer of endothelium attached to a basement membrane. Capillaries play a critically important role in the exchange of gases, nutrients, and wastes between tissues and blood.

The Veins

Veins and venules take blood from the capillary beds to the heart. First, the **venules** (small veins) drain blood from the capillaries and then join to form a vein. The walls of veins (and venules) have the same three layers as arteries, but there is less smooth muscle and connective tissue (see Fig. 12.1*c*). Therefore, the wall of a vein is thinner than that of an artery.

Veins often have **valves,** which allow blood to flow only toward the heart when open and prevent blood from flowing backward when closed. Valves are found in the veins that carry blood against the force of gravity, especially the veins of the lower limbs. Unlike blood flow in the arteries and arterioles, which is kept moving by the pumping of the heart, blood flow in veins is primarily due to skeletal muscle contraction.

Because the walls of veins are thinner, they can expand to a greater extent (see Fig. 12.1*d*). At any one time, about 70% of the blood is in the veins. In this way, the veins act as a blood reservoir. If blood is lost due to hemorrhaging, nervous stimulation causes the veins to constrict, providing more blood to the rest of the body. The largest veins in the body are the superior vena cava (20 mm wide) and the inferior vena cava (35 mm wide). These veins deliver O_2-poor blood into the heart.

Veins and venules return blood from the capillaries to the heart. Their walls are thinner than those of arteries and arterioles, and some of them have one-way valves. Veins and venules contain more blood than the arteries, arterioles, and capillaries combined.

12.2 The Human Heart

The **heart** is a cone-shaped, muscular organ about the size of a fist (Fig. 12.3). It is located between the lungs directly behind the sternum (breastbone) and is tilted so that the apex (the pointed end) is oriented to the body's left. The major portion of the heart, called the **myocardium,** consists largely of cardiac muscle tissue. The muscle fibers of the myocardium are branched and tightly joined to one another. The heart lies within the **pericardium,** a thick, serous membrane that secretes a small quantity of lubricating liquid. The inner surface of the heart is lined with endocardium, a membrane composed of connective tissue and endothelial tissue. The lining is continuous with the endothelium lining the blood vessels.

Internally, a wall called the septum separates the heart into a right side and a left side (Fig. 12.4). The heart has four chambers. The two upper, thin-walled atria (sing., **atrium**) are located above the two lower, thick-walled **ventricles.** The ventricles pump the blood to the lungs and the body.

The heart also has four valves that direct the flow of blood and prevent its backward movement. The two valves that lie between the atria and the ventricles are called the **atrioventricular valves.** These valves are supported by strong fibrous strings called **chordae tendineae.** The chordae, which are attached to muscular projections of the ventricular walls, support the valves and prevent them from inverting when the heart contracts. The atrioventricular valve on the body's right side is called the tricuspid valve because it has three flaps, or cusps. The valve on the left side is called the bicuspid (or mitral) valve because it has two flaps. The remaining two valves are the **semilunar valves,** whose flaps resemble half-moons, between the ventricles and their attached vessels. The pulmonary semilunar valve lies between the right ventricle and the pulmonary trunk. The aortic semilunar valve lies between the left ventricle and the aorta.

Humans have a four-chambered heart (two atria and two ventricles). A septum separates the right side from the left side.

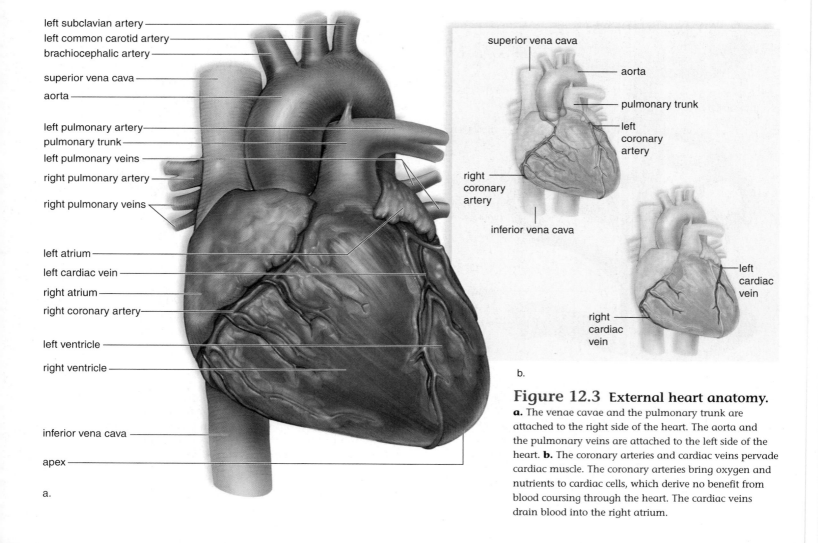

Figure 12.3 External heart anatomy.
a. The venae cavae and the pulmonary trunk are attached to the right side of the heart. The aorta and the pulmonary veins are attached to the left side of the heart. **b.** The coronary arteries and cardiac veins pervade cardiac muscle. The coronary arteries bring oxygen and nutrients to cardiac cells, which derive no benefit from blood coursing through the heart. The cardiac veins drain blood into the right atrium.

Path of Blood Through the Heart

By referring to Figure 12.4, we can trace the path of blood through the heart in the following manner:

- The superior vena cava and the inferior vena cava, which carry O_2-poor blood that is relatively high in carbon dioxide, enter the right atrium.
- The right atrium sends blood through an atrioventricular valve (the tricuspid valve) to the right ventricle.
- The right ventricle sends blood through the pulmonary semilunar valve into the pulmonary trunk and through the two **pulmonary arteries** to the lungs.
- Four **pulmonary veins,** which carry O_2-rich blood, enter the left atrium.
- The left atrium sends blood through an atrioventricular valve (the bicuspid valve) to the left ventricle.
- The left ventricle sends blood through the aortic semilunar valve into the aorta to the rest of the body.

From this description, it is obvious that O_2-poor blood never mixes with O_2-rich blood, and that blood must go through the lungs in order to pass from the right side to the left side of the heart. In fact, the heart is a double pump because the right ventricle sends blood into the lungs and the left ventricle sends blood into the rest of the body. Because the left ventricle has the harder job of pumping blood to the entire body, its walls are thicker than those of the right ventricle, which pumps blood a relatively short distance to the lungs. In a person with an average heart rate of 70 beats per minute, the output of the left ventricle is about 5.25 liters of blood per minute, which is about equal to the total amount of blood in the body.

The pumping of the heart sends blood out under pressure into the arteries. Because the left side of the heart is the stronger pump, blood pressure is greatest in the aorta. Blood pressure then decreases as the cross-sectional area of arteries and then arterioles increases. The **pulse** is a wave effect that passes down the walls of arteries when the aorta expands and then recoils with each ventricular contraction. The arterial pulse can be used to determine the heart rate, and a weak or "thready" pulse may indicate a weak heart or low blood pressure.

The right side of the heart pumps O_2-poor blood to the lungs and the left side of the heart pumps O_2-rich blood to the rest of the body.

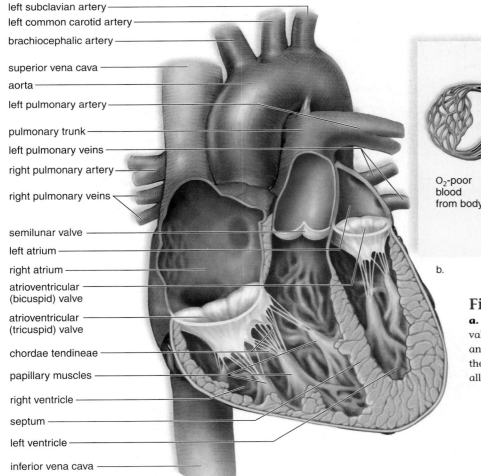

a.

b.

Figure 12.4 Internal view of the heart.

a. The heart has four valves. When the atrioventricular valves open, blood passes from the atria to the ventricles, and when the semilunar valves open, blood passes out of the heart. **b.** This diagrammatic representation of the heart allows you to trace the path of the blood through the heart.

The Heartbeat

Each heartbeat is called a **cardiac cycle** (Fig. 12.5). When the heart beats, first the two atria contract at the same time; then the two ventricles contract at the same time. Then all of the chambers relax. The word **systole** refers to contraction of heart muscle, and the word **diastole** refers to relaxation of heart muscle. The heart contracts, or beats, about 70 times a minute, and each heartbeat lasts about 0.85 second, apportioned as follows:

Cardiac Cycle		
Time	Atria	Ventricles
0.15 sec	Systole	Diastole
0.30 sec	Diastole	Systole
0.40 sec	Diastole	Diastole

A normal adult rate at rest can vary from 60 to 80 beats per minute, which adds up to about 2.5 million beats in a lifetime!

When the heart beats, the familiar "lub-dup" sound occurs. The longer and lower-pitched "lub" is caused by vibrations occurring when the atrioventricular valves close due to ventricular contraction. The shorter and sharper "dup" results when the semilunar valves close due to back pressure of blood in the arteries. A heart murmur, or a slight slush sound after the "lub," is often due to ineffective valves, which allow blood to pass back into the atria after the atrioventricular valves have closed. Congenital birth defects and rheumatic fever resulting from a bacterial infection are possible causes of faulty valves. Faulty valves can be surgically corrected or replaced.

Intrinsic Control of Heartbeat

The rhythmic contraction of the atria and ventricles is due to the intrinsic (or internal) conduction system of the heart that is made possible by the presence of nodal tissue, a unique type of cardiac muscle. Nodal tissue, which has both muscular and nervous characteristics, is located in two regions of the heart. The **SA (sinoatrial) node** is located in the upper dorsal wall of the right atrium; the **AV (atrioventricular) node** is located in the base of the right atrium very near the septum (Fig. 12.6*a*). The SA node initiates the heartbeat and sends out an excitation impulse every 0.85 second; this causes the atria to contract. When impulses reach the AV node, a slight delay allows the atria to finish contraction before the ventricles begin to contract. The signal for the ventricles to contract travels from the AV node through specialized cardiac muscle fibers called the **atrioventricular bundle** (AV bundle) before reaching the numerous and smaller **Purkinje fibers.**

The SA node is called the **cardiac pacemaker** because it usually keeps the heartbeat regular. If the SA node fails to

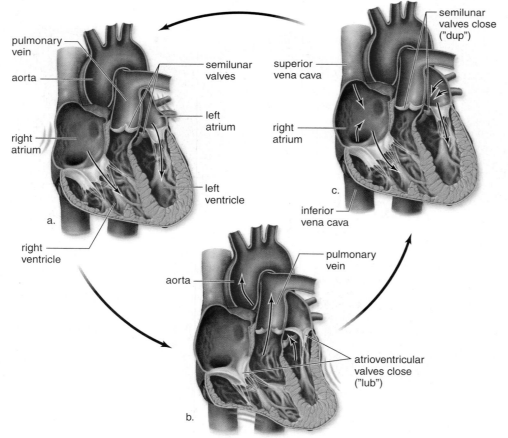

Figure 12.5 Generation of heart sounds during the cardiac cycle.
a. When the atria contract, the ventricles are relaxed and filling with blood. **b.** When the ventricles contract, the atrioventricular valves close, preventing blood from flowing back into the atria and producing the "lub" sound of the heartbeat. **c.** After the ventricles contract, the "dup" sound of the heartbeat results from the closing of the semilunar valves to prevent arterial blood from flowing back into the ventricles.

work properly, the heart still beats due to impulses generated by the AV node. But the beat is slower (40 to 60 beats per minute). To correct this condition, an artificial pacemaker may be implanted that gives an electrical stimulus to the heart every 0.85 second.

Extrinsic Control of Heartbeat

The body also has extrinsic (or external) ways to regulate the heartbeat. A cardiac control center in the medulla oblongata, a portion of the brain that controls internal organs, can alter the beat of the heart by way of the autonomic system, a portion of the nervous system. The autonomic system has two subdivisions: the parasympathetic division, which promotes functions of a resting state, and the sympathetic division, which brings about responses to increased activity or stress. The parasympathetic division decreases SA and AV nodal activity when we are inactive, and the sympathetic system increases these nodes' activity when we are active or excited.

The hormones epinephrine and norepinephrine, which are released by the adrenal medulla (inner portion of the adrenal gland), also stimulate the heart. During exercise, for example, the heart pumps faster and stronger due to sympathetic stimulation and because of the release of epinephrine and norepinephrine.

Each heartbeat produces sounds caused by the closing of the atrioventricular valves ("lub"), followed by the semilunar valves ("dup"). The intrinsic conduction system of the heart consists of the SA node, the AV node, the AV bundle, and the Purkinje fibers. The nervous and endocrine systems exert extrinsic control over the heartbeat.

The Electrocardiogram

An **electrocardiogram (ECG)** is a recording of the electrical changes that occur in the myocardium during a cardiac cycle. Body fluids contain ions that conduct electrical currents, and therefore the electrical changes in the myocardium can be detected on the skin's surface. Electrodes placed on the skin are connected by wires to an instrument that detects and records the myocardium's electrical changes. Figure 12.6b depicts the electrical changes during a normal cardiac cycle.

When the SA node triggers an impulse, the atrial fibers produce an electrical change called the P wave. The P wave indicates that the atria are about to contract. After that, the QRS complex signals that the ventricles are about to contract. The electrical changes that occur as the ventricular muscle fibers recover produce the T wave.

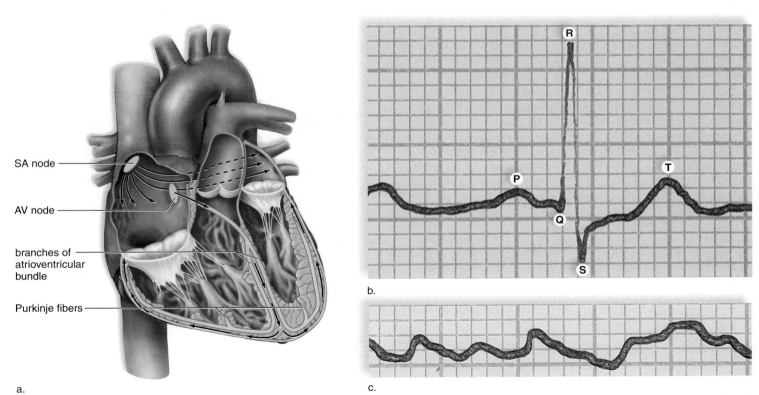

a.   b.   c.

Figure 12.6 Conduction system of the heart.
a. The SA node sends out a stimulus (black arrows), which causes the atria to contract. When this stimulus reaches the AV node, it signals the ventricles to contract. Impulses pass down the two branches of the atrioventricular bundle to the Purkinje fibers, and thereafter the ventricles contract. **b.** A normal ECG usually indicates that the heart is functioning properly. The P wave occurs just prior to atrial contraction; the QRS complex occurs just prior to ventricular contraction; and the T wave occurs when the ventricles are recovering from contraction. **c.** Ventricular fibrillation produces an irregular electrocardiogram due to irregular stimulation of the ventricles.

Various types of abnormalities can be detected by an electrocardiogram. One of these, called ventricular fibrillation, causes uncoordinated contraction of the ventricles (Fig. 12.6c). Ventricular fibrillation is of special interest because it can be caused by an injury or drug overdose. It is the most common cause of sudden cardiac death in a seemingly healthy person over age 35. Once the ventricles are fibrillating, they must be defibrillated by applying a strong electrical current for a short period of time. Then the SA node may be able to reestablish a coordinated beat.

> An electrocardiogram (ECG) displays the electrical changes that occur in the myocardium during the cardiac cycle. Each normal heartbeat results in a P wave, QRS complex, and T wave. Certain heart abnormalities can be detected with an ECG.

12.3 The Vascular Pathways

The cardiovascular system includes two circuits (Fig. 12.7). The **pulmonary circuit** circulates blood through the lungs, and the **systemic circuit** serves the needs of body tissues.

The Pulmonary Circuit

Blood from all regions of the body first collects in the right atrium and then passes into the right ventricle, which pumps it into the pulmonary trunk. The pulmonary trunk divides into the right and left pulmonary arteries, which branch as they approach the lungs. The arterioles take blood to the pulmonary capillaries, where gas exchange occurs. Blood then passes through the pulmonary venules, which lead to the four pulmonary veins that enter the left atrium. Since blood in the pulmonary arteries is O_2-poor but blood in the pulmonary veins is O_2-rich, it is not correct to say that all arteries carry blood high in oxygen and all veins carry blood low in oxygen. It is just the reverse in the pulmonary circuit.

> The pulmonary arteries take O_2-poor blood to the lungs, and the pulmonary veins return blood that is O_2-rich to the heart.

The Systemic Circuit

The systemic circuit includes all of the arteries and veins shown in Figure 12.8. The largest artery in the systemic circuit is the **aorta,** and the largest veins are the **superior vena cava** and the **inferior vena cava.** The superior vena cava collects blood from the head, the chest, and the arms, and the inferior vena cava collects blood from the lower body regions. Both

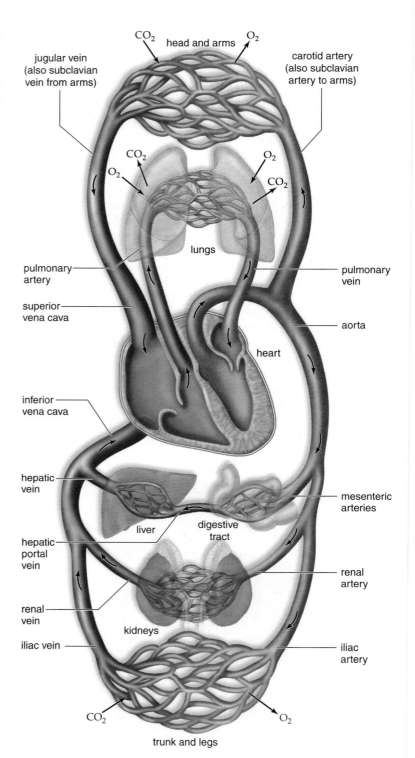

Figure 12.7 Path of blood.

Tracing blood from the right to the left side of the heart in the pulmonary circuit, involves the pulmonary vessels. Tracing blood from the digestive tract to the right atrium in the systemic circuit involves the hepatic portal vein, the hepatic vein, and the inferior vena cava. The blue colored vessels carry O_2-poor blood, and the red colored vessels carry O_2-rich blood; the arrows indicate the direction of blood flow.

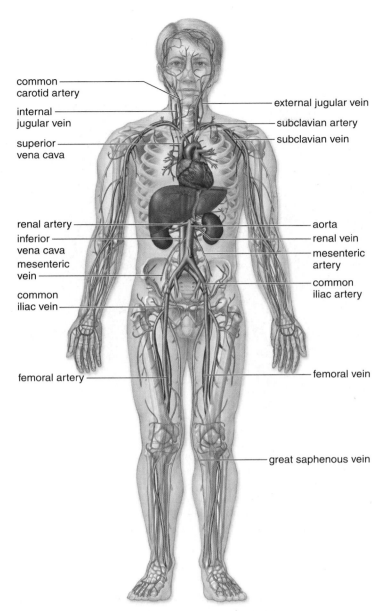

common carotid artery

internal jugular vein

superior vena cava

external jugular vein

subclavian artery

subclavian vein

renal artery

inferior vena cava

mesenteric vein

common iliac vein

aorta

renal vein

mesenteric artery

common iliac artery

femoral artery

femoral vein

great saphenous vein

Figure 12.8 Major arteries (red) and veins (blue) of the systemic circuit.
This representation of the major blood vessels of the systemic circuit shows how the systemic arteries and veins are arranged in the body. The superior and inferior venae cavae take their names from their relationship to which organ?

enter the right atrium. The aorta and the venae cavae serve as the major pathways for blood in the systemic circuit.

The path of systemic blood to any organ in the body begins in the left ventricle. For example, trace the path of blood to and from the legs in Figure 12.8:

left ventricle—aorta—common iliac artery—femoral artery—leg capillaries—femoral vein—common iliac vein—inferior vena cava—right atrium

Notice that, when tracing blood, you need only mention the aorta, the proper branch of the aorta, the region,

and the vein returning blood to the vena cava. In most instances, the artery and the vein that serve the same region are given the same name (Fig. 12.8).

The **coronary arteries** (see Fig. 12.3) serve the heart muscle itself. (The heart is not nourished by the blood in its own chambers.) The coronary arteries are the first branches off the aorta. They originate just above the aortic semilunar valve, and they lie on the exterior surface of the heart, where they divide into diverse arterioles. Because they have a very small diameter, the coronary arteries may become clogged, as discussed in section 12.5. The coronary capillary beds join to form venules. The venules converge to form the cardiac veins, which empty into the right atrium.

A *portal system* in blood circulation begins and ends in capillaries. One such system, the **hepatic portal system,** is associated with the liver. Capillaries that occur in the villi of the small intestine pass into venules that join to form the **hepatic portal vein.** This vein carries the blood to a set of capillaries in the liver, an organ that monitors the makeup of the blood (see Fig. 12.7). The **hepatic vein** leaves the liver and enters the inferior vena cava.

While Figure 12.7 is helpful in tracing the path of blood, remember that all parts of the body have both arteries and veins, as illustrated in Figure 12.8.

The systemic circuit takes blood from the left ventricle of the heart, through the body, and back to the right atrium. Coronary arteries serve the heart muscle itself; portal systems, such as the hepatic portal system, begin and end in capillaries.

Blood Pressure

When the left ventricle contracts, blood is forced into the aorta and then into other systemic arteries under pressure. **Systolic pressure** results from blood being forced into the arteries during ventricular systole, and **diastolic pressure** is the pressure in the arteries during ventricular diastole. Human **blood pressure** can be measured with a sphygmomanometer, which has a pressure cuff that determines the amount of pressure required to stop the flow of blood through an artery. Blood pressure is normally measured on the brachial artery, an artery in the upper arm. Today, digital manometers are often used to take one's blood pressure instead. Blood pressure is expressed in millimeters of mercury (mm Hg). A blood pressure reading consists of two numbers that represent systolic and diastolic pressures, respectively—for example, 120/80.

As blood flows from the aorta into the arteries and arterioles, blood pressure falls. Also, the difference between systolic and diastolic pressure gradually diminishes. In the

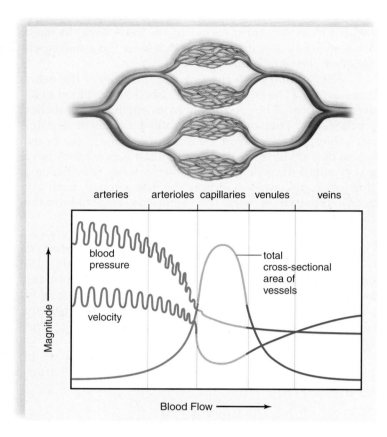

Figure 12.9 **Blood velocity and blood pressure.**
In capillaries, blood is under minimal pressure and has the least velocity. Blood pressure and velocity drop off because capillaries have a greater total cross-sectional area than arterioles.

capillaries, blood flow is slow and fairly even. This may be related to the very high total cross-sectional area of the capillaries (Fig. 12.9). It has been calculated that if all the blood vessels in a human were connected end to end, the total distance would reach around Earth at the equator two times! A large portion of this distance would be due to the quantity of capillaries.

Blood pressure in the veins is low and cannot move blood back to the heart, especially from the limbs. When skeletal muscles near veins contract, they put pressure on the veins and the blood they contain. Valves prevent the backward flow of blood in veins, and therefore muscle contraction is sufficient to move blood toward the heart (Fig. 12.10). **Varicose veins,** abnormal dilations in superficial veins, develop when the valves of the veins become weak and ineffective due to backward pressure from the blood. Varicose veins of the anal canal are known as hemorrhoids.

The beat of the heart keeps blood moving in arteries, while skeletal muscle contraction pushes blood in veins toward the heart. Valves prevent backward flow of blood in veins.

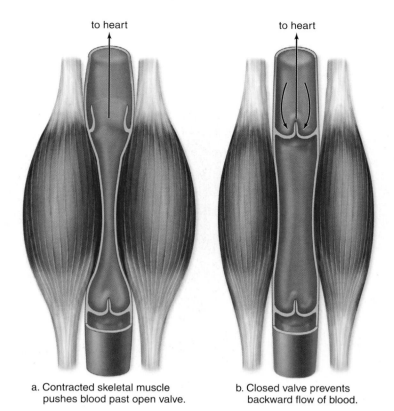

a. Contracted skeletal muscle pushes blood past open valve.

b. Closed valve prevents backward flow of blood.

Figure 12.10 **Cross section of a valve in a vein.**
a. Pressure on the walls of a vein, exerted by skeletal muscles, increases blood pressure within the vein and forces the valve open. **b.** When external pressure is no longer applied to the vein, blood pressure decreases, and back pressure forces the valve closed. Closure of the valve prevents the blood from flowing in the opposite direction.

12.4 Blood

If blood is collected from a person's vein into a test tube and prevented from clotting, it separates into two layers (Fig. 12.11a). The lower layer consists of the **formed elements,** and the upper layer is **plasma,** the liquid protein of blood. The formed elements consist of red blood cells, white blood cells, and blood platelets.

Blood has transport functions, regulatory functions, and protective functions. In addition to nutrients and wastes, blood also transports hormones. Blood helps regulate body temperature by dispersing body heat and helps regulate blood pressure because the plasma proteins contribute to the osmotic pressure of blood. Buffers in blood help maintain blood pH at about 7.4. Blood helps protect the body against invasion by disease-causing viruses and bacteria and against a potentially life-threatening loss of blood by clotting.

Plasma

Plasma contains a variety of inorganic and organic substances dissolved or suspended in water (Fig. 12.11b). Plasma proteins, which make up 7–8% of plasma, assist in

Figure 12.11 Composition

of blood. a. When blood is collected into a test tube containing an anticoagulant to prevent clotting and then centrifuged, it consists of three layers. The transparent straw-colored or yellow top layer is the plasma, the liquid portion of the blood. The thin, middle buffy coat layer consists of leukocytes and platelets. The bottom layer contains the erythrocytes. **b.** Breakdown of the components or plasma. **c.** Micrograph of the formed elements, which are also listed in (**d**).

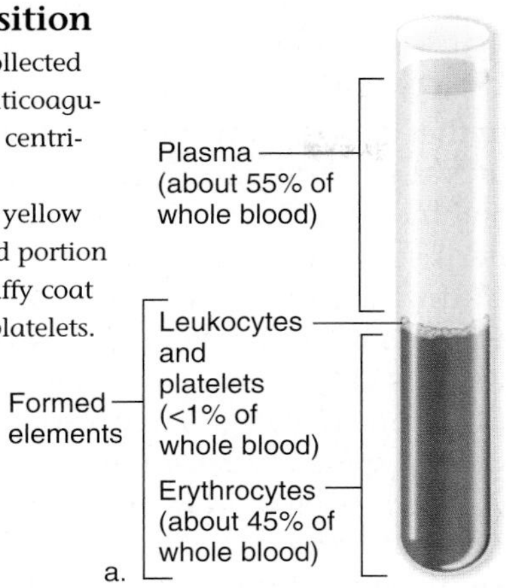

Plasma (about 55% of whole blood)

Leukocytes and platelets (<1% of whole blood)

Formed elements

Erythrocytes (about 45% of whole blood)

a.

Plasma		
Type	**Function**	**Source**
Water (90–92% of plasma)	Maintains blood volume; transports molecules	Absorbed from intestine
Plasma proteins (7–8% of plasma)	Maintain blood osmotic pressure and pH	
Albumin	Maintains blood volume and pressure	Liver
Globulins	Transport; fight infection	B lymphocytes
Fibrinogen	Clotting	Liver
Salts (less than 1% of plasma)	Maintain blood osmotic pressure and pH; aid metabolism	Absorbed from intestine
Gases		
Oxygen	Cellular respiration	Lungs
Carbon dioxide	End product of metabolism	Tissues
Nutrients		
Lipids	Food for cells	Absorbed from intestine
Glucose		
Amino acids		
Nitrogenous wastes	Excretion by kidneys	Liver
Urea		
Uric acid		
Other		
Hormones, vitamins, etc.	Aid metabolism	Varied

b.

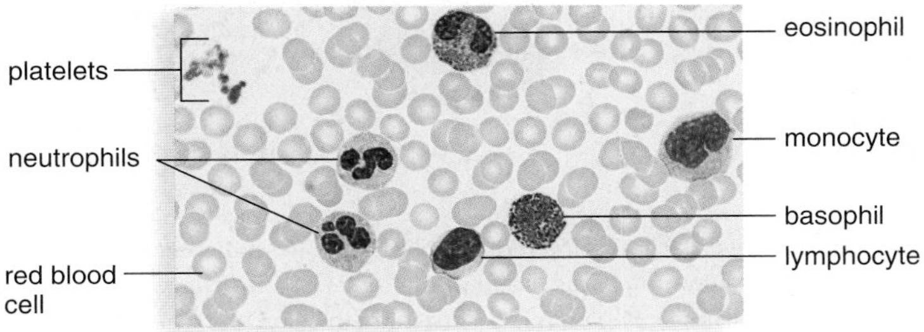

platelets

neutrophils

red blood cell

eosinophil

monocyte

basophil

lymphocyte

c.   250×

Formed Elements		
Type	**Function and Description**	**Source**
Red blood cells (erythrocytes)	Transport O_2 and help transport CO_2	Red bone marrow
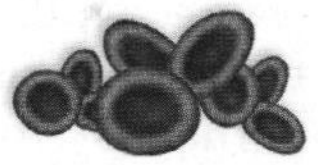 4 million–6 million per mm^3 blood	7–8 µm in diameter Bright-red to dark-purple biconcave disks without nuclei	
White blood cells (leukocytes) 5,000–11,000 per mm^3 blood	Fight infection	Red bone marrow
Neutrophils 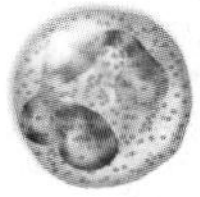 ★ 40–70%	10–14 µm in diameter Spherical cells with multilobed nuclei; fine, pink granules in cytoplasm; phagocytize pathogens	
Eosinophils 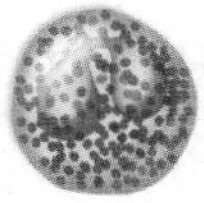 ★ 1–4%	10–14 µm in diameter Spherical cells with bilobed nuclei; coarse, deep-red, uniformly sized granules in cytoplasm; phagocytize antigen-antibody complexes and allergens	
Basophils 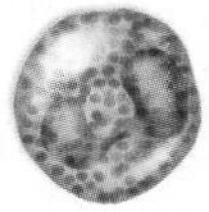 ★ 0–1%	10–12 µm in diameter Spherical cells with lobed nuclei; large, irregularly shaped, deep-blue granules in cytoplasm; release histamine, which promotes blood flow to injured tissues	
Lymphocytes  ★ 20–45%	5–17 µm in diameter (average 9–10 µm) Spherical cells with large, round nuclei; responsible for specific immunity	
Monocytes 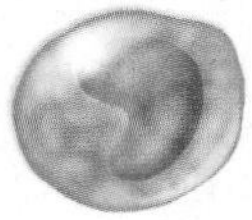 ★ 4–8%	10–24 µm in diameter Large spherical cells with kidney-shaped, round, or lobed nuclei; become macrophages that phagocytize pathogens and cellular debris	
Platelets (thrombocytes)	Aid clotting	Red bone marrow
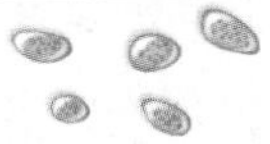 150,000–300,000 per mm^3 blood	2–4 µm in diameter Disk-shaped cell fragments with no nuclei; purple granules in cytoplasm	

Granular leukocytes

Agranular leukocytes

d.   ★ Appearance with Wright's stain.

transporting large organic molecules in blood. For example, **albumin** transports bilirubin, a breakdown product of hemoglobin. Lipoproteins transport cholesterol. Certain plasma proteins have specific functions. As discussed later in this section, fibrinogen is necessary to blood clotting, and immunoglobulins are antibodies that help fight infection. Plasma proteins also maintain blood volume because they are too large to leave capillaries. Therefore, capillaries always have lower water concentration than does tissue fluid, and water automatically diffuses into them.

> Plasma is the liquid part of blood. It contains various inorganic and organic substances that perform homeostatic functions, such as maintenance of blood pressure and pH, transport, clotting, and defense against pathogens.

The Red Blood Cells

Red blood cells (erythrocytes) are continuously manufactured in the red bone marrow of the skull, the ribs, the vertebrae, and the ends of the long bones. Normally, there are 4 to 6 million red blood cells per mm^3 of whole blood (see Fig. 12.11d).

Mature red blood cells don't have a nucleus and are biconcave disks (Fig. 12.12a, b). Their shape increases their flexibility for moving through capillary beds and their surface area for diffusion of gases. Red blood cells carry oxygen because they contain **hemoglobin,** the respiratory pigment. Because hemoglobin is a red pigment, the cells are red. A hemoglobin molecule contains four polypeptide chains (Fig. 12.12c). Each chain is associated with heme, a complex iron-containing group. The iron portion of hemoglobin acquires oxygen in the lungs and gives it up in the tissues. Hemoglobin also helps to carry carbon dioxide from tissues back to the lungs.

Carbon monoxide is a colorless and odorless air pollutant that comes primarily from the incomplete combustion of natural gas and gasoline. Unfortunately, carbon monoxide combines with hemoglobin more readily than does oxygen, and it stays combined for several hours, making hemoglobin unavailable for oxygen transport. Carbon monoxide detectors in homes can help prevent accidental death.

Possibly because they lack nuclei, red blood cells live only about 120 days. They are destroyed chiefly in the liver and the spleen, where they are engulfed by large phagocytic cells. The iron is mostly salvaged and reused, while the heme portion undergoes chemical degradation, and the liver excretes it into the bile as bile pigments.

When the body has an insufficient number of red blood cells or the red blood cells do not contain enough hemoglobin, an individual suffers from **anemia** and has a tired, rundown feeling. In the most common type of anemia, iron-deficiency anemia, the hemoglobin level is low, probably due to a diet that does not contain enough iron. Whenever arterial blood carries a reduced amount of oxygen, such as when an individual first takes up residence at a high altitude, the kidneys increase their production of a hormone called **erythropoietin,** which speeds the maturation of red blood cells in the bone marrow.

> Red blood cells are biconcave disks that lack a nucleus and contain hemoglobin. Hemoglobin carries oxygen to the tissues, and helps carry carbon dioxide back to the lungs. The condition of having an insufficient number of red blood cells or low hemoglobin is known as anemia.

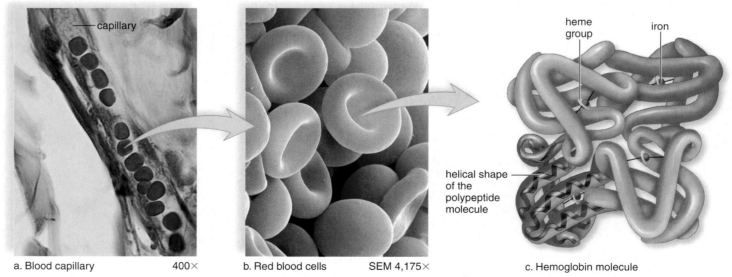

| a. Blood capillary | 400× | b. Red blood cells | SEM 4,175× | c. Hemoglobin molecule |

Figure 12.12 Physiology of red blood cells.
a. Red blood cells move in single file through the capillaries. **b.** Each red blood cell is a biconcave disk containing many molecules of hemoglobin, the respiratory pigment. **c.** Hemoglobin contains four polypeptide chains (purple). There is an iron-containing heme group in the center of each chain. Oxygen combines loosely with iron when hemoglobin is oxygenated. Oxyhemoglobin is bright red, and deoxyhemoglobin is a dark maroon color.

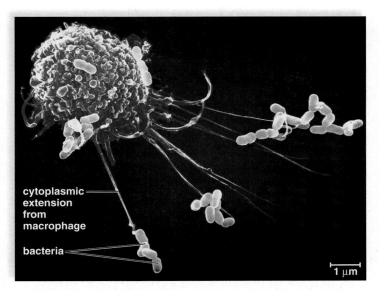

Figure 12.13 Macrophage (red) engulfing bacteria.

Monocyte-derived macrophages are the body's scavengers. They engulf microbes and debris in the body's fluids and tissues, as illustrated in this colorized scanning electron micrograph.

The White Blood Cells

White blood cells (leukocytes) differ from red blood cells in that they are usually larger, have a nucleus, lack hemoglobin, and without staining appear translucent (see Fig. 12.11c, d). White blood cells are not as numerous as red blood cells, with only 4,000–11,000 cells per mm³. White blood cells fight infection and play a role in the development of immunity, the ability to resist disease.

On the basis of structure, it is possible to divide white blood cells into **granular leukocytes** and **agranular leukocytes,** which are also called **mononuclear cells.** Granular leukocytes (neutrophils, basophils, and eosinophils) are filled with spheres that contain enzymes and proteins, which help white blood cells defend the body against microbes. **Neutrophils** are granular leukocytes with a multilobed nucleus joined by nuclear threads. They are the most abundant of the white blood cells and are able to phagocytize and digest bacteria. **Basophils** stain a deep blue and release histamine, which can be a problem for people with allergies. **Eosinophils** stain a deep red and are thought to fight parasitic worms, although they are also involved in some allergies. Agranular leukocytes (monocytes and lymphocytes) typically have a kidney-shaped or spherical nucleus.

Monocytes are the largest of the white blood cells, and they differentiate into phagocytic dendritic cells and macrophages. **Dendritic cells** are present in tissues that are in contact with the environment: skin, nose, lungs, and intestines. Once they have captured a microbe with their long, spiky arms, called dendrites, they stimulate other white blood cells to defend the body. **Macrophages,** well known as ferocious phagocytes (Fig. 12.13), play a similar role in

other organs, such as the liver, kidney, and spleen. The **lymphocytes** are of two types, B lymphocytes and T lymphocytes, and each type plays a specific role in immunity.

If the total number of white blood cells increases or decreases beyond normal, disease may be present. Sometimes an increase or decrease of only one type of white blood cell is a sign of infection. A person with infectious mononucleosis, caused by the Epstein-Barr virus, has an excessive number of lymphocytes of the B type. A person with AIDS, caused by an HIV infection, has an abnormally low number of T lymphocytes. Leukemia is a form of cancer characterized by uncontrolled production of abnormal white blood cells.

White blood cells live different lengths of time. Many live only a few days and are believed to die combating invading pathogens. Others live for months or even years.

> White blood cells, or leukocytes, are involved in fighting infection. The granular leukocytes include the neutrophils, basophils, and eosinophils. The agranular leukocytes, also called mononuclear cells, include the monocytes and lymphocytes.

The Platelets and Blood Clotting

Platelets (also called thrombocytes) result from fragmentation of certain large cells, called **megakaryocytes,** in the red bone marrow. Platelets are produced at a rate of 200 billion a day, and the blood contains 150,000–300,000 per mm³. These formed elements are involved in the process of blood **clotting,** or coagulation.

There are at least 12 clotting factors in the blood that participate with platelets in the formation of a blood clot. We will discuss the roles played by **fibrinogen** and **prothrombin,** which are proteins manufactured by the liver. Vitamin K, found in green vegetables and also formed by intestinal bacteria, is needed for prothrombin production. Vitamin K deficiency can result in clotting disorders.

Blood Clotting

When a blood vessel in the body is damaged, the process of clotting begins (Fig. 12.14a). First, platelets clump at the site of the puncture and partially seal the leak. Platelets and damaged tissue release **prothrombin activator,** which converts the plasma protein prothrombin to thrombin. This reaction requires calcium ions (Ca²⁺). **Thrombin,** in turn, acts as an enzyme that severs two short amino acid chains from each fibrinogen molecule, activating it. The activated fragments then join end to end, forming long threads of **fibrin.** Fibrin threads wind around the platelet plug in the damaged area of the blood vessel and provide the framework for the clot. Red blood cells are also

<table>

1. Blood vessel is punctured.

2. Platelets congregate and form a plug.

3. Platelets and damaged tissue cells release prothrombin activator, which initiates a cascade of enzymatic reactions.

</table>

Prothrombin activator

$$\text{Prothrombin} \xrightarrow[\text{Ca}^{2+}]{} \text{Thrombin}$$

$$\text{Fibrinogen} \xrightarrow[\text{Ca}^{2+}]{} \text{Fibrin threads}$$

4. Fibrin threads form and trap red blood cells.

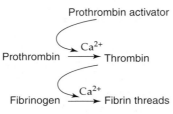

a. Blood-clotting process

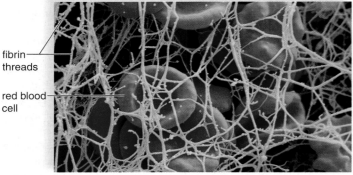

fibrin threads

red blood cell

b. Blood clot   4,400×

Figure 12.14 Blood clotting.

a. Platelets and damaged tissue cells release prothrombin activator, which acts on prothrombin in the presence of Ca^{2+} (calcium ions) to produce thrombin. Thrombin acts on fibrinogen in the presence of Ca^{2+} to form fibrin threads. **b.** A scanning electron micrograph of a blood clot shows red blood cells caught in the fibrin threads.

trapped within the fibrin threads; these cells make a clot appear red (Fig. 12.14*b*).

A fibrin clot is only temporary. As soon as blood vessel repair is initiated, an enzyme called plasmin destroys the fibrin network and restores the fluidity of the plasma. Sometimes it is desirable to inhibit the clotting process. This can be done with medications such as heparin, or even with medicinal leeches (see the Health Focus on page 236).

TABLE 12.1	Body Fluids Related to Blood
Name	**Composition**
Blood	Formed elements and plasma
Plasma	Liquid portion of blood
Serum	Plasma minus fibrinogen
Tissue fluid	Plasma minus most proteins
Lymph	Tissue fluid within lymphatic vessels

If blood is allowed to clot in a test tube, a yellowish fluid develops above the clotted material. This fluid is called **serum,** and it contains all the components of plasma except fibrinogen. Table 12.1 reviews the body fluids related to blood.

Hemophilia

Hemophilia is an inherited clotting disorder caused by a deficiency in a clotting factor. The slightest bump to an affected person can cause bleeding into the joints. Cartilage degeneration in the joints and resorption of underlying bone can follow. Bleeding into muscles can lead to nerve damage and muscular atrophy. Death can result from bleeding into the brain with accompanying neurological damage. Hemophiliacs usually require frequent blood transfusions, although they may also be treated with injections of the specific clotting factor they are lacking.

> **Platelets are cell fragments involved in blood clotting, a process that also involves the conversion of fibrinogen to fibrin by the enzymatic activity of thrombin. Clotting is necessary to minimize blood loss following an injury to a blood vessel.**

Bone Marrow Stem Cells

A **stem cell** is a cell that is ever capable of dividing and producing new cells that go on to differentiate into particular types of cells. It's been known for some time that the bone marrow has multipotent stem cells, which have the potential to give rise to other stem cells for the various formed elements (Fig. 12.15). Recent research has shown that bone marrow stem cells are also able to differentiate into other types of cells, including liver, bone, fat, cartilage, heart, and even neurons. The possibility exists that a patient's own bone marrow stem cells could be used for curing certain conditions that might develop, such as diabetes, heart disease, liver disease, or even brain disorders (e.g., Alzheimer disease or Parkinson disease). In one study, researchers examined the brains of women who had received bone marrow stem cells from male donors as a part of their treatment for leukemia. All recipients had transgender brain tissue—some of the neurons in their brains contained a Y chromosome! It seems that the bone marrow stem cells, in response to neurosignals, traveled to the brain, where they differentiated into neurons.

The use of a person's own stem cells does away with the problem of possible rejection. Among all the various types of *adult stem cells* in the body, bone marrow stem cells are particularly attractive because they are the most accessible. Some researchers prefer to work with *embryonic stem cells*, thinking that they are more likely to become any type of cell. Embryonic stem cells are available because many early-stage embryos remain unused in fertility clinics, although this issue has become controversial.

> **The bone marrow contains stem cells that can differentiate into the formed elements of blood. These stem cells can also apparently differentiate into cells that form many other types of tissues.**

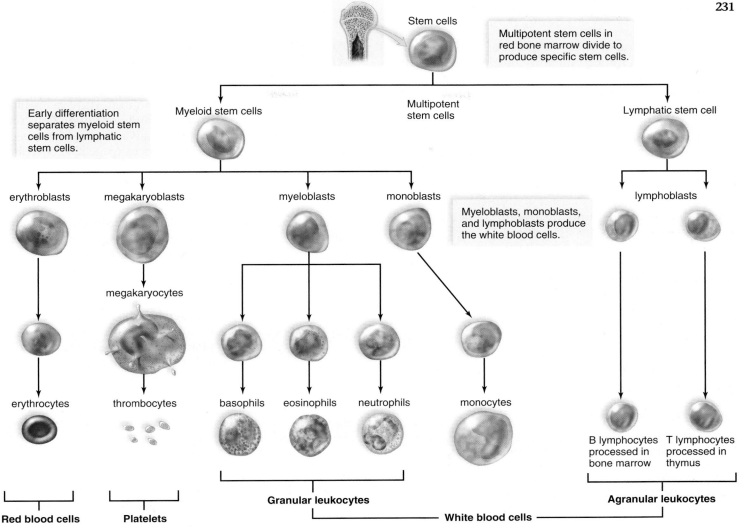

Figure 12.15 Blood cell formation in red bone marrow.
Multipotent stem cells give rise to two specialized types of stem cells. The myeloid stem cells give rise to still other cells, which become red blood cells, platelets, and all the white blood cells except lymphocytes. The lymphoid stem cells give rise to lymphoblasts, which become lymphocytes.

Capillary Exchange

Two forces primarily control movement of fluid through the capillary wall: osmotic pressure, created by salts and plasma proteins, which tends to cause water to move from the tissue fluid to the blood, and blood pressure, which tends to cause water to move in the opposite direction. At the arterial end of a capillary, blood pressure is higher than the osmotic pressure of blood (Fig. 12.16), so water exits a capillary at this end.

Midway along the capillary, where blood pressure is lower, the two forces essentially cancel each other, and there

Figure 12.16 Capillary exchange in the systemic circuit.
At the arterial end of a capillary (*left*) the blood pressure is higher than the osmotic pressure; therefore, water tends to leave the bloodstream. In the midsection, molecules, including oxygen and carbon dioxide, follow their concentration gradients. At the venous end of a capillary (*right*), the osmotic pressure is higher than the blood pressure; therefore, water tends to enter the bloodstream. Notice that the red blood cells and the plasma proteins are too large to exit a capillary.

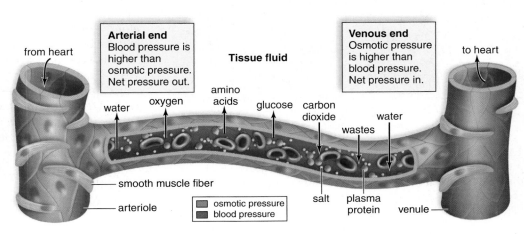

is no net movement of water. Solutes now diffuse according to their concentration gradient—nutrients (glucose and amino acids) and oxygen diffuse out of the capillary, and wastes (carbon dioxide) diffuse into the capillary. In the pulmonary circuit, where the O_2 concentration is higher in the lung tissues, and the CO_2 concentration is lower, the movement of these gases is reversed.

Red blood cells and almost all plasma proteins remain in the capillaries, but small substances leave, contributing to **tissue fluid,** the fluid between the body's cells. Tissue fluid tends to contain all the components of plasma but lesser amounts of protein, which generally stays in the capillaries.

At the venous end of a capillary, where blood pressure has fallen even more, osmotic pressure is greater than blood pressure, and water tends to move into the capillary. Almost the same amount of fluid that left the capillary returns to it, although some excess tissue fluid is always collected by the **lymphatic capillaries** (Fig. 12.17). Tissue fluid contained within lymphatic vessels is called **lymph.** Lymph is returned to the systemic venous blood when the major lymphatic vessels enter the subclavian veins in the shoulder region.

> Osmotic pressure and blood pressure are the main forces that control the movement of fluid through capillary walls. In the systemic circuit, oxygen exits, and carbon dioxide enters midway along a capillary. Excess tissue fluid, called lymph, is collected by lymphatic capillaries.

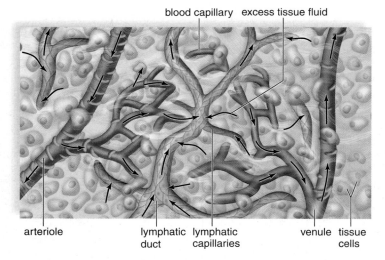

blood capillary   excess tissue fluid

arteriole   lymphatic duct   lymphatic capillaries   venule   tissue cells

Figure 12.17 Lymphatic capillaries.
A lymphatic capillary bed (shown here in green) lies near a blood capillary bed. When lymphatic capillaries take up excess tissue fluid, it becomes lymph.

12.5 Cardiovascular Disorders

Cardiovascular disease is the leading cause of untimely death in Western countries. Modern research efforts have resulted in improved diagnosis, treatment, and prevention. This section discusses some of the major advances that have been made in these areas. The Health Focus (page 234) describes ways to maintain cardiovascular health.

Atherosclerosis

Atherosclerosis is an accumulation of soft masses of fatty materials, particularly cholesterol, beneath the inner linings of arteries. Such deposits are called **plaque.** Plaque tends to protrude into the lumen of the vessel and interfere with the flow of blood (see Health Focus Fig. 12A). In most instances, atherosclerosis begins in early adulthood and develops progressively through middle age, but symptoms may not appear until an individual is 50 or older.

Plaque can cause platelets to adhere to the irregular arterial wall, forming a clot. As long as the clot remains stationary, it is called a **thrombus,** but if it dislodges and moves along with the blood, it is called an **embolus. Thromboembolism,** a clot that has been carried in the bloodstream but is now stationary, must be treated, or serious complications can arise.

> Atherosclerosis refers to the accumulation of fatty deposits beneath the inner linings of arteries. Platelets can adhere to these plaques, forming thrombi or emboli.

Stroke, Heart Attack, and Aneurysm

A cerebrovascular accident, also called a **stroke,** often results when a small cranial arteriole bursts or is blocked by an embolus. Lack of oxygen causes a portion of the brain to die, and paralysis or death can result. A person is sometimes forewarned of a stroke by a feeling of numbness in the hands or the face, difficulty in speaking, or temporary blindness in one eye. Cerebrovascular accidents become more common with age, but are certainly not limited to the elderly: on March 6, 2006, former Minnesota Twins baseball star Kirby Puckett died of a stroke at age 45.

If a coronary artery becomes partially blocked, the individual may suffer from **angina pectoris,** characterized by a squeezing or burning sensation in the chest. Nitroglycerin or related drugs dilate blood vessels and help relieve the pain. When a coronary artery is completely blocked, a portion of the heart muscle dies due to a lack of oxygen, and a myocardial infarction, or **heart attack,** occurs.

An **aneurysm** is the ballooning of a blood vessel, most often the abdominal artery or the arteries leading to the brain. Atherosclerosis and high blood pressure can weaken

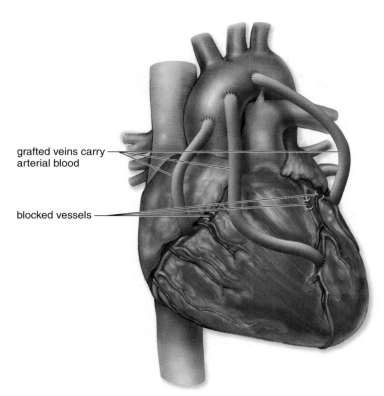

Figure 12.18 Coronary bypass operation.
During this operation, the surgeon grafts segments of another vessel, usually a small vein from the leg, between the aorta and the coronary vessels, bypassing areas of blockage. Patients who require surgery often receive two to five bypasses in a single operation.

grafted veins carry arterial blood

blocked vessels

the wall of an artery to the point that an aneurysm develops. If a major vessel such as the aorta should burst, death is likely. It is often possible to replace a damaged or diseased portion of a vessel, such as an artery, with a synthetic graft.

Coronary Bypass Operations

For many years, coronary bypass surgery has been a common way to treat an obstructed coronary artery. During this operation, a surgeon takes a segment from another blood vessel and stitches one end to the aorta and the other end to a coronary artery past the point of obstruction. Figure 12.18 shows a triple bypass in which three blood vessels have been used to allow blood to flow freely from the aorta to cardiac muscle by way of the coronary artery.

The results of some clinical trials have suggested that gene therapy may be another way to restore the blood supply to oxygen-starved cardiac muscle, without the need for major surgery. A harmless virus was engineered to contain a gene coding for a growth factor called VEGF, and then injected directly into affected areas of the heart. Although many patients appeared to show improvement, further testing will be required before such treatments can be judged safe enough for more widespread use.

Clearing Clogged Arteries

In **angioplasty,** a cardiologist threads a catheter into an artery in the groin or upper part of the arm and guides it

through a major blood vessel toward the heart. When the tube arrives at the region of plaque in an artery, a balloon attached to the end of the tube is inflated, forcing the vessel open. However, the artery may not always remain open because the trauma can cause smooth muscle cells in the wall of the artery to proliferate and close it.

More often, the same operation utilizes a cylinder of expandable metal mesh called a **stent.** Once the stent is in place, a balloon inside the stent is inflated, expanding it, and locking it in place (Fig. 12.19). The stent is most successful when it is coated with a drug called sirolimus, which seeps into the arterial lining and discourages cell growth.

Dissolving Blood Clots

Medical treatment for thromboembolism includes the use of tissue plasminogen activator (tPA). This drug converts plasminogen, a molecule found in blood, into plasmin, an enzyme that dissolves blood clots. In fact, tPA is the body's own way of converting plasminogen to plasmin. tPA is also being used for thrombolytic stroke patients but with limited success because some patients experience life-threatening bleeding in the brain. A better treatment might be new biotechnology drugs that act on the plasma membrane to prevent brain cells from releasing and/or receiving toxic chemicals caused by the stroke.

If a person has symptoms of angina or a stroke, aspirin may be prescribed. Aspirin reduces the stickiness of platelets, and thereby lowers the probability that a clot will form. Evidence indicates that aspirin protects against first heart attacks, but there is no clear support for taking aspirin every day to prevent strokes in symptom-free people. Even so, some physicians recommend taking a low dosage of aspirin every day.

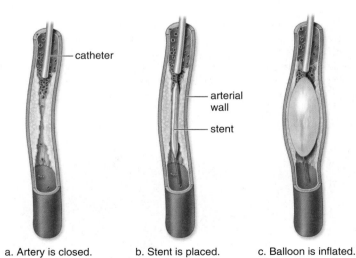

catheter

arterial wall

stent

a. Artery is closed.   b. Stent is placed.   c. Balloon is inflated.

Figure 12.19 Angioplasty with stent placement.
a. A plastic tube (catheter) is inserted into the coronary artery until it reaches the clogged area. **b.** A metal stent with a balloon inside it is pushed out the end of the plastic tube into the clogged area. **c.** When the balloon is inflated, the vessel opens, and the stent is left in place to keep the vessel open.

Prevention of Cardiovascular Disease

Certain genetic factors predispose an individual to cardiovascular disease, including family history of heart attack under age 55, male gender, and ethnicity (African Americans are at greater risk). Persons with one or more of these risk factors need not despair, however. It only means that they should pay particular attention to the following guidelines for a heart-healthy lifestyle.

The Don'ts

Smoking

When a person smokes, the drug nicotine, present in cigarette smoke, enters the bloodstream. Nicotine causes the arterioles to constrict and the blood pressure to rise. The heart has to pump harder to propel the blood through the lungs at a time when the blood's oxygen-carrying capacity is reduced by smoking.

Drug Abuse

Stimulants, such as cocaine and amphetamines, can cause an irregular heartbeat and lead to heart attacks even when using drugs for the first time. Intravenous drug use may also result in a cerebral blood clot and stroke.

Too much alcohol can destroy just about every organ in the body, the heart included. But investigators have discovered that people who take an occasional drink have a 20% lower risk of heart disease than do teetotalers. Two to four drinks a week for men and one to three drinks for women are sufficient.

Weight Gain

Hypertension (high blood pressure) is prevalent in persons who are more than 20% above the recommended weight for their height. More tissues require servicing, and the heart sends the extra blood out under greater pressure. It may be harder to lose weight once it is gained, and therefore, weight control should be a lifelong endeavor. Even a slight decrease in weight can bring with it a reduction in hypertension. Being overweight increases the risk of type 2 diabetes in which glucose damages blood vessels and makes them prone to the development of plaque, as discussed next.

The Do's

Healthy Diet

Diet influences the amount of cholesterol in the blood. Cholesterol is ferried by two types of plasma proteins, called LDL (low-density lipoprotein) and HDL (high-density lipoprotein). LDL (called "bad" lipoprotein) takes cholesterol from the liver to the tissues, and HDL (called "good" lipoprotein) transports cholesterol out of the tissues to the liver. When the LDL level in blood is high or the HDL level is abnormally low, plaque, which interferes with circulation, accumulates on arterial walls (Fig. 12A).

Eating foods high in saturated fat (red meat, cream, and butter) and foods containing trans-fats (most margarines, commercially baked goods, and deep-fried foods) raises the LDL-cholesterol level. Replacement of these harmful fats with healthier ones, such as monounsaturated fats (olive and canola oil) and polyunsaturated fats (corn, safflower, and soybean oil), is recommended. Cold-water fish (e.g., halibut, sardines, tuna, and salmon) contain polyunsaturated fatty acids and especially omega-3 polyunsaturated fatty acids that can reduce plaque.

Evidence is mounting to suggest a role for antioxidant vitamins (A, E, and C) in preventing cardiovascular disease.

Antioxidants protect the body from free radicals that oxidize cholesterol and damage the lining of an artery, leading to a blood clot that can block blood vessels. Nutritionists believe that consuming at least five servings of fruits and vegetables a day may protect against cardiovascular disease.

Cholesterol Profile

Starting at age 20, all adults are advised to have their cholesterol levels tested at least every five years. Even in healthy individuals, an LDL level above 160 mg/100 ml and an HDL level below 40 mg/100 ml are matters of concern. If a person has heart disease or is at risk for heart disease, an LDL level below 100 mg/100 ml is now recommended. Medications will, most likely, be prescribed for all those who do not meet these minimum guidelines.

Exercise

People who exercise are less apt to have cardiovascular disease. One study found that moderately active men who spent an average of 48 minutes a day on a leisure-time activity such as gardening, bowling, or dancing had one-third fewer heart attacks than peers who spent an average of only 16 minutes each day on those types of activities. Exercise not only helps keep weight under control, but may also help minimize stress and reduce hypertension.

Discussion Questions

1. Considering all the risk factors discussed in this reading, which, if any, could be affecting you right now? Do you think you are too young to worry about cardiovascular disease?

2. In addition to the factors discussed here, some scientists suspect that infections with certain types of bacteria and viruses may contribute to cardiovascular disease. What are some specific ways that infections might contribute to heart disease, for example? If this connection could be proven, can you think of any new treatment strategies that should be tested?

3. Stress is almost always mentioned as a contributing factor in heart disease. From what you know about the body's response to stress, how might chronic stress specifically cause damage to the cardiovascular system? Do you think there is such a thing as "good" stress, and if so, what differentiates good versus bad stress?

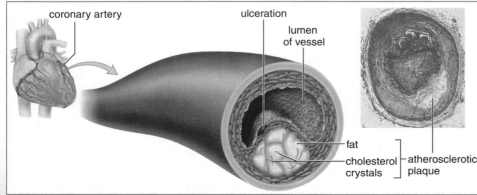

Figure 12A Coronary arteries and plaque.

Atherosclerotic plaque is an irregular accumulation of cholesterol and fat. When plaque is present in a coronary artery, a heart attack is more likely to occur because of restricted blood flow.

Figure 12.20
A total artificial heart (TAH).
The AbioCor replacement heart is designed to be implanted within the chest cavity.

Heart Transplants and Artificial Hearts

Heart transplants are usually successful today, but unfortunately, the need for hearts to transplant is greater than the supply because of a shortage of human donors. Today, a left ventricular assist device (LVAD), implanted in the abdomen, may help patients waiting for a transplant. A tube passes blood from the left ventricle to the device, which pumps it to the aorta. A cable passes from the device through the skin to an external battery the patient totes around.

Only a few patients have ever received a so-called **total artificial heart (TAH).** The model now available is called an AbioCor (Fig. 12.20). An internal battery and controller regulate the pumping speed, and an external battery powers the device by passing electricity through the skin. A rotating centrifugal pump moves silicon hydraulic fluid between left and right sacs to force blood out of the heart into the pulmonary trunk and the aorta. The exterior is made mainly of titanium, but the valves and membranes inside the ventricles are made of plastic, which has held up to beating 100,000 times a day for years. So far, all recipients have been near death and most have lived for only a short time after getting an AbioCor.

> **Medical treatment for strokes, heart attacks, and aneurysms may involve major surgery, angioplasty with or without stents, a drug called tPA, or a heart transplant.**

Hypertension

Hypertension, or high blood pressure, affects about 20% of all Americans. It is sometimes called "the silent killer" because it may not be detected until a stroke or heart attack occurs. Normal blood pressure values vary somewhat among different age groups, body sizes, and levels of athletic conditioning, but a general guideline is that hypertension is present in adult women whose blood pressure reading is 160/95 or above. For an adult male under age 45,

a reading above 130/90 is generally considered hypertensive, as is a reading above 140/95 in men older than 45.

Hypertension is most often due to a narrowing of the arteries. Consider trying to force a gallon of water through a straw, versus through a garden hose, in the same amount of time. You would need to apply much more force to get the water through the straw. In the same way, forcing blood through narrowed arteries over a prolonged period of time creates additional pressure on the cardiovascular system that can damage the blood vessels, heart, and other organs. Other factors that contribute to hypertension include obesity, smoking, chronic stress, and a high dietary salt intake (which tends to make the body retain more fluid, which in turn increases blood volume). Many medications can be used to treat high blood pressure, including diuretics (which reduce the blood volume by increasing urine output), vasodilators to dilate the blood vessels, and various drugs that improve heart function.

> **Hypertension is a major health problem in the United States, and it is often associated with narrowing of the arteries due to atherosclerosis. Some major factors that contribute to hypertension are obesity, smoking, stress, and high dietary salt intake.**

Summarizing the Concepts

12.1 The Blood Vessels
Blood vessels include arteries (and arterioles) that take blood away from the heart; capillaries, where exchange of substances with the tissues occurs; and veins (and venules) that take blood to the heart.

12.2 The Human Heart
The heart has a right and left side and four chambers. The venae cavae are connected to the right atrium; the pulmonary trunk is connected to the right ventricle. The pulmonary veins are connected to the left atria, and the aorta is connected to the left ventricle. Blood from the atria passes through atrioventricular valves into the ventricles. Blood leaving the ventricles passes through semilunar valves.

The right atrium receives O_2-poor blood from the body, and the right ventricle pumps it into the pulmonary circuit. On the left side, the left atrium receives O_2-rich blood from the lungs, and the left ventricle pumps it into the systemic circuit. During the cardiac cycle, the SA node (cardiac pacemaker) initiates the heartbeat by causing the atria to contract. The AV node conveys the stimulus to the ventricles, causing them to contract. The heart sounds, "lub-dup," are due to the closing of the atrioventricular valves, followed by the closing of the semilunar valves.

12.3 The Vascular Pathways
The cardiovascular system is divided into the pulmonary circuit and the systemic circuit. In the pulmonary circuit, the pulmonary trunk from the right ventricle of the heart and the two pulmonary

Health Focus

Although it may seem more like an episode of a popular TV show than a real-life medical treatment, the U.S. Food and Drug Administration has approved the use of leeches as medical "devices" for treating conditions involving poor blood supply to various tissues.

Leeches are bloodsucking, aquatic creatures, whose closest living relatives are earthworms (Fig. 12B). Prior to modern times, medical practitioners frequently applied leeches to patients, mainly in an attempt to remove the bad "humors" that they thought were responsible for many diseases. This practice was abandoned, thankfully, in the nineteenth century when it was realized that the "treatment" often harmed the patient.

True to their tenacious nature, however, leeches are making a comeback in twenty-first-century medicine. By applying leeches to tissues that have been injured by trauma or disease, blood supply can be improved. When reattaching a finger, for example, it is

Figure 12B Leeches.
Leeches can attach to human skin and suck blood out. Their use in medicine seems to be making a comeback.

easier to suture together the thicker-walled arteries than the veins. Poorly draining blood from the veins can pool in the appendage and threaten its survival. It turns out that leech saliva contains chemicals that dilate blood vessels and prevent blood from clotting by blocking the activity of thrombin. These effects can improve the circulation to the body part. Another substance in leech saliva actually anesthetizes the bite wound. In a natural setting, this allows the leech to feast on the blood supply of its victim undetected, but in a medical setting, it makes the

whole experience more tolerable, at least physically. Mentally, however, the application of leeches can still be a rather unsettling experience, and patient acceptance is a major factor limiting their more widespread use.

Discussion Questions

1. If you had an injury and your doctor said it might help, would you be willing to let leeches feast for a few minutes, say, on your hand? On your face?
2. Leeches were historically used for bloodletting, one of the oldest medical practices. Can you think of several reasons why, prior to the mid-nineteenth century, early physicians might have gotten the idea that removing blood could help cure various diseases?
3. The use of actual living organisms in medical treatments is sometimes called "biotherapy." Can you think of other situations in which living creatures might be used for human health benefits, either directly or indirectly?

---

arteries take O_2-poor blood to the lungs, and four pulmonary veins return O_2-rich blood to the left atrium of the heart.

To trace the path of blood in the systemic circuit, start with the aorta from the left ventricle. Follow its path until it branches to an artery going to a specific organ. It can be assumed that the artery divides into arterioles and capillaries and that the capillaries lead to venules which join to form a vein. The vein that takes blood to the vena cava most likely has the same name as the artery that delivered blood to the organ. In the adult systemic circuit, unlike the pulmonary circuit, the arteries carry O_2-rich blood, and the veins carry O_2-poor blood.

Portal systems begin and end in capillaries. The hepatic portal vein originates from intestinal capillaries and finishes in hepatic capillaries.

12.4 Blood

Blood has two main parts: the plasma and the formed elements. Plasma contains mostly water (90–92%) and proteins (7–8%), but it also contains nutrients and wastes.

The formed elements are red blood cells, white blood cells, and platelets. The red blood cells contain hemoglobin and function in oxygen transport. Defense against disease depends on the various types of white blood cells. Granular neutrophils and monocytes are phagocytic. Agranular lymphocytes, also called mononuclear cells, are involved in the development of immunity to disease. The platelets and a number of clotting factors are involved in blood clotting. The plasma proteins prothrombin and fibrinogen are

well known for their contribution to the clotting process, which requires a series of enzymatic reactions. When the process is over, fibrin has become fibrin threads.

When blood reaches a capillary, water moves out at the arterial end due to blood pressure. At the venule end, water moves in due to osmotic pressure. In between, nutrients diffuse out and wastes diffuse in.

12.5 Cardiovascular Disorders

Hypertension and atherosclerosis are two cardiovascular disorders that lead to stroke, heart attack, and aneurysm. Medical and surgical procedures are available to control cardiovascular disease, but the best policy is prevention by following a heart-healthy diet, getting regular exercise, maintaining a proper weight, and not smoking.

Testing Yourself

Choose the best answer for each question.

1. _____ lie between _____ and _____.
   a. Arteries, veins, capillaries
   b. Capillaries, arteries, veins
   c. Veins, arteries, capillaries
   d. None of these are correct.

2. The myocardium is made of
   a. muscle.
   b. epithelium.
   c. connective tissue.
   d. None of these are correct.

3. Gas (oxygen and carbon dioxide) exchange occurs across the
   _____ of the _____.
   a. veins, lungs
   b. capillaries, tissues
   c. arteries, tissues
   d. All of these are correct.

4. The average adult heart rate is about _____ beats per minute.
   a. 120   c. 45
   b. 100   d. 70

5. If the SA node fails and the AV node takes over, the heart
   rate will be
   a. slower.   c. the same.
   b. faster.   d. None these are correct.

6. Which association is incorrect?
   a. white blood cells—infection fighting
   b. red blood cells—blood clotting
   c. red blood cells—hemoglobin
   d. platelets—blood clotting

7. What is the type of cell that becomes any of the formed
   elements in blood?
   a. multipotent stem cells
   b. myeloid stem cells
   c. lymphoid stem cells
   d. All of these are correct.

8. In the tissues, the amount of carbon dioxide is _____ than in
   the capillaries.
   a. lower   c. neither higher nor lower
   b. higher   d. None of these are correct.

9. Lymph is formed from
   a. urine.   c. excess extracellular tissue fluid.
   b. whole plasma.   d. All of these are correct.

10. Anemia can result from
   a. an iron-deficient diet.
   b. a drop in red blood cell count.
   c. lack of hemoglobin production.
   d. All of these are correct.

*For questions 11–15, match the term with a function or
location in the key.*

Key:

   a. upper chambers   d. ventricular contraction
   b. atrial contraction   e. lower chambers
   c. initiates heartbeat

11. P wave

12. QRS complex

13. SA node

14. Atria

15. Ventricles

16. A decrease in lymphocytes would result in problems
   associated with
   a. clotting.
   b. immunity.
   c. oxygen transport.
   d. All of these are correct.

17. Label the following ECG wave chart.

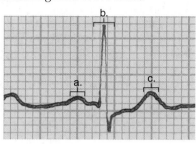

18. The best explanation for the slow movement of blood in
   capillaries is
   a. skeletal muscles press on veins, not capillaries.
   b. capillaries have much thinner walls than arteries.
   c. there are many more capillaries than arterioles.
   d. venules are not prepared to receive so much blood from
      the capillaries.
   e. All of these are correct.

19. Which of the following assists in the return of venous blood
   to the heart?
   a. valves
   b. skeletal muscle contraction
   c. respiratory movements
   d. reduction in cross-sectional area from venules to veins
   e. All of these are correct.

20. Label the following diagram of the systemic circuit.

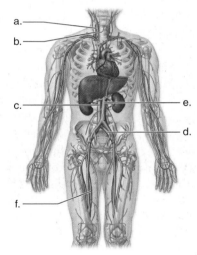

21. Water enters the venous side of capillaries because of
   a. active transport from tissue fluid.
   b. higher osmotic pressure than blood pressure.
   c. higher blood pressure on the venous side.
   d. higher blood pressure than osmotic pressure.
   e. higher red blood cell concentration on the venous side.

22. All arteries in the body contain O_2-rich blood with the
   exception of the
   a. aorta.
   b. pulmonary artery.
   c. renal artery.
   d. coronary arteries.

23. The "lub," or first heart sound, is produced by closing of the
   a. aortic semilunar valve.   c. tricuspid valve.
   b. pulmonary semilunar   d. bicuspid valve.
      valve.   e. both AV valves.

Understanding the Terms

agranular leukocyte 229
albumin 228
anemia 228
aneurysm 232
angina pectoris 232
angioplasty 233
aorta 224
arteriole 218
artery 218
atherosclerosis 232
atrioventricular bundle 222
atrioventricular valve 220
atrium 220
AV (atrioventricular)
   node 222
basophil 229
blood pressure 225
capillary 218
cardiac cycle 222
cardiac pacemaker 222
chordae tendineae 220
clotting 229
coronary artery 225
dendritic cell 229
diastole 222
diastolic pressure 225
electrocardiogram
   (ECG) 223
embolus 232
endothelium 218
eosinophil 229
erythropoietin 228
fibrin 229
fibrinogen 229
formed element 226
granular leukocyte 229
heart 220
heart attack 232
hemoglobin 228
hepatic portal system 225
hepatic portal vein 225
hepatic vein 225
hypertension 235
inferior vena cava 224
lymph 232

lymphatic capillaries 232
lymphocyte 229
macrophage 229
megakaryocyte 229
monocyte 229
mononuclear cell 229
myocardium 220
neutrophil 229
pericardium 220
plaque 232
plasma 226
platelet 229
prothrombin 229
prothrombin activator 229
pulmonary artery 221
pulmonary circuit 224
pulmonary vein 221
pulse 221
Purkinje fibers 222
red blood cell (erythrocyte)
   228
SA (sinoatrial) node 222
semilunar valve 220
serum 230
stem cell 230
stent 233
stroke 232
superior vena cava 224
systemic circuit 224
systole 222
systolic pressure 225
thrombin 229
thromboembolism 232
thrombus 232
tissue fluid 232
total artificial heart
   (TAH) 235
valve 219
varicose veins 226
vein 218
ventricle 220
venule 219
white blood cell
   (leukocyte) 229

Match the terms to these definitions:

a._____ Relaxation of a heart chamber.
b._____ Large systemic vein that returns blood
   to the right atrium of the heart.
c._____ Plasma protein that is converted into
   fibrin threads during blood clotting.
d._____ Iron-containing protein in red blood
   cells that combines with and transports oxygen.

Thinking Critically

1. A few specialized tissues do not contain any blood vessels, including capillaries. Can you think of two or three? How might these tissues survive without a direct blood supply?

2. Assume your heart rate is 70 beats per minute (bpm), and each minute your heart pumps 5.25 liters of blood to your body. Based on your age to the nearest year, about how many times has your heart beat so far, and how much volume has it pumped?

3. Examine the following abnormal ECG tracings. Compared to the normal tracing in Question 17 or Figure 12.6*b*, what is your best guess as to what type of problem each patient may have? What type of device can be implanted in a patient's heart to correct some abnormalities of the heart, and how does it generally work?

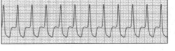   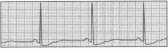

a.   b.

4. There are not enough living hearts available to meet the need for heart transplants. What are some of the major problems a manufacturer would have to overcome when attempting to design an artificial heart that will last for years inside a patient's body?

ARIS, the *Inquiry into Life* Website

ARIS, the website for *Inquiry into Life,* provides a wealth of information organized and integrated by chapter. You will find practice quizzes, interactive activities, labeling exercises, flashcards, and much more that will complement your learning and understanding of general biology.

www.aris.mhhe.com

Organs can fail to function *for various reasons. One of the most famous transplant patients in recent years was Mickey Mantle, whose liver was destroyed by years of alcohol abuse. In the photo below, a patient is undergoing dialysis because her kidneys are not functioning—perhaps they were damaged by cancer or an infection. In cases where an organ is destroyed beyond repair, transplantation of a healthy donor organ may be the only hope. However, the patient's own immune system may recognize certain molecules present on the donor organ, and this response must be overcome if the organ is not perfectly matched to the recipient. It is disappointing that the same immune system that defends us against invading pathogens can become an obstacle for a patient who needs an organ transplant to survive.*

The development of drugs to suppress the immune system has allowed organ transplantation to be performed routinely. In fact, the problem now isn't so much rejection of transplanted organs, but a shortage of suitable donors. Still, drugs that suppress the immune system are far from perfect. Sometimes they suppress the immune system too well, leaving transplant patients susceptible to infection. Fortunately, antibacterial and antifungal drugs can also help protect transplant patients from these infections. Research into the intricate workings of the immune system continues at a rapid pace, and this knowledge should lead to better ways of suppressing unwanted immune responses without depriving patients of the benefits of a normally functioning immune system.

chapter **13**

Lymphatic and Immune Systems

Chapter Concepts

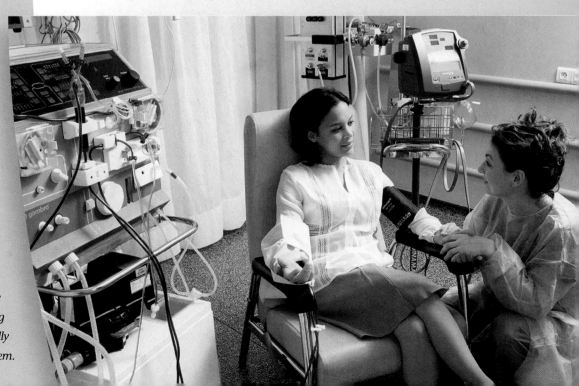

13.1   The Lymphatic System

The **lymphatic system** consists of lymphatic vessels and the lymphoid organs (Fig. 13.1). This system, which is closely associated with the cardiovascular system, has three main functions that contribute to homeostasis: (1) lymphatic capillaries take up excess tissue fluid and return it to the bloodstream; (2) small lymphatic capillaries absorb fats from the digestive tract and transport them to the bloodstream; and (3) the lymphatic system helps defend the body against disease. This last function is carried out by the white blood cells present in lymphatic vessels and lymphoid organs, as well as in the blood.

Lymphatic Vessels

Lymphatic vessels form a one-way system that begins with lymphatic capillaries. Most regions of the body are richly supplied with lymphatic capillaries, tiny, closed-ended vessels whose walls consist of simple squamous epithelium (Fig. 13.1). Lymphatic capillaries take up excess tissue fluid. Tissue fluid is mostly water, but it also contains solutes (i.e., nutrients, electrolytes, and oxygen) derived from plasma and cellular products (i.e., hormones, enzymes, and wastes) secreted by cells. The fluid inside lymphatic vessels is called **lymph.**

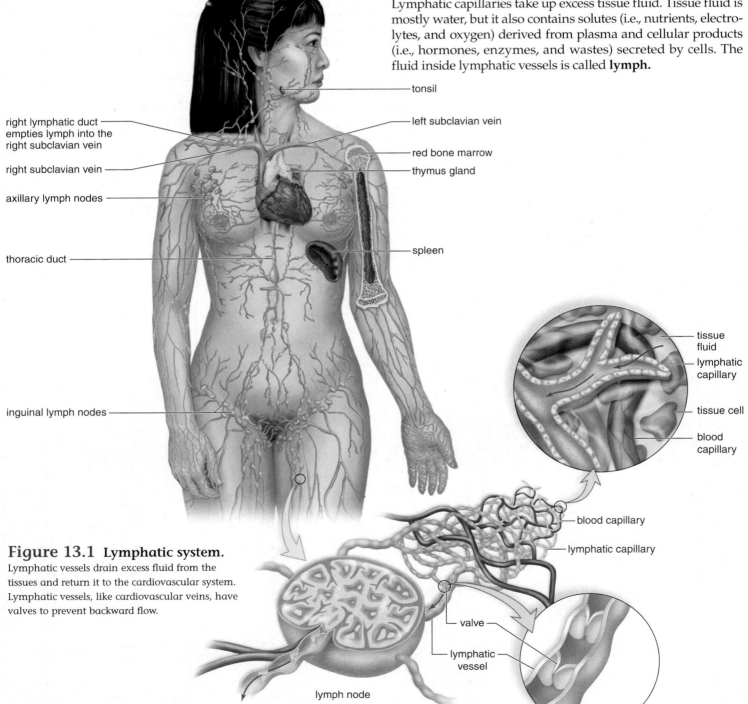

Figure 13.1 Lymphatic system.
Lymphatic vessels drain excess fluid from the tissues and return it to the cardiovascular system. Lymphatic vessels, like cardiovascular veins, have valves to prevent backward flow.

The lymphatic capillaries join to form lymphatic vessels that merge before entering one of two ducts: the thoracic duct or the right lymphatic duct. The larger, thoracic duct returns lymph collected from the body below the thorax and the left arm and left side of the head and neck into the left subclavian vein. The right lymphatic duct returns lymph from the right arm and right side of the head and neck into the right subclavian vein.

The construction of the larger lymphatic vessels is similar to that of cardiovascular veins, including the presence of valves. The movement of lymph within lymphatic capillaries is largely dependent upon skeletal muscle contraction. Lymph forced through lymphatic vessels as a result of muscular compression is prevented from flowing backward by one-way valves.

Edema is localized swelling caused by the accumulation of tissue fluid that has not been collected by the lymphatic system. It occurs if too much tissue fluid is made and/or if not enough of it is drained away. Edema can lead to tissue damage and eventual death, illustrating the importance of the function of the lymphatic system. The fat absorption and defense functions of the lymphatic system are equally important. Unfortunately, cancer cells sometimes enter lymphatic vessels and move undetected to other regions of the body where they produce secondary tumors. In this way, the lymphatic system sometimes assists metastasis, the spread of cancer far from its place of origin.

> Lymphatic capillaries in the tissues form lymphatic vessels that carry lymph from the tissues, eventually returning it to the blood. Edema is an accumulation of tissue fluid.

Lymphoid Organs

The primary **lymphoid organs** are red bone marrow and the thymus gland; the secondary lymphoid organs are the spleen and lymph nodes (Fig. 13.2). Lymphoid organs contain

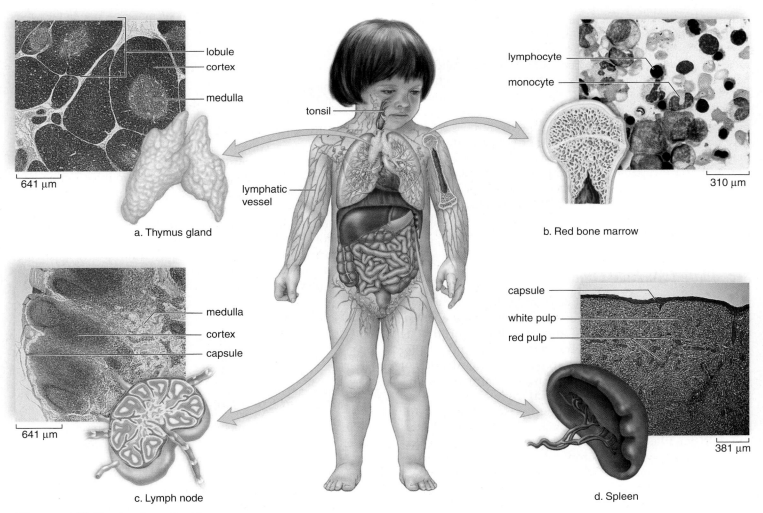

a. Thymus gland — lobule, cortex, medulla — 641 μm

b. Red bone marrow — lymphocyte, monocyte — 310 μm

c. Lymph node — medulla, cortex, capsule — 641 μm

d. Spleen — capsule, white pulp, red pulp — 381 μm

tonsil, lymphatic vessel

Figure 13.2 The lymphoid organs.
The thymus gland **(a)** and red bone marrow **(b)** are the primary lymphoid organs. Blood cells, including lymphocytes, are produced in red bone marrow. B cells mature in the bone marrow; T cells mature in the thymus. The lymph nodes **(c)** and the spleen **(d)** are secondary lymphoid organs. Lymph is cleansed in the nodes, and blood is cleansed in the spleen.

large numbers of white blood cells known as lymphocytes. There are two basic types of lymphocytes: B lymphocytes (B cells) and T lymphocytes (T cells). Lymphocytes develop and mature in the primary lymphoid organs, and some lymphocytes become activated in secondary lymphoid organs.

Primary Lymphoid Organs

Red bone marrow is the site of stem cells that are ever capable of dividing and producing blood cells. Some of these cells become the various types of white blood cells: neutrophils, eosinophils, basophils, lymphocytes, and monocytes (see Fig. 12.11).

In a child, most bones contain red bone marrow, but in an adult, it is present only in the bones of the skull, the sternum (breastbone), the ribs, the clavicle, the pelvic bones, and the vertebral column. The red bone marrow consists of a network of connective tissue fibers, which support the stem cells and their progeny. They are packed around thin-walled sinuses filled with venous blood. Differentiated blood cells enter the bloodstream at these sinuses. Bone marrow is not only the source of B cells, but also the place where B cells mature. T cells mature in the thymus.

The soft, bilobed **thymus gland** is located in the thoracic cavity between the trachea and the sternum above the heart. The thymus varies in size, but it is largest in children and shrinks as we get older. Connective tissue divides the thymus into lobules, which are filled with T cells and supporting cells. Immature T cells migrate from the bone marrow through the bloodstream to the thymus, where they mature. Only about 5% of these cells ever leave the thymus. Those that are capable of reacting to the body's own cells undergo a process of programmed cell death, or **apoptosis,** and only those that have the capability of reacting to foreign molecules or cells leave the thymus. It is said that the T cells remaining are capable of telling "self" from "nonself."

Secondary Lymphoid Organs

Lymphocytes frequently migrate from the bloodstream into the secondary lymphoid organs. Here, they may encounter foreign molecules or cells, after which they proliferate and become activated. These activated cells usually reenter the bloodstream, where they search for sites of infection or inflammation, like a squadron of highly trained military personnel seeking to destroy a specific enemy.

The **spleen** is located in the upper left side of the abdominal cavity behind the stomach. Most of the spleen is red pulp that filters the blood. Red pulp consists of blood vessels and sinuses where macrophages remove old and defective blood cells; lymphocytes cleanse the blood of foreign particles. The spleen also has white pulp that is inside the red pulp and consists of small areas of lymphatic tissue.

The spleen's outer capsule is relatively thin, and an infection or trauma can cause the spleen to burst. Although the spleen's functions are largely replaced by other organs, a person without a spleen is often slightly more susceptible to infections and may have to receive antibiotic therapy indefinitely.

Lymph nodes are small (about 1–25 mm in diameter), ovoid structures occurring along lymphatic vessels that cleanse lymph. Connective tissue divides the organ into nodules; each of these is packed with B and T cells and contains a sinus. As lymph courses through the many sinuses, resident macrophages engulf **pathogens,** which are disease-causing agents such as viruses and bacteria, and also debris.

Sometimes incorrectly called "glands," lymph nodes are named for their location. For example, inguinal nodes are in the groin, and axillary nodes are in the armpits. Physicians often feel for the presence of swollen, tender lymph nodes as evidence that the body is fighting an infection. This is a non-invasive, preliminary way to help make such a diagnosis.

> Lymphoid organs include primary organs (red bone marrow and the thymus gland), and secondary organs (the spleen and lymph nodes). Primary lymphoid organs are sites of lymphocyte development, and secondary lymphoid organs are sites where lymphocytes respond to foreign materials or cells.

13.2 Innate and Acquired Immunity

Immunity is the body's capability of removing or killing foreign substances, pathogens, and cancer cells. Mechanisms of **innate immunity** are fully functional without previous exposure to these substances, while **acquired immunity** is enhanced by exposure to specific antigens. An **antigen** is any molecule, usually a protein or carbohydrate, that stimulates an immune response. Mechanisms of innate immunity resemble a general police force, with members who are trained to deal with many different types of criminals. Acquired immunity is carried out mainly by the lymphocytes, which more closely resemble bounty hunters trained to seek out a very specific perpetrator.

Innate Immunity

Mechanisms of innate immunity can be divided into at least four types—physical and chemical barriers, inflammation, phagocytes and natural killer cells, and protective proteins.

Physical and Chemical Barriers

Skin and the mucous membranes lining the respiratory, digestive, and urinary tracts serve as mechanical barriers to entry of pathogens. The upper respiratory tract is lined by ciliated cells that sweep mucus and trapped particles up into the throat, where they can be swallowed or expelled.

The secretions of oil glands in the skin contain chemicals that weaken or kill certain bacteria on the skin. The stomach has an acid pH, which kills many types of bacteria or inhibits their growth. The various bacteria that normally

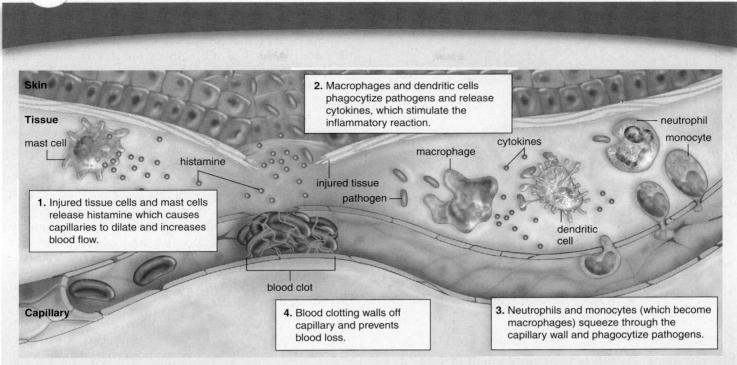

Skin

2. Macrophages and dendritic cells phagocytize pathogens and release cytokines, which stimulate the inflammatory reaction.

Tissue

mast cell

histamine

neutrophil

monocyte

cytokines

macrophage

injured tissue

pathogen

dendritic cell

1. Injured tissue cells and mast cells release histamine which causes capillaries to dilate and increases blood flow.

blood clot

Capillary

4. Blood clotting walls off capillary and prevents blood loss.

3. Neutrophils and monocytes (which become macrophages) squeeze through the capillary wall and phagocytize pathogens.

Figure 13.3 Inflammatory reaction.

Due to capillary changes in a damaged area and the release of chemical mediators, such as histamine by mast cells, an inflamed area exhibits redness, heat, swelling, and pain. The inflammatory reaction can be accompanied by other reactions to the injury. Macrophages and dendritic cells, present in the tissues, phagocytize pathogens, as do neutrophils, which squeeze through capillary walls from the blood. Macrophages and dendritic cells release cytokines, which stimulate the inflammatory and other immune reactions. A blood clot can form to seal a break in a blood vessel.

reside in the intestine and other areas, such as the vagina, take up nutrients and block binding sites that could potentially be used by pathogens.

Inflammation

Whenever tissue is damaged by physical or chemical agents or by pathogens, a series of events occurs that is known as the **inflammatory reaction.** Figure 13.3 illustrates the main participants in the inflammatory reaction.

An inflamed area has four signs: redness, heat, swelling, and pain. All of these signs are due to capillary changes in the damaged area. Chemicals such as **histamine,** released by damaged tissue cells as well as by **mast cells,** a type of immune cell found in tissues, cause nearby capillaries to dilate and become more permeable. This also results in increased blood flow, causing the skin to redden and feel warm. Increased capillary permeability allows fluids to escape into the tissues, resulting in swelling. The swollen area stimulates free nerve endings, causing the sensation of pain.

The inflammatory reaction can be accompanied by other responses to the injury. A blood clot can form to seal a break in a blood vessel. Phagocytes, discussed in the next section, are present. Antigens, chemical mediators, dendritic

cells, and macrophages move through the tissue fluid and lymph to the lymph nodes. Dendritic cells and macrophages cause T cells to mount a specific defense to the infection as described later in this section.

If the cause of inflammation cannot be eliminated, as occurs in tuberculosis and some types of arthritis, the inflammatory reaction may persist and become harmful rather than helpful. In these cases, anti-inflammatory medications such as aspirin, ibuprofen, or cortisone may be used to minimize the effects of various chemical mediators.

Phagocytes and Natural Killer Cells

At sites of inflammation, at least two types of **phagocytes** which literally means "eating cell," migrate through the walls of dilated capillaries. **Neutrophils** are usually the first cells to arrive at the scene, and they may accumulate to form pus, which has a whitish appearance mainly due to their presence. If the inflammatory reaction continues, more **monocytes** will migrate from the blood to the tissues, where they are then called **macrophages** (meaning, literally, "large eaters"). Macrophages are also important sources of **cytokines,** which are chemical messengers. Cytokines secreted by macrophages include colony-stimulating factors that cause the

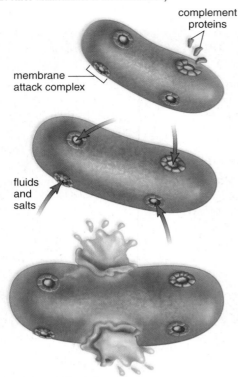

complement proteins

membrane attack complex

fluids and salts

Figure 13.4   Action of the complement system against a bacterium.
When complement proteins in the blood plasma are activated by an immune response, they form a membrane attack complex. This complex makes holes in bacterial cell walls and plasma membranes, allowing fluids and salts to enter until the cell eventually bursts.

bone marrow to produce more white cells, and others that act on the brain to induce a fever reponse. When a neutrophil or a macrophage encounters a pathogen, especially a bacterial cell, it will engulf the pathogen into an endocytic vesicle, which fuses with a lysosome inside the cell (Chapter 4). The lysosome has an acid pH that activates hydrolytic enzymes and various reactive oxygen compounds, which can usually destroy the pathogen. Also present in various tissues are phagocytic **dendritic cells,** which play an important role in interacting with T cells during acquired immune responses.

Natural killer (NK) cells are large, granular, lymphocyte-like cells that kill some virus-infected and cancer cells by cell-to-cell contact. Interestingly, NK cells and cytotoxic T cells use very similar mechanisms to kill their target cells, by inducing them to undergo apoptosis. NK cells, however, seek out and kill cells that lack a particular type of "self" molecule, called MHC-I (major histocompatibility class I), on their surface. Because some virus-infected and cancer cells may lack these MHC-I molecules, they often become susceptible to being killed by NK cells. However, since NK cells do not recognize specific viral or tumor antigens, and do not proliferate when exposed to a particular antigen, they are considered a part of the innate immune system.

Protective Proteins

The **complement system,** often simply called complement, is composed of a number of blood plasma proteins des-

ignated by the letter C and a subscript (for example, C_3). The complement proteins "complement" certain immune responses, which accounts for their name. For example, they are involved in and amplify the inflammatory reaction because certain complement proteins can bind to mast cells and trigger histamine release, and others can attract phagocytes to the scene. Some complement proteins bind to the surface of pathogens, as described later in this section, which ensures that the pathogens will be phagocytized by a neutrophil, dendritic cell, or macrophage.

Certain other complement proteins join to form a membrane attack complex that produces holes in the surface of bacteria and some viruses. Fluids and salts then enter the bacterial cell or virus to the point that they burst (Fig. 13.4).

Interferons are proteins produced by virus-infected cells as a warning to noninfected cells in the area. Interferons can bind to receptors of noninfected cells, causing them to prepare for possible attack by producing substances that interfere with viral replication. Interferons are used to treat certain viral infections, such as hepatitis C.

The body's innate immune system functions without previous exposure to antigens. Examples of innate immune mechanisms include physical and chemical barriers, inflammation, phagocytes and natural killer cells, and protective proteins.

Acquired Immunity

When innate defenses have failed to prevent an infection, acquired defenses come into play. Because acquired defenses do not ordinarily react to our own normal cells, it is said that the immune system is able to distinguish "self" from "nonself." Acquired defenses usually take five to seven days to become fully activated, and they may last for years; for example, once we recover from measles, we usually do not get measles a second time.

Acquired defenses primarily depend on the action of lymphocytes, which differentiate as either B cells or T cells. B cells and T cells are capable of recognizing antigens because they have specific antigen receptors—plasma membrane receptor proteins whose shape allows them to combine with particular antigens. Each lymphocyte has only one type of receptor. It is often said that the receptor and the antigen fit together like a lock and a key. Because we may encounter millions of different antigens during our lifetime, we need a diversity of B cells and T cells to protect us against them. Remarkably, diversification occurs to such an extent during the maturation process that there are specific B cells and/or T cells for almost any possible antigen.

B cells give rise to plasma cells, which produce antibodies that are capable of combining with and neutralizing a particular antigen. In contrast, T cells do not produce antibodies. Instead, they differentiate into either helper

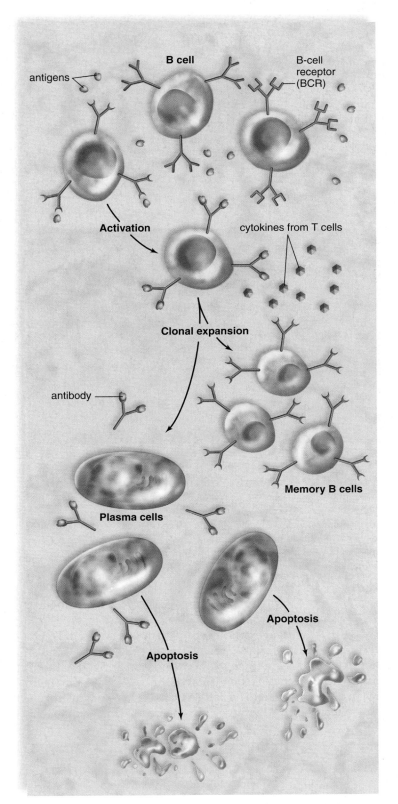

Figure 13.5 Clonal selection theory as it applies to B cells. Each B cell has a B-cell receptor (BCR) designated by shape that will combine with a specific antigen. Activation of a B cell occurs when its BCR can combine with an antigen (colored green). In the presence of cytokines, the B cell undergoes clonal expansion, producing many plasma cells and memory B cells. These plasma cells secrete antibodies specific to the antigen, and memory B cells immediately recognize the antigen in the future. After the infection passes, plasma cells undergo apoptosis, also called programmed cell death.

T cells, which release chemicals to regulate the immune response, or cytotoxic T cells, which attack and kill virus-infected cells and tumor cells.

B Cells and Antibody-Mediated Immunity

The receptor on a B cell is called a **B-cell receptor (BCR).** A B cell is activated in a lymph node or the spleen, when its BCRs bind to a specific antigen. Thereafter, the B cell divides by mitosis many times. In other words, it makes many copies of itself. Most of the resulting cells (clones) become plasma cells, which circulate in the blood and lymph. Plasma cells are larger than regular B cells because they have extensive rough endoplasmic reticulum for the mass production and secretion of antibodies to a specific antigen. Antibodies are the secreted form of the BCR of the B cell that was activated.

The **clonal selection theory** states that an antigen binds to the antigen receptor of only one type of B cell or T cell, and then this B cell or T cell divides, forming clones of itself. Note in Figure 13.5 that each B cell has a specific BCR represented by shape. Only the B cell with a BCR shape that fits the antigen (green circle) undergoes clonal expansion. During clonal expansion, cytokines secreted by helper T cells stimulate B cells to divide. Some cloned B cells become memory cells, which are the means by which long-term immunity is possible. If the same antigen enters the system again, memory B cells quickly divide and give rise to more plasma cells capable of rapidly producing the correct type of antibody.

Once the threat of an infection has passed, new plasma cells cease to develop, and those present undergo apoptosis, which involves a cascade of specific cellular events leading to the death and destruction of the cell (see Chapter 5).

Defense by B cells is called **antibody-mediated immunity** because the various types of activated B cells become plasma cells that produce antibodies. It is also called humoral immunity because these antibodies are present in blood and lymph. (Historically, a humor is any fluid normally occurring in the body.)

Characteristics of B Cells

- Provide antibody-mediated immunity against pathogens
- Produced and mature in bone marrow
- Reside in lymph nodes and spleen; circulate in blood and lymph
- Directly recognize antigen and then undergo clonal selection
- Clonal expansion produces antibody-secreting plasma cells as well as memory B cells

Structure of an Antibody Antibodies are also called **immunoglobulins (Igs).** They are typically Y-shaped molecules with two arms. Each arm has a "heavy" (long) polypeptide chain and a "light" (short) polypeptide chain. These chains have constant regions, where the sequence of amino acids is set, and variable regions, where the sequence of amino acids varies between antibodies (Fig. 13.6). The constant regions are almost the same within different classes of antibodies. The variable regions become hypervariable at their tips and form antigen-binding sites; their shape is specific for a particular antigen. It is the variable and hypervariable regions that allow the antibody to bind to a specific antigen. The antigen combines with the antibody at the antigen-binding site in a lock-and-key manner.

The antigen-antibody reaction can have several outcomes, but quite often the reaction produces complexes of antigens combined with antibodies. Such antigen-antibody complexes, sometimes called immune complexes, mark the antigens for destruction. For example, an antigen-antibody complex may be engulfed by neutrophils or macrophages, or it may activate complement. Complement makes pathogens more susceptible to phagocytosis, as discussed previously. Antibodies may also "neutralize" viruses or toxins by preventing their binding to cells.

Types of Antibodies   There are five major classes of antibodies (Table 13.1). IgG antibodies are the major type in blood, lymph, and tissue fluid. IgG antibodies bind to pathogens and their toxins and can activate the complement system. IgGs are the only antibodies that can cross the placenta. Along with IgAs, IgGs are in breast milk. IgM antibodies are pentamers, meaning that they contain five of the Y-shaped structures shown in Figure 13.6a. These antibodies appear in blood soon after a vaccination or infection and disappear relatively quickly. They are good activators of the complement system. IgMs are the antibodies in plasma that can cause red blood cells to clump. IgA antibodies are monomers in blood and lymph or dimers in tears, saliva, gastric juice, and mucous secretions. They are the main type of antibody found in body secretions. They bind to pathogens before they reach the bloodstream. IgD molecules appear on the surface of B cells when they are ready to be activated. IgE antibodies are important in the immune system's response to parasites (e.g., parasitic worms). They are best known, however, for immediate allergic responses, including anaphylactic shock (discussed in section 13.4).

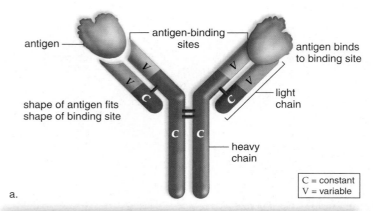
a.

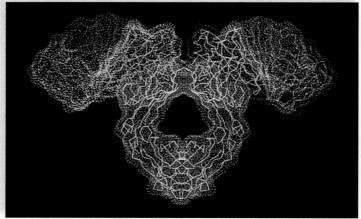
b.

Figure 13.6 Structure of an antibody.

a. An antibody contains two heavy (long) polypeptide chains and two light (short) chains arranged so that there are two variable regions, where a particular antigen is capable of binding with an antibody (*V* = variable region, *C* = constant region). The shape of the antigen fits the shape of the binding site. **b.** Computer model of an antibody molecule. The antigen combines with the two side branches.

T Cells and Cell-Mediated Immunity

When a T cell leaves the thymus, it has a unique **T-cell receptor (TCR)** similar to the BCR on B cells. Unlike B cells, however, T cells are unable to recognize an antigen without help. The antigen must be displayed, or "presented," to them by an **antigen-presenting cell (APC)** such as a dendritic cell or a macrophage. After ingesting and destroying a pathogen,

TABLE 13.1	Classes of Antibodies	
Class	**Presence**	**Function**
IgG	Main antibody type in circulation	Binds to pathogens, activates complement, and enhances phagocytosis
IgM	Antibody type found in circulation; largest antibody	Activates complement; clumps cells
IgA	Main antibody type in secretions such as saliva and milk	Prevents pathogens from attaching to epithelial cells in digestive and respiratory tract
IgD	Antibody type found on surface of immature B cells	Presence signifies readiness of B cell to respond to antigens
IgE	Antibody type found as antigen receptors on eosinophils in blood and on mast cells in tissues	Responsible for immediate allergic response and protection against certain parasitic worms

APCs travel to a lymph node or the spleen, where T cells also congregate. After breaking the pathogen apart with lysosomal enzymes, the APC presents a piece of the pathogen in the groove of an **MHC (major histocompatibility complex)** protein on its surface. Here, the antigen fragment/MHC combination can be recognized by the T cell. In addition to their role in antigen presentation to T cells, MHC proteins are also involved in recognition of foreign tissues. Therefore, prior to organ transplantation, medical personnel attempt to find an organ donor whose MHC molecules match those of the recipient as closely as possible.

There are two major types of T cells: **helper T cells** (T_H cells) and **cytotoxic T cells** (T_C cells, or CTLs). Each of these types has a TCR that can recognize an antigen fragment in combination with an MHC molecule. However, the major difference in antigen recognition by these two types of cells is that the T_H cells only recognize antigen presented by APCs with MHC class II molecules on their surface, while the T_C cells only recognize antigen presented by various cell types with MHC class I molecules on their surface (Fig. 13.7). After they are stimulated by APCs, T_H cells secrete various cytokines. The cytokines produced by T_H cells may help activate T_C cells, as well as B cells.

The T_C cells have storage vacuoles that contain many molecules of perforin and enzymes called granzymes. After an activated T_C cell binds to a virus-infected or cancer cell that is presenting foreign antigen on its MHC class I molecules, it releases perforin molecules, which form pores in the plasma membrane of the abnormal cell. This allows the granzymes to enter the target cell, which is induced to undergo apoptosis and die (Fig. 13.8a). Once T_C cells have delivered this lethal "hit," they are capable of moving on to another target cell (Fig. 13.8b).

Notice also in Figure 13.7 that, similar to B cells, some of the clonally expanded T cells become memory T cells. These may live for many years and can quickly jump-start an immune response to an antigen previously present in the body. Immunity mediated by T_H cells and T_C cells is sometimes referred to as **cell-mediated immunity.**

Because HIV, the virus that causes AIDS, infects helper T cells and other cells of the immune system, it suppresses many components of acquired immune responses and makes HIV-infected individuals susceptible to opportunistic infections (see Health Focus, "Opportunistic Infections and HIV"). Infected macrophages and dendritic cells also serve as reservoirs for HIV.

Characteristics of T Cells

- Produced in bone marrow; mature in thymus
- Function in cell-mediated immunity against virus-infected cells and cancer cells
- Recognize antigens presented in groove of an MHC molecule
- Cytotoxic T cells destroy nonself protein-bearing cells
- Helper T cells secrete cytokines that control the immune response

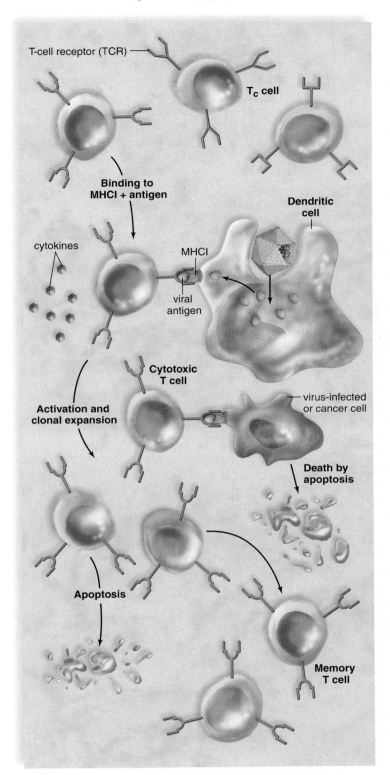

Figure 13.7 Clonal selection model as it applies to T_C cells. Each T cell has a T-cell receptor (TCR) designated by a shape that will combine only with a specific antigen. Activation of a T cell occurs when its TCR can combine with an antigen. A dendritic cell presents the antigen (colored green) in the groove of an MHC-I. Thereafter, the T_C cell undergoes clonal expansion, and many copies of the same type of T cell are produced. After the immune response has been successful, the majority of T cells undergo apoptosis, but a small number are memory T cells. Memory T cells provide protection should the same antigen enter the body again at a future time.

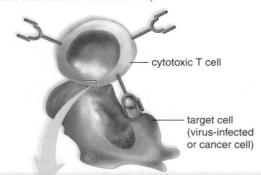

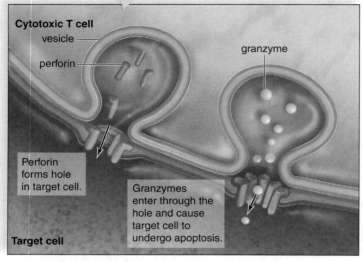

a.

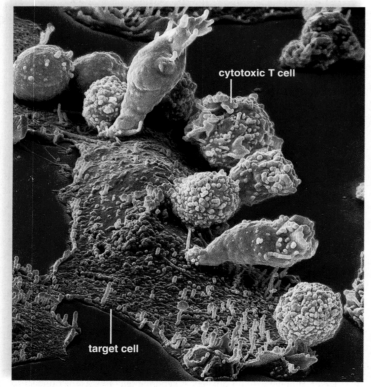

b.

Figure 13.8 Cell-mediated immunity.

a. How a T$_C$ cell destroys a virus-infected cell or cancer cell. **b.** The scanning electron micrograph shows cytotoxic T cells attacking and destroying a cancer cell (target cell).

Active vs. Passive Immunity

In general, acquired immune responses can be induced actively or passively. In **active immunity,** the individual alone produces an immune response against an antigen; in **passive immunity,** the individual is given prepared antibodies or cells via an injection.

Active Immunity

Active immunity usually develops naturally after a person is infected with a pathogen. However, active immunity can also be induced when a person is well, so that future infection will not take place. To prevent infections, people can be artificially immunized against them. The United States is committed to immunizing all children against the common types of childhood disease (Fig. 13.9a).

   Immunization involves the use of **vaccines,** substances that contain an antigen to which the immune system responds. Traditionally, vaccines are the pathogens themselves, or their products, that have been treated so they are no longer virulent (able to cause disease). Today, it is possible to genetically engineer bacteria to mass-produce a protein from pathogens, and this protein can be used as a vaccine. This method has now produced a vaccine against hepatitis B, a viral disease, and is being used to develop a vaccine against malaria, a protozoal disease.

   After a vaccine is given, it is possible to follow an immune response by determining the amount of antibody present in a sample of plasma; this is called the **antibody titer.** After the first exposure to a vaccine, a primary response occurs. For a period of several days, no antibodies are present; then the titer rises slowly, levels off, and gradually declines as the antibodies bind to the antigen or simply break down (Fig. 13.9b). After a second exposure to the vaccine, a secondary response occurs. The titer rises rapidly to a level much greater than before; then it slowly declines. The second exposure is called a "booster" because it boosts the immune response to a high level. The high levels of antigen-specific T cells and antibodies prevent disease symptoms even if the individual is exposed to the disease-causing agent.

   Active immunity is dependent upon the presence of memory B cells and memory T cells that are capable of responding to lower doses of antigen. Active immunity is usually long-lasting, although a booster may be required periodically. In recent years, the safety of certain vaccines has come into question (see the Bioethical Focus at the end of this chapter).

Active immunity can be induced by infection or by the use of vaccines. Active immunity is dependent upon the presence of memory B cells and memory T cells in the body.

Suggested Immunization Schedule

Vaccine	Age (months)	Age (years)
Hepatitis B	Birth, 1–2, 6–18	
Diphtheria, tetanus, pertussis (DTP)	2, 4, 6, 15–18	4–6
Tetanus only		11–12, 13–18
Haemophilus influenzae, type b	2, 4, 6, 12–15	
Polio (IPV)	2, 4, 6–18	4–6
Pneumococcal	2, 4, 6, 12–15	
Measles, mumps, rubella (MMR)	12–15	4–6, 11–12
Varicella (chicken pox)	12–18	2–18
Hepatitis A (in selected areas)	12–18	2–18
Human papilloma-virus, types 6, 11, 16, 18	–	11–12

a.

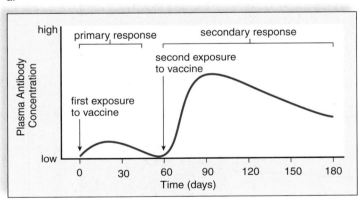

b.

Figure 13.9 Active immunity due to immunizations.

a. Suggested immunization schedule for infants and children. **b.** During immunization, the primary response, after the first exposure to a vaccine, is minimal, but the secondary response, which occurs after the second exposure, shows a dramatic rise in the amount of antibody present in plasma.

Passive Immunity

Passive immunity occurs when an individual is given prepared antibodies or immune cells to combat a disease. The passive transfer of antibodies is a common natural process. For example, newborn infants are passively immune to some diseases because IgG antibodies have crossed the placenta from the mother's blood (Fig. 13.10a). These antibodies soon disappear, however, so within a few months, infants become more susceptible to infections. Breast-feeding prolongs the natural passive

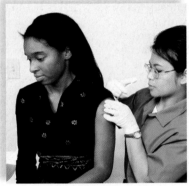

a. Antibodies (IgG) cross the placenta.

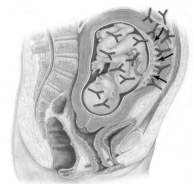

c. Antibodies can be injected by a physician.

b. Antibodies (IgG, IgA) are secreted into breast milk.

Figure 13.10
Passive immunity.
During passive immunity, antibodies are received by **(a)** crossing the placenta, **(b)** in breast milk, or **(c)** by injection. Because the body is not producing the antibodies, passive immunity is short lived.

immunity an infant receives from the mother because IgG and IgA antibodies are present in the mother's milk (Fig. 13.10b).

Even though passive immunity does not last, it is sometimes used to prevent illness in a patient who has been unexpectedly exposed to an infectious disease. Usually, the patient receives a gamma globulin injection (serum that contains antibodies), perhaps taken from individuals who have recovered from the illness (Fig. 13.10c). In the past, horses were immunized, and serum was taken from them to provide the needed antibodies against such diseases as diphtheria, botulism, and tetanus. Unfortunately, a patient who received these antibodies became ill about 50% of the time, because the serum contained some horse proteins that the individual's immune system recognized as foreign. This is called serum sickness. Passive immunity using horse antibodies against poisonous snake and spider venom is still used today. Antivenom antibodies are IgGs produced in horses.

Instead of antibodies, cells of the immune system may be passively transferred into a patient, although this practice is less common. The best example is a bone marrow transplant, in which stem cells that produce blood cells are replenished after a cancer patient's own bone marrow has been destroyed by radiation or chemotherapy.

> Passive immunity provides protection when an individual is in immediate danger of succumbing to an infectious disease. Immunity resulting from the passive transfer of antibodies is temporary because there are no memory cells.

Opportunistic Infections and HIV

AIDS (acquired immunodeficiency syndrome) is caused by the destruction of the immune system, especially the helper T cells, following an HIV (human immunodeficiency virus) infection. Then, the individual succumbs to many unusual types of infections that would not cause disease in a person with a healthy immune system. Such infections are known as opportunistic infections (OIs).

HIV not only kills helper T cells by directly infecting them; it also causes many uninfected T cells to die by a variety of mechanisms, including apoptosis. Many helper T cells are also killed by the person's own immune system as it tries to overcome the HIV infection. After initial infection with HIV, it may take up to ten years for an individual's helper T cells to become so depleted that the immune system can no longer organize a specific response to OIs (Fig. 13A). In a healthy individual, the number of helper T cells typically ranges from 800 to 1,000 cells per cubic millimeter (mm³) of blood. The appearance of the following specific OIs can be associated with the helper T-cell count:

- Shingles. Painful infection with varicella-zoster (chicken pox) virus. Helper T-cell count of less than 500/mm³.
- Candidiasis. Yeast infection of the mouth, throat, or vagina. Helper T-cell count of about 350/mm³.
- Pneumocystis pneumonia. Fungal infection causing the lungs to fill with fluid and debris. Helper T-cell count of less than 200/mm³.
- Kaposi sarcoma. Cancer of blood vessels due to human herpesvirus 8; gives rise to reddish-purple, coin-sized spots and lesions on the skin. Helper T-cell count of less than 200/mm³.
- Toxoplasmic encephalitis. Protozoan infection characterized by severe headaches, fever, seizures, and coma. Helper T-cell count of less than 100/mm³.
- *Mycobacterium avium* complex (MAC). Bacterial infection resulting in persistent fever, night sweats, fatigue, weight loss, and anemia. Helper T-cell count of less than 75/mm³.
- Cytomegalovirus. Viral infection that leads to blindness, inflammation of the brain, throat ulcerations. Helper T-cell count of less than 50/mm³.

Due to development of powerful drug therapies that slow the progression of AIDS, people infected with HIV in the United States are suffering a lower incidence of OIs than was common in the 1980s and 1990s.

Discussion Questions

1. People who don't have HIV also sometimes get OIs, such as shingles or candidiasis. How would you expect the outcome of, for example, a yeast infection to be different in a person without HIV compared to a person with HIV?
2. Kaposi sarcoma occurs relatively rarely in people who are infected with human herpesvirus 8, but are HIV-negative. How might an HIV infection increase the odds of getting Kaposi sarcoma?
3. Why do you suppose that the amount of HIV in the blood increases rapidly during the later stages of AIDS, as shown in Figure 13A?

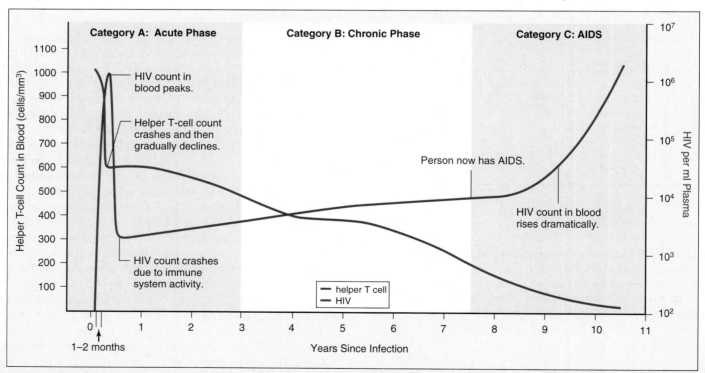

Figure 13A Progression of HIV infection during its three stages.
In category A, the individual may have no symptoms or very mild symptoms associated with the infection. By category B, opportunistic infections have begun to occur, such as candidiasis, shingles, and diarrhea. Category C is characterized by more severe opportunistic infections and is clinically described as AIDS.

Immune Therapies

Cytokines and Immunity

Cytokines are chemical messengers produced by T cells, macrophages, and other cells. Because cytokines regulate white blood cell formation and/or function, they are being investigated as possible adjunct therapy for cancer and AIDS. Both interferons and **interleukins,** which are cytokines produced by various white blood cells, have been used as immunotherapeutic drugs, particularly to enhance the ability of an individual's own T cells to fight cancer.

Because many cancer cells carry altered proteins on their cell surface, they should be attacked and destroyed by cytotoxic T cells. Whenever cancer develops, it is possible that cytotoxic T cells have not been activated. In that case, cytokines might awaken the immune system and lead to the destruction of the cancer. In one technique being investigated, researchers first withdraw T cells from the patient, present cancer cell antigens to them, and then activate the cells by culturing them in the presence of an interleukin. The T cells are reinjected into the patient, who is given doses of interleukin to maintain the killer activity of the T cells.

Scientists who are actively engaged in interleukin research believe that interleukins soon will be used as adjuncts for vaccines, for the treatment of chronic infectious disease, and perhaps for the treatment of cancer. Conversely, interleukin antagonists also may prove helpful in preventing skin and organ rejection, autoimmune diseases, and allergies.

Monoclonal Antibodies

Every plasma cell derived from the same B cell secretes antibodies against a specific antigen. These are **monoclonal antibodies** because all of them are the same type and because they are produced by plasma cells derived from the same B cell. One method of producing monoclonal antibodies in vitro (outside the body in glassware) is depicted in Figure 13.11. B cells are removed from an animal (usually mice are used) that has been exposed to a particular antigen. The resulting plasma cells are fused with myeloma cells (malignant plasma cells that live and divide indefinitely; they are immortal cells). The fused cells are called hybridomas—*hybrid*- because they result from the fusion of two different cells, and *-oma* because one of the cells is a cancer cell. Hybridomas producing the desired antibody are then isolated and purified.

At present, monoclonal antibodies are being used for quick and certain diagnosis of various conditions. For example, a particular hormone is present in the urine of a pregnant woman. A monoclonal antibody can be used to detect this hormone; if it is present, the woman is pregnant. Monoclonal antibodies are also used to identify infections. And because they can distinguish between cancerous and normal tissue cells, they are used to carry radioactive isotopes or toxic drugs to tumors, which can then be selectively destroyed. Herceptin is a monoclonal antibody used to treat breast cancer. It binds to a protein receptor on breast can-

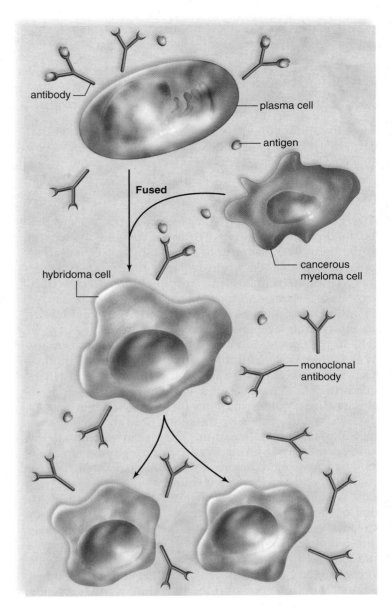

Figure 13.11 Production of monoclonal antibodies.
Plasma cells of the same type (derived from immunized mice) are fused with myeloma (cancerous) cells, producing hybridoma cells that are "immortal." Hybridoma cells divide and continue to produce the same type of antibody, called monoclonal antibodies.

cer cells and prevents the cancer cells from dividing so fast. Antibodies that bind to cancer cells also can activate complement, increase phagocytosis by macrophages and neutrophils, and attract NK cells.

Cytokines produced by immune cells may be useful in enhancing the immune system's response to diseases such as cancer, as well as to certain vaccines. Monoclonal antibodies, which react extremely specifically to a particular antigen, are being used in diagnostic tests as well as therapeutically.

Sometimes the immune system responds in a manner that harms the body, as when individuals develop allergies, receive an incompatible blood type, or suffer tissue rejection.

Allergies

Allergies are hypersensitivities to substances, such as pollen, food, or animal hair, that ordinarily would do no harm to the body. The response to these antigens, called **allergens,** usually includes some degree of tissue damage.

An **immediate allergic response** can occur within seconds of contact with the antigen. The response is caused by antibodies known as IgE (see Table 13.1). IgE antibodies are attached to receptors on the plasma membrane of mast cells in the tissues and also to eosinophils and basophils in the blood. When an allergen attaches to the IgE antibodies, these cells release histamine and other substances that bring about the allergic symptoms. When pollen is an allergen, histamine stimulates the mucous membranes of the nose and eyes, causing the runny nose and watery eyes typical of **hay fever.** In **asthma,** the airways leading to the lungs constrict, resulting in difficult breathing accompanied by wheezing. When food contains an allergen, nausea, vomiting, and diarrhea often result.

Anaphylactic shock is an immediate allergic response that occurs because the allergen has entered the bloodstream. Bee stings and penicillin shots are known to cause this reaction. Anaphylactic shock is characterized by a sudden and life-threatening drop in blood pressure due to increased permeability of the capillaries caused by histamine. Injecting epinephrine can counteract this reaction until medical help is available.

People with allergies can produce ten times more IgE than people without allergies. A new treatment using injections of monoclonal IgG antibodies against IgEs is currently being tested in individuals with severe food allergies. More routinely, injections of the allergen are given so that the body will build up high quantities of IgG antibodies (see Health Focus, "Immediate Allergic Responses"). The hope is that IgG's will combine with allergens before they have a chance to reach the IgE antibodies.

A **delayed allergic response** is initiated by memory T cells at the site of allergen contact in the body. The allergic response is regulated by the cytokines secreted by both T cells and macrophages. A classic example of a delayed allergic response is the skin test for tuberculosis (TB). When the test result is positive, the tissue where the antigen was injected becomes red and hardened, indicating prior exposure to tubercle bacilli, the cause of TB. Contact dermatitis, which occurs when a person is allergic to poison ivy and many other substances that touch the skin, is also an example of a delayed allergic response.

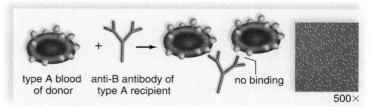

a. No agglutination

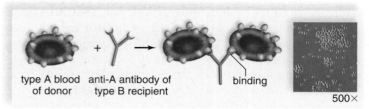

b. Agglutination

Figure 13.12 **Blood transfusions.**

No agglutination **(a)** versus agglutination **(b)** is determined by whether the recipient has antibodies in the plasma that can combine with antigens on the donor's red blood cells. The photos on the right show how the red blood cells appear under the microscope.

Blood-Type Reactions

Several blood typing systems are currently in use. The most important is the ABO system. People often carry information about their ABO type in their wallets in case an accident requires them to need blood.

ABO System

In the ABO system, the presence or absence of type A and type B antigens on red blood cells determines a person's blood type. For example, a person with type A blood has the A antigen on his or her red blood cells. Because it is considered "self," this molecule is not recognized as an antigen by this individual's body, although it can be an antigen to a recipient who does not have type A blood.

The simplified ABO system identifies four types of blood: A, B, AB, and O. A person's plasma contains antibodies to the antigens that are *not* present on the red blood cells. These antibodies are called anti-A and anti-B. This chart tells you what antibodies are present in the plasma of each blood type:

Blood Type	Antigen on Red Blood Cells	Antibody in Plasma
A	A	Anti-B
B	B	Anti-A
AB	A, B	None
O	None	Anti-A and anti-B

Because type A blood has anti-B and not anti-A antibodies in the plasma, a donor with type A blood can give blood to a recipient with type A blood (Fig. 13.12*a*). But giving type A blood to a type B recipient causes agglutination (Fig. 13.12*b*). **Agglutination,** or clumping of red blood cells, can stop blood from circulating in small blood vessels, lead-

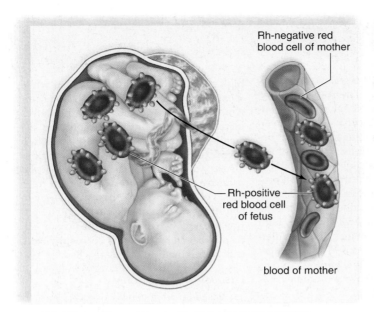

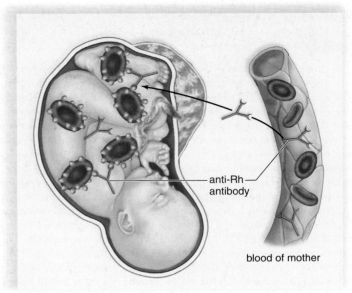

a. Fetal Rh-positive red blood cells leak across placenta into mother's bloodstream.

b. Mother forms anti-Rh antibodies that cross the placenta and attack fetal Rh-positive red blood cells.

Figure 13.13 Hemolytic disease of the newborn.

a. Due to a pregnancy in which the child is Rh positive, an Rh-negative mother can begin to produce antibodies against Rh-positive red blood cells. **b.** In the same, but more likely a subsequent pregnancy, these antibodies can cross the placenta and cause hemolysis of an Rh-positive child's red blood cells.

ing to organ damage. It is also followed by *hemolysis,* or bursting of red blood cells, which if extensive can cause the individual to die.

Theoretically, which blood type would be accepted by all other blood types? Type O blood has no antigens on the red blood cells and is sometimes called the universal donor. Which blood type could receive blood from any other blood type? Type AB blood has no anti-A or anti-B antibodies in the plasma and is sometimes called the universal recipient. In practice, however, it is not safe to rely solely on the ABO system when matching blood, and instead, samples of the two types of blood are physically mixed, and the result is microscopically examined before blood transfusions are done.

Today, blood transfusions are a matter of concern not only because blood types should match, but also because each person wants to receive blood that is free of infectious agents. Blood is tested for the more serious agents, such as those that cause AIDS, hepatitis, and syphilis. People with these infections can help protect the nation's blood supply by knowing when not to give blood.

Rh System

Another important antigen in matching blood types is the Rh factor. Eighty-five percent of the U.S. population have this particular antigen on their red blood cells and are Rh⁺ (Rh positive). Fifteen percent do not have this antigen and are Rh⁻ (Rh negative). Rh⁻ individuals normally do not have antibodies to the Rh factor, but they may make them when

exposed to the Rh factor. The designation of blood type usually also includes whether the person has or does not have the Rh factor on the red blood cells.

During pregnancy, if the mother is Rh⁻ and the father is Rh⁺, the child may be Rh⁺. The Rh⁺ red blood cells may leak across the placenta into the mother's cardiovascular system, especially as placental tissues normally break down before and at birth (Fig. 13.13*a*). Now, the mother produces anti-Rh antibodies. In this or a subsequent pregnancy with another Rh⁺ baby, these antibodies may cross the placenta and destroy the child's red blood cells (Fig. 13.13*b*). The resulting condition, called hemolytic disease of the newborn (HDN), can lead to brain damage and mental retardation or even death due to anemia and excess bilirubin in the blood.

The Rh problem is prevented by giving Rh⁻ women an Rh immunoglobulin injection, either towards the end of the pregnancy or no later than 72 hours after giving birth to an Rh⁺ child. This injection contains relatively low, safe levels of anti-Rh antibodies that attack any of the baby's red blood cells in the mother's blood before these cells can stimulate her immune system to produce her own antibodies. This injection must be given before the woman starts producing her own anti-Rh antibodies, so timing is important.

When hemolytic disease of the newborn occurs, antibodies produced by the mother have reacted to antigens on fetal red blood cells.

The runny nose and watery eyes of hay fever are often caused by an allergic reaction to the windblown pollen of trees, grasses, and ragweed, particularly in the spring and fall (Fig. 13B). Worse, if a person has asthma, the airways leading to the lungs constrict, resulting in difficult breathing characterized by wheezing. Most people can inhale pollen with no ill effects. But others have developed a hypersensitivity, meaning that their immune system responds in a deleterious manner. The problem stems from a type of antibody called immunoglobulin E (IgE) that causes the release of histamine from mast cells and also basophils whenever they are exposed to an allergen. Histamine causes the mucous membranes of the nose and eyes to release fluid as a defense against pathogen invasion. But in the case of allergies, copious fluid is released even though no real danger is present.

Many food allergies are also due to the presence of IgE antibodies, which usually bind to a protein in the food. The symptoms, such as nausea, vomiting, and diarrhea, are due to the mode of entry of the allergen. Skin symptoms may also occur, however. Adults are often allergic to shellfish, nuts, eggs, cows' milk, fish, and soybeans. Peanut allergy is a common food allergy in the United States, possibly because peanut butter is a staple in the diet. People seem to outgrow allergies to cows' milk and eggs more often than allergies to peanuts and soybeans.

Celiac disease occurs in people who are allergic to wheat, rye, barley, and sometimes oats—in short, any grain that contains gluten proteins. It is thought that the gluten proteins elicit a delayed cell-mediated immune response by T cells with the resultant production of cytokines. The symptoms of celiac disease include diarrhea, bloating, weight loss, anemia, bone pain, chronic fatigue, and weakness.

People can reduce their reactions to airborne and food allergens by avoiding the offending substances. Airlines are now required to provide a peanut-free zone in their planes for those who are allergic. Taking antihistamines can also be helpful.

If these precautions are inadequate, patients can be tested to measure their susceptibility to any number of possible allergens. A small quantity of a suspected allergen is injected just beneath the skin, and the strength of the subsequent reaction is noted. A wheal-and-flare response at the skin prick site demonstrates that IgE antibodies attached to mast cells have reacted to an allergen. In an immunotherapy called hyposensitization, ever-increasing doses of the allergen are periodically injected subcutaneously with the hope that the body will produce an IgG response against the allergen. IgG, in contrast to IgE, does not cause the release of histamine after it combines with the allergen. If IgG combines first, the allergic response does not occur. Patients know they are cured when the allergic symptoms go away. Therapy may have to continue for as long as two to three years.

Allergic-type reactions can occur without involving the immune system. Wasp and bee stings contain substances that cause swelling, even in those whose immune system is not sensitized to substances in the sting. Also, jellyfish tentacles and certain foods (e.g., fish that is not fresh and strawberries) contain histamine or closely related substances that can cause a reaction. Furthermore, immunotherapy is not possible in people who are allergic to penicillin and bee stings. Upon the first exposure high sensitivity has built up so that when reexposed, anaphylactic shock can occur. Among its many effects, histamine causes increased permeability of the capillaries, the smallest blood vessels. When this reaction occurs throughout the body, these individuals experience a drastic decrease in blood pressure that can be fatal within a few minutes. People who know they are allergic to bee stings can obtain a syringe of epinephrine to carry with them. This medication can counteract the symptoms of anaphylactic shock until medical help is reached.

Discussion Questions

1. Based on what you have read in this essay and the chapter, what are some factors that might determine a person's tendency to develop an immediate allergic response to an allergen?
2. An increase in allergic diseases has been reported in developed countries, compared to lower rates in less-developed countries. What are some possible reasons for this difference?
3. What are the specific effects of epinephrine that can save the life of a person who is in anaphylactic shock?

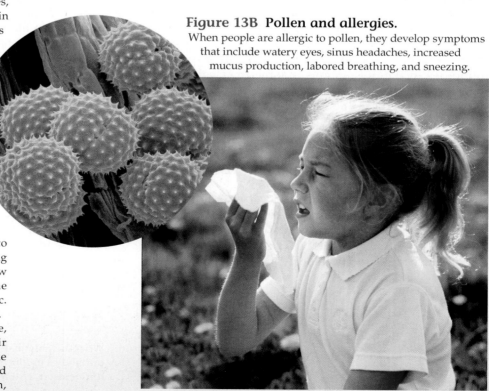

Figure 13B Pollen and allergies.
When people are allergic to pollen, they develop symptoms that include watery eyes, sinus headaches, increased mucus production, labored breathing, and sneezing.

Tissue Rejection

Certain organs, such as the skin, the heart, and the kidneys, could be transplanted easily from one person to another if the body did not attempt to reject them. Rejection of transplanted tissue results because the recipient's immune system recognizes that the transplanted tissue is not "self." Cytotoxic T cells respond by attacking the cells of the transplanted tissue.

It is best if the transplanted organ has the same type of MHC antigens as those of the recipient, because cytotoxic T cells recognize foreign MHC antigens. Organ rejection can also be controlled by carefully selecting the organ to be transplanted and administering **immunosuppressive** drugs. Two well-known immunosuppressive drugs, cyclosporine and tacrolimus, both act by inhibiting the production of certain cytokines by T cells.

Xenotransplantation is the use of animal organs instead of human organs in transplant patients. Scientists have chosen to use the pig because pigs are raised as a meat source and are prolific. Genetic engineering can make pig organs less antigenic by removing the MHC antigens. The ultimate goal is to make pig organs as widely accepted as type O blood. Although advances in xenotransplantation are encouraging, many researchers hope that tissue engineering, including the production of human organs that lack MHC antigens, will one day do away with the problem of rejection. Alternatively, scientists are learning how to grow organs from a patient's own tissues. In April of 2006, surgeons at Children's Hospital in Boston were able to use lab-grown urinary bladder tissue to rebuild defective bladders in human patients.

When tissue rejection occurs, the immune system has recognized and destroyed cells that bear foreign MHC antigens.

13.5 Disorders of the Immune System

When a person has an **autoimmune disease,** cytotoxic T cells or antibodies mistakenly attack the body's own cells as if they bear foreign antigens. Exactly what causes autoimmune diseases is not known. However, sometimes they occur after an individual has recovered from an infection.

In the autoimmune disease **myasthenia gravis,** antibodies attach to and interfere with neuromuscular junctions, and muscular weakness results. In **multiple sclerosis (MS),** T cells attack the myelin sheath of nerve fibers, and this causes various neuromuscular disorders. A person with **systemic lupus erythematosus (SLE)** has various symptoms prior to death due to kidney damage caused by the deposition of excessive antigen/antibody complexes. In **rheumatoid arthritis,** the joints are affected. Research suggests that heart

damage following rheumatic fever and type 1 diabetes are also autoimmune illnesses. As yet, there are no cures for autoimmune diseases, but they can sometimes be controlled with immunosuppressive drugs.

When a person has an immune deficiency, the immune system is unable to protect the body against disease. AIDS is an example of an acquired immune deficiency (see Health Focus, "Opportunistic Infections and HIV"). Immune deficiency may also be congenital (that is, inherited). Infrequently, a child may be born with an impaired immune system caused by a defect in lymphocyte development. In **severe combined immunodeficiency disease (SCID),** both antibody- and cell-mediated immunity are lacking or inadequate. Without treatment, even common infections can be fatal. Gene therapy has been successful in some SCID patients.

Diseases of the immune system include autoimmune or immune deficiency disorders.

Summarizing the Concepts

13.1 The Lymphatic System
The lymphatic system consists of lymphatic vessels and lymphoid organs. The lymphatic vessels absorb fats from the digestive tract and excess tissue fluid at blood capillaries and carry these to the bloodstream.

Lymphocytes are produced and accumulate in the lymphoid organs. Lymph is cleansed of pathogens and/or their toxins in lymph nodes, and blood is cleansed of pathogens and/or their toxins in the spleen. T cells mature in the thymus, while B cells mature in the red bone marrow where all blood cells are produced.

13.2 Innate and Acquired Immunity
Mechanisms of innate immunity include physical and chemical barriers, inflammation, phagocytes and natural killer cells, and protective proteins. Acquire immunity requires B cells and T cells. B cells undergo clonal selection with production of plasma cells and memory B cells after their antigen receptors combine with a specific antigen. Plasma cells secrete antibodies and eventually undergo apoptosis. Plasma cells are responsible for antibody-mediated immunity. The IgG antibody is a Y-shaped molecule that has two binding sites for a specific antigen. Memory B cells remain in the body and produce antibodies if the same antigen enters the body at a later date.

T cells are responsible for cell-mediated immunity. For a T cell to recognize an antigen, the antigen must be presented by an antigen-presenting cell (APC), such as a dendritic cell or macrophage, along with an MHC (major histocompatibility complex) molecule. Thereafter, the activated T cell undergoes clonal expansion until the illness has been stemmed. Then, most of the activated T cells undergo apoptosis. A few cells remain, however, as memory T cells.

Like B cells, each T cell bears antigen receptors. The two main types of T cells are cytotoxic T cells and helper T cells. Cytotoxic T cells kill virus-infected or cancer cells on contact because they

Bioethical Focus

Vaccination Requirements

In July 1999, a vaccine called Rotashield was removed from the U.S. market. An advisory committee to the U.S. government decided that the vaccine, designed to protect against serious diarrheal disease caused by rotavirus, was unsafe. Since the introduction of the vaccine less than a year earlier, 15 cases of serious, life-threatening bowel obstructions had been reported within 1–2 weeks following vaccination.

Similar concerns have been raised about the safety of other vaccines. In the late 1990s, some scientific studies seemed to establish a link between the common childhood measles-mumps-rubella (MMR) vaccine and an increasing incidence of autism, even though subsequent studies have failed to confirm this link. When the U.S. Army announced that it would require all military personnel to be vaccinated against anthrax, many filed lawsuits because they objected to being forced to take the vaccine. This kind of skepticism about vaccines can be traced all the way back to the early 1800s in England, when news first spread about Edward Jenner's demonstration that smallpox could be prevented by intentionally exposing a child to cowpox (which we now know is caused by a related virus). Newspapers of the time mocked Jenner, and published cartoons showing cows growing out of people's bodies after receiving Jenner's vaccine. With such a long tradition of doubt about the safety of vaccines, it is not surprising that some parents remain hesitant to vaccinate their children today.

The history of medicine has demonstrated without a doubt that since vaccines have been introduced, previously feared diseases such as smallpox, measles, polio, diphtheria, whooping cough, and tetanus have been reduced to rare occurrences. Most parents of young children today have never seen a case of these diseases, leading some of them to wonder why it is necessary to take even a small risk of making their child sick from a vaccine. Ironically, even parents who don't vaccinate their children benefit from the fact that a large percentage of children are vaccinated, and thus are less likely to transmit infectious diseases to their unvaccinated child. This concept of "herd immunity" has been shown to be relevant in recent outbreaks of measles and whooping cough in communities where increasing numbers of parents are not vaccinating their children.

The concept that vaccinating a high percentage of children can protect entire communities has led to immunization requirements. Most states require that a parent provide written proof of a child's immunizations prior to registering for school. If the child is not immunized, he or she will not be allowed to attend public school, unless the parent can demonstrate a medical reason, or a religious objection, that explains why the child is not vaccinated. With the rising popularity of the Internet, there is no shortage of websites voicing objections to any vaccine requirements, and going so far as to question the benefit of all vaccinations. The challenge for all potential parents, and for future health-care workers in particular, is to remain as informed as possible about the benefits and risks of vaccination, so that they can make wise choices for their patients and for their own children.

Form Your Own Opinion

1. Do you think the benefits of vaccination outweigh the risks? What level of risk would seem acceptable if you were having your own child vaccinated against polio, for example?
2. Do you think it is appropriate to allow people to forgo the required vaccinations for their children because of religious objections?
3. Suppose you are a nurse or a pediatrician and parents tell you they don't think the government should force them to vaccinate their child, because they read on a website that all vaccines are ineffective and dangerous. How would you respond?

---

bear a "nonself" protein on their MHC-I molecules. Helper T cells produce cytokines and stimulate other immune cells.

13.3 Active vs. Passive Immunity

Active immunity can be induced through infection or by vaccines when a person is well and in no immediate danger of contracting an infectious disease. Active immunity is dependent upon the presence of memory cells in the body.

Passive immunity is needed when an individual is in immediate danger of succumbing to an infectious disease. Passively acquired antibodies are short lived because the antibodies are administered to and not made by the individual.

Monoclonal antibodies, which are produced by the same plasma cell, are used for various purposes, from detecting infections to treating cancer.

13.4 Adverse Effects of Immune Responses

Allergic responses occur when the immune system reacts vigorously to substances not normally recognized as foreign. Delayed allergic responses, such as contact dermatitis, are due to the activity of T cells. Immune side effects also include blood-type reactions and tissue rejection.

13.5 Disorders of the Immune System

In autoimmune diseases, a person's own T cells or antibodies target the body's own cells or molecules. Examples are myasthenia gravis, multiple sclerosis, systemic lupus erythematosus, and rheumatoid arthritis.

Immunodeficiency diseases can be acquired, as occurs with AIDS, or inherited, as is seen in children with SCID, who are born unable to produce their own antibody- and cell-mediated immunity. Depending on the severity, immunodeficiencies can be fatal, usually as a result of other, common infections.

Testing Yourself

Choose the best answer for each question.

1. What is the term for localized swelling caused by fluid accumulation?
   a. node
   b. edema
   c. valve
   d. None of these are correct.

2. Which is a lymphoid organ?

   a. spleen
   b. lymph node
   c. thymus gland
   d. All of these are correct.

3. One-way flow of fluid in lymph vessels is aided by

   a. valves.
   b. skeletal muscle contractions.
   c. Both a and b are correct.
   d. None of these are correct.

4. Which of the following is not a function of the lymphatic system?

   a. produces red blood cells
   b. returns excess fluid to the blood
   c. transports lipids absorbed from the digestive system
   d. defends the body against pathogens

5. The _____ cleanses _____, and in its absence, problems associated with _____ may occur.

   a. spleen, lymph, oxygen transport
   b. liver, lymph, infection
   c. spleen, blood, infection
   d. spleen, blood, oxygen transport

6. Which cells phagocytize pathogens?

   a. neutrophils
   b. macrophages
   c. mast cells
   d. Both a and b are correct.

7. Complement

   a. is an innate defense mechanism.
   b. is involved in the inflammatory reaction.
   c. is a series of proteins present in the plasma.
   d. plays a role in destroying bacteria.
   e. All of these are correct.

8. _____ release histamines.

   a. Basophils
   b. Neutrophils
   c. Monocytes
   d. All of these are correct.

9. Which applies to T cells?

   a. mature in bone marrow
   b. mature in thymus gland
   c. Both a and b are correct.
   d. None of these are correct.

10. Antibody-mediated immunity is most directly associated with

   a. T cells.
   b. monocytes.
   c. basophils.
   d. B cells.

11. Plasma cells are

   a. the same as memory cells.
   b. formed from blood plasma.
   c. B cells that are actively secreting antibody.
   d. inactive T cells carried in the plasma.
   e. a type of red blood cell.

12. Vaccines are associated with

   a. active immunity.
   b. long-lasting immunity.
   c. passive immunity.
   d. Both a and b are correct.

13. MHC antigens play a role in

   a. active immunity.
   b. presentation of antigens to T cells.
   c. tissue transplantation.
   d. All of these are correct.

14. Spent plasma cells undergo

   a. activation.
   b. apoptosis.
   c. clonal selection.
   d. None of these are correct.

15. Type O blood has _____ antibodies in the plasma.

   a. A and B
   b. no
   c. self
   d. All of these are correct.

16. Label the four lymphoid organs, a–d. Which are primary, and which are secondary?

a. _____
b. _____
c. _____
d. _____

17. Label a–c on this IgG molecule using these terms: antigen-binding sites, light chain, heavy chain. What do V and C stand for in the diagram?

18. Which of the following is an innate defense of the body?

   a. immunoglobulin
   b. B cell
   c. T cell
   d. vaccine
   e. inflammation

19. Antibodies combine with antigens

   a. at variable regions.
   b. at constant regions.
   c. only if macrophages are present.
   d. Both a and c are correct.

20. Which one of these associations is mismatched?

   a. helper T cells—help complement react
   b. cytotoxic T cells—active in tissue rejection
   c. macrophages—activate T cells
   d. memory T cells—long-living line of T cells
   e. T cells—mature in thymus

21. An antigen-presenting cell (APC)

   a. presents antigens to T cells.
   b. secretes antibodies.
   c. marks each human cell as belonging to that particular person.
   d. secretes cytokines.

22. Which of the following would not be a participant in cell-mediated immune responses?

    a. helper T cells   d. cytotoxic T cells
    b. macrophages   e. plasma cells
    c. cytokines

23. Vaccines are

    a. the same as monoclonal antibodies.
    b. treated bacteria or viruses, or one of their proteins.
    c. short-lived.
    d. MHC proteins.
    e. All of these are correct.

24. Which of the following is not an example of an autoimmune disease?

    a. multiple sclerosis   c. hemolytic disease of the newborn
    b. rheumatic fever   d. systemic lupus erythematosus

Match the terms to these definitions:

a._____ Antigens prepared in such a way that they can promote active immunity without causing disease.

b._____ Fluid, derived from tissue fluid, that is carried in lymphatic vessels.

c._____ Foreign substance, usually a protein or a polysaccharide, that stimulates the immune system to react, for example, by producing antibodies.

d._____ Process of programmed cell death involving a cascade of specific cellular events leading to the death and destruction of the cell.

e._____ Lymphocyte that matures in the thymus and exists in three varieties, one of which kills antigen-bearing cells outright.

Understanding the Terms

acquired immunity 242
active immunity 248
agglutination 252
allergen 252
allergy 252
anaphylactic shock 252
antibody-mediated immunity 245
antibody titer 248
antigen 242
antigen-presenting cell (APC) 246
apoptosis 242
asthma 252
autoimmune disease 255
B-cell receptor (BCR) 245
cell-mediated immunity 247
clonal selection theory 245
complement system 244
cytokine 243
cytotoxic T cell 247
delayed allergic response 252
dendritic cell 244
edema 241
hay fever 252
helper T cell 247
histamine 243
immediate allergic response 252
immunity 242
immunization 248
immunoglobulin (Ig) 245
immunosuppressive 255

inflammatory reaction 243
innate immunity 242
interferon 244
interleukin 251
lymph 240
lymphatic system 240
lymphatic vessel 240
lymph node 242
lymphoid organ 241
macrophage 243
mast cell 243
major histocompatibility complex (MCH) 247
monoclonal antibody 251
monocyte 243
multiple sclerosis (MS) 255
myasthenia gravis 255
natural killer (NK) cell 244
neutrophil 243
passive immunity 248
pathogen 242
phagocyte 243
red bone marrow 242
rheumatoid arthritis 255
severe combined immunodeficiency disease (SCID) 255
spleen 242
systemic lupus erythematosus (SLE) 255
T-cell receptor (TCR) 246
thymus gland 242
vaccine 248

Thinking Critically

1. Many primitive organisms, such as invertebrates have no lymphocytes and thus lack an acquired immune system, but they have some components of an innate immune system, including phagocytes and certain protective proteins. What are some general features of innate immunity that make it very valuable to organisms lacking more specific antibody- and cell-mediated responses? What are some disadvantages to having only an innate immune system?

2. Compare the ways in which natural killer cells and cytotoxic T cells recognize the cells they are going to kill. Why does it make good biological sense for NK cells and cytotoxic T cells to recognize, for example, virus-infected cells in these different ways?

3. How do T cells and B cells differ in the form of antigens that they can recognize? Both cell types usually require the assistance of other cells in order to respond to an antigen. What cell types does each cell require?

4. Health-care practitioners sometimes test to see whether detectable levels of IgM have been produced against a specific pathogen. Why might this test be of more diagnostic value than a test for specific IgG or IgA?

5. Why is it that Rh incompatibility can be a serious problem when an Rh-negative mother is carrying an Rh-positive fetus, but ABO incompatibility between mother and fetus is usually no problem? That is, a type A mother can usually safely carry a type B fetus. (Hint: The antibodies produced by an Rh-negative mother against the Rh antigen are usually IgG, while the antibodies produced against the A or B antigens are IgM.)

ARIS, the *Inquiry into Life* Website

ARIS, the website for *Inquiry into Life,* provides a wealth of information organized and integrated by chapter. You will find practice quizzes, interactive activities, labeling exercises, flashcards, and much more that will complement your learning and understanding of general biology.

www.aris.mhhe.com

When you eat pizza,
taste buds in your mouth are stimulated by its tastes, and you begin to salivate. Salivary enzymes begin almost immediately to digest complex carbohydrates in the pizza's crust. You begin to chew, physically breaking down the larger pieces into smaller ones that are easier to swallow and more accessible to digestive enzymes. Your tongue moves the chewed food toward the pharynx in the back of the throat, and as you swallow, your epiglottis covers the opening of your airway, protecting you from choking. Coordinated muscular contractions then propel the food down your esophagus toward your stomach by a coordinated muscular contraction.

The stomach is where the action really begins. Muscular contractions of the stomach wall mix the food material with the strong acid and digestive enzymes secreted by glands in the inner stomach wall. After a few hours, the now partially digested food first enters the small intestine, where digestion is completed and absorption of small nutrient molecules into the body occurs. Then the remaining material enters the large intestine, where excess water is absorbed and the leftover, indigestible material known as feces can be eliminated from the body.

The process of digestion, complex as it is, is absolutely essential to life. Consumption of unhealthy food, too much food, or too little food can greatly affect our quality of life, and can even lead to life-threatening conditions. Since we truly "are what we eat," it is worthwhile to learn more about how our digestive system works and how to make healthy choices about our diet.

chapter **14**

Digestive System and Nutrition

Chapter Concepts

14.1 The Digestive Tract
- What is the general structure of the digestive tract? At what openings does the digestive tract begin and end? 260
- What are the two main types of digestion? 260
- What is the path of food, and what are the functions of the organs in that path? 260–266
- What happens to indigestible materials? 266

14.2 Three Accessory Organs
- What three main accessory organs assist the digestive process? 267
- How does each accessory organ contribute to the digestion of food? 267–268

14.3 Digestive Enzymes
- What nutrient molecules are absorbed following the digestion of carbohydrates? Of proteins? Of lipids? 268–269
- What are the main digestive enzymes, and what factors affect how they function? 268–270

14.4 Nutrition
- Why is proper nutrition important to good health? 270–279

14.5 Disorders of the Digestive System
- What are some common disorders of the digestive tract? Of the accessory organs? 279–281

14.1   The Digestive Tract

Digestion takes place within a tube called the digestive tract, which begins with the mouth and ends with the anus (Fig. 14.1). The functions of the digestive system are to ingest food, digest it to nutrients that can cross plasma membranes, absorb nutrients, and eliminate indigestible remains.

Digestion involves two main processes that occur simultaneously. During mechanical digestion, large pieces of food become smaller pieces, readying them for chemical digestion. Mechanical digestion begins with the chewing of food in the mouth and continues with the churning and mixing of food in the stomach. Parts of the digestive tract produce digestive enzymes. During chemical digestion, many different enzymes break down macromolecules to small organic molecules that can be absorbed. Each enzyme has a particular job to do.

The Mouth

The mouth, which receives food, is bounded externally by the lips and cheeks. The red portion of the lips is poorly keratinized, and this allows blood to show through.

Most people enjoy eating food largely because they like its texture and taste. Sensory receptors called taste buds occur primarily on the tongue, and when these are activated by the presence of food, nerve impulses travel by way of cranial nerves to the brain. The tongue is composed of skeletal muscle that contracts to change the shape of the tongue. Muscles exterior to the tongue enable it to move about. A fold of mucous membrane on the underside of the tongue attaches it to the floor of the mouth.

The roof of the mouth separates the nasal cavities from the mouth. The roof has two parts: an anterior (toward the front) **hard palate** and a posterior (toward the back) **soft palate**

Figure 14.1 The human digestive tract.

Trace the path of food from the mouth to the anus. The large intestine consists of the cecum, the colon (ascending, transverse, descending, and sigmoid colons), the rectum, and the anus. Note also the location of the accessory organs of digestion: the pancreas, the liver, and the gallbladder.

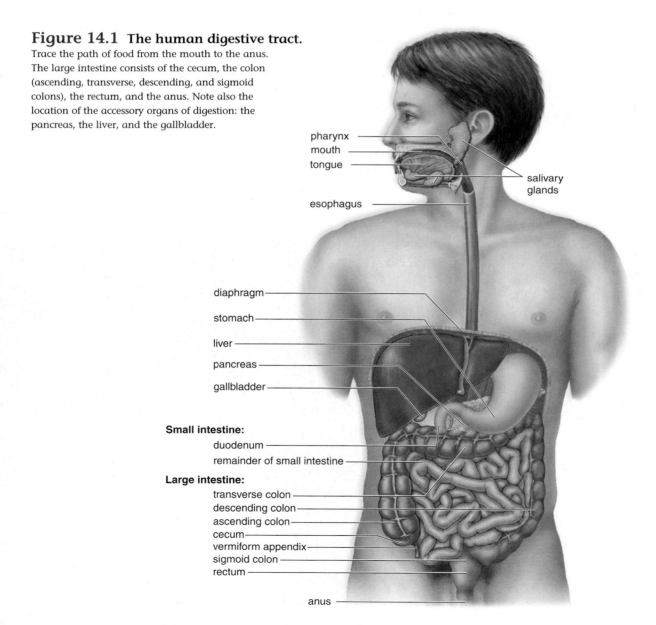

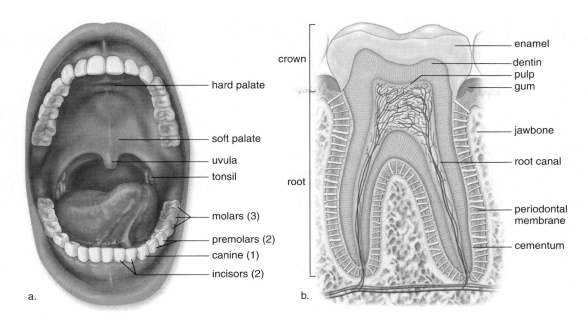

crown

hard palate

soft palate

uvula

tonsil

root

molars (3)

premolars (2)

canine (1)

incisors (2)

a.

enamel

dentin

pulp

gum

jawbone

root canal

periodontal membrane

cementum

b.

Figure 14.2 Adult mouth and teeth.

a. The chisel-shaped incisors bite; the pointed canines tear; the fairly flat premolars grind; and the flattened molars crush food. The last molar, called a wisdom tooth, may fail to erupt, or if it does, it is sometimes crooked and useless. Often dentists recommend the extraction of the wisdom teeth. **b.** Longitudinal section of a tooth. The crown is the portion that projects above the gum line and can be replaced by a dentist if damaged. When a "root canal" is done, the nerves are removed. When the periodontal membrane is inflamed, the teeth can loosen.

(Fig. 14.2*a*). The hard palate contains several bones, but the soft palate is composed entirely of muscle. The soft palate ends in a finger-shaped projection called the uvula. The tonsils are in the back of the mouth, on either side of the tongue, and in the nasopharynx, where they are called adenoids. The tonsils help protect the body against infections. If the tonsils become inflamed, the person has tonsillitis. The infection can spread to the middle ears. If tonsillitis recurs repeatedly, the tonsils may be surgically removed (called a tonsillectomy).

Three pairs of **salivary glands** (see Fig. 14.1) send juices (saliva) by way of ducts to the mouth. One pair of salivary glands lies at the sides of the face immediately below and in front of the ears. These glands swell when a person has the mumps, a disease caused by a viral infection. Salivary glands have ducts that open on the inner surface of the cheek at the location of the second upper molar. Another pair of salivary glands lies beneath the tongue, and still another pair lies beneath the floor of the mouth. The ducts from these salivary glands open under the tongue. You can locate the openings if you use your tongue to feel for small flaps on the inside of your cheek and under your tongue. Saliva contains an enzyme called **salivary amylase** that begins the process of digesting starch.

The Teeth

With our teeth, we chew food into pieces convenient for swallowing. During the first two years of life, the smaller 20 deciduous, or baby, teeth appear. These are eventually replaced by 32 adult teeth (Fig. 14.2*a*). The third pair of molars, called the wisdom teeth, sometimes fail to erupt. If they push on the other teeth and/or cause pain, they can be removed by a dentist or oral surgeon.

Each tooth has two main divisions, a crown and a root (Fig. 14.2*b*). The crown has a layer of enamel, an extremely hard outer covering of calcium compounds; dentin, a thick layer of bonelike material; and an inner pulp, which contains the nerves and the blood vessels. Dentin and pulp are also found in the root.

Tooth decay, called **dental caries,** or cavities, occurs when bacteria within the mouth metabolize sugar and give off acids, which erode teeth. Three measures can help to prevent tooth decay: eating a limited amount of sweets, daily brushing and flossing of teeth, and regular visits to the dentist. Fluoride treatments, particularly in children, can make the enamel stronger and more resistant to decay. Gum disease is more apt to occur with aging. Inflammation of the gums (gingivitis) can spread to the periodontal membrane, which lines the tooth socket. A person then has periodontitis, characterized by loss of bone and loosening of the teeth so that extensive dental work may be required. Stimulation of the gums in a manner advised by your dentist is helpful in controlling this condition. Medications are also available.

Over time, and especially with exposure to cigarette smoke, soft drinks, or coffee, the enamel layer can become stained. In recent years, various tooth-whitening treatments, most of which contain a bleaching agent such as hydrogen peroxide, have become a popular way to remove these stains.

The tongue mixes the chewed food with saliva. It then forms this mixture into a mass called a bolus in preparation for swallowing.

The salivary glands send saliva into the mouth, where the teeth chew the food and the tongue forms it into a bolus for swallowing.

TABLE 14.1	Path of Food		
Organ	**Function of Organ**	**Special Feature(s)**	**Function of Special Feature(s)**
Mouth	Receives food; starts digestion of starch	Teeth Tongue	Chew food Forms bolus
Pharynx	Passageway	—	—
Esophagus	Passageway	—	—
Stomach	Storage of food; acidity kills bacteria; starts digestion of protein	Gastric glands	Release gastric juices
Small intestine	Digestion of all foods; absorption of nutrients	Intestinal glands Villi	Release intestinal juices Absorb nutrients
Large intestine	Absorption of water; storage of indigestible remains	—	—

The Pharynx

The **pharynx** is a region that receives air from the nasal cavities and food from the mouth (see Fig. 14.1).

Table 14.1 traces the path of food. From the mouth, food (the bolus) passes through the pharynx and esophagus to the stomach, small intestine, and large intestine. The food passage and air passage cross in the pharynx because the trachea (windpipe) is ventral to (in front of) the esophagus, a long muscular tube that takes food to the stomach. Swallowing, a process that occurs in the pharynx (Fig. 14.3), is a **reflex action** performed automatically, without conscious thought. Usually, during swallowing, the soft palate moves back to close off the **nasopharynx,** and the trachea moves up under the **epiglottis** to cover the glottis. The **glottis** is the opening to the larynx (voice box), and therefore the air passage. The up-and-down movement of the Adam's apple, the front part of the larynx, is easy to observe when a person swallows. During swallowing, food normally enters the esophagus because the air passages are blocked. We do not breathe when we swallow.

Unfortunately, we have all had the unpleasant experience of having food "go the wrong way." The wrong way may be either into the nasal cavities or into the trachea. If it is the latter, coughing will most likely force the food up out of the trachea and into the pharynx again.

The air passage and food passage cross in the pharynx, which takes food to the esophagus. When swallowing, the soft palate closes off the nasopharynx, and the epiglottis covers the airway.

> The air passage and food passage cross in the pharynx which takes food to the esophagus. When we swallow, the soft palate closes off the nasopharynx, and the epiglottis covers the airway.

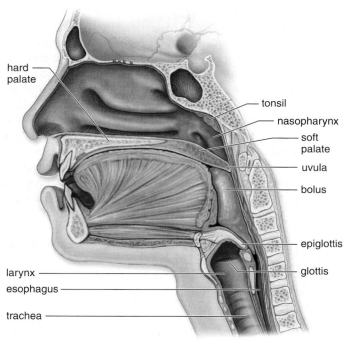

Figure 14.3 Swallowing.
When food is swallowed, the soft palate closes off the nasopharynx, and the epiglottis covers the glottis, forcing the bolus to pass down the esophagus. Therefore, a person does not breathe while swallowing.

The Esophagus

The **esophagus** is a muscular tube that passes from the pharynx, through the thoracic cavity and diaphragm, and into the abdominal cavity, where it joins the stomach. The esophagus is ordinarily collapsed, but it opens and receives the bolus when swallowing occurs.

A rhythmic contraction called **peristalsis** pushes the food along the digestive tract. Peristalsis begins in the esophagus and continues in all the organs of the digestive tract. Occasionally, peristalsis begins even though there is no food in the esophagus. This produces the sensation of a lump in the throat.

The esophagus plays no role in the chemical digestion of food. Its sole purpose is to conduct the food bolus from the mouth to the stomach. **Sphincters** are muscles that encircle tubes and act as valves; the tubes close when sphincters contract, and they open when sphincters relax. The entrance of the esophagus to the stomach is marked by a constriction that is often called a sphincter, although the muscle is not as developed as it would be in a true sphincter. Relaxation of the sphincter allows the bolus to pass into the stomach, while contraction prevents the acidic contents of the stomach from backing up into the esophagus.

Heartburn, which feels like a burning pain rising into the throat, occurs when some of the stomach contents escape into the esophagus. When vomiting occurs, contraction of the abdominal muscles and diaphragm propels the contents of the stomach upward through the esophagus.

The esophagus conducts the bolus from the pharynx to the stomach. Peristalsis begins in the esophagus and occurs along the entire length of the digestive tract.

The Wall of the Digestive Tract

The wall of the esophagus in the abdominal cavity is representative of that of the digestive tract. It is composed of four layers, shown in Figure 14.4 and listed here:

Mucosa (mucous membrane layer) A layer of epithelium supported by connective tissue and smooth muscle lines the lumen (central cavity) and contains glandular epithelial cells that secrete digestive enzymes and goblet cells that secrete mucus.

Submucosa (submucosal layer) A broad band of loose connective tissue that contains blood vessels lies beneath the mucosa. Lymph nodules, including some called Peyer's patches, are in the submucosa. Like the tonsils, they help protect us from disease.

Muscularis (smooth muscle layer) Two layers of smooth muscle make up this section. The inner, circular layer encircles the gut; the outer, longitudinal layer lies in the same direction as the gut. (The stomach also has oblique muscles.)

Serosa (serous membrane layer) Most of the digestive tract has a serosa, a very thin, outermost layer of squamous epithelium supported by connective tissue. The serosa secretes a serous fluid that keeps the outer surface of the intestines moist so that the organs of the abdominal cavity slide against one another.

The abdominal part of the esophagus, like most of the digestive tract, is composed of four layers: the mucosa, the submucosa, the muscularis, and the serosa.

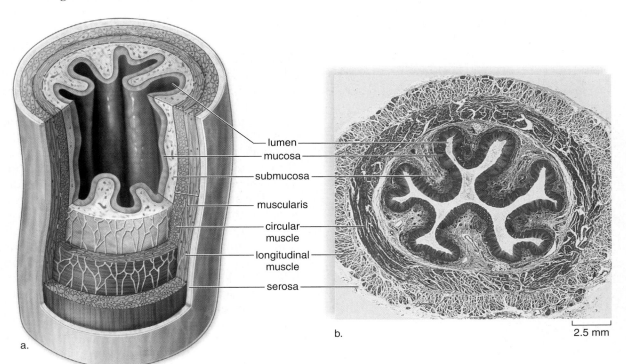

lumen
mucosa
submucosa
muscularis
circular muscle
longitudinal muscle
serosa

a.
b.
2.5 mm

Figure 14.4 Wall of the digestive tract.
a. Several different types of tissues are found in the wall of the digestive tract. Note the placement of circular muscle inside longitudinal muscle. **b.** Micrograph of the wall of the esophagus.

The Stomach

The **stomach** (Fig. 14.5*a*) is a thick-walled, J-shaped organ that lies on the left side of the abdominal cavity below the liver and diaphragm. The stomach is continuous with the esophagus above and the duodenum of the small intestine below. The stomach is about 25 cm (10 in.) long, regardless of the amount of food it holds, but the diameter varies, depending on how full it is. As the stomach expands, deep folds in its wall, called **rugae,** gradually disappear. When full, it can hold about 4 liters (1 gallon). The stomach receives food from the esophagus, stores food, starts the digestion of proteins, and moves food into the small intestine.

The columnar epithelium lining the stomach has millions of gastric pits, which lead into **gastric glands** (Fig. 14.5*b*). (The term "gastric" always refers to the stomach.) The gastric glands produce gastric juice, which contains pepsinogen, hydrochloric acid (HCl), and mucus. Pepsinogen becomes the enzyme **pepsin** when exposed to HCl. The HCl causes the stomach to have high acidity (a pH of about 2), and this is beneficial because it kills most of the bacteria and other microbes present in food. Although HCl does not digest food, it does break down the connective tissue of meat and activate pepsin.

The stomach acts both physically and chemically on food. Its wall contains three muscle layers: one layer is longitudinal and another is circular, as shown in Figure 14.4*a*; the third is obliquely arranged. This muscular wall not only moves the food along, but it also churns, mixing the food with gastric juice and breaking it down into small pieces.

Alcohol and other liquids are absorbed in the stomach, but food substances are not. Normally, the stomach empties in 2–6 hours. When food leaves the stomach, it is a thick, soupy liquid called **chyme.** Chyme enters the small intestine in squirts by way of the pyloric sphincter, which acts like a valve, repeatedly opening and closing.

> The stomach can expand to accommodate large amounts of food. When food is present, the stomach churns, mixing food with acidic gastric juice.

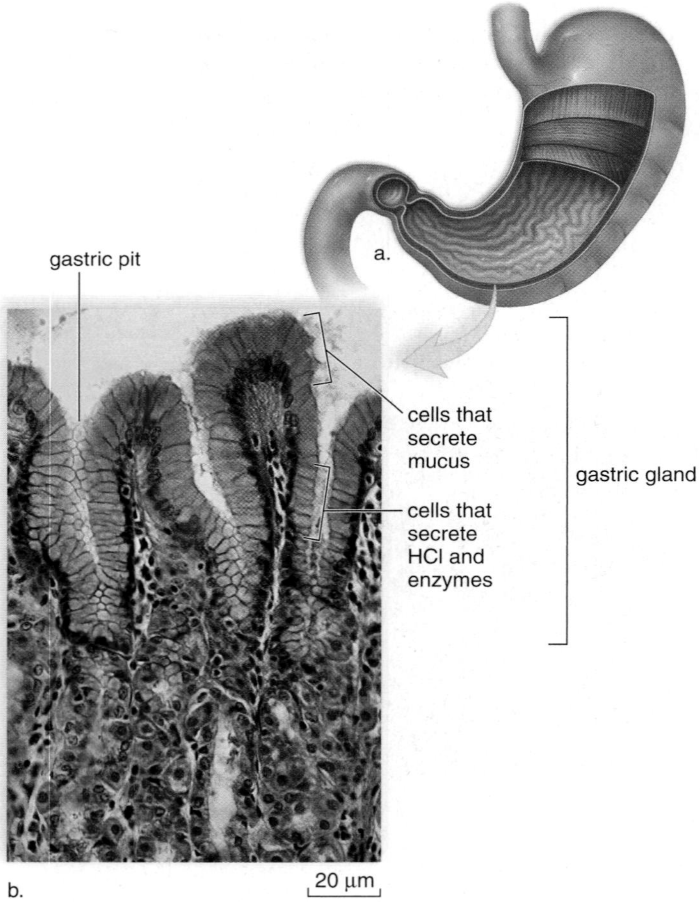

gastric pit

a.

cells that secrete mucus

gastric gland

cells that secrete HCl and enzymes

b.   20 μm

Figure 14.5 Anatomy of the stomach.

a. The stomach has a thick wall with folds that allow it to expand and fill with food. **b.** The mucosa contains gastric glands, which secrete gastric juice containing mucus and digestive enzymes.

The Small Intestine

The **small intestine** is named for its small diameter (compared to that of the large intestine), but perhaps it should be called the long intestine. The small intestine averages about 6 meters (18 ft) in length, compared to the large intestine, which is about 1.5 meters (4½ ft) in length.

The first 25 cm of the small intestine is called the **duodenum.** Ducts from the liver and pancreas join to form one duct that enters the duodenum (see Fig. 14.10). The small intestine receives bile from the liver and pancreatic juice from the pancreas via this duct. **Bile** emulsifies fat, meaning that it causes fat droplets to disperse in water. The intestine has a slightly basic pH because pancreatic juice contains sodium bicarbonate ($NaHCO_3$), which neutralizes the acid in chyme. The enzymes in pancreatic juice and enzymes produced by the intestinal wall complete the process of food digestion.

The surface area of the small intestine has been compared to that of a tennis court. What factors contribute to such a large surface area? The wall of the small intestine contains fingerlike projections called villi (sing. **villus**), which give the intestinal wall a soft, velvety appearance (Fig. 14.6). A villus has an outer layer of columnar epithelial cells, and each of these cells has thousands of microscopic extensions called microvilli. Collectively, in electron micrographs, microvilli give the villi a fuzzy border, known as a "brush border." The microvilli greatly increase the surface area of the villus for the absorption of nutrients.

Nutrients are absorbed into the vessels of a villus. A villus contains blood capillaries and a small lymphatic capillary, called a **lacteal.** The lymphatic system is an adjunct to the cardiovascular system; that is, its vessels carry a fluid called lymph to the cardiovascular veins. Sugars and amino

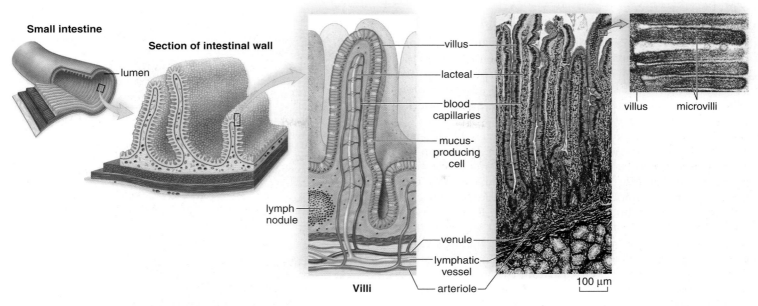

Small intestine

Section of intestinal wall

lumen

villus

lacteal

blood capillaries

mucus-producing cell

lymph nodule

venule

lymphatic vessel

Villi

arteriole

villus   microvilli

100 µm

Figure 14.6 Anatomy of the small intestine.

The wall of the small intestine has folds that bear fingerlike projections called villi. Microvilli, which project from the villi, absorb the products of digestion into the blood capillaries and the lacteals of the villi.

acids enter the blood capillaries of a villus. Glycerol and fatty acids (digested from fats) enter the epithelial cells of the villi, where they are joined and packaged as lipoprotein droplets that enter a lacteal. After nutrients are absorbed, the bloodstream eventually carries them to all the cells of the body.

> The large surface area of the small intestine facilitates absorption of nutrients into the cardiovascular system (sugars and amino acids) and the lymphatic system (fats).

Regulation of Digestive Secretions

The secretion of digestive juices is promoted by the nervous system and by hormones. A **hormone** is a substance produced by one set of cells that affects a different set of cells, the so-called target cells. Hormones are usually transported by the bloodstream. For example, when a person has eaten a meal particularly rich in protein, the stomach produces the hormone gastrin. Gastrin enters the bloodstream, and soon the stomach is churning, and the secretory activity of gastric glands is increasing. A hormone produced by the duodenal wall, GIP (gastric inhibitory peptide), works opposite to gastrin: it inhibits gastric gland secretion.

Cells of the duodenal wall produce two other hormones that are of particular interest—secretin and CCK (cholecystokinin). Acid, especially hydrochloric acid (HCl) present in chyme, stimulates the release of secretin, while partially digested protein and fat stimulate the release of CCK. Soon after these hormones enter the bloodstream,

the pancreas increases its output of pancreatic juice, which helps digest food, and the liver increases its output of bile. The gallbladder contracts to release bile. Figure 14.7 summarizes the actions of gastrin, secretin, and CCK.

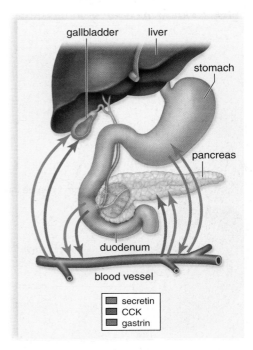

gallbladder   liver

stomach

pancreas

duodenum

blood vessel

secretin
CCK
gastrin

Figure 14.7 Hormonal control of digestive gland secretions. Gastrin (blue), produced by the lower part of the stomach, enters the bloodstream and thereafter stimulates the upper part of the stomach to produce more digestive juices. Secretin (green) and CCK (purple) produced by the duodenal wall, stimulate the pancreas to secrete its digestive juices and the gallbladder to release bile.

The Large Intestine

The **large intestine,** which includes the cecum, the colon, the rectum, and the anal canal, is larger in diameter than the small intestine (6.5 cm compared to 2.5 cm), but it is shorter in length (see Fig. 14.1). The large intestine absorbs water, salts, and some vitamins. It also stores indigestible material until it is eliminated at the anus.

The **cecum,** which lies below the junction with the small intestine, is the blind end of the large intestine. The cecum has a small projection called the **vermiform appendix** (*vermiform* means wormlike) (Fig. 14.8). In humans, the appendix also may play a role in fighting infection. This organ is subject to inflammation, a condition called appendicitis. If inflamed, the appendix should be removed before the fluid content rises to the point that the appendix bursts, a situation that may cause **peritonitis,** a generalized infection and inflammation of the lining of the abdominal cavity. Peritonitis can lead to death.

The **colon** includes the ascending colon, which goes up the right side of the body to the level of the liver; the transverse colon, which crosses the abdominal cavity just below the liver and the stomach; the descending colon, which passes down the left side of the body; and the sigmoid colon, which enters the **rectum,** the last 20 cm of the large intestine. The rectum opens at the **anus,** where **defecation,** the expulsion of feces, occurs. When feces are forced into the rectum by peristalsis, a defecation reflex occurs. The stretching of the rectal wall initiates nerve impulses to the spinal cord, and shortly thereafter, the rectal muscles contract, and the anal sphincters relax (Fig. 14.9). Ridding the body of indigestible remains is

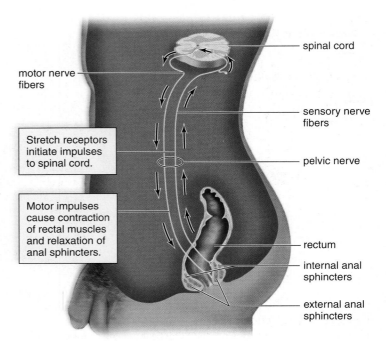

Figure 14.9 Defecation reflex.
The accumulation of feces in the rectum causes it to stretch, which initiates a reflex action resulting in rectal contraction and expulsion of the fecal material.

another way the digestive system helps maintain homeostasis. Feces are normally about three-quarters water and one-quarter solids. Bacteria, **fiber** (indigestible remains), and other indigestible materials are in the solid portion. Bacterial action on indigestible materials causes the odor of feces and also accounts for the presence of gas. A breakdown product of bilirubin (described in section 14.2) and the presence of oxidized iron cause the brown color of feces.

About 40% to 50% of the fecal mass consists of bacteria and other microbes; there are about 100 billion bacteria per gram of feces! For many years, scientists believed that facultative bacteria (bacteria that can live with or without oxygen), such as *Escherichia coli,* were the major inhabitants of the colon, but new culture methods show that over 99% of the colon bacteria are obligate anaerobes (bacteria that die in the presence of oxygen). Not only do the bacteria break down some indigestible material, but they also produce some vitamins and other molecules that our bodies can absorb and use. In this way, they perform a service for us, and we provide a good environment for them to live in.

Water is considered unsafe for swimming when the coliform (intestinal) bacterial count reaches a certain number. A high count indicates that a significant amount of feces has entered the water. The more feces present, the greater the possibility that disease-causing bacteria are also present.

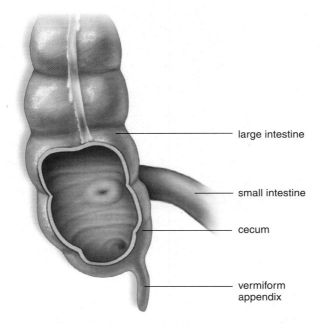

Figure 14.8 Junction of the small intestine and the large intestine.
The cecum is the blind end of the large intestine. The appendix is attached to the cecum.

The large intestine absorbs, water, salts, and some vitamins, and rids the body of indigestible wastes by defecation.

14.2 Three Accessory Organs

The pancreas, liver, and gallbladder are accessory digestive organs. Figure 14.10*a* shows how the pancreatic duct from the pancreas and the common bile duct from the liver and gallbladder enter the duodenum.

The Pancreas

The **pancreas** lies deep in the abdominal cavity, resting on the posterior abdominal wall. It is an elongated and somewhat flattened organ that has both an endocrine and an exocrine function. As an endocrine gland, it secretes insulin and glucagon, hormones that help keep the blood glucose level within normal limits. In this chapter, however, we are interested in

its exocrine function. Most pancreatic cells produce pancreatic juice, which contains sodium bicarbonate ($NaHCO_3$) to help neutralize the stomach acid, and digestive enzymes for all types of food. **Pancreatic amylase** digests starch, **trypsin** digests protein, and **lipase** digests fat.

> The pancreas has both endocrine and exocrine functions. It secretes pancreatic juice that contains sodium bicarbonate and digestive enzymes.

The Liver

The **liver,** which is the largest gland in the body, lies mainly in the upper right section of the abdominal cavity, under the diaphragm (see Fig. 14.1). The liver contains approximately 100,000 lobules that serve as its structural and functional units (Fig. 14.10*b*). Three structures are located between the lobules: a bile duct that takes bile away from the liver; a branch of the hepatic artery that brings O_2-rich blood to the liver; and a branch of the hepatic portal vein that transports nutrients from the intestines. Each lobule has a central vein, which enters a hepatic vein. In Figure 14.11, trace the path of blood from the intestines to the liver via the hepatic portal vein and from the liver to the inferior vena cava via the hepatic veins.

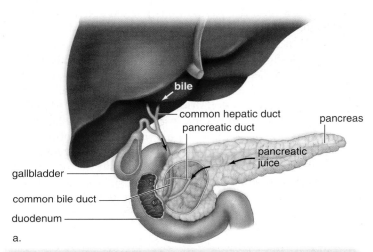

a.

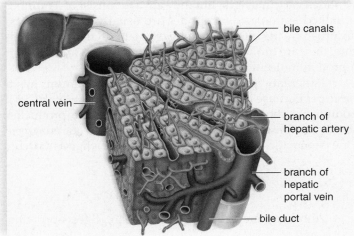

b.

Figure 14.10 Liver, gallbladder, and pancreas.

a. The liver makes bile, which is stored in the gallbladder and sent (black arrow) to the small intestine by way of the common bile duct. The pancreas produces digestive enzymes that are sent (black arrows) to the small intestine by way of the pancreatic duct. **b.** A hepatic lobule. The liver contains over 100,000 lobules, each lobule composed of many cells that perform the various functions of the liver. They remove and add materials to the blood and deposit bile in a duct.

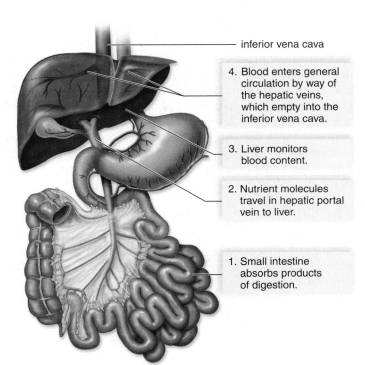

4. Blood enters general circulation by way of the hepatic veins, which empty into the inferior vena cava.

3. Liver monitors blood content.

2. Nutrient molecules travel in hepatic portal vein to liver.

1. Small intestine absorbs products of digestion.

Figure 14.11 Hepatic portal system.

The hepatic portal vein takes the products of digestion from the digestive system to the liver, where they are processed before entering a hepatic vein.

TABLE 14.2	Functions of the Liver

1. Detoxifies blood by removing and metabolizing poisonous substances
2. Stores iron (Fe^{2+}) and vitamins A, D, E, K, and B_{12}
3. Makes plasma proteins, such as albumins and fibrinogen, from amino acids
4. Stores glucose as glycogen after a meal, and breaks down glycogen to glucose to maintain the glucose concentration of blood between eating periods
5. Produces urea after breaking down amino acids
6. Removes bilirubin, a breakdown product of hemoglobin from the blood, and excretes it in bile, a liver product
7. Helps regulate blood cholesterol level, converting some to bile salts

In some ways, the liver acts as the gatekeeper to the blood (Table 14.2). As blood from the hepatic portal vein passes through the liver, it removes poisonous substances and detoxifies them. The liver also removes and stores iron and vitamins A, D, E, K, and B_{12}. The liver makes many of the plasma proteins and helps regulate the quantity of cholesterol in the blood.

The liver maintains the blood glucose level at 50–80 mg/100 ml, even though a person eats intermittently. When insulin is present, any excess glucose in the blood is removed and stored by the liver as glycogen. Between meals, glycogen is broken down to glucose, which enters the hepatic veins, and in this way, the blood glucose level remains constant.

If the supply of glycogen is depleted, the liver converts glycerol (from fats) and amino acids to glucose molecules. The conversion of amino acids to glucose necessitates deamination, the removal of amino groups. By a complex metabolic pathway, the liver then combines ammonia with carbon dioxide to form urea. Urea is the usual nitrogenous waste product from amino acid breakdown in humans.

The liver produces bile, which is stored in the gallbladder. Bile has a yellowish-green color because it contains the bile pigment bilirubin, derived from the breakdown of hemoglobin, the red pigment in red blood cells. Bile also contains bile salts. Bile salts are derived from cholesterol, and they emulsify fat in the small intestine. As you may recall from section 14.1, when fat is emulsified, it breaks up into droplets, providing a much larger surface area, which can be acted upon by a digestive enzyme from the pancreas.

The Gallbladder

The **gallbladder** is a pear-shaped, muscular sac attached to the surface of the liver (see Fig. 14.1). The liver produces about 400–800 ml of bile each day, and any excess is stored in the gallbladder. Water is reabsorbed by the gallbladder so that bile becomes a thick, mucuslike material. When needed, bile leaves the gallbladder and proceeds to the duodenum via the common bile duct.

The liver has many functions. Among these are the removal of toxic substances from the blood, the production of plasma proteins, the maintenance of blood sugar levels, and the production of bile, which is stored in the gallbladder.

14.3 Digestive Enzymes

Digestive enzymes, like other enzymes, are proteins with a particular shape that fits their substrate. They also function at an optimum pH, which maintains their shape, thereby enabling each enzyme to speed up its own specific reaction.

The various digestive enzymes present in the digestive juices, mentioned previously, help break down carbohydrates, proteins, nucleic acids, and fats, the major components of food. Starch is a polysaccharide, and its digestion begins in the mouth. Saliva from the salivary glands has a neutral pH and contains salivary amylase, the first enzyme to act on starch:

$$\text{starch} + H_2O \xrightarrow{\text{salivary amylase}} \text{maltose}$$

In this equation, salivary amylase is written above the arrow to indicate that it is neither a reactant nor a product in the reaction. It merely speeds the reaction in which its substrate, starch, is digested to many molecules of maltose, a disaccharide. Maltose molecules cannot be absorbed by the intestine; additional digestive action in the small intestine converts maltose to glucose, a monosaccharide. Glucose can be absorbed.

Protein digestion begins in the stomach. Gastric juice secreted by gastric glands has a very low pH—about 2—because it contains hydrochloric acid (HCl). Pepsinogen, a precursor that is converted to the enzyme pepsin when exposed to HCl, is also present in gastric juice. Pepsin acts on proteins, which are polymers of amino acids, to produce peptides:

$$\text{protein} + H_2O \xrightarrow{\text{pepsin}} \text{peptides}$$

Peptides vary in length, but they always consist of a number of linked amino acids. Peptides are usually too large to be absorbed by the intestinal lining, but later in the small intestine, they are broken down to amino acids, the monomer of a protein molecule.

Pancreatic juice, which enters the duodenum, has a basic pH because it contains sodium bicarbonate ($NaHCO_3$). Sodium bicarbonate neutralizes the acid in chyme, producing the slightly basic pH that is optimum for pancreatic enzymes. One pancreatic enzyme, pancreatic amylase, digests starch:

$$\text{starch} + H_2O \xrightarrow{\text{pancreatic amylase}} \text{maltose}$$

Another pancreatic enzyme, trypsin, digests protein:

$$\text{protein} + H_2O \xrightarrow{\text{trypsin}} \text{peptides}$$

Trypsin is secreted as trypsinogen, which is converted to trypsin in the duodenum.

Lipase, a third pancreatic enzyme, digests fat molecules in the fat droplets after they have been emulsified by bile salts:

$$\text{fat} \xrightarrow{\text{bile salts}} \text{fat droplets}$$
$$\text{fat droplets} + H_2O \xrightarrow{\text{lipase}} \text{glycerol} + 3 \text{ fatty acids}$$

TABLE 14.3	Major Digestive Enzymes			
Enzyme	**Produced By**	**Site of Action**	**Optimum pH**	**Digestion**
Salivary amylase	Salivary glands	Mouth	Neutral	Starch + $H_2O \longrightarrow$ maltose
Pancreatic amylase	Pancreas	Small intestine	Basic	Starch + $H_2O \longrightarrow$ maltose
Maltase	Small intestine	Small intestine	Basic	Maltose + $H_2O \longrightarrow$ glucose + glucose
Pepsin	Gastric glands	Stomach	Acidic	Protein + $H_2O \longrightarrow$ peptides
Trypsin	Pancreas	Small intestine	Basic	Protein + $H_2O \longrightarrow$ peptides
Peptidases	Small intestine	Small intestine	Basic	Peptide + $H_2O \longrightarrow$ amino acids
Nuclease	Pancreas	Small intestine	Basic	RNA and DNA + $H_2O \longrightarrow$ nucleotides
Nucleosidases	Small intestine	Small intestine	Basic	Nucleotide + $H_2O \longrightarrow$ base + sugar + phosphate
Lipase	Pancreas	Small intestine	Basic	Fat droplet + $H_2O \longrightarrow$ glycerol + fatty acids

Fats are triglycerides, which means that each one is composed of glycerol and three fatty acids. The end products of lipase digestion, glycerol and fatty acid molecules, are small enough to cross the cells of the intestinal villi, where absorption takes place. As mentioned previously, glycerol and fatty acids enter the cells of the villi, and within these cells, they are rejoined and packaged as lipoprotein droplets before entering the lacteals (see Fig. 14.6).

Peptidases and **maltase,** enzymes produced by the small intestine, complete the digestion of protein to amino acids and starch to glucose, respectively. Amino acids and glucose are small molecules that cross into the cells of the villi. Peptides, which result from the first step in protein digestion, are digested to amino acids by peptidases:

$$\text{peptidases}$$
$$\text{peptides} + H_2O \longrightarrow \text{amino acids}$$

Maltose, a disaccharide that results from the first step in starch digestion, is digested to glucose by maltase:

$$\text{maltase}$$
$$\text{maltose} + H_2O \longrightarrow \text{glucose} + \text{glucose}$$

Other disaccharides, each of which has its own enzyme, are digested in the small intestine. The absence of any one of these enzymes can cause illness. For example, as many as 75% of African Americans cannot digest lactose, the sugar found in milk, because they do not produce lactase, the enzyme that digests the sugar lactose. Consuming dairy products often gives these individuals the symptoms of **lactose intolerance** (diarrhea, gas, cramps), caused by a large quantity of nondigested lactose in the intestine. However, lactose-reduced products are available for most kinds of dairy products, including milk, cheese, and ice cream. It is important to include these foods that are high in calcium in the diet; otherwise, osteoporosis, a condition characterized by weak and fragile bones, may develop.

Table 14.3 lists some of the major digestive enzymes produced by the digestive tract, salivary glands, or the pancreas.

Digestive enzymes present in digestive juices help break down food to the nutrient molecules: glucose, amino acids, fatty acids, and glycerol. The first two are absorbed into the blood capillaries of the villi, and the last two re-form within epithelial cells before entering the lacteals.

Conditions for Digestion

Laboratory experiments can define the necessary conditions for digestion. For example, the four test tubes shown in Figure 14.12 can be prepared and observed for the digestion of egg white, a protein digested in the stomach by the enzyme pepsin.

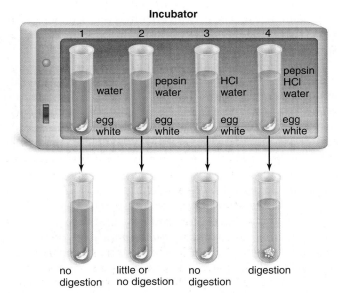

Figure 14.12 Digestion experiment.
This experiment is based on the optimum conditions for digestion by pepsin in the stomach. Knowing that the correct enzyme, optimum pH, optimum temperature, and correct substrate must be present for digestion to occur, explain the results of this experiment. Colors indicate pH of test tubes (blue, basic; pink, acidic).

After all tubes are placed in an incubator at body temperature for at least one hour, the results are as follows: Tube 1 is a control tube; no digestion has occurred in this tube because the enzyme and HCl are missing. (If a control gives a positive result, then the experiment is invalidated.) Tube 2 shows limited or no digestion because HCl is missing, and therefore, the pH is too high for pepsin to be effective. Tube 3 shows no digestion because although HCl is present, the enzyme is missing. Tube 4 shows the best digestive action because the enzyme is present and the presence of HCl has resulted in an optimum pH. This experiment supports the hypothesis that for digestion to occur, the substrate and enzyme must be present and the environmental conditions must be optimum. The optimal environmental conditions include a warm temperature and the correct pH.

14.4 Nutrition

The vigilance of your immune system, the strength of your muscles and bones, the ease with which your blood circulates—every aspect of your body's functioning depends on proper **nutrition,** the science of foods and nutrients. A **nutrient** is a component of food that performs a physiological function in the body. The six major classes of nutrients are carbohydrates, fats, proteins, vitamins, minerals, and water. The nutrients in our diet provide us with energy, promote growth and development, maintain the fluid balance and proper pH of blood, and regulate cellular metabolism. Carbohydrates and fats are the major sources of energy for the body. Proteins can be used for energy, but their primary function is to promote growth and development. Along with vitamins and minerals, proteins also help regulate metabolism, because nearly all enzymes are proteins. Water is necessary to metabolism because the content of a cell is about 70–80% water.

Food choices to fulfill your nutrient needs constitute the diet. Nutritionists often present dietary recommendations in a pyramid form. In 2005, the U.S. Department of Agriculture (USDA) released new dietary recommendations (Fig. 14.13). The daily diet should include about 6 oz of whole-grain breads, crackers, pasta, cereals, or rice. Fruits and vegetables, collectively, should be consumed in even greater quantities than grains. The USDA recommends limiting solid fats such as butter, stick margarine, lard, and shortening. However, the USDA does not shy away from dairy; it recommends that fat-free or low-fat milk or equivalent milk products be a part of the diet every day. Interestingly, the USDA says people should fulfill their protein needs by consuming lean meat such as poultry. Fish, beans, peas, nuts, and seeds are alternative sources of protein.

While food pyramids differ in various ways, most nutritionists can agree upon these simple guidelines:

1. To maintain your weight, balance your energy from food intake against your energy output.

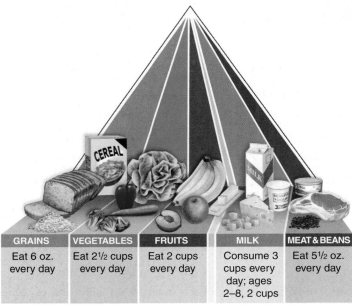

Source: U.S. Dept. of Agriculture

Figure 14.13 Food guide pyramid.
The United States Department of Agriculture (USDA) developed this pyramid as a guide to better health. The different widths of the food group bands suggest how much food a person should choose from each group. The six different colors illustrate that foods from all groups are needed each day for good health. The wider base is supposed to encourage the selection of foods with little or no solid fats or added sugars. The yellow band indicates fats and oils—eat only 1–2 tablespoons every day.

2. Eat a variety of foods, in order to acquire all needed nutrients.
3. A healthy diet
   - is moderate in total fat intake and low in saturated fat and cholesterol.
   - is rich in whole-grain products, legumes (e.g., beans and peas), and vegetables as sources of complex carbohydrates and fiber.
   - is low in refined carbohydrates, such as starches and sugars.
   - is low in salt and sodium intake.
   - contains adequate amounts of protein, largely from poultry, fish, and plant sources.
   - includes only moderate amounts of alcohol.
   - contains adequate amounts of minerals and vitamins, but avoids questionable food additives and supplements.

Carbohydrates

The quickest, most readily available source of energy for the body is glucose. Carbohydrates are digested to simple sugars, which are, or can be, converted to glucose. As mentioned in section 14.2, glucose is stored by the liver in the form of glycogen. Between eating periods, the blood glucose level is maintained at about 0.05% to 0.08% by the breakdown of glycogen or by the conversion of fat or amino acids to glucose. While body cells can utilize fatty acids as an energy source, brain cells require glucose. For this reason alone, it is necessary to include carbohydrates in the diet.

Complex sources of carbohydrates, such as the whole-grain foods featured in Figure 14.14, are recommended because they are digested to sugars gradually and contain fiber. Insoluble fiber, such as that found in wheat bran, has a laxative effect and may possibly guard against colon cancer by limiting the amount of time cancer-causing substances are in contact with the intestinal wall. Soluble fiber, such as that found in oat bran, combines with bile acids and cholesterol in the intestine and prevents them from being absorbed. The liver then removes cholesterol from the blood and changes it to bile acids, replacing the bile acids that were lost. While the diet should have an adequate amount of fiber, some evidence suggests that a diet too high in fiber can be detrimental, possibly impairing the body's ability to absorb iron, zinc, and calcium.

Simple sugars in foods such as candy and ice cream immediately enter the bloodstream, as do those derived from the digestion of starch within white bread and potatoes. These foods are said to have a high **glycemic index (GI)**, because the blood glucose response to these foods is high. When the blood glucose level rises rapidly, the pancreas produces an overload of insulin to bring the level under control. (Cells take up glucose in response to insulin.) Hunger quickly returns after a high glycemic meal. Investigators tell us that a chronically high insulin level also leads to insulin resistance and many harmful effects, such as increased fat deposition and a high blood fatty acid level. Over the years, the

TABLE 14.4	Reducing High Glycemic Index Carbohydrates

To reduce dietary sugar:

1. Eat fewer sweets, such as candy, soft drinks, ice cream, and pastry.

2. Eat fresh fruits or fruits canned without heavy syrup.

3. Use less sugar—white, brown, or raw—and less honey and syrups.

4. Avoid sweetened breakfast cereals.

5. Eat less jelly, jam, and preserves.

6. Drink pure fruit juices, not imitations.

7. When cooking, use spices, such as cinnamon, instead of sugar to flavor foods.

8. Do not put sugar in tea or coffee.

9. Avoid potatoes and processed foods made from refined carbohydrates, such as white bread, rice, and pasta.

body's cells can become insulin resistant, and type 2 diabetes can develop. Even if insulin resistance does not develop, researchers warn that a high blood fatty acid level can lead to an increased risk of coronary heart disease, hypertension, liver disease, and several types of cancer.

Table 14.4 gives suggestions on how to reduce your intake of high-GI carbohydrates.

> Complex carbohydrates, which contain fiber and a wide range of nutrients, are recommended, while high-GI carbohydrates are not recommended.

Figure 14.14 Complex carbohydrates.
To meet our energy needs, dietitians recommend consuming foods rich in complex carbohydrates, such as those shown here, rather than foods consisting of simple carbohydrates, such as candy and ice cream, or refined carbohydrates, such as white bread. Such carbohydrates provide monosaccharides, but few other types of nutrients.

Proteins

Foods rich in protein include red meat, fish, poultry, dairy products, legumes (i.e., peas and beans), nuts, and cereals. Following digestion of protein, amino acids enter the bloodstream and are transported to the tissues. Ordinarily, amino acids are not used as an energy source. Most are incorporated into structural proteins found in muscles, skin, hair, and nails. Others are used to synthesize such proteins as hemoglobin, plasma proteins, enzymes, and hormones.

Adequate protein formation requires 20 different types of amino acids. Of these, eight are required from the diet in adults (nine in children) because the body is unable to produce them. These are termed the **essential amino acids.** The body produces the other amino acids by simply transforming one type into another type.

Some protein sources, such as meat, milk, and eggs, are complete; they provide all 20 types of amino acids. Legumes (beans and peas), other types of vegetables, seeds, nuts, and also grains supply us with amino acids, but each of these alone is an incomplete protein source, because of a deficiency in at least one of the essential amino acids. The

TABLE 14.5	Complementary Proteins*		
Legumes	Seeds and Nuts	Grains	Vegetables
Green peas	Sunflower seeds	Wheat	Leafy green (e.g., spinach)
Navy beans	Sesame seeds	Rice	Broccoli
Soybeans	Macadamia nuts	Corn	Cauliflower
Black-eyed peas	Brazil nuts	Barley	Cabbage
Pinto beans	Peanuts	Oats	Artichoke hearts
Lima beans	Cashews	Rye	
Kidney beans	Hazelnuts		
Chickpeas	Almonds		
Black beans	Nut butter		

*Combine foods from any two or more columns to acquire all of the essential amino acids.

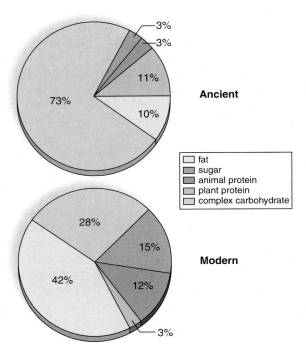

Figure 14.15 Ancient versus modern diet of native Hawaiians.

Among those Hawaiians who have switched back to an ancient diet that is rich in complex carbohydrates and plant protein and low in fat and sugar, the incidence of cardiovascular disease, cancer, and diabetes has dropped.

absence of one essential amino acid prevents utilization of the other 19 amino acids. Therefore, vegetarians are counseled to combine two or more incomplete types of plant products to acquire all the essential amino acids.

Table 14.5 lists complementary proteins—sources of protein whose amino acid content complement each other so that all the essential amino acids are present in the diet. Soybeans and tofu, which is made from soybeans, are rich in amino acids, but even so, you have to combine tofu with a complementary protein to acquire all the essential amino acids. Table 14.5 shows how to select various combinations of plant products to make sure the diet contains the essential amino acids when it does not contain meat.

Amino acids are not stored in the body, and a daily supply is needed. However, it does not take very much protein to meet the daily requirement. Two servings of meat a day (one serving is equal in size to a deck of cards) are usually plenty. While the body is harmed if the amount of protein in the diet is severely limited, it is also likely to be harmed by an overabundance of protein. Deamination of excess amino acids in the liver results in urea, our main nitrogenous excretion product. The water needed for excretion of urea can cause dehydration when a person is exercising and losing water by sweating. High-protein diets can also increase calcium loss in urine, especially those rich in animal proteins. Excretion of calcium can lead to kidney stones. The same comments apply to protein and amino acid supplements, which are generally not recommended by nutritionists.

Certain types of meat, especially red meat, are known to be high in saturated fats, while other sources, such as chicken, fish, and eggs, are more likely to be low in saturated fats. As discussed later in this section, an intake of saturated fats leads to cardiovascular disease. In one study, it was found that Hawaiians who follow the traditional native diet, which is rich in protein from plants, have a reduced

incidence of cardiovascular disease and cancer, compared to those who follow a modern diet, which is rich in fat and animal protein (Fig. 14.15).

Proteins are digested into amino acids, several of which are the essential amino acids required in the diet. Foods vary in their amino acid composition, which is why it is important to consume a variety of protein sources. Proteins from plant sources usually have less accompanying fat.

Lipids

Fats, oils, and cholesterol are lipids. Saturated fats, which are solids at room temperature, usually have an animal origin. Two well-known exceptions are palm oil and coconut oil, which contain mostly saturated fats and come from the plants mentioned. Butter and meats, such as marbled red meats and bacon, contain saturated fats. Saturated fats are particularly associated with cardiovascular disease. The worst offenders are those that contain trans fatty acids, which arise when unsaturated fatty acids are hydrogenated to produce a solid fat. Found largely in commercially produced products, trans fatty acids (often called trans-fats) may reduce the function of the plasma membrane receptors that clear cholesterol from the bloodstream.

Oils contain unsaturated fatty acids, which do not promote cardiovascular disease. Not only that, unsaturated oils, like whole grains and vegetables, have a low glycemic index and are more filling. Each type of oil has a percentage of monounsaturated and polyunsaturated fatty acids. Corn oil and safflower oil are high in polyunsaturated fatty acids. Polyunsaturated oils are nutritionally essential because they are the only type of fat that contains linoleic acid and linolenic acid, two fatty acids the body cannot make. The body needs these two polyunsaturated fatty acids to produce various hormones and the plasma membrane of cells. Since these fatty acids must be supplied by diet, they are called **essential fatty acids.**

Olive or canola oils contain a larger percentage of monounsaturated fatty acids than other types of cooking oils. Omega-3 fatty acids—characterized by a double bond in the third position—are believed to be especially protective against heart disease. Some cold-water fishes, such as salmon, sardines, and trout, are rich sources of omega-3, but they contain only about half as much as flaxseed oil, the best source from plants.

Fats That Cause Disease

Cardiovascular disease is often due to arteries blocked by **plaque,** which contains saturated fats and cholesterol. Cholesterol is carried in the blood by two types of lipoproteins: low-density lipoprotein (LDL) and high-density lipoprotein (HDL). LDL is thought of as "bad" because it carries cholesterol from the liver to the cells, while HDL is thought of as "good" because it carries cholesterol to the liver, which takes it up and converts it to bile salts. Saturated fats tend to raise LDL cholesterol levels, while unsaturated fats lower LDL cholesterol levels.

Controlled-feeding and statistical studies that have examined trans-fat intake, in relation to the risk of heart disease and diabetes, indicate that trans-fats are more prone to cause these diseases than saturated fats. Trans fatty acids are found in commercially packaged goods, such as cookies and crackers, commercially fried foods, such as french fries from some fast-food chains, and packaged snacks, such as microwave popcorn, as well as in vegetable shortening and some margarines. Any packaged goods that contain partially hydrogenated vegetable oils or "shortening" most likely contain trans-fat.

In addition to avoiding trans fatty acids and saturated fats, everyone should use diet and exercise to keep their cholesterol level within normal limits, rather than relying on medications for this purpose. Table 14.6 suggests ways to reduce dietary saturated fat and cholesterol. It is not a good idea to rely on commercially produced low-fat foods or those that contain fake fat (e.g., olestra) to reduce total fat intake. In some products, carbohydrates, namely sugars, have replaced the fat, and in others, protein is used. We have already discussed the dangers of simple sugars in the diet, and while replacement of fat by protein is nutritionally more sound, protein still contributes calories to the diet. Nutritionists also express concern about the consumption of fake fat, because high levels

TABLE 14.6	**Reducing Certain Lipids**

To reduce saturated fats and trans fats in the diet:

1. Choose poultry, fish, or dry beans and peas as a protein source.

2. Remove skin from poultry and trim fat from red meats before cooking and place on a rack so that fat drains off.

3. Broil, boil, or bake rather than fry.

4. Limit your intake of butter, cream, trans fats, shortenings, and tropical oils (coconut and palm oils).*

5. Use herbs and spices to season vegetables instead of butter, margarine, or sauces. Use lemon juice instead of salad dressing.

6. Drink skim milk instead of whole milk, and use skim milk in cooking and baking.

To reduce dietary cholesterol:

1. Avoid cheese, egg yolks, liver, and certain shellfish (shrimp and lobster). Preferably, eat white fish and poultry.

2. Substitute egg whites for egg yolks in both cooking and eating.

3. Include soluble fiber in the diet. Oat bran, oatmeal, beans, corn, and fruits, such as apples, citrus fruits, and cranberries are high in soluble fiber.

Although coconut and palm oils are from plant sources, they are mostly saturated fats.

of intake can lead to limited absorption of fat-soluble vitamins, diarrhea, and anal leakage. Products containing olestra often have extra fat-soluble vitamins added, however.

> The body requires certain essential fatty acids in the diet. An intake of saturated fats is associated with various health problems, while unsaturated fatty acids may protect against diseases.

Vitamins

Vitamins are organic compounds (other than carbohydrates, fats, and proteins) that the body uses for metabolic purposes but is unable to produce in adequate quantity. Many vitamins are portions of coenzymes, which are enzyme helpers. For example, niacin is part of the coenzyme NAD, and riboflavin is part of another dehydrogenase, FAD. Our bodies need coenzymes in only small amounts because each can be used over and over again. Not all vitamins are coenzymes; vitamin A, for example, is a precursor for the visual pigment that prevents night blindness. If vitamins are lacking in the diet, various symptoms develop (Fig. 14.16). Altogether, there are 13 vitamins, which are divided into those that are fat soluble (Table 14.7) and those that are water soluble (Table 14.8).

Antioxidants

Over the past 20 years, numerous statistical studies have been done to determine whether a diet rich in fruits and

TABLE 14.7	Fat-Soluble Vitamins			
Vitamin	**Functions**	**Food Sources**	**Conditions With**	
			Too Little	*Too Much*
Vitamin A	Antioxidant synthesized from beta-carotene; needed for healthy eyes, skin, hair, and mucous membranes, and for proper bone growth	Deep yellow/orange and leafy, dark green vegetables, fruits, cheese, whole milk, butter, eggs	Night blindness, impaired growth of bones and teeth	Headache, dizziness, nausea, hair loss, abnormal development of fetus
Vitamin D	A group of steroids needed for development and maintenance of bones and teeth	Milk fortified with vitamin D, fish liver oil; also made in the skin when exposed to sunlight	Rickets, bone decalcification and weakening	Calcification of soft tissues, diarrhea, possible renal damage
Vitamin E	Antioxidant that prevents oxidation of vitamin A and polyunsaturated fatty acids	Leafy green vegetables, fruits, vegetable oils, nuts, whole-grain breads and cereals	Unknown	Diarrhea, nausea, headaches, fatigue, muscle weakness
Vitamin K	Needed for synthesis of substances active in clotting of blood	Leafy green vegetables, cabbage, cauliflower	Easy bruising and bleeding	Can interfere with anticoagulant medication

vegetables can protect against cancer. Cellular metabolism generates free radicals, unstable molecules that carry an extra electron. The most common free radicals in cells are superoxide (O_2^-) and hydroxide (OH^-). In order to stabilize themselves, free radicals combine with DNA, proteins (including enzymes), or lipids, which are found in plasma membranes. Free radicals damage these cellular molecules and can thereby lead to disorders, perhaps even cancer.

Vitamins C, E, and A are believed to defend the body against free radicals, and therefore, they are termed antioxidants. These vitamins are especially abundant in fruits and vegetables. The dietary guidelines in Figure 14.13 suggest that we eat a minimum of four to five cups of fruits and vegetables a day. To achieve this goal, we should include salad greens, raw or cooked vegetables, dried fruit, and fruit juice, in addition to traditional apples and oranges.

Vitamin D

Skin cells contain a precursor cholesterol molecule that is converted to vitamin D after UV exposure. Vitamin D leaves the skin and is modified first in the kidneys and then in the liver until finally it becomes calcitriol. Calcitriol promotes the absorption of calcium by the intestines. The lack of vitamin D leads to rickets in children (Fig. 14.16a). Rickets, characterized by bowing of the legs, is caused by defective mineralization of the skeleton. Most milk today is fortified with vitamin D, which helps prevent the occurrence of rickets.

Vitamins are essential to cellular metabolism; many protect against identifiable illnesses and conditions.

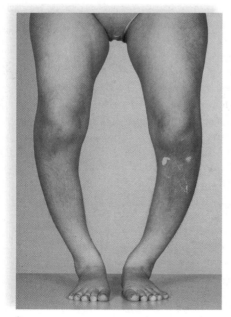

a.

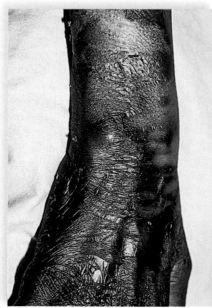

b.

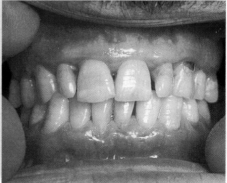

c.

Figure 14.16 Illnesses due to vitamin deficiency.

a. Bowing of bones (rickets) due to vitamin D deficiency. **b.** Dermatitis (pellagra) of areas exposed to light due to niacin (vitamin B_3) deficiency. **c.** Bleeding of gums (scurvy) due to vitamin C deficiency.

TABLE 14.8	Water-Soluble Vitamins			
Vitamin	**Functions**	**Food Sources**	**Conditions With**	
			Too Little	*Too Much*
Vitamin C	Antioxidant; needed for forming collagen; helps maintain capillaries, bones, and teeth	Citrus fruits, leafy green vegetables, tomatoes, potatoes, cabbage	Scurvy, delayed wound healing, infections	Gout, kidney stones, diarrhea, decreased copper
Thiamine (vitamin B_1)	Part of coenzyme needed for cellular respiration; also promotes activity of the nervous system	Whole-grain cereals, dried beans and peas, sunflower seeds, nuts	Beriberi, muscular weakness, enlarged heart	Can interfere with absorption of other vitamins
Riboflavin (vitamin B_2)	Part of coenzymes, such as FAD; aids cellular respiration, including oxidation of protein and fat	Nuts, dairy products, whole-grain cereals, poultry, leafy green vegetables	Dermatitis, blurred vision, growth failure	Unknown
Niacin (nicotinic acid)	Part of coenzymes NAD and NADP; needed for cellular respiration, including oxidation of protein and fat	Peanuts, poultry, whole-grain cereals, leafy green vegetables, beans	Pellagra, diarrhea, mental disorders	High blood sugar and uric acid, vasodilation, etc.
Folacin (folic acid)	Coenzyme needed for production of hemoglobin and formation of DNA	Dark leafy green vegetables, nuts, beans, whole-grain cereals	Megaloblastic anemia, spina bifida	May mask B_{12} deficiency
Vitamin B_6	Coenzyme needed for synthesis of hormones and hemoglobin; CNS control	Whole-grain cereals, bananas, beans, poultry, nuts, leafy green vegetables	Rarely, convulsions, vomiting, seborrhea, muscular weakness	Insomnia, neuropathy
Pantothenic acid	Part of coenzyme A needed for oxidation of carbohydrates and fats; aids in the formation of hormones and certain neurotransmitters	Nuts, beans, dark green vegetables, poultry, fruits, milk	Rarely, loss of appetite, mental depression, numbness	Unknown
Vitamin B_{12}	Complex, cobalt-containing compound; part of the coenzyme needed for synthesis of nucleic acids and myelin	Dairy products, fish, poultry, eggs, fortified cereals	Pernicious anemia	Unknown
Biotin	Coenzyme needed for metabolism of amino acids and fatty acids	Generally in foods, especially eggs	Skin rash, nausea, fatigue	Unknown

Dietary Supplements

Dietary supplements are nutrients and plant products (such as herbal teas) that are used to enhance health. The U.S. government does not require dietary supplements to undergo the same safety and effectiveness testing that new prescription drugs must complete before they are approved. Therefore, many of these products have not been tested scientifically to determine their benefits. Although people often think herbal products are safe because they are "natural", many plants, such as lobelia, comfrey, and kava kava, can be poisonous.

Supplements can be useful to correct a deficiency, but most people in the United States take them to greatly increase the dietary intake of a particular nutrient. Everyone should be aware, however, that ingesting high levels of certain nutrients can cause harm. Most fat-soluble vitamins are stored in the body and can accumulate to toxic levels, particularly vitamins A and D. Certain minerals can also be harmful, even deadly, when ingested in high doses.

Dietary supplements may provide a potential safeguard against cancer and cardiovascular disease, but nutritionists do not think people should take supplements instead of improving their intake of fruits and vegetables. There are many beneficial compounds in fruits that cannot be obtained from a vitamin pill. These compounds enhance each other's absorption or action and also perform independent biological functions.

> Most dietary supplements are not reviewed by government agencies. Some may be useful to correct deficiencies, but high levels of certain vitamins or minerals can be toxic.

Minerals

In addition to vitamins, the body requires various **minerals.** Minerals are divided into major minerals and trace minerals. The body contains more than 5 grams of each major mineral and less than 5 grams of each trace mineral (Fig. 14.17). The major minerals are constituents of cells and body fluids and are structural components of tissues. For example, calcium (present as Ca^{2+}) is needed for the construction of bones and teeth and for nerve conduction and muscle contraction. Phosphorus (present as PO_4^{3-}) is stored in the bones and teeth and is a part of phospholipids, ATP, and the nucleic acids. Potassium (K^+) is the major positive ion inside cells and is important in nerve conduction and muscle contraction, as is sodium (Na^+). Sodium also plays a major role in regulating the body's water balance, as does

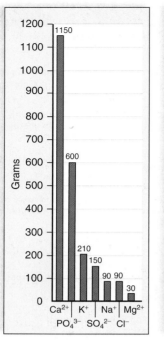
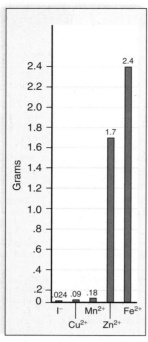

a. Major minerals   b. Trace minerals

Figure 14.17 Minerals in the body.
These charts show the usual amount of certain minerals in a 60-kilogram (135 lb) person. The functions of minerals are given in Table 14.9. **a.** The major minerals are present in amounts larger than 5 grams (about a teaspoon). **b.** Trace minerals are present in lesser amounts.

chloride (Cl^-). Magnesium (Mg^{2+}) is critical to the functioning of hundreds of enzymes.

The trace minerals are parts of larger molecules. For example, iron (Fe^{2+}) is present in hemoglobin, and iodine (I^-) is a part of thyroxine and triiodothyronine, hormones produced by the thyroid gland. Zinc (Zn^{2+}), copper (Cu^{2+}), and manganese (Mn^{2+}) are present in enzymes that catalyze a variety of reactions. Proteins called zinc-finger proteins, because of their characteristic shapes, bind to DNA when a particular gene is to be activated. As research continues, more and more elements are added to the list of trace minerals considered essential. During the past three decades, for example, very small amounts of selenium, molybdenum, chromium, nickel, vanadium, silicon, and even arsenic have been found to be essential to good health. Table 14.9 lists the functions of various minerals and gives their food sources and signs of deficiency and toxicity.

Occasionally, individuals do not receive enough iron, calcium, magnesium, or zinc in their diets. Adult females need more iron in the diet than males (18 mg compared to 10 mg) because they lose hemoglobin each month during menstruation. Stress can bring on a magnesium deficiency, and due to its high-fiber content, a vegetarian diet may make zinc less available to the body. However, a varied and complete diet usually supplies enough of each type of mineral.

Calcium

Many people take calcium supplements to counteract **osteoporosis,** a degenerative bone disease that afflicts an estimated one-fourth of older men and one-half of older women in the United States. Osteoporosis develops because bone-eating cells called osteoclasts become more active than bone-forming cells called osteoblasts. Therefore, the bones are porous, and they break easily because they lack sufficient calcium. Due to recent studies that show consuming more calcium does slow bone loss in elderly people, the guidelines have been revised. A calcium intake of 1,000 mg a day is recommended for men and for women who are premenopausal or who use estrogen replacement therapy; 1,300 mg a day is recommended for postmenopausal women who do not use estrogen replacement therapy. To achieve this amount, supplemental calcium is most likely necessary.

Vitamin D is an essential companion to calcium in preventing osteoporosis. Other vitamins may also be helpful; for example, magnesium has been found to suppress the cycle that leads to bone loss. Estrogen replacement therapy and exercise, in addition to adequate calcium and vitamin intake, also help prevent osteoporosis. Drinking more than nine cups of caffeinated coffee per day and smoking are risk factors for osteoporosis. Medications are also available that slow bone loss while increasing skeletal mass. These are still being studied for their effectiveness and possible side effects.

Sodium

The recommended amount of sodium intake per day is less than 2,400 mg, although the average American takes in 4,000–4,700 mg every day. In recent years, this imbalance has caused concern because high sodium intake has been linked to hypertension (high blood pressure) in some people. About one-third of the sodium we consume occurs naturally in foods; another one-third is added during commercial processing; and we add the last one-third either during home cooking or at the table in the form of table salt.

Clearly, it is possible for us to cut down on the amount of sodium in the diet. Table 14.10 gives recommendations for doing so.

Both major minerals and trace minerals play specific roles in the body. Calcium is needed for strong bones, for example. Excess sodium in the diet may lead to hypertension; therefore, excess sodium intake should be avoided.

Eating Disorders

Eating disorders are on the increase, and thus everyone needs to understand these illnesses. People with eating

TABLE 14.9	Minerals			
Mineral	**Functions**	**Food Sources**	**Conditions With**	
			Too Little	*Too Much*
Major minerals (more than 100 mg/day needed)				
Calcium (Ca^{2+})	Strong bones and teeth, nerve conduction, muscle contraction	Dairy products, leafy green vegetables	Stunted growth in children, low bone density in adults	Kidney stones; interferes with iron and zinc absorption
Phosphorus (PO_4^{3-})	Bone and soft tissue growth; part of phospholipids, ATP, and nucleic acids	Meat, dairy products, sunflower seeds, food additives	Weakness, confusion, pain in bones and joints	Low blood and bone calcium levels
Potassium (K^+)	Nerve conduction, muscle contraction	Many fruits and vegetables, bran	Paralysis, irregular heartbeat, eventual death	Vomiting, heart attack, death
Sodium (Na^+)	Nerve conduction, pH and water balance	Table salt	Lethargy, muscle cramps, loss of appetite	Edema, high blood pressure
Chloride (Cl^-)	Water balance	Table salt	Not likely	Vomiting, dehydration
Magnesium (Mg^{2+})	Part of various enzymes for nerve and muscle contraction, protein synthesis	Whole grains, leafy green vegetables	Muscle spasm, irregular heartbeat, convulsions, confusion, personality changes	Diarrhea
Trace minerals (less than 20 mg/day needed)				
Zinc (Zn^{2+})	Protein synthesis, wound healing, fetal development and growth, immune function	Meats, legumes, whole grains	Delayed wound healing, night blindness, diarrhea, mental lethargy	Anemia, diarrhea, vomiting, renal failure, abnormal cholesterol levels
Iron (Fe^{2+})	Hemoglobin synthesis	Whole grains, meats, prune juice	Anemia, physical and mental sluggishness	Iron toxicity disease, organ failure, eventual death
Copper (Cu^{2+})	Hemoglobin synthesis	Meat, nuts, legumes	Anemia, stunted growth in children	Damage to internal organs if not excreted
Iodine (I^-)	Thyroid hormone synthesis	Iodized table salt, seafood	Thyroid deficiency	Depressed thyroid function, anxiety
Selenium (SeO_4^{2-})	Part of antioxidant enzyme	Seafood, meats, eggs	Vascular collapse, possible cancer development	Hair and fingernail loss, discolored skin

disorders have attitudes and behaviors toward food that are outside the norm. Anyone, whether female or male and regardless of ethnicity, socioeconomic status, or intelligence level, can develop an eating disorder. However, some specific factors influence the development of eating disorders. The desire for slenderness in females or a manly physique in males may lead to dieting, and dieting is seen as a precursor to bulimia nervosa and anorexia nervosa. Neurochemicals have also been implicated in the development of these disorders. In any case, the individual may come to think of dieting, weight control, weight gain, and rigid self-control of appetite as moral issues.

The inability to produce an effective satiety hormone, called leptin, might be involved in obesity. It's difficult to determine if a person has a genetic tendency to be obese or learns eating and exercise habits that lead to weight gain. Food can also be used to lift a person's mood. Certainly, people are more sedentary than they used to be and they eat more processed foods, which can contain more calories than meals prepared from fresh foods at home.

It's also hard to know how to treat eating disorders. A combination of medication and personal/family counseling is the most common form of treatment.

TABLE 14.10	Reducing Dietary Sodium

To reduce dietary sodium:

1. Use spices instead of salt to flavor foods.

2. Add little or no salt to foods at the table, and add only small amounts of salt when you cook.

3. Eat unsalted crackers, pretzels, potato chips, nuts, and popcorn.

4. Avoid hot dogs, ham, bacon, luncheon meats, smoked salmon, sardines, and anchovies.

5. Avoid processed cheese and canned or dehydrated soups.

6. Avoid brine-soaked foods, such as pickles or olives.

7. Read labels to avoid high-salt products.

also determine how much body fat you have. The risk of heart disease is higher in obese individuals, and this alone tells us that excess body fat is not consistent with optimal health.

The treatment depends on the degree of obesity. Surgery to remove body fat, or even part of the stomach, may be required for those who are moderately or greatly obese. But for most people, a knowledge of good eating habits along with behavior modification may suffice, particularly if they combine a balanced diet with a sensible exercise program. A lifelong commitment to a properly planned program is the best way to prevent a cycle of weight gain followed by weight loss. Such a cycle is not conducive to good health.

> Persons with obesity
>
> - weigh 20% or more than is appropriate for their height.
> - have more body fat than is consistent with optimal health.
> - seldom exercise.

Figure 14.18
Recognizing obesity by its characteristics.

Obesity

As indicated in Figure 14.18, **obesity** is most often defined as a body weight 20% or more above the ideal weight for a person's height. By this standard, approximately 30% of adults in the United States are obese. Moderate obesity is 41–100% above ideal weight, and severe obesity is 100% or more above ideal weight.

Obesity is most likely caused by a combination of hormonal, metabolic, and social factors. It is known that obese individuals have more fat cells than normal, and when they lose weight, the fat cells simply get smaller; they don't disappear. The social factors that cause obesity include the eating habits of other family members. Consistently eating fatty foods, for example, will make you gain weight. Sedentary activities, such as watching television instead of exercising,

Bulimia Nervosa

Bulimia nervosa can coexist with either obesity or anorexia nervosa, which is discussed next. People with this condition have the habit of eating to excess (called binge eating) and then purging themselves by some artificial means, such as self-induced vomiting or use of a laxative. Bulimic individuals are overconcerned about their body shape and weight, and therefore, they may be on a very restrictive diet. A restrictive diet may bring on the desire to binge, and typically the person chooses to consume sweets, such as cakes, cookies, and ice cream (Fig. 14.19). The amount of food consumed is far beyond the normal number of calories for one meal, and the person keeps on eating until every bit is gone. Then, a feeling of guilt most likely brings on the next phase, which is a purging of all the calories that have been taken in.

Bulimia can be dangerous to a person's health. Blood composition is altered, leading to an abnormal heart rhythm, and damage to the kidneys can even result in death. At the very least, vomiting can lead to inflammation of the pharynx and esophagus, and stomach acids can cause the teeth to erode. The esophagus and stomach may even rupture and tear due to strong contractions during vomiting.

> Persons with bulimia nervosa have
>
> - recurrent episodes of binge eating: consuming a large amount of food in a short period and experiencing feelings of lack of control during the episode.
> - obsession about body shape and weight.
> - increase in fine body hair, halitosis, and gingivitis.
>
> Body weight is regulated by
>
> - a restrictive diet, excessive exercise.
> - purging (self-induced vomiting or misuse of laxatives).

Figure 14.19
Recognizing bulimia nervosa by its characteristics.

Persons with anorexia nervosa have	Body weight is kept too low by
• a morbid fear of gaining weight; body weight no more than 85% normal.	• a restrictive diet, often with excessive exercise.
• a distorted body image so that person feels fat even when emaciated.	• binge eating/purging (person engages in binge eating and then self-induces vomiting or misuses laxatives).
• in females, an absence of a menstrual cycle for at least three months.	

Figure 14.20
Recognizing anorexia nervosa by its characteristics.

The most important aspect of treatment is to get the patient on a sensible and consistent diet. Again, behavioral modification is helpful, and so perhaps is psychotherapy to help the patient understand the emotional causes of the behavior. Medications, including antidepressants, have sometimes helped to reduce the bulimic cycle and restore normal appetite.

Anorexia Nervosa

In **anorexia nervosa,** a morbid fear of gaining weight causes the person to be on a very restrictive diet. Athletes such as distance runners, wrestlers, and dancers are at risk of anorexia nervosa because they believe that being thin gives them a competitive edge. In addition to eating only low-calorie foods, the person may induce vomiting and use laxatives to bring about further weight loss. No matter how thin they have become, people with anorexia nervosa think they are overweight (Fig. 14.20). Such a distorted self-image may keep them from recognizing that they need medical help.

Actually, the person is starving and has all the symptoms of starvation, including low blood pressure, irregular heartbeat, constipation, and constant chilliness. Bone density decreases and stress fractures occur. The body begins to shut down; menstruation ceases in females; the internal organs, including the brain, don't function well; and the skin dries up. Impairment of the pancreas and digestive tract means that any food consumed does not provide nourishment. Death may be imminent. If so, the only recourse may be hospitalization and force-feeding. Eventually, it is necessary to use behavior therapy and psychotherapy to enlist the person's cooperation to eat properly. Family therapy may be necessary, because anorexia nervosa in children and teens is believed to be a way for them to gain some control over their lives.

Obesity, bulimia nervosa, and anorexia nervosa have complex causes and may be damaging to health. Therefore, they require competent medical and psychiatric attention.

14.5 Disorders of the Digestive System

Disorders of the digestive system can be grouped into two categories: disorders of the tract itself, and disorders of the accessory organs.

Disorders of the Digestive Tract

Stomach Ulcers

A thick layer of mucus normally protects the wall of the stomach. If this protective layer is broken down, the stomach wall can be damaged by the acidic pH of the stomach, resulting in an **ulcer,** or open sore in the wall caused by a gradual disintegration of the tissue. It now appears that most stomach ulcers are initiated by infection of the stomach by a bacterium called *Helicobacter pylori* (see Health Focus), which can impair the ability of the mucous cells to produce protective mucus. Thus, treatment of stomach ulcers now typically includes antibiotics to kill these bacteria, along with medication to reduce the amount of acid in the stomach. Other potential causes of stomach ulcers include certain viral infections, as well as the overuse of anti-inflammatory medications, which can have the side effect of damaging the stomach lining. Duodenal ulcers are also common because the duodenum receives acid chyme from the stomach.

Diarrhea and Constipation

Two common complaints associated with the large intestine are diarrhea and constipation. The major causes of diarrhea are infections of the lower intestinal tract and nervous stimulation. One type of infection is food poisoning caused by eating food contaminated with certain types of bacteria.

A New Culprit for Stomach Ulcers

A stomach (gastric) ulcer is a gaping, sometimes bloody, indentation in the stomach lining (Fig. 14A). Australian physician Barry Marshall didn't believe the accepted explanation that the major causes of stomach ulcers were stress or spicy foods. Instead, he and his colleague, Dr. Robin Warren, thought most stomach ulcers were due to infection with a bacterial species called *Helicobacter pylori*, which they had discovered in 1982. At that time, most physicians disagreed with their hypothesis, because they didn't believe any bacteria could survive in the very acidic conditions of the stomach. However, Marshall and Warren, a pathologist, had observed that nearly every biopsy of stomach ulcer tissue contained the corkscrew-shaped *H. pylori* bacteria.

To get the data they needed, Dr. Marshall decided to use himself as a guinea pig. One day in July 1984, he drank a culture containing *H. pylori*. About a week later, he started vomiting and suffering the painful symptoms of an ulcer. Medical tests confirmed that his stomach lining was teeming with bacteria. In a few weeks, his ulcer had disappeared, but the experiment had served its purpose. In recognition of his efforts, in 2005, Marshall and Warren were awarded the Nobel Prize in Physiology and Medicine

for their discovery of *H. pylori* and its role in ulcers and other stomach ailments.

Prior to Dr. Marshall's bold experiment, ulcers were usually treated mainly with drugs to reduce the acidity of the stomach. Patients got temporary relief, but the ulcers tended to return, and serious complications could arise. Some patients had bleeding ulcers and underwent an operation to remove a portion of the stomach. Thanks to the work of Marshall and Warren, in 1994 the National Institutes of Health endorsed antibiotics as a standard treatment for stomach ulcers. An international race is under way among scientists to develop a vaccine.

Discussion Questions

1. What kinds of adaptations would *H. pylori* need in order to survive in the acid pH of the stomach?
2. Why do you suppose that even after Dr. Marshall's experiment in 1984, it took ten years for his findings to be widely accepted by the medical community?
3. Do you think Dr. Marshall's experiment was ethical? If not, how would you have proposed to prove that *H. pylori* could cause ulcers? If so, could you conceive of a similar experiment that might be unethical?

— healthy gastric mucosa

— gastric ulcer

Figure 14A Stomach ulcer.
Drs. Barry Marshall and Robin Warren won a Nobel Prize in 2005 for discovering that most stomach ulcers are caused by a bacterial infection.

The intestinal wall becomes irritated by the infection, or by toxins produced by the organism, and peristalsis increases. Less water is absorbed, and the diarrhea that results helps rid the body of the infectious organisms. In nervous diarrhea, the nervous system stimulates the intestinal wall, and diarrhea results. Prolonged diarrhea can lead to dehydration because of water loss, and to disturbances in the heart's contraction, due to an imbalance of salts in the blood.

When a person is constipated, the feces are dry and hard. One reason for this condition is that some people have learned to inhibit the defecation reflex to the point that the urge to defecate is ignored. Chronic constipation can lead to the development of hemorrhoids, which are enlarged and inflamed blood vessels of the anus. Increasing the amount of water and fiber in the diet can help prevent constipation. Laxatives and enemas can also be used, but frequent use is discouraged because they can be irritating to the colon.

Polyps and Colon Cancer

The colon is subject to the development of **polyps,** which are small growths arising from the epithelial lining. They

may be benign or cancerous. Polyps are usually detected using a procedure known as colonoscopy, during which a long, flexible endoscope is inserted into the colon to visualize the wall of the large intestine. If colon cancer is detected and surgically removed while it is still confined to a polyp, a complete cure is expected.

Some investigators believe that dietary fat increases the likelihood of colon cancer because dietary fat causes an increase in bile secretion. Certain intestinal bacteria may convert bile salts to substances that promote the development of cancer. On the other hand, fiber in the diet seems to inhibit the development of colon cancer, possibly by diluting the concentration of bile salts and facilitating the movement of cancer-inducing substances through the intestine.

Disorders of the Accessory Organs

Disorders of the Pancreas

Pancreatitis is an inflammation of the pancreas. It can be caused by drinking too much alcohol, by gallstones that block the pancreatic duct, or by other unknown factors. In chronic pancreatitis, the digestive enzymes normally secreted by the pancreas can damage the pancreas and surrounding tissues. Eventually, chronic pancreatitis can decrease the ability of the pancreas to secrete insulin, potentially leading to diabetes.

Disorders of the Liver and Gallbladder

The liver performs many functions that are essential to life; therefore, diseases that affect the liver can be life threatening. A person who has a liver ailment may develop **jaundice,** which is a yellowish coloring in the whites of the eyes, as well as in the skin of light-pigmented persons. Jaundice can be caused by an abnormally large amount of bilirubin in the blood. Jaundice can also result from **hepatitis,** or inflammation of the liver. Hepatitis is most commonly caused by one of several viruses. Hepatitis A virus is usually acquired by consuming water or food that has been contaminated by sewage. Hepatitis B virus is usually spread by sexual contact, but can also be acquired through blood transfusions or contaminated needles. The hepatitis B virus is more contagious than the AIDS virus, which is spread in similar ways. Thankfully, however, a vaccine is available to prevent hepatitis B. Hepatitis C virus is spread by the same routes as hepatitis B, and it can lead to chronic hepatitis, liver cancer, and death. Unfortunately, no vaccine is available for hepatitis C.

Cirrhosis is another chronic disease of the liver. It is often seen in alcoholics due to malnutrition and to the toxic effects of consuming excessive amounts of alcohol, which acts as a toxin on the liver. In cirrhosis, the liver first becomes infiltrated with fat, and then the fatty liver tissue is replaced by fibrous scar tissue. Although the liver has amazing regenerative powers, if the rate of damage exceeds the rate of regeneration, there may not be enough time for the liver to repair itself. The preferred treatment in most cases of liver failure is liver transplantation, but the supply of livers currently cannot meet the demand. Artificial livers have been developed, but with very limited success so far.

In some individuals, the cholesterol present in bile can come out of solution and form crystals. If the crystals grow in size, they form gallstones. The passage of the stones from the gallbladder may block the common bile duct and cause obstructive jaundice. If this happens, the obstruction can sometimes be relieved; otherwise, the gallbladder must be removed.

> Common disorders of the digestive tract include stomach ulcers (most commonly caused by the bacterium *H. pylori*), diarrhea and constipation, and colon cancer. Important disorders of the accessory organs include pancreatitis, hepatitis, cirrhosis, and gallstones.

Summarizing the Concepts

14.1 The Digestive Tract
The digestive tract consists of the mouth, pharynx, esophagus, stomach, small intestine, and large intestine. Only these structures actually contain food, while the salivary glands, liver, and pancreas supply substances that aid in the digestion of food.

The salivary glands send saliva into the mouth, where the teeth chew the food and the tongue forms a bolus for swallowing. Saliva contains salivary amylase, an enzyme that begins the digestion of starch.

The air passage and food passage cross in the pharynx. When a person swallows, the air passage is usually blocked off, and food must enter the esophagus, where peristalsis begins.

The stomach expands and stores food. While food is in the stomach, the stomach churns, mixing food with the acidic gastric juice. Gastric juice contains pepsin, an enzyme that digests protein.

The duodenum of the small intestine receives bile from the liver and pancreatic juice from the pancreas. Bile, which is produced in the liver and stored in the gallbladder, emulsifies fat and readies it for digestion by lipase, an enzyme produced by the pancreas. The pancreas also produces enzymes that digest starch (pancreatic amylase) and protein (trypsin). The intestinal enzymes finish the process of chemical digestion.

The walls of the small intestine have fingerlike projections called villi where small nutrient molecules are absorbed. Amino acids and glucose enter the blood vessels of a villus. Glycerol and fatty acids are joined and packaged as lipoproteins before entering lymphatic vessels called lacteals in a villus.

The large intestine consists of the cecum, the colon (including the ascending, transverse, descending, and sigmoid colon), and the rectum, which ends at the anus. The large intestine absorbs water, salts, and some vitamins.

14.2 Three Accessory Organs
Three accessory organs of digestion—the pancreas, liver, and gallbladder—send secretions to the duodenum via ducts. The pancreas produces pancreatic juice, which contains digestive enzymes for carbohydrate, protein, and fat.

The liver produces bile, which is stored in the gallbladder. The liver receives blood from the small intestine by way of the hepatic portal vein. It has numerous important functions, and any malfunction of the liver is a matter of considerable concern.

14.3 Digestive Enzymes
Digestive enzymes are present in digestive juices and break down food into the nutrient molecules glucose, amino acids, fatty acids, and glycerol (see Table 14.3). Salivary amylase and pancreatic amylase begin the digestion of starch. Pepsin and trypsin digest protein to peptides. Lipase digests fat to glycerol and fatty acids. Intestinal enzymes finish the digestion of starch and protein.

Digestive enzymes have the usual enzymatic properties. They are specific to their substrate and speed up specific reactions at optimum body temperature and pH.

14.4 Nutrition
The nutrients released by the digestive process should provide us with an adequate amount of energy, essential amino acids and fatty acids, and all necessary vitamins and minerals.

Carbohydrates are necessary in the diet, but high glycemic index carbohydrates (simple sugars and refined starches) are not helpful because they cause a rapid release of insulin that can lead to health problems. Proteins supply us with essential amino acids, but it is wise to avoid meats that are fatty because fats from animal sources are saturated. While unsaturated fatty acids, particularly the omega-3 fatty acids, are protective against cardiovascular disease, the saturated fatty acids lead to plaque, which occludes blood vessels. Aside from carbohydrates, proteins, and fats, the body requires vitamins and minerals. The vitamins A, E, and C are antioxidants that protect cell contents from damage due to free radicals. The mineral calcium is needed for strong bones.

14.5 Disorders of the Digestive System

Various conditions can affect the function of the digestive tract, as well as the accessory organs. Most stomach ulcers are now thought to be caused by a bacterial infection, and thus can be cured with antibiotics. Diarrhea is frequently caused by intestinal infections or by excessive nervous stimulation of the gut. Pancreatitis, hepatitis, and cirrhosis are potentially serious diseases that affect the accessory organs.

Testing Yourself

Choose the best answer for each question.

1. The digestive system
   a. breaks down food into usable nutrients.
   b. absorbs nutrients.
   c. eliminates waste.
   d. All of these are correct.

2. Tooth decay is caused by bacteria metabolizing _____ and giving off _____.
   a. sugar, protein
   b. protein, sugar
   c. acids, sugar
   d. sugar, acids

3. _____ help(s) protect against infection.
   a. Teeth
   b. Uvula
   c. Tonsils
   d. None of these are correct.

4. Chemical digestion begins in the
   a. esophagus.
   b. stomach.
   c. pharynx.
   d. None of these are correct.

5. Which organ has both an exocrine and endocrine function?
   a. liver
   b. esophagus
   c. pancreas
   d. cecum

6. _____ is converted to _____ in the presence of _____.
   a. Pepsin, HCl, pepsinogen
   b. Amylase, HCl, pepsin
   c. Amylase, maltose, HCl
   d. Pepsinogen, pepsin, HCl

7. If trypsin is absent from pancreatic juice, _____ cannot be properly digested.
   a. starch
   b. glucose
   c. fat
   d. protein

8. The lack of _____ activity would result in failure to maintain water balance.
   a. small intestinal
   b. large intestinal
   c. gallbladder
   d. stomach

9. Which of these is not a function of the liver in adults?
   a. produces bile
   b. detoxifies alcohol
   c. stores glucose
   d. produces urea
   e. makes red blood cells

10. A good nonmeat source of protein is
    a. peas.
    b. yogurt.
    c. peanuts.
    d. All of these are correct.

For questions 11–15, match the ailment with the cause in the key.

Key:
   a. too little calcium
   b. too much calcium
   c. artificial fat
   d. lack of iron
   e. too little vitamin A

11. Kidney stones

12. Anemia

13. Osteoporosis

14. Night blindness

15. Anal leakage

16. Bile
    a. is an important enzyme for the digestion of fats.
    b. cannot be stored.
    c. is made by the gallbladder.
    d. emulsifies fat.
    e. All of these are correct.

17. Which transports nutrients from the small intestine to the liver?
    a. hepatic portal vein
    b. hepatic portal artery
    c. hepatic artery
    d. None of these are correct.

18. _____ increase surface area for nutrient absorption.
    a. Lacteals
    b. Microvilli
    c. Epithelial cells
    d. None of these are correct.

19. Label the following diagram of the large and small intestine junction.

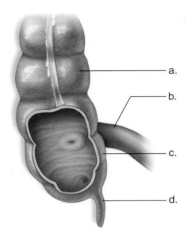

a.

b.

c.

d.

20. Which association is incorrect?

    a. mouth—starch digestion
    b. esophagus—protein digestion
    c. small intestine—starch, lipid, protein digestion
    d. stomach—food storage
    e. liver—production of bile

21. Label the following diagram of the hepatic lobule.

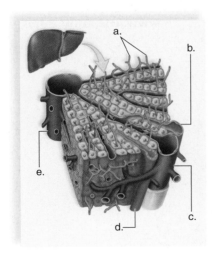

a.

b.

c.

d.

e.

22. Which of these substances could be absorbed directly without need of digestion?

    a. glucose   d. protein
    b. fat   e. nucleic acid
    c. polysaccharides

23. Peristalsis occurs

    a. from the mouth to the small intestine.
    b. from the beginning of the esophagus to the anus.
    c. only in the stomach.
    d. only in the small and large intestine.
    e. only in the esophagus and stomach.

24. Which association is incorrect?

    a. protein—trypsin   d. maltose—pepsin
    b. fat—bile   e. starch—amylase
    c. fat—lipase

25. Most of the products of digestion are absorbed across the

    a. squamous epithelium of the esophagus.
    b. striated walls of the trachea.
    c. convoluted walls of the stomach.
    d. fingerlike villi of the small intestine.
    e. smooth wall of the large intestine.

26. The amino acids that must be consumed in the diet are called essential. Nonessential amino acids

    a. can be produced by the body.
    b. are needed only occasionally.
    c. are stored in the body until needed.
    d. are required in smaller amounts.

27. Which of the following are often organic portions of important coenzymes?

    a. minerals
    b. vitamins
    c. proteins
    d. carbohydrates

28. Bulimia nervosa is not characterized by

    a. a restrictive diet often with excessive exercise.
    b. binge eating followed by purging.
    c. an obsession with body shape and weight.
    d. a distorted body image, so that the person feels fat even when emaciated.
    e. a health risk due to this complex.

29. Intestinal enzymes, such as maltase, are

    a. secreted by the intestine into the chyme.
    b. produced by the intestinal glands.
    c. produced by the pancreas.
    d. attached to the plasma membrane of microvilli in the epithelial cells of the mucosa.

Understanding the Terms

nasopharynx 262

nutrient 270

nutrition 270

obesity 278

osteoporosis 277

pancreas 267

pancreatic amylase 267

pancreatitis 280

pepsin 264

peptidase 269

peristalsis 262

peritonitis 266

pharynx 262

plaque 273

polyp 280

rectum 266

reflex action 262

rugae 264

salivary amylase 261

salivary gland 261

small intestine 264

soft palate 260

sphincter 263

stomach 264

trypsin 267

ulcer 279

vermiform appendix 266

villus 264

vitamin 273

Match the terms to these definitions:

a._____ First portion of the small intestine into which secretions from the liver and pancreas enter.

b._____ Muscle that surrounds a tube and closes or opens the tube by contracting and relaxing.

c._____ Discharge of feces from the rectum through the anus.

d._____ Fat-digesting enzyme secreted by the pancreas.

e._____ Saclike organ associated with the liver that stores and concentrates bile.

Thinking Critically

1. Suppose that, starting today, your body would lose the ability to perform either mechanical digestion or chemical digestion. Which type would you choose to keep and why?

2. Trace *all* of the types and locations of enzymes that might be involved in digesting a molecule of a complex carbohydrate such as starch, all the way through to the utilization of a glucose molecule by a body cell.

3. Suppose that, as you are swallowing a small piece of apple, your epiglottis malfunctions, and the apple "goes down the wrong pipe." If the piece is small enough to make it into one of your bronchial tubes, but doesn't obstruct your airway in any significant way, how long do you think it could stay there?

4. Pancreatic cancer is one of the most lethal cancers, with a survival rate of only about 5%. Why is the pancreas such a vital organ?

5. Suppose you are taking large doses of creatine, an amino acid supplement that is advertised for its ability to enhance muscle growth. Since your muscles can only grow at a certain rate, what do you suppose happens to all the creatine that is not used for synthesis of new muscle?

ARIS, the *Inquiry into Life* Website

ARIS, the website for *Inquiry into Life,* provides a wealth of information *Inquiry into Life* organized and integrated by chapter. You will find practice quizzes, interactive activities, labeling exercises, flashcards, and much more that will complement your learning and understanding of general biology.

www.aris.mhhe.com

Imagine that you are having lunch with some friends *when one of them suddenly seems unable to take a breath or talk. As he starts to turn red and points frantically to this throat, you realize it's no laughing matter. He begins to lose consciousness from lack of oxygen. By now you are out of your seat. You grab your friend from behind around his waist. You make a fist, with your thumb just below his ribcage, and grasp your fist with your other hand. Then you quickly pull both hands into his upper abdomen. After a couple of these thrusts, your friend expels the food that got caught in his windpipe and begins gasping for breath.*

Most of us learned about the Heimlich maneuver in grade school. Dr. Henry J. Heimlich published his initial report on this procedure in 1974, and a choking victim was saved by the method just one week later. Certainly thousands more potential fatalities due to choking have been averted using this method.

Because it is simple to perform and requires no special training, the Heimlich maneuver can be used even in the most remote locations. As documented in the 1998 Nova television program "Everest: The Death Zone," fellow climbers probably saved David Carter's life by performing the procedure on him when mucus generated by a respiratory infection began to obstruct his breathing at the extremely high altitude. The expedition leader, Ed Viesturs, performed several Heimlich maneuvers on Carter, which relieved the obstruction and allowed him to breathe.

Whether you are sitting in a restaurant in your hometown, or on the highest point on Earth, the exchange of gases by your respiratory system is absolutely essential to life.

chapter **15**

Respiratory System

C h a p t e r   C o n c e p t s

15.1   The Respiratory System

The primary function of the respiratory system is to allow oxygen from the air to enter the blood and carbon dioxide from the blood to exit into the air. During **inspiration,** or inhalation (breathing in), and **expiration,** or exhalation (breathing out), air is conducted toward or away from the lungs by a series of cavities, tubes, and openings (Fig. 15.1). **Ventilation,** another term for breathing, encompasses both inspiration and expiration.

The respiratory system also works with the cardiovascular system to accomplish these functions:

1. External respiration, the exchange of gases (oxygen and carbon dioxide) between air and the blood
2. Transport of gases to and from the lungs and the tissues.
3. Internal respiration, the exchange of gases between the blood and tissue fluid

As described in Chapter 7, cellular respiration, which produces ATP, uses the oxygen and gives off the carbon dioxide that makes gas exchange with the environment necessary. Without a continuous supply of ATP, the cells cease to function.

The Respiratory Tract

Table 15.1 traces the path of air from the nose to the lungs. As air moves in along the airways, it is cleansed, warmed, and moistened. Cleansing is accomplished by coarse hairs just inside the nostrils and by cilia and mucus in the nasal cavities and the other airways of the respiratory tract. In the nose, the hairs and the cilia act as screening devices. In the trachea and other airways, the cilia beat upward, carrying mucus, dust, and occasional bits of food that "went down the wrong way" into the pharynx, where the accumulation can be swallowed or expelled. The air is warmed by heat given off by the blood

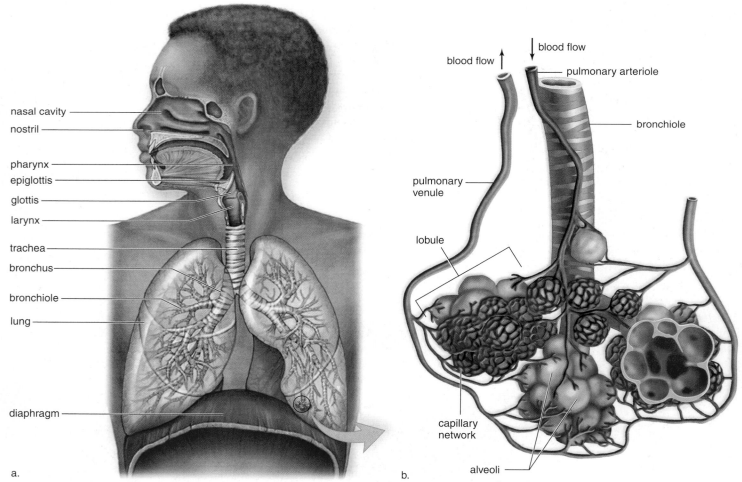

a.

b.

Figure 15.1 The respiratory tract.

a. The respiratory tract extends from the nose to the lungs, which are composed of air sacs called alveoli (arrow). **b.** Gas exchange occurs between air in the alveoli and blood within a capillary network that surrounds the alveoli. Notice that the pulmonary arteriole is colored blue—it carries O_2-poor blood away from the heart to the alveoli. Then carbon dioxide leaves the blood, and oxygen enters the blood. The pulmonary venule is colored red—it carries O_2-rich blood from alveoli toward the heart.

TABLE 15.1	Path of Air	
Structure	**Description**	**Function**
Upper Respiratory Tract		
Nasal cavities	Hollow spaces in nose	Filter, warm; and moisten air
Pharynx	Chamber posterior to oral cavity; lies between nasal cavity and larynx	Connection to surrounding regions
Glottis	Opening into larynx	Passage of air into larynx
Larynx	Cartilaginous organ that houses the vocal cords; voice box	Sound production
Lower Respiratory Tract		
Trachea	Flexible tube that connects larynx with bronchi	Passage of air to bronchi
Bronchi	Paired tubes inferior to the trachea that enter the lungs	Passage of air to lungs
Bronchioles	Branched tubes that lead from bronchi to alveoli	Passage of air to each alveolus
Lungs	Soft, cone-shaped organs that occupy lateral portions of thoracic cavity	Contain alveoli and blood vessels
Alveoli	Thin-walled microscopic air sacs in lungs	Gas exchange between air and blood

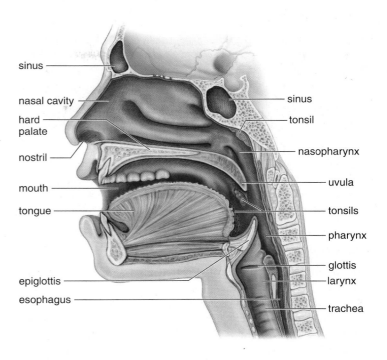

Figure 15.2 The path of air.
Air passes through the nasal cavities and mouth to and from the upper and lower respiratory tracts. The trachea is part of the lower respiratory tract; the other organs are in the upper respiratory tract.

vessels lying close to the surface of the lining of the airways, and it is moistened by the wet surface of these passages.

Conversely, as air moves out during expiration, it cools and loses its moisture. As the air cools, it deposits its moisture on the lining of the trachea and the nose, and the nose may even drip as a result of this condensation. The air still retains so much moisture, however, that upon expiration on a cold day, it condenses and forms a small cloud.

The Nose

The nose is a part of the upper respiratory tract, which contains the nasal cavities, the pharynx, and the larynx (Fig. 15.2). The nose, a prominent feature of the face, is the only external portion of the respiratory system. Air enters the nose through external openings called **nostrils.** The nose contains two **nasal cavities,** which are narrow canals separated from one another by a septum composed of bone and cartilage. Mucous membrane lines the nasal cavities. Bony ridges that project laterally into the nasal cavity increase the surface area for moistening and warming air during inhalation and for trapping water droplets during exhalation. Odor receptors are on the cilia of cells located high in the recesses of the nasal cavities.

The tear (lacrimal) glands drain into the nasal cavities by way of tear ducts. For this reason, crying produces a runny nose. The nasal cavities also communicate with **sinuses,** air-filled spaces that reduce the weight of the

skull and act as resonating chambers for the voice. The nasal cavities are separated from the mouth by a partition called the palate, which has two portions. Anteriorly, the hard palate is supported by bone, while posteriorly the soft palate is not supported by bone.

The Pharynx

The **pharynx** is a funnel-shaped passageway that connects the nasal and oral cavities to the larynx. Consequently, the pharynx, commonly referred to as the "throat," has three parts: the nasopharynx, where the nasal cavities open posterior to the soft palate; the oropharynx, where the mouth opens; and the laryngopharynx, which opens into the larynx. The soft palate has a soft extension called the uvula projecting into the oropharynx, which you can see by looking into your throat using a mirror.

The tonsils form a protective ring at the junction of the mouth and the pharynx. The tonsils are lymphatic tissue containing lymphocytes that protect against invasion of inhaled bacteria and viruses.

In the pharynx, the air passage and the food passage cross because the larynx, which receives air, lies above the esophagus, which receives food. The larynx leads to the trachea. Both the larynx and the trachea are normally open, allowing air to pass, but the esophagus is normally closed and opens only when a person swallows.

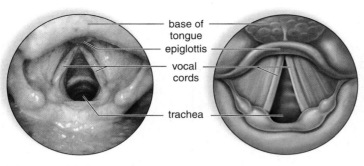

Figure 15.3 Placement of the vocal cords.
Viewed from above, the vocal cords stretch across the glottis, the opening
to the trachea. When air is expelled through the glottis, the vocal cords
vibrate, producing sound. The glottis is narrow when we make a high-
pitched sound, and it widens as the pitch deepens.

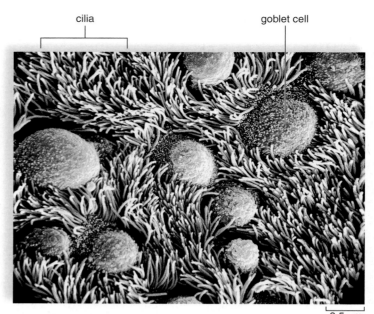

Figure 15.4 The surface of the trachea.
A scanning electron micrograph shows that the surface of the mucous
membrane lining the trachea consists of goblet cells and ciliated cells. The
cilia sweep mucus and the debris embedded in it toward the pharynx,
where they are swallowed or expectorated. Smoking causes the cilia to
disappear, allowing debris to enter the bronchi and lungs.

© Dr. Kessel & Dr. Kardon/Tissues & Organs/Visuals Unlimited

The Larynx

The **larynx** is a cartilaginous structure that serves as a passageway for air between the pharynx and the trachea. The
larynx can be pictured as a triangular box whose apex, the
Adam's apple, is at the front of the neck. The larynx is called
the voice box because it houses the vocal cords. The **vocal
cords** are mucosal folds supported by elastic ligaments, and
the slit between the vocal cords is an opening called the **glottis**
(Fig. 15.3). When air is expelled past the vocal cords through
the glottis, the vocal cords vibrate, producing sound. At the
time of puberty, the growth of the larynx and the vocal cords
is much more rapid and accentuated in the male than in the
female, causing the male to develop a more prominent Adam's
apple and a deeper voice. The voice "breaks" in the young
male due to his inability to control the longer vocal cords.

The high or low pitch of the voice is regulated when
speaking and singing by changing the tension on the vocal
cords. The greater the tension, as when the glottis becomes
more narrow, the higher the pitch. The loudness, or intensity, of the voice depends upon the amplitude of the vibrations—that is, the degree to which the vocal cords vibrate.

When food is swallowed, the larynx moves upward
against the **epiglottis,** a flap of tissue that prevents food
from passing into the larynx (Fig. 15.3). You can detect this
movement by placing your hand gently on your larynx and
swallowing.

The Trachea

The **trachea,** commonly called the windpipe, is a tube connecting the larynx to the primary bronchi. The trachea lies
above the esophagus and is held open by C-shaped cartilaginous rings. The open part of the C-shaped rings faces
the esophagus, and this allows the esophagus to expand
when swallowing. The mucosa that lines the trachea has a
layer of pseudostratified ciliated columnar epithelium (see
Chapter 11). The cilia that project from the epithelium keep
the lungs clean by sweeping mucus, produced by goblet
cells, and debris toward the pharynx, (Fig. 15.4). Smoking
is known to destroy these cilia, and consequently, the toxins

in cigarette smoke collect in the lungs. Smoking is discussed
more fully in the Health Focus later in this chapter.

The Bronchial Tree

The trachea divides into right and left primary bronchi
(sing., **bronchus**), which lead into the right and left lungs
(see Fig. 15.1). The bronchi branch into a great number of
secondary bronchi that eventually lead to **bronchioles.**
The bronchi resemble the trachea in structure, but as the
bronchial tubes divide and subdivide, their walls become
thinner, and the small rings of cartilage are no longer present. Each bronchiole leads to an elongated space enclosed
by a multitude of air pockets, or sacs, called alveoli (sing.,
alveolus). The components of the bronchial tree beyond the
primary bronchi compose the lungs.

The Lungs

The **lungs** are paired, cone-shaped organs that occupy the
thoracic cavity except for a central area that contains the trachea, the heart, and the esophagus. The right lung has three
lobes, and the left lung has two lobes, allowing room for
the heart, whose apex points left. A lobe is further divided
into lobules, and each lobule has a bronchiole serving many
alveoli. The apex of a lung is narrow, while the base is broad
and curves to fit the dome-shaped diaphragm, the muscle
that separates the thoracic cavity from the abdominal cavity. The other surfaces of the lungs follow the contours of
the ribs in the thoracic cavity.

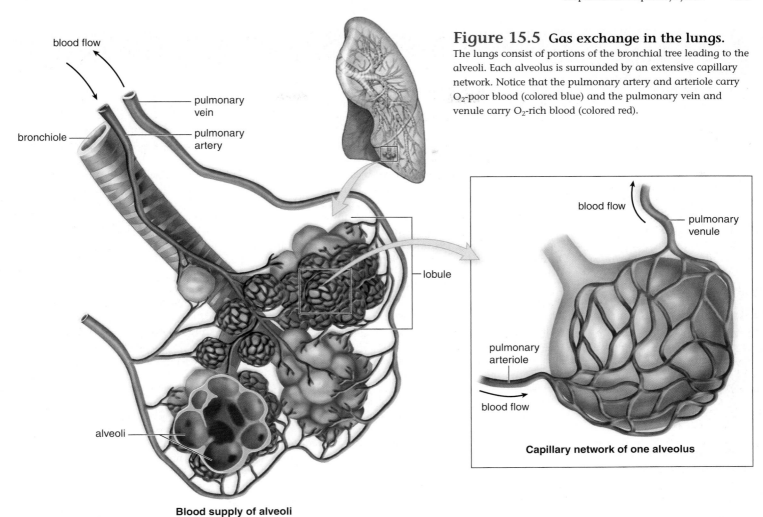

Figure 15.5 Gas exchange in the lungs.
The lungs consist of portions of the bronchial tree leading to the alveoli. Each alveolus is surrounded by an extensive capillary network. Notice that the pulmonary artery and arteriole carry O_2-poor blood (colored blue) and the pulmonary vein and venule carry O_2-rich blood (colored red).

blood flow

pulmonary vein

bronchiole

pulmonary artery

lobule

alveoli

blood flow

pulmonary venule

pulmonary arteriole

blood flow

Capillary network of one alveolus

Blood supply of alveoli

Each lung is covered by a very thin serous membrane called a **pleura** (see Fig. 15.7a). Another pleura covers the internal chest wall and diaphragm; both membranes produce a lubricating serous fluid that helps the pleurae slide freely against each other during inspiration and expiration. Surface tension is the tendency for water molecules to cling to each other due to hydrogen bonding between the molecules. Surface tension holds the two pleural layers together when the lungs recoil during expiration.

With each inhalation, air passes by way of the respiratory passageways to the alveoli. An alveolus is made up of simple squamous epithelium surrounded by blood capillaries. Gas exchange occurs between the air in an alveolus and the blood in the capillaries (Fig. 15.5). Oxygen diffuses across the alveolar and capillary walls to enter the bloodstream, while carbon dioxide diffuses from the blood across these walls to enter the alveoli.

If gas exchange is to occur, the alveoli must stay open to receive the inhaled air. Gas exchange takes place across moist cellular membranes, and yet the surface tension of water lining the alveoli is capable of causing them to close up. The alveoli are lined with **surfactant,** a film of lipoprotein that lowers the surface tension and prevents them from closing. The lungs collapse in some newborn babies, especially premature infants, who lack this film. The condition, called **infant respiratory distress syndrome,** is now treatable by surfactant replacement therapy.

The major function of the respiratory system is to facilitate the exchange of oxygen and carbon dioxide between the air and the blood. The respiratory tract consists of the nose, pharynx, larynx, trachea, bronchial tree, and lungs.

15.2 Mechanism of Breathing

During ventilation (breathing), air first moves into the lungs from the nose or mouth and then moves out of the lungs. A free flow of air is vitally important; therefore, a technique has been developed that allows physicians to detect any medical problem that prevents the lungs from filling with air and releasing it. An instrument called a spirometer records the volume of air exchanged during both

Most industrialized cities have photochemical smog, at least occasionally. Photochemical smog arises when primary pollutants react with one another under the influence of sunlight to form a more deadly combination of chemicals. For example, two primary pollutants, nitrogen oxides (NO_x) and volatile organic compounds (VOCs) including hydrocarbons, as well as alcohols, aldehydes, and ethers, react with one another in the presence of sunlight to produce nitrogen dioxide (NO_2), ozone (O_3), and PAN (peroxy-acetylnitrate). Ozone and PAN are commonly referred to as oxidants. Breathing oxidants affects the respiratory and nervous systems, resulting in respiratory distress, headache, and exhaustion.

Cities with warm, sunny climates that are large and industrialized, such as Los Angeles, Denver, and Salt Lake City in the United States, Sydney in Australia, Mexico City in Mexico, and Buenos Aires in Argentina, are particularly susceptible to photochemical smog. If the city is surrounded by hills, a thermal inversion may aggravate the situation. Normally, warm air near the ground rises, so that pollutants are dispersed and carried away by air currents. But sometimes during a thermal inversion, smog gets trapped near the Earth by a blanket of warm air (Fig. 15A). This may occur when a cold front brings in cold air, which settles beneath a warm layer. The trapped pollutants cannot disperse, and the results are dangerous to a person's respiratory health. Even healthy adults experience a reduction in lung capacity when exposed to photochemical smog for long periods or during vigorous outdoor activities. Repeated exposures to high concentrations of ozone are associated with respiratory problems, such as an increased rate of lung infections and permanent lung damage. Children, the elderly, asthmatics, and individuals with emphysema or other similar disorders are particularly at risk.

Even though federal legislation is in place to bring air pollution under control, more than half the people in the United States live in cities polluted by too much smog. In the long run, pollution prevention is usually easier and cheaper than pollution cleanup. Some prevention suggestions are as follows:

- Encourage use of public transportation and burn fuels that do not produce pollutants.
- Increase recycling in order to reduce the amount of waste that is incinerated.
- Reduce energy use so that power plants need to provide less.
- Use renewable energy sources, such as solar, wind, or water power.

- Require industries to meet clean-air standards.

Discussion Questions

1. One of the most significant sources of the pollutants that contribute to photochemical smog is exhaust from automobiles. The use of catalytic converters on car exhaust systems is one way to reduce the level of pollutants in the air. Suppose a car manufacturer found a way to completely eliminate these emissions from car exhaust. How much extra would you be willing to pay for a car with this device installed? $100? $1,000? Regardless of the cost, do you think these devices should be required on all new cars, even for people who live in less populated areas?
2. Why do you think exposure to ozone causes more serious respiratory problems in the young and the elderly than in healthy adults?
3. How often do you use public transportation? Whether you live in a big city or not, would you be willing to use public transportation if it added, for example, an hour to the total time you spend each day traveling to school or work? Why or why not?

a. Ground-level ozone formation

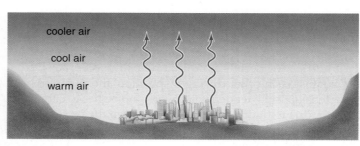

b. Normal pattern

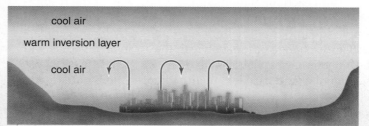

c. Thermal inversion

Figure 15A Thermal inversion.

a. Los Angeles is the "air pollution capital" of the world. Its millions of cars and thousands of factories make this city particularly susceptible to photochemical smog, which contains ozone due to the chemical reaction shown here. b. Normally, pollutants escape into the atmosphere when warm air rises. c. During a thermal inversion, a layer of warm air (warm inversion layer) overlies and traps pollutants in cool air below.

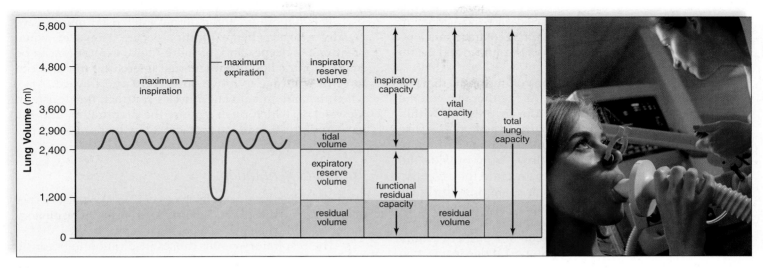

Figure 15.6 Measuring ventilation.

A spirometer measures the amount of air inhaled and exhaled with each breath. During inspiration, the pen moves up, and during expiration, the pen moves down. Vital capacity (red) is the maximum amount of air a person can exhale after taking the deepest inhalation possible.

normal and deep breathing. A spirogram shows the measurements recorded by a spirometer when a person breathes as directed by a technician (Fig. 15.6).

Respiratory Volumes

Normally when we are relaxed, only a small amount of air moves in and out with each breath. This amount of air, called the **tidal volume,** is only about 500 ml.

It is possible to increase the amount of air inhaled, and therefore the amount exhaled, by deep breathing. The maximum volume of air that can be moved in plus the maximum volume that can be moved out during a single breath is the **vital capacity.** It is called vital capacity because your life depends on breathing, and the more air you can move, the better off you are. A number of different illnesses discussed in section 15.4 can decrease vital capacity.

To increase your vital capacity, you must increase both the amount of air you breathe in and the amount you breathe out. You can increase inspiration (breathing in) by not only expanding the chest but also by lowering the diaphragm. Such forced inspiration usually increases the volume of air beyond the tidal volume by about 2,900 ml, and that amount is called the **inspiratory reserve volume.** You can increase the amount of air expired (breathed out) by contracting the abdominal and internal intercostal muscles. This so-called **expiratory reserve volume** is usually about 1,400 ml of air. You can see from Figure 15.6 that vital capacity is the sum of the tidal, inspiratory reserve, and expiratory reserve volumes.

In an average adult, only about 70% of the tidal volume actually reaches the alveoli; 30% remains in the airways. These passages are not used for gas exchange, and therefore, they are said to contain dead-space air. To ensure that a large portion of inhaled air reaches the lungs, it is better to breathe slowly and deeply. Also, note in Figure 15.6 that even after a very deep exhalation, some air (about 1,000

ml) remains in the lungs; this is called the **residual volume.** This air is not as useful for gas exchange because it has been depleted of oxygen. In some lung diseases, the residual volume builds up because the individual has difficulty emptying the lungs. This means that the vital capacity is reduced because the lungs have more residual volume.

> Normally, the amount of air that moves in and out of the lungs during each breath is much less than the maximum amount that can be inhaled and exhaled. There is always some air left in the lungs, even after maximum exhalation.

Inspiration and Expiration

To understand ventilation, the manner in which air enters and exits the lungs, it is necessary to remember the following facts:

1. Normally, there is a continuous column of air from the pharynx to the alveoli of the lungs.
2. The lungs lie within the sealed-off thoracic cavity. The rib cage, consisting of the ribs joined to the vertebral column posteriorly and to the sternum anteriorly, forms the top and sides of the thoracic cavity. The intercostal muscles lie between the ribs. The diaphragm and connective tissue form the floor of the thoracic cavity.
3. The lungs adhere to the thoracic wall by way of the pleura. Normally, any space between the two pleurae is minimal due to the surface tension of the fluid between them.

Inspiration

Inspiration is the active phase of ventilation because this is the phase in which the diaphragm and the external intercostal muscles contract (Fig. 15.7a). In its relaxed state, the diaphragm is dome shaped; during deep inspiration, it contracts and lowers. Also, the external intercostal muscles contract, and the rib cage moves upward and outward.

Following this contraction, the volume of the thoracic cavity is larger than it was before. As the thoracic volume increases, the lungs expand. Now the air pressure within the alveoli decreases, creating a partial vacuum. Because alveolar pressure is now less than atmospheric pressure outside the lungs, air naturally flows from outside the body into the respiratory passages and into the alveoli.

It is important to realize that air comes into the lungs because they have already opened up; air does not force the lungs open. This is why it is sometimes said that *humans inhale by negative pressure.* The creation of a partial vacuum in the alveoli causes air to enter the lungs. While inspiration is the active phase of breathing, the actual flow of air into the alveoli is passive.

Expiration

Usually, expiration is the passive phase of breathing, and no effort is required to bring it about. During expiration, the elastic properties of the thoracic wall and lungs cause them to recoil. In addition, the lungs recoil because the surface tension of the fluid lining the alveoli tends to draw them closed. During expiration, the abdominal organs press up against the diaphragm, and the rib cage moves down and inward (Fig. 15.7b).

The diaphragm and external intercostal muscles are usually relaxed when expiration occurs. However, when breathing is deeper and/or more rapid, expiration can be active. Contraction of the internal intercostal muscles can force the rib cage to move downward and inward. Also, when the abdominal wall muscles contract, they push on the viscera, which push against the diaphragm, and the increased pressure in the thoracic cavity helps expel air.

Control of Ventilation

Normally, adults have a breathing rate of 12 to 20 ventilations per minute. The rhythm of ventilation is controlled by a **respiratory center** located in the medulla oblongata of the brain.

The respiratory center causes inspiration to occur by automatically sending impulses to the diaphragm by way of the phrenic nerve, and to the intercostal muscles by way of the intercostal nerves (Fig. 15.8). When the respiratory center stops sending neuronal signals to the diaphragm and the rib cage, the diaphragm relaxes and resumes its dome shape. Now expiration occurs.

Although the respiratory center automatically controls the rate and depth of breathing, its activity can also be influenced by nervous input and chemical input. Following forced inhalation, stretch receptors in the alveolar walls initiate inhibitory nerve impulses that travel via the vagus nerve from the inflated lungs to the respira-

**Figure 15.7
Inspiration and expiration compared.**
a. During inspiration, the thoracic cavity and lungs expand so that air is drawn in.
b. During expiration, the thoracic cavity and lungs resume their original positions and pressures. Now, air is forced out.

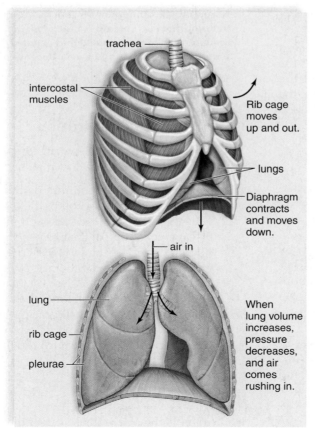

trachea

intercostal muscles

Rib cage moves up and out.

lungs

Diaphragm contracts and moves down.

air in

lung

rib cage

pleurae

When lung volume increases, pressure decreases, and air comes rushing in.

a. Inspiration

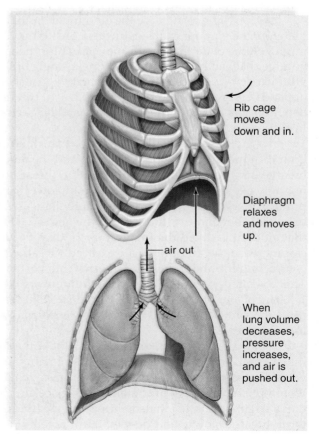

Rib cage moves down and in.

Diaphragm relaxes and moves up.

air out

When lung volume decreases, pressure increases, and air is pushed out.

b. Expiration

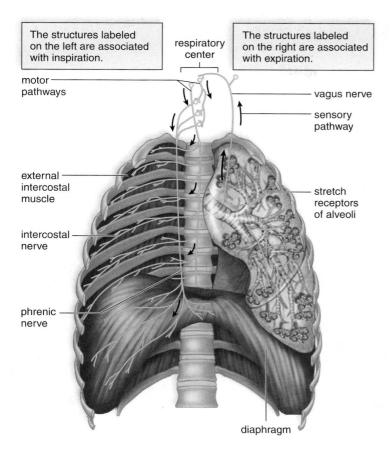

The structures labeled on the left are associated with inspiration.

respiratory center

The structures labeled on the right are associated with expiration.

motor pathways

vagus nerve

sensory pathway

external intercostal muscle

stretch receptors of alveoli

intercostal nerve

phrenic nerve

diaphragm

Figure 15.8 Nervous control of breathing.

The respiratory center automatically stimulates the external intercostal (rib) muscles and diaphragm to contract via the phrenic nerve. After forced inhalation, stretch receptors send inhibitory nerve impulses to the respiratory center via the vagus nerve. Usually, expiration automatically occurs due to lack of stimulation from the respiratory center to the diaphragm and intercostal muscles.

tory center. This stops the respiratory center from sending out nerve impulses.

Chemical Input The respiratory center is directly sensitive to the levels of carbon dioxide (CO_2) and hydrogen ions (H^+) in the blood. When they rise, the respiratory center increases the rate and depth of breathing. The center is not affected directly by low oxygen (O_2) levels. However, chemoreceptors in the **carotid bodies,** located in the carotid arteries, and in the **aortic bodies,** located in the aorta, are sensitive to the level of oxygen in the blood. When the concentration of oxygen decreases, these bodies communicate with the respiratory center, and the rate and depth of breathing increase.

During inspiration, due to nervous stimulation, the diaphragm lowers, and the rib cage lifts up and out.
During expiration, due to the lack of nervous stimulation, the diaphragm rises, and the rib cage lowers.

15.3 Gas Exchanges in the Body

The air passages of the respiratory system are simply conduits, and this system is absolutely essential to homeostasis, only because it allows gas exchange to occur. As mentioned previously, respiration includes not only the exchange of gases in the lungs (external respiration), but also the exchange of gases in the tissues (internal respiration) (Fig. 15.9). The principles of diffusion, alone, govern whether O_2 or CO_2 enters or leaves the blood in the lungs and in the tissues.

External Respiration

External respiration refers to the exchange of gases between air in the alveoli and blood in the pulmonary capillaries (see Fig. 15.5). Gases exert pressure, and the amount of pressure each gas exerts is called its partial pressure, symbolized as P_{O_2} and P_{CO_2}. Blood in the pulmonary capillaries has a higher P_{CO_2} than atmospheric air. Therefore, *CO_2 diffuses out of the plasma into the lungs.* Most of the CO_2 is carried as **bicarbonate ions** (HCO_3^-). As the little remaining free CO_2 begins to diffuse out, the following reaction is driven to the right:

$$H^+ + HCO_3^- \longrightarrow H_2CO_3 \xrightarrow{\text{carbonic anhydrase}} H_2O + CO_2$$

| hydrogen ion | bicarbonate ion | carbonic acid | water | carbon dioxide |

The enzyme **carbonic anhydrase,** present in red blood cells, speeds the breakdown of carbonic acid (H_2CO_3).

What happens if you hyperventilate (breathe at a high rate) and therefore push this reaction far to the right? The blood has fewer hydrogen ions, and alkalosis, a high blood pH, results. In that case, breathing will be inhibited, but in the meantime, you may suffer various symptoms, from dizziness to tetanic contractions of the muscles. What happens if you hypoventilate (breathe at a low rate) and this reaction does not occur? Hydrogen ions build up in the blood, and acidosis occurs. Buffers may compensate for the low pH, and breathing will most likely increase. Otherwise, you may become comatose and die.

The pressure pattern for O_2 during external respiration is the reverse of that for CO_2. Blood in the pulmonary capillaries is low in oxygen, and alveolar air contains a higher partial pressure of oxygen. Therefore, *O_2 diffuses into plasma and then into red blood cells in the lungs.* Hemoglobin takes up this oxygen and becomes **oxyhemoglobin** (HbO_2):

$$Hb + O_2 \longrightarrow HbO_2$$

| hemoglobin | oxygen | oxyhemoglobin |

Interestingly, only a relatively small percentage of the oxygen present in atmospheric air is utilized during normal external respiration. At sea level, air contains about

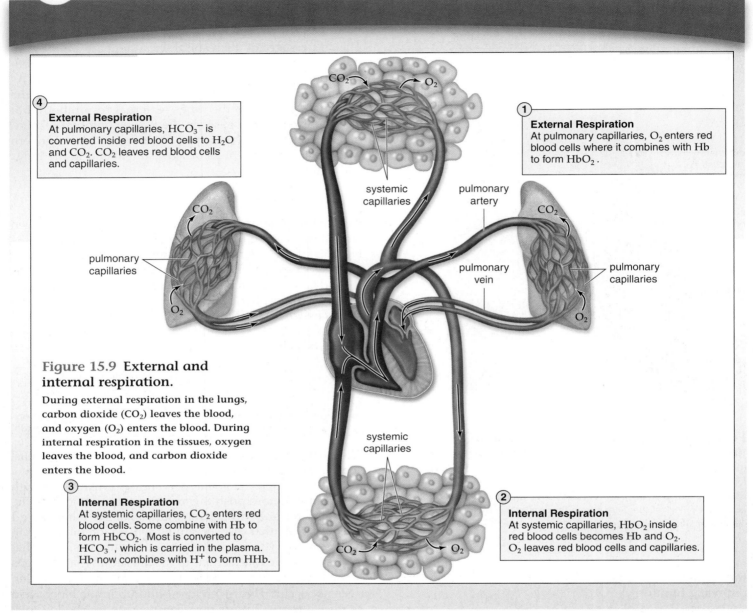

4 **External Respiration**
At pulmonary capillaries, HCO_3^- is converted inside red blood cells to H_2O and CO_2. CO_2 leaves red blood cells and capillaries.

1 **External Respiration**
At pulmonary capillaries, O_2 enters red blood cells where it combines with Hb to form HbO_2.

systemic capillaries

pulmonary artery

pulmonary capillaries

pulmonary vein

pulmonary capillaries

Figure 15.9 External and internal respiration.

During external respiration in the lungs, carbon dioxide (CO_2) leaves the blood, and oxygen (O_2) enters the blood. During internal respiration in the tissues, oxygen leaves the blood, and carbon dioxide enters the blood.

systemic capillaries

3 **Internal Respiration**
At systemic capillaries, CO_2 enters red blood cells. Some combine with Hb to form $HbCO_2$. Most is converted to HCO_3^-, which is carried in the plasma. Hb now combines with H^+ to form HHb.

2 **Internal Respiration**
At systemic capillaries, HbO_2 inside red blood cells becomes Hb and O_2. O_2 leaves red blood cells and capillaries.

21% O_2, and exhaled air still contains 16–17% O_2. While this system may not seem very efficient, it does explain how a person whose heart and lungs have failed can sometimes be revived by mouth-to-mouth resuscitation, which involves filling the person's lungs with exhaled air.

External respiration is the exchange of gases between the alveoli and the blood in the pulmonary capillaries. There is more CO_2 in the pulmonary capillaries than in the air, so CO_2 diffuses out of the plasma and into the lungs. There is less O_2 in the pulmonary capillaries, so O_2 diffuses into the blood where it binds to hemoglobin.

Internal Respiration

Internal respiration refers to the exchange of gases between the blood in systemic capillaries and the tissue fluid. Therefore, internal respiration services tissue cells and without internal respiration, cells could not continue to produce the ATP that allows them to exist. Blood in the systemic capillaries is a bright red color because red blood cells contain oxyhemoglobin. Oxyhemoglobin gives up O_2, which diffuses out of the blood into the tissues:

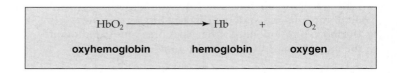

$$HbO_2 \longrightarrow Hb + O_2$$

oxyhemoglobin   **hemoglobin**   **oxygen**

Oxygen diffuses out of the blood into the tissues because the P_{O_2} of tissue fluid is lower than that of blood. The lower P_{O_2} is due to cells continuously using up oxygen in cellular respiration. *Carbon dioxide diffuses into the blood from the tissues* because the P_{CO_2} of tissue fluid is higher than that of blood. Carbon dioxide, produced continuously by cells, collects in tissue fluid.

After CO_2 diffuses into the blood, it enters the red blood cells, where a small amount is taken up by hemoglobin, forming **carbaminohemoglobin.** Most of the CO_2 combines with water, forming carbonic acid (H_2CO_3), which dissociates to hydrogen ions (H^+) and bicarbonate ions (HCO_3^-). The increased concentration of CO_2 in the blood drives the reaction to the right:

$$\underset{\substack{\text{carbon} \\ \text{dioxide}}}{CO_2} + \underset{\text{water}}{H_2O} \xrightarrow{\text{carbonic anhydrase}} \underset{\substack{\text{carbonic} \\ \text{acid}}}{H_2CO_3} \longrightarrow \underset{\substack{\text{hydrogen} \\ \text{ion}}}{H^+} + \underset{\substack{\text{bicarbonate} \\ \text{ion}}}{HCO_3^-}$$

The enzyme carbonic anhydrase, mentioned previously, speeds up the reaction. Bicarbonate ions diffuse out of red blood cells and are carried in the plasma. The globin portion of hemoglobin combines with excess hydrogen ions produced by the overall reaction, and Hb becomes HHb, called **reduced hemoglobin.** In this way, the pH of blood remains fairly constant. Blood that leaves the systemic capillaries is a dark maroon color because red blood cells contain reduced hemoglobin.

> **Internal respiration is the exchange of gases between the blood in the systemic capillaries and the tissues. Hemoglobin activity is essential for the transport of O_2, and also contributes to the transport of CO_2.**

15.4   Disorders of the Respiratory System

The respiratory tract is constantly exposed to the air in our environment, and thus it is susceptible to various infectious agents, as well as to pollution and, in some individuals, tobacco smoke. Disorders of the upper respiratory tract will be discussed before disorders of the lower respiratory tract.

Disorders of the Upper Respiratory Tract

The upper respiratory tract consists of the nasal cavities, the pharynx, and the larynx. Since it is responsible for filtering out many of the pathogens and other materials that may be present in the air, the upper respiratory tract is susceptible to a variety of viral and bacterial infections. Upper respiratory infections can also spread from these areas to the middle ear or the sinuses.

The Common Cold

Most "colds" are relatively mild infections of the upper respiratory tract characterized by sneezing, a runny nose, and perhaps a mild fever. Many different viruses can cause colds, the most common being a group called the rhinoviruses (*rhin* is Greek for nose). Most colds last from a few days to a week, when the immune response is able to eliminate the virus. Because colds are caused by viruses, antibiotics do not help, although decongestants and anti-inflammatory medications can ease the symptoms.

Pharyngitis, Tonsillitis, and Laryngitis

Pharyngitis is an inflammation of the throat, usually because of an infection. If the tonsils have not been removed previously, they often become involved as well. What we call "strep throat" is a bacterial infection caused by *Streptococcus pyogenes* that can lead to a generalized upper respiratory infection and even a systemic (affecting the body as a whole) infection. The symptoms of strep throat are severe sore throat, high fever, and white patches on a dark-red pharyngeal or tonsillar area (Fig. 15.10). Most cases of strep throat can be successfully treated with antibiotics.

Tonsillitis occurs when the **tonsils,** aggregates of lymphoid tissue in the pharynx, become inflamed and enlarged. The tonsils in the posterior wall of the nasopharynx are often called adenoids. If tonsillitis occurs frequently and the enlarged tonsils make breathing difficult, the tonsils can be removed surgically in a **tonsillectomy.** Fewer tonsillectomies are performed today than in the past because we now know that the tonsils help initiate immune responses to many of the pathogens that enter the pharynx; therefore, they are an important component of the body's immune system.

Laryngitis is an inflammation of the larynx with accompanying hoarseness, often leading to the inability to talk in an audible voice. Usually, laryngitis disappears after resting the vocal cords and treating any infection present.

Sinusitis

Sinusitis is an inflammation of the cranial sinuses, the cavities within the facial skeleton that drain into the nasal cavities. Sinusitis develops when nasal congestion blocks the tiny openings leading to the sinuses. Up to 10% of upper respiratory infections are accompanied by sinusitis, and allergies may also play a role. Symptoms may include postnasal discharge, headache, and facial pain that worsens when the patient bends forward. Successful treatment depends on restoring proper

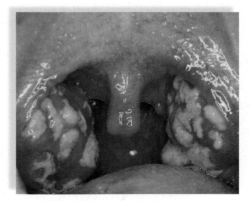

Figure 15.10
Strep throat.
Pharyngitis, often caused by the bacterium *S. pyogenes*, can lead to swollen, inflamed tonsils, sometimes with whitish patches.

drainage of the sinuses. Even a hot shower and sleeping with the head elevated can be helpful. Otherwise, spray decongestants are preferred over oral antihistamines, which may thicken rather than liquefy the material trapped in the sinuses.

Otitis Media

Otitis media is an inflammation of the middle ear. The middle ear is not a part of the respiratory tract, but this condition is considered here because, especially in children, nasal infections can spread to the ear by way of the **auditory (eustachian) tubes** that lead a from the nasopharynx to the middle ear. Pain is the primary symptom of otitis media. A sense of fullness, hearing loss, vertigo (dizziness), and fever may also be present. Antibiotics almost always bring about a full recovery, and recurrence is probably due to a new infection. Tubes (called tympanostomy tubes) are sometimes surgically placed in the eardrums of children with multiple recurrences to help prevent the buildup of pressure in the middle ear and the possibility of hearing loss (Fig. 15.11). Normally, the tubes fall out with time.

> The upper respiratory tract is prone to many different types of infections. Inflammations of several tissues of the upper respiratory tract, including the pharynx, tonsils, larynx, sinuses, and middle ears, comprise an important and common group of ailments.

Disorders of the Lower Respiratory Tract

Several disorders of the lower respiratory tract cause problems by obstructing normal air flow. Their causes range from a foreign object lodged in the trachea to excessive mucus produced in the bronchi. Other conditions tend to restrict the normal elasticity of the lung tissue itself (Fig. 15.12).

Disorders of the Trachea and Bronchi

One of the simplest, but most life-threatening disorders that affects the trachea is choking. As described in the story that opens this chapter, the best way for a person without extensive medical training to help someone who is choking is to perform the Heimlich maneuver. Trained medical personnel may also

Figure 15.11
A common treatment for middle ear infections.
Tympanostomy tubes may be surgically implanted in the eardrums to drain fluid from the middle ears of children with chronic otitis media.

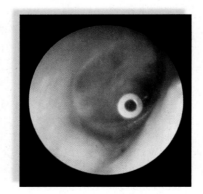

insert a breathing tube by way of an incision made in the trachea. The operation is called a tracheotomy, and the opening is a **tracheostomy.** Sometimes people whose larynx or trachea has been damaged or destroyed, usually as a result of smoking, must have a permanent tracheostomy tube installed.

Acute bronchitis is an inflammation of the primary and secondary bronchi. Usually it is preceded by a viral infection that has led to a secondary bacterial infection. Most likely, a nonproductive cough has become a deep cough that produces more mucus and perhaps pus (Fig. 15.12a). Typically this type of bacterial infection can be successfully treated with antibiotics.

In **chronic bronchitis,** the airways are inflamed and filled with mucus. A cough that brings up mucus is common. The bronchi have undergone degenerative changes, including the loss of cilia and their normal cleansing action. Under these conditions, an infection is more likely to occur. The most frequent cause of chronic bronchitis is smoking, although exposure to environmental pollutants can also be a contributing factor.

Asthma is a disease of the bronchi and bronchioles that is marked by wheezing, breathlessness, and sometimes a cough and expectoration of mucus. The airways are unusually sensitive to specific irritants, which can include a wide range of allergens such as pollen, animal dander, dust, cigarette smoke, and industrial fumes. Even cold air can be an irritant. When exposed to the irritant, the smooth muscle in the bronchioles undergoes spasms (Fig. 15.12b). Most asthma patients have some degree of bronchial inflammation that reduces the diameter of the airways and contributes to the seriousness of an attack. Asthma is not curable, but several types of drugs can either prevent or treat asthma attacks. One group of drugs, called the beta-agonists, works by dilating the bronchioles. Another group of drugs, the corticosteroids, can help control the inflammation and hopefully prevent an attack.

Diseases of the Lungs

Pneumonia is a viral or bacterial infection of the lungs in which the bronchi or alveoli fill with thick fluid (Fig. 15.12c). High fever and chills, with headache and chest pain, are symptoms of pneumonia. Rather than being a generalized lung infection, pneumonia may be localized in specific lobules of the lungs; obviously, the more lobules involved, the more serious is the infection. Pneumonia can be caused by several types of bacteria, viruses, and other infectious agents. Certain types of pneumonia mainly strike individuals with reduced immunity. For example, AIDS patients are subject to a particularly rare form of pneumonia caused by the fungus *Pneumocystis carinii.*

Pulmonary tuberculosis is caused by the bacterium *Mycobacterium tuberculosis.* When *M. tuberculosis* invade the lung tissue, the cells build a protective capsule around the invading bacteria, isolating them from the rest of the body. This tiny capsule is called a tubercle (Fig. 15.12d). If the body's resistance is high, the imprisoned organisms die, but if the resistance is low, the organisms eventually escape and spread. If a chest X ray detects active tubercles, the individual is put on appropriate drug therapy to ensure the localization of the disease and the eventual destruction of any live bacteria. It is

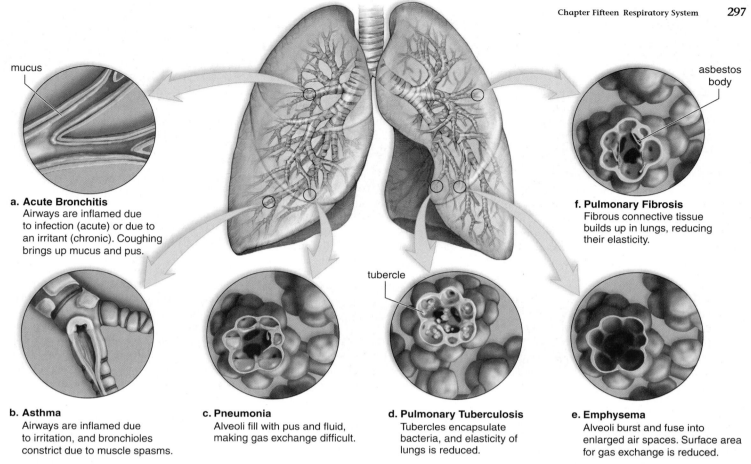

mucus

a. Acute Bronchitis
Airways are inflamed due
to infection (acute) or due to
an irritant (chronic). Coughing
brings up mucus and pus.

asbestos
body

f. Pulmonary Fibrosis
Fibrous connective tissue
builds up in lungs, reducing
their elasticity.

tubercle

b. Asthma
Airways are inflamed due
to irritation, and bronchioles
constrict due to muscle spasms.

c. Pneumonia
Alveoli fill with pus and fluid,
making gas exchange difficult.

d. Pulmonary Tuberculosis
Tubercles encapsulate
bacteria, and elasticity of
lungs is reduced.

e. Emphysema
Alveoli burst and fuse into
enlarged air spaces. Surface area
for gas exchange is reduced.

Figure 15.12 Common bronchial and pulmonary diseases.
Exposure to infectious pathogens and/or polluted air, including tobacco smoke, causes various diseases and disorders.

possible to tell if a person has ever been infected with tuberculosis bacteria with a TB skin test, in which a highly diluted extract of the bacteria is injected into the skin of the patient. A person who has never been infected with *M. tuberculosis* shows no reaction, but one who has had or is fighting an infection shows an area of inflammation that peaks in about 48 hours.

Emphysema is a chronic and incurable disorder in which the alveoli are distended and their walls damaged so that the surface area available for gas exchange is reduced (Fig. 15.12*e*). Emphysema is often preceded by chronic bronchitis. Air trapped in the lungs leads to alveolar damage and a noticeable ballooning of the chest. The elastic recoil of the lungs is reduced, so not only are the airways narrowed, but the driving force behind expiration is also reduced. The victim often feels out of breath and may have a cough. Because the surface area for gas exchange is reduced, less oxygen reaches the heart and the brain. Even so, the heart works furiously to force more blood through the lungs, and an increased workload on the heart can result. Because they are often seen in the same patient and tend to recur, emphysema, chronic bronchitis, and asthma are collectively called chronic obstructive pulmonary disease (COPD). COPD is the fourth leading cause of death in the United States, and is usually associated with smoking.

Cystic fibrosis (CF) is an example of a lung disease that is genetic, rather than infectious, although infections do play a role in the disease. One in 31 Americans carries the defective gene, but a child must inherit two copies of the faulty gene to have the disease. Still, CF is the most common inherited disease in the U.S. white population. The gene that is defective in CF codes for cystic fibrosis transmembrane regulator (CTFR), a protein needed for proper transport of Cl^- ions out of the epithelial cells of the lung. Because this also reduces the amount of water transported out of the lung cells, the mucus secretions become very sticky and can form plugs that interfere with breathing. Symptoms of CF include coughing and shortness of breath, and part of the treatment involves clearing mucus from the airways by vigorously clapping the patient on the back as well as by administering mucus-thinning drugs. None of these treatments is curative, however, and because the lungs can be severely affected, the median survival age for people with CF is only 30 years. Researchers are attempting to develop gene therapy strategies to replace the faulty CTFR gene.

Another common lung disease is **pulmonary fibrosis,** in which fibrous connective tissue builds up in the lungs, causing a loss of elasticity (Fig. 15.12*f*). This restricts the ability of the lungs to expand during inhalation, and so reduces the vital capacity and other lung volumes. Pulmonary fibrosis most commonly occurs in elderly persons, and the risk is increased by environmental exposure to silica (sand), various types of dust, and asbestos. Another environmental hazard that can lead to lung disease is photochemical smog (see Ecology Focus).

Health Focus

Is there a safe way to smoke?

No. All forms of tobacco can cause damage, and smoking even a small amount is dangerous. Tobacco is perhaps the only legal product whose advertised and intended use—that is, smoking it—will hurt the body.

Does smoking cause cancer?

Yes, and not only lung cancer. Besides lung cancer, smoking a pipe, cigarettes, or cigars is also a major cause of cancers of the mouth, larynx (voice box), and esophagus. In addition, smoking increases the risk of cancer of the bladder, kidney, pancreas, stomach, and uterine cervix.

What are the chances of being cured of lung cancer?

Very low; the five-year survival rate is only 13%. Fortunately, lung cancer is a largely preventable disease. In other words, by not smoking, it can probably be prevented.

Does smoking cause other lung diseases?

Yes. Smoking leads to chronic bronchitis, a disease in which the airways produce excess mucus, forcing the smoker to cough frequently. Smoking is also the major cause of emphysema, a disease that slowly destroys a person's ability to breathe.

Why do smokers have "smoker's cough"?

Normally, cilia (tiny hairlike formations that line the airways) beat outward and "sweep" harmful material out of the lungs. Smoke, however, decreases this sweeping action, so some of the poisons in the smoke remain in the lungs.

If you smoke but don't inhale, is there any danger?

Yes. Wherever smoke touches living cells, it does harm. So, even if smokers of pipes, cigarettes, and cigars don't inhale, they are at an increased risk for lip, mouth, and tongue cancer.

Does smoking affect the heart?

Yes. Smoking doubles the risk of heart disease, which is the United States' number-one killer. Smoking, high blood pressure, high cholesterol, and lack of exercise are all risk factors for heart disease.

Is there any risk for pregnant women and their babies?

Pregnant women who smoke endanger the health and lives of their unborn babies. When a pregnant woman smokes, she really is smoking for two because the nicotine, carbon monoxide, and other dangerous chemicals in smoke enter her bloodstream and then pass into the baby's body. Smoking mothers have more stillbirths and babies of low birth weight than nonsmoking mothers.

Does smoking cause any special health problems for women?

Yes. Women who smoke and use the birth control pill have an increased risk of stroke and blood clots in the legs. In addition, women who smoke increase their chances of getting cancer of the uterine cervix.

What are some of the short-term effects of smoking cigarettes?

Almost immediately, smoking can make it hard to breathe. Within a short time, it can also worsen asthma and allergies. Only seven seconds after a smoker takes a puff, nicotine reaches the brain, where it produces a morphinelike effect.

Are there any other risks to the smoker?

Yes, there are many more risks. Smoking is a cause of stroke, which is the third leading cause of death in the United States. Smokers are more likely to have and die from stomach ulcers than nonsmokers. Smokers have a higher incidence of cancer in general. If a person smokes and is exposed to radon or asbestos, the risk for lung cancer increases dramatically.

What are the dangers of passive smoking?

Passive smoking causes lung cancer in healthy nonsmokers. Children whose parents smoke are more likely to suffer from pneumonia or bronchitis in the first two years of life than children who come from smoke-free households. Passive smokers have a 30% greater risk of developing lung cancer than nonsmokers who live in a smoke-free house.

Are chewing tobacco and snuff safe alternatives to cigarette smoking?

No, they are not. Smokeless tobacco contains nicotine, the same addicting drug found in cigarettes and cigars. The juice from smokeless tobacco is absorbed through the lining of the mouth. There it can cause sores and white patches, which often lead to cancer of the mouth. Snuff dippers actually take in an average of over ten times more cancer-causing substances than cigarette smokers.

Discussion Questions

1. Before reading this Health Focus, were you aware of all the risks associated with smoking? If you are a smoker, how does knowing these risks affect the chances that you will quit? If you are a nonsmoker, do you think you might ever start?
2. Smoking is clearly addictive, because many people who want to quit are unable to do so. Whether you are a smoker or not (and it might be interesting to compare the responses of these two groups), what do you think is the percentage of smoking addiction that is physical versus psychological?
3. Suppose you became addicted to smoking two packs a day. If you decided to quit, how would you go about it? All at once, or a little at a time? Would you use any aids, such as nicotine gum or patches, or drugs? Perhaps a device that administers a mild electrical shock every time you smoke a cigarette? (Such a device exists!)

Lung Cancer Statistics

Lung cancer is the leading cause of death due to cancer for both men and women.

About 87% of lung cancer cases are linked to smoking. In the United States, an estimated 174,470 new cases and 162,460 deaths from lung cancer were expected in 2006.

This means that lung cancer will account for 28% of all cancer deaths. More people die of lung cancer than of colon, breast, and prostate cancers combined.

- Women who smoke are 1.5 times more likely to get lung cancer than are men who smoke.
- About 3,000 lung cancer deaths per year are attributed to secondhand smoke.
- Usually, 42% of people diagnosed with lung cancer die within the year.
- The five-year combined survival rate (the number of people alive after five years) after being diagnosed is only 13%.
- Even for those who are diagnosed early, the survival rate after five years is only about 49%.

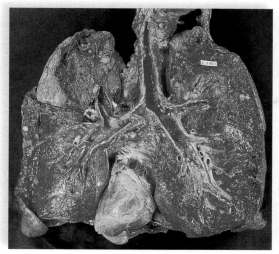

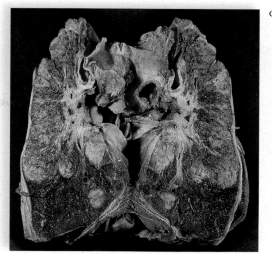

a.   b.

Figure 15.13
Normal and cancerous lungs compared.
a. Normal lungs with the heart in place. Note the healthy red color.
b. Lungs of a heavy smoker. Notice that the lungs are black except where cancerous tumors have formed.

Lung cancer is more prevalent in men than in women, but the incidence in women has increased recently, and lung cancer has surpassed breast cancer as a cause of death in women. This trend is directly correlated to increased numbers of women who smoke. Autopsies on smokers have revealed the progressive steps by which the most common form of lung cancer develops. The first step appears to be thickening of the cells lining the bronchi. Then cilia are lost, making it impossible to prevent dust and dirt from settling in the lungs. Following this, cells with atypical nuclei appear, followed by a tumor consisting of disordered cells with atypical nuclei. Normal lungs versus lungs with cancerous tumors are shown in Figure 15.13. A final step occurs when some of these cells break loose and penetrate other tissues, a process called metastasis. Now the cancer has spread. The original tumor may grow until a bronchus is blocked, cutting off the supply of air to that lung. The entire lung then collapses, the secretions trapped in the lung spaces become infected, and pneumonia or a lung abscess (localized area of pus) results. The only treatment that offers a possibility of cure is **pneumonectomy,** in which a lobe or the whole lung is removed before metastasis has had time to occur. If the cancer has spread, chemotherapy and radiation are also required.

The Health Focus lists the various illnesses, including cancer, that are apt to occur when a person smokes. Current research indicates that passive smoking—exposure to smoke exhaled by others who are smoking—can also cause lung cancer and other illnesses associated with smoking. The Bioethical Focus at the end of this chapter raises questions of fairness regarding attempts to cut back on passive smoking by issuing smoking bans. If a person stops voluntary smoking and avoids passive smoking, and if the body tissues are not already cancerous, they may return to normal over time.

The lower respiratory tract, including the delicate lung tissue, is susceptible to many infections and other diseases. Some diseases, such as chronic bronchitis and asthma, tend to obstruct the airways, while others, such as pulmonary fibrosis, restrict the lung's ability to expand normally.

Summarizing the Concepts

15.1 The Respiratory System
The respiratory tract consists of the nasal cavities (nose), the nasopharynx, the pharynx, the larynx (which contains the vocal cords), the trachea, the bronchi, the bronchioles, and the lungs. The bronchi, along with the pulmonary arteries and veins, enter the lungs, which contain many alveoli, air sacs surrounded by a capillary network.

15.2 Mechanism of Breathing
Inspiration begins when the respiratory center in the medulla oblongata sends excitatory nerve impulses to the diaphragm and the muscles of the rib cage. As they contract, the diaphragm lowers, and the rib cage moves upward and outward; the lungs expand, creating a partial vacuum, which causes air to rush in (inspiration). The respiratory center now stops sending impulses to the diaphragm and the muscles of the rib cage. As the diaphragm relaxes, it resumes its dome shape, and as the rib cage retracts, air is pushed out of the lungs (expiration).

15.3 Gas Exchanges in the Body
External respiration occurs in the lungs when CO_2 leaves the blood via the alveoli and O_2 enters the blood from the alveoli. Diffusion accounts for the passage of molecules in different directions: there is more carbon dioxide in pulmonary blood when it enters the lungs than in alveoli, and there is more oxygen in alveoli than in pulmonary blood when it enters the lungs. Because carbon dioxide is present in the blood as the bicarbonate ion (HCO_3^-), carbonic acid first forms and is broken down to carbon dioxide and water. Then carbon dioxide diffuses out of the blood. Oxygen is transported to the tissues in combination with hemoglobin as oxyhemoglobin (HbO_2).

Internal respiration occurs in the tissues when O_2 leaves and CO_2 enters the blood. When carbon dioxide enters the blood, carbonic acid forms and is broken down to the bicarbonate ion (HCO_3^-) and hydrogen ions. Carbon dioxide is mainly carried to the lungs within the plasma as the bicarbonate ion. Hemoglobin combines with hydrogen ions and becomes reduced (HHb).

15.4 Disorders of the Respiratory System
A number of illnesses are associated with the respiratory system. These disorders can be divided into those that affect the upper respiratory tract and those that affect the lower respiratory tract.

The common cold, pharyngitis, tonsillitis, and laryngitis are all conditions that most people experience at some time. In addition, these infections can spread into the sinuses and middle ears.

In 1964, the Surgeon General of the United States announced to the general public that smoking is hazardous to our health, and thereafter, a health warning was placed on packs of cigarettes. At that time, 40.4% of adults smoked, but by 1990, only about 26% of adults smoked. In the meantime, however, the public became aware that passive smoking—that is, just being in the vicinity of someone who is smoking—can also lead to cancer and other health problems. By now, many state and local governments have passed legislation that bans smoking in public places such as restaurants, elevators, public meeting rooms, and the workplace.

Is legislation that restricts the freedom to smoke ethical? Or is such legislation akin to racism and creating a population of second-class citizens who are segregated from the majority on the basis of a habit? Are the desires of nonsmokers being allowed to infringe on the rights of smokers? Or is this legislation one way to help smokers become nonsmokers? One study showed that workplace bans on smoking reduce the daily consumption of cigarettes among smokers by 10%.

Is legislation that disallows smoking in family-style restaurants fair, especially if bars and restaurants associated with casinos are not included in the ban on smoking? The selling of tobacco, and even the increased need for health care it generates, helps the economy. One smoker writes, "Smoking causes people to drink more, eat more, and leave larger tips. Smoking also powers the economy of Wall Street." Is this a reason to allow smoking to continue? Or should we simply require all places of business to put in improved air filtration systems? Would that do away with the dangers of passive smoking?

Does legislation that bans smoking in certain areas represent government invasion of our privacy? If yes, is reducing the chance of cancer a good enough reason to allow the government to invade our privacy? Some people are more prone to cancer than others. Should we all be regulated by the same legislation? Are we our brothers' keepers, meaning that we have to look out for one another?

Form Your Own Opinion

1. Besides the Surgeon General's warning, are there any other possible explanations for why the percentage of smoking adults declined between 1964 and 1990?

2. If you are a smoker, do you smoke in your house? Your car? If not, why not? If you are a nonsmoker, how would you respond if a good friend or a relative wanted to smoke in your house?

3. If there are smoking bans in your area, how have you personally been affected? If your area has no bans but passed one tomorrow, how might you be affected? Why?

The lower respiratory tract is subject to foreign bodies that can block the trachea and bronchi. In addition, infections can cause acute or chronic bronchitis, pneumonia, and pulmonary tuberculosis. Other common disorders include asthma, emphysema, cystic fibrosis, pulmonary fibrosis, and lung cancer. Smoking is a major risk factor associated with chronic bronchitis, emphysema, and lung cancer.

Testing Yourself

Choose the best answer for each question.

1. Label this diagram of the human respiratory system.

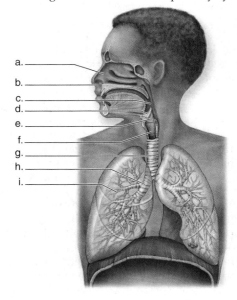

a.
b.
c.
d.
e.
f.
g.
h.
i.

2. The key component in internal, external, and cellular respiration is the transport and use of
   a. nitrogen.
   b. carbon monoxide.
   c. sodium.
   d. oxygen.

3. Cleansing inhaled air involves
   a. hairs.
   b. cilia.
   c. mucus.
   d. All of these are correct.

4. The pulmonary _____ are rich in oxygen.
   a. veins
   b. arteries
   c. Both a and b are correct.
   d. Neither a nor b is correct.

5. Food and air both travel through the
   a. lungs.
   b. pharynx.
   c. larynx.
   d. trachea.

6. Pulmonary fibrosis
   a. can be caused by inhaling asbestos.
   b. is most common in the elderly.
   c. leads to reduced vital capacity.
   d. All of these are correct.

7. Gas exchange occurs
   a. in the chambers of the heart.
   b. only in tissue capillaries.
   c. only in pulmonary capillaries.
   d. None of these are correct.

8. Most CO_2 is carried in the blood
   a. as HCO_3^-.
   b. on hemoglobin.
   c. dissolved in plasma.
   d. Both a and c are correct.

9. Reduced hemoglobin is carrying

   a. oxygen.
   b. CO_2.
   c. hydrogen ions.
   d. sodium.

10. The phrenic nerve stimulates the _____

    a. diaphragm
    b. lungs
    c. intestines
    d. All of these are correct.

11. A decrease in atmospheric pressure would probably

    a. make it more difficult to breathe.
    b. make it easier to breathe.
    c. have no effect on breathing.
    d. None of these are correct.

In questions 12–16, match the functions to the structures in the key.

Key:

   a. bronchioles
   b. glottis
   c. larynx
   d. bronchi
   e. lungs

12. Passage of air into the larynx

13. Passage of air into the lungs

14. Sound production

15. Gas exchange

16. Passage of air into the alveoli

17. The amount of air moved into and out of the lungs during regular breathing is known as the

    a. vital capacity.
    b. tidal volume.
    c. residual volume.
    d. None of these are correct.

18. Which of these statements is incorrect concerning inspiration?

    a. Rib cage moves up and out.
    b. Diaphragm contracts and moves down.
    c. Pressure in lungs decreases, and air comes rushing in.
    d. The lungs expand because air comes rushing in.

19. If the digestive and respiratory tracts were completely separate in humans, there would be no need for

    a. swallowing.
    b. a nose.
    c. an epiglottis.
    d. a diaphragm.
    e. All of these are correct.

20. Label the following diagram of the path of air.

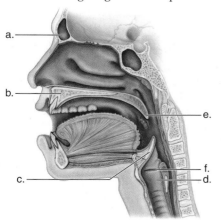

21. Which of these associations is correctly matched?

    a. expiration—pressure in lungs is less than atmospheric pressure
    b. inspiration—pressure in lungs is greater than atmospheric pressure
    c. expiration—pressure in lungs is greater than atmospheric pressure
    d. inspiration—pressure in lungs is the same as atmospheric pressure
    e. None of these are correct.

22. Lungs are

    a. paired, cone-shaped organs.
    b. each enclosed by a pleura.
    c. composed of alveoli.
    d. prevented from closing by surfactant.
    e. All of these are correct.

23. Label the respiratory volumes where indicated.

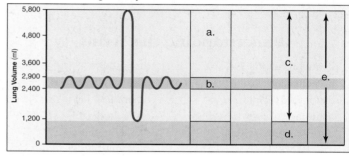

24. In tracing the path of air in humans, you would list the trachea

    a. directly after the nose.
    b. directly before the bronchi.
    c. before the pharynx.
    d. directly before the lungs.
    e. Both a and c are correct.

25. Carbon dioxide is carried in the plasma

    a. in combination with hemoglobin.
    b. as the bicarbonate ion.
    c. combined with carbonic anhydrase.
    d. only as a part of tissue fluid.
    e. All of these are correct.

26. Which of these statements is anatomically incorrect?

    a. The nose has two nasal cavities.
    b. The pharynx connects the nasal cavity and mouth to the larynx.
    c. The larynx contains the vocal cords.
    d. The trachea enters the lungs.
    e. The lungs contain many alveoli.

27. Internal respiration refers to

    a. the exchange of gases between the air and the blood in the lungs.
    b. the movement of air into the lungs.
    c. the exchange of gases between the blood and tissue fluid.
    d. cellular respiration, resulting in the production of ATP.

28. The presence of carbon dioxide in the blood

    a. affects the respiratory volumes.
    b. makes blood more acidic.
    c. makes respiratory exchanges more difficult.
    d. can increase the breathing rate.
    e. Both b and d are correct.

29. The enzyme carbonic anhydrase

    a. affects the heart rate.

    b. is found in red blood cells.

    c. is active once carbon dioxide enters the blood.

    d. attaches carbon dioxide to hemoglobin.

    e. Both b and c are correct.

30. Which of these statements correctly goes with expiration rather than inspiration?

    a. Rib cage moves up and out.

    b. Diaphragm relaxes and moves up.

    c. Pressure in lungs decreases, and air comes rushing out.

    d. Diaphragm contracts and lowers.

31. The respiratory center is directly sensitive to

    a. carbon dioxide and hydrogen ions.

    b. oxygen and carbon dioxide.

    c. the amount of air in the alveoli.

    d. the breathing rate.

Match the terms to these definitions:

    a._____ Common passageway for both food intake and air movement; located between the mouth and the esophagus.

    b._____ Amount of air remaining in the lungs after a forceful expiration.

    c._____ Form in which most of the carbon dioxide is transported in the bloodstream.

    d._____ Stage during breathing when air is pushed out of the lungs.

    e._____ Terminal, microscopic, grapelike air sac found in lungs.

Understanding the Terms

acute bronchitis 296	lungs 288
alveolus 288	nasal cavity 287
aortic body 293	nostril 287
asthma 296	otitis media 296
auditory (eustachian) tube 296	oxyhemoglobin 293
	pharyngitis 295
bicarbonate ion 293	pharynx 287
bronchiole 288	pleura 289
bronchus 288	pneumonectomy 299
carbaminohemoglobin 295	pneumonia 296
carbonic anhydrase 293	pulmonary fibrosis 297
carotid body 293	pulmonary tuberculosis 296
chronic bronchitis 296	
cystic fibrosis (CF) 297	reduced hemoglobin 295
emphysema 297	residual volume 291
epiglottis 288	respiratory center 292
expiration 286	sinus 287
expiratory reserve volume 291	sinusitis 295
	surfactant 289
external respiration 293	tidal volume 291
glottis 288	tonsillectomy 295
infant respiratory distress syndrome 289	tonsillitis 295
	tonsils 295
inspiration 286	trachea 288
inspiratory reserve volume 291	tracheostomy 296
	ventilation 286
internal respiration 294	vital capacity 291
laryngitis 295	vocal cord 288
larynx 288	
lung cancer 299	

Thinking Critically

1. The respiratory system of birds is quite different from that of mammals. Unlike mammalian lungs, the lungs of birds are relatively rigid and do not expand much during ventilation. Most birds also have thin-walled air sacs that fill most of the body cavity not occupied by other organs. Inspired air passes through the bird's lungs, into these air sacs, and back through the lungs again on expiration. In what ways might this system be more efficient than the mammalian lung?

2. The respiratory and cardiovascular systems work together to provide O_2 and remove CO_2 from the tissues. What are some specific ways that the activities of these two systems affect each other? What specifically happens to the other system, for example, if the activity of the cardiovascular system decreases, or the respiratory rate increases?

3. Carbon monoxide (CO) binds to hemoglobin 230–270 times more strongly than oxygen does, which means less O_2 is being delivered to tissues. What would be the specific cause of death if too much hemoglobin became bound to CO instead of to O_2?

4. As mentioned in the chapter, sometimes it is necessary to install a permanent tracheostomy—for example, in smokers who develop laryngeal cancer. Besides interfering with the ability to speak, what sorts of health problems would you expect to see in an individual with a permanent tracheostomy, and why?

5. People suffering from cystic fibrosis often have trouble when their pancreatic duct becomes obstructed due to thickening of the pancreatic secretions. What does this tell you about the function of *CFTR*? What kind of symptoms would you expect to experience if your pancreatic duct became blocked?

ARIS, the *Inquiry into Life* Website

ARIS, the website for *Inquiry into Life,* provides a wealth of information organized and integrated by chapter. You will find practice quizzes, interactive activities, labeling exercises, flashcards, and much more that will complement your learning and understanding of general biology.

www.aris.mhhe.com

Tom and Jennifer were excited about the impending birth of their second baby. Married for three years, they already had a healthy daughter, and they were happy that Megan would have a younger sister to play with. When Mary was born, however, it soon became clear that something was wrong. She only weighed 5 pounds, 4 ounces at birth.

The first time she urinated, there was an obvious tinge of blood in her urine. She also seemed to urinate much more frequently than normal. Her doctors used ultrasound to examine her abdominal organs. What they found was that Mary had signs of polycystic kidney disease (as shown in illustration), which meant that she might require hemodialysis, or even a kidney transplant, within a few years. Even worse was the knowledge that although both Tom and Jennifer were healthy, they were carriers of a faulty gene that they had unknowingly passed on to their new daughter.

There are two inherited forms of polycystic kidney disease (PKD): autosomal dominant and autosomal recessive. The autosomal dominant form is one of the most common inherited disorders in humans, and it is the most frequent genetic cause of kidney failure in the United States. However, the symptoms of this form of PKD usually don't show up until adulthood. Mary's doctors suspected that she had the autosomal recessive form, the symptoms of which can be much more severe in children, even while still in the womb. Although rare, this more severe form of PKD illustrates the critical functions of the urinary system.

chapter 16

Urinary System and Excretion

Chapter Concepts

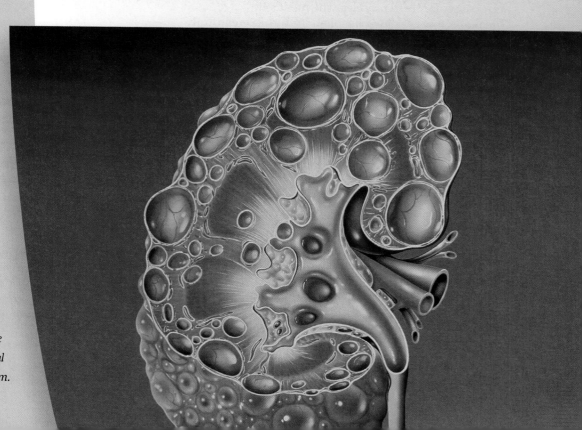

Urinary System

The kidneys are the primary organs of excretion. **Excretion** is the removal of metabolic wastes from the body. People sometimes confuse the terms excretion and defecation, but they do not refer to the same process. Defecation, the elimination of feces from the body, is a function of the digestive system. Excretion is the elimination of metabolic wastes, the products of metabolism. For example, the undigested food and bacteria that make up feces have never been a part of the functioning of the body, while the substances excreted in urine were once metabolites in the body.

Functions of the Urinary System

The urinary system produces urine and conducts it to outside the body. As the kidneys produce urine, they carry out the following four functions that contribute to homeostasis:

1. *Excretion of Metabolic Wastes* The kidneys excrete metabolic wastes, notably nitrogenous wastes. Urea is the primary nitrogenous end product of metabolism in human beings, but humans also excrete some ammonium, creatinine, and uric acid.

   Urea is a by-product of amino acid metabolism. The breakdown of amino acids in the liver releases ammonia, which the liver rapidly combines with carbon dioxide to produce urea. Ammonia is very toxic to cells, but urea is much less toxic. Some ammonia (NH_3) is also excreted as ammonium (NH_4^+).

   Creatine phosphate is a high-energy phosphate reserve molecule in muscles. The metabolic breakdown of creatine phosphate results in **creatinine.**

   The breakdown of nucleotides, such as those containing adenine and thymine, produces **uric acid.** Uric acid is rather insoluble. If too much uric acid is present in blood, crystals form and precipitate out. Crystals of uric acid sometimes collect in the joints, producing a painful ailment called **gout.**

2. *Maintenance of Water-Salt Balance* A principal function of the kidneys is to maintain the appropriate water-salt balance of the blood. Blood volume is intimately associated with the salt balance of the body. Salts, such as NaCl, have the ability to cause osmosis, the diffusion of water—in this case, into the blood. The more salts there are in the blood, the greater the blood volume and the greater the blood pressure. In this way, the kidneys are involved in regulating blood pressure.

   The kidneys also maintain the appropriate level of other ions, such as potassium ions (K^+), bicarbonate ions (HCO_3^-), and calcium ions (Ca^{2+}), in the blood.

3. *Maintenance of Acid-Base Balance* The kidneys regulate the acid-base balance of the blood. In order for a person to remain healthy, the blood pH should be just about 7.4. The kidneys monitor and help keep the blood pH at this level, mainly by excreting hydrogen ions (H^+) and reabsorbing the bicarbonate ions (HCO_3^-) as needed. Urine usually has a pH of 6 or lower because our diet often contains acidic foods.

4. *Secretion of Hormones* The kidneys assist the endocrine system in hormone secretion. The kidneys release renin, a substance that leads to the secretion of the hormone aldosterone from the adrenal cortex, the outer portion of the adrenal glands, which lie atop the kidneys. As described in section 16.3 aldosterone promotes the reabsorption of sodium ions (Na^+) by the kidneys.

   Whenever the oxygen-carrying capacity of the blood is reduced, the kidneys secrete the hormone **erythropoietin,** which stimulates red blood cell production.

   The kidneys also help activate vitamin D from the skin. Vitamin D is a hormone-like molecule that promotes calcium (Ca^{2+}) absorption from the digestive tract.

The kidneys are the primary organs of excretion, particularly of nitrogenous wastes. The kidneys are also major organs of homeostasis because they regulate the water-salt balance and the acid-base balance of the blood, as well as the secretion of certain hormones.

Organs of the Urinary System

The urinary system consists of the kidneys, ureters, urinary bladder, and urethra. Figure 16.1 shows these organs and also traces the path of urine.

Kidneys

The **kidneys** are paired organs located near the small of the back in the lumbar region on either side of the vertebral column. They lie in depressions dorsal to (behind) the peritoneum, where they receive some protection from the lower rib cage.

The kidneys are bean shaped and reddish-brown in color. The fist-sized organs are covered by a tough capsule of fibrous connective tissue called a renal capsule. Masses of adipose tissue adhere to each kidney. The concave side of a kidney has a depression called the hilum where a **renal artery** enters and a **renal vein** and a ureter exit the kidney.

Ureters

The **ureters** conduct urine from the kidneys to the bladder. They are small, muscular tubes about 25 cm long and 5 mm in diameter. Each descends behind the peritoneum, from the hilum of a kidney, to enter the bladder at its dorsal surface.

The wall of a ureter has three layers: an inner mucosa (mucous membrane), a smooth muscle layer, and an outer fibrous coat of connective tissue. Peristaltic contractions cause urine to enter the bladder even if a person is lying down. Urine enters the bladder in spurts at the rate of one to five per minute.

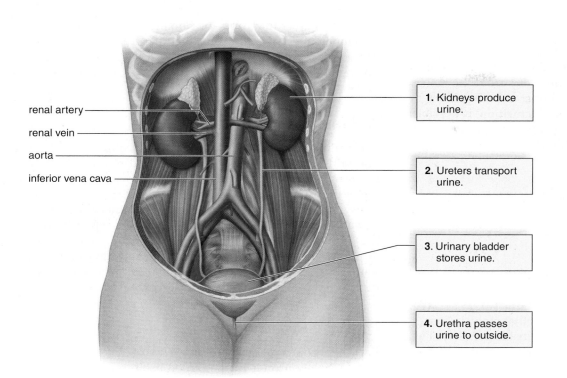

renal artery

renal vein

aorta

inferior vena cava

| 1. Kidneys produce urine. |

| 2. Ureters transport urine. |

| 3. Urinary bladder stores urine. |

| 4. Urethra passes urine to outside. |

Figure 16.1 **The urinary system.**
Urine is found only within the kidneys, the ureters, the urinary bladder, and the urethra. The kidneys are important organs of homeostasis because they excrete metabolic wastes and adjust the water-salt balance and the acid-base balance of the blood.

Urinary Bladder

The **urinary bladder** stores urine until it is expelled from the body. The bladder is located in the pelvic cavity, behind the pubic symphysis and behind the peritoneum. The urinary bladder has three openings (orifices)—two for the ureters and one for the urethra, which drains the bladder (Fig. 16.2).

The bladder wall is expandable because it contains a middle layer of circular fiber and two layers of longitudinal muscle. The transitional epithelium of the mucosa becomes thinner, and folds in the mucosa called *rugae* disappear as the bladder enlarges.

The bladder has other features that allow it to retain urine. After urine enters the bladder from a ureter, small folds of bladder mucosa act like a valve to prevent backward flow. Two sphincters lie in close proximity where the urethra exits the bladder. The internal sphincter occurs around the opening to the urethra. An external sphincter is composed of skeletal muscle that can be voluntarily controlled. Many people, especially the elderly, have some degree of **incontinence,** the involuntary loss of urine. Common causes of incontinence are age-related changes in the anatomy of the urinary bladder and nervous system diseases that affect the control of urination.

Urethra

The **urethra** is a small tube that extends from the urinary bladder to an external opening. Therefore, its function is to remove urine from the body. In females, the urethra is only about 4 cm long, while in males, the urethra averages 20 cm when the penis is flaccid (nonerect). As the urethra leaves the male urinary bladder, it is encircled by the prostate gland. In men over age 40, the prostate gland is subject to enlargement, as well as to prostate

cancer. Either of these conditions may restrict urination. Medical and surgical procedures can usually correct these disorders and restore a normal flow of urine (see section 21.5).

In females, the reproductive and urinary systems are not connected. In males, the urethra carries urine during urination and semen during ejaculation. This double function of the urethra in males does not alter the path of urine.

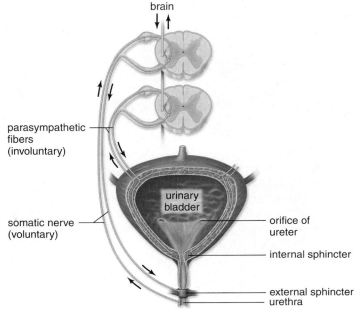

brain

parasympathetic fibers (involuntary)

somatic nerve (voluntary)

urinary bladder

orifice of ureter

internal sphincter

external sphincter
urethra

Figure 16.2 **Urination.**
As the bladder fills with urine, sensory impulses go to the spinal cord and then to the brain. When urination occurs, motor nerve impulses cause the bladder to contract and an internal sphincter to relax. Motor nerve impulses also cause an external sphincter to relax.

Urination

When the urinary bladder fills to about 250 ml with urine, stretch receptors send sensory nerve impulses to the spinal cord. Subsequently, motor nerve impulses from the spinal cord cause the urinary bladder to contract and the sphincters to relax so that urination, also called **micturition,** is possible (Fig. 16.2). In older children and adults, the brain controls this reflex, delaying urination until a suitable time.

Only the urinary system, consisting of the kidneys, ureters, urinary bladder, and urethra, contains urine.

16.2 Anatomy of the Kidney and Excretion

A lengthwise section of a kidney shows that many branches of the renal artery and renal vein reach inside a kidney (Fig. 16.3*a*). Removing the blood vessels shows that a kidney has three regions (Fig. 16.3*c*). The **renal cortex** is an outer, granulated layer that dips down in between a radially striated inner layer called the renal medulla. The **renal medulla** consists of cone-shaped tissue masses called renal pyramids. The **renal pelvis** is a central space, or cavity, that is continuous with the ureter.

Anatomy of a Nephron

Under higher magnification, the kidney is composed of over one million **nephrons,** sometimes called renal or kidney tubules (Fig. 16.3*b*). Each nephron has its own blood supply, including two capillary regions (Fig. 16.4). From the renal artery, an afferent arteriole leads to the **glomerulus,** a knot of capillaries inside the glomerular capsule. Blood leaving the glomerulus enters the efferent arteriole. The efferent arteriole takes blood to the **peritubular capillary network,** which surrounds the rest of the nephron. From there, the blood goes into a venule that joins the renal vein.

Parts of a Nephron

Each nephron is made up of several parts. First, the closed end of the nephron is pushed in on itself to form a cuplike

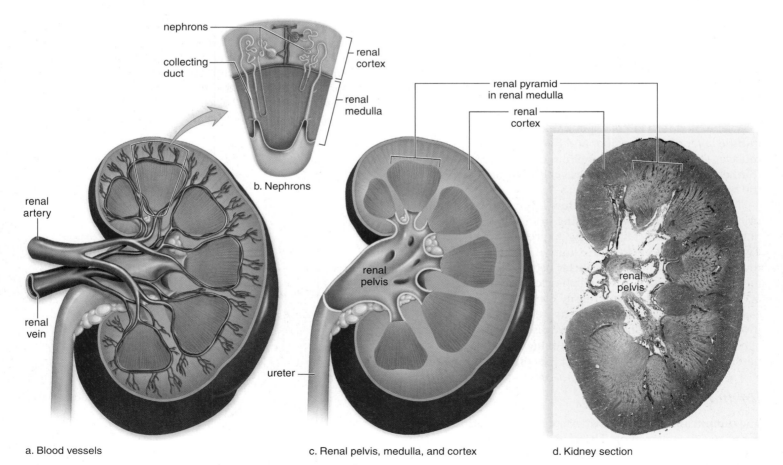

a. Blood vessels   c. Renal pelvis, medulla, and cortex   d. Kidney section

Figure 16.3 Anatomy of the kidney. a. A lengthwise section of the kidney showing the blood supply. Note that the renal artery divides into smaller arteries, and these divide into arterioles. Venules join to form small veins, which form the renal vein. **b.** An enlargement showing the placement of nephrons. **c.** The same section as "a" without the blood supply. Now it is easier to distinguish the renal cortex, medulla, and pelvis, which connects with the ureter. **d.** A lengthwise section of an actual kidney.

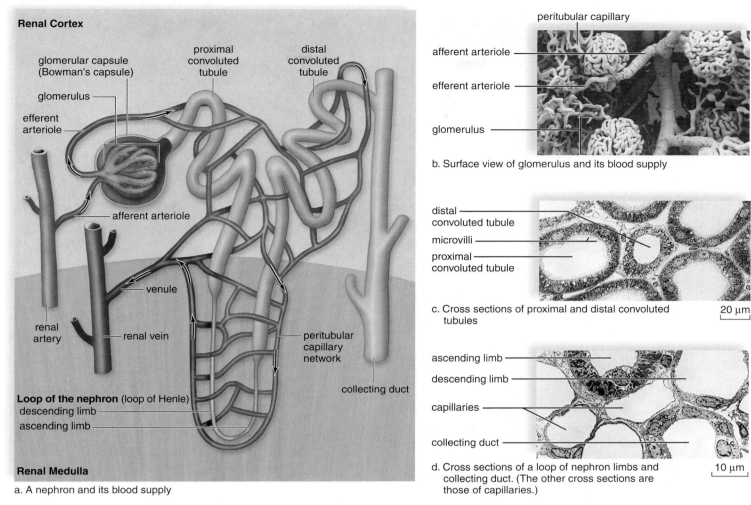

a. A nephron and its blood supply

Renal Cortex

glomerular capsule (Bowman's capsule)
glomerulus
efferent arteriole
afferent arteriole
proximal convoluted tubule
distal convoluted tubule
venule
renal artery
renal vein
peritubular capillary network
collecting duct
Loop of the nephron (loop of Henle)
descending limb
ascending limb
Renal Medulla

peritubular capillary
afferent arteriole
efferent arteriole
glomerulus

b. Surface view of glomerulus and its blood supply

distal convoluted tubule
microvilli
proximal convoluted tubule

c. Cross sections of proximal and distal convoluted tubules   20 μm

ascending limb
descending limb
capillaries
collecting duct

d. Cross sections of a loop of nephron limbs and collecting duct. (The other cross sections are those of capillaries.)   10 μm

Figure 16.4 Nephron anatomy.

a. You can trace the path of blood through a nephron by following the black arrows. **b, c, d.** A nephron is made up of a glomerular capsule, the proximal convoluted tubule, the loop of the nephron, the distal convoluted tubule, and the collecting duct.

©R. G. Kessel and R. H. Kardon, Tissues and Organs: A Text-Atlas of Scanning Electron Microscopy, 1979.

structure called the **glomerular capsule** (Bowman capsule). The inner layer of the glomerular capsule is composed of *podocytes* that have long cytoplasmic extensions (see Fig. 16.6). The podocytes cling to the capillary walls of the glomerulus and leave pores that allow easy passage of small molecules from the glomerulus to the inside of the glomerular capsule.

The glomerular capsule connects to a **proximal convoluted tubule (PCT).** The cuboidal epithelial cells lining this part of the nephron have numerous tightly packed microvilli that form a brush border, increasing the surface area for reabsorption. The tube then narrows and makes a U-turn called the **loop of the nephron** (loop of Henle), which is lined with simple squamous epithelium.

Following the loop, the tube becomes the **distal convoluted tubule (DCT),** which is composed of cuboidal epithelial cells that have numerous mitochondria but lack microvilli. This means the DCT is not specialized for reabsorption and is more likely to help move molecules from the blood into the tubule, a

process called tubular secretion. The distal convoluted tubules of several nephrons enter one collecting duct. Many **collecting ducts** carry urine to the renal pelvis.

As shown in Figure 16.4a, the glomerular capsule and the convoluted tubules always lie within the renal cortex. The loop of the nephron dips down into the renal medulla; a few nephrons have a very long loop of the nephron, which penetrates deep into the renal medulla. Collecting ducts are also located in the renal medulla, and they help give the renal pyramids their lined appearance (see Fig. 16.3d).

The kidney is composed mainly of nephrons, or renal tubules. A nephron is made up of a glomerular capsule, a proximal convoluted tubule, a loop of the nephron, a distal convoluted tubule and a collecting duct.

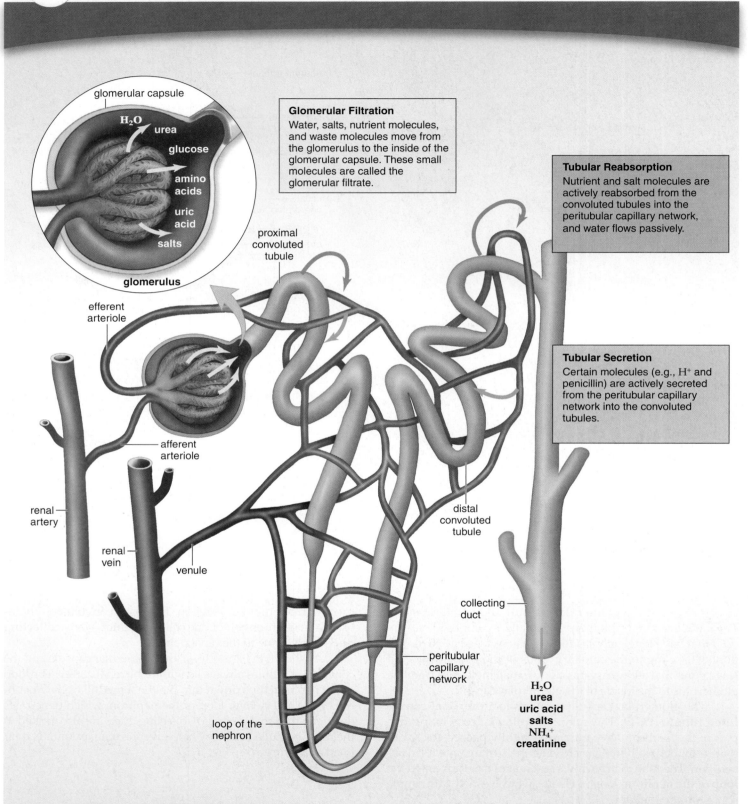

Glomerular Filtration
Water, salts, nutrient molecules, and waste molecules move from the glomerulus to the inside of the glomerular capsule. These small molecules are called the glomerular filtrate.

Tubular Reabsorption
Nutrient and salt molecules are actively reabsorbed from the convoluted tubules into the peritubular capillary network, and water flows passively.

Tubular Secretion
Certain molecules (e.g., H^+ and penicillin) are actively secreted from the peritubular capillary network into the convoluted tubules.

glomerular capsule

H_2O
urea
glucose
amino acids
uric acid
salts

glomerulus

proximal convoluted tubule

efferent arteriole

afferent arteriole

renal artery

renal vein

venule

distal convoluted tubule

collecting duct

peritubular capillary network

loop of the nephron

H_2O
urea
uric acid
salts
NH_4^+
creatinine

Figure 16.5 Processes in urine formation.

The three main processes in urine formation are described in boxes and color-coded to arrows that show the movement of molecules into or out of the nephron at specific locations. In the end, urine is composed of the substances within the collecting duct (see blue arrow).

Urine Formation

Figure 16.5 gives an overview of urine formation, which is divided into the following processes: glomerular filtration, tubular reabsorption, and tubular secretion.

Glomerular Filtration

Glomerular filtration occurs when whole blood enters the afferent arteriole and the glomerulus. Due to glomerular blood pressure, water and small molecules move from the glomerulus to the inside of the glomerular capsule. This is a filtration process because large molecules and formed elements are unable to pass through the capillary wall. In effect, then, blood in the glomerulus has two portions—the filterable components and the nonfilterable components:

Filterable Blood Components	Nonfilterable Blood Components
Water	Formed elements
Nitrogenous wastes	(blood cells and platelets)
Nutrients	Plasma proteins
Salts (ions)	

The **glomerular filtrate** contains small dissolved molecules in approximately the same concentration as plasma. Small molecules that escape being filtered and the nonfilterable components leave the glomerulus by way of the efferent arteriole.

As indicated in Table 16.1, nephrons in the kidneys filter 180 liters of water per day, along with a considerable amount of small molecules (such as glucose) and ions (such as sodium). If the composition of urine were the same as that of the glomerular filtrate, the body would continually lose water, salts, and nutrients. Therefore, we can conclude that the composition of the filtrate must be altered as this fluid passes through the remainder of the tubule.

Tubular Reabsorption

Tubular reabsorption occurs as molecules and ions are both passively and actively reabsorbed from the nephron into the blood of the peritubular capillary network. The osmolarity of the blood is maintained by the presence of both plasma proteins and salt. When sodium ions (Na^+) are actively reabsorbed, chloride ions (Cl^-) follow passively. The reabsorption of salt (NaCl) increases the osmolarity of the blood compared to the filtrate, and therefore water moves passively from the tubule into the blood. About 65% of Na^+ is reabsorbed at the proximal convoluted tubule.

Nutrients, such as glucose and amino acids, also return to the blood almost exclusively at the proximal convoluted tubule. This is a selective process because only molecules recognized by carrier proteins are actively reabsorbed. Glucose is an example of a molecule that ordinarily is completely reabsorbed because there is a plentiful supply of carrier proteins for it. However, every substance has a maximum rate of transport, and after all its carriers are in use, any excess in the filtrate will appear in the urine. For example, as reabsorbed levels of glucose approach 1.8–2 mg/ml plasma, the rest appears in the urine.

In diabetes mellitus, excess glucose appears in the blood, because the liver and muscles have failed to store glucose as glycogen. The kidneys cannot reabsorb all the glucose in the filtrate, and it appears in the urine. The presence of glucose in the filtrate increases its osmolarity, and therefore, less water is reabsorbed into the peritubular capillary network. The frequent urination and increased thirst experienced by untreated diabetics are due to the fact that less water is being reabsorbed.

We have seen that the filtrate that enters the proximal convoluted tubule is divided into two portions—components that are reabsorbed from the tubule into blood and components that are not reabsorbed and continue to pass through the nephron to be further processed into urine:

Reabsorbed Filtrate Components	Nonreabsorbed Filtrate Components
Most water	Some water
Nutrients	Much nitrogenous waste
Required salts (ions)	Excess salts (ions)

The substances that are not reabsorbed become the tubular fluid, which enters the loop of the nephron.

Tubular Secretion

Tubular secretion is a second way by which substances are removed from blood in the peritubular network and added to the tubular fluid. Hydrogen ions, potassium ions, creatinine, and drugs, such as penicillin, are some of the substances that are moved by active transport from the blood. In the end, urine contains (1) substances that have undergone glomerular filtration but have not been reabsorbed and (2) substances that have undergone tubular secretion.

TABLE 16.1	Reabsorption from Nephrons		
Substance	Amount Filtered (per day)	Amount Excreted (per day)	Reabsorption (%)
Water, L	180	1.8	99.0
Sodium, g	630	3.2	99.5
Glucose, g	180	0.0	100.0
Urea, g	54	30.0	44.0

L = liters, g = grams

Glomerular filtration and tubular secretion add molecules to urine; tubular reabsorption removes them.

16.3 Regulatory Functions of the Kidneys

The kidneys maintain the water-salt balance of the blood within normal limits. In this way, they also maintain the blood volume and blood pressure. Most of the water and salt (NaCl) present in the filtrate is reabsorbed across the wall of the proximal convoluted tubule.

Process of Water Reabsorption

The excretion of a hypertonic urine (one that is more concentrated than blood) is dependent upon the reabsorption of water from the loop of the nephron and the collecting duct. We can think of reabsorption of water as requiring (1) reabsorption of salt and (2) establishment of a solute gradient dependent on salt and urea before (3) water is reabsorbed.

Reabsorption of Salt

The kidneys regulate the salt balance in blood by controlling the excretion and the reabsorption of various ions. Sodium (Na^+) is an important ion in plasma that must be regulated, but the kidneys also excrete or reabsorb other ions, such as potassium ions (K^+), bicarbonate ions (HCO_3^-), and magnesium ions (Mg^{2+}), as needed.

Usually, more than 99% of sodium (Na^+) filtered at the glomerulus is returned to the blood. Most of this sodium (67%) is reabsorbed at the proximal tubule, and a sizable amount (25%) is reabsorbed by the ascending limb of the loop of the nephron. The rest is reabsorbed from the distal convoluted tubule and collecting duct.

Hormones regulate the reabsorption of sodium at the distal convoluted tubule. **Aldosterone,** a hormone secreted by the adrenal cortex, promotes the excretion of potassium ions (K^+) and the reabsorption of sodium ions (Na^+). The release of aldosterone is set in motion by the kidneys themselves. The **juxtaglomerular apparatus** is a region of contact between the afferent arteriole and the distal convoluted tubule (Fig. 16.6). When blood volume, and therefore blood pressure, is not sufficient to promote glomerular filtration, the juxtaglomerular apparatus secretes renin. **Renin** is an enzyme that changes angiotensinogen (a large plasma protein produced by the liver) into angiotensin I. Later, angiotensin I is converted to angiotensin II, a powerful vasoconstrictor that also stimulates the adrenal cortex to release aldosterone. The reabsorption of sodium ions is followed by the reabsorption of water. Therefore, blood volume and blood pressure increase.

Atrial natriuretic hormone (ANH) is a hormone secreted by the atria of the heart when cardiac cells are stretched due to increased blood volume. ANH inhibits the secretion of renin by the juxtaglomerular apparatus and the secretion of aldosterone by the adrenal cortex. Its effect, therefore, is to promote the excretion of Na^+, called natriuresis.

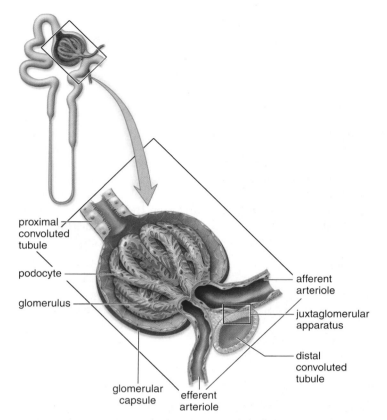

Figure 16.6 Juxtaglomerular apparatus.
The afferent arteriole and the distal convoluted tubule usually lie next to each other. The juxtaglomerular apparatus occurs where they touch. The juxtaglomerular apparatus secretes renin, a substance that leads to the release of aldosterone by the adrenal cortex. Reabsorption of sodium ions and water then occurs. Thereafter, blood volume and blood pressure increase.

Establishment of a Solute Gradient

The long loop of a nephron, which typically penetrates deep into the renal medulla, is made up of a descending limb and an ascending limb. Salt (NaCl) passively diffuses out of the lower portion of the ascending limb, but the upper, thick portion of the limb actively extrudes salt out into the tissue of the outer renal medulla (Fig. 16.7). Less and less salt is available for transport as fluid moves up the thick portion of the ascending limb. Because of these circumstances, there is an osmotic gradient within the tissues of the renal medulla: the concentration of salt is greater in the direction of the inner medulla. (Note that water cannot leave the ascending limb because the limb is impermeable to water.)

The large arrow at the far left in Figure 16.7 indicates that the innermost portion of the inner medulla has the highest concentration of solutes. This cannot be due to salt because active transport of salt does not start until fluid reaches the thick portion of the ascending limb. Urea is believed to leak from the lower portion of the collecting duct, and it is this molecule that contributes to the high solute concentration of the inner medulla.

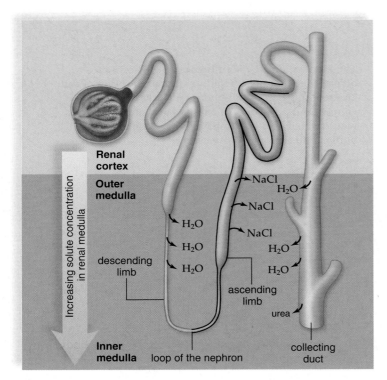

Renal cortex

Outer medulla

Increasing solute concentration in renal medulla

descending limb

H_2O

H_2O

H_2O

NaCl

H_2O

NaCl

NaCl

H_2O

H_2O

H_2O

ascending limb

urea

Inner medulla   loop of the nephron

collecting duct

Figure 16.7 Reabsorption of water at the loop of the nephron and the collecting duct.

Salt (NaCl) diffuses and is actively transported out of the ascending limb of the loop of the nephron into the renal medulla; also, urea is believed to leak from the collecting duct and to enter the tissues of the renal medulla. This creates a hypertonic environment, which draws water out of the descending limb and the collecting duct. This water is returned to the cardiovascular system. (The thick black outline of the ascending limb means that it is impermeable to water.)

Reabsorption of Water

Because of the osmotic gradient within the renal medulla, water leaves the descending limb along its entire length. This is a countercurrent mechanism: as water diffuses out of the descending limb, the remaining fluid within the limb encounters an even greater osmotic concentration of solute; therefore, water continues to leave the descending limb from the top to the bottom.

Fluid enters the collecting duct from the distal convoluted tubule. This fluid is hypotonic to the renal cortex, so to this point, the active transport of salt out of the ascending limbs has decreased the osmolarity of the filtrate (Fig. 16.7). This allows the production of urine that is hypotonic to the general body fluids—for example, when excess water needs to be excreted. When the urine needs to be hypertonic to body fluids (when the body is dehydrated, for example), the posterior pituitary gland releases **antidiuretic hormone (ADH).** In order to understand the action of this hormone, consider its name. *Diuresis* means an increased amount of urine, and *antidiuresis* means a decreased amount of urine. In the absence of ADH, the collecting duct is impermeable to water, and a dilute urine is produced. When ADH is present, the collecting duct becomes permeable to water. Since the filtrate within the collecting duct encounters the same osmotic gradient mentioned earlier, water can diffuse out

of the collecting duct into the renal medulla. Usually, more ADH is produced at night, presumably so that we don't have to interrupt our sleep as often to urinate. This also explains why the first urine of the day is more concentrated.

Diuretics

Diuretics are chemicals that increase the flow of urine. Drinking alcohol causes diuresis because it inhibits the secretion of ADH. The dehydration that follows is believed to contribute to the symptoms of a hangover. Caffeine is a diuretic because it increases the glomerular filtration rate and decreases the tubular reabsorption of Na^+. Diuretic drugs developed to counteract high blood pressure inhibit active transport of Na^+ at the loop of the nephron or at the distal convoluted tubule. A decrease in water reabsorption and a decrease in blood volume follow.

Acid-Base Balance

The pH scale, as discussed in Chapter 2, can be used to indicate the basicity (alkalinity) or the acidity of body fluids. A basic solution has a lesser hydrogen ion concentration $[H^+]$ than the neutral pH of 7.0, and an acidic solution has a greater $[H^+]$ than neutral pH. The normal pH for body fluids is about 7.4. This is the pH at which our proteins, such as cellular enzymes, function properly. If the blood pH rises above 7.4, a person is said to have **alkalosis,** and if the blood pH decreases below 7.4, a person is said to have **acidosis.** Alkalosis and acidosis are abnormal conditions that may need medical attention.

The foods we eat add basic or acidic substances to the blood, and so does metabolism. For example, cellular respiration adds carbon dioxide that combines with water to form carbonic acid, and fermentation adds lactic acid. The pH of body fluids stays at just about 7.4 via several mechanisms, primarily acid-base buffer systems, the respiratory center, and the kidneys.

Acid-Base Buffer Systems

The pH of the blood stays near 7.4 because the blood is buffered. A **buffer** is a chemical or a combination of chemicals that can take up excess hydrogen ions (H^+) or excess hydroxide ions (OH^-). One of the most important buffers in the blood is a combination of carbonic acid (H_2CO_3) and bicarbonate ions (HCO_3^-). When hydrogen ions (H^+) are added to blood, the following reaction occurs:

$$H^+ + HCO_3^- \longrightarrow H_2CO_3$$

When hydroxide ions (OH^-) are added to blood, this reaction occurs:

$$OH^- + H_2CO_3 \longrightarrow HCO_3^- + H_2O$$

These reactions temporarily prevent any significant change in blood pH. A blood buffer, however, can be overwhelmed unless some more permanent adjustment is made. The next adjustment to keep the pH of the blood constant occurs at pulmonary capillaries.

Respiratory Center

As discussed in Chapter 15, the respiratory center in the medulla oblongata increases the breathing rate if the hydrogen ion concentration of the blood rises. Increasing the breathing rate rids the body of hydrogen ions because the following reaction takes place in pulmonary capillaries:

$$H^+ + HCO_3^- \rightleftharpoons H_2CO_3 \rightleftharpoons H_2O + CO_2$$

In other words, when carbon dioxide is exhaled, this reaction shifts to the right, and the amount of hydrogen ions is reduced.

It is important to have the correct proportion of carbonic acid and bicarbonate ions in the blood. Breathing readjusts this proportion so that this particular acid-base buffer system can continue to absorb both H^+ and OH^- as needed.

The Kidneys

As powerful as the acid-base buffer and the respiratory center mechanisms are, only the kidneys can rid the body of a wide range of acidic and basic substances and otherwise adjust the pH. The kidneys are slower acting than the other two mechanisms, but they have a more powerful effect on pH. For the sake of simplicity, we can think of the kidneys as reabsorbing bicarbonate ions and excreting hydrogen ions as needed to maintain the normal pH of the blood (Fig. 16.8). If the blood is acidic, hydrogen ions are excreted, and bicarbonate ions are reabsorbed. If the blood is basic, hydrogen ions are not excreted, and bicarbonate ions are not reabsorbed. Because the urine is usually acidic, it follows that an excess of hydrogen ions is usually excreted. Ammonia (NH_3) provides another means of buffering and removing the hydrogen ions in urine: ($NH_3 + H^+ \rightarrow NH_4^+$). Ammonia (whose presence is quite obvious in the diaper pail or kitty litter box) is produced in tubule cells by the deamination of amino acids. Phosphate provides another means of buffering hydrogen ions in urine.

The importance of the kidneys' ultimate control over the pH of the blood cannot be overemphasized. As mentioned, the enzymes of cells cannot continue to function if the internal environment does not have near-normal pH.

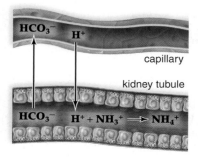

Figure 16.8 Acid-base balance. In the kidneys, bicarbonate ions (HCO_3^-) are reabsorbed and hydrogen ions (H^+) are excreted as needed to maintain the pH of the blood. Excess hydrogen ions are buffered, for example, by ammonia (NH_3), which is produced in tubule cells by the deamination of amino acids.

The kidneys, under the influence of the hormones aldosterone and ANH, regulate the reabsorption of salt. The reabsorption of salt and then water affects blood volume and pressure. The kidneys are more powerful regulators of acid-base (pH) balance than are blood buffers and the respiratory center.

16.4 Disorders of the Urinary System

Various disease conditions can affect the kidneys, ureters, urinary bladder, and urethra. The kidneys are essential for life, so if they are damaged sufficiently by disease, their function must somehow be restored if the patient is to survive.

Disorders of the Kidneys

Many major illnesses that affect other parts of the body, especially diabetes, hypertension, and certain autoimmune diseases, can also cause serious kidney disease. Most of these conditions tend to damage the glomeruli, resulting in a decreased glomerular filtration rate, and eventually kidney failure.

Infection of the kidneys is called **pyelonephritis.** Kidney infections almost always result from infections of the urinary bladder that spread via the ureter(s) to the kidney(s). Kidney infections are usually curable with antibiotics if diagnosed in time. However, an infection called hemolytic uremic syndrome, caused by various bacteria but most famously by a strain of *E. coli* known as O157:H7, can cause serious kidney damage in children even when antibiotics are administered.

Kidney stones are hard granules that can form in the renal pelvis. They may be composed of various materials, such as calcium, phosphate, uric acid, and protein. An estimated 5–15% of people will develop kidney stones at some time in their lives. Factors that contribute to the formation of these stones include too much animal protein in the diet, an imbalance in urinary pH, and/or urinary tract infection. Kidney stones may pass unnoticed in the urine flow. However, when a large kidney stone passes, strong contractions within a ureter can be excruciatingly painful. If a kidney stone grows large enough to block the renal pelvis or ureter, a reverse pressure builds up that may destroy the nephrons. To prevent this, it is sometimes necessary to break up a stone using a technique called lithotripsy, in which powerful ultrasound waves shatter the stone into smaller pieces that can be passed.

One of the first signs of kidney damage is the presence of albumin, white blood cells, and/or red blood cells in the urine. Once more than two-thirds of the nephrons have been destroyed by a disease process, urea and other waste products accumulate in the blood, a condition known as *uremia.* Although nitrogenous wastes can cause serious damage, the retention of water and salts by diseased kidneys is of even greater concern. The latter causes **edema,** the accumulation of fluid in body tissues. Imbalances in the ionic composition of body fluids can lead to loss of consciousness and heart failure.

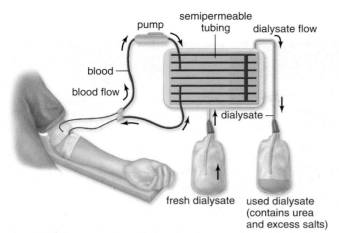

Figure 16.9 An artificial kidney machine.
As the patient's blood is pumped through dialysis tubing, the tubing is exposed to a dialysate (dialysis solution). Wastes exit the blood into the solution because of a preestablished concentration gradient. In this way, blood is not only cleansed, but its water-salt and acid-base balances are also adjusted.

Kidney Transplantation

Patients with renal failure may undergo a kidney transplant, which is the surgical replacement of a defective kidney with a functioning kidney from a donor. As with all organ transplants, organ rejection is a possibility (see Science Focus). Receiving a kidney from a close relative decreases the chances of rejection. The current one-year survival rate following a kidney transplant is 95–98%, but donor organs are in short supply. In the future, it may be possible to use kidneys from pigs or kidneys created in the laboratory (see Chapter 13).

> The kidneys can be affected by diseases of the entire body, such as diabetes, or by specific conditions, such as pyelonephritis and kidney stones. Patients with renal failure require either hemodialysis or a kidney transplant to survive.

Hemodialysis

Patients with renal failure can undergo **hemodialysis,** utilizing either an artificial kidney machine or continuous ambulatory peritoneal dialysis (CAPD). *Dialysis* is the diffusion of dissolved molecules through a semipermeable natural or synthetic membrane having pore sizes that allow only small molecules to pass through. In an artificial kidney machine (Fig. 16.9), the patient's blood is passed through a membranous tube, which is in contact with a dialysis solution, or **dialysate.** Substances more concentrated in the blood diffuse into the dialysate, and substances more common in the dialysate diffuse into the blood. The dialysate is continuously replaced to maintain favorable concentration gradients. In this way, the artificial kidney can be utilized either to extract substances from the blood, including waste products or toxic chemicals and drugs, or to add substances to the blood—for example, bicarbonate ions (HCO_3^-) if the blood is acidic. In the course of a 3- to 6-hour hemodialysis, 50 to 250 grams of urea can be removed from a patient, which greatly exceeds the amount excreted by normal kidneys. Therefore, a patient needs to undergo treatment only about two or three times a week.

Dialysis centers may not be available in rural areas, where CAPD is more commonly used. CAPD, or peritoneal dialysis, is so named because the peritoneum serves as the dialysis membrane. A fresh amount of dialysate is introduced directly into the abdominal cavity from a bag that is temporarily attached to a permanently implanted plastic tube. The dialysate flows into the peritoneal cavity by gravity. Waste and salt molecules pass from blood vessels in the abdominal wall into the dialysate before the fluid is collected several hours later. The solution is drained by gravity from the abdominal cavity into a bag, which is then discarded. One advantage of CAPD over an artificial kidney machine is that the individual can go about many of his or her normal activities during CAPD. A disadvantage, however, is an increased likelihood of infections.

Disorders of the Urinary Bladder and Urethra

Infections are probably the most common cause of problems in the urinary bladder and urethra. Urine leaving the kidneys is normally free of bacteria. The distal parts of the urethra, however, are normally colonized with bacteria. Most of the time, the flow of urine should be one-way from the bladder to the urethra. However, sometimes harmful bacteria from the urethra are able to gain access to the bladder, especially in females, whose urethra is shorter and broader than that of males. This helps explain why females are relatively more prone to bladder infections (Fig. 16.10). In the United States, approximately 25–40% of women aged 20–40 years have had a urinary tract infection. Infections of the

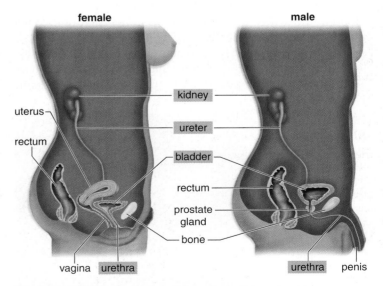

Figure 16.10 Female and male urinary tracts compared. Females have a short urinary tract compared to that of males. This makes it easier for bacteria to invade the urethra and helps explain why females are more likely than males to get a urinary tract infection.

Matching Organs for Transplantation

As we have seen in this chapter, the kidneys are essential for life. Since humans normally have two kidneys, but can function quite normally with only one, the kidney was an obvious choice as the first complex organ to be successfully transplanted (Fig. 16A). This occurred in 1954 at Peter Bent Brigham Hospital in Boston, where one of 23-year-old Ronald Herrick's kidneys was transplanted into his twin, Richard, who was dying of kidney disease. The operation was successful enough to allow Richard to live another eight years, while his brother lived with one kidney for over 50 years.

It was no accident, however, that the first successful transplant was performed between twins. Most cells and tissues of the body have molecules on their surface that can vary between individuals; that is, they are polymorphic. The major blood groups are

Figure 16A
Preparing for transplantation.
A donor kidney, kept moist and sterile inside a plastic bag, arrives in an operating room.

a familiar example; some individuals have only type A glycoproteins on their red blood cells, some have only type B, some have both (blood type AB), and some have neither (blood type O). Because an individual naturally has antibodies against the A or B glycoproteins their own cells lack, it is essential to make sure that donor blood is compatible before giving someone a blood transfusion.

Moreover, since other cells besides red blood cells also have these glycoproteins, the first requirement when attempting to match a potential organ donor with a recipient is to make sure the blood types match.

Another step in matching organs for transplantation involves determining whether the donor and the recipient share the same proteins, called human leukocyte antigens (HLA) or major histocompatibility complex (MHC) proteins. As the latter name implies, these are the major molecules (antigens) that are recognized by the recipient's immune system when transplanted tissue is rejected. There are two types of MHC molecules: class I and class II. Because they are found on almost all cells of the body, the class I molecules probably play a greater role in transplant rejection reactions. Each individual inherits from each parent three genes coding for these MHC-class I molecules,

---

bladder usually result in inflammation of the bladder, or **cystitis,** and infection of the urethra may result in **urethritis.** Almost all urinary tract infections can be cured with appropriate antibiotics.

Bladder stones (Fig. 16.11) can form in people of any age. They most commonly occur as a result of another condition that interferes with normal urine flow, such as a bladder infection with associated inflammation or prostate enlargement in men. They can also form from kidney stones that pass into the urinary bladder and become larger. The most common symptoms of bladder stones are pain, difficulty urinating or increased frequency of urination, and blood in the urine. Similar to kidney stones, bladder stones may be removed surgically or, preferably, by lithotripsy.

The most common type of cancer affecting the urinary system is cancer of the bladder. In the United States, bladder cancer is the fourth most common type of cancer in men and the tenth

most common in women. Because several toxic by-products of tobacco are secreted in the urine, it is estimated that smoking quintuples the risk of bladder cancer. Certain types of bladder cancer are very malignant; in these cases, the bladder must be removed. With no bladder, the ureters must either be diverted into the large intestine or attached to an opening (or stoma) that is created in the skin near the navel. In the latter case, the patient typically must wear a removable plastic bag to collect the urine. Recently, however, researchers have had success in growing entire human bladders in the lab, making the total replacement of human bladders a possibility (Fig. 16.12).

Problems that affect the urinary bladder and urethra include various infections, bladder stones, and cancer.

Figure 16.11
Bladder stones.
Hard granules such as those seen in this x-ray can form and lodge virtually anywhere in the urinary tract, including the kidneys, ureters, and urethra.

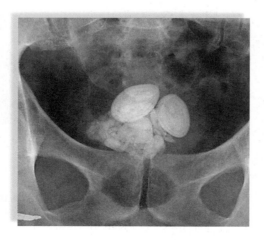

Figure 16.12
Organs grown in the lab. Researchers at Children's Hospital, Boston, have cultured bladder cells on biodegradable bladder-shaped molds to produce an organ that can be implanted in individuals with diseased or defective bladders.

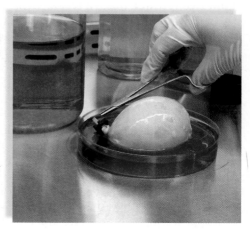

so most people have six different types of MHC-I molecules on their cells. Various types of tests can be done to determine what types of MHC-I molecules the donor cells and recipient cells have, but the donor organ is most compatible when all six match the recipient. This occurs 25% of the time between siblings who have the same mother and father, and also may occur by chance between unrelated individuals.

A final test that is usually performed prior to kidney transplantation is a cross-matching procedure in which the potential donor's cells are mixed with the blood (serum) of the recipient. This is to detect whether the recipient has any antibodies that could damage the cells of the donor kidney. Such antibodies could be present in patients who have been exposed to mismatched MHC molecules through a previous transplantation, blood transfusion, or even pregnancy. Even if the

donor and recipient seem closely matched, if the recipient has preformed antibodies that could react with the donor organ, the organ would probably go to another recipient.

Because a perfect match (such as between two identical twins) is rare, transplantations would not have become as commonplace as they are today without the development of immunosuppressive drugs that help prevent transplant rejection. Unfortunately, at the present time all drugs used for this purpose, such as azathioprine and cyclosporine A, cause a general suppression of the immune system, leaving transplant patients more susceptible to infections. Since transplant patients usually need to take these drugs for the rest of their lives, there is a definite need for new drugs that more specifically target the cells responsible for the transplant rejection reaction, but leave the rest of the immune system intact.

Discussion Questions:

1. Why do you suppose all humans with blood type A, for example, have antibodies against the B glycoprotein, even if they have never had a previous transplant, transfusion, or pregnancy?

2. Even if two individuals are tested and found to have a perfect match of all of their MHC-I and MHC-II molecules, a kidney transplanted between them still has a higher chance of failing compared to a transplant between identical twins. Can you explain why?

3. Besides the matching criteria described here, what other medical issues should be taken into consideration before a decision is made to transplant an organ into a recipient?

Summarizing the Concepts

16.1 Urinary System

The kidneys excrete nitrogenous wastes, including urea, uric acid, and creatinine. They maintain the normal water-salt balance and the acid-base balance of the blood, as well as assisting the endocrine system. The kidneys secrete erythropoietin, which stimulates red blood cell production and releases renin, which leads to the secretion of aldosterone by the adrenal cortex.

The kidneys produce urine, which is conducted by the ureters to the bladder where it is stored before being released by way of the urethra.

16.2 Anatomy of the Kidney and Excretion

Macroscopically, the kidneys are divided into the renal cortex, renal medulla, and renal pelvis. Microscopically, they contain the nephrons.

Each nephron has its own blood supply; the afferent arteriole approaches the glomerular capsule and divides to become the glomerulus, a capillary tuft. The efferent arteriole leaves the capsule and immediately branches into the peritubular capillary network.

A nephron has several parts: the glomerular capsule, the proximal convoluted tubule, the loop of the nephron, and the distal convoluted tubule. The spaces between the podocytes of the glomerular capsule allow small molecules to enter the capsule from the glomerulus. The cuboidal epithelial cells of the proximal convoluted tubule have many mitochondria and microvilli to carry out active transport (following passive transport) from the tubule to the blood. The cuboidal epithelial cells of the distal convoluted tubule have numerous mitochondria but lack microvilli. They carry out active transport from the blood to the tubule.

The steps in urine formation are glomerular filtration (takes place between glomerulus and glomerular capsule), tubular reabsorption (takes place between peritubular capillary and proximal convoluted tubule), and tubular secretion (takes place between peri-

tubular capillary and distal convoluted tubule). During glomerular filtration, water and other small molecules consisting of both nutrients and wastes move from the blood to the inside of the glomerular capsule. During tubular reabsorption, primarily water and nutrients are reabsorbed into the blood, mostly at the proximal convoluted tubule. During tubular secretion, certain substances (e.g., hydrogen ions) are transported into the tubule from the blood. Urine is composed primarily of nitrogenous waste products and salts in water.

16.3 Regulatory Functions of the Kidneys

Reabsorption of water and the production of a hypertonic urine involves three steps. (1) The reabsorption of salt increases blood volume and pressure because more water is also reabsorbed. Two hormones, aldosterone and ANH, control the kidneys' reabsorption of sodium (Na^+). (2) A solute gradient that increases toward the inner medulla is established. The ascending limb of the loop of the nephron establishes this gradient. (3) This gradient draws water from the descending limb of the loop of the nephron and also from the collecting duct. The permeability of the collecting duct is controlled by the hormone ADH.

The kidneys keep blood pH within normal limits. They reabsorb HCO_3^- and excrete H^+ as needed to maintain the pH at about 7.4. Ammonia also buffers H^+ in the urine.

16.4 Disorders of the Urinary System

Various types of medical conditions, including diabetes, kidney stones, and infections, can lead to renal failure, which necessitates undergoing hemodialysis by utilizing a kidney machine or CAPD, or kidney transplantation. The urinary bladder and urethra are also susceptible to infection, bladder stones, and cancer.

Testing Yourself

Choose the best answer for each question.

1. Kidneys control blood pH by excreting
   a. HCO_3^-.
   b. H^+.
   c. Both a and b are correct.
   d. Neither a nor b is correct.

2. Urea is a by-product of _____ metabolism.
    a. glucose   c. fat
    b. salt   d. amino acid
3. Urinary tract infections are usually caused by
    a. viruses.   c. fungi.
    b. bacteria.   d. None of these are correct.
4. Amino acid breakdown in the liver releases
    a. ammonia.   c. fats.
    b. glucose.   d. —COOH.
5. Breakdown of _____ produces _____, which can cause _____.
    a. proteins, fats, gout
    b. nucleotides, ammonia, high blood pressure
    c. nucleotides, uric acid, gout
    d. fats, acids, diabetes
6. An _____ artery exits the glomerulus.
    a. afferent   c. Both a and b are correct.
    b. efferent   d. Neither a nor b is correct.
7. _____ can act as a buffer in urine.
    a. Glucose   c. Amino acids
    b. Ammonia   d. NaOH
8. Which of these functions of the kidneys are mismatched?
    a. excretes metabolic wastes—rids the body of urea
    b. maintains the water-salt balance—helps regulate blood pressure
    c. maintains the acid-base balance—rids the body of uric acid
    d. secretes hormones—secretes erythropoietin
    e. All of these are properly matched.
9. Considering the parts of the nephron involved in urine formation from start to finish, which of these is listed out of order first?
    a. glomerular capsule
    b. proximal convoluted tubule
    c. distal convoluted tubule
    d. loop of the nephron
10. Which hormone causes a rise in blood pressure?
    a. aldosterone   c. erythropoietin
    b. oxytocin   d. atrial natriuretic hormone (ANH)
11. Excretion of a hypertonic urine in humans is most associated with the
    a. glomerular capsule and the tubules.
    b. proximal convoluted tubule only.
    c. loop of the nephron and collecting duct.
    d. distal convoluted tubule and peritubular capillary.
12. The presence of ADH (antidiuretic hormone) causes an individual to excrete
    a. sugars.   c. more water.
    b. less water.   d. Both a and c are correct.
13. In humans, water is
    a. found in the glomerular filtrate.
    b. reabsorbed from the nephron.
    c. in the urine.
    d. All of these are correct.
14. Filtration is associated with the
    a. glomerular capsule.   c. collecting duct.
    b. distal convoluted tubule. d. All of these are correct.
15. Which of the following is a structural difference between the urinary systems of males and females?
    a. Males have a longer urethra than females.
    b. In males, the urethra passes through the prostate.

    c. In males, the urethra serves both the urinary and reproductive systems.
    d. All of these are correct.
16. The function of erythropoietin is
    a. reabsorption of sodium ions.
    b. excretion of potassium ions.
    c. reabsorption of water.
    d. stimulation of red blood cell production.
    e. to increase blood pressure.
17. The purpose of the loop of the nephron in the process of urine formation is
    a. reabsorption of water.   c. reabsorption of solutes.
    b. production of filtrate. d. secretion of solutes.
18. Which of the following materials would not be filtered from the blood at the glomerulus?
    a. water   d. glucose
    b. urea   e. sodium ions
    c. protein
19. Which of the following materials would not be maximally reabsorbed from the filtrate?
    a. water   d. urea
    b. glucose   e. amino acids
    c. sodium ions
20. The renal medulla has a striped appearance due to the presence of which structures?
    a. loop of the nephron   c. peritubular capillaries
    b. collecting duct   d. Both a and b are correct.
21. By what transport process are most molecules secreted from the blood into the tubule?
    a. osmosis   c. active transport
    b. diffusion   d. facilitated diffusion
22. In males, the
    a. urethra carries semen during ejaculation.
    b. prostate gland plays a role in urine formation.
    c. sperm are diluted by urine in the urethra.
    d. posterior pituitary plays a role in urine formation.
23. The cells of the distal convoluted tubule
    a. have many mitochondria.
    b. lack microvilli.
    c. can move molecules from the blood to the tubules.
    d. All of these are correct.
24. H_2CO_3 can be converted to
    a. $H^+ + HCO_3^-$.   c. Both a and b are correct.
    b. $H_2O + CO_2$.   d. Neither a nor b is correct.
25. How much sodium filtered from the glomerulus is returned to the blood?
    a. 99%   c. 25%
    b. 75%   d. 10%
26. _____ reabsorption from the nephrons is, ordinarily, 100%.
    a. Sodium   c. Water
    b. Glucose   d. Urea
27. Urine transport follows which order?
    a. kidneys, urethra, bladder, ureter
    b. kidneys, ureter, bladder, urethra
    c. bladder, kidneys, ureter, urethra
    d. urethra, kidneys, ureter, bladder

28. _____ is the elimination of metabolic waste, and _____ is the elimination of undigested food.
    a. Excretion, absorption
    b. Defecation, excretion
    c. Excretion, defecation
    d. None of these are correct.

29. The use of an artificial kidney is known as
    a. filtration.
    b. excretion.
    c. hemodialysis.
    d. None of these are correct.

30. Label the following diagram of the urinary system and nearby blood vessels.

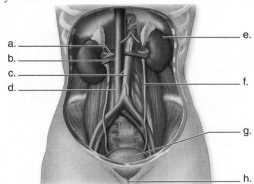

31. Label the following diagram of the kidneys.

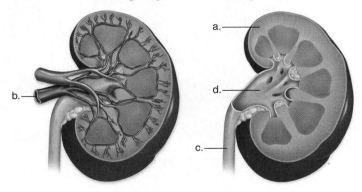

32. Which of these associations is mismatched?
    a. atrial natriuretic hormone (ANH)—secreted by the heart
    b. renin—secreted by the brain
    c. aldosterone—secreted by the adrenal cortex
    d. antidiuretic hormone (ADH)—secreted by the posterior pituitary

Understanding the Terms

acidosis 311
aldosterone 310
alkalosis 311
antidiuretic hormone (ADH) 311
atrial natriuretic hormone (ANH) 310
buffer 311
collecting duct 307
creatinine 304
cystitis 314
dialysate 313
distal convoluted tubule (DCT) 307
diuretic 311
edema 312
erythropoietin 304
excretion 304
glomerular capsule 307
glomerular filtrate 309
glomerular filtration 309
glomerulus 306
gout 304
hemodialysis 313
incontinence 305
juxtaglomerular apparatus 310
kidney 304
kidney stones 312
loop of the nephron 307
micturition 306
nephron 306
peritubular capillary network 306
proximal convoluted tubule (PCT) 307
pyelonephritis 312
renal artery 304
renal cortex 306
renal medulla 306
renal pelvis 306
renal vein 304
renin 310
tubular reabsorption 309
tubular secretion 309
urea 304
ureter 304
urethra 305
urethritis 314
uric acid 304
urinary bladder 305

Match the terms to these definitions:
    a._____ Portion of the nephron lying between the proximal convoluted tubule and the distal convoluted tubule that functions in water reabsorption.
    b._____ Movement of molecules from the contents of the nephron into blood primarily at the proximal convoluted tubule.
    c._____ Hormone secreted by the adrenal cortex that regulates the sodium and potassium balance of the blood.
    d._____ Outer portion of the kidney that appears granular.
    e._____ Surrounds a nephron and functions in reabsorption during urine formation.

Thinking Critically

1. The kidneys filter 180 liters of water every day, which is about 45 gallons. Therefore, it is essential that most of the water be reabsorbed. Can you think of any advantage to this type of system, versus a system that would be more dependent on specific excretion of waste materials without the need to lose and then reabsorb so much water?

2. How do the afferent and efferent arterioles play important, and somewhat opposite, roles in determining the glomerular filtration rate (GFR)?

3. The drug furosemide, brand name Lasix, is known as a "loop diuretic" because it works by inhibiting the transport of sodium ions in the loop of the nephron. How would this inhibition affect urine production, and why?

4. What would be the effect of a deficiency of antidiuretic hormone? Of too much aldosterone?

5. Suppose you developed kidney failure and needed hemodialysis, continuous ambulatory peritoneal dialysis, or a kidney transplant to survive. What would be some of the advantages and disadvantages to consider about each procedure?

ARIS, the *Inquiry into Life* Website

ARIS, the website for *Inquiry into Life*, provides a wealth of information organized and integrated by chapter. You will find practice quizzes, interactive activities, labeling exercises, flashcards, and much more that will complement your learning and understanding of general biology.

www.aris.mhhe.com

Integration and Control of the Human Body

As she sat in her car at the red light, Sarah noticed that the colors didn't seem quite right. "I must be stressed out," she thought to herself, but as she squinted at the traffic signal, the top light definitely seemed more orange than red. "Maybe there's something wrong with this light," she thought, but the next one was the same. At work later that day, she noticed that she was having trouble reading her email, and by the end of the day she had a splitting headache. She kept telling herself that she had just been working too hard. But even as she tried to remain calm, deep down she had a bad feeling. Within a few weeks, she was almost completely blind in one eye, and the sensations in her feet felt muffled, like they were wrapped in gauze. Her doctors diagnosed multiple sclerosis (MS), and were able to treat the inflammation in her optic nerves and spinal cord with high doses of immunosuppressive medications. Over the next few years, Sarah was able to keep her symptoms at bay by injecting herself daily with a drug called beta interferon, but she knew that someday she might need a wheelchair to get around.

MS is an inflammatory disease that affects the myelin sheaths, which wrap parts of some nerve cells like insulation around an electrical cord. As these sheaths deteriorate, the nerves no longer conduct impulses normally. For unknown reasons, MS often attacks the optic nerves first, before spreading to other areas of the brain. Most researchers think MS results from a misdirected attack on myelin by the body's immune system, although an infectious agent may also be involved. Like many diseases that affect the nervous system, there is no cure, so affected individuals must deal with their condition for the rest of their lives.

chapter 17

Nervous System

Chapter Concepts

17.1 Nervous Tissue

The nervous system has two major anatomical divisions. The **central nervous system (CNS)** consists of the brain and spinal cord, which are located in the midline of the body. The **peripheral nervous system (PNS)** consists of nerves that carry sensory messages to the CNS and motor commands from the CNS to the muscles and glands (Fig. 17.1). The division between the CNS and the PNS is arbitrary; the two systems work together and are connected to one another.

The nervous system contains two types of cells: neurons and neuroglia (neuroglial cells). **Neurons** are the cells that transmit nerve impulses between parts of the nervous system; **neuroglia** support and nourish neurons, maintain homeostasis, form myelin, and may aid in signal transmission. This section discusses the structure and function of neurons, the myelin sheath that surrounds cell extensions called axons, and how nerve impulses are transmitted through neurons to effectors.

Types of Neurons and Neuron Structure

Although the structure of the nervous system is extremely complex, the principles of its operation are simple. There are three classes of neurons: sensory neurons, interneurons, and motor neurons (Fig. 17.2). Their functions are best described in relation to the CNS. A **sensory neuron** takes messages to the CNS. Sensory neurons may be equipped with specialized endings called sensory receptors that detect changes in the environment. An **interneuron** lies entirely within the CNS. Interneurons can receive input from sensory neurons and also from other interneurons in the CNS. Thereafter, they sum up all the messages received from these neurons before they communicate with motor neurons. A **motor neuron** takes messages away from the CNS to an effector (an organ, muscle fiber, or gland). Effectors carry out our responses to environmental changes.

Neurons vary in appearance, but all of them have just three parts: a cell body, dendrites, and an axon. The **cell body** contains the nucleus, as well as other organelles. **Dendrites** are extensions leading toward the cell body that receive signals from other neurons and send them on to the cell body. An **axon** conducts nerve impulses away from the cell body toward other neurons or effectors.

All neurons have three parts: a cell body, dendrites, and an axon. Sensory neurons take messages to the CNS, and interneurons sum up sensory input before motor neurons take commands away from the CNS.

a.

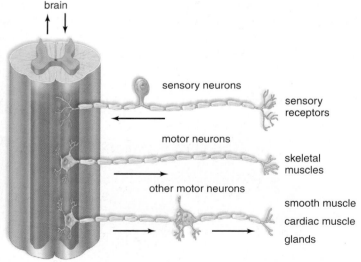

brain

sensory neurons

sensory receptors

motor neurons

skeletal muscles

other motor neurons

smooth muscle

cardiac muscle

glands

Central nervous system   **Peripheral nervous system**

b.

Figure 17.1 Organization of the nervous system.
a. In paraplegics, messages no longer flow between the body's lower limbs and the central nervous system (the spinal cord and brain). **b.** The sensory neurons of the peripheral nervous system take nerve impulses from sensory receptors to the central nervous system (CNS), and motor neurons take nerve impulses from the CNS to muscles and glands.

Myelin Sheath

Some axons are covered by a protective **myelin sheath.** In the PNS, this covering is formed by a type of neuroglia called **Schwann cells,** which contain the lipid substance myelin in their plasma membranes. The myelin sheath develops when Schwann cells wrap themselves around an axon as many as 100 times and in this way lay down many layers of plasma membrane. Since each Schwann cell myelinates only part of an axon, the myelin sheath is interrupted. The gaps where there is no myelin sheath are called **nodes of Ranvier** (Fig. 17.3). Each

Schwann cell only covers about one millimeter of an axon. Since a single axon may be several feet long (if it runs from the spinal cord to the foot, for example), several hundred or more Schwann cells may be required to myelinate a single axon.

In the PNS, myelin gives nerve fibers their white, glistening appearance and serves as an excellent insulator. The myelin sheath also plays an important role in nerve regeneration within the PNS. If an axon is accidentally severed, the myelin sheath may remain and serve as a passageway for new fiber growth. In the CNS, myelin is produced by the **oligodendroglial cells.** Unlike in the PNS, nerve regeneration does not seem to occur to any significant degree in the CNS.

The CNS is composed of two types of nervous tissue—gray matter and white matter. **Gray matter** is gray because it contains cell bodies and short, nonmyelinated fibers. **White matter** is white because it contains myelinated

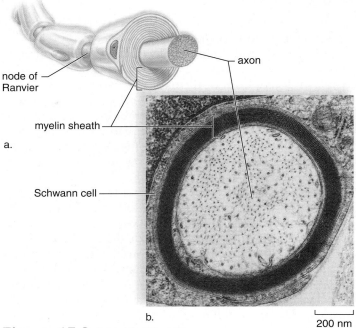

a.

b.   200 nm

Figure 17.3 Myelin sheath.
a. In the PNS, a myelin sheath forms when Schwann cells wrap themselves around an axon. **b.** Electron micrograph of a cross section of an axon surrounded by a myelin sheath.

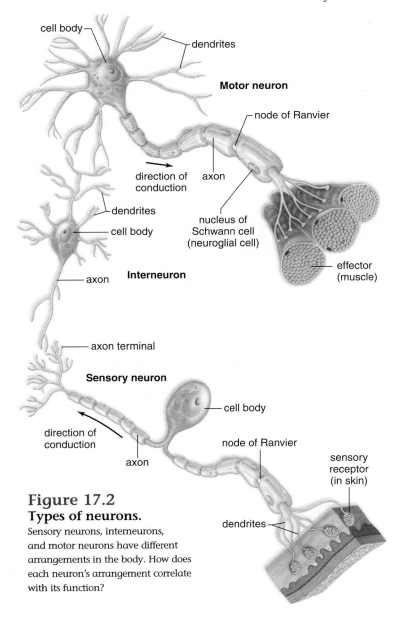

Figure 17.2
Types of neurons.

Sensory neurons, interneurons, and motor neurons have different arrangements in the body. How does each neuron's arrangement correlate with its function?

axons that run together in bundles called **tracts.** In the CNS, neurons with short axons make up the gray matter because no part of these neurons has a myelin sheath. However, some neurons in the CNS have long axons with a myelin sheath. These long axons, which make up the white matter of the brain, are carrying messages from one part of the CNS to another. The surface layer of the brain is gray matter, and the white matter lies deep within the gray matter. The central part of the spinal cord consists of gray matter, and the white matter surrounds the gray matter.

> Myelin sheaths cover the axons of most of the neurons in the PNS, and those that make up the white matter of the brain and spinal cord. Myelin is produced by Schwann cells in the PNS, and by oligodendroglial cells in the CNS.

The Nerve Impulse

The nervous system uses the **nerve impulse** to convey information. The nature of a nerve impulse has been studied by using excised axons and a voltmeter called an oscilloscope. Voltage, which is expressed in millivolts (mV), is a measure of the electrical potential difference between two points. In the case of a neuron, the two points are the inside and the outside of the axon. Voltage is displayed on the voltmeter screen as a trace, or pattern, over time.

Resting Potential

In the experimental setup shown in Figure 17.4a, a voltmeter is wired to two electrodes: one electrode is placed inside an axon, and the other electrode is placed outside. The axon is essentially a membranous tube filled with axoplasm (cytoplasm of the axon). When the axon is not conducting an impulse, the voltmeter records a potential difference across an axonal membrane (membrane of axon) equal to about –65 mV. This reading indicates that the inside of the axon is negative compared to the outside. This is called the **resting potential** because the axon is not conducting an impulse.

The existence of the polarity (charge difference) correlates with a difference in ion distribution on either side of the axonal membrane. As Figure 17.4a shows, the concentration of sodium ions (Na^+) is greater outside the axon than inside, and the concentration of potassium ions (K^+) is greater inside the axon than outside. The unequal distribution of these ions is due to the action of the **sodium-potassium pump,** a membrane protein that actively transports Na^+ out of and K^+ into the axon. The work of the pump maintains the unequal distribution of Na^+ and K^+ across the membrane.

The pump is always working because the membrane is somewhat permeable to these ions, and they tend to diffuse toward their lesser concentration. Since the membrane is more permeable to K^+ than to Na^+, there are always more positive ions outside the membrane than inside. This accounts for the polarity recorded by the voltmeter. Large, negatively charged organic ions in the axoplasm also contribute to the polarity across a resting axonal membrane.

> Because of the sodium-potassium pump, there is a greater concentration of Na^+ outside an axon and a greater concentration of K^+ inside an axon. An unequal distribution of ions causes the inside of an axon to be negative compared to the outside.

Action Potential

An **action potential** is a rapid change in polarity across an axonal membrane as the nerve impulse occurs. An action

Figure 17.4 Resting and action potentials of the axonal membrane.

a. Resting potential. A voltmeter that records voltage changes indicates the axonal membrane has a resting potential of –65 mV. There is a preponderance of Na^+ outside the axon and a preponderance of K^+ inside the axon. The permeability of the membrane to K^+ compared to Na^+ causes the inside to be negative compared to the outside. **b.** Action potential. Depolarization occurs when Na^+ gates open and Na^+ moves inside the axon. **c.** Repolarization occurs when K^+ gates open and K^+ moves outside the axon. **d.** Graph of the action potential.

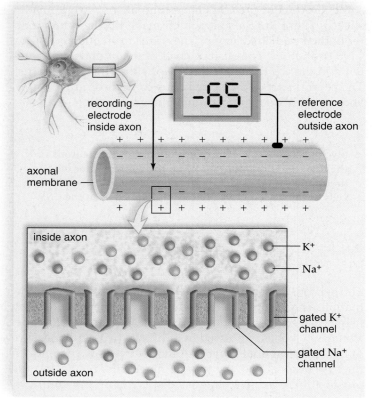

a. Resting potential: more Na^+ outside the axon and more K^+ inside the axon causes polarization.

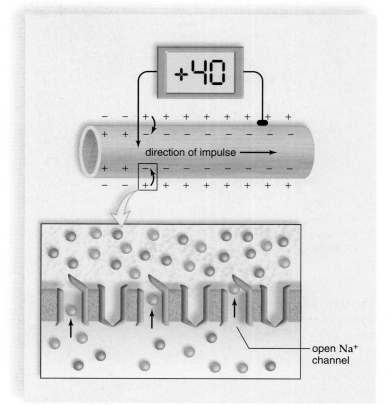

b. Action potential begins: depolarization occurs when Na^+ gates open and Na^+ moves inside the axon.

potential is an all-or-none phenomenon. If a stimulus causes the axonal membrane to depolarize to a certain level, called **threshold,** an action potential occurs. The strength of an action potential does not change, but an intense stimulus can cause an axon to fire (start an axon potential) more often in a given time interval than a weak stimulus.

The action potential requires two types of gated channel proteins in the membrane. One gated channel protein opens to allow Na+ to pass through the membrane to inside the cell, and the second channel opens to allow K+ to pass through the membrane to outside the cell (Fig. 17.4b,c).

Sodium Gates Open When an action potential begins, the gates of sodium channels open first, and Na+ flows down its concentration gradient into the axon. As Na+ moves inside the axon, the membrane potential changes from –65 mV to +40 mV. This is called a *depolarization* because the charge inside the axon changes from negative to positive (Fig. 17.4b).

Potassium Gates Open Second, the gates of potassium channels open, and K+ flows down its concentration gradient to outside the axon. As K+ moves outside the axon, the action potential changes from +40 mV back to –65 mV. This is a *repolarization* because the inside of the axon resumes a negative charge as K+ exits the axon (Fig. 17.4c). An action potential only takes two milliseconds. In order to visualize such rapid fluctuations in voltage across the axonal membrane, researchers generally find it useful to plot the voltage changes over time (Fig. 17.4d).

Conduction of an Action Potential In nonmyelinated axons, the action potential travels down an axon one small section at a time. As soon as an action potential has moved on, the previous section undergoes a **refractory period,** during which the sodium gates are unable to open. Notice, therefore, that the action potential cannot move backward and instead always moves down an axon toward its terminals. When the refractory period is over, the sodium-potassium pump has restored the previous ion distribution by pumping Na+ to outside the axon and K+ to inside the axon.

In myelinated axons, the gated ion channels that produce an action potential are concentrated at the nodes of Ranvier. Since ion exchange occurs only at the nodes, the action potential travels faster than in nonmyelinated axons. This is called saltatory conduction, meaning that the action potential "jumps" from node to node. Speeds of 200 meters per second (450 miles per hour) have been recorded.

An action potential is an electrochemical change that travels along the length of the axon as Na+ moves to the inside of the axon. During repolarization, K+ moves to the outside of the axon.

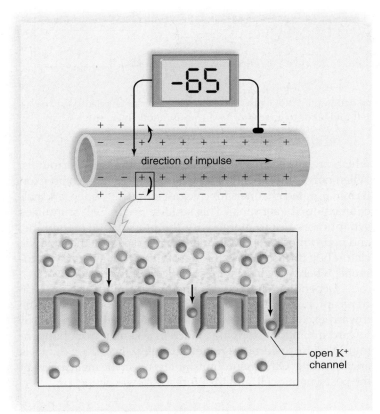

c. Action potential ends: repolarization occurs when K+ gates open and K+ moves outside the axon.

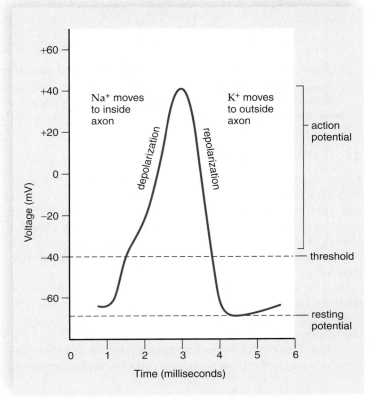

d. An action potential can be visualized if voltage changes are graphed over time.

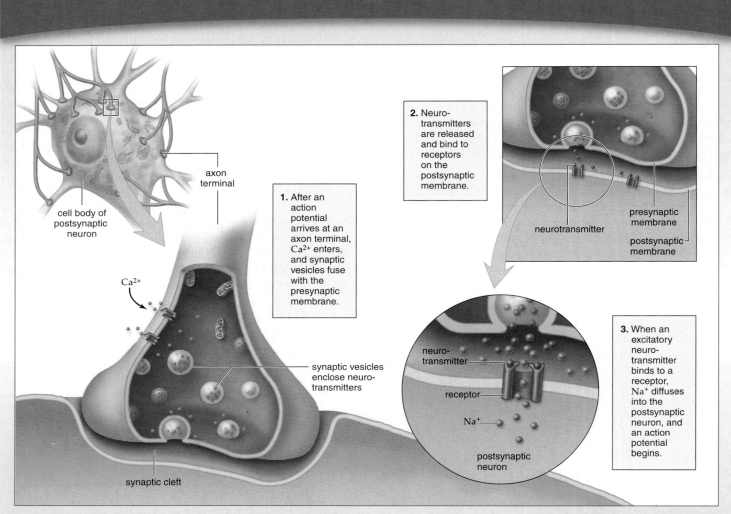

1. After an action potential arrives at an axon terminal, Ca^{2+} enters, and synaptic vesicles fuse with the presynaptic membrane.

2. Neurotransmitters are released and bind to receptors on the postsynaptic membrane.

3. When an excitatory neurotransmitter binds to a receptor, Na^+ diffuses into the postsynaptic neuron, and an action potential begins.

Figure 17.5 Chemical synapse structure and function.
Transmission across a synapse from one neuron to another occurs when an action potential causes a neurotransmitter to be released at the presynaptic membrane. The neurotransmitter diffuses across a synaptic cleft and binds to a receptor in the postsynaptic membrane. An action potential may begin.

Transmission Across a Synapse

Every axon branches into many fine endings, each tipped by a small swelling called an **axon terminal.** Each terminal lies very close to either the dendrite or the cell body of another neuron. This region of close proximity is called a **chemical synapse** (Fig. 17.5). It is important to note that the two neurons at a chemical synapse don't ever physically touch each other. Instead, they are separated by a tiny gap called the **synaptic cleft.** At a synapse, the membrane of the first neuron is called the *pre*synaptic membrane, and the membrane of the next neuron is called the *post*synaptic membrane.

Communication between the two neurons at a chemical synapse is carried out by molecules called **neurotransmitters,** which are stored in synaptic vesicles in the axon terminals. When nerve impulses traveling along an axon reach an axon terminal, gated channels for calcium ions (Ca^{2+}) open, and calcium enters the terminal. This sudden rise in Ca^{2+} stimulates synaptic vesicles to merge with the presynaptic membrane, and neurotransmitter molecules are released into the synaptic cleft. They diffuse across the cleft to the post-synaptic membrane, where they bind with specific receptor proteins.

Depending on the type of neurotransmitter and the type of receptor, the response of the postsynaptic neuron can be toward excitation (causing an action potential to happen) or toward inhibition (stopping an action potential from happening). Excitatory neurotransmitters that utilize gated ion channels are fast acting. Other neurotransmitters affect the metabolism of the postsynaptic cell, and therefore are slower acting.

Synaptic Integration

A single neuron may have many dendrites that carry signals to the cell body. Both dendrites and the cell body can have synapses with many other neurons. A neuron is on the receiving end of many excitatory and inhibitory signals. An excitatory neurotransmitter produces a potential change called a *signal* that drives the neuron closer to an action potential; an inhibitory neurotransmitter produces a signal that drives the neuron farther from an action potential. Excitatory signals have a depolarizing effect, and inhibitory signals usually have a hyperpolarizing effect; that is, they increase the potential difference across the axonal membrane.

Neurons integrate these incoming signals. **Synaptic integration** is the summing up of excitatory and inhibitory signals (Fig. 17.6). If a neuron receives many excitatory signals (either from different synapses or at a rapid rate from one synapse), the chances are that the neuron's axon will transmit a nerve impulse. On the other hand, if a neuron receives both inhibitory and excitatory signals, the summing up of these signals may prohibit the axon from firing.

> Synaptic integration is the summing up of inhibitory and excitatory signals received by a postsynaptic neuron.

Neurotransmitters

At least 25 different neurotransmitters have been identified, but two very well-known ones are **acetylcholine (ACh)** and **norepinephrine (NE).**

Once a neurotransmitter has been released into a synaptic cleft and has initiated a response, it is removed from the cleft. In some synapses, the postsynaptic membrane contains enzymes that rapidly inactivate the neurotransmitter. For example, the enzyme **acetylcholinesterase (AChE)** breaks down acetylcholine. In other synapses, the presynaptic membrane rapidly reabsorbs the neurotransmitter, possibly for repackaging in synaptic vesicles or for molecular breakdown. The short existence of neurotransmitters at a synapse prevents continuous stimulation (or inhibition) of postsynaptic membranes.

It is of interest to note here that many drugs that affect the nervous system act by interfering with or potentiating (enhancing) the action of neurotransmitters. As described in Figure 17.18, drugs can enhance or block the release of a neurotransmitter, mimic the action of a neurotransmitter or block the receptor, or interfere with the removal of a neurotransmitter from a synaptic cleft.

> Neurotransmitters diffuse across the synaptic cleft from one neuron to the next.

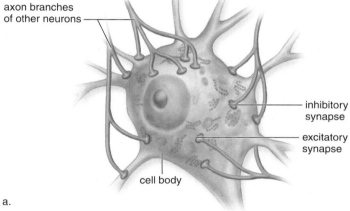

cell body of the neuron   axon terminals

axon branches of other neurons

inhibitory synapse

excitatory synapse

cell body

a.

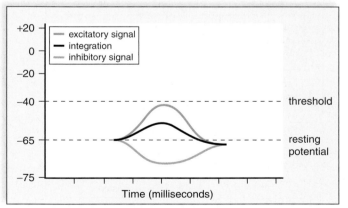

b.

Figure 17.6 Synaptic integration.

a. Inhibitory signals and excitatory signals are summed up in the dendrites and cell body of the postsynaptic neuron. Only if the combined signals cause the membrane potential to rise above threshold does an action potential occur. **b.** In this example, threshold was not reached.

17.2 The Central Nervous System

The spinal cord and the brain make up the central nervous system (CNS), where sensory information is received and motor control is initiated. Figure 17.7 illustrates how the CNS relates to the PNS. Both the spinal cord and the brain are protected by bone; the spinal cord is surrounded by vertebrae, and the brain is enclosed by the skull. Also, both the spinal cord and the brain are wrapped in protective membranes known as **meninges** (sing., meninx). The spaces between the meninges are filled with **cerebrospinal fluid (CSF),** which cushions and protects the CNS. A small amount of this fluid is sometimes withdrawn from around the cord for laboratory testing when a spinal tap (lumbar puncture) is performed to diagnose infections or other conditions that affect the brain and/or meninges.

The brain has hollow interconnecting cavities termed **ventricles,** which also connect with the hollow **central canal** of the spinal cord. The ventricles produce and serve as a reservoir for cerebrospinal fluid as does the central canal.

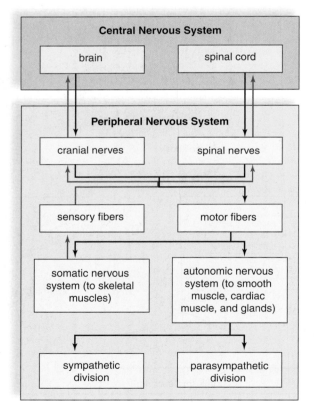

Figure 17.7 Organization of the nervous system.
The CNS is composed of the spinal cord and brain; the PNS is composed of the cranial and spinal nerves. Nerves contain both sensory and motor fibers. In the somatic system, nerves conduct impulses from sensory receptors to the CNS and motor impulses from the CNS to the skeletal muscles. In the autonomic system, consisting of the sympathetic and parasympathetic divisions, motor impulses travel to smooth muscle, cardiac muscle, and glands.

Normally, any excess cerebrospinal fluid drains away into the cardiovascular system. However, blockages can occur. In an infant, the brain and skull can enlarge due to cerebrospinal fluid accumulation, resulting in a condition called hydrocephalus ("water on the brain"). If cerebrospinal fluid collects in an adult, the brain cannot enlarge and instead is pushed against the skull. Regardless of age, the condition can cause brain damage if not corrected.

The CNS, which lies in the midline of the body and consists of the brain and the spinal cord, receives sensory information and initiates motor control.

The Spinal Cord

The **spinal cord** extends from the base of the brain through a large opening in the skull called the foramen magnum and into the vertebral canal formed by openings in the vertebrae.

Structure of the Spinal Cord

Figure 17.8*a* shows how an individual vertebra protects the spinal cord. The spinal nerves project from the cord between the vertebrae that make up the vertebral column. Fluid-filled **intervertebral disks** cushion and separate the vertebrae. If a disk ruptures, the vertebrae press on the spinal cord and spinal nerves, causing pain and loss of motor function.

A cross section of the spinal cord shows a central canal, gray matter, and white matter (Fig. 17.8*b,c*). The central canal contains cerebrospinal fluid, as do the meninges that protect the spinal cord. The gray matter is centrally located and shaped like the letter H. Portions of sensory neurons and motor neurons are found there, as are interneurons that communicate with these two types of neurons. The dorsal root of a spinal nerve contains sensory fibers entering the gray matter, and the ventral root of a spinal nerve contains motor fibers exiting the gray matter. The dorsal and ventral roots join before the spinal nerve leaves the vertebral canal. Spinal nerves are a part of the PNS.

The white matter of the spinal cord surrounds the gray matter. The white matter contains ascending tracts taking information to the brain (primarily located dorsally) and descending tracts taking information from the brain (primarily located ventrally). Because the tracts cross after they enter and exit the CNS, the left side of the brain controls the right side of the body, and the right side of the brain controls the left side of the body.

The spinal cord extends from the base of the brain into the vertebral canal formed by the vertebrae. A cross section shows that the spinal cord has a central canal, gray matter, and white matter.

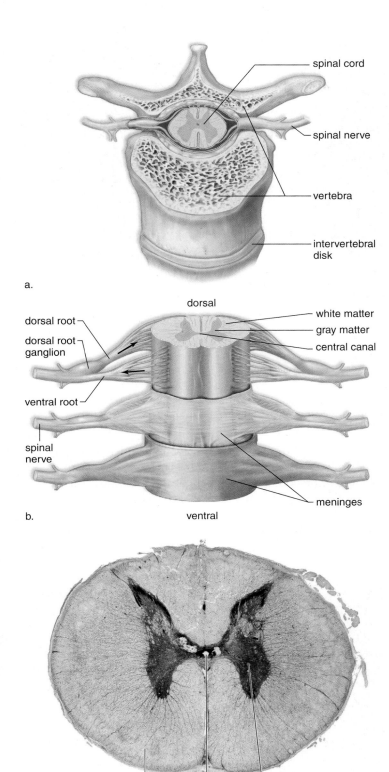

a.

b.

c.

Figure 17.8 Spinal cord.
a. The spinal cord passes through the vertebral canal formed by the vertebrae. **b.** The spinal cord has a central canal filled with cerebrospinal fluid, gray matter in an H-shaped configuration, and white matter. The white matter contains tracts that take nerve impulses to and from the brain. **c.** Photomicrograph of a cross section of the spinal cord.

Functions of the Spinal Cord

The spinal cord provides a means of communication between the brain and the peripheral nerves that leave the cord. When someone touches your hand, sensory receptors generate nerve impulses that pass through sensory fibers to the spinal cord and up ascending tracts to the brain. When we voluntarily move our limbs, motor impulses originating in the brain pass down descending tracts to the spinal cord and out to our muscles by way of motor fibers. Therefore, if the spinal cord is severed, we suffer a loss of sensation and a loss of voluntary control—that is, paralysis (see section 17.6).

We will see that the spinal cord is also the center for thousands of reflex arcs. A stimulus causes sensory receptors to generate nerve impulses that travel in sensory axons to the spinal cord. Interneurons integrate the incoming data and relay signals to motor neurons. The reflex is complete when motor axons cause skeletal muscles to contract or a gland or an organ to respond. Each interneuron in the spinal cord has synapses with many other neurons, and therefore, they send signals to several other interneurons and motor neurons.

The spinal cord plays a similar role for the internal organs. For example, when blood pressure falls, sensory receptors in the carotid arteries and aorta generate nerve impulses that pass through sensory fibers to the cord and then up an ascending tract to a cardiovascular center in the brain. Thereafter, nerve impulses pass down a descending tract to the spinal cord. Motor impulses then cause blood vessels to constrict so that the blood pressure rises.

The spinal cord serves as a means of communication between the brain and much of the body. The spinal cord is also a center for reflex actions.

The Brain

We will discuss the brain with reference to its four major parts: the cerebrum, the diencephalon, the cerebellum, and the brain stem. It will be helpful for you to associate these four regions with the brain's ventricles: the cerebrum with the two lateral ventricles, the diencephalon with the third ventricle, and the brain stem and the cerebellum with the fourth ventricle (Fig. 17.9a).

The Cerebrum

The **cerebrum** is the largest portion of the brain in humans. The cerebrum is the last center to receive sensory input and carry out integration before commanding voluntary motor responses. It communicates with and coordinates the activities of the other parts of the brain. The cerebrum carries out the higher thought processes required for learning and memory and for language and speech.

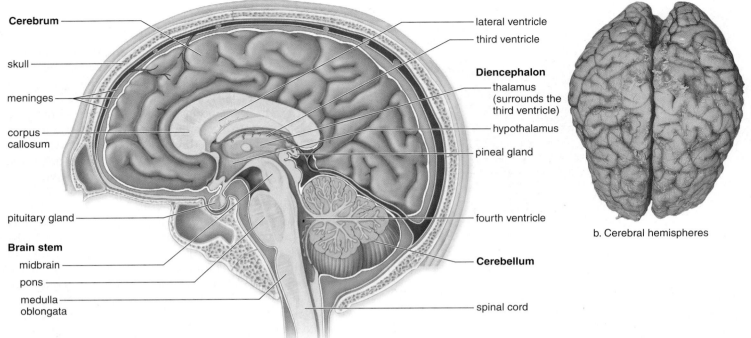

Cerebrum

skull

meninges

corpus
callosum

pituitary gland

Brain stem

midbrain

pons

medulla
oblongata

lateral ventricle

third ventricle

Diencephalon

thalamus
(surrounds the
third ventricle)

hypothalamus

pineal gland

fourth ventricle

Cerebellum

spinal cord

b. Cerebral hemispheres

a. Parts of brain

Figure 17.9 The human brain.

a. The cerebrum, seen here in longitudinal section, is the largest part of the brain in humans. The right cerebral hemisphere is shown here. **b.** Viewed from above, the cerebrum has left and right cerebral hemispheres. The hemispheres are connected by the corpus callosum.

The Cerebral Hemispheres   Just as the human body has two halves, so does the cerebrum. These halves are called the left and right **cerebral hemispheres** (Fig. 17.9b). A deep groove, the longitudinal *fissure,* divides the left and right cerebral hemispheres. Still, the two cerebral hemispheres are connected by a bridge of white matter within the **corpus callosum.**

Shallow grooves called *sulci* (sing., sulcus) divide each hemisphere into lobes (Fig. 17.10). The *frontal lobe* is the most ventral of the lobes (directly behind the forehead). The *parietal lobe* is dorsal to the frontal lobe. The *occipital lobe* is dorsal to the parietal lobe (at the rear of the head). The *temporal lobe* lies inferior to the frontal and parietal lobes (at the temple and the ear).

Figure 17.10
The lobes of a cerebral hemisphere.

Each cerebral hemisphere is divided into four lobes: frontal, parietal, temporal, and occipital. The frontal lobe contains centers for reasoning and movement, the parietal lobe for somatic sensing and taste, the temporal lobe for hearing, and the occipital lobe for vision.

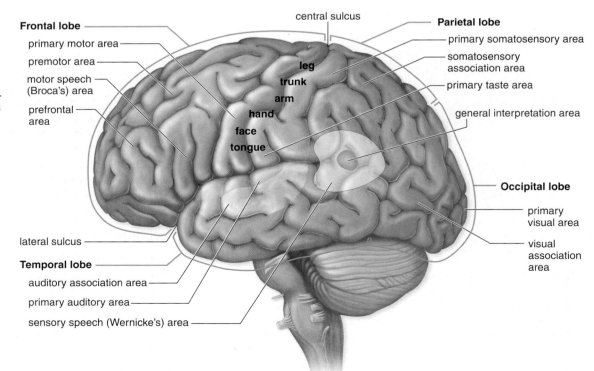

Frontal lobe

primary motor area

premotor area

motor speech
(Broca's) area

prefrontal
area

lateral sulcus

Temporal lobe

auditory association area

primary auditory area

sensory speech (Wernicke's) area

central sulcus

leg
trunk
arm
hand
face
tongue

Parietal lobe

primary somatosensory area

somatosensory
association area

primary taste area

general interpretation area

Occipital lobe

primary
visual area

visual
association
area

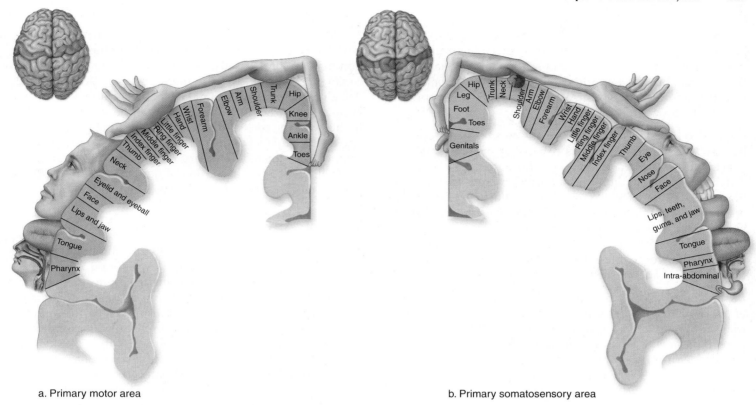

a. Primary motor area

b. Primary somatosensory area

Figure 17.11 The primary motor and somatosensory areas.
In these drawings, the size of the body part reflects the amount of cerebral cortex devoted to that body part. For example, the amount of primary motor cortex **(a)** and somatosensory cortex **(b)** devoted to the thumb, fingers, and hand is greater than that for the foot and toes.

The **cerebral cortex** is a thin but highly convoluted outer layer of gray matter that covers the cerebral hemispheres. Each fold, or convolution, in the cortex is called a gyrus. The cerebral cortex contains over one billion cell bodies and is the region of the brain that accounts for sensation, voluntary movement, and all the thought processes we associate with consciousness.

Primary Motor and Sensory Areas of the Cortex   The **primary motor area** is in the frontal lobe just ventral to the central sulcus. Voluntary commands to skeletal muscles begin in the primary motor area, and each part of the body is controlled by a certain section. Control of the versatile human hand takes up an especially large portion of the primary motor area (Fig. 17.11).

The **primary somatosensory area** is just dorsal to the central sulcus in the parietal lobe. Sensory information from the skin and skeletal muscles arrives here, where each part of the body is sequentially represented. A primary visual area in the occipital lobe receives information from our eyes, and a primary auditory area in the temporal lobe receives information from our ears. A primary taste area in the parietal lobe accounts for taste sensations, and a primary olfactory area for smell is located in the temporal lobe.

Association Areas   **Association areas** are places where integration occurs. Ventral to the primary motor area is a premotor area. The premotor area organizes motor functions for skilled motor activities, such as being able to walk and talk at the same time, and then the primary motor area sends signals to the cerebellum, which integrates them. Even a brief period of oxygen deprivation during birth can damage the motor areas of the cerebral cortex, which is one possible cause of cerebral palsy, a condition characterized by a spastic weakness of the arms and legs.

The somatosensory association area, located just dorsal to the primary somatosensory area, processes and analyzes sensory information from the skin and muscles. The visual association area associates new visual information with previously received visual information. It might "decide," for example, whether we have seen this face or scene or symbol in the past before. The auditory association area performs the same functions with regard to sounds.

Processing Centers   Processing centers of the cortex receive information from the other association areas and perform higher-level analytical functions. The **prefrontal area,** a processing center in the frontal lobe, receives information from the other association areas and uses it to reason and plan our actions. Integration in this center accounts for our most cherished human abilities to think critically and to formulate appropriate behaviors.

The unique ability of humans to speak is partially dependent upon two processing centers found only in the left cerebral cortex: **Wernicke's area** in the dorsal part of the left temporal lobe and **Broca's area** in the left frontal lobe. Broca's area is located just ventral to the portion of the primary motor area for speech musculature (lips, tongue, larynx, and so forth) (see also Fig. 17.14). Wernicke's area helps us understand both the written and spoken word and sends the information to Broca's area. Broca's area adds grammatical refinements and directs the primary motor area to stimulate the appropriate muscles for speaking.

Central White Matter Much of the rest of the cerebrum beneath the cerebral cortex is composed of white matter. Tracts within the cerebrum take information between the different sensory, motor, and association areas pictured in Figures 17.10 and 17.11. The corpus callosum, previously mentioned, contains tracts that join the two cerebral hemispheres. Descending tracts from the primary motor area communicate with various parts of the brain, and ascending tracts from lower brain centers send sensory information up to the primary somatosensory area.

Basal Nuclei While the bulk of the cerebrum is composed of tracts, there are masses of gray matter located deep within the white matter. These so-called **basal nuclei** (formerly termed basal ganglia) integrate motor commands, ensuring that proper muscle groups are activated or inhibited.

The Diencephalon

The hypothalamus and the thalamus are in the **diencephalon,** a region that encircles the third ventricle (see Fig. 17.9*a*). The **hypothalamus** forms the floor of the third ventricle. The hypothalamus is an integrating center that helps maintain homeostasis by regulating hunger, sleep, thirst, body temperature, and water balance. The hypothalamus manufactures hormones and controls the pituitary gland. Therefore, the hypothalamus serves as a link between the nervous and endocrine systems.

The **thalamus** consists of two masses of gray matter that form the sides and roof of the third ventricle. Most of the sensory input from the visual, auditory, taste, and somatosensory systems arrives at the thalamus via the cranial nerves and tracts from the spinal cord. The thalamus integrates this information and sends it on to the appropriate portions of the cerebrum. Most information from the olfactory system, however, goes directly to the olfactory part of the cortex without passing through the thalamus, perhaps because olfaction is an evolutionarily older system compared to the other senses. The thalamus is involved in arousal of the cerebrum, and it also participates in higher mental functions, such as memory and emotions.

The pineal gland, which secretes the hormone melatonin and regulates our body's daily rhythms, is located in the diencephalon (see Fig. 17.9*a*).

The Cerebellum

The **cerebellum** is separated from the brain stem by the fourth ventricle (see Fig. 17.9*a*). The cerebellum has two portions that are joined by a narrow median portion. Each portion is primarily composed of white matter, which, in longitudinal section, has a treelike pattern. Overlying the white matter is a thin layer of gray matter that forms a series of complex folds.

The cerebellum receives sensory input from the joints, muscles, and other sensory pathways about the present position of body parts. It also receives motor output from the cerebral cortex about where these parts should be located. After integrating this information, the cerebellum sends motor impulses, by way of the brain stem, to the skeletal muscles. In this way, the cerebellum maintains posture and balance. It also ensures that all of the muscles work together to produce smooth, coordinated voluntary movements. In addition, the cerebellum assists the learning of new motor skills, such as playing the piano or hitting a baseball.

The Brain Stem

The **brain stem** contains the midbrain, the pons, and the medulla oblongata (see Fig. 17.9*a*). The **midbrain** acts as a relay station for tracts passing between the cerebrum and the spinal cord or cerebellum. It also has reflex centers for visual, auditory, and tactile responses. The word *pons* means bridge in Latin, and true to its name, the **pons** contains bundles of axons traveling between the cerebellum and the rest of the CNS. In addition, the pons functions with the medulla oblongata to regulate the breathing rate and has reflex centers concerned with head movements in response to visual and auditory stimuli.

The **medulla oblongata** contains a number of reflex centers for regulating heartbeat, breathing, and vasoconstriction (blood pressure). It also contains the reflex centers for vomiting, coughing, sneezing, hiccuping, and swallowing. The medulla oblongata lies just superior to the spinal cord. It contains tracts that ascend or descend between the spinal cord and higher brain centers.

The electrical activity of the brain can be recorded in the form of an **electroencephalogram (EEG).** Electrodes are taped to different parts of the scalp, and an instrument records the so-called brain waves. The EEG is a diagnostic tool; for example, an irregular pattern can signify epilepsy or a brain tumor. A flat EEG signifies brain death, which is the subject of the Bioethical Focus at the end of this chapter.

The brain is made up of the cerebrum, the diencephalon, the cerebellum, and the brain stem.

17.3 The Limbic System and Higher Mental Functions

Emotions and higher mental functions are associated with the limbic system in the brain. The limbic system blends primitive emotions (rage, fear, joy, sadness) and higher mental functions (reason, memory) into a united whole. It accounts for why activities such as sexual behavior and eating seem pleasurable and also for why, say, mental stress can cause high blood pressure.

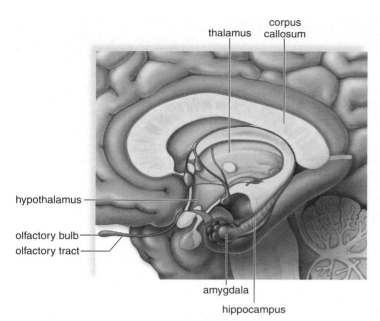

thalamus   corpus callosum

hypothalamus

olfactory bulb
olfactory tract

amygdala

hippocampus

Figure 17.12 The limbic system.
Structures deep within the cerebral hemispheres and surrounding the diencephalon join higher mental functions such as reasoning with more primitive feelings such as fear and pleasure.

Anatomy of the Limbic System

The **limbic system** is a complex network of tracts and nuclei that incorporates portions of the cerebral lobes, the basal nuclei, and the diencephalon. Two significant structures within the limbic system are the hippocampus and the amygdala, which are essential for learning and memory (Fig. 17.12). The **hippo-campus** is a seahorse-shaped structure deep in the temporal lobe that is well situated to communicate with the prefrontal area of the brain, which is also involved in learning and memory. The **amygdala** is an almond-shaped structure in the limbic system that allows us to respond to and display anger, avoidance, defensiveness, and fear. The amygdala prompts release of adrenaline and other hormones into the bloodstream. The inclusion of the frontal lobe in the limbic system means that we may be able to control strong feelings, despite the efforts of the amygdala to disrupt rational thought.

> The limbic system includes the hippocampus, which is involved in learning and memory, and the amygdala, which deals with emotions such as anger and fear.

Higher Mental Functions

As in other areas of biological study, brain research has progressed due to technological breakthroughs. Neuroscientists now have a wide range of techniques at their disposal for studying the human brain, including modern technologies that allow us to record its functioning (see Science Focus, Fig. 17A).

Memory and Learning

Just as the connecting tracts of the corpus callosum are evidence that the two cerebral hemispheres work together, so the limbic system indicates that cortical areas may work with lower centers to produce memory and learning. **Memory** is the ability to hold a thought in mind or to recall events from the past, ranging from a word we learned only yesterday to an early emotional experience that has shaped our lives. **Learning** takes place when we retain and utilize past memories.

Types of Memory   The last time you decided to try a new pizza restaurant, you may have looked up the number and tried to remember it for a short period of time. If you said you were trying to keep it in the forefront of your brain, you were exactly correct. The prefrontal area, which is active during **short-term memory,** lies just behind our forehead! On the other hand, there are probably some telephone numbers that you have memorized; these have gone into **long-term memory.** Think of the phone number of someone close to you whom you call frequently. Can you bring the number to mind without also thinking about the place or person associated with that number? Most likely you cannot, because long-term memory is typically a mixture of **semantic memory** (numbers, words, etc.) and **episodic memory** (persons, events, etc.). Due to brain damage, some people lose one type of memory but not the other. For example, without a working episodic memory, they can carry on a conversation but have no recollection of recent events. If you are talking to them and then leave the room, they don't remember you when you come back!

 Skill memory is another type of memory that can exist independent of episodic memory. Skill memory is involved in performing motor activities, such as riding a bike, playing ice hockey, or typing a letter using a computer keyboard. When a person first learns a skill, more areas of the cerebral cortex are involved than after the skill is perfected. In other words, you have to think about what you are doing when you learn a skill, but later the actions become automatic. Skill memory involves all the motor areas of the cerebrum below the level of consciousness.

Long-Term Memory Storage and Retrieval   The first step toward curing memory disorders is to know what parts of the brain are functioning when we remember something. Our long-term memories are stored in bits and pieces throughout the sensory association areas of the cerebral cortex. Visions are stored in the visual association area, sounds are stored in the auditory association area, and so forth. The hippocampus serves as a bridge between the sensory association areas and the prefrontal area. As discussed in the Science Focus, the prefrontal area communicates with the hippocampus when memories are stored and when these memories are brought to mind. Why are some memories so emotionally charged? The amygdala is responsible for fear conditioning and associating danger with sensory stimuli

Science Focus

How Memories Are Made

Neurobiologists want to understand higher mental functions, such as memory, from the behavioral to the molecular level. They believe that knowing how the healthy brain goes about its business is essential if we are ever to cure disorders such as Alzheimer disease and Parkinson disease, or even psychiatric disorders such as schizophrenia.

In times past, accidents, illnesses, or birth defects that selectively destroy a particular region of the brain have allowed investigators to assign functions to missing brain structures. Let's consider, for example, the hippocampus, a small, seahorse-shaped structure that lies in the medial temporal lobe. S. S. is a physicist whose hippocampus was destroyed after he contracted a herpes simplex infection. S. S. still has an IQ of 136, but he forgets a recent experience in just a few minutes. Although he is able to remember childhood events, he is unable to convert short-term memory into long-term memory. By observing S. S. and other patients like him, we can reason that the hippocampus must be involved in the storage of what are called episodic memories. Episodic memories pertain to episodes—that is, a series of occurrences.

Circumstantial data regarding the function of the hippocampus have been backed up by research in the laboratory. Monkeys can be taught to perform a delayed-response task, such as choosing a certain cup that always contains food even after a lapse of time between instruction and a subsequent testing period. In order to tell what parts of the brain are involved in a delayed-response task, investigators injected the brain with radioactive glucose, a source of molecular energy for neurons. The areas of the brain that are most active use the most energy. Therefore, those areas "light up" when they are active. After a monkey has performed a delayed-response task and its brain tissue is studied, radioactivity is particularly detected in the prefrontal area and the hippocampus. From this we know that the prefrontal area communicates with the hippocampus when an episodic memory is being stored.

Recently, the introduction of high-speed computers made it possible to view a functioning human brain. Data fed into the computers during positron emission tomography (PET) (Fig. 17A) and magnetic resonance imaging (MRI) produce brain scans—cross sections of the brain from all directions. Functional imaging occurs when the subjects undergoing PET or MRI are asked to perform a particular mental task, such as studying a few words and later selecting those words from a long list. In a recent study utilizing functional MRI, the hippocampus was maximally active when human subjects were asked to remember a word cued by a picture, while the so-called parahippocampal region was maximally active when subjects were asked to remember pictures for later recall.

Once it became accepted that the hippocampus and surrounding regions are needed to store and retrieve memories, investigators began to study the activity of individual neurons in the region. What exactly are neurons doing when we store memories and bring them back? Neurobiologists have discovered that a long-lasting effect, called long-term potentiation (LTP), occurs when synapses are active within the hippocampus. Long-term potentiation is carried out by glutamate, a type of neurotransmitter. Investigators at the Massachusetts Institute of Technology first showed that it was possible to teach mice to find their way around a maze. Then, by a dramatic feat of genetic engineering, they were able to produce mice that lacked the receptor for glutamate only in hippocampal cells. These defective mice were unable to learn to run the maze. These studies mark the first time it has been possible to study memory at all levels, from behavior to a molecular change in one part of the brain.

Discussion Questions

1. After reading the first part of this chapter, you have a fundamental understanding of how neurons work. Using this knowledge, how do you think memory might be stored in the brain—that is, what specific changes might take place at the neuronal level?

2. What are some possible advantages of storing different types of memories in different parts of the brain (e.g., short-term versus long-term)?

3. What is a "personality"? Is there a difference between our brains and our personalities? What region or regions of the brain do you think are most influential in determining our personality?

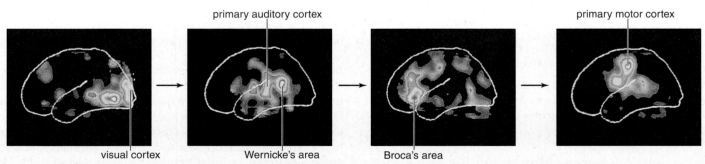

primary auditory cortex   primary motor cortex

visual cortex   Wernicke's area   Broca's area

1. The word is seen in the visual cortex.
2. Information concerning the word is interpreted in Wernicke's area.
3. Information from Wernicke's area is transferred to Broca's area.
4. Information is transferred from Broca's area to the primary motor area.

Figure 17A Cortical areas for language.

These PET images show the cortical pathway for reading words and then speaking them. Red indicates the most active areas of the brain, and blue indicates the least active areas.

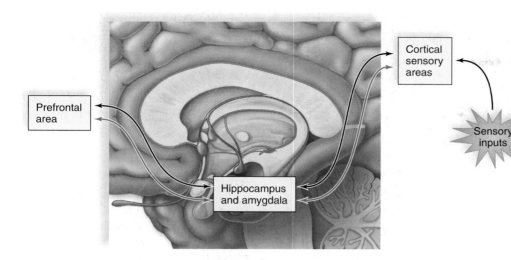

Figure 17.13 Long-term memory circuits.
The hippocampus and the amygdala are believed to be involved in the storage and retrieval of memories. Semantic memory (red arrows) and episodic memory (black arrows) are stored separately, and therefore, you can lose one without losing the other.

received from both the diencephalon and the cortical sensory areas. Figure 17.13 diagrammatically illustrates the long-term memory circuits of the brain.

Long-Term Potentiation While it is helpful to know the memory functions of various portions of the brain, an important step toward curing mental disorders is understanding memory on the cellular level. **Long-term potentiation (LTP)** is an enhanced response at synapses within the hippocampus. LTP is most likely essential to memory storage, but unfortunately, it sometimes causes a postsynaptic neuron to become so excited that it undergoes apoptosis, a form of cell death. This phenomenon, called excitotoxicity, may develop due to a mutation. The longer we live, the more likely it is that any particular mutation will occur.

Excitotoxicity is due to the action of the neurotransmitter glutamate, which is active in the hippocampus. When glutamate binds with a specific type of receptor in the postsynaptic membrane, calcium (Ca^{2+}) may rush in too fast, and the influx is lethal to the cell. A gradual extinction of brain cells in the hippocampus appears to be the underlying cause of Alzheimer disease (AD). Excitotoxicity may or may not be related to these abnormalities. Some researchers are trying to develop neuroprotective drugs that can guard brain cells against damage due to glutamate.

Higher mental functions of the brain include memory and learning. Different types of memory include short-term, long-term, semantic, episodic, and skill memory. Some of these functions are found in separate anatomical areas of the brain. An understanding of memory at the cellular level may lead to treatments for certain brain disorders.

Language and Speech

Language is obviously dependent upon semantic memory. Therefore, we would expect some of the same areas in the brain to be involved in both memory and language. Seeing words and hearing words depend on the primary visual cor-

tex in the occipital lobe and the primary auditory cortex in the temporal lobe, respectively. And speaking words depends on motor centers in the frontal lobe. Functional imaging by a technique known as PET (positron emission tomography) supports these suppositions (see Science Focus, Fig. 17A).

From studies of patients with speech disorders, it has been known for some time that damage to the motor speech (Broca's) area results in an inability to speak, while damage to the sensory speech (Wernicke's) area results in the inability to comprehend speech (Fig. 17.14). Actually, any disruption of pathways between various centers of the brain can contribute to an inability to comprehend our environment and use speech correctly. Indeed, damage to lower centers, especially the thalamus, can also cause speech difficulties. Remember that the thalamus passes on sensory information, and if the cerebral cortex does not receive the necessary sensory input, the motor areas cannot formulate the proper output.

One interesting aside pertaining to language and speech is the recognition that the left brain and right brain

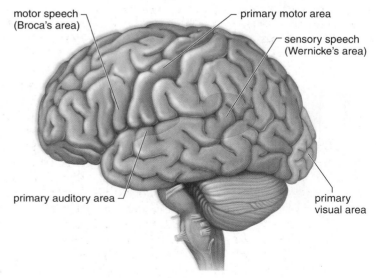

Figure 17.14 Language and speech.
Wernicke's area and Broca's area are thought to be involved in speech comprehension and use, as are the other labeled areas of the cerebral cortex.

have different functions. Only the left hemisphere, not the right, contains a Broca's area and Wernicke's area. In an attempt to cure epilepsy in the early 1940s, the corpus callosum was surgically severed in some patients. Later studies showed that these split-brain patients could only name objects seen by the left hemisphere. If objects were viewed only by the right hemisphere, a split-brain patient could choose the proper object for a particular use but was unable to name it. Based on these and various other studies, the popular idea developed that the left brain can be contrasted with the right brain along these lines:

Left Hemisphere	Right Hemisphere
Verbal	Nonverbal, visuo-spatial
Logical, analytical	Intuitive
Rational	Creative

Further, researchers generally came to believe that one hemisphere was dominant in each person, accounting in part for personality traits. However, recent studies suggest that the hemispheres simply process the same information differently. The right hemisphere is more global in its approach, whereas the left hemisphere is more specific. For example, if only the left cerebral cortex is functional, a person has difficulty communicating the emotions involved with language. Therefore, both sides of the brain play a role in language use.

Language is dependent upon memory; special centers in the left hemisphere help account for our ability to comprehend and use speech.

17.4   The Peripheral Nervous System

The peripheral nervous system (PNS) lies outside the central nervous system and is composed of nerves and ganglia. **Nerves** are bundles of axons; the axons that occur in nerves are called nerve fibers.

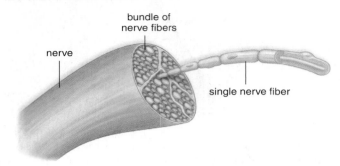

Sensory fibers carry information to the CNS, and motor fibers carry information away from the CNS. Ganglia (sing., **ganglion**) are swellings associated with nerves that contain collections of cell bodies.

Humans have 12 pairs of **cranial nerves** attached to the brain. By convention, the pairs of cranial nerves are referred to by Roman numerals (Fig. 17.15*a*). Some of the cranial nerves contain only sensory input fibers, while others contain only motor input fibers. The remaining are mixed nerves that contain both sensory and motor fibers. Cranial nerves are largely concerned with the head, neck, and facial regions of the body.

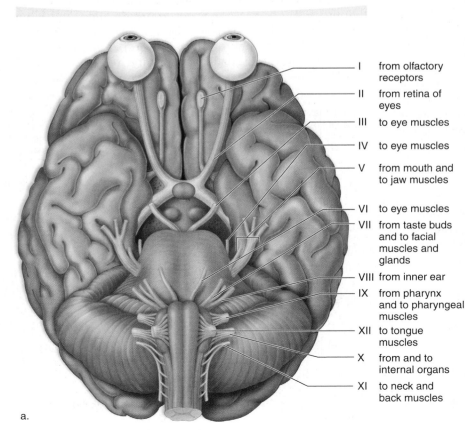

I	from olfactory receptors
II	from retina of eyes
III	to eye muscles
IV	to eye muscles
V	from mouth and to jaw muscles
VI	to eye muscles
VII	from taste buds and to facial muscles and glands
VIII	from inner ear
IX	from pharynx and to pharyngeal muscles
XII	to tongue muscles
X	from and to internal organs
XI	to neck and back muscles

a.

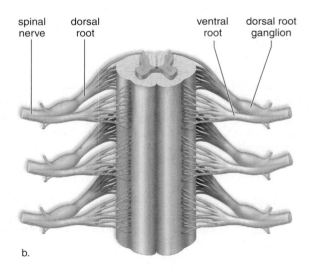

b.

Figure 17.15
Cranial and spinal nerves.

a. Ventral surface of the brain showing the attachments of the 12 pairs of cranial nerves. **b.** Cross section of the spinal cord showing three pairs of spinal nerves. The human body has 31 pairs of spinal nerves altogether, and each spinal nerve has a dorsal root and a ventral root attached to the spinal cord.

However, the vagus nerve (X) has branches not only to the pharynx and larynx, but also to most of the internal organs.

The **spinal nerves** of humans emerge in 31 pairs from between openings in the vertebral column of the spinal cord. Each spinal nerve originates when two short branches, or roots, join together (Fig. 17.15b). The dorsal root (at the back) contains sensory fibers that conduct impulses inward (toward the spinal cord) from sensory receptors. The cell body of a sensory neuron is in a **dorsal root ganglion.** The ventral root (at the front) contains motor fibers that conduct impulses outward (away from the cord) to effectors. Notice, then, that all spinal nerves are mixed nerves that contain many sensory and motor fibers. Each spinal nerve serves the particular region of the body in which it is located. For example, the intercostal muscles of the rib cage are innervated by thoracic nerves.

In the PNS, cranial nerves take impulses to and from the brain, and spinal nerves take impulses to and from the spinal cord.

Somatic System

The PNS is subdivided into the somatic system and the autonomic system. The **somatic system** serves the skin, skeletal muscles, and tendons. It includes nerves that take sensory information from external sensory receptors to the CNS and motor commands away from the CNS to the skeletal muscles. Some actions in the somatic system are due to **reflexes,** which are automatic responses to a stimulus. A reflex occurs quickly, without our even having to think about it. Other actions are voluntary, and these always originate in the cerebral cortex, as when we decide to move a limb.

The Reflex Arc

A reflex arc is a nerve pathway that carries out a reflex. Reflexes are programmed, built-in circuits that allow for protection and survival. They are present at birth and require no conscious thought to take place. Figure 17.16 illustrates the path of a withdrawal reflex that involves only the spinal cord. If your hand touches a sharp pin, sensory receptors in the skin generate nerve impulses that move along sensory fibers through the dorsal root ganglia toward the spinal cord. Sensory neurons that enter the cord dorsally pass signals on to many interneurons. Some of these interneurons synapse with motor neurons whose short dendrites and cell bodies are in the spinal cord. Nerve impulses travel along these motor fibers to

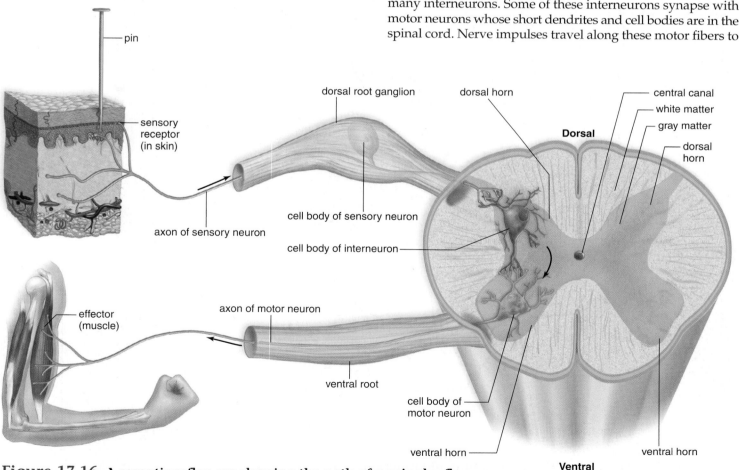

Figure 17.16 A somatic reflex arc showing the path of a spinal reflex.

A stimulus (e.g., sharp pin) causes sensory receptors in the skin to generate nerve impulses that travel in sensory axons to the spinal cord. Interneurons integrate data from sensory neurons and then relay signals to motor axons. Motor axons convey nerve impulses from the spinal cord to a skeletal muscle, which contracts. Movement of the hand away from the pin is the response to the stimulus.

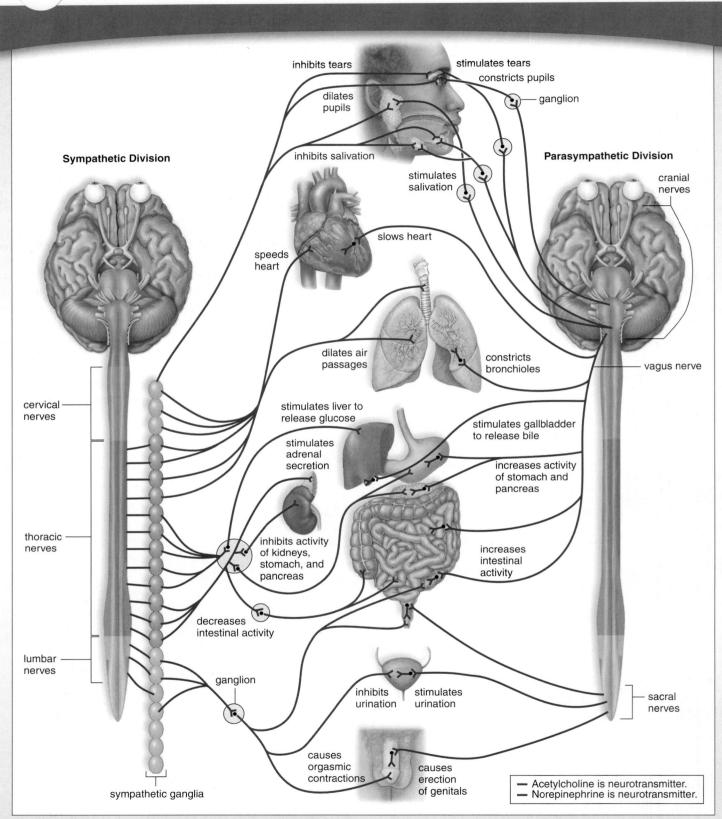

Figure 17.17 Autonomic system structure and function.

Sympathetic preganglionic fibers (*left*) arise from the cervical, thoracic, and lumbar portions of the spinal cord; parasympathetic preganglionic fibers (*right*) arise from the cranial and sacral portions of the spinal cord. Each system innervates the same organs but has contrary effects.

an effector, which brings about a response to the stimulus. In this case, the effector is a muscle, which contracts so that you withdraw your hand from the pin. Various other reactions are also possible—you will most likely look at the pin, wince, and cry out in pain. This whole series of responses occurs because some of the interneurons involved carry nerve impulses to the brain. The brain makes you aware of the stimulus and directs these other reactions to it. You do not feel the pain until your brain receives the information and interprets it.

In the somatic system, reflex arcs allow us to respond rapidly to external stimuli.

Autonomic System

The **autonomic system** of the PNS regulates the activity of cardiac and smooth muscle and glands. The system is composed of the sympathetic and parasympathetic divisions (Fig. 17.17). These two divisions have several features in common: (1) they function automatically and usually in an involuntary manner; (2) they innervate all internal organs; and (3) for each signal, they utilize two motor neurons that synapse at a ganglion. The first neuron has a cell body within the CNS, and its axon is called the preganglionic fiber. The second neuron has a cell body within the ganglion, and its axon is termed the postganglionic fiber.

Reflex actions, such as those that regulate the blood pressure and breathing rate, are especially important to the maintenance of homeostasis. These reflexes begin when the sensory neurons in contact with internal organs send messages to the CNS. They are completed by motor neurons within the autonomic system.

Sympathetic Division

The cell bodies of preganglionic fibers in the **sympathetic division** are located in the middle, or thoracic-lumbar, portion of the spinal cord. The fibers immediately terminate in ganglia that lie near the cord. Therefore, in this division, the preganglionic fiber is short, but the postganglionic fiber that makes contact with an organ is long:

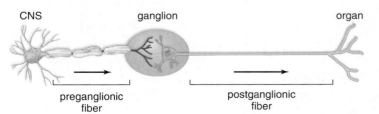

preganglionic
fiber

postganglionic
fiber

The sympathetic division is especially important during emergency situations when you might be required to fight back or run away. It accelerates the heartbeat and dilates the bronchi; active muscles, after all, require a ready supply of glucose and oxygen. On the other hand, the sym-

pathetic division inhibits the digestive tract—digestion is not an immediate necessity if you are under attack.

One of the functions of the sympathetic division is to activate the adrenal medulla to secrete the hormones epinephrine (adrenaline) and norepinephrine (NE) into the blood (also see Chapter 20). In fact, one special preganglionic sympathetic neuron goes right to the adrenal gland without stopping at a ganglion. The adrenaline and NE released by the adrenal medulla bind to receptors on various cell types, adding to the "fight or flight" response. Interestingly, the neurotransmitter released by sympathetic postganglionic axons is also NE. A small amount of the NE released by these axons spills over into the blood. During times of high sympathetic nerve activity, the amount of NE entering the blood from these neurons increases significantly.

Because they cause the heart to beat stronger and faster and constrict certain blood vessels, adrenaline and NE tend to increase blood pressure. Certain drugs, called beta blockers, can be used in patients with hypertension to block the activity of adrenaline and NE, which slows the heart and decreases blood pressure.

The sympathetic division brings about those responses we associate with "fight or flight."

Parasympathetic Division

The **parasympathetic division** includes a few cranial nerves (e.g., the vagus nerve), as well as fibers that arise from the sacral (bottom) portion of the spinal cord. Therefore, this division is often referred to as the craniosacral portion of the autonomic system. In the parasympathetic division, the preganglionic fiber is long, and the postganglionic fiber is short because the ganglia lie near or within the organ:

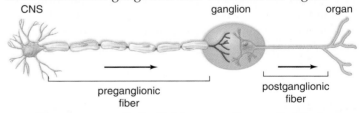

preganglionic
fiber

postganglionic
fiber

The parasympathetic division, sometimes called the housekeeper division, promotes all the internal responses we associate with "rest and digest." Opposite to the sympathetic system, it causes the pupil of the eye to contract, promotes digestion of food, and retards the heartbeat. The neurotransmitter utilized by the parasympathetic division is acetylcholine (ACh).

Table 17.1 compares and contrasts the characteristics of the somatic and autonomic systems.

The parasympathetic division brings about the responses we associate with a relaxed state.

TABLE 17.1	Comparison of Somatic Motor and Autonomic Motor Pathways		
	Somatic Motor Pathway	**Autonomic Motor Pathways**	
		Sympathetic	*Parasympathetic*
Type of control	Voluntary/involuntary	Involuntary	Involuntary
Number of neurons per message	One	Two (preganglionic shorter than postganglionic)	Two (preganglionic longer than postganglionic)
Location of motor fiber	Most cranial nerves and all spinal nerves	Thoracolumbar spinal nerves	Cranial (e.g., vagus) and sacral spinal nerves
Neurotransmitter	Acetylcholine	Norepinephrine	Acetylcholine
Effectors	Skeletal muscles	Smooth and cardiac muscle, glands	Smooth and cardiac muscle, glands

17.5   Drug Abuse

A wide variety of drugs affect the nervous system and can alter the mood and/or emotional state. In general, drugs either impact the limbic system or affect the action of a particular neurotransmitter at the synapse (Fig. 17.18). Stimulants are drugs that increase the likelihood of neuron excitation, and depressants decrease the likelihood of excitation. Increasingly, researchers believe that dopamine, among other neurotransmitters in the brain, is responsible for mood. Cocaine is known to potentiate the effects of dopamine by interfering with its uptake from synaptic clefts (Fig. 17.19). Many of the new medications developed to counter drug dependence and mental illness affect the release, reception, or breakdown of dopamine.

Drug abuse is apparent when a person takes a drug at a dose level and under circumstances that increase the potential for a harmful effect. A drug abuser often takes more of the drug than was intended. Drug abusers are apt to display a psychological and/or physical dependence on the drug. Dependence has developed when the person spends much time thinking about the drug or arranging to get it. With physical dependence, formerly called "addiction," the person has become tolerant to the drug—more of the drug is needed to get the same effect, and withdrawal symptoms occur when he or she stops taking the drug.

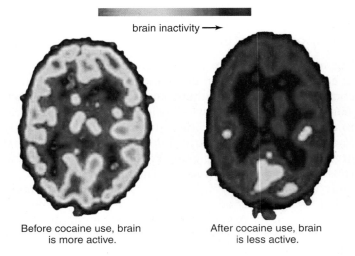

brain inactivity ⟶

Before cocaine use, brain is more active.

After cocaine use, brain is less active.

Figure 17.19 Drug abuse.
PET scans show that the usual activity of the brain is reduced and many areas become inactive after a person injects cocaine.

The misuse of drugs that affect the nervous system can lead to psychological and physical dependence.

Figure 17.18
Drug actions at a synapse.
A drug can ① cause the neurotransmitter (NT) to leak out of a synaptic vesicle into the axon terminal; ② prevent release of NT into the synaptic cleft; ③ promote release of NT into the synaptic cleft; ④ prevent reuptake of NT by the presynaptic membrane; ⑤ block the enzyme that causes breakdown of the NT; or ⑥ bind to a receptor, mimicking the action of an NT.

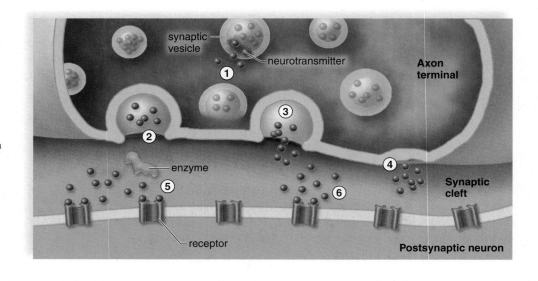

Some Specific Drugs of Abuse

Nicotine

In 2003, about 30% of the U.S. population aged 12 and older had used tobacco at least once in the month prior to being interviewed. Young adults between ages 18 and 25 reported the highest usage of tobacco, at 45%. When a person smokes a cigarette or cigar, or chews tobacco, nicotine is quickly distributed throughout the body. This causes a release of epinephrine from the adrenal cortex, which increases blood sugar levels and causes an initial feeling of stimulation. Then, as blood sugar falls, depression and fatigue set in, leading the smoker to seek more nicotine. In the CNS, nicotine causes neurons to release excess dopamine, which has a reinforcing effect that leads to dependence on the drug. Nicotine induces both physiological and psychological dependence, and once they start, many tobacco users find it extremely difficult to give up the habit. Withdrawal symptoms include headache, stomach pain, irritability, and insomnia. Pregnant women who smoke are more likely to have stillborn or premature infants, or babies born with other health problems.

Alcohol (Ethanol)

According to a 2003 survey by the Centers for Disease Control, 61% of adults had drunk alcohol in the past year, and 32% had consumed more than five drinks on a single day. When ingested, ethanol acts as a drug in the brain by influencing the action of GABA, an inhibitory neurotransmitter, or glutamate, an excitatory neurotransmitter. Alcohol is primarily metabolized in the liver, where it disrupts the normal workings of this organ so that fats cannot be broken down. Fat accumulation, the first stage of liver deterioration, begins after only a single night of heavy drinking. If heavy drinking continues, fibrous scar tissue is deposited during the next stage of liver damage. If heavy drinking stops at this point, the liver may still recover and become normal again. If not, the final and irrevocable stage, cirrhosis of the liver, occurs. Liver cells die, harden, and turn orange (cirrhosis means orange).

Alcohol is a carbohydrate and as such, can be converted into energy for the body to use. However, alcoholic beverages lack the vitamins, minerals, essential amino acids, and fatty acids the body needs. For this reason, many alcoholics are vitamin-deficient, undernourished, and prone to illness.

Based upon extensive research, the Surgeon General of the United States recommends that pregnant women drink no alcohol at all. Alcohol crosses the placenta freely and causes fetal alcohol syndrome in newborns, which is characterized by mental retardation and various physical defects.

Marijuana

Marijuana is the most commonly used illegal drug in the United States. Marijuana, or "pot," refers to the dried flowering tops, leaves, and stems of the Indian hemp plant *Cannabis sativa,* which contain and are covered by a resin that is rich in THC (tetrahydrocannabinol). Researchers have found that THC binds to a receptor in the brain for anandamide, a naturally occurring neurotransmitter that is important for short-term memory processing, and perhaps for feelings of contentment. Usually, marijuana is smoked in a cigarette form called a "joint," or in pipes or other devices, but it can also be consumed. The occasional user experiences a mild euphoria along with alterations in vision and judgment, which result in distortions of space and time. Motor incoordination, including the inability to speak coherently, takes place. Heavy use can result in hallucinations, anxiety, depression, body image distortions, paranoid reactions, and similar psychotic symptoms. Some researchers believe that long-term marijuana use leads to brain impairment. Fetal cannabis syndrome, which resembles fetal alcohol syndrome, has also been reported.

Cocaine and Crack

Cocaine is an alkaloid derived from the shrub *Erythroxylon coca.* It is sold in powder form and as crack, a more potent extract. Crack is cocaine that comes in a rocklike crystal that can be heated and its vapors inhaled. The term "crack" refers to the crackling sound heard when it is heated. Cocaine prevents the synaptic uptake of dopamine, and this causes the user to experience a rush sensation. The epinephrine-like effects of dopamine account for the state of arousal that lasts for several minutes after the rush experience.

A cocaine binge can go on for days, after which the user suffers a crash. During the binge period, the user is hyperactive and has little desire for food or sleep but an increased sex drive. During the crash period, the user is fatigued, depressed, and irritable, has memory and concentration problems, and displays no interest in sex. Indeed, men who use "coke" often become impotent.

Cocaine causes extreme physical dependence. With continued use, the body begins to make less dopamine to compensate for a seeming excess supply. The user subsequently experiences tolerance, withdrawal symptoms, and an intense craving for cocaine. Overdosing on cocaine can cause seizures and cardiac and respiratory arrest. It is possible that long-term cocaine abuse causes brain damage, and babies born to addicts suffer withdrawal symptoms and may have neurological and behavioral problems.

Heroin

Heroin is derived from morphine, an alkaloid of opium. Once heroin is injected into a vein, a feeling of euphoria, along with relief of any pain, occurs within three to six minutes. Heroin binds to receptors meant for the endorphins, naturally occurring neurotransmitters that kill pain and produce a feeling of tranquility. With repeated heroin use, the body's production of endorphins decreases. Tolerance develops so that the user needs to take more of the drug just to prevent withdrawal symptoms, and the original euphoria is no longer felt. Heroin withdrawal symptoms include perspiration, dilation of pupils, tremors, restlessness, abdominal cramps, vomiting, and increased blood pressure and respiratory rate. People who are excessively dependent may experience convulsions, respiratory failure, and death. Infants born to women who are physically dependent also experience those withdrawal symptoms.

"Club" Drugs

Ecstasy (MDMA), Rohypnol, and ketamine are examples of drugs that are abused by many teens and young adults who attend night-long dances called raves or trances. Of course, these drugs may also be abused by others as well. MDMA is chemically similar to methamphetamine. Many users say that "XTC" increases their feelings of well-being and friendliness, even love, for other people. However, it can cause many of the same side effects as other stimulants, such as increase in heart rate and blood pressure, muscle tension, and blurred vision. In high doses, MDMA can interfere with temperature regulation, leading to hyperthermia followed by damage to the liver, kidneys, and heart. Chronic MDMA use can also damage memory and lead to depression. Perhaps because of these side effects, several recent surveys have reported that MDMA use has been decreasing among youths aged 12 to 17.

Rohypnol is in the same class of drugs as valium, a popular sedative. When mixed with alcohol, Rohypnol, or "roofies," can render victims incapable of resisting, for example, sexual assault. It can also produce anterograde amnesia, which means victims may not remember events they experienced while under the influence of the drug. Ketamine is an anesthetic that veterinarians use to perform surgery on animals. When administered to humans, it typically induces a dreamlike state, sometimes with hallucinations. Because it can render the user unable to move, ketamine has also been used as a date rape drug. At higher doses, ketamine can cause dangerous reductions in heart and respiratory functions.

Methamphetamine

Methamphetamine, also called "meth" or "crank," is a powerful CNS stimulant. Meth is often synthesized or "cooked" in makeshift home laboratories, usually starting with either ephedrine or pseudoephedrine, common ingredients in many cold and asthma medicines. Many states have now passed laws making these medications more difficult to purchase. Meth is usually produced as a powder (speed) that can be snorted, or as crystals (crystal meth) that can be smoked. The most immediate effect is an initial "rush" of euphoria, due to high amounts of dopamine released in the brain. The user may also experience an increased energy level, alertness, and elevated mood. After the initial rush, a state of high agitation typically occurs that, in some individuals, leads to violent behavior. Chronic use can lead to paranoia, irritability, hallucinations, insomnia, tremors, hyperthermia, cardiovascular collapse, and death. Similar to cocaine, long-term meth abuse renders the brain less able to produce normal levels of dopamine; the user may have strong cravings for more meth, without being able to reach a satisfactory high.

Abused drugs often give users a temporary "rush," or feeling of euphoria, by altering the action of a particular neurotransmitter. Over time this feeling becomes increasingly difficult to obtain, and undesired effects increase.

17.6   Disorders of the Nervous System

A myriad of conditions can cause problems in the nervous system. Because of space considerations, we will consider only a few of the most serious, common, and/or interesting ones here.

Disorders of the Brain

Alzheimer disease (AD) is the most common cause of dementia, an impairment of brain function significant enough to interfere with the patient's ability to carry on daily activities. Signs of AD may appear before the age of 50, but the condition is usually seen in individuals past the age of 65. It is estimated that at least 40% of individuals older than 80 develop AD. One of the early symptoms is loss of memory, particularly for recent events. Characteristically, the person asks the same question repeatedly and becomes disoriented even in familiar places. Gradually, the person loses the ability to perform any type of daily activity and becomes bedridden. Death is often due to pneumonia resulting from the patient's debilitated state.

In AD patients, abnormal neurons are present throughout the brain but especially in the hippocampus and amygdala. These neurons have two abnormalities: (1) plaques, containing a protein called beta amyloid, envelop the axons, and (2) neurofibrillary tangles are in the axons and extend upward to surround the nucleus (Fig. 17.20). The neurofibrillary tangles arise because a protein called tau no longer has the correct shape. Tau holds microtubules in place so that they support the structure of neurons. When tau changes shape, it grabs onto other tau molecules, and the tangles result.

Researchers are working to discover the cause of beta amyloid plaques and neurofibrillary tangles. Several genes that predispose a person to AD have been identified. One of these, called $APOE_4$, is found in 65% of persons with AD. But it is

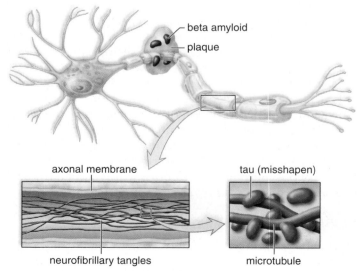

Figure 17.20 Alzheimer disease.
Some of the neurons of AD patients have beta amyloid plaques and neurofibrillary tangles. AD neurons are present throughout the brain but concentrated in the hippocampus and amygdala.

unknown why inheritance of this gene leads to AD. Moreover, although several treatments have been tried to date, none has been shown to prevent or cure AD. A variety of medications, however, can help ease symptoms such as anxiety, agitation, and depression.

Parkinson disease (PD) is characterized by a gradual loss of motor control, typically beginning between the ages of 50 and 60. Eventually the person develops a wide-eyed, unblinking expression, involuntary tremors of the fingers and thumbs, muscular rigidity, and a shuffling gait. Speaking and performing ordinary daily tasks becomes laborious. In Parkinson patients, the basal nuclei (see section 17.2) function improperly because of a degeneration of the dopamine-releasing neurons in the brain. Without dopamine, which is an inhibitory neurotransmitter, the excessive excitatory signals from the motor cortex and other brain areas result in the symptoms of PD.

Unfortunately, it is not possible to give PD patients dopamine directly because of the impermeability of the capillaries serving the brain. However, symptoms can be alleviated by giving patients L-dopa, a chemical that can be changed to dopamine in the body, until too few cells are left to do the job. Then patients must turn to a number of controversial surgical procedures. In 1998 the actor Michael J. Fox, who had been diagnosed with PD seven years earlier at the age of 30, had a surgical procedure called a thalamotomy, which has been successful in reducing the symptoms of some PD patients. However, in his biography *Lucky Man: A Memoir,* Fox admits that after a temporary improvement, his PD symptoms returned in full force.

Multiple sclerosis (MS) is the most common neurological disease that afflicts young adults. Nearly 350,000 people in the United States have MS, with approximately 10,000 new cases diagnosed each year. MS affects the myelin sheath of neurons in the white matter of the brain. Although the initiating cause remains unknown, MS is considered an autoimmune disease in which the patient's own white blood cells attack the myelin, oligodendrocytes, and eventually, neurons in the CNS. Specifically, MS is characterized by an infiltration of lymphocytes and macrophages into the brain, brain stem, optic nerves, and spinal cord. The most common symptoms of MS include fatigue, vision problems, limb weakness and/or pain, and abnormal sensations such as numbness or tingling. These symptoms, however, can be similar to those of several other brain disorders, so a technique called magnetic resonance imaging (MRI) can be very helpful in both diagnosing and monitoring the disease (Fig. 17.21). MS can also take sev-

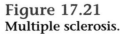

Figure 17.21
Multiple sclerosis.

This MRI scan shows the brain of an MS patient. The white areas, also known as MS plaques, are sites of active inflammation and myelin destruction.

eral clinical forms, from the milder relapsing-remitting form to more relentlessly progressive forms. Although MS is not generally considered a fatal disease, people with MS die an average of seven years earlier than the general population. No treatment has been shown to reverse the damage to the myelin of MS patients, but several medications are effective in slowing the progression of the disease, especially in the early stages.

A **stroke** results in disruption of the blood supply to the brain. There are two major forms of stroke, hemorrhagic and ischemic. In hemorrhagic stroke, bleeding occurs into the brain due to leakage from small arteries, often after years of damage by high blood pressure. Ischemic stroke occurs when there is a sudden loss of blood supply to an area of the brain, usually due to a thrombus or thromboembolism (see Chapter 12). In both types of stroke, the symptoms depend on the amount and specific area of brain tissue affected. However, some degree of paralysis and aphasia (loss of speech) are common signs. The risk of stroke increases with age, and African Americans have a higher incidence than other races in the United States. Physicians treating stroke patients must be careful to determine which type they are dealing with. For example, tissue plasminogen activator (tPA) can be used to help dissolve clots when treating ischemic stroke, but could have disastrous consequences if administered to a hemorrhagic stroke patient!

Meningitis is an infection of the meninges that surround the brain and spinal cord. Most cases of meningitis are caused by bacteria or viruses that have usually gained access to the meninges after first infecting the blood, middle ear, or sinuses. Certain viruses are also capable of infecting the brain and meninges by infecting peripheral nerve cells and being transported back to the brain. Since cells of the immune system have somewhat limited access to the brain, the infection may spread into the brain tissue. Bacterial meningitis is especially serious, even life-threatening if untreated. Physicians may diagnose meningitis based on clinical signs such as a stiff neck, fever, and headache, and may confirm the diagnosis by examining a sample of cerebrospinal fluid, usually obtained by a lumbar puncture. Most cases of bacterial meningitis are caused by organisms called *Haemophilus influenzae* or *Neisseria meningitidis.* Outbreaks of *N. meningitidis,* also called meningococcus, have occurred recently among college students. Fortunately, vaccines are available to prevent infection with both organisms.

Several brain diseases of humans and animals are caused by prions, which are infectious agents that most researchers believe are composed of protein only (see Chapter 28). Examples of human diseases include Creutzfeldt-Jakob disease and fatal familial insomnia; in animals, scrapie, mad cow disease, and chronic wasting disease are seen. Mad cow disease seems able to cross the species barrier and infect humans who consume infected tissues; over 100 human cases have been confirmed, mostly in the United Kingdom. All prion diseases are characterized by long incubation periods between initial exposure to the agent and onset of clinical signs, followed by rapid progress of the disease and, uniformly, death. Prion diseases are thought to result from the accumulation in the tissues of an abnormal form of a normal host (prion) protein. When this occurs in the brain, nerve cells die, forming areas that look like holes, which

Bioethical Focus

Declaration of Death

An electroencephalogram is a record of the brain's activity picked up from an array of electrodes on the forehead and scalp, as explained at the end of section 17.2. The complete and persistent absence of brain activity is often used as a clinical and legal criterion for brain death. Brain death can be present even though the body is still warm and moist. Indeed, the heart has been known to continue beating for more than a month after a declaration of brain death.

Organs that have never been deprived of oxygen and nutrients are more likely to be successfully transplanted. Even so, should it be permissible to remove organs for transplantation as soon as the person is declared legally dead on the basis of no brain activity?

As a safeguard, should we have to sign a card that permits organ removal if we have been declared legally dead on the basis of brain activity? Also, should organs be removed only if family members confirm that the person is legally dead and consent to their removal? What evidence should the family require?

On the other hand, perhaps it would be best if a society simply disallowed the removal of organs unless the person is no longer breathing and the heart has stopped beating. In that case, society would be denying some of its citizens the gift of life. Does the end justify the means—can we live with the remote possibility that organs will be removed from a living person because of the potential benefit to recipients?

Form Your Own Opinion

1. In 2005, the case of Terri Schiavo made national news. Schiavo suffered cardiac arrest in 1990 and was diagnosed as in a "persistent vegetative state." Her husband favored removing her from life support, but her parents believed she could recover. With whom did you tend to side, and why?
2. If you were declared legally brain dead, would you want your organs harvested for donation to a relative? A friend? A stranger?
3. Do you believe there are religious implications involved in removing organs from the body of a person who is, at least partially, still alive? If not, are you sympathetic to those who might object on religious grounds? Why or why not?

Figure 17.22
Effect of prions on the brain.

In this stained section of brain tissue from a patient with a prion disease, the spongiform changes (holes) in the tissue are due to accumulation of the abnormal prion protein.

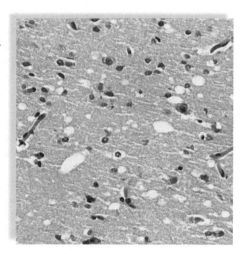

pathologists refer to as spongiform changes (Fig. 17.22). These diseases are also unique in that they can be inherited or can be spread by either eating infected tissues (as in mad cow disease) or through transplantation of certain organs or tissues (such as corneas) from an infected individual. Therefore, prion diseases are also called transmissible spongiform encephalopathies (TSEs). No treatment has yet been shown to delay the inevitably fatal progression of these diseases.

Disorders that affect the brain can be life-threatening and difficult to treat. Examples of serious brain diseases include Alzheimer disease, Parkinson disease, multiple sclerosis, and stroke. Infectious diseases of the brain include meningitis and prion diseases.

Disorders of the Spinal Cord

Spinal cord injuries may result from car accidents or other trauma. Since little or no nerve regeneration is possible in the CNS, any resulting disability is usually permanent. The cord may be completely cut across (called a transection) or only partially severed (a partial section). The location and extent of the damage produce a variety of effects, depending on the partial or complete stoppage of impulses passing up and down the spinal cord. If the spinal cord is completely transected, no sensations or somatic motor impulses are able to pass the point of injury. If the injury is between the first thoracic vertebra (T1) and the second lumbar vertebra (L2), paralysis of the lower body and legs occurs, a condition known as paraplegia. If the injury is in the neck region, the entire body below the neck is usually affected, a condition called quadriplegia. If the injury is above the second or third cervical vertebrae (C2 or C3), nervous control of respiration is also affected. In 1995, actor Christopher Reeve, best known for his role as "Superman," was thrown from his horse, crushing the top two cervical vertebrae. He immediately lost control of his arms and legs, and had to be placed on a respirator. Through an intense exercise and physical therapy program, Reeve regained the ability to move his left index finger and could take tiny steps while being held upright in a pool. Unfortunately, he died of cardiac arrest in 2004.

Amyotrophic lateral sclerosis (ALS), also known as Lou Gehrig's disease, is a devastating condition that affects the motor nerve cells of the spinal cord. ALS is incurable, and most people die within five years of being diagnosed, mainly due to failure of the respiratory muscles. Typically the first signs of ALS occur between the ages of 35 and 55 in the form of mild weakness or tremors in one or more limbs, starting in the fingers or toes and gradually involving the larger muscles

Figure 17.23
Physicist Steven Hawking.

Dr. Hawking has been a productive scientist and author for over 30 years, even though he has amyotrophic lateral sclerosis.

nearer to the body. There may be multiple causes of nerve cell death in ALS. For example, decreased levels of a protein called insulin-like growth factor-1 have been reported in ALS patients, and autoimmune reactions may play a role. Other than Lou Gehrig, the most famous person with ALS is probably physicist Steven Hawking, who has lived for over 30 years with the disease (Fig. 17.23). The fact that Hawking has remained an active scientist and author shows that ALS has very little effect on brain function.

> The outcome of spinal cord injury varies depending on the severity and location of the injury, but the damage is usually permanent. ALS also affects the spinal cord, with uniformly fatal results.

Disorders of the Peripheral Nerves

Guillain-Barré syndrome (GBS) is an inflammatory disease that causes demyelination of peripheral nerve axons. It is thought to result from an abnormal immune reaction to one of several types of infectious agents. In GBS patients, antibodies (see Chapter 13) formed against these bacteria or viruses seem to cross-react with the myelin. Symptoms usually begin with weakness or unsteadiness two to four weeks after an illness or immunization. The muscle weakness typically starts in the lower limbs and ascends to the upper limbs over a period of days to weeks. Patients typically lose their reflex responses in affected areas, such as the patellar reflex (upward movement of the leg after the tendon below the knee is tapped with an instrument). Interestingly, sensory nerve function usually remains normal. The disease process may eventually affect the respiratory muscles, so that

mechanical ventilation is required. In most cases, however, the paralysis begins to improve in a few weeks, and most patients make a full recovery within six to twelve months.

Myasthenia gravis (MG) is an autoimmune disorder in which antibodies are formed that react against the acetylcholine receptor (AchR) at the neuromuscular junction of the skeletal muscles. Normally, when the action potential arrives at the presynaptic terminal of peripheral motor nerves, acetylcholine (ACh) is released into the synaptic cleft and binds to the AchR on the muscle cell, stimulating contraction. In MG, antibodies bind to the AchR and block binding of ACh, preventing muscle stimulation.

Interestingly, the most common symptom of MG is weakness of the muscles of the eye or eyelid, although over time the weakness usually spreads to the face, trunk, and then the limbs. In severe cases, the respiratory muscles can be affected, leading to death if untreated. Even though there is no cure, MG patients often respond very well to treatment with immunosuppressive drugs and inhibitors of acetylcholinesterase, an enzyme that normally destroys the ACh after it is released. A technique called plasmapheresis can also be used in both GBS and MG to remove the harmful antibodies from the patient's blood.

> Guillain-Barré syndrome (GBS) and myasthenia gravis (MG) are autoimmune disorders that attack the peripheral nerves. In GBS the target is myelin; in MG it is the acetylcholine receptor on muscle cells.

Summarizing the Concepts

17.1 Nervous Tissue

There are three types of neurons. Sensory neurons take information from sensory receptors to the CNS; interneurons occur within the CNS; and motor neurons take information from the CNS to effectors (muscles or glands). A neuron is composed of dendrites, a cell body, and an axon. Long axons are covered by a myelin sheath.

When an axon is not conducting a nerve impulse, the inside of the axon is negative (−65 mV) compared to the outside. The sodium-potassium pump actively transports Na^+ out of an axon and K^+ to the inside of an axon. The resting potential is due to the leakage of K^+ to the outside of the neuron. When an axon is conducting a nerve impulse (action potential), Na^+ first moves into the axoplasm, and then K^+ moves out of the axoplasm.

Transmission of a nerve impulse from one neuron to another takes place when a neurotransmitter molecule is released into a synaptic cleft. The binding of the neurotransmitter to receptors in the postsynaptic membrane causes excitation or inhibition. Integration is the summing of excitatory and inhibitory signals.

17.2 The Central Nervous System

The CNS consists of the spinal cord and brain, which are both protected by bone. The CNS receives and integrates sensory input and formulates motor output. The gray matter of the spinal cord contains neuron cell bodies; the white matter consists of myelinated axons that occur in bundles called tracts. The spinal cord sends sensory information to the brain, receives motor output from the brain, and carries out reflex actions.

In the brain, the cerebrum has two cerebral hemispheres connected by the corpus callosum. Sensation, reasoning, learning and memory, and language and speech take place in the cerebrum. The cerebral cortex is a thin layer of gray matter covering the cerebrum. The cerebral cortex of each cerebral hemisphere has four lobes: a frontal, parietal, occipital, and temporal lobe. The primary motor area, in the frontal lobe, sends out motor commands to lower brain centers, which pass them on to motor neurons. The primary somatosensory area, in the parietal lobe, receives sensory information from lower brain centers in communication with sensory neurons. Association areas are located in all the lobes. The cortex also contains processing centers for reasoning and speech.

The brain has a number of other regions. The hypothalamus controls homeostasis, and the thalamus specializes in sending sensory input on to the cerebrum. The cerebellum primarily coordinates skeletal muscle contractions. The medulla oblongata and the pons have centers for vital functions, such as breathing and the heartbeat.

17.3 The Limbic System and Higher Mental Functions
The limbic system connects portions of the cerebral cortex with the hypothalamus, the thalamus, and basal nuclei. In the limbic system, the hippocampus acts as a conduit for sending messages to long-term memory and retrieving them once again for the prefrontal area. The amygdala adds emotional overtones to memories.

17.4 The Peripheral Nervous System
The peripheral nervous system contains only nerves and ganglia. Voluntary actions always involve the cerebrum, but reflexes are automatic, and some do not require involvement of the brain. In the somatic system, for example, a stimulus causes sensory receptors to generate nerve impulses that are then conducted by sensory fibers to interneurons in the spinal cord. Interneurons signal motor neurons, which conduct nerve impulses to a skeletal muscle that contracts, producing the response to the stimulus.

The autonomic (involuntary) system controls the smooth muscle of the internal organs and glands. The sympathetic division is associated with responses that occur during times of stress, and the parasympathetic system is associated with responses that occur during times of relaxation.

17.5 Drug Abuse
Although neurological drugs are quite varied, each type has been found to either promote or prevent the action of a particular neurotransmitter. Drug abusers often take more of a drug than is intended and develop psychological and/or physical dependence on the drug.

17.6 Disorders of the Nervous System
The precise cause of many of the disorders affecting the CNS remains unclear. Alzheimer disease seems to be due to an accumulation of abnormal proteins in the brain. Parkinson disease is due to a degeneration of dopamine-secreting neurons. Multiple sclerosis is the result of destruction of myelin and myelin-producing cells. Stroke, whether hemorrhagic or ischemic, is the result of a disruption of the blood supply to the brain. Infections of the brain include meningitis and prion diseases.

Injuries to the spinal cord can cause a spectrum of problems, from weakness in one limb to complete paralysis. Amyotrophic lateral sclerosis is a fatal disease resulting from a loss of motor neurons in the spinal cord. Guillain-Barré syndrome and myasthenia gravis are both autoimmune disorders that attack the peripheral nerves.

Testing Yourself

Choose the best answer for each question.

1. In the PNS, myelin is formed by
   a. oligodendroglial cells.
   b. Schwann cells.
   c. axons.
   d. motor neurons.

2. During the refractory period,
   a. sodium gates are open.
   b. sodium gates are closed.
   c. sodium and potassium gates are open.
   d. None of these are correct.

3. Membranes surrounding the CNS are collectively called
   a. meninges.
   b. myelin.
   c. vesicles.
   d. None of these are correct.

4. Gray matter is found on the _____ of the brain and the _____ of the spinal cord.
   a. inside, inside
   b. outside, outside
   c. outside, inside
   d. inside, outside

5. The cerebellum
   a. coordinates skeletal muscle movements.
   b. receives sensory input from the joints and muscles.
   c. receives motor input from the cerebral cortex.
   d. All of these are correct.

6. An interneuron can relay information
   a. from a sensory neuron to a motor neuron.
   b. only from a motor neuron to another motor neuron.
   c. only from a sensory neuron to another sensory neuron.
   d. None of these are correct.

7. The sympathetic division of the autonomic system will
   a. increase heart rate and digestive activity.
   b. decrease heart rate and digestive activity.
   c. cause pupils to constrict.
   d. None of these are correct.

8. Somatic is to skeletal muscle as autonomic is to
   a. cardiac muscle.
   b. smooth muscle.
   c. gland.
   d. All of these are correct.

9. Effects of drugs can include
   a. prevention of neurotransmitter release.
   b. prevention of reuptake by the presynaptic membrane.
   c. blockages to a receptor.
   d. All of these are correct.
   e. None of these are correct.

10. Damage to the Wernicke's area of the brain could cause problems in
    a. vision.
    b. taste.
    c. smell.
    d. speech.

11. A damaged hippocampus can cause problems associated with
    a. long-term memory.
    b. episodic memory.
    c. consolidating short-term into long-term memory.
    d. All of these are correct.

12. Homeostasis is dependent on which of these?
    a. hypothalamus
    b. pons
    c. medulla oblongata
    d. All of these are correct.

13. The summing up of inhibitory and excitatory signals is
    a. excitation.
    b. integration.
    c. depolarization.
    d. All of these are correct.

14. Which are the first and last elements in a spinal reflex?
    a. axon and dendrite
    b. sensory receptor and muscle effector
    c. ventral horn and dorsal horn
    d. brain and skeletal muscle
    e. motor neuron and sensory neuron

15. A spinal nerve takes nerve impulses
    a. to the CNS.
    b. away from the CNS.
    c. both to and away from the CNS.
    d. from the CNS to the spinal cord.

16. Which of these correctly describes the distribution of ions on either side of an axon when it is not conducting a nerve impulse?
    a. more sodium ions (Na^+) outside and more potassium ions (K^+) inside
    b. more K^+ outside and less Na^+ inside
    c. charged protein outside; Na^+ and K^+ inside
    d. Na^+ and K^+ outside and water only inside
    e. chlorine ions (Cl^-) on outside and K^+ and Na^+ on inside

17. When the action potential begins, sodium gates open, allowing Na^+ to cross the membrane. Now the polarity changes to
    a. negative outside and positive inside.
    b. positive outside and negative inside.
    c. There is no difference in charge between outside and inside.
    d. neutral outside and positive inside.

18. Transmission of the nerve impulse across a synapse is accomplished by the
    a. movement of Na^+ and K^+.
    b. release of a neurotransmitter by a dendrite.
    c. release of a neurotransmitter by an axon.
    d. release of a neurotransmitter by a cell body.
    e. All of these are correct.

19. The autonomic system has two divisions, called the
    a. CNS and PNS.
    b. somatic and skeletal.
    c. efferent and afferent.
    d. sympathetic and parasympathetic.

20. Synaptic vesicles are
    a. at the ends of dendrites and axons.
    b. at the ends of axons only.
    c. along the length of all long fibers.
    d. All of these are correct.

21. Which two parts of the brain are least likely to work together?
    a. thalamus and cerebrum
    b. cerebrum and cerebellum
    c. hypothalamus and medulla oblongata
    d. cerebellum and medulla oblongata
    e. hypothalamus and amygdala

22. The limbic system
    a. involves portions of the cerebral lobes, the basal nuclei, and the diencephalon.
    b. is responsible for our deepest emotions, including pleasure, rage, and fear.
    c. is a system necessary to memory storage.
    d. All of these are correct.

23. Label this diagram of the brain.

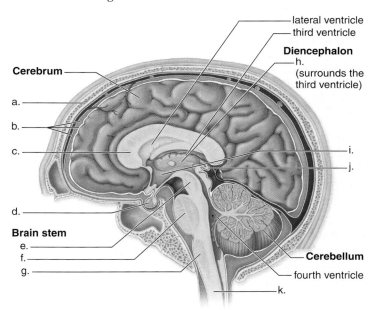

24. Label this diagram of the action potential.

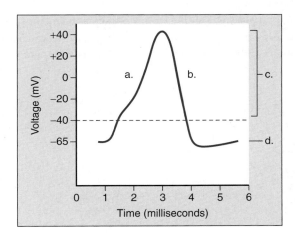

Understanding the Terms

acetylcholine (ACh) 325
acetylcholinesterase (AChE) 325
action potential 322
Alzheimer disease (AD) 340
amygdala 331
association area 329
autonomic system 337
axon 320
axon terminal 324
basal nuclei 330
brain stem 330
Broca's area 329
cell body 320
central canal 326
central nervous system (CNS) 320
cerebellum 330
cerebral cortex 329
cerebral hemisphere 328
cerebrospinal fluid (CSF) 326
cerebrum 327
chemical synapse 324
corpus callosum 328
cranial nerve 334
dendrite 320
diencephalon 330
dorsal root ganglion 335
electroencephalogram (EEG) 330
episodic memory 331
ganglion 334
gray matter 321
hippocampus 331
hypothalamus 330
interneuron 320
intervertebral disk 326

learning 331
limbic system 331
long-term memory 331
long-term potentiation (LTP) 333
medulla oblongata 330
memory 331
meninges (sing., meninx) 326
meningitis 341
midbrain 330
motor neuron 320
multiple sclerosis (MS) 341
myelin sheath 320
nerve 334
nerve impulse 321
neuroglia 320
neuron 320
neurotransmitter 324
node of Ranvier 320
norepinephrine (NE) 325
oligodendroglial cells 321
parasympathetic division 337
Parkinson disease (PD) 341
peripheral nervous system (PNS) 320
pons 330
prefrontal area 329
primary motor area 329
primary somatosensory area 329
reflex 335
refractory period 323
resting potential 322
Schwann cell 320
semantic memory 331
sensory neuron 320
short-term memory 331

skill memory 331
sodium-potassium pump 322
somatic system 335
spinal cord 326
spinal nerve 335
stroke 341
sympathetic division 337

synaptic cleft 324
synaptic integration 325
thalamus 330
threshold 323
tract 321
ventricle 326
Wernicke's area 329
white matter 321

Match the terms to these definitions:

a._____ Action potential (an electrochemical change) traveling along an axon.

b._____ One of the large paired structures making up the cerebrum of the brain.

c._____ The summing of excitatory and inhibitory signals received at synapses by a neuron.

d._____ Neuron that takes information from the central nervous system to effectors.

e._____ Part of the cerebral cortex that functions in memory storage and retrieval.

Thinking Critically

1. How does myelin specifically speed up an action potential?

2. What major difference between the structure of the nerves in the central nervous system and the peripheral nervous system might explain why nerves can often regenerate in the PNS, but not in the CNS?

3. How do you suppose scientists and physicians have determined which areas of the brain are responsible for different activities, as shown in Figures 17.10, 17.11, and 17.14?

4. Dopamine is a key neurotransmitter in the brain. Parkinson patients suffer because of a lack of this chemical, while abusers of drugs such as nicotine, cocaine, and meth enjoy an enhancement of dopamine activity. Would these drugs be a possible treatment, or even a cure, for Parkinson disease? Why or why not?

5. What are some of the factors that make brain diseases such as Alzheimer, Parkinson, and MS so difficult to treat?

ARIS, the *Inquiry into Life* Website

ARIS, the website for *Inquiry into Life,* provides a wealth of information organized and integrated by chapter. You will find practice quizzes, interactive activities, labeling exercises, flashcards, and much more that will complement your learning and understanding of general biology.

www.aris.mhhe.com

Drew Carey is hardly recognizable *without the customary round-rimmed glasses nestled on top of his nose. Frustrated by being denied the joys of scuba diving, Carey decided to undergo corrective eye surgery. You may not be planning frequent scuba trips, but what would you do to throw away your glasses or contact lenses? The hassle and expense of corrective lenses can be draining, and can even be a hindrance for athletes and entertainers. It is no wonder that a growing number of celebrities have decided to toss their specs and give laser eye surgery a try.*

The most common type of corrective laser eye surgery is known as LASIK (laser in-situ keratomileusis). The surgeon first cuts a hinged flap of thin tissue off the outer layer of the eyeball (cornea) and then lifts the flap out of the way. Next, a laser is used to contour the shape of the cornea to the specific needs of the individual. The outer layer is then laid back in place, and the eye usually heals within a few days. Many individuals have reported near-perfect vision within 24 hours after surgery. Still, there are risks, including the possibility of over- or under-correction, infection, and even a decrease in vision that can't be corrected with glasses or contacts. The initial costs, although steadily decreasing, can be inhibitory, and most insurance policies deny such cosmetic procedures. Not all individuals are candidates; factors such as age and health history can affect eligibility.

Despite these potential drawbacks, for many people the possible benefits of LASIK are enticing. The initial clinical trials of the procedure in the United States began in 1996, with FDA approval in 1999. Today it is estimated that more than a million LASIK procedures are done in the United States annually.

chapter 18

Senses

Chapter Concepts

18.1 Sensory Receptors and Sensations

Sensory receptors are specialized cells that detect certain types of stimuli (sing., **stimulus**). Based on the source of the stimulus, sensory receptors can be classified as interoceptors or exteroceptors.

Interoceptors receive stimuli from inside the body, such as changes in blood pressure, blood volume, and the pH of the blood. Interoceptors are directly involved in homeostasis and are regulated by a negative feedback mechanism. For example, when blood pressure rises, pressoreceptors signal a regulatory center in the brain, which then sends out nerve impulses to the arterial walls, causing them to relax so that the blood pressure falls. Once the pressoreceptors are no longer stimulated, the system shuts down.

Exteroceptors are sensory receptors that detect stimuli from outside the body, such as those that result in taste, smell, vision, hearing, and equilibrium (Table 18.1). Their function is to send messages to the central nervous system to report changes in environmental conditions.

Types of Sensory Receptors

Based on the nature of the stimulus, interoceptors and exteroceptors can be further classified into four categories: chemoreceptors, photoreceptors, mechanoreceptors, and thermoreceptors.

Chemoreceptors respond to chemical substances in the immediate vicinity. As Table 18.1 indicates, taste and smell are dependent on this type of sensory receptor, but certain chemoreceptors in various other organs are sensitive to internal conditions. For example, chemoreceptors that monitor blood pH are located in the carotid arteries and aorta. If the pH lowers, the breathing rate increases. As more carbon dioxide is expired, the blood pH rises.

Photoreceptors respond to light energy. Our eyes contain photoreceptors that are sensitive to light rays, and thereby provide us with the sense of vision. Stimulation of the photoreceptors known as rod cells results in black-and-white vision, while stimulation of the photoreceptors known as cone cells results in color vision (see section 18.4).

Mechanoreceptors are stimulated by mechanical forces, which most often result in pressure of some sort.

For example, when we hear, airborne sound waves are converted to fluid-borne pressure waves that can be detected by mechanoreceptors in the inner ear. Similarly, mechanoreceptors are responding to fluid-borne pressure waves when we detect changes in gravity and motion, helping us keep our balance. These receptors are in the vestibule and semicircular canals of the inner ear, respectively (see section 18.5).

The sense of touch is dependent on pressure receptors in the skin and tongue that are sensitive to either strong or slight pressures. Pressoreceptors located in certain arteries detect changes in blood pressure, and stretch receptors in the lungs detect the degree of lung inflation. Proprioceptors, which respond to the stretching of muscle fibers, tendons, joints, and ligaments, make us aware of the position of our limbs.

Thermoreceptors, located in the hypothalamus and skin, are stimulated by changes in temperature. Those that respond when temperatures rise are called warmth receptors, and those that respond when temperatures lower are called cold receptors.

Both interoceptors and exteroceptors are categorized as chemoreceptors, photoreceptors, mechanoreceptors, and thermoreceptors.

How Sensation Occurs

Sensory receptors respond to environmental stimuli by generating nerve impulses. **Detection** occurs when environmental changes, such as pressure to the fingertips or light to the eye, stimulate sensory receptors. **Sensation**, which is conscious perception of stimuli, occurs when nerve impulses arrive at the cerebral cortex of the brain.

As we discussed in Chapter 17, sensory receptors are the first element in a reflex arc. We are only aware of a reflex action when sensory information reaches the brain. At that time, the brain integrates this information with other information received from other sensory receptors. After all, if you burn yourself and quickly remove your hand from a hot stove, the brain receives information not only from your skin, but also from your eyes, nose, and all sorts of sensory receptors.

TABLE 18.1	Exteroceptors			
Sensory Receptor	**Stimulus**	**Category**	**Sense**	**Sensory Organ**
Taste cells	Chemicals	Chemoreceptor	Taste	Taste buds
Olfactory cells	Chemicals	Chemoreceptor	Smell	Olfactory epithelium
Rod cells and cone cells in retina	Light rays	Photoreceptor	Vision	Eye
Hair cells in spiral organ	Sound waves	Mechanoreceptor	Hearing	Ear
Hair cells in semicircular canals	Motion	Mechanoreceptor	Rotational equilibrium	Ear
Hair cells in vestibule	Gravity	Mechanoreceptor	Gravitational equilibrium	Ear

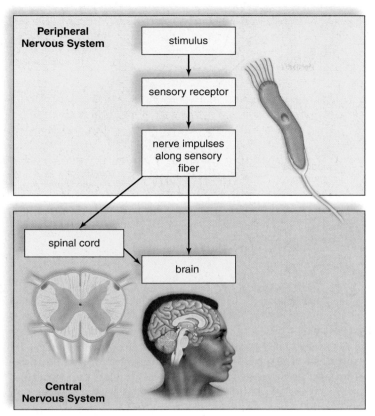

Figure 18.1 Sensation.
The stimulus is received by a sensory receptor, which generates nerve impulses (action potentials). Nerve impulses are conducted to the CNS by sensory nerve fibers within the PNS, and only those impulses that reach the cerebral cortex result in sensation and perception.

Some sensory receptors are free nerve endings or encapsulated nerve endings, while others are specialized cells closely associated with neurons. The plasma membrane of a sensory receptor contains receptor proteins that react to the stimulus. For example, the receptor proteins in the plasma membrane of chemoreceptors bind to certain molecules. When this happens, ion channels open, and ions flow across the plasma membrane. If the stimulus is sufficient, nerve impulses begin and are carried by a sensory nerve fiber within the PNS to the CNS (Fig. 18.1). In this process, known as *sensory transduction,* energy from the chemical or physical stimulus is transduced by the sensory receptors into an electrical signal. This electrical signal is in the form of nerve impulses that travel along the sensory nerve fiber. The stronger the stimulus, the greater the frequency of nerve impulses. Nerve impulses that reach the spinal cord first are conveyed to the brain by ascending tracts. If nerve impulses finally reach the cerebral cortex, sensation and perception occur.

All sensory receptors initiate nerve impulses; the sensation that results depends on the part of the brain receiving the nerve impulses. Nerve impulses that begin in the optic nerve reach the visual cortex, and thereafter, we see objects. Nerve impulses that begin in the auditory nerve reach the auditory cortex, and thereafter, we hear sounds.

Before sensory receptors initiate nerve impulses, they carry out **integration,** the summing up of signals. One type of integration is called **sensory adaptation,** a decrease in the response to a stimulus. For example, as you sit reading this, you are probably not aware of the feeling of the chair beneath you, until now as you are currently reminded. Some authorities believe that when sensory adaptation occurs, sensory receptors have stopped sending impulses to the brain. Others believe that the thalamus has filtered out the ongoing stimuli. You will recall from Chapter 17 that most sensory information is conveyed from the brain stem through the thalamus to the cerebral cortex. The thalamus acts as a gatekeeper and only passes on information of immediate importance. Just as we gradually become unaware of particular environmental stimuli, we can suddenly become aware of stimuli that may have been present for some time. This can be attributed to the workings of the thalamus, which has synapses with all the major ascending sensory tracts.

The functioning of our sensory receptors makes a significant contribution to homeostasis. Without sensory input, we would not receive information about our internal and external environment. This information leads to appropriate reflex and voluntary actions to keep the internal environment constant.

> Detection occurs when environmental changes stimulate sensory receptors. Sensation, the conscious perception of stimuli, occurs when nerve impulses reach the cerebral cortex of the brain.

18.2 Somatic Senses

It is logical to consider those senses whose receptors are associated with the skin, muscles, joints, and viscera together as the somatic senses. Some of these receptors, such as those that sense the position of our limbs, are interoceptors, while others, like those that detect changing conditions at body surfaces, are exteroceptors. Regardless, these receptors send nerve impulses via the spinal cord to the somatosensory areas of the cerebral cortex (see Fig. 17.10). Somatic sensory receptors can be categorized into three types: proprioceptors, cutaneous receptors, and pain receptors.

Proprioceptors

Proprioceptors are mechanoreceptors involved in reflex actions that maintain muscle tone, and thereby the body's equilibrium and posture. They help us know the position of our limbs in space by detecting the degree of muscle relaxation, the stretch of tendons, and the movement of ligaments. Muscle spindles act to increase the degree of muscle contraction, and sensory receptors, called Golgi tendon organs, act to decrease it. The result is a muscle that has the proper length and tension, or muscle tone.

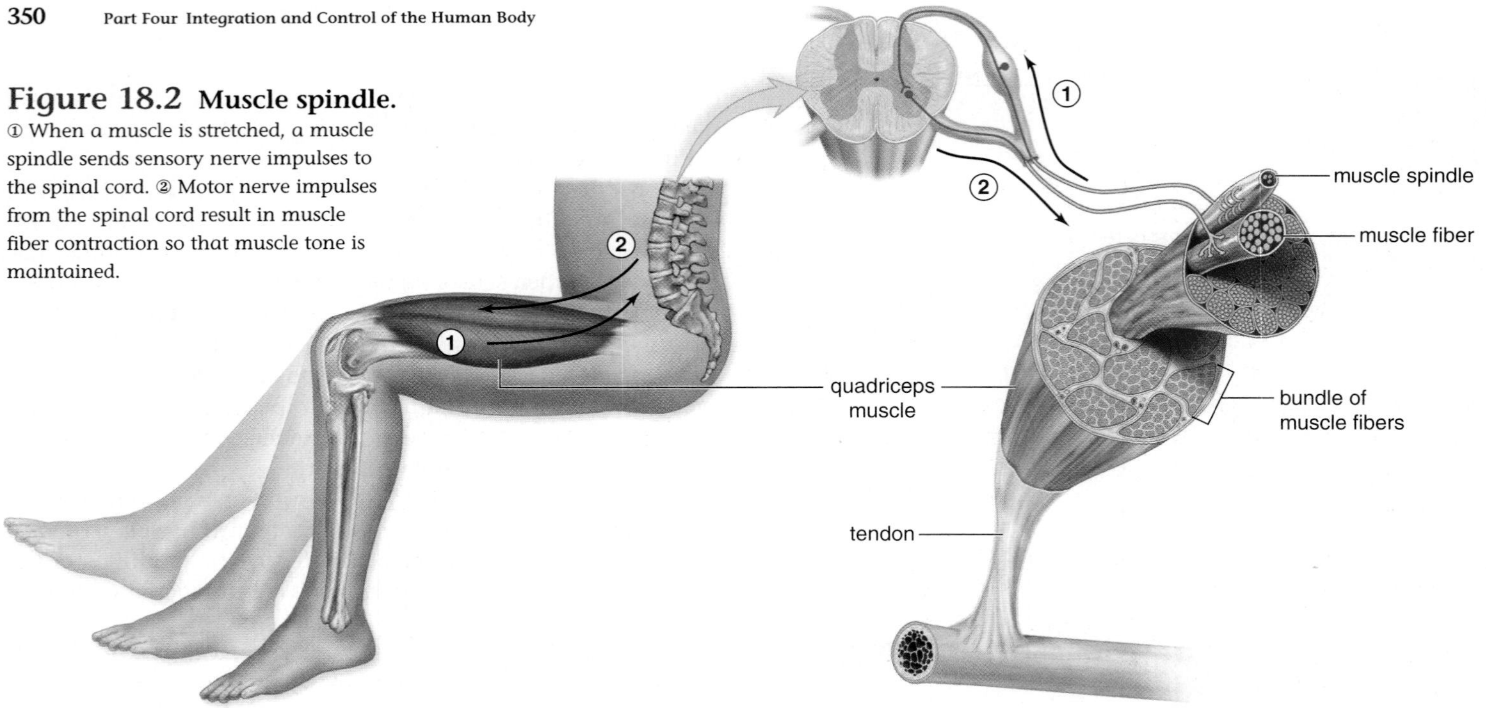

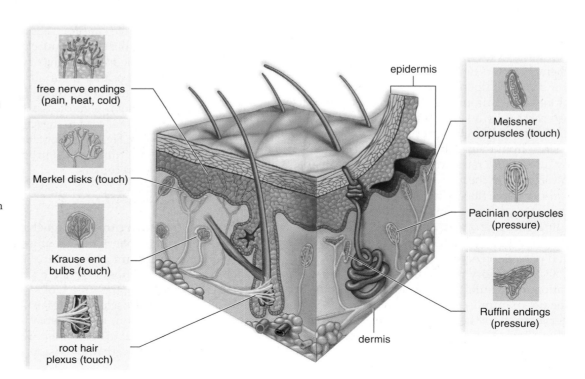

Figure 18.2 Muscle spindle.
① When a muscle is stretched, a muscle spindle sends sensory nerve impulses to the spinal cord. ② Motor nerve impulses from the spinal cord result in muscle fiber contraction so that muscle tone is maintained.

Figure 18.2 illustrates the activity of a muscle spindle. In a muscle spindle, sensory nerve endings are wrapped around thin muscle cells within a connective tissue sheath. When the muscle relaxes and undue stretching of the muscle spindle occurs, nerve impulses are generated. The rapidity of the nerve impulses generated by the muscle spindle is proportional to the stretching of a muscle. A reflex action then occurs, which results in contraction of muscle fibers adjoining the muscle spindle. The knee-jerk reflex, which involves muscle spindles, offers an opportunity for physicians to test a reflex action. The information sent by muscle spindles to the CNS is used to maintain the body's equilibrium and posture, despite the force of gravity always acting upon the skeleton and muscles.

Cutaneous Receptors

The skin is composed of two layers: the epidermis and the dermis. In Figure 18.3, the artist has dramatically indicated these two layers by separating the epidermis from the dermis in one location. The epidermis is stratified squamous epithelium in which cells become keratinized as they rise to

Figure 18.3 Cutaneous receptors in human skin.
The classical view is that each cutaneous receptor has the main function shown here. However, investigators report that matters are not so clear-cut. For example, microscopic examination of the skin of the ear shows only free nerve endings (pain receptors), and yet the skin of the ear is sensitive to all sensations. Therefore, it appears that the receptors of the skin are somewhat, but not completely, specialized.

the surface, where they are sloughed off. The dermis is a thick connective tissue layer. The dermis contains **cutaneous receptors,** which make the skin sensitive to touch, pressure, pain, and temperature (warmth and cold). The dermis contains a mosaic of these tiny receptors, as you can determine by slowly passing a metal probe over your skin. At certain points, you will feel touch or pressure, and at others, you will feel heat or cold (depending on the probe's temperature).

Four types of cutaneous receptors are sensitive to fine touch. *Meissner corpuscles* and *Krause end bulbs* are concentrated in the fingertips, the palms, the lips, the tongue, the nipples, the penis, and the clitoris. *Merkel disks* are found where the epidermis meets the dermis. A free nerve ending called a *root hair plexus* winds around the base of a hair follicle and fires if the hair is touched.

The two different types of cutaneous receptors that are sensitive to pressure are Pacinian corpuscles and Ruffini endings. *Pacinian corpuscles* are onion-shaped sensory receptors that lie deep inside the dermis. *Ruffini endings* are encapsulated by sheaths of connective tissue and contain lacy networks of nerve fibers.

Temperature receptors are simply free nerve endings in the epidermis. Some free nerve endings are responsive to cold; others are responsive to warmth. Cold receptors are far more numerous than warmth receptors, but the two types have no known structural differences.

Pain Receptors

The skin and many internal organs and tissues have pain receptors (free nerve endings, also called *nociceptors*), which are sensitive to chemicals released by damaged cells. When damage occurs due to mechanical, thermal, or electrical stimulation, or a toxic substance, cells release chemicals that stimulate pain receptors. Aspirin and ibuprofen reduce pain by inhibiting the synthesis of one class of these chemicals.

Sometimes, stimulation of internal pain receptors is felt as pain from the skin, as well as the internal organs. This is called **referred pain.** Some internal organs have a referred pain relationship with areas located in the skin of the back, groin, and abdomen; for example, pain from the heart is typically felt in the left shoulder and arm. This most likely happens when nerve impulses from the pain receptors of internal organs travel to the spinal cord and synapse with neurons also receiving impulses from the skin. Pain receptors are protective because they alert us to possible danger. For example, without the pain of appendicitis, we might never seek the medical help needed to avoid a ruptured appendix.

Proprioceptors help maintain our posture; sensory receptors in the skin are sensitive to touch, pressure, pain, and temperature. Pain receptors are a type of chemoreceptor that respond to chemicals released by damaged tissues.

18.3 Senses of Taste and Smell

Taste and smell are called chemical senses because their receptors are sensitive to molecules in the food we eat and the air we breathe. Taste cells and olfactory cells are classified as chemoreceptors.

Sense of Taste

The sensory receptors for the sense of taste, the taste cells, are located in **taste buds.** About 10,000 taste buds are embedded in epithelium, primarily on the tongue (Fig. 18.4). Many lie along the walls of the papillae, the small elevations on the tongue that

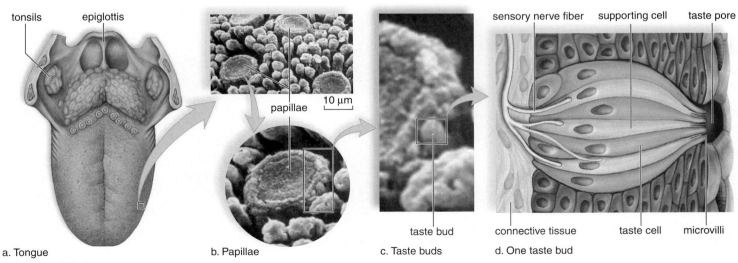

Figure 18.4 Taste buds in humans.
a., b. Papillae on the tongue contain taste buds that are sensitive to sweet, sour, salty, bitter, and perhaps umami. **c.** Taste buds occur along the walls of the papillae. **d.** Taste cells end in microvilli that bear receptor proteins for certain molecules. When molecules bind to the receptor proteins, nerve impulses are generated and go to the brain, where the sensation of taste occurs.

are visible to the naked eye. Isolated taste buds are also present on the hard palate, the pharynx, and the epiglottis. Different taste cells can detect at least four primary types of taste: salty, sour, bitter, and sweet. However, there is evidence that more receptors do exist. For example, a receptor capable of detecting certain amino acids has been described. In particular, this receptor detects the amino acid glutamate, which is part of the flavor enhancer monosodium glutamate (MSG). These receptors have been called "umami" receptors, after the Japanese word meaning delicious. Taste buds containing sensory receptors for each type of taste are located throughout the tongue.

How the Brain Receives Taste Information

Taste buds open at a taste pore. They have supporting cells and a number of elongated taste cells that end in microvilli. When molecules bind to receptor proteins of the microvilli, nerve impulses are generated in sensory nerve fibers that go to the brain. When they reach the gustatory (taste) cortex, they are interpreted as particular tastes.

Since we can respond to a range of salty, sour, bitter, sweet, and perhaps other tastes, the brain appears to survey the overall pattern of incoming sensory impulses and to take a "weighted average" of their taste messages as the perceived taste. Note that even though our senses are dependent on sensory receptors, the cortex integrates the incoming information and gives us our sense perceptions.

> The microvilli of taste cells have receptor proteins for molecules that cause the brain to distinguish between different types of taste.

Sense of Smell

Approximately 80–90% of what we perceive as "taste" is actually due to the sense of smell, which explains why food tastes dull when we have a head cold or a stuffed-up nose. Our sense of smell is dependent on **olfactory cells** located within olfactory epithelium high in the roof of the nasal cavity (Fig. 18.5). Olfactory cells are modified neurons. Each cell ends in a tuft of about five olfactory cilia, which bear receptor proteins for odor molecules.

How the Brain Receives Odor Information

Each olfactory cell has only one out of about 1,000 different types of receptor proteins. Nerve fibers from like olfactory

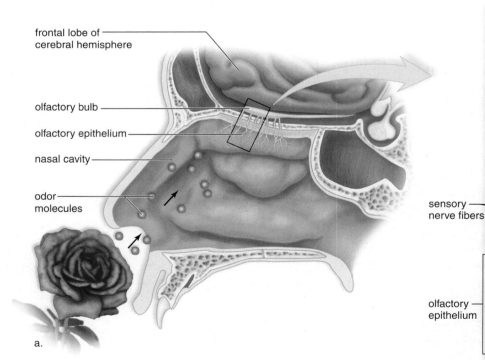

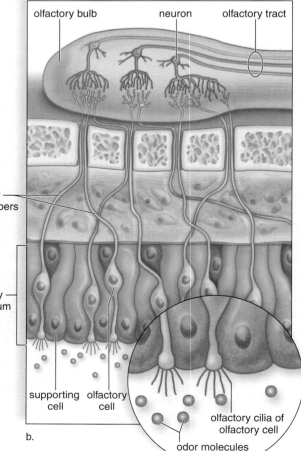

Figure 18.5 Olfactory cell location and anatomy.
a. The olfactory epithelium in humans is located high in the nasal cavity. **b.** Olfactory cells end in cilia that bear receptor proteins for specific odor molecules. The cilia of each olfactory cell can bind to only one type of odor molecule (signified here by color). For example, if a rose causes olfactory cells sensitive to "blue" and "green" odor molecules to be stimulated, then neurons designated by blue and green in the olfactory bulb are activated. The primary olfactory area of the cerebral cortex interprets the pattern of stimulation as the scent of a rose.

cells lead to the same neuron in the olfactory bulb, an extension of the brain. An odor contains many odor molecules, which activate a characteristic combination of receptor proteins. For example, a rose might stimulate olfactory cells, designated by blue and green in Figure 18.5*b*, while a hyacinth might stimulate a different combination. An odor's signature in the olfactory bulb is determined by which neurons are stimulated. The neurons communicate this information via the olfactory tract to the brain; when the information reaches the olfactory cortex, we know we have smelled a rose or a hyacinth.

Have you ever noticed that a certain aroma vividly brings to mind a certain person or place? A person's perfume may remind you of someone else, or the smell of boxwood may remind you of your grandfather's farm. The olfactory bulbs have direct connections to the limbic system and its centers for emotions and memory (see Chapter 17). One investigator showed that when subjects smelled an orange while viewing a painting, they not only remembered the painting when asked about it later, but they also had many deep feelings about it.

Olfactory epithelium contains olfactory cells. The cilia of olfactory cells have receptor proteins for odor molecules that cause the brain to distinguish odors.

18.4 Sense of Vision

Vision requires the work of the eyes and the brain. As we shall see, much processing of stimuli occurs in the eyes before nerve impulses are sent to the brain. Still, researchers estimate that at least a third of the cerebral cortex takes part in processing visual information.

Anatomy and Physiology of the Eye

The eyeball, an elongated sphere about 2.5 cm in diameter, has three layers, or coats: the sclera, the choroid, and the retina (Fig. 18.6). Only the retina contains photoreceptors for light energy. Table 18.2 lists the functions of the parts of the eye.

The outer layer, called the **sclera,** is white and fibrous except for the **cornea,** which is made of transparent collagen fibers. The cornea is the window of the eye. The middle, thin, darkly pigmented layer, the **choroid,** is vascular and absorbs stray light rays that photoreceptors have not absorbed. Toward the front, the choroid becomes the donut-shaped **iris.** The iris contains smooth muscle that regulates the size of the **pupil,** a hole in the center of the iris through which light enters the eyeball. The color of the iris (and therefore, the color of your eyes) correlates with its pigmentation. Heavily pigmented eyes are brown, while lightly pigmented eyes are green or blue. Behind the iris,

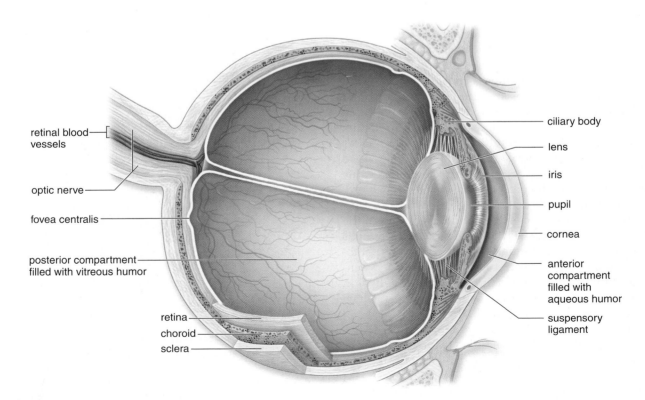

retinal blood vessels
optic nerve
fovea centralis
posterior compartment filled with vitreous humor
retina
choroid
sclera
ciliary body
lens
iris
pupil
cornea
anterior compartment filled with aqueous humor
suspensory ligament

Figure 18.6 Anatomy of the human eye.
Notice that the sclera (the outer layer of the eye) becomes the cornea and that the choroid (the middle layer) is continuous with the ciliary body and the iris. The retina (the inner layer) contains the photoreceptors for vision. The fovea centralis is the region where vision is most acute.

TABLE 18.2	Functions of the Parts of the Eye
Part	**Function**
Sclera	Protects and supports eyeball
Cornea	Refracts light rays
Choroid	Absorbs stray light
Ciliary body	Holds lens in place; aids accommodation
Iris	Regulates light entrance
Pupil	Admits light
Retina	Contains sensory receptors for sight
Rods	Make black-and-white vision possible
Cones	Make color vision possible
Fovea centralis	Makes acute vision possible
Other	
Lens	Refracts and focuses light rays
Humors	Transmit light rays and support eyeball
Optic nerve	Transmits impulse to brain

the choroid thickens and forms the circular ciliary body. The **ciliary body** contains the **ciliary muscle,** which controls the shape of the lens for near and far vision.

The **lens,** attached to the ciliary body by ligaments, divides the eye into two compartments; the one in front of the lens is the anterior compartment, and the one behind the lens is the posterior compartment. The anterior compartment is filled with a clear, watery fluid called the **aqueous humor.** A small amount of aqueous humor is continually produced each day. Normally, it leaves the anterior compartment by way of tiny ducts. When a person has **glaucoma,** these drainage ducts are blocked, and aqueous humor builds up. If glaucoma is not treated, the resulting pressure compresses the arteries that serve the nerve fibers of the retina, where photoreceptors are located. The nerve fibers begin to die due to lack of nutrients, and the person becomes partially blind. Eventually, total blindness can result (see section 18.7).

The third layer of the eye, the **retina,** is located in the posterior compartment, which is filled with a clear, gelatinous material called the **vitreous humor.** The retina contains photoreceptors called rod cells and cone cells. The rods are very sensitive to light, but they do not see color; therefore, at night or in a darkened room, we see only shades of gray. The cones, which require bright light, are sensitive to different wavelengths of light, and therefore, we have the ability to distinguish colors. The retina has a very special region called the **fovea centralis,** where cone cells are densely packed. Light is normally focused on the fovea when we look directly at an object. This is helpful because vision is most acute in the fovea centralis. Sensory nerve fibers from the retina form the **optic nerve,** which takes nerve impulses to the visual cortex.

The eye has three layers: the outer sclera, the middle choroid, and the inner retina. Only the retina contains photoreceptors for light energy.

Function of the Lens

The lens, assisted by the cornea and the humors, focuses images on the retina (Fig. 18.7). Focusing starts with the cornea and continues as the rays pass through the lens and the humors. The image produced is much smaller than the object because light rays are bent (refracted) when they are brought into **focus.** If the eyeball is too long or too short, the person may need corrective lenses to bring the image into focus.

Visual accommodation occurs for close vision. During visual accommodation, the lens rounds up in order to bring the image to focus on the retina. The shape of the lens is controlled by the ciliary muscle within the ciliary body. When we view a distant object, the ciliary muscle is relaxed, causing the suspensory ligaments attached to the ciliary body to be taut; therefore, the lens remains relatively flat (Fig. 18.7a). When we view a near object, the ciliary muscle contracts, releasing the tension on the suspensory ligaments, and the lens rounds up due to its natural

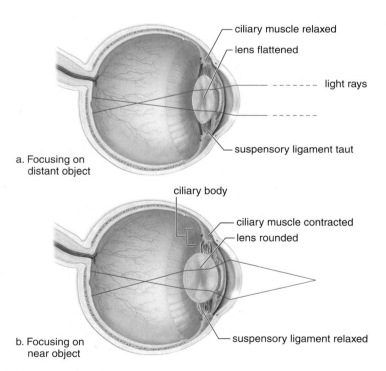

a. Focusing on distant object

b. Focusing on near object

Figure 18.7 Focusing the human eye.
Light rays from each point on an object are bent by the cornea and the lens in such a way that an inverted and reversed image forms on the retina. **a.** When focusing on a distant object, the lens is flat because the ciliary muscle is relaxed and the suspensory ligament is taut. **b.** When focusing on a near object, the lens accommodates; that is, it becomes rounded because the ciliary muscle contracts, causing the suspensory ligament to relax.

elasticity (Fig. 18.7*b*). Now the image is focused on the retina. Because close work requires contraction of the ciliary muscle, it very often causes muscle fatigue, known as eyestrain. Usually after the age of 40, the lens loses some of its elasticity and is unable to accommodate. Bifocal lenses may then be necessary for those who already have corrective lenses.

> **The lens, assisted by the cornea and the humors, focuses images on the retina. When viewing a close object, the lens rounds up in order to bring the image to focus on the retina.**

Visual Pathway to the Brain

The pathway for vision begins once light has been focused on the photoreceptors in the retina. Some integration occurs in the retina, where nerve impulses begin before the optic nerve transmits them to the brain.

Function of Photoreceptors   The photoreceptors in the eye are of two types—**rod cells** and **cone cells;** Figure 18.8 illustrates their structure. Both rods and cones have an outer segment joined to an inner segment by a stalk. Pigment molecules are embedded in the membrane of the many disks present in the outer segment. Synaptic vesicles are located at the synaptic endings of the inner segment.

The visual pigment in rods is a deep purple pigment called rhodopsin. **Rhodopsin** is a complex molecule made up of the protein opsin and a light-absorbing molecule called **retinal,** which is a derivative of vitamin A. When a rod absorbs light, rhodopsin splits into opsin and retinal, leading to a cascade of reactions and the closure of ion channels in the rod cell's plasma membrane. The release of inhibitory transmitter molecules from the rod's synaptic vesicles ceases. Thereafter, signals go to other neurons in the retina. Rods are very sensitive to light, and are therefore suited to night vision. (Because carrots are rich in vitamin A, it is true that eating carrots can improve your night vision.) With the exception of the fovea centralis, rod cells are plentiful throughout the entire retina; therefore, they also provide us with peripheral vision and perception of motion.

The cones, on the other hand, are located primarily in the fovea centralis and are activated by bright light. They allow us to detect the fine detail and the color of an object. **Color vision** depends on three different kinds of cones, which contain pigments called the B (blue), G (green), and R (red) pigments. Each pigment is an iodopsin made up of retinal and an opsin. There is a slight difference in the opsin structure of each, which accounts for the ability of the pigments to absorb different wavelengths (colors) of visible light. Various combinations of cones are believed to be stimulated by in-between shades of color.

> **The receptors for sight are the rods and the cones. The rods permit vision in dim light at night, and the cones permit vision in the bright light needed for color vision.**

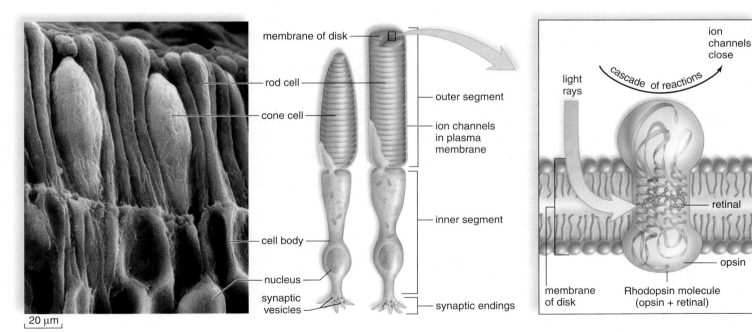

Figure 18.8 Photoreceptors in the eye.

The outer segment of rods and cones contains stacks of membranous disks that contain visual pigments. In rods, the membrane of each disk contains rhodopsin, a complex molecule composed of the protein opsin and the pigment retinal. When rhodopsin absorbs light energy, it splits, releasing opsin, which sets in motion a cascade of reactions that cause ion channels in the plasma membrane to close. Thereafter, nerve impulses go to the brain.

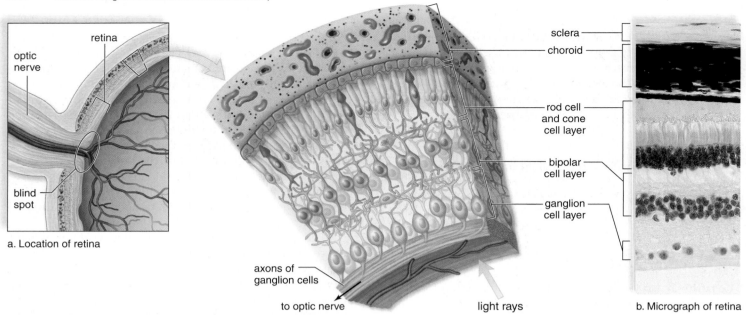

a. Location of retina

sclera
choroid
rod cell and cone cell layer
bipolar cell layer
ganglion cell layer

optic nerve
retina
blind spot

axons of ganglion cells
to optic nerve
light rays

b. Micrograph of retina

Figure 18.9 Structure and function of the retina.

a. The retina is the inner layer of the eyeball. Rod cells and cone cells, located at the back of the retina nearest the choroid, synapse with bipolar cells, which synapse with ganglion cells. Integration of signals occurs at these synapses; therefore, much processing occurs in bipolar and ganglion cells. Notice also that many rod cells share one bipolar cell, but cone cells do not. Certain cone cells synapse with only one ganglion cell. Cone cells, in general, distinguish more detail than do rod cells. **b.** Micrograph shows that the sclera and choroid are relatively thin compared to the retina, which has several layers of cells.

Function of the Retina   The retina has three layers of neurons (Fig. 18.9). The layer closest to the choroid contains the rod cells and cone cells; the middle layer contains bipolar cells; and the innermost layer contains ganglion cells, whose sensory fibers become the optic nerve. Only the rod cells and the cone cells are sensitive to light, and therefore, light must penetrate to the back of the retina before they are stimulated.

The rod cells and the cone cells synapse with the bipolar cells, which in turn synapse with ganglion cells whose axons become the optic nerve. Notice in Figure 18.9 that there are many more rod cells and cone cells than ganglion cells. In fact, the retina has as many as 150 million rod cells and 6 million cone cells but only one million ganglion cells. The sensitivity of cones versus rods is due in part to how directly they connect to ganglion cells. As many as 150 rods may activate the same ganglion cell. No wonder stimulation of rods results in vision that is blurred and indistinct. In contrast, some cone cells in the fovea centralis activate only one ganglion cell. This explains why cones, especially in the fovea, provide us with a sharper, more detailed image of an object.

As signals pass to bipolar cells and ganglion cells, synaptic integration occurs. Each ganglion cell receives signals from rod cells covering about one square millimeter of retina (about the size of a thumbtack hole). This region is the ganglion cell's receptive field. Some time ago, scientists discovered that a ganglion cell is stimulated only by signals received from the center of its receptive field; otherwise, it is inhibited. If all the rod cells in the receptive field receive light, the ganglion cell responds in a neutral way—that is, it reacts only weakly or perhaps not at all. This supports the hypothesis that considerable processing occurs in the retina before ganglion cells generate nerve impulses, which are carried in the optic nerve to the visual cortex. Additional integration occurs in the visual cortex.

Synaptic integration and processing begin in the retina before nerve impulses are sent to the brain.

Blind Spot   Figure 18.9*a* provides an opportunity to point out that there are no rods and cones where the optic nerve exits the retina. Therefore, no vision is possible in this area. You can prove this to yourself by putting a dot to the right of center on a piece of paper. Use your right hand to move the paper slowly toward your right eye, while you look straight ahead. The dot will disappear at one point—this is your right eye's **blind spot.**

From the Retina to the Visual Cortex   As stated previously, the axons of ganglion cells in the retina assemble to form the optic nerves. The optic nerves carry nerve impulses from the eyes to the optic chiasma. The **optic chiasma** has an X shape, formed by a crossing-over of optic nerve fibers. Fibers from the right half of each retina converge and continue on together in the *right optic tract,* and fibers from the left half of each retina converge and continue on together in the *left optic tract* (Fig. 18.10).

The optic tracts sweep around the hypothalamus, and most fibers synapse with neurons in nuclei (masses of neuron cell bodies) within the thalamus. Axons from the thalamic nuclei form optic radiations that take nerve impulses to the *visual area* within the occipital lobe. Notice in Figure 18.10 that the image arriving at the thalamus, and therefore the visual area, has been split because the left optic tract carries information about the right portion of the visual field and the right optic tract carries information about the left portion of the visual field. Therefore, the right and left visual areas must communicate with each other for us to see the entire visual field. Also,

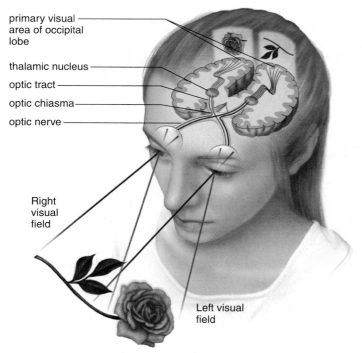

Figure 18.10 Optic chiasma.
Both eyes "see" the entire visual field. Because of the optic chiasma, data from the right half of each retina go to the right visual cortex, and data from the left half of each retina go to the left visual cortex. These data are then combined to allow us to see the entire visual field. Note that the visual pathway to the brain includes the thalamus, which has the ability to filter sensory stimuli.

because the image is inverted and reversed, it must be righted in the brain for us to correctly perceive the visual field.

The most surprising finding has been that the primary visual area acts like a post office, parceling out information regarding color, form, motion, and possibly other attributes to different portions of the adjoining visual association area. Therefore, the brain has taken the visual field apart, even though we see a unified visual field. The visual association areas are believed to rebuild the field and give us an understanding of it at the same time.

> The visual pathway begins in the retina and passes through the thalamus before reaching the primary visual area in the occipital lobe of the brain.

18.5 Sense of Hearing

The ear has two sensory functions: hearing and balance (equilibrium). The sensory receptors for both of these are located in the inner ear, and each consists of **hair cells** with stereocilia (long microvilli) that are sensitive to mechanical stimulation. They are mechanoreceptors.

Anatomy of the Ear

Figure 18.11 shows that the ear has three divisions: outer, middle, and inner. The **outer ear** consists of the **pinna** (external flap), which collects and funnels sound into the **auditory canal.** The opening of the auditory canal is lined with fine hairs and sweat glands. Modified sweat glands are located in the upper wall of the canal; they secrete earwax, a substance that helps guard the ear against the entrance of foreign materials, such as air pollutants and microorganisms.

The **middle ear** begins at the **tympanic membrane** (eardrum) and ends at a bony wall containing two small openings covered by membranes. These openings are called the **oval window** and the **round window.** Between the tympanic

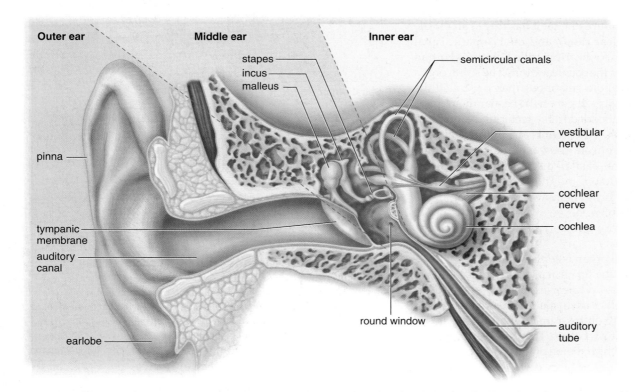

**Figure 18.11
Anatomy of the human ear.**
In the middle ear, the malleus (hammer), the incus (anvil), and the stapes (stirrup) amplify sound waves. In the inner ear, the mechanoreceptors for equilibrium are in the semicircular canals and the vestibule, and the mechanoreceptors for hearing are in the cochlea.

membrane and the oval window are three small bones, collectively called the **ossicles.** Individually, they are the **malleus** (hammer), the **incus** (anvil), and the **stapes** (stirrup) because their shapes resemble these objects. The malleus adheres to the tympanic membrane, and the stapes touches the oval window. An **auditory tube** (eustachian tube), which extends from the middle ear to the nasopharynx, permits equalization of air pressure. Chewing gum, yawning, and swallowing in elevators and airplanes help move air through the auditory tubes upon ascent and descent. As this occurs, we often hear the ears "pop."

Whereas the outer ear and the middle ear contain air, the inner ear is filled with fluid. Anatomically speaking, the **inner ear** has three areas: the **semicircular canals** and the **vestibule** function in equilibrium; the **cochlea** functions in hearing. The cochlea resembles the shell of a snail because it spirals.

Auditory Pathway to the Brain

Hearing requires the ear, the cochlear nerve, and the auditory cortex of the brain. The sound pathway begins with the auditory canal.

Through the Auditory Canal and Middle Ear The process of hearing begins when sound waves enter the auditory canal. Just as ripples travel across the surface of a pond, sound waves travel by the successive vibrations of molecules. When these vibrations impinge upon the tympanic membrane, it begins to vibrate at the same frequency as the sound waves. The malleus then takes the pressure from the inner surface of the tympanic membrane and passes it, by means of the incus, to the stapes in such a way that the pressure is multiplied about 20 times as it moves. The stapes strikes the membrane of the oval window, causing it to vibrate, and in this way, the pressure is passed to the fluid within the cochlea.

From the Cochlea to the Auditory Cortex Examining part of the cochlea in cross section reveals that it has three canals: the vestibular canal, the **cochlear canal,** and the tympanic canal (Fig. 18.12). The sense organ for hearing, called the **spiral organ** (organ of Corti), is located in the cochlear canal. The spiral organ consists of little hair cells and a gelatinous material called the **tectorial membrane.** The hair cells sit on the basilar membrane, and their stereocilia are embedded in the tectorial membrane.

When the stapes strikes the membrane of the oval window, pressure waves move from the vestibular canal to the tympanic canal across the basilar membrane. The basilar membrane moves up and down, and the stereocilia of the hair cells embedded in the tectorial membrane bend. Then nerve impulses begin in the cochlear nerve and travel to the brain. When they reach the auditory cortex in the temporal lobe, they are interpreted as a sound.

Each part of the spiral organ is sensitive to different wave frequencies, or pitch. Near the tip, the spiral organ responds to low pitches, such as a tuba, and near the base, it responds to higher pitches, such as a bell or a whistle. The nerve fibers from each region along the length of the spiral organ lead to slightly different areas in the auditory cortex. The pitch sensation we experience depends upon which region of the basilar membrane vibrates and which area of the auditory cortex is stimulated.

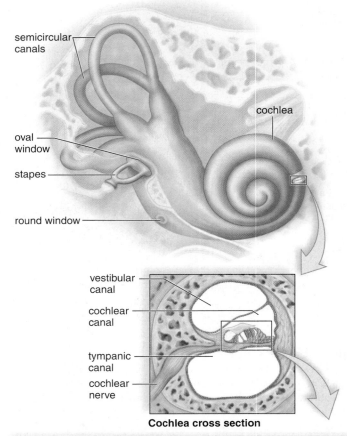

Cochlea cross section

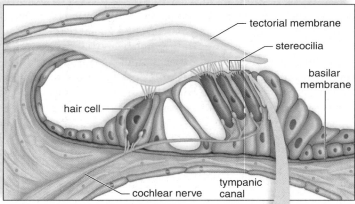

Spiral organ

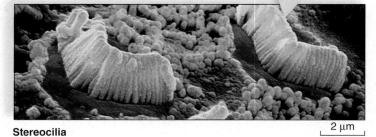

Stereocilia 2 μm

Figure 18.12 **Mechanoreceptors for hearing.**

The spiral organ is located within the cochlea. In the uncoiled cochlea, note that the spiral organ consists of hair cells resting on the basilar membrane, with the tectorial membrane above. Pressure waves move from the vestibular canal to the tympanic canal, causing the basilar membrane to vibrate. This causes the stereocilia (or at least a portion of the more than 20,000 hair cells) embedded in the tectorial membrane to bend. Nerve impulses traveling in the cochlear nerve to the auditory cortex result in hearing.

Health Focus

Preventing a Loss of Hearing

Some degree of age-associated hearing loss is very common (see section 18.7), but deafness due to stereocilia damage from exposure to loud noises is preventable. Hospitals are now aware that even the ears of newborns need to be protected from noise and are taking steps to make sure neonatal intensive care units and nurseries are as quiet as possible.

In today's society, exposure to the types of noises listed in Table 18A is common. Noise is measured in decibels, and any noise above a level of 80 decibels could result in damage to the hair cells of the spiral organ. Hearing loss occurs once 25% to 30% of these cells are damaged. Eventually, the stereocilia and then the hair cells disappear completely (Fig. 18A). If listening to city traffic for extended periods can damage hearing, it stands to reason that frequent attendance at rock concerts, constantly playing a stereo loudly, or using earphones at high volume is also damaging to hearing.

Legendary rock music performer Pete Townsend of The Who recently admitted that long-term exposure to loud music has caused him to experience hearing loss. Digital music players, such as the iPod, may pose an additional risk because they can hold thousands of songs and play for hours without recharging. Like all the sensory systems, the ears can become desensitized to loud music, so that it seems less loud. The response of many music enthusiasts, of course, is to turn up the volume.

The first hint of danger could be temporary hearing loss, a "full" feeling in the ears, muffled hearing, or tinnitus (e.g., ringing in the ears). If you have any of these symptoms, modify your listening habits immediately to prevent further damage. If exposure to noise is unavoidable, specially designed noise-reduction earmuffs are available, and it is also possible to purchase earplugs made from a

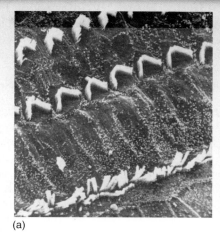
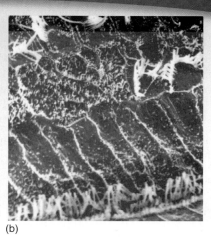

(a)   (b)

Figure 18A Hearing loss.
a. Normal hair cells in the spiral organ of a guinea pig. **b.** Damaged hair cells in the spiral organ of a guinea pig. This damage occurred after 24-hour exposure to a noise-level typical of a rock concert.

compressible, spongelike material at the drugstore or sporting-goods store. These earplugs are not the same as those worn for swimming, and they should not be used interchangeably.

In addition to loud music, noisy indoor or outdoor equipment can also be harmful to hearing if hearing protection is not used. Even motorcycles and recreational vehicles such as snowmobiles and motocross bikes can contribute to a gradual loss of hearing. Exposure to intense sounds of short duration, such as a burst of gunfire, can result in an immediate hearing loss. Hunters may have a significant hearing reduction in the ear opposite the shoulder where the gun is held. (The butt of the rifle offers some protection to the ear nearest the gun when it is shot.)

Finally, people need to be aware that some medicines are ototoxic. Anticancer drugs, most notably cisplatin, and certain antibiotics (e.g., streptomycin, kanamycin, and gentamicin) make ears especially sus-

ceptible to a hearing loss. Anyone taking such medications needs to be especially careful to protect the ears from loud noises.

Discussion Questions

1. Rank the following senses, from 1 to 7, according to which you consider most valuable: balance, hearing, pain, smell, taste, touch, vision.
2. If you listen to loud music or are exposed to other noise, are you concerned that the hair cells in the spiral organs of your inner ears may become damaged? Why or why not?
3. Assume that you suddenly became deaf today, and list three to five of your everyday activities that would be most affected. If your deafness could not be corrected medically, what specific steps could you take to help you communicate with hearing people? With other deaf people?

TABLE 18A	Noises that Affect Hearing	
Type of Noise	**Sound Level (decibels)**	**Effect**
Jet engine, shotgun, rock concert	Over 125	Beyond threshold of pain; potential for hearing loss high
Nightclub, "boom box," thunderclap	Over 120	Hearing loss likely
Chain saw, pneumatic drill, jackhammer, snowmobile, garbage truck, cement mixer	100–200	Regular exposure of more than one minute risks permanent hearing loss
Farm tractor, newspaper press, subway, motorcycle	90–100	Fifteen minutes of unprotected exposure potentially harmful
Lawn mower, food blender	85–90	Continuous daily exposure for more than eight hours can cause hearing damage
Diesel truck, average city traffic noise	80–85	Annoying; constant exposure may cause hearing damage
Source: National Institute on Deafness and Other Communication Disorders, National Institute of Health.		

Volume is a function of the amplitude of sound waves. Loud noises cause the fluid within the vestibular canal to exert more pressure and the basilar membrane to vibrate to a greater extent. The brain interprets the resulting increased stimulation as volume. Researchers believe that the brain interprets the tone of a sound based on the distribution of the hair cells stimulated. Exposure to loud noises over a long period of time is one of the causes of hearing loss (see Health Focus).

> **The mechanoreceptors for hearing are hair cells on the basilar membrane of the spiral organ.**

18.6 Sense of Equilibrium

Mechanoreceptors in the semicircular canals detect rotational and/or angular movement of the head (**rotational equilibrium**), while mechanoreceptors in the utricle and saccule detect movement of the head in the vertical or horizontal planes (**gravitational equilibrium**) (Fig. 18.13).

Through their communication with the brain, these mechanoreceptors help us achieve equilibrium, but other structures in the body are also involved. For example, we already mentioned that proprioceptors are necessary for maintaining our equilibrium. Vision, if available, also provides extremely helpful input the brain can act upon.

Rotational Equilibrium Pathway

Rotational equilibrium involves the three semicircular canals, which are arranged so that there is one in each dimension of space. The base of each of the three canals, called an **ampulla**, is slightly enlarged. Little hair cells, whose stereocilia are embedded within a gelatinous material called a cupula, are found within the ampullae. Because of the way the semicircular canals are arranged, each ampulla responds to head rotation in a different plane of space. As fluid within a semicircular canal flows over and displaces a cupula, the stereocilia of the hair cells bend, and the pattern of impulses carried by the vestibular nerve to the brain stem and cerebellum changes. The brain uses information from the hair cells within the ampullae of the semicircular canals to maintain rotational equilibrium through appropriate motor output to various skeletal muscles that can right our present position in space as need be.

Gravitational Equilibrium Pathway

Gravitational equilibrium depends on the **utricle** and the **saccule**, two membranous sacs located in the vestibule. Both of these sacs contain little hair cells, whose stereocilia are embedded within a gelatinous material called an otolithic membrane. Calcium carbonate ($CaCO_3$) granules, or **otoliths**, rest on this membrane. The utricle is especially sensitive to horizontal (back-and-forth) movements and the bending of the head, while the saccule responds best to vertical (up-and-down) movements.

When the body is still, the otoliths in the utricle and the saccule rest on the otolithic membrane above the hair cells. When the head bends or the body moves in the horizontal and vertical planes, the otoliths are displaced and the otolithic membrane sags, bending the stereocilia of the hair cells beneath. If the stereocilia move toward the largest stereocilium, called the kinocilium, nerve impulses increase in the vestibular nerve. If the stereocilia move away from the kinocilium, nerve impulses decrease in the vestibular nerve. If you are upside down, nerve impulses in the vestibular nerve cease. These data reach the vestibular cortex, which uses them to determine the direction of the movement of the head at the moment. The brain uses this information to maintain gravitational equilibrium through appropriate motor output to various skeletal muscles that can right our present position in space as need be.

Table 18.3 summarizes how the various parts of the ear function in both hearing and equilibrium.

> **The semicircular canals contribute to the sense of rotational equilibrium, and the utricle and the saccule function in gravitational equilibrium.**

TABLE 18.3	Functions of the Parts of the Ear		
Part	**Medium**	**Function**	**Mechanoreceptor**
Outer Ear	*Air*		
Pinna		*Collects sound waves*	—
Auditory canal		Filters air	—
Middle Ear	*Air*		
Tympanic membrane and ossicles		Amplify sound waves	—
Auditory tube		Equalizes air pressure	—
Inner Ear	*Fluid*		
Semicircular canals		Rotational equilibrium	Stereocilia embedded in cupula
Vestibule (contains utricle and saccule)		Gravitational equilibrium	Stereocilia embedded in otolithic membrane
Cochlea (contains spiral organ)		Hearing	Stereocilia embedded in tectorial membrane

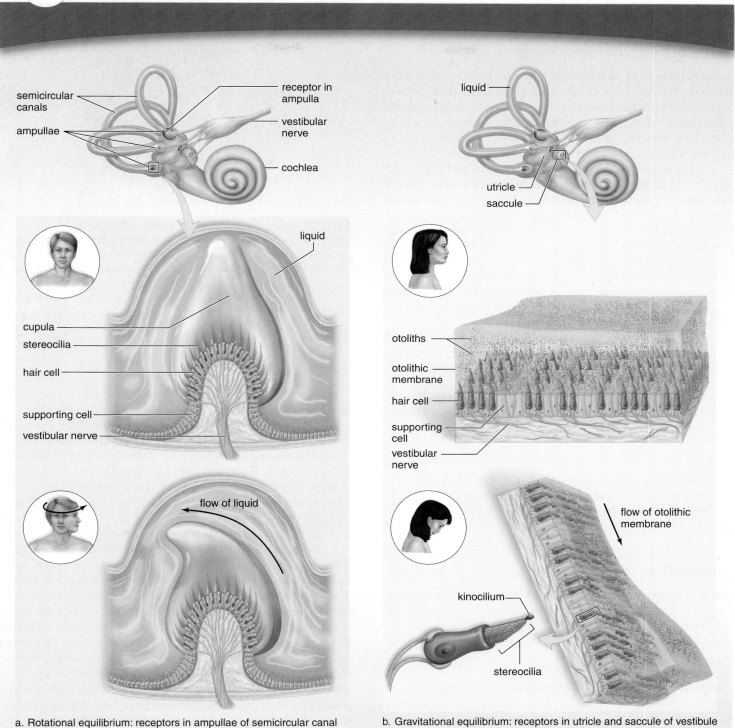

a. Rotational equilibrium: receptors in ampullae of semicircular canal

b. Gravitational equilibrium: receptors in utricle and saccule of vestibule

Figure 18.13 Mechanoreceptors for equilibrium.

a. Rotational equilibrium. The ampullae of the semicircular canals contain hair cells with stereocilia embedded in a cupula. When the head rotates, the cupula is displaced, bending the stereocilia. Thereafter, nerve impulses travel in the vestibular nerve to the brain.
b. Gravitational equilibrium. The utricle and the saccule contain hair cells with stereocilia embedded in an otolithic membrane. When the head bends, otoliths are displaced, causing the membrane to sag and the stereocilia to bend. If the stereocilia bend toward the kinocilium, the longest of the stereocilia, nerve impulses increase in the vestibular nerve. If the stereocilia bend away from the kinocilium, nerve impulses decrease in the vestibular nerve. This difference tells the brain in which direction the head moved.

18.7 Disorders that Affect the Senses

Disorders of Taste and Smell

It is hard for most of us to imagine our world without the ability to taste and smell. Disorders that affect these senses may not sound very serious, but these senses contribute substantially to our enjoyment of life. In addition, unpleasant smells and tastes can warn us about dangers such as fire, poisonous fumes, and spoiled food.

In most people, the sense of smell begins to decline after age 60, and a large proportion of elderly people lose some of their olfactory sense. Some people are born with a poor sense of smell or taste, and others completely lack a sense of smell, a condition known as anosmia. However, the most common conditions that affect our sense of taste and smell are upper respiratory infections, allergies, exposure to certain drugs and chemicals, and certain types of brain tumors. Brain injury due to trauma can also cause smell or taste problems, and smoking impairs the ability to identify odors and diminishes the sense of taste.

The extent of a loss of smell or taste can be tested using the lowest concentration of a chemical that a person can detect and recognize. Scientists have developed an easily administered "scratch-and-sniff" test to evaluate the sense of smell. For taste, patients are asked to evaluate different chemical concentrations, sometimes applied directly to specific areas of the tongue.

> **The most frequent causes of disorders of taste and smell are infections, allergies, exposure to certain drugs and chemicals, and brain tumors.**

Disorders of the Eye

Color blindness and problems with visual focus are two common abnormalities of the eye. More serious disorders can result in blindness.

Color Blindness

Complete color blindness is extremely rare. Instead, in most instances a particular type of cone is lacking or deficient in number. The most common mutation is a lack of either red or green cones, resulting in various types of color blindness (Fig. 18.14). If the eye lacks red cones, the green colors are accentuated, and vice versa. Most genetic mutations affecting color vision are X-linked, recessive traits. Since females have two copies of the X chromosome, they are more likely to have a normal copy of the gene. Therefore, color blindness is much more common in males. Approximately 5% to 8% of males and 0.5% of females have some type of deficiency in color vision.

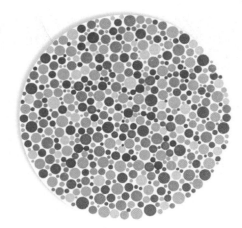

Figure 18.14 Testing for color blindness.
This is one type of test plate used to diagnose color blindness. There are many different types of color blindness, and thus many different tests are used. In this case, you should be able to see the number "16" in green among the dots.

Reproduced from Ishihara's Tests for Colour Deficiency published by KANEHARA TRADING INC., located at Tokyo in Japan. But tests for color deficiency cannot be conducted with this material. For accurate testing, the original plates should be used.

Visual Focus

The majority of people can see what is designated as a size 20 letter 20 feet away, and so are said to have 20/20 vision. Persons who can see close objects but cannot see the letter from this distance are said to be **nearsighted.** These individuals usually have an elongated eyeball, so that when they attempt to look at a distant object, the lens of the eye brings the image into focus in front of the retina (Fig. 18.15a). They can see close objects because the lens can compensate for the long eyeball. To see distant objects, these people can wear concave lenses, which diverge the light rays so that the image focuses on the retina.

Persons who can easily see the optometrist's chart but cannot see close objects well are **farsighted;** these individuals can see distant objects better than they can see close objects. They have a shortened eyeball, and when they try to see close objects, the image is focused behind the retina. When the object is distant, the lens can compensate for the short eyeball. When the object is close, these persons must wear convex lenses to increase the bending of light rays so that the image can be focused on the retina (Fig. 18.15b). People over the age of 45 usually begin to have trouble seeing objects close up due to a condition called presbyopia, in which the lens of the eye begins to lose its elasticity and thus cannot focus on close objects.

When the cornea or lens is uneven, the image is fuzzy. The light rays cannot be evenly focused on the retina. This condition, known as **astigmatism,** can be corrected by wearing an unevenly ground lens to compensate for the uneven cornea (Fig. 18.15c). Rather than wearing glasses or contact lenses, many people with problems of visual focus are now choosing to undergo a laser surgery procedure called LASIK (see the introduction to this chapter).

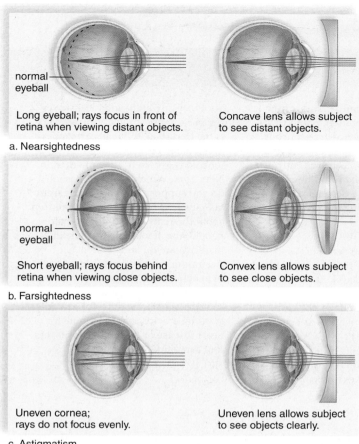

Long eyeball; rays focus in front of retina when viewing distant objects.

Concave lens allows subject to see distant objects.

a. Nearsightedness

normal eyeball

Short eyeball; rays focus behind retina when viewing close objects.

Convex lens allows subject to see close objects.

b. Farsightedness

normal eyeball

Uneven cornea; rays do not focus evenly.

Uneven lens allows subject to see objects clearly.

c. Astigmatism

Figure 18.15 Common abnormalities of the eye and possible corrective lenses.
a. A concave lens in nearsighted persons focuses light rays on the retina.
b. A convex lens in farsighted persons focuses light rays on the retina. **c.** An uneven lens in persons with astigmatism focuses light rays on the retina.

Common Causes of Blindness

The most frequent causes of blindness in adults are retinal disorders, glaucoma, and cataracts, in that order. Retinal disorders are varied. In diabetic retinopathy, the leading cause of blindness in the U.S. for people under the age of 65, capillaries to the retina become damaged, and hemorrhages and blocked vessels can occur. Surgical lasers can be used to seal the leaking blood vessels, but eventually scarring can lead to blindness. Careful regulation of blood glucose levels in these patients may be protective. A common condition in people over 50 years of age is macular degeneration, in which the cones are destroyed because thickened choroid vessels no longer function as they should. The retina itself can also become torn or detached, often following trauma to the eye or head. If treated quickly enough, the retina can usually be reattached with a laser.

Glaucoma occurs when the drainage system of the eyes fails, so fluid builds up and destroys the nerve fibers responsible for peripheral vision. Eye doctors usually check for glaucoma during routine exams, but it is advisable for everyone to be aware of the disorder because it can come on quickly, and damage to the nerves can be permanent. Those who have experienced acute glaucoma report that the eyeball feels as heavy as a stone. Rarely, the abnormalities that lead to glaucoma can be present at birth, even though the disease symptoms may not appear for a few months or years. Regular visits to an eye-care specialist, especially by the elderly, are a necessity in order to catch conditions such as glaucoma early enough to allow effective treatment.

With aging, the eye is increasingly subject to cataracts. Cataracts are cloudy spots on the lens of the eye that eventually pervade the whole lens, which takes on a milky-white to yellowish color and becomes incapable of transmitting light (Fig. 18.16). Factors that increase the risk of cataracts include exposure to ultraviolet or other types of radiation, diabetes, and heavy alcohol consumption. In addition, the risk of cataracts doubles in men who smoke 20 or more cigarettes a day and in women who smoke 30 or more a day. A possible reason is that smoking reduces the delivery of blood, and therefore nutrients, to the lens. In most cases vision can be restored by surgical removal of the lens and replacement with a clear plastic artificial lens. Cataract surgery can be very successful in restoring vision. In fact, it is the most frequently performed surgery in the United States, with over 1.5 million cataract surgeries done each year. (The Bioethical Focus explores the question of whether we should rely on cataract surgery to solve this problem or take preventive steps to keep cataracts from developing in the first place.)

Since several disorders of the eye have been associated with exposure to ultraviolet irradiation, it is recommended that everyone, especially those who live in sunny climates or work outdoors, wear sunglasses that absorb ultraviolet light. Large lenses worn close to the eyes offer the best protection. The Sunglass Association of America has adopted the following system for categorizing sunglasses:

- Cosmetic lenses, which absorb at least 70% of UV-B (the more harmful type of ultraviolet radiation), 20%

Figure 18.16 Cataract in a human eye.
A cataract is a lens that has become opaque or cloudy. Risk factors that increase the chances of cataract formation include UV exposure, diabetes, and smoking. Cataracts can usually be surgically removed and replaced with an artificial lens.

A cataract is a cloudiness of the lens that usually occurs in the elderly. A dense, centrally placed cataract causes severe blurring of vision. Certain factors have been identified as contributing to the chances of having a cataract:

- Smoking 20 or more cigarettes a day doubles the risk of cataracts in men, and smoking more than 30 cigarettes a day increases the chances of cataracts in women.
- Exposure to the ultraviolet radiation in sunlight can more than double the risk of cataracts.
- As many as one-third of cataracts may be caused by being overweight. Diet, rather than exercise, seems to reduce cataract formation.

Most cataract operations today are performed on an outpatient basis, with minimal postoperative discomfort and a high expectation of sight restoration. Should we simply rely on medical science to restore our eyesight? Or should we act responsibly and take all possible steps to keep from developing cataracts—such as by refraining from smoking, wearing sunglasses and a wide-brimmed hat, and watching our weight?

Form Your Own Opinion

1. Suppose you lose your sense of hearing after years of listening to loud music on your MP3 player. Should you be eligible for insurance benefits to help pay for thousands of dollars of hearing aids or other devices to help you function normally?
2. Some state governments collect relatively large amounts of revenue by taxing cigarettes. For example, New York City currently adds $3 in taxes per pack. Interestingly, Virginia's tax is the lowest, at 2.5 cents per pack, presumably because much more tobacco is grown in Virginia than in New York. Which state's philosophy do you support? What if the cigarette tax money is used to educate children about the risks of smoking, to help people stop smoking, or to subsidize medical care for smokers?
3. Assuming the money would be used in beneficial ways, would you support higher taxes on any other products that can damage our health—for example, trans fatty acids that contribute to heart disease or digital music players that can damage your hearing?

---

of UV-A, and 60% of visible light. Such lenses are worn for comfort or appearance rather than protection.
- General-purpose lenses that absorb at least 95% of UV-B, 60% of UV-A, and 60–92% of visible light. These lenses are good for outdoor activities in temperate regions.
- Special-purpose lenses, which block at least 99% of UV-B, 60% of UV-A, and 60–92% of visible light. They are good for bright sun combined with sand, snow, or water.

An abnormally shaped eyeball may prevent the lens from focusing an image on the retina. If the eyeball is too long, a person is nearsighted. If the eyeball is too short, the person is farsighted. Common causes of blindness include retinal disorders, glaucoma, and cataracts.

Disorders of Hearing and Equilibrium

Hearing loss can develop gradually or suddenly and has many potential causes. Especially when we are young, the middle ear is subject to infections that can lead to hearing impairment if not treated promptly by a physician (see section 15.4). Hearing loss is also one of the most common conditions of older adults, affecting one in three people over age 60. Age-associated hearing loss usually develops gradually; the first sign may be problems understanding conversations when there is noise in the background. Actually, in most individuals age-related hearing loss probably begins at around age 20, with the ability to hear high-pitched sounds affected first. With age, the mobility of ossicles decreases, and in otosclerosis, new filamentous bone grows over the stirrup, impeding its movement. Surgical treatment is the only remedy for this type of conduction deafness. Some types of hearing loss seem to run in families, but most can be attributed to the effect of years of frequent exposure to loud noise (see *Health Focus*).

Sudden deafness, which may occur all at once or over a period of up to three days, most commonly affects people between the ages of 30 and 50, usually in only one ear. Potential causes include infections, head trauma, and the side effects of certain drugs. Depending on the cause, sudden deafness often resolves itself within a few days.

Deafness can also be present at birth. Several genetic disorders are associated with deafness at birth, as are infections with German measles (rubella) and mumps virus when contracted by a woman during her pregnancy. For this reason, every female should be immunized against these viruses before she reaches childbearing age.

Disorders of equilibrium often manifest as **vertigo,** the feeling that a person or the environment is moving when no motion is occurring. It is possible to simulate a feeling of vertigo by spinning your body rapidly and then stopping suddenly. Vertigo can be caused by problems in the brain as well as the inner ear. An estimated 20% of patients who experience symptoms of vertigo have benign positional vertigo (BPV), which may result from the formation of particles in the semicircular canals. When individuals with BPV move their heads suddenly, especially when lying down, these particles shift like pebbles inside a tire. After the head movement, there is an initial delay, but when the movement stops, the particles tumble down with gravity, stimulating the stereocilia and resulting in the sensation of movement. Interestingly, most cases of BPV either resolve on their own or can be cured using a series of precise head movements designed to cause the particles to settle into a position where they can no longer stimulate the stereocilia.

Because the senses of hearing and equilibrium are anatomically linked, certain disorders can affect both. An example of this phenomenon is Meniere's disease, which is usually characterized by vertigo, a feeling of fullness in the affected ear(s), tinnitus (ringing in the ears), and hearing loss. In about 80% of patients, only one ear is affected, and the disease usually strikes between the ages of 20 and 50. The exact cause of Meniere's disease is unknown, but it seems related to an increased volume of fluid in the semicircular canals, vestibule, and/or cochlea; for this reason, it has been called "glaucoma of the ear." There is no cure for Meniere's disease, but the vertigo and feeling of pressure can often be managed by adhering to a low-salt diet, which is thought to decrease the volume of fluid produced. Any hearing loss that occurs, however, is usually permanent.

Hearing loss can affect people of all ages. Several genetic disorders cause deafness at birth, while certain infections can lead to deafness in newborns and young children. Some degree of hearing loss is very common in older adults, and is usually associated with prolonged exposure to loud noise. A common symptom of disorders of equilibrium is vertigo, the feeling that the person or his or her environment is moving.

Summarizing the Concepts

18.1 Sensory Receptors and Sensations
Sensory receptors can be classified as interoceptors or exteroceptors, depending on whether they receive stimuli from inside or outside the body. Both types can also be categorized as chemoreceptors, photoreceptors, mechanoreceptors, or thermoreceptors, according to the nature of the stimulus they receive. Detection occurs when environmental changes stimulate sensory receptors, generating nerve impulses. Sensation occurs when these nerve impulses reach the cerebral cortex. Perception is an interpretation of the meaning of sensations.

18.2 Somatic Senses
Proprioception is illustrated by the action of muscle spindles, which are stimulated when muscle fibers stretch. Proprioception helps maintain equilibrium and posture.

The skin contains sensory receptors, called cutaneous receptors, for touch, pressure, pain, and temperature (warmth and cold). The pain of internal organs is sometimes felt in the skin and is called referred pain.

18.3 Senses of Taste and Smell
Taste and smell are due to chemoreceptors that are stimulated by molecules in the environment. After molecules bind to plasma membrane receptor proteins on the microvilli of taste cells and the cilia of olfactory cells, nerve impulses eventually reach the cerebral cortex, which determines the taste and odor according to the pattern of stimulation. The gustatory (taste) cortex responds to a range of salty, sour, bitter, sweet, and perhaps "umami" tastes.

18.4 Sense of Vision
Vision requires the eye, the optic nerves, and the visual cortex in the occipital lobe. The eye has three layers. The outer layer, called the sclera, can be seen as the white of the eye; it also becomes the transparent bulge in the front of the eye called the cornea. The middle pigmented layer, called the choroid, absorbs stray light rays. The rod cells (sensory receptors for dim light) and the cone cells (sensory receptors for bright light and color) are located in the retina, the inner layer of the eyeball. The cornea, the humors, and especially the lens bring the light rays to focus on the retina. To see a close object, accommodation occurs as the lens rounds up.

The visual pathway begins when light strikes rhodopsin within the membranous disks of rod cells; rhodopsin splits into opsin and retinal. A cascade of reactions leads to the closing of ion channels in a rod cell's plasma membrane. Inhibitory transmitter molecules are no longer released, and signals are passed to other neurons in the retina.

Integration occurs in the retina, which is composed of three layers of cells: the rod and cone layer, the bipolar cell layer, and the ganglion cell layer. The axons of ganglion cells become the optic nerve, which takes nerve impulses to the brain. The visual field is taken apart by the optic chiasma before reaching the visual cortex in the occipital lobe. The primary visual area parcels out signals for color, form, and motion to the visual association area. Then the cortex rebuilds the field.

18.5 Sense of Hearing
Hearing is dependent on the ear, the cochlear nerve, and the auditory cortex of the brain.

The ear is divided into three parts: outer, middle, and inner. The outer ear consists of the pinna and the auditory canal, which direct sound waves to the middle ear. The middle ear begins with the tympanic membrane and contains the ossicles (malleus, incus, and stapes). The malleus is attached to the tympanic membrane, and the stapes is attached to the oval window, which is covered by a membrane. The inner ear contains the cochlea and the semicircular canals, plus the utricle and the saccule.

The auditory pathway begins when the outer ear receives and the middle ear amplifies the sound waves, which then strike the oval window membrane. Its vibrations set up pressure waves across the cochlear canal, which contains the spiral organ, consisting of hair cells whose stereocilia are embedded within the tectorial membrane. When the basilar membrane vibrates, the stereocilia of the hair cells bend. Nerve impulses begin in the cochlear nerve and are carried to the primary auditory area in the temporal lobe of the cerebral cortex.

18.6 Sense of Equilibrium
The ear also contains mechanoreceptors for our sense of equilibrium. Rotational equilibrium is dependent on the stimulation of hair cells within the ampullae of the semicircular canals. Gravitational equilibrium relies on the stimulation of hair cells within the utricle and the saccule.

18.7 Disorders that Affect the Senses
Many conditions can affect the various sensory organs. Most are not life-threatening, but all certainly can affect the quality of life. Although certain disorders of the senses can be present at birth, most affect older individuals more commonly.

Testing Yourself

Choose the best answer for each question.

1. Chemoreceptors are involved in
   a. hearing.   d. vision.
   b. taste.   e. Both b and c are correct.
   c. smell.

2. Sensing both pressure and temperature would be a function of the
   a. eyes.   c. muscles.
   b. ears.   d. skin.

3. A problem with vision could be related to the
   a. retina.   d. cochlea.
   b. cornea.   e. All except d are correct.
   c. lens.

4. A color-blind person has (an) abnormal
   a. rods.   d. cones.
   b. cochlea.   e. None of these are correct.
   c. cornea.

For questions 5–9, match structures to the functions in the key.

Key:
   a. pressure receptor   d. pain receptors
   b. absorb stray light rays   e. proprioception
   c. hearing

5. Choroid

6. Cochlea

7. Muscle spindle

8. Pacinian corpuscle

9. Nociceptors

10. Free nerve endings sense temperature and
    a. pressure.   c. vision.
    b. pain.   d. hearing.

11. Tasting "sweet" versus "salty" is a result of
    a. activating different sensory receptors.
    b. activating many versus few sensory receptors.
    c. activating no sensory receptors.
    d. None of these are correct.

12. Our perception of pitch is dependent upon the region of the
    _____ vibrated and the regions of the _____ stimulated.
    a. cochlea, spinal cord
    b. basilar membrane, spinal cord
    c. basilar membrane, auditory cortex
    d. cochlea, cerebellum

13. The ossicle that articulates with the tympanic membrane is the
    a. malleus.   c. incus.
    b. stapes.   d. All of these are correct.

14. The pupil
    a. is made of connective tissue.
    b. is an opening.

c. is regulated by the iris.
d. allows for the passage of light to the lens.
e. All are correct except(a).

15. Label the following diagram of the optic chiasma.

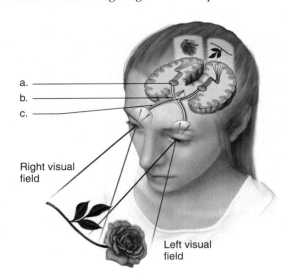

a. _____
b. _____
c. _____

Right visual field

Left visual field

16. Label the following diagram of the human ear.

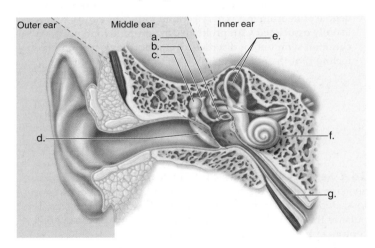

Outer ear   Middle ear   Inner ear

a.
b.
c.
e.
d.
f.
g.

17. Which of these is not a proper contrast between olfactory receptors and equilibrium receptors?

	Olfactory receptors	Equilibrium receptors
a.	located in nasal cavities	located in the inner ear
b.	chemoreceptors	mechanoreceptors
c.	respond to molecules in air	respond to movements of the body
d.	communicate with brain via a tract	communicate with brain via vestibular nerve
e.	All of these contrasts are correct.	

18. Label the following diagram of the human eye.

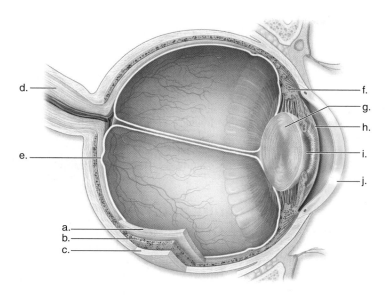

19. Which of these sequences is the correct path for light rays entering the human eye?

    a. sclera, retina, choroid, lens, cornea
    b. fovea centralis, pupil, aqueous humor, lens
    c. cornea, pupil, lens, vitreous humor, retina
    d. cornea, fovea centralis, lens, choroid, rods
    e. optic nerve, sclera, choroid, retina, humors

20. Which association matches an incorrect function to a structure?

    a. lens—focusing
    b. cones—color vision
    c. iris—regulation of amount of light
    d. choroid—location of cones
    e. sclera—protection

21. Which is the correct location of the spiral organ?

    a. between the tympanic membrane and the oval window in the inner ear
    b. in the utricle and saccule within the vestibule
    c. between the tectorial membrane and the basilar membrane in the cochlear canal
    d. between the nasal cavities and the throat
    e. between the outer and inner ear within the semicircular canals

22. Which of these associations is mismatched?

    a. semicircular canals—inner ear
    b. utricle and saccule—outer ear
    c. auditory canal—outer ear
    d. cochlea—inner ear
    e. ossicles—middle ear

23. Retinal is

    a. a derivative of vitamin A.   d. found in both rods and cones.
    b. sensitive to light energy.   e. All of these are correct.
    c. a part of rhodopsin.

24. Both olfactory receptors and sound receptors have cilia, and both

    a. are chemoreceptors.   d. initiate nerve impulses.
    b. are a part of the brain.   e. All of these are correct.
    c. are mechanoreceptors.

25. Which pair of terms is properly matched?

    a. nearsightedness—light rays focus in front of the retina
    b. nearsightedness—light rays focus behind the retina
    c. farsightedness—light rays focus in front of the retina
    d. None of these are correct.

Understanding the Terms

Match the terms to these definitions:

a._____ Portion of the inner ear that resembles a snail's shell and contains the spiral organ, the sense organ for hearing.

b._____ Sensory receptor that responds to chemical stimulation—for example, receptors for taste and smell.

c._____ Visual pigment found in the rods, whose activation by light energy leads to vision.

d._____ Awareness of a stimulus; occurs only in the cerebral cortex.

e._____ Type of receptor that gives rise to nerve impulses when stereocilia are bent or tilted.

Thinking Critically

1. Some sensory receptors, such as those for taste, smell, and pressure, readily undergo the process of sensory adaptation, or decreased response to a stimulus. In contrast, receptors for pain are less prone to adaptation. Why does this make good biological sense? What do you think happens to children who are born without the ability to feel pain normally?

2. How would you summarize the role of the thalamus in the delivery of sensory information to the brain?

3. Trace the path of a photon of light through all the layers of the retina discussed in this chapter. Then trace the nerve impulse from a photoreceptor located on the right side of the left eyeball through the visual cortex.

4. Suppose you find yourself in a room with very little light, and you are having trouble finding the light switch. You notice that you can see better with your peripheral vision than in the center of your visual field. Why is this the case?

5. Some animals have more powerful senses than humans do. For example, dogs are able to sense odors at concentrations nearly 100 million times lower, cats can see in light that is about one-sixth as dim, and eagles can see four to seven times better than humans. What are three or four possible explanations for the more acute senses of some animals?

ARIS, the *Inquiry into Life* Website

ARIS, the website for *Inquiry into Life*, provides a wealth of information organized and integrated by chapter. You will find practice quizzes, interactive activities, labeling exercises, flashcards, and much more that will complement your learning and understanding of general biology.

www.aris.mhhe.com

Arnold Schwarzenegger *is a seven-time Mr. Olympia, a movie star, and the current governor of California. He has also admitted that he used muscle-building anabolic steroids during his bodybuilding career in the 1970s. Because the performance-enhancing drugs were legal at the time, Schwarzenegger once stated that he had no regrets about taking them under a doctor's supervision. In recent years, however, he has backed away from these comments, suggesting that drug use sends the wrong message to children.*

Anabolic steroids, which are drugs or hormones chemically related to testosterone, promote muscle growth. The Anabolic Steroid Control Act passed in 1990 criminalized the use of these drugs for "athletic purposes" in the United States. However, steroid use among athletes, from high schools to the professional ranks, is increasingly hotly debated. Recent reports have estimated that somewhere between 4% and 12% of male high school seniors, and about 1% of their female counterparts, have used anabolic steroids. The potential side effects of these drugs (discussed in more detail in Chapter 20) range from impotence and reduced sperm count in men to masculinization and menstrual cycle changes in women.

Despite these potential dangers, the list of professional athletes who have admitted to using steroids to build muscles and improve athletic performance continues to grow. Governor Schwarzenegger seems to have escaped unscathed, but for the rest of us, are performance-enhancing drugs worth the risk to our overall health?

chapter **19**

Musculoskeletal System

19.1 Anatomy and Physiology of Bone

The skeleton provides attachment sites for the muscles, whose contraction makes the bones move so that we can work out, play tennis, type papers, and perform all manner of other activities. Together, the bones and muscles comprise the *musculoskeletal system*. In this chapter, we discuss the structures and functions of first the skeleton and then the muscles. Bones and the tissues associated with them are connective tissues, while muscles are, of course, composed of muscular tissue.

Organization of Tissues in the Skeleton

Bones are classified by their shape: long, short, flat, or irregular. Long bones do not have to be terribly long—the bones in your fingers are classified as long. But a long bone must be longer than it is wide. The bones of the arms and legs are long bones except for the kneecap and the wrist and ankle bones.

A long bone of the arm or leg can be used to illustrate general principles of organization of tissues (Fig. 19.1). A bone is enclosed by a tough, fibrous, connective tissue covering called the **periosteum,** which is continuous with the ligaments that go across a joint. The periosteum contains blood vessels that enter the bone and give off branches that service osteocytes, located in lacunae. Each expanded end of a long bone is termed an epiphysis; the portion between the epiphyses is called the diaphysis.

Splitting a bone open, as in Figure 19.1, shows that the diaphysis is not solid but has a medullary cavity containing yellow bone marrow. Yellow bone marrow contains a large amount of fat. The medullary cavity is bounded at the sides by compact bone. The epiphyses of a long bone are composed largely of spongy bone. Even in adults, the epiphyses contain red bone marrow, where all the types of blood cells are formed. Beyond the spongy bone, there is a thin shell of compact bone and, finally, a layer of hyaline cartilage, called **articular cartilage** because it occurs at a joint.

Figure 19.1 shows how the spongy bone of an epiphysis is connected to the compact bone of the diaphysis, and micrographs illustrate the structure of compact bone and hyaline cartilage. Notice that the blood vessel coming from outside the bone penetrates the periosteum and enters the central canals of compact bone before continuing on into spongy bone.

> Bones are enclosed by a tough periosteum. The diaphysis, or shaft, of a bone houses a medullary cavity that usually contains yellow bone marrow in adults. The epiphyses, or ends, of a bone usually contain red bone marrow.

Structure of Bone and Associated Tissues

The primary connective tissues of the skeleton are bone, cartilage, and dense fibrous connective tissue. All connective tissues contain cells separated by a matrix that contains fibers.

Bone

Bone tissue is strong because the matrix contains mineral salts, notably calcium phosphate. **Compact bone** is highly organized and composed of tubular units called **osteons.** Bone cells called **osteocytes** lie in lacunae, or tiny chambers, arranged in concentric circles around a central canal (Fig. 19.1). The mineralized extracellular substance (*matrix*), which also contains collagen fibers, fills the spaces between the lacunae. Central canals contain blood vessels, lymphatic vessels, and nerves. Tiny canals called canaliculi run through the matrix, connecting the lacunae with each other and with the central canal. The canaliculi contain extensions of osteocytes surrounded by tissue fluid. In this way, the canaliculi bring nutrients from the blood vessel in the central canal to the cells in the lacunae.

Compared to compact bone, **spongy bone** has an unorganized appearance (Fig. 19.1). Osteocytes are found in trabeculae, which are numerous thin plates separated by unequal spaces. Although the spaces make spongy bone lighter than compact bone, spongy bone is still designed for strength. Just as braces are used for support in buildings, the plates of spongy bone follow lines of stress. The spaces of spongy bone are often filled with **red bone marrow,** a specialized tissue that produces all types of blood cells.

In infants, red bone marrow is found in the cavities of most bones. In adults, red bone marrow occurs in a more limited number of bones (see section 19.2).

Cartilage

Cartilage is not as strong as bone, but it is more flexible because the matrix is gel-like and contains many collagenous and elastic fibers. The cells lie within lacunae, which are irregularly grouped. Cartilage has no blood vessels, and therefore, injured cartilage is slow to heal.

All three types of cartilage—hyaline cartilage, fibrocartilage, and elastic cartilage—are associated with bones. Hyaline cartilage is firm and somewhat flexible. The matrix appears uniform and glassy (Fig. 19.1), but actually it contains a generous supply of collagen fibers. Hyaline cartilage is found at the ends of long bones and in the nose, at the ends of the ribs, and in the larynx and trachea.

Fibrocartilage is stronger than hyaline cartilage because the matrix contains wide rows of thick collagen fibers. Fibrocartilage is able to withstand both tension and pressure. This type of cartilage is found where support is of prime importance—in the disks between the vertebrae and in the cartilaginous disks between the bones of the knee.

Elastic cartilage is more flexible than hyaline cartilage because the matrix contains mostly elastin fibers. This type of cartilage is found in the ear flaps and epiglottis.

Dense Fibrous Connective Tissue

Dense fibrous connective tissue contains rows of cells called fibroblasts separated by bundles of collagen fibers. Dense

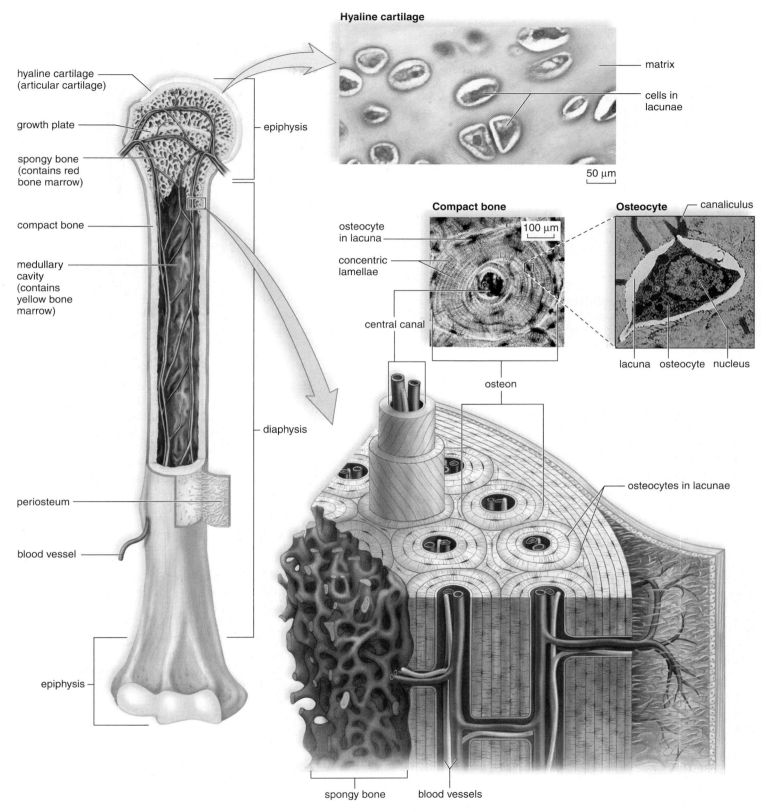

Figure 19.1 Anatomy of a bone from the macroscopic to the microscopic level.
A long bone is encased by fibrous membrane (periosteum) except where it is covered at the ends by hyaline cartilage, shown on the micrograph (*right, top*). Spongy bone, located beneath the cartilage, may contain red bone marrow. The central shaft contains yellow bone marrow and is bordered by compact bone, which is shown in the enlargement and micrograph (*right, center*).

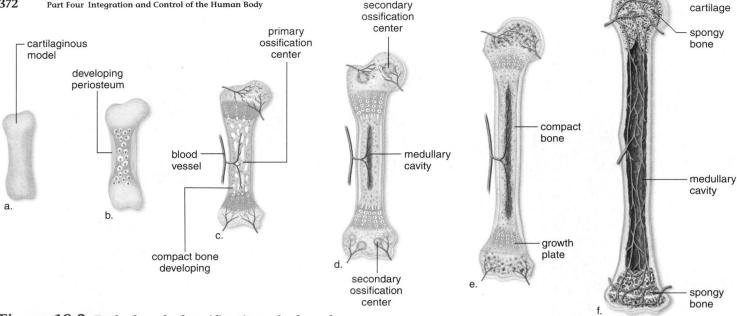

Figure 19.2 Endochondral ossification of a long bone.
a. A cartilaginous model develops during embryonic development. **b.** A periosteum develops. **c.** A primary ossification center contains spongy bone surrounded by compact bone. **d.** The medullary cavity forms in the shaft, and secondary ossification centers develop at the ends of the long bone. **e.** Growth is still possible as long as cartilage remains at the growth plate. **f.** When the bone is fully formed, the growth plate disappears.

fibrous connective tissue forms the flared sides of the nose. **Ligaments,** which bind bone to bone, are dense fibrous connective tissue, and so are **tendons,** which connect muscle to bone at **joints.** Joints are also called articulations.

> The two main types of bone are compact bone and spongy bone. All three types of cartilage—hyaline, elastic, and fibrocartilage—are found in association with bones. Ligaments connect bones to bones, and tendons connect muscles to bones.

Bone Growth and Remodeling

Bones are composed of living tissues, as exemplified by their ability to undergo growth and remodeling.

Bone Development and Growth

The bones of the human skeleton, except those of the skull, first appear during embryonic development as hyaline cartilage. These cartilaginous structures are then gradually replaced by bone, a process called endochondral ossification (Fig. 19.2).

During endochondral ossification, the cartilage begins to break down in the center of a long bone, which is now covered by a periosteum. **Osteoblasts** invade the region and begin to lay down spongy bone in what is called a primary ossification center. Other osteoblasts lay down compact bone beneath the periosteum. As the compact bone thickens, osteoclasts break down the spongy bone, and the cavity created becomes the medullary cavity.

The ends of developing bone continue to grow, but soon, secondary ossification centers appear in these regions. Here, spongy bone forms and does not break down. Also, a band of cartilage called a **growth plate** remains between the primary ossification center and each secondary center. The limbs keep increasing in length as long as the growth plates are present. The rate of growth is mainly controlled by growth hormone and sex hormones. Eventually, the growth plates become ossified, and the bone stops growing.

Remodeling of Bones

In the adult, bone is continually being broken down and built up again, a process called remodeling. **Osteoclasts,** which are derived from a red bone marrow stem cell, break down bone, remove worn-out cells, and deposit calcium in the blood. Simultaneously, the bone is repaired by the work of osteoblasts. As they form new bone, osteoblasts take calcium from the blood. Eventually, some of these cells get caught in the matrix they secrete and are converted to osteocytes, the cells found within the lacunae of osteons. Because of continual remodeling, the thickness of bones can change due to excessive use or disuse. Hormones such as calcitonin and parathyroid hormone (see Chapter 20) can also affect the thickness of bones. When women reach menopause, the likelihood of a bone disease called osteoporosis increases, in part due to reduced estrogen levels (see section 19.6).

> Bone is living tissue that develops, grows, and remodels itself. In all these processes, osteoclasts break down bone, and osteoblasts build bone.

19.2 Bones of the Skeleton

The functions of the skeleton pertain to particular bones:

The skeleton supports the body. The bones of the lower limbs (the
femur in particular and also the tibia) support the entire
body when we are standing, and the coxal bones of the
pelvic girdle support the abdominal cavity.

The skeleton protects soft body parts. The bones of the skull protect
the brain; the rib cage, composed of the ribs, thoracic
vertebrae, and sternum, protects the heart and lungs.

The skeleton produces blood cells. All bones in the fetus have
spongy bone containing red bone marrow that
produces blood cells. In the adult, the flat bones of the
skull, ribs, sternum, clavicles, and also the vertebrae and
pelvis produce blood cells.

The skeleton stores minerals and fat. All bones have an
extracellular substance that contains calcium phosphate.
When bones are remodeled, osteoclasts break down
bone and return calcium ions and phosphate ions to the
bloodstream. Fat is stored in the yellow bone marrow
(see Fig. 19.1).

*The skeleton, along with the muscles, permits flexible body
movement.* While articulations (joints) occur between all
the bones, we associate body movement, in particular,
with the bones of the lower limbs (especially the
femur and tibia) and the feet (tarsals, metatarsals, and
phalanges) because we use them when walking.

Classification of the Bones

The approximately 206 bones of the skeleton are classified
according to whether they occur in the axial skeleton or the
appendicular skeleton. The **axial skeleton** is in the midline
of the body, and the **appendicular skeleton** consists of the
limbs along with their girdles (Fig. 19.3).

As mentioned, long bones, exemplified by the humerus
and femur, are longer than they are wide. In contrast, short
bones, such as the carpals and tarsals, are cube shaped—that
is, their lengths and widths are about equal. Flat bones, such as
those of the skull, are platelike, with broad surfaces. Irregular
bones, such as the vertebrae and facial bones, have varied
shapes that permit connections with other bones. Some authorities recognize another category, the round bones, which are
circular in shape and usually embedded in a tendon. A good
example of a round bone is the patella, or kneecap.

None of the bones of the skeleton are smooth; they
have articulating depressions and protuberances at various
joints. They also have projections, often called processes,
where the muscles attach, as well as openings for nerves
and/or blood vessels to pass through.

**The skeleton is divided into the axial and appendicular
skeletons. Bones have different shapes, with protuberances
at joints and processes where the muscles attach.**

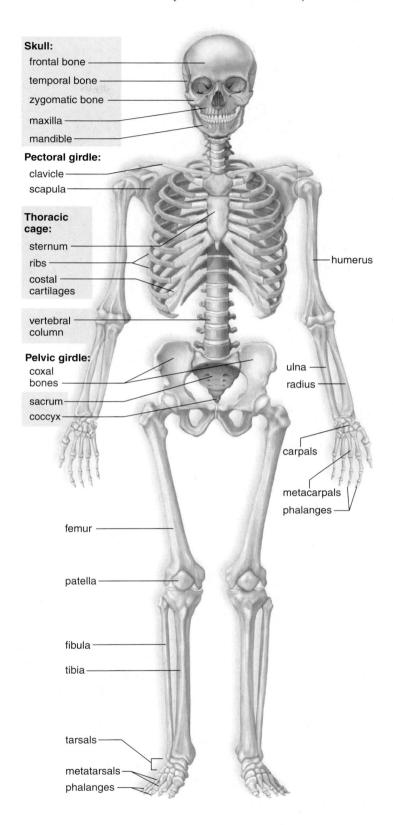

Figure 19.3 The skeleton.
The skeleton of a human adult contains bones that belong to the axial skeleton
(shaded in blue) and those that belong to the appendicular skeleton (unshaded).
The bones of the axial skeleton are located along the body's axis, and the bones
of the appendicular skeleton are located in the girdles and appendages.

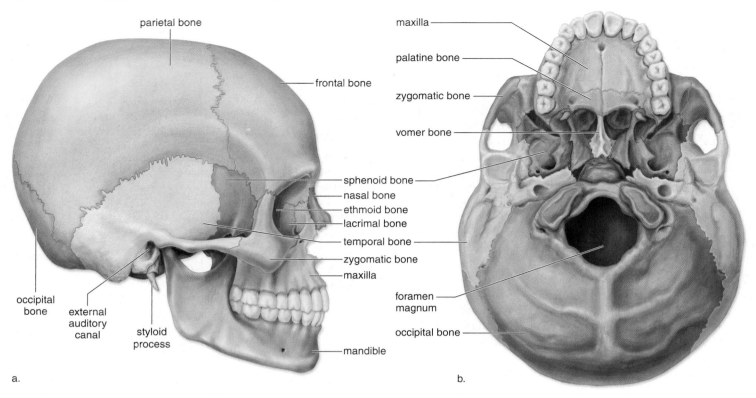

Figure 19.4 Bones of the skull.
a. Lateral view. **b.** Inferior view.

The Axial Skeleton

The axial skeleton lies in the midline of the body and consists of the skull, hyoid bone, vertebral column, rib cage, and ossicles.

The Skull

The **skull** is formed by the cranium (braincase) and the facial bones. It should be noted, however, that some cranial bones also help to form the face.

The Cranium The cranium protects the brain. In adults, it is composed of eight bones fitted tightly together. In newborns, certain cranial bones are not completely formed and instead are joined by membranous regions called **fontanels.** The fontanels usually close by the age of 24 months by the process of intramembranous ossification.

Some of the bones of the cranium contain the **sinuses,** air spaces lined by mucous membrane. The sinuses reduce the weight of the skull and give a resonant sound to the voice. Two sinuses, called the mastoid sinuses, drain into the middle ear. Mastoiditis, a condition that can lead to deafness, is an inflammation of these sinuses.

The major bones of the cranium have the same names as the lobes of the brain: frontal, parietal, occipital, and temporal. On the top of the cranium (Fig. 19.4a), the **frontal bone** forms the forehead, the **parietal bones** extend to the sides, and the **occipital bone** curves to form the base of the skull. At the base, there is a large opening, the **foramen magnum** (Fig. 19.4b), through which the spinal cord passes and connects

with the brain stem. Below the much larger parietal bones, each **temporal bone** has an opening (external auditory canal) that leads to the middle ear.

The **sphenoid bone,** which is shaped like a bat with wings outstretched, extends across the floor of the cranium from one side to the other. The sphenoid is the keystone of the cranial bones because all the other bones articulate with it. The sphenoid completes the sides of the skull and also helps form the eye sockets. The eye sockets are called orbits because of our ability to rotate our eyes. The **ethmoid bone,** which lies in front of the sphenoid, also helps form the orbits and the nasal septum. The orbits are completed by various facial bones.

> The cranium contains eight bones: the frontal, two parietal, the occipital, two temporal, the sphenoid, and the ethmoid.

The Facial Bones The most prominent facial bones are the mandible, the maxillae (maxillary bones), the zygomatic bones, and the nasal bones.

The **mandible,** or lower jaw, is the only movable portion of the skull (Fig. 19.5a), and its action permits us to chew our food. It also forms the chin. Tooth sockets are located on the mandible and on the two **maxillae,** which form the upper jaw and also the anterior portion of the hard palate. The palatine bones make up the posterior portion of the hard palate and the floor of the nose (see Fig. 19.4b).

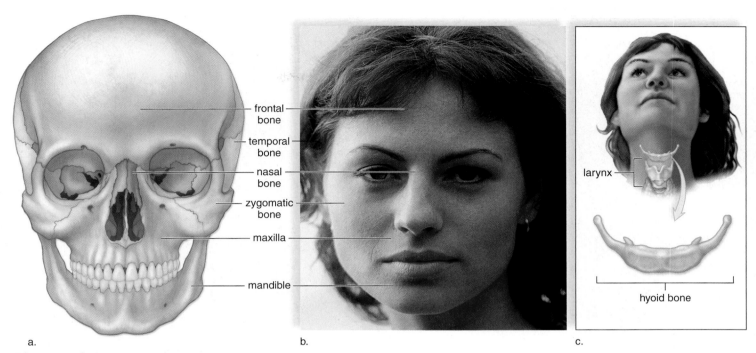

a.   b.   c.

Figure 19.5 Bones of the face and the hyoid bone.

a. The frontal bone forms the forehead and eyebrow ridges, and the zygomatic bones form the cheekbones. The maxillae have numerous functions: assisting in the formation of the eye sockets and the nasal cavity; forming the upper jaw; and containing sockets for the upper teeth. The mandible is the lower jaw with sockets for the lower teeth. The mandible has a projection we call the chin. **b.** The maxillae, frontal bones, and nasal bones help form the external nose. **c.** The hyoid bone attaches to the larynx as shown.

The lips and cheeks have a core of skeletal muscle. The **zygomatic bones** are the cheekbone prominences, and the **nasal bones** form the bridge of the nose. Other bones (e.g., ethmoid and vomer) are a part of the nasal septum, which divides the interior of the nose into two nasal cavities. The lacrimal bone (see Fig. 19.4a) contains the opening for the nasolacrimal canal, which brings tears from the eyes to the nose.

The temporal and frontal bones are cranial bones that contribute to the face. The temporal bones account for the flattened areas we call the temples. The frontal bone forms the forehead and has supraorbital ridges where the eyebrows are located. Glasses sit where the frontal bone joins the nasal bones (Fig. 19.5b).

While the ears are formed only by cartilage and not by bone, the nose is a mixture of bones, cartilages, and connective tissues. The cartilages complete the tip of the nose, and fibrous connective tissue forms the flared sides of the nose.

> Among the facial bones, the mandible is the lower jaw, where the chin is located, the two maxillae form the upper jaw, the two zygomatic bones are the cheekbones, and the two nasal bones form the bridge of the nose.

The Hyoid Bone

Although the **hyoid bone** is not part of the skull, we include it here because it is part of the axial skeleton. The hyoid is the only bone in the body that does not articulate with another bone. It is attached to processes of the temporal bones by muscles and ligaments and to the larynx by a membrane (Fig. 19.5 c). The larynx is the voice box at the top of the trachea in the neck region. The hyoid bone anchors the tongue and serves as the site of attachment for the muscles associated with swallowing.

The Vertebral Column

The **vertebral column** consists of 33 vertebrae (Fig. 19.6). Normally, the vertebral column has four curvatures that provide resilience and strength for an upright posture. **Scoliosis** is an abnormal lateral (sideways) curvature of the spine. Two other well-known abnormal curvatures are kyphosis, an abnormal posterior curvature that often results in a hunchback, and lordosis, an abnormal anterior curvature resulting in a swayback.

The vertebral column forms when the vertebrae join. The spinal cord, which passes through the vertebral canal, gives off the spinal nerves at the intervertebral foramina. Among other functions, spinal nerves control skeletal muscle contraction. The spinous processes of the vertebrae can be felt as bony projections along the midline of the back. The spinous processes and also the transverse processes, which extend laterally, serve as attachment sites for the muscles that move the vertebral column.

Types of Vertebrae   The various vertebrae are named according to their location in the vertebral column. The cervical vertebrae are in the neck. The first cervical vertebra, called the **atlas** holds

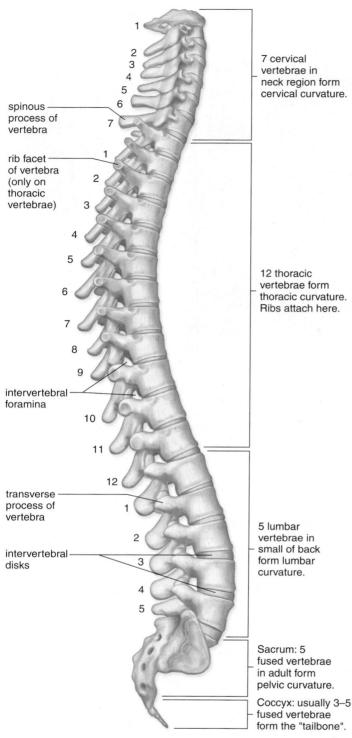

Figure 19.6 **The vertebral column.**

1
2
3
4
5
6
7

7 cervical vertebrae in neck region form cervical curvature.

spinous process of vertebra

rib facet of vertebra (only on thoracic vertebrae)

1
2
3
4
5
6
7
8
9
10
11
12

12 thoracic vertebrae form thoracic curvature. Ribs attach here.

intervertebral foramina

transverse process of vertebra

1
2
3
4
5

5 lumbar vertebrae in small of back form lumbar curvature.

intervertebral disks

Sacrum: 5 fused vertebrae in adult form pelvic curvature.

Coccyx: usually 3–5 fused vertebrae form the "tailbone".

The vertebral column is flexible because the vertebrae are separated by intervertebral disks. The vertebrae are named for their location in the vertebral column. For example, the thoracic vertebrae are in the thorax. Note that humans have a coccyx, which is also called a tailbone.

up the head. It is so named because Atlas, of Greek mythology, held up the world. Movement of the atlas permits the "yes" motion of the head. It also allows the head to tilt from side to side. The second cervical vertebra is called the **axis** because it allows a degree of rotation, as when we shake the head "no." The thoracic vertebrae have long, thin, spinous processes and

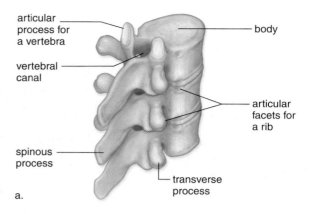

articular process for a vertebra

vertebral canal

spinous process

body

articular facets for a rib

transverse process

a.

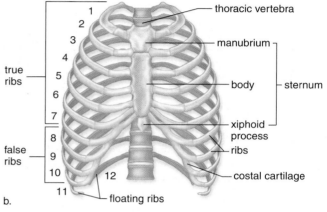

1
2
3
4
5
6
7

thoracic vertebra

manubrium

true ribs

body

sternum

xiphoid process

ribs

false ribs

8
9
10
11

12

costal cartilage

floating ribs

b.

Figure 19.7 **Thoracic vertebrae and the rib cage.**
a. The thoracic vertebrae articulate with each other at the articular processes and with the ribs at the articular facets. A thoracic vertebra has two facets for articulation with a rib; one is on the body, and the other is on the transverse process. **b.** The rib cage consists of the 12 thoracic vertebrae, the 12 pairs of ribs, the costal cartilages, and the sternum. The rib cage protects the lungs and the heart.

articular facets for the attachment of the ribs (Fig. 19.7a). Lumbar vertebrae have a large body and thick processes. The five sacral vertebrae are fused together in the sacrum. The coccyx, or tailbone, is usually composed of three to five fused vertebrae.

Intervertebral Disks Between the vertebrae are **intervertebral disks** composed of fibrocartilage that act as a kind of padding. They prevent the vertebrae from grinding against one another and absorb shock caused by movements such as running, jumping, and even walking. The presence of the disks allows the vertebrae to move as we bend forward, backward, and from side to side. Unfortunately, these disks become weakened with age and can even herniate and rupture. If a disk presses against the spinal cord and/or spinal nerves, pain results, and surgical removal of the disk may be necessary to relieve the pain.

The vertebral column consists of 33 joined vertebrae. It supports the head and trunk, protects the spinal cord, and serves as a site for muscle attachment. The intervertebral disks provide padding and account for the flexibility of the vertebral column.

The Rib Cage

The rib cage, also called the thoracic cage, is composed of the thoracic vertebrae, the ribs and their associated cartilages, and the sternum (Fig. 19.7b). The rib cage is part of the axial skeleton.

The rib cage demonstrates how the skeleton is protective but also flexible. The rib cage protects the heart and lungs; yet it swings outward and upward upon inspiration, and then downward and inward upon expiration. Because of their proximity to vital organs, however, the ribs also have the potential to cause harm. A fractured rib can puncture a lung.

The Ribs

There are twelve pairs of ribs, and all twelve connect directly to the thoracic vertebrae in the back. Each rib is a flattened bone that originates at a particular thoracic vertebra and proceeds toward the ventral thoracic wall. A rib articulates with the body and transverse process of its corresponding thoracic vertebra. It curves outward and then forward and downward.

The upper seven pairs of ribs connect directly to the sternum by means of costal cartilages. These are called the "true ribs." The next three pairs of ribs do not connect directly to the sternum, and they are called the "false ribs." They attach to the sternum by means of a common cartilage. The last two pairs are called "floating ribs" because they do not attach to the sternum at all.

The Sternum

The **sternum,** or breastbone, lies in the midline of the body. Along with the ribs, it helps protect the heart and lungs. The sternum is a flat bone that is shaped like a knife.

The sword-shaped sternum is composed of three bones that fuse during fetal development. These bones are the manubrium (the handle), the body (the blade), and the xiphoid process (the point of blade). The manubrium articulates with the clavicles of the appendicular skeleton and the first pair of ribs. The manubrium joins with the body of the sternum at an angle. This is an important anatomical landmark because it occurs at the level of the second rib, and therefore allows the ribs to be counted. Medical personnel sometimes count the ribs to locate the apex of a patient's heart, which is usually between the fifth and sixth ribs.

The xiphoid process is the third part of the sternum. Composed of hyaline cartilage in the child, it becomes ossified in the adult. The variably shaped xiphoid process serves as an attachment site for the diaphragm, which divides the thoracic cavity from the abdominal cavity.

The rib cage, consisting of the thoracic vertebrae, the ribs, costal cartilages, and the sternum, protects the heart and lungs in the thoracic cavity.

The Appendicular Skeleton

The appendicular skeleton consists of the bones within the pectoral and pelvic girdles and their attached limbs. The two (left and right) pectoral girdles and upper limbs are specialized for flexibility; the pelvic girdle and lower limbs are specialized for strength.

The Pectoral Girdle and Upper Limb

A **pectoral girdle** (shoulder girdle) consists of a scapula (shoulder blade) and a clavicle (collarbone) (Fig. 19.8). The **clavicle** extends across the top of the thorax; it articulates

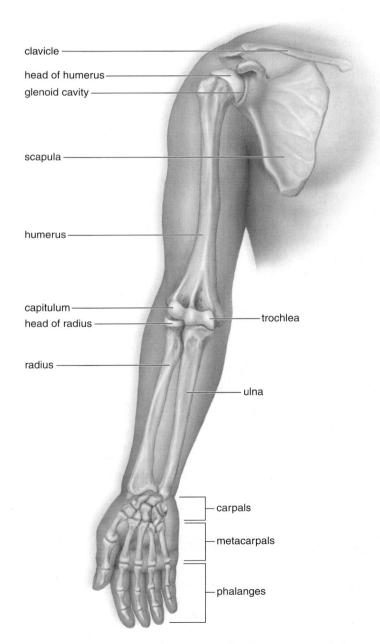

Figure 19.8 Bones of the pectoral girdle and upper limb (arm, forearm, and hand).

with (joins with) the sternum and the acromion process of the **scapula,** a bone external to the ribs in the back. The muscles of the arm and chest attach to the coracoid process of the scapula. The **glenoid cavity** of the scapula articulates with, and is much smaller than, the head of the humerus, the single long bone in the arm. This joint allows the arm to move in almost any direction, but has reduced stability and thus is the joint that is most apt to be dislocated. Tendons that encircle and help form a socket for the humerus are collectively called the **rotator cuff.** Vigorous circular movements of the arm can lead to rotator cuff injuries.

The components of a pectoral girdle follow freely the movements of the upper limb, which consists of the humerus in the arm and the radius and ulna in the forearm. The **humerus** has a smoothly rounded head that fits into the glenoid cavity of the scapula, as mentioned. The shaft of the humerus has a tuberosity (protuberance) where the deltoid, a prominent muscle of the shoulder, attaches. Enlargement of this tuberosity occurs in people who do a lot of heavy lifting.

The far end of the humerus has two protuberances, called the capitulum and the trochlea, which articulate respectively with the **radius** and the **ulna** at the elbow. The bump at the back of the elbow is the olecranon process of the ulna.

When the arm is held with the palm turned forward, the radius and ulna are about parallel to one another. When the arm is turned so that the palm faces backward, the radius crosses in front of the ulna, a feature that contributes to the easy twisting motion of the forearm.

The hand has many bones, and this increases its flexibility. The wrist has eight **carpal** bones, which look like small pebbles. From these, five **metacarpal** bones fan out to form a framework for the palm. The metacarpal bone that leads to the thumb is opposable to the other digits. (The term **digits** refers to either fingers or toes.) The knuckles are the enlarged distal ends of the metacarpals. Beyond the metacarpals are the **phalanges,** the bones of the fingers and the thumb. The phalanges of the hand are long, slender, and lightweight.

> The pectoral girdles and arms are specialized for flexible movement. A pectoral girdle consists of a scapula and clavicle. The scapula articulates with the humerus, which articulates with the radius and ulna. The wrist is composed of carpal bones, and the hand contains metacarpal bones and phalanges.

The Pelvic Girdle and Lower Limb

Figure 19.9 shows how the lower limb is attached to the pelvic girdle. The **pelvic girdle** (hip girdle) consists of two heavy, large coxal bones (hipbones). The **pelvis** is a basin composed of the pelvic girdle, sacrum, and coccyx. The pelvis bears the weight of the body, protects the organs within the pelvic cavity, and serves as the site of attachment for the lower limbs.

Each **coxal bone** has three parts: the ilium, the ischium, and the pubis, which are fused in the adult (Fig. 19.9). The hip

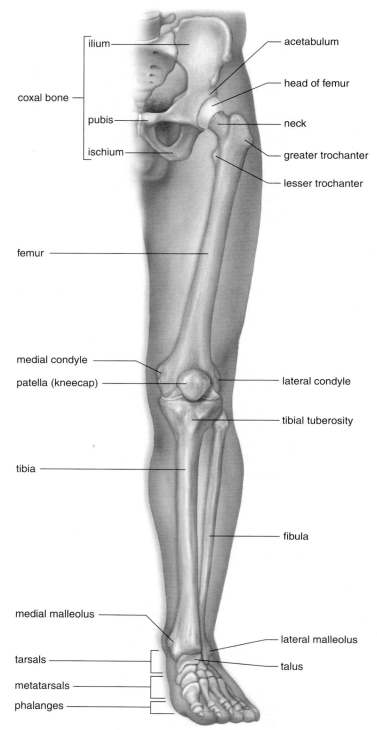

Figure 19.9 Bones of the pelvic girdle and lower limb.

socket, called the acetabulum, occurs where these three bones meet. The ilium is the largest part of the coxal bones, and our hips occur where it flares out. We sit on the ischium, which has a posterior spine called the ischial spine. The pubis, from which the term pubic hair is derived, is the anterior part of a coxal bone. The two pubic bones are joined together by a fibrocartilage disk at the pubic symphysis.

The male and female pelves differ from one another. In the female, the ilia are more flared, the pelvic cavity is broader, and the outlet is wider. These adaptations facilitate giving birth.

The **femur** (thighbone) is the longest and strongest bone in the body. The head of the femur articulates with the coxal bones at the acetabulum, and the short neck better positions the legs for walking. The femur has two large processes, the greater and lesser trochanters, which are places of attachment for the thigh muscles and the muscles of the buttocks. At its distal end, the femur has medial and lateral condyles that articulate with the **tibia.** This is the region of the knee and the **patella,** or kneecap. The patella is held in place by the quadriceps tendon, which continues as a ligament that attaches to the tibial tuberosity. At the distal end, the medial malleolus of the tibia causes the inner bulge of the ankle. The **fibula** is the more slender bone in the leg. The fibula has a head that articulates with the tibia and a prominence (lateral malleolus) that forms the outer bulge of the ankle.

Each foot has an ankle, an instep, and five toes. The many bones of the foot give it considerable flexibility, especially on rough surfaces. The ankle contains seven **tarsal** bones, one of which (the talus) can move freely where it joins the tibia and fibula. Strange to say, the calcaneus, or heel bone, is also considered part of the ankle. The talus and calcaneus support the weight of the body.

The instep has five elongated **metatarsal** bones. The distal end of the metatarsals forms the ball of the foot. If the ligaments that bind the metatarsals together become weakened, flat feet are apt to result. The bones of the toes are called phalanges, just like those of the fingers, but in the foot, the phalanges are stout and extremely sturdy.

> The pelvic girdle and lower limbs are adapted to support the weight of the body. The pelvic girdle consists of two coxal bones, each of which is composed of an ilium, ischium and pubis. The leg is formed by the femur, tibia and fibula, tarsal bones, metatarsal bones, and phalanges. The femur is the longest and strongest bone in the body.

Articulations

Bones are joined at the joints, which are classified as fibrous, cartilaginous, or synovial based on their structure and their ability to move. Most fibrous joints, such as the **sutures** between the cranial bones, are immovable. Cartilaginous joints are connected by hyaline cartilage, as in the costal cartilages that join the ribs to the sternum, or by fibrocartilage, as in the intervertebral disks. Cartilaginous joints tend to be slightly movable. Synovial joints are freely movable.

In **synovial joints,** the two bones are separated by a cavity. Ligaments hold the two bones in place as they form a capsule. Tendons also help stabilize the joint. The joint capsule is lined by a **synovial membrane,** which produces synovial fluid, a lubricant for the joint. Major synovial joints include the shoulder, elbow, hip, and knee. Aside from articular cartilage, the knee contains menisci (sing., **meniscus**), crescent-shaped pieces of hyaline cartilage between the bones (Fig. 19.10). These give added stability and act as shock absorbers. Unfortunately,

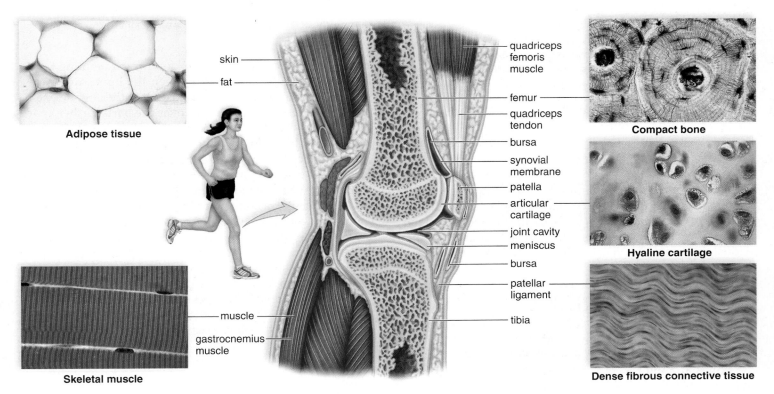

Adipose tissue

skin

fat

Skeletal muscle

muscle

gastrocnemius muscle

quadriceps femoris muscle

femur

quadriceps tendon

bursa

synovial membrane

patella

articular cartilage

joint cavity

meniscus

bursa

patellar ligament

tibia

Compact bone

Hyaline cartilage

Dense fibrous connective tissue

Figure 19.10 Knee joint.

The knee joint is a synovial joint. Notice the cavity between the bones, which is encased by ligaments and lined by synovial membrane. The patella (kneecap) serves to guide the quadriceps tendon over the joint when flexion or extension occurs.

TABLE 19.1	Examples of Movements at Synovial Joints	
Type	**Example**	
Flexion	Forearm toward the arm	
Extension	Forearm away from the arm	
Abduction	Arms sideways, away from body	
Adduction	Arms back to the body	
Rotation	Head to answer "no"	

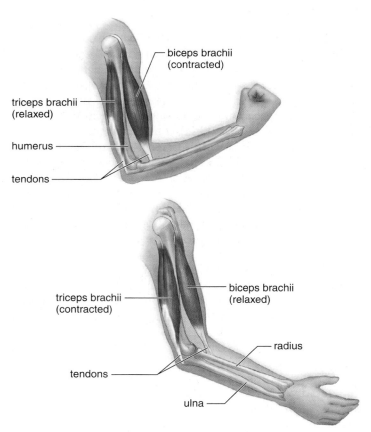

Figure 19.11 Antagonistic muscles.
Muscles can exert force only by shortening; therefore, they often work as antagonistic pairs. The biceps and triceps brachii exemplify an antagonistic pair of muscles that act opposite to one another. The biceps brachii raises the forearm, and the triceps brachii lowers the forearm.

athletes often suffer injury to the menisci, known as torn cartilage. The anterior and posterior ligaments, which help stabilize the knee by attaching the femur to the tibia, are also prone to being torn (sometimes called "ruptured"). The knee joint also contains 13 fluid-filled sacs called bursae (sing., **bursa**), which ease friction between the tendons and ligaments. Inflammation of the bursae is called **bursitis;** tennis elbow is a form of bursitis.

There are different types of synovial joints. The knee and elbow joints are **hinge joints** because, like a hinged door, they largely permit movement in one direction only. The joint between the radius and ulna is a **pivot joint** in which only rotation is possible. More movable are the **ball-and-socket joints** which allow movement in all planes, even rotational movement. For example, the ball of the femur fits into a socket on the hipbone. Some of the movements at synovial joints are listed in Table 19.1.

> Joints are regions of articulations between bones. Synovial joints are freely movable and can be categorized by the type of movement they allow.

19.3 Skeletal Muscles

Humans have three types of muscle tissue: smooth, cardiac, and skeletal (see Fig. 11.5). Skeletal muscle makes up the greatest percentage of muscle tissue in the body. Skeletal muscle is voluntary because its contraction can be consciously stimulated and controlled by the nervous system.

Skeletal Muscles Work in Pairs

Skeletal muscles are covered by several layers of fibrous connective tissue called fascia, which extends beyond the muscle to become its tendon. Tendons are attached to the skeleton, and the contraction of skeletal muscle causes the bones at a joint to move. When skeletal muscles contract, one bone remains fairly stationary, and the other one moves. The **origin** of a muscle is on the stationary bone, and the **insertion** of a muscle is on the bone that moves. Table 19.1 lists some terms that are used to describe these move-

ments, and gives examples. Flexion means bending bones at a joint so that the angle between the bones decreases, and the involved bones move closer together, while extension increases that angle and moves the parts farther apart. Abduction at a joint moves the involved part (usually a limb) away from the middle of the body, while adduction brings it back toward the midline. Rotation moves the part around an axis, due to a twisting motion at the joint. Note that not all of these movements can occur at all joints.

Frequently, a body part is moved by a group of muscles working together. Even so, one muscle does most of the work and is called the prime mover. The assisting muscles are called the synergists. When muscles contract, they shorten. Therefore, muscles can only pull; they cannot push.

Most muscles have antagonists, and antagonistic pairs bring about movement in opposite directions. For example, the biceps brachii and the triceps brachii are antagonists; one flexes the forearm, and the other extends the forearm (Fig. 19.11).

> Skeletal muscles are voluntary. When skeletal muscles contract, some act as prime movers, others are synergists, and still others are antagonists.

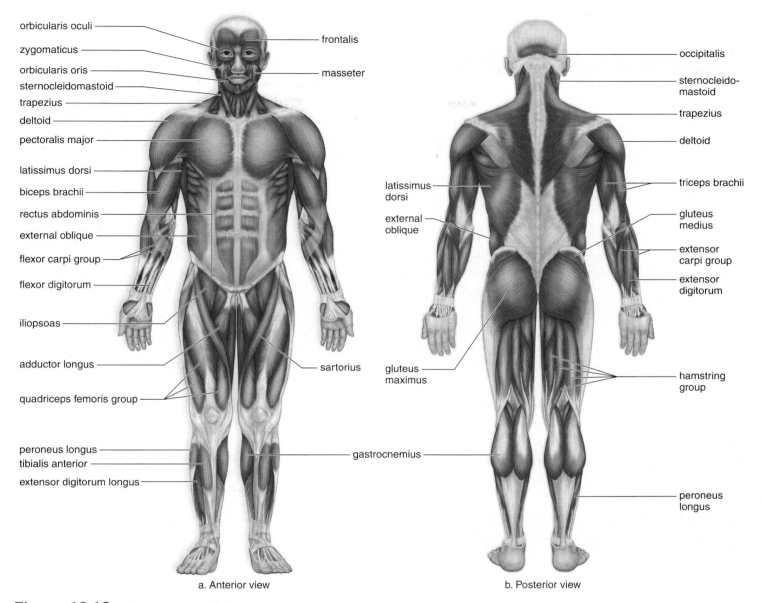

Figure 19.12 Human musculature.
Skeletal muscles located near the surface in **(a)** anterior and **(b)** posterior views.

Major Skeletal Muscles

There are approximately 650 skeletal muscles in the human body; the major ones are shown in Figure 19.12. As listed in Tables 19.2 and 19.3, the muscles of the head are involved in facial expression and chewing; those of the neck participate in head movements; the muscles of the trunk and upper limb move the arm, the forearm, and the fingers; and the muscles of the buttocks and lower limb move the thigh, the leg, and the toes.

Nomenclature

Skeletal muscles are named based on the following characteristics:

1. Size. The gluteus maximus that makes up the buttocks is the largest muscle (*maximus* means greatest). Other terms used to indicate size are vastus (huge) and longus (long).

2. Shape. The deltoid is shaped like a triangle. (The Greek letter delta has this appearance: Δ.) The trapezius is shaped like a trapezoid. Another term used to indicate shape is latissimus (wide).

3. Location. The frontalis overlies the frontal bone. The external obliques are located outside the internal obliques. Another term used to indicate location is pectoralis (chest).

4. Direction of muscle fibers. The rectus abdominis is a longitudinal muscle of the abdomen (*rectus* means straight). The orbicularis oculi is a circular muscle around the eye.

5. Number of attachments. The biceps brachii has two attachments, or origins (*bi-* means two).

6. Action. The extensor digitorum extends the fingers (digits). The adductor longus is a long muscle that adducts the thigh. Another term used to indicate action is masseter (to chew).

TABLE 19.2	Muscles (Anterior View)
Name	**Action**
Head and neck	
Frontalis	Wrinkles forehead and lifts eyebrows
Orbicularis oculi	Closes eye (winking)
Zygomaticus	Raises corner of mouth (smiling)
Masseter	Closes jaw, chewing
Orbicularis oris	Closes and protrudes lips (kissing)
Upper limb and trunk	
External oblique	Compresses abdomen; rotates trunk
Rectus abdominis	Flexes spine; compresses abdomen
Pectoralis major	Flexes and adducts arm ventrally (pulls arm across chest)
Deltoid	Abducts and moves arm up and down in front
Biceps brachii	Flexes forearm and rotates hand outward
Lower limb	
Adductor longus	Adducts and flexes thigh
Iliopsoas	Flexes thigh at hip joint
Sartorius	Raises and rotates thigh and leg
Quadriceps femoris group	Extends leg at knee; flexes thigh
Peroneus longus	Everts foot
Tibialis anterior	Dorsiflexes and inverts foot
Flexor digitorum longus	Flexes toes
Extensor digitorum longus	Extends toes

TABLE 19.3	Muscles (Posterior View)
Name	**Action**
Head and neck	
Occipitalis	Moves scalp backward
Sternocleidomastoid	Turns head to side; flexes neck and head
Trapezius	Extends head; raises scapula as when shrugging shoulders
Upper limb and trunk	
Latissimus dorsi	Extends and adducts arm dorsally (pulls arm across back)
Deltoid	Abducts and moves arm up and down in front
External oblique	Rotates trunk
Triceps brachii	Extends forearm
Flexor carpi group	Flexes hand
Extensor carpi group	Extends hand
Flexor digitorum	Flexes fingers
Extensor digitorum	Extends fingers
Buttocks and lower limb	
Gluteus medius	Abducts thigh
Gluteus maximus	Extends thigh back (forms buttocks)
Hamstring group	Flexes leg and extends thigh
Gastrocnemius	Plantar flexes foot (tiptoeing)

19.4 Mechanism of Muscle Fiber Contraction

We have already examined the structure of skeletal muscle as seen with the light microscope. Skeletal muscle tissue has alternating light and dark bands, giving it a striated appearance. These bands are due to the arrangement of myofilaments in a muscle fiber.

Muscle Fiber

A muscle fiber is a cell containing the usual cellular components, but special names have been assigned to some of these components (Table 19.4). Figure 19.13 shows the microscopic structure of a muscle fiber. The plasma membrane is called the **sarcolemma;** the cytoplasm is the sarcoplasm; and the endoplasmic reticulum is the **sarcoplasmic reticulum.** A muscle fiber also has some unique anatomical characteristics. One feature is its T (for transverse) system; the sarcolemma forms **T (transverse) tubules** that penetrate, or dip down, into the cell so that they come in contact—but do not fuse—with expanded portions of the sarcoplasmic reticulum. The expanded portions of the sarcoplasmic reticulum are calcium storage sites. Calcium ions (Ca^{2+}), as we shall see, are essential for muscle contraction.

The sarcoplasmic reticulum encases hundreds and sometimes even thousands of **myofibrils,** each about 1 μm in diameter, which are the contractile portions of the muscle fibers. Any other organelles, such as mitochondria, are located in the sarcoplasm between the myofibrils. The sarcoplasm also contains glycogen, which provides stored energy for muscle contraction, and the red pigment myoglobin, which binds oxygen until it is needed for muscle contraction.

Myofibrils and Sarcomeres

Myofibrils are cylindrical in shape and run the length of the muscle fiber. The light microscope shows that skeletal

TABLE 19.4	Microscopic Anatomy of a Muscle
Name	**Function**
Sarcolemma	Plasma membrane of a muscle fiber that forms T tubules
Sarcoplasm	Cytoplasm of a muscle fiber that contains the organelles, including myofibrils
Glycogen	A polysaccharide that stores energy for muscle contraction
Myoglobin	A red pigment that stores oxygen for muscle contraction
T tubule	Extension of the sarcolemma that extends into the muscle fiber and conveys impulses that cause Ca^{2+} to be released from the sarcoplasmic reticulum
Sarcoplasmic reticulum	The smooth ER of a muscle fiber that stores Ca^{2+}
Myofibril	A bundle of myofilaments that contracts
Myofilament	Actin filaments and myosin filaments whose structure and functions account for muscle striations and contractions

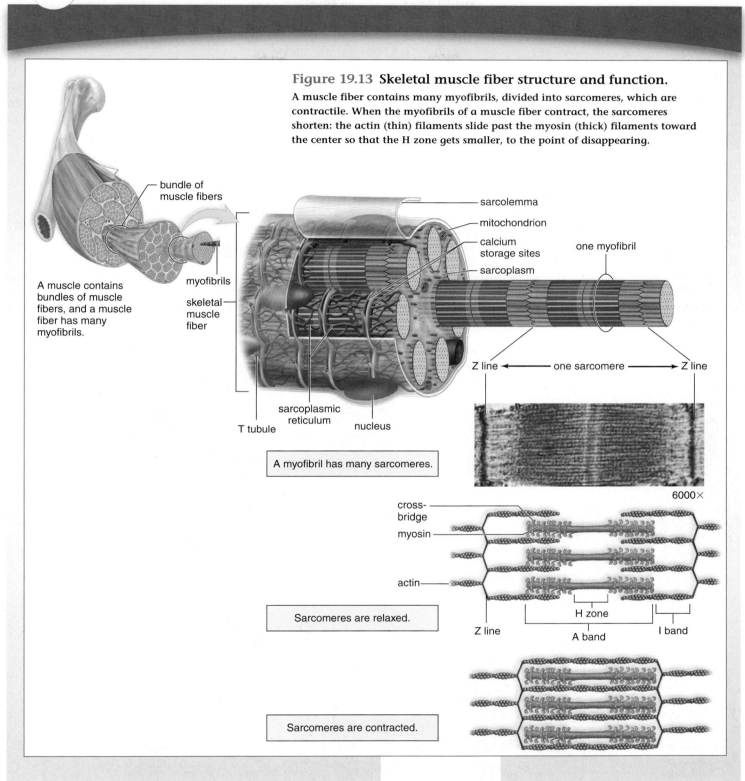

Figure 19.13 **Skeletal muscle fiber structure and function.**

A muscle fiber contains many myofibrils, divided into sarcomeres, which are contractile. When the myofibrils of a muscle fiber contract, the sarcomeres shorten: the actin (thin) filaments slide past the myosin (thick) filaments toward the center so that the H zone gets smaller, to the point of disappearing.

bundle of muscle fibers

A muscle contains bundles of muscle fibers, and a muscle fiber has many myofibrils.

myofibrils

skeletal muscle fiber

sarcolemma

mitochondrion

calcium storage sites

sarcoplasm

one myofibril

sarcoplasmic reticulum

T tubule

nucleus

Z line ← one sarcomere → Z line

A myofibril has many sarcomeres.

6000×

cross-bridge

myosin

actin

Sarcomeres are relaxed.

H zone

Z line

A band

I band

Sarcomeres are contracted.

muscle fibers have light and dark bands called striations. The electron microscope shows that the striations of skeletal muscle fibers are formed by the placement of myofilaments within units of myofibrils called **sarcomeres.** A sarcomere extends between two dark lines, called the Z lines. A sarcomere contains two types of protein myofilaments. The thick filaments are made up of a protein called myosin, and the thin filaments are made up of a protein called actin. Other proteins are also present. The I band is light colored because it contains only actin filaments attached to a Z line. The dark regions of the A band contain overlapping actin and myosin filaments, and its H zone has only myosin filaments.

Myofilaments

The thick and thin filaments differ in the following ways:

Thick Filaments   A thick filament is composed of several hundred molecules of the protein **myosin.** Each myosin molecule is shaped like a golf club, with the straight portion of the molecule ending in a double globular head. The myosin heads occur on each side of a sarcomere but not in the middle.

Thin Filaments   A thin filament primarily consists of two intertwining strands of the protein **actin.** Two other proteins, called tropomyosin and troponin, also play a role, as we will discuss later in this section.

Sliding Filaments   We will also see that when muscles are stimulated, impulses travel down a T tubule, and calcium is released from the sarcoplasmic reticulum. Now the muscle fiber contracts as the sarcomeres within the myofibrils shorten. When a sarcomere shortens, the actin (thin) filaments slide past the myosin (thick) filaments and approach one another. This causes the I band to shorten and the H zone to almost

or completely disappear. The movement of actin filaments in relation to myosin filaments is called the **sliding filament theory** of muscle contraction. During the sliding process, the sarcomere shortens, even though the filaments themselves remain the same length. ATP supplies the energy for muscle contraction. Although the actin filaments slide past the myosin filaments, it is the myosin filaments that do the work. Myosin filaments break down ATP and form cross-bridges that pull the actin filaments toward the center of the sarcomere.

> When muscle fibers are stimulated to contract, myofilaments slide past one another, causing sarcomeres to shorten.

Skeletal Muscle Contraction

Muscle fibers are stimulated to contract by motor neurons whose axons are in nerves (Fig. 19.14). The axon of one motor neuron can stimulate from a few to several muscle fibers of a

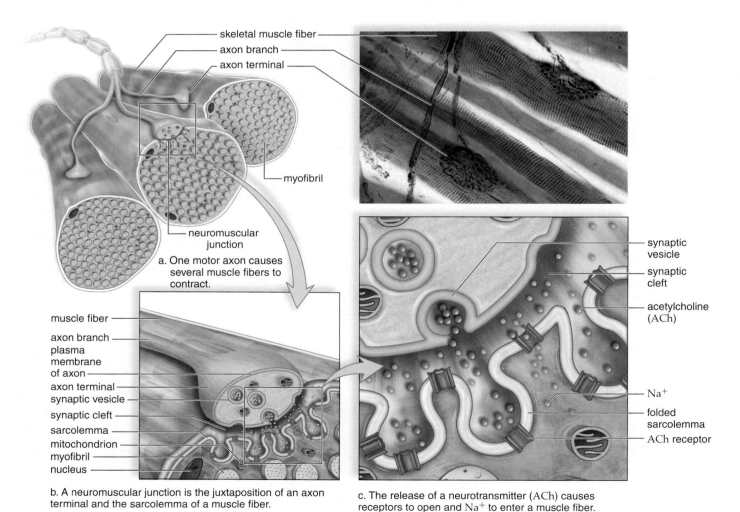

a. One motor axon causes several muscle fibers to contract.

b. A neuromuscular junction is the juxtaposition of an axon terminal and the sarcolemma of a muscle fiber.

c. The release of a neurotransmitter (ACh) causes receptors to open and Na$^+$ to enter a muscle fiber.

Figure 19.14 Neuromuscular junction.
a. The branch of a motor nerve fiber ends in an axon terminal that meets but does not touch a muscle fiber. **b.** A synaptic cleft separates the axon terminal from the sarcolemma of the muscle fiber. **c.** Nerve impulses traveling down a motor fiber cause synaptic vesicles to discharge a neurotransmitter (ACh) that diffuses across the synaptic cleft. When the neurotransmitter is received by the sarcolemma of a muscle fiber, impulses begin that lead to muscle fiber contraction.

muscle because each axon has several branches. A branch of an axon ends in an axon terminal that lies in close proximity to the sarcolemma of a muscle fiber. A small gap, called a synaptic cleft, separates the axon terminal from the sarcolemma. This entire region is called a **neuromuscular junction** (Fig. 19.14).

Axon terminals contain synaptic vesicles that are filled with the neurotransmitter acetylcholine (ACh). When nerve impulses traveling down a motor neuron arrive at an axon terminal, the synaptic vesicles release ACh into the synaptic cleft. ACh quickly diffuses across the cleft and binds to receptors in the sarcolemma. Now the sarcolemma generates impulses that spread over itself and down T tubules to the sarcoplasmic reticulum. The release of Ca^{2+} ions from the sarcoplasmic reticulum causes the filaments within the sarcomeres to slide past one another. Sarcomere contraction causes myofibril contraction, which in turn results in the contraction of a muscle fiber and, eventually, a whole muscle.

Botox is a trade name for botulinum toxin A, a neurotoxin produced by a bacterium. Botox, which can be injected by a physician to prevent wrinkling of the brow and skin around the eyes, prevents the release of ACh, and therefore, the contraction of muscles in these areas.

> At a neuromuscular junction, nerve impulses bring about the release of neurotransmitter molecules, which then signal a muscle fiber to contract.

The Molecular Mechanism of Contraction

As shown in Figure 19.15, two other proteins are associated with actin filaments. Threads of *tropomyosin* wind about an actin filament, and *troponin* occurs at intervals along the threads. When Ca^{2+} ions are released from the sarcoplasmic reticulum, they combine with troponin, and this causes the tropomyosin threads to shift their position.

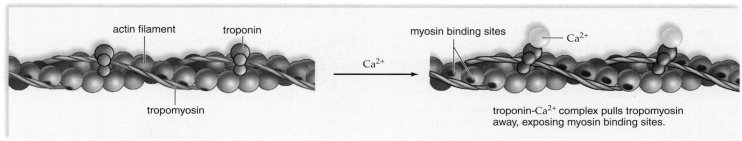

a. Function of Ca^{2+} ions in muscle contraction

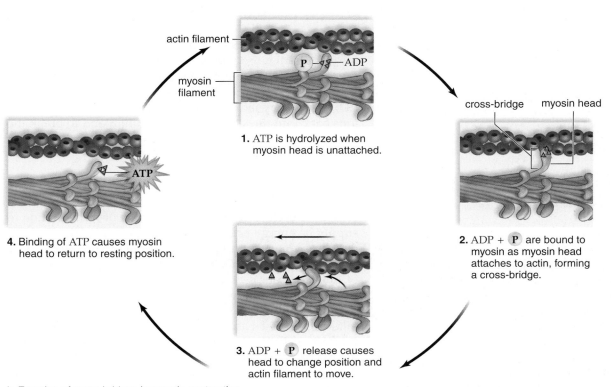

b. Function of cross-bridges in muscle contraction

1. ATP is hydrolyzed when myosin head is unattached.

2. ADP + P are bound to myosin as myosin head attaches to actin, forming a cross-bridge.

3. ADP + P release causes head to change position and actin filament to move.

4. Binding of ATP causes myosin head to return to resting position.

**Figure 19.15
The role of specific proteins and calcium in muscle contraction.**

a. Upon release, calcium binds to troponin, exposing myosin binding sites. **b.** After breaking down ATP (1), myosin heads bind to an actin filament (2), and later, a power stroke causes the actin filament to move (3). When another ATP binds to myosin, the head detaches from actin (4), and the cycle begins again. Although only one myosin head is featured, many heads are active at the same time.

The double globular heads of a myosin filament have ATP binding sites, where an ATPase splits ATP into ADP and ⓟ. The ADP and ⓟ remain on the myosin heads until the heads attach to an actin filament, forming cross-bridges. Now, ADP and ⓟ are released, and the cross-bridges change their positions. This is the power stroke that pulls the actin filament toward the center of the sarcomere. When ATP molecules again bind to the myosin heads, the cross-bridges are broken, and heads detach from the actin filament. Actin filaments move nearer the center of the sarcomere each time the cycle is repeated. When nerve impulses cease, the sarcoplasmic reticulum actively transports Ca^{2+} ions back into the sarcoplasmic reticulum, and the muscle relaxes.

> After Ca^{2+} ions bind to troponin, myosin filaments break down ATP and attach to actin filaments, forming cross-bridges that pull actin filaments to the center of a sarcomere.

Energy for Muscle Contraction

ATP produced previous to strenuous exercise lasts a few seconds, and then muscles acquire new ATP in three different ways: creatine phosphate breakdown, cellular respiration, and fermentation. Creatine phosphate breakdown and fermentation are anaerobic, meaning that they do not require oxygen. Creatine phosphate breakdown, used first, is a way to acquire ATP before oxygen starts entering mitochondria. Cellular respiration is aerobic and only takes place when oxygen is available. If exercise is vigorous to the point that oxygen cannot be delivered fast enough to working muscles, fermentation occurs. Fermentation brings on oxygen debt.

Creatine Phosphate Breakdown

Creatine phosphate is a high-energy compound built up when a muscle is resting. Creatine phosphate cannot participate directly in muscle contraction. Instead, it can regenerate ATP by the following reaction:

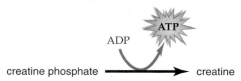

This reaction occurs in the midst of sliding filaments, and therefore is the speediest way to make ATP available to muscles. Creatine phosphate provides enough energy for only about eight seconds of intense activity, and then it is spent. Creatine phosphate is rebuilt when a muscle is resting by transferring a phosphate group from ATP to creatine.

Cellular Respiration

Cellular respiration, completed in mitochondria, usually provides most of a muscle's ATP. Glycogen and fat are stored in muscle cells. Therefore, a muscle cell can use glucose (from glycogen) and fatty acids (from fat) as fuel to produce ATP if oxygen is available:

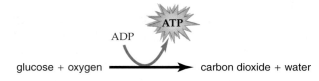

Myoglobin, an oxygen carrier similar to hemoglobin, is synthesized in muscle cells, and its presence accounts for the reddish-brown color of skeletal muscle fibers. Myoglobin has a higher affinity for oxygen than does hemoglobin. Therefore, myoglobin can pull oxygen out of blood and make it available to muscle mitochondria that are carrying on cellular respiration. Then, too, the ability of myoglobin to temporarily store oxygen reduces a muscle's immediate need for oxygen when cellular respiration begins. The end products (carbon dioxide and water) are usually no problem. Carbon dioxide leaves the body at the lungs, and water simply enters the extracellular space. The by-product, heat, keeps the entire body warm.

Fermentation

Like creatine phosphate breakdown, fermentation supplies ATP without consuming oxygen. During fermentation, glucose is broken down to lactate (lactic acid):

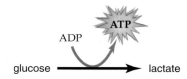

The accumulation of lactate in a muscle fiber makes the cytoplasm more acidic, and eventually enzymes cease to function well. If fermentation continues longer than two or three minutes, cramping and fatigue set in. Cramping seems to be caused by lack of the ATP needed to pump calcium ions back into the sarcoplasmic reticulum and to break the linkages between the actin and myosin filaments so that muscle fibers can relax.

Oxygen Debt

When a muscle uses creatine phosphate or fermentation to supply its energy needs, it incurs an **oxygen debt.** Oxygen debt is obvious when a person continues to breathe heavily after exercising. The ability to run up an oxygen debt is one of muscle tissue's greatest assets. Without oxygen, brain tissue cannot last nearly as long as muscles can.

In people who exercise regularly, the number of muscle mitochondria increases, and so fermentation is not needed to produce ATP. Their bodies' mitochondria can start consuming oxygen as soon as the ADP concentration begins to rise during muscle contraction. Because mitochondria can break down fatty acids instead of glucose, blood glucose is spared for the activity of the brain. (The brain, unlike other organs, can only utilize glucose to produce ATP.) Because less lactate is produced in people who train, the pH of the blood remains steady, and there is less of an oxygen debt.

Repaying an oxygen debt requires replenishing creatine phosphate supplies and disposing of lactate. Lactate can be changed back to pyruvate and metabolized completely in mitochondria, or it can be sent to the liver to reconstruct glycogen. A marathon runner who has just crossed the finish line is not usually exhausted due to oxygen debt. Instead, the runner has used up all the muscles', and probably the liver's, glycogen supply. It takes about two days to replace glycogen stores on a high-carbohydrate diet.

> Working muscles require a supply of ATP. Anaerobic creatine phosphate breakdown and fermentation can quickly generate ATP. Cellular respiration in mitochondria is best for sustained exercise.

19.5 Whole Muscle Contraction

Researchers sometimes study muscles in the laboratory in an effort to understand whole muscle contraction in the body.

In the Laboratory

When a muscle fiber is isolated, placed on a microscope slide, and provided with ATP plus various salts, it contracts completely along its entire length. This observation has resulted in the *all-or-none law,* which states that a muscle fiber contracts completely or not at all. In contrast, a whole muscle shows degrees of contraction. To study whole muscle contraction in the laboratory, an isolated muscle is stimulated electrically, and the mechanical force of contraction is recorded as a visual pattern called a **myogram.** When the strength of the stimulus is above a threshold level, the muscle contracts and then relaxes. This action—a single contraction that lasts only a fraction of a second—is called a **muscle twitch.**

Figure 19.16*a* is a myogram of a muscle twitch, which is customarily divided into three stages: the *latent period,* or the period of time between stimulation and initiation of contraction; the *contraction period,* when the muscle shortens; and the *relaxation period,* when the muscle returns to its former length. It's interesting to use our knowledge of muscle fiber contraction to understand these events. From our study thus far, we know that a muscle fiber in an intact muscle contracts when calcium leaves storage sacs and relaxes when calcium returns to storage sacs.

But unlike the contraction of a muscle fiber, a whole muscle has degrees of contraction, and a twitch can vary in height (strength) depending on the degree of stimulation. Why? Obviously, a stronger stimulation causes more individual fibers to contract than before.

If a whole muscle is given a rapid series of stimuli, it can respond to the next stimulus without relaxing completely. Summation is increased muscle contraction until maximal sustained contraction, called **tetanus,** is achieved (Fig. 19.16*b*). The myogram no longer shows individual twitches; rather, the twitches are fused and blended completely into a straight line. Tetanus continues until the muscle becomes fatigued due to depletion of energy reserves. Fatigue is apparent when a muscle relaxes even though stimulation continues.

In the Body

In the body, nerves cause muscles to contract. As mentioned, each axon within a nerve stimulates a number of muscle fibers. A nerve fiber, together with all of the muscle fibers it innervates, is called a **motor unit.** A motor unit obeys the all-or-none law. Why? Because all the muscle fibers in a motor unit are stimulated at once, and they all either contract or do not contract. A variable of interest is the number of muscle fibers within a motor unit. For example, in the ocular muscles that move the eyes, the innervation ratio is one motor axon per 23 muscle fibers, while in the gastrocnemius muscle of the leg, the ratio is about one motor axon per 1,000 muscle fibers. Thus, moving the eyes requires finer control than moving the legs.

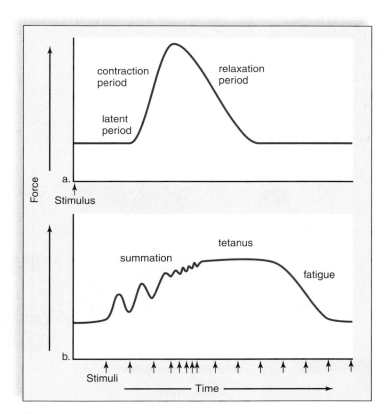

Figure 19.16 Physiology of skeletal muscle contraction.
Stimulation of a muscle dissected from a frog resulted in these myograms. **a.** A simple muscle twitch has three periods: latent, contraction, and relaxation. **b.** Summation and tetanus. When the muscle is not permitted to relax completely between stimuli, the contraction gradually increases in intensity. The muscle becomes maximally contracted until it fatigues.

Tetanic contractions ordinarily occur in the body because, as the intensity of nervous stimulation increases, more and more motor units are activated. This phenomenon, known as recruitment, results in stronger and stronger muscle contractions. But while some muscle fibers are contracting, others are relaxing. Because of this, intact muscles rarely fatigue completely. Even when muscles appear to be at rest, they exhibit **muscle tone,** in which some of their fibers are always contracting. Muscle tone is particularly important in maintaining posture. If all the fibers within the muscles of the neck, trunk, and legs were to suddenly relax, the body would collapse.

> Observation of muscle contraction in the laboratory helps explain how muscles contract in the body. A contracting muscle exhibits degrees of contraction dependent on recruitment, and a muscle at rest exhibits tone.

Athletics and Muscle Contraction

Athletes who excel in a particular sport, and much of the general public as well, are interested in staying fit by exercising. The Health Focus gives suggestions for exercise programs according to age.

Exercise and Size of Muscles   Muscles that are not used or that are used for only very weak contractions decrease in size, or **atrophy.** Atrophy can occur when a limb is placed in a cast or when the nerve serving a muscle is damaged. If nerve stimulation is not restored, muscle fibers are gradually replaced by fat and fibrous tissue. Unfortunately,

atrophy can cause muscle fibers to shorten progressively, leaving body parts contracted in contorted positions.

Forceful muscular activity over a prolonged period causes muscle to increase in size as the number of myofibrils within the muscle fibers increases. Increase in muscle size, called **hypertrophy,** occurs only if the muscle contracts to at least 75% of its maximum tension.

Slow-Twitch and Fast-Twitch Muscle Fibers   We have seen that all muscle fibers metabolize both aerobically and anaerobically. Some muscle fibers, however, utilize one method more than the other to provide myofibrils with ATP. Slow-twitch fibers tend to be aerobic, and fast-twitch fibers tend to be anaerobic (Fig. 19.17).

Slow-twitch fibers have a steadier tug and more endurance, despite having motor units with a smaller number of fibers. These muscle fibers are most helpful in sports such as long-distance running, biking, jogging, and swimming. Because they produce most of their energy aerobically, they tire only when their fuel supply is gone. Slow-twitch fibers have many mitochondria and are dark in color because they contain myoglobin, the respiratory pigment found in muscles. They are also surrounded by dense capillary beds and draw more blood and oxygen than fast-twitch fibers. Slow-twitch fibers have a low maximum tension, which develops slowly, but these muscle fibers are highly resistant to fatigue. Because slow-twitch fibers have a substantial reserve of glycogen and fat, their abundant mitochondria can maintain a steady, prolonged production of ATP when oxygen is available.

Fast-twitch fibers tend to be anaerobic and seem designed for strength because their motor units contain many fibers. They provide explosions of energy and are most helpful in sports activities such as sprinting, weight lifting, swinging a golf club, or throwing a shot. Fast-twitch fibers are light in color because

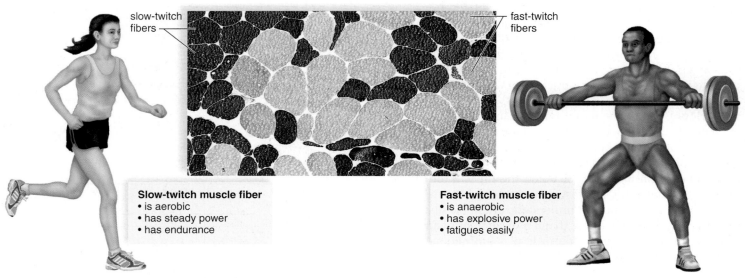

slow-twitch fibers

fast-twitch fibers

Slow-twitch muscle fiber
• is aerobic
• has steady power
• has endurance

Fast-twitch muscle fiber
• is anaerobic
• has explosive power
• fatigues easily

Figure 19.17 Slow- and fast-twitch muscle fibers.
If your muscles contain many slow-twitch fibers (dark color), you probably do better at a sport such as cross-country running. But if your muscles contain many fast-twitch fibers (light color), you probably do better at a sport such as weight lifting.

Health Focus

Exercise, Exercise, Exercise

Sensible exercise programs improve muscular strength, muscular endurance, and flexibility for people of all ages (Table 19A). Muscular strength is the force a muscle group (or muscle) can exert against a resistance in one maximal effort. Muscular endurance is judged by the ability of a muscle to contract repeatedly or sustain a contraction for an extended period. Flexibility is tested by observing the range of motion at a joint.

Exercise also improves cardiorespiratory endurance. The heart rate and capacity increase, and the air passages dilate so that the heart and lungs are able to support prolonged muscular activity. The blood level of high-density lipoprotein (HDL), the molecule that prevents the development of plaque in blood vessels, increases. Also, body composition, the proportion of protein to fat, changes favorably when you exercise.

Exercise reduces the risk of diabetes mellitus, especially type 2 or insulin-resistant diabetes, which is a large and growing health problem in many countries around the world. Being overweight and inactive increases the chances of developing type 2 diabetes. Studies have shown that healthy eating and regular exercise are more effective at reducing the risk of diabetes than medication alone. Exercise seems to help prevent certain kinds of cancer. Cancer prevention involves eating properly, not smoking, avoiding cancer-causing chemicals and radiation, undergoing appropriate medical screening tests, and knowing the early warning signs of cancer. However, studies show that people who exercise are less likely to develop colon, breast, cervical, uterine, and ovarian cancers.

Exercise promotes the activity of osteoblasts in young people as well as older people. Physical training with weights can improve bone density and also muscular strength and endurance in all adults, regardless of age. Even men and women in their 80s and 90s make substantial gains in bone and muscle strength, which can help them lead more independent lives. Exercise also helps prevent osteoporosis, a condition in which the bones are weak and tend to break. The stronger the bones when a person is young, the lower the chance of osteoporosis as a person ages.

Exercise helps prevent weight gain, not only because the level of activity increases, but also because muscles metabolize faster than other tissues. As a person becomes more muscular, fat is less likely to accumulate.

Exercise relieves depression and enhances the mood. Some report that exercise actually makes them feel more energetic, and after exercising, particularly in the late afternoon, they sleep better that night. Self-esteem rises, not only because of improved appearance, but also due to other factors that are not well understood. However, it is known that vigorous exercise releases endorphins, hormonelike chemicals that alleviate pain and provide a feeling of tranquility.

A sensible exercise program is one that provides all the benefits without the detriments of a too strenuous program. Overexertion can actually be harmful to the body and might result in a lasting sports injury, such as a bad back or bad knees.

Dr. Arthur Leon at the University of Minnesota performed a study involving 12,000 men, and the results showed that only moderate exercise is needed to lower the risk of a heart attack by one-third. In another study conducted by the Institute for Aerobics Research in Dallas, Texas, which included 10,000 men and more than 3,000 women, even a little exercise was found to lower the risk of death from circulatory diseases and cancer. Simply increasing daily activity by walking to the corner store instead of driving and by taking the stairs instead of the elevator can improve your health.

Discussion Questions

1. What type(s) of muscle activity (strength, endurance, flexibility) are most important for the following sports?
   a. football
   b. golf
   c. gymnastics
   d. long-distance running
   e. sprinting
2. Provide a specific, reasonable mechanism by which exercise could decrease your chances of developing the following: diabetes, hypertension, cancer.
3. Since there are so many documented heath benefits to exercising, what are some major factors that keep people from starting and sticking to a regular exercise program?

TABLE 19A	A Checklist for Staying Fit		
Children, 7–12	**Teenagers, 13–18**	**Adults, 19–55**	**Seniors, 55 and up**
Vigorous activity 1–2 hours daily	Vigorous activity 1 hour 3–5 days a week; otherwise, 1/2 hour daily moderate activity	Vigorous activity 1 hour 3 days a week; otherwise, 1/2 hour daily moderate activity	Moderate exercise 1 hour daily 3 days a week; otherwise, 1/2 hour daily moderate activity
Free play	Build muscle with calisthenics	Exercise to prevent lower back pain: aerobics, stretching, yoga	Plan a daily walk
Build motor skills through team sports, dance, swimming	Plan aerobic exercise to control buildup of fat cells	Take active vacations: hike, bicycle, cross-country ski	Daily stretching exercise
Encourage more exercise outside of physical education classes	Pursue tennis, swimming, horseback riding—sports that can be enjoyed for a lifetime	Find exercise partners: join a running club, bicycle club, outing group	Learn a new sport or activity: golf, fishing, ballroom dancing
Initiate family outings: bowling, boating, camping, hiking	Continue team sports, dancing, hiking, swimming		Try low-impact aerobics, boating. Before undertaking new exercises, consult your doctor

they have fewer mitochondria, little or no myoglobin, and fewer blood vessels than slow-twitch fibers do. Fast-twitch fibers can develop maximum tension more rapidly than slow-twitch fibers can, and their maximum tension is greater. However, their dependence on anaerobic energy leaves them vulnerable to an accumulation of lactate that causes them to fatigue quickly.

> Success in a particular sport is determined, in part, by a person's proportion of slow-twitch and fast-twitch muscle fibers.

19.6 Disorders of the Musculoskeletal System

Disorders of the Skeleton and Joints

Because bones are rigid, they are susceptible to being broken, or fractured. The most common cause of bone fractures is trauma, although pathological fractures can also occur during normal everyday activities if the bones are diseased. If the skin is punctured by the injury, the fracture is generally termed an "open" fracture; otherwise, it is "closed." Some common types of fractures include greenstick fractures, which do not pass all the way through the bone; comminuted fractures, which result in three or more bone fragments (Fig. 19.18); and stress fractures, one of the most common injuries in athletes. Stress fractures are small cracks in the bone that occur most commonly on the lower leg when muscles become fatigued and unable to absorb the stress of athletic activities. These small fractures usually heal with rest—that is, by avoiding the activity that produced them. Other types of uncomplicated fractures may be treated with casts or splints. Compound and comminuted fractures, however, often require surgery to realign bones or replace bone fragments. In children, the growth plate is the last part of the bone to ossify, leaving it more susceptible to fracture. Fracture of the growth

Figure 19.18 A comminuted fracture.

This X ray shows a tibia that has fractured into at least three pieces. The fibula (left) is also fractured.

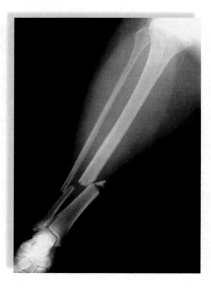

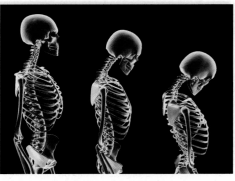

Figure 19.19 Osteoporosis.

Osteoporosis can eventually result in the slumped posture and loss of height characteristically seen in some elderly people.

plate may cause the affected bone to stop growing, resulting in limbs that are crooked or uneven in length.

Osteoporosis is a condition in which bone loses mass and mineral content. This leads to an increased risk of fractures, especially of the pelvis, vertebrae, and wrist. As the bone softens affected individuals often develop a characteristic curvature of the spine (Fig. 19.19). Eighty percent of those affected by osteoporosis are women. About one in four women will have an osteoporosis-related fracture in their lifetime. Women generally have about 30% less bone mass than men to begin with, and after menopause women typically lose about 2% of their bone mass per year. Estrogen replacement therapy (ERT) has been shown to increase bone mass and reduce fractures, but because it also can increase the risk of cardiovascular disease and certain types of cancer, patients must be well-informed before beginning ERT. Other factors that decrease the risk of osteoporosis for everyone include consuming 1,000–1,500 mg of calcium per day and engaging in regular physical exercise.

There are many different types of **arthritis,** or inflammation of the joints. The Arthritis Foundation estimates that nearly one in three adult Americans has some degree of chronic joint pain. Arthritis is second only to heart disease as a cause of work disability in the United States. The most common form of arthritis is **osteoarthritis (OA),** or degenerative joint disease, which results from the deterioration of the cartilage in one or more synovial joints. As the cartilage continues to soften, crack, and wear away, the resulting bone-on-bone contact causes pain and a decreased range of motion. Symptoms of OA, such as pain or stiffness in the joints after periods of inactivity, typically begin after age 40 and progress slowly. The hands, hips, knees, lower back, and neck are most often affected (Fig. 19.20). **Rheumatoid arthritis (RA)** is considered an autoimmune disease, in which the body's immune system attacks the joints as well as other tissues. RA affects more women than men, and often first strikes people in their 20s and 30s. The initiating cause of RA remains to be discovered, but genetic factors may play a major role. Over-the-counter anti-inflammatory drugs such as ibuprofen, or more powerful prescription drugs such as corticosteroids, can reduce the pain of most types of arthritis. Some studies have shown

Bioethical Focus

Performance-Enhancing Drugs

Athletes cannot receive Olympic medals if they have taken certain performance-enhancing drugs. There is no doubt that regular use of drugs such as anabolic steroids leads to kidney disease, liver dysfunction, hypertension, and a myriad of other undesirable side effects. Even so, shouldn't individuals be allowed to take these drugs if they want to? Anabolic steroids are synthetic forms of the male sex hormone testosterone. Taking large doses, along with strength training, leads to much larger muscles than otherwise. Extra strength and endurance can give an athlete an advantage in sports such as racing, swimming, and weight lifting.

Is the Olympic committee justified in outlawing the use of anabolic steroids, and if so, on what basis? "Unfair advantage" is controversial because some athletes naturally have an unfair advantage over other athletes. Should these drugs be outlawed on the basis of health reasons? Excessive practice alone and a purposeful decrease or increase in weight to better perform in a sport can also injure a person's health. In other words, how can you justify allowing some behaviors that enhance performance and not others?

Form Your Own Opinion

1. If anabolic steroids should be banned or restricted, what about nutritional "supplements" such as androstenedione, which is thought to be converted to testosterone in the body?
2. Are there any similarities between taking anabolic steroids and taking legal drugs, such as aspirin, to reduce inflammation in the body? What about injecting anti-inflammatory drugs directly into sore joints?
3. Suppose a new pay-per-view sports network allowed athletes to take any performance-enhancing drug they wanted to take, in order to see what the human body could achieve. What factors would then limit the athlete's performance?

Figure 19.20
The effects of arthritis.
Arthritis often affects the hands, resulting in swelling, pain, and loss of the ability to perform normal functions.

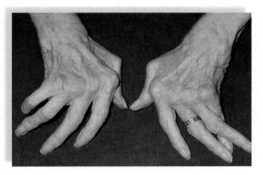

that dietary supplements containing glucosamine and chondroitin sulfate are also beneficial. If necessary, the damaged joint can be surgically replaced with a prosthesis (artificial substitute).

Disorders of the Muscles

Muscle cramps and twitches don't usually rise to the level of a "disorder," but almost everyone has experienced them. A true muscle cramp doesn't originate in the muscle itself, but rather is caused by abnormal nerve activity. This type of cramp is more common at night or after exercise, and can usually be ended by stretching the muscle. Most of us have also experienced occasional muscle twitches, or "fasciculations." These also result from abnormal nerve activity, but only involve a small part of a muscle. Twitches most commonly occur in the eyelids, between the thumb and forefinger, and in the feet. They may occur more frequently due to stress, caffeine, or lack of sleep.

Fibromyalgia is a disorder that causes chronic pain in the muscles and ligaments. It affects about 4 million Americans, mostly women in their mid-30s to late-50s. The pain is usually most severe in the neck, shoulders, back, and hips. Although the cause of fibromyalgia is unknown, it may be due to low

levels of serotonin or other neurotransmitters involved in pain perception and sleep regulation. As a result, patients with fibromyalgia may be more sensitive to pain, and they usually are unable to get enough deep sleep. As a result, chronic fatigue is also a major symptom. There are no specific treatments, but most people can reduce their symptoms by combining medications to reduce pain and improve sleep with regular exercise and a healthy diet.

Muscular dystrophies (MD) are a group of genetic diseases that affect the muscles. Most people are probably most familiar with these diseases from the annual Jerry Lewis Labor Day telethon sponsored by the Muscular Dystrophy Association. These disorders vary considerably in their degree of severity. The more severe forms may cause problems beginning at a very young age, while other forms may not cause problems until middle age or later. The most common type of MD, called Duchenne MD, is due to an abnormality in the gene coding for a muscle-associated protein called dystrophin. Because it is an X-linked recessive disorder, Duchenne MD mainly affects boys. Muscle weakness is usually noticeable by age 3 to 5. By age 12, most affected children are unable to walk, and will eventually need a respirator to breathe. Corticosteroids have been shown to slow the degenerative process of Duchenne MD, but these drugs often have side effects. Gene therapy, in which the correct form of the faulty gene is inserted into the patient's cells, is being investigated as a potential cure for MD.

Three common disorders affecting the skeletal system are fractures, osteoporosis, and arthritis. Fewer diseases affect the muscles specifically, but important examples are fibromyalgia and muscular dystrophy.

Summarizing the Concepts

19.1 Anatomy and Physiology of Bone

Bone and the associated tissues (cartilage and dense fibrous connective tissue) make up the skeleton. In a long bone, hyaline (articular) cartilage covers the ends, while periosteum (fibrous connective tissue) covers the rest. Spongy bone, which contains red bone marrow, is in the ends of long bones, and yellow bone marrow is in the medullary cavity of the shaft. The wall of the shaft is compact bone.

Bone tissue contains mineral salts, which increase its strength. Compact bone is highly organized. It is made up of units called osteons, which consist of a central canal surrounded by tiny canals called canaliculi, and chambers called lacunae. Bone cells called osteocytes dwell in these lacunae. Spongy bone is less organized, although still very strong. Cartilage is not as strong as bone, but is more flexible because it contains many collagenous and elastic fibers.

Bone is a living tissue that can grow and repair itself. The prenatal human skeleton is at first cartilaginous, but is later replaced by a bony skeleton.

19.2 Bones of the Skeleton

The skeleton supports and protects the body, produces blood cells, serves as a storehouse for fats and mineral salts, particularly calcium phosphate, and permits flexible movement.

The axial skeleton lies in the midline of the body and consists of the skull, the hyoid bone, the vertebral column, and the rib cage. The skull contains the cranium, which protects the brain, and the facial bones.

The appendicular skeleton consists of the bones of the pectoral girdles, upper limbs, pelvic girdle, and lower limbs. The pectoral girdles and upper limbs are adapted for flexibility. The pelvic girdle and lower limbs are adapted for strength: the femur is the strongest bone in the body.

There are three types of joints: fibrous joints, such as the sutures of the cranium, are immovable; cartilaginous joints, such as those between the ribs and sternum and the pubic symphysis, are slightly movable; and the synovial joints, consisting of a membrane-lined (synovial membrane) capsule, are freely movable.

19.3 Skeletal Muscles

Whole muscles work in antagonistic pairs; for example, the biceps flexes the forearm, and the triceps extends it. Muscles are grouped by location and named for various characteristics.

19.4 Mechanism of Muscle Fiber Contraction

The sarcolemma of a muscle fiber forms T tubules that extend into the fiber and almost touch the sarcoplasmic reticulum, which stores calcium ions. When calcium ions (Ca^{2+}) are released into muscle fibers, actin filaments slide past myosin filaments within the sarcomeres of a myofibril.

At a neuromuscular junction, synaptic vesicles release acetylcholine (ACh), which binds to protein receptors on the sarcolemma, causing impulses to travel down T tubules and calcium to leave the sarcoplasmic reticulum. Calcium ions bind to troponin and cause tropomyosin threads to shift their position, revealing myosin binding sites on actin. Myosin is an ATPase, and once it breaks down ATP, the myosin head is ready to attach to actin. The release of ADP + Ⓟ causes the head to change its position. This is the power stroke that causes the actin filament to slide toward the center of a sarcomere.

19.5 Whole Muscle Contraction

In the laboratory, muscle contraction is described in terms of a muscle twitch, summation, and tetanus. In the body, muscles exhibit tone, in which tetanic contraction involving a number of muscle fibers is the rule. A muscle fiber has three ways to acquire ATP after muscle contraction begins: (1) creatine phosphate, built up when a muscle is resting, donates phosphates to ADP, forming ATP; (2) oxygen-dependent cellular respiration occurs within mitochondria; and (3) fermentation with the concomitant accumulation of lactate quickly produces ATP. Fermentation can result in oxygen debt because oxygen is needed to complete the metabolism of lactate.

Certain sports, such as running and swimming, can be associated with slow-twitch fibers, which are dark in color and have a plentiful supply of mitochondria and myoglobin. Other sports, such as weight lifting, can be associated with fast-twitch fibers, which rely on an anaerobic means of acquiring ATP. They have few mitochondria and less myoglobin, and their motor units contain more muscle fibers. Fast-twitch fibers are known for their explosive power, but they fatigue quickly.

19.6 Disorders of the Musculoskeletal System

Major disorders affecting the skeletal system include fractures, osteoporosis, and arthritis. Fractures can be open or closed, and some specific types are greenstick, comminuted, and stress fractures. Osteoporosis is a softening of bone due to loss of mass and mineral content, which affects women more commonly than men. Among the many types of arthritis are osteoarthritis, or degenerative joint disease, and rheumatoid arthritis, which is an autoimmune disease.

Muscles can be affected by cramps and twitches, which are very common but usually just an annoyance. More serious diseases of the muscles include fibromyalgia, which results in intense pain and may be due to low levels of serotonin, and muscular dystrophy, a group of genetic disorders that affect muscle development.

Testing Yourself

Choose the best answer for each question.

1. Bone is classified as _____ tissue.
   a. epithelial
   b. connective
   c. the same category as muscle
   d. Both b and c are correct.

2. _____ connect bone to bone, and _____ connect muscle to bone.
   a. Ligaments, ligaments   c. Ligaments, tendons
   b. Tendons, ligaments   d. None of these are correct.

3. _____ are involved in the breakdown of bone.
   a. Monocytes   c. Lymphocytes
   b. Osteoblasts   d. Osteoclasts

4. _____ occupies the _____.
   a. Cartilage, medullary cavity
   b. Marrow, foramen magnum
   c. Marrow, medullary cavity
   d. None of these are correct.

5. The only bone that does not articulate with any other bone is the
   a. temporal bone.   d. maxilla.
   b. mandible.   e. palatine bone.
   c. hyoid bone.

6. The ribs articulate with the _____ vertebrae.
   a. cervical   d. sacral
   b. thoracic   e. All of these are correct.
   c. lumbar

7. The knee joint contains

   a. cartilage.
   b. cruciate ligaments.
   c. synovial fluid.
   d. bursae.
   e. All of these are correct.

8. The _____ of the muscle attaches to the bone that is stationary during movement.

   a. origin
   b. insertion
   c. belly
   d. None of these are correct.

9. The biceps and triceps are considered

   a. synergists.
   b. antagonists.
   c. protagonists.
   d. None of these are correct.

For questions 10–14, match each term with the correct characteristic given in the key.

Key:

   a. closing the angle at a joint
   b. movement in all planes
   c. made of cartilage
   d. movement toward the body
   e. fluid-filled sac

10. Ball-and-socket joints

11. Flexion

12. Bursa

13. Meniscus

14. Adduction

15. Muscles can be named based on

   a. size.
   b. shape.
   c. location.
   d. action.
   e. All of these are correct.

16. Lack of calcium in muscles will

   a. result in no contraction.
   b. cause weak contraction.
   c. cause strong contraction.
   d. have no effect.
   e. None of these are correct.

17. To increase the force of muscle contraction,

   a. individual muscle cells have to contract with greater force.
   b. motor units have to contract with greater force.
   c. more motor units need to be recruited.
   d. All of these are correct.
   e. None of these are correct.

18. Myoglobin content is higher in _____ -twitch fibers respiring _____.

   a. slow, aerobically
   b. fast, aerobically
   c. slow, anaerobically
   d. None of these are correct.

19. Label the following diagram of the skull.

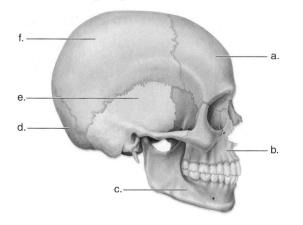

20. When muscles contract,

   a. sarcomeres increase in length.
   b. actin breaks down ATP.
   c. myosin slides past actin.
   d. the H zone disappears.
   e. calcium is taken up by the sarcoplasmic reticulum.

For questions 21–25, match the names of the muscles in the key to the following diagram.

Key:

   a. deltoid
   b. sartorius
   c. gastrocnemius
   d. triceps brachii
   e. biceps brachii

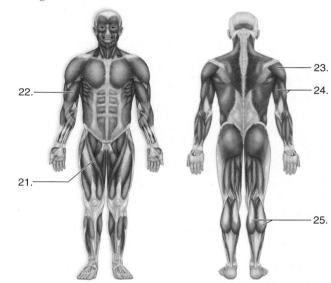

For questions 26–29, match each bone with the correct location given in the key.

Key:

   a. arm
   b. forearm
   c. pectoral girdle
   d. pelvic girdle
   e. thigh
   f. leg

26. Ulna

27. Tibia

28. Clavicle

29. Femur

For questions 30–35, label the diagram of a muscle fiber, using these terms: myofibril, Z line, T tubule, sarcomere, sarcolemma, sarcoplasmic reticulum.

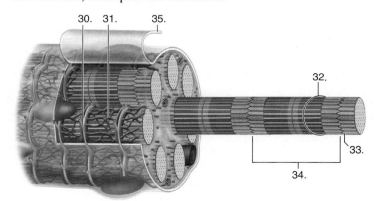

36. In a muscle fiber,
    a. the sarcolemma is connective tissue holding the myofibrils together.
    b. the T system consists of tubules.
    c. both actin and myosin filaments have cross-bridges.
    d. there is no endoplasmic reticulum.
    e. All of these are correct.
37. Nervous stimulation of muscles
    a. occurs at a neuromuscular junction.
    b. results in an action potential that travels down the T system.
    c. causes calcium to be released from expanded regions of the sarcoplasmic reticulum.
    d. All of these are correct.

Understanding the Terms

actin 384
appendicular skeleton 373
arthritis 390
articular cartilage 370
atrophy 388
axial skeleton 373
ball-and-socket joint 380
bursa 380
bursitis 380
compact bone 370
creatine phosphate 386
digits 378
fontanel 374
foramen magnum 374
growth plate 372
hinge joint 380
hypertrophy 388
insertion 380
intervertebral disk 376
joint 372
ligament 372
meniscus 379
motor unit 387
muscle tone 388
muscle twitch 387
myofibril 382
myogram 387
myosin 384
neuromuscular junction 385
origin 380

osteoarthritis (OA) 390
osteoblast 372
osteoclast 372
osteocyte 370
osteon 370
osteoporosis 390
oxygen debt 386
pectoral girdle 377
pelvic girdle 378
periosteum 370
pivot joint 380
red bone marrow 370
rheumatoid arthritis (RA) 390
sarcolemma 382
sarcomere 383
sarcoplasmic reticulum 382
scoliosis 375
sinus 374
skull 374
sliding filament theory 384
spongy bone 370
suture 379
synovial joint 379
synovial membrane 379
tendon 372
tetanus 387
T (transverse) tubule 382
vertebral column 375

Match the terms to these definitions:

a._____ Region where an axon terminal approaches a muscle fiber and the stimulus for muscle contraction is relayed.
b._____ Fibrous connective tissue that joins bone to bone at a joint.
c._____ Cartilaginous wedges that separate the surfaces of bones in synovial joints.
d._____ Muscle protein making up the thick filaments that attach to and pull the thin filaments to shorten a sarcomere.
e._____ Blood-cell-forming tissue located in the spaces within spongy bone.

Thinking Critically

1. When cancer patients receive chemotherapy or radiation to destroy the cancer cells, it is sometimes necessary to replace the bone marrow with a transplant. However, the bone itself is very resistant to these treatments. What is the difference between bone and bone marrow?

2. In section 19.1 we mentioned that hormones such as calcitonin and parathyroid hormone are involved in bone metabolism. Since these hormones are produced by far-away organs (the thyroid and parathyroid glands, respectively), how can these endocrine organs "know" whether bones should be built up or broken down, and what specific cells in the bone respond to these hormones?

3. Determine whether each of the following joints is capable of these movements: flexion, extension, abduction, adduction, rotation. You may "cheat" by using your own body.
   a. elbow   d. phalanges
   b. neck   e. hip
   c. knee

4. Similar to the activity in our muscle cells, certain bacteria can utilize sugars by aerobic cellular respiration (which requires oxygen) or by fermentation (which does not). However, like muscle cells, these bacteria only use fermentation when oxygen is not available. Why is aerobic respiration preferable? What is the specific role of oxygen in this process? (You may want to refer back to Chapter 7.)

5. Most authorities emphasize the importance of calcium consumption for the prevention of osteoporosis. However, surveys have shown that osteoporosis is rare in some countries, such as China, where calcium intake is lower than in the United States. These studies often point to the higher level of animal protein consumed in the U.S. as having a detrimental effect on bone mass. If this were true, what physiological mechanism might explain it? Can you think of any other differences between the lifestyles in China and the U.S. that also might explain the difference?

ARIS, the *Inquiry into Life* Website

ARIS, the website for *Inquiry into Life,* provides a wealth of information organized and integrated by chapter. You will find practice quizzes, interactive activities, labeling exercises, flashcards, and much more that will complement your learning and understanding of general biology.

www.aris.mhhe.com

Hank is only in second grade, *but if you were around him for very long you would probably notice that is he has some extra responsibilities compared to most other kids in his class. Several times a day, Hank gets a small black kit out of his backpack, loads a plastic-covered needle onto an instrument, places it against his finger, and pushes a button. Once a drop of blood appears, he neatly and efficiently adds it to a strip, which he then inserts into a reader. If the number is too high, Hank knows it is time to go to the school nurse for a dose of insulin. If it is too low, he needs to eat a snack to restore his blood glucose level. Because Hank has juvenile diabetes, he doesn't remember a time when he didn't have to perform these extra steps throughout his day.*

Other advances in the care of diabetes, aside from those that allow a second-grader to monitor his blood sugar level, include the production of insulin. Prior to the development of recombinant DNA technology, which now allows human insulin to be produced in large quantities, insulin was derived from the pancreases of pigs or cows. This required laborious purification, and because the animal insulins were not identical to the human form, sometimes immunological reactions occurred. Another recent advance is the insulin pump, a device a little bigger than a cell phone, which can deliver precise amounts of insulin via a small plastic catheter, more accurately mimicking the pancreas's natural release of the correct amount of insulin needed. Currently about 150,000 people in the United States are using an insulin pump. In the future, it may be possible to implant a device into diabetes patients that will not only monitor the blood sugar level, but will also provide the appropriate doses of insulin. Until then, Hank and many others like him will keep checking and adjusting their blood glucose levels several times each day.

Endocrine System

Chapter Concepts

20.1 Endocrine Glands and Hormones
- What are hormones, how are they distributed in the body, and how is their effect regulated? 396–398
- How does the mechanism of action of most peptide hormones differ from that of steroid hormones? 398–399
- What is meant by a "first messenger"? A "second messenger"? 398

20.2 Hypothalamus and Pituitary Gland
- What is the relationship between the hypothalamus and the posterior pituitary? The anterior pituitary? 399–400
- What hormones are produced by the posterior and anterior pituitary, respectively? What is the function of each hormone? 400–401

20.3 Thyroid and Parathyroid Glands
- Where are the thyroid and parathyroid glands located? 400
- What hormones are produced by the thyroid gland? What is the function of each? 402
- What hormone is produced by the parathyroid gland? What is the function of this hormone? 402

20.4 Adrenal Glands
- Where are the adrenal glands located? 403
- What hormones are produced by the adrenal glands? What is the function of each? 403–404

20.5 Pancreas
- What does it mean to say that the pancreas has both exocrine and endocrine functions? 405
- What hormones are produced by the pancreas? How does each affect the blood glucose level? 405

20.6 Other Endocrine Glands
- What other body organs and tissues function as endocrine glands? 405–408
- What is a growth factor? A local hormone? 408

20.7 Pheromones
- What is a pheromone? 408

20.8 Disorders of the Endocrine System
- What are some common disorders involving abnormal function of the pituitary gland? The thyroid gland? The adrenal glands? The pancreas? 408–412

20.1 Endocrine Glands and Hormones

The endocrine system consists of glands and tissues that secrete hormones. **Hormones** are chemicals that affect the behavior of other glands or tissues, which can be located far away from the sites of hormone production. Hormones influence the metabolism of cells, the growth and development of body parts, and homeostasis. Notice in Table 20.1 that several hormones directly affect the blood glucose, calcium, and sodium levels. Other hormones are involved in the function of various organs, including the reproductive organs. Hormones that promote cell division and mitosis are also sometimes called growth factors.

Not all hormones act between body parts. Local hormones are not carried in the bloodstream; instead, they affect neighbing cells. As we shall see, prostaglandins, which medi-ate pain and inflammation, are good examples of local hormones.

Endocrine glands have no ducts; instead, they secrete their hormones into tissue fluid. From there, hormones diffuse into the bloodstream for distribution throughout the body. Figure 20.1 depicts the locations of the major endocrine glands in the body, and Table 20.1 lists the hormones they release. In contrast, exocrine glands secrete their products through ducts. For example, the salivary glands send saliva into the mouth by way of the salivary ducts.

Hormones and Homeostasis

Homeostasis requires close cooperation between the endocrine and nervous systems. Like the nervous system, the endocrine system is intimately involved in homeostasis.

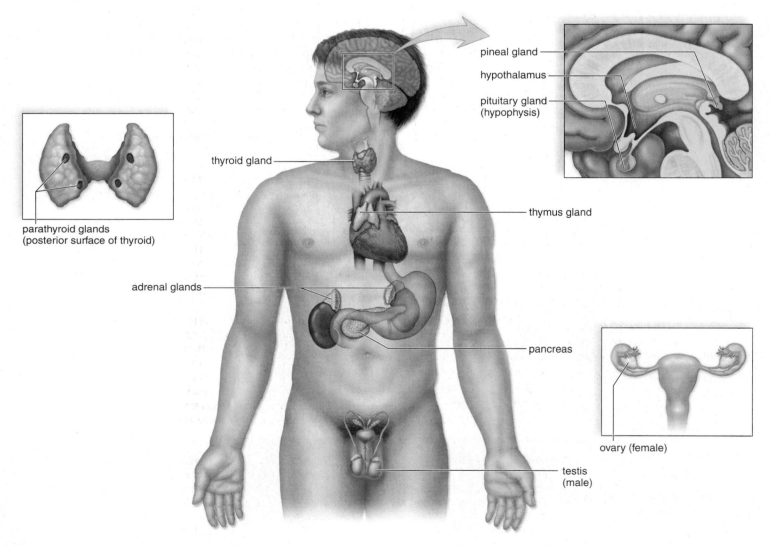

Figure 20.1 The endocrine system.
Anatomical location of major endocrine glands in the body. The hypothalamus and pituitary gland are in the brain, the thyroid and parathyroids are in the neck, and the adrenal glands and pancreas are in the abdominal cavity. The gonads include the ovaries in females, located in the pelvic cavity, and the testes in males, located outside this cavity in the scrotum. Also shown are the pineal gland, located in the brain, and the thymus gland, which lies ventral to the thoracic cavity.

TABLE 20.1	Principal Endocrine Glands and Hormones			
Endocrine Gland	**Hormone Released**	**Chemical Class**	**Target Tissues/Organs**	**Chief Function(s) of Hormone**
Hypothalamus	Hypothalamic-releasing and -inhibiting hormones	Peptide	Anterior pituitary	Regulate anterior pituitary hormones
Pituitary gland				
Posterior pituitary	Antidiuretic (ADH)	Peptide	Kidneys	Stimulates water reabsorption by kidneys
	Oxytocin	Peptide	Uterus, mammary glands	Stimulates uterine muscle contraction, release of milk by mammary glands
Anterior pituitary	Thyroid-stimulating (TSH)	Glycoprotein	Thyroid	Stimulates thyroid
	Adrenocorticotropic (ACTH)	Peptide	Adrenal cortex	Stimulates adrenal cortex
	Gonadotropic (FHS, LH)	Glycoprotein	Gonads	Egg and sperm production; sex hormone production
	Prolactin (PRL)	Protein	Mammary glands	Milk production
	Growth (GH)	Protein	Soft tissues, bones	Cell division, protein synthesis, and bone growth
	Melanocyte-stimulating (MSH)	Peptide	Melanocytes in skin	Unknown function in humans; regulates skin color in lower vertebrates
Thyroid	Thyroxine (T_4) and triiodothyronine (T_3)	Iodinated amino acid	All tissues	Increases metabolic rate; regulates growth and development
	Calcitonin	Peptide	Bones, kidneys, intestine	Lowers blood calcium level
Parathyroids	Parathyroid (PTH)	Peptide	Bones, kidneys, intestine	Raises blood calcium level
Adrenal gland				
Adrenal cortex	Glucocorticoids (cortisol)	Steroid	All tissues	Raise blood glucose level; stimulate breakdown of protein
	Mineralocorticoids (aldosterone)	Steroid	Kidneys	Reabsorb sodium and excrete potassium
	Sex hormones	Steroid	Gonads, skin, muscles, bones	Stimulate reproductive organs and bring about sex characteristics
Adrenal medulla	Epinephrine and norepinephrine	Modified amino acid	Cardiac and other muscles	Released in emergency situations; raise blood glucose level
Pancreas	Insulin	Protein	Liver, muscles, adipose tissue	Lowers blood glucose level; promotes formation of glycogen
	Glucagon	Protein	Liver, muscles, adipose tissue	Raises blood glucose level
Gonads				
Testes	Androgens (testosterone)	Steroid	Gonads, skin, muscles, bones	Stimulate male sex characteristics
Ovaries	Estrogens and progesterone	Steroid	Gonads, skin, muscles, bones	Stimulate female sex characteristics
Thymus	Thymosins	Peptide	T lymphocytes	Stimulate production and maturation of T lymphocytes
Pineal gland	Melatonin	Modified amino acid	Brain	Controls circadian and circannual rhythms; possibly involved in maturation of sexual organs

However, while the nervous system is organized to respond rapidly to stimuli by transmitting nerve impulses, the endocrine system typically secretes its hormones into the blood, resulting in a slower but more prolonged response. The blood concentration of a substance often prompts an endocrine gland to secrete its hormone. For example, the parathyroid glands secrete a hormone when the blood calcium level falls below normal. This hormone causes osteo-

clasts to slowly release calcium from bone. It takes time for the cells to respond, but the effect is longer lasting.

The production of hormones is usually controlled in two ways: by negative feedback, and by the action of other hormones. When controlled by **negative feedback,** an endocrine gland can be sensitive to either the condition it is regulating or the blood level of the hormone it is producing. For example, when the blood glucose level rises, the pancreas produces insulin. Insulin causes the liver to store glucose, and glucose is removed from the blood for liver storage. The stimulus for the production of insulin is, thereby, inhibited, and the pancreas stops producing insulin. Or, the anterior pituitary produces a thyroid-stimulating hormone that causes the thyroid to produce two hormones. When these hormones rise, the anterior pituitary stops producing thyroid-stimulating hormone.

The effect of a hormone also can be controlled by the release of an antagonistic hormone. The effect of insulin, for example, is offset by the production of glucagon by the pancreas. Insulin lowers the blood sugar level, while glucagon raises it. Similarly, a hormone produced by the thyroid gland lowers the blood calcium level, but another from the parathyroid gland increases blood calcium.

> The production of hormones is often regulated by negative feedback, in which the level of the hormone itself, or of some chemical induced by the hormone, inhibits the further secretion of the hormone. Both the production and the effects of hormones can also be influenced by the presence of other hormones.

The Action of Hormones

Hormones have a wide range of effects on cells. Some of these effects induce a target cell to increase its uptake of particular substances (e.g., glucose) or ions (e.g., calcium). Other hormones bring about an alteration of the target cell's structure in some way.

Hormones fall into two basic chemical classes: (1) **peptide hormones** are either peptides, proteins, glycoproteins, or modified amino acids; (2) **steroid hormones** always have the same complex of four carbon rings, but each specific hormone has different side chains. Because their effect is amplified through cellular mechanisms, both classes of hormones can function at extremely low concentrations. The chemical composition of a hormone also determines how it must be administered as a medication. Protein hormones, such as insulin, must be administered by injection. If these hormones were taken orally, they would be digested in the stomach and small intestine. Steroid hormones, such as those in birth control pills, can be taken orally.

Hormones are a type of chemical signal. **Chemical signals** are a means of communication between cells, between body parts, or even between individuals. They typically affect the metabolism of target cells that have receptors to receive them (Fig. 20.2). In the case of peptide hormones, these receptors are usually found on the cell surface (Fig. 20.3), while receptors for steroid hormones are usually in the nucleus but sometimes in the cytoplasm (Fig. 20.4). Epinephrine is a typical peptide hormone, which binds to a receptor on muscle cells,

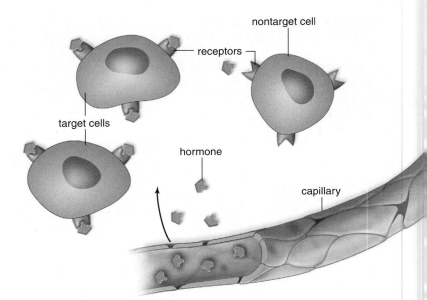

Figure 20.2 Target cell concept.
Most hormones are distributed by the bloodstream to target cells. Target cells have receptors for the hormone, and the hormone combines with the receptor as a key fits a lock.

as shown in Figure 20.3. This leads to a conversion of ATP to cyclic AMP (adenosine monophosphate), or cAMP. In this chemical signaling process, the peptide hormone is called the **first messenger,** and cAMP is called the **second messenger.**

The second messenger typically sets in motion an enzymatic pathway sometimes termed an enzyme cascade, because each enzyme, in turn, activates another. Because enzymes work over and over, every step in an enzyme cascade leads to more reactions—the binding of even a single peptide hormone molecule can result in a thousandfold response. In muscle cells, the activation of these cellular signaling pathways by epinephrine leads to the breakdown of glycogen to glucose.

To better understand the terms "first messenger" and "second messenger," imagine that the adrenal medulla, which produces epinephrine, is the home office that sends out a courier (i.e., the first messenger, epinephrine) to a factory (the cell). The courier doesn't have clearance to enter the factory, so when he arrives there, he tells a supervisor through a screen door that the home office wants the factory to produce a particular product. The supervisor (i.e., cAMP, the second messenger) walks over and flips a switch (the enzymatic pathway), and a product is made.

Steroid hormones act in a somewhat different way. After diffusing through the plasma membrane, these hormones typically bind to receptors in the cytoplasm or nucleus (see Fig. 20.4). Either way, the hormone-receptor complex usually binds to DNA and activates transcription of certain genes. Translation of messenger RNA (mRNA) transcripts results in enzymes and other proteins that can carry out a response to the hormonal signal. Since it takes longer to synthesize new proteins than to activate enzymes already present in cells,

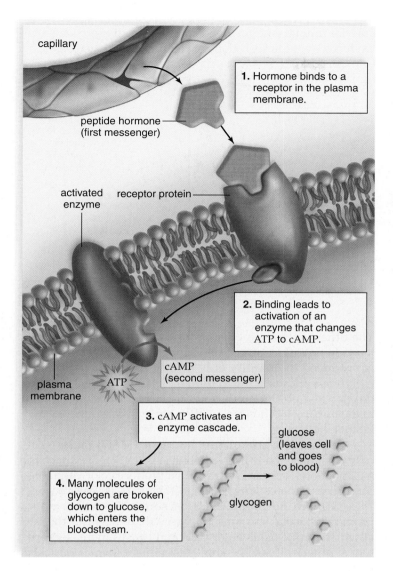

Figure 20.3 Peptide hormone action.

The peptide hormone (first messenger) binds to a receptor in the plasma membrane. Thereafter, a second messenger (cyclic AMP in this case) forms and activates an enzyme cascade.

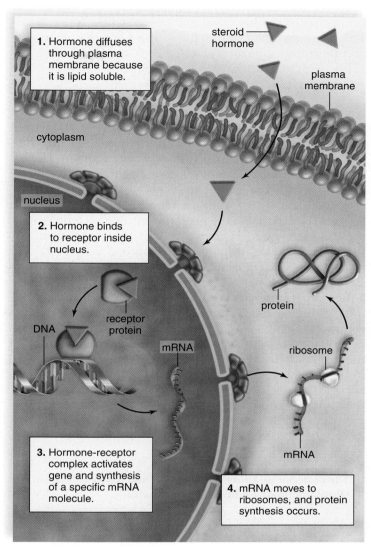

Figure 20.4 Steroid hormone action.

A steroid hormone passes directly through the target cell's plasma membrane before binding to a receptor in the nucleus or cytoplasm. The hormone-receptor complex binds to DNA, and gene expression follows.

steroid hormones generally act more slowly than peptide hormones. Their actions usually last longer, however.

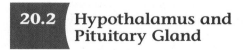

Hormones bind to receptor proteins on or in target cells. Peptide hormones (e.g., epinephrine) usually bind to a receptor on the plasma membrane, while steroid hormones diffuse through the plasma membrane and bind to a receptor inside the cell.

20.2 Hypothalamus and Pituitary Gland

The **hypothalamus** helps regulate the internal environment in two ways. Through the autonomic system, it modulates heartbeat, blood pressure, hunger and appetite, body temperature, and water balance. It also controls the glandular secretions of the **pituitary gland** (hypophysis). The pituitary, a small gland about 1 cm in diameter, is connected to the hypothalamus by a stalklike structure. The pituitary has two portions: the posterior pituitary and the anterior pituitary.

Posterior Pituitary

Neurons in the hypothalamus called neurosecretory cells produce the hormones **antidiuretic hormone (ADH)** and oxytocin (Fig. 20.5, *left*). These hormones pass through axons into the **posterior pituitary** where they are stored in axon terminals. Certain neurons in the hypothalamus are sensitive to the water-salt balance of the blood. When these cells determine that the blood is too concentrated, ADH is released from the posterior pituitary. Upon reaching the kidneys, ADH causes water to be reabsorbed. As the blood becomes diluted by reabsorbed water, ADH is no longer released.

Oxytocin, the other hormone made in the hypothalamus, causes uterine contraction during childbirth and milk letdown when a baby is nursing. The more the uterus contracts during labor, the more nerve impulses reach the hypothalamus, causing oxytocin to be released. Similarly, the more a baby suckles, the more oxytocin is released. In both instances, the release of oxytocin from the posterior pituitary is controlled by **positive feedback**—that is, the stimulus continues to bring about an effect that ever increases in intensity.

> The posterior pituitary releases ADH and oxytocin, both of which are produced by the hypothalamus.

Anterior Pituitary

A portal system, consisting of two capillary networks connected by a vein, lies between the hypothalamus and the anterior pituitary (Fig. 20.5, *right*). The hypothalamus controls the anterior pituitary by producing **hypothalamic-releasing hormones** and, in some instances, **hypothalamic-inhibiting hormones.** For example, one hypothalamic-releasing hormone stimulates the anterior pituitary to secrete a thyroid-stimulating hormone, and a particular hypothalamic-inhibitory hormone prevents the anterior pituitary from secreting prolactin.

Three of the six hormones produced by the **anterior pituitary** have an effect on other glands: **thyroid-stimulating hormone (TSH)** stimulates the thyroid to produce the thyroid hormones; **adrenocorticotropic hormone (ACTH)** stimulates the adrenal cortex to produce cortisol; and **gonadotropic hormones** (FSH and LH) stimulate the gonads—the testes in males and the ovaries in females—to produce gametes and sex hormones. In each instance, the blood level of the last hormone in the hypothalamus-

anterior pituitary-target gland control system exerts negative feedback over the secretions of the first two structures:

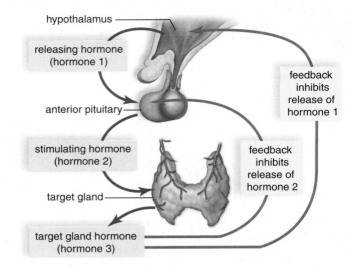

The other three hormones produced by the anterior pituitary do not affect other endocrine glands. **Prolactin (PRL)** is produced in quantity only during pregnancy and after childbirth. It causes the mammary glands in the breasts to develop and produce milk and suppress ovulation for a time in the nursing mother. It also plays a role in carbohydrate and fat metabolism.

Melanocyte-stimulating hormone (MSH) causes skin-color changes in many fishes, amphibians, and reptiles that have melanophores, special skin cells that produce color variations. The concentration of this hormone in humans is very low.

Growth hormone (GH), or somatotropic hormone, promotes skeletal and muscular growth. It increases the rate at which amino acids enter cells and causes increased protein synthesis. It also promotes fat metabolism as opposed to glucose metabolism.

The anterior pituitary normally produces higher levels of GH during childhood and adolescence, when most body growth is occurring. The amount of growth hormone produced during childhood is clearly one factor that influences height, although other factors, such as the genetic predisposition to attain a certain height, are also important (see Section 20.8).

> The hypothalamus controls the anterior pituitary by secreting releasing hormones and inhibitory hormones into a portal system. The anterior pituitary produces TSH, ACTH, gonadotropic hormones, prolactin, MSH, and GH.

20.3 Thyroid and Parathyroid Glands

The **thyroid gland** is a large gland located in the neck, where it is attached to the trachea just below the larynx (see Fig. 20.1). The parathyroid glands are embedded in the posterior surface of the thyroid gland.

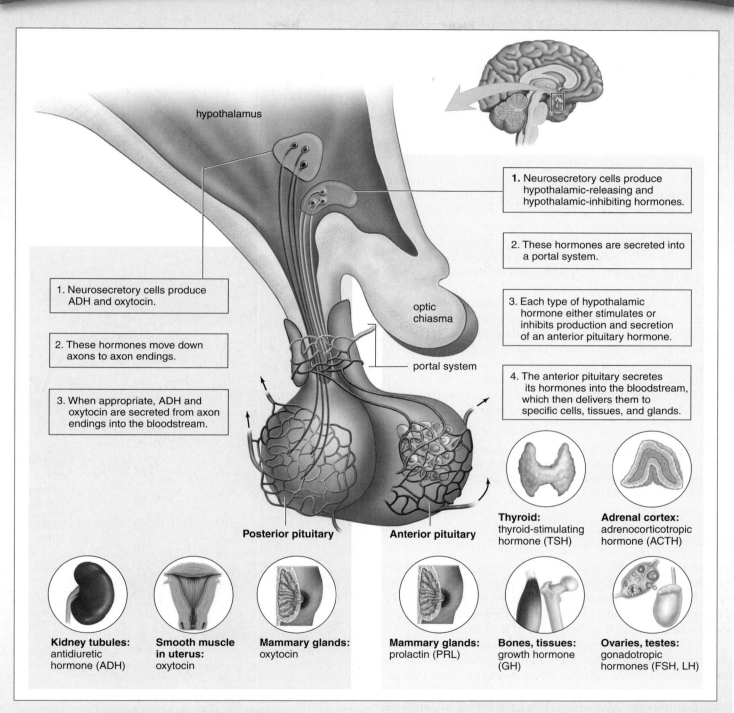

1. Neurosecretory cells produce hypothalamic-releasing and hypothalamic-inhibiting hormones.

2. These hormones are secreted into a portal system.

3. Each type of hypothalamic hormone either stimulates or inhibits production and secretion of an anterior pituitary hormone.

4. The anterior pituitary secretes its hormones into the bloodstream, which then delivers them to specific cells, tissues, and glands.

1. Neurosecretory cells produce ADH and oxytocin.

2. These hormones move down axons to axon endings.

3. When appropriate, ADH and oxytocin are secreted from axon endings into the bloodstream.

hypothalamus

optic chiasma

portal system

Posterior pituitary

Anterior pituitary

Thyroid: thyroid-stimulating hormone (TSH)

Adrenal cortex: adrenocorticotropic hormone (ACTH)

Kidney tubules: antidiuretic hormone (ADH)

Smooth muscle in uterus: oxytocin

Mammary glands: oxytocin

Mammary glands: prolactin (PRL)

Bones, tissues: growth hormone (GH)

Ovaries, testes: gonadotropic hormones (FSH, LH)

Figure 20.5 Hypothalamus and the pituitary.

Left: The hypothalamus produces two hormones, ADH and oxytocin, which are stored and secreted by the posterior pituitary. *Right:* The hypothalamus controls the secretions of the anterior pituitary, and the anterior pituitary controls the secretions of the thyroid gland, adrenal cortex, and gonads, which are also endocrine glands. These activities of the hypothalamus illustrate that the nervous system exerts control over the endocrine system and is intimately involved in maintaining the constancy of the internal environment.

Thyroid Gland

The thyroid gland contains a large number of follicles, each a small spherical structure made of thyroid cells that produce triiodothyronine (T_3), which contains three iodine atoms, and **thyroxine (T_4),** which contains four iodine atoms. To produce T_3 and T_4, the thyroid gland actively acquires iodine from the bloodstream. The concentration of iodine in the thyroid gland is approximately 25 times that found in the blood. The primary source for iodine in our diets is iodized salt.

The thyroid hormones T_3 and T_4 increase the metabolic rate. They do not have a single target organ; instead, they stimulate most cells of the body to metabolize more glucose and utilize more energy. Although the thyroid releases about ten times more T_4 than T_3, T_3 has a much more potent activity. Fortunately, the liver is able to convert most of the T_4 produced by the thyroid gland to T_3. Interestingly, even though T_3 and T_4 are considered peptide hormones because they are synthesized from the amino acid tyrosine, their receptor is actually located inside cells, more like a steroid hormone receptor.

The thyroid gland also produces a hormone called **calcitonin,** which helps control blood calcium levels (Fig. 20.6). Calcium (Ca^{2+}) plays a significant role in many processes, including nerve conduction, muscle contraction, and blood clotting. The thyroid gland secretes calcitonin when the blood calcium level rises. The primary effect of calcitonin is to bring about the deposition of calcium in the bones. It does this by temporarily reducing the activity and number of osteoclasts, the cells that break down bone (see Chapter 19). When the blood calcium level lowers to normal, the thyroid stops releasing calcitonin.

> The thyroid gland produces the hormones triiodothyronine (T_3), thyroxine (T_4), and calcitonin. T_3 and T_4 are involved in regulating overall body metabolism, and calcitonin lowers the blood calcium level.

Parathyroid Glands

Many years ago, the four parathyroid glands were sometimes mistakenly removed during thyroid surgery because they are so small. **Parathyroid hormone (PTH),** the hormone produced by the **parathyroid glands,** causes the blood phosphate (HPO_4^{2-}) level to decrease and the blood calcium level to increase.

PTH promotes the activity of osteoclasts and the release of calcium from the bones. PTH also promotes the reabsorption of calcium by the kidneys, where it helps activate vitamin D. Vitamin D, in turn, stimulates the absorption of calcium from the intestine. These effects bring the blood calcium level back to the normal range. As soon as blood calcium is back to normal, the parathyroid glands no longer secrete PTH.

> Parathyroid hormone promotes the activity of osteoclasts, causing a release of calcium from the bones. It also increases reabsorption of calcium by the kidneys and helps activate vitamin D.

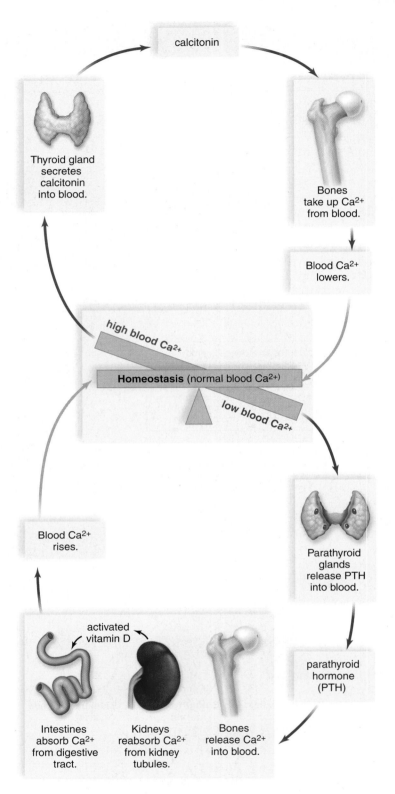

Figure 20.6 Regulation of blood calcium level.
Top: When the blood calcium (Ca^{2+}) level is high, the thyroid gland secretes calcitonin. Calcitonin promotes the uptake of Ca^{2+} by the bones, and therefore the blood Ca^{2+} level returns to normal. *Bottom:* When the blood Ca^{2+} level is low, the parathyroid glands release parathyroid hormone (PTH). PTH causes the bones to release Ca^{2+} and the kidneys to reabsorb Ca^{2+} and activate vitamin D; thereafter, the intestines absorb Ca^{2+}. Therefore, the blood Ca^{2+} level returns to normal.

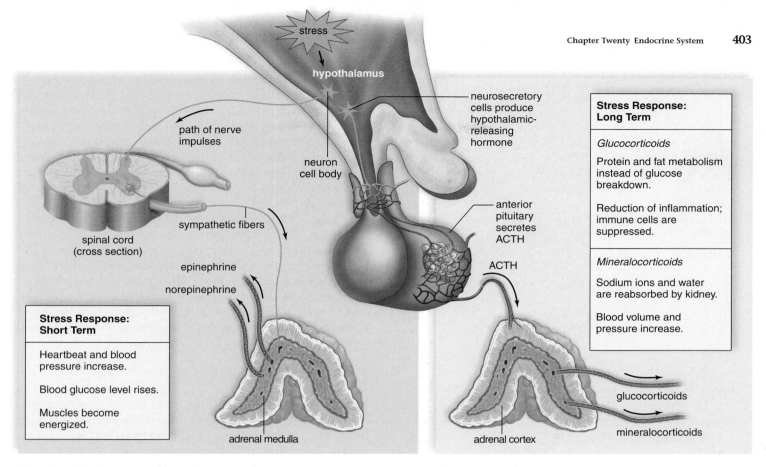

Figure 20.7 The stress response.
Both the adrenal cortex and the adrenal medulla are under the control of the hypothalamus when they help us respond to stress. *Left:* Nervous stimulation causes the adrenal medulla to provide a rapid, but short-term stress response. *Right:* The adrenal cortex provides a slower, but long-term stress response. ACTH causes the adrenal cortex to release glucocorticoids. Independently, the adrenal cortex releases mineralocorticoids.

20.4 Adrenal Glands

The **adrenal glands** sit atop the kidneys (see Fig. 20.1). Each adrenal gland consists of an inner portion called the **adrenal medulla** and an outer portion called the **adrenal cortex.** These portions, like the anterior pituitary and the posterior pituitary, are two functionally distinct endocrine glands. The adrenal medulla is under nervous control, and a portion of the adrenal cortex is under the control of adrenocorticotropic hormone (ACTH), an anterior pituitary hormone. Stress of all types, including emotional and physical trauma, prompts the hypothalamus to stimulate a portion of the adrenal glands (Fig. 20.7).

Adrenal Medulla

The hypothalamus initiates nerve impulses that travel by way of the brain stem, spinal cord, and sympathetic nerve fibers to the adrenal medulla, which then secretes its hormones.

Epinephrine (adrenaline) and **norepinephrine** (NE) produced by the adrenal medulla rapidly bring about all the body changes that occur when an individual reacts to an emergency situation in a fight-or-flight manner (Fig. 20.7, *left*) The effects of these hormones provide a short-term response to stress.

Adrenal Cortex

In contrast, the hormones produced by the adrenal cortex provide a longer-term response to stress (Fig. 20.7, *right*). The two major types of hormones produced by the adrenal cortex are the mineralocorticoids and the glucocorticoids. The **mineralocorticoids** regulate salt and water balance, leading to increases in blood volume and blood pressure. The **glucocorticoids,** under the control of ACTH, regulate carbohydrate, protein, and fat metabolism, leading to an increase in the blood glucose level. Glucocorticoids also suppress the body's inflammatory response. Cortisone, the medication often administered for inflammation of joints, is a glucocorticoid.

The adrenal cortex also secretes a small amount of male sex hormones and a small amount of female sex hormones in both sexes. That is, in the male, both male and female sex hormones are produced by the adrenal cortex, and in the female, both male and female sex hormones are also produced by the adrenal cortex.

Glucocorticoids

ACTH stimulates the portion of the adrenal cortex that secretes the glucocorticoids. **Cortisol** is a biologically significant glucocorticoid produced by the adrenal cortex.

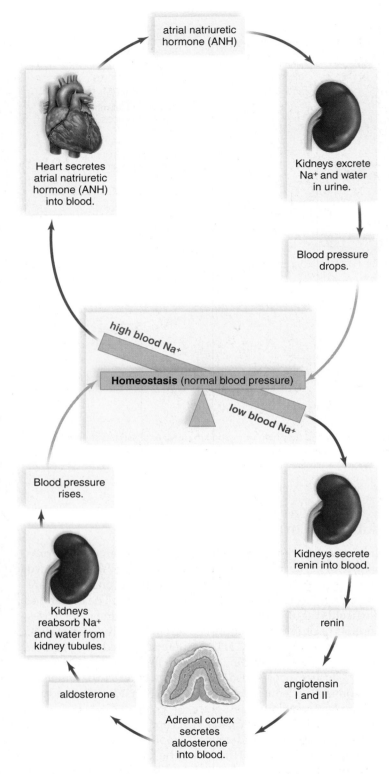

Figure 20.8 Regulation of blood pressure and volume.
Top: When the blood Na⁺ is high, high blood volume causes the heart to secrete atrial natriuretic hormone (ANH). ANH causes the kidneys to excrete Na⁺, and water follows. The blood volume and pressure then return to normal. *Bottom:* When the blood sodium (Na⁺) level is low, low blood pressure causes the kidneys to secrete renin. Renin leads to the secretion of aldosterone from the adrenal cortex. Aldosterone causes the kidneys to reabsorb Na⁺, and water follows, so that blood volume and pressure return to normal.

Cortisol raises the blood glucose level in at least two ways: (1) it promotes the breakdown of muscle proteins to amino acids, which are taken up by the liver from the bloodstream. The liver then converts these excess amino acids into glucose, which enters the blood. (2) Cortisol promotes the metabolism of fatty acids, rather than carbohydrates, and this spares glucose. The rise in blood glucose is beneficial to a person under stress because glucose is the preferred energy source for neurons.

Cortisol also counteracts the inflammatory response that leads to the pain and swelling of joints in arthritis and bursitis. Very high levels of glucocorticoids in the blood can suppress the body's defense system, including the inflammatory response that occurs at infection sites. Cortisone and other glucocorticoids can relieve swelling and pain from inflammation, but by suppressing pain and immunity, they can also make a person more susceptible to injury and infection.

Mineralocorticoids

Aldosterone is the most important of the mineralocorticoids. Aldosterone primarily targets the kidney, where it promotes renal absorption of sodium (Na⁺) and renal excretion of potassium (K⁺), and thereby helps regulate blood volume and blood pressure (Fig. 20.8).

The secretion of mineralocorticoids is not controlled by the anterior pituitary. When the blood sodium level and, therefore, blood pressure are low, the kidneys secrete **renin.** Renin is an enzyme that converts the plasma protein angiotensinogen to angiotensin I, which is changed to angiotensin II by a converting enzyme found in lung capillaries. Angiotensin II stimulates the adrenal cortex to release aldosterone. The effect of this system, called the renin-angiotensin-aldosterone system, is to raise blood pressure in two ways: angiotensin II constricts the arterioles, and aldosterone causes the kidneys to reabsorb sodium. When the blood sodium level rises, water is reabsorbed, in part because the hypothalamus secretes ADH (see Section 20.2). Then blood pressure increases to normal.

When the atria of the heart are stretched due to a great increase in blood volume, cardiac cells release a hormone called **atrial natriuretic hormone (ANH),** which inhibits the secretion of aldosterone from the adrenal cortex. The effect of this hormone is to cause the excretion of sodium—that is, *natriuresis.* When sodium is excreted, so is water, and therefore, blood pressure lowers to normal.

The adrenal medulla secretes epinephrine and norepinephrine; the adrenal cortex secretes glucocorticoids and mineralocorticoids. Glucocorticoids such as cortisol increase blood glucose and suppress inflammatory responses. Mineralocorticoids such as aldosterone cause the kidneys to reabsorb sodium, which raises the blood pressure.

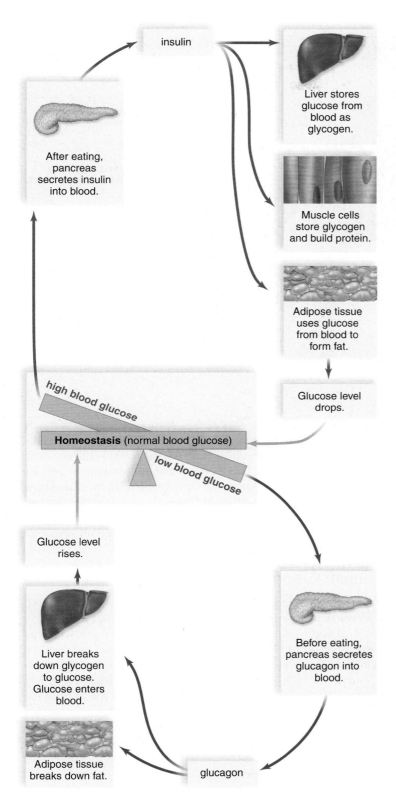

Figure 20.9 Regulation of blood glucose level.

Top: When the blood glucose level is high, the pancreas secretes insulin. Insulin promotes the storage of glucose as glycogen and the synthesis of proteins and fats (as opposed to their use as energy sources). Therefore, insulin lowers the blood glucose level to normal. *Bottom:* When the blood glucose level is low, the pancreas secretes glucagon. Glucagon acts opposite to insulin; therefore, glucagon raises the blood glucose level to normal.

20.5 Pancreas

The **pancreas** is a long organ that lies transversely in the abdomen between the kidneys and near the duodenum of the small intestine (see Fig. 20.1). It is composed of two types of tissue. Exocrine tissue produces and secretes digestive juices that go by way of ducts to the small intestine. The **pancreatic islets** (islets of Langerhans) are clusters of endocrine cells that produce and secrete the hormones **insulin** and **glucagon** directly into the blood, and thereby help regulate the blood glucose level (Fig. 20.9).

Insulin is secreted when the blood glucose level is high, which usually occurs just after eating. Insulin stimulates the uptake of glucose by cells, especially liver cells, muscle cells, and adipose tissue cells. In liver and muscle cells, glucose is then stored as glycogen. In muscle cells, the breakdown of glucose supplies energy for protein metabolism, and in fat cells, the breakdown of glucose supplies glycerol for the formation of fat. In these various ways, insulin lowers the blood glucose level.

Glucagon is secreted from the pancreas, usually before eating, when the blood glucose level is low. The major target tissues of glucagon are the liver and adipose tissue. Glucagon stimulates the liver to break down glycogen to glucose. Fat and protein are also used as energy sources by the liver, thus sparing glucose. Adipose tissue cells break down fat to glycerol and fatty acids. The liver takes these up and uses them as substrates for glucose formation. In these various ways, glucagon raises the blood glucose level.

The two antagonistic hormones insulin and glucagon, both produced by the pancreas, maintain the normal level of glucose in the blood.

20.6 Other Endocrine Glands

The **gonads** are the testes in males and the ovaries in females. The gonads are endocrine glands. Other lesser known glands and some tissues also produce hormones.

Testes and Ovaries

The **testes** are located in the scrotum, and the **ovaries** are located in the pelvic cavity. Under the influence of the gonadotropic hormones secreted by the anterior pituitary, the testes produce sperm and **androgens** (e.g., **testosterone**), which are the male sex hormones. Under the influence of the gonadotropic hormones, the ovaries produce eggs and also estrogen and progesterone, the female sex hormones. The hypothalamus and the pituitary gland control the hormonal secretions of these organs in a similar manner as previously described for the thyroid gland.

Greatly increased testosterone secretion at the time of puberty stimulates the growth of the penis and the testes. Testosterone also brings about and maintains the male second-

balding in men and women; hair on face and chest in women

deepening of voice in women

breast enlargement in men and breast reduction in women

liver dysfunction and cancer

kidney disease and retention of fluids, called "steroid bloat"

reduced testicular size, low sperm count, and impotency

'roid mania– delusions and hallucinations; depression upon withdrawal

severe acne

high blood cholesterol and atherosclerosis; high blood pressure and damage to heart

in women, increased size of ovaries; cessation of ovulation and menstruation

stunted growth in youngsters by prematurely halting fusion of the growth plates

Figure 20.10 **The side effects of anabolic steroid use.**

ary sex characteristics that develop during puberty, such as the growth of a beard, axillary (underarm) hair, and pubic hair. Testosterone also prompts the larynx and the vocal cords to enlarge, causing the voice to change. It is partially responsible for the muscular strength of males. This is the reason some athletes take supplemental amounts of illegal **anabolic steroids,** which are either testosterone or related chemicals. The side effects of taking anabolic steroids are listed in Figure 20.10. Testosterone also stimulates oil and sweat glands in the skin; therefore, it is largely responsible for acne and body odor. Another side effect of testosterone is baldness. Genes for baldness are probably inherited by both sexes, but baldness occurs more often in males because of the presence of testosterone.

The female sex hormones, **estrogen** and **progesterone,** have many effects on the body. In particular, estrogen secreted at the time of puberty stimulates the growth of the uterus and the vagina. Estrogen is necessary for egg maturation and is largely responsible for the secondary sex characteristics in females, including female body hair and fat distribution. In general, females have a more rounded appearance than males because of a greater accumulation of fat beneath the skin. Also, the pelvic girdle is wider in females than in males, resulting in a larger pelvic cavity. Both estrogen and progesterone are required for breast development and regulation of the uterine cycle, which includes monthly menstruation (discharge of blood and mucosal tissues from the uterus).

Thymus Gland

The **thymus gland** is a lobular gland that lies just beneath the sternum (see Fig. 20.1). This organ reaches its largest size and

is most active during childhood. With aging, the organ gets smaller and becomes fatty. Lymphocytes that originate in the bone marrow and then pass through the thymus differentiate into T lymphocytes. The lobules of the thymus are lined by epithelial cells that secrete hormones called thymosins. These hormones aid in the differentiation of lymphocytes packed inside the lobules. There is some evidence that these hormones would enhance T-lymphocyte function if injected into AIDS or cancer patients.

Pineal Gland

The **pineal gland,** which is located in the brain (see Fig. 20.1) produces the hormone **melatonin,** primarily at night. Melatonin is involved in our daily sleep-wake cycle; normally, we grow sleepy at night when melatonin levels increase and awaken once daylight returns and melatonin levels are low. Daily 24-hour cycles such as this are called **circadian rhythms,** and they are controlled by a biological clock located in the hypothalamus, as discussed in the Health Focus.

Based on animal research, it appears that melatonin also regulates sexual development. It has been noted that children whose pineal gland has been destroyed due to a brain tumor experience early puberty.

The gonads, thymus gland, and pineal gland are also endocrine organs. The gonads secrete the sex hormones, the thymus gland secretes thymosins, and the pineal gland secretes melatonin.

The hormone melatonin is now sold as a nutritional supplement. The popular press promotes its use in pill form for sleep, aging, cancer treatment, sexuality, and more. At best, melatonin may have some benefits in certain sleep disorders. But most physicians do not yet recommend it for that use because so little is known about its dosage requirements and possible side effects.

Melatonin is produced by the pineal gland in greatest quantity at night and smallest quantity during the day. Notice in **Figure 20A** that melatonin's production cycle accompanies our natural sleep-wake cycle. Rhythms with a period of about 24 hours are called circadian ("around the day") rhythms. All circadian rhythms seem to be controlled by an internal biological clock because they are free-running—that is, they have a regular cycle even in the absence of environmental cues. In scientific experiments, humans have lived in underground bunkers where they never see the light of day. In a few people, the sleep-wake cycle drifts badly, but in most, the daily activity schedule is just about 25 hours.

An individual's internal biological clock is reset each day by the environmental day-night cycle. Characteristically, biological clocks that control circadian rhythms are reset by environmental cues, or else they drift out of phase with the environmental day-night cycle.

Recent research suggests that our biological clock lies in a cluster of neurons within the hypothalamus called the suprachiasmatic nucleus (SCN). The SCN undergoes spontaneous cyclical changes in activity, and therefore, it can act as a pacemaker for circadian rhythms. Neural connections between the retina and the SCN indicate that reception of light by the eyes most likely resets the SCN and keeps our biological rhythms on a 24-hour cycle. Some people suffer from seasonal affective disorder, or SAD. As the days get darker and darker during the fall and winter, they become depressed, sometimes severely. They find it difficult to keep up because their biological clock has fallen behind without early morning light to reset it. If so, a half-hour dose of simulated daylight from a portable light box first thing in the morning makes them feel operational again. The SCN also controls the secretion of melatonin by the pineal gland, and, in turn, melatonin may quiet the operation of the neurons in the SCN.

Research is still going forward to see if melatonin will be effective for circadian rhythm disorders such as SAD, jet lag, sleep phase problems, recurrent insomnia in the totally blind, and some other less common disorders. So-called jet lag occurs when you travel across several time zones and your biological clock is out of phase with local time. Jet-lag symptoms gradually disappear as your biological clock adjusts to the environmental signals of the local time zone.

Many young people have a sleep phase problem because their circadian cycle lasts 25 to 26 hours. As they lengthen their day, they get out of sync with normal times for sleep and activity.

Clinical trials revealed that melatonin could shift circadian rhythms and reset our biological clock. Melatonin given in the afternoon shifts rhythms earlier, while melatonin given in the morning shifts rhythms later. For most people, the process was gradual: the average rate of change was about an hour a day. Before you try melatonin, however, you might want to consider that melatonin is known to affect reproductive behavior in other mammals. It's a matter of deciding whether any potential side effect from melatonin use is worth the possible benefits.

Discussion Questions

1. Have you ever experienced any degree of seasonal affective disorder? If so, do you think it was at least partly psychological, or did you feel it was a "purely" chemical disorder over which you had no control? Is it possible that both explanations could be true?

2. Are you more of a "morning person" or a "night person"? Can you think of any way that variations in melatonin secretion could account for how different people feel about, for example, getting up early in the morning?

3. As with melatonin, some Internet sites proclaim the benefits of using growth hormone for weight loss, increased energy, and the like. However, there are also documented cases of diabetes and other problems resulting from its use, and some of the products have been contaminated or have been found to contain little or no growth hormone. Assuming you have reached a normal height, would you ever take a supplement containing growth hormone to try to improve your overall health? Why or why not?

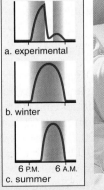

Figure 20A
Melatonin production.
Melatonin production is greatest at night when we are sleeping. Light suppresses melatonin production **(a),** so its duration is longer in the winter **(b)** than in the summer **(c).**

a. experimental

b. winter

6 P.M.   6 A.M.
c. summer

Hormones from Other Tissues

Some organs that are usually not considered endocrine glands do indeed secrete hormones. We have already mentioned that the heart produces atrial natriuretic hormone. And you will recall that the stomach and the small intestine produce peptide hormones that regulate digestive secretions. A number of other types of tissues also produce hormones.

Leptin

Leptin is a protein hormone produced by adipose tissue. Leptin acts on the hypothalamus, where it signals satiety—

that the individual has had enough to eat. Paradoxically, the blood of obese individuals may be rich in leptin. It is possible that the leptin they produce is ineffective because of a genetic mutation, or else their hypothalamic cells lack a suitable number of receptors for leptin.

Growth Factors

A number of different types of organs and cells produce peptide **growth factors,** which stimulate cell division and mitosis. Like hormones, they act on cell types with specific receptors to receive them. Some, such as lymphokines, are released into the blood; others diffuse to nearby cells. Growth factors of particular interest include:

Granulocyte and macrophage colony-stimulating factor (GM-CSF) is secreted by many different tissues. GM-CSF causes bone marrow stem cells to form either granulocyte or macrophage cells, depending on the concentration of these two cell types.

Platelet-derived growth factor is released from platelets and from many other cell types. It helps in wound healing and causes an increase in the number of fibroblasts, smooth muscle cells, and certain cells of the nervous system.

Epidermal growth factor and nerve growth factor stimulate the cells indicated by their names, as well as many others. These growth factors are also important in wound healing.

Tumor angiogenesis factor stimulates the formation of capillary networks and is released by tumor cells. One treatment for cancer is to prevent the activity of this growth factor.

Prostaglandins

Prostaglandins are potent chemical signals produced within cells from arachidonate, a fatty acid. Prostaglandins are not distributed in the blood; instead, they act locally, quite close to where they were produced. In the uterus, prostaglandins cause muscles to contract; therefore, they are implicated in the pain and discomfort of menstruation in some women. Also, prostaglandins mediate the effects of pyrogens, chemicals that are believed to reset the temperature regulatory center in the brain. Aspirin reduces body temperature and controls pain because of its effect on prostaglandins.

Certain prostaglandins reduce gastric secretions and have been used to treat ulcers; others lower blood pressure and have been used to treat hypertension; and still others inhibit platelet aggregation and have been used to prevent thrombosis. However, different prostaglandins have contrary effects, and it has been very difficult to successfully standardize their use. Therefore, prostaglandin therapy is still considered experimental.

Many tissues aside from the traditional endocrine glands produce hormones. Some of these enter the bloodstream, and some act only locally.

20.7 Pheromones

Chemical signals that act between individuals are called **pheromones.** The effects of pheromones are better documented in other animals than in humans. For example, female moths release a sex attractant that is received by male moth antennae even several miles away. Humans do produce a rich supply of airborne chemicals from a variety of areas, including the scalp, oral cavity, armpit (axilla), genital areas, and feet. Studies suggest that women prefer the axillary odors of men who are a different MHC type from themselves. (MHC molecules are plasma membrane proteins involved in immunity.) Choosing a mate of a different MHC type could conceivably improve the immune response of offspring. Several studies indicate that axillary secretions can affect the menstrual cycle. Women who live in the same household often have menstrual cycles in synchrony. Researchers have found that a woman's axillary extract can alter another woman's cycle by a few days.

Pheromones are chemical signals that act between individuals. Although pheromones are clearly important in some animal species, their relevance in humans is less certain.

20.8 Disorders of The Endocrine System

The endocrine glands play a major role in regulating the development and function of many body systems. Therefore, an increase or decrease in the production of most hormones can cause significant disease. Increased production of a hormone is often due to cancer affecting the endocrine gland, while decreased production can be due to various conditions that result in destruction of the gland.

Disorders of the Pituitary Gland

Because it controls the secretions of several other endocrine glands, the pituitary has been called the "master gland" of the endocrine system. As such, disorders of the pituitary gland can have dramatic effects on the body. As discussed in Section 20.2, the posterior pituitary produces antidiuretic hormone (ADH), which reduces the amount of urine formed by increasing water reabsorption by the collecting ducts of the kidney (see Section 16.3). If too little ADH is secreted, a condition known as **diabetes insipidus (DI)** results. Patients with DI are usually very thirsty, produce large volumes of urine, and can become severely dehydrated if the condition is untreated. Another type of DI occurs when the kidneys are unable to respond to ADH produced by the pituitary. Consumption of alcohol also inhibits ADH production, such that the symptoms of a hangover are, in part, caused by dehydration.

The anterior pituitary produces several important hormones, including growth hormone (GH) and adreno-

Figure 20.11   Effect of growth hormone on height.
Too much growth hormone can lead to giantism, while an insufficient amount results in limited stature and even pituitary dwarfism.

corticotropic hormone (ACTH). If too little GH is produced during childhood, the individual will have **pituitary dwarfism,** characterized by normal proportions but small stature. Conversely, if too much GH is secreted during childhood, a person may become a giant (Fig. 20.11). Giants usually have poor health, primarily because GH has a secondary effect on the blood glucose-level, promoting diabetes mellitus (discussed later in this section).

On occasion, GH is overproduced in the adult, and a condition known as **acromegaly** results. Since long bone growth is no longer possible in adults, only the feet, hands, and face (particularly the chin, nose, and eyebrow ridges) can respond, and these portions of the body become overly large (Fig. 20.12)

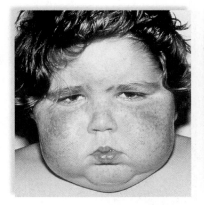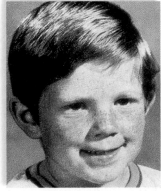

Figure 20.13   Cushing syndrome.
Cushing syndrome results from hypersecretion of hormones due to an adrenal cortex tumor. *Left:* Patient first diagnosed with Cushing syndrome. *Right:* Four months later, after therapy.

Some pituitary tumors produce large amounts of ACTH, which is a common cause of **Cushing syndrome.** The excess ACTH stimulates the adrenal cortex to overproduce cortisol. The increased amount of cortisol causes muscle protein to be metabolized and subcutaneous fat to be deposited in the midsection. The result is a swollen "moon" face and an obese trunk (Fig. 20.13). Depending on the cause and duration of Cushing syndrome, some people may experience more dramatic changes, including masculinization, high blood pressure, and weight gain. Cushing syndrome can also be caused by tumors of the adrenal gland or by taking high or prolonged doses of cortisone or similar drugs.

> Disorders of the pituitary gland can affect the production of many hormones. Decreased production of antidiuretic hormone results in diabetes insipidus. A deficiency of growth hormone in childhood causes pituitary dwarfism, while an excess may cause giantism. Acromegaly results from excess GH production in an adult. Secretion of too much ACTH can overstimulate the adrenal cortex, resulting in Cushing syndrome.

**Figure 20.12
Acromegaly.**
Acromegaly is caused by overproduction of GH in the adult. It is characterized by enlargement of the bones in the face, the hands, and the feet as a person ages.

Age 9   Age 16   Age 33   Age 52

a. Cretinism

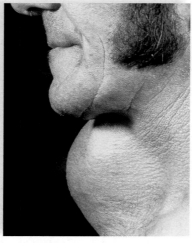
b. Simple goiter

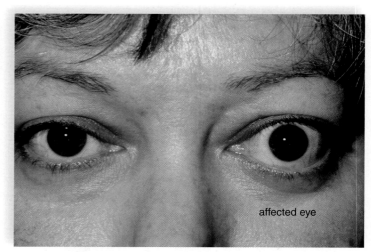
affected eye

c. Exophthalmia

Figure 20.14 Abnormalities of the thyroid.
a. Individuals with cretinism have hypothyroidism; from infancy or childhood on, they do not grow and develop as others do. **b.** An enlarged thyroid gland is often caused by a lack of iodine in the diet. Without iodine, the thyroid is unable to produce its hormones, and continued anterior pituitary stimulation causes the gland to enlarge and form a simple goiter. **c.** In hyperthyroidism, accumulation of fluid behind the eyes can cause the eyes to protrude (exophthalmia). Only this individual's left eye was affected.

Disorders of the Thyroid, Parathyroid, and Adrenal Glands

If the thyroid gland does not produce enough thyroid hormones for any reason, **hypothyroidism** occurs. Failure of the thyroid to develop properly in childhood results in a condition called **cretinism** (Fig. 20.14a). Individuals with this condition are short and stocky. Thyroid hormone therapy can initiate growth, but unless treatment is begun within the first two months of life, mental retardation results. Hypothyroidism can also occur in adults, most often when the immune system produces antibodies that destroy the gland, a condition known as Hashimoto thyroiditis. Untreated hypothyroidism may produce a group of clinical symptoms called **myxedema,** which is characterized by lethargy, weight gain, hair loss, constipation, slow heart rate, and thickened or puffy skin. The administration of adequate doses of thyroid hormones restores normal body functions and appearance.

If iodine is lacking in the diet, the thyroid gland is unable to produce sufficient amounts of T_3 and T_4. In response to constant stimulation by TSH released by the anterior pituitary, the thyroid gland enlarges, resulting in a **simple goiter** (Fig. 20.14b). The incidence of simple goiters was much higher in the United States prior to the addition of iodine to most salt.

Hyperthyroidism results from the oversecretion of thyroid hormones. In Graves disease, antibodies are produced that react with the TSH receptor on thyroid follicular cells, mimicking the effect of TSH and causing the production of too much T_3 and T_4. One typical sign of Graves disease is **exophthalmia,** or excessive protrusion of the eyes (Fig. 20.14c). The eyes protrude because of the edema in eye socket tissues and swelling of the muscles that move the eyes. The patient with Graves disease usually becomes hyperactive, nervous, and irritable, and may suffer from insomnia, diarrhea, and abnormal heart rhythms. Drugs that inhibit the synthesis or release of thyroid hormones, or their effects on body tissues, are used to manage Graves disease. Hyperthyroidism can also be caused by a thyroid tumor, which is usually detected as a lump during physical examination. Removal or destruction of a portion of the thyroid gland, either by surgery or by administration of radioactive iodine, is often curative.

When insufficient parathyroid hormone production leads to a dramatic drop in the blood calcium level, tetany results. In **tetany,** the body shakes from continuous muscle contraction, brought about by increased excitability of the nerves.

As mentioned previously, an increased production of cortisol by the adrenal glands is known as Cushing syndrome. In contrast, adrenal gland insufficiency is called **Addison disease.** Perhaps the most famous person to suffer from Addison disease was President John F. Kennedy. The most common cause of Addison disease in the United States is destruction of the adrenal cortex by the immune system. Clinical symptoms do not appear until about 90% of both adrenal cortexes have been destroyed. These symptoms are fairly nonspecific and may include weakness, weight loss, abdominal pain, mood disturbances, and hyperpigmentation of the skin (Fig. 20.15). Because the decreased production of mineralocorticoids can affect the balance of sodium and potassium, which in turn can have dramatic effects on the heart, Addison disease can be fatal unless properly treated by replacing the missing hormones.

Disorders of the thyroid and adrenal glands cause significant diseases. Untreated hypothyroidism causes cretinism in children and myxedema in adults. A simple goiter can result from insufficient dietary iodine intake. Graves disease is one type of hyperthyroidism, and adrenal gland insufficiency is also known as Addison disease.

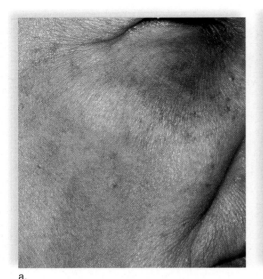
a.

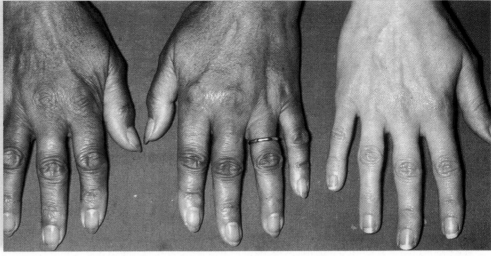
b.

Figure 20.15 Addison disease.

Addison disease is characterized by a peculiar bronzing of the skin, particularly noticeable in light-skinned individuals. Note the color of (**a**) the face and (**b**) the hands compared to the hand of an individual without the disease.

Diabetes Mellitus

Approximately 16 million Americans have **diabetes mellitus,** often referred to simply as diabetes, a condition that affects their ability to regulate their glucose metabolism. People with diabetes either do not produce enough insulin (type 1) or cannot properly use the insulin they produce (type 2). In either case, although blood glucose levels rise, and cellular famine exists in the midst of plenty. Some of the excess glucose in the blood is excreted into the urine, and water follows, causing the volume of urine to increase. Since the glucose in the blood cannot be used, the body turns to the metabolism of fat, which leads to the buildup of ketones in the blood. In turn, these ketones are metabolized to form various acids, which can build up in the blood (acidosis) and lead to coma and death.

The glucose tolerance test is often used to diagnose diabetes mellitus. After a person ingests a known amount of glucose, the blood glucose concentration is measured at intervals. In a diabetic, the blood glucose level usually rises greatly and remains elevated for hours (Fig. 20.16). In the meantime, glucose appears in the urine. In a nondiabetic, the blood glucose level rises somewhat, and then returns to normal after about two hours.

About 10% of diabetics in the U.S. have type 1 diabetes, in which the pancreas does not produce enough insulin. This condition usually begins in childhood or adolescence, and thus it is sometimes called juvenile-onset diabetes. However, type 1 diabetes can also occur later in life, following exposure to a virus, an autoimmune reaction, or an environmental agent that leads to destruction of the pancreatic islets.

Since individuals with type 1 diabetes are suffering from an insulin shortage, treatment simply consists of providing the needed insulin through daily injections. These injections control the diabetic symptoms but can still cause inconveniences, since either an overdose of insulin or missing a meal can result in hypoglycemia (low blood sugar). Symptoms of hypoglycemia include perspiration, pale skin, shallow breathing, and anxiety. Because the brain requires a constant supply of glucose, unconsciousness can result. The treatment is quite simple: immediate ingestion of a sugary snack or fruit juice can very quickly counteract hypoglycemia. Better control of glucose levels can sometimes be achieved with an insulin pump, a small device worn outside

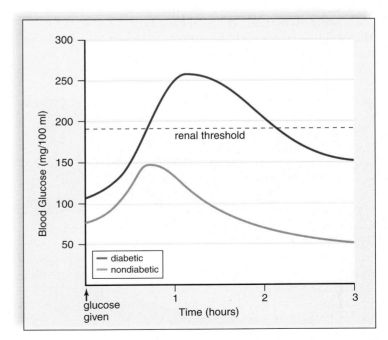

Figure 20.16 Glucose tolerance test.

Following the administration of 100 g of glucose, the blood glucose level rises dramatically in the diabetic and glucose appears in the urine. Also, the blood glucose level at 2 hours is equal to or more than 200 mg/100 ml.

Higher-order multiple births (triplets or more) in the United States increased 19% between 1980 and 1994. During these years, it became customary to use fertility drugs (gonadotropic hormones) to stimulate the ovaries to produce several eggs at once. Fertilization of multiple eggs leads to multiple births. The risks for premature delivery, low birth weight, and developmental abnormalities rise sharply for higher-order multiple births. And the physical and emotional burden placed on the parents is extraordinary. They face endless everyday chores and find it difficult to maintain normal social relationships, if only because they get insufficient sleep. Finances are strained in order to provide for the children's needs, including housing and child-care assistance. About one-third report that they received no help from relatives, friends, or neighbors in the first year after the birth. Trips to the hospital for accidental injury are more frequent because a parent with only two arms and two legs cannot keep so many children safe at one time.

Many clinicians are now urging that all possible steps be taken to ensure that the chance of higher-order multiple births is reduced. However, none of the ethical choices to bring this about are attractive. Fertility drugs could be outlawed, but if that were done, some couples might be denied the possibility of ever having a child. A higher-order multiple pregnancy could be terminated, but many people object to that practice. Another option might be selective reduction, in which one or more of the fetuses is killed by an injection of potassium chloride. However, selective reduction could very well result in psychological and social complications for the mother and surviving children. Finally, the parents could opt to utilize in vitro fertilization (in which the eggs are fertilized in the lab), with the intent that only one or two zygotes would be placed in the woman's womb. But then any leftover zygotes might never have an opportunity to continue development.

Form Your Own Opinion

1. If you or your "significant other" wanted children but were unable to conceive, would either of you take fertility drugs? Why or why not?
2. Suppose a woman who has taken fertility drugs is found to be carrying ten fetuses, and it is certain that most or all will not survive to term. Which would be the best option: termination of all fetuses, selective reduction, or intervening only to save the mother's life? What factors might influence your opinion?
3. Assuming that more zygotes will be created by in vitro fertilization than will ever be implanted, would it be better use their cells for research and possible cures for various diseases, to leave them frozen indefinitely, or to destroy them?

---

the body that is connected to a plastic catheter inserted under the skin (Fig. 20.17). It is also possible to transplant a working pancreas, or even fetal pancreatic islet cells, into patients with type 1 diabetes.

Most diabetics in the U.S. have type 2 diabetes. Often, the patient is overweight or obese—adipose tissue produces a substance that impairs insulin receptor function. Since type 2 diabetes usually doesn't occur before age 40, it was formerly called adult-onset diabetes. However, because of the increasing prevalence of obesity in children, type 2 diabetes is occurring at younger ages. Normally, the binding of insulin to its receptor on cell surfaces causes the number of glucose transporters to increase in the plasma membrane, but not in the type 2 diabetic. Treatment usually involves weight loss, which can sometimes control symptoms in the type 2 diabetic. However, since many type 2 diabetics also have low insulin levels, they may require injections of insulin, along with medications to increase the effectiveness of the insulin they produce.

Long-term complications of both types of diabetes are blindness, kidney disease, and cardiovascular disorders, including reduced circulation. The latter can lead to gangrene in the arms and legs. Diabetic women who become pregnant have an increased risk of diabetic coma, and the child of a diabetic is more likely to be stillborn or to die shortly after birth. These complications are uncommon, however, if the mother's blood glucose level is carefully regulated during pregnancy.

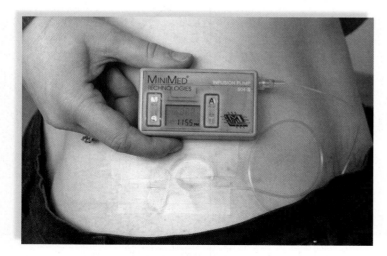

Figure 20.17 An insulin pump.
Insulin pumps administer insulin through an implanted catheter in small, preprogrammed doses throughout the day. Most insulin pumps can be worn under clothing.

Diabetes mellitus is a fairly common disease affecting the ability to regulate the uptake of glucose by cells. In type 1 diabetes, the pancreas produces insufficient insulin; in type 2, cells cannot respond properly to insulin.

Summarizing the Concepts

20.1 Endocrine Glands and Hormones

The major endocrine glands and hormones are listed in Table 20.1. Negative feedback and antagonistic hormonal actions control the secretion of hormones.

Hormones are chemical signals that act at a distance. Hormones are either peptides or steroids. Reception of a peptide hormone at the plasma membrane activates an enzyme cascade inside the cell. Steroid hormones combine with a receptor in the cell, and the complex attaches to and activates DNA. Protein synthesis follows.

20.2 Hypothalamus and Pituitary Gland

The hypothalamus produces antidiuretic hormone (ADH) and oxytocin, which are stored in the posterior pituitary until they are released.

The hypothalamus produces hypothalamic-releasing and, in some instances, hypothalamic-inhibiting hormones, which pass to the anterior pituitary by way of a portal system. The anterior pituitary produces at least six types of hormones, and some of these stimulate other hormonal glands to secrete hormones. If the latter, the hypothalamus, anterior pituitary, and target gland are involved in a negative feedback control system.

20.3 Thyroid and Parathyroid Glands

The thyroid gland requires iodine to produce thyroxine and triiodothyronine, which increase the metabolic rate. The thyroid gland also produces calcitonin, which helps lower the blood calcium level. The parathyroid glands secrete parathyroid hormone, which raises the blood calcium and decreases the blood phosphate levels.

20.4 Adrenal Glands

The adrenal glands respond to stress: immediately, the adrenal medulla secretes epinephrine and norepinephrine, which bring about responses we associate with emergency situations. On a long-term basis, the adrenal cortex produces the glucocorticoids (e.g., cortisol) and the mineralocorticoids (e.g., aldosterone). Cortisol stimulates the conversion of amino acids to glucose, raising the blood glucose level. Aldosterone causes the kidneys to reabsorb sodium ions (Na^+) and to excrete potassium ions (K^+).

20.5 Pancreas

The pancreas has both exocrine and endocrine functions. The pancreatic islets secrete insulin and glucagon. Insulin lowers the blood glucose level by stimulating the uptake of glucose by various types of cells. Glucagon raises the blood glucose level.

20.6 Other Endocrine Glands

The gonads produce the sex hormones; the thymus secretes thymosins, which stimulate T-lymphocyte production and maturation; and the pineal gland produces melatonin, which may be involved in circadian rhythms and the development of the reproductive organs.

Tissues also produce hormones. Adipose tissue produces leptin, which acts on the hypothalamus, and various tissues produce growth factors. Prostaglandins are produced and act locally.

20.7 Pheromones

Pheromones are chemical signals that act between individuals, sometimes at great distances. There is some evidence for the activity of pheromone-like chemicals in humans.

20.8 Disorders of the Endocrine System

Increases or decreases in the amounts of hormones synthesized by endocrine glands cause several common disorders. Disorders caused by pituitary gland dysfunction include diabetes insipidus, growth hormone disorders, and Cushing syndrome. Insufficient thyroid gland function results in cretinism in childhood and myxedema in adults, while Graves disease is a type of hyperthyroidism. Addison disease is a serious condition in which the adrenal cortexes have been destroyed, often by an autoimmune process.

Testing Yourself

Choose the best answer for each question.

1. Hormones are never
   a. steroids.
   b. amino acids.
   c. glycoproteins.
   d. fats (triglycerides).

2. _____ glands are ductless.
   a. Exocrine
   b. Endocrine
   c. Both a and b are correct.
   d. Neither a nor b is correct.

3. _____ is released through positive feedback and causes _____
   a. Insulin, stomach contractions
   b. Oxytocin, stomach contractions
   c. Oxytocin, uterine contractions
   d. None of these are correct.

4. Growth hormone is produced by the
   a. posterior adrenal gland.
   b. posterior pituitary gland.
   c. anterior pituitary gland.
   d. kidneys.
   e. None of these are correct.

5. The body's response to stress includes
   a. water reabsorption by the kidneys.
   b. blood pressure increase.
   c. increase in the blood glucose level.
   d. heart rate increase.
   e. All of these are correct.

6. Glucagon causes
   a. use of fat for energy.
   b. glycogen to be converted to glucose.
   c. use of amino acids to form fats.
   d. Both a and b are correct.
   e. None of these are correct.

7. Long-term complications of diabetes include
   a. blindness.
   b. kidney disease.
   c. circulatory disorders.
   d. All of these are correct.

8. Which hormones typically cross the plasma membrane?
   a. peptide hormones
   b. steroid hormones
   c. Both a and b are correct.
   d. Neither a nor b is correct.

Match the action in questions 9-13 to the correct hormone in the key.

Key:

  a. glucagon   d. insulin
  b. prostaglandins   e. leptin
  c. melatonin

9. Raises blood glucose levels

10. Conversion of glucose to glycogen

11. Hunger control

12. Controls circadian rhythms

13. Causes uterine contractions

14. Anabolic steroid use can cause

  a. liver damage.   d. reduced testicular size.
  b. severe acne.   e. All of these are correct.
  c. balding.

15. PTH causes the blood levels of _____ to increase and _____ to decrease.

  a. calcium, sodium   d. phosphate, calcium
  b. calcium, phosphate   e. None of these are correct.
  c. phosphate, sodium

16. Lack of aldosterone will cause a blood imbalance of

  a. sodium.   d. All of these are correct.
  b. potassium.   e. None of these are correct.
  c. water.

17. Name the endocrine glands in the following diagram.

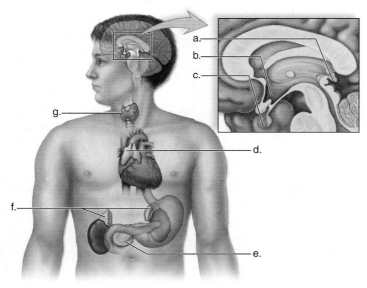

18. The anterior pituitary controls the secretion(s) of

  a. both the adrenal medulla and the adrenal cortex.
  b. both the thyroid gland and the adrenal cortex.
  c. both the ovaries and the testes.
  d. Both b and c are correct.

19. Diabetes mellitus is associated with

  a. too much insulin in the blood.
  b. a blood glucose level that is too high.
  c. blood that is too dilute.
  d. All of these are correct.

20. Complete the following diagram by filling in blanks a–e.

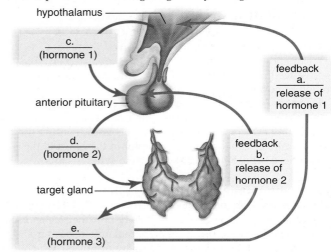

Match the hormones in questions 21–25 to the correct gland in the key.

Key:

  a. pancreas   d. thyroid gland
  b. anterior pituitary   e. adrenal medulla
  c. posterior pituitary   f. adrenal cortex

21. Cortisol

22. Growth hormone (GH)

23. Oxytocin storage

24. Insulin

25. Epinephrine

26. Name the hormones in the following diagram.

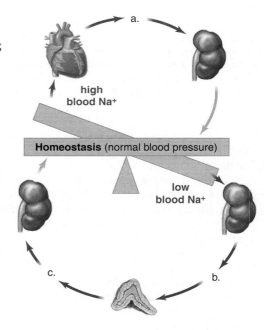

27. Which of these is not a pair of antagonistic hormones?

  a. insulin—glucagon
  b. calcitonin—parathyroid hormone
  c. cortisol—epinephrine
  d. aldosterone—atrial natriuretic hormone (ANH)
  e. thyroxine—growth hormone

28. Which hormone and condition are mismatched?
    a. growth hormone—acromegaly
    b. thyroxine—goiter
    c. parathyroid hormone—tetany
    d. cortisol—myxedema
    e. insulin—diabetes

29. Tropic hormones are hormones that affect other endocrine tissues. Which of the following would be considered a tropic hormone?
    a. calcitonin
    b. oxytocin
    c. glucagon
    d. melatonin
    e. follicle-stimulating hormone

Understanding the Terms

acromegaly 409
Addison disease 410
adrenal cortex 403
adrenal gland 403
adrenal medulla 403
adrenocorticotropic hormone (ACTH) 400
aldosterone 404
anabolic steroid 406
androgen 405
anterior pituitary 400
antidiuretic hormone (ADH) 400
atrial natriuretic hormone (ANH) 404
calcitonin 402
chemical signal 398
circadian rhythm 406
cortisol 403
cretinism 410
Cushing syndrome 409
diabetes insipidus (DI) 408
diabetes mellitus 411
endocrine gland 396
epinephrine 403
estrogen 406
exophthalmia 410
first messenger 398
glucagon 405
glucocorticoid 403
gonad 405
gonadotropic hormone 400
growth factor 408
growth hormone (GH) 400
hormone 396
hyperthyroidism 410
hypothalamic-inhibiting hormone 400
hypothalamic-releasing hormone 400
hypothalamus 399

hypothyroidism 410
insulin 405
leptin 407
melanocyte-stimulating hormone (MSH) 400
melatonin 406
mineralocorticoid 403
myxedema 410
negative feedback 397
norepinephrine (NE) 403
ovary 405
oxytocin 400
pancreas 405
pancreatic islets (of Langerhans) 405
parathyroid gland 402
parathyroid hormone (PTH) 402
peptide hormone 398
pheromone 408
pineal gland 406
pituitary dwarfism 409
pituitary gland 399
positive feedback 400
posterior pituitary 400
progesterone 406
prolactin (PRL) 400
prostaglandin (PG) 408
renin 404
second messenger 398
simple goiter 410
steroid hormone 398
testes 405
testosterone 405
tetany 410
thymus gland 406
thyroid gland 400
thyroid-stimulating hormone (TSH) 400
thyroxine (T_4) 402

Match the terms to these definitions:

a._____ Organ in the neck that secretes several important hormones, including thyroxine and calcitonin.

b._____ Condition characterized by a high blood glucose level and the appearance of glucose in the urine.

c._____ Hormone secreted by the anterior pituitary that stimulates the adrenal cortex to release glucocorticoid.

d._____ Hormone secreted by the pancreas that causes the liver to break down glycogen and raises the blood glucose level.

e._____ Hormone released by the posterior pituitary that causes contraction of the uterus and milk letdown.

Thinking Critically

1. Since some of their functions overlap, why is it necessary to have both a nervous system and an endocrine system?

2. Considering the mechanism involved in each of the following processes, explain which would have the fastest or slowest effect when your body is responding to a physical threat, and why:
    a. a steroid hormone, such as cortisol, raising blood sugar
    b. muscle contraction enabling you to run away
    c. a peptide hormone such as epinephrine increasing blood supply to the muscles

3. Certain endocrine disorders, such as Cushing syndrome, can be caused by excessive secretion of a hormone (in this case, ACTH) by the pituitary gland, or by a problem with the adrenal gland itself. If you were able to look at the adrenal glands of a Cushing patient, how could you tell the difference between a pituitary problem and a primary adrenal gland problem?

4. In diagnosing diabetes mellitus, the glucose tolerance test is considered more useful than direct measurement of blood insulin concentration. Why not just measure the amount of insulin in the blood, assuming a diabetic would have less insulin than a nondiabetic?

5. In animals, pheromones can affect many different behaviors. Since humans do produce a number of airborne chemicals, what potential human behaviors could be affected by these chemicals? How would these pheromones be received, and would we necessarily be conscious of their effects on us?

ARIS, the *Inquiry into Life* Website

ARIS, the website for *Inquiry into Life,* provides a wealth of information organized and integrated by chapter. You will find practice quizzes, interactive activities, labeling exercises, flashcards, and much more that will complement your learning and understanding of general biology.

www.aris.mhhe.com

Continuance of the Species

part V

21

22

23

24

25

26

James and Lisa already have three healthy boys. However, Lisa has always wanted a girl, and figures this is her last chance to have one. "I don't want to have to keep trying," she told her physician at a routine checkup as she and Tom began to think about having a fourth baby. She was surprised when her doctor told her that, in fact, it is now possible to greatly increase the odds of having a baby of a particular sex. Unlike previous gender selection methods, modern methods can virtually guarantee a male or female baby. For example, the website of the Genetics and IVF Institute, based in Fairfax, Virginia, describes the use of Microsort® technology, which can separate sperm carrying two X chromosomes from those carrying one X and one Y. For a $3,765 fee, this company will sort one semen sample and implant the sorted sperm into a woman's uterus. So far, the company's data suggest a 91% success rate when girls were desired and a 76% success rate for boys. Other more invasive (and expensive) methods offer even higher success rates.

Is it ethical to select the gender of your baby? Perhaps—but in certain countries, gender selection has had some troubling consequences. In India, for example, even though the practice is against the law, doctors have used amniocentesis and, more recently, ultrasound technology to detect female babies, which are often then aborted. The blame for this practice falls mostly on traditional cultural attitudes toward females. In one sense, gender selection might be advantageous by preventing baby girls from being aborted, but will this also perpetuate cultural practices that devalue females?

chapter **21**

Reproductive System

Chapter Concepts

21.1 Male Reproductive System

The male reproductive system includes the organs depicted in Figure 21.1 and listed in Table 21.1. The male gonads are paired **testes** (sing., testis), which are suspended within the sac-like **scrotum.**

Genital Tract

Sperm produced by the testes mature within the epididymides (sing., **epididymis**), which are tightly coiled ducts lying outside the testes. Maturation seems to be required in order for sperm to swim to the egg. When sperm leave an epididymis, they enter a **vas deferens** (pl., vasa deferentia), where they may also be stored for a time. Each vas deferens passes into the abdominal cavity, where it curves around the urinary bladder and empties into an ejaculatory duct. The ejaculatory ducts connect to the **urethra.**

At the time of ejaculation, sperm leave the penis in a fluid called **semen** (seminal fluid). The seminal vesicles, the prostate gland, and the bulbourethral glands (Cowper glands) add secretions to semen. The pair of **seminal vesicles** lie at the base of the bladder, and each has a duct that joins with a vas deferens. The **prostate gland** is a single, donut-shaped gland that surrounds the upper portion of the urethra just below the urinary bladder. **Bulbourethral**

TABLE 21.1	Male Reproductive Organs
Organ	**Function**
Testes	Produce sperm and sex hormones
Epididymides	Ducts where sperm mature and are stored
Vasa deferentia	Conduct and also store sperm
Seminal vesicles	Contribute nutrients and fluid to semen
Prostate gland	Contributes basic fluid to semen
Urethra	Conducts sperm
Bulbourethral glands	Contribute viscous fluid to semen
Penis	Organ of sexual intercourse

glands are pea-sized organs that lie underneath the prostate on either side of the urethra.

Each component of semen seems to have a particular function. Sperm are more viable in a basic solution, and semen, which is milky in appearance, has a slightly basic pH (about 7.5). Swimming sperm require energy, and semen contains the sugar fructose, which presumably serves as an energy source. Semen also contains prostaglandins, chemicals that cause the uterus to contract. Some investigators believe that uterine contractions help propel the sperm toward the egg.

The **penis** (Fig. 21.2) is the male organ of sexual intercourse. The penis has a long shaft and an enlarged tip called

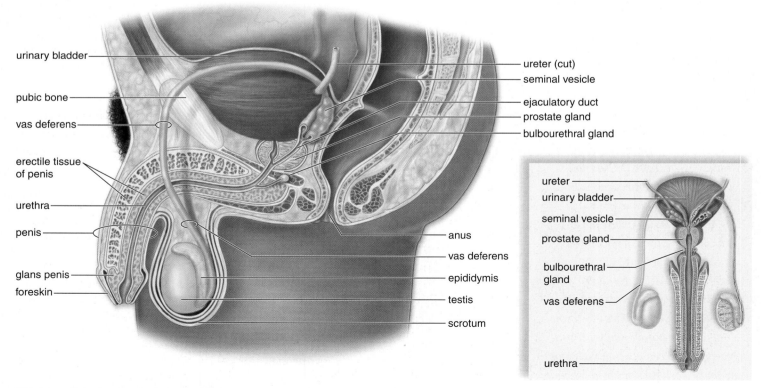

Figure 21.1 The male reproductive system.

The testes produce sperm. The seminal vesicles, the prostate gland, and the bulbourethral glands provide a fluid medium for the sperm, which move from the vasa deferentia through the ejaculatory ducts to the urethra in the penis. The foreskin (prepuce) is removed when a penis is circumcised.

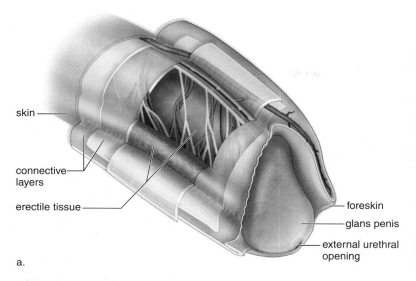

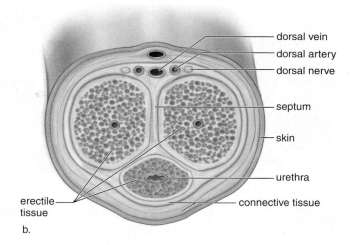

Figure 21.2 **Penis anatomy.**

a. Beneath the skin and the connective tissue lies the urethra, surrounded by erectile tissue. This tissue expands to form the glans penis, which in uncircumcised males is partially covered by the foreskin (prepuce). **b.** Two other columns of erectile tissue in the penis are located dorsally.

the glans penis. The glans penis is normally covered by a layer of skin called the foreskin (prepuce). Circumcision, the surgical removal of the foreskin, is usually done for religious reasons or perceived health benefits, although as yet science has found no overall positive or negative impact on a man's health.

> Sperm mature in the epididymides and are stored in the vasa deferentia before entering the urethra just prior to ejaculation. The accessory glands (seminal vesicles, prostate gland, and bulbourethral glands) produce semen. The penis is the male organ of sexual intercourse.

Orgasm in Males

Spongy, erectile tissue containing distensible blood spaces extends through the shaft of the penis. When a man is sexually excited, the arteries in the penis relax and widen. Increased blood flow causes the penis to become hard and erect. Also, the veins that normally carry blood away from the penis get compressed, and this maintains an erection.

When sexual stimulation intensifies, sperm enter the urethra from the vasa deferentia, and the accessory glands contribute secretions to the semen. Once semen is in the urethra, rhythmic muscle contractions cause it to be expelled from the penis in spurts (ejaculation). During ejaculation, a sphincter closes off the urinary bladder so that no urine enters the urethra, and no semen enters the bladder. (Notice that the urethra carries urine or semen at different times.)

The contractions that expel semen from the penis are a part of male orgasm, the physiological and psychological sensations that occur at the climax of sexual stimulation. The psychological sensation of pleasure is centered in the brain, but the physiological reactions involve the genital (reproductive) organs and associated muscles, as well as the entire body. Marked muscular tension is followed by contraction and relaxation.

Following ejaculation and/or loss of sexual arousal, the penis returns to its normal flaccid state. After ejaculation, a male typically experiences a period of time, called the refractory period, during which stimulation does not bring about an erection. The length of the refractory period increases with age.

There may be in excess of 400 million sperm in the approximately 2–6 ml of semen expelled during ejaculation. However, the sperm count can be much lower than this, and fertilization of the egg by a sperm still can take place.

> Semen is ejaculated from the penis at the time of male orgasm.

Male Gonads, the Testes

The testes, which produce sperm and also the male sex hormones, lie outside the abdominal cavity of the human male, within the scrotum. The testes begin their development inside the abdominal cavity but descend into the scrotal sacs during the last two months of fetal development. If, by chance, the testes do not descend and the male is not treated or operated on to place the testes in the scrotum, sterility—the inability to produce offspring—usually follows. This is because the internal temperature of the body is too high to produce viable sperm. The scrotum helps regulate the temperature of the testes by holding them closer or farther away from the body. Inactivity may be contributing to the growing problem of infertility in males. Sitting for extended periods of time—whether in the car, in front of a computer, or on the couch—increases the scrotal temperature and may cause low sperm production.

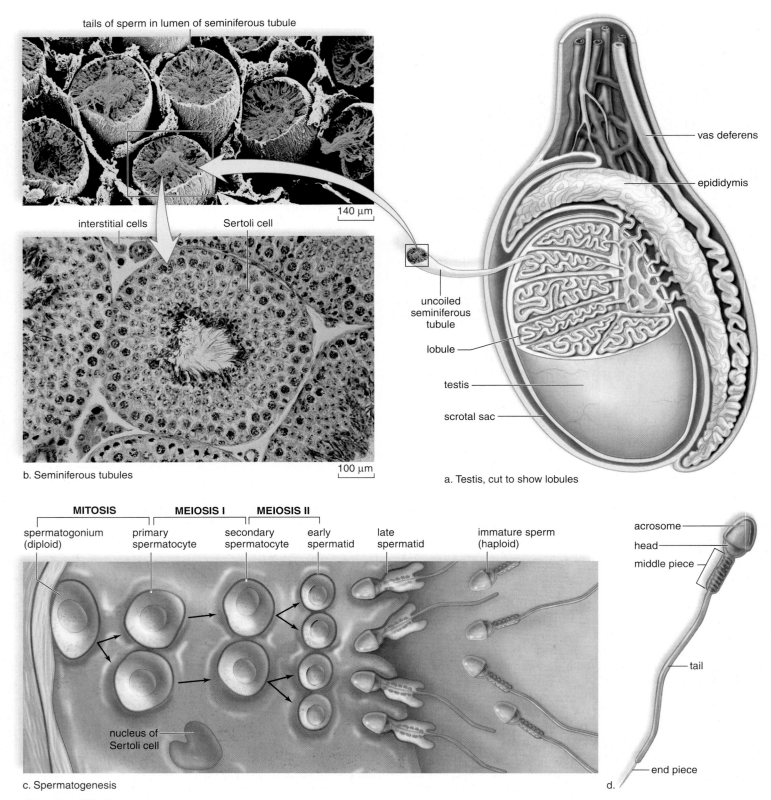

tails of sperm in lumen of seminiferous tubule

140 μm

interstitial cells   Sertoli cell

b. Seminiferous tubules

100 μm

vas deferens

epididymis

uncoiled
seminiferous
tubule

lobule

testis

scrotal sac

a. Testis, cut to show lobules

MITOSIS   MEIOSIS I   MEIOSIS II

spermatogonium   primary   secondary   early   late   immature sperm
(diploid)   spermatocyte   spermatocyte spermatid spermatid   (haploid)

nucleus of
Sertoli cell

c. Spermatogenesis

acrosome

head

middle piece

tail

end piece

d.

Figure 21.3 Testis and sperm.

a. The lobules of a testis contain seminiferous tubules. **b.** Light micrographs of a cross section of the seminiferous tubules, where spermatogenesis occurs. Note the location of interstitial cells in clumps among the seminiferous tubules. **c.** Diagrammatic representation of spermatogenesis, which occurs in the walls of the tubules. **d.** A sperm has a head, a middle piece, tail, and an end piece. The nucleus is in the head, which is capped by the enzyme-containing acrosome.

Seminiferous Tubules

A longitudinal section of a testis shows that it is composed of compartments called lobules, each of which contains one to three tightly coiled **seminiferous tubules** (Fig. 21.3*a*). Altogether, these tubules have a combined length of approximately 250 meters. A microscopic cross section of a seminiferous tubule reveals that it is packed with cells undergoing **spermatogenesis** (Fig. 21.3*b, c*), the production of sperm. Newly formed spermatogonia (sing., spermatogonium) move away from the outer wall and become primary spermatocytes that undergo meiosis I to produce secondary spermatocytes with 23 chromosomes. Secondary spermatocytes undergo meiosis II to produce four spermatids that are also haploid. Spermatids then differentiate into sperm. Also present are Sertoli cells, (sustentacular cells) which support, nourish, and regulate the spermatogenic cells.

Mature **sperm,** or spermatozoa, have three main parts: a head, a middle piece, and a tail (Fig. 21.3*d*). Mitochondria in the middle piece provide energy for the movement of the tail, which has the structure of a flagellum (see Fig. 3.14). The head contains a nucleus covered by a cap called the **acrosome,** which stores enzymes needed to penetrate the egg. The ejaculated semen of a normal human male contains several hundred million sperm, but only one sperm normally enters an egg. Sperm usually do not live more than 48 hours in the female genital tract.

Interstitial Cells

The male sex hormones, the androgens, are secreted by cells that lie between the seminiferous tubules. Therefore, they are called **interstitial cells** (Fig. 21.3*b*). The most important of the androgens is testosterone, whose functions are discussed next.

Hormonal Regulation in Males

The hypothalamus has ultimate control of the testes' sexual function because it secretes a hormone called **gonadotropin-releasing hormone (GnRH)** that stimulates the anterior pituitary to secrete the gonadotropic hormones. There are two gonadotropic hormones—**follicle-stimulating hormone (FSH)** and **luteinizing hormone (LH)**—in both males and females. In males, FSH promotes the production of sperm in the seminiferous tubules, which also release the hormone inhibin. Inhibin, in turn, inhibits further FSH synthesis.

LH in males is sometimes given the name **interstitial cell-stimulating hormone (ICSH)** because it controls the production of testosterone by the interstitial cells. All these hormones are involved in a negative feedback relationship that maintains the fairly constant production of sperm and testosterone (Fig. 21.4).

Testosterone, the main sex hormone in males, is essential for the normal development and functioning of the organs listed in Table 21.1. Testosterone also brings about and maintains the male secondary sex characteristics that develop at the time of puberty. Males are generally taller than females and have broader shoulders and longer legs

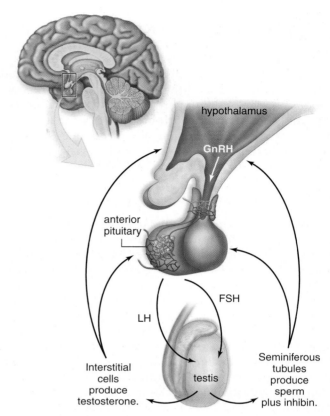

Figure 21.4 Hormonal control of testes.
GnRH (gonadotropin-releasing hormone) stimulates the anterior pituitary to produce FSH and LH. FSH stimulates the testes to produce sperm, and LH stimulates the testes to produce testosterone. Testosterone from interstitial cells and inhibin from the seminiferous tubules exert negative feedback control over the hypothalamus and the anterior pituitary, and this ultimately regulates the level of testosterone in the blood.

relative to trunk length. The deeper voices of males compared to those of females are due to a larger larynx with longer vocal cords. Since the so-called Adam's apple is a part of the larynx, it is usually more prominent in males than in females. Testosterone causes males to develop noticeable hair on the face, chest, and occasionally other regions of the body, such as the back. Testosterone also leads to the receding hairline and pattern baldness that occur in males.

Testosterone is responsible for the greater muscular development of males. Knowing this, both males and females sometimes take anabolic steroids, which are either testosterone or related steroid hormones resembling testosterone. Health problems involving the kidneys, the cardiovascular system, and hormonal imbalances can arise from such use. The testes shrink in size, and feminization of other male traits occurs (see Fig. 20.10).

The gonads in males are the testes, which produce sperm as well as testosterone, the most significant male sex hormone.

21.2   Female Reproductive System

The female reproductive system includes the organs depicted in Figure 21.5 and listed in Table 21.2. The female gonads are paired **ovaries** that lie in shallow depressions, one on each side of the upper pelvic cavity. **Oogenesis** is the production of an egg, or **oocyte,** the female gamete. The ovaries usually alternate in producing one oocyte per month. **Ovulation** is the process by which an oocyte bursts from an ovary and usually enters an oviduct.

TABLE 21.2	Female Reproductive Organs
Organ	**Function**
Ovaries	Produce oocyte and sex hormones
Oviducts (uterine or fallopian tubes)	Conduct oocyte; location of fertilization; transport early zygote
Uterus (womb)	Houses developing fetus
Cervix	Contains opening to uterus
Vagina	Receives penis during sexual intercourse; serves as birth canal and as the exit for menstrual flow

The Genital Tract

The **oviducts,** also called uterine or fallopian tubes, extend from the uterus to the ovaries; however, the oviducts are not attached to the ovaries. Instead, they have fingerlike projections called fimbriae (sing., **fimbria**) that sweep over the ovaries. When an oocyte bursts from an ovary during ovulation, it usually is swept into an oviduct by the combined action of the fimbriae and the beating of cilia that line the oviducts.

Once in the oviduct, the oocyte is propelled slowly by ciliary movement and tubular muscle contraction toward the uterus. An oocyte lives only approximately 6 to 24 hours unless fertilization occurs. Fertilization, and therefore formation of a **zygote,** usually takes place in the oviduct. The developing embryo normally arrives at the uterus after several days and then embeds, or **implants,** itself in the uterine lining, which has been prepared to receive it.

The **uterus** is a thick-walled, muscular organ about the size and shape of an inverted pear. Normally, it lies above and is tipped over the urinary bladder. The oviducts join the uterus at its upper end, while at its lower end, the **cervix** connects with the vagina nearly at a right angle. The cervix normally contains a mucus plug that helps to keep bacteria out of the uterus.

Development of the embryo normally takes place in the uterus. This organ, sometimes called the womb, is approximately 5 cm wide in its usual state but is capable of stretching to over 30 cm wide to accommodate a growing

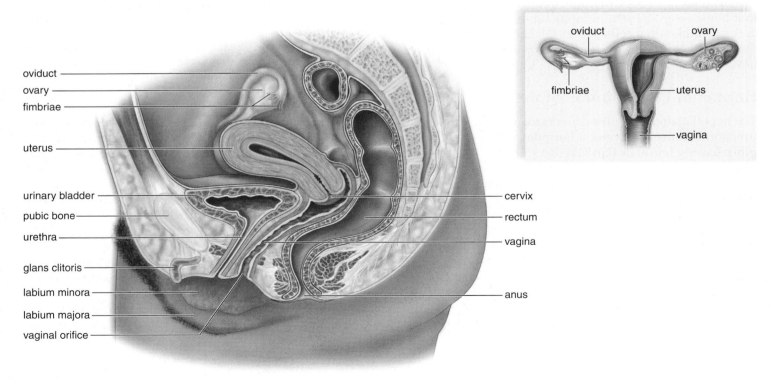

Figure 21.5 The female reproductive system.
The ovaries release one egg a month; fertilization occurs in the oviduct, and development occurs in the uterus. The vagina is the birth canal, as well as the organ of sexual intercourse and the outlet for menstrual flow.

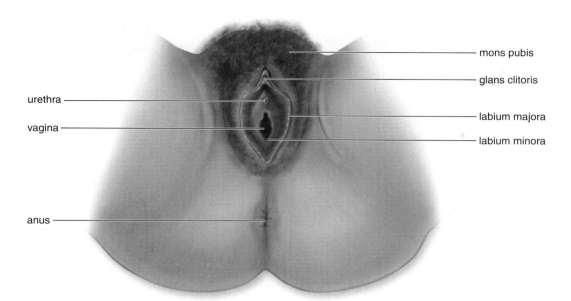

urethra

vagina

anus

mons pubis

glans clitoris

labium majora

labium minora

Figure 21.6 External genitals of the female.
The opening of the vagina is partially blocked by a membrane called the hymen. Physical activities and sexual intercourse disrupt the hymen.

baby. The lining of the uterus, called the **endometrium,** participates in the formation of the placenta (see Section 21.3), which supplies nutrients needed for embryonic and fetal development. The endometrium has two layers—a basal layer and an inner, functional layer. In the nonpregnant female, the functional layer of the endometrium varies in thickness according to a monthly cycle of events called the uterine cycle.

A small opening in the cervix leads to the vaginal canal. The **vagina** is a tube that lies at a 45° angle to the small of the back. The mucosal lining of the vagina lies in folds and can extend. This is especially important when the vagina serves as the birth canal, and it facilitates sexual intercourse, when the vagina receives the penis. The vagina also acts as the exit for menstrual flow.

External Genitals

The external genital organs of the female are known collectively as the **vulva** (Fig. 21.6). The vulva includes two large, hair-covered folds of skin called the labia majora (sing., labium major). The labia majora extend backward from the mons pubis, a fatty prominence underlying the pubic hair. The labia minora are two small folds lying just inside the labia majora. They extend forward from the vaginal opening to encircle and form a foreskin for the glans clitoris. The glans clitoris is the organ of sexual arousal in females and, like the penis, contains a shaft of erectile tissue that becomes engorged with blood during sexual stimulation.

The cleft between the labia minora contains the openings of the urethra and the vagina. The vagina may be partially closed by a ring of tissue called the hymen. The hymen is ordinarily ruptured by sexual intercourse or by other types of physical activity. If remnants of the hymen persist after sexual intercourse, they can be surgically removed.

Notice that the urinary and reproductive systems in the female are entirely separate. For example, the urethra carries only urine, and the vagina serves only as the birth canal and the organ for sexual intercourse.

Orgasm in Females

Upon sexual stimulation, the labia minora, the vaginal wall, and the clitoris become engorged with blood. The breasts also swell, and the nipples become erect. The labia majora enlarge, redden, and spread away from the vaginal opening.

The vagina expands and elongates. Blood vessels in the vaginal wall release small droplets of fluid that seep into the vagina and lubricate it. Mucus-secreting glands beneath the labia minora on either side of the vagina also provide lubrication for entry of the penis into the vagina. Although the vagina is the organ of sexual intercourse in females, the clitoris plays a significant role in the female sexual response. The extremely sensitive clitoris can swell to two or three times its usual size. The thrusting of the penis and the pressure of the public symphyses of the partners act to stimulate the clitoris.

Orgasm occurs at the height of the sexual response. Blood pressure and pulse rate rise, breathing quickens, and the walls of the vagina, uterus, and oviducts contract rhythmically. A sensation of intense pleasure is followed by relaxation when organs return to their normal size. Females have little or no refractory period between orgasms, and multiple orgasms can occur during a single sexual experience.

The female genital tract—oviducts, uterus, vagina—ends at the external genitals. The clitoris plays an active role in the sexual response of females, which also involves vaginal contractions and culminates in uterine and oviduct contractions.

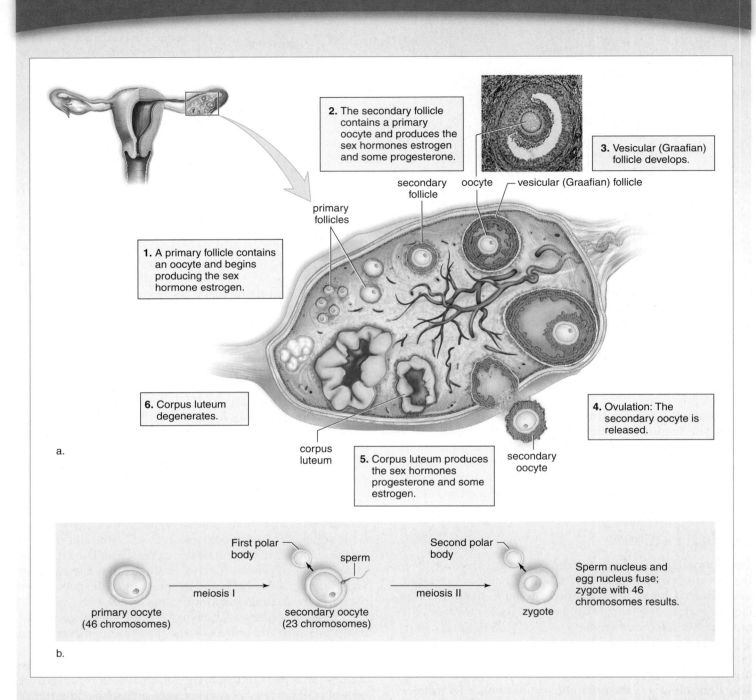

2. The secondary follicle contains a primary oocyte and produces the sex hormones estrogen and some progesterone.

3. Vesicular (Graafian) follicle develops.

secondary follicle   oocyte   vesicular (Graafian) follicle

primary follicles

1. A primary follicle contains an oocyte and begins producing the sex hormone estrogen.

6. Corpus luteum degenerates.

4. Ovulation: The secondary oocyte is released.

a.

corpus luteum

5. Corpus luteum produces the sex hormones progesterone and some estrogen.

secondary oocyte

First polar body

sperm

Second polar body

Sperm nucleus and egg nucleus fuse; zygote with 46 chromosomes results.

primary oocyte (46 chromosomes)

meiosis I

secondary oocyte (23 chromosomes)

meiosis II

zygote

b.

Figure 21.7 Ovarian cycle.
a. A single follicle goes through 6 stages in one place within the ovary. As a follicle matures, layers of follicle cells surround the developing oocyte. Eventually, the mature follicle ruptures, and the secondary oocyte is released. The follicle then becomes the corpus luteum, which eventually disintegrates. **b.** During oogenesis, the chromosome number is reduced from 46 to 23. Fertilization restores the full number of chromosomes.

21.3 Female Hormone Levels

Hormone levels cycle in the female on a monthly basis, and the ovarian cycle drives the uterine cycle, as discussed in this section.

The Ovarian Cycle

A longitudinal section through an ovary shows that it is made up of an outer cortex and an inner medulla (Fig. 21.7a). In the cortex are many **follicles,** and each one contains an immature oocyte. A female is born with as many as 2 million follicles, but the number is reduced to 300,000–400,000 by the time of puberty. Only a small number of follicles (about 400) ever mature because a female usually produces only one oocyte per month during her reproductive years. Oocytes, present at birth, age as the woman ages. This may be one reason older women are more likely to produce children with genetic defects.

The **ovarian cycle** occurs as a follicle changes from a primary to a secondary to a vesicular (Graafian) follicle (Fig. 21.7a). Epithelial cells of a primary follicle surround a primary oocyte. Pools of follicular fluid surround the oocyte in a secondary follicle. In a vesicular follicle, a fluid-filled cavity increases to the point that the follicle wall balloons out on the surface of the ovary.

As a follicle matures, oogenesis, depicted in Figure 21.7b, is initiated and continues. The primary oocyte divides, producing two haploid cells. One cell is a secondary oocyte, and the other is a polar body. The vesicular follicle bursts, releasing the secondary oocyte. This process is referred to as ovulation. Once a vesicular follicle has lost the secondary oocyte, it develops into a **corpus luteum,** a glandlike structure that produces progesterone.

The secondary oocyte enters an oviduct. If a sperm enters the secondary oocyte, fertilization occurs, which completes meiosis. An egg with 23 chromosomes and a second polar body result. When the sperm nucleus unites with the egg nucleus, a zygote with 46 chromosomes is produced. If zygote formation and pregnancy do not occur, the corpus luteum begins to degenerate after about ten days.

Phases of the Ovarian Cycle

The ovarian cycle is commonly divided into two phases. The first half of the cycle is called the follicular phase, and the second half is the luteal phase. During the *follicular phase,* follicle-stimulating hormone (FSH), produced by the anterior pituitary, promotes the development of a follicle in the ovary, which secretes estrogen and some progesterone (Fig. 21.8). As the estrogen level in the blood rises, it exerts feedback control over the anterior pituitary secretion of FSH so that the follicular phase comes to an end.

Presumably, an estrogen spike causes a sudden secretion of a large amount of GnRH from the hypothalamus. This

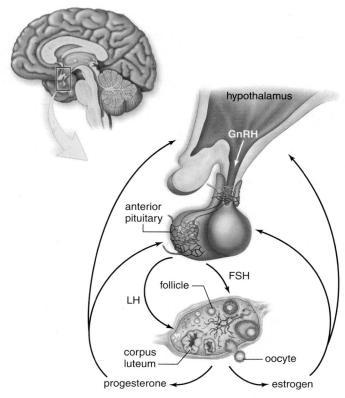

Figure 21.8 Hormonal control of ovaries.

The hypothalamus produces GnRH (gonadotropin-releasing hormone). GnRH stimulates the anterior pituitary to produce FSH (follicle-stimulating hormone) and LH (luteinizing hormone). FSH stimulates the follicle to produce primarily estrogen, and LH stimulates the corpus luteum to produce primarily progesterone. Estrogen and progesterone maintain the sex organs (e.g., uterus) and the secondary sex characteristics, and they exert feedback control over the hypothalamus and the anterior pituitary. Feedback control regulates the relative amounts of estrogen and progesterone in the blood.

leads to a surge of LH production by the anterior pituitary and to ovulation at about the 14th day of a 28-day cycle.

Now, the *luteal phase* begins. During the luteal phase of the ovarian cycle, LH promotes the development of the corpus luteum, which secretes progesterone and some estrogen. As the blood level of progesterone rises, it exerts feedback control over the anterior pituitary secretion of LH so that the corpus luteum in the ovary begins to degenerate. As the luteal phase comes to an end, the low levels of progesterone and estrogen in the body cause menstruation to begin, as discussed next.

One ovarian follicle per month produces a secondary oocyte. The oocyte enters an oviduct, where it can possibly be fertilized. Following ovulation, the follicle develops into the corpus luteum.

The Uterine Cycle

The female sex hormones, **estrogen** and **progesterone,** have numerous functions. One of their functions is to affect the endometrium, causing the uterus to undergo a cyclical series of events known as the **uterine cycle** (Fig. 21.9, *bottom*). Twenty-eight-day cycles are divided as follows:

During *days 1–5,* a low level of female sex hormones in the body causes the endometrium to disintegrate and its blood vessels to rupture. On day 1 of the cycle, a flow of blood and tissues passes out of the vagina during **menstruation,** also called the menstrual period.

During *days 6–13,* increased production of estrogen by a new ovarian follicle in the ovary causes the endometrium to thicken and become vascular and glandular. This is called the *proliferative phase* of the uterine cycle.

On *day 14* of a 28-day cycle, ovulation usually occurs.

During *days 15–28,* increased production of progesterone by the corpus luteum in the ovary causes the endometrium of the uterus to double or triple in thickness (from 1 mm to 2–3 mm) and the uterine glands to mature, producing a thick mucous secretion. This is called the *secretory phase* of the uterine cycle. The endometrium is now prepared

to receive the developing embryo. If this does not occur, the corpus luteum in the ovary degenerates, and the low level of sex hormones in the female body results in the endometrium breaking down during menstruation.

Table 21.3 and Figure 21.9 compare the stages of the uterine cycle with those of the ovarian cycle.

Menstruation

During menstruation, arteries that supply the uterine lining constrict and the capillaries weaken. Blood spilling from the damaged vessels detaches layers of the lining, not all at once, but in random patches. Endometrium, mucus, and blood descend from the uterus, through the vagina, creating menstrual flow. Fibrinolysin, an enzyme released by dying cells, prevents the blood from clotting. Menstruation lasts from 3 to 10 days, as the uterus sloughs the thick lining that was three weeks in the making.

During the uterine cycle, the endometrium builds up and then is broken down during menstruation.

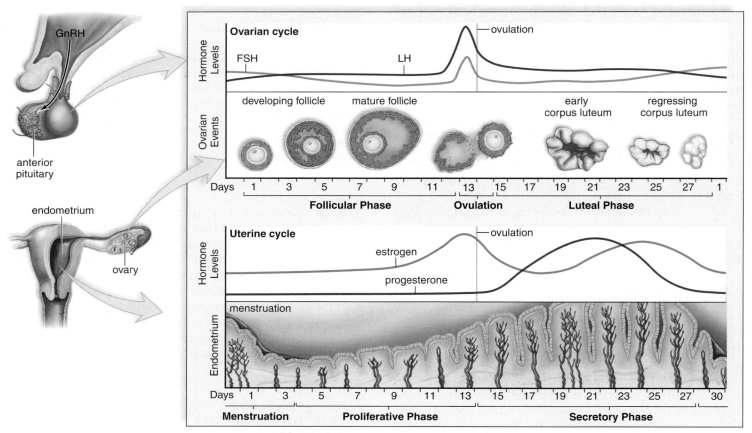

Figure 21.9 Female hormone levels during the ovarian and uterine cycles.
During the follicular phase of the ovarian cycle *(top),* FSH released by the anterior pituitary promotes the maturation of a follicle in the ovary. The ovarian follicle produces increasing levels of estrogen, which causes the endometrium to thicken during the proliferative phase of the uterine cycle *(bottom).* After ovulation and during the luteal phase of the ovarian cycle, LH promotes the development of the corpus luteum. This structure produces increasing levels of progesterone, which causes the endometrium to become secretory. Menstruation and the proliferative phase begin when progesterone production declines to a low level.

TABLE 21.3		Ovarian and Uterine Cycles		
Ovarian Cycle	**Events**	**Uterine Cycle**	**Events**	
Follicular phase—Days 1–13	FSH secretion begins.	Menstruation—Days 1–5	Endometrium breaks down.	
	Follicle maturation occurs. Estrogen secretion is prominent.	Proliferative phase—Days 6–13	Endometrium rebuilds.	
Ovulation—Day 14*	LH spike occurs.			
Luteal phase—Days 15–28	LH secretion continues. Corpus luteum forms. Progesterone secretion is prominent.	Secretory phase—Days 15–28	Endometrium thickens, and glands are secretory.	

*Assuming a 28-day cycle.

Fertilization and Pregnancy

If fertilization does occur, an embryo begins development even as it travels down the oviduct to the uterus. The endometrium is now prepared to receive the developing embryo, which becomes implanted in the lining several days following fertilization (Fig. 21.10). Pregnancy has now begun; an abortion is the removal of an implanted embryo.

The **placenta,** which sustains the developing embryo and later fetus, originates from both maternal and fetal tissues. It is the region of exchange of molecules between fetal and maternal blood, although the two rarely mix. At first, the placenta produces **human chorionic gonadotropin (hCG),** which maintains the corpus luteum in the ovary until the placenta begins its own production of progesterone and estrogen. A pregnancy test detects the presence of hCG in the blood or urine. If this molecule is present, a woman is pregnant.

Progesterone and estrogen produced by the placenta have two effects: they shut down the anterior pituitary so that no new follicle in the ovaries matures, and they maintain the endometrium so that the corpus luteum in the ovary is no longer needed. Usually, no menstruation occurs during pregnancy.

Estrogen and Progesterone

Estrogen and progesterone affect not only the uterus but other parts of the body as well. Estrogen is largely responsible for the secondary sex characteristics in females, including body hair and fat distribution. In general, females have a more rounded appearance than males because of a greater accumulation of fat beneath the skin. Like males, females develop axillary and pubic hair during puberty. In females, the upper border of pubic hair is horizontal, but in males, it tapers toward the navel. Both estrogen and progesterone are also required for breast development. Other hormones are involved in milk production following pregnancy and milk letdown when a baby begins to nurse.

The pelvic girdle is wider and deeper in females, so the pelvic cavity usually has a larger relative size compared to that of males. This means that females have wider hips than males and that their thighs converge at a greater angle toward the knees. Because the female pelvis tilts forward, females tend to have an abdominal bulge, protruding buttocks, and more of a lower back curve than males.

Menopause

Menopause, the period in a woman's life during which the ovarian and uterine cycles cease, is likely to occur between ages 45 and 55. The ovaries are no longer responsive to the gonadotropic hormones produced by the anterior pituitary, and the ovaries no longer secrete estrogen or progesterone. At the onset of menopause, the uterine cycle becomes irregular, but as long as menstruation occurs, it is still possible for a woman to conceive. Therefore, a woman is usually not considered to have completed menopause until menstruation has been absent for a year.

The hormonal changes during menopause often produce physical symptoms, such as "hot flashes" (caused by circulatory irregularities), dizziness, headaches, insomnia,

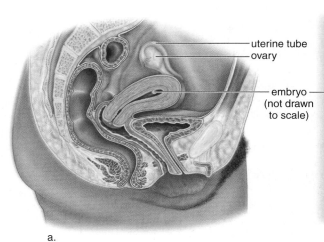

uterine tube
ovary

embryo
(not drawn
to scale)

a.

endometrium

b.

**Figure 21.10
Implantation.**
a. Site of implantation of an embryo in the uterine wall. **b.** A scanning electron micrograph showing an embryo implanted in the endometrium on day 12 following fertilization.

sleepiness, and depression. These symptoms may be mild or even absent. Until recently, many women took combined estrogen-progestin drugs to ease menopausal symptoms. However, a new study conducted by the Women's Health Initiative (WHI) found that long-term use of the combined drugs by postmenopausal women caused increases in breast cancer, heart attacks, strokes, and blood clots. For most women, those risks outweigh the drugs' actual benefits—a small decrease in hip fractures and a decrease in cases of colorectal cancer.

> Estrogen and progesterone produced by the ovaries are the female sex hormones. They foster the development of the reproductive organs, maintain the uterine cycle, and bring about the secondary sex characteristics in females.

21.4   Control of Reproduction

Several means are available to dampen or enhance our reproductive potential. Birth control methods are used to regulate the number of children an individual or couple will have.

Birth Control Methods

The most reliable method of birth control is abstinence—that is, not engaging in sexual contact. This form of birth control has the added advantage of preventing the spread of sexually transmitted diseases. Table 21.4 lists other means of birth control used in the United States, and rates their effectiveness. For example, with the birth control pill, we expect nearly 100% effectiveness, meaning that no sexually active women will get pregnant within the year. On the other hand, with natural family planning, one of the least effective methods, we expect that 70% will not get pregnant and 30% will get pregnant within the year.

Figure 21.11 features some of the most effective and commonly used means of birth control. **Contraceptives** are medications and devices that reduce the chance of pregnancy. Oral contraception (**birth control pills**) often involves taking a combination of estrogen and progesterone on a daily basis (Fig. 21.11a). The estrogen and progesterone in the birth control pill effectively shut down the pituitary production of both FSH and LH so that no follicle begins to develop in the ovary; since ovulation does not occur, pregnancy cannot take place. Because of possible side effects, women taking birth control pills should see a physician regularly.

TABLE 21.4	Common Birth Control Methods			
Name	**Procedure**	**Methodology**	**Effectiveness**	**Risk**
Abstinence	Refrain from sexual intercourse	No sperm in vagina	100%	None
Vasectomy	Vasa deferentia are cut and tied	No sperm in semen	Almost 100%	Irreversible sterility
Tubal ligation	Oviducts cut and tied	No oocytes in oviduct	Almost 100%	Irreversible sterility
Oral contraception	Hormone medication taken daily	Anterior pituitary does not release FSH and LH	Almost 100%	Thromboembolism, especially in smokers
Contraceptive implants	Tubes of progestin (form of progesterone) implanted under skin	Anterior pituitary does not release FSH and LH	More than 90%	Presently none known
Contraceptive injections	Injections of hormones	Anterior pituitary does not release FSH and LH	About 99%	Breast cancer? Osteoporosis?
Intrauterine device (IUD)	Plastic coil inserted into uterus by physician	Prevents implantation	More than 90%	Infection (pelvic inflammatory disease)
Diaphragm	Latex cup inserted into vagina to cover cervix before intercourse	Blocks entrance of sperm to uterus	With jelly, about 90%	Latex, spermicide allergy
Cervical cap	Latex cap held by suction over cervix	Delivers spermicide near cervix	Almost 85%	UTI;[1] latex or spermicide allergy
Male condom	Latex sheath fitted over erect penis	Traps sperm and prevents STDs	About 85%	Latex allergy
Female condom	Polyurethane liner fitted inside vagina	Blocks entrance of sperm to uterus and prevents STDs	About 85%	Latex allergy
Coitus interruptus	Penis withdrawn before ejaculation	Prevents sperm from entering	About 75%	Presently none known
Jellies, creams, foams	These spermicidal products are inserted before intercourse	Kill a large number of sperm	About 75%	UTI,[1] vaginitis, allergy
Natural family planning	Day of ovulation determined by record keeping, various methods of testing	Intercourse avoided on certain days of the month	About 70%	Presently none known
Douche	Vagina cleansed after intercourse	Washes out sperm	Less than 70%	Presently none known

[1] = urinary tract infection

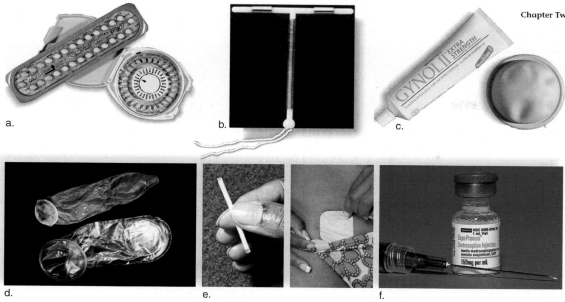

Figure 21.11
Various methods of birth control.
a. Oral contraception (birth control pills). **b.** Intrauterine device. **c.** Spermicidal jelly and diaphragm. **d.** Male and female condoms. **e.** Contraceptive implant and patch. **f.** Depo-Provera, a progesterone injection.

An **intrauterine device (IUD)** is a small piece of molded plastic that is inserted into the uterus by a physician. IUDs are believed to alter the environment of the uterus and oviducts so that fertilization probably will not occur—but if fertilization should occur, implantation cannot take place. The type of IUD featured in Figure 21.11*b* has copper wire wrapped around the plastic.

The **diaphragm** is a soft latex cup with a flexible rim that lodges behind the pubic bone and fits over the cervix (Fig. 21.11*c*). Each woman must be properly fitted by a physician, and the diaphragm can be inserted into the vagina no more than two hours before sexual relations. Also, it must be used with spermicidal jelly or cream and should be left in place for at least six hours after sexual relations. The cervical cap is a minidiaphragm.

There has been renewed interest in barrier methods of birth control, because these methods offer some protection against sexually transmitted diseases (Fig. 21.11*d*). A **female condom,** now available, consists of a large polyurethane tube with a flexible ring that fits onto the cervix. The open end of the tube has a ring that covers the external genitals. A **male condom** is most often a latex sheath that fits over the erect penis. The ejaculate is trapped inside the sheath, and thus does not enter the vagina. When used in conjunction with a spermicide, the protection is better than with the condom alone.

Contraceptive implants (Fig. 21.11*e*) utilize synthetic progesterone to prevent ovulation by disrupting the ovarian cycle. The newest version consists of a single capsule that remains effective for about three years. A **contraceptive patch** is also available. Known by the brand name Ortho Evra, the patch is applied to the skin once a week for three weeks, then no patch is worn during the menstrual period. Since women who use the patch are exposed to higher levels of estrogen than those who use oral contraceptives, the risk of side effects is somewhat higher with the patch.

Contraceptive injections are available as progesterone only (Fig. 21.11*f*) or as a combination of estrogen and progesterone. The length of time between injections can vary from three months to a few weeks.

Contraceptive vaccines are now being developed. For example, a vaccine intended to immunize women against hCG, the hormone so necessary to maintaining the implantation of the embryo, was successful in a limited clinical trial. Since hCG is not normally present in the body, no autoimmune reaction is expected, but the immunization does wear off with time. It may also be possible to develop a safe antisperm vaccine for use in women.

Morning-After Pills

A morning-after pill, or emergency contraception, is a medication that can prevent pregnancy after unprotected intercourse. The expression "morning-after pill" is a misnomer in that a woman can begin taking the medication one to several days after unprotected intercourse.

One type, a kit called Preven, contains four synthetic progesterone pills; two are taken up to 72 hours after unprotected intercourse, and two more are taken 12 hours later. The medication upsets the normal uterine cycle, making it difficult for an embryo to implant itself in the endometrium. A recent study estimated that the medication was 85% effective in preventing unintended pregnancies.

Mifepristone, better known as RU-486, is a pill that causes the loss of an implanted embryo by blocking the progesterone receptor proteins of endometrial cells. Without functioning receptors for progesterone, the endometrium sloughs, carrying the embryo with it. When taken in conjunction with a prostaglandin to induce uterine contractions, RU-486 is 95% effective. In August 2006, the United States Food and Drug Administration approved another drug called Plan B, which is up to 89% effective in preventing pregnancy if taken within 72 hours after unprotected sex. It is available to women age 18 and older without a prescription.

The birth control methods and devices now available vary in effectiveness. New methods of birth control are expected to be developed.

Ecology Focus

Endocrine-Disrupting Contaminants

Rachel Carson's book *Silent Spring,* published in 1962, predicted that pesticides would have a deleterious effect on animal life. Soon thereafter, it was found that pesticides caused the egg shells of bald eagles to become so thin that their eggs broke and the chicks died. Additionally, populations of terns, gulls, cormorants, and lake trout declined after they ate fish contaminated by high levels of environmental toxins. The concern was so great that the United States Environmental Protection Agency (EPA) came into existence. This agency together with civilian environmental groups have brought about a reduction in pollution release and a cleaning up of emissions. Even so, we are now aware of more subtle effects of pollutants.

Hormones influence nearly all aspects of physiology and behavior in animals. Therefore, when wildlife in contaminated areas began to exhibit certain abnormalities, researchers began to think that certain pollutants can affect the endocrine system. In England, male fish exposed to sewage developed ovarian tissue and produced a metabolite normally found only in females during egg formation. In California, western gulls displayed abnormalities in gonad structure and nesting behaviors. Hatchling alligators in Florida possessed abnormal gonads and hormone concentrations linked to nesting.

At first, such effects seemed to involve only the female hormone estrogen, and researchers therefore called the contaminants ecoestrogens. Many of the contaminants interact with hormone receptors, and in that way cause developmental effects. Others bind directly with sex hormones, such as testosterone and estradiol (a potent estrogen). Still others alter the physiology of growth hormones and neurotransmitters responsible for brain development and behavior. Therefore, the preferred term today for these pollutants is endocrine-disrupting contaminants (EDCs).

Many EDCs are chemicals used as pesticides and herbicides in agriculture, and some are associated with the manufacture of various other organic molecules, such as PCBs (polychlorinated biphenyls). Some chemicals shown to influence hormones are found in plastics, food additives, and personal hygiene products. In mice, phthalate esters, which are plastic components, affect neonatal development when present in the part-per-trillion range. It is of great concern that EDCs have been found at levels comparable to functional hormone levels in the human body. Therefore, it is not surprising that EDCs are affecting the endocrine systems of a wide range of organisms (Fig. 21A).

Scientists and representatives of industrial manufacturers continue to debate whether EDCs pose a health risk to humans. Some suspect that EDCs lower sperm counts, reduce male and female fertility, and increase rates of certain cancers (breast, ovarian, testicular, and prostate). Additionally, some studies suggest that EDCs contribute to learning deficits and behavioral problems in children.

Laboratory and field research continues to identify chemicals that have the ability to influence the endocrine system. Millions of tons of potential EDCs are produced annually in the United States, and the EPA is under pressure to certify these compounds as safe. The European Economic Community has already restricted the use of certain EDCs and has banned the production of specific plastic components intended for use by children. Only through continued scientific research and the cooperation of industry can we identify the risks the EDCs pose to the environment, wildlife, and humans.

Discussion Questions

1. There are undeniable benefits to the use of pesticides, one of the major ones being more attractive, insect-free food. How would you personally weigh these benefits compared to the risks of ingesting endocrine-disrupting contaminants?

2. How might endocrine-disrupting chemicals specifically affect cells? (You may need to review how peptide and steroid hormones work in Chapter 20.)

3. What would be some implications for agricultural and manufacturing industries if the use of all potential endocrine-disrupting chemicals were banned? How much extra would you be willing to pay for your monthly food (and other products) if their costs increased?

Figure 21A Exposure to endocrine-disrupting contaminants.
Various types of wildlife, as well as humans, are exposed to endocrine-disrupting contaminants that can seriously affect their health and reproductive abilities.

21.5 Disorders of the Reproductive System

In this section, we first discuss diseases that can be transmitted by sexual contact, such as AIDS and hepatitis, but that do not necessarily affect the reproductive organs themselves. Then we describe some of the most common conditions affecting the reproductive system.

Sexually Transmitted Diseases

Many diseases are transmitted by sexual contact and are therefore called sexually transmitted diseases (STDs). This discussion centers on those that are most prevalent. AIDS, genital herpes, genital warts, and hepatitis B are caused by viruses; therefore, they do not respond to traditional antibiotics. Antiviral drugs have been developed to treat many of these diseases, but in many cases infection is lifelong. Chlamydia, gonorrhea, and syphilis are bacterial diseases treatable (and usually curable) with appropriate antibiotic therapy if diagnosed early enough. However, some of these bacteria are becoming resistant to antibiotics. Unfortunately, the human body does not generally become immune to the infectious agents that cause STDs, making it difficult to develop effective vaccines.

AIDS

Acquired immunodeficiency syndrome (AIDS) is caused by a virus known as HIV (human immunodeficiency virus). AIDS is considered a pandemic because the disease is prevalent in the entire human population around the globe. As of June 2006, the World Health Organization reports that AIDS has killed more than 25 million people worldwide, and about 38.6 million more are currently estimated to be infected with HIV. The epidemic has been particularly devastating in sub-Saharan Africa, where 30% to 40% of the population is infected in some countries. It is also estimated that AIDS could kill 31 million people in India and 18 million in China by the year 2025. In the United States, an estimated one million people are infected, and over 500,000 have died. Homosexual men make up the largest proportion of people with AIDS in the U.S., but the fastest rate of increase is now seen among heterosexuals, and over half of all new HIV infections are occurring in people under the age of 25.

HIV is transmitted by sexual contact with an infected person, including vaginal or anal intercourse and oral/genital contact. Also, needle-sharing among intravenous drug users is a very efficient way to transmit HIV. The Health Focus suggests ways to protect yourself against HIV and other STDs. Babies born to HIV-infected mothers may become infected before or during birth or through breast-feeding after birth.

The primary host cells infected by HIV are the helper T lymphocytes (helper T cells), macrophages, and dendritic cells. Helper T cells, also called CD4⁺ T cells, are the type of white blood cell that stimulates B cells to produce antibodies, and cytotoxic T cells to kill virus-infected cells (see Chapter 13). One of the hallmarks of an HIV infection that is progressing

to AIDS is a decline in the patient's number of CD4⁺ T cells. HIV is thought to kill these cells directly by infecting them, and also indirectly by inducing apoptosis, even of uninfected cells. Cytotoxic T cells are also thought to kill large numbers of HIV-infected cells. Eventually (usually years after the initial infection), the number of CD4⁺ T cells falls so low that acquired immune responses become severely crippled, and the individual usually succumbs to infection by microorganisms that are normally harmless (see Health Focus in Chapter 13.).

Stages of an HIV Infection   When a person initially becomes infected with HIV, the virus replicates at a high level in the macrophages and CD4⁺ T cells, and spreads throughout the lymphoid tissues of the body. Even though the person's blood contains a high "viral load" at this point, most do not experience any symptoms at all. A few do have mononucleosis-like symptoms such as fever, chills, aches, and swollen lymph nodes, but soon these symptoms disappear. Even if a person had an HIV test at this time, it would probably be negative because it tests for the presence of anti-HIV antibodies, which may not have been produced yet. This means that early in the infection, a person can be highly infectious, even though the HIV blood test is negative.

Usually no other HIV-related symptoms occur for several years. But within a few weeks after infection, the immune system does produce antibodies against the virus, and the HIV blood test becomes positive. During this stage of infection, the body is able to maintain a reasonably healthy immune system by producing as many as one to two billion new CD4⁺ helper T cells per day. As long as the body's production of new CD4⁺ T cells is able to keep pace with the loss of CD4⁺ T cells, the person stays relatively healthy. However, even during this asymptomatic stage, evidence indicates that the virus continues to replicate at a high rate (Fig. 21.12). Eventually, the number of CD4⁺ T cells begins to decline, and symptoms appear.

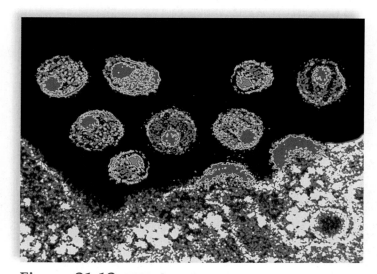

Figure 21.12 HIV, the AIDS virus.
False-colored micrographs show HIV particles budding from an infected helper T cell. These viruses can infect helper T cells and also macrophages, which work with helper T cells to stem the infection.

In order to help physicians assess the prognosis of individuals at different stages of HIV infection, the U.S. Centers for Disease Control and Prevention have defined three categories of HIV infection, based mainly on the type of clinical disease seen in the patient (Fig. 21.13). Within each category, the number of CD4$^+$ T cells is also monitored. Patients in *category A* usually have CD4$^+$ T-cell counts of 500 or greater per mm^3 of blood, and either lack symptoms completely or have persistently enlarged lymph nodes. Patients in *category B* have symptoms that aren't as severe as those of category C, but are indicative of an unhealthy immune system; their CD4$^+$ T-cell count is usually between 200 and 499 per mm^3 of blood. Examples of category B symptoms include thrush, a yeast infection of the oral cavity, and repeated episodes of shingles, a reactivation of a childhood chickenpox virus infection. Patients in *category C* usually have CD4$^+$ T-cell counts less than 200 per mm^3 of blood, and have one or more AIDS-indicator conditions, which are mainly opportunistic bacterial or fungal infections. An **opportunistic infection** is one that has the *opportunity* to occur only because the immune system is weakened. Examples of opportunistic infections that kill most AIDS patients include bacterial pneumonia, invasive Candida (yeast) infections, Kaposi's sarcoma (caused by a human herpesvirus), various mycobacterial infections, Pneumocystis pneumonia, and toxoplasmosis of the brain.

Although not composing an official category of HIV infection, another small, yet very interesting group of HIV-infected people remain healthy for at least ten years, even without any antiviral treatments. Often called long-term nonprogressors, these people have been very interesting for HIV researchers to study, because their viral loads remain low and their CD4$^+$ T-cell counts usually remain normal. As of this writing, no single factor has been identified that explains how some people are able to remain relatively healthy in the face of HIV infection, but their existence continues to provide hope that effective host resistance to HIV is possible.

Treatments for HIV Infection   Prior to the development of antiviral drugs, most HIV-infected individuals eventually progressed to category C or full-blown AIDS, which was almost inevitably fatal. There is still no cure for HIV infection, but since 1995 an increasing number of drugs have become available to help control the replication and spread of HIV, and the life span of many HIV-infected patients has increased to the point where HIV longer is no longer considered an automatic death sentence. Unfortunately, this is not the case in developing nations where most people cannot afford these drugs.

In order to understand how antiretroviral drugs work, a brief review of the life cycle of HIV is necessary (Fig. 21.14). HIV is a type of RNA virus called a retrovirus. HIV has a protein on its outer surface called gp120, which attaches to the CD4 molecule on a host T cell, macrophage, or dendritic cell. This attachment causes *fusion* of HIV with the host plasma membrane, releasing HIV's inner core inside the cell. This core contains the viral RNA, plus several viral enzymes. One such enzyme, called *reverse transcriptase*, makes a DNA copy of the viral RNA, called cDNA. Duplication occurs and the result is double-stranded DNA (dsDNA). During *integration*, the viral enzyme integrase inserts dsDNA into host DNA, where it will remain for the life of the cell. Following integration, viral DNA serves as a template for production of more viral RNA. Some of the viral RNA proceeds to the ribosomes, where it brings about the synthesis of long polypeptides that need to be cleaved into smaller pieces before they can be assembled into new viruses. This cleavage process depends on a third viral enzyme, called *protease*. Finally, the new virus particles are assembled and then released from the host plasma membrane by a process called budding.

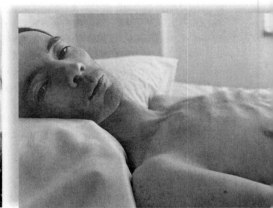

a. AIDS patient Tom Moran, at diagnosis   b. AIDS patient Tom Moran, 6 months later   c. AIDS patient Tom Moran, in terminal stage of AIDS

Figure 21.13 The course of an AIDS infection.
These photos show the effect of an HIV infection in one individual who progressed through all the phases of AIDS. At first, the body can fight off the effects of an HIV infection, but then the infection takes over as the number of T cells drops from thousands to hundreds. Eventually, an AIDS patient will die from the opportunistic diseases that flourish because the immune system has been destroyed.

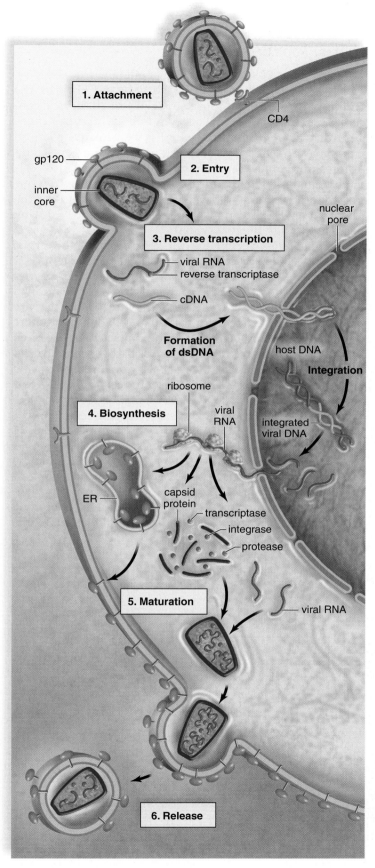

1. Attachment

CD4

gp120

inner core

2. Entry

nuclear pore

3. Reverse transcription

viral RNA
reverse transcriptase

cDNA

Formation of dsDNA

host DNA

Integration

ribosome

4. Biosynthesis

viral RNA

integrated viral DNA

ER

capsid protein

transcriptase

integrase

protease

5. Maturation

viral RNA

6. Release

Figure 21.14 Reproduction of the retrovirus HIV.
HIV uses reverse transcription to produce cDNA (DNA copy of RNA genes); double-stranded DNA (dsDNA) integrates into the cell's chromosomes before the virus reproduces and buds from the cell.

Currently, three classes of antiretroviral drugs are available: (1) reverse transcriptase inhibitors, which block the production of viral DNA from RNA; (2) protease inhibitors, which block the viral enzyme that cleaves long viral polypeptides prior to assembly; and (3) fusion inhibitors, which interfere with HIV's ability to fuse with the host plasma membrane during initial infection. A fourth category, integrase inhibitors, which inhibit the insertion of viral DNA into host DNA, are currently in clinical trials. When the first antiretroviral drug, known as AZT, became available in 1995, it was initially very effective at reducing the viral load in many patients, but the virus tended to become resistant to the drug very quickly. This is because the HIV reverse transcriptase enzyme is very error-prone, which means the virus mutates at a very high rate, resulting in drug-resistant mutants. However, today HIV-infected patients are typically treated with three or four antiretroviral drugs at the same time, a combination called highly active antiretroviral therapy (HAART). Because it is unlikely that a particular mutant virus will be resistant to several drugs all at once, HAART is usually able to inhibit HIV replication to the point that no virus is detectable in the blood. It is important to note, however, that these patients presumably still have large numbers of HIV-infected cells, and thus they can still transmit the virus. Other potential problems with HAART include various side effects, which can be severe, and the complexity of dosing schedules for the different medications.

Investigators have found that when HAART is discontinued, the virus rebounds; therefore, therapy is usually continued indefinitely. An HIV-positive pregnant woman who takes reverse transcriptase inhibitors during her pregnancy reduces the chances of HIV transmission to her newborn. If possible, drug therapy should be delayed until the 10th to 12th weeks of her pregnancy to minimize any adverse effects of AZT on fetal development.

HIV Vaccines It is unlikely that the global HIV pandemic will ever be controlled unless an effective vaccine against HIV is developed. Such a vaccine will likely need to bring about a twofold immune response: antibodies that can bind to the virus and prevent it from entering cells, and cytotoxic T cells that can kill any virus-infected cells. Some of the obstacles to developing an effective vaccine include the existence of many strains of HIV, the tendency of HIV to mutate its proteins, thus making it a "moving target" for the immune system, and the ability of HIV to "hide" inside cells. Currently in 2006, over 30 small-scale studies of HIV vaccines are in progress, as are several larger-scale studies. One type of vaccine that has been studied is a recombinant vaccine, in which a piece of DNA encoding an HIV protein is inserted into another organism (usually another type of virus), which can then infect cells and elicit a cytotoxic T-cell response, without any danger of causing an infection with HIV. Other types of vaccines focus more on inducing an antibody response, especially targeted against the gp120 protein.

Unfortunately, most of these trials have yielded disappointing results, and most investigators now agree that a combination of various types of vaccines may be the best strategy.

Genital Herpes

Genital herpes is caused by the herpes simplex virus, of which there are two types: type 1, which usually causes cold sores and fever blisters, and type 2, which more often causes genital herpes. Crossover infections do occur, however. That is, type 1 has been known to cause genital infections, while type 2 has been known to cause cold sores and fever blisters.

Genital herpes is one of the more prevalent sexually transmitted diseases today. At any one time, millions of persons may be having recurring symptoms. After infection, the first symptoms usually include a tingling or itching sensation before blisters appear at the infected site within 2–20 days (Fig. 21.15). Once the blisters rupture, they leave painful ulcers, which take from five days to three weeks to heal. These symptoms can be accompanied by fever, pain upon urination, and swollen lymph nodes.

After the blisters heal, the virus is only dormant, and blisters can recur repeatedly at variable intervals. Sunlight, sexual intercourse, menstruation, and stress seem to cause the symptoms of genital herpes to recur. While the virus is dormant, it primarily resides in the ganglia of the sensory nerves associated with the affected skin.

Infection of the newborn can occur if the child comes in contact with a lesion in the birth canal. Within 1–3 weeks, the infant may be gravely ill and can become blind, have neurological disorders, including brain damage, or die. Birth by cesarean section prevents these adverse developments.

Human Papillomaviruses

There are over 100 types of human papillomaviruses (HPV). Vaccines are expected soon for HPV 16 and HPV 18, the strains of the virus that cause about 70% of all cervical cancers. Most strains cause warts, and about 30 types cause **genital warts,** which are sexually transmitted. Genital warts may appear as flat or raised warts on the penis and/or foreskin of males, and on the vulva, vagina, and/or cervix of females. Note that if the wart(s) are only on the cervix, there may be no outward signs or symptoms of the disease. Newborns can also be infected with HPV during passage through the birth canal.

About ten types of HPV can cause cancer of the cervix, the second leading cause of cancer death in women in the U.S. (approximately 500,000 deaths per year). These HPV types produce a viral protein that inactivates a host protein called p53, which normally acts as a "brake" on cell division. Once p53 has been inactivated in a particular cell, that cell is more prone to the uncontrolled cell division characteristic of cancer. Early detection of cervical cancer is possible by means of a **Pap smear,** in which a few cells are removed from the region of the cervix for microscopic examination. It the cells are cancerous, a hysterectomy (removal of the uterus) may be recommended. Teenagers who have had multiple sex partners seem particularly susceptible to HPV infection. More and more cases of cancer of the cervix are being diagnosed in this age group.

Currently, there is no cure for an HPV infection, but the warts can be treated effectively by surgery, freezing, application of an acid, or laser burning. However, even after treatment, the virus can sometimes be transmitted. Therefore, once someone has been diagnosed with genital warts, absti-

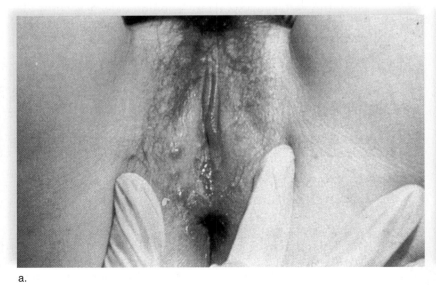

a.

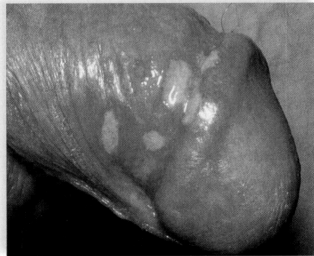

b.

Figure 21.15 Genital herpes.

There are several types of herpes viruses, and usually the ones called herpes simplex-2 cause genital herpes. About one million persons become infected each year, most of them teens and young adults. Symptoms of genital herpes are due to an outbreak of blisters, which can be present on the labia of females **(a)** or on the penis of males **(b).**

nence or the use of a condom is recommended to prevent transmission of the virus.

In June 2006, the U.S. Food and Drug Administration licensed an HPV vaccine that is effective against the four most common types of HPV found in the U.S., including the two types that cause about 70% of cervical cancers. Since the vaccine doesn't protect those who are already infected, ideally children should be vaccinated before they become sexually active.

Hepatitis B

Of the several types of viruses that can cause hepatitis, or inflammation of the liver, hepatitis B is the most likely to be transmitted sexually. As mentioned in Chapter 14, hepatitis B virus (HBV) is more contagious than HIV, and HBV is transmitted by similar routes. About 50% of persons infected with HBV develop flu-like symptoms, including fatigue, fever, headache, nausea, vomiting, and muscle aches. As the virus replicates in the liver, a dull pain in the upper right half of the abdomen may occur, as may jaundice, a yellowish cast to the skin. Some persons have an acute infection that only lasts a few weeks, while others develop a chronic form of the disease that leads to liver failure and the need for a liver transplant. A safe and effective vaccine is available for the prevention of HBV infection. This vaccine is now on the list of recommended immunizations for children.

Chlamydia

Chlamydia is named for the tiny bacterium that causes it, *Chlamydia trachomatis*. New chlamydial infections are more numerous than any other sexually transmitted disease. Some estimate that the actual incidence could be as high as 6 million new cases per year. For every reported case in men, more than five cases are detected in women. The low rates in men suggest that many of the sex partners of women with chlamydia are not diagnosed or reported.

Chlamydial infections of the lower reproductive tract are usually mild or asymptomatic. About 8–21 days after infection, men experience a mild burning sensation on urination and a mucous discharge. The infection can spread to the prostate gland and epididymides in males.

Women may have a vaginal discharge, along with the symptoms of a urinary tract infection. If not properly treated, the infection can eventually cause **pelvic inflammatory disease (PID).** This condition is characterized by inflamed oviducts that become partially or completely blocked by scar tissue. As a result, the woman may become sterile or subject to ectopic pregnancy. Some believe that, in any case, chlamydial infections increase the possibility of premature and still-born births. If a newborn is exposed to chlamydia during delivery, pneumonia or inflammation of the eyes can result (Fig. 21.16).

Detection and Treatment of Chlamydia New and faster laboratory tests are now available for detecting a chlamydial infection. Their expense sometimes prevents public clinics from testing for chlamydia. Thus, the following criteria have

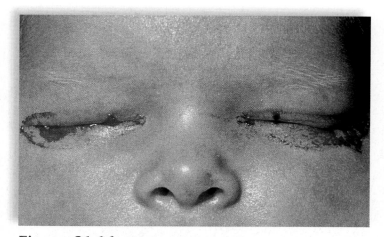

Figure 21.16 **Chlamydia eye infection.**
This newborn's eyes were infected after passing through the birth canal of an infected mother.

been suggested to help physicians decide which women should be tested: no more than 24 years old; a new sex partner within the preceding two months; cervical discharge; bleeding during parts of the vaginal exam; and use of a nonbarrier method of contraception. Some doctors, however, are routinely prescribing additional antibiotics appropriate to treating chlamydia for anyone who has gonorrhea, because 40% of females and 20% of males with gonorrhea also have chlamydia.

Gonorrhea

Gonorrhea is caused by the bacterium *Neisseria gonorrhoeae*. Although as many as 20% of males are asymptomatic, usually they complain of pain at urination and have a thick, greenish-yellow urethral discharge 3–5 days after contact with an infected partner. Unfortunately, 60–80% of females are asymptomatic until they develop severe PID. Similarly, scarring of each vas deferens may occur in untreated males.

Gonorrhea proctitis, or infection of the anus, with symptoms that may include pain in the anus and blood or pus in the feces, is spread by anal intercourse. Oral sex can cause infection of the throat and the tonsils. Gonorrhea can also spread to other parts of the body, causing heart damage or arthritis. If, by chance, the person touches infected genitals and then his or her eyes, a severe eye infection can result.

Eye infection leading to blindness can occur as a baby passes through the birth canal. Because of this, all vaginally delivered infants receive eyedrops containing antibacterial agents, such as silver nitrate, tetracycline, or penicillin, as a protective measure.

Reported gonorrhea rates declined steadily until the late 1990s, and since then, they have increased by 9%. Gonorrhea rates among African Americans are 30 times greater than among Caucasians. Also, women using the birth control pill have a greater risk of contracting gonorrhea because hormonal contraceptives cause the genital tract to be more receptive to certain pathogens. As with AIDS, condoms protect against gonorrheal and chlamydial infections.

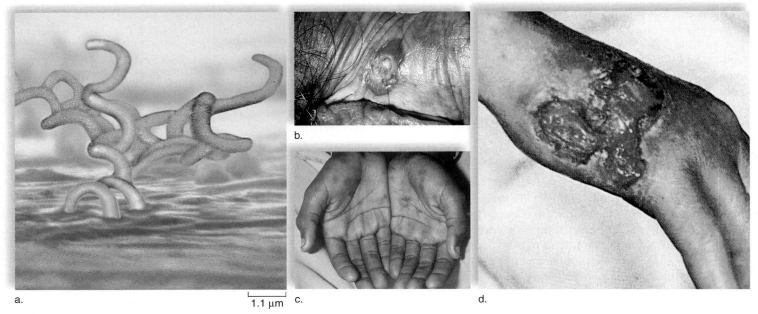

a.   1.1 μm c.   d.

Figure 21.17 Syphilis.

a. Scanning electron micrograph of *Treponema pallidum,* the cause of syphilis. **b.** The primary stage of syphilis is a chancre at the site where the bacterium enters the body. **c.** The secondary stage is a body rash that occurs even on the palms of the hands and soles of the feet. **d.** In the tertiary stage, gummas may appear on the skin or internal organs.

Syphilis

Syphilis is caused by the bacterium *Treponema pallidum,* an actively motile corkscrewlike organism (Fig. 21.17*a*). Syphilis has three stages, which can be separated by intervening latent periods during which the bacteria are not multiplying. During the *primary stage,* a hard chancre (ulcerated sore with hard edges) indicates the site of infection (Fig. 21.17*b*). The chancre can go unnoticed, especially since it usually heals spontaneously, leaving little scarring. During the *secondary stage,* the individual breaks out in a rash, proof that bacteria have invaded and spread throughout the body. Curiously, the rash does not itch and is even seen on the palms of the hands and the soles of the feet (Fig. 21.17*c*). Hair loss can occur, and infectious gray patches may develop on the mucous membranes, including the mouth. These symptoms disappear of their own accord.

Not all cases of secondary syphilis go on to the tertiary stage. During the *tertiary stage,* which lasts until the patient dies, syphilis may affect the cardiovascular system; weakened arterial walls (aneurysms) develop, particularly in the aorta. In other instances, the disease may affect the nervous system so that the infected person shows psychological disturbances. In still another variety of the tertiary stage, gummas, large destructive ulcers, may develop on the skin or within the internal organs (Fig. 21.17*d*).

Congenital syphilis, which is present at birth, is caused by syphilitic bacteria crossing the placenta. The child is born blind and/or with numerous anatomical malformations.

As with many other bacterial diseases, penicillin has been effective in treating syphilis. Its control depends on prompt and adequate treatment of all new cases; therefore, it is crucial for all sexual contacts to be traced so they can be treated. Diagnosis of syphilis can be made through blood tests or microscopic examination of fluids from lesions.

> Important STDs caused by viruses include AIDS, genital herpes, human papillomavirus infections, and hepatitis B. Bacterial STDs include chlamydia, gonorrhea, and syphilis.

Common Conditions Affecting the Male Reproductive System

Erectile Dysfunction

It is estimated that about 50% of men aged 40 to 75 have experienced some degree of **erectile dysfunction (ED),** or impotence, which is the inability to produce or maintain an erection sufficient to perform sexual intercourse. In recent years it has become unusual to watch a sporting event on TV without seeing commercials for medications used to treat ED. All of these medications, (e.g., trade names Viagra, Levitra, and Cialis) work by increasing the blood flow to the penis during sexual arousal. Specifically, they inhibit an enzyme called PDE-5, found mainly in the penis, which normally breaks down some of the chemicals responsible for an erection. Surveys have shown these medications to be helpful in about 65% of men who are experiencing ED.

ED can have many causes. Since a normal erection is dependent on blood flow, almost any condition that affects the cardiovascular system, such as atherosclerosis, high blood pressure, or diabetes mellitus, can lead to ED. It is very likely that changes in the cardiovascular system associated with aging alone can also produce some degree of ED. Certain medications can cause ED, as can smoking, alcohol, and some illicit drugs. Psychological factors, especially depression, can also be very important contributors to ED. Depending on the cause, other therapies besides the Viagra-type drugs may be indicated. For example, chemicals that dilate blood vessels can be injected directly into the penis, applied as a gel or a patch, or inserted into the urethra as a suppository. If chemical treatments do not work, it is possible to surgically implant devices into the penis that either cause it to remain semi-rigid all the time or that must be inflated with a small pump, which is usually implanted in the scrotum. Obviously, such devices are usually a last resort in men with ED for whom other approaches have been unsuccessful.

Disorders of the Prostate

Beginning around age 40, most men have some enlargement of the prostate gland, which may eventually grow from its normal size of a walnut to that of a lime or even a lemon. This condition is called **benign prostatic hyperplasia (BPH).** Since the prostate gland wraps around the urethra, this enlargement initially may force the bladder muscle to work harder to empty the urethra, leading to irritation of the bladder and an urge to urinate more frequently (Fig. 21.18). As the

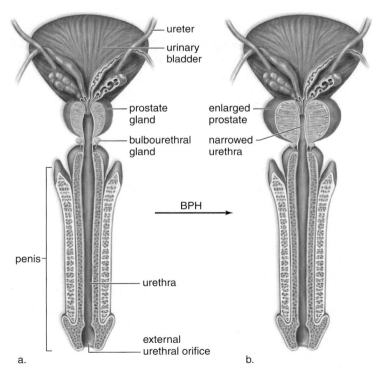

Figure 21.18 Benign prostatic hyperplasia (BPH).
a. This longitudinal section shows how the prostate gland normally surrounds the urethra. **b.** When the prostate gland enlarges, it can constrict the urethra as it exits the bladder, making it difficult to fully empty the bladder.

prostate grows larger, this increased pressure may result in bladder or kidney damage, and may even completely block urination, a condition that is fatal if untreated.

Early signs of an enlarged prostate include a more frequent urge to urinate, a weak stream of urine, and/or a sense of not fully emptying the bladder after urination. The presence of an enlarged prostate can usually be confirmed by a simple digital exam of the prostate, which a physician performs by inserting the finger of a gloved hand into the rectum. Depending on the severity of the symptoms, treatments for BPH range from avoiding liquids in the evening (to reduce the need to urinate during the night), to taking medications that shrink the prostate and/or improve urine flow, to having surgery.

Prostate enlargement is often due to an enzyme (5-alpha reductase) that acts on the male sex hormone testosterone, converting it to a substance that promotes prostate growth. That growth is needed during puberty, but continued growth in the adult is undesirable. Several chemical substances have been shown to interfere with the action of this enzyme. Saw palmetto, which is sold in tablet form as an over-the-counter nutritional supplement, should not be taken unless BPH is confirmed by a physician, but it has been shown to be effective, particularly in the early stages of prostate enlargement. Prescription drugs such as Finasteride are more powerful inhibitors of the enzyme, but men taking these drugs may experience erectile dysfunction and loss of libido. The other drugs used to treat BPH are the alpha blockers, a major class of drugs also used to treat high blood pressure because they relax the smooth muscle in arterial walls. Although these drugs probably do not affect the overall size of the prostate, they do relax the smooth muscle in the gland and thus relieve symptoms.

If drugs don't control symptoms or the symptoms are severe, surgery may be necessary to reduce the pressure on the urethra and/or reduce the size of the prostate. In the most common operation, called transurethral resection of the prostate, the surgeon scrapes away the innermost core of the gland using a small instrument inserted into the urethra. In a more limited operation, called transurethral incision of the prostate, the doctor simply makes one or two small cuts in the prostate to relieve pressure. Alternatively, a newer procedure called transurethral microwave thermotherapy uses microwaves to produce heat that destroys, and hopefully shrinks, the enlarged gland.

Cancer of the prostate is the most commonly diagnosed cancer in American males, with one in ten men expected to develop the disease in their lifetime. Many men are concerned that BPH may be associated with prostate cancer, but the two conditions are not necessarily related. BPH tends to occur in the inner zone of the prostate, while cancer tends to develop in the outer area. If prostate cancer is detected early, usually by a blood test for prostate-specific antigen (PSA), it can usually be successfully treated. However, prostate cancer is the second leading cancer-related cause of death in men in the U.S., because when not detected early, it has a tendency to metastasize and spread throughout the body.

Testicular Cancer

As mentioned in the story that opened Chapter 11, Lance Armstrong is probably the most famous testicular cancer survivor. **Testicular cancer** is the most common type of cancer in males aged 15–35. There are several types of testicular cancer, and some of them have a greater tendency to metastasize to other tissues. If discovered and treated early, testicular cancer is one of the most curable of all cancers. Therefore, experts recommend that men perform a monthly testicular self-examination after a hot shower or bath, when the scrotal skin is looser. Men should examine each testicle for any lumps or abnormal tenderness and for any change in the size of one testicle compared to the other. Any such findings should be reported to a doctor as soon as possible. If testicular cancer is confirmed, treatment usually includes surgery to remove the affected testicle, plus radiation or chemotherapy for types of cancer that have a high risk of spreading. However, if diagnosed and treated before it has spread, the cure rate for testicular cancer approaches 100%.

> Erectile dysfunction and benign prostatic hyperplasia are common conditions in men aged 40 and over. Effective medications are available to treat either condition. Testicular cancer tends to occur in younger men, and it too can be effectively treated if diagnosed early.

Common Conditions Affecting the Female Reproductive System

Endometriosis

Endometriosis is the most common cause of chronic pelvic pain in women. Estimates indicate that between 10% and 33% of women aged 25–35 years will have some degree of this condition, although it can also occur earlier or later in life. **Endometriosis** is the presence of endometrial-like tissue at locations outside the uterine cavity, such as the ovaries, the oviducts, the outside of the uterus, on virtually any other abdominal organ. The presence of this abnormal tissue can induce a chronic inflammatory reaction, which commonly results in pain, abnormal menstrual cycles, or infertility. The exact cause of endometriosis is unclear, but various theories have been suggested. During menstruation, it is possible that live cells shed from the endometrium actually flow backwards, through the oviducts and into the abdomen, where they are able to form endometrial tissue. Another hypothesis suggests that under certain hormonal conditions, the cells that normally line the abdominal organs can become endometrial tissue.

Endometriosis is often diagnosed when a surgeon sees the abnormal tissue during a surgical procedure to investigate the cause of abdominal pain (Fig. 21.19). The initial

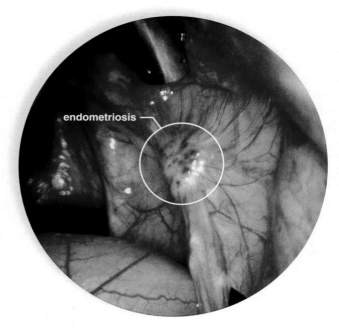

Figure 21.19 Endometriosis.
Endometriosis is a common cause of pelvic pain, abnormal menstruation, and infertility in women. The condition occurs when endometrial tissue, which is often darkly pigmented, grows outside the uterus (usually in the abdominal cavity).

treatment usually focuses on controlling the pain, but may also involve drugs, such as oral contraceptives, which inhibit ovulation, and thus reduce the normal hormonally induced changes in the endometrial tissue. If these approaches do not work, surgery may be performed to either remove the abnormal endometrial tissue or to remove the uterus and ovaries in women who do not wish to have children.

Ovarian Cancer

Ovarian cancer accounts for about 4% of all cancers in the United States. Unlike testicular cancer, ovarian cancer typically occurs in individuals over the age of 40. However, it also has a worse prognosis than testicular cancer, with only 50% of women surviving five years after diagnosis. The main reason for this may be that ovarian cancer is often "silent," showing no obvious signs or symptoms until late in its development. The most common sign is enlargement of the abdomen due to accumulation of fluid. Rarely, abnormal vaginal bleeding occurs. In women over 40, vague digestive disturbances (stomach discomfort, gas, distension) that persist and cannot be explained by any other cause may indicate the need for a thorough evaluation for ovarian cancer.

Women over age 40 should have a cancer-related checkup every year. Early detection requires periodic, thorough pelvic examinations. The Pap smear, useful in detecting cervical cancer, does not reveal ovarian cancer. Testing for the level of tumor marker CA-125, a protein antigen, is helpful. Surgery, radiation therapy, and drug therapy are

Health Focus

Preventing Transmission of STDs

Being aware of how STDs are spread (Fig. 21B) and then observing the following guidelines will greatly help prevent the transmission of STDs.

Sexual Activities Transmit STDs

Abstain from sexual intercourse or develop a long-term monogamous (always the same partner) *sexual relationship with a partner who is free of STDs.*

Refrain from multiple sex partners or having relations with someone who has multiple sex partners. If you have sex with two other people and each of these has sex with two people and so forth, the number of people having sexual contact is quite large.

Remember that, although the prevalence of AIDS is presently higher among homosexuals and bisexuals, the highest rate of increase is now occurring among heterosexuals. The lining of the uterus is only one cell thick, and thus it can allow infected cells from a sexual partner to enter the blood.

Be aware that having relations with an intravenous drug user is risky because the behavior of this group risks hepatitis and an HIV infection. Be aware that anyone who already has another sexually transmitted disease is more susceptible to an HIV infection.

Avoid anal-rectal intercourse (in which the penis is inserted into the rectum) because the lining of the rectum is thin and cells infected with HIV can easily enter the body there.

Unsafe Sexual Practices Transmit STDs

Always use a latex condom during sexual intercourse if you do not know for certain that your partner has been free of STDs for some time. Be sure to follow the directions supplied by the manufacturer.

Avoid fellatio (kissing and insertion of the penis into a partner's mouth) *and cunnilingus* (kissing and insertion of the tongue into the vagina) because they may be a means of transmission. The mouth and gums often have cuts and sores that facilitate the entrance of infected cells.

Be cautious about using alcohol or any drug that may prevent you from being able to control your behavior.

Drug Use Transmits Hepatitis and HIV

Stop, if necessary, or do not start the habit of injecting drugs into your veins. Be aware that hepatitis and HIV can be spread by blood-to-blood contact.

Always use a new, sterile needle for injection or one that has been cleaned in bleach if you are a drug user and cannot stop your behavior.

Discussion Questions

1. How can you be certain that your partner doesn't have any STDs? Do people necessarily know they have an STD?
2. What does it mean if a person tells you he or she just had an AIDS test, and it was negative? What does this test detect in the blood?
3. Which methods of birth control are most effective in preventing transmission of HIV? Which are least effective?

a.   b.

Figure 21B Transmission of STDs.

a. Sexual activities transmit STDs. **b.** Sharing needles transmits STDs. In the United States, more than 65 million people are living with an incurable STD. An additional 15 million people become infected with one or more STDs each year. Approximately one-fourth of these new infections occur in teenagers. Despite the fact that STDs are quite widespread, many people remain unaware of the risks and the sometimes deadly consequences of becoming infected.

treatment options. Surgery usually involves removing one or both ovaries, the uterus, and the oviducts. With some very early tumors, only the involved ovary is removed, especially in young women who would like to preserve their ability to become pregnant. In advanced disease, surgeons attempt to remove all intra-abdominal cancerous tissue to improve the patient's chances of survival.

Women who have never had children are twice as likely to develop ovarian cancer as those who have. Early age at first pregnancy, early menopause, and the use of oral contraceptives, which reduces ovulation frequency, appear to protect against ovarian cancer. If a woman has had breast cancer, her chances of developing ovarian cancer double.

Disorders of Menstruation

Most menstrual cycles occur every 21–35 days, with 3–10 days of bleeding during which 30–40 milliliters of blood are lost. Abnormalities include too little bleeding (or a complete lack of menstruation) or too much bleeding. The former usually indicates a problem with the control of menstruation by the hypothalamus or pituitary gland. This may be seen, for example, in young women who are undergoing strenuous athletic training, and/or who have very low levels of body fat. Excessive bleeding may be due to a large number of other disorders, including various STDs, certain medications, or problems with blood clotting.

The most common disorder of menstruation is dysmenorrhea, or painful menstruation. About half of all women experience this condition at some point, which includes such symptoms as agonizing cramps, headaches, backaches, and nausea. Painful menstruation is the leading cause of lost time from school and work among young women. There are many potential causes of dysmenorrhea. One common factor is a high level of prostaglandins released by the endometrium during the luteal and menstrual phases of the cycle. Research has shown that prostaglandin levels are four or five times higher than normal in women who experience painful menstruation. Since prostaglandins are known to cause uterine contractions, investigators suspect that these chemicals are responsible for much of the pain these women experience.

Other potential causes of painful menstruation include endometriosis (discussed previously) and various types of ovarian cysts. These fluid-filled sacs can develop on the ovaries of females of all ages, but are more common in the childbearing years. Two major types of ovarian cysts are follicular cysts, usually caused by the failure of a developing follicle to burst, or corpus luteum cysts, which occur when the corpus luteum doesn't break down normally. On rare occasions, surgery may be needed to remove ovarian cysts.

Premenstrual syndrome (PMS) is a group of symptoms related to the menstrual cycle. Anywhere from two weeks to a few days before menstruation, a significant number of women experience such symptoms as irrational mood swings, headaches, joint pain, digestive upsets, and sore breasts. The mood swings can be particularly troublesome if they lead to behaviors uncharacteristic of the individual.

Self-help measures for painful menstruation include applying a heating pad to the lower abdomen, drinking warm beverages, and taking warm showers or baths. Over-the-counter anti-inflammatory medications may be helpful; if not, a physician can prescribe other medications. For PMS, avoiding salt, sugar, caffeine, and alcohol may also help, especially when symptoms are present. A physician should be consulted for severe symptoms.

Common disorders of the female reproductive system include endometriosis, ovarian cancer, and difficulties with menstruation, which include dysmenorrhea and PMS.

Infertility

Infertility is the failure of a couple to achieve pregnancy after one year of regular, unprotected intercourse. The American Medical Association estimates that 15% of all couples are infertile. The cause of infertility can be attributed to the male (40%), the female (40%), or both (20%).

Causes of Infertility

The most frequent cause of infertility in males is low sperm count and/or a large proportion of abnormal sperm, which can be due to environmental influences. The public is particularly concerned about endocrine-disrupting contaminants, which are discussed in the Ecology Focus. But thus far, it appears that a sedentary lifestyle coupled with smoking and alcohol consumption is the most common cause of male infertility. When males spend most of the day sitting in front of a computer or the TV or driving, the testes temperature remains too high for adequate sperm production.

Body weight appears to be one of the most significant factors in causing female infertility. In women of normal weight, fat cells produce a hormone called leptin that stimulates the hypothalamus to release GnRH. In overweight women, the ovaries often contain many small follicles, and the woman fails to ovulate. Other causes of infertility in females are blocked oviducts due to pelvic inflammatory disease and endometriosis, as discussed earlier in this section.

Sometimes the causes of infertility can be corrected by medical intervention so that couples can have children. If no obstruction is apparent and body weight is normal, it is possible to give females fertility drugs, which are gonadotropic hormones that stimulate the ovaries and bring about ovulation. Such hormone treatments may cause multiple ovulations and multiple births, as discussed in the Bioethical Focus on page 412.

When reproduction does not occur in the usual manner, many couples adopt a child. Others sometimes try one of the assisted reproductive technologies discussed in the following paragraphs.

Bioethical Focus

Assisted Reproductive Technologies

The dizzying array of assisted reproductive technologies has progressed from simple in vitro fertilization to the ability to freeze eggs or sperm or even embryos for future use. Older women who never had the opportunity to freeze their eggs can now still have children if they use donated eggs—perhaps harvested from a fetus.

Legal complications associated with assisted reproductive technologies abound, ranging from which mother has first claim to the child—the surrogate mother, the woman who donated the egg, or the primary caregiver—to which partner has first claim to frozen embryos following a divorce. Legal issues concerning who has the right to use what techniques have rarely been discussed, much less decided upon. Some reproductive clinics will help anyone, male or female, no

questions asked as long as the person has the ability to pay. As scientific advances continue, it may soon be possible to guarantee the sex of the child and to make sure the child will be free from a particular genetic disorder. It would not be surprising if, in the future, zygotes could be engineered to have any particular trait desired by the parents.

Even today, eugenic (good gene) goals are evidenced by the fact that reproductive clinics advertise for egg and sperm donors, primarily in elite college newspapers. The question becomes, "Is it too late for us as a society to make ethical decisions about reproductive issues?" Should we come to a consensus about what techniques should be allowed and who should be able to use them? Should a woman investigate a couple before she has a child for them?

Form Your Own Opinion

1. Assuming you and your partner wanted to have a child but could not, how far would you go to have one? Sperm sorting with intrauterine fertilization? Artificial insemination by a donor? In vitro fertilization?

2. Many assisted reproductive technologies are expensive. Assuming your annual household income were $60,000, how much would you be willing to spend to be guaranteed a healthy child? $10,000? $30,000?

3. Assuming you and your partner would like to have a baby, would you like to select the gender? How about the height? Athletic ability?

Assisted Reproductive Technologies

Assisted reproductive technologies (ART) are techniques that increase the chances of pregnancy. Often, sperm and/or eggs are retrieved from the testes and ovaries, and fertilization takes place in a clinical or laboratory setting. The Bioethical Focus on this page raises some of the legal and ethical questions associated with ART.

Artificial Insemination by Donor (AID)   During artificial insemination, sperm are placed in the vagina by a physician. Sometimes a woman is artificially inseminated by her partner's sperm. This is especially helpful if the partner has a low sperm count, because the sperm can be collected over a period of time and concentrated so that the sperm count is sufficient to result in fertilization. Often, however, a woman is inseminated by sperm acquired from a donor who is a complete stranger to her. At times, a combination of partner and donor sperm is used.

A variation of AID is *intrauterine insemination (IUI)*. In IUI, fertility drugs are given to stimulate the ovaries, and then the donor's sperm is placed in the uterus, rather than in the vagina.

If the prospective parents wish, sperm can be sorted into those that are believed to be x-bearing or y-bearing to increase the chances of having a child of the desired sex.

In Vitro Fertilization (IVF)   During IVF, conception occurs in laboratory glassware. Ultrasound machines can now spot follicles in the ovaries that hold immature oocytes; therefore, the latest method is to forgo the administration of fertility drugs and retrieve immature oocytes by using a needle. The immature oocytes are then brought to maturity in glassware before concentrated sperm are

added. After about two to four days, the embryos are ready to be transferred to the uterus of the woman, who is now in the secretory phase of her uterine cycle. If desired, the embryos can be tested for a genetic disease, so that only those free of disease are used. If implantation is successful, development is normal and continues to term.

Gamete Intrafallopian Transfer (GIFT)   The term **gamete** refers to a sex cell, either a sperm or an oocyte. Gamete intrafallopian transfer was devised to overcome the low success rate (15–20%) of in vitro fertilization. The method is exactly the same as in vitro fertilization, except the oocytes and the sperm are placed in the oviducts immediately after they have been brought together. GIFT has the advantage of being a one-step procedure for the woman—the oocytes are removed and reintroduced all in the same time period. A variation on this procedure is to fertilize the oocytes in the laboratory and then place the zygotes in the oviducts.

Surrogate Mothers   In some instances, women are contracted and paid to have babies. These women are called surrogate mothers. The sperm and even the oocyte can be contributed by the contracting parents.

Intracytoplasmic Sperm Injection (ICSI)   In this highly sophisticated procedure, a single sperm is injected into an oocyte. It is used effectively when a man has severe infertility problems.

When corrective procedures fail to reverse infertility, assisted reproductive technologies may be considered.

Summarizing the Concepts

21.1 Male Reproductive System

Spermatogenesis, occurring in the seminiferous tubules of the testes, produces sperm that mature and are largely stored in the epididymides. Sperm enter the vasa deferentia before entering the urethra along with secretions produced by the seminal vesicles, prostate gland, and bulbourethral glands. Sperm and these secretions are called semen.

The external genitals of males are the penis, the organ of sexual intercourse, and the scrotum, which contains the testes. Orgasm in males is a physical and emotional climax during sexual intercourse that results in ejaculation of semen from the penis.

Secretions from the hypothalamus, the anterior pituitary, and the testes maintain testosterone levels produced by the interstitial cells of the testes. FSH from the anterior pituitary promotes spermatogenesis in seminiferous tubules; LH promotes testosterone production by the interstitial cells.

21.2 Female Reproductive System

Oogenesis within the ovaries typically occurs in one follicle each month. This follicle balloons out of the ovary and bursts, releasing an oocyte, which enters an oviduct. The oviducts lead to the uterus, where implantation and development occur. The external genitals include the vaginal opening, the clitoris, the labia minora, and the labia majora.

The vagina is the organ of sexual intercourse, the birth canal, and the exit for menstrual fluids. The external genitals, especially the clitoris, play an active role in orgasm, which culminates in vaginal, uterine, and oviduct contractions.

21.3 Female Hormone Levels

The ovarian cycle is under the control of the hypothalamus and anterior pituitary. FSH from the anterior pituitary causes maturation of a follicle that secretes estrogen and some progesterone. After ovulation and during the second half of the cycle, LH from the anterior pituitary converts the follicle into the corpus luteum, which secretes progesterone and some estrogen.

In the uterine cycle, estrogen produced by a follicle causes the endometrium to rebuild. A surge of LH causes ovulation to occur, usually on day 14 of a 28-day cycle. Progesterone, produced by the corpus luteum, causes the endometrium to thicken and become secretory. Then a low level of hormones causes the endometrium to break down as menstruation occurs.

If fertilization takes place, the embryo implants itself in the thickened endometrium. The corpus luteum in the ovary is maintained for a while because of hCG production by the placenta. Later, the placenta produces both estrogen and progesterone. Menstruation usually does not occur during pregnancy.

21.4 Control of Reproduction

Numerous birth control methods and devices, including the birth control pill, diaphragm, and condom, are available for those who wish to prevent pregnancy. Several morning-after pills are now available.

21.5 Disorders of the Reproductive System

Sexually transmitted diseases include AIDS, a pandemic disease that devastates the immune system; hepatitis B, which can lead to liver failure; genital herpes, which flares up repeatedly; human papilloma viruses, the cause of genital warts and cervical cancer; chlamydia and gonorrhea, which can lead to pelvic inflammatory disease; and syphilis, which may cause cardiovascular and neurological complications if untreated.

Common disorders of the male reproductive tract include erectile dysfunction, prostate problems, and testicular cancer. Common disorders of the female reproductive tract include endometriosis, ovarian cancer, and various conditions that affect menstruation. Some couples are infertile, and if so, they may use assisted reproductive technologies in order to have a child. Artificial insemination and in vitro fertilization have been followed by more sophisticated techniques, such as intracytoplasmic sperm injection.

Testing Yourself

Choose the best answer for each question.

1. Which of these structures contributes fluid to semen?
   a. prostate
   b. seminal vesicles
   c. bulbourethral gland
   d. All of these are correct.
   e. None of these are correct.

2. Sperm cells mature in the
   a. urethra.
   b. urinary bladder.
   c. epididymides.
   d. prostate.
   e. None of these are correct.

3. The _____ contains enzymes needed to penetrate the oocyte
   a. urethra
   b. acrosome
   c. prostate
   d. vas deferens
   e. None of these are correct.

4. Production of testosterone in the interstitial cells is controlled by
   a. LH.
   b. ICSH.
   c. GnRH.
   d. All of these are correct.
   e. Both a and b are correct.

5. Assuming a 28-day menstrual cycle, LH levels should be highest around day
   a. 1.
   b. 10.
   c. 14.
   d. 20.
   e. 25.

For questions 6–10, match the hormones with the function in the key.

Key:
   a. development of corpus luteum
   b. promote sperm production
   c. stimulate hormone release from pituitary
   d. muscle development
   e. thickening of endometrium

6. FSH

7. Testosterone

8. LH

9. Progesterone

10. GnRH

11. A rise in estrogen levels in the blood causes the follicular phase to
    a. begin.
    b. end.
    c. It has no effect.
    d. None of these are correct.

12. Low sperm count is usually caused by
    a. radiation.
    b. disease.
    c. high testes temperature.
    d. All of these are correct.
    e. None of these are correct.

13. Which of these diseases is caused by a virus?

    a. gonorrhea
    b. genital warts
    c. chlamydia
    d. syphilis.
    e. None of these are correct.

14. Late detection of gonorrhea can cause

    a. chlamydia.
    b. hepatitis.
    c. PID.
    d. syphilis.
    e. All of these are correct.

15. In the category B stage of HIV infection, patients usually exhibit

    a. more T cells than category C, but certain other infections are common.
    b. a reduced amount of T cells, but often no symptoms of an HIV infection.
    c. opportunistic infections and likelihood of immediate death.
    d. a positive blood test.
    e. Both a and d are correct.

16. Label the anatomical structures in this diagram of the female reproductive system.

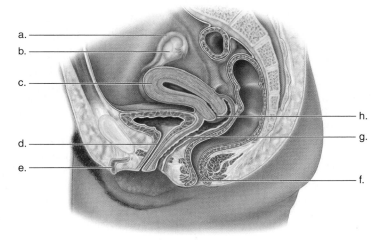

17. Label the anatomical structures in this diagram of the male reproductive system.

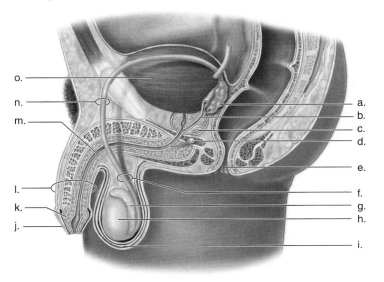

18. In tracing the path of sperm, you would mention the vasa deferentia before the

    a. testes.   c. urethra.
    b. epididymides.   d. uterus.

19. An oocyte is fertilized in the

    a. vagina.   c. oviduct.
    b. uterus.   d. ovary.

20. For nine months, pregnancy is maintained by the

    a. anterior pituitary.   c. corpus luteum.
    b. ovaries.   d. placenta.

21. Home pregnancy tests are based on the presence of

    a. estrogen.   c. follicle-stimulating hormone.
    b. progesterone.   d. human chorionic gonadotropin.

For questions 22–28, match the terms in the key with the following graph of the female reproductive cycles.

Key:

    a. estrogen   e. follicular phase
    b. ovulation   f. luteal phase
    c. menstruation   g. progesterone
    d. LH

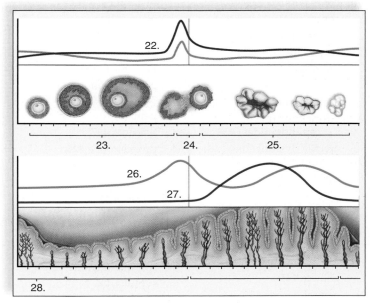

Match the method of protection in questions 29–31 with the means of birth control in the key.

Key:

    a. vasectomy   d. diaphragm
    b. oral contraception   e. male condom
    c. intrauterine device (IUD)   f. coitus interruptus

29. Blocks entrance of sperm to uterus

30. Traps sperm and also prevents STDs

31. Prevents implantation of an embryo

32. Female oral contraceptives prevent pregnancy because

    a. the pill inhibits the release of luteinizing hormone.
    b. oral contraceptives prevent the release of an egg.
    c. follicle-stimulating hormone is not released.
    d. All of these are correct.

33. During pregnancy,
    a. the ovarian and uterine cycles occur more quickly than before.
    b. GnRH is produced at a higher level than before.
    c. the ovarian and uterine cycles do not occur.
    d. the female secondary sex characteristics are not maintained.

Understanding the Terms

acquired immunodeficiency syndrome (AIDS) 431
acrosome 421
benign prostatic hyperplasia (BPH) 437
birth control pill 428
bulbourethral gland 418
cervix 422
chlamydia 435
contraceptive 428
contraceptive implant 429
contraceptive injection 429
contraceptive patch 429
contraceptive vaccine 429
corpus luteum 425
diaphragm 429
endometriosis 438
endometrium 423
epididymis (pl., epididymides) 418
erectile dysfunction (ED) 436
estrogen 426
female condom 429
fimbria 422
follicle 425
follicle-stimulating hormone (FSH) 421
gamete 441
genital herpes 434
genital warts 434
gonadotropin-releasing hormone (GnRH) 421
gonorrhea 435
human chorionic gonadotropin (hCG) 427
implant 422
infertility 440
interstitial cell 421
interstitial cell-stimulating hormone (ICSH) 421

intrauterine device (IUD) 429
luteinizing hormone (LH) 421
male condom 429
menopause 427
menstruation 426
oocyte 422
oogenesis 422
opportunistic infection 432
ovarian cancer 438
ovarian cycle 425
ovary 422
oviduct 422
ovulation 422
Pap smear 434
pelvic inflammatory disease (PID) 435
penis 418
placenta 427
progesterone 426
prostate gland 418
scrotum 418
semen 418
seminal vesicle 418
seminiferous tubule 421
sperm 421
spermatogenesis 421
syphilis 436
testes 418
testicular cancer 438
testosterone 421
urethra 418
uterine cycle 426
uterus 422
vagina 423
vas deferens (pl., vasa deferentia) 418
vulva 423
zygote 422

Match the terms to these definitions:

a._____ Organ that leads from the uterus to the external genitals and serves as the birth canal and organ of sexual intercourse in females.

b._____ Lining of the uterus, which becomes thickened and vascular during the uterine cycle.

c._____ Gland located around the male urethra below the urinary bladder; adds secretions to semen.

d._____ Hormone secreted by the anterior pituitary gland that stimulates the development of an ovarian follicle in a female or the production of sperm in a male.

e._____ External genitals of the female that surround the opening of the vagina.

Thinking Critically

1. Compare the anatomy of the male and female reproductive tracts. Which organs serve similar functions? Which organs serve a unique function, with no analogous function in the other gender?

2. In the animal kingdom, only primates menstruate. Other mammals come into season, or "heat," during certain times of the year, while still others only ovulate after having sex. What would be some possible advantages of a monthly menstrual cycle?

3. Is it possible for some HIV-infected people to have relatively normal numbers of CD4+ T cells in their blood (< 500 per mm^3), but still be in category C because they have one or more AIDS-defining opportunistic infections? Why or why not?

4. Suppose you have a friend who has HIV and has been on HAART for several years. Recently she has begun to develop some opportunistic infections, and lab tests have shown that her CD4+ T-cell numbers are dropping, indicative that the HIV in her body is developing resistance. Are there any possible options for her, or will she progress to full-blown AIDS soon?

5. The condition called benign prostatic hyperplasia is usually not life threatening, but prostate cancer can be. Since the prostate gland is typically enlarged in both conditions, why is one condition benign and the other potentially life threatening?

ARIS, the *Inquiry into Life* Website

ARIS, the website for *Inquiry into Life,* provides a wealth of information organized and integrated by chapter. You will find practice quizzes, interactive activities, labeling exercises, flashcards, and much more that will complement your learning and understanding of general biology.

www.aris.mhhe.com

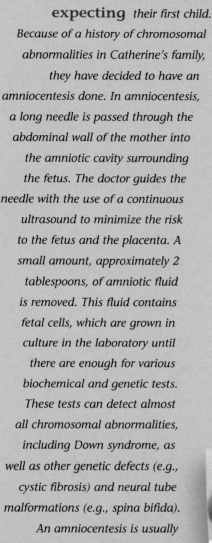

Tim and Catherine are expecting *their first child. Because of a history of chromosomal abnormalities in Catherine's family, they have decided to have an amniocentesis done. In amniocentesis, a long needle is passed through the abdominal wall of the mother into the amniotic cavity surrounding the fetus. The doctor guides the needle with the use of a continuous ultrasound to minimize the risk to the fetus and the placenta. A small amount, approximately 2 tablespoons, of amniotic fluid is removed. This fluid contains fetal cells, which are grown in culture in the laboratory until there are enough for various biochemical and genetic tests. These tests can detect almost all chromosomal abnormalities, including Down syndrome, as well as other genetic defects (e.g., cystic fibrosis) and neural tube malformations (e.g., spina bifida).*

An amniocentesis is usually done between the fifteenth and twentieth week of the pregnancy. Since the procedure carries a risk of miscarriage of about 1 in 370, it is not done unless a problem is suspected. In this case, Catherine has a brother with a genetic abnormality, so her pregnancy is considered a high-risk pregnancy. Tim and Catherine hope this test will enable them to prepare appropriately for the arrival of this child.

chapter **22**

Development and Aging

Chapter Concepts

22.1 Principles of Animal Development
- What prevents several sperm from fertilizing one oocyte? 446
- What are the cellular stages, the tissue stages, and the organ stages of development? 447–449
- How does the development of the nervous system proceed from the tissue stage to the organ stage? 449
- What are two processes of animal development, and how do they occur? 450–453

22.2 Human Embryonic and Fetal Development
- What is the purpose of extraembryonic membranes in human development? 454
- What are the two phases of human development, and what events occur during each stage? 453–460
- How does fetal circulation differ from adult circulation? 458–459
- What is the function of a placenta during pregnancy and fetal development? Do maternal blood and fetal blood ever mix? 460
- What are the stages of birth? 461, 464

22.3 Human Development After Birth
- What is aging, and what are the hypotheses about why we age? 465–467

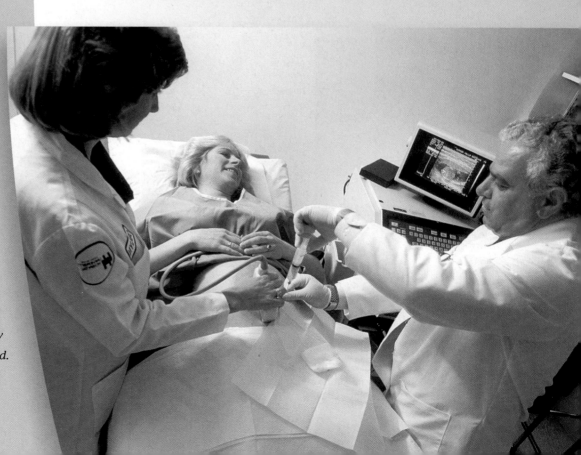

22.1 Principles of Animal Development

Animal development begins with a single cell that multiplies and changes to form a complete organism.

Fertilization

Fertilization, which results in a zygote, requires that the sperm and oocyte interact. Recall that a sperm has three distinct parts: a head, a middle piece, and a tail. The head contains a nucleus and is capped by a membrane-bounded acrosome.

In humans, the plasma membrane of the oocyte is surrounded by an extracellular matrix termed the zona pellucida (Fig. 22.1). In turn, the zona pellucida is surrounded by a few layers of adhering follicle cells. These cells nourished the oocyte when it was in a follicle of the ovary. During fertilization, a sperm travels from the interior of the oviduct to the inside of the oocyte. During this journey, (1) the sperm squeezes through the follicle cells; (2) the sperm releases acrosomal enzymes that allow it to penetrate the zona pellucida; (3) the plasma membrane of the sperm head fuses with the plasma membrane of the oocyte; (4) the sperm enters the oocyte and its nucleus is released; (5) the sperm nucleus and the oocyte nucleus fuse.

Several sperm attempt to make this journey, but only one completes it. It could be that the acrosomal enzymes of several sperm are needed to forge a pathway through the zona pellucida, and in this way the other sperm assist the one sperm that goes on to enter the oocyte. If more than one sperm enters an oocyte, an event called polyspermy, the zygote would have too many chromosomes, and development would be abnormal. Prevention of polyspermy depends on changes in the oocyte's plasma membrane and in the zona pellucida. As soon as a sperm touches an oocyte, the oocyte's plasma membrane depolarizes (from -65 mV to 10 mV), and this prevents the binding of any other sperm. Then the oocyte releases substances that lead to lifting of the zona pellucida away from the surface of the oocyte. Now, sperm cannot bind to the zona pellucida either.

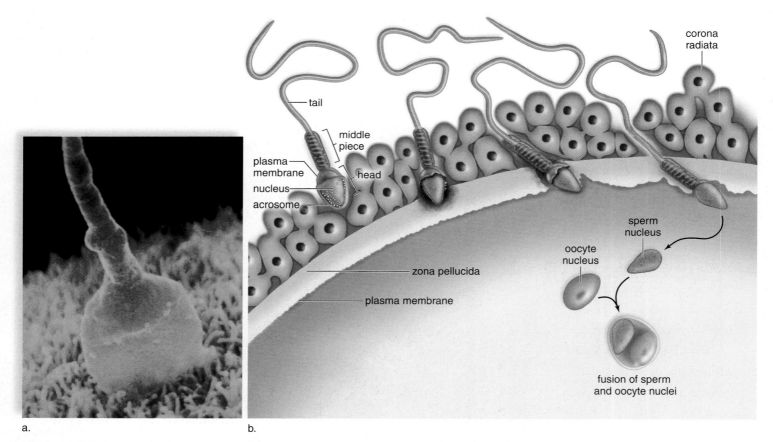

a.   b.

Figure 22.1 Fertilization.

a. During fertilization, a single sperm enters the oocyte. **b.** A sperm makes its way through the follicle cells surrounding the oocyte. The head of a sperm has a membrane-bounded acrosome filled with enzymes. When released, these enzymes digest a pathway for the sperm through the zona pellucida. After it binds to the plasma membrane of the oocyte a sperm enters the oocyte. When the sperm nucleus fuses with the oocyte nucleus, fertilization is complete.

Early Stages of Animal Development

The early stages of animal development occur at the cellular, tissue, and organ levels of organization.

Cellular Stages of Development

The cellular stages of development are (1) cleavage resulting in a multicellular embryo, and (2) formation of the blastula. **Cleavage** is cell division without growth. DNA replication and mitotic cell division occur repeatedly, and the cells get smaller with each division. In other words, cleavage increases only the number of cells; it does not change the original volume of the egg cytoplasm.

As shown in Figure 22.2, cleavage in a lancelet, a primitive chordate, is equal (cells of uniform size) and results in a **morula,** which is a ball of cells. The next cellular stage in lancelet development is formation of a **blastula,** which is a hollow ball of cells having a fluid-filled cavity called a **blastocoel.** The blastocoel forms when the cells of the morula pump Na^+ into extracellular spaces and water follows by osmosis. The water collects in the center, and the result is a hollow ball of cells.

The zygotes of other animals, such as a frog, chick, or human, which are vertebrates, also undergo cleavage and form a morula. In frogs, cleavage is not equal because of the presence of yolk, a dense nutrient material. When yolk is present, the zygote and embryo exhibit polarity, and the embryo has an animal pole (smaller cells) and a vegetal pole (larger cells containing yolk). The animal pole of a frog embryo has a deep gray color because the cells contain melanin granules, and the vegetal pole has a yellow color because the cells contain yolk.

Similarly, all vertebrates have a blastula stage, but the appearance of the blastula can be different from that of a lancelet. A chicken is a vertebrate animal that develops on land and lays a hard-shelled egg containing a large amount of yolk. In the developing chick, the yolk does not participate in cleavage. The blastula forms as a layer of cells that spreads out over the yolk. The blastocoel is a space that separates these cells from the yolk:

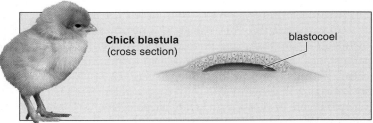

The blastula of humans resembles that of the chick embryo, but this resemblance cannot be related to the amount of yolk because the human egg contains little yolk. Rather, the evolutionary history of these two animals can explain this similarity. Since both birds and mammals are related to reptiles, all three groups develop similarly, despite a difference in the amount of yolk in their eggs.

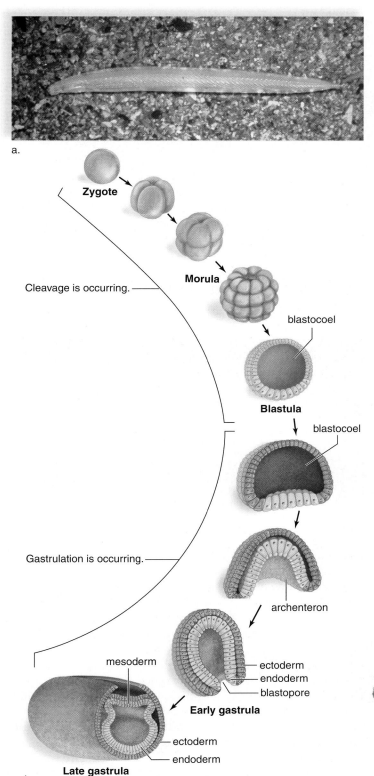

Figure 22.2 Lancelet early development.

a. A lancelet. **b.** The early stages of development are exemplified in the lancelet. Cleavage produces a number of cells that form a cavity, the blastocoel. Invagination during gastrulation produces the germ layers ectoderm and endoderm. Then the mesoderm arises.

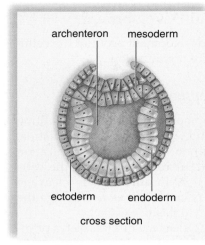

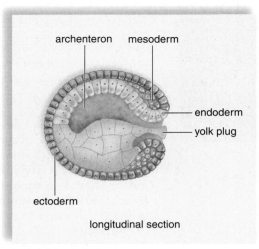

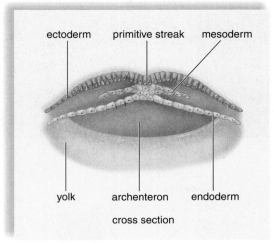

a. Lancelet late gastrula   b. Frog late gastrula   c. Chick late gastrula

Figure 22.3 Comparative development of mesoderm.
a. In the lancelet, mesoderm forms by an outpocketing of the archenteron. **b.** In the frog, mesoderm forms by migration of cells between the ectoderm and endoderm. **c.** In the chick, mesoderm also forms by invagination of cells.

Tissue Stages of Development

The tissue stages of development are: (1) the early gastrula and (2) the late gastrula. The early **gastrula** stage begins when certain cells begin to push, or invaginate, into the blastocoel, creating a double layer of cells (see Fig. 22.2). Cells migrate during this and other stages of development, sometimes traveling quite a distance before reaching a destination, where they continue developing. As cells migrate, they "feel their way" by changing their pattern of adhering to extracellular proteins and cytoskeletal elements.

An early gastrula has two layers of cells. The outer layer of cells is called the **ectoderm,** and the inner layer is called the **endoderm.** The endoderm borders the gut, which at this point is termed either the archenteron or the primitive gut. The pore, or hole, created by invagination is the **blastopore,** and in a lancelet, the blastopore eventually becomes the anus. **Gastrulation** is not complete until three layers of cells that will develop into adult organs are produced. In addition to ectoderm and endoderm, the late gastrula has a middle layer of cells called the **mesoderm.**

Figure 22.2 illustrates gastrulation in a lancelet, and Figure 22.3 compares the lancelet, frog, and chick late gastrula stages. In the lancelet, mesoderm formation begins as outpocketings from the archenteron (Fig. 22.3*a*). These outpocketings will grow in size until they meet and fuse, forming two layers of mesoderm. The space between them is the coelom. The coelom is a body cavity lined by mesoderm that contains internal organs. (In humans, as you know, the coelom becomes the thoracic and abdominal cavities.)

In the frog, the cells containing yolk do not participate in gastrulation, and therefore, they do not invaginate. Instead, a slitlike blastopore is formed when the animal pole cells begin to invaginate from above, forming endoderm. Animal pole cells also move down over the yolk, to invaginate from below. Some yolk cells, which remain temporarily in the region of the

TABLE 22.1	Embryonic Germ Layers
Embryonic Germ Layer	**Vertebrate Adult Structures**
Ectoderm (outer layer)	Nervous system; epidermis of skin; epithelial lining of oral cavity and rectum
Mesoderm (middle layer)	Musculoskeletal system; dermis of skin; cardio-vascular system; urinary system; reproductive system—including most epithelial linings; outer layers of respiratory and digestive systems
Endoderm (inner layer)	Epithelial lining of digestive tract and respiratory tract, associated glands of these systems, epithelial lining of urinary bladder

blastopore, are called the yolk plug. Mesoderm forms when cells migrate between the ectoderm and endoderm (Fig. 22.3*b*). Later, a splitting of the mesoderm creates the coelom.

The chicken egg contains so much yolk that endoderm formation does not occur by invagination. Instead, an upper layer of cells becomes ectoderm, and a lower layer becomes endoderm. Mesoderm arises by an invagination of cells along the edges of a longitudinal furrow in the midline of the embryo. Because of its appearance, this furrow is called the primitive streak (Fig. 22.3*c*). Later, the newly formed mesoderm splits to produce a coelomic cavity.

Ectoderm, mesoderm, and endoderm are called the embryonic **germ layers.** No matter how gastrulation takes place, the result is the same: three germ layers are formed. It is possible to relate the development of future organs to these germ layers (Table 22.1).

A mature gastrula has three germ layers: ectoderm, endoderm, and mesoderm. Each germ layer develops into specific organs.

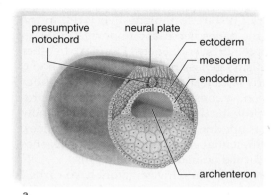
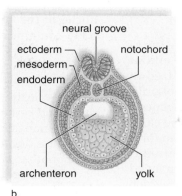
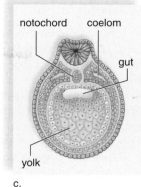
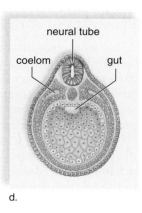

a.   b.   c.   d.

Figure 22.4 Development of neural tube and coelom in a frog embryo.
a. Ectodermal cells that lie above the future notochord (called presumptive notochord) thicken to form a neural plate. **b.** The neural groove and folds are noticeable as the neural tube begins to form. **c.** Splitting of the mesoderm produces a coelom, which is completely lined by mesoderm. **d.** A neural tube and a coelom have now developed.

Organ Stages of Development

The organs of an animal's body develop from the three embryonic germ layers. Scientists have devoted much study to how the nervous system develops.

The newly formed mesoderm cells lying along the main longitudinal axis of the animal coalesce to form a dorsal supporting rod called the **notochord.** The notochord persists in lancelets, but in frogs, chicks, and humans, it is later replaced by the vertebral column. Therefore, these animals are called vertebrates.

The nervous system develops from midline ectoderm located just above the notochord. At first, a thickening of cells, called the **neural plate,** is seen along the dorsal surface of the embryo. Then, neural folds develop on either side of a neural groove, which becomes the **neural tube** when these folds fuse. Figure 22.4 shows cross sections of frog development to illustrate the formation of the neural tube. At this point, the embryo is called a **neurula.** Later, the anterior end of the neural tube develops into the brain, and the rest becomes the spinal cord. In addition, the neural crest is a band of cells that develops where the neural tube pinches off from the ectoderm. Neural crest cells migrate to various locations, where they contribute to formation of skin and muscles, in addition to the adrenal medulla and the ganglia of the peripheral nervous system.

Midline mesoderm cells that did not contribute to the formation of the notochord now become two longitudinal masses of tissue. These two masses become blocked off into somites, which are serially arranged along both sides along the length of the notochord. Somites give rise to muscles associated with the axial skeleton and to the vertebrae. The serial origin of axial muscles and the vertebrae testifies that vertebrates are segmented animals. Lateral to the somites,

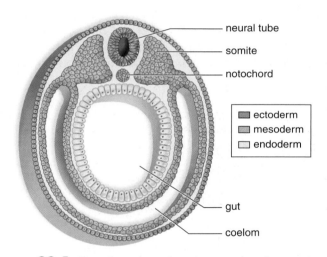

Figure 22.5 Vertebrate embryo, cross section.
At the neurula stage, each of the germ layers, indicated by color (see key), can be associated with the later development of particular parts. The somites give rise to the muscles of each segment and to the vertebrae, which replace the notochord in vertebrates.

the mesoderm splits, forming the mesodermal lining of the coelom (Fig. 22.5).

A primitive gut tube is formed by endoderm as the body itself folds into a tube. The heart, too, begins as a simple tubular pump. Organ formation continues until the germ layers have given rise to the specific organs listed in Table 22.1. Figure 22.5 will help you relate the formation of vertebrate structures and organs to the three embryonic layers of cells: the ectoderm, the mesoderm, and the endoderm.

During neurulation, the neural tube develops just above the notochord. At the neurula stage of development, cross sections of all chordate embryos are similar in appearance.

Processes of Animal Development

Aside from growth, the process of development requires (1) cellular differentiation and (2) morphogenesis. **Cellular differentiation** occurs when cells become specialized in structure and function; for example, a muscle cell looks different and acts differently than a nerve cell. **Morphogenesis** produces the shape and form of the body. One of the earliest indications of morphogenesis is cell migration to form the germ layers. Later, morphogenesis includes **pattern formation,** which means how tissues and organs are arranged in the body. Apoptosis, or programmed cell death, which was first discussed in Chapter 5, is an important part of pattern formation.

Cellular Differentiation

At one time, investigators mistakenly believed that irreversible genetic changes must account for cellular differentiation. Perhaps the genes are parceled out as development occurs, they said, and that's why cells of the body have different structures and functions. However, as we will discuss in Chapter 25 the process of cloning shows that all the nuclei of the body's cells are totipotent. Every cell in the body contains all the genes necessary to develop into an organism (see Chapter 25 Science Focus, Fig. 25A). In other words, all an organism's genes are in each cell of the body.

The process by which different tissues come about becomes clearer when we consider that specialized cells produce only certain proteins. In other words, we now know that specialization is not due to the parceling out of genes; rather, it is due to differential gene expression. Certain genes and not others are turned on in differentiated cells.

Therefore, in recent years, investigators have turned their attention to discovering the mechanisms that lead to differential gene expression during development. Differentiation must begin long before we can recognize specialized types of cells. Ectodermal, endodermal, and mesodermal cells in the gastrula look quite similar, but they must be different because they develop into different organs.

Investigations by early researchers showed that the cytoplasm of a frog's egg is not uniform. It is polar and has both an anterior/posterior axis and a dorsal/ventral axis, which can be correlated with the **gray crescent,** a gray area that appears after the sperm fertilizes the egg (Fig. 22.6a).

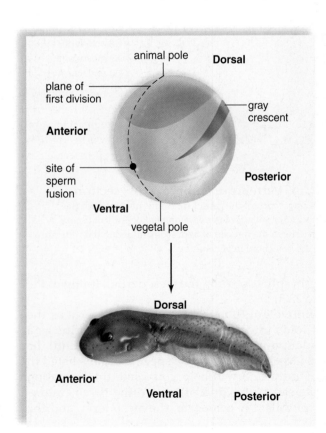

a. Frog's egg is polar and has axes.

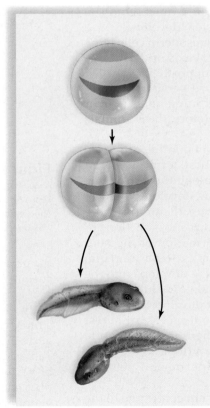

b. Each cell receives a part of the gray crescent.

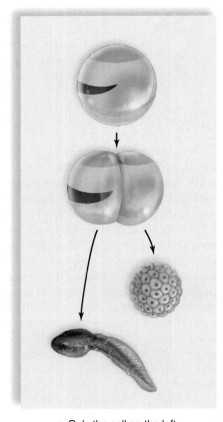

c. Only the cell on the left receives the gray crescent.

Figure 22.6 Experimental determination of cytoplasmic influence on development.
a. A frog's egg has anterior/posterior and dorsal/ventral axes that correlate with the position of the gray crescent. **b.** When researchers cut the fertilized egg such that the gray crescent is divided in half, each daughter cell is capable of developing into a complete tadpole. **c.** If the fertilized egg is cut such that only one daughter cell receives the gray crescent, then only that cell can become a complete embryo.

When researchers in the lab divided the fertilized egg such that each half contained gray crescent, each experimentally separated daughter cell developed into a complete embryo (Fig. 22.6*b*). However, if the egg was divided so that only one daughter cell received the gray crescent, only that cell became a complete embryo (Fig. 22.6*c*). This experiment allows us to hypothesize that the gray crescent must contain particular chemical signals that are needed for development to proceed normally.

The oocyte is now known to contain substances called maternal determinants that influence the course of development. These determinants consist of RNAs and proteins synthesized from the maternal genome and stored in the egg. **Cytoplasmic segregation** is the parceling out of maternal determinants as mitosis occurs:

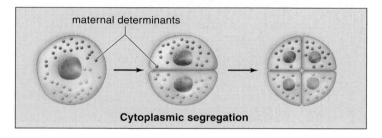

maternal determinants

Cytoplasmic segregation

Cytoplasmic segregation helps determine how the various cells of the morula will later develop. As development proceeds, specialization of cells is influenced by maternal determinants and the signals given off by neighboring cells.

Induction is the ability of one embryonic tissue to influence the development of another tissue by the use of signals called *inducers.* Inducers are chemical signals that alter the metabolism of the receiving cell and activate particular genes.

Induction and Frog Experiments   A frog embryo's gray crescent becomes the dorsal lip of the blastopore, where gastrulation begins. Since this region is necessary for com-

plete development, early investigators called the dorsal lip of the blastopore the primary organizer. The cells closest to the primary organizer become endoderm, those farther away become mesoderm, and those farthest away become ectoderm. This suggests that a molecular concentration gradient may act as a chemical signal (inducer) to cause germ layer differentiation.

The gray crescent of a frog's egg marks the dorsal side of the embryo where the mesoderm becomes notochord and ectoderm becomes nervous system. Experiments have shown that presumptive (potential) notochord tissue induces the formation of the nervous system. If presumptive nervous system tissue, located just above the presumptive notochord, is cut out and transplanted to another region of an embryo, nervous tissue does not develop in that region. The presumptive nervous system is missing some sort of signal to differentiate. On the other hand, if presumptive notochord tissue is cut out and transplanted beneath what would be ectoderm in another region, this other ectoderm does differentiate into nervous tissue. This experiment shows that signals from the presumptive notochord system cause ectoderm to differentiate into nervous tissue.

A well-known series of inductions accounts for the development of the vertebrate eye. An optic vesicle, which is a lateral outgrowth of a developing brain, induces the overlying ectoderm to thicken and become the lens of the eye. The developing lens, in turn, induces an optic vesicle to form an optic cup, where the retina forms.

Induction and Roundworm Experiments   More recently, work with the roundworm *Caenorhabditis elegans* has also shown that induction is necessary to the process of differentiation. The study of this organism, and also the fruit fly (*Drosophila melanogaster*) and the mouse (*Mus musculus*) has contributed much to modern-day developmental genetics (Fig. 22.7). These organisms are called model organisms because the study of their development produced concepts that help us understand development in general.

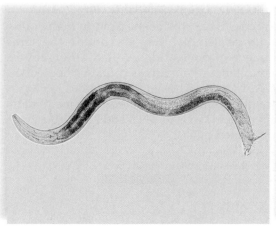

a.   b.   c.

Figure 22.7 Model organisms.
The study of (**a**) *Caenorhabditis elegans,* (**b**) *Drosophila melanogaster,* and (**c**) *Mus musculus* has furthered our understanding of animal development.

C. elegans is only 1 mm long, and vast numbers can be raised in the laboratory in either petri dishes or a liquid medium. Development of *C. elegans* takes only three days, and the adult worm contains only 959 cells. Investigators have been able to watch the developmental process from beginning to end, especially since the worm is transparent. Many modern genetic studies have been done utilizing this worm. Its entire genome has been sequenced. Individual genes have been altered and cloned, and their products have been injected into cells or extracellular fluid.

Fate maps, diagrams that trace the differentiation of developing cells, have been developed that show the destiny of each cell as it arises following successive cell divisions. Some investigators have studied in detail the development of the vulva, a pore through which eggs are laid. A cell called the anchor cell induces the vulva to form. The cell closest to the anchor cell receives the most inducer and becomes the inner vulva. This cell, in turn, produces another inducer, which acts on its two neighboring cells, and they become the outer vulva. Work with *C. elegans* has shown that a series of inductions occurs as development proceeds.

Morphogenesis

Pattern formation is the ultimate in morphogenesis. To understand the concept of pattern formation, think about the pattern of your own body. Your vertebral column runs along the main axis, and your arms and legs occur in certain locations. Pattern formation refers to how this arrangement of body parts comes about during development. Fruit fly experiments, in particular, have contributed to our knowledge of pattern formation. A fruit fly may be larger than a roundworm, but a few pairs can produce hundreds of offspring in a couple of weeks, all within a small bottle kept on a laboratory bench.

The Fruit Fly Experiments   In fruit flies, investigators have discovered certain genes, now called **morphogen genes,** that determine the relationship of individual parts. For example, some genes control which end of the animal will be the head and which the tail, and others determine how many segments the animal will have.

Each of these genes codes for a protein that is present in a gradient—cells at the start of the gradient contain high levels of the protein, and cells at the end of the gradient contain low levels of the protein. These gradients are called morphogen gradients because they determine the shape of the organism (Fig. 22.8). A morphogen gradient is efficient because it has a range of effects, depending on the particular concentration in a portion of the animal. Sequential sets of master genes code for morphogen gradients that activate the next set of master genes in turn, and that's one reason development turns out to be so orderly.

Homeotic genes act by controlling the identity of each segment. They encode master regulatory proteins that control the expression of other genes, which in turn are responsible for the development of segment-specific structures.

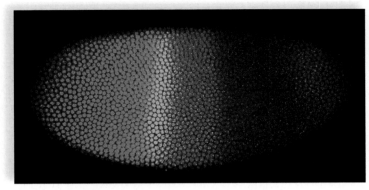
a.

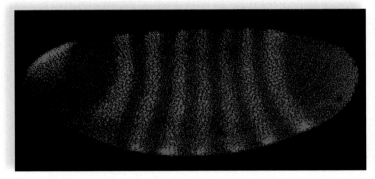
b.

Figure 22.8 Morphogen gradients in the fruit fly.
Different morphogen gradients appear as development proceeds. **a.** This early gradient determines which end is the head and which the tail. The different colors show the concentration gradients of two different proteins. **b.** Another gradient determines the number of segments.

Mutations in homeotic genes cause particular segments to develop as if they were located elsewhere in the body. For example, the second thoracic segment of the fruit fly should develop a pair of wings. However, a mutation in the homeotic gene that controls the development of the third thoracic segment allows that segment to develop wings when it should not, resulting in a four-winged fly (Fig. 22.9*a*).

Homeotic genes have now been found in many other organisms, and surprisingly, they all contain the same particular sequence of nucleotides, called a **homeobox.** (Because homeotic genes in mammals contain a homeobox, they are called *Hox* genes.) The homeobox codes for a particular sequence of 60 amino acids called a homeodomain: A homeodomain protein binds to DNA and helps determine which genes are turned on:

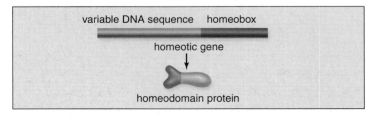

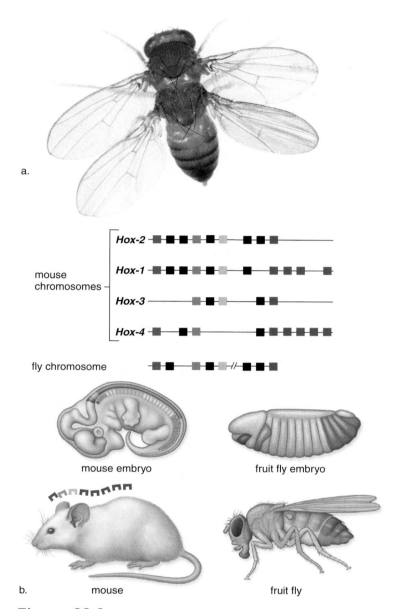

Figure 22.9 Pattern formation in *Drosophila*.
Homeotic genes control pattern formation, an aspect of morphogenesis.
a. If homeotic genes are activated at inappropriate times, abnormalities such as a fly with four wings occur. **b.** The green, blue, yellow, and red colors show that homologous homeotic genes occur on four mouse chromosomes and on a fly chromosome in the same order. These genes are color-coded to the region of the embryo, and therefore the adult, where they regulate pattern formation. The black boxes are homeotic genes that are not identical between the two animals. In mammals, homeotic genes are called *Hox* genes.

In *Drosophila*, homeotic genes are located on a single chromosome. In mice and also humans, the same four clusters of homeotic genes are located on four different chromosomes (Fig. 22.9*b*). In all three types of animals, homeotic genes are expressed from anterior to posterior in the same order. The first clusters determine the final development of

anterior segments of the embryo, while those later in the sequence determine the final development of posterior segments of the embryo.

Since the homeotic genes of so many different organisms contain the same homeodomain, we know that this nucleotide sequence arose early in the history of life and that it has been largely conserved as evolution occurred. In general, it has been very surprising to learn that developmental genetics is similar in organisms ranging from yeasts to plants to a wide variety of animals. Certainly, the genetic mechanisms of development appear to be quite similar in all animals.

Apoptosis   We have already discussed the importance of apoptosis (programmed cell death) in the normal day-to-day operation of the body (see Fig. 5.2). Apoptosis is also an important part of pattern formation in all organisms. In humans, for example, we know that apoptosis is necessary for the shaping of the hands and feet; if it does not occur, the child is born with webbing between the fingers and toes.

The fate maps of *C. elegans* indicate that apoptosis occurs in 131 cells as development takes place. When a cell-death signal is received, an inhibiting protein becomes inactive, allowing a cell-death cascade to proceed, which ends in enzymes destroying the cell.

> The processes of development include cellular differentiation and morphogenesis. Cellular differentiation includes induction, and morphogenesis (e.g., pattern formation) includes apoptosis.

22.2 Human Embryonic and Fetal Development

In humans, the length of the time from conception (fertilization followed by implantation) to birth (parturition) is approximately nine months. It is customary to calculate the time of birth by adding 280 days to the start of the last menstrual period because this date is usually known, whereas the day of fertilization is usually unknown. Because the time of birth is influenced by so many variables, only about 5% of babies actually arrive on the forecasted date.

Human development before birth is often divided into embryonic development (months 1 and 2) and fetal development (months 3–9). **Embryonic development** consists of early formation of the major organs, and fetal development is the refinement of these structures. Today, developmental biology encompasses the study of all stages of the human life cycle, from the embryonic period through infancy, childhood,

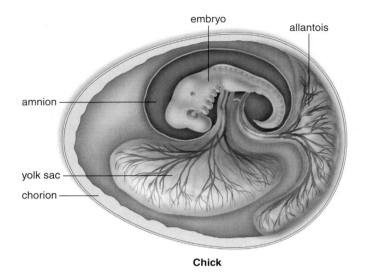

Chick

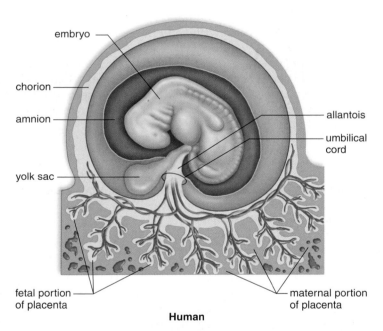

Human

Figure 22.10 Extraembryonic membranes.
Extraembryonic membranes, which are not part of the embryo, are found during the development of chicks and humans. Each has a specific function.

adolescence, and adulthood. During these stages, stem cells in particular continue to produce new cells, which undergo growth, differentiation, and morphogenesis.

Extraembryonic Membranes

Before we consider human development chronologically, we must understand the placement of **extraembryonic membranes,** which are best understood by considering their function in reptiles and birds. In reptiles, these membranes made development on land possible. If an embryo develops in the water, the water supplies oxygen for the embryo and takes away waste products. The surrounding water prevents desiccation, or drying out, and provides a protective cushion. For an embryo that develops on land, all these functions are performed by the extraembryonic membranes.

In the chick, the extraembryonic membranes develop from extensions of the germ layers, which spread out over the yolk. Figure 22.10 (*top*) shows the chick surrounded by the membranes. The **chorion** lies next to the shell and carries on gas exchange. The **amnion** contains the protective amniotic fluid, which bathes the developing embryo. The **allantois** collects nitrogenous wastes, and the **yolk sac** surrounds the remaining yolk, which provides nourishment.

Humans (and other mammals) also have these extraembryonic membranes (Fig. 22.10, *bottom*) The chorion develops into the fetal half of the placenta; the yolk sac, which has little yolk, is the first site of blood cell formation; the allantoic blood vessels become the umbilical blood vessels; and the amnion contains fluid to cushion and protect the embryo, which develops into a fetus. Therefore, the functions of the extraembryonic membranes in humans have been modified to suit internal development, but their very presence indicates our relationship to birds and to reptiles. It is interesting to note that all chordate animals develop in water, either in bodies of water or within amniotic fluid.

> The presence of extraembryonic membranes in reptiles made development on land possible. Humans also have these membranes, but their functions have been modified for internal development.

Embryonic Development

Embryonic development encompasses the first two months of development following fertilization.

The First Week

Fertilization occurs in the distal third (closest to the ovary) of an oviduct (Fig. 22.11), and cleavage begins even as the embryo passes down this duct to the uterus. By the time the embryo reaches the uterus on the third day, it is a morula. The morula is not much larger than the zygote because, even though multiple cell divisions have occurred, the newly formed cells do not grow. By about the fifth day, the morula is transformed into the blastocyst. The **blastocyst** has a fluid-filled cavity, a single layer of outer cells called the **trophoblast,** and an inner cell mass. Later, the trophoblast, reinforced by a layer of mesoderm, gives rise to the chorion, one of the extraembryonic membranes (see Fig. 22.10). The inner cell mass eventually becomes the embryo, which develops into a fetus. Embryonic stem cells are derived from this inner cell mass.

The Second Week

At the end of the first week, the embryo begins the process of implanting in the wall of the uterus. The trophoblast secretes enzymes to digest away some of the tissue and blood vessels of the endometrium of the uterus (Fig. 22.11). The embryo is now about the size of the period at the end of this sentence. The trophoblast begins to secrete the hormone **human chorionic gonadotropin (hCG),** which is the basis for the pregnancy test and serves to maintain the corpus luteum past the time it normally disintegrates (see Chapter 21). Because of this, the endometrium is maintained and menstruation does not occur.

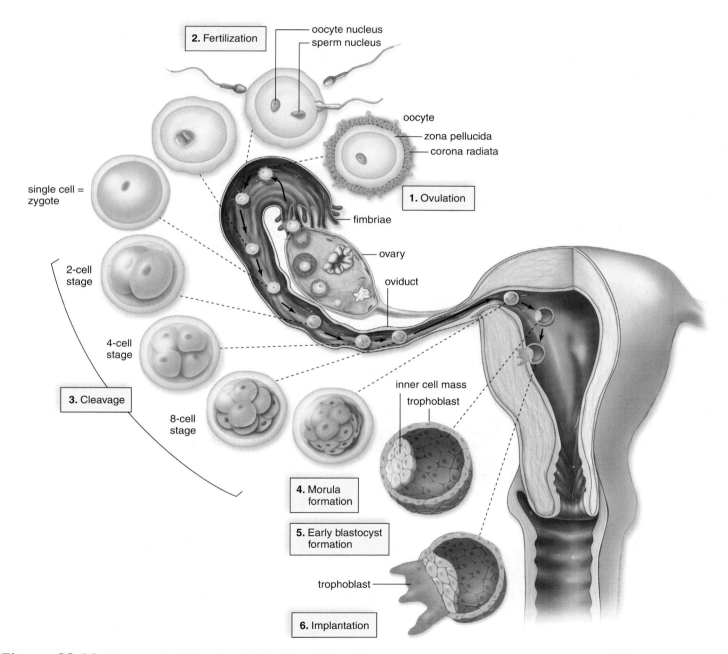

Figure 22.11 Human development before implantation.
Structures and events proceed counterclockwise. (1) At ovulation, the secondary oocyte leaves the ovary. A single sperm nucleus enters the oocyte and (2) fertilization occurs in the oviduct. As the zygote moves along the oviduct, it undergoes (3) cleavage to produce (4) a morula. (5) The blastocyst forms and (6) implants itself in the uterine lining.

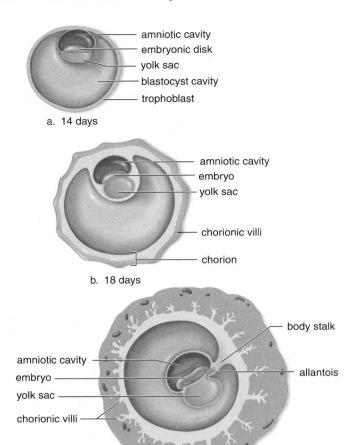

a. 14 days

- amniotic cavity
- embryonic disk
- yolk sac
- blastocyst cavity
- trophoblast

b. 18 days

- amniotic cavity
- embryo
- yolk sac
- chorionic villi
- chorion

c. 21 days

- body stalk
- amniotic cavity
- embryo
- yolk sac
- chorionic villi
- allantois

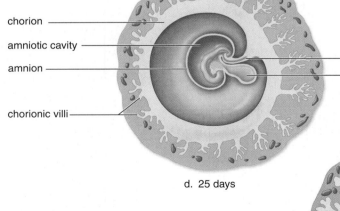

d. 25 days

- chorion
- amniotic cavity
- amnion
- chorionic villi
- allantois
- yolk sac

Figure 22.12 Human embryonic development.
a. At first, the embryo contains no organs, only tissues. The amniotic cavity is above the embryonic disk, and the yolk sac is below. **b.** The chorion develops villi, the structures so important to the exchange between mother and child. **c, d.** The allantois and yolk sac, two more extraembryonic membranes, are positioned inside the body stalk as it becomes the umbilical cord. **e.** At 35+ days, the embryo has a head region and a tail region. The umbilical cord takes blood vessels between the embryo and the chorion (placenta).

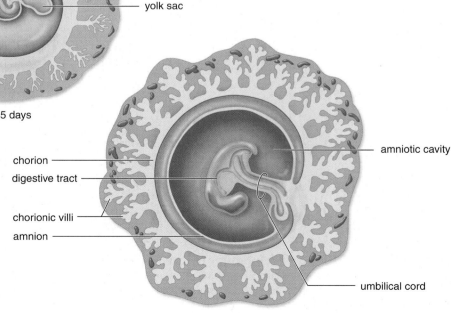

- chorion
- digestive tract
- chorionic villi
- amnion
- amniotic cavity
- umbilical cord

e. 35+ days

As the week progresses, the inner cell mass detaches itself from the trophoblast, and two more extraembryonic membranes form (Fig. 22.12*a*). The yolk sac, which forms below the embryonic disk, has no nutritive function as it does in chicks, but it is the first site of blood cell formation. The amnion has a cavity where the embryo (and then the fetus) develops. In humans, amniotic fluid acts as an insulator against cold and heat and also absorbs shock, such as that caused by the mother exercising.

Gastrulation occurs during the second week. The inner cell mass now has flattened into the **embryonic disk,** composed of two layers of cells: ectoderm above and endoderm below. Once the embryonic disk elongates to form the so-called primitive streak, the third germ layer, mesoderm, forms by invagination of cells along the streak. The trophoblast is re-inforced by mesoderm and becomes the chorion (Fig. 22.12*b*).

It is possible to relate the development of future organs to these germ layers (see Table 22.1).

The Third Week

Two important organ systems appear during the third week. The nervous system is the first organ system to be visually evident. At first, a thickening appears along the entire dorsal length of the embryo, and then invagination occurs as neural folds appear. When the neural folds meet

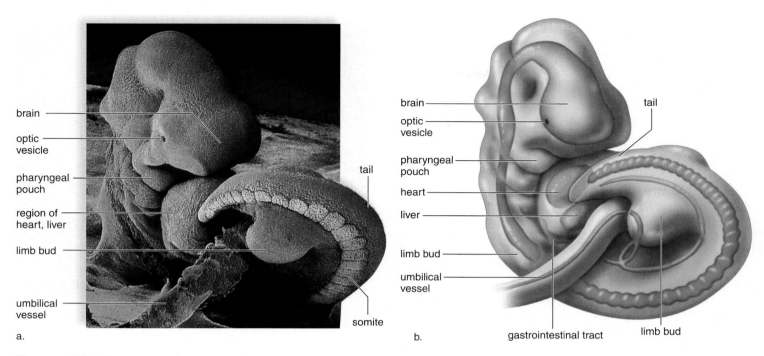

Figure 22.13 Human embryo at beginning of fifth week.
a. Scanning electron micrograph. **b.** The embryo is curled so that the head touches the heart and liver, the two organs whose development is farther along than the rest of the body. The organs of the gastrointestinal tract are forming, and the arms and the legs develop from the bulges that are called limb buds. The tail is an evolutionary remnant; its bones regress and become those of the coccyx (tailbone). The pharyngeal pouches become functioning gills only in fishes and amphibian larvae; in humans, the first pair of pharyngeal pouches becomes the auditory tubes. The second pair becomes the tonsils, while the third and fourth pairs become the thymus gland and the parathyroid glands.

at the midline, the neural tube, which later develops into the brain and the nerve cord, is formed (see Fig. 22.4). After the notochord is replaced by the vertebral column, the nerve cord is called the spinal cord.

Development of the heart begins in the third week and continues into the fourth week. At first, there are right and left heart tubes; when these fuse, the heart begins pumping blood, even though its chambers are not fully formed. The veins enter posteriorly and the arteries exit anteriorly from this largely tubular heart, but later the heart twists so that all major blood vessels are located anteriorly.

The Fourth and Fifth Weeks

At four weeks, the embryo is barely larger than the height of this print. A bridge of mesoderm called the body stalk connects the caudal (tail) end of the embryo with the chorion, which has treelike projections called **chorionic villi** (Fig. 22.12*c*, *d*). The fourth extraembryonic membrane, the allantois, is contained within this stalk, and its blood vessels become the umbilical blood vessels. The head and the tail then lift up, and the body stalk moves anteriorly by constriction. Once this process is complete, the **umbilical cord,** which connects the developing embryo to the placenta, is fully formed (Fig. 22.12*e*).

Little flippers called limb buds appear (Fig. 22.13); later, the arms and the legs develop from the limb buds, and even the hands and the feet become apparent. At the same time—during the fifth week—the head enlarges, and the sense organs become more prominent. It is possible to make out the developing eyes, ears, and even the nose.

The Sixth Through Eighth Weeks

A remarkable change in external appearance takes place during the sixth through eighth weeks of development. The embryo changes from a form that is difficult to recognize as human to one easily recognized as human. Concurrent with brain development, the head achieves its normal relationship with the body as a neck region develops. The nervous system is developed well enough to permit reflex actions, such as a startle response to touch. At the end of this period, the embryo is about 38 mm (1.5 inches) long and weighs no more than an aspirin tablet, even though all organ systems have been established.

During embryonic development all major organs start to develop.

Fetal Development and Birth

Fetal development includes the third through ninth months of development. At this time, the fetus looks human (Fig. 22.14).

The Third and Fourth Months

At the beginning of the third month, the fetal head is still very large, the nose is flat, the eyes are far apart, and the ears are well formed. Head growth now begins to slow down as the rest of the body increases in length. Epidermal refinements, such as eyelashes, eyebrows, hair on head, fingernails, and nipples, appear.

Cartilage begins to be replaced by bone as ossification centers appear in most of the bones. Cartilage remains at the ends of the long bones, and ossification is not complete until age 18 or 20 years. The skull has six large membranous areas called **fontanels,** which permit a certain amount of flexibility as the head passes through the birth canal and allow rapid growth of the brain during infancy. Progressive fusion of the skull bones causes the fontanels to usually close by 16 months of age.

Sometime during the third month, it is possible to distinguish males from females. Researchers have discovered a series of genes on the X and Y chromosomes that cause the differentiation of gonads into testes or ovaries. Once these have differentiated, they produce the sex hormones that influence the differentiation of the genital tract.

At this time, either testes or ovaries are located within the abdominal cavity, but later, in the last trimester of fetal development, the testes descend into the scrotal sacs as discussed in Chapter 21. Sometimes the testes fail to descend, and in that case, an operation may be done later to place them in their proper location.

During the fourth month, the fetal heartbeat is loud enough to be heard when a physician applies a stethoscope to the mother's abdomen. By the end of this month, the fetus is about 152 mm (6 inches) in length and weighs about 171 grams (6 oz).

> During the third and fourth months, it is obvious that the skeleton is becoming ossified. The sex of the individual can now be distinguished.

The Fifth Through Seventh Months

During the fifth through seventh months, the mother begins to feel movement. At first, there is only a fluttering sensation, but as the fetal legs grow and develop, kicks and jabs are felt. The fetus, though, is in the fetal position, with the head bent down and in contact with the flexed knees.

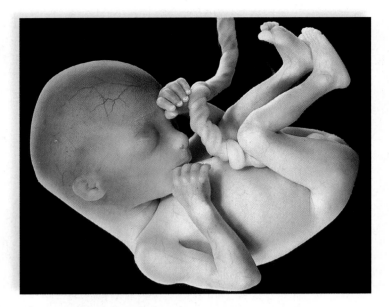

Figure 22.14 Three- to four-month-old fetus.
At this stage, the fetus looks human, with its face, hands, and fingers well defined.

The wrinkled, translucent skin is covered by a fine down called **lanugo.** This, in turn, is coated with a white, greasy, cheeselike substance called **vernix caseosa,** which probably protects the delicate skin from the amniotic fluid. The eyelids are now fully open.

At the end of this period, the length has increased to about 300 mm (12 inches), and the weight is about 1,380 grams (3 lb). It is possible that, if born now, the baby will survive.

Fetal Circulation

The fetus has circulatory features that are not present in the adult circulation (Fig. 22.15). All of these features are necessary because the fetus does not use its lungs for gas exchange. For example, much of the blood entering the right atrium is shunted into the left atrium through the **oval opening** (foramen ovale) between the two atria. Also, any blood that does enter the right ventricle and is pumped into the pulmonary trunk is shunted into the aorta by way of the **arterial duct** (ductus arteriosus).

Blood within the aorta travels to the various branches, including the iliac arteries, which connect to the **umbilical arteries** leading to the placenta. Exchange of gases and nutrients between maternal blood and fetal blood takes place at the placenta. The **umbilical vein** carries blood rich in nutrients and oxygen to the fetus. The umbilical vein enters the liver and then joins the **venous duct,** which merges with the inferior vena cava, a vessel that returns blood to the heart. It is interesting to note that the umbilical arteries and vein run alongside one another in the umbilical cord, which is cut at birth, leaving only the umbilicus (navel).

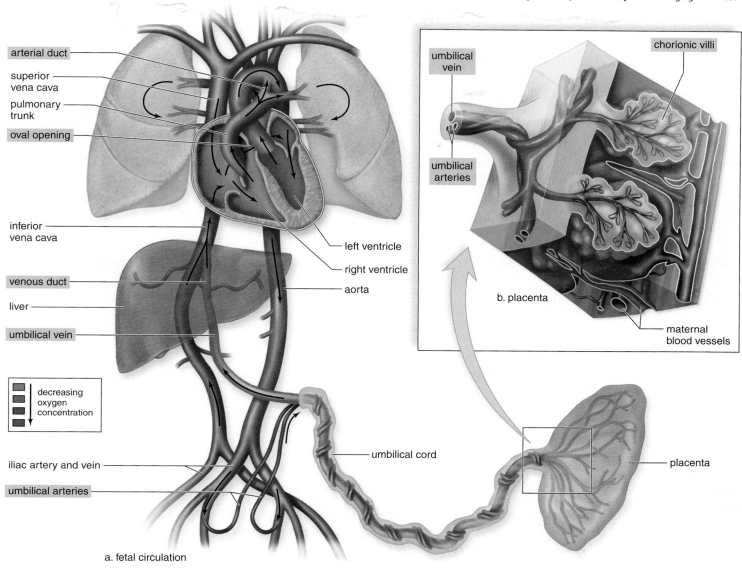

a. fetal circulation

arterial duct
superior vena cava
pulmonary trunk
oval opening
inferior vena cava
venous duct
liver
umbilical vein

decreasing oxygen concentration

iliac artery and vein
umbilical arteries

left ventricle
right ventricle
aorta

umbilical cord

umbilical vein
umbilical arteries
chorionic villi
b. placenta
maternal blood vessels

placenta

Figure 22.15 Fetal circulation and the placenta.

a. The lungs are not functional in the fetus, and the blood passes directly from the right atrium to the left atrium or from the right ventricle to the aorta. The umbilical arteries take fetal blood to the placenta, and the umbilical vein returns fetal blood from the placenta. **b.** At the placenta, an exchange of molecules between fetal and maternal blood takes place across the walls of the chorionic villi. Oxygen and nutrient molecules diffuse into the fetal blood, and carbon dioxide and urea diffuse out of the fetal blood. Labels for blood vessels unique to a fetus are highlighted in red.

The most common of all cardiac defects in the newborn is the persistence of the oval opening. Once the umbilical cord has been tied and the lungs have expanded, blood enters the lungs in quantity. The return of this blood to the left side of the heart usually causes a flap to cover the opening. Incomplete closure occurs in nearly one out of four individuals, but even so, passage of the blood from the right atrium to the left atrium rarely occurs because either the opening is small or it closes when the atria contract. In a small number of cases, the passage of impure blood from the right side to the left side of the heart is sufficient to cause a "blue baby." Such a condition can now be corrected by open heart surgery.

The arterial duct closes because endothelial cells divide and block off this duct. Remains of the arterial duct and parts of the umbilical arteries and vein are later transformed into connective tissue.

In the fetus, blood is shunted away from the lungs and moves to and away from the placenta through the umbilical blood vessels located in the umbilical cord. Exchange of substances between fetal blood and maternal blood takes place at the placenta, which forms from the chorion and uterine tissue.

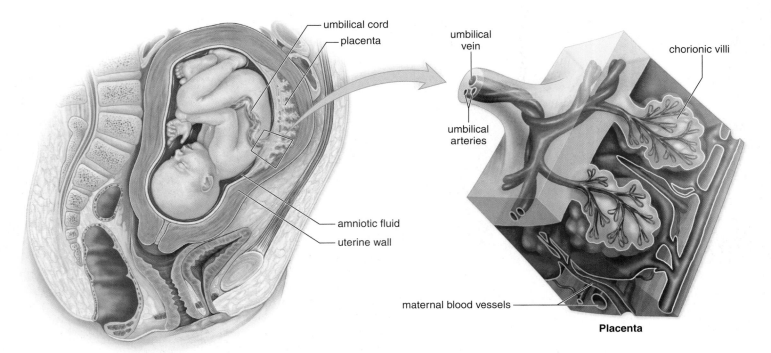

Figure 22.16 Anatomy of the placenta in a fetus at six to seven months.

The placenta is composed of both fetal and maternal tissues. Chorionic villi penetrate the uterine lining and are surrounded by maternal blood. Exchange of molecules between fetal and maternal blood takes place across the walls of the chorionic villi.

Structure and Function of the Placenta   Human beings belong to the group of mammals called placental mammals. The **placenta** is firmly attached to the uterine wall by the allantois and by fingerlike projections called the chorionic villi. The placenta is clearly a structure that functions only before birth, and when the child is born, the placenta becomes part of the afterbirth, as described later in this section.

The placenta functions in gas, nutrient, and waste exchange between the embryonic (later fetal) and maternal circulatory systems. The placenta begins to form once the embryo is fully implanted. At first, the entire chorion has chorionic villi that project into the endometrium. Later, these disappear in all areas except where the placenta develops. By the tenth week, the placenta is fully formed and is producing progesterone and estrogen. These hormones have two effects: (1) due to their negative feedback control of the hypothalamus and the anterior pituitary, they prevent any new follicles from maturing, and (2) they maintain the lining of the uterus so that the corpus luteum is no longer needed. Usually no menstruation occurs during pregnancy.

The placenta has a fetal side contributed by the chorion and a maternal side consisting of uterine tissues. Notice in Figure 22.16 how the chorionic villi are surrounded by maternal blood; yet, maternal and fetal blood never mix, since exchange always takes place across the walls of the chorionic villi. Carbon dioxide and other wastes move from the fetal side to the maternal side of the placenta, and nutrients and oxygen move from the maternal side to the fetal side.

The umbilical cord stretches between the placenta and the fetus. Although the umbilical cord may seem to travel from the placenta to the intestine, actually it is simply taking fetal blood to and from the placenta. The umbilical cord is the lifeline of the fetus because it contains the umbilical arteries and vein, which transport waste molecules (carbon dioxide and urea) to the placenta for disposal and take oxygen and nutrient molecules from the placenta to the rest of the fetal circulatory system.

As discussed in the Health Focus, harmful chemicals can also cross the placenta. This is of particular concern during the embryonic period, when various structures are first forming. Each organ or part seems to have a sensitive period during which a substance can alter its normal development. For example, if a woman takes the drug thalidomide, a tranquilizer, between days 27 and 40 of her pregnancy, the infant is likely to be born with deformed limbs. If thalidomide is taken after day 40, however, the infant is born with normal limbs.

During mammalian development, the embryo and later the fetus depends on the placenta for gas exchange, for acquiring nutrients, and for ridding the body of wastes.

Birth

The uterus has contractions during the last trimester of the pregnancy. At first, these are light, lasting about 20–30 seconds and occurring every 15–20 minutes. Near the end of pregnancy, the contractions may become stronger and more frequent; however, the onset of true labor is marked by uterine contractions that occur regularly every 10–15 minutes and last for 40 seconds or longer. These contractions are generally called "labor pains," although, when they first appear, they are not usually painful and only become painful at a later stage.

A positive feedback mechanism regulates the onset and continuation of labor. Uterine contractions are induced by stretching of the cervix, which also brings about the release of oxytocin from the posterior pituitary. Oxytocin stimulates uterine contractions, which push the fetus downward, and the cervix stretches even more. This cycle keeps repeating itself until the baby is born. Three events, in any order, indicate that delivery will soon occur: (1) The uterine contractions are occurring about every five minutes and becoming stronger. (2) The amnion, which contains amniotic fluid, ruptures, causing water to flow out of the vagina.

This event is sometimes referred to as "breaking water." (3) A plug of mucus from the cervical canal leaves the vagina. This plug prevents bacteria and sperm from entering the uterus during pregnancy. Its expulsion may be the least noticeable of the events, unless it is mixed with blood.

Stage 1

Parturition, the process of giving birth to an offspring, can be divided into three stages (Fig. 22.17). Prior to the first stage, there can be a "bloody show" caused by expulsion of the mucous plug. At first, the uterine contractions of labor occur in such a way that the cervical canal slowly disappears as the lower part of the uterus is pulled upward toward the baby's head (Fig. 22.17*b*). This process is called effacement, or "taking up the cervix." With further contractions, the baby's head acts as a wedge to assist cervical dilation. If the amniotic membrane has not already ruptured, it is apt to do so during this stage, releasing the amniotic fluid, which leaks out the vagina. The first stage of parturition ends once the cervix is dilated completely.

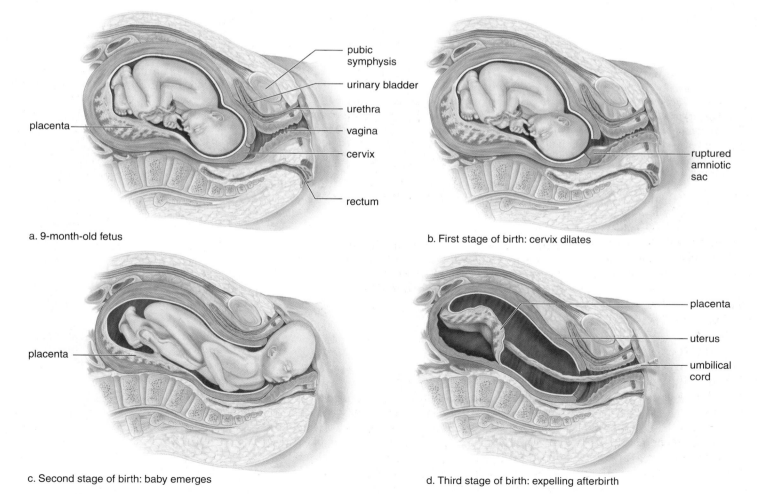

a. 9-month-old fetus

b. First stage of birth: cervix dilates

c. Second stage of birth: baby emerges

d. Third stage of birth: expelling afterbirth

Figure 22.17 Three stages of parturition (birth).
a. Position of fetus just before birth begins. **b.** Dilation of cervix. **c.** Birth of baby. **d.** Expulsion of afterbirth.

While some birth defects are not preventable, others are, and therefore all females of childbearing age are advised to take everyday precautions to protect any future and/or presently developing embryos and fetuses from these defects. For example, it is best if a woman has a physical exam even before she becomes pregnant. At that time, it can be determined if she has been immunized against rubella (German measles). Depending on exactly when a pregnant woman has the disease, rubella can cause blindness, deafness, mental retardation, heart malformations, and other serious problems in an unborn child. A vaccine to prevent the disease can be given to a woman before she gets pregnant but not to a woman who is already pregnant because it contains live viruses. Also, the presence of HIV (the causative agent for AIDS) should be tested for because preventive therapies are available to improve maternal and infant health.

Good health habits are a must during pregnancy, including proper nutrition, adequate rest, and exercise. Moderate exercise can usually continue throughout pregnancy and hopefully will contribute to ease of delivery. Basic nutrients are required in adequate amounts to meet the demands of both fetus and mother. A growing number of studies confirm that small, thin newborns are more likely to develop certain chronic diseases, such as diabetes and high blood pressure, when they become adults than are babies who are born heavier.

An increased amount of minerals, such as calcium for bone growth and iron for red blood cell formation, and certain vitamins, such as vitamin B_6 for proper metabolism and folate (folic acid), are required. A pregnant woman needs more folate per day to meet an increased rate of cell division and DNA synthesis in her own body and in that of the developing child. A maternal deficiency of folate has been linked to development of neural tube defects in the fetus, which include spina bifida (spinal cord or spinal fluid bulge through the back) and anencephaly (absence of a brain). Perhaps as many as 75% of these defects could be avoided by adequate folate intake even before pregnancy occurs. Consuming fortified breakfast cereals is a good way to meet

TABLE 22A	Behaviors Harmful to the Unborn

Drinking alcohol

Smoking cigarettes

Taking illegal drugs

Taking any medication not approved by a physician

Exposure to environmental toxins and radiation

folate needs, because they contain a more absorbable form of folate.

As shown on Figure 22.16, the placenta is the region for exchange between the fetal and maternal circulatory systems. Good health habits include avoiding the substances listed in Table 22A, which can cross the placenta and harm the fetus. Cigarette smoke poses a serious threat to the health of a fetus because it contains not only carbon monoxide but also other fetotoxic chemicals. Children born to smoking mothers have a greater chance of a cleft lip or palate, increased incidence of respiratory diseases, and later on, more reading disorders than those born to mothers who did not smoke during pregnancy.

Alcohol easily crosses the placenta, and only one drink a day appears to increase the chance of a spontaneous abortion. The more alcohol consumed, the greater the chance of physical abnormalities if the pregnancy continues. Heavy consumption of alcohol puts a fetus at risk of a mental defect because alcohol enters the brain of the fetus. Babies born to heavy drinkers are apt to undergo delirium tremens after birth—shaking, vomiting, and extreme irritability—and to have fetal alcohol syndrome (FAS). Children with FAS have decreased weight, height, and head size, with malformation of the head and face (Fig. 22A). Later, mental retardation is common, as are numerous other physical malformations.

Certainly, pregnant women should completely avoid illegal drugs, such as marijuana, cocaine, and heroin. "Cocaine babies" now make up 60% of drug-affected babies. Severe fluctuations in blood pressure produced by the use of cocaine temporarily deprive the developing brain of oxygen. Cocaine babies have visual problems, lack coordination, and are mentally retarded.

Children born to women who received X-ray treatment during pregnancy for, say, cancer are apt to have birth defects and/or to develop leukemia later. Even low X-ray levels are apt to cause mutations in the developing embryo or fetus. Dental and other diagnostic X rays that result in only a small amount of radiation are probably safe. Still, a woman should be sure her physician knows that she is or may be pregnant. Similarly, toxic chemicals, such as pesticides, and many organic industrial chemicals, such as vinyl chloride, formaldehyde, asbestos, and benzenes, are mutagenic and can cross the placenta, resulting in abnormalities. Lead circulating in a pregnant woman's blood can cause a child to be mentally retarded.

A woman has to be very careful about taking medications while pregnant. Excessive vitamin A, sometimes used to treat acne, may damage an embryo. In the 1950s and 1960s, DES (diethylstilbestrol), a synthetic hormone related to the natural female hormone estrogen, was given to pregnant women to prevent cramps, bleeding, and threatened miscarriage. But in the 1970s and 1980s, some adolescent girls and young women

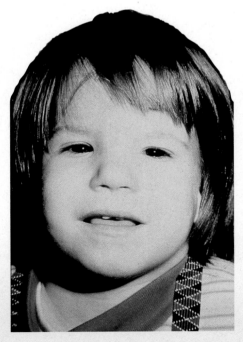

Figure 22A A child with fetal alcohol syndrome.

whose mothers had been treated with DES showed various abnormalities of the reproductive organs and an increased tendency toward cervical cancer. Other sex hormones, including birth control pills, can possibly cause abnormal fetal development, including abnormalities of the sex organs.

The drug thalidomide was a popular tranquilizer during the 1950s and 1960s in many European countries and to a degree in the United States. The drug, which was taken to prevent nausea in pregnant women, arrested the development of arms and legs in some children and also damaged heart and blood vessels, ears, and the digestive tract. Some mothers of affected children report that they took the drug for only a few days. Because of such experiences, physicians are generally very cautious about prescribing drugs during pregnancy, and no pregnant woman should take any drug—even an ordinary cold remedy—without first checking with her physician.

Unfortunately, immunization for sexually transmitted diseases is not possible. The AIDS virus can cross the placenta and cause mental retardation. As mentioned, proper medication can greatly reduce the chance of this happening. When a mother has herpes, gonorrhea, or chlamydia, newborns can become infected as they pass through the birth canal. Blindness and other physical and mental defects may develop. Birth by cesarean section could prevent these occurrences.

Rh-negative women who have given birth to Rh-positive children should receive an Rh immunoglobulin injection within 72 hours to prevent the mother's body from producing Rh antibodies. She will start producing these antibodies when some of the child's Rh-positive red blood cells enter her bloodstream, possibly before but particularly at birth. Rh antibodies can cause nervous system and heart defects in a fetus. The first Rh-positive baby is not usually affected. But in subsequent pregnancies, antibodies created at the time of the first birth cross the placenta and begin to destroy the blood cells of the fetus, causing anemia and other complications.

The birth defects we have been discussing are particularly preventable because they are not due to inheritance of an abnormal number of chromosomes or any other genetic abnormality. More women are having babies after the age of 35, and first births among women older than 40 have increased by 50% since 1980. The chance of an older woman bearing a child with a birth defect unrelated to genetic inheritance is no greater than that of a younger woman. However, as discussed in Chapter 26 there is a greater risk of an older woman having a child with a chromosomal abnormality, leading to premature delivery, cesarean section, low birth weight, and certain syndromes. Chapter 26 discusses how to detect chromosomal and other genetic defects in utero so that therapy for these disorders can begin as soon as possible.

Now that physicians and laypeople are aware of the various ways birth defects can be prevented, it is hoped that the incidence of birth defects will decrease in the future (Fig. 22B).

Discussion Questions

1. Should a woman be prosecuted for causing damage to her unborn child by drinking or illegal drug use?
2. Why does good medical care during pregnancy only reduce the *risk* of birth defects, but not guarantee a healthy child?
3. Since folate, which has been shown to reduce the risk of neural tube defects, is needed by the embryo during the early weeks of development, when should a woman begin to take extra folate?

Figure 22B Child care.
Proper care of a child begins before birth in order to prevent birth defects.

Stage 2

During the second stage of parturition, the uterine contractions occur every 1–2 minutes and last about one minute each. They are accompanied by a desire to push, or bear down. As the baby's head gradually descends into the vagina, the desire to push becomes greater. When the baby's head reaches the exterior, it turns so that the back of the head is uppermost (Fig. 22.17c). If the vaginal orifice does not expand enough to allow passage of the head, an **episiotomy** may be performed. This incision, which enlarges the opening, is sewn together later. As soon as the head is delivered, the baby's shoulders rotate so that the baby faces either to the right or the left. At this time, the physician may hold the head and guide it downward, while one shoulder and then the other emerges. The rest of the baby follows easily.

Once the baby is breathing normally, the umbilical cord is cut and tied, severing the child from the placenta. The stump of the cord shrivels and leaves a scar, which is the umbilicus.

Stage 3

The placenta, or **afterbirth,** is delivered during the third stage of parturition (Fig. 22.17d). About 15 minutes after delivery of the baby, uterine muscular contractions shrink the uterus and dislodge the placenta. The placenta is then expelled into the vagina. As soon as the placenta and its membranes are delivered, the third stage of parturition is complete.

> **During the first stage of parturition, the cervix dilates; during the second stage, the child is born; and during the third stage, the afterbirth is expelled.**

Female Breast and Lactation

A female breast contains 15 to 25 lobules, each with a milk duct. Each milk duct begins at the nipple and divides into numerous other ducts that end in blind sacs called alveoli (Fig. 22.18).

During pregnancy, the breasts enlarge as the ducts and alveoli increase in number and size. The same hormones that affect the mother's breasts can also affect the child's. Some newborns, including males, even secrete a small amount of milk for a few days.

Usually, no milk is produced during pregnancy. The hormone prolactin is needed for lactation to begin, and the production of this hormone is suppressed because of the feedback control that the increased amount of estrogen and progesterone during pregnancy has on the pituitary. Once the baby is delivered, however, the pituitary begins secreting prolactin. It takes a couple of days for milk production to begin, and in the meantime, the breasts produce **colostrum,** a thin, yellow, milky fluid rich in protein, including antibodies.

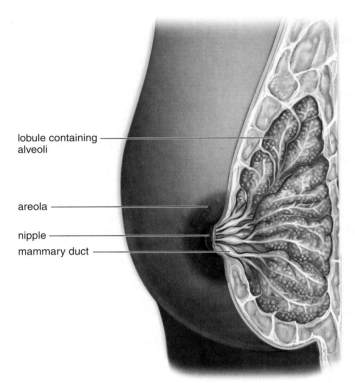

lobule containing alveoli

areola

nipple

mammary duct

Figure 22.18 Female breast anatomy.
The female breast contains lobules consisting of ducts and alveoli. The alveoli are lined by milk-producing cells in the lactating (milk-producing) breast. Milk droplets congregate in the ducts that enter sinuses just behind the nipple. When a baby suckles, the sinuses are squeezed.

The continued production of milk requires a suckling child. When a breast is suckled, the nerve endings in the areola are stimulated, and a nerve impulse travels along neural pathways from the nipples to the hypothalamus, which directs the pituitary gland to release the hormone oxytocin. When this hormone arrives at the breast, it causes contraction of the lobules so that milk flows into the ducts (called milk letdown), where it may be drawn out of the nipple by the suckling child.

Whether to breast-feed or not is a private decision based in part on a woman's particular circumstances. However, it is well known that breast milk contains antibodies produced by the mother than can help a baby survive. Babies have immature immune systems, less stomach acid to destroy foreign antigens, and also unsanitary habits. Breast-fed babies are less likely to develop stomach and intestinal illnesses, including diarrhea, during the first 13 weeks of life. Breast-feeding also has physiological benefits for the mother. Suckling causes uterine contractions that can help the uterus return to its normal size, and breast-feeding uses up calories and thus can help a woman return to her normal weight.

> **Following delivery of the baby, the breast tissue produces colostrum, followed by milk.**

22.3   Human Development After Birth

Development does not cease once birth has occurred but continues throughout the stages of life: infancy, childhood, adolescence, and adulthood. **Aging** encompasses these progressive changes, which contribute to an increased risk of infirmity, disease, and death (Fig. 22.19).

Today, there is great interest in **gerontology,** the study of aging, because our society includes more older individuals than ever before, and the number is expected to rise even more. In the next half-century, the number of people over age 65 will increase by 147%. The human life span is judged to be a maximum of 120–125 years. The present goal of gerontology is not necessarily to increase the life span, but to increase the health span, the number of years an individual enjoys the full functions of all body parts and processes.

Hypotheses About Aging

Of the many hypotheses about what causes aging, three are considered here.

Genetic in Origin

Several lines of evidence indicate that aging has a genetic basis: (1) The number of times a cell divides is species-specific. Perhaps as we grow older, more and more cells are unable to divide, and instead, they undergo degenerative changes and die. (2) Some cell lines may become nonfunctional long before the maximum number of divisions has occurred. Whenever DNA replicates, mutations can occur, and this can lead to the production of nonfunctional proteins. Eventually, the number of inadequately functioning cells can build up, which contributes to the aging process. (3) The children of long-lived parents tend to live longer than those of short-lived parents. Recent work suggests that when an animal produces fewer free radicals, it lives longer. Free radicals are unstable molecules that have an unpaired electron. In order to stabilize themselves, free radicals react with another molecule, such as DNA or proteins (e.g., enzymes) or lipids found in plasma membranes. Eventually, these molecules are unable to function, and the cell is destroyed. Certain genes code for antioxidant enzymes that detoxify free radicals. Research suggests that animals with particular forms of these genes—and therefore more efficient antioxidant enzymes—live longer.

Whole-Body Process

A decline in the hormonal system can affect many different organs of the body. For example, type 2 diabetes is common in older individuals. The pancreas makes insulin, but the cells lack the receptors that enable them to respond. Menopause in women occurs for a similar reason. There is plenty of follicle-stimulating hormone in the bloodstream, but the ovaries do not respond. Perhaps aging results from the loss of hormonal activities and a decline in the functions they control.

Figure 22.19 Aging.

Aging is a slow process during which the body undergoes changes that eventually bring about death, even if no marked disease or disorder is present. Medical science is trying to extend the human life span as well as the health span, the length of time the body functions normally.

The immune system, too, no longer performs as it once did, and this can affect the body as a whole. The thymus gland gradually decreases in size, and eventually most of it is replaced by fat and connective tissue. The incidence of cancer increases among the elderly, which may signify that the immune system is no longer functioning as it should. This idea is also substantiated by the increased incidence of autoimmune diseases in older individuals.

It is possible, though, that aging is not due to the failure of a particular system that can affect the body as a whole, but to a specific type of tissue change that affects all of the organs and even the genes. It has been noticed for some time that proteins—such as the collagen fibers present in many support tissues—become increasingly cross-linked as people age. Undoubtedly, this cross-linking contributes to the stiffening and loss of elasticity characteristic of aging tendons and ligaments. It may also account for the inability of such organs as the blood vessels, the heart, and the lungs to function as they once did. Some researchers have now found that glucose has the tendency to attach to any type of protein, which is the first step in a cross-linking process. They are presently experimenting with drugs that can prevent cross-linking.

Extrinsic Factors

The current data about the effects of aging are often based on comparisons of the elderly to younger age groups. But perhaps today's elderly were not as aware when they were younger of the importance, for example, of diet and exercise to general health. It is possible, then, that much of what we attribute to aging is instead due to years of poor health habits.

Consider, for example, osteoporosis. This condition is associated with a progressive decline in bone density in both males and females so that fractures are more likely to occur after only minor trauma. Osteoporosis is common in the elderly—by age 65, one-third of women will have vertebral fractures, and by age 81, one-third of women and one-sixth of men will have suffered a hip fracture. While there is no denying that a decline in bone mass occurs as a result of aging, certain extrinsic factors are also important. The occurrence of osteoporosis itself is associated with cigarette smoking, heavy alcohol intake, and inadequate calcium intake. Not only is it possible to eliminate these negative factors by personal choice, but it is also possible to add a positive factor. A moderate exercise program has been found to slow the progressive loss of bone mass.

Even more important, a sensible exercise program and a proper diet that includes at least five servings of fruits and vegetables a day will most likely help eliminate cardiovascular disease. Experts no longer believe that the cardiovascular system necessarily suffers a large decrease in functioning ability with age. Persons 65 years of age and older can have well-functioning hearts and open coronary arteries if their health habits are good and they continue to exercise regularly.

Effect of Age on Body Systems

Data about how aging affects body systems are necessarily based on past events. It is possible that, in the future, age will not have these effects or at least not to the same degree as those described here.

Skin

As aging occurs, skin becomes thinner and less elastic because the number of elastic fibers decreases and the collagen fibers undergo cross-linking, as discussed previously. Also, there is less adipose tissue in the subcutaneous layer; therefore, older people are more likely to feel cold. The loss of thickness partially accounts for sagging and wrinkling of the skin.

Homeostatic adjustment to heat is also limited because there are fewer sweat glands for sweating to occur. There are fewer hair follicles, so the hair on the scalp and the limbs thins out. The number of oil (sebaceous) glands is reduced, and the skin tends to crack. Older people also experience a decrease in the number of melanocytes, making their hair gray and their skin pale. In contrast, some of the remaining pigment cells are larger, and pigmented blotches appear on the skin.

Processing and Transporting

Cardiovascular disorders are the leading cause of death today. The heart shrinks because of a reduction in cardiac muscle cell size. This leads to loss of cardiac muscle strength and reduced cardiac output. Still, in the absence of disease, the heart is able to meet the demands of increased activity. It can double its rate or triple the amount of blood pumped each minute even though the maximum possible output declines.

Because the middle layer of arteries contains elastic fibers, which are most likely subject to cross-linking, the arteries become more rigid with time. The internal diameter of arteries is further reduced by plaque, a buildup of fatty material. Therefore, blood pressure readings gradually rise with age. Such changes are common in individuals living in Western industrialized countries but not in agricultural societies. A diet low in cholesterol and saturated fatty acid has been suggested as a way to control degenerative changes in the cardiovascular system.

Blood flow to the liver is reduced, and this organ does not metabolize drugs as efficiently as before. This means that, as a person gets older, less medication is needed to maintain the same level in the bloodstream.

Cardiovascular problems are often accompanied by respiratory disorders, and vice versa. Growing inelasticity of lung tissue means that ventilation is reduced. Because we rarely use the entire vital capacity, these effects are not noticed unless the demand for oxygen increases.

Blood supply to the kidneys is also reduced. The kidneys become smaller and less efficient at filtering wastes. Salt and water balance are difficult to maintain, and the elderly dehydrate faster than young people. Difficulties involving urination include incontinence (lack of bladder control) and the inability to urinate. In men, the prostate gland may enlarge and reduce the diameter of the urethra, making urination so difficult that surgery may be needed.

The loss of teeth, which is frequently seen in elderly people, is more apt to be the result of long-term neglect than aging. The digestive tract loses tone, and secretion of saliva and gastric juice is reduced, but there is no indication of reduced absorption. Therefore, an adequate diet, rather than vitamin and mineral supplements, is recommended. Elderly people commonly complain of constipation, increased gas, and heartburn; gastritis, ulcers, and cancer can also occur.

Integration and Coordination

While most tissues of the body regularly replace their cells, some at a faster rate than others, the brain and the muscles ordinarily do not. However, contrary to previous opinion, recent studies show that few neural cells of the cerebral cortex are lost during the normal aging process. This means that cognitive skills remain unchanged even though a loss in short-term memory characteristically occurs. Although the elderly learn more slowly than the young, they can acquire and remember new material. It has been noted that when more time is given for the subject to respond, age differences in learning decrease.

Neurons are extremely sensitive to oxygen deficiency, and if neuron death does occur, it may be due not to aging itself but to reduced blood flow in narrowed blood vessels. Specific disorders, such as depression, Parkinson disease, and Alzheimer disease, are sometimes seen, but they are not common. Reaction time, however, does slow, and more stimulation is needed for hearing, taste, and smell receptors to function as before. After age 50, the ability to hear tones at higher frequencies decreases gradually, and this can make it difficult to identify individual voices and to understand conversation in a group. The lens of the eye does not accommodate as well and also may develop a cataract. Glaucoma, the buildup of pressure due to increased fluid, is more likely to develop because of a reduction in the size of the anterior cavity of the eye.

Loss of skeletal muscle mass is not uncommon, but it can be controlled by a regular exercise program. The capacity to do heavy labor decreases, but routine physical work should be no problem. A decrease in the strength of the respiratory muscles and inflexibility of the rib cage contribute to the inability of the lungs to expand as before, and reduced muscularity of the urinary bladder contributes to an inability to empty the bladder completely, and therefore to the occurrence of urinary infections.

As noted before, aging is accompanied by a decline in bone density. Osteoporosis, characterized by a loss of calcium and mineral from bone, is not uncommon, but evidence indicates that proper health habits can prevent its occurrence. Arthritis, which causes pain upon movement of a joint, is also seen.

Weight gain occurs because the basal metabolism decreases and inactivity increases. Muscle mass is replaced by stored fat and retained water.

The Reproductive System

Females undergo menopause, and thereafter, the level of female sex hormones in the blood falls markedly. The uterus and the cervix are reduced in size, and the walls of the oviducts and the vagina become thinner. The external genitals become less pronounced. In males, the level of androgens falls gradually over the age span of 50–90, but sperm production continues until death. Sexual activity need not decline with age, however, due to hormone therapy and other treatments.

It is of interest that, as a group, females live longer than males. Although their health habits may be better, it is also possible that the female sex hormone estrogen offers

Figure 22.20 Remaining active.
The aim of gerontology is to allow seniors to enjoy living.

women some protection against cardiovascular disorders when they are younger. Males suffer a marked increase in heart disease in their forties, but an increase is not noted in females until after menopause, when women lead men in the incidence of stroke. Men are still more likely than women to have a heart attack, however.

Conclusion

We have listed many adverse effects of aging, but it is important to emphasize that such effects are not inevitable (Fig. 22.20). We must discover any extrinsic factors that precipitate these adverse effects and guard against them. Just as it is wise to make the proper preparations to remain financially independent when older, it is also wise to realize that biologically successful old age begins with the health habits developed when we are younger. The *Bioethical Focus* at the end of this chapter suggests other ways that aging people can plan ahead.

The deterioration of organ systems associated with aging may be prevented in part by practicing good health habits.

In 2003, 12% of the U.S. population was over 65 years of age. According to U.S. Census Bureau projections, as the "baby boomers" age, the over-65 age group is expected to make up 20% of the population by 2030. As the population of the United States ages, more and more people will be dealing with "end-of-life" issues. These are the practical, legal, medical, and even spiritual decisions that must be made when a person is expected to die within approximately six months. You may be involved in making such decisions with your parents. Legal decisions involve financial matters, and a will should be made to protect these decisions. Medical decisions relate to the degree of family involvement, the type of care desired, and whether this care will be delivered in the dying person's own home, in the home of a family member, or in an institutional care setting. Often, family members or physicians end up making these medical end-of-life decisions because the patients have

waited until they are too sick to communicate their desires. For some, this is due to the discomfort of discussing the possibility of their death with family members. Often the medical terminology is difficult to comprehend, and a patient may not understand the choices available. For many, cultural or religious beliefs influence the type of medical treatment they want or do not want, including the use of feeding tubes, ventilators, chemotherapy, and drastic life-saving procedures.

People can plan their end-of-life medical treatment in what is termed an *advance directive*. The two types of advance directives are a living will and a durable power of attorney for health care. A living will expresses the patient's desires concerning the type and degree of medical intervention. A durable power of attorney names another person who will make only health-care decisions for the person. A durable power of attorney is especially critical if the patient wants a non-

family member to make these decisions, as in the case of an unmarried couple or same-sex partners. The American Psychological Association suggests that patients plan ahead while they are not ill and have time to find answers to all of their questions, write down their choices, complete the legal paperwork, and share their decisions with their family members and physicians.

Form Your Own Opinion

1. Should all Americans be required to complete an advance directive for their medical care? If so, by what age?
2. In the absence of an advance directive, how should decisions be made concerning medical treatment for an incapacitated patient if family members disagree?
3. How would you feel if asked to discuss these issues with your parents while they are still healthy?

Summarizing the Concepts

22.1 Principles of Animal Development
Development begins at fertilization. Only one sperm actually enters the oocyte and this sperm's nucleus fuses with the oocyte nucleus.

The early developmental stages in animals proceed from the cellular level to the tissue level to the organ level. During the cellular stages, cleavage (cell division) occurs, but there is no overall growth. The result is a morula, which becomes the blastula when an internal cavity (the blastocoel) appears.

During the tissue stages, gastrulation (invagination of cells into the blastocoel) results in formation of the germ layers: ectoderm, mesoderm, and endoderm. Both the cellular and tissue stages can be affected by the amount of yolk.

Organ formation can be related to germ layers. For example, during neurulation, the nervous system develops from midline ectoderm, just above the notochord.

Cellular differentiation begins with cytoplasmic segregation in the egg. After the first cleavage of a frog embryo, only a daughter cell that receives a portion of the gray crescent is able to develop into a complete embryo. Therefore, cytoplasmic segregation of maternal determinants occurs during the early development of a frog. Induction is also part of cellular differentiation. In *C. elegans,* investigators have shown that induction is an ongoing process in which one tissue after the other regulates the development of another tissue through chemical signals coded for by particular genes.

Work with *Drosophila* has allowed researchers to identify morphogen genes that determine the shape and form of the body. During development, sequential sets of morphogen genes code for morphogen gradients that activate the next set of morphogen

genes, in turn. Homeotic genes control which appendages are on each *Drosophila* segment by coding for proteins that turn on particular genes. Homeotic genes code for proteins that contain a homeodomain, a particular sequence of 60 amino acids. Homologous homeotic genes have been found in a wide variety of organisms, and therefore, they must have arisen early in the history of life and been conserved.

22.2 Human Embryonic and Fetal Development
Human development before birth can be divided into embryonic development (months 1 and 2) and fetal development (months 3–9).

The extraembryonic membranes appear early in human development. The trophoblast of the blastocyst is the first sign of the chorion, which goes on to become the fetal part of the placenta. Exchange of gases, nutrients, and wastes occurs between fetal and maternal blood at the placenta. The amnion contains amniotic fluid, which cushions and protects the embryo. The yolk sac and allantois are also present.

Fertilization occurs in the oviduct, and cleavage occurs as the embryo moves toward the uterus. The morula becomes the blastocyst before implanting in the endometrium of the uterus. Organ development begins with neural tube and heart formation. There follows a steady progression of organ formation during embryonic development. During fetal development, features are refined, and the fetus adds weight. Birth, which requires three stages, occurs about 280 days after the start of the mother's last menstruation.

22.3 Human Development After Birth
Development after birth consists of infancy, childhood, adolescence, and adulthood. Aging may be due to cellular repair changes, which are genetic in origin. Other factors that may affect aging are changes in body processes and certain extrinsic factors.

Testing Yourself

Choose the best answer for each question.

1. Which occurs in the egg to prevent fertilization by multiple sperm cells?

   a. Plasma membrane depolarizes.
   b. Zona pellucida moves away from the egg.
   c. Killer enzymes are secreted.
   d. Both a and b are correct.

2. The inner lining of the digestive tract forms from which germ layer?

   a. endoderm   c. mesoderm
   b. ectoderm   d. blastula

3. Similarities in the development of birds, mammals, and reptiles indicate

   a. evolutionary relationships.
   b. a coincidence.
   c. same amount of yolk.
   d. None of these are correct.

4. In some animals, the notochord is replaced by the

   a. neural plate.   c. neurula.
   b. vertebral column.   d. None of these are correct.

5. Evidence suggests that _____ may act as a chemical signal to induce germ layer differentiation.

   a. temperature   d. a concentration gradient
   b. embryo size   e. None of these are correct.
   c. embryo shape

6. Programmed cell death is called

   a. apoptosis.   c. reduction.
   b. induction.   d. None of these are correct.

7. The _____ is part of the umbilical cord.

   a. yolk sac   d. All of these are correct.
   b. amnion   e. None of these are correct.
   c. allantois

8. The umbilical vein carries blood rich in

   a. nutrients.   c. Both a and b are correct.
   b. oxygen.   d. Neither a nor b is correct.

9. The archenteron will develop into the

   a. heart.   c. gut.
   b. vertebral column.   d. None of these are correct.

10. The outer cell layer in the blastocyst is called the

    a. ectoderm.   d. embryo.
    b. endoderm.   e. None of these are correct.
    c. trophoblast.

11. Implantation occurs at the _____ stage.

    a. gastrula   c. morula
    b. blastula   d. None of these are correct.

12. The cavity in the gastrula is the

    a. blastocoel.   d. embryo cavity.
    b. gastrocoel.   e. None of these are correct.
    c. archenteron.

13. The ability to experimentally cause a second neural plate to develop above notochord donor tissue is due to

    a. segregation.   d. gastrulation.
    b. induction.   e. None of these are correct.
    c. reduction.

14. Identify the stages of animal development in the following diagram.

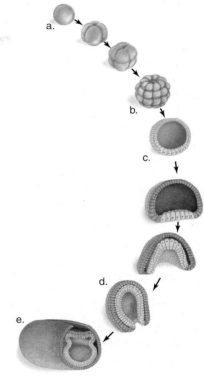

15. Which of these pairs is mismatched?

    a. brain—ectoderm
    b. gut—endoderm
    c. bone—mesoderm
    d. lens—endoderm
    e. heart—mesoderm

16. In humans, the fetus

    a. has four extraembryonic membranes.
    b. has developed organs and is recognizably human.
    c. is dependent upon the placenta for excretion of wastes and acquisition of nutrients.
    d. All of these are correct.

17. Identify the regions of a cross section of an animal embryo in the following diagram.

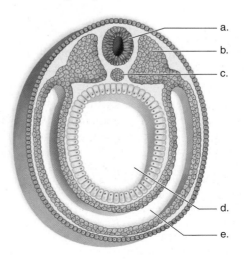

18. In human development, which part of the blastocyst develops into a fetus?

    a. morula
    b. trophoblast
    c. inner cell mass
    d. chorion
    e. yolk sac

19. Genes that control the structural pattern of an animal are

    a. homeotic genes.
    b. morphogen genes.
    c. induction genes.
    d. All of these are correct.
    e. Both a and b are correct.

20. Morphogenesis is associated with

    a. protein gradients.
    b. induction.
    c. homeotic genes.
    d. All of these are correct.

21. In many embryos, differentiation begins at what stage?

    a. cleavage
    b. blastula
    c. gastrula
    d. neurula
    e. after the completion of stages a–d

22. Which of the germ layers is best associated with development of the heart?

    a. ectoderm
    b. mesoderm
    c. endoderm
    d. neurula
    e. All of these are correct.

23. Which of these pairs is mismatched?

    a. cleavage—cell division
    b. blastula—gut formation
    c. gastrula—three germ layers
    d. neurula—nervous system
    e. Both b and c are mismatched.

24. In the following sequence (a–e), which stage is out of order first?

    a. cleavage
    b. blastula
    c. morula
    d. gastrula
    e. neurula

Understanding the Terms

afterbirth 464	embryonic disk 456
aging 465	endoderm 448
allantois 454	episiotomy 464
amnion 454	extraembryonic
arterial duct 458	membrane 454
blastocoel 447	fate map 452
blastocyst 454	fertilization 446
blastopore 448	fontanel 458
blastula 447	gastrula 448
cellular differentiation 450	gastrulation 448
chorion 454	germ layer 448
chorionic villi 457	gerontology 465
cleavage 447	gray crescent 450
colostrum 464	homeobox 452
cytoplasmic	homeotic gene 452
segregation 451	human chorionic
ectoderm 448	gonadotropin
embryonic development 453	(hCG) 455

induction 451	parturition 461
lanugo 458	pattern formation 450
mesoderm 448	placenta 460
morphogenesis 450	trophoblast 454
morphogen gene 452	umbilical artery 458
morula 447	umbilical cord 457
neural plate 449	umbilical vein 458
neural tube 449	venous duct 458
neurula 449	vernix caseosa 458
notochord 449	yolk sac 454
oval opening 458	

Match the terms to these definitions:

a._____ Process by which a chemical or a tissue influences the development of another tissue.

b._____ Primary tissue layer of a vertebrate embryo; namely ectoderm, mesoderm, or endoderm.

c._____ Movement of early embryonic cells to establish body outline and form.

d._____ Structure formed during the development of placental mammals from the chorion and the uterine wall; allows the embryo, and then the fetus, to acquire nutrients and rid itself of wastes.

e._____ Cell division without cytoplasmic addition or enlargement; occurs during first stage of development.

Thinking Critically

1. Mitochondria contain their own genetic material. Mitochondrial mutations are inherited from mother to child, without any contribution from the father. Why?

2. Are the maternal determinants found in the egg also morphogens? Why or why not?

3. In one form of prenatal testing for fetal genetic abnormalities, chorionic villi samples are taken and analyzed. Why is this possible? Why doesn't this harm the fetus?

4. The Health Focus in this chapter suggests that women of childbearing age begin taking precautions to prevent birth defects (good health habits, rubella vaccination, etc.). Why should these precautions begin before a woman is pregnant?

5. What can you do now to prevent some of the aging-related changes described in section 22.3?

ARIS, the *Inquiry into Life* Website

ARIS, the website for *Inquiry into Life,* provides a wealth of information organized and integrated by chapter. You will find practice quizzes, interactive activities, labeling exercises, flashcards, and much more that will complement your learning and understanding of general biology.

www.aris.mhhe.com

Actress Jane Seymour and her daughter, *Katie Flynn and son, Sean Flynn, look very much alike. Children often look very much like their parents.*

When a new baby is born, relatives often comment that the baby has "the mother's eyes" or "the father's mouth." In reality, what the baby inherited was the parents' genes, half from the mother and half from the father. Due to meiosis, each sperm and each oocyte contains only half the chromosomes found in the rest of the cells of the body. When a sperm and oocyte unite, the full chromosome number is restored.

We now know that there are many hundreds of genes located on each of the chromosomes. The physical and cellular characteristics of an individual, such as hairline or blood type, are all controlled by the genes. The combination of your traits is dependent on which genes you inherited from your mother and father. And because each of your siblings inherited a different combination of genes from your parents, you do not look exactly like a sibling either, unless you have an identical twin. In this chapter, we examine the chances of you inheriting a particular gene from your parents.

Patterns of Inheritance

Chapter Concepts

23.1   Mendel's Laws

Today, most people know that DNA is the genetic material, and they may have heard that scientists have just sequenced all the bases in the DNA of human cells. In contrast, they may never have heard of Gregor Mendel, an Austrian monk who, in 1860, developed certain laws of heredity after doing crosses between garden pea plants (Fig. 23.1). Mendel investigated genetics at the organism level, and this is still the level that intrigues most of us on a daily basis. We observe, for example, that facial and other features run in families, and we would like some convenient way of explaining this observation. And so it is appropriate to begin our study of genetics at the organism level and learn to use Mendel's laws of heredity.

Gregor Mendel

Mendel's parents were farmers, so he no doubt acquired the practical experience he needed to grow pea plants during childhood. Mendel was also a mathematician; he kept careful and complete records, even though he crossed and catalogued some 24,034 plants through several generations. He concluded that the plants transmitted distinct factors (now called genes found on chromosomes) to offspring. In pea plants and in humans, the chromosomes come in pairs, called *homologous pairs* of chromosomes. We can recognize

Figure 23.1   Gregor Mendel, 1822–84.
Mendel grew and tended the pea plants he used for his experiments. For each experiment, he observed as many offspring as possible. For a cross that required him to count the number of round seeds to wrinkled seeds, he observed and counted a total of 7,324 peas!

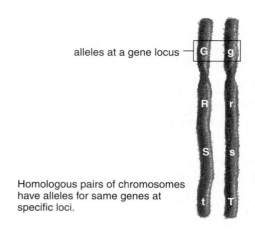

alleles at a gene locus

Homologous pairs of chromosomes have alleles for same genes at specific loci.

Figure 23.2   Homologous chromosomes.
The letters represent alleles—that is, alternate forms of a gene. Each allelic pair, such as *Gg* or *Tt*, is located on homologous pairs of chromosomes at a particular gene locus.

a homologous pair because both of its members have the same length and centromere location. They also carry genes for the same traits in the same order. For example, one trait in humans is finger length. One member of the homologous pair of chromosomes could contain the gene for long fingers, while the other member could contain the gene for short fingers. Alternate forms of a gene for the same trait are called **alleles.** Alleles are always at the same spot, or **locus,** on each member of the homologous pair. In Figure 23.2, the letters on the homologous chromosomes stand for alleles. *G* is an allele of *g,* and vice versa; *R* is an allele of *r,* and vice versa. *G* could never be an allele of *R* because *G* and *R* are at different loci and govern different traits.

Mendel knew nothing about homologous pairs of chromosomes, but his work with pea plants (described in the Science Focus) caused him to decide that pea plants have two factors for every trait, such as stem length. He observed that one of the factors controlling the same trait can be dominant over the other, which is recessive. Therefore, a tall pea plant could have one factor for long stem and another for short stem. While most of the offspring from two such tall pea plants were tall, some were short. Mendel reasoned that this would be possible if the gametes (i.e., sperm and oocyte) contained only one factor for each trait. Only in that way could an offspring receive two factors for short stem. On the basis of such studies, Mendel formulated his first law, the **law of segregation.**

The law of segregation states the following:

■   Each individual has two factors for each trait.

■   The factors segregate (separate) during the formation of the gametes.

■   Each gamete contains only one factor from each pair of factors.

■   Fertilization gives each new individual two factors for each trait.

a. Widow's peak

b. Straight hairline

Figure 23.3 Widow's peak.

In humans, **(a)** widow's peak is dominant over **(b)** straight hairline. Therefore, the man in (a) could have the genotype *Ww* or *WW*. The woman's genotype has to be *ww* because only this genotype can result in straight hairline. Suppose this man reproduces with this woman. If the man's genotype is *Ww*, then they could have a child with a straight hairline.

The Inheritance of a Single Trait

The **phenotype** of an individual refers to what the person actually looks like. Phenotype may describe either an individual's physical characteristics, or microscopic and metabolic characteristics. For example, a widow's peak and a straight hairline are considered phenotypes (Fig. 23.3). The **genotype** refers to the alleles the chromosomes carry that are responsible for that trait. Since each chromosome is found in a homologous pair in diploid organisms, there are two alleles for each trait, one on each member of the homologous pair. Mendel suggested using letters to indicate these factors, which we now call alleles. A capital letter indicates a **dominant allele** and a lowercase letter indicates a **recessive allele.** The word *dominant* is not meant to imply that the dominant allele is better or stronger than the recessive allele. Dominant means that this allele will mask the expression of the recessive allele when they are together in the same organism.

When working with the alleles that determine hairline, we will use this key:

W = Widow's peak (dominant allele)

w = Straight hairline (recessive allele)

The key tells us which letter of the alphabet to use for the gene in a particular problem. It also tells which allele is dominant, since a capital letter signifies dominance.

TABLE 23.1	Genotype Related to Phenotype	
Genotype	**Genotype**	**Phenotype**
WW	Homozygous dominant	Widow's peak
Ww	Heterozygous	Widow's peak
ww	Homozygous recessive	Straight hairline

In the case of a single trait, such as hairline, there are three possible combinations of the two alleles: *WW*, *Ww* and *ww*. These are considered the possible genotypes for this trait. If the two members of the allelic pair are the same (*homo*), the organism is said to be **homozygous.** Both *WW* and *ww* are homozygous, with *WW* homozygous dominant, and *ww* homozygous recessive. If the two members of the allelic pair are different (*hetero*), the organism is said to be **heterozygous.** Only *Ww* is a heterozygous genotype. But what about the phenotype? When an organism has a *W,* it exhibits the dominant phenotype regardless of whether the other allele is *W* or *w.* Both *WW* and *Ww* exhibit the dominant phenotype, a widow's peak. The recessive allele *w* is masked by the expression of the dominant allele *W* in the heterozygous genotype. The only time the recessive phenotype, straight hairline, is expressed, is when the genotype is homozygous recessive, or *ww.* The phenotypes and genotypes for this trait are summarized in Table 23.1.

Gamete Formation

Whereas the genotype has two alleles for each trait, the gametes (i.e., sperm and egg) have only one allele for each trait in accordance with Mendel's law of segregation. During gamete formation, the homologous chromosomes separate, and there is only one member of each pair of chromosomes in each gamete. Therefore, there is only one allele for each trait, such as type of hairline, in each gamete. (You may want to review meiosis in Chapter 5 at this time.)

In the simplest terms, *no two letters in a gamete can be the same letter of the alphabet.* If the genotype of an individual is *Ww*, the gametes for this individual will contain either a *W* or a *w* but not both letters. If the genotype is *WwLl*, all combinations of any two different letters can be present.

Science Focus

The Investigations of Gregor Mendel

Virtually every culture in history has attempted to explain inheritance patterns. An understanding of these patterns has always been important to agriculture and animal husbandry, the science of breeding animals. However, it was not until the 1860s that the Austrian monk Gregor Mendel developed the fundamental laws of heredity after performing a series of ingenious experiments. Previously, he had studied science and mathematics at the University of Vienna. When Mendel began his work, most plant and ani-

mal breeders acknowledged that both sexes contribute equally to a new individual. They thought that parents of contrasting appearance always produced offspring of intermediate appearance. This concept, called the *blending concept of inheritance,* meant that a cross between plants with red flowers and plants with white flowers would yield only plants with pink flowers. When red and white flowers reappeared in future generations, the breeders mistakenly attributed this to instability in the genetic material.

Mendel chose to work with the garden pea, *Pisum sativum* (Fig. 23A). The garden pea was a good choice for several reasons. The plants were easy to cultivate and had a short generation time. Although peas normally self-pollinate (pollen only goes to the same flower), they could be cross-pollinated by hand by transferring pollen from an anther to a stigma. Many varieties of peas were available, and Mendel chose 22 for his experiments. When these varieties self-pollinated, they were *true-breeding*—meaning

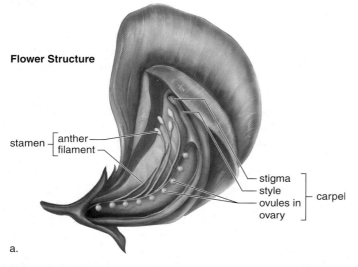

Flower Structure

stamen — anther — filament

stigma
style
ovules in
ovary
— carpel

a.

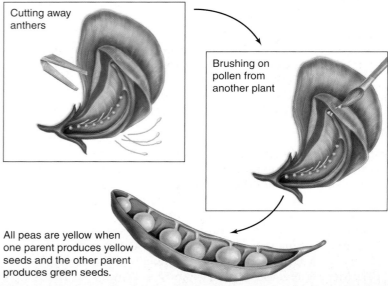

Cutting away anthers

Brushing on pollen from another plant

All peas are yellow when one parent produces yellow seeds and the other parent produces green seeds.

Figure 23A Garden pea anatomy and a few traits.

a. In the garden pea, *Pisum sativum,* pollen grains produced in the anther contain sperm, and ovules in the ovary contain oocytes. When Mendel performed crosses, he brushed pollen from one plant onto the stigma of another plant. After sperm fertilized oocytes, the ovules developed into seeds (peas). The open pod shows the results of a cross between plants with yellow seeds and plants with green seeds. **b.** Mendel selected traits like these for study. He made sure his parent (P generation) plants bred true, and then he cross-pollinated the plants. The offspring called F_1 (first filial) generation always resembled the parent with the dominant characteristic (*left*). Mendel then allowed the F_1 plants to self-pollinate. In the F_2 (second filial) generation, he always achieved a 3:1 (dominant to recessive) ratio.

Practice Problems 1*

1. **For each of the following genotypes, give all possible gametes.**

   a.   *WW*
   b.   *WWSs*
   c.   *Tt*
   d.   *Ttgg*
   e.   *AaBb*

2. **For each of the following, state whether a genotype or a gamete is represented.**

   a.   *D*
   b.   *Ll*
   c.   *Pw*
   d.   *LlGg*

*Answers to Practice Problems appear in Appendix A.

One-Trait Crosses

It is now possible for us to consider a particular cross. If a homozygous man with a widow's peak (Fig. 23.3*a*) reproduces with a woman with a straight hairline (Fig. 23.3*b*), what kind of hairline will their children have? To solve this problem, (1) use the key already established (Table 23.1) to indicate the genotype of each parent; (2) determine the possible gametes from each parent; (3) combine all possible gametes; and (4) determine the genotypes and the phenotypes of all the offspring. (When writing a heterozygous genotype, *always put the capital letter first* to avoid confusion.)

In the following diagram, the letters in the first row are the genotypes of the parents. Each parent has only one type of gamete in regard to hairline, and therefore all that the offspring were like the parent plants and like each other. In contrast to his predecessors, Mendel studied the inheritance of relatively simple and discrete traits, such as seed shape, seed color, and flower color.

Mendel studied both one-trait and two-trait crosses. He cross-pollinated the plants for the F_1 (first filial) generation. None of the offspring of this generation were intermediate between the parents. All of the offspring exhibited the trait of one or the other parent, which Mendel termed the dominant trait. The trait that disappeared in the offspring he called the recessive trait. He then allowed the F_1 plants to self-pollinate to obtain an F_2 generation. He counted many plants and always found approximately the same ratios of plants exhibiting dominant and recessive traits in this generation.

Most likely, Mendel's background in mathematics prompted him to use a statistical basis for his breeding experiments. As Mendel followed the inheritance of individual traits, he kept careful records, and he used his understanding of the mathematical laws of probability to interpret his results. He arrived at a theory that is now called a *particulate theory of inheritance* because it is based on the existence of minute particles or hereditary units, which we call genes. Inheritance involves reshuffling the same genes from generation to generation.

Mendel achieved his success in genetics by studying large numbers of offspring, keeping careful records, and treating his data quantitatively. However, a certain amount of luck was involved in the traits Mendel chose. None of the traits he analyzed exhibited the more complicated patterns of inheritance.

Discussion Questions

1. Would the genotype of a true-breeding plant be homozygous or heterozygous? Why?
2. Besides the reasons listed above, what other reasons would make the pea plant a better object for genetic study than human beings?
3. What "more complicated patterns of inheritance" could have confused Mendel's results?

Trait	Characteristics				F_2 Results*	
	Dominant		Recessive		Dominant	Recessive
Stem length	Tall		Short		787	277
Seed shape	Round		Wrinkled		5,474	1,850
Seed color	Yellow		Green		6,022	2,001
Flower color	Purple		White		705	224

*All of these produce approximately a 3:1 ratio. For example, $\frac{787}{277} \cong \frac{3}{1}$.

b.

the offspring have similar genotypes and phenotypes. The children are heterozygous (Ww) and have a widow's peak:

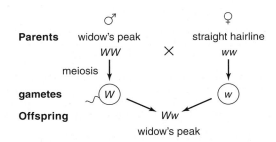

These children are **monohybrids;** that is, they are heterozygous for only one pair of alleles. If they reproduce with someone else of the same genotype, what type of hairline will their children have? In this problem ($Ww \times Ww$), each parent has two possible types of gametes (W or w), and we must ensure that all types of sperm have an equal chance to fertilize all possible types of oocytes. One way to do this is to use a **Punnett square** (Fig. 23.4), in which all possible types of sperm are lined up vertically and all possible types of oocytes are lined up horizontally (or vice versa), and every possible combination of gametes occurs within the squares.

After we determine the genotypes and the phenotypes of the offspring, we can first determine the genotypic and then the phenotypic ratio. The genotypic ratio is 1 WW: 2 Ww: 1 ww, or simply 1:2:1, but the phenotypic ratio is 3:1. Why? Because three individuals have a widow's peak and one has a straight hairline. This 3:1 phenotypic ratio is always expected for a monohybrid cross when one allele is completely dominant over the other.

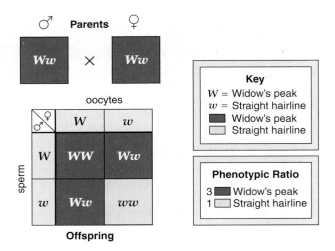

Figure 23.4 Monohybrid cross.
A Punnett square diagrams the results of a cross. When the parents are heterozygous, each child has a 75% chance of having the dominant phenotype and a 25% chance of having the recessive phenotype.

The exact ratio is more likely to be observed if a large number of matings take place and if a large number of offspring result. Only then do all possible kinds of sperm have an equal chance of fertilizing all possible kinds of eggs. Naturally, in humans, we do not routinely observe hundreds of offspring from a single type of cross. The best interpretation of Figure 23.4 in humans is to say that each child has three chances out of four to have a widow's peak, or one chance out of four to have a straight hairline. It is important to realize that *chance has no memory*; for example, if two heterozygous parents already have three children with a widow's peak and are expecting a fourth child, this child still has a 75% chance of having a widow's peak and a 25% chance of having a straight hairline.

> When solving a genetics problem, it is assumed that all possible types of sperm fertilize all possible types of eggs. The results can be expressed as a probable phenotypic ratio. It is also possible to state the chance of an offspring showing a particular phenotype.

One-Trait Crosses and Probability

Another method of calculating the expected ratios uses the rules of probability or chance. First, we must know the *product rule of probability*, which states that the chance of two or more independent events occurring together is the product (multiplication) of their chance of occurring separately. For example, if you flip a coin, the chance of getting "heads" is ½ (one head out of two possibilities, or ½). If you flip two coins, the chance of getting two heads is ½ × ½, or ¼. You must multiply together the probability of each head occurring separately (½).

In the cross just considered ($Ww \times Ww$), what is the chance of obtaining either a W or a w from a parent?

The chance of W = ½, and the chance of w = ½.

Therefore, the probability of having these genotypes is as follows:

1. The chance of WW = ½ × ½ = ¼
2. The chance of Ww = ½ × ½ = ¼
3. The chance of wW = ½ × ½ = ¼
4. The chance of ww = ½ × ½ = ¼

Now we must consider the *sum rule of probability*, which states that the chance of an event that can occur in more than one way is the sum (addition) of the individual chances. Therefore, to calculate the chance of an offspring having a widow's peak, add the chances of WW, Ww, or wW from the preceding list to arrive at ¾, or 75%. The chance of offspring with straight hairline (only ww from the preceding) is ¼, or 25%.

The One-Trait Testcross

It is not possible to tell by observation if an individual expressing a dominant allele is homozygous dominant or heterozygous. Therefore, breeders sometimes do a so-called **testcross**—they cross an animal showing the dominant phenotype with one showing the recessive phenotype. The recessive phenotype is used because it has a *known* genotype. The results of a testcross should indicate the genotype of an animal with the dominant phenotype (Fig. 23.5).

The results of crosses between people also tell us the genotype of the parents. For example, Figure 23.5 shows two possible results when a man with a widow's peak reproduces with a woman who has a straight hairline. If the man is homozygous dominant, all his children will have a widow's peak. If the man is heterozygous, each child has a 50% chance of having a straight hairline. The birth of just one child with a straight hairline indicates that the man is heterozygous.

> ## Practice Problems 2*
> 1. **Both a man and a woman are heterozygous for freckles. Freckles are dominant over no freckles. What is the chance that their child will have freckles?**
>
> 2. **Both you and your sister or brother have attached earlobes, but your parents have unattached earlobes. Unattached earlobes (E) are dominant over attached earlobes (e). What are the genotypes of your parents?**
>
> 3. **A father has dimples, the mother does not have dimples, and all five of their children have dimples. Dimples (D) are dominant over no dimples (d). Give the probable genotypes of all persons concerned.**
>
> *Answers to Practice Problems appear in Appendix A.*

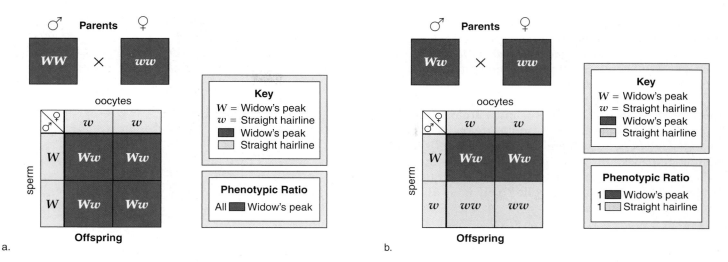

Figure 23.5 One-trait testcross.

A testcross determines if an individual with the dominant phenotype is homozygous or heterozygous. **a.** Because all offspring show the dominant characteristic, the individual is most likely homozygous, as shown. **b.** Because the offspring show a 1:1 phenotypic ratio, the individual is heterozygous, as shown.

The Inheritance of Two Traits

Figure 23.6 represents two pairs of homologues undergoing meiosis. The homologues are distinguished by length—that is, one pair of homologues is short and the other is long. (The color signifies that we inherit chromosomes from our parents; one homologue of each pair is the "paternal" chromosome, and the other is the "maternal" chromosome.) When the homologues separate (segregate) during meiosis, each gamete receives one member from each pair of homologues. The homologues separate

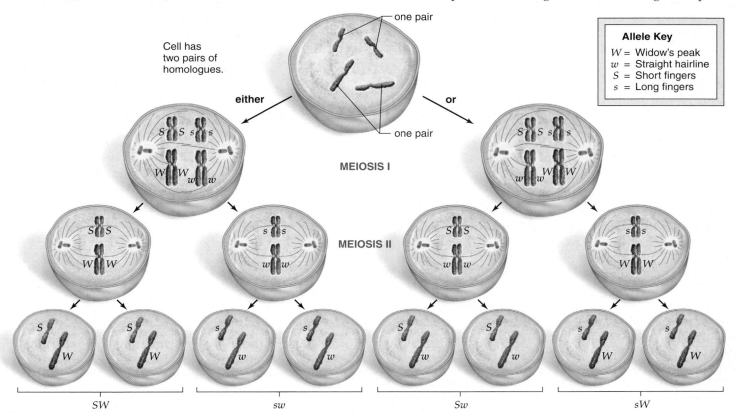

Figure 23.6 Homologous pairs of chromosomes during meiosis.

A cell has two pairs of homologous chromosomes (homologues) distinguished here by length, not color. The short pair of homologues carries alleles for finger length, and the long pair of homologues carries alleles for type of hairline. The homologues, and the alleles they carry, segregate independently during meiosis I. This means that either a red or blue chromosome from each pair can go into either of the daughter cells. Therefore, all possible combinations of chromosomes and alleles occur in the gametes. The last line of the illustration lists the possible allele combinations for this particular cross.

independently; it does not matter which member of a pair goes into which gamete. In the simplest terms, a gamete in Figure 23.6 will *receive one short and one long chromosome of either color.* Therefore, all possible combinations of chromosomes are in the gametes.

We will assume that the alleles for two genes are on these homologues. The alleles *S* and *s* are on one pair of homologues, and the alleles *W* and *w* are on the other pair of homologues. Because there are no restrictions as to which homologue goes into which gamete, a gamete can receive either an *S* or an *s* and either a *W* or a *w* in any combination. In the end, the gametes will collectively have all possible combinations of alleles.

Independent Assortment

Mendel had no knowledge of meiosis, but he could see that his results were attainable only if the sperm and eggs contain every possible combination of factors. This caused him to formulate his second law, the **law of independent assortment.**

The law of independent assortment states the following:

■ Each pair of factors separates independently (without regard to how the others separate).

■ All possible combinations of factors can occur in the gametes.

Figure 23.6 illustrates the law of segregation and the law of independent assortment. Because it does not matter which homologue of a pair faces which spindle pole, a daughter cell can receive either homologue and either allele.

Two-Trait Crosses

In the two-trait cross depicted in Figure 23.7, a person homozygous for widow's peak and short fingers (*WWSS*) reproduces with one who has a straight hairline and long fingers (*wwss*). The law of segregation tells us that the gametes for the *WWSS* grandparent have to be *WS* and the gametes for the *wwss* grandparent have to be *ws*. Therefore, their offspring will all have the genotype *WwSs* and the same phenotype (widow's peak with short fingers). This genotype is called a **dihybrid** because the individual is heterozygous in two regards: hairline and fingers.

When the dihybrid *WwSs* reproduces with another dihybrid that is *WwSs,* what gametes are possible? The law of segregation tells us that each gamete can have only one letter of each kind, and the law of independent assortment tells us that all combinations are possible. Therefore, these are the gametes for both dihybrids: *WS, Ws, wS,* and *ws.*

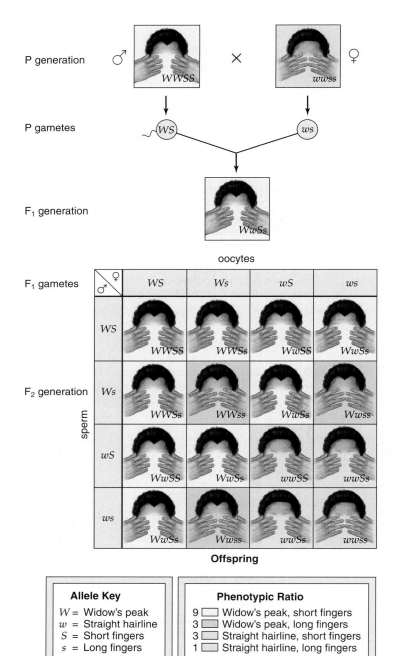

Figure 23.7 Dihybrid cross.
Since each dihybrid can form four possible types of gametes, four different phenotypes occur among the offspring in the proportions shown.

A Punnett square represents all possible sperm fertilizing all possible oocytes, so following are the expected phenotypic results for a dihybrid cross:

9 widow's peak and short fingers:
3 widow's peak and long fingers:
3 straight hairline and short fingers:
1 straight hairline and long fingers.

This 9:3:3:1 phenotypic ratio is always expected for a dihybrid cross when simple dominance is present. We can use this expected ratio to predict the chances of each child receiving a certain phenotype. For example, the chance of getting the two dominant phenotypes together is 9 out of 16, and the chance of getting the two recessive phenotypes together is 1 out of 16.

Two-Trait Crosses and Probability

Instead of using a Punnett square, it is also possible to use the product rule and the sum rule of probability to predict the results of a dihybrid cross. For example, we know that the probable results for two separate monohybrid crosses are as follows:

Probability of widow's peak = ¾

Probability of short fingers = ¾

Probability of straight hairline = ¼

Probability of long fingers = ¼

Using the product rule, we can calculate the probable outcome of a dihybrid cross as follows:

Probability of widow's peak and short fingers =
    ¾ × ¾ = 9⁄16

Probability of widow's peak and long fingers =
    ¾ × ¼ = 3⁄16

Probability of straight hairline and short fingers =
    ¼ × ¾ = 3⁄16

Probability of straight hairline and long fingers =
    ¼ × ¼ = 1⁄16

In this way, the rules of probability tell us that the expected phenotypic ratio when all possible sperm fertilize all possible oocytes is 9:3:3:1.

Practice Problems 3*

1. **Attached earlobes are recessive. What genotype do children have if one parent is homozygous recessive for earlobes and homozygous dominant for hairline, and the other is homozygous dominant for unattached earlobes and homozygous recessive for hairline?**

2. **If an individual from this cross reproduces with another of the same genotype, what are the chances that they will have a child with a straight hairline and attached earlobes?**

3. **A child who does not have dimples or freckles is born to a man who has dimples and freckles (both dominant) and a woman who does not. What are the genotypes of all persons concerned?**

Answers to Practice Problems appear in Appendix A:

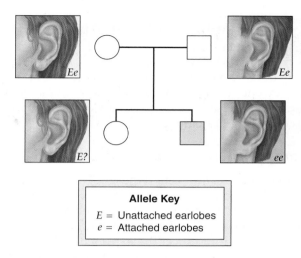

Allele Key
E = Unattached earlobes
e = Attached earlobes

Figure 23.8 A human pedigree.
In a pedigree, females are represented by circles and males by squares. An affected individual is shaded.

Pedigrees

In humans, it is impossible to carry out some of the crosses that can be done in plants, such as a test cross or self-pollination. Instead, human inheritance is diagrammed in a pedigree. A **pedigree** is a chart of a family's history with regard to a particular genetic trait. In the chart, males are designated by squares and females by circles. A line between a square and a circle represents a mating. A vertical line going downward leads directly to a single child; if there are more children, they are placed off a horizontal line. If the purpose of the pedigree is to follow the inheritance of a particular trait, those individuals with the trait, called the affected individuals, are usually shaded. A pedigree is often used to determine the inheritance of a genetic disease, as we will see in Chapter 26.

Figure 23.8 shows a pedigree involving the trait attached earlobes. The two parents have free earlobes, and the firstborn child, a girl, also has free earlobes. The second child, a son, has attached earlobes, and so the square is shaded. How is this trait inherited? Having free earlobes (*E*) is dominant to attached earlobes (*e*). That means the son must have the genotype *ee*. He inherited the *e* allele from each of his parents, which means that his parents both must have the genotype *Ee*. What about his sister? Since she has free earlobes, she must have an *E*. But it is impossible to tell whether she has inherited another *E* or an *e*. Since her genotype is unknown, this is usually written as *E?*.

In humans, pedigrees are used to follow a particular genetic trait.

23.2　Beyond Simple Inheritance Patterns

Certain traits, such as those studied in section 23.1, are controlled by one set of alleles that follows a simple dominant or recessive inheritance. But we now know of many other types of inheritance patterns.

Incomplete Dominance and Codominance

Incomplete dominance occurs when the heterozygote is intermediate between the two homozygotes. For example, when a curly-haired Caucasian reproduces with a straight-haired Caucasian, their children have wavy hair. When two wavy-haired persons reproduce, the expected phenotypic ratio among the offspring is 1:2:1—that is, one curly-haired child to two with wavy hair to one with straight hair (Fig. 23.9). We can explain incomplete dominance by assuming that only one allele codes for a product and the single dose of the product gives the intermediate result.

　　Codominance occurs when alleles are equally expressed in a heterozygote. A familiar example is the human blood type AB, in which the red blood cells have the characteristics of both type A and type B blood. We can explain codominance by assuming that both genes code for a product, and we observe the results of both products being present. Blood type inheritance is also said to be an example of multiple alleles.

Multiple Allele Inheritance

When a trait is controlled by **multiple alleles,** the gene exists in several allelic forms. But each person has only two of the possible alleles.

ABO Blood Types

Three alleles for the same gene control the inheritance of ABO blood types. These alleles determine the presence or absence of antigens on red blood cells:

I^A = A antigen on red blood cells
I^B = B antigen on red blood cells
i = Neither A nor B antigen on red blood cells

　　Each person has only two of the three possible alleles, and both I^A and I^B are dominant over i. Therefore, there are two possible genotypes for type A blood and two possible genotypes for type B blood. On the other hand, I^A and I^B are fully expressed in the presence of the other. Therefore, if a person inherits one of each of these alleles, that person will have type AB blood. Type O blood can result only from the inheritance of two i alleles.

　　The possible genotypes and phenotypes for blood type are as follows:

Phenotype	Genotype
A	$I^A I^A$, $I^A i$
B	$I^B I^B$, $I^B i$
AB	$I^A I^B$
O	ii

　　Blood typing can sometimes aid in paternity suits. However, the results of a blood test can only suggest that the tested individual *might* be the father, not that he definitely *is* the father. For example, it is possible, but not definite, that a man with type A blood (genotype $I^A i$) is the father of a child with type O blood. On the other hand, a blood test can sometimes definitely prove that a man is *not* the father. For example, a man with type AB blood cannot possibly be the father of a child with type O blood. Therefore, blood tests can be used in legal cases only to exclude a man from possible paternity.

　　As a point of interest, the Rh factor is inherited separately from A, B, AB, or O blood types. When you are Rh positive, your red blood cells have a particular antigen, and when you are Rh negative, that antigen is absent. There are multiple recessive alleles for Rh⁻, but they are all recessive to Rh⁺.

　　Figure 23.10 shows that matings between certain genotypes can have surprising results in terms of blood type.

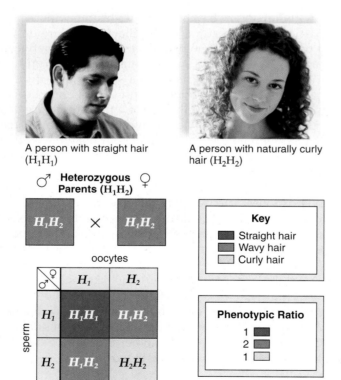

A person with straight hair (H_1H_1)

A person with naturally curly hair (H_2H_2)

Heterozygous Parents (H_1H_2)

H_1H_2　×　H_1H_2

oocytes

Key
■ Straight hair
■ Wavy hair
□ Curly hair

♂\♀	H_1	H_2
H_1	H_1H_1	H_1H_2
H_2	H_1H_2	H_2H_2

sperm

Phenotypic Ratio
1 ■
2 ■
1 □

Offspring

Figure 23.9　Incomplete dominance.

Among Caucasians, neither straight nor curly hair is dominant. When two wavy-haired individuals reproduce, each offspring has a 25% chance of having either straight or curly hair and a 50% chance of having wavy hair, the intermediate phenotype.

In incomplete dominance a blending of traits occurs. In codominance both traits are visible. Inheritance by multiple alleles occurs when a gene exists in more than two allelic forms. However, each individual inherits only two alleles for these genes.

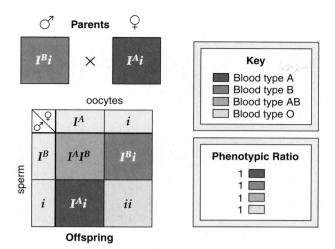

Figure 23.10 Inheritance of blood types.
Blood type exemplifies multiple allele inheritance. A mating between individuals with type A blood and type B blood can result in any one of the four blood types. Why? Because the parents are $I^A i$ and $I^B i$. If both parents had type AB blood, predict the blood types of their potential offspring.

Practice Problems 4*

1. If a person with straight hair marries someone with wavy hair, can they have a child with curly hair?

2. A child with type O blood is born to a mother with type A blood. What is the genotype of the child? The mother? What are the possible genotypes of the father?

3. From the following blood types, determine which baby belongs to which parents:

   Baby 1, type O   Mrs. Doe, type A

   Baby 2, type B   Mr. Doe, type A

                           Mrs. Jones, type A

                           Mr. Jones, type AB

*Answers to Practice Problems appear in Appendix A.

Sex-Linked Inheritance

Normally, both males and females have 23 pairs of chromosomes; 22 pairs are called **autosomes,** and one pair is the sex chromosomes. The **sex chromosomes** are so named because they differ between the sexes. In humans, males have the sex chromosomes X and Y, and females have two X chromosomes.

Traits controlled by genes on the sex chromosomes are said to be **sex-linked;** an allele on an X chromosome is **X-linked,** and an allele on the Y chromosome is Y-linked. Most sex-linked genes are only on the X chromosomes, and the Y chromosome is blank for these. Very few alleles have been found on the much smaller Y chromosome.

It would be logical to suppose that a sex-linked trait is passed from father to son or from mother to daughter, but this is not the case. A male always receives an X-linked allele from his mother, from whom he inherited an X chro-

mosome. *The Y chromosome from the father does not carry an allele for the trait.* Usually a sex-linked genetic disorder is recessive; therefore, a female must receive two alleles, one from each parent, before she has the condition.

Sex-Linked Alleles

When considering X-linked traits, the allele on the X chromosome is shown as a letter attached to the X chromosome. For example, following is the key for red-green color blindness, a well-known X-linked recessive disorder:

$$X^B = \text{normal vision}$$

$$X^b = \text{color blindness}$$

The possible genotypes and phenotypes in both males and females are

Genotypes	Phenotypes
$X^B X^B$	Female who has normal color vision
$X^B X^b$	Carrier female who has normal color vision
$X^b X^b$	Female who is color blind
$X^B Y$	Male who has normal vision
$X^b Y$	Male who is color blind

The second genotype is a carrier female because, although a female with this genotype appears normal, she is capable of passing on an allele for color blindness. Color-blind females are rare because they must receive the allele from both parents; color-blind males are more common because they need only one recessive allele to be color blind. The allele for color blindness has to be inherited from their mother because it is on the X chromosome; males only inherit the Y chromosome from their father.

Now let us consider a mating between a man with normal vision and a heterozygous woman (Fig. 23.11).

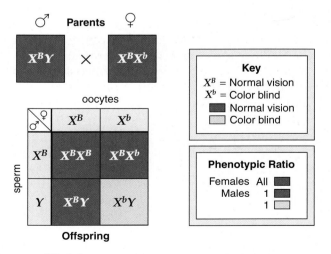

Figure 23.11 Cross involving an X-linked allele.
The male parent is normal, but the female parent is a carrier—that is, an allele for color blindness is located on one of her X chromosomes. Therefore, each son has a 50% chance of being color blind. The daughters will appear normal, but each one has a 50% chance of being a carrier.

What is the chance that this couple will have a color-blind daughter? A color-blind son? All daughters will have normal color vision because they all receive an X^B from their father. The sons, however, have a 50% chance of being color blind, depending on whether they receive an X^B or an X^b from their mother. The inheritance of a Y chromosome from their father cannot offset the inheritance of an X^b from their mother. Because the Y chromosome doesn't have an allele for the trait, it can't possibly prevent color blindness in a son. Notice in Figure 23.11 that the phenotypic results for sex-linked traits are given separately for males and females.

> The X chromosome carries alleles that are not on the Y chromosome. Therefore, a recessive allele on the X chromosome that was inherited from his mother is expressed in males.

Polygenic Inheritance

Many traits, such as size or height, shape, weight, skin color, metabolic rate, and behavior, are governed by several sets of alleles. **Polygenic inheritance** occurs when a trait is governed by two or more sets of alleles. The individual has a copy of all allelic pairs, possibly located on many different pairs of chromosomes. Each dominant allele codes for a product, and therefore the dominant alleles have a quantitative effect on the phenotype, and these effects are additive. The result is a *continuous variation* of phenotypes, resulting in a distribution of these phenotypes that resembles a bell-shaped curve. The more genes involved, the more continuous are the variations and distribution of the phenotypes.

Skin Color

Skin color is an example of a polygenic trait that is likely controlled by many pairs of alleles. Even so, we will use the simplest model and assume that skin has only two pairs of alleles (*Aa* and *Bb*) and that each capital letter contributes pigment to the skin. When a very dark person reproduces with a very light person, the children have medium-brown skin. When two people with the genotype *AaBb* reproduce with one another, the children may range in skin color from very dark to very light:

Genotypes	Phenotypes
AABB	Very dark
AABb or *AaBB*	Dark
AaBb or *AAbb* or *aaBB*	Medium brown
Aabb or *aaBb*	Light
aabb	Very light

Notice again that a range of phenotypes exists and several possible phenotypes fall between the two extremes. Therefore, the distribution of these phenotypes is expected to follow a bell-shaped curve, meaning that few people have the extreme phenotypes and most people have the phenotype that lies in the middle (Fig. 23.12).

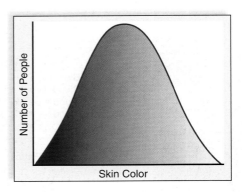

Figure 23.12   Polygenic inheritance.
Skin color is controlled by many pairs of alleles which results in a range of phenotypes. The vast majority of people have skin colors in the middle range while fewer people have skin colors in the extreme range.

> Polygenic inheritance results in a continuous variation of phenotypes.

Practice Problems 5*

1. Both the mother and the father of a color-blind male appear to have normal vision. From whom did the son inherit the allele for color blindness? What are the genotypes of the mother, the father, and the son?

2. A woman is color blind. What are the chances that her sons will be color blind? If she is married to a man with normal vision, what are the chances that her daughters will be color blind? Will be carriers?

3. A husband and wife both have normal vision. The wife gives birth to a color-blind daughter. Is it more likely that the father had normal vision or that he was color blind? What does this lead you to deduce about the girl's parentage?

4. A certain polygenic trait is controlled by three pairs of alleles: *A* vs *a*, *B* vs *b*, and *C* vs *c*. What are the two extreme genotypes for this trait?

*Answers to Practice Problems appear in Appendix A.

23.3   Environmental Influences

Environmental factors, such as nutrition or temperature, can also influence the expression of genetic traits. Polygenic traits seem to be particularly influenced by the environment. For example, in the case of height, differences in nutrition are one of the factors that bring about a bell-shaped curve (Fig. 23.13).

Temperature can also affect the phenotypes of plants and animals. For example, primroses have white flowers when grown above 32°C and red flowers when grown at 24°C. The coats of Siamese cats and Himalayan rabbits are darker in color at the ears, nose, paws, and tail. Himalayan

Figure 23.13 Polygenic inheritance and environmental effects.

When you record the heights of a large group of people chosen at random, the values follow a bell-shaped curve. Such a continuous distribution is due to control of a trait by several sets of alleles. Environmental effects are also involved.

rabbits are known to be homozygous for the allele *ch,* which is involved in the production of melanin. Experimental evidence suggests that the enzyme coded for by this gene is active only at low temperatures, and therefore black fur occurs only at the extremities, where body heat is lost to the environment (Fig. 23.14).

Figure 23.14 Coat color in Himalayan rabbits.

The influence of the environment on the phenotype has been demonstrated by plucking out the fur from one area and applying an ice pack. The new fur that grew in was black instead of white, showing that the enzyme that produces melanin (a dark pigment) in the rabbit is active only at low temperatures.

Investigators try to determine what percentage of various human traits is due to nature (inheritance) and what percentage is due to nurture (the environment).

Some studies use identical and fraternal twins who have been separated since birth and thus have been raised in different environments. The supposition is that, if identical twins in different environments share the same trait, that trait is most likely inherited. These studies have found that identical twins are more similar in their intellectual talents, personality traits, and levels of lifelong happiness than are fraternal twins when both groups have been separated since birth. Biologists conclude that all behavioral traits are partly inheritable and that genes exert their effects by acting together in complex combinations susceptible to environmental influences.

Scientists have also learned how to manipulate the environment in order to prevent the damaging effects of some genetic disorders. For example, the bodies of people who have the genetic disorder phenylketonuria (PKU) lack an enzyme that breaks down the amino acid phenylalanine. The accumulation of phenylalanine leads to neurological damage in these patients. But scientists have discovered that if these patients modify their diet so that they consume very little protein containing phenylalanine, they can avoid many of the damaging effects of this disease.

Environmental agents, such as nutrition and temperature, can affect the expression of certain genes. Traits that are polygenic seem to be particularly affected by environmental influences.

Figure 23.15 Linkage group.

In this individual, alleles *A* and *B* are on one member of a homologous pair, and alleles *a* and *b* are on the other member. Therefore, the individual is a dihybrid. **a.** When linkage is complete, this dihybrid produces only two types of gametes in equal proportion. **b.** When linkage is incomplete, this dihybrid produces four types of gametes because crossing-over has occurred. The recombinant gametes occur in reduced proportion because crossing-over occurs infrequently.

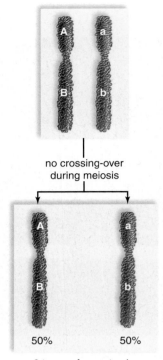

no crossing-over
during meiosis

50%   50%

2 types of gametes in
equal proportions

a. Complete linkage

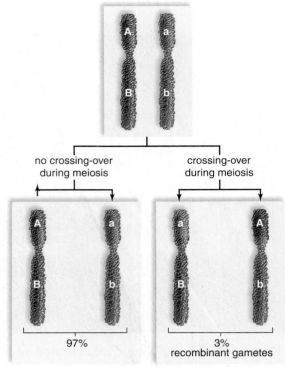

no crossing-over
during meiosis

crossing-over
during meiosis

97%

3%
recombinant gametes

4 types of gametes in unequal proportions

b. Incomplete linkage

23.4   Inheritance of Linked Genes

A chromosome doesn't contain just one or two alleles; it contains a long series of alleles in a definite sequence. The sequence is fixed because each allele has its own particular locus on a chromosome. All the alleles on one chromosome form a **linkage group** because they tend to be inherited together. Figure 23.15*a* shows the results of a cross when linkage is complete: a dihybrid produces only two types of gametes in equal proportion.

During meiosis (see Chapter 5) crossing-over sometimes occurs between the nonsister chromatids of a tetrad. During crossing-over, the chromatids exchange genetic material, and therefore genes. If crossing-over occurs between the two alleles of interest, a dihybrid produces four types of gametes instead of two (Fig. 23.15*b*). Recombinant gametes (with recombined alleles) occur in reduced number because crossing-over is infrequent.

The occurrence of crossing-over can help tell the sequence of genes on a chromosome because crossing-over occurs more often between distant genes than between genes that are close together on a chromosome. For example, consider these homologous chromosomes:

Z   S   R   G
z   s   r   g

pair of homologous chromosomes

We expect recombinant gametes to include *G* and *z* more often than *R* and *s*. In keeping with this observation, investigators formerly used recombination frequencies to map the chromosomes of, say, fruit flies, but also humans

to a degree. Each 1% of crossing-over is equivalent to one map unit between genes.

The possibility of using linkage data to map human chromosomes is limited because we can work only with matings that have occurred by chance. This, coupled with the fact that humans tend not to have numerous offspring, means that additional methods must be used to sequence the genes on human chromosomes. Today, it is customary to rely on biochemical methods coupled with the use of a computer to map the human chromosomes, as we shall discuss in Chapter 26.

> The presence of linkage groups changes the expected results of genetic crosses. The frequency of recombinant gametes that occurs due to the process of crossing-over has been used to map chromosomes.

Practice Problems 6[*]

1. Suppose that individuals with the genotype *AaBb* reproduce, and the phenotypic ratio that results is just about 3:1. What ratio would you have expected? What may have caused the observed ratio?

2. The genes for ABO blood type and for the presence of fingernails are on the same homologous pair of chromosomes. In an actual family, 45% of the offspring have type B blood and no fingernails; 45% have type O blood and fingernails; 5% have type B blood and fingernails; and 5% have type O blood with no fingernails. What process accounts for the recombinant phenotypes?

*Answer to Practice Problems appear in Appendix A.

Bioethical Focus

Selecting Children

Do you approve of selecting a baby's gender even before it is conceived? As you know, the sex of a child is dependent upon whether an X-bearing sperm or a Y-bearing sperm enters the egg. A new technique has been developed that can separate X-bearing sperm from Y-bearing sperm. First, the sperm are dosed with a DNA-staining chemical. Because the X chromosome has slightly more DNA than the Y chromosome, it takes up more dye. When a laser beam shines on the sperm, the X-bearing sperm shine a little more brightly than the Y-bearing sperm. A machine sorts the sperm into two groups on this basis. The results are not perfect. Following artificial insemination, there's about an 91% success rate for a girl and about a 76% success rate for a boy.

Selecting children is a real possibility today. First, the parents must donate sperm and oocytes to the process of in vitro fertilization. Then, selection of an embryo on the basis of genes is accomplished by testing one of the cells of an 8-cell embryo for the desired alleles. The other seven cells go on developing in the usual manner. The embryos that will have the desired trait (s) are either intro-

duced into the woman's uterus or saved in case they are needed for a second try. Hopefully, one embryo will implant in the uterine wall and after nine months a child will be born with the desired traits.

Some might argue that while it is acceptable to use vaccines to prevent illnesses or to give someone a heart transplant, it goes against nature to choose gender. But what if the mother is a carrier of an X-linked genetic disorder, such as hemophilia, a blood disorder, or Duchenne muscular dystrophy, a debilitating disorder? Is it acceptable to bring a child into the world with a genetic disorder that may cause an early death? Would it be better to select sperm for a girl, who at worst would be a carrier like her mother? Previously, a pregnant woman with these concerns had to wait for the results of an amniocentesis test and then decide whether to abort the pregnancy if the child was a boy. Is it better to increase the chances of having a girl to begin with?

Some authorities do not find gender selection acceptable for any reason. Even if it doesn't lead to a society with far more mem-

bers of one sex than another, problems could result. Once you separate reproduction from the sex act, it might open the door to genetically designing children.

The process of selecting children can also be wasteful because many more embryos may be produced than are needed. Do you feel the end justifies the means? That is, it is all right to sacrifice some embryos in order to produce a healthy child?

Form Your Own Opinion

1. Do you think it is acceptable to choose the gender of a baby—even if it requires artificial insemination at a clinic? Why or why not?
2. Do you see any difference between choosing gender and choosing eggs or embryos free of a genetic disease for reproduction purposes? Explain.
3. Do you think it is acceptable to genetically select children before they are born? For example, would it be all right to select an embryo that could, as a newborn, donate umbilical cord stem cells to cure its father of a serious disease?

Summarizing the Concepts

23.1 Mendel's Laws

The genes are on the chromosomes; each gene has a minimum of two alternative forms, called alleles. It is customary to use letters to represent the genotypes of individuals. Homozygous dominant is indicated by two capital letters, and homozygous recessive is indicated by two lowercase letters. Heterozygous is indicated by a capital letter and a lowercase letter. Whereas an individual has two alleles for each trait, the gametes have one allele for each trait. Therefore, in keeping with Mendel's law of segregation, each heterozygous individual can form two types of gametes.

When performing an actual cross, it is assumed that all possible types of sperm fertilize all possible types of oocytes. Use of the Punnett square helps carry out this assumption. The results may be expressed as a probable phenotypic ratio; it is also possible to state the chance of an offspring having a particular phenotype. When a heterozygote reproduces with a heterozygote (called a monohybrid cross), the phenotypic ratio is 3:1; that is, each child has a 75% chance of having the dominant phenotype and a 25% chance of having the recessive phenotype.

Testcrosses are used to determine if an individual with the dominant phenotype is homozygous or heterozygous. If an individual expressing the dominant allele reproduces with an individual expressing the recessive allele and an offspring with the recessive phenotype results, we know that the individual with the dominant phenotype is heterozygous. Notice that when a heterozygote reproduces with a homozygous recessive, the

phenotypic ratio is 1:1; each child has a 50% chance of having the recessive phenotype and a 50% chance of having the dominant phenotype.

In keeping with Mendel's law of independent assortment, an individual heterozygous for two traits can form four types of gametes. Therefore, the phenotypic ratio for a dihybrid cross is 9:3:3:1; in other words, $\frac{9}{16}$ of the offspring have the two dominant traits, $\frac{3}{16}$ have one of the dominant traits with one of the recessive traits, $\frac{3}{16}$ have the other dominant trait with the other recessive trait, and $\frac{1}{16}$ have both recessive traits.

23.2 Beyond Simple Inheritance Patterns

There are many exceptions to Mendel's laws, including degrees of dominance (curly hair) and multiple alleles (ABO blood types). Some traits are sex-linked, meaning that although they do not determine gender, they are carried on the sex chromosomes. Most of the alleles for these traits are carried on the X chromosome, while the Y is blank. The phenotypic results of crosses are given for females and males separately. For polygenic traits such as skin color and height, several genes each contribute to the overall phenotype in equal, small degrees. The environment plays a role in the continuously varying expression that follows a bell-shaped curve.

23.3 Environmental Influence

Environmental factors, such as nutrition and temperature, affect the expression of certain traits. In humans, nutrition helps determine height. In Siamese cats and Himalayan rabbits, the effect of temperature is apparent in the distribution of fur color. Humans can manipulate the environment to lessen the damaging effects of genetic defects.

23.4 Inheritance of Linked Genes

All the genes on one chromosome form a linkage group, which is broken only when crossing-over occurs. Genes that are linked tend to go together into the same gamete. If crossing-over occurs, a dihybrid cross yields all possible phenotypes among the offspring, but the expected ratio is greatly changed because recombinant phenotypes are reduced in number.

Crossing-over data are used to map the chromosomes of animals, such as fruit flies, but it is not possible in humans to prearrange matings in order to map the chromosomes.

Testing Yourself

Choose the best answer for each question.

1. According to the law of segregation,
    a. each gamete contains two copies of each factor for each trait.
    b. each individual has one copy of each factor for each trait.
    c. fertilization gives each new individual one factor for each trait.
    d. All of these are correct.
    e. None of these are correct.

2. Peas are good for genetic studies because they
    a. cannot self-pollinate.
    b. have a long generation time.
    c. are easy to grow.
    d. are more genetically variable than most plants.

3. To test whether F_1 plants were Yy, Mendel testcrossed them and observed an offspring ratio of
    a. 1 yellow : 1 green.
    b. 3 yellow : 1 green.
    c. 3 green : 1 yellow.
    d. all yellow.
    e. all green.

4. Which of the following genotypes indicates a heterozygous individual?
    a. *AB*   c. *Aa*
    b. *AA*   d. *aa*

5. List the gametes produced by *AaBb*.
    a. *Aa, Bb*
    b. *A, a, B, b*
    c. *AB, ab*
    d. *AB, Ab, aB, ab*

6. Homologous chromosomes contain identical
    a. genes.
    b. alleles.
    c. DNA sequences.
    d. All of these are correct.

7. The genotypic ratio from the F_2 of a monohybrid cross is
    a. 1:1.   d. 9:3:3:1.
    b. 3:1.   e. 1:1:1:1.
    c. 1:2:1.

8. If two parents with short fingers (dominant), have a child with long fingers, what is the chance their next child will have long fingers?
    a. no chance   d. $1/16$
    b. $1/2$ e. $3/16$
    c. $1/4$

9. Using the product rule, determine the probability that an *Aa* individual will be produced from an $Aa \times Aa$ cross.
    a. 50%   d. 100%
    b. 25%   e. 0%
    c. 75%

10. A testcross is a cross between
    a. a recessive individual of unknown genotype and a homozygous dominant.
    b. two dominant individuals of unknown genotype.
    c. a dominant individual of unknown genotype and a homozygous recessive.
    d. a dominant individual of unknown genotype and a homozygous dominant.
    e. a recessive individual of unknown genotype and a homozygous recessive.

11. Using the following diagram, show how four types of gametes (*SW, sw, Sw, sW*) are produced from the dihybrid by attaching the correct letter to each chromosome.

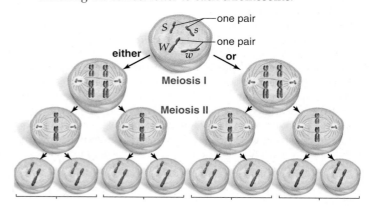

12. According to the law of independent assortment,
    a. all possible combinations of factors can occur in the gametes.
    b. only the parental combinations of gametes can occur in the gametes.
    c. only the nonparental combinations of gametes can occur in the gametes.

13. Using the product rule, what is the probability that a dihybrid cross will produce a homozygous recessive in both traits?
    a. $9/16$
    b. $3/16$
    c. $1/4$
    d. $1/8$
    e. $1/16$

14. Which of the following is not a feature of polygenic inheritance?
    a. Effects of dominant alleles are additive.
    b. Genes affecting the trait may be on multiple chromosomes.
    c. Environment influences phenotype.
    d. Recessive alleles are harmful.

15. The ABO blood system exhibits
    a. codominance.
    b. dominance.
    c. multiple alleles.
    d. All of these are correct.
    e. None of these are correct.

16. In the following diagram, indicate the alleles in the gametes and the genotypes of the offspring. In addition, determine the phenotypic ratio of the second generation. **Key:** W = widow's peak, w = straight hairline, S = short fingers, s = long fingers.

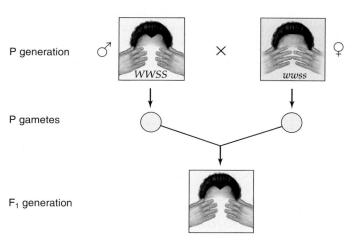

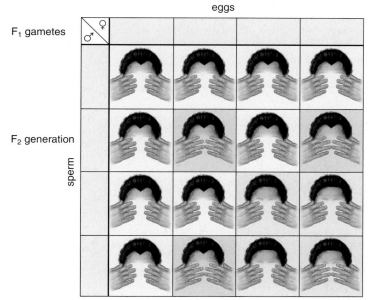

Offspring

For questions 17–22, match each type of cross with one of the expected ratios in the key. Answers can be used more than once.

Key:

a. 3:1
b. 1:1
c. 9:3:3:1
d. 1:1:1:1
e. 1:2:1

17. Phenotypic ratio from a dihybrid cross; both genes express dominance

18. Genotypic ratio from a monohybrid cross; the gene exhibits incomplete dominance

19. Genotypic ratio from a monohybrid cross; the gene exhibits complete dominance

20. Phenotypic ratio from a monohybrid cross; the gene exhibits complete dominance

21. Genotypic ratio from a monohybrid cross; the gene exhibits codominance

22. Phenotypic ratio from a testcross to a monohybrid; the gene exhibits dominance

23. Tay-Sachs disease is caused by a single gene and results when a person cannot produce the enzyme hexosaminidase. Heterozygous individuals produce about half the amount of enzyme found in homozygous normal people. At the biochemical level, this trait exhibits

    a. complete dominance.   c. codominance.
    b. incomplete dominance.

24. Assume that two parents with normal vision have a color-blind son. Which parent is responsible for the son's color blindness?

    a. the mother
    b. the father
    c. either parent
    d. None of these are correct because two normal parents cannot have a color-blind son.

25. All the genes on one chromosome are said to form a

    a. chromosomal group.
    b. recombination group.
    c. linkage group.
    d. crossing-over group.

26. If alleles R and s are linked on one chromosome and r and S are linked on the homologous chromosome, what gametes will be produced? (Assume that no crossing-over occurs.)

    a. RS, Rs, rS, rs
    b. RS, rs
    c. Rs, rS
    d. R, S, r, s

Additional Genetic Problems

1. How many genetically different types of gametes would be produced by an $AaBBCc$ individual?

    a. 1   d. 6
    b. 2   e. 8
    c. 4

2. In sheep, coat color is determined by a single, dominant gene. A white ram and a black ewe produce a black lamb. Which color is dominant?

    a. white   c. impossible to determine
    b. black

3. Watermelon fruit are either short (A?) or long (aa), and green (G?) or striped (gg). Genes A and G are on separate chromosomes. What proportion of the progeny would be short and green when crossing dihybrids?

    a. $\frac{1}{2}$ d. $\frac{3}{4}$
    b. $\frac{9}{16}$ e. $\frac{1}{16}$
    c. $\frac{1}{4}$

4. If a child has type O blood and the mother has type A, which of the following could be the blood type of the child's father?

    a. A only   d. A or O
    b. B only   e. A, B, or O
    c. O only

5. Dimpled chin is a single-gene dominant trait. A man without a dimpled chin marries a woman with a dimpled chin whose mother lacked the dimple. What proportion of their children would lack the dimple?

a. ¼

b. ½

c. ¾

d. All of these are correct.

e. None of these are correct.

Understanding the Terms

allele 472	locus 472
autosome 481	monohybrid 475
codominance 480	multiple alleles 480
dihybrid 478	pedigree 479
dominant allele 473	phenotype 473
genotype 473	polygenic inheritance 482
heterozygous 473	Punnett square 475
homozygous 473	recessive allele 473
incomplete dominance 480	sex chromosome 481
law of independent	sex-linked 481
assortment 478	testcross 476
law of segregation 472	X-linked 481
linkage group 484	

Match the terms to these definitions:

a._____ Gridlike device used to calculate the expected results of a simple genetic cross.

b._____ Alternate forms of a gene that occur at the same locus on homologous chromosomes.

c._____ Allele that exerts its phenotypic effect in the heterozygote; it masks the expression of the recessive allele.

d._____ Cross between an individual with the dominant phenotype and an individual with the recessive phenotype to see if the individual with the dominant phenotype is homozygous or heterozygous.

e._____ Genes of an individual for a particular trait or traits; for example, *BB* or *Aa*.

Thinking Critically

1. Explain at the molecular level why blood types A and B are dominant to blood type O.

2. When can the genotype be readily determined from the phenotype in the case of complete dominance? In the case of incomplete dominance? In the case of codominance?

3. What ratios would be obtained from a dihybrid testcross when the dominant expressing parent has the genotype *AaBb*?

4. Based on the information in Figure 23.6, if two pairs of chromosomes give rise to four different types of gametes, how many different types of gametes can arise from 23 pairs of chromosomes?

5. If identical twins have exactly the same genetic makeup, why are there any differences at all between them?

ARIS, the *Inquiry into Life* Website

ARIS, the website for *Inquiry into Life,* provides a wealth of information organized and integrated by chapter. You will find practice quizzes, interactive activities, labeling exercises, flashcards, and much more that will complement your learning and understanding of general biology.

www.aris.mhhe.com

Today, it's possible for scientists to create genetically modified organisms ("GMOs"), which may include plants, bacteria, or animals. Through the process of genetic engineering, GMOs are often created by taking genes from one organism and inserting them into another. GMOs and genetic engineering have had great benefits to society—from producing higher yields of important crops to engineering bacteria to produce medical products such as human insulin! However, some concerns exist about GMOs, because their long-term health effects are unknown. Currently, there is debate over whether GMO foods should be labeled so that consumers can make informed decisions.

The amazing technological advances that allow us to create GMOs would not have been possible without the discovery of DNA (deoxyribonucleic acid). By the mid-1950s, scientists had conducted experiments to determine that DNA was the genetic material that offspring inherit from parents. In the 50 years since then, we have made great strides in understanding the structure of DNA, how DNA codes for proteins, and the cause of many human genetic disorders. In fact, we now have the complete DNA sequence of the human genome! Now scientists can ask questions such as: How do changes in DNA structure lead to genetic diseases like cystic fibrosis and Huntington disease? How do differences in DNA sequences relate to risk of cancer? We can identify, treat, and potentially even cure many human diseases that were incurable before DNA was discovered.

chapter 24

DNA Biology and Technology

Chapter Concepts

24.1 DNA Structure and Replication

24.2 Gene Expression

24.3 DNA Technology

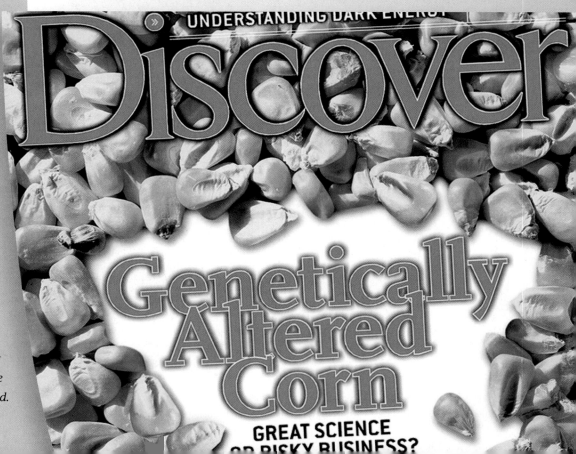

24.1 DNA Structure and Replication

We now know that **DNA (deoxyribonucleic acid)** makes up our genes, but that is a relatively recent discovery. A little more than 50 years ago, scientists were not sure whether the genetic material was composed of DNA or protein. In 1952, Alfred Hershey and Martha Chase used experiments with viruses to answer this question. Hershey and Chase thought that if they could determine which part of the virus—the DNA in the viral core, or the protein in the viral coat, or capsid—enters a bacterium and directs the production of more viruses, they would know whether genes were made of DNA or protein.

Hershey and Chase performed two experiments (Fig. 24.1). In the first experiment, viral DNA was labeled with radioactive ^{32}P (represented in yellow). The viruses were allowed to attach to and inject their genetic material into the bacterium *Escherichia coli (E. coli)*. Then the culture was agitated in a kitchen blender to remove whatever remained of the viruses on the outside of the bacteria. Finally, the culture was centrifuged (spun at high speed) so that the bacteria collected as a pellet at the bottom of the centrifuge tube. In this experiment investigators found most of the ^{32}P-labeled DNA (yellow) in the bacteria and not in the liquid medium. Why? Because the DNA had entered the bacteria.

In the second experiment, viral protein in capsids was labeled with radioactive ^{35}S (yellow). The viruses were allowed to attach to and inject their genetic material into *E. coli*, then the same procedure was followed to remove viruses and collect a pellet of bacteria. In this experiment scientists found ^{35}S-labeled protein (yellow) in the liquid medium and not in the bacteria. Why? Because the radioactive capsids remained on the outside of the bacteria.

These results indicated that the DNA of a virus (not a protein) enters the host, where viral reproduction takes place. Therefore, DNA is the genetic material and transmits the genetic information needed to make new viruses.

Experiments with viruses that infect bacteria showed that DNA, not protein, is the genetic material.

Structure of DNA

The structure of DNA was determined by James Watson and Francis Crick in the early 1950s. The data they used and

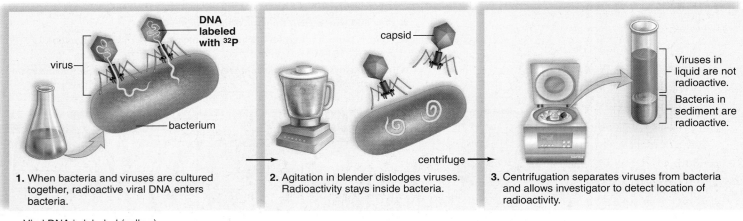

a. Viral DNA is labeled (yellow).

1. When bacteria and viruses are cultured together, radioactive viral DNA enters bacteria.
2. Agitation in blender dislodges viruses. Radioactivity stays inside bacteria.
3. Centrifugation separates viruses from bacteria and allows investigator to detect location of radioactivity.

Viruses in liquid are not radioactive.
Bacteria in sediment are radioactive.

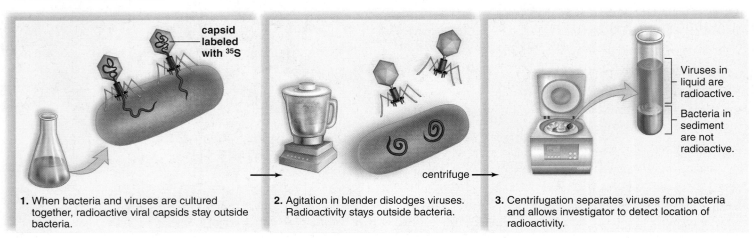

b. Viral capsid is labeled (yellow).

1. When bacteria and viruses are cultured together, radioactive viral capsids stay outside bacteria.
2. Agitation in blender dislodges viruses. Radioactivity stays outside bacteria.
3. Centrifugation separates viruses from bacteria and allows investigator to detect location of radioactivity.

Viruses in liquid are radioactive.
Bacteria in sediment are not radioactive.

Figure 24.1 Hershey—Chase experiments.

These experiments concluded that viral DNA, not protein, was responsible for directing the production of new viruses.

Science Focus

Finding the Structure of DNA

In 1953, James Watson, an American biologist, began an internship at the University of Cambridge, England. There he met Francis Crick, a British physicist, who was interested in molecular structures. Together, they set out to determine the structure of DNA and to build a model that would explain how DNA, the genetic material, can vary from species to species and even from individual to individual. They also discovered how DNA replicates (makes a copy of itself) so that daughter cells receive an identical copy.

The bits and pieces of data available to Watson and Crick were like puzzle pieces they had to fit together. *This is what they knew from the research of others:*

1. DNA is a polymer of nucleotides, each one having a phosphate group, the sugar deoxyribose, and a nitrogen-containing base. There are four types of nucleotides because there are four different bases: adenine (A) and guanine (G) are purines, while cytosine (C) and thymine (T) are pyrimidines.
2. A chemist, Erwin Chargaff, had determined in the late 1940s that regardless of the species under consideration, the number of purines in DNA always equals the number of pyrimidines. Further, the amount of adenine equals the amount of thymine (A = T), and the amount of guanine equals the amount of cytosine (G = C). These findings came to be known as Chargaff's rules.
3. Rosalind Franklin and Maurice Wilkins, working at King's College, London, had just prepared an X-ray diffraction photograph of DNA. It showed that DNA is a double helix of constant diameter and that the bases are regularly stacked on top of one another (Fig. 24A).

Using these data, Watson and Crick deduced that DNA has a twisted, ladder-like structure; the sugar-phosphate molecules make up the sides of the ladder, and the bases make up the rungs. Further, they determined that if A is normally hydrogen-bonded with T, and G is normally hydrogen-bonded with C (in keeping with Chargaff's rules), then the rungs always have a constant width, consistent with the X-ray photograph.

Watson and Crick built an actual model of DNA out of wire and tin. This double-helix model does indeed allow for differences in DNA structure between species because the base pairs can be in any order. Also, the model suggests that complementary base pairing plays a role in the replication of DNA. As Watson and Crick pointed out in their original paper, "It has not escaped our notice that the specific pairing we have postulated immediately suggests a possible copying mechanism for the genetic material."

Discussion Questions

1. Based on Chargaff's rules, if a segment of DNA is composed of 20% adenine (A) bases, what is the percentage of guanine (G)?
2. Watson and Crick's discovery of DNA is clearly one of the most important biological discoveries in the last century. Speculate on a world today without knowledge of DNA. Can you think of some human diseases, specifically, for which the discovery of DNA has allowed detection, treatment, and cures?
3. We are now getting to the point where we may be able to produce "smart drugs" that allow treatment of illnesses such as heart disease based on the genetic structure of the DNA of the individual being affected. That is, instead of one drug to treat arterial blockage around the heart, there may be as many as six or more, depending on the individual's genotype and genetic family history. What are the pros and cons of such drugs?

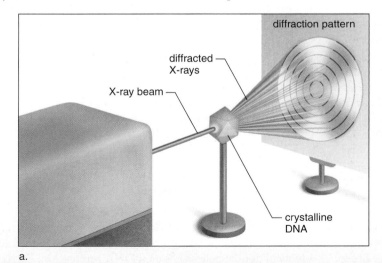

a.

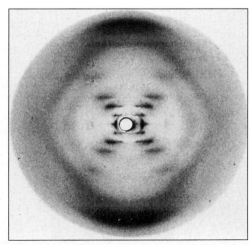

b.

Figure 24A X-ray diffraction of DNA.
a. When a crystal is X-rayed, the way in which the beam is diffracted reflects the pattern of the molecules in the crystal. The closer together two repeating structures are in the crystal, the farther from the center the beam is diffracted. **b.** The diffraction pattern of DNA produced by Rosalind Franklin. The crossed (X) pattern in the center told investigators that DNA is a helix, and the dark portions at the top and the bottom told them that some feature is repeated over and over. Watson and Crick determined that this feature was the hydrogen-bonded bases.

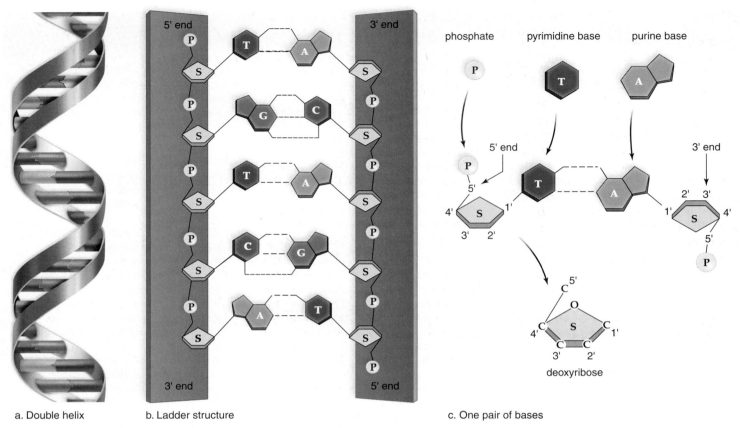

a. Double helix   b. Ladder structure   c. One pair of bases

Figure 24.2 Overview of DNA structure.
a. DNA double helix. **b.** Unwinding the helix reveals a ladder configuration in which the uprights are composed of sugar and phosphate molecules and the rungs are complementary bases. The bases in DNA pair in such a way that the phosphate-sugar backbones are oriented in different directions. **c.** Notice that 3' and 5' are part of system for numbering the carbon atoms that make up the sugar.

how they interpreted the data to deduce DNA's structure are reviewed in the Science Focus.

DNA is a chain of nucleotides. Each nucleotide is a complex of three subunits—phosphoric acid (phosphate), a pentose sugar (deoxyribose), and a nitrogen-containing base. There are four possible bases: two are **purines** with a double ring, and two are **pyrimidines** with a single ring. Adenine (A) and guanine (G) are purines; thymine (T) and cytosine (C) are pyrimidines.

A DNA polynucleotide *strand* has a backbone made up of alternating phosphate and sugar molecules. The bases are attached to the sugar but project to one side. DNA has two such strands, and the two strands twist about one another in the form of a **double helix** (Fig. 24.2a). The strands are held together by hydrogen bonding between the bases: A always pairs with T by forming two hydrogen bonds, and G always pairs with C by forming three hydrogen bonds. Notice that a purine is always bonded to a pyrimidine. This is called **complementary base pairing.** When the DNA helix unwinds, it resembles a ladder (Fig. 24.2b). The sides of the ladder are the phosphate-sugar backbones, and the rungs of the ladder are the complementary paired bases.

The two DNA strands are antiparallel—that is, they run in opposite directions, which you can verify by noticing that the sugar molecules are oriented differently. The carbon atoms in a sugar molecule are numbered, and the fifth carbon atom (5') is uppermost in the strand on the left, while the third carbon atom (3') is uppermost in the strand on the right (Fig. 24.2c).

> DNA is a double helix with phosphate-sugar backbones on the outside and paired bases on the inside. In complementary base pairing, adenine (A) pairs with thymine (T), and guanine (G) pairs with cytosine (C).

Replication of DNA

When the body grows or heals itself, cells divide. Each new cell requires an exact copy of the DNA contained in the chromosomes. The process of copying one DNA double helix into two identical double helices is called **DNA replication.** Each original strand serves as a template for the formation of a complementary new strand. DNA replication is termed *semiconservative* because a new double helix has one conserved old strand and one new strand. Replication results in two DNA helices that are identical to each other and to the original molecule (Fig. 24.3).

Figure 24.4 shows how complementary nucleotides pair in the daughter strand.

1. Before replication begins, the two strands that make up parental DNA are hydrogen-bonded to one another.
2. The enzyme *DNA helicase* unwinds and "unzips" the double-stranded DNA (i.e., the weak hydrogen bonds between the paired bases break).
3. New complementary DNA nucleotides, always present in the nucleus, fit into place by the process of complementary base pairing. These are positioned and joined by the enzyme *DNA polymerase*.
4. To complete replication, the enzyme *DNA ligase* seals any breaks in the sugar-phosphate backbone.
5. The two double helix molecules are identical to each other and to the original DNA molecule.

Chemotherapeutic drugs for cancer treatment stop replication and, therefore, cell division. Some chemotherapeutic drugs are analogs that have a similar, but not identical, structure to the four nucleotides in DNA. When these are mistakenly used by the cancer cells to synthesize DNA, replication stops, and the cancer cells die off.

> **During DNA replication, DNA unwinds and unzips, and new strands that are complementary to the original strands form.**

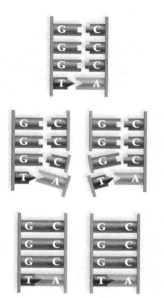

Parental DNA molecule contains so-called old strands hydrogen-bonded by complementary base pairing.

Region of replication. Parental DNA is unwound and unzipped. New nucleotides are pairing with those in old strands.

Replication is complete. Each double helix is composed of an old (parental) strand and a new (daughter) strand.

Figure 24.4 Ladder configuration and DNA replication.
The ladder configuration best illustrates how complementary nucleotides, available in the cell, pair with those of each old strand before they are joined together to form a daughter strand.

24.2 Gene Expression

A **gene** is a segment of DNA found on a chromosome that specifies the amino acid sequence of a polypeptide; sometimes either a single polypeptide or multiple polypeptides form proteins. Note that not all DNA is considered "coding" DNA, or DNA that makes proteins. Some stretches of DNA are associated with the proper functioning of the genes, and other stretches of DNA are "non-coding." In genes, DNA stores the information necessary for forming polypeptides in the form of a nucleotide code.

Genetic disorders are often caused by a missing or malfunctioning enzymatic or structural protein in the cell. Persons with Duchenne muscular dystrophy, for example, are missing the protein called dystrophin. Lack of

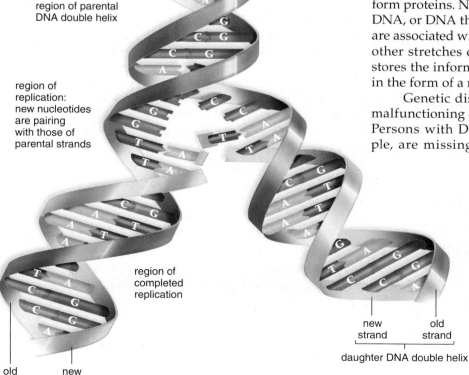

region of parental DNA double helix

region of replication: new nucleotides are pairing with those of parental strands

region of completed replication

old strand   new strand

daughter DNA double helix

new strand   old strand

daughter DNA double helix

**Figure 24.3
Overview of DNA replication.**
Replication is called semiconservative because each new double helix is composed of an old (parental) strand and a new (daughter) strand.

dystrophin can cause weakening, paralysis, and even death. In the genetic disorder Huntington disease, the protein huntingtin is altered and unable to carry out its usual functions. This causes degeneration of neurons in the brain and symptoms such as uncontrolled movements, depression, and loss of intellectual faculties.

In the next sections, we will learn how DNA codes for proteins. However, DNA does not directly control protein synthesis; instead, its information is transcribed into RNA, which is more directly involved in protein synthesis.

RNA

Like DNA, **RNA (ribonucleic acid)** is a polynucleotide. However, the nucleotides in RNA contain the sugar ribose, not deoxyribose. Also, the bases in RNA are adenine (A), cytosine (C), guanine (G), and uracil (U). In other words, the base uracil replaces the thymine found in DNA. Finally, RNA is generally single stranded, while DNA is double stranded (Fig. 24.5). Table 24.1 summarizes the similarities and differences between DNA and RNA.

There are three major classes of RNA, each with specific functions in protein synthesis:

Messenger RNA (mRNA) takes a message from DNA to the ribosomes. In eukaryotic cells, DNA is in the nucleus, and the ribosomes are in the cytoplasm.

Ribosomal RNA (rRNA), along with proteins, makes up the ribosomes, where proteins are synthesized.

Transfer RNA (tRNA) transfers amino acids to the ribosomes.

Gene expression (production of a protein) requires two steps, known as transcription and translation. During **transcription,** one type of polynucleotide (DNA) is transcribed letter by letter into another type of polynucleotide (RNA). Next, during **translation,** the RNA transcript directs the sequence of amino acids in a polypeptide. With the help of RNA, a gene specifies the sequence of amino acids in a protein. In this way, genes control the structure and processes that occur in cells.

> A gene specifies the sequence of amino acids in a polypeptide through the processes of transcription and translation. In this way, genes direct the formation of proteins necessary for many of life's functions.

Transcription

During transcription, a segment of the DNA serves as a template for the production of an RNA molecule. Although all three classes of RNA are formed by transcription, we will focus on transcription to form mRNA, the first step in protein synthesis.

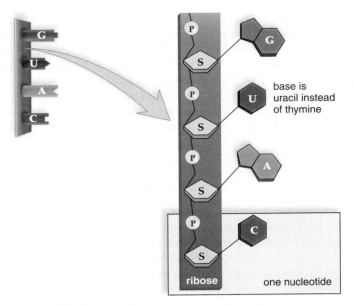

base is uracil instead of thymine

ribose   one nucleotide

Figure 24.5 Structure of RNA.
Like DNA, RNA is a polymer of nucleotides. In an RNA nucleotide, the sugar ribose is attached to a phosphate molecule and to a base, either G, U, A, or C. Notice that in RNA, the base uracil replaces thymine as one of the pyrimidine bases. RNA is generally single stranded, whereas DNA is double stranded.

Messenger RNA

Transcription begins when the enzyme **RNA polymerase** binds tightly to a **promoter,** a region of DNA that contains a special sequence of nucleotides. This enzyme opens up the DNA helix just in front of it so that complementary base pairing can occur in the same way as in DNA replication. Then, RNA polymerase joins the RNA nucleotides, and an mRNA molecule results. When mRNA forms, it has a sequence of bases complementary to that of the DNA; wherever A, T, G, or C are present in the DNA template, U, A, C, or G, respectively, are incorporated into the mRNA molecule (Fig. 24.6). Now, mRNA is a faithful copy of the sequence of bases in DNA.

Processing of mRNA After the mRNA is transcribed in eukaryotic cells, it must be processed before entering the cytoplasm.

The newly synthesized **primary mRNA** molecule becomes a *mature mRNA* molecule after processing. Most

TABLE 24.1	DNA Structure Compared to RNA Structure	
	DNA	**RNA**
Sugar	Deoxyribose	Ribose
Bases	Adenine, guanine, thymine, cytosine	Adenine, guanine, uracil, cytosine
Strands	Double stranded	Single stranded
Helix	Yes	No

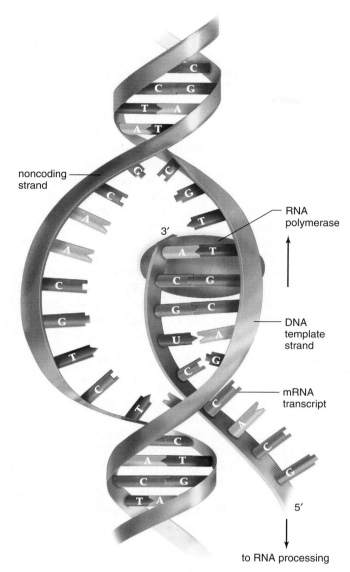

Figure 24.6 Transcription of DNA to form mRNA.
During transcription, complementary RNA is made from a DNA template. At the point of attachment of RNA polymerase, the DNA helix unwinds and unzips, and complementary RNA nucleotides are joined together. After RNA polymerase has passed by, the DNA strands rejoin and the mRNA transcript dangles to the side.

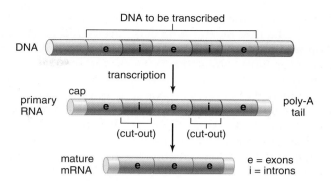

Figure 24.7 mRNA processing.
During processing, a cap and tail are added to mRNA, and the introns (i) are removed so that only exons (e) remain.

Surprisingly, RNA, not the protein, is the enzyme, and so it is called a **ribozyme.**

Ordinarily, processing brings together all the exons of a gene. In some instances, cells use only certain exons to form a mature RNA transcript. The result can be a different protein product in each cell. Alternate mRNA splicing may account for white blood cells' ability to produce a specific antibody for each type of bacteria and virus we encounter on a daily basis.

> During transcription, the enzyme RNA polymerase copies a strand of the DNA into a complementary RNA molecule. In eukaryotic cells, the primary mRNA molecule is processed by modification of the 5′ and 3′ ends and removal of the introns. The mature mRNA leaves the nucleus.

Translation

Translation is the second step by which gene expression leads to protein synthesis. Translation requires several enzymes, mRNA, and two other types of RNA: transfer RNA and ribosomal RNA.

The Genetic Code

The sequence of bases in DNA is transcribed into mRNA, which ultimately codes for a particular sequence of amino acids to form a polypeptide. Can four mRNA bases (A, C, G, U) provide enough combinations to code for 20 amino acids? If only one base stood for an amino acid (i.e., a "singlet code"), then only 4 amino acids would be possible. If two bases stood for one amino acid, there would only be 16 possible combinations (4×4). If the code is a triplet, then there are 64 possible triplets of the four bases ($4 \times 4 \times 4$). It should come as no surprise then to learn that the genetic code is a **triplet code** (each triplet is also called

genes in humans are interrupted by segments of DNA that are not part of the gene. These portions are called **introns** because they are intragene segments. The other portions of the gene are called **exons** because they are ultimately expressed. Only exons result in a protein product.

Primary mRNA contains bases that are complementary to both exons and introns, but during processing, (1) one end of the mRNA is modified by the addition of a cap, composed of an altered guanine nucleotide, and the other end is modified by the addition of a poly-A tail, a series of adenosine nucleotides. (2) The introns are removed, and the exons are joined to form a mature mRNA molecule consisting of continuous exons. This splicing of mRNA is done by a complex composed of both RNA and protein(Fig. 24.7).

a **codon**) and that this code is *conserved*. That is, notice in Figure 24.8 that most amino acids are coded for by more than one codon. For example, leucine has six codons and serine has four codons; this conservation offers some protection against possibly harmful mutations that change the sequence of bases. For example, if a mutation caused a DNA triplet to change from AAA to AAG, the corresponding mRNA codon would change from UUU to UUC, but the amino acid would remain the same—phenylalanine. Of the 64 codons, 61 code for amino acids, while the remaining three are *stop codons* (UAA, UGA, UAG), codons that do not code for amino acids but instead signal polypeptide termination.

The genetic code is just about universal in all living things. This means that a codon in a fruit fly codes for the same amino acid as in a bird, a fern, or a human. It also suggests that the code dates back to the very first organisms on Earth and that all living things are related.

Transfer RNA

Transfer RNA (tRNA) molecules bring amino acids to the ribosomes, the site of protein synthesis. Each tRNA molecule is a single-stranded polynucleotide that doubles back on itself such that complementary base pairing creates a bootlike shape. On one end is an amino acid, and on the other end is an **anticodon,** a triplet of three bases complementary to a specific codon of mRNA (Fig. 24.9). Because there are 64 codons, at least one tRNA molecule is possible for each of the 20 amino acids found in proteins.

When a tRNA–amino acid complex comes to the ribosome, its anticodon pairs with an mRNA codon. For example, if the codon is CGG, what is the anticodon, and what amino acid will be attached to the tRNA molecule? Based on Figure 24.8, the answer to this question is as follows:

Codon (mRNA)	Anticodon (tRNA)	Amino Acid (protein)
CGG	GCC	Arginine

The order of the codons of the mRNA determines the order that tRNA–amino acids come to a ribosome and, therefore, the final sequence of amino acids in a protein.

Figure 24.10 shows gene expression. During transcription, the base sequence in DNA is copied into a sequence of bases in mRNA. During translation, tRNAs bring amino acids to the ribosomes in the order dictated by the base sequence of mRNA. The sequence of amino acids forms a polypeptide chain, or complete protein.

First Base	Second Base				Third Base
	U	**C**	**A**	**G**	
U	UUU phenylalanine	UCU serine	UAU tyrosine	UGU cysteine	U
	UUC phenylalanine	UCC serine	UAC tyrosine	UGC cysteine	C
	UUA leucine	UCA serine	UAA *stop*	UGA *stop*	A
	UUG leucine	UCG serine	UAG *stop*	UGG tryptophan	G
C	CUU leucine	CCU proline	CAU histidine	CGU arginine	U
	CUC leucine	CCC proline	CAC histidine	CGC arginine	C
	CUA leucine	CCA proline	CAA glutamine	CGA arginine	A
	CUG leucine	CCG proline	CAG glutamine	CGG arginine	G
A	AUU isoleucine	ACU threonine	AAU asparagine	AGU serine	U
	AUC isoleucine	ACC threonine	AAC asparagine	AGC serine	C
	AUA isoleucine	ACA threonine	AAA lysine	AGA arginine	A
	AUG (start) methionine	ACG threonine	AAG lysine	AGG arginine	G
G	GUU valine	GCU alanine	GAU aspartate	GGU glycine	U
	GUC valine	GCC alanine	GAC aspartate	GGC glycine	C
	GUA valine	GCA alanine	GAA glutamate	GGA glycine	A
	GUG valine	GCG alanine	GAG glutamate	GGG glycine	G

Figure 24.8 Messenger RNA codons.
Notice that in this chart, each of the codons (in boxes) is composed of three letters representing the first base, second base, and third base. For example, find the box where C for the first base and A for the second base intersect. You will see that U, C, A, or G can be the third base. The bases CAU and CAC are codons for histidine; the bases CAA and CAG are codons for glutamine.

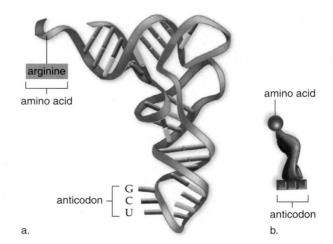

anticodon ⎡ G
           ⎢ C
           ⎣ U

a.

amino acid

amino acid

anticodon

b.

Figure 24.9 Transfer RNA: amino acid carrier.
a. A tRNA is a polynucleotide that folds into a bootlike shape because of complementary base pairing. At one end of the molecule is its specific anticodon—in this case, GCU. At the other end, an amino acid attaches that corresponds to this anticodon—in this case, arginine. **b.** tRNA is represented like this in the illustrations that follow.

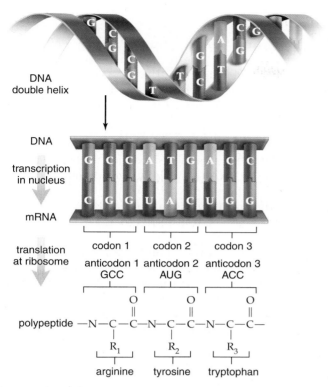

Figure 24.10 Overview of gene expression.
One strand of DNA acts as a template for mRNA synthesis, and the sequence of bases in mRNA determines the sequence of amino acids in a polypeptide.

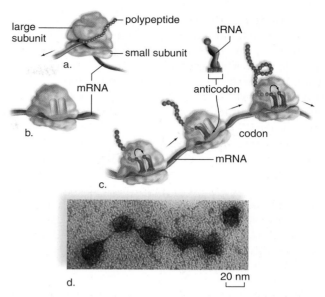

Figure 24.11 Polyribosome structure and function.
a. Side view of a ribosome shows positioning of mRNA and growing protein. **b.** Frontal view of a ribosome, shows tRNA binding sites. **c.** Several ribosomes, collectively called a polyribosome, move along an mRNA at one time. Therefore, several polypeptides can be made at the same time **d.** Electron micrograph of a polyribosome.

Ribosomes and ribosomal RNA

Ribosomes are small structural bodies found in the cytoplasm and on the endoplasmic reticulum where translation occurs. Ribosomes are composed of many proteins and several ribosomal RNAs (rRNAs). In eukaryotic cells, rRNA is produced in a nucleolus within the nucleus. There it joins with proteins manufactured in the cytoplasm to form two ribosomal subunits, one large and one small. The subunits leave the nucleus and join together in the cytoplasm to form a ribosome just as protein synthesis begins.

A ribosome has a binding site for mRNA as well as binding sites for two tRNA molecules. These binding sites facilitate complementary base pairing between tRNA anticodons and mRNA codons. As the ribosome moves down the mRNA molecule, new tRNAs arrive, and a polypeptide forms and grows longer. Translation terminates once the polypeptide is fully formed and a mRNA stop codon is reached. The ribosome then dissociates into its two subunits and falls off the mRNA molecule.

As soon as the initial portion of mRNA has been translated by one ribosome and the ribosome has begun to move down the mRNA, another ribosome attaches to the same mRNA. Therefore, several ribosomes are often attached to and translating a single mRNA, thus forming several copies of a polypeptide simultaneously. The entire complex is called a **polyribosome** (Fig. 24.11).

During translation, the sequence of bases in mRNA determines the order that tRNA–amino acid complexes come to a ribosome and, therefore, the order of amino acids in a particular polypeptide.

Translation Requires Three Steps

During translation, the codons of an mRNA base-pair with the anticodons of tRNA molecules carrying specific amino acids. The order of the codons determines the order of the tRNA molecules at a ribosome and the sequence of amino acids in a polypeptide. The process of translation must be extremely orderly so that the amino acids of a polypeptide are sequenced correctly.

Protein synthesis involves three steps: initiation, elongation, and termination. Enzymes are required for each of the three steps to function properly. The first two steps, initiation and elongation, require energy.

Initiation

Initiation is the step that brings all the translation components together. Proteins called initiation factors are required to assemble the small ribosomal subunit, mRNA, initiator tRNA, and the large ribosomal subunit for the start of protein synthesis.

Initiation is shown in Figure 24.12. In prokaryotes, a small ribosomal subunit attaches to the mRNA in the vicinity of the *start codon* (AUG). The first or initiator tRNA pairs with this codon because its anticodon is UAC. Then, a large ribosomal subunit joins to the small subunit (Fig. 24.12). Although similar in many ways, initiation in eukaryotes is much more complicated and will not be discussed here.

As already discussed, a ribosome has three binding sites for tRNAs. One of these is called the E (for exit) site, the second is the P (for peptide) site, and the third is the A (for amino acid) site. The initiator tRNA happens to be capable of binding to the P site, even though it carries only the amino acid methionine (see Fig. 24.8). The A site is for tRNA carrying the next amino acid, and the E site is for any tRNAs that are leaving a ribosome.

> **Following initiation, mRNA and the first tRNA are at a ribosome.**

Elongation

Elongation is the protein synthesis step in which a polypeptide increases in length one amino acid at a time. In addition to the participation of tRNAs, elongation requires elongation factors, which facilitate the binding of tRNA anticodons to mRNA codons at a ribosome.

Elongation is shown in Figure 24.13 as a series of four steps:

1. A tRNA with an attached peptide is already at the P site, and a tRNA carrying the next amino acid in the chain is just arriving at the A site.
2. Once the next tRNA is in place at the A site, the peptide chain will be transferred to this tRNA.
3. Needed to bring about this transfer are energy and a ribozyme, which is part of the ribosomal subunit. The energy contributes to peptide bond formation, which makes the peptide one amino acid longer by adding the peptide from the A site.
4. Next, translocation occurs – the mRNA moves forward one codon length, and the peptide-bearing

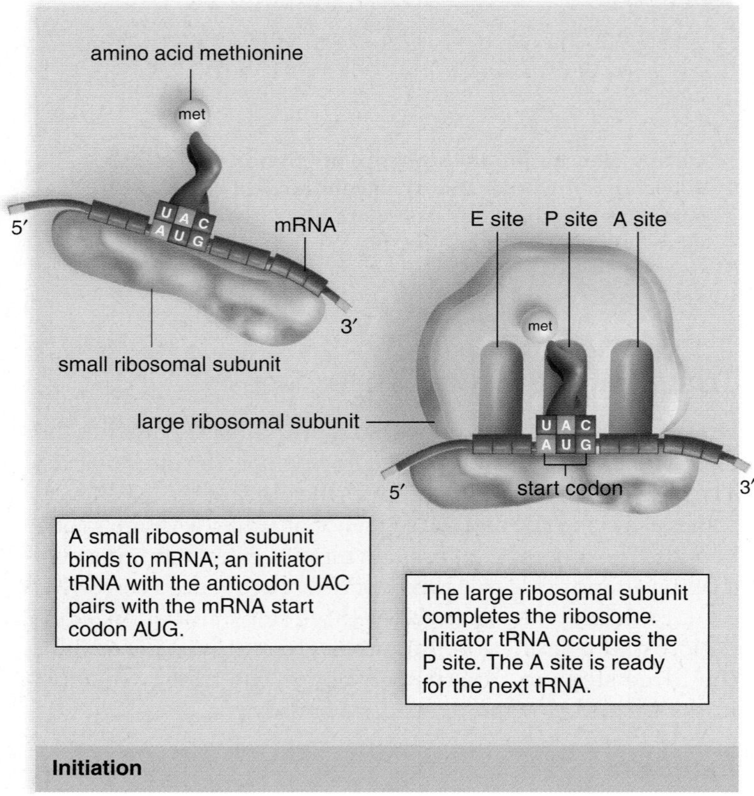

Initiation

A small ribosomal subunit binds to mRNA; an initiator tRNA with the anticodon UAC pairs with the mRNA start codon AUG.

The large ribosomal subunit completes the ribosome. Initiator tRNA occupies the P site. The A site is ready for the next tRNA.

Figure 24.12 Initiation.
During initiation, participants in the translation process assemble as shown. The start codon, AUG also codes for the first amino acid methionine.

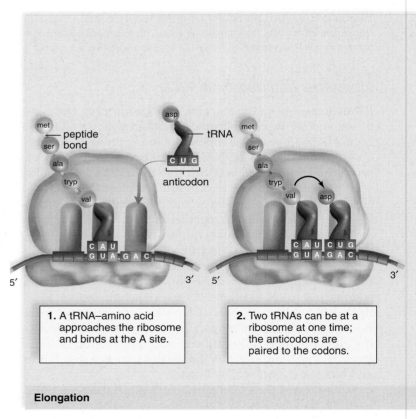

Elongation

1. A tRNA–amino acid approaches the ribosome and binds at the A site.

2. Two tRNAs can be at a ribosome at one time; the anticodons are paired to the codons.

Figure 24.13 Elongation.
Note that a polypeptide is already at the P site. During elongation, polypeptide synthesis occurs as amino acids are added one at a time to the growing chain.

tRNA is now at the ribosome P site. The "spent" tRNA is now at the E site and it then exits. The new codon is at the A site and is ready to receive the next complementary tRNA.

The complete cycle in steps 1–4, above, is repeated at a rapid rate (e.g., about 15 times each second in *E. coli*).

During elongation, amino acids are added one at a time to the growing polypeptide.

Termination

Termination is the final step in protein synthesis. During termination, as shown in Figure 24.14, the polypeptide and the assembled components that carried out protein synthesis are separated from one another.

Termination of polypeptide synthesis occurs at a stop codon. Termination requires a protein called a release factor, which cleaves the polypeptide from the last tRNA. After this occurs, the polypeptide is set free and begins to take on its three-dimensional shape. The ribosome dissociates into its two subunits.

During termination, the ribosome separates into its two subunits, and the polypeptide is released.

Properly functioning proteins are of paramount importance to the cell and to the organism. For example, if an organism inherits a faulty gene, the result can be a genetic disorder (such as Huntington disease) caused by a malfunctioning protein or a propensity toward cancer. Proteins are the link between genotype and phenotype. The DNA sequence underlying these proteins distinguishes different types of organisms. In addition to accounting for the difference between cell types, proteins account for the differences between organisms.

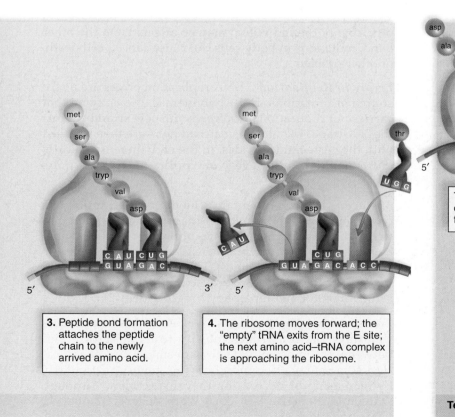

3. Peptide bond formation attaches the peptide chain to the newly arrived amino acid.

4. The ribosome moves forward; the "empty" tRNA exits from the E site; the next amino acid–tRNA complex is approaching the ribosome.

The ribosome comes to a stop codon on the mRNA. A release factor binds to the site.

The release factor hydrolyzes the bond between the last tRNA at the P site and the polypeptide, releasing them. The ribosomal subunits dissociate.

Termination

Figure 24.14 Termination.
During termination, the finished polypeptide is released, as are the mRNA and the last tRNA.

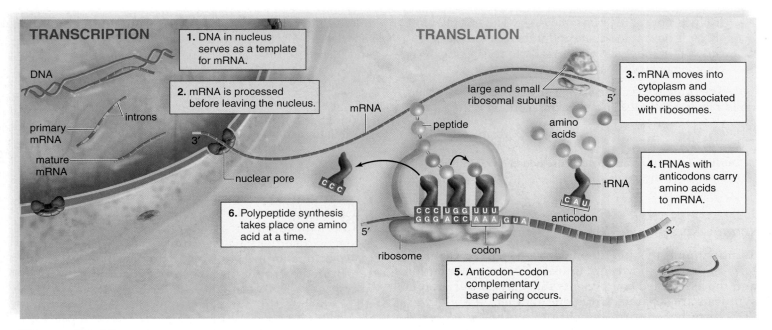

Figure 24.15 Summary of gene expression.
Messenger RNA is produced and processed in the nucleii during transcription, and protein synthesis occurs in the cytoplasm at the ribosomes during translation.

Review of Gene Expression

Genes are made up of DNA in the nucleus that contains a triplet code. Gene expression involves transcription and translation (Fig. 24.15). During transcription, a segment of a DNA strand serves as a template for the formation of messenger RNA (mRNA). The bases in mRNA are complementary to those in DNA; every three mRNA bases is a *codon* (a triplet code) for a certain amino acid. Messenger RNA is processed before it leaves the nucleus, during which time the introns are removed and the ends are modified. Messenger RNA carries a sequence of codons to the *ribosomes*. A transfer RNA (tRNA) molecule then bonds to a particular amino acid and has an *anticodon* that pairs with a codon in mRNA. During translation, tRNAs and their attached amino acids arrive at the ribosomes, where the linear sequence of codons of mRNA determines the order in which amino acids form a chain and become incorporated into a protein.

Genes and Gene Mutations

Whereas early geneticists thought of genes as sections of a chromosome, in molecular genetics, a gene is a sequence of DNA bases that code for a product, most often a protein. Therefore, it follows that a **gene mutation** is a change in the sequence of bases within a gene. Gene mutations can lead to malfunctioning proteins in cells.

Causes of Mutations

Three causes of mutations are errors in replication, mutagens, and transposons. If a mutation occurs in a gamete, the offspring of the individual may be affected. On the other hand, mutations in body cells can cause cancer, cell death, or other problems.

Errors in Replication DNA replication errors are a rare source of mutations. DNA polymerase, the enzyme that carries out replication, proofreads the new strand against the old strand. Usually mismatched pairs are then replaced with the correct nucleotides. In the end, there is typically only one mistake for every one billion nucleotide pairs replicated.

Mutagens Environmental influences called **mutagens** cause mutations in humans. Mutagens include radiation (e.g., radioactive elements, X rays, ultraviolet [UV] radiation) and certain organic chemicals (e.g., chemicals in cigarette smoke and certain pesticides). The rate of mutations resulting from mutagens is low because *DNA repair enzymes* constantly monitor and repair any irregularities.

Transposons **Transposons** are specific DNA sequences that have the remarkable ability to move within and between chromosomes. Their movement to a new location sometimes alters neighboring genes, particularly by increasing or decreasing their expression. Although "movable elements" in corn were described over 40 years ago, their significance was only realized recently. So-called *jumping genes* have now been discovered in bacteria, fruit flies, and humans, and it is likely that all organisms have such elements. The presence of white kernels in corn is due to a transposon located within a gene coding for a pigment-producing enzyme (Fig. 24.16*a, b*). So-called "Indian corn" displays a variety of

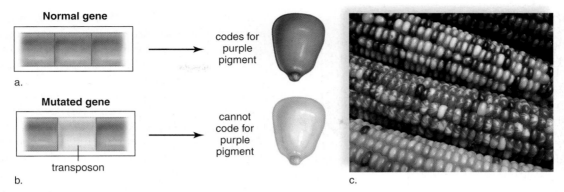

Figure 24.16 Transposon.

a. A purple coding gene ordinarily codes for a purple pigment **b.** A transposon "jumps" into the purple-coding gene. This mutated gene is unable to code for purple pigment and a white kernel results. **c.** Indian corn displays a variety of colors and patterns. Transposons are responsible for the speckled or striped patterns.

colors and patterns because of transposons (Fig. 24.16c). In a rare human neurological disorder called Charcot-Marie-Tooth disease, a transposon called Mariner causes the muscles and nerves of the legs and feet to gradually wither away. Believe it or not, the Mariner transposon is also found in fruit flies!

Types of Mutations

Two types of mutations are frameshift mutations and point mutations.

Frameshift Mutations

A sequence of codons constitutes a *reading frame* because the codons are read starting at a specific location. When one or more nucleotides are either inserted or deleted from DNA, the result can be a polypeptide that codes for the wrong sequence of amino acids. Such a mutation is called a *frameshift mutation*. Consider this sentence: THE CAT ATE THE RAT. If the letter C is deleted from this sentence, the sentence will read like this: THE ATA TET HER AT—something that doesn't make sense.

Point Mutations

Point mutations involve a substitution of one nucleotide for another. The result of a point mutation can be variable, so there are subcategories of point mutations. For example, if UAC is changed to UAU, the sequence of amino acids will not change because both of these codons code for tyrosine (see Fig. 24.8). Therefore, this is called a *silent mutation*. On the other hand, if UAC is changed to UAG, the result could be disastrous, because UAG is a stop codon, and a shorter or dysfunctional protein could result. This is, therefore, called a *nonsense mutation*. Finally, a *missense mutation* affects the shape of the protein. For example, when valine, instead of glutamate, occurs in a hemoglobin chain at one location, sickle cell disease results. Abnormal hemoglobin stacks up inside cells, and their sickle

shape makes them clog small vessels. Hemorrhaging results and leads to damage and pain in internal organs and joints.

> Mutations (a change in DNA base sequence) can lead to proteins that function poorly or not at all. Frameshift and point mutations can arise due to errors in DNA replication, mutagens, or transposons.

24.3 DNA Technology

Knowledge of DNA biology has led to our ability to manipulate the genes of organisms. We can clone genes and then use them to alter the genomes of viruses and cells, whether bacterial, plant, or animal cells. This practice, called **genetic engineering**, has innumerable uses, from producing a product to treating cancer and genetic disorders. Typically, these applications begin with gene cloning.

The Cloning of a Gene

In biology, **cloning** is the production of identical copies of an organism through some asexual means. The members of a bacterial colony on a petri dish are clones because they all came from the division of the same original cell. Human identical twins are also clones, because the first two cells of the embryo separated and each became a complete individual.

Gene cloning is the production of many identical copies of a single gene. Biologists clone genes for a number of reasons. They might want to determine the difference in base sequence between a normal gene and a mutated gene, learn how a cloned gene codes for a particular protein, or use the genes to alter the phenotypes of other organisms in a beneficial way. Scientists typically use one of two ways to clone the gene: recombinant DNA technology or the polymerase chain reaction.

Using Recombinant DNA Technology

Figure 24.17 shows the steps in cloning a human gene, which begin with the production of recombinant DNA. **Recombinant DNA (rDNA)** contains DNA from two or more different sources, such as the human cell and the bacterial cell in Figure 24.17. To make rDNA, a researcher needs a **vector,** a piece of DNA that can be manipulated,

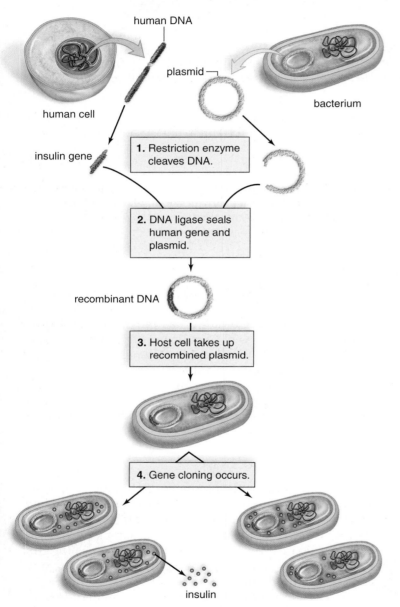

human DNA

plasmid

human cell

bacterium

insulin gene

1. Restriction enzyme cleaves DNA.

2. DNA ligase seals human gene and plasmid.

recombinant DNA

3. Host cell takes up recombined plasmid.

4. Gene cloning occurs.

insulin

Figure 24.17 Cloning a human gene.
Human DNA and plasmid DNA are cleaved by a specific type of restriction enzyme. For example, human DNA containing the insulin gene is spliced into a plasmid by the enzyme DNA ligase. Gene cloning is achieved after a bacterium takes up the plasmid. If the gene functions normally as expected the product (e.g., insulin) may also be retrieved.

in order to add foreign DNA to it. One common vector is a plasmid. **Plasmids** are small accessory rings of DNA from bacteria that are not part of the bacterial chromosome and are capable of self-replicating. Plasmids were discovered by investigators studying the bacterium *E. coli.*

Two enzymes are needed to introduce foreign DNA into vector DNA: (1) a **restriction enzyme** to cleave DNA and (2) **DNA ligase** to seal DNA into an opening created by the restriction enzyme. Hundreds of restriction enzymes occur naturally in bacteria, where they act as a primitive immune system by cutting up any viral DNA that enters the cell. They are called restriction enzymes because they *restrict* the growth of viruses, but they can also be used as molecular scissors to cut double-stranded DNA at a specific site. For example, the restriction enzyme called *Eco*RI always recognizes and cuts double-stranded DNA in the following manner when DNA has the sequence of bases GAATTC:

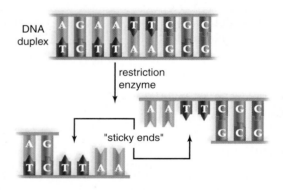

DNA duplex

restriction enzyme

"sticky ends"

Notice that a gap now exists into which a piece of foreign DNA can be placed if it ends in bases complementary to those exposed by the restriction enzyme. To ensure this, it is only necessary to cleave the foreign DNA with the same type of restriction enzyme. The single-stranded, but complementary, ends of the two DNA molecules are called "sticky ends" because they can bind a piece of foreign DNA by complementary base pairing. Sticky ends facilitate the insertion of foreign DNA into vector DNA.

DNA ligase, an enzyme that functions in DNA replication, is then used to seal the foreign piece of DNA into the vector. Bacterial cells take up recombinant plasmids, especially if the cells are treated to make them more permeable. Thereafter, as the plasmid replicates, so does the foreign DNA and thus the gene is cloned.

Using the Polymerase Chain Reaction

The **polymerase chain reaction (PCR)** can create millions of copies of a segment of DNA very quickly in a test tube without the use of a vector or a host cell. The original sample for PCR is usually just a portion of the entire genome. PCR is very specific—it *amplifies* (makes copies of) a targeted DNA sequence that can be less than one part in a million of the total DNA sample!

PCR requires the use of DNA polymerase, the enzyme that carries out DNA replication; a set of primers specific to the targeted DNA sequence; and a supply of nucleotides for the new DNA strands. Primers are single-stranded DNA sequences that start the replication process on each strand. PCR is a chain reaction because the targeted DNA is repeatedly replicated as long as the process continues (Fig. 24.18). The amount of DNA doubles with each replication cycle: after one cycle, there are two copies of the targeted DNA sequence; after two cycles, there are four copies; and so forth.

PCR has been in use for several years, and now almost every laboratory has automated PCR machines to conduct the procedure. Automation became possible after a temperature-insensitive (thermostable) DNA polymerase was extracted from the bacterium *Thermus aquaticus,* which lives in hot springs. The enzyme, called *Taq* polymerase, substitutes for the DNA polymerase used during cellular DNA replication. During the PCR cycle, the mixture of DNA and primers is heated to 95°C to separate the two strands of the double helix so that the primers can bind to single-stranded

DNA. The *Taq* enzyme can withstand the high temperature used to separate double-stranded DNA; therefore, replication does not have to be interrupted by the need to add more enzyme.

DNA amplified by PCR is often analyzed for various purposes. For example, mitochondrial DNA base sequences have been used to decipher the evolutionary history of human populations. Since so little DNA is required for PCR to be effective, it has even been possible to sequence DNA taken from a 76,000-year-old mummified human brain.

DNA Fingerprinting

To identify an individual in a criminal investigation or a paternity case, DNA can be subjected to **DNA fingerprinting.** DNA fingerprinting is based on the differing sequences of DNA nucleotides that exist between individuals. Originally, these differences were discovered using a restriction enzyme to digest DNA and cut it into fragments of different lengths. Each individual has a unique fragment pattern. It is possible to visualize fragment lengths by using a technique called gel electrophoresis. During gel electrophoresis, DNA fragments move through a jellylike, porous material (the gel) while an electrical current is applied. DNA is negatively charged and thus runs through a gel

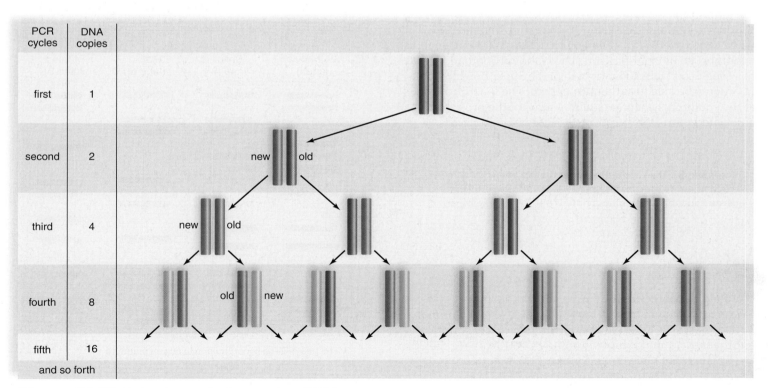

Figure 24.18 Polymerase chain reaction (PCR).
PCR allows the production of many identical copies of DNA in a laboratory setting. Assuming you start with only one copy, you can create millions of copies with only 20–25 cycles. For this reason only a tiny DNA sample is necessary for forensic genetics.

toward the positively charged electrode (Fig. 24.19). Shorter fragments migrate farther through the gel than longer fragments, thus separating them by their lengths.

Today, DNA fingerprinting, now called "DNA profiling" is most often carried out by detecting how many short tandem repeats of DNA are in a particular portion of the genome. Short tandem repeats are a repeated sequence, usually 2–6 bases in length (e.g., CACACA . . .), and are also called **microsatellites.** People differ by how many repeats they have at different locations in their genome, creating DNA fragments of different lengths. Thus, these fragments can also be visualized on a gel. However, because these microsatellites are in particular locations in the genome, PCR is used to amplify only that portion of the genome. It is customary to test the number of repeats at several locations to further ensure that an individual is correctly identified. While reasonably large amounts of DNA were needed for the original DNA fingerprinting technique, only a tiny sample is needed for microsatellite analysis because PCR is used to make many copies of the target DNA sequence.

DNA fingerprinting has many uses. Medically, it can be used to identify the presence of a viral infection or a mutated gene that could predispose someone to cancer. In forensics, DNA fingerprinting from a single sperm is enough to identify a suspected rapist because the DNA can first be amplified by PCR. It can identify the parents of a child or the remains of someone who died, such as the victims of the World Trade Center disaster. In the latter instance, all that was needed were skin cells left on a personal item, such as a toothbrush or cigarette butt. The National Football League has even used DNA to mark each of the Super Bowl footballs, in order to be able to authenticate them. The Bioethical Focus at the end of this chapter explores some of the pros and cons associated with DNA profiling.

Recombinant DNA technology and the polymerase chain reaction are two ways to clone DNA. DNA fingerprinting permits identification of individuals and/or their relatives.

Biotechnology

The field of **biotechnology** uses natural biological systems to create a product or to achieve a goal desired by humans. Today, bacteria, plants, and animals can be genetically engineered to make biotechnology products (Fig. 24.20). Organisms that have had a foreign gene inserted into their DNA are called **transgenic organisms.**

Transgenic Bacteria

Recombinant DNA technology is used to produce transgenic bacteria, which are grown in huge vats called bioreactors. The bacteria express the cloned gene, and the gene product is

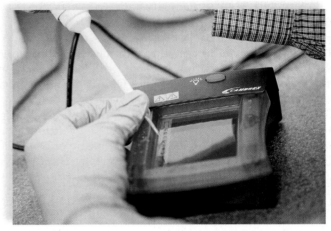

1. Digest DNA samples with restriction enzymes

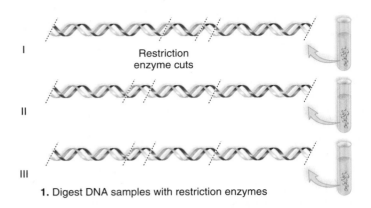

2. Apply samples to gel, and perform electrophoresis

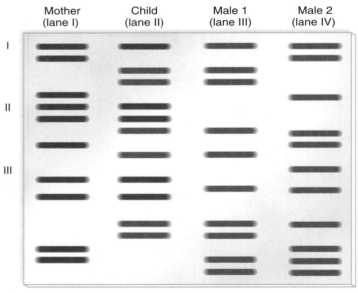

3. The use of DNA fingerprints to establish paternity

Figure 24.19 DNA fingerprints.
Traditionally, DNA fingerprinting was performed like this. First, DNA samples were digested to fragments. The fragments were separated by gel electrophoresis and stained; then the resulting fragment length pattern was observed. In this example, the sample in lane I is from a mother and the DNA in lane II from her child. Then, in a paternity case, the sample from lane III is more likely to be from the child's father than is the sample in lane IV.

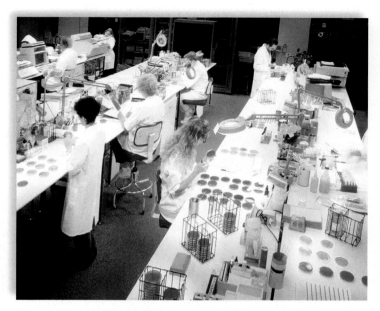

Figure 24.20 Biotechnology products.
Products such as clotting factor VIII, which is administered to hemophiliacs, can be made by transgenic bacteria, plants, or animals. After being processed and packaged, they are sold as commercial products.

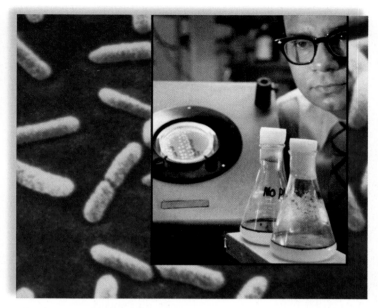

Figure 24.21 Bioremediation.
These bacteria, which are capable of decomposing oil, were engineered and patented by the investigator Dr. Chakrabarty. The flask toward the rear contains oil and no bacteria; the flask toward the front contains the engineered bacteria and is almost clear of oil.

often collected from the medium the bacteria are grown in. Several biotechnology products produced by bacteria are now on the market, including insulin, human growth hormone, tPA (tissue plasminogen activator), and hepatitis B vaccine.

Transgenic bacteria have many other uses as well. Some have been produced to promote the health of plants. For example, "frost plus" bacteria that normally live on plants and encourage the formation of ice crystals have been changed to frost-minus. Also, a bacterium that normally colonizes the roots of corn plants has now been endowed with genes (from another bacterium) that code for an insect toxin. The toxin protects the roots from insects.

Bacteria can be selected for their ability to degrade a particular substance, and this ability can then be enhanced by genetic engineering. For instance, naturally occurring bacteria that eat oil can be genetically engineered to do an even better job of cleaning up beaches after oil spills (Fig. 24.21). Bacteria can also remove sulfur from coal before it is burned and help clean up toxic waste dumps. One such strain was given genes that allowed it to clean up levels of toxins that would have killed other strains. Further, these bacteria were given "suicide" genes that caused them to self-destruct when their job was done.

Organic chemicals are often synthesized by having catalysts act on precursor molecules or by using bacteria to carry out the synthesis. Today, it is possible to go one step further and manipulate the genes that code for these enzymes. For instance, biochemists discovered a strain of bacteria that is especially good at producing phenylalanine, an organic chemical needed to make aspartame, better known as NutraSweet®. They isolated, altered, and cloned the appropriate genes so that various bacteria could be genetically engineered to produce phenylalanine.

Bacteria are being genetically altered to perform all sorts of tasks, in both the laboratory and the environment.

Transgenic Plants

Techniques have been developed to introduce foreign genes into immature plant embryos or into plant cells called **protoplasts** whose cell walls have been removed. It is possible to treat protoplasts with an electric current while they are suspended in a liquid containing foreign DNA. The electric current makes tiny, self-sealing holes in the plasma membrane through which genetic material can enter. Protoplasts go on to develop into mature plants containing and expressing the foreign DNA.

Foreign genes transferred to cotton, corn, and potato strains have made these plants resistant to pests because their cells now produce an insect toxin. Similarly, soybeans have been made resistant to a common herbicide that is sprayed to kill weeds that compete with soybean growth. Some corn and cotton plants are both pest- and herbicide-resistant. These and other genetically engineered crops that are expected to have increased yields are now sold commercially. However, as discussed in the Health Focus, the public is concerned about the possible effect of genetically modified organisms, also called GMOs, on their health and the environment.

Scientists hope that plants can one day be engineered to have leaves that boost their intake of carbon dioxide or cut down on water loss. Also, yields would increase if wheat and rice could be engineered to utilize C_4 instead of C_3 photosynthesis. Already, it is possible to water modified tomato plants with salt water instead of fresh water, because they have been genetically altered to sequester the salt in their leaves.

Plant pharming is the use of engineered plants to produce human proteins, such as hormones, clotting factors, and antibodies, in their seeds. One type of antibody made by corn can deliver radioisotopes to tumor cells to control cancer growth, and another made by soybeans can be used to treat genital herpes.

Transgenic Animals

Techniques have been developed to insert genes into the eggs of animals. It is possible to microinject foreign genes into eggs by hand, but another method uses vortex mixing. The eggs are placed in an agitator with DNA and silicon-carbide needles. The needles make tiny holes in the eggs through which the DNA can enter. When these eggs are fertilized, the resulting offspring are transgenic animals. Using this technique, many types of animal eggs have acquired the gene for bovine growth hormone (bGH). The procedure has been used to produce larger fishes, cows, pigs, rabbits, and sheep.

Animal pharming, the use of transgenic farm animals to produce pharmaceuticals, is being pursued by a number of firms. Figure 24.22 outlines the procedure for producing transgenic animals. DNA containing the gene of interest is injected into donor eggs. Following in vitro fertilization, the zygotes are placed in host females, where they develop. After female offspring mature, the product is secreted in their milk. Plans are under way to produce drugs for the treatment of cystic fibrosis, cancer, blood diseases, and other disorders by animal pharming.

> **Plants are genetically engineered to improve yield, and animals are engineered to increase their size. Animal pharming, the use of transgenic organisms to produce pharmaceuticals, is now a reality.**

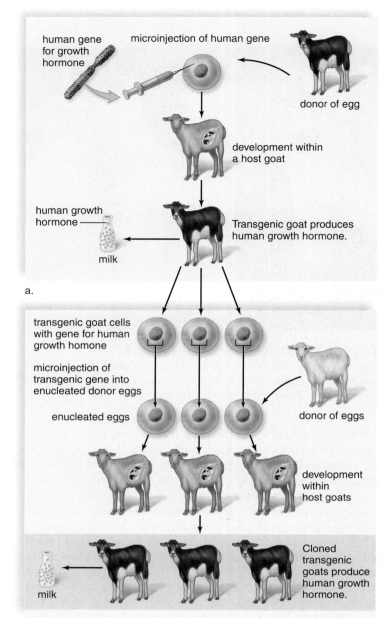

a.

b.

Figure 24.22 Transgenic animals.
a. A genetically engineered egg develops in a host to create a transgenic goat that produces a biotechnology product in its milk. **b.** Nuclei from the transgenic goat are transferred into donor eggs, which develop into cloned transgenic goats.

Summarizing the Concepts

24.1 DNA Structure and Replication
In two separate experiments, investigators labeled the protein coat of a T$_2$ virus with ^{35}S and the DNA with ^{32}P. They then showed that the radioactive P alone is largely taken up by the bacterial host and is followed by reproduction of viruses. These are convincing data that DNA is the genetic material.

DNA is a double helix with two sugar-phosphate backbones and paired nitrogen-containing bases. A (adenine) is paired with T (thymine), and G (guanine) is paired with C (cytosine). During replication, DNA "unzips," and then a complementary strand forms opposite to each original strand. This means that replication is semiconservative because each double helix contains one old strand and one new strand.

24.2 Gene Expression
RNA is a single-stranded nucleic acid in which A pairs with U (uracil), while G still pairs with C. DNA specifies the synthesis of proteins using a triplet code: every three bases is transcribed

into an mRNA codon that stands for one amino acid. This is the information DNA passes from generation to generation.

Protein synthesis requires transcription and translation. During transcription, mRNA complementary to one of the DNA strands is produced. The mRNA moves to the cytoplasm, where it becomes associated with the ribosomes. During translation, tRNA molecules, attached to their own particular amino acid, travel to a ribosome, and through complementary base pairing between codons of tRNA and codons of mRNA, the tRNAs, and the amino acids they carry, are sequenced in a predetermined way to form a polypeptide chain.

Are Genetically Engineered Foods Safe?

A series of focus groups conducted by the Food and Drug Administration (FDA) in 2000 showed that although most participants believed that genetically engineered foods, now also called GMOs, might offer benefits, they also feared possible unknown long-term health consequences. In Canada, Conrad G. Brunk, a bioethicist at the University of Waterloo in Ontario, has said, "When it comes to human and environmental safety, there should be clear evidence of the absence of risks. The mere absence of evidence is not enough."

The discovery by activists that a type of genetically engineered corn called StarLink had inadvertently made it into the food supply triggered the recall of taco shells, tortillas, and many other corn-based foodstuffs from supermarkets. Further, the makers of StarLink were forced to buy back StarLink from farmers and to compensate food producers at an estimated cost of several hundred million dollars.

StarLink is a type of "BT" corn. It contains a foreign gene taken from a common soil organism, *Bacillus thuringiensis,* which makes a protein that is toxic to many insect pests. About a dozen BT varieties, including corn, potato, and even a tomato, have now been approved for human consumption. These strains contain a gene for an insecticidal protein called CryIA. Instead, StarLink contained a gene for a related protein called Cry9C, which researchers thought might slow down the chances of pest resistance to BT corn. In order to get FDA approval for use in foods, the makers of StarLink performed the required tests. Like the other now-approved strains, StarLink wasn't poisonous to rodents, and its biochemical structure is not similar to those of most chemicals in food that commonly cause allergic reactions in humans (called allergens). But the Cry9C protein resisted digestion

longer than the other BT proteins when it was put in simulated stomach acid and subjected to heat. Because most food allergens resist digestion in a similar fashion, StarLink was not approved for human consumption.

The scientific community is now trying to devise more tests for allergens because it has not been possible to determine conclusively whether Cry9C is or is not an allergen. Also, at this point, it is unclear how resistant to digestion a protein must be in order to be an allergen, and it is also unclear what degree of amino acid sequence similarity a potential allergen must have to a known allergen to raise concern. Dean D. Metcalfe, chief of the Laboratory of Allergic Diseases at the National Institute of Allergy and Infectious Diseases, said, "We need to understand thresholds for sensitization to food allergens and thresholds for elicitation of a reaction with food allergens."

Other scientists are concerned about the following potential drawbacks to the planting of BT corn: (1) resistance among populations of the target pest, (2) exchange of genetic material between the transgenic crop and related plant species, and (3) BT crops' impact on nontarget species. They feel that many more studies are needed before stating for certain that BT corn has no ecological drawbacks.

Despite controversies, the planting of Genetically engineered corn increased from 2001–2004. The USDA reports that U.S. farmers planted genetically engineered corn on 45% of all corn acres, up from 26% in 2001. In all, U.S. farmers planted at least 100 million acres with mostly genetically engineered corn, soybeans, and cotton (Fig. 24B). Some groups advocate that GMOs should be labeled as such, but this may not be easy to accomplish because, for example, most cornmeal is derived from

both conventional and genetically engineered corn. So far, there has been no attempt to sort out one type of food product from the other. However, at many health food stores foods that *do not* contain GMOs are now labeled.

Discussion Questions

1. Do you think genetically modified organisms (GMOs) should be labeled? Construct an argument both for and against labeling GMOs.
2. Small groups of activists are strongly advocating the complete removal of GMOs from the market. A few people, called ecoterrorists, are taking such drastic actions as burning crops or even setting biotechnology labs on fire. Do you think GMOs should be removed from the market? Why or why not? What further information would you need to make your decision?
3. Rice is a staple in the diet of millions of people worldwide, many of them living in less-developed countries. In some of those same countries, vitamin A deficiency is a major cause of blindness in small, malnourished children. Scientists have developed a new form of rice, called "golden rice," which is genetically modified to assist with the metabolism of vitamin A and might prevent millions of cases of blindness. The scientists who created golden rice claim it is safe for consumption, although critics say the health effects are not yet fully understood. Given its nutritional potential, should golden rice be planted and distributed on a wide scale? What information would you need to make your decision?

a.

b.

c.

Figure 24B Genetically engineered crops. Genetically engineered (a) corn, (b) cotton crops, and (c) soybeans are increasingly being planted by today's farmers.

DNA Fingerprinting and the Criminal Justice System

Traditional fingerprinting has been used for years to identify criminals and to exonerate those wrongly accused of crimes. However, sometimes fingerprints are not left at the crime scene, while at other times fingerprints may be present but not connected to the particular crime at all. DNA profiling, on the other hand, does not depend on fingerprints. It can be used to identify suspects based on the unique DNA sequences in their genes, as obtained from biological samples found at the crime scene. One advantage of DNA profiling is that only a tiny sample is needed because PCR is used. So, the sample can come from a drop of blood, semen from a rape victim, or even a single hair root!

In DNA profiling, target sequences of DNA containing microsatellites are amplified using PCR and separated using gel electrophoresis to determine the number of tandem DNA repeats, as described in section 24.3. Modern technology uses automated equipment that can determine the length of the fragments because the DNA is labeled with fluorescent dyes that are picked up by a laser beam as they move through a gel contained in capillaries. A person could have a single fragment at a microsatellite locus, indicating he or she is a homozygote, or two fragments, indicating he or she is a heterozygote. The suspect's genotype is then compared to that found at the crime scene. A mismatch at one microsatellite locus is enough to exclude a suspect. However, a match at a single locus is not enough to prove that the DNA sample at the crime scene belongs to a particular suspect. Why? The particular genotype is compared to the frequency of that genotype in the FBI database. Let's say that a particular genotype is present in 10% of the population of U.S. males, and that a suspect in a rape case has DNA that matches that in semen collected from the crime scene. This means there is a 10% chance of *accidental match*. Is this guilt beyond a reasonable doubt? Surely not. Therefore, the microsatellite genotyping procedure is done over several different locations in the genome (usually 7–8) to ensure that the probability of accidental match is incredibly low. Thus, advocates of DNA profiling claim that identification is "beyond a reasonable doubt."

Opponents of this technology, however, point out that it is not without its problems. Police or laboratory negligence can invalidate the evidence. For example, during the O.J. Simpson trial, the defense claimed that the DNA evidence was inadmissible because it could not be proven that the police had not "planted" O.J.'s blood at the crime scene. Problems involving sloppy laboratory procedures and the credibility of forensic experts have also been reported.

In addition to identifying criminals, DNA fingerprinting can be used to establish paternity and maternity, determine nationality for immigration purposes, and identify victims of a national disaster, such as 9/11.

Considering the usefulness of DNA fingerprints, perhaps everyone should be required to contribute blood to create a national DNA fingerprint databank. Some say, however, this would constitute "search without cause," which is unconstitutional.

Form Your Own Opinion

1. Would you be willing to provide your DNA for a national DNA databank? Why or why not? What types of privacy restrictions would you want on your DNA?
2. If not everyone, do you think that convicted felons, at least, should be required to provide DNA for a databank?
3. Should all defendants have access to DNA fingerprinting (at government expense) to prove they didn't commit a crime? Should this include those already convicted of crimes who want to reopen their cases using new DNA evidence?

A ribosome has a binding site for two tRNAs at a time. The tRNA at the first binding site passes a peptide to a newly arrived tRNA–amino acid at the second binding site. Then translocation occurs: the ribosome moves, and the tRNA with the peptide is at the first binding site. In this way, an amino acid grows one amino acid at a time until a polypeptide is formed.

In molecular genetics, a gene is a sequence of DNA bases that code for a product. A gene mutation is a change in the sequence of bases within a gene. The result can be a malfunctioning protein. Mutations can arise due to DNA replication errors, mutagens, or transposons.

24.3 DNA Technology

Genetic engineering includes techniques that allow us to manipulate genes.

It is possible to clone a gene by utilizing recombinant DNA technology or the polymerase chain reaction (PCR). Recombinant DNA contains DNA from two different sources. A restriction enzyme is used to cleave both bacterial plasmid DNA and foreign DNA. The "sticky ends" produced facilitate the insertion of foreign DNA into vector DNA. The foreign gene is sealed into the vector DNA by DNA ligase. Both bacterial plasmids and viruses can serve as vectors to carry foreign genes into bacterial host cells.

PCR uses the enzyme *taq* polymerase to quickly make multiple copies of a specific piece (target) of DNA. PCR is a chain reaction because the targeted DNA is repeatedly replicated as long as the process continues. Following PCR, it is possible to carry out DNA profiling.

Biotechnology uses natural biological systems to create a product or to achieve an end desired by human beings. Transgenic organisms have had a foreign gene inserted into them. Genetically engineered bacteria, agricultural plants, and farm animals now produce commercial products such as hormones and vaccines. Bacteria secrete the product. The seeds of plants and the milk of animals, or other animal products, contain the product of interest.

Transgenic bacteria have been produced to promote the health of plants, perform bioremediation, and produce chemicals. Transgenic crops that have been engineered to resist herbicides and pests are commercially available. Transgenic animals have been given various genes, in particular the one for bovine growth hormone (bGH). Cloning of animals is now possible.

Testing Yourself

Choose the best answer for each question.

1. In the Hershey-Chase experiments, radioactive phosphorus was found
   a. outside the bacterial cells.
   b. inside the bacterial cells.
   c. both inside and outside the bacterial cells.

2. If Hershey and Chase had used an isotope of nitrogen to label viruses in their experiment, they would have detected the isotope

   a. outside the bacterial cells.
   b. inside the bacterial cells.
   c. both inside and outside the bacterial cells.

3. According to Chargaff's rules, in DNA, the amount of

   a. A = T and G = C.
   b. A = G and T = C.
   c. A = C and T = G.
   d. A = T = C = G.

4. The X-ray diffraction work of Franklin and Wilkins showed that DNA

   a. contains four types of bases.
   b. is composed of nucleotides.
   c. is a double helix.
   d. contains hydrogen bonds between bases.

5. The two strands of a DNA molecule run in opposite directions, so they are said to be

   a. complementary.   c. semidiscontinuous.
   b. antiparallel.   d. semiconservative.
   e. inversely parallel.

6. During DNA replication, the parental strand ATTGGC would code for the daughter strand

   a. ATTGGC.   c. TAACCG.
   b. CGGTTA.   d. GCCAAT.

For questions 7–9, match the function in DNA replication with the enzyme in the key.

Key:

   a. DNA helicase   c. DNA ligase
   b. DNA polymerase

7. Addition of new complementary DNA nucleotides to the daughter strand

8. Unwinds and unzips DNA

9. Seals breaks in the sugar-phosphate backbone

10. In the following diagram, indicate the components of the DNA molecule. Include hydrogen bonds, phosphate, sugar, and the four bases (adenine, cytosine, guanine, thymine). Two hydrogen bonds exist between A–T base pairs and three exist between G–C pairs. Also, remember that G and A are double-ring structures, while C and T are single rings.

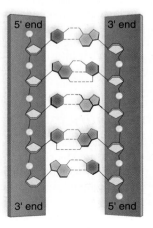

For questions 11–13, match the function with the type of RNA in the key.

Key:

   a. messenger RNA   c. transfer RNA
   b. ribosomal RNA

11. Carries amino acids to ribosomes

12. Is a structural component of the organelle responsible for protein synthesis

13. Carries information from genes to the translation machinery

14. In contrast to DNA, RNA

   a. is a helix.   c. is double stranded.
   b. contains uracil.   d. contains deoxyribose sugar.

15. Which of the following is not a function of RNA polymerase?

   a. breaking hydrogen bonds in DNA
   b. recognizing promoter sequences
   c. proofreading
   d. building RNA

16. Using the information in Figure 24.8, determine the sequence of amino acids that would be produced following transcription and translation of the following DNA template strand:

   TACGTTCCAACT

   a. tyrosine - valine - proline - threonine
   b. methionine - glutamine - glycine
   c. isoleucine - valine - proline
   d. methionine - glutamine - glycine - threonine

17. If a cell could not form polyribosomes,

   a. transcription and translation would not occur.
   b. several similar polypeptides would not be made as quickly.
   c. mistakes would happen during translation, and mutations would occur.
   d. transfer RNAs could not deliver amino acids, and translocation would not occur.

18. Which elements come together during initiation?

   a. mRNA and ribosome
   b. mRNA, ribosome, and initiator tRNA
   c. ribosome and initiator tRNA
   d. mRNA, small ribosomal subunit, and initiator tRNA
   e. small ribosomal subunit and initiator tRNA

19. If a cell could not carry out translocation during translation,

   a. the translation components could not come together.
   b. the chain would not fall off the ribosome at the end of translation.
   c. the chain would not elongate.
   d. tRNA would not base-pair with mRNA.

20. Mutation rates are typically low due to

   a. proofreading by DNA polymerase.
   b. DNA repair enzymes.
   c. silent mutations.
   d. None of these are correct.
   e. All of these are correct.

21. Transposable elements cause mutations by

   a. altering DNA nucleotides.
   b. removing segments of genes.
   c. inserting themselves into genes.
   d. breaking up RNA molecules.

22. These enzymes are needed to introduce foreign DNA into a vector.
    a. DNA gyrase and DNA ligase
    b. DNA ligase and DNA polymerase
    c. DNA gyrase and DNA polymerase
    d. restriction enzyme and DNA gyrase
    e. restriction enzyme and DNA ligase

23. During the PCR reaction, the DNA sample is heated in order to
    a. separate it into single strands.
    b. allow primers to bind.
    c. allow DNA polymerase to work.
    d. synthesize RNA.

24. Gel electrophoresis separates DNA fragments according to
    a. the sequence of their bases.
    b. differences in electrical charge.
    c. their length.
    d. the number of mutations they carry.

25. Which of the following is not true about mRNA?
    a. Exons are spliced together by ribozymes.
    b. When transcribed, mRNA is a copy of sequential nucleotides complementary to those of DNA.
    c. Exons are removed from the mRNA molecule before it leaves the nucleus.
    d. mRNA is formed in the nucleus by RNA polymerase.

26. The process of converting the information contained in the nucleotide sequence of RNA into a sequence of amino acids is called
    a. transcription.   c. translocation.
    b. translation.   d. replication.

27. During protein synthesis, an anticodon of a transfer RNA (tRNA) pairs with
    a. amino acids in the polypeptide.
    b. DNA nucleotide bases.
    c. ribosomal RNA (rRNA) nucleotide bases.
    d. messenger RNA (mRNA) nucleotide bases.
    e. other tRNA nucleotide bases.

Understanding the Terms

anticodon 496
biotechnology 504
cloning 501
codon 496
complementary base
    pairing 492
DNA (deoxyribonucleic
    acid) 490
DNA fingerprinting 503
DNA ligase 502
DNA replication 492
double helix 492
elongation 498
exon 495
gene 493
gene cloning 501
gene mutation 500
genetic engineering 501
initiation 497

intron 495
messenger RNA
    (mRNA) 494
microsatellites 504
mutagen 500
plasmid 502
polymerase chain reaction
    (PCR) 502
polyribosome 497
primary mRNA 494
promoter 494
protoplasts 505
purine 492
pyrimidine 492
recombinant DNA
    (rDNA) 502
restriction enzyme 502
ribosomal RNA (rRNA) 494
ribosome 497

ribozyme 495
RNA (ribonucleic acid) 494
RNA polymerase 494
termination 499
transcription 494
transfer RNA (tRNA) 494

transgenic organism 504
translation 494
transposon 500
triplet code 495
vector 502

Match the terms to these definitions:

a._____ Agent, such as radiation or a chemical, that brings about a mutation.

b._____ Enzyme that speeds the formation of RNA from a DNA template.

c._____ Process whereby a DNA strand serves as a template for the formation of mRNA.

d._____ Process whereby the sequence of codons in mRNA determines (is translated into) the sequence of amino acids in a polypeptide.

e._____ Three nucleotides on a tRNA molecule attracted to a complementary codon on mRNA.

Thinking Critically

1. Use what you have learned in this chapter about the scientific experiments conducted to determine that DNA is the genetic material and how scientists deduced the structure of DNA.

2. In this chapter, you learned about gene cloning. Now, as you have likely heard in the news, we can clone whole organisms—most recently, dogs, cats, and mules. Soon, we may even be able to clone human beings! Should cloning of pets be allowed (for a cost of about $30,000 to the consumer)? What about humans? What are the scientific arguments for and against cloning?

3. In the same way that viruses were used to infect bacteria in the Hershey-Chase experiments, we can use transgenic viruses to infect humans and help treat genetic disorders. That is, the viruses are genetically modified to contain "normal" human genes to try to replace functional human genes. Using this type of gene therapy, a person is infected with a particular virus, which then delivers the "normal" human gene to cells by infecting them. What are some pros and cons of viral gene therapy?

4. Forensic genetics is now used in many court cases. While it is easy to prove innocence using genetics (by identifying a mismatch at a single microsatellite locus), it is more difficult to prove guilt. Suppose DNA found at a crime scene does not belong to the victim and matches the DNA of a suspect. Is the suspect guilty? What information would you need to prove guilt "beyond a reasonable doubt"?

ARIS, the *Inquiry into Life* Website

ARIS, the website for *Inquiry into Life,* provides a wealth of information organized and integrated by chapter. You will find practice quizzes, interactive activities, labeling exercises, flashcards, and much more that will complement your learning and understanding of general biology.

www.aris.mhhe.com

If there's one thing 25-year-old Emily loves, *it's lying in the sun. "Baking on the beach" is what she calls it. A little baby oil, a radio, some sunglasses. She does it for hours. Unfortunately, while Emily rubs in the oil, she doesn't notice the small mole on the back of her right arm. Dark and curvy, the little mole looks like a splotch of spilled ink. Doctors call it melanoma.*

If Emily is lucky, she or her physician will see the mole before the cancer cells break away and spread. If not, the cancer cells may travel through Emily's cardiovascular system, making their way to her lungs, liver, brain, and ovaries. Then, surgery alone will not cure her.

About 32,000 cases of melanoma are diagnosed each year. One in five persons diagnosed dies within five years. For most people, it's easy to avoid the disease. Most important, don't "bake on the beach" and don't visit tanning salons. Protective clothing is a must, and wearing sunscreen will also help.

The melanoma that developed on Emily's right arm went through certain phases. First, a single cell underwent a mutation that caused it to begin to divide repeatedly. As it divided, other mutations occurred until a cell developed that had the ability to invade surrounding tissues and begin tumors in other organs. Cancer cells may break away and spread, a process called metastasis.

chapter 25

Control of Gene Expression and Cancer

Chapter Concepts

25.1 Control of Gene Expression

- How do bacteria regulate gene expression? 512, 514
- How do reproductive and therapeutic cloning differ? 513
- At what levels and by what methods do eukaryotes regulate gene expression? 514–516
- Why and how do eukaryotic cells signal one another? 516

25.2 Cancer: A Failure of Genetic Control

- What are the general characteristics of cancer cells? 517–518
- What two types of genes, in particular, malfunction in cancer cells? What do these genes do in normal cells? 518–519
- How is the risk for cancer increased by heredity? 519
- What environmental carcinogens, in particular, are known to play a role in the development of cancer? 519
- What types of routine screening tests are available to detect and diagnose cancer? 519–520
- What three types of therapy are presently the standard ways to treat cancer? 520
- What types of therapies are expected in the future? 520
- What protective steps can you take to reduce your risk of cancer? 521

25.1 Control of Gene Expression

Each of us started out as a single cell. Through DNA replication and the cell cycle, each of the trillions of cells in our bodies received a copy of all the genes present in the original cell. Early in mammalian development, the cells of the developing embryo can be separated from each other, and each will develop into a normal organism. Nature does this sometimes when it forms identical twins. Later in development, the cells of the embryo differentiate from each other into hundreds of different cell types with specialized functions. However, each cell is still *totipotent*, meaning that it has the complete potential to develop into an organism, because it still contains all of the genes that were present in the original cell. So if the genes are not different, what makes a skin cell different from a red blood cell? Differences in gene expression account for the specialization of the various types of cells. Genes are turned on and off at different times such that each gene's product is made only when and where it is needed. For example, skin cells turn on the genes to make keratin, while red blood cell precursors turn on the genes to make hemoglobin.

Many steps are required for gene expression, and regulation can occur at any of these steps. Both positive controls (increasing the final product) and negative controls (decreasing the final product) can be found, and most genes use several different modes of regulation. We first take a look at gene regulation in bacteria because it is simpler than in eukaryotes.

Gene Expression in Bacteria

The bacterium *Escherichia coli* that lives in your intestine can use various sugars as a source of energy and carbon. This organism can quickly adjust its gene expression to match your diet. The enzymes that are needed to break down lactose, the sugar present in milk, are found encoded together in an operon in the bacterial DNA. An **operon** is a cluster of genes usually coding for proteins related to a particular metabolic pathway, along with the short DNA sequences that coordinately control their transcription. The control sequences consist of a promoter, a sequence of DNA where RNA polymerase first attaches to begin transcription, and an operator, a sequence of DNA where a repressor protein binds (Fig. 25.1).

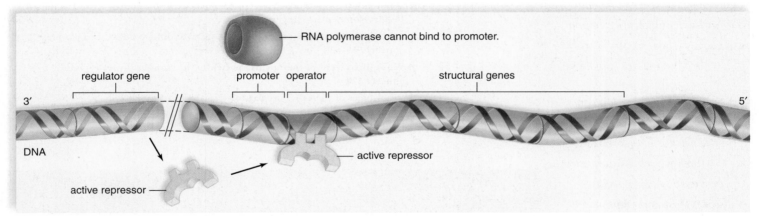

a. **Lactose absent.** Enzymes needed to metabolize lactose are not produced.

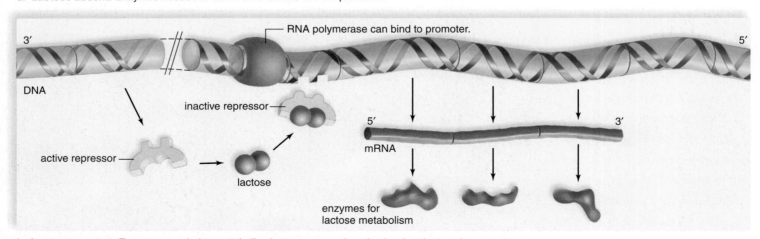

b. **Lactose present.** Enzymes needed to metabolize lactose are produced only when lactose is present.

Figure 25.1 The *lac* operon.

a. When lactose is absent, the regulator gene codes for a repressor that is normally active. When it binds to the operator, RNA polymerase cannot attach to the promoter, and structural genes are not expressed. **b.** When lactose is present, it binds to the repressor, changing its shape so that it is inactive and cannot bind to the operator. Now, RNA polymerase binds to the promoter, and the structural genes are expressed.

ScienceFocus

Reproductive and Therapeutic Cloning

Cloning has aroused much interest because of its scientific and ethical implications. In humans, both reproductive and therapeutic cloning are controversial.

In **reproductive cloning,** the desired end is an individual that is genetically identical to the original individual. The cloning of plants has been done for quite some time (see section 10.3.). However, the cloning of adult animals was thought to be impossible because investigators found it difficult to induce a differentiated cell to return to its pre-differentiated state. Although it contained the same genes as before, differentiation somehow seemed to alter the ability of those genes to be expressed.

In March 1997, Scottish scientists announced they had cloned a sheep, which they named Dolly. To do this, donor cells were taken from the mammary gland of an adult ewe and implanted into another egg that had its own nucleus removed (an enucleated egg cell). Although this procedure had been attempted before with no success, these researchers tried something new. They starved the donor cells, which caused them

to stop dividing and go into a resting stage. When the nuclei from the donor cells were then implanted into the enucleated eggs, they were more sensitive to the cytoplasmic developmental signals of the egg (Fig. 25A). Even so, only one out of 29 clones was successful. Now it is common practice to clone various farm animals with desirable traits and even to clone rare animals that otherwise might become extinct. However, the cloning of humans raises many ethical concerns. In the United States, no federal funds can be used on experiments to clone human beings. There have been claims of human cloning, but none of these have been validated by scientists.

In **therapeutic cloning,** the desired end is not an individual, but mature cells of various types. The purpose of therapeutic cloning is to learn more about how specialization of cells occurs and to provide cells and tissue that could be used to treat various human illnesses, such as diabetes, spinal cord injuries, and Parkinson disease.

There are two possible ways to carry out therapeutic cloning. The first uses the same procedure as for reproductive cloning,

except that embryonic cells, called *embryonic stem cells,* are used. This is controversial because the embryonic cells could have developed into an individual. The second method uses *adult stem cells,* which are found in many organs of the body. For example, the bone marrow has stem cells that produce new blood cells, as does the umbilical cord of newborns. It has already been possible to use stem cells from the brain to regenerate nerve tissue for the treatment of Parkinson disease. However, the goal is to develop techniques that allow scientists to turn any adult stem cell into any type of specialized cell.

Discussion Questions

1. Why is the cloning of a human being so controversial when the cloning of other animals, such as farm animals, is not?
2. How would cells grown in a laboratory be used to treat someone with a spinal cord injury or Parkinson disease?
3. It is now possible to freeze and store umbilical cord blood at birth. Why might you want to pay for having your child's blood preserved in this way?

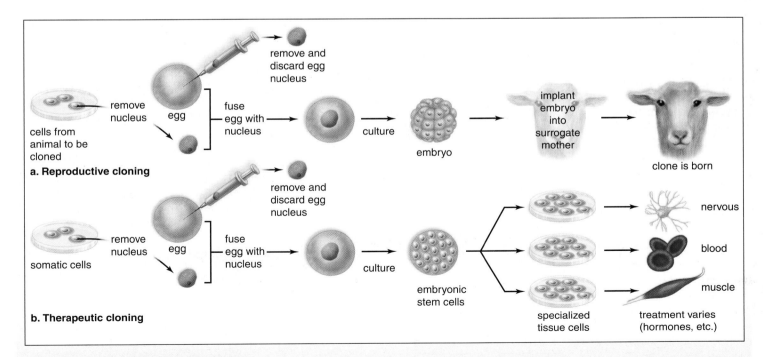

Figure 25A Two types of cloning.
a. The purpose of reproductive cloning is to produce an individual that is genetically identical to the one that donated a 2n nucleus. The 2n nucleus is placed in an enucleated egg, and after several divisions, the embryo comes to term in a surrogate mother. **b.** The purpose of therapeutic cloning is to produce specialized tissue cells. A 2n nucleus is placed in an enucleated egg, and after several divisions, the embryonic cells (called embryonic stem cells) are separated and treated to become specialized cells.

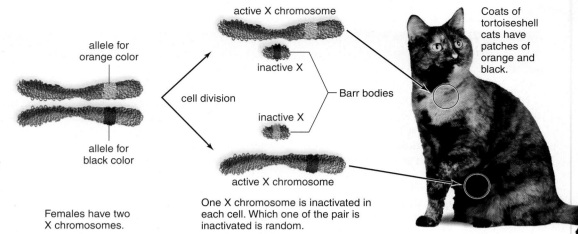

Figure 25.2 X-inactivation.

In cats, the alleles for black or orange coat color are carried on the X chromosome. Random X-inactivation occurs in females. Therefore, in heterozygous females, some of the cells express the allele for black coat color while other cells express the allele for orange coat color. If the X-inactivation happened early in development, resulting in large patches, a calico pattern is found. If it occurred later in development, producing smaller patches of color, a tortoiseshell pattern is found. The white color on calico cats is provided by another gene.

In the *lac* operon, the structural genes for three enzymes that are needed for lactose metabolism are under the control of one promoter/operator complex. If lactose is absent, a protein called a **repressor** binds to the operator. When the repressor is bound to the operator, RNA polymerase cannot transcribe the three structural genes of the operon. The *lac* repressor is encoded by a **regulator gene** located outside of the operon.

When lactose is present, the lactose binds with the *lac* repressor so that the repressor is unable to bind to the operator. Then RNA polymerase is able to transcribe the structural genes into a single mRNA, which is then translated into the three different enzymes. In this way, the enzymes needed to break down lactose are only synthesized when lactose is present.

The *lac* operon is considered an inducible operon; it is only activated when lactose induces its expression. Other bacterial operons are repressible operons; they are usually active until a repressor turns them off.

Gene Expression in Eukaryotes

Many eukaryotic genes encode essential metabolic enzymes or cellular components that are present in all cells. These so-called "housekeeping genes" are not finely regulated because they are always needed, at least to some degree. So while a red blood cell makes hemoglobin, a muscle cell makes myosin, and a pancreatic cell makes insulin as part of their specialized functions, they all express a set of genes necessary to maintain general cellular functions such as transport and acquiring energy:

Cell type	Red blood	Muscle	Pancreatic
Gene type			
Housekeeping	▭	▭	▭
Hemoglobin	▭	☐	☐
Insulin	☐	☐	▭
Myosin	☐	▭	☐

Levels of Gene Control

Eukaryotic genes exhibit control of gene expression at five different levels: (1) chromatin structure, (2) transcriptional control, (3) posttranscriptional control, (4) translational control, and (5) posttranslational control.

Chromatin Structure   Eukaryotes use chromatin packing as a way to keep genes turned off. Genes within darkly staining, highly condensed portions of chromatin, called *heterochromatin*, are inactive. A dramatic example of this occurs with the X chromosome in mammalian females. Females have two X chromosomes, while males have only one X chromosome. This inequality is balanced by the inactivation of one X chromosome in every cell of the female body. Each inactivated X chromosome, called a Barr body in honor of its discoverer, can be seen as a small, darkly staining mass of condensed chromatin along the inner edge of the nuclear envelope. During early prenatal development, one or the other X chromosome in a cell is randomly shut off by chromatin condensation into heterochromatin. All of the cells that form from division of that cell will have the same X chromosome inactivated; therefore, females have patches of tissue that differ in which X chromosome is being expressed. If that female is heterozygous for a particular X-linked gene, she will be a mosaic, containing patches of cells expressing different alleles. This can be clearly seen in calico and tortoiseshell cats (Fig. 25.2).

Active genes in eukaryotic cells are associated with more loosely packed chromatin, called *euchromatin*. However, even euchromatin must be "unpacked" before it can be transcribed. The presence of nucleosomes limits access to the DNA by the transcription machinery. A chromatin remodeling complex pushes the nucleosomes aside to open up sections of DNA for expression:

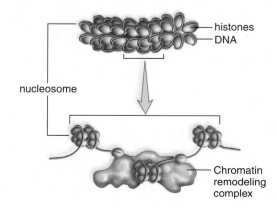

Transcriptional Control   Although eukaryotic cells use various other levels of gene expression control, as depicted in Figure 25.3, the transcriptional level is the most important. As in prokaryotes, transcriptional control is dependent on the interaction of proteins with particular DNA sequences. The proteins are called transcription factors and activators, while the DNA sequences are called enhancers and promoters. Transcription factors and activators are discussed in further detail later in this section. In eukaryotes, each gene has its own promoter.

Posttranscriptional Control   Following transcription, messenger RNA (mRNA) is processed before it leaves the nucleus and passes into the cytoplasm. The primary mRNA is converted to the mature mRNA by the addition of a poly-A tail and a guanine cap, and by the removal of the introns and splicing back together of the exons. The same mRNA can be spliced in different ways to make slightly different products in different tissues. For example, both the hypothalamus and the thyroid gland produce the hormone calcitonin, but the calcitonin mRNA that exits the nucleus contains different combinations of exons in the two tissues.

The speed of transport of mRNA from the nucleus into the cytoplasm can ultimately affect the amount of gene product following transcription. There is a difference in the length of time it takes various mRNA molecules to pass through a nuclear pore.

Translational Control   The longer an mRNA remains in the cytoplasm before it is broken down, the more gene product can be translated. Differences in the poly-A tail or the guanine cap can determine how long a particular transcript remains active before it is destroyed by a ribonuclease associated with ribosomes. Hormones can cause the stabilization of certain mRNA transcripts. For example, the mRNA for vitelline, an egg membrane protein, can persist for three weeks if it is exposed to estrogen, as opposed to 15 hours without estrogen.

Posttranslational Control   Some proteins are not active immediately after synthesis. After translation, insulin is folded into a three-dimensional structure that is inactive. Then a sequence of about 30 amino acids is enzymatically removed from the middle of the molecule, leaving two polypeptide chains that are bonded together by disulfide (S–S) bonds. This activates the protein. Other modifications, such as phosphorylation, also affect the activity of a protein. Many proteins only function a short time before they are degraded or destroyed by the cell.

Transcription Factors and Activators

In eukaryotes, **transcription factors** are proteins that help RNA polymerase bind to a promoter. Several transcription factors per gene form a *transcription initiation complex* that also helps pull double-stranded DNA apart and even acts

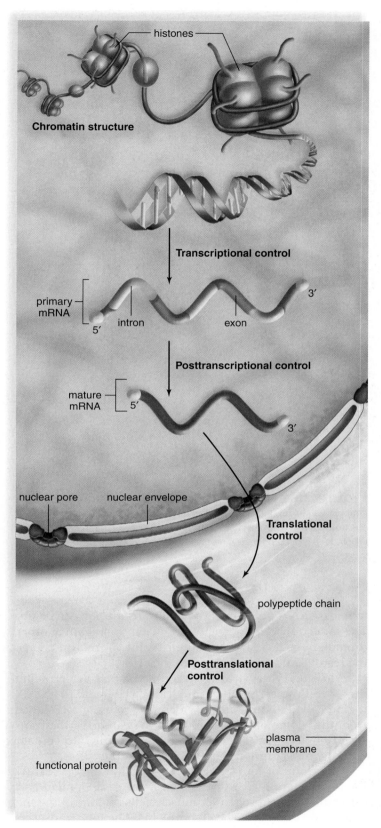

Figure 25.3 Levels at which control of gene expression occurs in eukaryotic cells.

The five levels of control are: (1) chromatin structure, (2) transcriptional control, and (3) posttranscriptional control, which occur in the nucleus; and (4) translational and (5) posttranslational control, which occur in the cytoplasm.

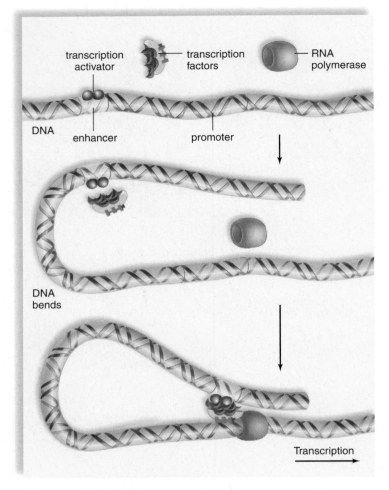

Figure 25.4 Transcription factors and activators.
Transcription factors facilitate the binding of RNA polymerase to a promoter. The transcription activators bind to enhancers that speed transcription when a DNA loop brings all the regulator gene proteins together.

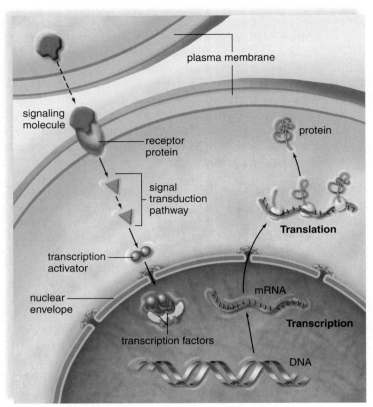

Figure 25.5 Signal transduction pathway.
A signaling molecule is received at a specific receptor protein. A signal transduction pathway terminates when a transcription activator enhances the transcription of a specific gene. Translation of the mRNA transcript follows, and a protein product results.

to release RNA polymerase so that transcription can begin. The same transcription factors are used over again at other promoters, so it is easy to imagine that if one malfunctions, the result could be disastrous to the cell.

In eukaryotes, **transcription activators** are proteins that speed transcription dramatically. They bind to a DNA region called an **enhancer,** which can be quite a distance from the promoter. A loop in the DNA can bring the transcription activator attached to the enhancer into contact with the transcription initiation complex (Fig. 25.4).

A group of four related transcription factors, called the Myo-D family, is responsible for the development and differentiation of progenitor muscle cells by controlling the expression of muscle-specific proteins. The four transcription factors appear to act at different times during development of the muscle cells, leading the cells down a path to become fully differentiated muscle cells.

Signaling Between Cells

Cells are in constant communication with one another. Typically, a cell produces a signaling molecule that binds to a receptor protein in a target cell's plasma membrane. The molecule causes the receptor protein to initiate a series of reactions that will change the receiving cell's behavior. If the cell is a nerve cell, it may transmit a nerve impulse; if it is a pancreatic cell, it may produce insulin.

The series of reactions is a **signal transduction pathway.** In this instance, the end result of the signal transduction pathway is stimulation of a transcription activator. The transcription activator will then help turn on a gene. In this way, specialization of cells comes about and is coordinated so that cells usually assume a structure and function suitable to their location in the body (Fig. 25.5).

Control of gene expression occurs at a number of levels in eukaryotes. Activation of chromatin and use of transcription factors and activators controls gene expression. Signaling between cells turns on particular genes.

25.2 | Cancer: A Failure of Genetic Control

Abnormal cells that defy the normal regulation of the cell cycle and have the ability to invade and colonize other areas are called cancer cells. Normal cells exhibit *contact inhibition*—in other words, when they come in contact with a neighboring cell, they stop dividing. Cells that begin to proliferate abnormally lose contact inhibition and form tumors. They pile on top of one another and grow in multiple layers. As long as the tumor stays clustered in a single mass, it is considered benign, or noncancerous. However, when the cells gain the ability to invade surrounding tissues, they are said to be cancerous. Cancer cells are able to travel through the blood or lymph and start secondary tumors elsewhere in the body (Fig. 25.6). These tumors are called **metastatic** tumors, and the cancer is said to have metastasized. Metastatic cancer is much more difficult to fight and makes recovery doubtful.

Characteristics of Cancer Cells

The primary characteristics of cancer cells are as follows:

Cancer cells are genetically unstable. Generation of cancer cells appears to be linked to mutagenesis. A cell acquires a mutation that allows it to continue to divide. Eventually one of the progeny cells will acquire another mutation and gain the ability to form a tumor. Further mutations occur, and the most aggressive cell becomes the dominant cell of the tumor. Metastatic tumor cells undergo multiple mutations and also tend to have chromosomal aberrations and rearrangements.

Cancer cells do not correctly regulate the cell cycle. Cancer cells continue to cycle through the cell cycle. The normal controls of the cell cycle do not operate to stop the cycle and allow the cells to differentiate. Because of that, cancer cells tend to be nonspecialized. Both the rate of cell division and the number of cells increase. The genes involved in cancer that control the cell cycle will be described later in this discussion.

Cancer cells escape the signals for cell death. A cell that has genetic damage or problems with the cell cycle will initiate apoptosis, or programmed cell death. However, cancer cells do not respond to internal signals to die, and they continue to divide even with genetic damage. Cells from the immune system, when they detect an abnormal cell, will send signals to that cell, inducing apoptosis. Cancer cells also ignore these signals.

Most normal cells have a built-in limit to the number of times they can divide before they die. One of the reasons normal cells stop entering the cell cycle

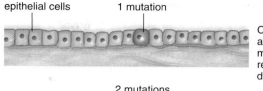

epithelial cells   1 mutation

Cell (dark pink) acquires a mutation for repeated cell division.

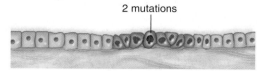

2 mutations

New mutations arise, and one cell (green) has the ability to start a tumor.

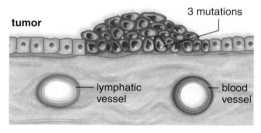

tumor   3 mutations

lymphatic vessel   blood vessel

The tumor is at its place of origin. One cell (purple) mutates further.

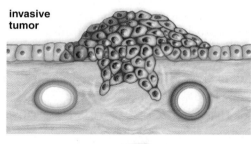

invasive tumor

Cells have gained the ability to invade underlying tissues by producing a proteinase enzyme.

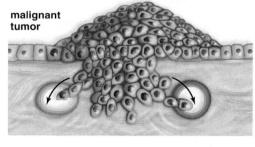

malignant tumor

Cancer cells now have the ability to invade lymphatic and blood vessels.

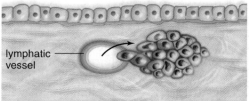

distant tumor

lymphatic vessel

New metastatic tumors are found some distance from the tumor.

Figure 25.6 Development of cancer.
Several mutations contribute to the development of a tumor. A single abnormal cell begins the process, and the most aggressive cell, thereafter, becomes the one that divides the most and forms the tumor. Eventually, cancer cells gain the ability to invade underlying tissue and travel to other parts of the body, where they develop new tumors.

is that the telomeres become shortened. **Telomeres** are sequences at the ends of the chromosomes that keep them from fusing with each other. With each cell division, the telomeres shorten, eventually becoming short enough to signal apoptosis. Cancer cells turn on the gene that encodes the enzyme telomerase, which is capable of rebuilding and lengthening the telomeres. Cancer cells thus show characteristics of "immortality" in that they can enter the cell cycle repeatedly.

Cancer cells can survive and proliferate elsewhere in the body. Many of the changes that must occur for cancer cells to metastasize and form tumors elsewhere in the body are not understood. The cells apparently disrupt the normal adhesive mechanism and move to another place within the body. They travel through the blood and lymphatic vessels and then invade new tissues, where they form tumors. As a tumor grows, it must increase its blood supply by forming new blood vessels, a process called **angiogenesis.** Tumor cells switch on genes that code for the production of growth factors encouraging blood vessel formation. These new blood vessels supply the tumor cells with the nutrients and oxygen they require for rapid growth, but they also rob normal tissues of nutrients and oxygen.

Proto-oncogenes and Tumor Suppressor Genes

Carcinogenesis, the formation of cancer, requires multiple mutations. When these mutations are examined, many are found to occur in two different categories of genes: proto-oncogenes and tumor suppressor genes. **Proto-oncogenes** encode proteins that promote the cell cycle and prevent apoptosis. They are often likened to the gas pedal of a car because they cause acceleration of the cell cycle. **Tumor suppressor genes** encode proteins that inhibit the cell cycle and promote apoptosis. They are often likened to the brakes of a car because they inhibit acceleration through the cell cycle.

Proto-oncogenes

When proto-oncogenes mutate, they become cancer-causing genes, called **oncogenes.** These are called "gain-of-function" mutations because when the product is present, the effect is seen. For example, the Ras protein functions in a signal transduction pathway. When a **growth factor** binds to a receptor on the cell surface, the Ras protein is activated and in turn, activates other proteins that eventually cause changes in gene transcription (Fig. 25.7). This initiates cell division. Therefore, when activated, Ras sends the signal for the cell to divide. When Ras is altered, it is always activated. No growth factor needs to bind to the cell for the altered Ras to send the signal to divide. An altered Ras protein is found

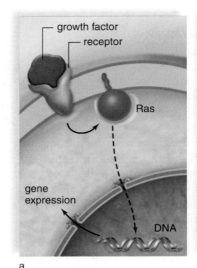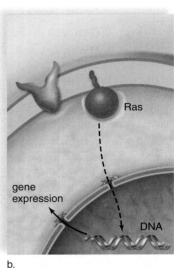

a.   b.

Figure 25.7 Activity of Ras protein.
a. When a growth factor binds to its receptor on the cell surface, Ras is activated and sends a signal to the nucleus to express genes for cell division. **b.** When Ras is mutated, it sends the signals for cell division without the presence of a growth factor.

in approximately 25% of all tumors, including **leukemias** (cancer of blood-forming cells) and **lymphomas** (cancers of lymphatic tissue).

Tumor Suppressor Genes

When tumor suppressor genes mutate, their products no longer inhibit the cell cycle. These are called "loss-of-function" mutations because when the product is absent, the effect is seen. The retinoblastoma (RB) protein acts as one of the main brakes on the progress through the cell cycle. When the gene is mutated or missing, this allows the cell to repeatedly enter the cell cycle. p53 is another protein encoded by a tumor suppressor gene (Fig. 25.8). The normal p53 protein plays a role

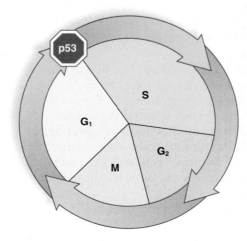

Figure 25.8 Activity of p53 tumor suppressor.
The p53 protein prevents the cell from entering the S phase of the cell cycle if there is DNA damage. If the damage cannot be repaired, it forces the cell to undergo apoptosis. If p53 is absent, the cell continues to cycle.

in cell cycle control, apoptosis, and the genetic stability of the cell. If the DNA of the cell is damaged, p53 halts the cell cycle until repairs can be made. If the damage is severe, p53 forces the cell into apoptosis. It is such an important protein that it is easy to see why almost half of all human cancers have a mutation in this gene.

Causes of Cancer

Two significant causes of cancer are heredity and environmental factors. Particular types of cancer seem to run in families, providing evidence that heredity does play a role in the development of cancer (Fig. 25.9).

There is both a hereditary and a nonhereditary form of the disease retinoblastoma, which forms eye tumors. In the hereditary form, one copy of the gene encoding the retinoblastoma protein is damaged because of a chromosomal aberration or mutation. The individual thus inherits one copy of a normal retinoblastoma gene and one copy of a "bad" retinoblastoma gene. Since the retinoblastoma gene is a tumor suppressor gene, as long as the normal gene produces RB protein, cancer will not develop. But, if the normal gene is somehow mutated and becomes nonfunctional, the person will most likely develop cancer. This explains why cancer, by itself, is not inherited; rather, it is the increased potential for cancer that is inherited. There are still many factors in carcinogenesis that are not understood. In hereditary retinoblastoma, cancers develop earlier than in nonhereditary retinoblastoma. The reason is that individuals with nonhereditary retinoblastoma inherit two normal genes. Both genes must mutate and become nonfunctional in order for that person to develop cancer. This usually takes longer.

Environmental factors include chemical **carcinogens,** ionizing radiation, and viruses. Each of these factors causes mutations in the DNA. Cigarette smoke contains deadly chemical carcinogens. Smoking is the major cause of cancer of the lungs, larynx, oral cavity, and esophagus. It has also been linked to increased death rates for cancers of the bladder, kidney, and pancreas. UV light exposure from natural sunlight or from tanning beds increases the risk of skin cancer. Several viruses have been associated with cancer as well. Epstein Barr virus is found in nearly all nasopharyngeal cancer specimens, and certain human papilloma viruses (HPV) are linked to cervical cancer. The Health Focus describes ways to protect yourself from these environmental factors.

Diagnosis of Cancer

Routine screening tests help detect cancer. For example, the Pap smear only requires that a physician take a sample of cells from the cervix, which is then examined microscopically for signs of abnormality. A mammogram can reveal a breast tumor that is too small to be felt and at a time when the cancer is still highly curable. A colonoscopy detects polyps while they are small enough to be destroyed by laser therapy.

Tumor marker tests are blood tests for possible cancer in the body. These "markers" are generally normal proteins that are produced in very low amounts. Cancer cells, however, produce these proteins in excess. If cancer is present, the level of the marker may rise. For example, the PSA (prostate-specific antigen) test detects prostate cancer. PSA is normally produced by the prostate and found in the blood. When PSA levels rise in the blood, a problem with the prostate is suspected. This test does not differentiate between benign conditions and cancer of the prostate, so further testing is required. Physicians also can use tumor marker tests to determine if a cancer is responding to therapy or if a cancer has returned.

Various **genetic tests** are also available. Tests for genetic mutations in proto-oncogenes and tumor suppressor genes are making it possible to detect the likelihood that cancer is present or will develop in the near future. Tests are available that signal the possibility of colon, bladder, breast, and thyroid cancers, as well as melanoma.

For example, a genetic test is available for the presence of mutations in the *BRCA1* and *BRCA2* (*breast ca*ncer 1 and 2) genes. Mutations in these genes are involved in many cases of both breast and ovarian cancer. Women with mutations in one of these genes are three to seven times more likely to develop breast cancer than women with normal *BRCA1* and *BRCA2* genes. The test involves a blood sample that is sent to a genetics laboratory. A person who tests positive for a mutation in one of these genes has an increased *risk* of developing cancer, but not all people who inherit an altered gene develop the disease. On the other hand, anyone who tests positive should be carefully monitored for the development of breast and ovarian cancer. Cancers that are caught early are easier to treat successfully. The Bioethical Focus at the end of this chapter describes some of the advantages and disadvantages of genetic testing for cancer genes.

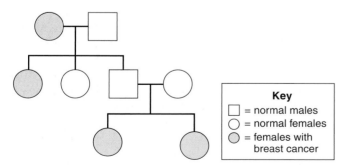

Key
- ☐ = normal males
- ◯ = normal females
- ◉ = females with breast cancer

Figure 25.9 A family pedigree of breast cancer.
In this family, breast cancer has affected three generations. Note that the male in the second generation passed the mutant gene on to his daughters. He has one sister with breast cancer and one without.

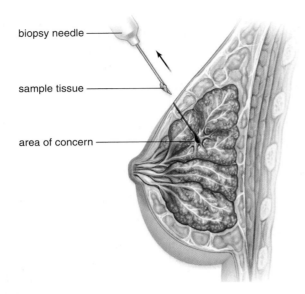

biopsy needle

sample tissue

area of concern

Figure 25.10 Needle biopsy of the breast.
A sample of tissue can be taken with a biopsy needle to help a physician
determine treatment options.

Confirming the Diagnosis

A diagnosis of cancer can be confirmed without major surgery. For example, biopsies allow removal of cells for examination (Fig. 25.10), and sophisticated techniques, such as laparoscopy, permit viewing of body parts. Various imaging techniques are also available. A radioactive scan, obtained after a radioactive isotope is administered, can reveal any abnormal isotope accumulation due to a tumor. During ultrasound, echoes of high-frequency sound waves directed at a part of the body can reveal the size, shape, and location of tumors.

Treatment of Cancer

Surgery, radiation, and chemotherapy are the standard methods of cancer therapy. Surgery alone is sufficient for tumors that are contained. But because there is always the danger that some cancer cells were left behind, surgery is often preceded by and/or followed by radiation therapy or chemotherapy. Because radiation is mutagenic, dividing cells such as cancer cells are more susceptible to its effects than other cells. Thus, the theory is that radiation will cause cancer cells to mutate and undergo apoptosis. Powerful X rays or gamma rays can be administered through an externally applied beam or, in some instances, by implanting tiny radioactive sources into the patient's body. Cancer of the cervix and larynx, early stages of prostate cancer, and Hodgkin disease are often treated with radiation therapy alone.

Chemotherapy is a way to catch cancer cells that have spread throughout the body. Most chemotherapeutic drugs kill cells by damaging their DNA or interfering with DNA synthesis. The hope is that all cancer cells will be killed, while leaving enough normal cells untouched to allow the body to keep functioning. Particularly in breast and colon cancer, chemotherapy can reduce the chance of recurrence following surgical removal of the tumor. One drug, taxol, extracted from the bark of the Pacific yew tree, was found to be particularly effective against advanced ovarian cancers, as well as breast, head, and neck tumors. Taxol interferes with microtubules needed for cell division. Now chemists have synthesized a family of related drugs, called taxoids, which may be more powerful and have fewer side effects than taxol itself.

Certain types of cancer, such as leukemias, lymphomas, and testicular cancer, are now successfully treated by a combination of chemotherapeutic agents. The survival rate for children with childhood leukemia is 80%. Hodgkin disease, a lymphoma, once killed two out of three patients. Now, combination therapy using four different drugs can wipe out the disease in a matter of months in three out of four patients, even when the cancer is not diagnosed immediately.

The red bone marrow contains large populations of dividing cells; therefore, red bone marrow is particularly prone to destruction by chemotherapeutic drugs. In bone marrow autotransplantation, a patient's stem cells are harvested and stored before chemotherapy begins. Quite high doses of radiation or chemotherapeutic drugs are then given within a relatively short period of time. This prevents multidrug resistance from occurring, and the treatment is more likely to catch each and every cancer cell. Then the stored stem cells, which are needed to produce blood cells, are returned to the patient by injection. They automatically make their way to bony cavities, where they initiate blood cell formation.

Future Therapies

Future therapies include cancer vaccines, *p53* gene therapy, and inhibitory drugs for angiogenesis and tumors. Cancer vaccines elicit an immune response against a tumor protein, allowing the body to destroy the tumor. The gene for the p53 protein can be injected directly into the tumor cells of patients.

Antiangiogenic drugs, which confine and reduce tumors by breaking up the network of new capillaries in the vicinity of a tumor, should be routine therapy soon.

Cancer cells display several characteristics not found in normal cells. Mutations in proto-oncogenes and tumor suppressor genes promote cancer development. Both heredity and environmental factors play a role in cancer.

Health Focus

Prevention of Cancer

Protective Behaviors

Be tested for cancer.
Do a shower test for breast cancer or testicular cancer. Women should get a Pap smear for cervical cancer annually if they are over 21 or are sexually active. Have a friend or physician check your skin annually for any unusual moles. Have other exams done regularly by a physician.

Be aware of occupational hazards.
Exposure to several different industrial agents (nickel, chromate, asbestos, vinyl chloride, etc.) and/or radiation increases the risk of various cancers. Your employer should notify you of potential hazards in your workplace. The risk of several of these cancers is increased when combined with cigarette smoking.

Use sunscreen.
Almost all cases of basal-cell and squamous-cell skin cancers are sun-related. Use a sunscreen of at least SPF 15 (Fig. 25B) and wear protective clothing if you are going to be out during the brightest part of the day. Don't sunbathe on the beach or in a tanning salon.

Check your home for radon.
Excessive radon exposure in homes increases the risk of lung cancer, especially in cigarette smokers. It is best to test your home and take the proper remedial actions.

Avoid unnecessary X rays.
Even though most medical and dental X rays are adjusted to deliver the lowest dose possible, unnecessary X rays should be avoided. Sensitive areas of the body that are not being X-rayed should be protected with lead screens.

Practice safe sex.
Human papilloma viruses (HPV) cause genital warts and are associated with cervical cancer, as well as tumors of the vulva, vagina, anus, penis, and mouth. HPVs may be involved in 90–95% of all cases of cervical cancer, and 20 million people in the United States have an infection that can be transmitted to others.

Carefully consider hormone therapy.
Estrogen therapy to control menopausal symptoms increases the risk of endometrial cancer. However, including progesterone in

Figure 25B UV Protection.
Use of sunscreen with SPF of 15 or higher can help prevent skin cancer, as can limiting midday exposure.

estrogen replacement therapy helps minimize this risk.

Maintain a healthy weight.
The risk of cancer (especially colon, breast, and uterine cancers) is 55% greater among obese women and 33% greater among obese men, compared to people of normal weight. Eating a low-fat, healthy diet helps maintain your weight as well as reducing your risk of cancer.

Exercise regularly.
Regular activity can reduce the incidence of certain cancers, especially colon and breast cancer.

Increase consumption of foods that are rich in vitamins A and C.
Beta-carotene, a precursor of vitamin A, is found in dark-green, leafy vegetables, carrots, and various fruits. Vitamin C is present in citrus fruits. These vitamins are called antioxidants because they prevent the formation of chemicals called free radicals that can damage the DNA in the cell. Vitamin C also prevents the conversion of nitrates and nitrites into carcinogenic nitrosamines in the digestive tract.

Include vegetables from the cabbage family in your diet.
The cabbage family includes cabbage, broccoli, Brussels sprouts, kohlrabi, and cauliflower. These vegetables may reduce the risk of gastrointestinal and respiratory tract cancers.

Limit consumption of salt-cured, smoked, or nitrite-cured foods.
Salt-cured or pickled foods may increase the risk of stomach and esophageal cancer.

Smoked foods, such as ham and sausage, contain chemical carcinogens similar to those in tobacco smoke. Nitrites are sometimes added to processed meats (e.g., hot dogs and cold cuts) and other foods to protect them from spoilage. These are converted to carcinogenic nitrosamines in the digestive tract.

Be moderate in the consumption of alcohol.
Cancers of the mouth, throat, esophagus, larynx, and liver occur more frequently among heavy drinkers, especially when accompanied by smoking cigarettes or chewing tobacco.

Don't smoke.
Cigarette smoking accounts for about 30% of all cancer deaths. Smoking is responsible for 90% of lung cancer cases among men and 79% among women—about 87% altogether. People who smoke two or more packs of cigarettes a day have lung cancer mortality rates 13 to 23 times greater than those of nonsmokers. Cigars and smokeless tobacco (chewing tobacco or snuff) increase the risk of cancers of the mouth, larynx, throat, and esophagus.

Discussion Questions
1. Which of these cancer-preventive measures are you currently practicing?
2. Why do you think tobacco use increases the risk of other types of cancer besides lung cancer?
3. Why does the use of tanning beds also increase the incidence of skin cancer?

Bioethical Focus

Genetic Testing for Cancer Genes

Over the past decade, genetic tests have become available for certain cancer genes. If women test positive for defective *BRCA1* and *BRCA2* genes, they have an increased risk for early-onset breast and ovarian cancer. If individuals test positive for the *APC* gene, they are at greater risk for the development of colon cancer. Other genetic tests exist for rare cancers, including retinoblastoma and Wilms tumor.

Advocates of genetic testing say that it can alert those who test positive for these mutated genes to undergo more frequent mammograms or colonoscopies. Early detection of cancer clearly offers the best chance for successful treatment.

Others feel that genetic testing is unnecessary because nothing can presently be done to prevent the disease. Perhaps it is enough for those who have a family history of cancer to schedule more frequent check-ups, beginning at a younger age.

People opposed to genetic testing worry that being predisposed to cancer might threaten one's job or health insurance. They suggest that genetic testing be confined to a research setting, especially since it is not known which particular mutations in the genes predispose a person to cancer. They are afraid, for example, that a woman with a defective *BRCA1* or *BRCA2* gene might make the unnecessary decision to have a radical mastectomy. The lack of proper counseling also concerns many. In a study of 177 patients who underwent *APC* gene testing for susceptibility to colon cancer, less than 20% received counseling before the test. Moreover, physicians misinterpreted the test results in nearly one-third of the cases.

Another concern is that testing negative for a particular genetic mutation may give people the false impression that they are not at risk for cancer. Such a false sense of security can prevent them from having routine cancer screening. Regular testing and avoiding known causes of cancer—such as smoking, a high-fat diet, or too much sunlight—are important for everyone.

Form Your Own Opinion

1. Should genetic testing for cancer be available for everyone, or should genetic testing be confined to a research setting? Explain.
2. If genetic testing for cancer were offered to you, would you take advantage of it? Why or why not?
3. Do you feel that everyone should do all they can to avoid having cancer, such as by not smoking, or is it the individual's choice to take that risk? Explain.

Summarizing the Concepts

25.1 Control of Gene Expression

Each cell in the human body is totipotent, meaning that it has the complete potential to develop into an organism. Specialization of various types of cells is due to differences in gene expression.

In bacteria, genes are clustered into operons. An operon contains a group of genes, usually coding for proteins related to a particular metabolic pathway, along with the promoter, a DNA sequence where RNA polymerase binds to begin transcription, and an operator, where a repressor protein binds. In the case of the *lac* operon, the structural genes code for enzymes that metabolize lactose. When lactose is absent, the repressor binds to the operator and shuts off the operon. When lactose is present, it binds to the repressor so that it can no longer bind to the operator, and this allows RNA polymerase to transcribe the structural genes. Some bacterial operons are repressible instead of inducible.

Control of gene expression is more complicated in eukaryotes. Eukaryotes have five levels of control: chromatin structure, transcriptional control, posttranscriptional control, translational control, and posttranslational control. Genes that are found within highly condensed heterochromatin are inactive. A dramatic example of this is inactivated X chromosomes, called Barr bodies. Even genes that are found in looser-packed DNA, called euchromatin, must have the nucleosomes shifted in order to be expressed. The transcriptional level of control is the most important level. In posttranscriptional control, differences in mRNA processing affect gene expression. Translational and posttranslational controls occur in the cytoplasm and involve the length of time the mRNA and protein are functional. Proteins can be activated by a variety of methods.

Transcription factors and activators bind to DNA and increase transcription rates. Signaling between cells involves a signal transduction pathway that results in changes in gene expression.

25.2 Cancer: A Failure of Genetic Control

Cancer cells defy the normal regulation of the cell cycle and can invade and colonize other areas of the body. Cancer cells do not exhibit contact inhibition and thus form tumors. When tumor cells gain the ability to invade surrounding tissues, they are said to be malignant. When they are able to spread to other areas of the body, they are called metastatic.

In terms of their primary characteristics, cancer cells are genetically unstable, do not correctly regulate the cell cycle, escape the signals for cell death, and can survive and proliferate elsewhere in the body. Multiple mutations and chromosomal aberrations are present in cancer cells. They continue to cycle through the cell cycle and tend to be nonspecialized. They do not respond to either internal or external signals for apoptosis, and they induce the expression of telomerase to repair their telomeres.

Many of the mutations found in cancers occur either in proto-oncogenes or in tumor suppressor genes. Proto-oncogenes promote the cell cycle and prevent apoptosis. A mutated Ras protein sends signals for the cell to divide even in the absence of an external growth factor. Tumor suppressor genes inhibit the cell cycle and promote apoptosis. The genes encoding RB and p53 are examples of these.

Two significant causes of cancer are heredity and environmental factors. In the disease retinoblastoma, a mutated copy of the retinoblastoma gene is inherited. Environmental factors include carcinogens, ionizing radiation, and viruses.

Routine screening tests can help detect cancer. Various tumor marker and genetic tests are available. Biopsies can confirm the diagnosis of cancer. Surgery, radiation, and chemotherapy are the standard methods of cancer treatment.

Testing Yourself

Choose the best answer for each question.

1. An operon is a short sequence of DNA
   a. that prevents RNA polymerase from binding to the promoter.
   b. that prevents transcription from occurring.
   c. and the sequences that control its transcription.
   d. that codes for the repressor protein.
   e. that functions to prevent the repressor from binding to the operator.

2. Which of these correctly describes the function of a regulator gene for the *lac* operon?
   a. prevents transcription from occurring
   b. a sequence of DNA that codes for the repressor
   c. prevents the repressor from binding to the operator
   d. keeps the operon off until lactose is present
   e. Both b and d are correct.

3. Which of the following correctly describes reproductive cloning?
   a. The desired end is an individual that is genetically identical to the original individual.
   b. The desired end is mature cells of various cell types.
   c. An embryonic nucleus is placed in an enucleated and starved embryonic donor cell.
   d. Ethical concerns surround the embryo, which could become an individual, if allowed.

4. The product of therapeutic cloning differs from the product of reproductive cloning in that it is
   a. composed of mature cells of various types, not an individual.
   b. genetically identical to the somatic cells obtained from the donor.
   c. composed of mature cells of various types with the ability to become an individual.
   d. an individual that is genetically identical to the donor of the nucleus.

5. Which of these associations is mismatched?
   a. loosely packed chromatin—gene can be active
   b. transcription factors—gene is inactivated
   c. mRNA—translation can begin
   d. heterochromatin—Barr body

6. Label the following diagram of X-inactivation.

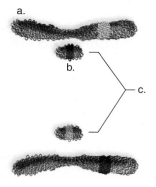

7. What is a Barr body?
   a. an example of gene regulation using chromatin packing
   b. an inactive X chromosome
   c. an X chromosome not producing gene products
   d. the reason that tortoiseshell cats have patches of both black and orange on their coats
   e. All of these are correct.

8. How is transcription directly controlled in eukaryotic cells?
   a. through the use of signal transduction pathways
   b. by means of cell signaling
   c. using transcription factors and activators
   d. when chromatin is packed to keep genes turned on
   e. None of these are correct.

9. Label the following diagram of transcription factors and activators.

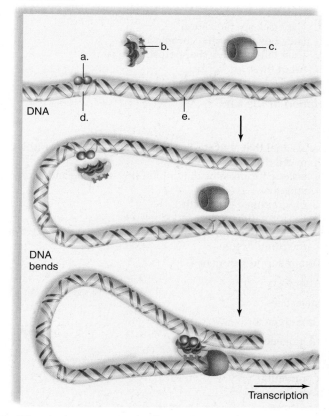

10. What is the purpose of cell signaling in eukaryotes?
    a. to change the behavior of the receiving cell
    b. to initiate a set of reactions through the receptor protein
    c. to turn on a gene
    d. to transmit a nerve impulse, for example
    e. All of these are correct.

11. Eukaryotic cells signal one another through the use of
    a. repressor proteins acting upon operons.
    b. molecules that bind to a receptor protein in the membrane of a target cell.
    c. transcription factors and activators acting upon RNA polymerase.
    d. RNA polymerase acting upon transcription promoters and enhancers.
    e. nerve impulses, which cause a receiving cell to change its behavior.

12. Sequence the events that could lead to the development of cancer:
    (1) Cells have gained the ability to invade underlying tissue using a proteinase enzyme.
    (2) New metastatic tumors are found some distance from the primary tumor.
    (3) A cell acquires a mutation that allows for repeated cell division.
    (4) A tumor develops, and cells mutate further.
    (5) New mutations arise, and one cell has the ability to start a tumor.

    a. 5, 4, 3, 2, 1
    b. 3, 4, 5, 2, 1
    c. 3, 5, 4, 1, 2
    d. 1, 2, 3, 4, 5
    e. 4, 3, 5, 1, 2

13. Which of these is a characteristic of cancer cells?
    a. They do not exhibit contact inhibition.
    b. They lack specialization.
    c. They have abnormal chromosomes.
    d. They fail to undergo apoptosis.
    e. All of these are correct.

For questions 14–18, match the cancer-related word to its description in the key.

Key:
    a. a signal that activates a cell-signaling pathway resulting in cell division
    b. encode for proteins that inhibit the cell cycle and promote apoptosis
    c. programmed cell death
    d. the formation of new blood vessels
    e. the process of establishing new tumors at a distance from the primary tumor

14. Tumor suppressor genes

15. Apoptosis

16. Metastasis

17. Growth factor

18. Angiogenesis

19. Which of these is not a possible cause of cancer?
    a. radiation
    b. metastasis
    c. viruses
    d. genes
    e. oncogenes

20. Which of these is not a possible behavior that could help prevent cancer?
    a. maintaining a healthy weight
    b. eating more dark-green, leafy vegetables, carrots, and various fruits
    c. not smoking
    d. consuming large amounts of smoked and pickled foods
    e. consuming alcohol only in moderation

21. Cancers can be diagnosed using which of the following methods?
    a. physical examination
    b. tumor marker test
    c. genetic mutation test
    d. X ray or ultrasound
    e. All of these are correct.

Understanding the Terms

angiogenesis 518
carcinogen 519
chemotherapy 520
enhancer 516
genetic test 519
growth factor 518
leukemia 518
lymphoma 518
metastatic 517
oncogene 518
operon 512
proto-oncogene 518

regulator gene 514
repressor 514
reproductive cloning 513
signal transduction pathway 516
telomere 518
therapeutic cloning 513
transcription activator 516
transcription factor 515
tumor marker test 519
tumor suppressor gene 518

Match the terms to these definitions:

a._____ A group of bacterial genes that are regulated as a unit.
b._____ Chemical signal that regulates mitosis and differentiation of cells that have receptors for it; sometimes a contributing factor in cancer.
c._____ Spread of cancer from the place of origin throughout the body; caused by the ability of cancer cells to migrate and invade tissues.
d._____ Protein that helps initiate transcription by binding to an enhancer.
e._____ Cancer-causing gene.

Thinking Critically

1. Why do mutations in both proto-oncogenes and tumor suppressor genes end up causing the same disease?

2. What kind of genes do you think would be included in the category of "housekeeping" genes?

3. Fragile X syndrome is a genetic disease in which part of the X chromosome is fragile and may break off. Females experience varying degrees of disability, from no symptoms to very severe symptoms. Why are there varying degrees of disability?

4. When the enzyme telomerase was first discovered, some people thought it might be "the fountain of youth" since it could immortalize cells. Why was this not found to be true?

5. Why is it only the *risk* for cancer that is inherited?

ARIS, the *Inquiry into Life* Website

ARIS, the website for *Inquiry into Life,* provides a wealth of information organized and integrated by chapter. You will find practice quizzes, interactive activities, labeling exercises, flashcards, and much more that will complement your learning and understanding of general biology.

www.aris.mhhe.com

Charise lives with sickle cell anemia. *She is one of an estimated 60,000 Americans who suffer from this chronic, often debilitating genetic disease. The distinctive feature of sickle cell anemia is excruciatingly painful episodes called sickle cell crises. A crisis can be initiated by many factors, including dehydration, fever, overexertion, or exposure to cold. In a person with sickle cell anemia, the normally spherical red blood cells can become sickle-shaped and very stiff. They then block narrow blood vessels in the lungs, liver, bone, muscles, brain, eyes, spleen, and kidneys, causing tissue injury and extreme pain. The immune system is also weakened, and so people with sickle cell anemia are very susceptible to bacterial infections.*

As a child, Charise was often very tired and experienced pain in her joints. She was always in some pain, but then, without warning, she would experience crises of intense pain. She also had many, many infections, and so she got further and further behind in school. As a teenager, she is on pain medication to handle both the chronic pain and the intense pain of a crisis. She is very careful with her health and visits her doctor regularly. For Charise, this disease also takes a toll on her mental health. She suffers from depression and worries about having a stroke. She would like to have children but wants to make sure that she does not pass this disease on. Neither of her parents have sickle cell anemia, but both were carriers of the disease. She plans to undergo genetic counseling before she starts a family.

chapter 26

Genetic Counseling

Chapter Concepts

26.1 Counseling for Chromosomal Disorders

Now that potential parents are becoming aware that many illnesses are caused by abnormal chromosomal inheritance or by faulty genes, more couples are seeking **genetic counseling** to determine the risk of a chromosomal mutation or genetic mutation in their family. For example, they might be prompted to seek counseling after several miscarriages, if several relatives have a particular medical condition, or if they already have a child with a genetic defect. The counselor will help the couple understand the mode of inheritance and the medical consequences of a particular genetic disorder, and will inform them of the possible decisions they might wish to make.

Various human disorders result from abnormal chromosome number or structure. Such disorders often result in a **syndrome,** which is a group of symptoms that always occur together. Table 26.1 lists several syndromes that are due to an abnormal chromosome number. If expectant parents are concerned that their unborn child might have a chromosomal defect, the counselor might recommend karyotyping the fetus's chromosomes.

Karyotyping

A **karyotype** is a visual display of the chromosomes, arranged by size, shape, and banding pattern. Any cell in the body, except red blood cells, which lack a nucleus, can be a source of chromosomes for karyotyping. In adults, it is easiest to use white blood cells separated from a blood sample for this purpose. For fetuses, cells can be obtained by either amniocentesis or chorionic villi sampling. These cells can be used for karyotype analysis or for DNA or protein analysis, as described later.

Amniocentesis is a procedure for obtaining a sample of amniotic fluid from the uterus of a pregnant woman. Blood tests and the mother's age are used to determine whether the procedure should be done. The risk of spontaneous abortion increases by about 0.25–0.5% due to amniocentesis, and doctors use the procedure only if it is medically warranted.

Amniocentesis is not usually performed until after the fifteenth week of pregnancy. A long needle is passed through the abdominal and uterine walls to withdraw a small amount of fluid, which also contains fetal cells (Fig. 26.1a). The amniotic fluid is then tested. Karyotyping of the chromosomes may be delayed as long as four weeks so that the cells can be cultured to increase their number. DNA tests for other genetic disorders can also be done.

Chorionic villi sampling (CVS) is a procedure for obtaining chorionic cells in the region where the placenta will develop. This procedure is usually done between the tenth and fourteenth weeks of pregnancy. A long, thin suction tube is inserted through the vagina into the uterus (Fig. 26.1b). Ultrasound, which gives a picture of the uterine contents, is used to place the tube between the uterine lining and the chorionic villi. Then a sampling of chorionic cells is obtained by suction. Results are available within one to two weeks. But testing amniotic fluid is not possible because no amniotic fluid is collected. Also, CVS carries a greater risk of spontaneous abortion than amniocentesis—0.5–1% compared to 0.25–0.5%. The advantage of CVS is getting the results of karyotyping at an earlier date.

The Karyotype

After a cell sample has been obtained, the cells are stimulated to divide in a culture medium. A chemical is used to stop mitosis during metaphase when chromosomes are the most highly compacted and condensed. The cells are then killed, spread on a microscope slide, and dried. Stains are applied to the slides, and the cells are photographed. Staining causes the chromosomes to have dark and light cross-bands of varying widths, and these can be used, in addition to size and shape, to pair up the chromosomes. Today, a computer is used to arrange the chromosomes in pairs (Fig. 26.1c). The karyotype of a person who has Down syndrome usually has three chromosomes 21, instead of the usual two (Fig. 26.1d, e).

TABLE 26.1		Syndromes from Abnormal Chromosome Numbers			
Syndrome	**Sex**	**Chromosomes**	**Chromosome Number**	**Frequency**	
				Spontaneous Abortions	*Live Births*
Down	M or F	Trisomy 21	47	1/40	1/800
Poly-X	F	XXX (or XXXX)	47 or 48	0	1/1,500
Klinefelter	M	XXY (or XXXY)	47 or 48	1/300	1/800
Jacobs	M	XYY	47	?	1/1,000
Turner	F	X	45	1/18	1/2,500

Figure 26.1 **Human karyotype preparation.**

A karyotype is an arrangement of an individual's chromosomes into numbered pairs according to their size, shape, and banding pattern. **a.** Amniocentesis and **b.** chorionic villi sampling provide cells for karyotyping to determine if the unborn child has a chromosomal abnormality. **c.** After cells are treated, a computer constructs the karyotype. **d.** Karyotype of a normal male. **e.** Karyotype of a male with Down syndrome, showing three chromosomes 21.

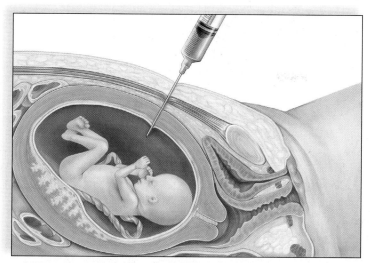

a. During amniocentesis, a long needle is used to withdraw amniotic fluid containing fetal cells.

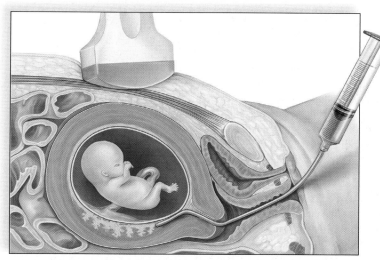

b. During chorionic villi sampling, a suction tube is used to remove cells from the chorion, where the placenta will develop.

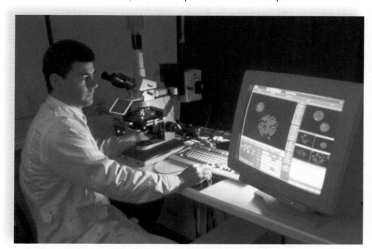

c. Cells are microscopically examined and photographed. A computer is used to arrange chromosomes by pairs.

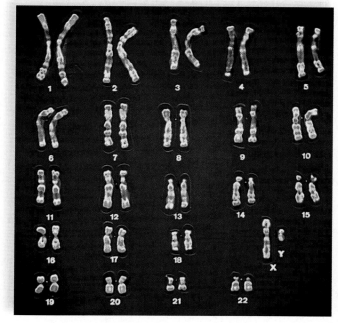

d. Normal male karyotype with 46 chromosomes

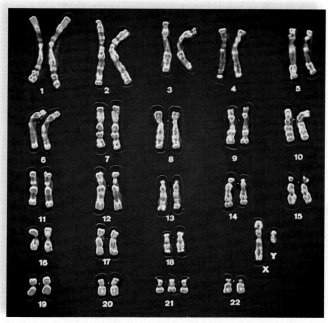

e. Down syndrome karyotype with an extra chromosome 21

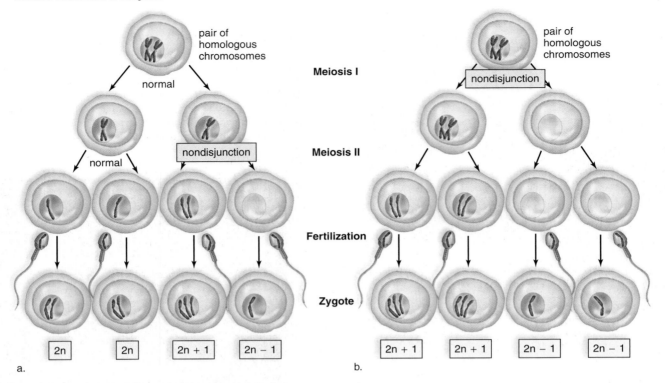

Figure 26.2 Nondisjunction of chromosomes during oogenesis.

a. Nondisjunction can occur during meiosis II if the sister chromatids separate but the resulting chromosomes go into the same daughter cell. Then the egg will have one more or one less than the usual number of chromosomes. Fertilization of these abnormal eggs with normal sperm produces an abnormal zygote with abnormal chromosome numbers. **b.** Nondisjunction can also occur during meiosis I and result in abnormal eggs that also have one more or one less than the normal number of chromosomes. Fertilization of these abnormal eggs with normal sperm results in a zygote with an abnormal chromosome number. In humans (2n), the normal diploid number of chromosomes is 46. An individual who is 2n + 1 would have 47 chromosomes, while an individual who is 2n − 1 would have 45 chromosomes.

Changes in Chromosome Number

Normally, a human receives 22 pairs of autosomes and two sex chromosomes. Sometimes individuals are born with either too many or too few autosomes or sex chromosomes, most likely due to nondisjunction during meiosis. **Nondisjunction** occurs during meiosis I, when both members of a homologous pair go into the same daughter cell, or during meiosis II, when the sister chromatids fail to separate and both daughter chromosomes go into the same gamete. Figure 26.2 assumes that nondisjunction has occurred during oogenesis; some abnormal eggs have 24 chromosomes, while others have only 22 chromosomes. If an egg with 24 chromosomes is fertilized with a normal sperm, the result is a **trisomy,** so called because one type of chromosome is present in three copies. If an egg with 22 chromosomes is fertilized with a normal sperm, the result is a **monosomy,** so called because one type of chromosome is present in a single copy.

Normal development depends on the presence of exactly two of each kind of chromosome. Too many chromosomes are tolerated better than a deficiency of chromosomes, and several trisomies are known to occur in humans. Among autosomal trisomies, only trisomy 21 (Down syndrome) has a reasonable chance of survival after birth. This is probably due to the fact that chromosome 21 is the smallest of the chromosomes.

The chances of survival are greater when trisomy or monosomy involves the sex chromosomes. In normal XX females, one of the X chromosomes becomes a darkly staining mass of chromatin called a **Barr body** (after the person who discovered it). A Barr body is an inactive X chromosome; therefore, we now know that the cells of females function with a single X chromosome just as those of males do. This is most likely the reason that a zygote with one X chromosome (Turner syndrome) can survive. Then, too, all extra X chromosomes beyond a single one become Barr bodies, and this explains why poly-X females and XXY males are seen fairly frequently. An extra Y chromosome, called Jacobs syndrome, is tolerated in humans, most likely because the Y chromosome carries few genes. Jacobs syndrome (XYY) is due to nondisjunction during meiosis II of spermatogenesis. We know this because two Ys are present only during meiosis II in males.

Nondisjunction changes the chromosome number in gametes. Offspring sometimes inherit an extra chromosome (trisomy) or are missing a chromosome (monosomy).

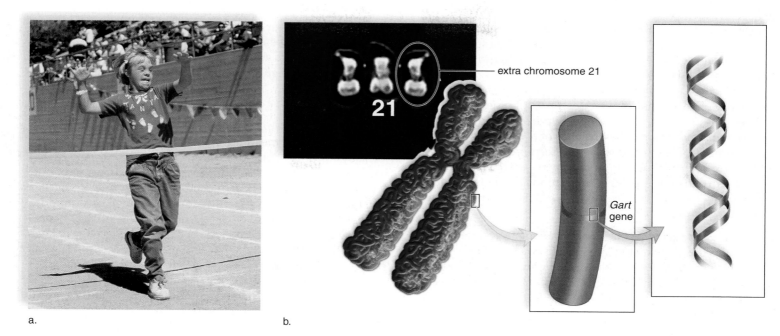

a.   b.

Figure 26.3 Abnormal autosomal chromosome number.

Persons with Down syndrome have an extra chromosome 21. **a.** Common characteristics of the syndrome include a wide, rounded face and a fold on the upper eyelids. Mental retardation, along with an enlarged tongue, makes it difficult for a person with Down syndrome to speak distinctly. **b.** Karyotype of an individual with Down syndrome shows an extra chromosome 21. More sophisticated technologies allow investigators to pinpoint the location of specific genes associated with the syndrome. An extra copy of the *Gart* gene, which leads to a high level of purines in the blood, may account for the mental retardation seen in persons with Down syndrome.

Down Syndrome

The most common autosomal trisomy seen among humans is trisomy 21, also called Down syndrome (Fig. 26.3). This syndrome is easily recognized by the following characteristics: short stature, an eyelid fold, a flat face, stubby fingers, a wide gap between the first and second toes, a large, fissured tongue, a round head, a palm crease (the so-called simian line), and mental retardation, which can sometimes be severe.

Persons with Down syndrome usually have three copies of chromosome 21 because the egg had two copies instead of one. (In 23% of the cases studied, however, the sperm had the extra chromosome 21.) The chances of a woman having a Down syndrome child increase rapidly with age, starting at about age 40, and the reasons for this are still being determined.

Although an older woman is more likely to have a Down syndrome child, most babies with Down syndrome are born to women younger than age 40 because this is the age group having the most babies. Karyotyping can detect a Down syndrome child. However, young women are not routinely encouraged to undergo the procedures necessary to get a sample of fetal cells (i.e., amniocentesis or chorionic villi sampling) because the risk of complications is greater than the risk of having a Down syndrome child. Fortunately, a test based on substances in maternal blood can help identify fetuses who may need to be karyotyped.

The genes that cause Down syndrome are located on the long arm of chromosome 21 (Fig. 26.3b), and extensive investigative work has been directed toward discovering the specific genes responsible for the characteristics of the syndrome. Thus far, investigators have discovered several genes that may account for various conditions seen in persons with Down syndrome. For example, they have located genes most likely responsible for the increased tendency toward leukemia, cataracts, accelerated rate of aging, and mental retardation. The gene for mental retardation, dubbed the *Gart* gene, causes an increased level of purines in the blood, a finding associated with mental retardation. One day, it may be possible to control the expression of the *Gart* gene even before birth so that at least this symptom of Down syndrome does not appear.

Changes in Sex Chromosome Number

An abnormal sex chromosome number is the result of inheriting too many or too few X or Y chromosomes. Figure 26.2 can be used to illustrate nondisjunction of the sex chromosomes during oogenesis, if you assume that the chromosomes shown represent X chromosomes. Nondisjunction during oogenesis or spermatogenesis can result in gametes that have too few or too many X or Y chromosomes. After fertilization, the syndromes listed in Table 26.1 (other than Down syndrome, which is autosomal) are possibilities.

A person with only one sex chromosome, an X (Turner syndrome), is a female, and a person with more than one X chromosome plus a Y (Klinefelter syndrome) is a male. This shows that in humans, the presence of a Y chromosome, not the number of X chromosomes, determines maleness. The *SRY* (sex-determining *r*egion of *Y*) gene, on the short arm of the Y chromosome, produces a hormone called testis-determining factor, which plays a critical role in the development of male genitals.

Turner Syndrome   Females with Turner syndrome have only one sex chromosome, an X, as shown in Figure 26.4*a*. They are usually short and may have malformed features, such as a webbed neck, high palate, and small jaw. Many have congenital heart and kidney defects. Most have ovarian failure and do not undergo puberty or menstruate with-

out taking sex hormone replacement therapy. However, pregnancy has been achieved through in vitro fertilization using donor eggs. Women with Turner syndrome have a normal range of intelligence but often have a nonverbal learning disability. They can lead very successful, fulfilling lives if they receive appropriate care.

Klinefelter Syndrome   About 1 in 650 males are born with two X chromosomes and one Y chromosome (Fig. 26.4*b*). The symptoms of this condition (referred to as "47, XXY") are often so subtle that only 25% are ever diagnosed, and those are usually not diagnosed until after age 15. Earlier diagnosis opens the possibility for educational accommodations and other interventions that can help mitigate common symptoms, which include speech and language delays. Those 47, XXY males who develop more severe symptoms as adults are referred to as having "Klinefelter syndrome." All 47, XXY adults will require assisted reproduction in order to father children. Affected individuals commonly receive testosterone supplementation beginning at puberty.

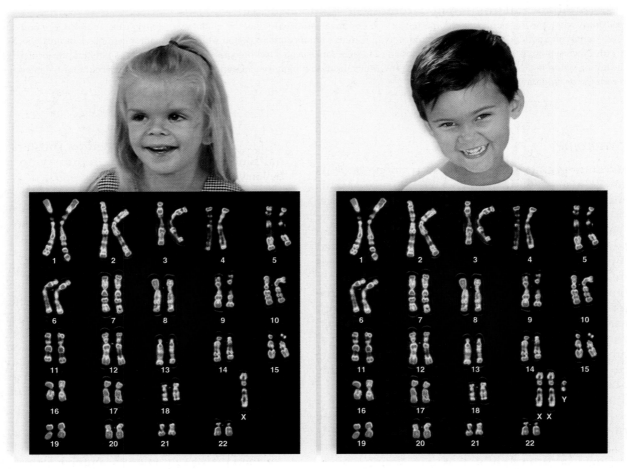

a. Turner syndrome   b. Klinefelter syndrome

Figure 26.4   Abnormal sex chromosome number.
a. People with Turner syndrome have only one sex chromosome, an X, as shown on the karyotype. **b.** People with Klinefelter syndrome have more than one X chromosome plus a Y chromosome, as shown. Both groups can look relatively normal (especially as children) and can lead relatively normal lives.

Health Focus

Living with Klinefelter Syndrome

In 1996, at the age of 25, I was diagnosed with Klinefelter syndrome (KS). Being diagnosed has changed my life for the better.

I was a happy baby, but when I was still very young, my parents began to believe that there was something wrong with me. I knew something was different about me, too, as early on as five years old. I was very shy and had trouble making friends. One minute I'd be well behaved, and the next I'd be picking fights and flying into a rage. Many psychologists, therapists, and doctors tested me because of school and social problems and severe mood changes. Their only diagnosis was "learning disabilities" in such areas as reading comprehension, abstract thinking, word retrieval, and auditory processing. No one could figure out what the real problem was, and I hated the tutoring sessions I had. In the seventh grade, a psychologist told me that I was stupid and lazy, I would probably live at home for the rest of my life, and I would never amount to anything. For the next five years, he was basically right, and I barely graduated from high school.

I believe, though, that I have succeeded because I was told that I would fail. I quit the tutoring sessions when I enrolled at a community college; I decided I could figure things out on my own. I received an associate degree there, then transferred to a small liberal arts college. I never told anyone about my learning disabilities and never sought special help. However, I never had a semester below a 3.0, and I graduated with two B.S. degrees. I was accepted into a graduate program but decided instead to accept a job as a software engineer even though I did not have an educational background in this field. As I later learned, many KS'ers excel in computer skills. I had been using a computer for many years and had learned everything I needed to know on my own, through trial and error.

Around the time I started the computer job, I went to my physician for a physical. He sent me for blood tests because he noticed that my testes were smaller than usual. The results were conclusive: Klinefelter syndrome with sex chromosomes XXY. I initially felt denial, depression, and anger, even though I now had an explanation for many of the problems I had experienced all my life. But then I decided to learn as much as I could about the condition and treatments available. I now give myself a testosterone injection once every two weeks, and it has made me a different person, with improved learning abilities and stronger thought processes in addition to a more outgoing personality.

I found, though, that the best possible path I could take was to help others live with the condition. I attended my first support group meeting four months after I was diagnosed. By spring 1997, I had developed an interest in KS that was more than just a part-time hobby. I wanted to be able to work with this condition and help people forever. I have been very involved in KS conferences and have helped to start support groups in the U.S., Spain, and Australia.

Since my diagnosis, it has been my dream to have a son with KS, although when I was diagnosed, I found out it was unlikely that I could have biological children. Through my work with KS, I had the opportunity to meet my fiancee Chris. She has two wonderful children: a daughter, and a son who has the same condition that I do. There are a lot of similarities between my stepson and me, and I am happy I will be able to help him get the head start in coping with KS that I never had. I also look forward to many more years of helping other people seek diagnosis and live a good life with Klinefelter syndrome.

Stefan Schwarz
stefan13@cox.net
http://klinefeltersyndrome.org

Discussion Questions

1. If you had a child with Stefan's behavioral problems would you have looked for a genetic explanation?
2. Why do testosterone injections improve the characteristic problems of Klinefelter Syndrome?
3. If there were no treatments available, would you still want to know if you had a genetic disease? Why or why not?

The Heath Focus, "Living with Klinefelter Syndrome," suggests that it is best for parents to know right away that they have a child with this disorder because much can be done to help the child lead a normal life.

Poly-X Females   A poly-X female has more than two X chromosomes and thus extra Barr bodies in the nucleus. Females with three X chromosomes have no distinctive phenotype, aside from a tendency to be tall and thin. Although some have delayed motor and language development, most poly-X females are not mentally retarded. Some may have menstrual difficulties, but many menstruate regularly and are fertile. Their children usually have a normal karyotype.

Females with more than three X chromosomes occur rarely. Unlike XXX females, XXXX females are tall and more likely to be retarded. They exhibit various physical abnormalities, but may menstruate normally.

Jacobs Syndrome   XYY males, who have Jacobs syndrome, can only result from nondisjunction during spermatogenesis. Affected males are usually taller than average, suffer from persistent acne, and tend to have speech and reading problems. At one time, it was suggested that these men were likely to be criminally aggressive, but more recent evidence indicates that the incidence of such behavior among them may be no greater than among XY males.

Changes in Chromosomal Structure

A mutation is a permanent genetic change. A change in the structure of a chromosome that can be detected microscopically is a **chromosomal mutation.** A skilled technician can detect differences in chromosomal structure based on a change in the normal banding patterns (see Fig. 26.1*d,e*).

Chromosomal mutations occur when chromosomes break. Various environmental agents—radiation, certain

a.

b.

Figure 26.5 Deletion.

a. When chromosome 7 loses an end piece, the result is Williams syndrome. **b.** These children, although unrelated, have similar appearance, health, and behavioral problems.

organic chemicals, or even viruses—can cause chromosomes to break apart. Ordinarily, when breaks occur, the segments reunite to give the same sequence of genes. But their failure to reunite correctly can result in one of several types of mutations: deletion, duplication, translocation, or inversion. Chromosomal mutations can occur during meiosis, and if the offspring inherits the abnormal chromosome, a syndrome may develop.

Deletions and Duplications

A **deletion** occurs when a single break causes a chromosome to lose an end piece or when two simultaneous breaks lead to the loss of an internal chromosomal segment. An individual who inherits a normal chromosome from one parent and a chromosome with a deletion from the other parent no longer has a pair of alleles for each trait, and a syndrome can result.

Williams syndrome occurs when chromosome 7 loses a tiny end piece (Fig. 26.5). Children who have this syndrome look like pixies because they have turned-up noses, wide mouths, small chins and large ears. Although their academic skills are poor, they exhibit excellent verbal and musical abilities. The gene that governs the production of the protein elastin is missing, which affects the health of the cardiovascular system and causes their skin to age prematurely. Such individuals are very friendly but need an ordered life, perhaps because of the loss of a gene for a protein that is normally active in the brain.

Cri du chat (cat's cry) syndrome is seen when chromosome 5 is missing an end piece. The affected individual has a small head, is mentally retarded, and has facial abnormalities. Abnormal development of the glottis and larynx results in the most characteristic symptom-the infant's cry resembles that of a cat.

In a **duplication,** a chromosomal segment is repeated in the same chromosome or in a nonhomologous chromosome. In any case, the individual has more than two alleles for certain traits. An inverted duplication is known to occur in chromosome 15. **Inversion** means that a segment joins in the direction opposite from normal. Children with this syndrome, called inv dup 15 syndrome, have poor muscle tone, mental retardation, seizures, a curved spine, and autistic characteristics, including poor speech, hand flapping, and lack of eye contact (Fig. 26.6).

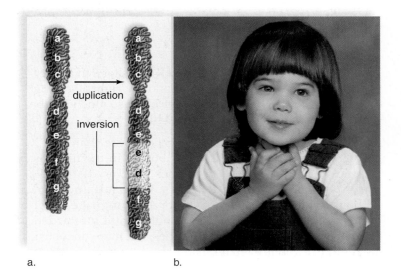

a.

b.

Figure 26.6 Duplication.

a. When a piece of chromosome 15 is duplicated and inverted, (**b**) a syndrome results that is characterized by poor muscle tone and autistic characteristics.

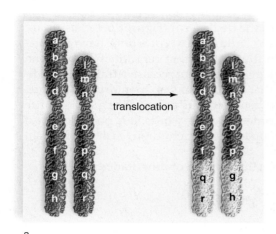

a.

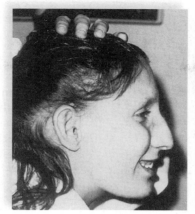

b.

Figure 26.7
Translocation.
a. When chromosomes 2 and 20 exchange segments, (**b**) Alagille syndrome, characterized by distinctive facial features, sometimes results because the translocation disrupts an allele on chromosome 20.

Translocation

A **translocation** is the exchange of chromosomal segments between two nonhomologous chromosomes. A person who has both of the involved chromosomes has the normal amount of genetic material and is healthy, unless the chromosome exchange breaks an allele into two pieces. The person who inherits only one of the translocated chromosomes will no doubt have only one copy of certain alleles and three copies of certain other alleles. A genetic counselor begins to suspect a translocation has occurred when spontaneous abortions are commonplace and family members suffer from various syndromes. A special microscopic technique allows a technician to determine that a translocation has occurred.

In 5% of cases, a translocation that occurred in a previous generation between chromosomes 21 and 14 is the cause of one type of Down syndrome. The affected person inherits two normal chromosomes 21 and an abnormal chromosome 14 that contains a segment of chromosome 21. In these cases, Down syndrome is not related to the age of the mother, but instead tends to run in the family of either the father or the mother.

Figure 26.7 shows a father and daughter who have a translocation between chromosomes 2 and 20. Although they have the normal amount of genetic material, they have the distinctive face, abnormalities of the eyes and internal organs, and severe itching characteristic of Alagille syndrome. People with this syndrome ordinarily have a deletion on chromosome 20; therefore, it can be deduced that the translocation disrupted an allele on chromosome 20 in the father. The symptoms of Alagille syndrome range from mild to severe, so some people may not be aware they have the syndrome. This father did not realize it until he had a child with the syndrome.

Inversion

An inversion occurs when a segment of a chromosome is turned 180 degrees. You might think this is not a problem because the same genes are present, but the reverse sequence of alleles can lead to altered gene activity.

Crossing-over between an inverted chromosome and the noninverted homologue can lead to recombinant chromosomes that have both duplicated and deleted segments. This happens because alignment between the two homologues is only possible when the inverted chromosome forms a loop (Fig. 26.8).

Chromosomal mutations can lead to various syndromes among offspring when the mutation produces chromosomes that have deleted, duplicated, translocated, or inverted segments.

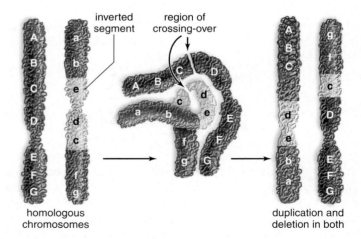

inverted segment
region of crossing-over

homologous chromosomes

duplication and deletion in both

Figure 26.8 Inversion.

Left: A segment of one homologue is inverted. Notice that in the inverted segment *edc* occurs instead of *cde. Middle:* The two homologues can pair only when the inverted sequence forms an internal loop. After crossing-over, a duplication and a deletion can occur. *Right:* The homologue on the left has *AB* and *ba* sequences and neither *gf* nor *FG* genes. The homologue on the right has *gf* and *FG* sequences and neither *AB* nor *ba* genes.

26.2 Counseling for Genetic Disorders

Even if no chromosomal abnormality is likely, amniocentesis still might be done because biochemical tests can now detect over 400 different genetic disorders in the fetus. The genetic counselor determines ahead of time what tests might be warranted. To do this, the counselor needs to know the medical history of the family in order to construct a pedigree.

Family Pedigrees

A pedigree is a chart that shows the inheritance of a particular genetic trait through successive generations of a family. In the pedigree, males are represented by squares and females by circles. A line connecting a male and female represents a mating of those individuals resulting in children, which are shown in order of birth on a horizontal line connected to and below the parents' line. An individual whose symbol is colored displays the genetic trait under consideration; in the context of genetic disorders, a shaded individual has the genetic disease.

Based on the counselor's knowledge of genetic disorders, he or she might already know a trait's pattern of inheritance—that is, whether it is autosomal dominant, autosomal recessive, or X-linked recessive. If not, the pedigree might help the counselor determine the pattern of inheritance. The counselor can then calculate the chances that any child born to the couple will have the abnormal phenotype.

Pedigrees for Autosomal Disorders

Studying a pedigree allows a genetic counselor to decide whether an inherited condition is due to an autosomal dominant or an autosomal recessive allele. Does the following pattern of inheritance represent an autosomal dominant or an autosomal recessive characteristic?

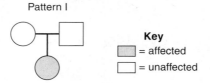

Pattern I

Key
■ = affected
□ = unaffected

In this pattern, the child is affected, but neither parent is; this can happen only if the disorder is recessive and the parents are *Aa.* Notice that the parents are **carriers** because they are unaffected but are capable of having a child with the genetic disorder. If the family pedigree suggests that the parents are carriers for an autosomal recessive disorder, the counselor might suggest confirming this by doing the appropriate genetic test. Then, if the parents so desire, it would be possible to do prenatal testing for the genetic disorder, as described later in this section.

Notice that in the pedigree in Figure 26.9, cousins are the parents of three children, two of whom have the disorder. Aside from illustrating that reproduction between cousins is more likely to bring out recessive traits, this pedigree also shows that "chance has no memory"; therefore, each child born to heterozygous parents has a 25% chance of having the disorder. In other words, it is possible that if a heterozygous couple has four children, each child might have the condition.

Now consider this pattern of inheritance:

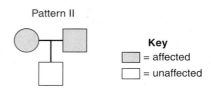

Pattern II

Key
■ = affected
□ = unaffected

In this pattern, the child is unaffected, but the parents are both affected. This can happen if the condition is autosomal dominant and the parents are *Aa.* Figure 26.10 lists other ways to recognize an autosomal dominant pattern of inheritance.

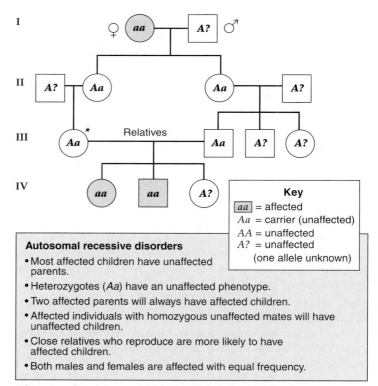

Autosomal recessive disorders
- Most affected children have unaffected parents.
- Heterozygotes (*Aa*) have an unaffected phenotype.
- Two affected parents will always have affected children.
- Affected individuals with homozygous unaffected mates will have unaffected children.
- Close relatives who reproduce are more likely to have affected children.
- Both males and females are affected with equal frequency.

Figure 26.9 Autosomal recessive pedigree.
The list gives ways to recognize an autosomal recessive disorder. How would you know the individual at the asterisk is heterozygous?

*See Appendix A for answers.

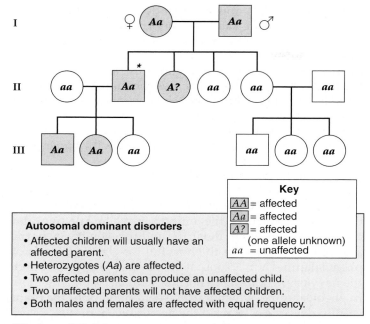

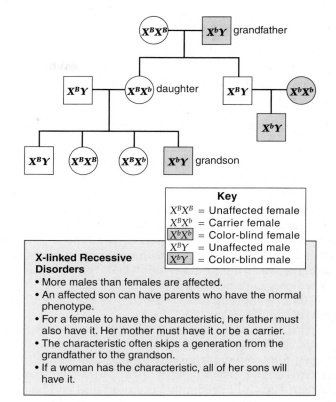

Figure 26.10 Autosomal dominant pedigree.
The list gives ways to recognize an autosomal dominant disorder. How would you know the individual at the asterisk is heterozygous?

* See Appendix A for answers.

This pedigree also illustrates that when both parents are unaffected, all their children are unaffected. Why? Because neither parent has a dominant gene that passes the condition on.

Pedigrees for Sex-Linked Disorders

Sex-linked disorders can be carried on the X or Y chromosome.

X-Linked Disorders   Figure 26.11 gives a pedigree for an X-linked recessive disorder. Recall that sons inherit an X-linked recessive allele from their mothers because their fathers gave them a Y chromosome. More males than females have the disorder because recessive alleles on the X chromosome are always expressed in males (since the Y chromosome lacks an allele for the disorder). Females who have the condition inherited the allele from both their mother and their father, and all the sons of this female will have the condition. If a male has an X-linked recessive condition, his daughters are carriers, even if their mother is normal. Therefore, X-linked recessive conditions often pass from grandfather to grandson. Figure 26.11 lists other ways to recognize a recessive X-linked disorder.

Only a few known traits are X-linked dominant traits. If a disorder is X-linked dominant, affected males pass the trait *only* to daughters, who have a 100% chance of having the condition. Females can pass an X-linked dominant allele to both sons and daughters. If a female is heterozygous and

Figure 26.11 X-linked recessive pedigree.
This pedigree for color blindness exemplifies the inheritance pattern of an X-linked recessive disorder. The list gives various ways of recognizing the X-linked recessive pattern of inheritance.

her partner is normal, each child has a 50% chance of escaping an X-linked dominant disorder, depending on which maternal X chromosome is inherited.

Y-Linked Disorders   A few genetic disorders are carried on the Y chromosome; one that is well known is involved in determining gender during development. Others code for membrane proteins, including an enzyme that regulates the movement of ADP into and ATP out of mitochondria!

How would a counselor recognize a Y-linked pattern of inheritance? Y-linked disorders are present only in males and are passed directly from father to *all* sons because males must inherit a Y chromosome from their fathers.

Unusual Inheritance Patterns

No doubt a genetic counselor will occasionally come across some rather unusual inheritance patterns. For example, a mother (and not a father) passes mutated mitochondrial genes to her offspring through the cytoplasm of the egg. (The sperm contains little or no cytoplasm.) Genetic disorders due to mutated mitochondrial DNA usually manifest themselves in defective energy metabolism.

Genetic Disorders of Interest

The field of medical genetics has traditionally focused on disorders caused by single-gene mutations, and we will discuss a few of the better-known disorders.

Autosomal Recessive Disorders

Inheritance of two recessive alleles is required before an autosomal recessive disorder will appear.

Tay-Sachs disease is a well-known autosomal recessive disorder that usually occurs among Jewish people in the United States, most of whom are of central and eastern European descent. Tay-Sachs disease results from a lack of the enzyme hexosaminidase A (Hex A) and the subsequent storage of its substrate, a glycosphingolipid, in lysosomes. Lysosomes build up in many body cells, but the primary sites of storage are the cells of the brain, which accounts for the onset of symptoms and the progressive deterioration of psychomotor functions (Fig. 26.12a).

At first, it is not apparent that a baby has Tay-Sachs disease. However, development begins to slow down between four and eight months of age, and neurological impairment and psychomotor difficulties then become apparent. The child gradually becomes blind and helpless, develops uncontrollable seizures, and eventually becomes paralyzed.

Cystic fibrosis (CF) is an autosomal recessive disorder that occurs among all ethnic groups, but it is the most common lethal genetic disorder among Caucasians in the United States. Research has demonstrated that chloride ions (Cl^-) fail to pass through a plasma membrane channel protein in the cells of these patients (Fig. 26.12b). Ordinarily, after chloride ions have passed through the membrane, sodium ions (Na^+) and water follow. It is believed that lack of water is the cause of abnormally thick mucus in bronchial tubes and pancreatic ducts. In these children, this mucus is so thick and viscous that it interferes with the function of the lungs and pancreas. To ease breathing, the thick mucus in the lungs must be manually loosened periodically, but the lungs still become infected frequently. Clogged pancreatic ducts prevent digestive enzymes from reaching the small intestine, and to improve digestion, CF patients take digestive enzymes mixed with applesauce before every meal.

Phenylketonuria (PKU) is an autosomal recessive metabolic disorder that affects nervous system development. Affected individuals lack the enzyme needed for the normal metabolism of the amino acid phenylalanine, and therefore, it appears in the urine and the blood. Newborns are routinely tested in the hospital for elevated levels of phenylalanine in the blood. If an elevated level is detected, the newborn will develop normally if placed on a diet low in phenylalanine, which must be continued until the brain is fully developed, around the age of seven, or else severe mental retardation will develop. Some doctors recommend that the diet continue for life, but in any case, a pregnant woman with phenylketonuria must be on the diet in order to protect her unborn child from harm.

Sickle cell disease is an autosomal recessive disorder in which the red blood cells are not biconcave disks like normal red blood cells; rather, they are irregular, and often sickle shaped (Fig. 26.13a). This defect is caused by an abnormal type of hemoglobin that differs from normal hemoglobin by one amino acid in the protein globin. The single amino acid change causes hemoglobin molecules to stack up and form insoluble rods, and the red blood cells become sickle shaped.

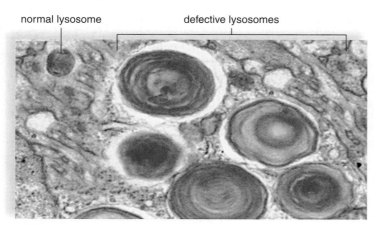

a. Malfunctioning lysosomes in Tay-Sachs disease.

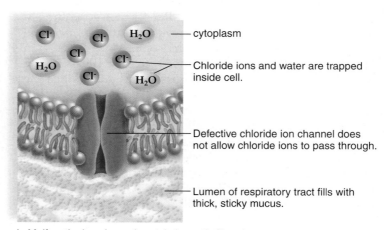

b. Malfunctioning channel protein in cystic fibrosis.

Figure 26.12 Genetic disorders.
a. A neuron in Tay-Sachs disease contains many defective lysosomes because of a deficient lysosomal enzyme. **b.** Cystic fibrosis is due to a faulty protein that is supposed to regulate the flow of chloride ions into and out of cells through a channel protein.

Because sickle-shaped cells can't pass along narrow capillary passageways as disk-shaped cells can, they clog the vessels and break down. This is why persons with sickle cell disease experience poor circulation, anemia, and low resistance to infection, as described in the introduction to this chapter. Internal hemorrhaging leads to further complications, such as jaundice, episodic pain in the abdomen and joints, and damage to internal organs.

Sickle cell heterozygotes have sickle cell traits in which the blood cells are normal unless they experience dehydration or mild oxygen deprivation. Still, at present, most experts believe that persons with the sickle cell trait do not need to restrict their physical activity.

Autosomal Dominant Disorders

Inheritance of only one dominant allele is necessary for an autosomal dominant genetic disorder to appear. We discuss only two of the many autosomal dominant disorders here.

Marfan syndrome, an autosomal dominant disorder, is caused by a defect in an elastic connective tissue protein called fibrillin. This protein is normally abundant in the lens of the eye, the bones of the limbs, fingers, and ribs, and also in the wall of the aorta. The affected person often has a dislocated lens, long limbs and fingers, and a caved-in chest. The aorta wall is weak and may burst without warning. A tissue graft can strengthen the aorta, but people with Marfan syndrome still should not overexert themselves.

Huntington disease is a neurological disorder that leads to progressive degeneration of brain cells (Fig. 26.13*b*). The disease is caused by a mutated copy of the gene for a protein called huntingtin as described in the Health Focus on page 541. Most patients appear normal until they are of middle age and have already had children, who may later also be stricken. Occasionally, the first sign of the disease appears during the teen years or even earlier. There is no effective treatment, and death comes 10 to 15 years after the onset of symptoms.

Several years ago, researchers found that the gene for Huntington disease was located on chromosome 4. They developed a test to detect the presence of the gene; however, few people want to know they have inherited the gene because there is no cure. At least now we know that the disease stems from a mutation that causes the huntingtin protein to have too many copies of the amino acid glutamine. The normal version of huntingtin has stretches of between 10 and 25 glutamines. If huntingtin has more than 36 glutamines, it changes shape and forms large clumps inside neurons. Even worse, it attracts and causes other proteins to clump with it. One of these proteins, called CBP, helps nerve cells survive. Researchers hope to combat the disease by boosting CBP levels.

Incompletely Dominant Disorders

The prognosis in **familial hypercholesterolemia (FH)** parallels the number of LDL-cholesterol receptor proteins in the plasma membrane. A person with two mutated alleles lacks LDL-cholesterol receptors; a person with only one mutated allele has half the normal number of receptors; and a person with two normal alleles has the usual number of receptors.

The presence of excessive cholesterol in the blood causes cardiovascular disease. Therefore, people with no receptors die of cardiovascular disease before 30 years of age. These individuals develop cholesterol deposits in the

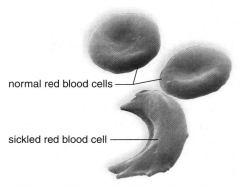

normal red blood cells

sickled red blood cell

Abnormally shaped red blood cells in sickle cell disease.
a.

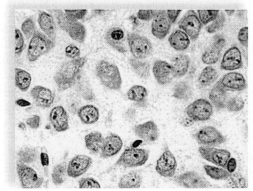

Many neurons in normal brain.
b.

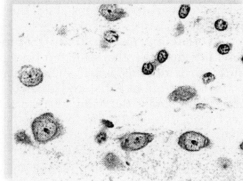

Loss of neurons in Huntington brain.

Figure 26.13 Genetic disorders.

a. Persons with sickle cell disease have sickle-shaped red blood cells because of an abnormal hemoglobin A molecule. **b.** Huntington disease is characterized by increasingly serious psychomotor and mental disturbances because of the loss of nerve cells.

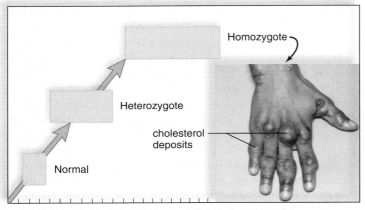

a. Cholesterol levels in familial hypercholesterolemia (FH)

b. Abnormal muscle in muscular dystrophy

Figure 26.14 Genetic disorders.

a. Familial hypercholesterolemia is incompletely dominant. Persons with one mutated allele (heterozygotes) have an abnormally high level of cholesterol in the blood, and those with two mutated alleles (homozygotes) have a higher level still. In homozygotes, the cholesterol level is sometimes so high that the excess forms visible deposits in the skin. **b.** In Duchenne muscular dystrophy, an X-linked recessive disorder, the calves of the legs enlarge because fibrous tissue develops as muscles waste away due to lack of the protein dystrophin.

skin, tendons, and cornea. Individuals with one-half the number of receptors may die while still young or after they have reached middle age. People with the full number of receptors do not have familial hypercholesterolemia (Fig. 26.14*a*).

X-Linked Recessive Disorders

Most X-linked disorders are recessive; therefore, more males than females are affected. **Color blindness,** an X-linked recessive disorder, does not prevent males from leading a normal life. About 8% of Caucasian men have red-green color blindness. Most of them see brighter greens as tans, olive greens as browns, and reds as reddish browns. A few cannot tell reds from greens at all; they see only yellows, blues, blacks, whites, and grays.

Duchenne muscular dystrophy is an X-linked recessive disorder characterized by wasting away of the muscles. Symptoms, which include a waddling gait, toe walking, frequent falls, and difficulty in rising, may appear as soon as the child starts to walk. Muscle weakness intensifies until the individual is confined to a wheelchair. Death usually occurs by age 20; therefore, affected males are rarely fathers. The recessive allele remains in the population by passing from carrier mother to carrier daughter.

The absence of a protein, now called dystrophin, is the cause of Duchenne muscular dystrophy. Much investigative work determined that the lack of dystrophin causes calcium to leak into the cell, which promotes the action of an enzyme that dissolves muscle fibers. When the body attempts to repair the tissue, fibrous tissue forms (Fig. 26.14*b*), and this cuts off the blood supply so that more and more cells die. Immature muscle cells can be injected into

muscles, but it takes 100,000 cells for dystrophin production to increase by 30–40%.

Another X-linked recessive disorder is **hemophilia,** of which there are two common types. Hemophilia A is due to the absence or minimal presence of a clotting factor known as factor VIII, and hemophilia B is due to the absence of clotting factor IX. Hemophilia is called the bleeder's disease because the affected person's blood either does not clot or clots very slowly. Although hemophiliacs bleed externally after an injury, they also bleed internally, particularly around joints. Hemorrhages can be stopped with transfusions of fresh blood (or plasma) or concentrates of the missing clotting protein. Factors VIII and IX are also now available as biotechnology products.

At the turn of the century, hemophilia was prevalent among the royal families of Europe, and all of the affected males could trace their ancestry to Queen Victoria of England. Of Queen Victoria's 26 grandchildren, four grandsons had hemophilia, and four granddaughters were carriers. Because none of Queen Victoria's relatives were affected, it seems that the faulty allele she carried arose by mutation, either in Victoria or in one of her parents. Her carrier daughters, Alice and Beatrice, introduced the allele into the ruling houses of Russia and Spain, respectively. Alexis, the last heir to the Russian throne before the Russian Revolution, was a hemophiliac. There are no hemophiliacs in the present British royal family because Victoria's eldest son, King Edward VII, did not receive the allele.

Inherited disorders can be autosomal recessive, autosomal dominant, incompletely dominant, or X-linked recessive.

Testing for Genetic Disorders

Prospective parents know if either of them has an autosomal dominant disorder because the person will show it. However, genetic testing is required to detect if either is a carrier for an autosomal recessive disorder. In any case, a genetic counselor can explain to a couple the chances of a child having the disorder. If a woman is already pregnant, the parents may want to know if the unborn child has the disorder. If the woman is not pregnant, the parents may opt for testing of an embryo or egg before she does become pregnant.

The method of testing depends on the particular genetic disorder. In some instances, it is appropriate to test for a specific protein, and in others to test for the mutated gene.

Testing for a Protein

Some genetic mutations lead to disorders caused by a missing enzyme. As mentioned previously, babies with Tay-Sachs disease lack an enzyme called hexosaminidase A (hex A). It is possible to test for the quantity of hex A in a sample of cells and determine from that whether the individual is homozygous normal, a carrier, or a Tay-Sachs patient. If the parents are carriers, each child has a 25% chance of having Tay-Sachs disease. This knowledge may lead prospective parents to have an egg or embryo tested, as discussed later in this section.

In the case of PKU, an enzyme is missing, but the test looks for an excess amount of the enzyme's substrate, namely phenylalanine. A lab technician places a medium containing the newborn's blood on a bacterial culture, and if the bacteria grow, the newborn has PKU.

Testing the DNA

Two types of DNA testing are possible: testing for a genetic marker and using a DNA probe.

Genetic Markers   Testing for a genetic marker is similar to the traditional procedure for DNA fingerprinting (see Chapter 24). As an example, consider that individuals with Huntington disease have an abnormality in the sequence of their bases at a particular location on a chromosome. This abnormality in sequence is a **genetic marker.** As you know, restriction enzymes cleave DNA at particular base sequences. Therefore, fragments that result from the use of a restriction enzyme will be differently sized according to whether the individual has or does not have Huntington disease (Fig. 26.15).

DNA Probes   A **DNA probe** is a single-stranded piece of DNA that will bind to complementary DNA. For the purpose of genetic testing, the DNA probe bears a genetic mutation of interest. A DNA chip is a new technology that consists of a very small glass square containing several rows of DNA probes. The chip allows testing for many genetic disorders at one time. First, DNA from an individual is cut into fragments using restriction enzymes. The fragments are then tagged with a fluorescent dye and converted to single DNA strands before being applied to the chip (Fig. 26.16). Fragments that contain a mutated gene bind to one of the probes, and binding is detected by a laser scanner. Therefore, the results tell if an individual has particular mutated genes. The use of a DNA chip to discover all the particular mutations of an individual is called **genetic profiling.** Genetic profiling might become a regular part of a medical checkup in the future.

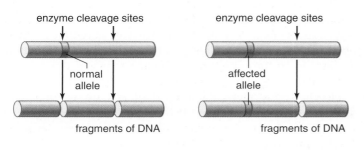

a. Normal fragmentation pattern

b. Genetic disorder fragmentation pattern

Figure 26.15 Use of a genetic marker to test for a genetic mutation.
a. In this example, DNA from a normal individual has certain restriction enzyme cleavage sites. **b.** DNA from another individual lacks one of the cleavage sites, and this loss indicates that the person has a mutated gene. In heterozygotes, half of their DNA would have the cleavage site, and half would not have it. (In other instances, the gain in a cleavage site could be an indication of a mutation.)

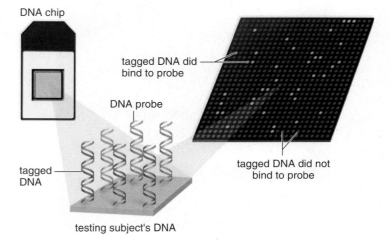

testing subject's DNA

Figure 26.16 Use of a DNA chip to test for mutated genes.
This DNA chip contains rows of DNA sequences for mutations that indicate the presence of particular genetic disorders. If DNA fragments derived from an individual's DNA bind to a sequence, the individual has the mutation.

Testing the Fetus, Embryo, or Egg

If a woman is already pregnant, some serious fetal abnormalities can be detected by ultrasound. DNA of fetal cells, the embryo, and the egg can also be obtained and tested for genetic defects.

Ultrasound   An ultrasound probe scans the mother's abdomen, while a transducer transmits high-frequency sound waves that are transformed into a picture on a video screen. This picture shows the fetus inside the uterus (Fig. 26.17). Ultrasound can tell if a baby has a serious condition, such as spina bifida, the most common neural tube defect. Spina bifida results when the spine fails to close properly during the first month of pregnancy. In severe cases, the spinal cord protrudes through the back and may be covered by skin or a thin membrane.

Testing Fetal Cells   For testing purposes, fetal cells may be acquired through these means:

Amniocentesis. As mentioned in section 26.1, amniocentesis is not usually performed until a woman is at least 15 weeks pregnant. The chance of losing the fetus by this means is slightly less than for chorionic villi sampling (CVS). However, the wait for genetic testing of cells (about 20 days) is longer than for CVS. Because amniotic fluid is collected, it can be tested for alpha fetoprotein (AFP), a chemical that is present when a fetus has a neural tube defect.

Chorionic villi sampling (CVS). As mentioned previously, this procedure is done earlier than amniocentesis. There is a slightly higher risk of miscarriage, and

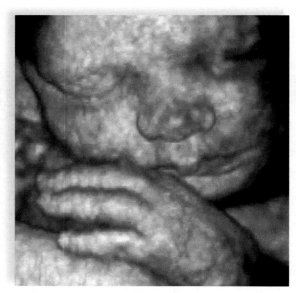

Figure 26.17   Ultrasound.
An ultrasound exam produces images of the fetus that can allow physicians to detect serious defects, such as a neural tube defect. Today, ultrasound can produce images that look three-dimensional.

facial and limb abnormalities induced by the procedure have been reported but not confirmed. Because no amniotic fluid is collected, the AFP test cannot be done. The wait for genetic test results (about 10 days) is less than for amniocentesis.

Fetal cells in mother's blood. As early as nine weeks into the pregnancy, a small number of fetal cells can be isolated from the mother's blood using a cell sorter. It is best to gather nucleated fetal red blood cells, because they are short-lived and, therefore, must be related to the present pregnancy.

Only about 1 in 70,000 of the cells in a mother's blood is a fetal cell, and therefore, PCR (see Chapter 24) must be used to amplify the DNA from the few cells collected. Only the DNA tests using genetic markers or DNA probes can be done, but the procedure poses no risk whatsoever to the fetus.

Testing the Embryo   If prospective parents are carriers for a genetic disorder, they may want assurance that their offspring will be free of the disorder. Testing the embryo will provide this assurance.

Following in vitro fertilization (IVF), it is possible to test the embryo. Once cell division has produced a six- to eight-celled embryo, removal of one of these cells for testing purposes has no effect on normal development. Only embryos that test negative for the genetic disorders of interest are placed in the uterus to continue developing.

So far, about 1,000 children free of alleles for genetic disorders that run in their families have been born worldwide following embryo testing. In the future, it's possible that embryos that test positive for a disorder could be treated by gene therapy so that they, too, would be allowed to continue to term.

Testing the Egg   Recall that meiosis in females results in a single egg and at least two polar bodies (see Fig. 5.16). Polar bodies, which later disintegrate, receive very little cytoplasm, but they do receive a haploid number of chromosomes. When a woman is heterozygous for a recessive genetic disorder, about half the polar bodies will have received the mutated allele, and in these instances, the egg received the normal allele. Therefore, if a polar body tests positive for a mutated allele, the egg received the normal allele. Only normal eggs are then used for IVF. Even if the sperm should happen to carry the mutation, the zygote will, at worst, be heterozygous. But the phenotype will appear normal.

Today, genetic counseling is most appropriate for genetic defects caused by single-gene defects, whose pattern of inheritance is shown by a family pedigree. Genetic testing may involve protein and DNA testing or testing of the fetus, embryo, or egg.

Trinucleotide Repeat Expansion Disorders (TRED)

Several of the genetic diseases we have studied result from a single change in the DNA that results in a single change in an amino acid in a particular protein. For example, the most common form of sickle cell disease results from a one-nucleotide change in the DNA that causes a one-amino-acid difference in the protein hemoglobin. However, in at least 15 human diseases, including Huntington disease and fragile X syndrome, the problem lies in a DNA sequence that is repeated too many times, called a *trinucleotide repeat.*

As described in the text, Huntington disease is a neurological disorder in which the huntingtin protein is altered. The trinucleotide sequence CAG normally appears several times at the beginning of the gene that encodes the huntingtin protein. The presence of multiple CAGs results in multiple glutamines in the protein. People with 10–35 copies of the CAG display no signs of Huntington disease; however, people with more than 40 copies have the disease.

Fragile X syndrome is one of the most common genetic causes of mental retardation. As children, fragile X individuals may be hyperactive or autistic; their speech is delayed in development and often repetitive in nature. As adults, males have large testes and big, usually protruding ears. They are short in stature, but the jaw is prominent, and the face is long and narrow (Fig. 26A*a*). The term *fragile X syndrome* originated because diagnosis used to be dependent upon observation of an X chromosome whose tip is attached to the rest of the chromosome by only a thin thread (Fig. 26A*b*). This disease results from repeats of the trinucleotide CGG. People with 6–50 copies appear normal, while those with over 230 copies have the disease. Those with between 50 and 230 copies are carriers of fragile X, with either no symptoms or very mild symptoms.

People who have the intermediate number of repeats are said to have a premutation. Premutations can lead to full-blown mutations in future generations by expansion of the number of repeats. This is sometimes called genetic anticipation. The number of repeats increases with each generation, as does the severity of the disease (Fig. 26A*c*). It is not known why the repeats expand in number. Using the current molecular technology, it is possible to accurately determine the number of trinucleotide repeats present in an individual.

Discussion Questions

1. The repeat expansion appear to occur during meiosis but not mitosis. Why might this be true?
2. If a male has fragile X syndrome, is his mother or father more likely to be a carrier for the disease? Why?
3. Why is fragile X but not Huntington disease considered a syndrome?

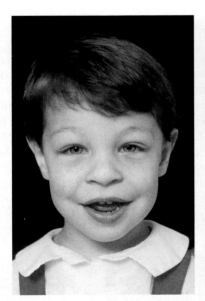

a. Young male with fragile X syndrome

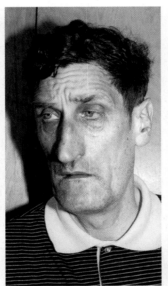

Same individual when mature

b. Fragile X chromosome

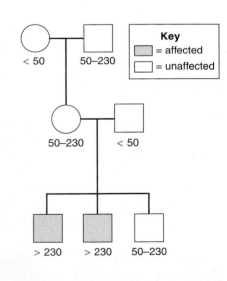

c. Inheritance pattern for fragile X syndrome

Figure 26A **TRED in fragile X syndrome.**

a. A young male with fragile X syndrome appears normal but with age develops an elongated face with a prominent jaw and ears that noticeably protrude. **b.** An arrow points out the fragile site of this fragile X chromosome. **c.** The number of base-pair repeats at the fragile site are given for each person, showing that the incidence and severity increase with each succeeding generation. A grandfather who has a premutation with 50–230 repeats has no symptoms but transmits the condition to his grandsons through his daughter. Two grandsons have full-blown mutations with more than 230 base repeats.

26.3 | Genomics

In the previous century, researchers discovered the structure of DNA, how DNA replicates, and how protein synthesis occurs. Genetics in the twenty-first century concerns **genomics,** the study of genomes—our genes and the genes of other organisms. The enormity of the task can be appreciated by knowing that, at the very least, humans have 20–25,000 genes that code for proteins, and many organisms have even more encoding genes than that.

Sequencing the Bases

We now have a working draft of our **genome,** the sequence of all the base pairs in all the DNA in all our chromosomes. This feat was accomplished by the **Human Genome Project,** a 13-year effort that involved university and private laboratories around the world. How did they do it? First, investigators developed a laboratory procedure that would allow them to decipher a short sequence of base pairs, and then they devised an instrument that would carry out this procedure automatically. Over the span of the project DNA sequencers were constantly improved, so that today's instruments can automatically analyze up to 2 million base pairs of DNA in a 24-hour period.

Genome Comparisons

The genomes of many other organisms, including a common bacterium, a yeast, and a mouse, are also in the final-draft stage (Table 26.2). Researchers have identified many similarities between the sequence of human bases and those of other organisms whose DNA sequences are also known. From this, we can conclude that we share a large number of genes with much simpler organisms, such as bacteria. However, with a genome size of 3 billion base pairs, there are also many differences.

In one study, researchers compared the human genome with that of chromosome 22 in chimpanzees. Among the many genes that differed in sequence were three types of particular interest: a gene for proper speech development, several for hearing, and several for smell (Fig. 26.18). The gene necessary for proper speech development is thought to have played an important role in human evolution. You can

TABLE 26.2	Sequenced Genomes
Year Sequenced	**Organisms**
2004	*Rattus norvegicus*, the brown Norway rat
2003	*Homo sapiens*, the human being
2002	*Mus musculus*, the mouse; *Fugu rubripes*, the Japanese puffer fish
2000	*Drosophila melanogaster*, the fruit fly
1998	*Caenorhabditis elegans*, a form of roundworm
1997	*Escherichia coli*, a bacterium
1996	*Methanococcus jannaschii*, an archaea
1995	*Saccaromyces cerevisiae*, a form of yeast

suppose that changes in hearing may have facilitated using language for communication between people. Changes in smell genes are a little more problematic. The investigators speculated that the olfaction genes may have affected dietary changes or sexual selection. Or, they may have been involved in traits other than just smell.

The researchers were surprised to find that many of the other genes they located and studied are known to cause human diseases, if abnormal. They wondered if comparing genomes would be a way of finding genes that are associated with human diseases. Investigators are taking all sorts of avenues to link human base sequence differences to illnesses.

The HapMap Project

The HapMap project is a new undertaking whose goal is to catalog common sequence differences that occur in human beings. People have been found to inherit patterns of sequence differences, now called haplotypes. For example, if one haplotype of a person has an A rather than a G at a particular location in a chromosome, there are probably other particular base differences near the A. To discover the most common haplotypes, genetic data from African, Asian, and European populations will be analyzed (Fig. 26.19). The goal of the project is to link haplotypes to the risk for specific illnesses, hoping this knowledge will lead to new methods of preventing, diagnosing, and treating disease.

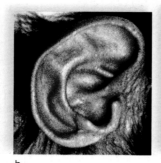
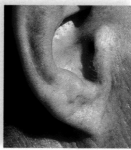
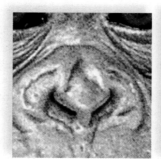

a.   b.   c.

Figure 26.18 **Studying genomic differences between chimpanzees and humans.**
Did changes in the genes for (**a**) speech, (**b**) hearing, and (**c**) smell influence the evolution of humans?

Figure 26.19 Studying genomic differences among people.
The HapMap project will compare sequences among African, Asian, and European populations.

Proteomics and Bioinformatics

Proteomics is the study of the structure, function, and inter-action of cellular proteins. Recall that the known sequence of bases in the human genome predicts that at least 20–25,000 genes are translated into proteins. The translation of all these genes results in a collection of proteins called the human proteome.

Computer modeling of the three-dimensional shape of these proteins is an important part of proteomics. If the primary structure of these proteins is now known, it should be possible to predict their final three-dimensional shape. The study of protein function is also essential to the discovery of better drugs. One day, it may be possible to correlate drug treatment to the particular genome of the individual to increase efficiency and decrease side effects.

Bioinformatics is the application of computer technologies to the study of the genome. Genomics and proteomics produce raw data, and these fields depend upon computer analysis to find significant patterns in the data. As a result of bioinformatics, scientists hope to find cause-and-effect relationships between various genetic profiles and genetic disorders caused by multifactor genes. Such a program is discussed in the Bioethical Focus.

Also, the current genome sequence contains many gene "deserts," regions of genes that have no known function. Bioinformatics might identify the functions of these regions by correlating any sequence changes with resulting phenotypes. New computational tools will most likely be needed to accomplish these goals.

Expanding Present-Day Genomics

Investigators now believe that the human genome is three-dimensional. The sections of DNA that encode proteins are but one dimension—what are the other two? The second layer is composed of DNA sequences that interrupt and separate the genes. These sections have been preserved, which suggests that they may have important functions. Could these sequences contribute to what makes one person different from another? Doesn't it seem strange that fruit flies have fewer coding genes than roundworms, and that rice plants have more than humans? Our protein-coding sequences account for less than 2% of the DNA in the human chromosomes. The rest has been dismissed as evolutionary junk. But is it? Some investigators have found so far that much of the "junk" codes for RNA, and they are trying to discover what this RNA does.

The third layer of our genome lies outside DNA! It's called the epigenic layer because it consists of the proteins, such as histones and other chemicals, that surround and adhere to DNA. The epigenic layer seems to play an important role in growth, aging, and cancer, and might explain the unusual patterns of inheritance of certain diseases. It's possible the epigenic layer will be easier for us to modify than the other layers of the genome.

Sequencing the base pairs of the human genome has led to several in-depth studies of the genome and the possibility of finding new medicines.

Gene Therapy

Gene therapy is the insertion of genetic material into human cells for the treatment of genetic disorders and various other human illnesses, such as cardiovascular disease and cancer. Gene therapy includes both ex vivo (outside the body) and in vivo (inside the body) therapy.

Ex Vivo Gene Therapy

Figure 26.20 describes an ex vivo methodology for treating children who have SCID (severe combined immunodeficiency). These children lack the enzyme ADA (adenosine deaminase), which is involved in the maturation of T and B cells. In order to carry out gene therapy, bone marrow stem cells are removed from the blood and infected with an RNA retrovirus that carries a normal gene for the enzyme. Then the cells are returned to the patient. Bone marrow stem cells are preferred for this procedure because they divide to produce more cells with the same genes. Patients who have undergone this procedure show significantly improved immune function associated with a sustained rise in the level of ADA enzyme activity in the blood.

As discussed in section 26.2, in one form of familial hypercholesterolemia, the liver cells lack receptor proteins for removing cholesterol from the blood. The resulting high levels of blood cholesterol make the patient subject to fatal heart attacks at a young age. Using ex vivo gene therapy, a small portion of the liver is surgically excised and then infected with a retrovirus containing a normal gene for the receptor before being returned to the patient. Several patients have experienced lowered serum cholesterol levels following this procedure.

In Vivo Gene Therapy

Cystic fibrosis patients lack a gene that codes for the transmembrane carrier of the chloride ion (see section 26.2). They often die due to numerous infections of the respiratory tract. In gene therapy trials, the gene needed to cure cystic fibrosis is sprayed into the nose or delivered to the lower respiratory tract by an adenovirus vector or by using liposomes, microscopic vesicles made of lipoproteins. Investigators are trying to improve uptake and are also hypothesizing that a combination of different vectors might be more successful.

Genes are being used to treat medical conditions, such as poor coronary circulation. It has been known for some time that VEGF (vascular endothelial growth factor) can cause the growth of new blood vessels. The gene that codes for this growth factor can be injected into the heart, alone or within a virus, to stimulate branching of coronary blood vessels. Patients report that they have less chest pain and can run longer on a treadmill.

Gene therapy is increasingly relied upon as a part of cancer treatment. Genes are being used to make healthy cells more tolerant of chemotherapy, while making tumor cells more sensitive. Knowing that the tumor suppressor gene *p53* brings about apoptosis (cell death), researchers are interested in finding a way to selectively introduce *p53* into cancer cells, and in that way, kill them.

> The Human Genome Project has spawned new fields, including genomics, proteomics, and bioinformatics, and it may lead to more types of gene therapy.

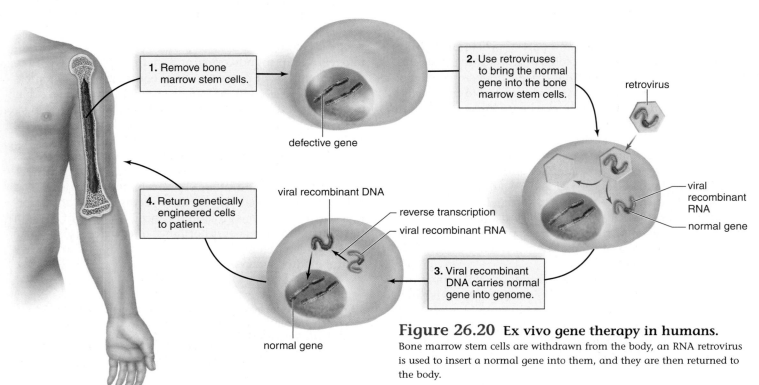

1. Remove bone marrow stem cells.

defective gene

2. Use retroviruses to bring the normal gene into the bone marrow stem cells.

retrovirus

viral recombinant RNA

normal gene

4. Return genetically engineered cells to patient.

viral recombinant DNA

reverse transcription

viral recombinant RNA

3. Viral recombinant DNA carries normal gene into genome.

normal gene

Figure 26.20 Ex vivo gene therapy in humans.
Bone marrow stem cells are withdrawn from the body, an RNA retrovirus is used to insert a normal gene into them, and they are then returned to the body.

Bioethical Focus

Bioinformatics in Iceland

The population of Iceland, approximately 275,000 people, is small and homogeneous. Very little immigration has occurred since the Vikings and Celts first settled there almost 1,200 years ago. The country also has extensive health data and family trees dating back to 1915. In December 1998, Iceland's parliament granted an exclusive license to the health, genealogy, and genetic information of its citizens to the biotechnology company deCODE Genetics for $200 million. The company created the DNA database for the country and correlated their findings with the health and genealogy data. The purpose of such a national health database was to uncover the genetic causes of various diseases and to thrust Iceland into the biotechnology forefront. So far, several disease genes, including one for osteoarthritis, have been identified. One of the benefits for the people of Iceland is that any medications developed

from information obtained from the database will be made available without cost. Other countries, such as Estonia and Tonga, have followed suit in assembling national genetic databases.

From the beginning, opposition to the Iceland plan has been fierce. The Iceland Medical Association opposed the measure for fear that employers and insurance companies would have access to the data and use it to discriminate against individuals. Although the data are encrypted, opponents believe there is enough information to make it personally identifiable. The database is also able to gather information without the expressed consent of the patient, because the Icelandic government set up the program as a required program, unless a person specifically "opts out." This is a special problem for dead Icelanders, who cannot "opt out." In 2003, the Icelandic Supreme Court declared

the Health Sector Database Act unconstitutional as a result of a case brought by a woman on behalf of her dead father whose data had been automatically included in the database. Finally, deCODE Genetics has a monopoly on the data. They can sell it for a fee to other interested parties, including pharmaceutical and insurance companies.

Form Your Own Opinion

1. Who has the right to collect, access, and benefit from your personal genetic information?
2. Should people be compelled to provide genetic information if it benefits the population as a whole?
3. What measures should be in place to protect individuals from genetic discrimination?

Summarizing the Concepts

26.1 Counseling for Chromosomal Disorders

It is possible to karyotype the chromosomes of a cell arrested in metaphase of mitosis. Amniocentesis and chorionic villi sampling can provide fetal cells for the karyotyping of chromosomes when a chromosomal abnormality is suspected.

Nondisjunction during meiosis can result in an abnormal number of autosomes or sex chromosomes in the gametes. Down syndrome results when an individual inherits three copies of chromosome 21. Females who are XO have Turner syndrome, and those who are XXX are poly-X females. Males with Klinefelter syndrome are XXY. Males who are XYY have Jacobs syndrome.

Changes in chromosomal structure also affect the phenotype. Chromosomal mutations include deletions, duplications, translocations, and inversions. Translocations do not necessarily cause any difficulties, if the person has inherited both translocated chromosomes. However, the translocation can disrupt a particular allele, and then a syndrome will follow. An inversion can lead to chromosomes that have a deletion and a duplication when the inverted piece loops back to align with the noninverted homologue, and crossing-over follows between the nonsister chromatids.

26.2 Counseling for Genetic Disorders

A family pedigree is a visual representation of the history of a genetic disorder in a family. Constructing a pedigree helps a genetic counselor decide whether a genetic disorder is autosomal recessive (see Fig. 26.9) or dominant (see Fig. 26.10), x-linked (see Fig. 26.11), or follows some other pattern of inheritance. After deciding the pattern of inheritance, a counselor can tell prospective parents the chances that any child born to them will have the genetic disorder.

Today, we have been most successful in identifying genetic disorders caused by single-gene mutations. Tay-Sachs disease (a

lysosomal storage disease), cystic fibrosis (faulty regulator of chloride channel), phenylketonuria (inability to metabolize phenylalanine), and sickle cell disease (sickle-shaped red blood cells) are all autosomal recessive disorders. Marfan syndrome (defective elastic connective tissue) and Huntington disease (abnormal huntingtin protein) are both autosomal dominant disorders. Familial hypercholesterolemia (liver cells lack cholesterol receptors) is an incompletely dominant autosomal disorder. Among X-linked disorders are muscular dystrophy (absence of dystrophin leads to muscle weakness) and hemophilia (inability of blood to clot).

Testing a protein or testing the DNA can reveal the presence of certain genetic disorders. For example, persons with Tay-Sachs disease lack the enzyme hex A, and those with PKU have excessive phenylalanine in their blood. To test the DNA for mutations, it is possible to use genetic markers or DNA chips.

The fetus, embryo, or egg can be tested to identify a genetic defect before birth. An ultrasound allows for the detection of severe abnormalities, such as spina bifida. Fetal cells can be obtained by amniocentesis or chorionic villi sampling, or by sorting fetal cells from the mother's blood. Following in vitro fertilization, it is possible to test the embryo. A cell is removed from an eight-celled embryo, and if it is found to be genetically healthy, the embryo is implanted in the uterus, where it develops to term. Before IVF, a polar body can also be tested. If the woman is heterozygous and the polar body has the genetic defect, the egg does not have it. Following IVF, the embryo is implanted in the uterus.

26.3 Genomics

The Human Genome Project has succeeded in obtaining a working draft of the sequence of all the base pairs in all the DNA of the human chromosomes. The genomes of many other organisms are also in final-draft stage. Comparison of the DNA sequences of different organisms is revealing information concerning the relatedness and evolution of these organisms. The HapMap project is cataloging common sequence differences that occur in human beings. Proteomics, the study of the structure, function, and interaction of cellular proteins, employs computer modeling of the three-dimensional shape of these proteins. Bioinformatics is the application of computer technologies to the study of the genome. The use of these tools is expanding our understanding of the various dimensions of the human genome.

During gene therapy, a genetic defect or other type of illness is treated by giving the patient a foreign gene. During ex vivo therapy, cells are removed from the patient, treated, and returned to the patient. In vivo therapy consists of directly giving the patient a foreign gene that will improve his/her health.

Testing Yourself

Choose the best answer for each question.

1. A karyotype is prepared
   a. for an unborn child through amniocentesis or chorionic villi sampling.
   b. by arranging chromosomes into pairs by size, shape, and banding pattern.
   c. as a means of diagnosing an abnormal chromosomal disorder.
   d. using a photograph of a cell sample arrested during metaphase.
   e. All of these are correct.

2. Fetal chromosomes are most easily acquired for study using
   a. amniotic fluid.   c. a blood sample.
   b. a saliva sample.   d. a hair sample.

3. An individual can have too many or too few chromosomes as a result of
   a. nondisjunction.   d. amniocentesis.
   b. Barr bodies.   e. monosomy.
   c. trisomy.

For questions 4–8, match the chromosome disorder to its description in the key.

Key:
   a. female with undeveloped ovaries and uterus, unable to undergo puberty, normal intelligence, can live normally with hormone replacement
   b. XXY male, can inherit more than two X chromosomes
   c. male or female, mentally retarded, short stature, flat face, stubby fingers, large tongue, simian palm crease
   d. XXX or XXXX female
   e. caused by nondisjunction during spermatogenesis

4. Klinefelter syndrome

5. Poly-X female

6. Down syndrome

7. Turner syndrome

8. Jacobs syndrome

For questions 9–12, match the chromosome mutation to its description in the key.

Key:
   a. turned-up nose, wide mouth, small chin, large ears, poor academic skills, excellent verbal and musical abilities, prematurely aging cardiovascular system
   b. deletion in chromosome 5
   c. poor muscle tone, mental retardation, seizures, curved spine, autistic characteristics, poor speech, hand flapping, lack of eye contact
   d. translocation between chromosomes 2 and 20

9. Alagille syndrome

10. Inv dup 15 syndrome

11. Williams syndrome

12. Cri du chat syndrome

13. The following pedigree pertains to color blindness. Using the letter B for the normal allele, the genotype of the starred individual is _____.

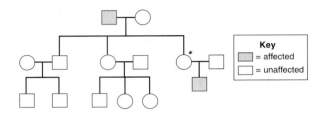

14. Which of the following is incorrect regarding genetic testing? Genetic testing
   a. is beneficial in determining if parents are carriers for autosomal recessive disorders.
   b. can determine if the enzyme hexosaminidase A is present in an unborn fetus.
   c. can be carried out on an embryo prior to in vitro fertilization.
   d. indirectly using polar bodies can provide information about eggs.
   e. is routinely used to determine genetic defects caused by multiple genes.

15. Proteomics
   a. is the application of computer technologies to the study of the genome.
   b. is the study of the structure, function, and interaction of cellular proteins.
   c. is the study of the human genome to better understand how genes work.
   d. is all the genes that occur in a cell.
   e. is the study of a person's complete genotype, or genetic profile.

16. Parents who do not have Tay-Sachs disease (recessive) produce a child who has Tay-Sachs. What are the chances that each child born to this couple will have Tay-Sachs disease?
   a. 100%   d. 0%
   b. 75%   e. None of these are correct.
   c. 25%

17. Which is a late-onset neuromuscular genetic disorder?
   a. cystic fibrosis   c. phenylketonuria
   b. Huntington disease   d. Tay-Sachs disease

18. Two affected parents have an unaffected child. The trait involved is
   a. autosomal recessive.   c. controlled by multiple alleles.
   b. incompletely dominant.   d. autosomal dominant.

19. Because of the Human Genome Project, we know
   a. humans have gene deserts with no known function.
   b. humans have 20–25,000 genes.
   c. our genome size.
   d. All of these are correct.
   e. Only a and c are correct.

20. Gene therapy has been used to treat which of the following disorders?
    a. cystic fibrosis
    b. hypercholesterolemia
    c. severe combined immunodeficiency
    d. All of these are correct.
21. If a boy is color blind,
    a. his father must be color blind.
    b. his mother must be color blind or a carrier.
    c. his father must be normal or a carrier.
    d. his siblings are also color blind.
    e. Both b and d are correct.

Additional Genetics Problems

1. A color-blind female and a male with normal vision have children.
   a. What are the chances their son will be color blind?
   b. What are the chances their daughter will be color blind?
2. A man has Duchenne muscular dystrophy.
   a. What are the possible genotypes of his mother?
   b. If he lived to produce a child with a homozygous normal female, what are the chances a daughter would have the disease?
3. A woman has Marfan syndrome, inherited from her father. She has a young son who also shows signs of the syndrome.
   a. What are the possible genotypes of the child?
   b. What are the possible genotypes of his father?
4. A 25-year-old man believes he has Huntington disease and does not want to pass this trait down to his children.
   a. If he does have the disease, is it possible for him to father a normal son? Explain.
   b. If he does have the disease, is it possible for him to father a normal daughter? Explain.
5. A child with familial hypercholesterolemia has half the normal number of receptors for cholesterol.
   a. What are the possible genotypes of his parents? (Keep in mind that they lived long enough to have children.)
   b. Could he live long enough to have children? Explain, using his genotype.
6. A woman has relatives with cystic fibrosis and does not want to have a child that suffers from the disease. She and her spouse go in for genetic testing and counseling. What are the chances of their having a child with the disease?

Understanding the Terms

amniocentesis 526
Barr body 528
bioinformatics 543
carrier 534
chorionic villi sampling (CVS) 526
chromosomal mutation 531
color blindness 538
cystic fibrosis (CF) 536
deletion 532
DNA probe 539
Duchenne muscular dystrophy 538
duplication 532
familial hypercholesterolemia (FH) 537
gene therapy 544
genetic counseling 526
genetic marker 539
genetic profiling 539
genome 542
genomics 542
hemophilia 538
Human Genome Project 542
Huntington disease 537
inversion 532
karyotype 526
Marfan syndrome 537
monosomy 528
nondisjunction 528
phenylketonuria (PKU) 536
proteomics 543
sickle cell disease 536
syndrome 526
Tay-Sachs disease 536
translocation 533
trisomy 528

Match the terms to these definitions:
a. _____ Abnormality in chromosome number or structure.
b. _____ One more chromosome than usual.
c. _____ Dark-staining body in the nuclei of female mammals that contains a condensed, inactive X chromosome.
d. _____ Change in chromosome structure in which a segment of a chromosome is turned around 180°.
e. _____ Change in chromosome structure in which a particular segment is present more than once in the same chromosome.

Thinking Critically

1. Why would expectant parents want to undergo fetal genetic testing if they know that regardless of the findings they plan to have the baby?
2. Why are there no viable trisomies of chromosome 1?
3. Why can a person carrying a translocation be normal except for the inability to have children?
4. Certain genetic disorders involve a change in a gene's DNA, but the amino acid sequence of the protein encoded by the gene is unchanged. How can this result in a genetic disorder?
5. In a genomic comparison between humans and yeast, what genes would you expect to be similar?

ARIS, the *Inquiry into Life* Website

ARIS, the website for *Inquiry into Life*, provides a wealth of information organized and integrated by chapter. You will find practice quizzes, interactive activities, labeling exercises, flashcards, and much more that will complement your learning and understanding of general biology.

www.aris.mhhe.com

Evolution and Diversity

We tend to think of evolution as happening over long timescales. *However, we have recently learned that human activities can accelerate the process of evolution quite rapidly—even over a few years! For example, farmers are continuously challenged by the fact that insects evolve pesticide resistance (see the Health Focus at the end of this chapter). That is, when new pesticides are sprayed, a certain (usually small) proportion of insect populations are resistant to the pesticide. These individuals survive and reproduce, while those that are susceptible die. Thus, over time, the proportion of resistant individuals increases to the point that the pesticide is no longer effective and crop damage increases. Then, new pesticides have to be used or developed. Similar problems have arisen with antibiotics; much in the same way, bacteria evolve resistance to antibiotics, making some ineffective for treating certain infections.*

The good news is that our understanding of evolutionary biology has helped change human behavior to deal with the evolution of resistance in pests and bacteria. For example, many farmers now use "integrated pest management," which involves rotating pesticide types as well as using natural enemies (predators) to fight crop pests. Similarly, doctors no longer prescribe antibiotics unless they are relatively certain a patient has a bacterial infection, and they tend to prescribe the minimum strength antibiotic to do the job. These are examples of why evolution is important in people's everyday lives.

In this chapter, you will learn about evidence that indicates evolution has occurred and about the way the evolutionary process works.

chapter 27

Evolution of Life

Chapter Concepts

27.1 Origin of Life
- What type of evolution preceded biological evolution? Explain. 550–551
- What conditions were needed for a true cell to come into being? 551–552

27.2 Evidence of Evolution
- What types of evidence show that common descent with modification has occurred? 552–558

27.3 The Process of Evolution
- What type of change occurs when a population evolves? Do individual organisms evolve? 559–562
- What five agents lead to evolutionary changes? 562–566
- Which agent allows populations to become better adapted to their environment? 564

27.4 Speciation
- How does speciation occur? 566–569
- What is meant by the phrase "the pace of speciation"? 567

27.5 Classification
- What are the basic categories of classification? 569
- What does evolution have to do with classification of organisms? 569
- What are the five kingdoms and the three domains? What kingdoms are in Domain Eukarya? 569–570
- How do evolutionary trees of traditionalists differ from cladograms done by phylogeneticists? 571

Colorado potato beetles,
Leptinotarsa decimlineata

27.1 Origin of Life

In order to understand evolution, it is first important to understand how life began. The common ancestor for all living things was the first cell or cells. The planet Earth was in existence a long time before the first cell arose—over a billion years, in fact. Earth is 4.6 billion years old, and the earliest fossils of prokaryotes are 3.5 billion years old. A billion years is about 13 million human life spans, assuming humans live 75 years. The origin of the first cell is an event of low probability, because a complex series of events would have had to occur—but this length of time is long enough for an event of low probability to have occurred.

Today we do not believe that life arises spontaneously from nonlife, and we say that "life comes only from life." However, the very first living thing had to have come from nonliving chemicals. Under the conditions of early Earth, it is possible that a chemical reaction produced the first cell(s). A particular mix of inorganic chemicals could have reacted to produce small organic molecules such as glucose, amino acids, and nucleotides. Then these would have polymerized into macromolecules. Once a plasma membrane formed, a structure called a **protocell** could have come into existence (Fig. 27.1).

Evolution of Small Organic Molecules

Most chemical reactions take place in water, and the first protocell undoubtedly arose in the ocean. But where? One possibility is that the protocell formed on the surface of the seas and in seaside pools where much energy was available. Ultraviolet radiation was intense in these areas because there was no ozone shield, the layer of O_3 that today blocks much of the ultraviolet radiation coming from the sun. Oxygen molecules later reacted with one another to produce the ozone shield.

In 1953, Stanley Miller and Harold Urey performed an experiment (known as the Miller-Urey experiment) that supports the hypothesis that small organic molecules were formed at the ocean's surface. In the early Earth, volcanoes erupted constantly, and the first atmospheric gases would have consequently contained methane (CH_4), ammonia (NH_3), and hydrogen (H_2). These gases could then have been washed into the ocean by the first rains. Fierce lightning and unabated ultraviolet radiation would have allowed them to react and produce the first organic molecules.

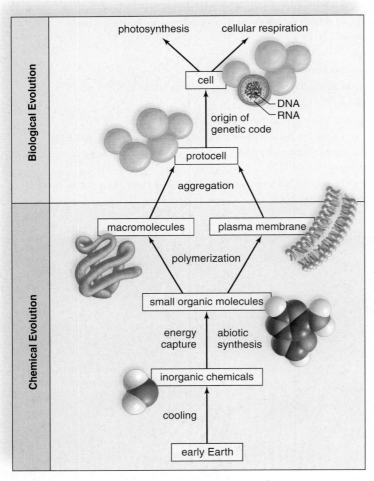

Figure 27.1 Origin of the first cell(s).
A chemical evolution may have produced the first cell. First, inorganic chemicals reacted to produce small organic molecules, which polymerized to form macromolecules. With the origination of the plasma membrane, the first primitive cell (a protocell) evolved, and once this cell could replicate, life began.

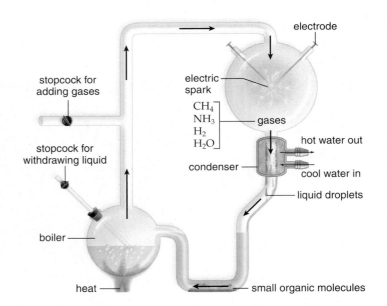

Figure 27.2 Miller and Urey's apparatus and experiment.
Gases that were thought to be present in the early Earth's atmosphere were admitted to the apparatus, circulated past an energy source (electric spark), and cooled to produce a liquid that could be withdrawn. Upon chemical analysis, the liquid was found to contain various small organic molecules.

Figure 27.3 Chemical evolution at hydrothermal vents.
Minerals that form at deep-sea hydrothermal vents like this one can catalyze the formation of ammonia and even organic molecules.

To test the hypothesis of chemical evolution, Miller placed the inorganic chemicals mentioned in a closed system, heated the mixture, and circulated it past an electric spark (simulating lightning). After a week, the solution contained a variety of amino acids and organic acids (Fig. 27.2). This and other similar experiments support the hypothesis that inorganic chemicals in the absence of oxygen (O_2) and in the presence of a strong energy source can result in organic molecules.

Is there any other place in the oceans where life could have evolved? Scientists have discovered mid-oceanic ridges within the depths of the sea. Hydrothermal (hot water) vents (openings) occur in the region of mid-oceanic ridges. A vent can be huge, measuring 10–15 meters wide with sides about 15–20 meters high. Hot water spewing out of these vents contains a mix of iron-nickel sulfides. Amazingly, scientists have discovered communities of organisms, including tube worms and giant clams, living in the regions of hydrothermal vents. It is possible that in the past, the right combination of conditions occurred at these vents to initiate life (Fig. 27.3).

The formation of small organic molecules is thought to be the first step toward the origination of a cell, from which other life-forms evolved.

Macromolecules

Once formed, the first small organic molecules gave rise to still larger molecules and then macromolecules. There are three primary hypotheses concerning this stage in the origin of life. One is the **RNA-first hypothesis,** which suggests that only the macromolecule RNA (ribonucleic acid) was needed at this time to progress toward formation of the first cell or cells. Scientists formulated this hypothesis after discovering that RNA can sometimes be both a genetic substrate and an enzyme. Such RNA molecules are called ribozymes. The first genes and enzymes could thus both have been composed of RNA, since we now know that ribozymes exist. Scientists who support this hypothesis say it was an "RNA world" some 4 billion years ago.

Another hypothesis is termed the **protein-first hypothesis.** Sidney Fox has shown that amino acids polymerize abiotically (without life) when exposed to dry heat. He suggests that amino acids collected in shallow puddles along the rocky shore, and the heat of the sun caused them to form **proteinoids,** small polypeptides that have some catalytic properties. When proteinoids are returned to water, they form **microspheres,** structures composed only of protein that have many of the properties of a cell. Some of these proteins could have had enzymatic properties. This hypothesis assumes that DNA genes came after protein enzymes arose.

The third hypothesis is put forth by Graham Cairns-Smith. He believes that clay was especially helpful in causing the polymerization of both proteins and nucleic acids at the same time. Clay attracts small organic molecules and contains iron and zinc, which may have served as inorganic catalysts for polypeptide formation. In addition, clay tends to collect energy from radioactive decay and then discharge it when the temperature or humidity changes, possibly providing a source of energy for polymerization. Cairns-Smith suggests that RNA nucleotides and amino acids became associated in such a way that polypeptides were ordered by, and helped synthesize, RNA.

Chemical reactions likely produced the macromolecules we associate with living things.

The Protocell

After macromolecules formed, something akin to a modern plasma membrane was needed to separate them from the environment. Thus, before the first true cell arose, there would likely have been a protocell, a structure that had a lipid-protein membrane and carried on energy metabolism. Fox has shown that if lipids are made available to microspheres, the two tend to become associated, producing a lipid-protein membrane (Fig. 27.4*a*).

Figure 27.4 Protocell components.
a. Microspheres, which are composed only of protein, have a number of cellular characteristics and could have evolved into the protocell. **b.** Liposomes form automatically when phospholipid molecules are put into water. Plasma membranes may have evolved similarly.

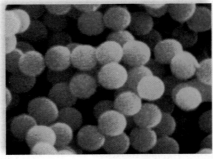

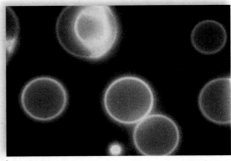

a.   b.

Aleksandr Oparin, a Soviet biochemist, showed that under appropriate conditions of temperature, ionic composition, and pH, concentrated mixtures of macromolecules can give rise to complex units called coacervate droplets. Coacervate droplets have a tendency to absorb and incorporate various substances from the surrounding solution. Eventually, a semipermeable boundary may form around the droplet. In a liquid environment, phospholipid molecules automatically form droplets called **liposomes** (Fig. 27.4b). Perhaps the first plasma-like membrane formed in this manner. However it happened, development of the plasma membrane was key because it separated the genetic material from the outside environment.

The Heterotroph Hypothesis

It has been suggested that the protocell likely was a **heterotroph,** an organism that takes in preformed food. During the early evolution of life, the ocean contained abundant nutrition in the form of small organic molecules. This suggests that heterotrophs preceded **autotrophs,** organisms that make their own food.

At first, the protocell may have used preformed ATP, but as this supply dwindled, cells that could extract energy from carbohydrates to transform ADP to ATP were favored. Glycolysis is a common metabolic pathway in living things, and this testifies to its early evolution in the history of life. Since there was no free oxygen, we can assume that the protocell carried on a form of fermentation. It seems logical that the protocell at first had limited ability to break down organic molecules and that it took millions of years for glycolysis to evolve completely.

The True Cell

A true cell is a membrane-bounded structure that can carry on protein synthesis to produce the enzymes that allow DNA to replicate. The central concept of genetics states that DNA directs protein synthesis and that information flows from DNA to RNA to protein. It is possible that this sequence developed in stages.

According to the RNA-first hypothesis, RNA would have been the first genetic material to evolve, and the first true cell would have had RNA genes. These genes would have directed and enzymatically carried out protein synthesis, as in ribozymes. Also, today we know that some viruses have RNA as their genetic material. These viruses have a protein enzyme

called reverse transcriptase that uses RNA as a template to form DNA, which then undergoes protein formation. Perhaps, with time, reverse transcription gave rise to DNA genes. Once DNA genes existed, they may have specified proteins.

According to the protein-first hypothesis, proteins, or at least polypeptides, were the first of the three molecules (i.e., DNA, RNA, and protein) to arise. Only after the protocell developed sophisticated enzymes did it have the ability to synthesize DNA and RNA from small molecules provided by the ocean. Researchers point out that because nucleic acids are very complicated molecules, the likelihood that RNA arose on its own is minimal.

Cairns-Smith proposes that polypeptides and RNA evolved simultaneously. Therefore, the first true cell would have contained RNA genes that could have replicated because of the presence of proteins. This eliminates the baffling chicken-and-egg paradox: which came first, proteins or RNA? It does mean, however, that two unlikely events would have had to happen at the same time.

Once the protocells acquired genes that could replicate, they became cells capable of reproducing, and biological evolution began.

The hypothesis that the origin of life followed a transition from small organic molecules to macromolecules to protocells to true cells is currently widely favored by scientists. However, recently some scientists have argued that early cells may have come from asteroids from Mars that hit the Earth. Recent expeditions to Mars have shown that water possibly existed there in the past. Nonetheless, it is thought that cells, the basic building block of life, arose at some point from nonliving matter.

> Once the protocell was capable of reproduction, it became a true cell, and biological evolution began.

27.2 Evidence of Evolution

Evolution is all the changes that have occurred in living things since the beginning of life due to differential reproductive success. That is, some individuals reproduce more than others because they are better "fit" to their environment. Table 27.1 indicates that Earth is about 4.6 billion years old and that prokaryotes, probably the first living organisms, evolved about 3.5 billion years ago. The eukaryotic cell arose about 2.1 billion years ago, but multicellularity didn't begin until perhaps 700 million years ago. This means that only unicellular organisms were present for 80% of the time that life has existed on Earth. Most evolutionary events we will be discussing in future chapters occurred in less than 20% of the history of life!

Evolution is defined as "common descent." Because of descent with modification, all living things share the same fundamental characteristics: they are made of cells, take chemicals and energy from the environment, respond to external stimuli, and reproduce. Living things are diverse because individual organisms exist in the many environments throughout the Earth, and the features that enable them to survive in those environments are quite diverse.

Many fields of biology provide evidence that evolution through descent with modification occurred in the past and is still occurring. Let us look at the various types of evidence for evolution.

Fossil Evidence

Fossils are the remains and traces of past life or any other direct evidence of past life. Most fossils consist only of hard parts of organisms, such as shells, bones, or teeth, because these are usually preserved after death. The soft parts of a dead organism are often consumed by scavengers or decomposed by bacteria. Occasionally, however, an organism is buried quickly and in such a way that decomposition is never completed or is completed so slowly that the soft parts leave an imprint of their structure. Traces include trails, footprints, burrows, worm casts, or even preserved droppings.

The great majority of fossils are found embedded in sedimentary rock. Sedimentation, a process that has been going on since Earth formed, can take place on land or in bodies of water. The weathering and erosion of rocks produces particles that vary in size and are called sediment. As such particles accumulate, sediment becomes a stratum (pl., strata), a recognizable layer of rock. Any given stratum is older than the one above it and younger than the one immediately below it, so that the relative age of fossils can be determined based on their depth.

Paleontologists are biologists who study the fossil record and from it draw conclusions about the history of life. Particularly interesting are the fossils that serve as **transitional links** between groups. For example, the famous fossils of *Archaeopteryx* are intermediate between reptiles and birds (Fig. 27.5). The dinosaur-like skeleton of this fossil has reptilian features, including jaws with teeth and a long, jointed tail, but *Archaeopteryx* also had feathers and wings, all suggesting that reptiles evolved from birds. Other transitional links among fossil vertebrates suggest that fishes evolved before amphibians, which evolved before reptiles, which evolved before both birds and mammals in the history of life.

Figure 27.5 Transitional fossils.

a. *Archaeopteryx* was a transitional link between reptiles and birds. Fossils indicate it had feathers and wing claws. Most likely, it was a poor flier. Perhaps it ran over the ground on strong legs and climbed into trees with the assistance of these claws. **b.** *Archaeopteryx* also had a feather-covered, reptilian tail that shows up well in this artist's representation. (Orange labels = reptilian characteristics; green label = bird characteristic.)

TABLE 27.1 The Geological Timescale: Major Divisions of Geological Time and Some of the Major Evolutionary Events of Each Time Period

Era	Period	Epoch	Millions of Years Ago	Plant Life	Animal Life
Cenozoic*		Holocene	0.01	Human influence on plant life	Age of *Homo sapiens*
			Significant Mammalian Extinction		
	Quaternary	Pleistocene	(0.01–2)	Herbaceous plants spread and diversify.	Presence of ice age mammals. Modern humans appear.
		Pliocene	(5–1.8)	Herbaceous angiosperms flourish.	First hominids appear.
		Miocene	(23–25)	Grasslands spread as forests contract.	Apelike mammals and grazing mammals flourish; insects flourish.
	Tertiary	Oligocene	(36–23)	Many modern families of flowering plants evolve.	Browsing mammals and monkeylike primates appear.
		Eocene	(57–36)	Subtropical forests with heavy rainfall thrive.	All modern orders of mammals are represented.
		Paleocene	(65–57)	Flowering plants continue to diversify.	Primitive primates, herbivores, carnivores, and insectivores appear.
			Mass Extinction: Dinosaurs and Most Reptiles		
Mesozoic	Cretaceous		(144–65)	Flowering plants spread; conifers persist.	Placental mammals appear; modern insect groups appear.
	Jurassic		(231–144)	Flowering plants appear.	Dinosaurs flourish; birds appear.
			Mass Extinction		
	Triassic		(248–118)	Forests of conifers and cycads dominate.	First mammals appear; first dinosaurs appear; corals and molluscs dominate seas.
			Mass Extinction		
Paleozoic	Permian		(280–248)	Gymnosperms diversify.	Reptiles diversify; amphibians decline.
	Carboniferous		(360–280)	Age of great coal-forming forests: ferns, club mosses, and horsetails flourish.	Amphibians diversify; first reptiles appear; first great radiation of insects.
			Mass Extinction		
	Devonian		(408–360)	First seed plants appear. Seedless vascular plants diversify.	Jawed fishes diversify and dominate the seas; first insects and first amphibians appear.
	Silurian		(438–408)	Seedless vascular plants appear.	First jawed fishes appear.
			Mass Extinction		
	Ordovician		488.3	Nonvascular land plants appear. Marine algae flourish.	Invertebrates spread and diversify; jawless fishes (first vertebrates) appear.
	Cambrian		542	First plants appear on land. Marine algae flourish.	All invertebrate phyla present; first **chordates** appear.
Precambrian Time			600	Oldest soft-bodied invertebrate fossils	
			1,400–700	Protists evolve and diversify.	
			2,200	Oldest eukaryotic fossils	
			2,700	O_2 accumulates in atmosphere.	
			3,500	Oldest known fossils (prokaryotes)	
			4,600	Earth forms.	

*Many authorities divide the Cenozoic era into the Paleogene period (contains the Paleocene, Eocene, and Oligocene epochs) and the Neogene period (contains the Miocene, Pliocene, Pleistocene, and Holocene epochs).

Figure 27.6 Dinosaurs of the late Cretaceous period.

Parasaurolophus walkeri, although not as large as other dinosaurs, was one of the largest plant-eaters of the late Cretaceous period. The crest atop its head was about 2 meters long and was used to make booming calls. Also living at this time were the rhino-like dinosaurs represented here by *Triceratops* (left), another herbivore.

Geological Timescale

As a result of studying strata, scientists have divided Earth's history into eras, and then periods and epochs (Table 27.1). The fossil record has helped determine the dates given in the table. There are two ways to date fossils. The relative dating method determines the relative order of fossils and strata depending on the layer of rock in which they were found, but it does not determine the actual date they were formed.

The absolute dating method relies on radioactive dating techniques to assign an actual date to a fossil. All radioactive isotopes have a particular half-life, the length of time it takes for half of the radioactive isotope to change into another stable element. Carbon 14 (^{14}C) is the only radioactive isotope in organic matter. Assuming a fossil contains organic matter, half of the ^{14}C will have changed to nitrogen 14 (^{14}N) in 5,730 years. To estimate how much ^{14}C was in the organism to begin with, it is assumed that organic matter always begins with the same amount of ^{14}C. Scientists compare the ^{14}C radioactivity of the fossil to that of a modern sample of organic matter. For example, if a fossil has one-fourth the amount of radioactive ^{14}C as a modern sample, then the fossil is approximately 11,460 years old (2 half-lives). After 50,000 years, however, the amount of ^{14}C radioactivity is so low that it cannot be used to measure the age of a fossil accurately. In that event, certain other ratios of isotopes with longer half-lives can be used to date rocks even billions of years old, and then the age of a fossil contained in the rock can be inferred in a similar way. Using both relative and absolute dating methods, we can learn from fossils about the various organisms and environments that existed across the planet during any time period.

Mass Extinctions

Extinction is the death of every member of a species. During mass extinctions, a large percentage of species become extinct within a relatively short period of time. So far, there have been five major mass extinctions. These occurred at the ends of the Ordovician, Devonian, Permian, Triassic, and Cretaceous periods (see Table 27.1), and a sixth is likely occurring now, probably as a result of human activities (discussed in Chapter 36). Following mass extinctions, the remaining groups of organisms are likely to spread out and fill the habitats vacated by those that have become extinct.

It was proposed in 1977 that the Cretaceous extinction (or "Cretaceous crisis") was due to an asteroid that exploded, producing meteorites that fell to Earth. A large meteorite striking Earth could have produced a cloud of dust that mushroomed into the atmosphere, blocking out the sun and causing plants to freeze and die. A huge crater that could have been caused by a meteorite involved in the Cretaceous extinction was found in the Caribbean–Gulf of Mexico region on the Yucatán Peninsula. During the Cretaceous period, great herds of dinosaurs roamed the plains, as did *Parasaurolophus walkeri* and *Triceratops* (Fig. 27.6), but all dinosaur species went extinct near the end of the Cretaceous period.

In 1984, paleontologists found that marine animals have a mass extinction about every 26 million years, and surprisingly, astronomers can offer an explanation. Our sun moves up and down as it orbits in the Milky Way, a starry galaxy. Astronomers predict that when this vertical movement causes our solar system to approach certain other members of the Milky Way, an unstable situation develops that could lead to a meteorite striking Earth. This evidence suggests that mass extinctions can be associated with extraterrestrial events, but these events are not necessarily the only cause of mass extinctions.

Fossils allowed scientists to construct the geological timescale that traces the history of life. Several mass extinctions have occurred in the past, possibly due to extraterrestrial events.

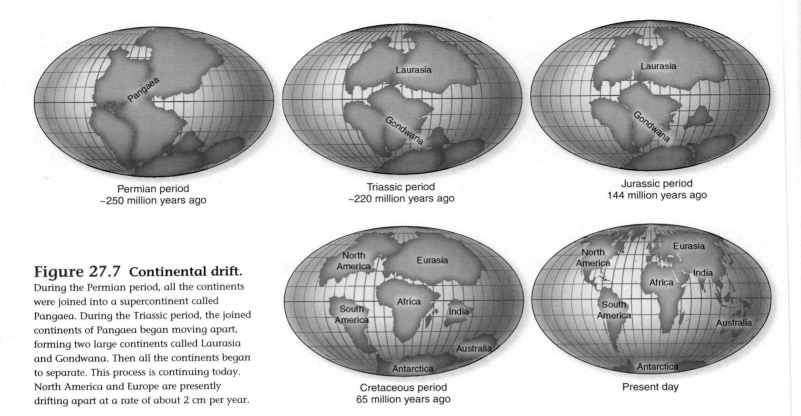

Permian period
~250 million years ago

Triassic period
~220 million years ago

Jurassic period
144 million years ago

Figure 27.7 Continental drift.
During the Permian period, all the continents were joined into a supercontinent called Pangaea. During the Triassic period, the joined continents of Pangaea began moving apart, forming two large continents called Laurasia and Gondwana. Then all the continents began to separate. This process is continuing today. North America and Europe are presently drifting apart at a rate of about 2 cm per year.

Cretaceous period
65 million years ago

Present day

Biogeographical Evidence

Another type of evidence that supports evolution through descent with modification is found in the field of **biogeography,** the study of the distribution of species throughout the world. The world's six biogeographical regions each have their own distinctive mix of living things. For example, the mammals and flowering plants of North America are different from those in Africa, even though parts of the two continents have similar environmental conditions. If you want to see zebras and lions, you have to go to Africa, not to the midwestern United States. Similarly, cactuses flourish in the deserts of North America, but euphorbias, not cactuses, occupy similar arid habitats of Africa. What is the best explanation for this phenomenon? Different mammals and flowering plants evolved separately in each biogeographical region, and barriers such as mountain ranges and oceans prevented them from migrating to other regions.

Many of these barriers arose through a process called **continental drift.** That is, the continents have never been fixed; rather, their positions and the positions of the oceans have changed over time (Fig. 27.7). During the Permian period, all the present landmasses belonged to one continent and then later drifted apart. As evidence of this, fossils of one species of seed fern (*Glossopteris*) have been found on all the southern continents separated by oceans. This species' presence on Antarctica is evidence that this continent was not always frozen. In contrast, many Australian species are restricted to that continent, including the majority of marsupials (pouched mammals such as the kangaroo). What is the

explanation for these distributions? Some organisms must have evolved and spread out before the continents broke up; then they became extinct.

> The distribution of many organisms on Earth is explainable by knowing when they evolved, either before or after the continents moved apart.

Anatomical Evidence

The fact that anatomical similarities exist among organisms provides further support for evolution via descent with modification. Vertebrate forelimbs are used for flight (birds and bats), orientation during swimming (whales and seals), running (horses), climbing (arboreal lizards), or swinging from tree branches (monkeys). Yet all vertebrate forelimbs contain the same sets of bones organized in similar ways, despite their dissimilar functions. The most plausible explanation for this unity is that the basic forelimb plan belonged to a common ancestor, and then the plan was modified in the succeeding groups as each continued along its own evolutionary pathway. Structures that are anatomically similar because they are inherited from a common ancestor are called **homologous structures** (Fig. 27.8). In contrast, **analogous structures** serve the same function, but are not constructed similarly, nor do they share a common ancestry. The wings of birds and insects and the eyes of octopi and humans are analogous structures and are similar due to a common environment, not common ancestry. The presence of homology, not analogy, is evidence that organisms are related.

Vestigial structures are anatomical features that are fully developed in one group of organisms but that are reduced and may have no function in similar groups. Most birds, for example, have well-developed wings for flight. However, some bird species (e.g., ostrich) have greatly reduced wings and do not fly. Similarly, snakes have no use for hindlimbs, and yet some have remnants of hindlimbs in a pelvic girdle and legs. The presence of vestigial structures can be explained by common descent. Vestigial structures occur because organisms inherit their anatomy from their ancestors; they are traces of an organism's evolutionary history.

The homology shared by vertebrates extends to their embryological development (Fig. 27.9). At some time during development, all vertebrates have a postanal tail and exhibit paired pharyngeal pouches. In fishes and amphibian larvae, these pouches develop into functioning gills. In humans, the first pair of pouches becomes the cavity of the middle ear and the auditory tube. The second pair becomes the tonsils, while the third and fourth pairs become the thymus and parathyroid glands. Why should terrestrial vertebrates develop and then modify structures like pharyngeal pouches that have lost their original function? The most likely explanation is that fishes are ancestral to other vertebrate groups.

In 1859, Charles Darwin (See Science Focus, p. 560) speculated that whales evolved from a land mammal. His hypothesis has now been substantiated. In recent years the fossil record has yielded an incredible parade of fossils that link modern whales and dolphins to land ancestors (Fig. 27.10). The presence of a vestigial pelvic girdle and legs in modern whales is also significant evidence.

> Organisms that share homologous structures are closely related and have a common ancestry. Studies of comparative anatomy and embryological development reveal homologous structures.

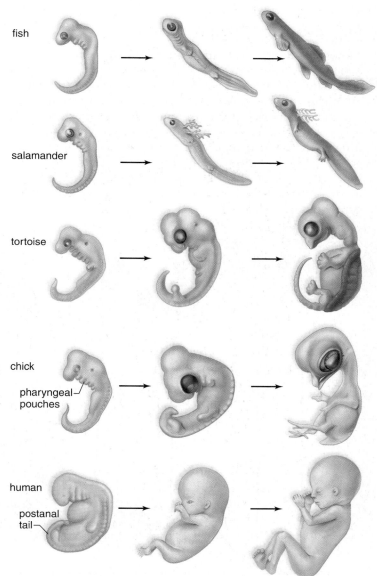

Figure 27.9 Significance of developmental similarities.
At these comparable developmental stages, vertebrate embryos have many features in common, which suggests they evolved from a common ancestor. (These embryos are not drawn to scale).

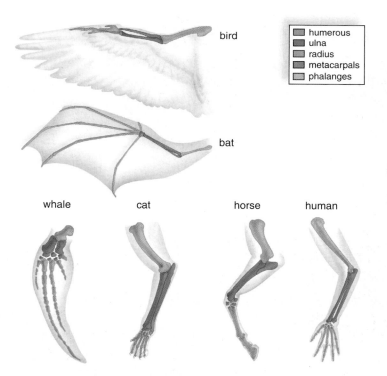

Figure 27.8 Significance of homologous structures.
Although the specific design details of vertebrate forelimbs are different, the same bones are present (note color-coding). Homologous structures provide evidence of a common ancestor.

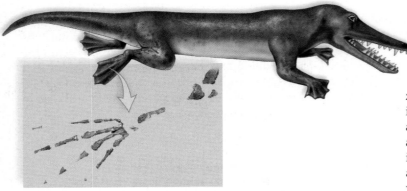

Figure 27.10 Ancestor to whales.

Ambulocetus, an ancestor to modern whales, dated 50 MYA. The presence of limbs is evidence that land-based mammals gave rise to whales.

Biochemical Evidence

Almost all living organisms use the same basic biochemical molecules, including DNA, ATP (adenosine triphosphate), and many identical or nearly identical enzymes. Further, organisms use the same DNA triplet code for the same 20 amino acids in their proteins. Since the sequences of DNA bases in the genomes of many organisms are now known, it has become clear that humans share a large number of genes with much simpler organisms. It appears that life's vast diversity has come about by only a slight difference in many of the same genes. The result has been widely divergent types of bodies.

When the degree of similarity in DNA nucleotide sequences or the degree of similarity in amino acid sequences of proteins is examined, the more similar the DNA sequences are, generally the more closely related the organisms are. For example, humans and chimpanzees are about 99% similar! Cytochrome *c* is a molecule that is used in the electron transport chain of all the organisms appearing in Figure 27.11. Data regarding differences in the amino acid sequence of cytochrome *c* show that the sequence in a human differs from that in a monkey by only one amino acid, from that in a duck by 11 amino acids, and from that in a yeast by 51 amino acids. These data are consistent with other data regarding the anatomical similarities of these organisms and, therefore, how closely they are related.

Evolution is no longer considered a hypothesis. It is one of the great unifying theories of biology. In science, the word *theory* is reserved for those conceptual schemes that are supported by a large number of observations and scientific experiments. The theory of evolution has the same status in biology that the germ theory of disease has in medicine.

> Many lines of evidence support the theory of evolution by descent with modification. Recently, biochemical evidence has also been found to support evolution. A hypothesis is strengthened when it is supported by many different lines of evidence.

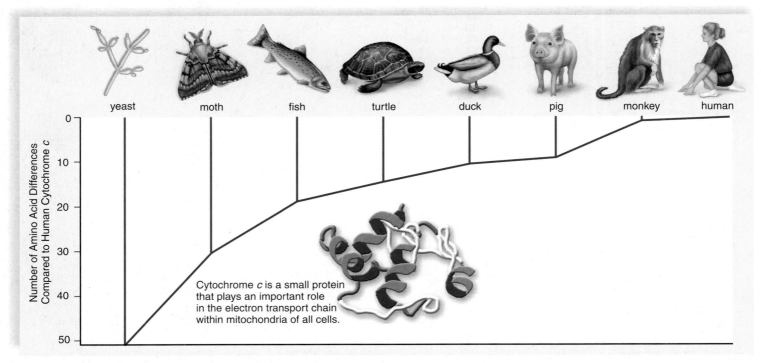

Cytochrome *c* is a small protein that plays an important role in the electron transport chain within mitochondria of all cells.

Figure 27.11 Significance of biochemical differences.

The branch points in this diagram indicate the number of amino acids that differ between human cytochrome *c* and the organisms depicted. These biochemical data are consistent with those provided by a study of the fossil record and comparative anatomy.

27.3   The Process of Evolution

Some people have the misconception that individuals evolve; however, evolution occurs at the population level. As evolution takes place, genetic changes occur within a population, and over generations, these lead to phenotypic changes that are commonly seen in that population. In this section we will consider a change in gene frequencies within a population over time, defined as **microevolution.** Microevolution can be studied using population genetics, which investigates changes in gene frequencies.

Population Genetics

A **population** is all the members of a single species that occupy a particular area at the same time and that interbreed and exchange genes. A population could be all the green frogs in a frog pond, all the field mice on a farm, or all the English daisies on a hill. The members of a population reproduce with one another to produce the next generation.

Each member of a population is assumed to be free to reproduce with any other member, and when reproduc-

tion occurs, the genes of one generation are passed on in the manner described by Mendel's laws. Therefore, in this so-called Mendelian population (as discussed in Ch. 23) of sexually reproducing individuals, the total number of alleles at all the gene loci in all the members make up a **gene pool** for the population. It is customary to describe this gene pool in terms of allele frequencies for the various genes. Using this methodology, two investigators, G. H. Hardy, an English mathematician, and W. Weinberg, a German physician, discovered a principle that now bears their names.

Hardy and Weinberg decided to use the binomial equation $p^2 + 2pq + q^2$ to calculate the genotype and allele frequencies of a population. Figure 27.12 shows how this is done. Once you know the allele frequencies, you can calculate the ratio of genotypes in the next generation using a Punnett square. The data from Figure 27.12 reveal that the next generation will have exactly the same ratio of genotypes as before:

		eggs		Genotype frequencies:
		0.6 L	0.4 l	0.36 LL + 0.48 Ll + 0.16 ll = 1
sperm	0.6 L	0.36 LL	0.24 Ll	or
	0.4 l	0.24 Ll	0.16 ll	$L^2 + 2\,Ll + l^2$

It is important to realize that the sperm and eggs represented in this Punnett square are actually the frequencies of alleles L and l in an entire population, not gametes produced by individuals.

The Hardy-Weinberg Principle

The Hardy-Weinberg principle states that allele frequencies in a gene pool will remain at equilibrium, and thus constant, after one generation of random mating in a large, sexually reproducing population as long as five conditions are met:

1. *No mutations.* Genetic mutations are an alteration in an allele, due to a change in DNA composition. Under Hardy-Weinberg assumptions, allele changes do not occur, or changes in one direction are balanced by changes in the opposite direction.
2. *No genetic drift.* Genetic drift is random changes in allele frequencies by chance. If a population is very large, changes in allele frequencies due to chance alone are insignificant.
3. *No gene flow.* Gene flow is the sharing of alleles between two populations through interbreeding. If there is no gene flow, migration of individuals, and therefore their genes, into or out of the population does not occur.
4. *Random mating.* Random mating occurs when individuals pair by chance, not according to their genotypes or phenotypes.
5. *No selection.* Often, the environment selects certain phenotypes to reproduce and have more offspring than other phenotypes. If selection does not occur, no phenotype is favored over another to reproduce.

$$p^2 + 2\,pq + q^2$$

p^2 = frequency of homozygous dominant individuals *(AA)*

p = frequency of dominant allele *(A)*

q^2 = frequency of homozygous recessive individuals *(aa)*

q = frequency of recessive allele *(a)*

$2\,pq$ = frequency of heterozygous individuals *(Aa)*

Realize that $p + q = 1$ (There are only 2 alleles.)

$p^2 + 2\,pq + q^2 = 1$ (These are the only genotypes so the total frequency of the genotypes in a population must add up to 1, or 100%.)

Example: An investigator has determined by inspection that 16% of a human population has a recessive trait. Using this information, we can complete all the genotype and allele frequencies for this population.

Given: $q^2 = 16\% = 0.16$ are homozygous recessive individuals

Therefore, $q = \sqrt{0.16} = 0.4$ = frequency of recessive allele
$p = 1.0 - 0.4 = 0.6$ = frequency of dominant allele
$p^2 = (0.6)(0.6) = 0.36 = 36\%$ are homozygous dominant individuals
$2\,pq = 2(0.6)(0.4) = 0.48 = 48\%$ are heterozygous individuals

or

$2\,pq = 1.00 - 0.52 = 0.48$ 84% have the dominant phenotype

Figure 27.12 Calculating gene pool frequencies using the Hardy-Weinberg equation.

a.

b.

Figure 27.13 Microevolution.

Both dark-colored and light-colored individuals occur in populations of the peppered moth, *Biston betularia*. **a.** When tree trunks are light, dark-colored moths are seen and eaten by predatory birds, and the light-colored moths increase in number. **b.** When tree trunks are dark due to pollution, light-colored moths are seen and eaten by predatory birds, and the dark-colored moths increase in number.

In real life, these conditions are rarely, if ever, met, and allele frequencies in the gene pool of a population do change from one generation to the next. Because a change in allele frequencies is our definition of *micro*evolution, then evolution has occurred. A significance of the Hardy-Weinberg principle is that microevolution can be detected by noting deviations from a Hardy-Weinberg equilibrium of allele frequencies in the gene pool of a population.

Such deviations suggest that one or more of the five conditions is occurring in a population. Figure 27.13 gives an example of microevolution due to selection in a population of peppered moths. Peppered moths can be dark colored or light colored, and the percentage of each in the population can vary. Predatory birds are the selective agent that causes the makeup of the population to vary. When dark-colored moths rest on light trunks in a nonpolluted area, they are seen and eaten by these birds. With pollution, the trunks of trees darken, so light-colored moths stand out and are eaten more than dark-colored moths. We know that evolution has occurred in Figure 27.13 because the population changes from 10% dark-colored phenotype to 80% dark-colored phenotype over time. In this example, evolution has occurred because a selective force (predatory birds) favored one genotype over another.

> The Hardy-Weinberg principle predicts that allele frequencies in a population will remain constant generation after generation, and this provides a baseline by which to judge whether evolution has occurred. A change of allele frequencies in the gene pool of a population signifies that evolution has occurred.

Five Agents of Evolutionary Change

The list of conditions for genetic equilibrium stated previously implies that the opposite conditions can cause evolutionary change. These conditions are mutations, genetic drift, gene flow, nonrandom mating, and natural selection.

Mutations

Mutations are genetic changes that provide the raw material for evolutionary change; mutations create new alleles. For example, a mutation can result in a nucleotide change in a gene. A mutation can be "silent" if the nucleotide change does not result in an amino acid change or if it is recessive and masked by a dominant allele in a diploid organism. If a mutation does, however, affect protein function, it can be harmful to an organism. Mutations are random and are most often thought to result in no change or a negative effect on an individual's reproductive success.

In a changing environment, however, even a seemingly harmful mutation that results in a phenotypic change can be a source of an adaptive variation. For example, the water flea *Daphnia* ordinarily thrives at temperatures around 20°C, but there is a mutation that requires *Daphnia* to live at temperatures between 25°C and 30°C. The adaptive value of this mutation is entirely dependent on environmental conditions.

Genetic Drift

Genetic drift refers to changes in the allele frequencies of a gene pool due to chance, as illustrated by the green and brown frogs in Figure 27.14. As you can imagine, genetic drift has greater effects in smaller populations. For example, the chance death of one individual in a population of a million will not have an appreciable effect on allele frequencies, but the chance death of one individual in a population of ten could change the frequency of an allele by 10% or even cause its loss altogether (if that individual was the only one with that allele). In nature, two situations, called founder effect and bottleneck effect, lead to small populations whereby genetic drift can drastically affect allele frequencies in a gene pool.

The **founder effect** occurs when a few individuals form a new colony, and only a fraction of the total genetic diversity of the original gene pool is represented in these individuals. The particular alleles carried by the founders is dictated by chance alone. The Amish population of Lancaster, Pennsylvania, is an isolated religious sect descended from a few German founders. Today, as many as one in 14 individuals in this group carries a recessive allele that causes an unusual form of dwarfism (it affects only the lower arms and legs) and polydactylism (extra fingers) (Fig. 27.15). Genetic drift has caused this proportion to be much higher in the Amish than the non-Amish, where the allele is found in only one in 1,000 people.

Sometimes a population is subjected to near extinction because of a natural disaster (e.g., earthquake or fire) or because of human interference. The disaster acts as a bottleneck, preventing the majority of genotypes from breeding to form the next generation. For example, a large genetic similarity found in cheetahs is believed to be due to a **bottleneck effect.** In a skin grafting study, most cheetahs failed to reject skin grafts from unrelated cheetahs because they were so genetically similar.

Gene Flow

Gene flow is the movement of alleles between populations, as occurs when individuals migrate from one population to another and breed in that new population. For example, adult plants are not able to migrate, but their gametes are often either blown by the wind or carried by insects. The wind, in particular, can carry pollen for long distances and can therefore be a factor in gene flow among plant populations.

Gene flow among populations keeps their gene pools similar. It also prevents close adaptation to a local environment.

Nonrandom Mating

Nonrandom mating occurs when individuals pair up, not by chance, but according to their genotypes or phenotypes. Inbreeding, or mating between relatives to a greater extent than by chance, is an example of nonrandom mating. Inbreeding decreases the proportion of heterozygotes and increases the proportions of homozygotes at all gene loci. In a human population, inbreeding increases the frequency of recessive abnormalities (see Fig. 27.15).

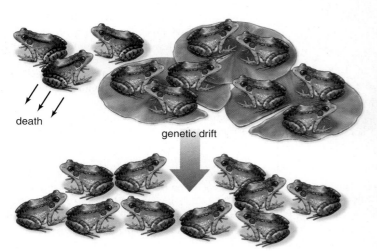

Figure 27.14 Genetic drift.
Genetic drift occurs when by chance only certain members of a population (in this case, green frogs) reproduce and pass on their alleles to the next generation. The allele frequencies of the next generation's gene pool may be markedly different from those of the previous generation, particularly in small populations.

Figure 27.15 Founder effect.
A member of the founding population of Amish in Pennsylvania had a recessive allele for a rare kind of dwarfism linked with polydactylism. The percentage of the Amish population now carrying this allele is much higher compared to that of the general population.

Although Charles Darwin is often credited as the first to believe in descent with modification, biologists before him had slowly begun to accept the idea of evolution. Jean-Baptiste de Lamarck (1744–1829), a predecessor of Darwin, concluded after studying the succession of life-forms in geological strata, that more complex organisms are descended from less complex organisms. To explain the process of adaptation to the environment, Lamarck proposed of inheritance of acquired characteristics—that the environment can bring about inherited change. One example he gave—and the one for which he is most famous—is that the long neck of a giraffe developed over time because animals stretched their necks to reach food high in trees and then passed gradually longer necks to their offspring (Fig. 27A).

This hypothesis for the inheritance of acquired characteristics has never been substantiated. The molecular mechanism of inheritance explains why. Phenotypic changes acquired during an organism's lifetime do not result in genetic changes that can be passed to subsequent generations. As an example, consider tail cropping in Doberman pincers. All Doberman puppies are born with tails, even though their parents' tails are most often cropped. That is, tail cropping is a phenotypic change that is not inherited in the DNA. We now know that Lamarck's ideas, although important for advancing ideas about evolution, were incorrect.

Charles Darwin (1809–1882) came to a different conclusion from Lamarck's after going on a five-year trip as a naturalist aboard the ship the HMS *Beagle*. He read a book by Charles Lyell, a geologist who suggested the world is very old and has been undergoing gradual changes for many many years. This meant that there was time for evolution to occur.

Because the ship sailed in the tropics of the Southern Hemisphere, Darwin encountered different living things that were more abundant and varied than those found in his native England. When Darwin compared the animals of Africa to those of South America, he noted that the African

Early giraffes probably had short necks that they stretched to reach food.

Their offspring had longer necks that they stretched to reach food.

Eventually, the continued stretching of the neck resulted in today's giraffe.

Figure 27A Jean-Baptiste de Lamarck's proposal of acquired characteristics.

ostrich and the South American rhea, although similar in appearance, were actually different animals. He reasoned that they had a different line of descent because they were on different continents. When he arrived at the Galápagos Islands, he began to study the diversity of finches (see Fig. 27.20), whose adaptations could best be explained by assuming they had diverged from a common ancestor. He found such a hypothetical ancestor on the mainland of South America, supporting his theory. With this type of evidence, Darwin concluded that species evolve (change) with time.

When Darwin returned home, he spent the next 20 years gathering data to support the principle of biological evolution. His most significant contribution was his theory of natural selection, which explains how populations of a species become adapted to their environment. This theory is explained in Figure 27B. Before formulating the theory, Darwin read an essay on human population growth written by Thomas Malthus. Malthus observed that although humans have a great reproductive potential, many environmental variables, such as availability of food and living space, tend to keep the human population in check with factors such as disease and famine. Darwin applied these ideas to all populations of organisms. A population is all the members of a species living in one particular place. Darwin calculated that a single pair of elephants could have 19 million descendants in 750 years. He realized that other organisms have an even greater reproductive potential than this pair of elephants, yet usually population sizes remain about the same. Darwin decided there is a constant struggle for existence, whereby only certain members of a population survive to reproduce. Those individuals best adapted to their environment produce the greatest number of offspring, and it is their traits that increase in frequency in a population in successive generations. This so-called "survival of the fittest" causes the next generation to be better adapted to the environment than the previous generation.

Darwin's theory of natural selection was nonteleological, meaning that there is no design or purpose in the works or processes of nature. However, rather than believing that organisms strive to adapt themselves to the environment, Darwin concluded that the *environment acts on individual phenotypes to select those individuals that are best adapted.* These individuals have been "naturally selected" to pass on their characteristics to the next generation. In contrast, the Lamarckian explanation for the long neck of the giraffe was incorrect because ancestors of the modern giraffe were "trying" to reach into the trees to browse on high-growing vegetation. Lamarck's proposal is teleological because, according to him, the outcome (longer necks) is predetermined. Darwin's theory of evolution, rather than being progressive or "forward looking" implies that the changing environment does not move toward any predetermined outcome.

The critical elements of Darwin's theory are as follows:

- **Variations.** Individual members of a population vary in physical characteristics. To be affected by natural selection, physical variations must be inherited from generation to generation by reproduction rather than being environmentally induced. If there is no variation in a trait in a population, natural selection cannot act.
- **Overproduction and struggle for existence.** The members of all populations compete with each other for limited resources. Certain members are able to capture or utilize these resources better than others.
- **Survival of the fittest.** Just as humans carry on artificial breeding programs to select which plants and animals will reproduce, natural selection by the environment determines which members of a population survive and reproduce. While Darwin emphasized the importance of survival, modern evolutionists emphasize the importance of unequal reproduction. That is, certain members of the population produce more

Early giraffes probably had necks of various lengths.

Natural selection due to competition led to survival of the longer-necked giraffes and their offspring.

Eventually, only long-necked giraffes survived the competition.

Figure 27B Charles Darwin's theory of natural selection.

offspring than others simply because they happen to have a variation or variations that make them better suited to the environment. In a biological sense, *fitness* is the number of fertile offspring an individual produces throughout its lifetime.

- **Adaptation.** The result of natural selection is that populations come to resemble the "best types"—those individuals that produce the most offspring because they are better adapted to the environment.

Darwin was prompted to publish his findings only after he received a letter from another naturalist, Alfred Russel Wallace, who had come to the same conclusions about evolution. Although both scientists subsequently presented their ideas at the same meeting of the famed Royal Society in London in 1858, only Darwin had outlined his reasoning for the theory in a draft of *The Origin of Species by Means of Natural Selection,* which he had completed 16 years earlier and eventually published in 1859. This book is still studied by many biologists today.

Can natural selection account for the origin of new species and for the great diversity of life? Yes, Darwinian selection is the only accepted scientific theory for the diversity of life.

Discussion Questions

1. Currently, a debate is in progress regarding the teaching of intelligent design alongside evolution as a theory for the diversity of life. Explain why the intelligent design idea is teleological.
2. Explain why variation in a trait must be present in order for natural selection to operate.
3. Explain why Lamarck's idea of "inheritance of acquired characteristics" is incorrect.

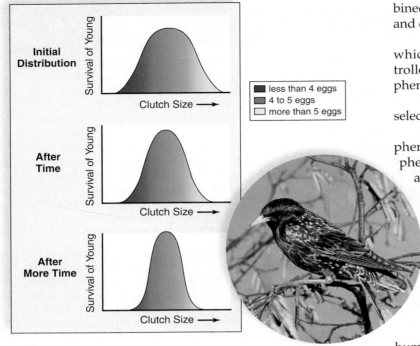

Figure 27.16 Stabilizing selection.
Stabilizing selection occurs when natural selection favors the intermediate phenotype over the extremes. For example, Swiss starlings that lay four to five eggs (usual clutch size) have more surviving young than birds that lay fewer than four eggs or more than five eggs.

Natural Selection

Natural selection is the process by which some individuals produce more offspring than others. The Science Focus outlines how Charles Darwin explained evolution by natural selection. Here, we restate these steps in the context of modern evolutionary theory. Evolution by natural selection requires:

1. **Individual variation.** The members of a population differ from one another.
2. **Inheritance.** Many of these differences are heritable genetic differences.
3. **Overproduction.** Individuals in a population are engaged in a struggle for existence because breeding individuals in a population tend to produce more off-spring than the environment can support.
4. **Differential reproductive success.** Individuals that are better adapted to their environment produce more offspring than those that are not as well adapted, and consequently, their fertile offspring will make up a greater proportion of the next generation.

In biology, the **fitness** of an individual is measured by the number of fertile offspring produced throughout its lifetime. Gene mutations are the ultimate source of variation because they provide new alleles. However, in sexually reproducing organisms, genetic variation can also result from crossing-over and independent assortment of chromosomes during meiosis and also fertilization when gametes are com-

bined. A different combination of alleles can lead to a new and different phenotype.

In this context, consider that most of the traits on which natural selection acts are polygenic and thus controlled by more than one gene. Such traits have a range of phenotypes that follow a bell-shaped curve.

The three main types of natural selection are stabilizing selection, directional selection, and disruptive selection.

Stabilizing selection occurs when an intermediate phenotype is favored. With stabilizing selection, extreme phenotypes are selected against, and individuals near the average are favored. Stabilizing selection can improve adaptation of the population to those aspects of the environment that remain constant. As an example, consider that when Swiss starlings lay four to five eggs, more young survive than when the female lays more or less than this number (Fig. 27.16). Genes determining physiological characteristics, such as the production of yolk, and behavioral characteristics, such as how long the female will mate, are involved in determining clutch size.

Through the years, hospital data have shown that human infants born with an intermediate birth weight (3–4 kg) have a better chance of survival than those born with an extreme birth weight—either higher or lower than that range. Stabilizing selection serves to reduce the variability in birth weight in human populations.

Directional Selection **Directional selection** occurs when an extreme phenotype is favored and the distribution curve shifts in that direction (Fig. 27.17). This changes the average phenotype in a population. Such a shift can occur when a population is adapting to a changing environment. For example, the gradual increase in the size of the modern horse, *Equus*, can be correlated with a change in the environment from forest conditions to grassland conditions. *Hyracotherium*, the ancestor of the modern horse, was about the size of a dog and was adapted to the forestlike environment of the Eocene epoch of the Paleogene period. This animal could have hidden among the trees for protection, and its low-crowned teeth would have been appropriate for browsing on leaves. Later, in the Miocene and Pliocene epochs, grasslands began to replace the forests. Then the ancestors of *Equus* were subject to selective pressure for the development of strength, intelligence, speed, and durable grinding teeth. A larger size provided the strength needed for combat, elongated legs ending in hooves gave speed for escaping from enemies, and the durable grinding teeth enabled the animals to feed efficiently on grasses. Nevertheless, the evolution of the horse should not be viewed as a straight line of descent; there were many side branches that became extinct. The evolution of peppered moths discussed previously is another good example of directional selection.

Disruptive Selection In **disruptive selection,** two or more extreme phenotypes are favored over any intermediate phenotype (Fig. 27.18). For example, British land snails

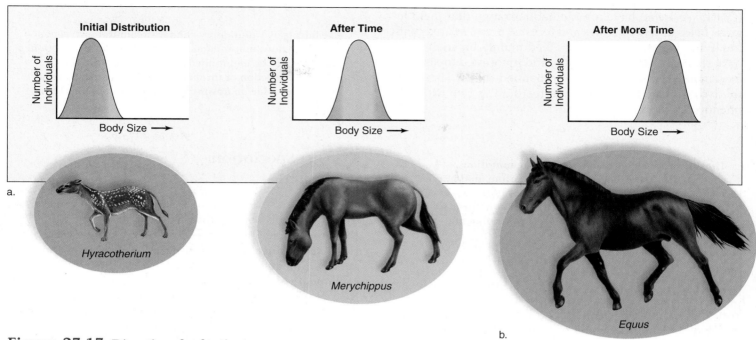

Figure 27.17 Directional selection.

a. Directional selection occurs when natural selection favors one extreme phenotype, resulting in a shift in the distribution curve. **b.** For example, *Equus*, the modern-day horse, which is adapted to a grassland habitat, is much larger than its ancestor, *Hyracotherium,* which was adapted to a forest habitat.

Figure 27.18 Disruptive selection.

a. Disruptive selection favors two or more extreme phenotypes. **b.** Today, British land snails comprise mainly two different phenotypes, each adapted to a different habitat. Snails with dark shells are more prevalent in forested areas, and light-banded snails are more prevalent in areas with low-lying vegetation.

(*Cepaeanemoralis*) have a wide habitat range that includes grass fields and hedgerows and forested areas. In areas with low-lying vegetation, thrushes feed mainly on snails with dark shells that lack light bands, and in forested areas, they feed mainly on snails with light-banded shells. Therefore, the two different habitats have resulted in two different phenotypes in the population.

> The agents of evolutionary change are mutations, genetic drift, gene flow, nonrandom mating, and natural selection. These processes cause changes in the allele frequencies of a population. Of these, only natural selection results in adaptation to the environment.

Maintenance of Variation

You might think that genetic variation, particularly due to deleterious alleles, would eventually disappear because natural selection tends to remove those alleles from a population. But sickle cell disease exemplifies how genetic variation is sometimes maintained in a population. Persons homozygous for the allele that causes sickle cell disease have sickle-shaped red blood cells, which can clog blood vessels and deprive the body of oxygen. Therefore, you would expect this condition to be selected against and eliminated from a population. However, heterozygotes for the sickle cell allele have some sickle-shaped cells, and are also resistant to malaria, a disease caused by a parasite that lives in red blood cells. Malaria is a leading killer in many parts of the world. As a result, the allele for sickle cell disease is maintained in relatively high frequency in regions where there is a high incidence of malaria. A study of the three genotypes and phenotypes involved shows why:

Genotype	Phenotype	Result
$Hb^A Hb^A$	Normal	Dies due to malarial infection
$Hb^A Hb^S$	Some sickle cells	Lives due to protection from malaria
$Hb^S Hb^S$	Sickle cell disease[1]	Dies due to sickle cell disease

[1]All red blood cells sickle shaped

The frequency of the sickle cell allele in some parts of Africa is 0.40, while among African Americans, it is only 0.05 due to lowered incidence of malaria in the United States. In Africa, the favored heterozygote keeps the two homozygotes equally present in the population. Maintenance of the same ratio of two or more phenotypes in each generation is called balanced polymorphism.

> Five agents of evolutionary change are: mutations, genetic drift, gene flow, non-random mating and natural selection. Variation can be maintained in populations through balancing selection or through alleles that cause diseases that may provide an advantage in the heterozygous form.

27.4 Speciation

Usually, a species occupies a certain geographical range, within which several subpopulations exist. For our present discussion, **species** is defined as a group of subpopulations that are capable of interbreeding and are isolated reproductively from other species. The subpopulations of the same species can exchange genes, but different species do not exchange genes. Reproductive isolation of similar species is accomplished by the isolating mechanisms listed in Table 27.2. **Prezygotic isolating mechanisms** are in place before fertilization, and thus reproduction is never attempted. **Postzygotic isolating mechanisms** are in place after fertilization, so reproduction may take place, but it does not produce fertile offspring.

The Process of Speciation

Speciation has occurred when one species gives rise to two species, each of which continues on its own evolutionary pathway. How can we recognize speciation? Whenever reproductive isolation develops between two formerly interbreeding groups of populations, speciation has occurred. One type of speciation, called **allopatric speciation,** usually occurs when populations become separated by a geographic barrier and gene flow is no longer possible. Figure 27.19 illustrates an

TABLE 27.2	Reproductive Isolating Mechanisms
Isolating Mechanism	**Example**
Prezygotic	
Habitat isolation	Species at same locale occupy different habitats
Temporal isolation	Species reproduce at different seasons or different times of day
Behavioral isolation	In animals, courtship behavior differs, or they respond to different songs, calls, pheromones, or other signals
Mechanical isolation	Genitalia unsuitable for one another
Postzygotic	
Gamete isolation	Sperm cannot reach or fertilize egg
Zygote mortality	Fertilization occurs, but zygote does not survive
Hybrid sterility	Hybrid survives but is sterile and cannot reproduce
F_2 fitness	Hybrid is fertile, but F_2 hybrid has reduced fitness

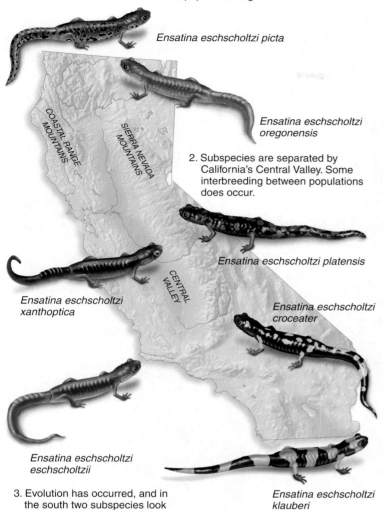

1. Members of a northern ancestral population migrated southward.

Ensatina eschscholtzi picta

COASTAL RANGE MOUNTAINS

SIERRA NEVADA MOUNTAINS

Ensatina eschscholtzi oregonensis

2. Subspecies are separated by California's Central Valley. Some interbreeding between populations does occur.

Ensatina eschscholtzi platensis

CENTRAL VALLEY

Ensatina eschscholtzi xanthoptica

Ensatina eschscholtzi croceater

Ensatina eschscholtzi eschscholtzii

3. Evolution has occurred, and in the south two subspecies look quite different from one another.

Ensatina eschscholtzi klauberi

Figure 27.19 Allopatric speciation.

In this example of allopatric speciation, the Central Valley of California is separating a range of populations descended from the same northern ancestral species. Those to the west along the coastal mountains and those to the east along the Sierra Nevada mountains experience gene flow, but gene flow is limited between the eastern populations and the western populations. Members of the most southerly eastern and western populations are quite different in color pattern.

example of allopatric speciation that has been extensively studied in California. Apparently, members of an ancestral population of *Ensatina* salamanders existing in the Pacific Northwest migrated southward, establishing a series of populations. Each population was exposed to its own selective pressures along the coastal and Sierra Nevada mountains. Due to the presence of the Central Valley of California, which is largely dry and thus unsuitable habitat for amphibians, gene flow rarely occurs between eastern and western populations of *Ensatina*. Genetic differences also increased from north to south, resulting in two distinct forms of *Ensatina* salamanders in Southern California that differ dramatically in color.

It is also possible that a single population could suddenly divide into two reproductively isolated groups without being geographically isolated. The best evidence for this type of speciation, called **sympatric speciation,** is found among plants, where multiplication of the chromosome number in one plant prevents it from successfully reproducing with others of its kind. Self-reproduction can maintain such a new plant species.

> Speciation is the origin of a new species. This usually requires geographic isolation followed by reproductive isolation.

Adaptive Radiation

One of the best examples of "allopatric" speciation is provided by the finches on the Galápagos Islands, located 600 miles west of Ecuador, South America. The 13 species of finches found there are often called Darwin's finches because Darwin first realized their significance as an example of how evolution works. These species are believed to be descended from mainland finches that migrated to one of the islands. We can imagine that after the original species on a single island increased, some individuals dispersed to other islands.

The islands are ecologically different enough to have promoted divergent feeding habits. This is apparent because, although the birds physically resemble each other in many respects, they have different beaks, each adapted to gathering and eating a different type of food (Fig. 27.20). There are seed-eating ground finches, cactus-eating ground finches, insect-eating tree finches, also with different-sized beaks; and a warbler-type tree finch, with a beak adapted to eating insects and gathering nectar. Among the tree finches is a woodpecker type, which lacks the long tongue of a true woodpecker but makes up for this by using a cactus spine or a twig to ferret out insects. Remarkably, each of these types is found on islands where its beak matches the abundant food type. Therefore, Darwin's finches are an example of **adaptive radiation,** or the proliferation of a species by adaptation to different ways of life.

The Pace of Speciation

Currently, there are two hypotheses about the pace of speciation and, therefore, evolution. One hypothesis is called the phyletic gradualism model, and the other is called the punctuated equilibrium model. Each model gives a different answer to the question of why so few transitional links are found in the fossil record.

Traditionally, evolutionists have supported a model called **phyletic gradualism,** which states that change is very slow but steady within a lineage before and after a divergence

Figure 27.20 The Galápagos finches.

Each of these finches is adapted to gathering and eating a different type of food. Note the different sizes and shapes of the beaks in the different species. Tree finches have beaks largely adapted to eating insects and, at times, plants. The woodpecker finch, a tool-user, uses a cactus spine or twig to probe in the bark of a tree for insects. Ground finches have beaks adapted to eating prickly-pear cactus or different-sized seeds.

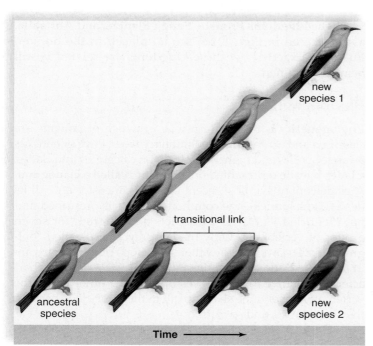

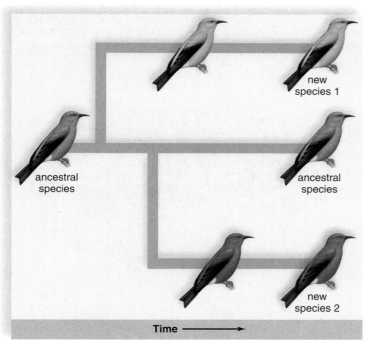

b.

Figure 27.21 Phyletic gradualism compared to punctuated equilibrium.

a. Supporters of the phyletic gradualism model believe that speciation takes place gradually and many transitional links occur. **b.** Supporters of the more recent, punctuated equilibrium model believe that speciation occurs rapidly, with no transitional links.

(splitting of the line of descent) (Fig. 27.21a). Therefore, it is not surprising that few transitional links such as *Archaeopteryx* (see Fig. 27.5) have been found. Indeed, the fossil record, even if it were complete, might be unable to show when speciation

has occurred. Why? Because a new species comes about after reproductive isolation, and reproductive isolation cannot be detected in the fossil record! Only when a new species evolves and displaces the existing species is the new species likely to show up in the fossil record.

A model of evolution called **punctuated equilibrium** has also been proposed (Fig. 27.21b). It says that long periods of stasis, or no visible change, are followed by rapid periods of speciation. With reference to the length of the fossil record (about 3.5 billion years), speciation occurs relatively rapidly, and this can explain why few transitional links are found. Mass extinction events are often followed by rapid (relative to the age of the Earth) periods of speciation.

Adaptive radiation is an example of allopatric speciation that is easily observable. Whether speciation occurs slowly or rapidly is being debated.

27.5 Classification

Recall that each type of organism is given a scientific name and the scientific name for modern humans is *Homo sapiens*. The first word, *Homo,* is the genus, a classification category that contains many species. The second word is the species name, which may describe the organism. The word *sapiens* refers to a large brain. When species are classified, they are placed in a hierarchy of categories: **species, genus, family, order, class, phylum,** and **kingdom.** This text uses one further classification category, called a **domain.**

Phylogeny is the evolutionary relationship among organisms. Ideally, classification reflects phylogeny in that it tells how organisms are related through evolution and common ancestry. Species in the same genus are more closely related than species in separate genera and so forth as we proceed from genus to domain. Each higher classification category is more inclusive than the one below it; therefore, there are many more species within a kingdom than in a phylum, for example.

Five-Kingdom System

For many years, most biologists favored a five-kingdom classification system consisting of Plantae, Animalia, Fungi, Protista, and Monera. Organisms were placed into these kingdoms on the basis of type of cell (prokaryotic or eukaryotic), level of organization (unicellular or multicellular), and mode of nutrition. In this system, the organisms in the kingdom Monera are distinguished by their structure—they are prokaryotic (lack a membrane-bounded nucleus)—whereas the organisms in the other kingdoms are eukaryotic (have a membrane-bounded nucleus). An evolutionary tree, also called a phylogenetic tree, depicts the relationships among organisms based on how they are classified. Figure 27.22 is an evolutionary tree depicting the five-kingdom system of classification. The tree suggests that protists evolved from monerans and that fungi, plants, and animals evolved from protists via three separate lines of evolution. In an evolutionary tree, two or more groups that separate from the same juncture share the same **common ancestor.** The five-kingdom system of classification suggests that fungi, plants, and animals share the same ancestor, presumably an extinct protist known only from the fossil record.

Three-Domain System

Within the past ten years, new information has called into question the five-kingdom system of classification. The molecule rRNA probably changes slowly during evolution and, indeed, may change when there is a major evolutionary event. Molecular data based on the sequencing of rRNA suggest that there are three domains: **Bacteria, Archaea,** and **Eukarya.**

Cellular data also support the three-domain system. Bacteria and archaea are both unicellular prokaryotes that lack a membrane-bounded nucleus. However, bacteria and archaea are distinguishable from each other on the basis of lipid and cell wall biochemistry. The biochemical attributes of many archaea allow them to live in very hostile environments, including anaerobic swamps, salty bodies of water, and even hot, acidic environments, such as hot springs and geysers. Molecular and cellular data also suggest that the archaea and eukarya are more closely related to each other than either is to the bacteria. The evolutionary tree depicted in Figure 27.23 reflects this evolutionary relationship. It shows that the archaea and the eukarya share a more recent common ancestor than do all three domains.

The **kingdoms Protista, Fungi, Plantae,** and **Animalia,** as illustrated in Figure 1.5, are all placed in the domain Eukarya. Exactly how these kingdoms are related is still being determined.

Phylogenetics

Phylogenetics is the modern way in which organisms are classified and arranged in evolutionary trees. Phylogeneticists arrange species and higher classification categories into clades. **Clades** may be represented on a diagram called a **cladogram.** A clade contains a most recent common ancestor and all its descendant species—the common ancestor is presumed and not identified. Figure 27.24 depicts a cladogram for seven groups of vertebrates. Only the lamprey, the so-called "outgroup," lacks jaws, but the other six groups of vertebrates are in the same clade because they all have jaws, a derived characteristic relative to their ancestors. On the other hand, the vertebrates beyond the shark are all in the same clade because they have lungs, and so forth.

Figure 27.24 is somewhat misleading because, although single traits are noted on the tree, phylogeneticists use much more data to arrange groups of organisms into clades. Phylogeneticists are aided in their endeavor by the computer and any and all available data, including morphological data and DNA sequences. In making decisions, they are often guided

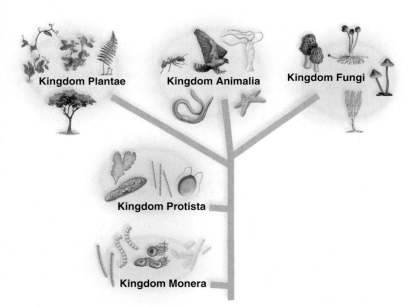

Figure 27.22 Five-kingdom system of classification.
All prokaryotes are in the kingdom Monera. The eukaryotes are in kingdoms Protista, Fungi, Plantae, and Animalia. The evolutionary tree shows the lines of descent. Note that the three domain system of classification (Fig. 27.23) is currently preferred.

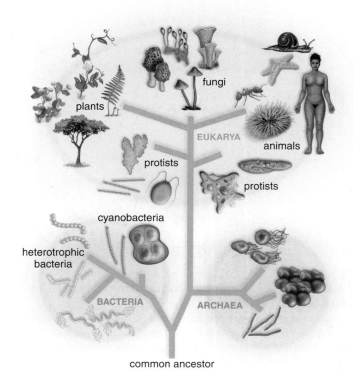

Figure 27.23 The three-domain system of classification.
Representatives of each domain are depicted. The phylogenetic tree shows that domain Archaea is more closely related to domain Eukarya than either is to domain Bacteria.

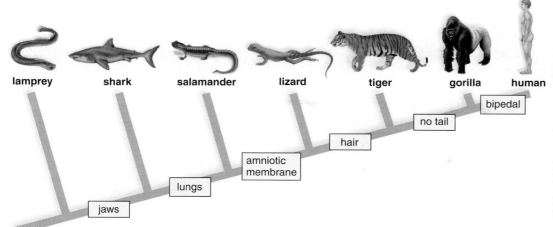

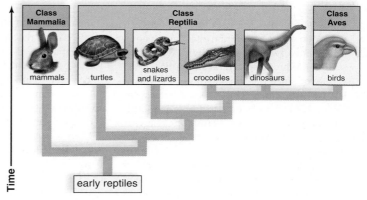

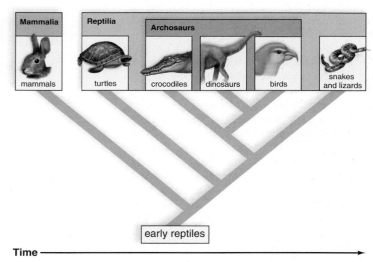

Figure 27.24 Cladogram.
A cladogram gives comparative information about relationships. Organisms in the same clade share the same derived characteristics. Humans and all the other vertebrates shown, except lampreys, are in the same clade as sharks because they all have jaws. However, humans are also alone in a clade because only they are bipedal. This cladogram is simplified because phylogeneticists actually use a great deal more data to construct cladograms.

by the principle of parsimony, which states that the pattern that requires the fewest evolutionary changes is the most likely.

Like phylogeneticists, another group of scientists called "traditionalists" consider descent from a common ancestor when grouping organisms, but they also consider the amount of adaptive evolutionary change. For example, traditionalists place crocodiles (class Reptilia) and birds (class Aves) in separate classes (Fig. 27.25a) because of the adaptive advantage of feathers, despite the fact that they agree with phylogeneticists that these groups share a recent common ancestor. However, phylogeneticists place crocodiles and birds in the same group (Archosaurs, Fig. 27.25b) because of the many traits they share.

Organisms are classified into groups based on their evolutionary relationships. Currently a three domain system replaces the previously preferred five kingdom system of classification. Phylogenetics is used to classify groups of closely related organisms, called clades, into evolutionary trees.

Summarizing the Concepts

27.1 Origin of Life
Chemical reactions are believed to have led to the formation of the first true cell(s). Inorganic chemicals, probably derived from the primitive atmosphere, reacted to form small organic molecules. These reactions occurred in the ocean, either on the surface or in the region of hydrothermal vents deep within.

After small organic molecules such as glucose, amino acids, and nucleotides arose, they polymerized to form the macromolecules. Amino acids joined to form proteins, and nucleotides joined to form nucleic acids. Perhaps RNA was the first nucleic acid. The RNA-first hypothesis is supported by the discovery of ribozymes, RNA enzymes. The protein-first hypothesis is supported by the observation that amino acids polymerize abiotically when exposed to dry heat.

Once a plasma membrane developed, the protocell came into being. Eventually, the DNA → RNA → protein system evolved, and a true cell came into being.

27.2 Evidence of Evolution
The fossil record and biogeography, as well as studies of comparative anatomy, development, and biochemistry, all provide evidence of evo-

Figure 27.25 Traditional versus cladistic view of reptilian phylogeny.
a. According to traditionalists, crocodiles and birds are in separate classes.
b. According to phylogeneticists, crocodiles and birds share a recent common ancestor and should be in the same clade.

lution. The fossil record gives clues about the history of life in general and allows us to trace the descent of a particular group. Biogeography shows that the distribution of organisms on Earth can be influenced by a combination of evolutionary and geological processes. Comparing the anatomy and the development of organisms reveals homologous

HealthFocus

Through Darwin's theory of natural selection, we have come to understand why bacteria become resistant to the antibiotics we use to treat patients. Some people refer to the use of antibiotics as "artificial selection," because humans are involved. Nonetheless, the process is the same as natural selection—antibiotics kill bacteria that are susceptible, but bacterial populations within a single patient are usually so large that resistant individuals are likely to survive and reproduce. Through time, the frequency of resistant individuals increases in the population to the point that a certain antibiotic may no longer be effective. Keep in mind that a patient infected with a resistant strain might require several antibiotics.

Ever since the introduction of antibiotics, resistance has often evolved soon after (Table 27A). An extreme example is methicillin, an antibiotic that it only took one year for bacteria to evolve resistance against! This is the type of "accelerated evolution" you learned about at the beginning of this chapter. Anitbacterial resistance creates the need for even newer antibiotics. The development of a single new antibiotic is estimated to cost between $400 and $500 million. Antibiotic resistance adds $30 billion to annual medical costs in the United States alone!

Some strains of tuberculosis (or TB), a disease caused by bacteria in the genus *Mycobacterium*, are resistant to multiple antibiotics. TB is spread though the air from one person to another, usually when an infected person coughs or sneezes. TB is the most common infectious disease today, infecting about one-third of the world's population, or about 2 billion people. You may not realize it, but TB kills 2–3 million people worldwide each year—right up there with AIDS and malaria! In extreme cases, infections can spread through the lungs, causing lesions and even holes, and eventually leading to death in untreated patients.

When a patient tests positive for TB, he or she is usually put on a six-month course of the antibiotic isoniazid. Many patients feel better after a few weeks on the drug, and some discontinue its use. This selects for resistant strains of TB. Our understanding of evolutionary biology has helped doctors treat patients with TB. In New York City, for example, patients are treated with what is called "direct observation therapy," in which a doctor or nurse actually watches a patient take the medicine. In addition, the initial diagnosis of TB includes testing whether the strain is drug resistant. This way, doctors can treat patients with an effective drug regimen, instead of allowing their infection to persist and spread to other people.

In patients with strains that are resistant to isoniazid, four drugs are recommended for treatment. Multidrug-resistant strains of TB can be very difficult and costly to treat; usually an 18-month course of multiple antibiotics is necessary, and treatment for strains of TB resistant to methicillin alone can cost $50,000 per person!

Discussion Questions

1. In New York City, state health officials have the power to quarantine TB patients who do not take their medicine. That is, they can essentially lock them up for as long as needed (often up to 18 months) to treat their illness. However, some of the medications can have serious side effects. What do you think about this policy?
2. Many people in less-developed countries die from TB, not because their disease is incurable, but simply because they do not have health insurance and cannot afford the medications. Should we in the United States pay more for our medications so that pharmaceutical companies can provide them to lower-income people at a reduced cost or for free?
3. Antibiotics kill bacteria only, not viruses. Knowing what you know about antibiotic resistance and natural selection, when would you prescribe antibiotics if you were a doctor?

TABLE 27A	Dates of Antibiotic Discovery and Resistance	
Antibiotic	**Discovery/ Introduction**	**Resistance**
Penicillin	1928/1943	1946
Sulfonamides	1930s	1940s
Streptomycin	1943/1945	1959
Chloramphenicol	1947	1959
Tetracycline	1948	1953
Erythromycin	1952	1988
Vancomycin	1956	1988/1993
Methicillin	1960	1961
Ampicillin	1961	1972
Cefotaxime/ceftazidime	1981/1985	1983/1984/1988

structures among those that share common ancestry. All organisms have certain biochemical molecules in common, and these chemical similarities indicate the degree of relatedness.

27.3 The Process of Evolution

Microevolution is a process that involves a change in allele frequencies within the gene pool of a sexually reproducing population. The Hardy-Weinberg principle states that gene pool frequencies arrive at an equilibrium that is maintained generation after generation unless disrupted by mutations, genetic drift, gene flow, nonrandom mating, or natural selection. Any change from the

initial allele frequencies in the gene pool of a population signifies that evolution has occurred.

27.4 Speciation

Speciation is the origin of new species. This usually requires geographic isolation, followed by reproductive isolation. The evolution of several species of finches on the Galápagos Islands is an example of speciation caused by adaptive radiation because each one has a different way of life.

Currently, there are two hypotheses about the pace of speciation. Traditionalists support phyletic gradualism—slow, steady change leading to speciation. In contrast, a more recent model, called punctuated equilibrium, proposes that long periods of stasis are interrupted by rapid speciation.

27.5 Classification

Classification involves assigning species to a hierarchy of categories: kingdom, phylum, class, order, family, genus, and species, and in this text, domain. The five-kingdom system of classification recognizes these kingdoms: Monera (the bacteria), Protista (algae, protozoans), Fungi, Plantae, and Animalia. The more recent which is three-domain system (Bacteria, Archaea, and Eukarya), based on molecular data, is currently preferred. Both bacteria and archaea are prokaryotes. Members of the kingdoms Protista, Fungi, Plantae, and Animalia are eukaryotes.

Phylogeneticists classify and diagram the evolutionary relationships among organisms. They use as many characteristics as possible to put species in clades, which are represented on portions of a diagram called a cladogram. A clade contains a most recent common ancestor and all its descendant species, which share the same derived characteristics relative to their ancestors.

Testing Yourself

Choose the best answer for each question.

1. The atmosphere in which life arose lacked
   a. carbon.
   b. nitrogen.
   c. oxygen.
   d. hydrogen.

2. The RNA-first hypothesis for the origin of cells is supported by the discovery of
   a. ribozymes.
   b. proteinoids.
   c. polypeptides.
   d. nucleic acid polymerization.

3. Protocells probably obtained energy as
   a. photosynthetic autotrophs.
   b. chemoautotrophs.
   c. heterotrophs.
   d. None of these are correct.

4. All true cells are able to
   a. replicate DNA.
   b. synthesize sugars.
   c. absorb nutrients.
   d. export minerals.

5. DNA genes may have arisen from RNA genes via
   a. DNA polymerase.
   b. RNA polymerase.
   c. reverse transcriptase.
   d. DNA ligase.

6. Fossils that serve as transitional links allow scientists to
   a. determine how prehistoric animals interacted with each other.
   b. deduce the order in which various groups of animals arose.
   c. relate climate change to evolutionary trends.
   d. determine why evolutionary changes occur.

7. Carbon dating cannot be used to determine the age of dinosaur fossils because
   a. levels of atmospheric carbon were very low when dinosaurs were alive.
   b. dinosaurs contained very low levels of carbon.
   c. dinosaur fossils contain very low levels of carbon.
   d. the half-life of radioactive carbon is too short.

8. Marine animals experience mass extinction approximately every 26 million years. Scientists believe this pattern is due to meteorites that reach Earth because
   a. our solar system shifts its location in the Milky Way in a 26-million-year pattern.
   b. the sun moves far away from Earth every 26 million years.
   c. Earth moves into an asteroid belt every 26 million years.
   d. the moon moves close to Earth every 26 million years.

9. Which of the following groups of organisms might be found on multiple continents? See Table 27.1, and remember that the continents began to move apart about 250 million years ago.
   a. primitive primates
   b. reptiles
   c. birds
   d. Both b and c are correct.
   e. All of these are correct.

10. The flipper of a dolphin and the fin of a tuna are
    a. homologous structures.
    b. homogeneous structures.
    c. analogous structures.
    d. reciprocal structures.

11. Which of the following is not an example of a vestigial structure?
    a. human tailbone
    b. ostrich wings
    c. pelvic girdle in snakes
    d. dog kidney

12. Which of the following is sure to be a population?
    a. grizzly bears of the Rocky Mountains
    b. mosquitoes of the United States
    c. barracuda in the Caribbean Sea
    d. sequoia grove in Sequoia National Park
    e. dandelions of Pennsylvania

13. The frequency of a rare disorder expressed as an autosomal recessive trait is 0.0064. Using the Hardy-Weinberg principle, determine the frequency of carriers for the disease in this population.
    a. 0.147
    b. 0.080
    c. 0.920
    d. 0.020
    e. 0.846

14. Which of the following generally results in a gain in genetic variability?
    a. genetic drift
    b. mutation
    c. founder effect
    d. bottleneck

For questions 15–19, match the description with the appropriate term in the key.

Key:
   a. mutation
   b. natural selection
   c. founder effect
   d. bottleneck
   e. gene flow
   f. nonrandom mating

15. The Northern elephant seal went through a severe population decline as a result of hunting in the late 1800s. As a result of a hunting ban, the population has rebounded but is now homozygous for nearly every gene studied.

16. A small, reproductively isolated religious sect called the Dunkers was established by 27 families that came to the United States from Germany 200 years ago. The frequencies for blood group alleles in this population differ significantly from those in the general U.S. population.

17. Turtles on a small island tend to mate with relatives more often than turtles on the mainland.

18. Within a population, plants that produce an insect toxin are more likely to survive and reproduce than plants that do not produce the toxin.

19. The gene pool of a population of bighorn sheep in the southwest U.S. is altered when several animals cross over a mountain pass and join the population.

20. People who are heterozygous for the cystic fibrosis gene are more likely than others to survive a cholera epidemic. This heterozygote advantage seems to explain why homozygotes are maintained in the human population, and is an example of

    a. disruptive selection.
    b. balanced polymorphism.
    c. high mutation rate.
    d. nonrandom mating.

21. The creation of new species due to geographic barriers is called

    a. isolation speciation.
    b. allopatric speciation.
    c. allelomorphic speciation.
    d. sympatric speciation.
    e. symbiotic speciation.

22. Which of the following models is supported by the observation that few transitional links are found in the fossil record?

    a. phyletic gradualism
    b. punctuated equilibrium
    c. Both a and b are correct.
    d. None of these are correct.

23. Organisms have been placed in the five-kingdom classification system based on

    a. level of organization and mode of nutrition.
    b. genetic diversity and mode of nutrition.
    c. genetic diversity and level of organization.
    d. reproductive traits and mode of nutrition.
    e. reproductive traits and genetic diversity.

24. The three-domain classification system has recently been developed based on

    a. mitochondrial biochemistry and plasma membrane structure.
    b. cellular and rRNA sequence data.
    c. plasma membrane and cell wall structure.
    d. rRNA sequence data and plasma membrane structure.
    e. nuclear and mitochondrial biochemistry.

Understanding the Terms

adaptive radiation 567
allopatric speciation 566
analogous structure 556
Archaea 570
autotroph 552
Bacteria 570
biogeography 556
bottleneck effect 563
clade 570
cladogram 570
class 569
common ancestor 570
continental drift 556
directional selection 564
disruptive selection 564
domain 569
Eukarya 570
evolution 552
family 569
fitness 564
fossil 553
founder effect 563
gene flow 563

gene pool 559
genetic drift 562
genus 569
heterotroph 552
homologous structure 556
kingdom 569
kingdom Animalia 570
kingdom Fungi 570
kingdom Plantae 570
kingdom Protista 570
liposome 552
microevolution 559
microsphere 551
mutation 562
natural selection 564
nonrandom mating 563
order 569
phyletic gradualism 567
phylogenetics 570
phylogeny 569
phylum 569
population 559

postzygotic isolating mechanism 566
prezygotic isolating mechanism 566
protein-first hypothesis 551
proteinoid 551
protocell 550
punctuated equilibrium 569

RNA-first hypothesis 551
speciation 566
species 566, 569
stabilizing selection 564
sympatric speciation 567
transitional link 553
vestigial structure 557

Match the terms to these definitions:

a._____ Process by which populations become adapted to their environment.

b._____ Type of natural selection in which an extreme phenotype is favored, usually in a changing environment.

c._____ An evolutionary model that proposes periods of rapid change dependent on speciation followed by long periods of stasis.

d._____ Structure that is similar in two or more species because of common ancestry.

e._____ Movement of genes from one population to another via sexual reproduction between members of the populations.

Thinking Critically

1. Viruses such as HIV are rapidly replicated and have very high mutation rates; thus, evolution of the virus can be observed in a single infected person. Using what you have learned in this chapter, explain why HIV is so hard to treat, even though multiple drugs to treat HIV have been developed.

2. If the conditions for the Hardy-Weinberg principle are rarely, if ever, met in nature, why is it such an important idea?

3. You observe a wasting disease in cattle that you know is genetically caused and thus heritable. The disease is fatal in young cattle. The allele frequency for the gene that causes the disease is 0.05 in the United States, but 0.35 in South America. Explain why such a difference in allele frequencies might exist.

4. Why are homologous structures, as opposed to analogous ones, used to determine the evolutionary relationships of species and to reconstruct phylogenies?

5. Why do scientists continue to devise experimental systems that mimic the conditions of the early Earth, despite the fact that it is difficult to do and there is no way to know for sure what the conditions on Earth were billions of years ago? Are you aware of any such experiments and their results?

ARIS, the *Inquiry into Life* Website

ARIS, the website for *Inquiry into Life,* provides a wealth of information organized and integrated by chapter. You will find practice quizzes, interactive activities, labeling exercises, flashcards, and much more that will complement your learning and understanding of general biology.

www.aris.mhhe.com

What do Doc Holliday (legendary gunslinger), Vivian Leigh (Scarlett O'Hara in Gone with the Wind), and Jimmie Rodgers (the "Father of Country Music") have in common? They all died of tuberculosis (TB), an infectious disease caused by the bacterium Mycobacterium tuberculosis. *Prior to the development of effective drugs to kill this microscopic organism, TB was also know as "consumption" because it seemed to consume people from within as they coughed up bloody sputum, struggled to breathe, and wasted away. During that time, about 75% of people with the disease died within five years of becoming infected. Today the disease can usually be treated successfully, if appropriate antibacterial drugs are available. However, TB remains one of the most common infectious diseases in the world, and approximately 10% of the M. tuberculosis bacteria isolated from these patients have become resistant to the antibiotics most commonly used to treat it. Moreover, because the human immunodeficiency virus attacks the very cells of the human immune system that are most essential for controlling TB, the HIV epidemic has fueled a resurgence of TB in many parts of the world.*

In this chapter you will learn about the bacteria, protozoa, fungi, and viruses that occupy the vast microbial world. Some, like M. tuberculosis (see micrograph) and HIV, are best known for causing deadly diseases. The great majority, however, are beneficial, and even essential to life on Earth.

chapter 28

Microbiology

Chapter Concepts

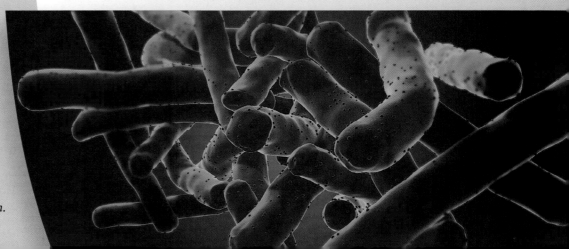

28.1   The Microbial World

Antonie van Leeuwenhoek (1632–1723) was a Dutch tradesman and scientist. He was apparently very skilled at working with glass lenses, which enabled him to greatly improve the microscopes that already existed in the 1600s. Using these instruments, he was among the first to view microscopic life forms in a drop of water, which he called "animalcules" and described in this way:

> [They] were incredibly small, nay so small, in my sight, that I judged that even if 100 of these very wee animals lay stretched out one against another, they could not reach to the length of a grain of coarse Sand.

Leeuwenhoek and others after him believed that the "wee animals" he had observed could arise spontaneously from inanimate matter. For about 200 years, scientists carried out various experiments to determine the origin of microscopic organisms in laboratory cultures. Finally, in about 1859, Louis Pasteur devised the experiment shown in Figure 28.1. In the first experiment, when flasks containing sterilized broth were exposed to either outdoor or indoor air, they often became contaminated with microbial growth. In the second experiment, however, if the neck of the flask was curved so that microbes could not enter the broth from the air, no growth occurred. Thus, the bacteria were not arising spontaneously in the broth, but rather were coming from the air. In 1884, Pasteur also suggested that something even smaller than a bacterium was the cause of rabies, and it was he who chose the word *virus* from a Latin word meaning poison. Pasteur went on to develop a vaccine for rabies, which he used to save the life of a young French boy who had been bitten by a rabid dog.

Microbiology is the study of microbes, a term that includes the viruses and also the microorganisms called bacteria, archaea, protists, and some fungi. Most, but not all, of these organisms are so small that they require a microscope to be seen. Today we know that bacteria and other microbes are incredibly numerous in air, water, soil, and on objects. A single spoonful of soil can contain 10^{10} bacteria, and the total number of bacteria on Earth has been estimated at around 10^{30}, which clearly exceeds the numbers of any other type of organism on Earth. It is a good thing bacteria are microscopic—if they were the size of beetles, the Earth would be covered in a layer of bacteria several miles deep!

We only need to turn on the nightly TV news to hear about diseases caused by microbes, which in this case are often called *pathogens*. AIDS, SARS, and bird flu are recent examples of "emerging" infectious diseases (see Health Focus on page 599). However, many microbes have beneficial—and sometimes essential—roles to play in human

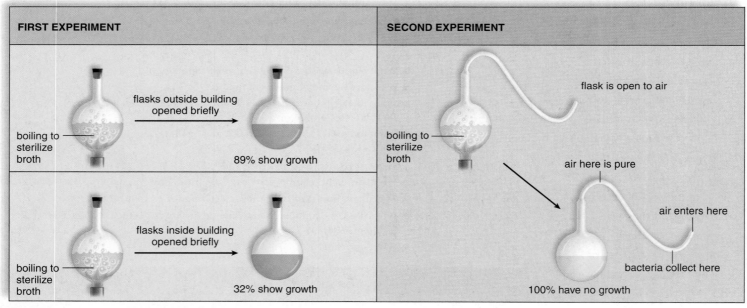

Figure 28.1 Pasteur's experiments.
Pasteur disproved the theory of spontaneous generation of microbes by performing these types of experiments.

health as well as in the proper functioning of the biosphere. For example, you have approximately 10^{14} bacteria living on or in your body right now, even though your body is composed of about 10^{13} human cells! In other words, you have ten times as many bacterial cells as human cells (although the bacterial cells are much smaller). Most of these microbes, also known as the **normal microflora,** or microbiota, have beneficial effects. Bacteria that live on your skin help crowd out harmful microbes that might grow in those areas, and bacteria in the intestines aid in digestion and synthesize vitamin K and vitamin B_{12}. Our bodies are actually a finely balanced ecosystem for microbes.

Bacteria and fungi are the **decomposers** that play essential roles in various nutrient cycles on Earth by breaking down organic and inorganic materials that can be reused by plants and animals. Some bacteria perform photosynthesis, often in ways completely different from plant photosynthesis. We have also learned to use bacteria to clean up our environment. Wastewater that leaves your bathroom and kitchen is probably cleaned at a treatment plant by processes that rely on the activity of bacteria. It seems that bacteria can consume virtually any substance found on Earth. Soil contaminated by oil spills or just about any other toxic compounds can be cleaned by encouraging the growth of bacteria that eat these compounds.

Not only do bacteria play essential roles in the environment, but they are also valuable for industrial processes, particularly food processing. Also, most antibiotics known today were first discovered in soil bacteria or fungi. Genetic techniques can be used to alter the products generated by bacterial cultures. Biotechnology is the development and exploitation of these organisms. A variety of valuable products can be made in this way, including insulin and vaccines. Overall, although it might not seem so to an individual suffering from tuberculosis, AIDS, or some other serious infectious disease, most microbes are far more beneficial than harmful.

> Microbes include the viruses and also the microorganisms called bacteria, archaea, protists, and some fungi. Most, but not all, microbes are microscopic, and some are the most numerous organisms on Earth. Some microbes are pathogens that cause disease, but many more are beneficial to humans and to the environment.

28.2 Bacteria and Archaea

Both bacteria (domain Bacteria) and archaea (domain Archaea) are prokaryotes, but each is placed in its own domain because of molecular and cellular differences. Prokaryotes do not have nuclei or the membrane-bound cytoplasmic organelles found in eukaryotic cells.

Biology of Bacteria

Bacteria have three basic shapes: rod (bacillus, pl., bacilli); spherical (coccus, pl., cocci); and spiral or helical (spirillum, pl., spirilla) (Fig. 28.2). A bacillus or coccus can occur singly or may occur in particular arrangements. For example, when cocci form a cluster, they are called staphylococci, while cocci that form chains are called streptococci.

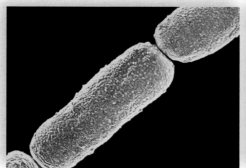

a. Bacilli (rod):   SEM 35,000×
   Bacillus anthracis

b. Cocci (spherical):   SEM 3,520×
   Streptococcus thermophilus

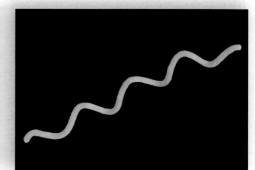

c. Spirillum (curved):   SEM 6,250×
   Spirillum volutans

Figure 28.2 Typical shapes of bacteria.
a. Bacilli, rod-shaped bacteria. **b.** Cocci, round bacteria. **c.** Spirillum, a spiral-shaped bacterium.

The typical structure of a bacterium is shown in Figure 28.3. All bacterial cells have a plasma membrane, which is a lipid bilayer, similar to the plasma membrane in plant and animal cells. Most bacterial cells are further protected by a cell wall that contains the unique molecule peptidoglycan. Bacteria can be classified by differences in their cell walls, which are detected using a staining procedure devised more than 100 years ago by Hans Christian Gram. When you go to the doctor for a bacterial infection, one of the most common tests performed is the **Gram stain,** because different antibiotics tend to be more effective against Gram-positive or Gram-negative bacteria. Cell walls that have a thick layer of peptidoglycan outside the plasma membrane stain purple with the Gram stain procedure, and are called Gram-positive bacteria. If the peptidoglycan layer is either thin or lacking altogether, the cells stain pink and are considered Gram-negative. In addition to their plasma membrane, Gram-negative bacteria have an outer membrane that contains molecules called **lipopolysaccharides (LPS).** When these Gram-negative cells are killed by your immune system, LPS molecules are released, stimulating inflammation and fever. Beyond the cell wall, some bacteria have a slimy polysaccharide capsule that can protect the cell from dehydration and the immune system.

Motile bacteria have **flagella** for locomotion but never cilia. The flagellum is a stiff, curved filament that rotates like a propeller. Bacterial flagella are structurally distinct from eukaryotic flagella. Some bacteria have fimbriae that bind to specific surface receptors of cells. For example, bacteria that cause urinary tract infections can bind to urinary tract cells. Drinking cranberry juice seems to inhibit this binding.

Most bacteria have a single circular chromosome, which is located in a **nucleoid** region, rather than in a membrane-bounded nucleus. Many bacteria also harbor accessory rings of DNA called plasmids, which can carry genes for antibiotic resistance, among other things, and are commonly used to carry foreign DNA into other bacteria during genetic engineering (see Chapter 24). Bacterial cells also contain abundant ribosomes, as well as various types of granules that store nutrients such as glycogen and lipids. Notice that bacteria do not have an endoplasmic reticulum, a Golgi apparatus, mitochondria, or chloroplasts. This diagram summarizes the major structural features of bacteria:

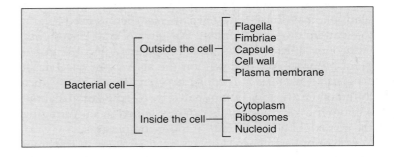

Bacterial Reproduction

Bacteria reproduce asexually. After a period of sufficient growth, the bacterial cell simply divides into two new cells, with each cell getting a copy of the genome and approximately half of the cytoplasm. This process is known as **binary fission** (Fig. 28.4). Each daughter cell is a *clone,* or exact copy, of the parent cell. When bacteria are spread out and grown on agar plates, each individual cell can give rise to a colony of millions of cells. Each of these daughter cells is a clone of the initial cell. Some bacteria need only 20 minutes to reproduce, while others grow more slowly, with generation times of a day or more.

Under unfavorable conditions, some bacteria (mainly Gram-positives) can produce resistant structures called **endospores,** thick-walled, dehydrated structures capable of surviving the harshest conditions, perhaps even for thousands of years. Endospores are *not for reproduction,* but simply a way to survive unfavorable conditions.

Sexual reproduction does not occur among prokaryotes, but at least three means of genetic recombination have been observed. **Conjugation** takes place when the so-called male cell passes DNA to the female cell by way of a sex pilus. **Transformation** occurs when a bacterium takes up DNA released into the medium by dead bacteria. During **transduction,** viruses carry portions of bacterial DNA from one bacterium to another.

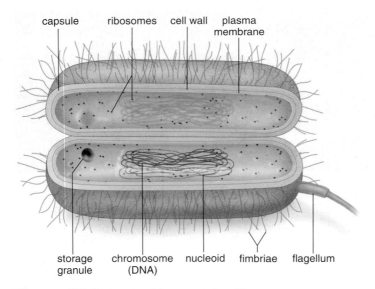

Figure 28.3 Typical bacterial cell.
Bacteria are prokaryotes and thus have no membrane-bound nuclei or organelles.

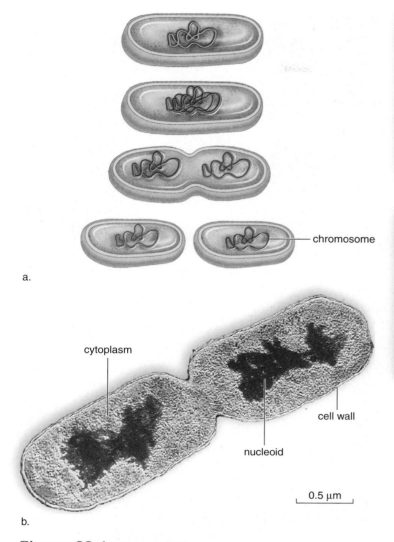

a.

b.

0.5 μm

Figure 28.4 Binary fission.

a. A diagram and (**b**) a micrograph illustrate binary fission, a process that occurs when a bacterium reproduces. Bacteria grow and the chromosome is duplicated before binary fission occurs.

Bacterial Metabolism

Although most bacteria are structurally similar, they demonstrate a remarkable range of metabolic abilities. Most bacteria are **heterotrophic** and require an outside source of organic compounds in the same way that animals do. Some heterotrophic bacteria are anaerobic and cannot use oxygen to capture electrons at the end of their electron transport chain (see Chapter 7). Instead, they use a variety of substances as final electron acceptors. Sulfate-reducing bacteria transfer the electrons to sulfate, producing hydrogen sulfide, which smells like rotten eggs. Denitrifying bacteria use nitrate, while others use minerals, such as iron or manganese, as electron receivers.

Other bacteria are **chemoautotrophs.** They reduce carbon dioxide to an organic compound by using ener-

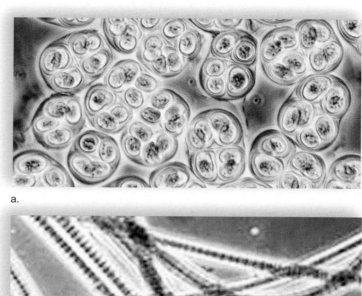

a.

b.

Figure 28.5 Cyanobacteria.

a. In *Chroococcus*, single cells are grouped in a common gelatinous sheath. **b.** Filaments of cells occur in *Oscillatoria*, a common inhabitant in freshwater environments.

getic electrons derived from chemicals, such as ammonia, hydrogen gas, and hydrogen sulfide. Electrons can also be extracted from certain minerals, such as iron.

Some bacteria are photosynthesizers that use solar energy to produce their own food. **Cyanobacteria** are believed to have arisen some 3.8 billion years ago and to have produced much of the oxygen in the atmosphere at that time. Also called blue-green algae, these organisms contain green chlorophyll and have other pigments that give them a bluish-green color (Fig. 28.5). Other bacterial photosynthesizers split hydrogen sulfide, instead of water, in anaerobic environments. Therefore, they produce sulfur rather than oxygen as a by-product of photosynthesis.

Aside from producing oxygen, cyanobacteria are often the first colonizers of rocks. Many are capable of both carbon fixation and nitrogen fixation, needing only minerals, air, sunlight, and water for growth. Cyanobacteria are also notorious for forming toxic "blue-green algae" blooms in waters enriched by nutrients, which may require that some lakes be closed to human use. Cyanobacteria (and in some cases eukaryotic algae) form a symbiotic relationship with fungi in a lichen. The fungi provide a place for the cyanobacteria to grow and obtain water and mineral nutrients. Some cyanobacteria even appear to live in a symbiotic relationship in the fur of sloths, providing the animals with a form of camouflage in their dense green forest home!

> Bacteria are prokaryotes, meaning they lack a true nucleus and membrane-bound organelles. Most have a single chromosome. Although they reproduce asexually, many can exchange DNA through genetic recombination. Although they are structurally similar, bacteria are metabolically very diverse.

Bacterial Diseases in Humans

Most types of bacteria do not cause disease, but a significant number do. Why does one microbe cause disease, while another, closely related species is completely harmless? Pathogenic microbes often carry genes that code for specific virulence factors that determine the type and extent of illness they are capable of causing. For example, right now you have billions of *E. coli* living in your large intestine, without causing any problems. However, some strains of *E. coli* have acquired genes that make them dangerous invasive pathogens. The strain of *E. coli* called O157:H7 has the ability to generate a toxin that damages the lining of the intestine, resulting in a bloody diarrhea. It has also obtained other virulence factors that help it stick to the lining of the gut more efficiently. In 2006, around 200 people became ill and at least three died from *E. coli* O157:H7 contamination in bagged spinach.

By acquiring genes coding for virulence factors, a "good" bacterium can be converted into a "bad" bacterium. The genes coding for virulence factors frequently move between individual bacteria, sometimes even when they are of a different species or genus. Antibiotic resistance genes often pass between organisms in the same ways.

Here we discuss a few of the most common types of bacterial diseases in humans.

Streptococcus Infections

More different types of human disease are caused by bacteria from the genus *Streptococcus* than by any other type of bacterium. *Streptococcus pneumoniae* can cause pneumonia, meningitis, and middle ear infections. Up to 70% of apparently healthy adults are carriers of this potentially harmful bacterium, which causes disease mainly in children and the elderly. *Streptococcus mutans* is found on the teeth and contributes to tooth decay and the formation of dental caries.

The most common illness caused by streptococci is pharyngitis, commonly called strep throat, which is usually due to infection by *Streptococcus pyogenes* (Fig. 28.6a). *S. pyogenes* also causes relatively mild skin diseases, such as impetigo in infants (Fig. 28.6b). A number of more serious and invasive diseases can occur after or during infections with *S. pyogenes*. The nature and severity of these diseases depend on which

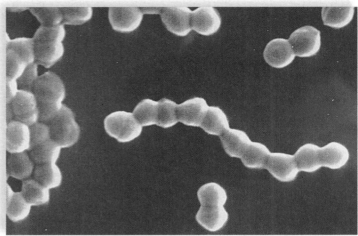

a. *Streptococcus pyogenes*   0.5 μm

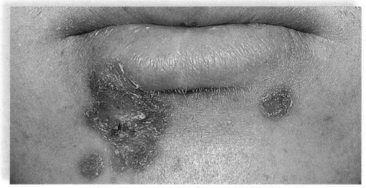

b. Impetigo

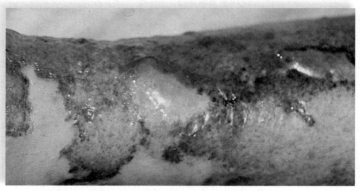

c. Flesh-eating disease

Figure 28.6 *Streptococcus pyogenes*.
a. *Streptococcus pyogenes*, shown here in an electron micrograph, causes more types of human disease than any other. **b.** Impetigo is a common mild skin infection. **c.** Flesh-eating disease is rare and life-threatening.

virulence factors the pathogen carries. Scarlet fever is a strep infection caused by a *Streptococcus* strain that produces a toxin causing a red rash. An intense immune response to the infection can lead to rheumatic fever, which is characterized by a high temperature, swollen joints that form nodules, and heart damage. Although rarely seen in the United States today, rheumatic fever killed more school-age children than all other diseases combined in the early twentieth century.

By releasing enzymes that destroy connective and muscle tissues and kill cells, *Streptococcus pyogenes* infection can lead to large-scale cell lysis and tissue destruction. So-called "flesh-eating" bacteria cause necrotizing fasciitis (Fig. 28.6c), in which rapid tissue damage can require amputation of infected limbs or, in 40% of cases, rapid death.

Tuberculosis

Tuberculosis (TB) is one of the leading worldwide causes of death due to infectious disease. When first described by the father of medical microbiology, Robert Koch, in 1882, TB caused one of every seven deaths in Europe, and one-third of all deaths of young adults. Today, it is estimated that one-third of the world's population has TB, causing approximately 2 million deaths each year. Tuberculosis is a chronic disease caused by *Mycobacterium tuberculosis,* a pathogen closely related to the causative agent of leprosy. Generally, *M. tuberculosis* infection occurs in the lungs, but it can occur elsewhere as well.

M. tuberculosis is very slow growing, and the extent of the disease is determined by host susceptibility. An immune response results in inflammation of the lungs. Active lesions can produce dense structures called **tubercles** (Fig. 28.7). These can persist for years, causing symptoms such as coughing and spreading the bacteria to other areas of the lungs and body as well as to other individuals. Eventually, the damaged lung tissue hardens and calcifies, leaving characteristic spots that can be observed with

chest X rays. In recent years, the incidence of tuberculosis is on the rise in the United States, mainly because the bacteria have developed drug resistance. Worldwide, the AIDS pandemic is probably the most important factor, since HIV damages the helper T cells that are important in fighting the TB bacterium. An estimated one-third of the nearly 40 million people infected with HIV worldwide are also infected with *M. tuberculosis.*

Food Poisoning

Whether the source was mishandled food at a salad bar, or potato salad that sat in the sun too long at a picnic, all of us have probably experienced the intestinal discomfort known as food poisoning. Two basic types of bacteria cause food poisoning: those that cause infections once they are in the intestine, and those that produce toxins while they are growing in food.

Salmonella is a classic example of a food poisoning agent that does not produce gastroenteritis until it has reproduced in the intestines for several days. On the other hand, *Staphylococcus* acts more quickly because it produces toxins that taint food. *Clostridium botulinum,* the causative agent of *botulism,* produces, perhaps, the most toxic substance on Earth. When people are canning and don't heat foods to a high enough temperature, and under sufficient pressure, *Clostridium* can produce endospores that survive the canning process. When the spores germinate in the airless environment of the can or bottle, they become toxin-producing cells.

Chlamydia Infections

Chlamydia trachomatis is a small intracellular parasite that causes a variety of human diseases, including trachoma, the leading cause of preventable blindness in the world. Over 20 million people each year become blind because of trachoma. Infections occur by direct contact or by contact with items such as clothing and towels. Repeated infection leads to chronic conjunctivitis and scarring. Left untreated, the eyelid curls inward, allowing the eyelashes to scrape the cornea covering the eyeball. Damage and scarring of the cornea lead to blindness. Chlamydia is also one of the most common sexually transmitted diseases in the United States. As many as 18% of American women have vaginal chlamydial infections, most of which go undetected. In some cases, the infection can develop into life-threatening pelvic inflammatory disease.

Drug Control of Bacterial Diseases

Antibiotics kill or inhibit bacteria by interfering with their unique metabolic pathways. Therefore, they are not expected to harm human cells. Some antibiotics, such as erythromycin and tetracyclines, inhibit bacterial protein synthesis by taking advantage of differences between bacterial and human ribosomes. Others, such as penicillins and cephalosporins, inhibit bacterial cell wall biosynthesis.

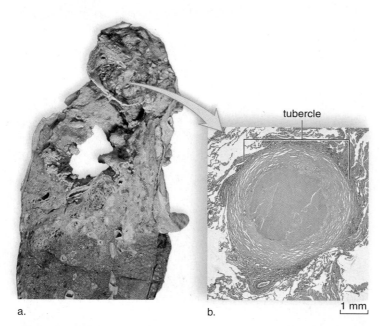

Figure 28.7 Tuberculosis.
Tuberculosis, an infection caused by the bacterium *Mycobacterium tuberculosis,* usually settles in the lungs. **a.** Lung tissue with a large cavity surrounded by tubercles, which are hard, calcified nodules where the bacteria are trapped. **b.** Photomicrograph shows a cross section of one large tubercle, with a diseased center.

There are problems associated with antibiotic therapy. Some individuals are allergic to certain antibiotics, and the reaction may even be fatal. Antibiotics not only kill off disease-causing bacteria, but they may also reduce the number of beneficial bacteria in the intestinal tract and vagina. This can lead to vaginal yeast infections, or even to the overgrowth of yeast in the oral cavity, a condition known as thrush.

Most important perhaps is the growing resistance of bacteria to antibiotics, as discussed on page 572. Antibiotics were introduced in the 1940s, and for several decades they worked so well it appeared that infectious diseases had been brought under control. However, we now know that bacterial strains can mutate and become resistant to a particular antibiotic. Worse yet, when bacteria swap bits of DNA, resistance can pass between different bacteria. Penicillin and tetracycline now have a failure rate of more than 20% against certain strains of gonococci, which cause gonorrhea. Bacteria with multiple drug resistances are a serious threat in hospital settings, where they thrive, particularly by causing wound infections in surgical patients.

Until recently, most pharmaceutical companies had concluded that the dozens of antibiotics in their arsenal were sufficient, and stopped focusing on developing new antibiotics. With the rise of antibiotic resistance, however, many companies are now working to develop innovative kinds of antibacterial therapies. The need for new, potent antibiotics is especially critical since 9/11. (See the Bioethical Focus at the end of the chapter.)

> **Bacterial diseases of humans include various streptococcal infections, tuberculosis, food poisoning, and chlamydial infections. Antibiotics are drugs that selectively kill or inhibit bacteria; however, the development of antibiotic resistance is becoming an increasingly serious problem.**

Biology of Archaea

Archaea and bacteria are not close relatives, even though both are prokaryotes; therefore, they are placed in separate domains. Based on a number of criteria, including nucleic acid similarities, archaea may be more closely related to

TABLE 28.1	Comparison of Domains Archaea and Eukarya	
	DOMAIN	
Feature	**Archaea**	**Eukarya**
Nucleus	No	Yes
Organelles	No	Yes
Introns	Sometimes	Yes
Histones	Yes	Yes
RNA polymerase	Several types	Several types
Methionine is at start of protein synthesis	Yes	Yes

a.

b.

c.

Figure 28.8 Extreme habitats.

a. Halophilic archaea can live in salt lakes. **b.** Thermophilic archaea can live in the hot springs of Yellowstone National Park. **c.** Methanogens live in swamps and in the guts of animals.

eukarya, even though the prokaryotic cell is much different from the eukaryotic cell (Table 28.1).

Most of the **archaea** studied to date live in extreme environments (Fig. 28.8). It has been suggested that archaea are more like ancient forms of life than are bacteria, due to the ability of archaea to live in thermal and anaerobic environments that mimic conditions on early Earth. Archaea have remarkable strategies for thriving in hyperthermal areas, acidic environments, hypersaline habitats, and deeply anaerobic locations. These abilities are reflected in the major groups of archaea: *thermoacidophiles* (high temperature and low pH environments); *methanogens* (anaerobic environments); and *halophiles* (salty environments). Nevertheless, in recent years, it has become evident that archaea are also found in large numbers in less extreme environments. Archaea have been isolated from the human colon; however, no archaea have been proven to be associated with any human disease.

Archaeal Structure

Archaea tend to be 0.1 to 15 microns in size. Their DNA genome is a single, closed, circular molecule, often smaller than a bacterial genome. Archaea that stain Gram-positive have a thick polysaccharide cell wall; those that stain Gram-negative have a protein or glycoprotein surface layer. The plasma membrane of archaea differs markedly from those of bacteria and eukaryotes. Rather than a lipid bilayer, archaea often have a monolayer of lipids with branched side chains. Their chemical characteristics make archaea tolerant to acid and heat. Like bacteria, archaea reproduce asexually by binary fission.

Archaeal Metabolism

Some archaea are heterotrophs, while others are autotrophs. Many halophiles perform a unique type of photosynthesis by utilizing a pigment similar to retinal in human eyes. When the pigment absorbs solar energy, it moves a hydrogen ion to outside the plasma membrane. This establishes an H^+ gradient that promotes ATP sythesis.

Among archaea, the **methanogens** can reduce carbon dioxide to methane. Methanogens are found in anaerobic environments such as swamps, lake sediments, rice paddies, and the intestines of animals. Cows, which have large populations of methanogens in their digestive tracts, release a significant amount of methane into the environment. Since methane is second only to carbon dioxide as a greenhouse gas that contributes to global warming, some scientists are suggesting that the amount of methane produced by cattle should be reduced by changing the animals' diets, adding substances to their feed to alter the microbial populations, or developing a vaccine to specifically inhibit the growth of methanogens. However, the relative contribution of cattle, compared to other potential sources of methane, remains controversial.

The archaea are prokaryotes, but they may be more closely related to the eukarya than to bacteria. Many archaea live in extreme environments, similar to those that may have been found on early Earth. Archaea have a unique plasma membrane structure that helps them resist acid and heat. The methanogenic archaea may contribute significantly to global warming by producing methane, a greenhouse gas.

28.3 Protists

Protists (domain Eukarya, kingdom Protista) are eukaryotes, and they have various organelles. For the most part, protists are unicellular and microscopic, which sets them apart from the multicellular plants, animals, and fungi that are also in domain Eukarya.

Unlike the microscopic prokaryotes, protists are structurally diverse, having many complex shapes. Their diversity also extends to various lifestyles and modes of reproduction. Most carry out asexual reproduction, with sexual reproduction as an option, particularly under unfavorable conditions. In protists, sexual reproduction often results in a spore or cyst that can survive until environmental conditions improve.

Protists offer us a glimpse into the past because they are most likely related to the first eukaryotic cell to have evolved. According to the endosymbiotic hypothesis, mitochondria may have resulted when a nucleated cell engulfed aerobic bacteria, and chloroplasts may have originated when a nucleated cell engulfed cyanobacteria (see Fig. 3.15). They also bridge the gap between the first eukaryotic cells and multicellular organisms. Some protists are not only multicellular but also have differentiated tissues.

It is difficult to tell how the various types of protists are related, and some scientists classify them into several different kingdoms, instead of just one as we will do. For the sake of discussion, we will group protists according to their modes of nutrition: the algae are photosynthetic, the protozoans are heterotrophic by ingestion, and the water molds are heterotrophic by absorption.

Biology and Diversity of Algae

Algae can be unicellular or form colonies or filaments. The unicellular and colonial forms may have flagella. Some types of algae are multicellular seaweeds, such as kelp. Unicellular types can be as small as bacteria (about 1 micron), while multicellular forms can grow as large as hundreds of feet. In aqueous environments, small algae are often a component of suspended, photosynthesizing organisms termed **phytoplankton.** Algae are important in terrestrial ecosystems as well, being found in soils, on

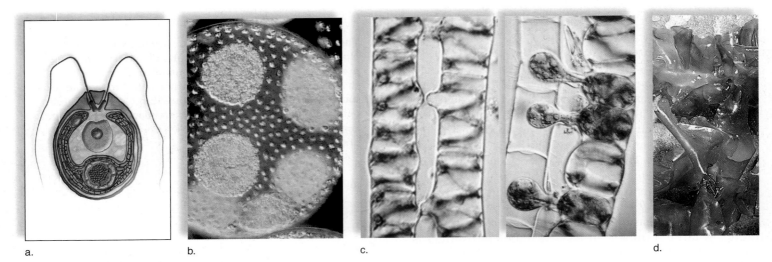

a.   b.   c.   d.

Figure 28.9 Representative green algae.

a. *Chlamydomonas* is a motile, unicellular green alga. **b.** *Volvox* is a colony of flagellated cells. **c.** *Spirogyra* is a filamentous green alga in which each cell has a ribbonlike chloroplast. During conjugation, the cell contents of one filament enter the cells of another filament. Zygote formation follows. **d.** *Ulva* is a multicellular green alga known as sea lettuce.

rocks, and on trees. In addition, algae contribute to the formation of coral reefs and partner with fungi in lichens (see Fig. 28.27).

Algae have chloroplasts that contain chlorophyll and sometimes other pigments. Algae perform plantlike photosynthesis. Oxygen is released from water, and electrons energized by the sun are used to reduce carbon dioxide to carbohydrate. Thus, algae are important primary producers of food for heterotrophs and provide oxygen to the atmosphere. **Pyrenoids** are organelles found in algae that are active in starch storage and metabolism.

Algae generally have a rigid cell wall, sometimes made of cellulose. Beyond the cell wall, many algae produce a slime layer that can be harvested and used in foods.

Green Algae

Green algae are believed to be closely related to the first plants, because both groups have some of the same characteristics: (1) a cell wall that contains cellulose, (2) chlorophylls *a* and *b,* and (3) storage of reserve food as starch inside the chloroplast. Green algae are symbiotic with other organisms, particularly fungi and lichens.

The many different types of green algae include those that are unicellular, colonial, or multicellular (Fig. 28.9). *Chlamydomonas* is a unicellular green alga with a cup-shaped chloroplast and two whiplike flagella. *Volvox* is a colony (loose association of cells) in which thousands of flagellated cells are arranged in a single layer surrounding a watery interior. Each cell resembles a *Chlamydomonas. Spirogyra* is filamentous, consisting of end-to-end chains of cells. A ribbonlike chloroplast is arranged in a spiral within each cell. *Spirogyra* can repro-

duce by conjugation, a process whereby physical contact between cells allows for the transmission of genetic material. Multicellular *Ulva* is commonly called sea lettuce because of its leafy appearance.

Diatoms

Diatoms are the most numerous unicellular algae in the oceans, and they are also common in fresh water. In addition to chlorophyll, they utilize a brown pigment to capture solar energy. Diatoms constitute a large fraction of the phytoplankton and are important primary producers at the base of marine food chains. Diatoms are easy to recognize under the microscope because they have a wide variety of elaborate shells made of silica (Fig. 28.10). The two overlapping shells have intricately shaped depressions, pores, and passageways that bring the diatom's plasma membrane in contact with the environment. The fossilized remains of diatoms, also called diatomaceous earth, accumulate on the ocean floor and can be mined for use as filtering agents and abrasives.

Dinoflagellates

Dinoflagellates are best known for the **red tide** they cause when they greatly increase in number, an event called an algal *bloom. Gonyaulax,* one species implicated in red tides, produces a very potent toxin (Fig. 28.11). This is harmful by itself, but it also accumulates in shellfish. The shellfish are not damaged, but people can become quite ill when they eat them.

Despite some harmful effects, dinoflagellates are important members of the phytoplankton in marine and freshwater ecosystems. Generally, dinoflagellates are photosynthetic,

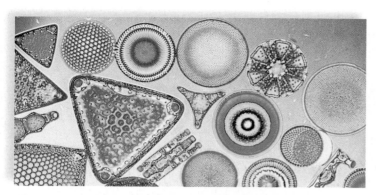

Figure 28.10 Assorted fossilized diatoms.
The overlapping shells of a diatom are made of silica. Scientists use their delicate markings to identify the particular species.

although colorless heterotrophs are known to live as symbionts inside other organisms. Corals, which build coral reefs, contain large numbers of symbiotic dinoflagellates.

Many dinoflagellates have protective cellulose plates that become encrusted with silica, turning them into hard shells. Dinoflagellates have two flagella. One is located in a groove that encircles the protist, and the other is in a longitudinal groove and has a free end. The arrangement of the flagella makes a dinoflagellate whirl as it moves; in fact, the name dinoflagellate is derived from the Greek *dino-*, for whirling.

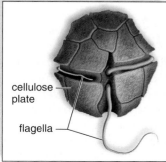

cellulose plate

flagella

Figure 28.11 Dinoflagellates.
Dinoflagellates have cellulose plates; these belong to *Gonyaulax*, a dinoflagellate that contains a reddish-brown pigment and is responsible for occasional "red tides."

It is interesting to note that luminous dinoflagellates produce a twinkling light that can give seas a phosphorescent glow at night, particularly in the tropics.

Red Algae

Although unicellular forms are known, **red algae** are mainly multicellular, ranging from simple filaments to leafy structures. As seaweeds, red algae often resemble the brown algae seaweeds, although red algae can be more delicate (Fig. 28.12).

In addition to chlorophyll, the red algae contain red and blue pigments that give them characteristic colors. Coralline algae have calcium carbonate in their cell walls and contribute to the formation of coral reefs.

Red algae produce a number of useful gelling agents. Agar is commonly used in microbiology laboratories to solidify culture media and commercially to encapsulate vitamins or drugs. Agar is also a gelatin substitute for vegetarians, an antidrying agent in baked goods, and an additive in cosmetics, jellies, and desserts. Carageenan is a related product used in cosmetics and chocolate manufacture. *Porphyra,* a red seaweed, is popular as a sushi wrap in Japan.

Brown Algae

Brown algae are the conspicuous multicellular seaweeds that dominate rocky shores along cold and temperate coasts. The color of brown algae is due to accessory pigments that actually range from pale beige to yellow-brown to almost black. These pigments allow the brown algae to extend their range down into deeper waters because the pigments are more efficient than green chlorophyll in absorbing the sunlight away from the ocean surface. The alga produces a slimy matrix that retains water when the tide is out and the seaweed is exposed. This gelatinous material, algin, is used in ice cream, cream cheese, and some cosmetics.

Figure 28.12 Red alga.
Red algae, represented by *Chondrus crispus*, are smaller and more delicate than brown algae.

Figure 28.13 Brown alga.
Brown algae form multicellular seaweeds. This is the giant marine kelp *Macrocystis pyrifera*.

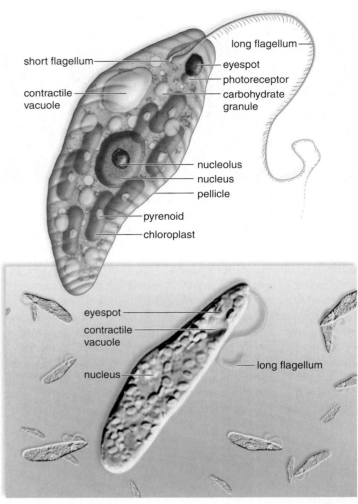

LM 200×

Figure 28.14 *Euglena*.
In *Euglena*, a very long flagellum propels the body, which is enveloped by a flexible pellicle.

The Sargasso Sea in the North Atlantic Ocean commonly has large floating mats of brown algae called sargassum. The most familiar brown algae are kelp (genus *Laminaria*) in coastal regions and giant marine kelp (genus *Macrocystis*) in deeper waters (Fig. 28.13). Tissue differentiation in kelp results in blades, stalks, and holdfasts, which are analogous to the leaves, stems, and roots of plants.

Euglenoids

Euglenoids are freshwater unicellular organisms that typify the problem of classifying protists. Many have chloroplasts, but some do not. Those that lack chloroplasts ingest or absorb their food. Those that have chloroplasts are believed to have originally acquired them by ingestion and subsequent endosymbiosis of a green algal cell. Three, rather than two, membranes surround these chloroplasts. The outermost membrane is believed to represent the plasma membrane of an original host cell that engulfed a green alga.

Euglenoids have two flagella, one of which is typically much longer than the other and projects out of the anterior, vase-shaped invagination (Fig. 28.14). Near the base of this flagellum is an eyespot, which shades a photoreceptor for detecting light. Because euglenoids are bounded by a flexible pellicle composed of protein strips lying side by side, they can assume different shapes as the underlying cytoplasm undulates and contracts.

Algae are diverse photosynthetic protists that serve as producers in aquatic ecosystems. The seaweeds are commercially important for the multipurpose gelatinous substances they produce.

Biology and Diversity of Protozoans

A **protozoan** is a usually motile, eukaryotic, unicellular protist. In addition, this text groups the protozoans as all heterotrophic by ingestion, as are animals. However, some scientists place the dinoflagellates and the euglenoids among the protozoans on the basis of motility and genetic data.

One common way to divide protozoans is based on mode of locomotion; one group is nonmotile, and the other groups use flagella, cilia, or pseudopods for getting about. Protozoans are distributed in great number in many habitats; for example, in aquatic environments, they are part of the **zooplankton,** microscopic suspended organisms that feed on other organisms. A number of protozoans are parasitic, often causing diseases of the blood. In many cases, their complex life cycles hinder the development of suitable treatment.

Protozoans are unicellular, but their cells are very complex. Some protozoans have more than one nucleus. Those in the genus *Paramecium* have a large macronucleus and a small micronucleus. The macronucleus produces mRNA and directs metabolic functions. The micronucleus is important during sexual reproduction. However, reproduction in protozoans is usually asexual by cell division.

Protozoans have the usual complement of eukaryotic organelles. Also, their organelles perform many of the functions we normally associate with the organs of multicellular organisms. For example, when protozoans ingest food, phagocytic vacuoles act as a "stomach" where digestion occurs. **Contractile vacuoles** are responsible for osmoregulation, particularly in freshwater environments. They pump out the excess water that tends to enter the cell.

Many protozoans produce cysts that protect them during difficult environmental conditions and promote dispersal to areas better suited for growth. Encystation involves the development of a cell with a thick cell wall and low metabolic rate, analogous to the endospores produced by some bacteria. The cyst germinates when better environmental conditions return.

Ciliates

Ciliates are the largest group of protozoans. All of them have cilia, hairlike structures that rhythmically beat, moving the cell forward or in reverse. Typically, the cell rotates as it moves. Cilia also help capture prey and particles and then move them toward the mouthparts. After phagocytic vacuoles engulf food, they, combine with lysosomes, which supply the enzymes needed for digestion.

Some ciliates can be 3 mm long, which is large enough to be seen with the naked eye. Most are freely motile, but some can be anchored such as *Stentor,* which forms a stalk, attaching itself to a surface and collecting food with its cilia (Fig. 28.15*a*). *Paramecium* is the most widely known ciliate, and it is commonly used for research and teaching. It is shaped like a slipper and has visible contractile vacuoles (Fig. 28.15*b*). *Paramecium* has been important for studying ciliate sexual reproduction, which involves *conjugation,* with interactions between micronuclei and macronuclei (Fig. 28.15*c*).

Amoeboids

Amoeboids move by pseudopods, processes that form when cytoplasm streams forward in a particular direction (Fig. 28.16*a*). They usually live in aquatic environments, such as oceans and freshwater lakes and ponds, where they are a part of the zooplankton. When amoeboids feed, their pseudopods surround and engulf their prey, which may

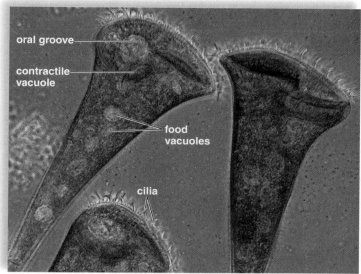

a. *Stentor*   200 µm

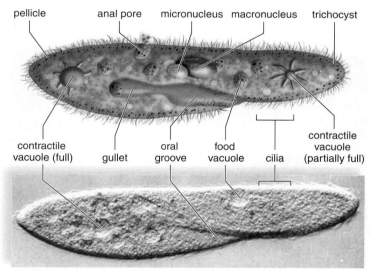

b. *Paramecium*

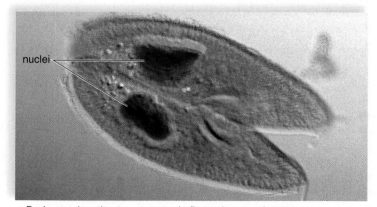

c. During conjugation two paramecia first unite at oral areas

Figure 28.15 Ciliates.

Ciliates are the most complex of the protists. **a.** *Stentor,* a large, vase-shaped, freshwater ciliate. **b.** Structure of a *Paramecium,* adjacent to an electron micrograph. Note the oral groove and the gullet and anal pore. **c.** A form of sexual reproduction called conjugation occurs periodically in paramecia.

be algae, bacteria, or other protists. Digestion then occurs within a food vacuole.

Parasitic amoeboids in the genus *Entamoeba* cause amoebic dysentery. Complications arise when the parasite invades the intestinal lining and reproduces there. If the parasite enters the body proper, liver and brain involvement can be fatal.

Foraminiferans and radiolarians are related amoeboid groups that have a skeleton called a test. In **foraminiferans**, pseudopods extend through openings in the test, which covers the plasma membrane. Because each geological time period has a distinctive form of foraminiferan, they can be used to date sedimentary rock. Depositions of millions of years followed by geological upheaval formed the White Cliffs of Dover along the southern coast of England (Fig. 28.16b). Also, the great Egyptian pyramids are built of foraminiferan limestone. In radiolarians, the test is internal (Fig. 28.16c). The tests of dead foraminiferans and radiolarians form a deep layer of sediment on the ocean floor. Their presence is used as an indicator of oil deposits on land or sea.

Zooflagellates

Zooflagellates are heterotrophic protozoans that propel themselves using one or more flagella. Most are symbiotic, and many are parasitic. *Trypanosoma brucei*, the cause of African sleeping sickness, is transmitted by the tsetse fly (Fig. 28.17). It attacks the patient's blood, causing inflammation that decreases oxygen flow to the brain. Chagas disease, caused by *T. cruzi*, is transmitted by the kissing bug, so called because it tends to bite lips. *Giardia lamblia* is well known as a human pathogen that can cause epidemics of diarrhea, known as giardiasis. The cysts of *Giardia* can pass through water treatment plants because they are not well retained by sand filters and are impervious to chlorine disinfection. *Giardia* is common in environments contaminated with fecal materials, such as day-care centers, and can be found in natural surface waters.

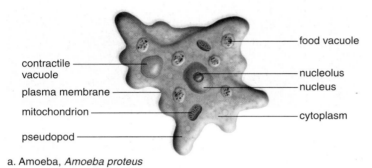

a. Amoeba, *Amoeba proteus*

contractile vacuole
plasma membrane
mitochondrion
pseudopod
food vacuole
nucleolus
nucleus
cytoplasm

250 µm

b. Foraminiferan, *Globigerina*, and the White Cliffs of Dover, England

c. Radiolarian tests   SEM 200×

Figure 28.16 Amoeboids.

a. Structure of *Amoeba proteus*, an amoeboid common in freshwater ponds.
b. Pseudopods of a live foraminiferan project through holes in the calcium carbonate shell. Fossilized shells were so numerous they became a large part of the White Cliffs of Dover when a geologic upheaval occurred.
c. Skeletal tests of radiolarians. © Dr. Richard Kessel & Dr. Gene Shih/Visuals Unlimited.

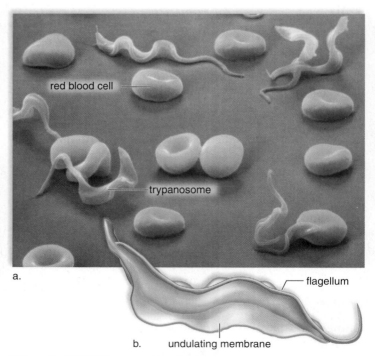

red blood cell
trypanosome
a.
flagellum
b.   undulating membrane

Figure 28.17 Zooflagellates.

a. Micrograph shows *Trypanosoma brucei*, a causal agent of African sleeping sickness, among red blood cells. **b.** The drawing shows the general structure of *T. brucei*.

Sporozoans

The apicomplexa are commonly called **sporozoans** because they produce spores. All phases of the life cycle are generally nonmotile, with the exception of male gametes and zygotes. Sporozoans are either intercellular or extracellular parasites. An apical complex composed of fibrils, microtubules, and organelles is found at one end of the single cell. Enzymes for attacking host cells are secreted through a pore at the end of the apical complex.

Malaria **Malaria** is the most widespread and dangerous sporozoan disease. It is endemic in approximately 50% of the habitable surface of the Earth. Some 300 million people are infected with malaria worldwide. Two to four million people die each year from malaria, a million of these under age five. The malaria parasite, one of several *Plasmodium* species, has a complex life cycle that involves transmission by a mosquito vector (Fig. 28.18).

The sexual reproduction phase in the mosquito ends with the migration of a developmental stage called sporozoites into the salivary glands. These sporozoites are then transmitted by the mosquito's bite to a human, where they reproduce asexually in the liver and red blood cells, forming merozoites. The infected red blood cells frequently rupture, releasing merozoites and toxins, and causing the person to experience the chills and fever that are characteristic of malaria. Some of these merozoites become gametocytes, which, if ingested by another mosquito, can start the sexual reproduction phase again. People who survive malaria gain limited immunity to reinfection, and no vaccine is available. Interestingly, the sickle cell trait common in Africans provides some protection against malaria. Efforts to control mosquito populations have been successful in the United States, where malaria is rarely seen. The control of mosquitoes worldwide has now been hampered by the rise of resistance to DDT.

Other Sporozoan Diseases A related protozoan, *Toxoplasma,* is commonly transmitted by cat feces. Toxoplasma infection generally causes no appreciable

Sexual phase in mosquito

female gamete — male gamete — food canal — zygote — sporozoite — salivary glands

1. In the gut of a female *Anopheles* mosquito, gametes fuse, and the zygote undergoes many divisions to produce sporozoites, which migrate to her salivary gland.

2. When the mosquito bites a human, the sporozoites pass from the mosquito salivary glands into the bloodstream and then the liver of the host.

Figure 28.18 Life cycle of *Plasmodium vivax,* a species that causes malaria. Asexual reproduction of this sporozoan occurs in humans, while sexual reproduction takes place within the *Anopheles* mosquito.

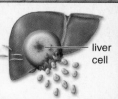

3. Asexual spores (merozoites) produced in liver cells enter the bloodstream and then the red blood cells. — liver cell

6. Some merozoites become gametocytes, which enter the bloodstream. If taken up by a mosquito, they become gametes.   gametocytes

Asexual phase in humans

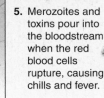

4. When the red blood cells rupture, merozoites invade and reproduce asexually inside new red blood cells.

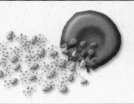

5. Merozoites and toxins pour into the bloodstream when the red blood cells rupture, causing chills and fever.

symptoms, but the parasite can be harmful to a developing fetus. Thus, pregnant women are advised not to interact with cats and to avoid working in gardens or other locations where cats may defecate. Another related protozoal disease is caused by *Cryptosporidium*. The organism and its cysts are common in surface waters and in the feces of animals and birds, and can pass through sand filters in drinking-water plants, while being unaffected by chlorine treatment. *Cryptosporidium* infection usually causes a self-limiting gastroenteritis, but can lead to a fatal watery diarrhea in some cases.

Molds as Protists

We tend to think of molds as fungi, but the water molds and slime molds are classified as protists. Unlike fungi, they both have flagellated cells.

Water Molds

Most **water molds** are saprotrophic, meaning that they feed on dead organic matter. They usually live in water, where they decompose remains and form furry growths when they parasitize fish. In spite of their common name, some water molds live on land and parasitize insects and plants. The water mold *Phytophthora* was responsible for the 1840s potato famine in Ireland.

Water molds have a filamentous body, as do fungi, but the cell walls of water molds are largely composed of cellulose, whereas fungi have cell walls of chitin. During asexual reproduction, water molds produce flagellated spores. During sexual reproduction, they produce eggs and sperm. Their phylum name, Oomycota, refers to the enlarged tips (oogonia), where eggs are produced.

Slime Molds

In forests and woodlands, slime molds feed on dead plant material. They also feed on bacteria, keeping their population under control. The vegetative cells of slime molds are mobile and amoeboid. They ingest their food by phagocytosis.

Usually, **plasmodial (acellular) slime molds** exist as a plasmodium, a diploid, multinucleated, cytoplasmic mass enveloped by a slime sheath, that creeps along, phagocytizing decaying plant material in a forest or agricultural field (Fig. 28.19). At times unfavorable to growth, such as during a drought, the plasmodium develops many sporangia, reproductive structures that produce spores. The spores produced by a sporangium can survive until moisture is sufficient for them to germinate. In plasmodial slime molds, spores release a haploid flagellated cell or an amoeboid cell. Eventually, two of these fuse to form a zygote that feeds and grows, producing a multinucleated plasmodium once again.

Plasmodium, *Physarum*

Sporangia, *Hemitrichia* 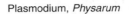 1 mm

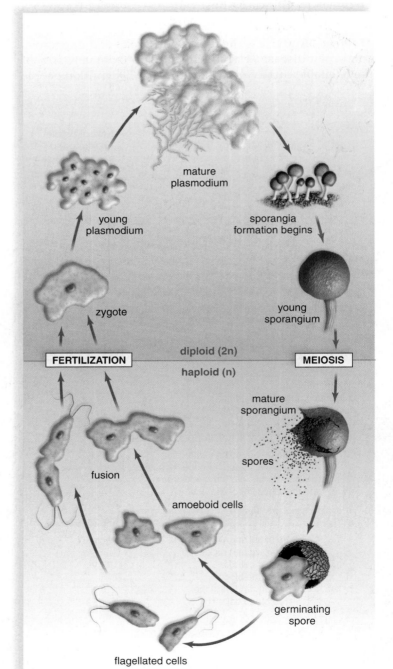

Figure 28.19 Plasmodial slime molds.
The diploid adult forms sporangia during sexual reproduction, when conditions are unfavorable to growth. Haploid spores germinate, releasing haploid amoeboid or flagellated cells that fuse.

Cellular Slime Molds

Cellular slime molds exist as individual amoeboid cells. They are common in soil, where they feed on bacteria and yeasts. *Dictyostelium* is the commonly researched example. As food supplies dwindle, the cells release a chemical that causes them to aggregate into a sluglike pseudoplasmodium, which eventually gives rise to a fruiting body that produces spores. When favorable conditions return, the spores germinate, releasing haploid amoeboid cells, and this cycle, which is asexual, begins again. A sexual cycle is known to occur under very moist conditions.

> **All protists are eukaryotic; most are unicellular and microscopic. Three major groups of protists are the algae, the protozoans, and the water molds/slime molds. Algae contain chloroplasts and thus are photosynthetic. Most protozoans are motile, and ingest their food. A few protozoans cause important human diseases, such as malaria and sleeping sickness.**

28.4 Fungi

Fungi (domain Eukarya, kingdom Fungi) are a structurally diverse group of eukaryotes that are strict heterotrophs. Unlike animals, fungi (sing., fungus) release digestive enzymes into the external environment and digest their food outside the body, while animals ingest their food and digest it internally. Most fungi are saprotrophs; they decompose the corpses of plants, animals, and microbes. Along with bacteria, fungi play an important role in ecosystems by breaking down complex organic molecules and returning inorganic nutrients to producers of food—that is, photosynthesizers. Fungi can degrade even cellulose and lignin in the woody parts of trees. It is common to see fungi (brown rot or white rot) on the trunks of fallen trees. The body of a fungus can become large enough to cover acres of land.

Biology of Fungi

The body of a fungus is composed of a mass of individual filaments called hyphae (sing., **hypha**); collectively, the mass of filaments is called a **mycelium** (pl., mycelia) (Fig. 28.20). Some fungi have cross walls that divide a hypha into a chain of cells. These hyphae are termed septate. Septa have pores that allow

cytoplasm and even organelles to pass from one cell to the other along the length of the hypha. Nonseptate fungi have no cross walls, and their hyphae are multinucleated.

Hyphae give the mycelium quite a large surface area per volume of cytoplasm, which facilitates absorption of nutrients into the body of a fungus. Hyphae grow at their tips, and the mycelium absorbs and then passes nutrients on to the growing tips. Individual hyphae can grow quite rapidly, as much as 18 feet per day. Altogether, a single mycelium may add as much as a kilometer of new hyphae in a single day!

Fungal cells are quite different from plant cells, not only because they lack chloroplasts, but also because their cell walls contain chitin, not cellulose. Chitin, like cellulose, is a polymer of glucose, but in chitin, each glucose molecule has a nitrogenous-containing amino group attached to it. Chitin is also the major structural component of the exoskeleton of arthropods, such as insects, lobsters, and crabs. Unlike plants, the energy reserve of fungi is not starch, but glycogen, as in animals. Fungi are nonmotile and do not have flagella at any stage in their life cycle. They move toward a food source by growing toward it.

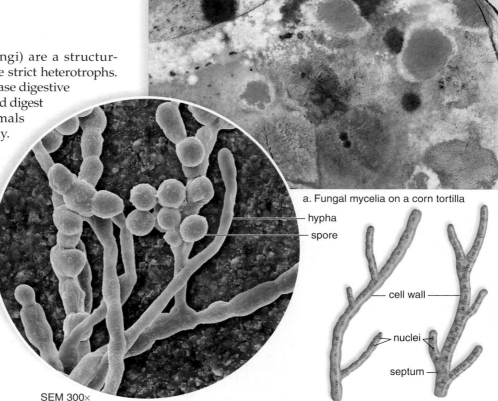

a. Fungal mycelia on a corn tortilla

SEM 300×
b. Specialized fungal hyphae that bear spores

c. nonseptate hypha   d. septate hypha

Figure 28.20 Fungal mycelia and hyphae.
a. A colorful variety of fungal mycelia growing on a corn tortilla.
b. Scanning electron micrograph of specialized aerial fungal hyphae that bear spores. Hyphae are either (**c**) nonseptate (do not have cross walls) or (**d**) septate (have cross walls).

a.

b.

Figure 28.21 Dispersal of spores.

Fungi reproduce by spore formation. **a.** As with this earthstar fungus, hordes of spores are released into the air. **b.** The microscopic appearance of the spores.

Source image: From C.Y. Shih and R.G. Kessel, Living Images. Sciences Books International, 1982, Boston.

Fungi are adapted to life on land by producing wind-blown spores during both asexual and sexual reproduction (Fig. 28.21). A **spore** is a haploid reproductive cell that develops into a new organism without the need to fuse with another reproductive cell. In fungi, spores germinate into new mycelia. Sexual reproduction in fungi involves conjugation of hyphae from two different mating types (usually designated + and −). Often, the haploid nuclei from the two hyphae do not immediately fuse to form a zygote. The hyphae contain + and − nuclei for long periods of time.

Figure 28.22 Black bread mold, *Rhizopus stolonifer.*

The mycelium of this mold utilizes sporangia to produce windblown spores. A zygospore forms during the sexual life cycle.

Eventually, the nuclei fuse to form a zygote that undergoes meiosis, followed by spore formation.

Diversity of Fungi

Fungi are differentiated on the basis of their mode of sexual reproduction.

Zygospore Fungi

The **zygospore fungi** (phylum Zygomycota) are mainly saprotrophs, but some are parasites of small soil protists or worms and even insects, such as the housefly. *Rhizopus stolonifer* (Fig. 28.22) is well known to many of us as the mold that appears on old bread, even when it has been refrigerated. In *Rhizopus,* the hyphae are specialized: some are horizontal and exist on the surface of the bread; others grow into the bread to anchor the mycelium and carry out digestion; and still others are stalks that bear sporangia. A **sporangium** is a capsule that produces spores.

When *Rhizopus* reproduces sexually, the ends of + and − hyphae join, haploid nuclei fuse, and a thick-walled zygospore results. The zygospore undergoes a period of dormancy before meiosis and germination take place. Following germination, aerial hyphae, with sporangia at their tips, produce many spores. The spores are dispersed by air currents and give rise to new mycelia.

Sac Fungi

The **sac fungi** (phylum Ascomycota) are named for their characteristic cuplike sexual reproductive structure called an ascocarp. Many sac fungi reproduce by producing chains

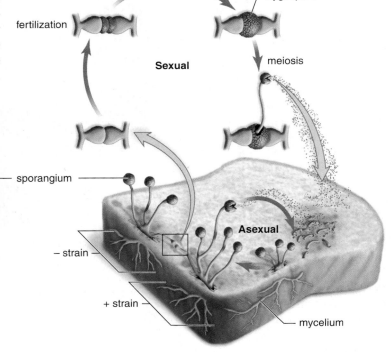

of sexual spores called conidia (sing., **conidium**). Cup fungi, morels, and truffles have conspicuous ascocarps (Fig. 28.23). Truffles, underground symbionts of hazelnut and oak trees, are highly prized as gourmet delights. Pigs and dogs are trained to sniff out truffles in the woods, but they are also cultivated on the roots of seedlings.

Other sac fungi are plant pathogens. Powdery mildews grow on leaves, while chestnut blight and Dutch elm disease destroy trees. *Ergot,* a parasitic sac fungus that infects rye, produces hallucinogenic compounds similar to LSD. Psychoses resulting from ingesting grain contaminated with rye ergot may have led to the Salem, Massachusetts, witch trials in 1692.

Some of the most familiar fungi were formerly considered "imperfect fungi," because their life cycles are incompletely understood. However, based on DNA sequencing and other recent information, these fungi are now classified as sac fungi. Examples include *Penicillium,* the original source of penicillin, a breakthrough antibiotic that led to the important class of "cillin" antibiotics that have saved millions of lives. Other species of *Penicillium, roquefortii* and *camemberti,* are necessary to the production of blue cheeses (Fig. 28.24). *Aspergillus* is used widely for its ability to produce citric acid and various enzymes; certain species also cause serious infections in people with compromised immune systems.

Yeasts   The term **yeasts** is generally applied to unicellular fungi, and many of these organisms are sac fungi. *Saccharomyces cerevisiae,* brewer's yeast, is representative of budding yeasts. When unequal binary fission occurs, a small cell gets pinched off and then grows to full size. Sexual reproduction occurs when the food supply runs out, producing spores (Fig. 28.25).

Figure 28.24
Blue cheese.
The blue color of this aromatic cheese is due to the presence of fungal conidia.

When some yeasts ferment, they produce ethanol and carbon dioxide. In the wild, yeasts grow on fruits, and historically, the yeasts already present on grapes were used to produce wine. Today, selected yeasts are added to relatively sterile grape juice in order to make wine. Also, yeasts are added to grains to make beer and liquor. Both the ethanol and carbon dioxide are retained in beers and sparkling wines, while carbon dioxide is released from wines. In breadmaking, the carbon dioxide produced by yeasts causes the dough to rise; the gas pockets are preserved as the bread bakes.

Club Fungi

Club fungi (phylum Basidiomycota) are named for their characteristic sexual reproductive structure called a basidium. Basidia are enclosed within a basidiocarp, which develops after + and – hyphae join. The union of + and – nuclei occurs in the basidium, which produces spores by meiosis. We recognize most club fungi by their basidiocarps (Fig. 28.26). When you eat a mushroom, you are consuming a basidiocarp. Shelf or bracket fungi found on dead trees are also basidiocarps. Less well known are puffballs, bird's nest fungi, and stinkhorns. In puffballs, spores are produced inside parchmentlike membranes, and the spores are released through a pore or when the membrane breaks down. Stinkhorns resemble a mushroom, with a spongy stalk and a compact, slimy cap. Stinkhorns emit an incredibly disagreeable odor. Flies are attracted by the odor; when they linger to feed on the sweet jelly, the flies pick up spores and later distribute them. Club fungi generally reproduce sexually, but can also make asexual spores.

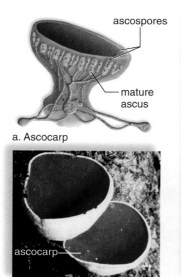

a. Ascocarp

b. Cup fungi

c. Morels

Figure 28.23 Sexual reproduction in sac fungi.
Some sac fungi are known to us by their ascocarp, the structure that produces spores as a part of sexual reproduction. **a.** Diagram of an ascocarp containing asci, where spores are produced. **b.** The ascocarps of cup fungi. **c.** The ascocarp of a morel, which is a prized delicacy.

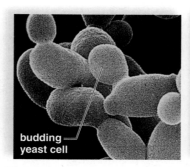

a.

b.

Figure 28.25 Asexual reproduction in sac fungi.
a. Yeasts, unique among fungi, reproduce by budding. **b.** The sac fungi usually reproduce asexually by producing spores called conidia.

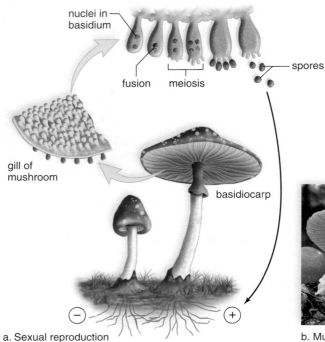

nuclei in basidium

fusion   meiosis   spores

gill of mushroom

basidiocarp

(−)   (+)

a. Sexual reproduction

Figure 28.26 Sexual reproduction in club fungi.

Most club fungi are known to us by their basidiocarp, the structure that produces spores as a part of sexual reproduction. **a.** Sexual life cycle of a basidiocarp and a basidium (pl., basidia) where spores are produced. Also shown are basidiocarps of (**b**) scarlet hood mushroom, (**c**) chicken-of-the-woods shelf fungi, (**d**) and a giant puffball. Puffballs contain many spores, and their basidiocarps tend to be rounded. When they are mature, any pressure from outside, such as a raindrop or the kick of a shoe, ejects the spores through a hole as a cloud of dust.

b. Mushroom   c. Shelf fungi   d. Giant puffball

Environmental Aspects of Fungi

Fungi often have symbiotic relationships with other organisms.

Fungi and Photosynthesizers

Lichens are associations between fungi and cyanobacteria or green algae. The different lichen species are identified according to the fungal partner. Lichens are efficient at acquiring nutrients and moisture, and therefore, they can survive in poor soils, as well as on rocks with no soil. These primary colonizers produce organic matter and create new soil, allowing plants to invade the area. Lichens exhibit three structures: compact *crustose* lichens, often seen on bare rocks or tree bark; shrublike *fruticose* lichens; and leaflike *foliose* lichens (Fig. 28.27).

The body of a lichen has three layers. The fungus forms a thin, tough upper layer and a loosely packed lower layer; these shield the photosynthetic cells in the middle layer. Specialized fungal hyphae, which penetrate or envelop the photosynthetic cells, transfer organic nutrients directly to the fungus. The fungus protects the algae from predation and desiccation and provides them with minerals and water. Lichens can reproduce asexually by releasing fragments that

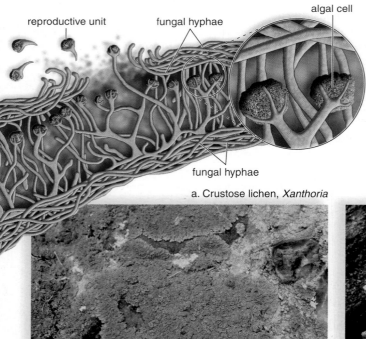

reproductive unit   fungal hyphae   algal cell

fungal hyphae

Figure 28.27 Lichen morphology.

a. A section of a compact crustose lichen shows the placement of the algal cells and the fungal hyphae, which encircle and penetrate the algal cells. **b.** Fruticose lichens are shrublike. **c.** Foliose lichens are leaflike.

a. Crustose lichen, *Xanthoria*   b. Fruticose lichen, *Cladonia*   c. Foliose lichen, *Xanthoparmelia*

sac fungi reproductive cups

contain hyphae and an algal cell. As with many symbioses, the relationship between fungi and algae was likely a pathogen-and-host interaction originally, but became mutually beneficial over evolutionary time. Still, how to confirm this conclusion remains an issue for debate.

Mycorrhizal fungi form mutualistic relationships with the roots of most plants, helping them grow more successfully in dry or poor soils, particularly those deficient in phosphates. It is important to encourage the growth of mycorrhizal fungi when restoring lands damaged by strip mining or chemical pollution. The relationship is very ancient, as it is seen in early plant fossils. Perhaps it helped plants adapt to life on dry land. Mycorrhizal fungi generally go unnoticed, except for the truffle, discussed previously.

Mycorrhizal fungi may live on the outside of roots, enter the cortex of roots, or penetrate root cells. The fungus and plant cells can easily exchange nutrients, with the plant providing organic nutrients to the fungus and the fungus bringing water and minerals to the plant. The fungal hyphae greatly increase the surface area from which the plant can absorb water and nutrients.

Fungal Diseases of Plants   We have already mentioned that various types of fungi parasitize plants. Many fungal pathogens of plants enter through the stomata (breathing pores in leaves) or through a wound. The spores of rice blast disease (genus *Pyricularia*) attach to the leaf and germinate, sending out hyphae and forming a mycelium. The end of a hypha begins to enlarge, and hydrostatic pressure helps force a penetration peg through the leaf cuticle. After that, hyphae can grow within the leaf, causing lesions.

Smuts and rusts are club fungi that parasitize cereal crops, such as corn, wheat, oats, and rye (Fig. 28.28). They are of great economic importance because of the crops they destroy each year. The corn smut mycelium grows between corn kernels and secretes substances that promote tumor development. The life cycle of rusts may be particularly complex in that it often requires two different plant host species to complete the cycle. Black stem rust of wheat uses barberry bushes as an alternate host. Eradication of barberry bushes in areas where wheat is grown helps control this rust. Wheat rust can also be controlled by producing new and resistant strains of wheat.

Fungal Diseases of Humans

Mycoses, fungal diseases of humans, vary in levels of seriousness. *Cutaneous mycoses* affect only the epidermis, *subcutaneous mycoses* affect deeper skin layers, and *systemic mycoses* spread their effects throughout the body. Many fungal diseases are acquired from the environment. Ringworm comes from soil fungi, rose gardener's disease from thorns, Chicago's disease from old buildings, and basketweaver's disease from grass cuts.

Moldlike fungi cause infections of the skin called **tineas.** Athlete's foot, caused by a species of Trichophyton,

a. Corn smut, *Ustilago*

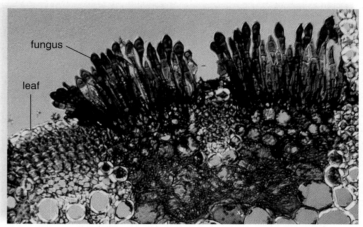
b. Wheat rust, *Puccinia*

Figure 28.28 Smuts and rusts.
a. Corn smut. **b.** Wheat rust. Both are important fungal diseases affecting agricultural crops in the United States and other countries.

is a tinea characterized by itching and peeling of the skin between the toes (Fig. 28.29a). In ringworm, which can be caused by several different fungi, the fungus releases enzymes that degrade keratin and collagen in skin. The area of infection becomes red and inflamed, and the fungal colony grows outward, forming a ring of inflammation. As the center of the lesion begins to heal, ringworm acquires its characteristic appearance, a red ring surrounding an area of healed skin (Fig. 28.29b). Tinea infections of the scalp are most common among children 4–14 years of age, although adults can also be infected. Some people can harbor and spread the fungus, but show no signs of disease. At any one time, an estimated 3–8% of the U.S. population is infected with scalp ringworm.

The vast majority of people living in the Midwest have been infected with *Histoplasma capsulatum*. This common soil fungus, often associated with bird droppings, leads in most

Figure 28.29
Human fungal
diseases.

a. Athlete's foot and
(b) ringworm are caused
by *Tinea* species. **c.** Thrush,
or oral candidiasis, is
characterized by the
formation of white patches
on the tongue.

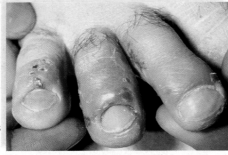

a.

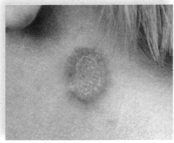

b.

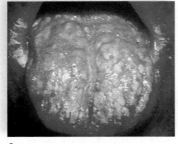

c.

cases to a mild "fungal flu." Over 40 million people have been infected in the United States, with 500,000 new cases annually. Less than half notice any symptoms, but approximately 3,000 show severe disease, called histoplasmosis, and about 50 die each year. The fungal pathogen lives and grows within cells of various tissues and causes systemic illness. Lesions may form in the lungs and leave calcifications that resemble those of tuberculosis when viewed in X-ray images.

Candida albicans causes the widest variety of fungal infections. Disease occurs when the normal microflora balance in an organ is disturbed. For example, the bacterium *Lactobacillus* normally produces organic acids that lower the pH of the vagina, and this inhibits *Candida,* a normal inhabitant of the vagina, from proliferating. When *Lactobacilli* are killed off by antibiotics, *Candida* can proliferate, and the result is inflammation, itching, and discharge. Oral thrush, a *Candida* infection of the mouth, is common in newborns and AIDS patients (Fig. 28.29*c*). In immunocompromised individuals, *Candida* can move throughout the body, causing a systemic infection that can damage the heart, the brain, and other organs.

Control of Fungi

Fungi are more closely related to animals than to bacteria. Fungal cells contain a nucleus, organelles, and ribosomes that are like those of human cells. The strong similarities between fungal cells and human cells make it more difficult to design antimicrobials against fungi that do not also harm humans than it is to design them against bacteria. Researchers exploit any biochemical differences they can discover. The biosynthesis of steroids in fungi differs somewhat from the same pathways in humans. A variety of fungicides are directed against steroid biosynthesis, including some that are applied to fields of grain. Fungicides based

on heavy metals are applied to the seeds of sorghum and other crops. For humans, topical agents are available to treat yeast infections and tineas, and systemic medications are available for systemic fungal infections.

> Fungi include the familiar molds and mushrooms, and some of them can cause significant diseases in humans and plants. The dispersal agents of fungi are windblown spores, which they produce during asexual and sexual reproduction. Because fungi are decomposers of animal and plant remains, they assist ecological cycles by returning inorganic nutrients to photosynthesizers.

28.5   Viruses, Viroids, and Prions

As we learned in Chapter 1, all living things are composed of cells. **Viruses** are not composed of cells; they are acellular. Also, viruses are obligate parasites, meaning that they can only reproduce inside a living cell (called the host cell) by utilizing at least some of the machinery (ribosomes, certain enzymes, etc.) of that cell. Therefore, the question arises, "Are viruses alive?"

Scientists and philosophers have long argued this question. Some say that viruses are not alive. After all, not only are they acellular, but some have been synthesized in the lab from chemicals! Moreover, when viruses are outside a host cell, they are totally quiescent, exhibiting no metabolic activity. Others argue that viruses are alive because they have a genome that directs their reproduction when they are inside the host cell.

If viruses are not considered alive, then certainly the simpler viroids and prions are not alive either. Viroids are strands of RNA that can reproduce inside a cell, and prions are protein molecules that cause other proteins to become prions.

Biology of Viruses

The structure of viruses and the precise way they reproduce varies considerably for different viruses.

Viral Structure

Most viruses are much smaller than bacteria. Viruses are usually less than 0.2 micron in size, while most bacteria measure at least 1 micron. Interestingly, the largest virus discovered so far, called the Mimivirus, is about 0.4 micron in size, which is actually bigger than the smallest known bacterium.

Viruses come in a variety of shapes, including helical, spheres, polyhedrons, and more complex forms. Even so, a virus always has at least two parts—an outer **capsid** composed of protein subunits and an inner core of nucleic acid—which can be either DNA or RNA (Fig. 28.30). The viral genome can be single- or double-stranded DNA, or single- or

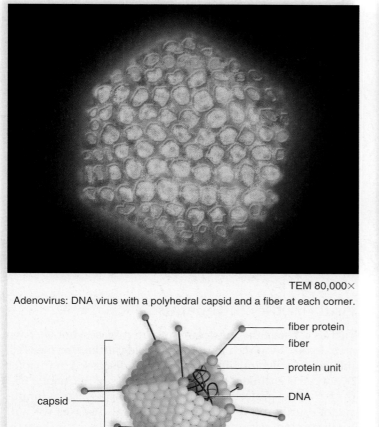

TEM 80,000×

Adenovirus: DNA virus with a polyhedral capsid and a fiber at each corner.

- fiber protein
- fiber
- protein unit
- DNA

capsid

a.

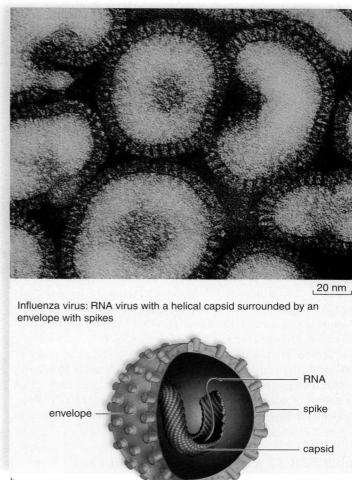

20 nm

Influenza virus: RNA virus with a helical capsid surrounded by an envelope with spikes

- RNA
- spike
- capsid

envelope

b.

Figure 28.30 Viruses.

a. Despite their diversity, all viruses have an outer capsid composed of protein subunits and a nucleic acid core that is composed of either DNA or RNA, but not both. **b.** Some types of viruses also have a membranous envelope.

double-stranded RNA. This diversity in genetic material is different from all cellular organisms, which always have a double-stranded DNA genome. Special viral enzymes that help a virus reproduce can also be inside the capsid.

In some viruses, especially those that infect animals, the capsid is surrounded by a membrane called an **envelope.** This envelope is made of lipid, and is usually derived from the host cell's plasma membrane when the virus buds from the host cell. Viral glycoproteins called spikes often extend from the envelope. These spikes are critical to viral infection because they help the virus bind to the surface of the host cell before entering it.

Viral Reproduction

Viruses are specific to a particular host. Viruses called bacteriophages infect only bacterial cells, usually just a particular group or species of bacteria. Some viruses only infect plants, while others only infect animals. Some viruses can cause

disease only in humans. The specificity of a virus for a host occurs because the spike of a virus and a receptor molecule on a cell's plasma membrane fit together like a hand fits a glove, or else the virus cannot bind to and enter the cell.

Figure 28.31 illustrates the life cycle of a typical enveloped animal RNA virus. This life cycle has six steps:

1. During *attachment,* the spikes of the virus bind to a specific receptor molecule on the surface of a host cell. The host cell normally uses this receptor for another purpose, but it is effectively "hijacked" by the virus for its own purposes.
2. During *entry,* also called penetration, the viral envelope fuses with the host's plasma membrane, and the rest of the virus (capsid and viral genome) enters the cell. The genome is freed when cellular enzymes remove the capsid, a process called uncoating.
3. *Replication* occurs when a viral enzyme makes complementary copies of the genome, which is RNA in this case.

4. During *biosynthesis,* some of these RNA molecules serve as mRNA for the production of more capsid and spike proteins, using host ribosomes.
5. At *assembly,* a mature capsid forms around a copy of the viral genome.
6. During *budding,* new viruses are released from the cell surface. During this process, they acquire a portion of the host cell's plasma membrane and spikes, which were specified by viral genes during biosynthesis. The enveloped viruses are now free to spread the infection to other cells.

Latency   Some animal viruses can become latent (hidden) inside the host cell. Sometimes the viral DNA genome can actually integrate itself into the host cell genome. An integrated viral genome is called a **provirus.** During latency, new viruses are not produced, but the viral genome is reproduced along with the host cell. Environmental stresses, such as ultraviolet radiation, can induce the latent virus to enter the biosynthesis stage, leading to the production of new virus particles. In some cases of latency, the viral genome simply remains in the cell and doesn't integrate itself into the host DNA.

Retroviruses add an interesting twist to the story of viral reproduction. The genome of a retrovirus is RNA, but these viruses are able to convert their genome into DNA because they contain an enzyme called reverse transcriptase. This enzyme, which is not found in host cells, is so named

because it catalyzes a process that is the reverse of normal transcription, which goes from DNA to RNA. The DNA copy (cDNA) of the retroviral genome can be integrated into the host DNA. So a retrovirus, such as HIV that causes AIDS, is able to exist as a provirus (see Fig. 21.14). Not only is the integrated provirus resistant to antiviral medications taken by the host, but it is also able to escape detection by the host immune system.

> Viruses are acellular, obligate parasites. They always contain a genome composed of either DNA or RNA, which is surrounded by a capsid made of proteins. Some viruses also have an envelope made of lipid. A typical viral life cycle includes six steps: attachment, entry, replication, biosynthesis, assembly, and budding.

Viral Diseases in Humans

Viruses cause many important human diseases, but we have room to discuss only a few of them: colds, influenza, measles, and four herpesviral infections. AIDS is discussed in Chapter 21, and viruses linked to cancer are discussed in Chapter 25.

The best protection against most viral diseases is immunization utilizing a vaccine. Of the viral diseases we will be discussing, vaccines are readily available for influenza, measles, and chickenpox. Vaccines against the herpes simplex viruses, as well as AIDS, may still be years away. Vaccination against some viral diseases provides protection for the life of an individual, but others must be administered every year.

Figure 28.31 **Life cycle of an animal virus.**
The influenza virus, like many others that infect animal cells, has an RNA genome. Notice also that the mode of entry requires uncoating and that the virus acquires an envelope and spikes when it buds from the host cell.

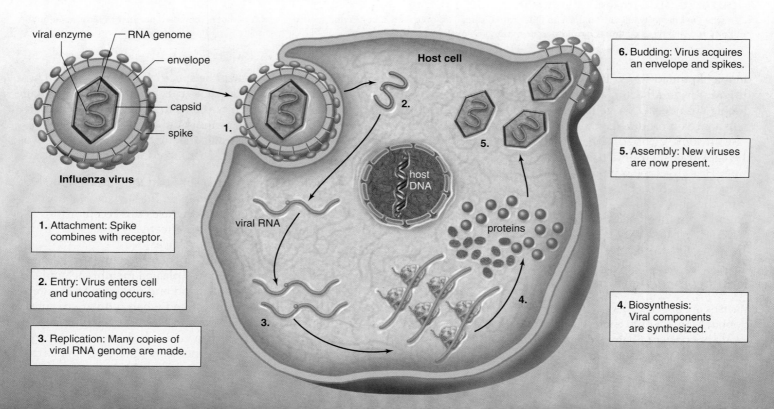

viral enzyme — RNA genome
envelope
capsid
spike
1.
Influenza virus

Host cell

2.

host DNA

viral RNA

proteins

5.
4.
3.

6. Budding: Virus acquires an envelope and spikes.

5. Assembly: New viruses are now present.

4. Biosynthesis: Viral components are synthesized.

1. Attachment: Spike combines with receptor.

2. Entry: Virus enters cell and uncoating occurs.

3. Replication: Many copies of viral RNA genome are made.

Emerging Viral Pathogens

You hear about it on the evening news—a new viral pathogen has emerged. West Nile virus. SARS. Hantavirus. HIV. Where do these apparently new pathogens come from? How do they cause such widespread diseases? A best-seller from a decade ago, *The Hot Zone,* told the true story of the Drs. Jaax, who were working on viruses for the military in the Washington, D.C., area. An outbreak of Ebola virus in a colony of monkeys had to be confirmed and contained. Ebola is one of a group of viruses that cause hemorrhagic fevers. These terrible pathogens quickly cause extensive tissue damage, leading to internal bleeding, multiple organ failure, and death. The Ebola the doctors encountered was called the Reston strain, for Reston, Virginia, where the animals were housed. There are also Sudan and Zaire strains of Ebola, named for outbreaks in those places. Lassa virus, which can cause a hemorrhagic fever, was named after Lassa, Nigeria.

Figure 28A shows the areas of the world where viral diseases have recently emerged. Such diseases seem to emerge in several different ways. In some cases, the virus is simply transported from one location to another. Flu strains move from Southeast Asia to the United States each year. The West Nile virus is making headlines because it has changed its range, being transported into the United States and taking hold in bird and mosquito populations. Severe Acute Respiratory Syndrome (SARS) was clearly transported from Southeast Asia to Toronto, Canada. A world where you can wake up in Bangkok and sleep in Los Angeles is a world where disease can spread at unprecedented rates.

Some viruses infect both humans and animals. These may cause a more serious disease in humans, while the animal population maintains the disease, acting as a reservoir. Diseases with animal reservoirs are hard to control. Smallpox, on the other hand, was successfully eradicated because it infects only humans.

Many viral diseases are transmitted by vectors, usually insects, that carry disease from an infected individual to a healthy one. Diseases spread by insects may require massive efforts to control the vector population. A disease can emerge when a mutation allows a virus to use a different insect vector. If a virus transmitted by a rare species of mosquito becomes able to be transmitted by a more common species, the disease can spread more easily. This illustrates that the mode of transmission can be very important. What if the HIV virus were transmitted in the same way as the common cold?

Viruses can emerge by acquiring new surface antigens. The immune system, unable to recognize the new "face" of the virus, allows it to cause disease. Viruses that continually change their "faces," such as HIV, are far more difficult to eradicate. Currently, there is concern about the "bird flu," which arose in Southeast Asia in a region where markets are overcrowded with humans and animals. The infecting virus carries surface antigens from a human flu virus and a bird flu virus. The bird flu has a relatively high mortality rate, has been encountered infrequently by the human immune system, and may cause a pandemic, if there is no time to prepare sufficient vaccine. Luckily, the disease doesn't transmit among humans very well.

Discussion Questions

1. During the Ebola outbreak that occurred at the primate facility in Reston, Virginia, several people produced antibodies against the virus, indicating that they had been infected. However, none developed disease symptoms characteristic of other Ebola virus infections in humans. What are some possible reasons for this?

2. It is commonly observed that viral diseases that require close contact to be transmitted, such as herpesviruses or even the common cold, typically don't produce acute, life-threatening symptoms (at least not right away). However, many viruses transmitted by insects, such as yellow fever or West Nile virus, cause very acute, life-threatening diseases. How might the mode of transmission be related to the type of disease caused?

3. The strain of "bird flu" that is currently in the news likely arose by a genetic reassortment. Specifically, what is the mechanism of this genetic reassortment—in other words, how does it happen?

Figure 28A
Emerging Viral diseases.

Emerging diseases, such as those noted here according to their country of origin, are new or demonstrate increased prevalence. The disease-causing agents may have acquired new virulence factors, or environmental factors may have encouraged their spread to an increased number of hosts. Avian influenza (bird flu) is a new emerging disease.

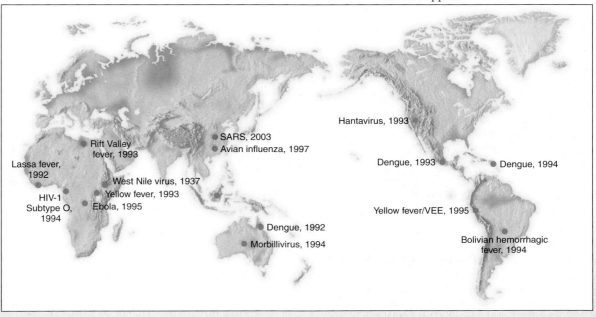

Hantavirus, 1993
SARS, 2003
Avian influenza, 1997
Rift Valley fever, 1993
Dengue, 1993
Dengue, 1994
Lassa fever, 1992
West Nile virus, 1937
HIV-1 Subtype O, 1994
Yellow fever, 1993
Ebola, 1995
Yellow fever/VEE, 1995
Dengue, 1992
Morbillivirus, 1994
Bolivian hemorrhagic fever, 1994

The Common Cold and Influenza

Colds are most commonly caused by rhinoviruses, and the symptoms usually include a runny nose, mild fever, and fatigue. The flu, caused by the influenza virus, is characterized by more severe symptoms, such as a high fever, chills, body aches, and severe fatigue. Cold symptoms tend to run their course within a week, but the flu may last for two or three weeks, and can result in death, especially in elderly patients or those with weakened immune systems. While most common cold viruses are endemic (always present) in the human population, flu outbreaks tend to be epidemic, affecting large numbers of people in more limited geographic areas at any one time. Why can you get several colds or the flu year after year? There are over 100 different strains of rhinoviruses, plus several other viruses that cause very similar symptoms. In the case of colds, you become immune only to those strains you have contracted before. However, the reason you may get the flu each year is that the influenza virus can change rapidly. Small changes called **antigenic drift,** especially those affecting the surface spikes, may be enough to make the virus capable of temporarily evading the immune response of individuals who were immune to the original virus (Fig. 28.32a). Therefore, manufacturers of influenza vaccines try to incorporate the strains that are predicted to cause the highest number of cases each year; still, antigenic drift can lead to local epidemics.

The genome of the influenza virus is composed of eight segments of RNA. When two different influenza viruses infect the same cell, these RNA segments can get mixed up as the viruses reproduce. When new virus particles are assembled that contain RNA from both original viruses, this *reassortment* event, called an **antigenic shift,** leads to large changes in the surface spikes (Fig. 28.32b). Because the human population has not previously been exposed to this combination of antigens, the result can be a pandemic, or worldwide epidemic. Unfortunately, it is impossible to predict what reassortment events may occur, so vaccine manufacturers must play catch-up whenever a new virus evolves.

Measles

Measles is one of the most contagious human diseases. An unvaccinated person can become infected simply by breathing the air in a room where an infected person has been, even two hours later. Measles is a major killer worldwide, but the number of deaths is decreasing in many areas due to efforts to vaccinate children, who still make up the majority of measles fatalities. In the United States, the number of measles cases dropped from five million each year to a few thousand after the introduction of the measles vaccine in 1963. In 2005,

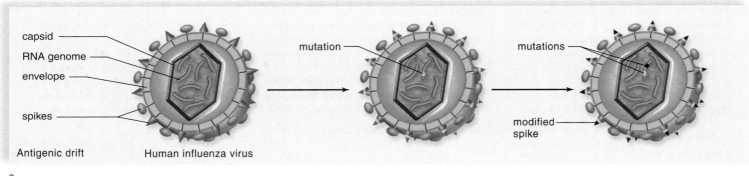

a.

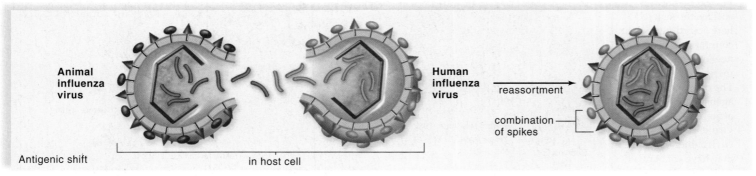

b.

Figure 28.32 Antigenic drift and shift.

a. In antigenic drift, small mutations gradually change surface antigens so that antibodies to the original virus become less effective. As time goes by, people are more likely to become ill upon exposure to the virus. **b.** In antigenic shift, even more serious major changes take place on surface antigens as genome segments are reassorted between two influenza viruses. Now, antibodies are sure to be ineffective, and most people will become ill when they are exposed to the virus.

the World Health Organization announced that worldwide measles deaths had declined from 873,000 in 1999 to 530,000 in 2005. The vaccine against measles is usually given to infants in combination with vaccines against mumps and rubella, collectively called the MMR vaccine (see Fig. 13.9).

The virus is spread, by way of the respiratory route, as airborne particles. After exposure, there is a seven- to twelve-day incubation period before the onset of flu-like symptoms, including fever. A red rash develops on the face and moves to the trunk and limbs. The rash lasts for several days. In the more-developed nations, the fatality rate is about 1 in 3,000; however, in the less-developed countries, the fatality rate is 10–15%.

Herpesviruses

Herpesviruses cause chronic infections that remain latent for much of the time. Eight herpesviruses capable of infecting humans have been described. Some of these infect epithelial cells and neurons, while others infect blood cells. Some herpesviruses do not seem to cause any pathology, but we will discuss four herpesviruses that cause disease in humans.

Herpes simplex virus type 1 (HSV-1) is usually associated with cold sores and fever blisters around the mouth. It is estimated that 70–90% of the adult population is infected with HSV-1. Herpes infections of the genitals are generally caused by HSV-2 and are transmitted through sexual contact. Painful blisters fill with clear fluid that is very infectious (see Fig. 21.15). Genital herpes infections are widespread in the United States, affecting 20–25% of the adult population. About 88% of individuals infected with genital herpes will experience recurrences of symptoms induced by various stresses.

Another herpesvirus that infects humans is the Varicella-Zoster virus that causes chickenpox (Fig. 28.33). Later in life, usually after age 60, the latent virus can reemerge as a related disease called shingles. Painful blisters form in an area innervated by a single sensory neuron, often on the upper chest or face. The symptoms may last for several weeks or, in some cases, months. The U.S. Food and Drug Administration approved a vaccine for the prevention of chickenpox in 1995, and many states are now requiring that children receive this vaccine. In May 2006, the FDA approved the first vaccine for prevention of shingles; it is for use in people over age 60.

Epstein-Barr virus (EBV) is the herpesvirus associated with infectious mononucleosis, or "mono." Like HSV-1, EBV is very common—as many as 95% of adults aged 35 to 40 have been infected. When infected with EBV as children, most people have no symptoms. In contrast, those infected as teenagers and young adults often develop infectious mononucleosis. The disease is named after the tendency of the cells infected by the virus, which are also known as mononuclear cells, to become more numerous in the blood. The symptoms of mono usually include fever, fatigue, and swollen lymph nodes. Like all herpesviruses, EBV establishes a lifelong latent infection. EBV has also been

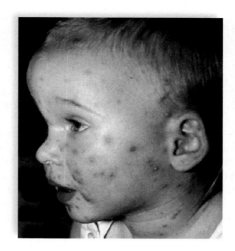

Figure 28.33 Chickenpox.
The lesions characteristic of chickenpox are pus-filled blisters that break and crust. The chickenpox virus remains latent in nerve cells and can produce shingles decades later.

associated with other conditions, such as chronic fatigue syndrome and certain rare cancers.

Antiviral Drugs

Because viruses use the machinery of host cells for viral replication, it is difficult to develop drugs that affect viral replication without harming host cells. As noted earlier, however, many viruses use their own enzymes to copy their genetic material, and many antiviral drugs inhibit these enzymes. As discussed in Chapter 21, a variety of antiretroviral agents have been developed to inhibit the reverse transcriptase and protease enzymes used by HIV, which can delay the onset of AIDS. Other drugs, such as acyclovir and valacyclovir (or Valtrex), are commonly used to control recurrences of HSV-1 and -2 infections. These compounds mimic nucleotides, and thus can inhibit viral genome synthesis. Other antiviral drugs may affect virus attachment, entry, or assembly. If no effective antiviral drugs are available for a particular viral infection, patients are often told to simply let the virus run its course, because antibiotics are not effective against viral infections.

Many viruses can cause disease in humans. Examples of human diseases caused by viruses include the common cold, influenza, measles, cold sores, genital herpes, chickenpox, and infectious mononucleosis. The latter four conditions are all caused by herpesviruses. Many antiviral drugs inhibit the enzymes viruses use to replicate their genetic material.

Viroids and Prions

Viroids and prions are also acellular pathogens. A **viroid** consists of a circular piece of naked RNA. The viroid RNA is ten times shorter than that of most viral genomes, and apparently it does not code for any proteins. Viroid replication causes diseases in plants, the only known hosts. The mechanism of viroid diseases, such as potato spindle tuber and apple scar skin, are not known.

Bioethical Focus

The events of September 11, 2001, caused scientists and ethicists to consider the possibility of bioterrorism—the use of a biological agent, such as a microbe, to kill thousands of people at once. After several people died when anthrax spores were sent through the mail, bioterrorism became a reality to the U.S. public as well. Bioterrorism has been employed throughout the history of the civilized world, so much so, that the United States, the United Kingdom, and the USSR signed the 1972 Biological Weapons Convention banning research and the production of offensive biological agents.

The possibility of using microbes and inexpensive nerve agents as weapons is particularly attractive to countries that lack the means to produce conventional weapons of war. When a group of Japanese extremists used a nerve agent to kill many innocent civilians in the Tokyo subway system, it became apparent that individuals, as well as hostile countries, can obtain and use these agents as weapons.

Numerous ethical questions surround the development of biological agents that can kill people. For example, should scientists be allowed to perfect and publish their findings regarding such agents? Many biologists say not publishing these findings would be detrimental to the progress of science. "How do doctors talk about research if we don't

publish it?" asks Dr. Ariella Rosengard of the University of Pennsylvania. She recently led a team that synthesized a protein from smallpox virus that is 100 times more effective at inhibiting immune system proteins called complement (see Section 13.2) than is a similar protein from a related virus. This knowledge may help to explain why smallpox virus is so lethal to humans, but it could also be useful to bioterrorists attempting to produce even more lethal forms of the virus. While it may seem that potentially harmful information like this shouldn't be published, such research is important for developing suitable vaccines.

Other scientists feel the need to be more socially responsible. "When you wield power equivalent to nuclear weapons, you need some oversight," says John Steinbruner of the Center for International and Security Studies in Maryland. This organization and others are attempting to define what types of research should be allowed, and what types should possibly be banned. The main difficulty is that a better understanding of what makes these biological agents so deadly could save lives, as well as destroy them.

Form Your Own Opinion

1. Should the United States maintain an active research program to develop

weapons of biological warfare? Why or why not? Is the use of these kinds of weapons more objectionable than conventional warfare? Than nuclear warfare?

2. Many experts believe that Russia still possesses large stockpiles of smallpox virus produced during the Cold War. The United States also retains some smallpox virus, reportedly for research purposes. Ideally, if both countries could agree, would it be best to destroy all samples of this deadly virus? What if other countries also have the virus?

3. In response to September 11, 2001, as well as to the mailing of anthrax spores, the U.S. government has increased funding for research into infectious agents that bioterrorists could use. However, this also means that more scientists will be working with these agents and developing knowledge that could be used for detrimental purposes. Should scientists working, for example, with *Bacillus anthracis*, the bacterium that causes anthrax, be required to submit to extensive background checks? To periodic questioning by the FBI? If so, might this discourage people with good intentions from doing this work?

Prions are proteinaceous infectious particles that cause degenerative diseases of the nervous system in humans and other animals. They are derived from normal proteins of unknown function in the brains of healthy individuals. Disease occurs when the normal proteins change into the abnormal, prion shape. This forms more prion proteins, which go on to convert other normal proteins into the wrong shape. These abnormal proteins seem to build up in the brain, causing a loss of neurons and what appear to be holes in the tissue (see Fig. 17.22).

The first prion disease to be described was scrapie, which occurs in sheep. It is called scrapie because, as the disease affects the animal's brain, a sheep begins to scrape off much of its wool. Scrapie is not capable of causing disease in humans, but another prion disease in cattle, called bovine spongiform encephalopathy, or mad cow disease, does appear capable of infecting humans and has caused over 100 human deaths in Great Britain and a few other countries. This disease is now called variant Creutzfeldt-Jakob disease (vCJD), to distinguish it from the "classic" form of CJD that occurs spontaneously in a very low percentage of people. Most cases of vCJD have occurred in people who ate meat

from cows that had mad cow disease. As of June 2006, however, no human cases of vCJD have been documented in the United States. Chronic wasting disease is another prion disease that occurs in deer, elk, and moose, but there have been no documented cases of transmission to humans. One prion disease that passes directly from human to human is kuru, which was transmitted among a cannibalistic tribe in New Guinea due to their traditional practice of consuming their dead relatives. Even though prion diseases are as frightening as they are fascinating, their incidence in humans remains very low; a person is statistically more likely to die in a soccer accident than of a prion disease.

Viroids and prions are acellular infectious agents. Viroids are plant pathogens composed entirely of RNA. Prions are misfolded proteins that cause other, normal proteins to become misfolded, which affects the brain. Examples of human prion diseases are Creutzfeldt-Jakob disease (CJD), variant CJD, and kuru.

Summarizing the Concepts

28.1 The Microbial World

Microbiology is primarily the study of bacteria, archaea, protists, fungi, and viruses. Most microbes are microscopic, and some are extremely numerous. Two of the most significant contributors to the early discipline of microbiology were Antonie van Leeuwenhoek, who was the first to visualize microbes using his microscopes, and Louis Pasteur, whose many accomplishments include disproving the idea that microbes could arise spontaneously.

Microbes are perhaps best known for causing infectious diseases, which can range from the merely annoying to the swiftly fatal. However, most microorganisms are far more beneficial than harmful, not only to humans, but to the Earth's ecosystem.

28.2 Bacteria and Archaea

Bacteria are unicellular prokaryotic organisms that lack nuclei and membrane-bounded organelles. The circular DNA genome comprises the nucleoid. Bacteria reproduce asexually by binary fission. Most bacteria are heterotrophs, requiring organic carbon foods. Some bacteria require oxygen, while others are obligate anaerobes. Some bacteria are autotrophs, either by photosynthesis or chemosynthesis.

Archaea share many cellular features with bacteria. They are prokaryotic, thus devoid of membrane-bounded organelles, and they contain a circular genome. Nucleic acid sequences and biochemical differences distinguish the archaea from the bacteria. The plasma membrane of archaea has unique lipids that allow them to survive in extreme environments. Archaea can be heterotrophs or autotrophs. Some perform photosynthesis using a unique pigment. Others are methanogenic, producing methane. Archaea are often associated with extreme habitats, but actually are widespread in the environment.

28.3 Protists

Protists are a diverse group of eukaryotes widespread in moist environments. Protists are generally unicellular, but still they are quite complex because they (1) have a combination of characteristics not seen in the other eukaryotic kingdoms; (2) have complicated life cycles that include the ability to withstand hostile environments; and (3) sometimes have organelles not seen in other types of eukaryotic cells.

Green algae possess chlorophylls *a* and *b*, store reserve food as starch, and have cell walls of cellulose, as do plants. Green algae include those that are unicellular (*Chlamydomonas*), filamentous (*Spirogyra*), multicellular (*Ulva*), and colonial (*Volvox*). The life cycle varies among the green algae. In most, the zygote undergoes meiosis, and the adult is haploid.

Diatoms, which have an outer layer of silica, are extremely numerous in both marine and freshwater ecosystems. Dinoflagellates usually have cellulose plates and two flagella, one at a right angle to the other. They are extremely numerous in the ocean and, on occasion, produce a neurotoxin when they form red tides. The large red and brown algae, called seaweeds, are economically important. Red algae are often more delicate than brown algae and are usually found in warmer waters. Giant kelp grows under the surface of the sea in dense groves. Euglenoids are flagellated cells with a pellicle instead of a cell wall. Their chloroplasts are most likely derived from a green alga, through endosymbiosis.

Protozoans are grouped according to mode of locomotion. The ciliates move by means of their many cilia. They are remarkably diverse in form, and as exemplified by *Paramecium,* they show how complex a protist can be, despite being a single cell. Amoeboids move and feed by forming pseudopods. Parasitic amoeboids, in the genus *Entamoeba,* cause amoebic dysentery. Zooflagellates, which move by flagella, are often symbiotic. Trypanosomes are zooflagellates that cause African sleeping sickness in humans. The sporozoans are nonmotile parasites that form spores; *Plasmodium* causes malaria. Water molds, which ordinarily are saprotrophs, are sometimes parasitic. Slime molds are amoeboid and ingest food by phagocytosis.

28.4 Fungi

Fungi are unicellular or multicellular eukaryotes that are strict heterotrophs. After external digestion, they absorb the resulting nutrient molecules. Most fungi act as saprotrophic decomposers that aid the cycling of chemicals in ecosystems by decomposing dead remains. Some fungi are parasitic, especially on plants, and others are mutualistic with plant roots and algae.

The body of a fungus is composed of thin filaments, called hyphae, that form masses termed mycelia. The cell wall contains chitin. Fungi are not motile at any stage in their life cycle. Nonseptate hyphae have no cross walls, forming multinucleate cells. Septate hyphae have cross walls, but pores allow for exchange of cytoplasm.

Fungi produce nonmotile, and often windblown, spores during both asexual and sexual reproduction. During sexual reproduction, hyphae tips fuse so that hyphae containing both + and − nuclei usually result. Following nuclear fusion, meiosis occurs during the production of the sexual spores. Fungi are divided into the zygospore fungi, which form zygospores; the sac fungi, which form ascocarps; and the club fungi, which form basidiocarps.

Fungi often have symbiotic relationships with other organisms. Lichens are symbiotic associations between fungi and cyanobacteria, or green algae. These are often primary colonizers in poor soils or on rocks. The algae provide organic nutrients to the fungi, while the fungi provide protection and enhanced absorption of water and minerals. Mycorrhizal fungi enjoy a mutualistic relationship with plant roots, such that the fungus helps the plant absorb minerals and water, while the plant supplies the fungus with organic nutrients. Most plants grow much better when colonized by mycorrhizal fungi.

Fungal diseases of humans include ringworm and athlete's foot. *Candida,* a yeast, is well known for causing thrush and vaginal infections.

28.5 Viruses, Viroids, and Prions

Viruses are minute, acellular pathogens that reproduce as obligate intracellular parasites. Viruses always have an outer capsid composed of protein and an inner core of nucleic acid. Many animal viruses also have an envelope with spikes. The viral genome can be single- or double-stranded DNA or RNA.

The life cycle of an animal virus typically has six steps: attachment, entry, genome replication, biosynthesis, assembly, and budding. During budding, the viruses acquire a portion of the host cell's plasma membrane. Retroviruses (e.g., HIV) have RNA genomes that are converted to DNA by reverse transcriptase, before integrating in the host cell's chromosome. Viruses cause such diseases as influenza, measles, and chickenpox.

Viroids are obligate, intracellular plant pathogens that are autonomously replicating short RNA molecules. They contain no protein. Prions are proteinaceous, infectious particles that damage the brain, causing dementia. They contain no nucleic acid. Prions cause scrapie and mad cow disease.

Testing Yourself

Choose the best answer for each question.

1. Which is not true of prokaryotes? They
   a. are living cells.
   b. lack a nucleus.
   c. all are parasitic.
   d. include both archaea and bacteria.
   e. evolved early in the history of life.

2. Cyanobacteria, unique among bacteria, photosynthesize and
   a. give off oxygen.
   b. do not have chlorophyll.
   c. do not have a cell wall.
   d. need a fungal partner.

3. Archaea differ from bacteria in that they
   a. have a nucleus.
   b. have membrane-bound organelles.
   c. have peptidoglycan in their cell walls.
   d. are often photosynthetic.
   e. None of the above are correct.

4. Bacterial endospores function in
   a. reproduction.
   b. survival.
   c. protein synthesis.
   d. storage.

5. Which of the following is a place where microbes can be found?
   a. on inanimate objects
   b. in the air
   c. in the human body
   d. All of these are correct.

6. Decomposers
   a. break down dead organic matter in the environment by secreting digestive enzymes.
   b. break down living organic matter by secreting digestive enzymes.
   c. destroy living cells and then break them down with digestive enzymes.
   d. live in close association with another species.

7. Which bacterium is spherical in shape?
   a. bacillus
   b. coccus
   c. spirillum

8. Which bacterium is curved and spiral in shape?
   a. bacillus
   b. coccus
   c. spirillum

9. Bacteria
   a. are acellular in nature.
   b. lack a nucleus
   c. lack a plasma membrane.
   d. are eukaryotic.

10. Which of the following statements is false?
    a. Slime molds and water molds are protists.
    b. Some algae have flagella.
    c. Amoeboids have pseudopods.
    d. Among protists, only green algae ever have a sexual life cycle.
    e. Conjugation occurs among some green algae.

11. Which pair is properly matched?
    a. water mold—flagellate
    b. trypanosome—protozoan
    c. *Plasmodium vivax*—mold
    d. amoeboid—algae

12. Which of the following statements is incorrect?
    a. Unicellular protists can be quite complex.
    b. Euglenoids are motile but have chloroplasts.
    c. Plasmodial slime molds are amoeboid but have sporangia.
    d. *Volvox* is colonial but is box-shaped.
    e. Both b and d are incorrect.

13. Dinoflagellates
    a. usually reproduce sexually.
    b. have protective cellulose plates.
    c. are insignificant producers of food and oxygen.
    d. have cilia instead of flagella.
    e. tend to be larger than brown algae.

14. Ciliates
    a. can move by pseudopods.
    b. are not as varied as other protists.
    c. have a gullet for food gathering.
    d. are closely related to the radiolarians.

15. Which feature is best associated with hyphae?
    a. strong, impermeable walls
    b. rapid growth
    c. large surface area
    d. pigmented cells
    e. Both b and c are correct.

16. A fungal spore
    a. contains an embryonic organism.
    b. germinates directly into an organism.
    c. is always windblown.
    d. is most often diploid.
    e. Both b and c are correct.

17. In the life cycle of black bread mold, the zygospore
    a. undergoes meiosis and produces spores.
    b. produces spores as a part of sexual reproduction.
    c. is a thick-walled dormant stage.
    d. is equivalent to asci and basidia.
    e. All of these are correct.

18. Conidia are formed
    a. asexually at the tips of special hyphae.
    b. during sexual reproduction.
    c. by all types of fungi except water molds.
    d. only when it is windy and dry.
    e. as a way to survive a harsh environment.

19. Lichens
    a. cannot reproduce.
    b. need a nitrogen source to live.
    c. are parasitic on trees.
    d. are able to live in extreme environments.

20. Mycorrhizal fungi
   a. are a type of lichen.
   b. are in a mutualistic relationship.
   c. help plants gather solar energy.
   d. help plants gather inorganic nutrients.
   e. Both b and d are correct.

21. Tinea infections are generally found
   a. in the eye.   c. on the skin.
   b. in cartilage.   d. in the vagina.

22. Which of the following diseases is not caused by *Candida?*
   a. oral thrush
   b. tineas
   c. vaginal yeast infection
   d. candidiasis

23. Chemoautotrophic bacteria acquire energy
   a. by parasitizing other organisms.
   b. from the sun.
   c. by oxidizing inorganic compounds.
   d. by anaerobic photosynthesis.
   e. Both c and d are correct.

24. Which pair is mismatched?
   a. red algae—multicellular, delicate, seaweed
   b. diatoms—silica shell, boxlike, golden brown
   c. euglenoids—flagella, pellicle, eyespot
   d. kelp—green alga, seaweed, microscopic
   e. *Paramecium*—cilia, calcium carbonate shell, gullet

25. Which is found in slime molds but not in fungi?
   a. nonmotile spores
   b. amoeboid vegetative cells
   c. zygote formation
   d. photosynthesis
   e. All of these are correct.

26. Which pair is mismatched?
   a. water mold—potato famine
   b. trypanosome—African sleeping sickness
   c. *Plasmodium vivax*—malaria
   d. amoeboid—severe diarrhea
   e. AIDS—*Giardia lamblia*

27. Fungi are classified according to
   a. sexual reproductive structures.
   b. shape of the sporocarp.
   c. mode of nutrition.
   d. type of cell wall.
   e. All of these are correct.

28. When sac fungi and club fungi reproduce sexually, they produce
   a. ascocarps and basidiocarps.
   b. spores.
   c. conidiospores.
   d. hyphae.
   e. Both a and b are correct.

29. Which association is mismatched?
   a. protozoans and algae—both protists
   b. fungi and bacteria—both prokaryotes
   c. protozoans and bacteria—both usually heterotrophic
   d. dinoflagellates and cyanobacteria—both photosynthesizers

30. Label the life cycle of an RNA animal virus in the following diagram.

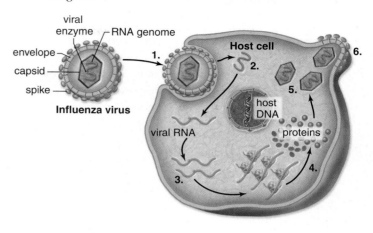

31. Some scientists consider viruses nonliving because
   a. they do not locomote.
   b. they cannot reproduce independently.
   c. their nucleic acid does not code for protein.
   d. they are acellular.
   e. Both b and d are correct.

32. The envelope of an animal virus is usually derived from the _____ of its host cell.
   a. cell wall   c. capsule
   b. plasma membrane   d. receptors

33. Retroviruses
   a. only parasitize plant cells.
   b. have a reverse life cycle.
   c. include HIV.
   d. carry on anaerobic cellular metabolism.
   e. All of these are correct.

34. Capsid proteins are synthesized during which phase of viral replication?
   a. replication
   b. biosynthesis
   c. assembly
   d. proteination
   e. All of these are correct.

35. Small changes in influenza surface antigens lead to
   a. antigenic drift.   c. antigenic schism.
   b. antigenic shift.   d. antigenic sift.

36. Antigenic shift requires
   a. reassortment of viral genomes.
   b. point mutations in viral genomes.
   c. large changes to host cells.
   d. incorporation of protein subunits.

37. Shingles is a disease observed mainly in the elderly that is a reactivation of which of the following?
   a. herpes   c. chickenpox
   b. polio   d. AIDS

38. RNA retroviruses have a special enzyme that
   a. disintegrates host DNA.   d. translates host DNA.
   b. polymerizes host DNA   e. repairs viral DNA.
   c. transcribes viral RNA to cDNA.

Understanding the Terms

algae 583	malaria 589
amoeboid 587	methanogen 583
antibiotic 581	mycelium 591
antigenic drift 600	mycorrhizal fungi 595
antigenic shift 600	mycoses 595
archaea 583	normal microflora 577
bacteria 577	nucleoid 578
binary fission 578	phytoplankton 582
brown alga 585	plasmodial (acellular)
capsid 596	slime mold 590
cellular slime mold 591	prion 602
chemoautotroph 579	protist 583
ciliate 587	protozoan 586
club fungi 593	provirus 598
conidium (pl., conidia) 593	pyrenoid 584
conjugation 578	red alga 585
contractile vacuole 587	red tide 584
cyanobacteria 579	retrovirus 598
decomposer 577	sac fungi 592
diatom 584	sporangium 592
dinoflagellate 584	spore 592
endospore 578	sporozoan 589
envelope 597	tinea 595
euglenoid 586	transduction 578
flagella 578	transformation 578
fungus 591	tubercle 581
Gram stain 578	viroid 601
green alga 584	virus 596
heterotrophic 579	water mold 590
hypha 591	yeast 593
lichen 594	zooflagellate 588
lipopolysaccharide	zooplankton 587
(LPS) 578	zygospore fungi 592

Match the terms to these definitions:

a._____ Infectious strand of RNA devoid of a capsid and much smaller than a virus.

b._____ Differential staining procedure that divides bacteria into those that are positive and those that are negative.

c._____ Diseases caused by fungi.

d._____ Organelle that pumps water out of a protist.

e._____ Symbiotic relationship between certain fungi and algae in which the fungi possibly provide inorganic food or water and the algae provide organic food.

Thinking Critically

1. In the chapter we mentioned that although a few large bacteria have been discovered, most are very small. Why does there seem to be a limit on bacterial size? Would you expect the largest bacteria to be able to grow and reproduce faster or slower than the smallest bacteria?

2. Even though archaea and bacteria are both prokaryotic, archaea are more closely related to eukaryotic cells (including plants, animals, fungi, and protists) than they are to bacteria. Since archaea and bacteria are both prokaryotic, what kind of evidence suggests that they are more closely related to humans than to, say, an *E. coli* bacterium? (See Table 28.1.)

3. Sickle cell disease occurs mostly in people of African ancestry. When a person inherits two copies of the sickle cell gene, his or her hemoglobin is abnormal, causing red blood cells to be more fragile and sticky than usual. This leads to anemia and circulatory problems. About 70,000 Americans have sickle cell disease, and about 2 million carry one copy of the gene. Even though it can cause a serious disease, the sickle cell gene is thought to have been favored by natural selection, because it gives some protection against malaria. Considering the type of cells infected by *Plasmodium* species, how might this protection work? What do you think may happen to the prevalence of the sickle cell anemia gene if an effective vaccine against malaria is ever developed and widely used?

4. Many viruses contain their own enzymes for replicating their genetic material, while others use the host cell's enzymes. Would DNA viruses or RNA viruses be more likely to produce their own enzyme(s) for this purpose, and why?

5. Why is it easier to create drugs to treat bacterial infections than protozoal or fungal infections? Specifically, why do drugs such as penicillin have a very low incidence of side effects?

ARIS, the *Inquiry into Life* Website

ARIS, the website for *Inquiry into Life,* provides a wealth of information organized and integrated by chapter. You will find practice quizzes, interactive activities, labeling exercises, flashcards, and much more that will complement your learning and understanding of general biology.

www.aris.mhhe.com

chapter **29**

Plants

If all the plants on Earth were to suddenly die *all animals, including humans, would soon perish. Yet, if animals were to die, plants would still survive and flourish. The fossil record indicates that plants colonized land before animals. This means that they were able to perform the incredible feat of moving onto dry land from an aquatic environment. The first plants to conquer land were the bryophytes (mosses, liverworts, and hornworts), which evolved around 450 million years ago (MYA). Their lack of vascular tissue and the inability to conduct water and nutrients keeps the bryophytes rather diminutive in size. The seedless vascular plants, of which ferns are the most notable, evolved about 350 MYA. Vascular tissue allowed ferns and their allies to grow very tall. If you could travel back to this time, you would see steamy jungles and swamps dominated by seedless vascular plants, many much larger than the ones in existence today. Gymnosperms were the first seed plants to evolve around 365 MYA and the angiosperms followed around 245 MYA. The ability to produce seeds is a highly successful trait. Gymnosperms dominated the land during the Mesozoic era when the dinosaurs ruled. Like the dinosaurs, the gymnosperms of that day, which include the cycads and evergreen trees, were also generally quite large. In fact, botanists like to call the Mesozoic the "age of the cycads" instead of the "age of the dinosaurs" because the cycads were much larger and more numerous during this era than they are today. As the reign of the dinosaurs waned, the angiosperms evolved. Today, angiosperms are the most widespread of the plants because of their ability to produce flowers and seeds enclosed by fruit. On every continent, plants display an astonishing diversity as well as amazing abilities to adapt to varied environments.*

Chapter Concepts

29.1   Evolutionary History of Plants

Plants (domain Eukarya, kingdom Plantae) are vital to human survival. Our dependence on them is nothing less than absolute (see Ecology Focus). Due to their present great variety, plants serve us in many different ways. Like the algae, plants are photosynthetic eukaryotes. Plants are believed to have evolved from the green alga, over 500 million years ago, because plants and green algae share several important characteristics: they contain chlorophylls *a* and *b* and various accessory pigments, store excess carbohydrates as starch, and have cellulose in their cell walls. In recent years, biochemists have compared the sequences of DNA bases coding for ribosomal RNA between organisms. The results suggest that plants are most closely related to a group of green algae known as stoneworts, perhaps those in the genus *Chara*:

Chara

The evolution of plants is marked by four events, which can be conveniently associated with the four major groups of plants living today (Fig. 29.1). The nonvascular plants (e.g., mosses), and all other groups of plants, nourish a multicellular embryo within the body of the female plant. This feature distinguishes plants from green algae. Since green algae do not protect the embryo as all plants do, protection of the multicellular embryo may be the first evolved feature that separated plants from green algae during the Paleozoic era.

The seedless vascular plants (e.g., ferns) and the other two groups of plants (gymnosperms and angiosperms) have **vascular tissue.** Vascular tissue is specialized for the transport of water and solutes throughout the body of a plant. The fossil record indicates that vascular plants evolved about 430 million years ago, during the Silurian period.

The gymnosperms (e.g., conifers), which are primarily cone-bearing plants, and the angiosperms (the flowering plants) produce seeds. A **seed** contains an embryo and stored organic nutrients within a protective coat. When you plant a seed, a plant of the next generation emerges; in other words, seeds disperse offspring. Seeds are highly resistant structures well suited to protect a plant embryo from drought, and to some extent from predators, until conditions are favorable for germination. Gymnosperms appear in the fossil record about 400 million years ago during the carboniferous period.

The fourth evolutionary event was the advent of the **flower,** a reproductive structure. Flowers attract pollinators, such as insects, and they also give rise to fruits that contain the seeds. Plants with flowers may have evolved about 200 million years ago during the cretaceous period.

Alternation of Generations

All plants have a life cycle that includes **alternation of generations.** In this life cycle, two multicellular individuals alternate,

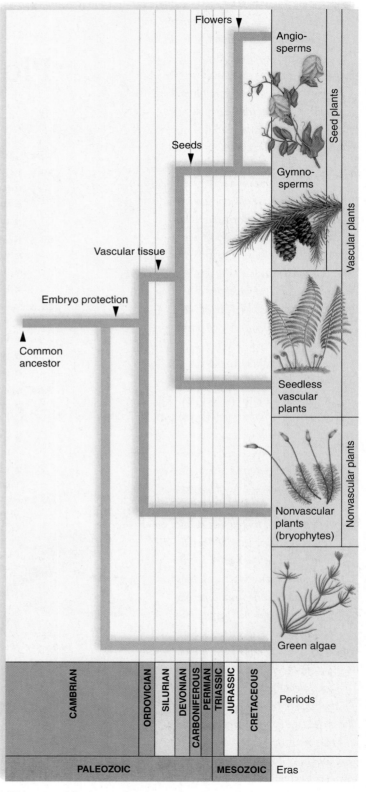

Figure 29.1 Evolutionary history of plants.
The evolution of plants involves four significant innovations: (1) Protection of a multicellular embryo was seen in the first plants to live on land. (2) The evolution of vascular tissue was another important adaptation for life on land. (3) The evolution of the seed increased the chance of survival for the next generation. (4) The evolution of the flower fostered the use of animals as pollinators and the use of fruits to aid in the dispersal of seeds.

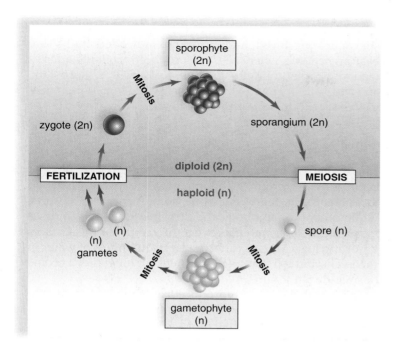

Figure 29.2 Alternation of generations.

each producing the other (Fig. 29.2). The two individuals are (1) a sporophyte, which represents the diploid generation, and (2) a gametophyte, which represents the haploid generation.

The **sporophyte** (2n) is so named for its production of spores by meiosis. A **spore** is a haploid reproductive cell that develops into a new organism without needing to fuse with another reproductive cell. In the plant life cycle, a spore undergoes mitosis and becomes a gametophyte.

The **gametophyte** (n) is so named for its production of gametes. In plants, eggs and sperm are produced by mitotic cell division. A sperm and egg fuse, forming a diploid zygote that undergoes mitosis and becomes the sporophyte.

Two observations are in order. First, meiosis produces haploid spores. This is consistent with the realization that the sporophyte is the diploid generation and spores are hap-

loid reproductive cells. Second, mitosis occurs as a spore becomes a gametophyte, and mitosis also occurs as a zygote becomes a sporophyte. Indeed, it is the occurrence of mitosis at these times that results in two generations.

Plants differ as to which generation is dominant—that is, more conspicuous. In nonvascular plants, the gametophyte is dominant, but in the other three groups of plants, the sporophyte is dominant. In the history of plants, only the sporophyte evolves vascular tissue; therefore, the shift to sporophyte dominance is an adaptation to life on land. Notice that as the sporophyte gains in dominance, the gametophyte becomes microscopic (Fig. 29.3). It also becomes dependent upon the sporophyte.

All plants have an alternation of generations life cycle, but the appearance of the generations varies widely. In ferns, a gametophyte is a small, independent, heart-shaped structure. Archegonia are female reproductive organs that produce eggs. Antheridia are male reproductive organs that produce motile sperm. The eggs are fertilized by flagellated sperm, which swim to the archegonia in a film of water. In contrast, the female gametophyte of an angiosperm, called the embryo sac, is retained within the body of the sporophyte plant and consists of seven cells within a structure called an ovule. Following fertilization, the ovule becomes a seed. In seed plants, pollen grains are mature, sperm-bearing male gametophytes. Pollen grains are transported by wind, insects, bats, or birds, and therefore, they do not need free water to reach the egg. In seed plants, reproductive cells are protected from drying out in the land environment.

Plants evolved from a group of green algae around 500 million years ago. Evolution of a protected multicellular embryo, vascular tissue, seeds, and flowers occurred in the different groups of plants. Plants exhibit alternation of generations during their life cycles, meaning that two multicellular individuals alternate, each producing the other.

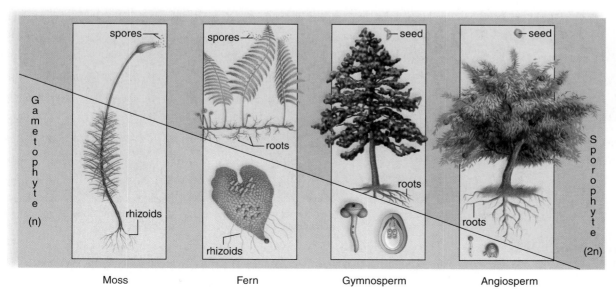

Moss Fern Gymnosperm Angiosperm

Figure 29.3 Reduction in the size of the gametophyte.
As plants became adapted to life on land, the size of the gametophyte decreased, and the size of the sporophyte increased, as evidenced by these representatives of today's plants. In the moss and fern, spores disperse the gametophyte. In gymnosperms and angiosperms, seeds disperse the sporophyte.

29.2 Nonvascular Plants

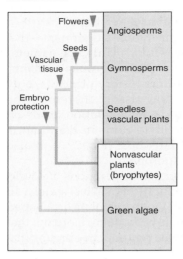

The **nonvascular plants** lack vascular tissue. Although the nonvascular plants often have a "leafy" appearance, they do not have true roots, stems, and leaves—which, by definition, must contain true vascular tissue. Therefore, the nonvascular plants are said to have rootlike, stemlike, and leaflike structures.

The gametophyte is the dominant generation in nonvascular plants—that is, it is the generation we recognize as the plant. Further, flagellated sperm swim in a continuous film of water to the vicinity of the egg. The sporophyte, which develops from the zygote, is attached to, and derives its nourishment from, the gametophyte shoot.

The nonvascular plants consist of three divisions—one each for hornworts, liverworts, and mosses. The placement of these plants in separate divisions reflects current thinking that they are not closely related. In this section, we discuss the liverworts and mosses.

Liverworts

Liverworts exist in two types: those that have a flat, lobed thallus (body; pl., thalli) are the most familiar, while those that are leafy are more numerous. *Marchantia*, a familiar liverwort (Fig. 29.4), has a smooth upper surface. The lower surface bears numerous *rhizoids* (rootlike hairs) that project into the soil. *Marchantia* reproduces both asexually and sexually. Gemma cups on the upper surface of the thallus contain gemmae, groups of cells that detach from the thallus and can asexually start a new plant. Sexual reproduction depends on disk-headed stalks that bear antheridia, where flagellated sperm are produced, and on umbrella-headed stalks that bear archegonia, where eggs are produced. Following fertilization, tiny sporophytes composed of a foot, a short stalk, and a capsule begin growing within archegonia. Windblown spores are produced within the capsule.

Mosses

Mosses can be found from the Arctic through the tropics to parts of the Antarctic. Although most live in damp, shaded locations in the temperate zone, some survive in deserts, while others inhabit bogs and streams. In forests, mosses frequently form a mat that covers the ground or the surfaces of rotting logs. Mosses can store large quantities of water in their cells, but if a dry spell continues for long, they become dormant until it rains.

Most mosses can reproduce asexually by fragmentation. Just about any part of the plant is able to grow and eventually produce leaflike thalli. Figure 29.5 describes the life cycle of a typical temperate-zone moss. The gametophyte of mosses has two stages. First, there is the algalike *protonema*, a branching filament of cells. Then, after about three days of favorable growing conditions, upright leafy thalli appear at intervals along the protonema. Rhizoids anchor the thalli, which bear antheridia and archegonia. An **antheridium** consists of a short stalk, an outer layer of sterile cells, and an inner mass of cells that become the flagellated sperm. An **archegonium,** which looks like a vase with a long neck, has a single egg located inside its base.

The dependent sporophyte consists of a *foot*, which grows down into the gametophyte tissue; a *stalk;* and an upper *capsule*, or *sporangium*, where windblown spores are produced. At first, the sporophyte is green and photosynthetic; at maturity, it is brown and nonphotosynthetic. Since the gametophyte is the dominant generation, it seems consistent for spores to be dispersal agents—that is, when the haploid spores germinate, the gametophyte is in a new location.

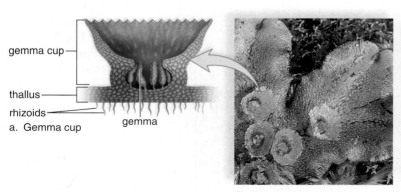

gemma cup

thallus

rhizoids

gemma

a. Gemma cup

Thallus with gemmae cups

male gametophyte

b. Male gametophytes bear antheridia

female gametophyte

c. Female gametophytes bear archegonia

Figure 29.4 Liverwort, *Marchantia.*

a. A gemma is a group of cells that can detach and start a new plant. **b.** Antheridia are present in disk-shaped stalks. **c.** Archegonia are present in umbrella-shaped stalks.

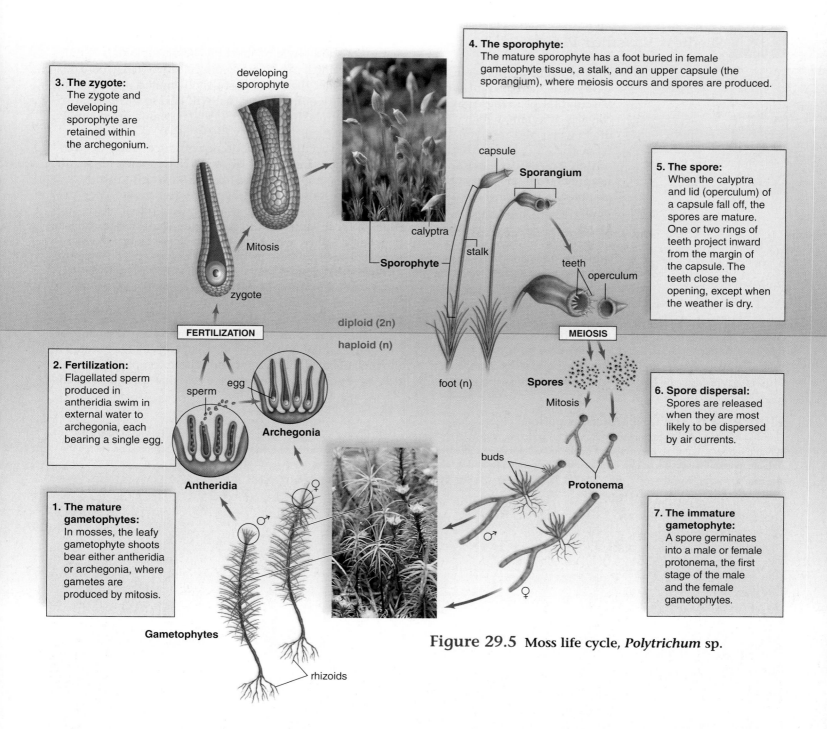

3. The zygote:
The zygote and developing sporophyte are retained within the archegonium.

developing sporophyte

4. The sporophyte:
The mature sporophyte has a foot buried in female gametophyte tissue, a stalk, and an upper capsule (the sporangium), where meiosis occurs and spores are produced.

Mitosis

capsule

Sporangium

calyptra

5. The spore:
When the calyptra and lid (operculum) of a capsule fall off, the spores are mature. One or two rings of teeth project inward from the margin of the capsule. The teeth close the opening, except when the weather is dry.

Sporophyte

stalk

teeth

operculum

zygote

diploid (2n)

FERTILIZATION

haploid (n)

MEIOSIS

2. Fertilization:
Flagellated sperm produced in antheridia swim in external water to archegonia, each bearing a single egg.

sperm

egg

Archegonia

foot (n)

Spores

Mitosis

6. Spore dispersal:
Spores are released when they are most likely to be dispersed by air currents.

Antheridia

buds

Protonema

1. The mature gametophytes:
In mosses, the leafy gametophyte shoots bear either antheridia or archegonia, where gametes are produced by mitosis.

♂

♀

7. The immature gametophyte:
A spore germinates into a male or female protonema, the first stage of the male and the female gametophytes.

♂

♀

Gametophytes

rhizoids

Figure 29.5 Moss life cycle, *Polytrichum* sp.

Adaptations and Uses of Nonvascular Plants

Although mosses are nonvascular, they are better than flowering plants at living on stone walls and on fences and even in the shady cracks of hot, exposed rocks. For these particular microhabitats, being small and simple seems to be a selective advantage. When bryophytes colonize bare rock, they help convert the rocks to soil that can be used for the growth of other organisms.

In areas such as bogs, where the ground is wet and acidic, dead mosses, especially *Sphagnum,* do not decay. The accumulated moss, called peat or bog moss, can be used as fuel. Peat moss is also commercially important because it has special nonliving cells that can absorb moisture. Thus, peat moss is often used in gardens to improve the water-holding capacity of the soil.

The nonvascular plants include the liverworts and the mosses, plants that have a dominant gametophyte. Fertilization requires an outside source of moisture; windblown spores are dispersal agents.

29.3   Seedless Vascular Plants

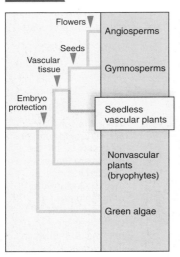

All the other plants we will study are **vascular plants.** Vascular tissue in these plants consists of **xylem,** which conducts water and minerals up from the soil, and **phloem,** which transports organic nutrients from one part of the plant to another. The vascular plants usually have true roots, stems, and leaves. The roots absorb water from the soil, and the stem conducts water to the leaves. Xylem, with its strong-walled cells, supports the body of the plant against the pull of gravity. The leaves are fully covered by a waxy cuticle except where it is interrupted by stomata, little pores whose size can be regulated to control water loss.

The sporophyte is the dominant generation in vascular plants. This is advantageous because the sporophyte is the genera-tion with vascular tissue. Another advantage of having a dominant sporophyte relates to its being diploid. If a faulty gene is present, it can be masked by a functional gene. Then, too, the greater the amount of genetic material, the greater the possibility of mutations that will lead to increased variety and complexity. Indeed, vascular plants are complex, extremely varied, and widely distributed.

Some vascular plants do not produce seeds. Seedless vascular plants include whisk ferns, club mosses, horsetails, and ferns, which disperse their offspring by producing windblown spores. When the spores germinate, they produce a small gametophyte that is independent of the sporophyte for its nutrition. In these plants, antheridia release flagellated sperm, which swim in a film of external water to the archegonia, where fertilization occurs. Because spores are dispersal agents and the nonvascular gametophyte is independent of the sporophyte, these plants cannot wholly benefit from the adaptations of the sporophyte to a terrestrial environment.

The seedless vascular plants formed the great swamp forests of the Carboniferous period (Fig. 29.6). A large number of these plants died but did not decompose completely. Instead, they were compressed to form the coal that we still mine and

Figure 29.6 The Carboniferous period.

Growing in the swamp forests of the Carboniferous period were treelike club mosses (*left*), treelike horsetails (*right*), and shorter, fernlike foliage (*lower left*). When the trees fell, they were covered by water and did not decompose well. Sediment built up and turned to rock, and the resulting compression caused the organic material to become coal, a fossil fuel that helps run our industrialized society today.

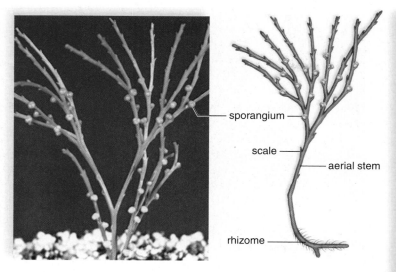

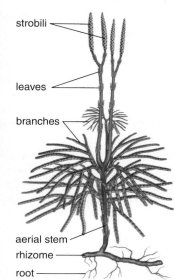

Figure 29.7 Whisk fern, *Psilotum.*
Whisk ferns have no roots or leaves—the branches carry on photosynthesis. The sporangia are yellow.

Figure 29.8 Club moss, *Lycopodium.*
Green photosynthetic stems are covered by scalelike leaves, and the sporangia are on sporophylls arranged into strobili (cones).

burn today. (Oil has a similar origin, but it most likely formed in marine sedimentary rocks and included animal remains.)

Whisk Ferns

The **psilotophytes** are represented by *Psilotum,* a whisk fern (Fig. 29.7). Whisk ferns, named for their resemblance to whisk brooms, are found in Arizona, Texas, Louisiana, and Florida, as well as Hawaii and Puerto Rico. *Psilotum* looks like a rhyniophyte, an ancient vascular plant that is known only from the fossil record. Its aerial stem forks repeatedly and is attached to a *rhizome,* a fleshy horizontal stem that lies underground. Whisk ferns have no leaves, so the branches carry on photosynthesis. Sporangia, located at the ends of short branches, produce spores that are dispersal agents.

The *Psilotum* gametophyte is independent, small (less than a centimeter), and found underground associated with mutualistic mycorrhizal fungi. The resemblance of its life cycle to that of a fern suggests that *Psilotum* is actually a fern devoid of leaves and roots. If so, it could be said to be a vestigial fern.

Club Mosses

The **club mosses** are common in moist woodlands of the temperate zone, where they are known as ground pines. Typically, a branching rhizome sends up aerial stems less than 30 cm tall. Tightly packed, evergreen, scalelike leaves cover stems and branches, giving the plant a mossy look (Fig. 29.8). In club mosses, the sporangia are borne on terminal clusters of leaves, called *strobili,* which are club-shaped. The spores are sometimes harvested and sold as lycopodium powder, or vegetable sulfur, which is used in pharmaceuticals and in fireworks because it is highly flammable.

The majority of club mosses live in the tropics and subtropics, where many of them are epiphytes—plants that live on, but do not parasitize, trees. The closely related spike mosses (*Selaginella*) are extremely varied and include the resurrection plant, which curls up into a tight ball when dry but unfurls, as if by magic, when moistened.

Horsetails

Horsetails, which thrive in moist habitats around the globe, are represented by *Equisetum,* the only genus in existence today (Fig. 29.9). A perennial rhizome produces roots and photosynthetic aerial stems that can stand about 1.3 meters

Figure 29.9 Horsetail, *Equisetum.*
Whorls of branches and tiny leaves are at the nodes of the stem. Spore-producing sporangia are borne in strobili (cones).

high. In some species, the whorls of slender green side branches at the joints (nodes) of the stem bear a fanciful resemblance to a horse's tail. The small, scalelike leaves are whorled at the nodes and are nonphotosynthetic. Many horsetails have strobili at the tips of branch-bearing stems; others send up special buff-colored, naked stems that bear the strobili.

The stems of horsetails are tough and rigid because of silica deposited in their cell walls. Early Americans, in particular, used horsetails to scour pots and called them "scouring rushes." Today, they are still used as ingredients in a few abrasive powders. *Equisetum* is sometimes considered a weed in the garden, and it is difficult to control because any portion of the rhizome left in the soil will regenerate.

Ferns

Ferns are the largest group of plants other than the flowering plants, and they display great diversity in form and habitat. Ferns are most abundant in warm, moist, tropical regions, but they are also found in northern regions and in relatively dry, rocky, places. They range in size from low-growing plants resembling mosses to tall trees. The *fronds* (leaves) that grow from a rhizome can vary. The royal fern has fronds that stand 6 feet tall; those of the maidenhair fern are branched, with broad leaflets; and those of the hart's tongue fern are straplike and leathery. Figure 29.10 gives examples of fern diversity.

Life Cycle, Adaptations, and Uses of Ferns Ferns have true roots, stems, and leaves; that is, they contain vascular tissue. The well-developed leaves fan out, capture solar energy, and photosynthesize. The life cycle of a typical temperate fern is shown in Figure 29.11. The dominant sporophyte produces wind-blown spores. When the spores germinate, a tiny green and independent gametophyte develops. The gametophyte is water dependent because it lacks vascular tissue. Also, flagellated sperm produced within antheridia require an outside source of moisture to swim to the eggs in the archegonia. Upon fertilization the zygote develops into a sporophyte. In nearly all ferns, the leaves of the sporophyte first appear in a curled-up form called a fiddlehead, which unrolls as it grows. Once established, some ferns, such as the bracken fern *Pteridium aquilinum,* can spread into drier areas by means of vegetative (asexual) reproduction. For example, ferns can spread when their rhizomes grow horizontally in the soil, producing fiddleheads that become new fronds.

In terms of economic value, people use ferns in decorative bouquets and as ornamental plants in the home and garden. Wood from tropical tree ferns often serves as a building material because it resists decay, particularly by termites. Ferns, especially the ostrich fern, are eaten as food,

Cinnamon fern, *Osmunda cinnamomea*

Hart's tongue fern, *Campyloneurum scolopendrium*

Maidenhair fern, *Adiantum pedatum*

Figure 29.10 Fern diversity.
The many different types of ferns all grow in places that offer moisture. These photos show the dominant sporophyte. The separate, delicate, and almost microscopic gametophyte is water dependent and produces sperm that require moisture in order to swim to the egg.

and in the Northeast, many restaurants feature fiddleheads as a special treat, although studies have shown that some are carcinogenic.

The seedless vascular plants include ferns, whisk ferns, club mosses, and horsetails. The sporophyte is dominant, and the offspring are dispersed by spores. The independent, nonvascular gametophyte produces flagellated sperm.

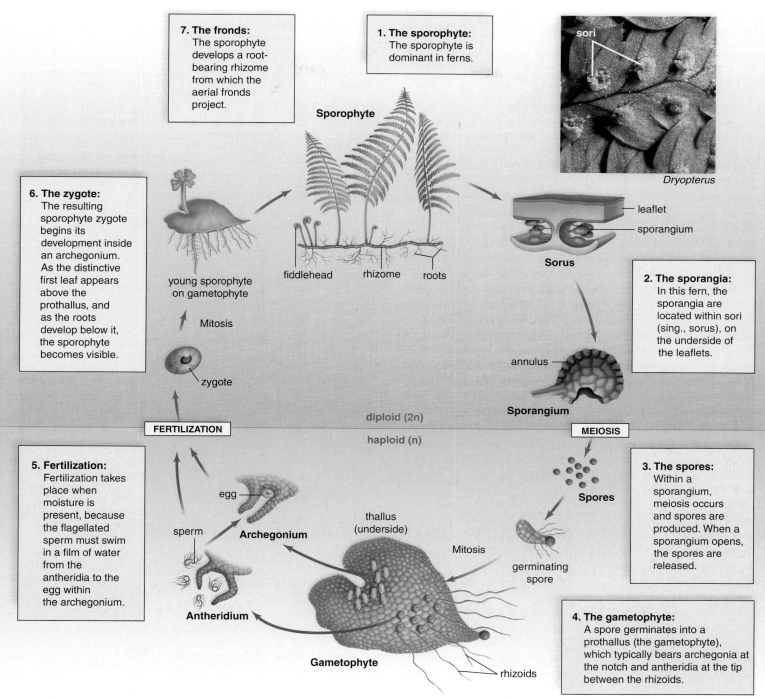

7. The fronds: The sporophyte develops a root-bearing rhizome from which the aerial fronds project.

1. The sporophyte: The sporophyte is dominant in ferns.

sori

Dryopterus

Sporophyte

leaflet

sporangium

6. The zygote: The resulting sporophyte zygote begins its development inside an archegonium. As the distinctive first leaf appears above the prothallus, and as the roots develop below it, the sporophyte becomes visible.

young sporophyte on gametophyte

fiddlehead   rhizome   roots

Sorus

2. The sporangia: In this fern, the sporangia are located within sori (sing., sorus), on the underside of the leaflets.

Mitosis

zygote

annulus

FERTILIZATION

diploid (2n)

Sporangium

haploid (n)

MEIOSIS

5. Fertilization: Fertilization takes place when moisture is present, because the flagellated sperm must swim in a film of water from the antheridia to the egg within the archegonium.

egg

sperm

Archegonium

Antheridium

Spores

3. The spores: Within a sporangium, meiosis occurs and spores are produced. When a sporangium opens, the spores are released.

thallus (underside)

Mitosis

germinating spore

Gametophyte

rhizoids

4. The gametophyte: A spore germinates into a prothallus (the gametophyte), which typically bears archegonia at the notch and antheridia at the tip between the rhizoids.

Figure 29.11 Fern life cycle.

29.4 Seed Plants

The gymnosperms and angiosperms are seed plants, the most plentiful plants in the biosphere today. Seeds contain a sporophyte embryo and stored food within a protective seed coat. The seed coat and stored food allow an embryo to survive harsh conditions during long periods of dormancy (arrested state) until environmental conditions become favorable for growth. Seeds can even remain dormant for hundreds of years. When a seed germinates, the stored food is a source of nutrients for the growing seedling. The survival value of seeds largely accounts for the dominance of seed plants today.

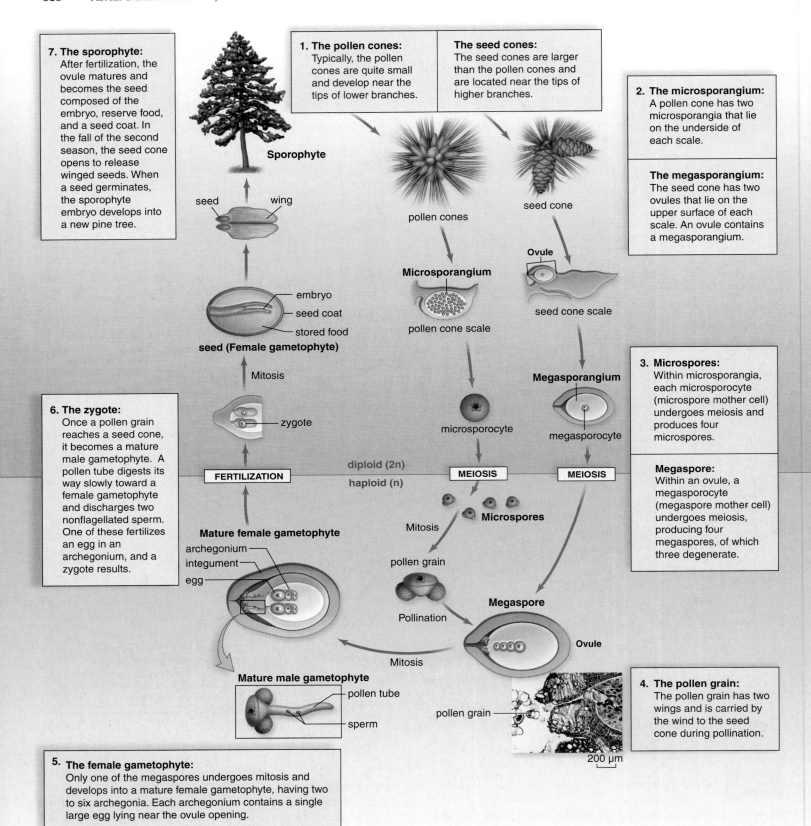

7. The sporophyte:
After fertilization, the ovule matures and becomes the seed composed of the embryo, reserve food, and a seed coat. In the fall of the second season, the seed cone opens to release winged seeds. When a seed germinates, the sporophyte embryo develops into a new pine tree.

1. The pollen cones:
Typically, the pollen cones are quite small and develop near the tips of lower branches.

The seed cones:
The seed cones are larger than the pollen cones and are located near the tips of higher branches.

2. The microsporangium:
A pollen cone has two microsporangia that lie on the underside of each scale.

The megasporangium:
The seed cone has two ovules that lie on the upper surface of each scale. An ovule contains a megasporangium.

3. Microspores:
Within microsporangia, each microsporocyte (microspore mother cell) undergoes meiosis and produces four microspores.

Megaspore:
Within an ovule, a megasporocyte (megaspore mother cell) undergoes meiosis, producing four megaspores, of which three degenerate.

4. The pollen grain:
The pollen grain has two wings and is carried by the wind to the seed cone during pollination.

6. The zygote:
Once a pollen grain reaches a seed cone, it becomes a mature male gametophyte. A pollen tube digests its way slowly toward a female gametophyte and discharges two nonflagellated sperm. One of these fertilizes an egg in an archegonium, and a zygote results.

5. The female gametophyte:
Only one of the megaspores undergoes mitosis and develops into a mature female gametophyte, having two to six archegonia. Each archegonium contains a single large egg lying near the ovule opening.

Sporophyte

seed · wing

embryo · seed coat · stored food
seed (Female gametophyte)

Mitosis

zygote

FERTILIZATION

Mature female gametophyte

archegonium · integument · egg

Mature male gametophyte
pollen tube · sperm

pollen cones

Microsporangium
pollen cone scale

microsporocyte

seed cone

Ovule

seed cone scale

Megasporangium

megasporocyte

diploid (2n)
haploid (n)

MEIOSIS   **MEIOSIS**

Microspores

Mitosis

pollen grain

Pollination

Megaspore

Ovule

Mitosis

pollen grain

200 μm

Figure 29.12 Pine life cycle.
The sporocyte is dominant, and its sporangia are borne in cones. There are two types of cones: pollen cones (microstrobili) and seed cones (megastrobili).

Figure 29.13 Conifers.
a. Conifers are adapted to living in cold climates. **b.** In pine trees, the seed cones become woody, but the pollen cones do not. **c.** The seed cones of a juniper have fleshy scales that fuse into a berrylike structure.

a. A northern coniferous forest

b. Seed cones and pollen cones of lodgepole pine

c. Juniper seed cones are fleshy

In the life cycle of the pine, a gymnosperm (Fig. 29.12), seed plants are *heterosporous,* meaning that they have two types of spores, and they produce two kinds of gametophytes—male and female. **Pollen grains,** which resist drying out, become the multicellular male gametophyte. **Pollination** occurs when a pollen grain is brought to the vicinity of a female gametophyte by wind or, in the case of flowering plants, by either wind or a pollinator. A **pollinator** is an animal, usually an insect, that carries pollen from flower to flower. After a pollen grain germinates, sperm move toward the female gametophyte through a growing **pollen tube.** *No external water is needed to accomplish fertilization.* The female gametophyte develops within an **ovule,** which eventually becomes a seed. The seed disperses offspring.

Gymnosperms

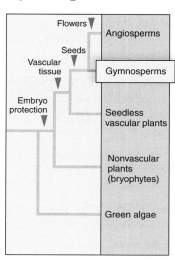

Most **gymnosperms** are cone-bearing plants. On the surfaces of their cone scales are ovules, which later become seeds. These ovules are said to be naked because they are not completely enclosed by diploid tissue (as the ovules in angiosperms are). This characteristic gives the gymnosperms their name; *gymnos* is a Greek word meaning naked, and *sperma* means seed. Botanists now divide the gymnosperms into four groups: conifers, cycads, ginkgoes, and gnetophytes.

Conifers

The better-known gymnosperms are evergreen, cone-bearing trees called **conifers.** The 575 species of conifers include the familiar pine, spruce, fir, cedar, hemlock, redwood, and cypress. Perhaps the oldest and largest trees in the world are conifers. Bristlecone pines in the Nevada mountains are known to be more than 4,900 years old, and a number of redwood trees in California are 2,000 years old and more than 100 meters tall.

Adaptations and Uses of Conifers Conifers are adapted to cold, dry weather. Thus, vast areas of northern temperate regions are covered in coniferous forests (Fig. 29.13a). The tough, needlelike leaves of pines conserve water because they have a thick cuticle and recessed stomata.

Note in the life cycle of the pine (see Fig. 29.12) that the dominant sporophyte produces two kinds of cones. **Pollen cones** (microstrobili) produce windblown pollen, and **seed cones** (megastrobili) produce windblown seeds (Fig. 29.13b, c). These cones are adaptations to a land environment.

Conifers supply much of the wood used to construct buildings and to manufacture paper. They also produce many valuable chemicals, such as those extracted from resin (pitch), a substance that protects conifers from attack by fungi and insects, and is used for waterproofing and in varnish and paints.

Other Gymnosperms

Cycads have large, finely divided leaves growing in clusters at the top of the stem. Therefore, they resemble palms or ferns, depending on their height. Pollen cones or seed cones, which grow at the top of the stem surrounded by

pollen cone   seed cones

a.

b.

c.

Figure 29.14 Other gymnosperms.

a. Cycad cones. **b.** The female Ginkgo trees are known for their malodorous seeds. **c.** *Welwitschia* is known for its enormous straplike leaves.

the leaves, can be huge—more than a meter long and weighing 40 kg (Fig. 29.14*a*). Cycads were very plentiful in the Mesozoic era at the time of the dinosaurs, and it's likely that dinosaurs fed on cycad seeds. Now, cycads are in danger of extinction because they grow very slowly, a distinct disadvantage.

Ginkgoes, although plentiful in the fossil record, are represented today by only one surviving species, *Ginkgo biloba,* the maidenhair tree, so named because its leaves resemble those of the southern maidenhair fern, *Adiantum capillus-veneris.* Female trees produce fleshy seeds, which ripen in the fall, and give off such a foul odor that male trees are usually preferred for planting (Fig. 29.14*b*). Ginkgo trees are resistant to pollution and do well along city streets and in city parks.

The three living genera of **gnetophytes** don't resemble one another. *Gnetum,* which occurs in the tropics, consists of trees or climbing vines with broad, leathery leaves arranged in pairs. *Ephedra,* occurring only in southwestern North America and southeast Asia, is a shrub with small, scalelike leaves. *Welwitschia,* living in the deserts of southwestern Africa, has only two enormous, straplike leaves, which split lengthwise as the plant ages (Fig. 29.14*c*).

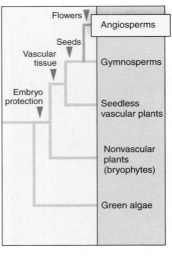

Angiosperms

Angiosperms, the flowering plants, are an exceptionally large and successful group of plants. There are 240,000 known species of angiosperms—six times the number of species of all other plant groups combined. Angiosperms live in all sorts of habitats, from fresh water to desert, and from the frigid north to the torrid tropics. They range in size from the tiny, almost microscopic duck weed to *Eucalyptus* trees over 100 meters tall. It would be impossible to exaggerate the importance of angiosperms in our everyday lives. As discussed in the Ecology Focus, they provide us with clothing, food, medicines, and other commercially valuable products.

The name angiosperm is derived from the Greek words *angio,* meaning vessel, and *sperma,* meaning seed. The seed develops from an ovule within an ovary, which becomes a fruit. Therefore, angiosperms produce covered seeds (in contrast to the exposed seeds of gymnosperms). Although the oldest fossils of angiosperms date back no more than about 135 million years, the angiosperms probably arose much earlier. Indirect evidence suggests the possible ancestors of angiosperms may have originated as long as 200 million years ago. Scientists believe that angiosperms coevolved with the insects that act as pollinators. As angiosperms became diverse, flying insects also enjoyed an adaptive radiation.

Monocots and Eudicots

Most flowering plants belong to one of two classes. The **Monocotyledones** (Liliopsida), often shortened to **monocots,** are composed of about 65,000 species; the **Eudicotyledones,** shortened to **eudicots** are composed of about 175,000 species. Table 29.1 lists the differences between monocots and eudicots.

TABLE 29.1	Monocots and Eudicots
Monocots	**Eudicots**
One cotyledon	Two cotyledons
Flower parts in threes or multiples of three	Flower parts in fours or fives or multiples of four or five
Usually herbaceous	Woody or herbaceous
Usually parallel venation	Usually net venation
Scattered bundles in stem	Vascular bundles in a ring
Fibrous root system	Taproot system

Beavertail cactus

Water lily

Louisiana iris

Dogwood flowers

Daisy

Butterfly weed flowers

Figure 29.15
Flower diversity.
Regardless of size and shape, flowers share certain features.

One of the chief differences between monocots and eudicots involves their seeds. Monocots are so called because they have only one cotyledon (seed leaf) in their seeds. Eudicots are so called because they have two cotyledons in their seeds. Cotyledons are the seed leaves—they contain nutrients that nourish the developing embryo.

The Flower

Although flowers vary widely in appearance (Fig. 29.15), most have certain structures in common. The flower stalk expands slightly at the tip into a **receptacle,** which bears the other flower parts. These parts, called **sepals, petals, stamens,** and **carpels,** are attached to the receptacle in whorls (circles) (Fig. 29.16). The sepals, collectively called the **calyx,** protect the flower bud before it opens. The sepals may drop off or may be colored like the petals. Usually, however, sepals are green and remain attached to the receptacle. The petals, collectively called the **corolla,** are quite diverse in size, shape, and color. The corolla often attracts a particular pollinator.

Each stamen consists of a saclike **anther,** where pollen is produced, and a stalk called a **filament.** In most flowers, the anther is positioned where the pollen can be carried away by wind or a pollinator. One or more carpels (sometimes called pistils) is at the center of a flower. A carpel has three major regions: ovary, style, and stigma. The swollen base is the **ovary,** which contains from one to hundreds of ovules. The **style** elevates the **stigma,** which is sticky or otherwise adapted to receive pollen grains. Glands in the region of the ovary produce nectar, a nutrient that pollinators gather as they go from flower to flower. The flower explains the success of the angiosperms.

anther

filament

stamens

stigma

pollen tube

style

ovary

ovule

carpel (pistil)

receptacle

petals (corolla)

sepals (calyx)

Figure 29.16 Generalized flower.
A flower has four main parts: sepals, petals, stamens, and carpels. Each stamen has an anther and a filament. A carpel has a stigma, a style, and an ovary. The ovary contains ovules.

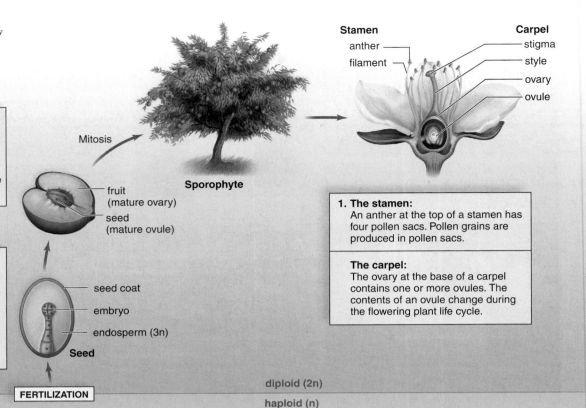

7. The sporophyte:
The embryo within a seed is the immature sporophyte. When a seed germinates, growth and differentiation produce the mature sporophyte of a flowering plant.

Mitosis

fruit (mature ovary)
seed (mature ovule)

Sporophyte

Stamen
anther
filament

Carpel
stigma
style
ovary
ovule

1. The stamen:
An anther at the top of a stamen has four pollen sacs. Pollen grains are produced in pollen sacs.

The carpel:
The ovary at the base of a carpel contains one or more ovules. The contents of an ovule change during the flowering plant life cycle.

6. The seed:
The ovule now develops into the seed, which contains an embryo and food enclosed by a protective seed coat. The wall of the ovary and sometimes adjacent parts develop into a fruit that surrounds the seed(s).

seed coat
embryo
endosperm (3n)

Seed

FERTILIZATION

diploid (2n)
haploid (n)

pollen grain
tube cell nucleus
generative cell

(mature male gametophyte)

Pollination

5. Double fertilization:
On reaching the ovule, the pollen tube discharges the sperm. One of the two sperm migrates to and fertilizes the egg, forming a zygote; the other unites with the two polar nuclei, producing a 3n (triploid) endosperm nucleus. The endosperm nucleus divides to form endosperm, food for the developing plant.

integument
polar nuclei
sperm
egg
pollen tube

pollen tube
sperm

antipodals
polar nuclei
egg
synergids

Embryo sac
(mature female gametophyte)

Figure 29.17
Flowering plant life cycle.
The parts of the flower involved in reproduction are the stamens and the carpel. Reproduction has been divided into significant stages: female gametophyte development, male gametophyte development, and sporophyte development.

4. The mature male gametophyte:
A pollen grain that lands on the carpel of the same type of plant germinates and produces a pollen tube, which grows within the style until it reaches an ovule in the ovary. Inside the pollen tube, the generative cell nucleus divides and produces two nonflagellated sperm. A fully germinated pollen grain is the mature male gametophyte.

The mature female gametophyte:
The ovule now contains the mature female gametophyte (embryo sac), which typically consists of eight haploid nuclei embedded in a mass of cytoplasm. The cytoplasm differentiates into cells, one of which is an egg and another of which contains two polar nuclei.

Life Cycle, Adaptations, and Uses of Flowering Plants
Figure 29.17 depicts the life cycle of a typical flowering plant. Sexual reproduction in flowering plants is dependent on the flower, which produces both pollen and seeds. A gymnosperm, such as a pine tree, you will recall, depends wholly on the wind to disperse both pollen and seeds. In contrast, among angiosperms, some species have windblown pollen, and others rely on a pollinator to carry pollen to another member of the same species. Bees, wasps, flies, butterflies, moths, beetles,

and even bats are pollinators. The flower provides the pollinator with nectar, a sugary substance that, in the case of bees, will become honey in the hive. Several species of bees (such as honeybees and bumblebees) also have two "pollen baskets," one on each of their hind legs. When bees make a brief stop on a flower, they suck up nectar with specialized mouthparts and gather pollen grains, which are stored in their pollen baskets. Honey and pollen are the main diets of immature bees (called larvae). When a bee visits another flower, some of the pollen

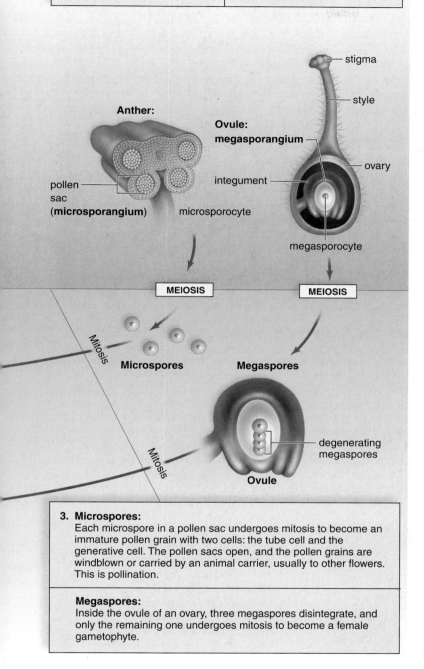

2. The microsporangia:
Pollen sacs of the anther are microsporangia, where each microsporocyte undergoes meiosis to produce four microspores.

The megasporangium:
First, an ovule within an ovary contains a megasporangium, where a megasporocyte undergoes meiosis to produce four megaspores.

3. Microspores:
Each microspore in a pollen sac undergoes mitosis to become an immature pollen grain with two cells: the tube cell and the generative cell. The pollen sacs open, and the pollen grains are windblown or carried by an animal carrier, usually to other flowers. This is pollination.

Megaspores:
Inside the ovule of an ovary, three megaspores disintegrate, and only the remaining one undergoes mitosis to become a female gametophyte.

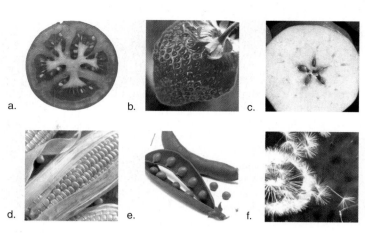

Figure 29.18 Fruits.

Angiosperms have diverse fruits. **a.** A tomato is technically a berry. **b.** A strawberry is an aggregate fruit formed from many carpels within a single flower. **c.** Apples have seeds inside a fleshy fruit. **d.** A corn cob carries many fruits (kernels). **e.** Peas have seeds in pods. **f.** Dandelion fruits are carried by wind.

ultraviolet shadings that lead the pollinator to the location of nectar at the base of the flower. Night-blooming flowers are usually aromatic and white or cream-colored, so that their smell alone, rather than color, can attract nocturnal pollinators such as bats.

Fruits, the final product of a flower, aid in the dispersal of seeds (Fig. 29.18). The production of fruit is another advantage angiosperms have over gymnosperms. Dry fruits, such as pods, assist the dispersal of windblown seeds. Mature pods sometimes explode and shoot out the seeds; other pods, such as those of peas and beans, simply break open and then the seeds are scattered. The infamous dandelion produces a one-seeded fruit with an umbrella-like crown of intricately branched hairs. The slightest gust of wind catches the hairs, raising and propelling the fruit into the air like a parachute.

Fruits also help disperse angiosperm seeds in two other ways. One means is exemplified by the coconut, a fruit that can drift for thousands of kilometers across seas and oceans. A second method of seed dispersal occurs because the seeds of some fruits have a fleshy covering that animals eat as a source of food. Only the fleshy part is digested, so the stones or seeds pass through the animal's digestive system and are deposited perhaps far away from the parent plant. Still other fruits, like those of the cocklebur, have seeds with hooks, which catch on the fur of animals and are carried away.

grains are rubbed off, and in this way the bee delivers pollen from flower to flower. Other types of pollinators also carry pollen between flowers of the same species, because pollinators and flowers are adapted to one another.

Flowering plants that have windblown pollen are usually not showy, whereas insect-pollinated flowers and bird-pollinated flowers are often colorful. Many flowers are adapted to attract quite specific pollinators. For example, bee-pollinated flowers are usually blue or yellow and have

Gymnosperms include the conifers, cycads, ginkgoes, and gnetophytes. Pollen grains, produced in pollen cones, carry the sperm to the egg, which is located in the ovule of a seed cone. After fertilization, the ovule develops into the seed, which is then released from the cone and dispersed by wind. Angiosperms are the flowering plants. The flower attracts animals (e.g., insects), which aid in pollination, and produces seeds enclosed by fruit, which aids in the seeds' dispersal.

EcologyFocus

Plants: Could We Do Without Them?

Humans derive most of their nourishment from three flowering plants: wheat, corn, and rice. All three of these plants are members of the grass family and are collectively, along with other species, called grains (Fig. 29A). Wheat is commonly used in the United States to produce flour and bread. It was first cultivated in the Near East about 8000 B.C.; hence, it is thought to be one of the earliest cultivated plants. Early settlers brought wheat to North America, where many varieties of corn, more properly called maize, were already in existence. Rice originated in southeastern Asia several thousand years ago, where it grew in swamps. Brown rice results when the seeds are threshed to remove the hulls but the seed coat and complete embryo remain. In white rice, the seed coat and embryo have been removed, leaving only the starchy endosperm.

Many foods of both plant and animal origin are bland or tasteless without spices. The Americas were discovered when Columbus was seeking a new route to the Far East to acquire spices, such as nutmeg, oregano, rosemary, and sage. Do you have an "addiction" to sugar? This simple carbohydrate comes almost exclusively from two plants—sugarcane (grown in South America, Africa, Asia, and the Caribbean) and sugar beets (grown mostly in Europe and North America). Each provides about 50% of the world's sugar.

Our most popular drinks—coffee, tea, and cola—also come from flowering plants. Coffee originated in Ethiopia, while tea is thought to have been first used somewhere in central Asia. The coca plant is the source of both the cola in the original Coca-Cola and cocaine.

Cotton and rubber are two plants that still have many uses today (Fig. 29B). Until a few decades ago, cotton and other natural fibers were our only source of clothing. Interestingly, when Levi Strauss wanted to make a tough pair of jeans, he needed a stronger fiber than cotton, so

he used hemp. Hemp (marijuana) is now known primarily as a hallucinogenic drug. Rubber had its origin in Brazil from the thick, white sap of the rubber tree. To produce a stronger rubber, such as that in tires, sulfur is added, and the sap is heated in a process called vulcanization. This produces a flexible material less sensitive to temperature changes. Today, though, much rubber is synthetically produced.

Plants have also been used for centuries for a number of important household items, including the house itself. We are most familiar with lumber as the major structural portion in buildings. This wood comes mainly from a variety of conifers: pine, fir, and spruce, among others. In the tropics, trees and even herbs provide important components for houses. In rural parts of Central and South America, palm leaves are preferable to tin for roofs, since they last as long as ten years and are quieter during rainstorms. In the Near East, numerous houses along rivers are made entirely of reeds.

An actively researched area of plant use today is that of medicinal plants. Currently, about 50% of all pharmaceutical drugs originate from plants. The treatment of cancers appears to rest in the discovery of miracle plants. Indeed, the National Cancer Institute and most pharmaceutical companies have spent millions of dollars to send botanists out to collect and test plant samples from around the world. Tribal medicine men, or shamans, of South America and Africa have already been of great importance in developing numerous drugs. Many plant extracts continue to be misused for their hallucinogenic or other effects on the human body; examples are coca, the source of cocaine and crack, and opium poppy, the source of heroin.

In addition to all these uses of plants, we should not forget their aesthetic value. Flowers brighten any yard, ornamental plants accent landscaping, and trees provide cooling shade during the summer as well as shelter from harsh winds during the winter.

Discussion Questions

1. Besides cotton, can you think of other plants from which we derive fabric?
2. Why do you think plants are such a good source of drugs for human use? What advantage does this give the plants?
3. With over 250,000 species of angiosperms alone, why do you think we get most of our nourishment from only three plants (wheat, corn, and rice)?

Rubber

Cotton

Figure 29B Plants used commercially.

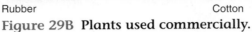

grain head

Wheat plants, *Triticum*

Figure 29A Plants used for food.

ear

Corn plants, *Zea*

grain head

Rice plants, *Oryza*

Summarizing the Concepts

29.1 Evolutionary History of Plants

Plants are multicellular, photosynthetic organisms adapted to a land existence. Their most significant evolved features are a protected embryo, vascular tissue, seeds, and flowers. Plants have an alternation of generations life cycle, but some have a dominant gametophyte (haploid generation), while others have a dominant sporophyte (diploid generation).

Plants are divided into nonvascular plants, seedless vascular plants, and seed plants based on the presence or absence of vascular tissue and the type of structure that disperses the plant.

29.2 Nonvascular Plants

The nonvascular plants, which include the liverworts and the mosses, do not have vascular tissue, and therefore, they lack true roots, stems, and leaves.

In nonvascular plants, the gametophyte is dominant; antheridia produce swimming sperm that use external water to reach the eggs in the archegonia. Following fertilization, the dependent moss sporophyte consists of a foot, a stalk, and a capsule within which windblown spores are produced by meiosis. Each spore germinates to produce a gametophyte. The use of windblown spores to disperse the gametophyte is the nonvascular plants' chief adaptation to reproducing on land.

29.3 Seedless Vascular Plants

Vascular plants have vascular tissue—that is, xylem and phloem. Xylem transports water and minerals and also supports plants against the pull of gravity. Phloem transports organic nutrients from one part of the plant to another.

In the life cycle of vascular plants, the sporophyte is dominant. Whisk ferns, club mosses, horsetails, and ferns are seedless vascular plants that were prominent in swamp forests during the Carboniferous period. In the life cycle of seedless vascular plants, spores disperse the gametophyte, and the separate gametophyte produces flagellated sperm. Vegetative (asexual) reproduction is sometimes used to disperse ferns in dry habitats.

In seedless vascular plants, the dominant sporophyte produces windblown spores. The separate gametophyte produces flagellated sperm within antheridia and eggs within archegonia.

29.4 Seed Plants

Seed plants have a life cycle that is fully adapted to the land environment. The spores develop inside the body of the sporophyte and are of two types. The male gametophyte is the pollen grain, which produces nonflagellated sperm, and the female gametophyte, which is located within an ovule, produces an egg. The pollen grain is windblown or carried by an animal, whereas flagellated sperm of seedless vascular plants require an outside source of moisture to reach an egg.

Gymnosperms, which include the conifers, produce seeds that are uncovered (naked). In gymnosperms, the male gametophytes develop in pollen cones. The female gametophyte develops within an ovule located on the scales of seed cones. Following pollination and fertilization, the ovule becomes a winged seed that is dispersed by wind. Seeds contain the next sporophyte generation. Angiosperms, also called the flowering plants, produce seeds that are covered by fruits. The petals of flowers attract pollinators, and the ovary develops into a fruit, which aids seed dispersal. Angiosperms provide most of the food that sustains terrestrial animals, and they are the source of many products used by humans.

Testing Yourself

Choose the best answer for each question.

1. Unlike green algae, plants
   a. contain chloroplasts.
   b. nourish their embryos within the female tissue.
   c. store carbohydrates as starch.
   d. contain cellulose in their cell walls.

2. The spores produced by a plant are
   a. haploid and genetically different from each other.
   b. haploid and genetically identical to each other.
   c. diploid and genetically different from each other.
   d. diploid and genetically identical to each other.

3. The gametophyte is the dominant generation in
   a. ferns.   c. gymnosperms.
   b. mosses.   d. angiosperms.

4. Label the stages in the following diagram of the plant life cycle.

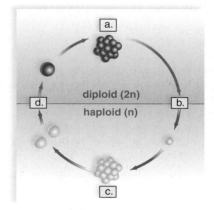

5. In mosses, meiosis occurs in
   a. antheridia.   c. capsules.
   b. archegonia.   d. protonema.

6. A fern sporophyte will develop on which region of the gametophyte?
   a. near the notch, where the archegonia are located
   b. near the tip, where the antheridia are located
   c. anywhere on the gametophyte
   d. All of these are correct.

7. The gymnosperm that is a common street tree and represented by only one living species is the
   a. pine.   c. cycad.
   b. ginkgo.   d. ephedra.

For questions 8–13, match the description with the appropriate plant group in the key. Some answers can be used more than once. Some questions will have more than one answer.

Key:
   a. nonvascular plants   c. gymnosperms
   b. seedless vascular plants d. angiosperms

8. Contain xylem and phloem

9. Produce seeds

10. Produce flagellated sperm

11. Gametophyte is separate from sporophyte

12. Produce seeds on scales

13. Produce flowers

14. In contrast to seed plants, seedless vascular plants
    a. are not heterosporous.
    b. produce gametophytes.
    c. have male gametophytes that travel to the female gametophytes.
    d. produce ovules.

15. Which statement about the conifer life cycle is false?
    a. Meiosis produces spores.
    b. The gametophyte is the dominant generation.
    c. The seed is a dispersal agent.
    d. Male gametophytes are carried by the wind.

16. Cycads are in danger of extinction because
    a. they occupy very specialized niches.
    b. they have unusual mineral nutrient requirements.
    c. they grow very slowly.
    d. their natural habitats are becoming too warm.

17. Flowering plants are called angiosperms because they
    a. produce seeds within fruits.
    b. have a dominant sporophyte generation.
    c. produce flowers.
    d. are more diverse than any other group of plants.

18. Label the parts of the flower in the following diagram.

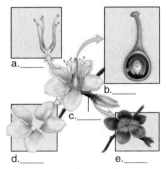

19. Unlike eudicots, monocots have
    a. woody tissue.
    b. two seed leaves.
    c. scattered vascular bundles in their stems.
    d. flower parts in multiples of fours or fives.

20. A seed contains a mature
    a. embryo.   c. ovary.
    b. ovule.   d. pollen grain.

21. Endosperm is produced by the union of the
    a. egg and polar nuclei.   c. polar and sperm nuclei.
    b. egg and sperm nuclei.   d. polar and egg nuclei.

Understanding the Terms

alternation of generations 608
angiosperm 618
anther 619
antheridium 610
archegonium 610
calyx 619
carpel 619
club moss 613
conifer 617
corolla 619
cycad 617
eudicot (Eudicotyledones) 618
fern 614
filament 619
flower 608
fruit 621
gametophyte 609
ginkgo 618
gnetophyte 618
gymnosperm 617
horsetail 613
liverwort 610
monocot (Monocotyledones) 618
moss 610
nonvascular plant 610
ovary 619
ovule 617
petal 619
phloem 612
pollen cone 617
pollen grain 617
pollen tube 617
pollination 617
pollinator 617
psilotophyte 613
receptacle 619
seed 608
seed cone 617
sepal 619
spore 609
sporophyte 609
stamen 619
stigma 619
style 619
vascular plant 612
vascular tissue 608
xylem 612

Match the terms to these definitions:

a._____ Egg-producing structure, as in the moss life cycle.
b._____ Seed plant with uncovered (naked) seeds; examples are conifers and cycads, which bear cones.
c._____ In the life cycle of a plant, the diploid generation that produces spores by meiosis.
d._____ Mature ovule that contains a sporophyte embryo with stored food enclosed by a protective coat.
e._____ Structure in a flower that consists of a stigma, a style, and an ovary.

Thinking Critically

1. What characteristics of angiosperms have allowed this group of plants to dominate Earth?

2. What advantages do being small with no roots give the bryophytes?

3. Compare and contrast the plant alternation of generation life cycle with the human life cycle.

4. What characteristics of gymnosperms have allowed some members of this group to have a life span much longer than that of most angiosperms?

5. Why are pollen cones usually located on the lower branches, whereas seed cones are located on the upper branches of the same tree? (What does this type of arrangement prevent?)

ARIS, the *Inquiry into Life* Website

ARIS, the website for *Inquiry into Life,* provides a wealth of information organized and integrated by chapter. You will find practice quizzes, interactive activities, labeling exercises, flashcards, and much more that will complement your learning and understanding of general biology.

www.aris.mhhe.com

A monarch butterfly emerges from its cocoon. *We recognize the beauty of a butterfly, but we tend to think of many other invertebrates as pests. However, most biologists would argue that "invertebrates rule the world," since invertebrate species far outnumber vertebrates. Insects, in particular, number in the millions of species, and upon close inspection, their lifestyles are fascinating.*

Take, for example, the termite, a colonial insect. Although termites are generally thought of as wood-eating pests, there are over 3,000 species, most of which are not pests! As a group, termites are over 200 million years old, and some species are capable of building mounds that are the strongest and tallest (20 feet, equivalent to a 2,000-foot building) nonhuman structures on Earth. Within their mound, they build ventilation shafts that function like air conditioning units, keeping the mound at a constant temperature. There are typically three social classes of termites (all born from the same queen, which can lay 3 million eggs per year): Workers build, soldiers defend, and harvesters help feed everyone. Harvesters can recycle nutrients to improve soil fertility. Within the mound, harvesters actually cultivate fungi, which serve as food for the nest and may also have antibacterial properties that prevent infection.

Termites are an example of the fascinating diversity of invertebrates throughout the world. In this chapter, you will learn about the lifestyles and features of some of the major phyla of invertebrates.

chapter **30**

Animals: Part I

Chapter Concepts

30.1 Evolutionary Trends Among Animals

Animals are members of the domain Eukarya and the kingdom Animalia. Animals are extremely diverse (Fig. 30.1), but in general they share the following characteristics:

1. Heterotrophic; usually acquire food by ingestion followed by digestion
2. Typically have the power of movement or locomotion by means of muscle fibers
3. Multicellular; most have specialized cells that form tissues and organs
4. Have a life cycle in which the adult is typically diploid
5. Usually undergo sexual reproduction and produce an embryo that goes through developmental stages

The more than 30 animal phyla are believed to have evolved from a protistan ancestor approximately 700 million years ago. All but one of these phyla are considered **invertebrates,** which are animals without an endoskeleton of bone or cartilage. Chordata is the only phylum whose members have an endoskeleton of bone or cartilage. Most chordates are **vertebrates,** in which the vertebral column replaces the notochord of early chordates.

TABLE 30.1	Animal Phyla, Part I*

Multicellular organisms with well-developed tissues; usually motile; heterotrophic by ingestion, generally in a digestive cavity; adults are diploid.

Phylum Porifera: sponges

Asymmetrical without well-developed tissues; body wall perforated by pores; inner cavity lined by collar cells. Example: sponges

Phylum Cnidaria: cnidarians

Two tissue layers; radial symmetry; sac body plan; tentacles having stinging cells with nematocysts. Examples: hydra, jellyfish, corals, anemones

Phylum Platyhelminthes: flatworms

Bilateral symmetry; three germ layers; solid (no body cavity); digestive tract with one opening. Examples: Planarian, tapeworms, liver flukes

Phylum Nematoda: roundworms

Pseudocoelomate with tube-within-a-tube plan. Examples: some important animal parasites

Phylum Mollusca: molluscs

Soft-bodied coelomate with body divided into three parts: foot, visceral mass, and mantle; many have shells and a rasping tongue called a radula. Examples: clams, squid, snails, octopuses

Phylum Annelida: annelids

Segmented worms; coelomate having a digestive tract with specialized parts; bristles called setae on each segment. Examples: earthworms, marine worms, leeches

Phylum Arthropoda: arthropods

Exoskeleton covers segmented body; jointed appendages; most successful of all animal phyla. Examples: crustaceans (shrimps, crabs, lobsters, barnacles, copepods, crayfish), insects (flies, ants, bees, grasshoppers, cockroaches, beetles), arachnids (scorpions, spiders, ticks), horseshoe crabs

*Animal phyla continued in Table 31.1, page 652.

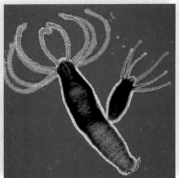
Hydra, *Hydra*

Green crayfish, *Barbi*

Figure 30.1
Animal diversity.

Hydras, crayfish, and humans are all multicellular heterotrophic animals, but they differ in certain fundamental ways. In terms of level of organization, hydras have only tissues, but crayfish and humans have organ systems. Hydras do not have an identifiable head, whereas crayfish and humans do. Their body plans also differ—hydras do not have a complete digestive system, while crayfish and humans have a complete digestive tract surrounded by body wall. Crayfish and humans share other features, even though one is an invertebrate and the other is a vertebrate. They both have a skeleton with jointed appendages, but the crayfish skeleton is external, while the human skeleton is internal.

Human being, *Homo*

This chapter discusses examples of several invertebrate phyla with regard to several key evolutionary innovations (Fig. 30.2). Although the exact history of animal evolution is not known, an examination of these innovations does provide insight into the possible evolutionary history of animals. Table 30.1 previews the animal groups we will discuss in this chapter.

Level of Organization

Three levels of organization exist in the animal kingdom: the *cellular, tissue,* and *organ* levels. At the cellular level, an animal has no true tissues; sponges are a good example of animals that have the cellular level of organization. Three germ layers are possible in animals: ectoderm, endoderm, and mesoderm. Animals that have only an ectoderm and endoderm, exemplified by *Hydra,* have the tissue level of organization, which is more recently evolved than the cellular level. Animals that have all three germ layers have the organ level of organization, which is the most recently evolved. This includes the vast majority of animals; therefore, crayfish and humans have this level of organization (see Fig. 30.1).

Visual Focus

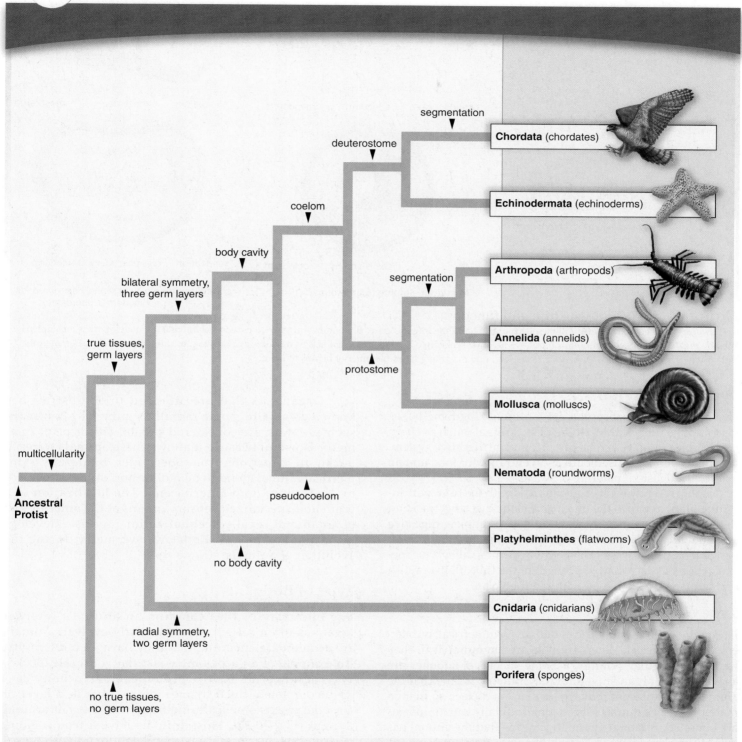

Figure 30.2 Evolutionary tree.

All animals are believed to be descended from a protist; the Porifera (sponges), with the cellular level of organization, may have evolved separately.

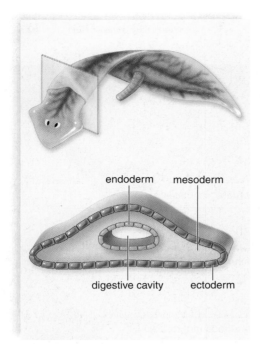

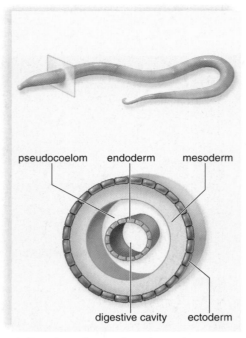

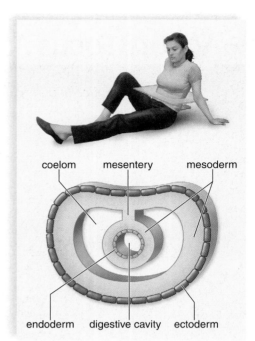

| endoderm | mesoderm | | pseudocoelom | endoderm | mesoderm | | coelom | mesentery | mesoderm |

digestive cavity   ectoderm

digestive cavity   ectoderm

endoderm   digestive cavity   ectoderm

a. **Acoelomate** (flatworms)   b. **Pseudocoelomate** (roundworms)   c. **Coelomate** (molluscs, annelids, arthropods, echinoderms, chordates)

Figure 30.3 Coelom structure and function.

a. Flatworms have no body cavity, and mesodermal tissue fills the interior space. **b.** Roundworms have a pseudocoelom, and the body cavity is incompletely lined by mesodermal tissue. In animals that have no other skeleton, a fluid-filled coelom acts as a hydrostatic skeleton. **c.** Humans are true coelomates; the body cavity is completely lined by mesodermal tissue, and mesentery holds the internal organs in place.

Type of Body Plan

Two body plans are observed in the animal kingdom: the sac plan and the tube-within-a-tube plan. Animals such as *Hydra* have the **sac plan,** which has an incomplete digestive system. Its single opening is used both as an entrance for food and an exit for undigested material, or waste.

More recently evolved are animals with the **tube-within-a-tube plan,** in which the inner tube is the digestive tract and the outer tube is the body wall. This is a complete digestive system with a separate entrance for food and an exit for undigested material, or waste. Having two openings allows for specialization of parts along the length of the tube. Both crayfish and humans have the tube-within-a-tube plan.

Type of Symmetry

Animals can be asymmetrical, radially symmetrical, or bilaterally symmetrical. When animals are **asymmetrical,** they have no particular symmetry. Some species of sponges are asymmetrical organisms. **Radial symmetry** means that the animal is organized circularly, similar to a wheel, so that no matter how the animal is sliced longitudinally, mirror images are obtained. **Bilateral symmetry** means that the animal has definite right and left halves; only a longitudinal cut down the center of the animal will produce a mirror image.

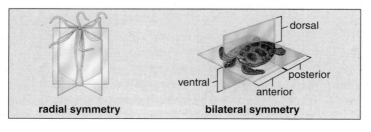

radial symmetry   bilateral symmetry

Organisms that are attached to a substrate are known as **sessile.** Some radially symmetrical animals, such as corals, are considered sessile. This type of symmetry is useful because it allows these animals to reach out in all directions from one center. Bilaterally symmetrical animals tend to be mobile and to move forward by leading with an anterior end. The localization of a brain and specialized sensory organs at the anterior end of an animal is called **cephalization,** and is a development that accompanied bilateral symmetry during the evolution of animals.

Type of Body Cavity

The body cavity that contains an animal's internal organs is often called a **coelom.** Early-evolved animals are **acoelomate,** meaning that they have no body cavity. An example of an acoelomate is a flatworm (Fig. 30.3a). The roundworms have a **pseudocoelom,** a body cavity incompletely lined by mesoderm—that is, a layer of mesoderm exists beneath the body wall but not around the gut (Fig. 30.3b). Most animals have a true coelom, meaning the body cavity is completely lined with mesoderm. Earthworms, crayfish, and humans are all coelomates (Fig. 30.3c).

Based on their embryonic development, scientists recognize two major groups of coelomates. When the first embryonic opening becomes the mouth, the animal is a **protostome.** Clams, earthworms, and crayfish are protostomes. When the second opening becomes the mouth, the animal is a **deuterostome.** Starfishes and humans are deuterostomes.

Segmentation

Segmentation is the repetition of body parts along the length of the body. Animals can be nonsegmented or segmented. Among coelomates, molluscs and echinoderms are nonsegmented, while annelids, arthropods, and chordates are segmented. Segmentation in these animals leads to specialization of parts because the various segments can become differentiated for specific purposes, as is particularly evident in arthropods and chordates.

In vertebrates, muscles develop from repeated blocks of mesoderm called somates that occur in the embryo. Segmentation is also seen in the series of bones called vertebrae that make up the vertebral column.

Jointed Appendages

In arthropods, an exoskeleton protects and covers soft tissues, and the jointed appendages provide a means of locomotion that lends itself to living on land as well as in the water. As we will learn in Chapter 31, vertebrates have a jointed endoskeleton that protects their internal organs and still permits body movement on both land and water.

> The evolution of animals is marked by certain key innovations in body plan, such as type of symmetry, type of body cavity, and whether or not they are segmented.

30.2 Introducing the Invertebrates

The sponges, cnidarians, flatworms, and nematodes (roundworms) demonstrate several key evolutionary trends (see Fig. 30.2).

Sponges: Multicellular

Sponges are placed in phylum Porifera (5,000 species) because their saclike bodies are perforated by many pores. Sponges are aquatic, largely marine animals that vary greatly in size, shape, and color. Sponges are multicellular but lack organized tissues. Therefore, *sponges have the cellular level of organization.* Evolutionists believe

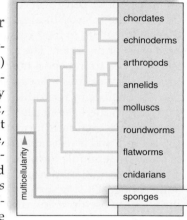

that sponges are out of the mainstream of animal evolution and represent an evolutionary dead end.

In sponges, the outer layer of the body wall contains flattened epidermal cells, some of which have contractile fibers; the middle layer is a semifluid matrix with wandering amoeboid cells; and the inner layer is composed of flagellated cells called **collar cells,** or choanocytes (Fig. 30.4). The beating of the flagella produces water currents that flow through the pores into the central cavity and out through the osculum, the upper opening of the body. Even a simple sponge only 10 cm

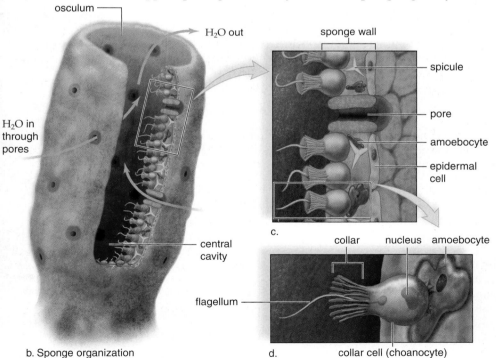

a. Yellow tube sponge, *Aplysina fistularis* b. Sponge organization c. d. collar cell (choanocyte)

Figure 30.4 Simple sponge anatomy.

a. A simple sponge. **b.** The wall of a sponge contains two layers of cells, the outer epidermal cells and the inner collar cells as shown in **(c).** The collar cells **(d,** enlarged) have flagella that beat, moving the water through pores as indicated by the blue arrows in **(b).** Food particles in the water are trapped by the collar cells and digested within their food vacuoles. Amoebocytes transport nutrients from cell to cell; spicules form an internal skeleton in some sponges.

tall is estimated to filter as much as 100 liters of water each day. It takes this much water to supply the needs of the sponge. A sponge is a sessile **filter feeder,** an organism that filters its food from the water by means of a straining device—in this case, the pores of the walls and the microvilli making up the collar of collar cells. Microscopic food particles that pass between the microvilli are engulfed by the collar cells and digested by them in food vacuoles or are passed to the amoeboid cells for digestion. The amoeboid cells also act as a circulatory device to transport nutrients from cell to cell, and they produce the sex cells (the egg and the sperm) and spicules.

Sponges can reproduce both asexually and sexually. They reproduce asexually by fragmentation followed by regeneration, gemmule formation, or budding. A gemmule is like a spore and is resistant to drying out, freezing, or the lack of oxygen. Gemmule formation occurs in freshwater sponges, and when conditions are favorable, a gemmule gives rise to an adult sponge. During sexual reproduction, sperm are released through the osculum and drawn into other sponges through the pores, where they fertilize eggs within the body. Most sponges are **hermaphroditic,** meaning they possess both male and female sex organs; however, they usually do not self-fertilize. After fertilization, the zygote develops into a flagellated larva that may swim to a new location. Sponges are capable of regeneration—if the cells of a sponge are separated, they will reassemble into a complete and functioning organism!

Sponges are classified on the basis of their skeleton. Some sponges have an internal skeleton composed of **spicules,** small needle-shaped structures with one to six rays. Chalk sponges have spicules made of calcium carbonate; glass sponges have spicules that contain silica. Most sponges have a skeleton composed of fibers of spongin, a soft protein. A bath sponge is the dried spongin skeleton from which all living tissue has been removed. Today, however, commercial "sponges" are usually synthetic. Note that the popular "loofah" is not really a sponge; it is actually a spongy plant related to cucumbers.

Sponges have a cellular level of organization and most likely evolved independently from protozoans. They are the only animals that digest their food strictly within cells.

Cnidarians: True Tissues

Cnidarians (phylum Cnidaria, about 9,000 species) are a large and ancient group of invertebrates with a rich fossil record. They are multicellular, tubular, or bell-shaped animals that reside mainly in shallow coastal waters, except for the oceanic jellyfishes. Cnidarians are radially symmetrical.

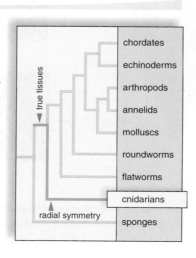

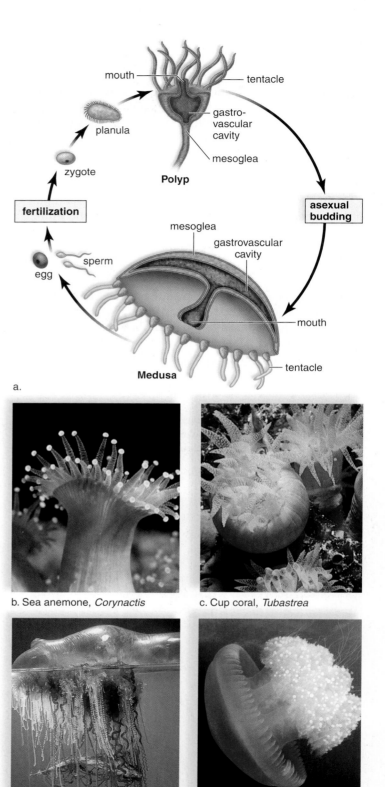

b. Sea anemone, *Corynactis*   c. Cup coral, *Tubastrea*

d. Portuguese man-of-war, *Physalia*   e. Jellyfish, *Crambione*

Figure 30.5 Cnidarian diversity.
a. The life cycle of a cnidarian. Some cnidarians have both a polyp stage and a medusa stage; in others, one stage may be dominant or absent altogether. **b.** The anemone, which is sometimes called the flower of the sea, is a solitary polyp. **c.** Corals are colonial polyps residing in a calcium carbonate or proteinaceous skeleton. **d.** The Portuguese man-of-war is a colony of modified polyps and medusae. **e.** True jellyfishes undergo the complete life cycle; this is the medusa stage. The polyp is small.

During development, cnidarians have two germ layers (ectoderm and endoderm), and as adults *cnidarians have the tissue level of organization.* Unique to cnidarians are specialized stinging cells, called **cnidocytes,** which give the phylum its name. Each cnidocyte has a capsule called a **nematocyst,** which contains a long, spirally coiled, hollow thread. When the trigger of the cnidocyte is touched, the nematocyst is discharged. Some threads merely trap a prey or predator; others have spines that penetrate and inject paralyzing toxins. These can give a nasty "sting" that some people experience when swimming; the box jellyfish from Australia has a sting so powerful it can be deadly to humans!

Cnidarians have two basic body forms—the polyp and the medusa. The mouth of a **polyp** is directed upward from the substrate, while the mouth of a jellyfish of the **medusan** is directed downward. A medusa has more jellylike packing material called mesoglea than a polyp, and its tentacles are concentrated on the margin of the bell. During its life cycle, a cnidarian typically alternates between the sessile polyp stage and the sexual, motile, medusa stage (Fig. 30.5a). A medusa produces eggs and sperm, and the zygote develops into a ciliated larva capable of dispersal. In many of today's cnidarians, one stage is dominant and the other is reduced; in some species, one form is absent altogether. However, at one time in the evolutionary history of these animals, their life cycle may have always included these two forms.

Cnidarians are quite diverse (Fig. 30.5b–e). Sea anemones (class Anthozoa) are solitary polyps, often very large and with thick walls. They can be brightly colored, resembling beautiful flowers. In the same class are **corals,** from which coral reefs take their name. Corals are similar to sea anemones, but they have calcium carbonate skeletons. Some corals are solitary, but most are colonial, forming either flat and rounded or upright and branched colonies. The slow accumulation of coral skeletons forms coral reefs, which are areas of biological abundance in warmer waters (see the Bioethical Focus at the end of this chapter). The Portuguese man-of-war (class Hydrozoa) (Fig. 30.5d) is a colony of polyp and medusa types of individuals. One polyp becomes a gas-filled float, and the other polyps are specialized for feeding. The medusae are specialized for reproduction. In jellyfishes (class Scyphozoa) (Fig. 30.5e), the medusa is the primary stage of the life cycle, and the polyp remains quite small and inconspicuous.

Hydra

Hydra (class Hydrozoa) is a freshwater cnidarian commonly found attached to underwater plants or rocks in most lakes and ponds. The body is a small, tubular polyp about 7.5 mm in length (Fig. 30.6). Like all cnidarians, a hydra has the sac body plan; that is, a single opening serves as both mouth and anus. The outer tissue layer is a protective epidermis derived from ectoderm. The inner tissue layer, derived from endoderm, is called a gastrodermis. The two tissue layers are separated by mesoglea. There are both circular and longitudinal muscle fibers. Nerve cells located below the epidermis, near the mesoglea, interconnect and form a nerve net that communicates with sensory cells throughout

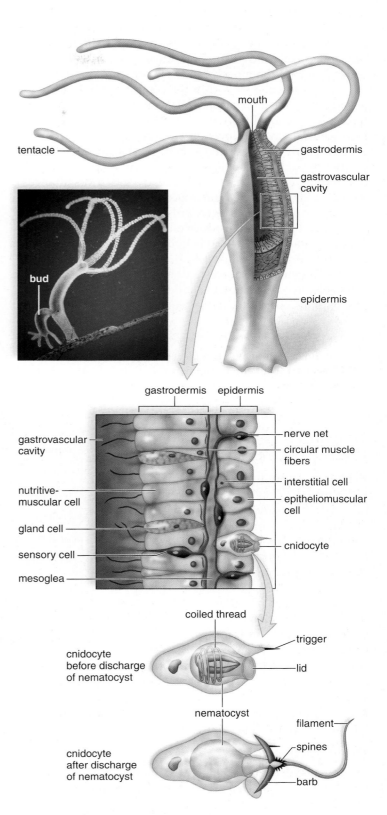

Figure 30.6 Anatomy of *Hydra*.
Top: The body of *Hydra* is a small, tubular polyp whose wall contains two tissue layers. *Hydra* reproduces asexually by forming outgrowths called buds (see photo) that develop into a complete animal. *Middle:* Various types of cells in the body wall. *Bottom:* Cnidocytes are cells that contain nematocysts.

the body. The nerve net allows transmission of impulses in several directions at once. Because they have both muscle fibers and nerve fibers, cnidarians are capable of directional movement; the body can contract or extend, and the tentacles that ring the mouth can reach out and grasp prey.

Digestion begins within the central cavity but is completed within the food vacuoles of gastrodermal cells. Nutrient molecules pass by diffusion to the other cells of the body. The large central cavity allows gastrodermal cells to exchange gases directly with a watery medium. Because the central cavity carries on digestion and acts as a circulatory system by distributing food and gases, it is called a **gastrovascular cavity.** All cnidarians have a gastrovascular cavity.

Although hydras exist only as polyps (there are no medusae), they can reproduce either sexually or asexually. When sexual reproduction is going to occur, an ovary or a testis develops in the body wall. Like sponges, cnidarians can also regenerate from a small piece. When conditions are favorable, hydras produce small outgrowths, or buds, that pinch off and begin to live independently.

> Cnidarians exhibit the tissue level of organization. Their life cycle typically alternates between a medusa stage and a polyp stage.

Flatworms: Bilateral Symmetry

Flatworms (phylum Platy-helminthes, 20,000 species) have *bilateral symmetry*. Like all the other animal phyla we will study, they also have three germ layers. However, the flatworms have no coelom—they are acoelomates. The presence of mesoderm, in addition to ectoderm and endoderm, gives bulk to the animal and leads to greater

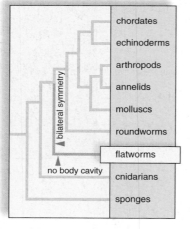

complexity. Free-living flatworms have muscles and excretory, reproductive, and digestive systems. The worms lack respiratory and circulatory systems; because the body is flat and thin, diffusion alone can pass needed oxygen and other substances from cell to cell.

Planarians

Freshwater planarians (class Turbellaria), shown in Figure 30.7, are small (several millimeters to several centimeters), literally flat, worms. Some tend to be colorless; others have brown or black pigmentation. Planarians live in lakes, ponds, streams, and springs, where they feed on small living or dead organisms, such as worms and crustaceans.

Planarians have an excretory system that consists of a network of interconnecting canals extending through much of the

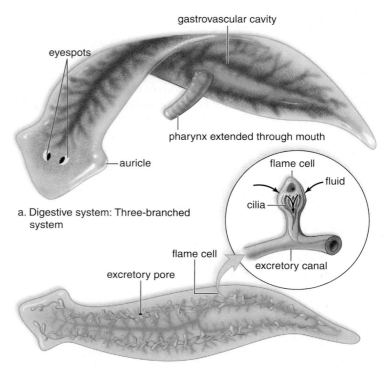

a. Digestive system: Three-branched system

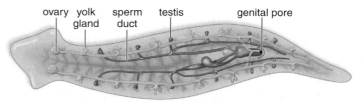

b. Excretory system: Flame-cell system

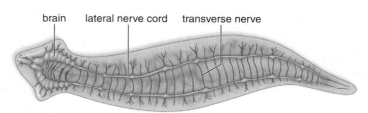

c. Reproductive system: Hermaphroditic system

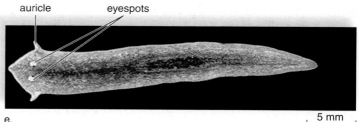

d. Nervous system: Ladder-style system

e.   5 mm

Figure 30.7 Planarian anatomy.

a. A planarian extends its pharynx to suck food into a gastrovascular cavity that branches throughout the body. **b.** The excretory system includes flame cells. **c.** The reproductive system (shown in pink and blue) has both male and female organs. **d.** The nervous system has a ladderlike appearance. **e.** A flatworm, such as *Dugesia*, is bilaterally symmetrical and has a head region with eyespots.

body. The beating of cilia in the **flame cells** (so named because the beating of the cilia reminded early investigators of the flickering of a flame) keeps the water moving toward the excretory pores.

Planarians have a **ladderlike nervous system.** A small anterior brain and two lateral nerve cords are joined by cross-branches. Planarians exhibit cephalization; in addition to a brain, they have light-sensitive organs (the eyespots) and chemosensitive organs located on the auricles. Their three muscle layers—outer circular, inner longitudinal, and diagonal—allow for varied movement. A ciliated epidermis allows planarians to glide along a film of mucus.

The animal captures food by wrapping itself around the prey, entangling it in slime, and pinning it down. Then the planarian extends a muscular pharynx, and by a sucking motion, tears up and swallows the food. The pharynx leads into a three-branched gastrovascular cavity in which digestion is completed. The digestive tract is considered incomplete because it has only one opening.

Planarians can reproduce both sexually and asexually. Although *hermaphroditic,* the worms practice cross-fertilization, in which the penis of one is inserted into the genital pore of the other (and vice versa), and reciprocal transfer of sperm takes place. Fertilized eggs hatch in 2–3 weeks as tiny worms. Asexual reproduction occurs by regeneration—if you slice a planarian in half, two new planarians will grow!

Parasitic Flatworms

The parasitic flatworms belong to two classes: the tapeworms (class Cestoda) and the flukes (class Trematoda).

Tapeworms As adults, tapeworms are endoparasites (internal parasites) of various vertebrates, including humans. They vary in length from a few millimeters to nearly 20 meters.

Tapeworms have a tough integument, a specialized body covering resistant to the host's digestive juices. Their excretory, muscular, and nervous systems are similar to those of other flatworms. Tapeworms have a well-developed anterior region, called the **scolex,** that bears hooks for attachment to the intestinal wall of the host and suckers for feeding. Behind the scolex are **proglottids,** a series of reproductive units with a full set of male and female sex organs. Each proglottid fertilizes its own eggs, which number in the thousands. Immature proglottids are located closer to the scolex, while mature (or gravid) proglottids are further away. The gravids, which contain fertilized eggs, break away and are eliminated in the host's feces. A secondary host must ingest the eggs for the life cycle to continue. Figure 30.8 illustrates the life cycle of the pork tapeworm, *Taenia solium,* where the human is the primary host and the pig is the secondary host. The larvae burrow through the intestinal wall and travel in the bloodstream to finally lodge and encyst in muscle. The cyst is a small, hard-walled structure that contains a larva called a bladder worm. When humans eat infected meat that has not been thoroughly cooked, the bladder worm breaks out of the cyst, attaches itself to the intestinal wall, and grows to adulthood. Then the life cycle begins again.

Flukes Flukes are all endoparasites of various vertebrates. Their flattened and oval-to-elongated body is covered by a nonciliated integument. At the anterior end of these animals, an oral sucker is surrounded by sensory papillae, and there is at least one other sucker used for attachment to the host. Although the digestive system is reduced compared to that of free-living flatworms, the alimentary canal is well-developed. The excretory and muscular systems are similar to those of free-living flatworms, but the nervous system is reduced, with poorly developed sense organs. Most flukes are hermaphroditic.

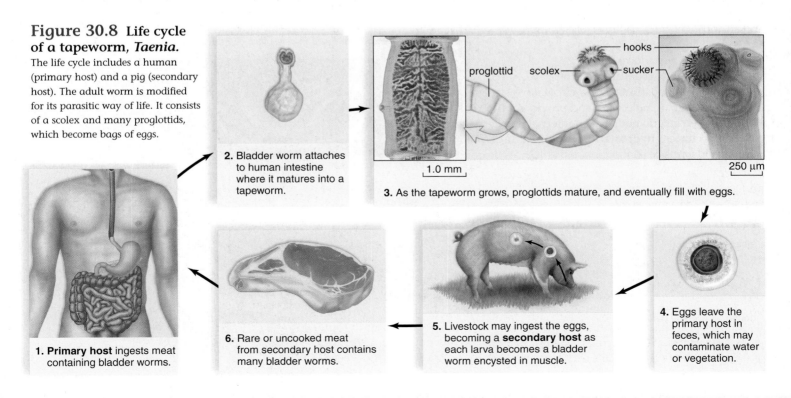

Figure 30.8 **Life cycle of a tapeworm, *Taenia*.**
The life cycle includes a human (primary host) and a pig (secondary host). The adult worm is modified for its parasitic way of life. It consists of a scolex and many proglottids, which become bags of eggs.

proglottid   scolex   hooks   sucker

1.0 mm   250 μm

2. Bladder worm attaches to human intestine where it matures into a tapeworm.

3. As the tapeworm grows, proglottids mature, and eventually fill with eggs.

1. Primary host ingests meat containing bladder worms.

6. Rare or uncooked meat from secondary host contains many bladder worms.

5. Livestock may ingest the eggs, becoming a **secondary host** as each larva becomes a bladder worm encysted in muscle.

4. Eggs leave the primary host in feces, which may contaminate water or vegetation.

Flukes are usually named for the type of vertebrate organ they inhabit; for example, there are blood, liver, and lung flukes. The **blood fluke** (*Schistosoma* spp.) occurs predominantly in the Middle East, Asia, and Africa. Nearly 800,000 infected persons die each year from schistosomiasis. Adult flukes are small (approximately 2.5 cm long) and may live for years in their human hosts. The Chinese liver fluke, *Clonorchis sinensis,* is a major parasite of humans, cats, dogs, and pigs. This 20-mm-long fluke is commonly found in many regions of the Orient. It requires two intermediate hosts, a snail and a fish. Eggs are shed into the water in feces of the human or other mammal host and enter the body of a snail, where they undergo development. Larvae escape into the water and bore into the muscles of a fish. When humans eat infected fish, the juveniles migrate into the bile duct, where they mature. A heavy infection can cause destruction of the liver and death.

> Flatworms are bilaterally symmetrical and many species, such as planaria, have excretory systems and nervous systems. Some parasitic flatworms cause diseases in humans.

Roundworms: Pseudocoelomates

Roundworms (phylum Nematoda, 90,000 species) possess two anatomical features not seen in animals discussed previously: *a tube-within-a-tube body plan* and *a body cavity.* With a tube-within-a-tube body plan, the digestive tract is considered complete; there is both a mouth and an anus. The body cavity is a **pseudocoelom** (see Fig. 30.3*b*). The fluid-filled pseudocoelom provides space for the development of organs, substitutes for a circulatory system by allowing easy passage of molecules, and provides a type of skeleton. Worms, in general, do not have an internal or external skeleton, but they do have a **hydrostatic skeleton,** a fluid-filled interior that supports muscle contraction and enhances flexibility.

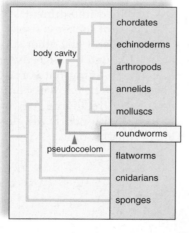

Roundworms are **nonsegmented,** meaning that they have a smooth outside body wall. These worms, which are generally colorless and less than 5 cm in length, occur almost anywhere—in the sea, in fresh water, and in the soil—typically in large numbers. Many roundworms are beneficial; for example, *Caenorhabditis elegans,* a free-living nematode, is a model animal used in genetics and developmental biology. Roundworms can also have negative effects. Parasitic roundworms live anaerobically in many plants and every type of animal, including humans.

Ascaris

In the roundworm, *Ascaris lumbricoides* (Fig. 30.9), females tend to be larger (20–35 cm in length) than males. Both sexes move by a characteristic whiplike motion because they have only longitudinal muscles and no circular muscles next to the body wall. *Ascaris* species are most commonly parasites of humans and pigs.

A female *Ascaris* is very reproductively prolific, producing over 200,000 eggs daily. The eggs are eliminated with host feces and, under the right conditions, can develop into a worm within two weeks. The eggs enter the host's body via uncooked vegetables, soiled fingers, or ingested fecal material and hatch in the intestines. The juveniles make their way into the veins and lymphatic vessels and are carried to the heart and lungs. From the lungs, the larvae travel up the trachea, where they are swallowed and eventually reach the intestines. There, the larvae mature and begin feeding on intestinal contents. The symptoms of an *Ascaris* infection depend upon the site and the stage of infection.

Figure 30.9 Roundworm anatomy.

a. The roundworm *Ascaris.* **b.** Roundworms such as *Ascaris* have a pseudocoelom and a complete digestive tract with a mouth and an anus. **c.** The larvae of the roundworm *Trichinella* penetrate striated muscle fibers, where they coil in a cyst formed from the muscle fiber.

a. *Ascaris*

c. *Trichinella*   SEM 400×

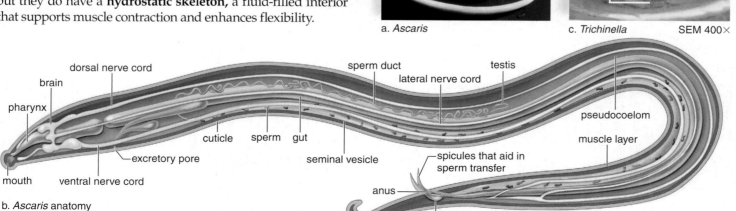

b. *Ascaris* anatomy

Other Roundworms

Trichinosis is a fairly serious infection caused by *Trichinella spiralis,* a roundworm that rarely infects humans in the United States. Humans contract the disease when they eat pork that is not fully cooked and contains encysted larvae. After maturation, the female adult burrows into the wall of the host's small intestine, where she deposits live larvae that are then carried by the bloodstream and encyst in the skeletal muscles (Fig. 30.9c). When the adults are in the small intestine, digestive disorders, fatigue, and fever occur. When the larvae encyst, the symptoms include aching joints, muscle pain, and itchy skin.

Elephantiasis is caused by a roundworm called the filarial worm, which utilizes the mosquito as an intermediate host. When a mosquito bites an infected person, it transports larvae to a new host. Because the adult worms reside in lymphatic vessels, fluid return is impeded, and the limbs of an infected human can swell to an enormous size, even resembling those of an elephant.

Other roundworm infections are more common in the United States. Children frequently acquire a pinworm infection, and hookworm is seen in the southern states, as well as worldwide. Good hygiene, proper disposal of sewage, and cooking meat thoroughly usually protect people from parasitic roundworms.

> Roundworms have bilateral symmetry, a tube-within-a-tube body plan, three germ layers, the organ level of organization, and a pseudocoelom. The rest of the animal phyla to be discussed have these features and also a true coelom.

30.3 Molluscs: Coelomates

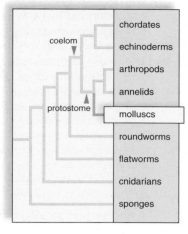

The **molluscs** (phylum Mollusca, 110,000 species) and all the other species we will study are coelomates. All coelomates have bilateral symmetry, three germ layers, the organ level of organization, and the tube-within-a-tube body plan. The advantages of having a coelom include: freer body movements, space for development of complex organs, greater surface area for absorption of nutrients, and protection of internal organs from damage. On the basis of embryological development, molluscs, annelids, and arthropods are protostomes, and echinoderms and chordates are deuterostomes.

Although almost everyone enjoys looking at the intricate patterns and beauty of seashells, few people are aware of the tremendous diversity found in the phylum Mollusca (Fig. 30.10). The molluscs make up the second largest animal phylum.

Unique Characteristics of Molluscs

Despite being a very large and diversified group, all molluscs have a body composed of at least three distinct parts:

1. The **visceral mass** is the soft-bodied portion that contains internal organs, including a highly specialized

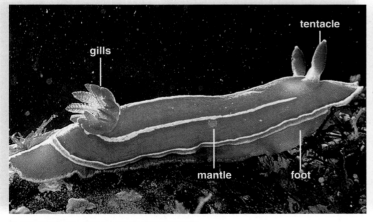

c. Nudibranch, *Glossodoris macfarlandi*

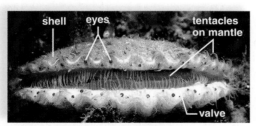

a. Chiton, *Tonicella*

b. Scallop, *Pecten* sp.

Figure 30.10 Diversity of molluscs.
a. A chiton is believed to be most like the ancestral mollusc that gave rise to all the other types. Recent evidence suggests that ancestral molluscs such as the chiton were segmented, but that this segmentation was lost throughout much of the phylum. **b.** Like clams, a scallop is a bivalve. **c.** Like snails, a nudibranch is a gastropod. **d.** Like squids, an octopus is a cephalopod.

d. Two-spotted octopus, *Octopus bimaculatus*

digestive tract, paired kidneys, and reproductive organs. The nervous system of a mollusc consists of several ganglia connected by nerve cords. The amount of cephalization and sensory organs varies from nonexistent in clams to complex in squid and octopuses.

2. The **foot** is a strong, muscular portion used for locomotion. Molluscan groups can be distinguished by modifications of the foot. Molluscs exhibit varying amounts of mobility. Oysters are sessile; snails are extremely slow moving; and squid are fast-moving, active predators.

3. The **mantle,** a membranous or sometimes muscular covering, envelops but does not completely enclose the visceral mass. The *mantle cavity* is the space between the two folds of the mantle. The mantle may secrete a *shell,* which is an exoskeleton.

Another feature often present is a rasping, tongue-like *radula,* an organ that bears many rows of teeth and is used to obtain food.

Gastropods

The **gastropods** (class Gastropoda) include nudibranchs, conchs, and snails. In gastropods, whose name means "stomach-footed," the foot is ventrally flattened, and the animal moves by muscle contractions that pass along the foot. Many are herbivores that use their radulas to scrape food from surfaces. Others are carnivores, using their radulas to bore through surfaces, such as bivalve shells, to obtain food.

While nudibranchs, also called sea slugs, lack a shell, conchs and snails have a univalve coiled shell in which the visceral mass spirals. Land snails, such as *Helix aspera,* have a head with two pairs of tentacles; one pair bears eyes at the tips (Fig. 30.11*a*). The shell not only offers protection, but also prevents desiccation (drying out). While aquatic gastropods have gills, land snails have a mantle that is richly supplied with blood vessels and

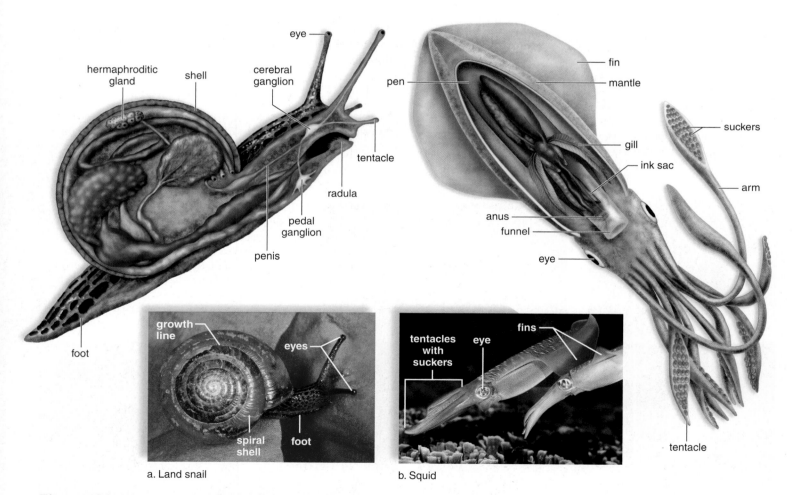

a. Land snail

b. Squid

Figure 30.11 Gastropod and cephalopod anatomy.
a. Snails have a long muscular foot that allows them to creep along slowly by way of muscular contraction. Note the absence of lungs in this land snail.
b. Squids are torpedo-shaped and adapted for fast swimming by jet propulsion.

functions as a lung. Reproduction is also adapted to a land existence. Land snails are hermaphroditic; when two snails meet, each inserts its penis into the vagina of the other, and following fertilization and the deposit of eggs externally, development proceeds directly without formation of swimming larvae.

Cephalopods

In **cephalopods** (class Cephalopoda, meaning head-footed), including octopuses, squid, and nautiluses, the foot has evolved into tentacles about the head (Fig. 30.11*b*). Aside from the tentacles, which seize prey, cephalopods have a powerful beak and a radula (toothy tongue) to tear prey apart. Cephalization is apparent. The eyes are superficially similar to those of vertebrates—they have a lens and a retina with photoreceptors. However, the eye is constructed so differently from the vertebrate eye that the so-called **camera-type eye** must have evolved independently in both the molluscs and in the vertebrates. In cephalopods, the brain is formed from a fusion of ganglia, and nerves leaving the brain supply various parts of the body. An especially large pair of nerves controls the rapid contraction of the mantle, allowing these animals to move quickly by a jet propulsion of water. Rapid movement and the secretion of a dark ink help cephalopods escape their enemies. Octopuses have no shell, and squid have only a remnant of a shell concealed beneath the skin. Octopuses, like some other species of cephalopods, are thought to be among the most intelligent invertebrates and are even capable of learning!

Bivalves

Clams, mussels, oysters, and scallops are called **bivalves** (class Bivalvia) because their shells have two parts. In a clam, the shell, which is secreted by the mantle, is composed of protein and calcium carbonate, with an inner layer of mother-of-pearl. If a foreign body is placed between the mantle and the shell, pearls form as concentric layers of shell are deposited about the particle.

Figure 30.12 shows the internal anatomy of the freshwater clam, *Anodonta*. The adductor muscles hold the valves of the shell together. Within the mantle cavity, the gills, which are the organs for gas exchange in aquatic forms, hang down on either side of the visceral mass, which lies above the foot.

The clam is a filter feeder. Food particles and water enter the mantle cavity by way of the incurrent siphon, a posterior opening between the two valves. Mucous secretions cause smaller particles to adhere to the gills, and ciliary action sweeps them toward the mouth.

The Visceral Mass   The heart of a clam lies just below the hump of the shell within the pericardial cavity, the

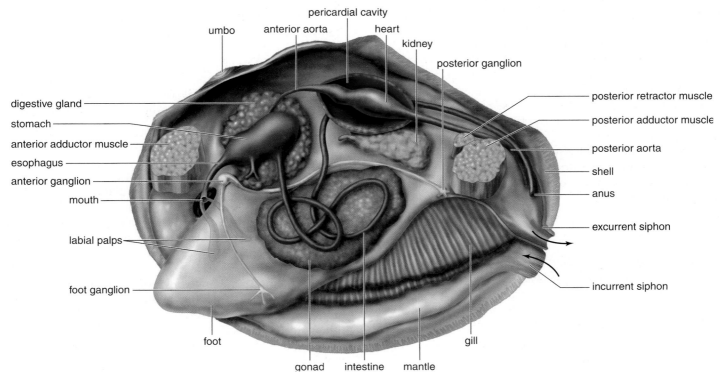

Figure 30.12 Clam, *Anodonta*.
In this drawing of a clam, one of the bivalve shells and the mantle have been removed from one side. Follow the path of food from the incurrent siphon to the gills, the mouth, the stomach, the intestine, the anus, and the excurrent siphon. Locate the three ganglia: anterior, foot, and posterior. The heart lies in the reduced coelom.

TABLE 30.2 Comparison of Clam, Squid, and Land Snail

Feature	Clam	Squid	Land Snail
Food gathering	Filter feeder	Active predator	Herbivore
Skeleton	Heavy shell for protection	No external skeleton	Shell protects visceral mass
Circulation	Open	Closed	Open
Cephalization	None	Marked	Present
Locomotion	"Hatchet" foot	Jet propulsion	Flat foot
Nervous System	3 separate ganglia	Brain and nerves	Cerebral ganglion

only remains of the reduced coelom. The heart pumps blood into a dorsal aorta that leads to the various organs of the body. Within the organs, however, blood flows through spaces, or sinuses, rather than through vessels. This is called an **open circulatory system** because the blood is not entirely contained within blood vessels. An open circulatory system would be inefficient in a fast-moving animal.

The **nervous system** of a clam is composed of three pairs of ganglia (anterior, foot, and posterior), which are all connected by nerves. Clams lack cephalization. The "hatchet" foot projects anteriorly from the shell, and by expanding the tip of the foot with blood and pulling the body after it, the clam moves forward.

The **digestive system** of the clam includes a mouth with labial palps, an esophagus, a stomach, and an intestine, which coils about in the visceral mass and then is surrounded by the heart as it extends to the anus. Two excretory kidneys remove waste from the pericardial cavity for excretion into the mantle cavity.

The sexes are usually separate. The gonad (i.e., ovary or testis) is located around the coils of the intestine. All clams have some type of larval stage, and marine clams have a trochophore larva.

Comparison

Table 30.2 compares the clam and land snail, which are adapted to less active lifestyles, with the squid, a more active animal.

Evolutionarily, molluscs are the first phylum to have a coelom. Gastropods, cephalopods, and bivalves represent the diversity of molluscs, all of which have the following unique characteristics: a visceral mass; a strong, muscular foot; and a mantle.

30.4 Annelids: Segmented Worms

Annelids (phylum Annelida, about 12,000 species) are *segmented,* as evidenced by the rings that encircle their bodies on the outside. Segmentation is also seen in arthropods and chordates.

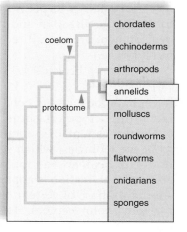

Internally, the segments of annelids are partitioned by septa. The well-developed coelom is fluid-filled and serves as a supportive hydroskeleton. Along with the partitioning of the coelom, this hydroskeleton permits each body segment to move independently. Locomotion occurs by contraction and expansion of each body segment, propelling the animal forward. Thus, instead of just burrowing in the mud, a terrestrial annelid is also capable of crawling on the surface.

The most familiar members of this phylum include marine worms, leeches, and earthworms. The majority of annelids live in a marine environment. Annelids vary in size from microscopic to tropical earthworms as much as 4 m long.

In annelids, the tube-within-a-tube body plan has led to specialization of the digestive tract. For example, the digestive system may include a pharynx, esophagus, crop, gizzard, intestine, and accessory glands. Annelids have an extensive **closed circulatory system** with blood vessels that run the length of the body and branch to every segment. The nervous system consists of a brain connected to a **ventral solid nerve cord,** with ganglia in each segment. The excretory system consists of nephridia in most segments. A **nephridium** is a tubule that collects waste material and excretes it through an opening in the body wall.

Polychaetes

Most annelids are marine; their class name, Polychaeta, refers to the presence of many setae. **Setae** are bristles that anchor the worm or help it move. The setae are in bundles on parapodia, which are paddlelike appendages found on most segments. These are used in swimming, but also as respiratory organs. Ragworms, such as *Nereis* (Fig. 30.13a), prey on crustaceans and other small animals, which they capture using a pair of strong, chitinous jaws that extend with the pharynx when they feed. Associated with its way of life, *Nereis* has a well-defined head region with eyes and other sense organs (Fig. 30.13b).

Other polychaetes are sessile tube worms, with tentacles that form a funnel-shaped fan (Fig. 30.13c). They are filter feeders; water currents, created by the action of cilia, trap food particles that are directed toward the mouth.

Polychaetes have breeding seasons, and only during these times do the worms have functional sex organs.

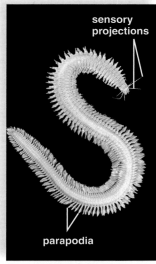

a. Ragworm, *Nereis*

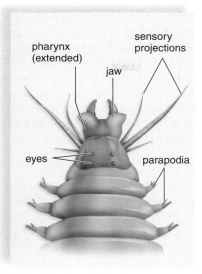

b. Head region of *Nereis*

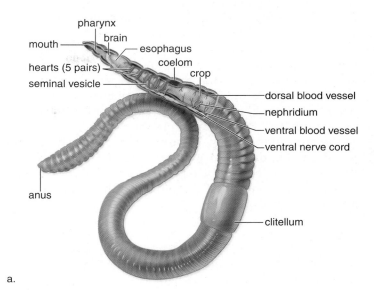

a.

c. Christmas tree worms, *Spirobranchus*

Figure 30.13
Polychaete diversity.
a. Ragworms are predatory polychaetes that have a well-defined head region (**b**). Note also the parapodia, which are used for swimming and as respiratory organs. **c.** Christmas tree worms (a type of tube worm) are sessile filter feeders whose ciliated tentacles spiral.

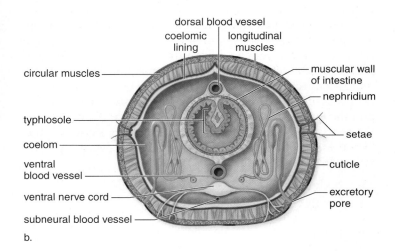

b.

In *Nereis,* many worms concurrently shed a portion of their bodies containing either eggs or sperm, and these float to the surface, where fertilization takes place. The zygote rapidly develops into a trochophore larva, just as in marine clams. The existence of this larval form in both the annelids and molluscs is evidence that these two groups of animals are related.

Polychaetes are marine worms with bundles of setae attached to parapodia.

Oligochaetes

The oligochaetes (class Oligochaeta), which include earthworms, have few setae per segment. Earthworms (e.g., *Lumbricus*) do not have a well-developed head or parapodia (Fig. 30.14). Their setae protrude in pairs directly from the surface of the body. Locomotion, which is accomplished section by section, utilizes muscle contraction and the setae. When longitudinal muscles contract, segments bulge, and their setae

c.

Figure 30.14 Earthworm, *Lumbricus*.
a. Internal anatomy of the anterior part of an earthworm. Notice that each body segment bears a pair of setae and that internal septa divide the coelom into compartments. **b.** Cross section of an earthworm. **c.** When earthworms mate, they are held in place by a mucus secreted by the clitellum. The worms are hermaphroditic, and when mating, sperm pass from the seminal vesicles of each to the seminal receptacles of the other.

protrude into the soil. Then, when circular muscles contract, the setae are withdrawn, and these segments move forward in sequence, pushing the whole animal forward.

Earthworms reside in soil where there is adequate moisture to keep the body wall moist for gas exchange. They are scavengers that do not have an obvious head and feed on leaves or any other organic matter, living or dead, which they can conveniently take into their mouths along with dirt. Food drawn into the mouth by the action of the muscular pharynx is stored in a crop and ground up in a thick, muscular gizzard. Digestion and absorption occur in a long intestine, whose dorsal surface has an expanded region, called a typhlosole, that increases the surface for absorption.

Segmentation

Earthworm segmentation, which is obvious externally, is also internally evidenced by the presence of septa. The long, ventral solid nerve cord leading from the brain has ganglionic swellings and lateral nerves in each segment. The paired **nephridia** in most segments have two openings: one is a ciliated funnel that collects coelomic fluid, and the other is an exit in the body wall. Between the two openings is a convoluted region where waste material is removed from the blood vessels about the tubule. Red blood moves anteriorly in the dorsal blood vessel, which connects to the ventral blood vessel by five pairs of connectives called "hearts." Pulsations of the dorsal blood vessel and the five pairs of hearts are responsible for blood flow. As the ventral vessel takes the blood toward the posterior regions of the worm's body, it gives off branches in every segment. Altogether, segmentation is evidenced by:

- body rings
- coelom divided by septa
- setae on most segments
- ganglia and lateral nerves in each segment
- nephridia in most segments
- branch blood vessels in each segment

Reproduction

Earthworms are hermaphroditic: the male organs are the testes, the seminal vesicles, and the sperm ducts, and the female organs are the ovaries, the oviducts, and the seminal receptacles. When mating, two worms lie parallel to each other facing in opposite directions. The fused midbody segment, called a clitellum, secretes mucus that protects the sperm from drying out as they pass between the worms. After the worms separate, the clitellum of each produces a slime tube, which is moved along over the anterior end by muscular contractions. As it passes, eggs and the sperm received earlier are deposited, and fertilization occurs. The slime tube then forms a cocoon to protect the worms as they develop. There is no larval stage.

Comparison with Clam Worm

Comparing the anatomy of marine clam worms (class Polychaeta) to that of terrestrial earthworms (class Oligochaeta) highlights the manner in which earthworms are adapted to life on land. A lack of cephalization is seen in the nonpredatory earthworms that extract organic remains from the soil they eat. Their lack of parapodia helps reduce the possibility of water loss and facilitates burrowing in soil. The clam worm makes use of external water, while the earthworm provides a mucous secretion to aid fertilization. It is the aquatic form that has the swimming, or **trochophore** larva, not the land form.

> Earthworms, which burrow in the soil, lack obvious cephalization and parapodia. They are hermaphroditic and have no larval stage.

Leeches

Leeches (class Hirudinea) are usually found in fresh water, but some are marine or even terrestrial. They have the same body plan as other annelids, but they have no setae, and each body ring has several transverse grooves. Most leeches are only 2–6 cm in length, but some, including the medicinal leech, are as long as 20 cm.

Among their modifications are two **suckers,** a small one around the mouth and a large, posterior one. While some leeches are free-living, feeding on plant material or invertebrates, most are fluid feeders that attach themselves to open wounds. Some bloodsuckers, such as the medicinal leech, can cut through tissue. Leeches are able to keep blood flowing and prevent clotting by means of a substance in their saliva known as **hirudin,** a powerful anticoagulant.

> Leeches are modified in a way that lends itself to the parasitic way of life. Some are external parasites known as bloodsuckers.

30.5   Arthropods: Jointed Appendages

Arthropods (phylum Arthropoda) are extremely diverse (Fig. 30.15). Over one million species have been discovered and described, but some experts suggest that as many as 30 million arthropods could exist—most of them insects.

What characteristics account for the success of arthropods? Arthropod literally means "jointed foot," but actually they have freely

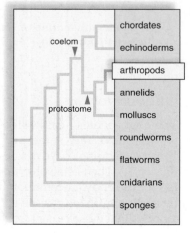

a. Flat-backed millipede, *Sigmoria*

b. Tarantula, *Aphonopelma*

c. Dungeness crab, *Cancer*

d. Paper wasp, *Polistes*

e. Stone centipede, *Lithobius*

Figure 30.15 **Arthropod diversity.**
a. A millipede has only one pair of antennae, and the head is followed by a series of segments, each with two pairs of appendages. **b.** The hairy tarantulas of the genus *Aphonopelma* are dark in color and sluggish in movement. Their bite is harmless to people. **c.** A crab is a crustacean with a calcified exoskeleton, one pair of claws, and four other pairs of walking legs. **d.** A wasp is an insect with two pairs of wings, both used for flying, and three pairs of walking legs. **e.** A centipede has only one pair of antennae, and the head is followed by a series of segments, each with a single pair of appendages.

movable *jointed appendages.* The exoskeleton of arthropods is composed primarily of **chitin,** a strong, flexible, nitrogenous polysaccharide. The jointed exoskeleton serves many functions, including protection, attachment for muscles, locomotion by jointed appendages, and prevention of desiccation. However, because this exoskeleton is hard and nonexpandable, arthropods must **molt,** or shed their exoskeleton, as they grow larger. Before molting, the body secretes a new, larger exoskeleton, which is soft and wrinkled, underneath the old one. After enzymes partially dissolve and weaken the old exoskeleton, the animal breaks it open and wriggles out. The new exoskeleton then quickly expands and hardens.

Arthropods are segmented, but some segments are fused into regions such as a head, a thorax, and an abdomen. Trilobites, an early and now extinct arthropod, had a pair of appendages on each body segment. In modern arthropods, appendages are specialized for such functions as walking, swimming, reproducing, eating, and sensory reception. These modifications account for much of the diversity of arthropods.

Arthropods have a well-developed nervous system that includes a brain and a ventral solid nerve cord. The head bears various types of sense organs, including anten-

nae (or feelers) and two types of eyes—compound and simple. The compound eye (when present) is composed of many complete visual units, each of which operates independently. The lens of each visual unit focuses an image on the light-sensitive membranes of a small number of photoreceptors within that unit. Vision is not acute, but it is much better for detecting details of movement than is possible with human eyes.

Crustaceans

Crustaceans (subphylum Crustacea) are a group of largely marine arthropods that include barnacles, shrimps, lobsters, and crabs. There are also some freshwater crustaceans, including the crayfish, and some terrestrial ones, including the sowbug, or pillbug.

Crustaceans are named for their hard shells; the exoskeleton is calcified to a greater degree in some forms than in others. Although crustacean anatomy is extremely diverse, the head usually bears a pair of compound eyes and five pairs of appendages. The first two pairs, called antennae, lie in front of the mouth and have sensory functions. The other three pairs are mouthparts used in feeding.

In crayfish, such as *Cambarus,* the thorax bears five pairs of **walking legs.** The first walking leg is a pinching claw (Fig. 30.16). The gills are situated above the walking legs. The head and thorax are fused into a cephalothorax, which is covered on the top and sides by a nonsegmented carapace. The abdominal segments, which contain much musculature, are equipped with swimmerets, small paddlelike structures. The last two segments bear the uropods and the telson, which make up a fan-shaped tail to propel the crayfish backward.

Internal Organs

The digestive system includes a stomach, which is divided into two main regions. The anterior portion consists of a gas-

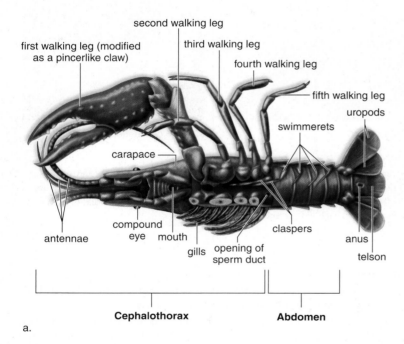

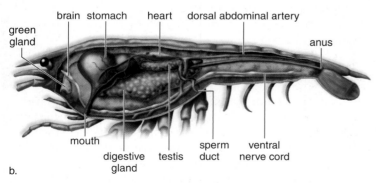

Figure 30.16 Male crayfish, *Cambarus.*
a. Externally, it is possible to observe the crayfish's jointed appendages, including the swimmerets, and the walking legs, which include claws. These appendages, plus a portion of the carapace, have been removed from the right side so that the gills are visible. **b.** An internal view shows the parts of the digestive and circulatory systems. Note the ventral nerve cord.

tric mill, a special grinding apparatus equipped with chitinous teeth. The posterior region acts as a filter to prevent coarse particles from entering the digestive glands where absorption takes place. **Green glands** lying in the head region, anterior to the esophagus, excrete metabolic wastes through a duct that opens externally at the base of the antennae. The coelom is reduced to a space around the reproductive system. A heart within a pericardial cavity pumps blood containing the respiratory pigment hemocyanin into a **hemocoel** consisting of sinuses (open spaces), where the hemolymph flows about the organs. (Whereas hemoglobin is a red, iron-containing pigment, hemocyanin is a blue, copper-containing pigment.) Thus, the blood is not contained within blood vessels, constituting an open circulatory system.

The nervous system of the crayfish is very similar to that of the earthworm. The crayfish has a brain and a ventral nerve cord that passes posteriorly. Along the length of the nerve cord, segmental ganglia give off 8 to 19 paired lateral nerves.

The sexes are separate in the crayfish, and the gonads are located just ventral to the pericardial cavity. In the male, a coiled sperm duct opens to the outside at the base of the fifth walking leg. Sperm transfer is accomplished by the first two pairs of swimmerets, which are enlarged and quite strong. In the female, the ovaries open at the bases of the third walking legs. A stiff fold between the bases of the fourth and fifth pairs of walking legs serves as a seminal receptacle. Following fertilization, the eggs are attached to the swimmerets of the female.

Insects

Insects (subphylum Uniramia) are so numerous (likely over one million species) and so diverse (Fig. 30.17) that the study of this one group is a major specialty in biology called **entomology.** Some insects show remarkable behavioral adaptations, as exemplified by the social systems of bees, ants, termites, and other colonial insects.

Insects have certain features in common. The body is divided into a head, a thorax, and an abdomen. The head usually bears a pair of sensory antennae, a pair of compound eyes, and several simple eyes. The mouthparts are adapted to the specific insect's way of life; for example, a grasshopper has mouthparts that chew, and a butterfly has a long tube for siphoning the nectar of flowers. The abdomen contains most of the internal organs; the thorax bears three pairs of legs and the wings—either one or two pairs or none. **Wings,** if present, enhance an insect's ability to survive by providing a way of escaping enemies, finding food, facilitating mating, and dispersing the offspring. The exoskeleton of an insect is lighter and contains less chitin than that of many other arthropods.

Here, we will discuss the features of a grasshopper as a representative example of insect form and function. In the grasshopper (Fig. 30.18), the third pair of legs is suited

a. Housefly, *Musca*

b. Walking stick, *Diapheromera*

c. Bee, *Apis*

Figure 30.17
Insect diversity.

a. Flies have a single pair of wings and lapping mouthparts. **b.** Walking sticks are herbivorous, with biting and chewing mouthparts. **c.** Bees have four translucent wings and a thorax separated from the abdomen by a narrow waist. **d.** Dragonflies have two pairs of similar wings. They catch and eat other insects while flying. **e.** Butterflies have forewings larger than their hindwings. Their mouthparts form a long tube for siphoning up nectar from flowers.

d. Dragonfly, *Aeshna*

e. American copper butterfly, *Lycaena*

to jumping. There are two pairs of wings. The forewings are tough and leathery, and when folded back at rest, they protect the broad, thin hindwings. On the lateral surface, the first abdominal segment bears a large tympanum on each side for the reception of sound waves.

Internal Organs

The digestive system of a grasshopper is suitable for a herbivorous diet. The mouthparts chew the food, which is temporarily stored in the crop before passing into a gastric mill, where it is finely ground before digestion is completed in the stomach. Nutrients are absorbed into the hemocoel from outpockets called gastric ceca. The stomach leads into an intestine and a rectum, which empties by way of an anus. The excretory system consists of **Malpighian tubules,** which extend into the hemocoel and collect nitrogenous wastes that are excreted into the

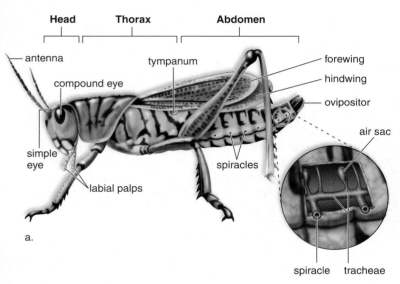

a.

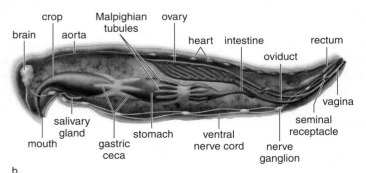

b.

Figure 30.18 Female grasshopper, *Romalea*.

a. Externally, the body of a grasshopper is divided into three sections and has three pairs of legs. The tympanum receives sound waves, and the jumping legs and the wings are for locomotion. **b.** Internally, the digestive system is specialized. The Malpighian tubules excrete a solid nitrogenous waste (uric acid). A seminal receptacle receives sperm from the male, which has a penis.

ScienceFocus

A fresh human corpse is lying on the floor. Within 10–15 minutes, blowflies with blue and green metallic wings arrive. An animal corpse is a perfect site for egg production. The flies lay eggs in the mouth, nostrils, ears, eyes, and any open wounds. Depending on the temperature, the eggs will hatch in the next 12–24 hours. The small hatchling larval flies, more commonly known as maggots, feed on fat and collagen in the corpse in order to grow. Their development occurs in three stages called instars, during which the larvae shed skin and grow. When the third instar grows, it leaves the corpse and forms a pupa that takes about 14 days (depending on the temperature) to emerge as an adult blowfly. The adult has to wait a day or two until it can fly because the wings need to expand and the body has to harden. Once capable of flight, the adult leaves to find a mate as well as another corpse in which to lay eggs, thereby completing the life cycle.

What do blowflies have to do with crime scene investigation? If the corpse involved is a human, the timing of the various stages of the life cycle of the flies (e.g., newly hatched eggs, maggot size, maggot weight, or pupa remains) can help forensic scientists determine the postmortem interval (or PMI) since the time of death. Forensic entomology uses insects such as blowflies to determine the time of death. A forensic entomologist can even determine if a human corpse has been moved based on the fact that different species of blowflies prefer to lay eggs in different environments. For example, some species lay their eggs in shade, while others lay eggs in sunny habitats; similarly, some species inhabit urban areas, while others inhabit rural areas. Thus, the combination of life stage(s) and species of blowfly found on a corpse can help a forensic entomologist determine the time and possible location of a person's death.

Discussion Questions

1. Given that the hatching of eggs and the growth rate of blowfly larvae depend on temperature, how sure do you think forensic entomologists can be about time of death?
2. As a juror, how would you feel about forensic entomology evidence entered into court?
3. When might forensic entomology evidence be useful in a murder case?

---

digestive tract. The formation of a solid nitrogenous waste, namely uric acid, conserves water.

The respiratory system begins with openings in the exoskeleton called spiracles. From here, the air enters small tubules called **tracheae** (Fig. 30.18*a*). The tracheae branch and rebranch until they end intracellularly, where the actual exchange of gases takes place. The movement of air through this complex of tubules is not passive. Rather, air is actively pumped by alternate contraction and relaxation of the body wall. Breathing by tracheae is suitable to the small size of insects (most are less than 60 mm in length), since the tracheae would be crushed by any significant amount of weight.

The circulatory system contains a slender, tubular heart that lies against the dorsal wall of the abdominal exoskeleton and pumps hemolymph into an aorta that leads to a hemocoel, where it circulates before returning to the heart again. The hemolymph is colorless and lacks a respiratory pigment because the tracheal system is responsible for gas exchange.

Reproduction and Development

Grasshopper reproduction is adapted to life on land. The male grasshopper has a penis, and sperm passed to the female are stored in a seminal receptacle. Internal fertilization protects both gametes and zygotes from drying out. The female deposits the fertilized eggs in the ground with her ovipositor.

Metamorphosis is a change in form and physiology that occurs as an immature stage, called a larva, becomes an adult. Grasshoppers undergo **gradual metamorphosis** as they mature. The immature grasshopper, called a nymph, looks like an adult grasshopper, even though it differs somewhat in shape and form. Other insects, such as butterflies, undergo complete metamorphosis, involving drastic changes in form. At first, the animal is a wormlike larva (caterpillar) with chewing mouthparts. It then forms a case, or cocoon, about itself and becomes a pupa. During this stage, the body parts are completely reorganized; the adult then emerges from the cocoon. This life cycle allows the larvae and adults to use different food sources. Most eating by insects occurs during the larval stage. The field of forensic science has found a way to put our knowledge of insect life cycle stages to good use (see Science Focus).

Comparison with Crayfish

The grasshopper and the crayfish share a common ancestor, but they have diverged in their morphology to adapt to aquatic versus terrestrial environments. In crayfish, gills take up oxygen from water, while in the grasshopper, tracheae allow oxygen-laden air to enter the body. Appropriately, the crayfish has an oxygen-carrying pigment, but a grasshopper has no such pigment in its blood. The crayfish excretes a liquid nitrogenous waste (ammonia), while the grasshopper excretes a solid nitrogenous waste (uric acid). Only grasshoppers have: (1) a tympanum

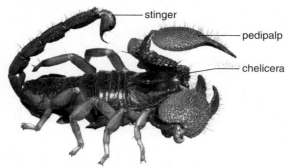

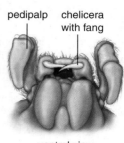

a. Kenyan giant scorpion, *Pandinus*

b. Goldenrod spider, *Misumena*

ventral view
of mouthparts

c. Wood ticks, *Ixodes*

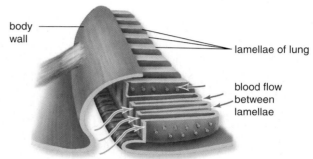

air flowing in through spiracle

d. Book lung anatomy

Figure 30.19 Arachnid diversity.
a. A scorpion has pincerlike chelicerae and pedipalps; its long abdomen ends with a stinger that contains venom. **b.** Goldenrod spiders can change colors from yellow to white to camouflage with color of the flower they are found on. **c.** In the western United States, the wood tick carries a debilitating disease called Rocky Mountain spotted fever. **d.** Arachnids breathe by means of book lungs, in which the "pages" are double sheets of thin tissue (lamellae).

for the reception of sound waves; and (2) a penis in males for passing sperm to females without possible desiccation and an ovipositor in females for laying eggs in soil. Crayfish utilize their uropods when they swim; a grasshopper has legs for hopping and wings for flying.

Arachnids

The **arachnids** (subphylum Chelicerata) include terrestrial scorpions, spiders, ticks, and mites (Fig. 30.19). In this group, the cephalothorax bears six pairs of appendages: the chelicerae and the pedipalps and four pairs of walking legs. The cephalothorax is followed by an abdomen that contains internal organs.

Scorpions are the oldest terrestrial arthropods. Today, they occur in the tropics, subtropics, and temperate regions of North America. They are nocturnal and spend most of the day hidden under a log or a rock. In scorpions, the pedipalps are large pincers, and the long abdomen ends with a stinger that contains venom.

Ticks and mites are parasites. Ticks suck the blood of vertebrates and sometimes transmit diseases, such as Rocky Mountain spotted fever or Lyme disease. Chiggers, the larvae of certain mites, feed on the skin of vertebrates.

Spiders, the most familiar arachnids, have a narrow waist that separates the cephalothorax from the abdomen. Each chelicera consists of a basal segment and a fang that delivers poison to paralyze or kill prey. Spiders use silk threads for all sorts of purposes, from lining their nests to catching prey. Biologists have used web-building behavior to discover how spider families are related.

The internal organs of spiders also show how they are adapted to a terrestrial way of life. Malpighian tubules work in conjunction with rectal glands to reabsorb ions and water before a relatively dry nitrogenous waste (uric acid) is excreted. Invaginations of the inner body wall form the lamellae ("pages") of spiders' so-called book lungs. Air flowing into the folded lamellae on one side exchanges gases with blood flowing in the opposite direction on the other side.

The arthropods are a diverse group of animals, that are adapted to either an aquatic or a terrestrial way of life. Jointed appendages and a water-repellent exoskeleton assist locomotion and lessen the threat of drying out on land.

Conservation of Coral Reefs

Coral reefs are among the most biologically diverse and productive communities on Earth. Coral reefs tend to be found in warm, clear, and shallow tropical waters worldwide and are typically formed by reef-building corals, which are cnidarians. Aside from being beautiful and giving shelter to many colorful species of fishes, coral reefs help generate economic income from tourism, protect ocean shores from erosion and may serve as the source of medicines derived from the antimicrobial compounds that reef-dwelling organisms produce.

However, coral reefs around the globe are being destroyed for a variety of reasons, most of them linked to human development. Deforestation, for example, causes tons of soil to settle on the top of coral reefs. This sediment prevents photosynthesis of symbiotic algae that provide food for the corals. When the algae die, so do the corals, which then turn white. This so-called coral "bleaching" has been seen in the Pacific Ocean and the Caribbean Sea. Recent evidence suggests that global warming could be linked to coral bleaching and death because corals can only tolerate a narrow range of temperatures. As global temperatures rise, so do water temperatures, and corals can die as a result. Increases in aquatic nutrients, such as those that come from fertilizers and wash into the ocean, can make corals more susceptible to diseases, which can also kill them.

Marine scientist Edgardo Gomez estimates that 90% of coral reefs in the Philippines are dead or deteriorating due to human activities such as pollution and, especially, overfishing. Fishing methods that employ dynamite or cyanide to kill or stun the fish for food or the pet trade can easily kill corals. Paleobiologist Jeremy Jackson of the Smithsonian Tropical Research Institute near Panama City, Panama, estimates that we may lose 60% of the world's coral reefs by the year 2050.

Figure 30B Yellow coral polyps, *Parazoanthus gracilis.*

Form Your Own Opinion

1. How can we make the public care more about the loss of coral reefs?
2. Considering what is causing the loss of coral reefs, is it possible to save them? How?
3. Can you think of any circumstances under which dire predictions can be used to help conserve the environment?

Summarizing the Concepts

30.1 Evolutionary Trends Among Animals
Animals are motile, multicellular heterotrophs that ingest their food. An evolutionary tree based on these features depicts evolutionary trends among the animals.

30.2 Introducing the Invertebrates
Table 30.3 contrasts selected anatomical features used to help classify animals. Sponges are asymmetrical, and cnidarians have radial symmetry. All other phyla contain bilaterally symmetrical animals. Flatworms have three germ layers but no coelom. Round-worms have a pseudocoelom and a tube-within-a-tube body plan.

30.3 Molluscs: Coelomates
Molluscs, along with annelids and arthropods, are protostomes because the first embryonic opening becomes the mouth. The body of a mollusc typically contains a visceral mass, a mantle, and a foot. Predaceous squid display marked cephalization, move rapidly by jet propulsion, and have a closed circulatory system. Bivalves (e.g., clams), which have a hatchet foot, are filter feeders.

30.4 Annelids: Segmented Worms
Annelids are segmented worms; segmentation is seen both externally and internally. Polychaetes are marine worms that have parapodia. A clam worm is a predaceous marine worm with a defined head region. Earthworms are oligochaetes that scavenge for food in the soil and do not have a well-defined head region.

30.5 Arthropods: Jointed Appendages
Arthropods are the most varied and numerous of animals. Their success is largely attributable to a flexible exoskeleton and specialized body regions.

Like many other arthropods, crustaceans have a head that bears compound eyes, antennae, and mouthparts. The crayfish, a crustacean, also illustrates such features as an open circulatory system, respiration by gills, and a ventral solid nerve cord.

Like most other insects, grasshoppers have wings and three pairs of legs attached to the thorax. Grasshoppers also have many other features that illustrate adaptation to a terrestrial life, including respiration by tracheae.

Spiders are arachnids with chelicerae, pedipalps, and four pairs of walking legs attached to a cephalothorax. Spiders, too, are adapted to life on land, and they spin silk that is used in various ways.

TABLE 30.3	Comparison of Invertebrates				
	Sponges	**Cnidarians**	**Flatworms**	**Roundworms**	**Other Phyla***
Level of organization	Cell	Tissue	Organs	Organs	Organs
Type of body plan	—	Sac	Sac	Tube-within-a-tube	Tube-within-a-tube
Type of symmetry	Radial or none	Radial	Bilateral	Bilateral	Bilateral except echinoderms
Type of body cavity	—	—	None	Pseudocoelom	Coelom
Segmentation	—	—	—	—	Only annelids, arthropods, and chordates
Jointed appendages	—	—	—	—	Only arthropods and chordates

*Molluscs, annelids, arthropods, echinoderms, chordates

Testing Yourself

Choose the best answer for each question.

1. Animals with a cellular level of organization produce
   a. cells only.
   c. cells and organs.
   b. cells and tissues.
   d. cells, tissues, and organs.

2. Which of the following has a sac type of body plan?
   a. earthworm
   d. ant
   b. hydra
   e. mosquito
   c. leech

For problems 3–7, indicate the type of symmetry exhibited by the animal listed. Each answer can be used more than once.

Key:
   a. asymmetry
   c. bilateral symmetry
   b. radial symmetry

3. Lobster

4. Jellyfish

5. Sponge

6. Planarian

7. Octopus

8. Nonsegmented body parts are characteristic of
   a. leeches.
   c. spiders.
   b. clams.
   d. humans.

9. Which of the following animals lacks true tissues?
   a. snail
   d. squid
   b. beetle
   e. coral
   c. sponge

10. Cnidarians are considered to be organized at the tissue level because they contain
    a. ectoderm and endoderm.
    d. endoderm and mesoderm.
    b. ectoderm.
    e. mesoderm.
    c. ectoderm and mesoderm.

11. A cnidocyte is
    a. a digestive cell.
    c. a stinging cell.
    b. a reproductive cell.
    d. an excretory cell.

12. Digestion in hydra is carried out in the
    a. gastrovascular cavity.
    c. medusa.
    b. polyp.
    d. pharynx.

For problems 13–16, match the feature with the type of flatworm. Each answer may be used more than once. Each problem may have more than one answer.

Key:
   a. planarian
   c. fluke
   b. tapeworm

13. Endoparasite

14. Light-sensitive organs

15. Digestive tract lacks an anus

16. Hermaphroditic

17. Unlike flatworms, roundworms have:
    a. an internal skeleton and a tube-within-a-tube body plan.
    b. an external skeleton and a tube-within-a-tube body plan.
    c. an internal skeleton and a body cavity.
    d. an external skeleton and a body cavity.
    e. a tube-within-a-tube body plan and a body cavity.

18. Compared to an animal species that has a coelom, one that lacks a coelom
    a. is more flexible.
    b. has more complex organs.
    c. is more likely to tolerate temperature variations.
    d. is less able to store excretory wastes.

19. Which of the following is not a feature of most coelomates?
    a. radial symmetry
    b. three germ layers
    c. tube-within-a-tube body plan
    d. organ level of organization

20. A mollusc shell is secreted by the
    a. foot.
    c. visceral mass.
    b. head.
    d. mantle.

21. The type of mollusc that produces tentacles is a
    a. gastropod.
    c. univalve.
    b. bivalve.
    d. cephalopod.

22. A feature of annelids is
    a. segmented body.
    c. sac body plan.
    b. acoelomate.
    d. radial symmetry.

23. Label the parts of an earthworm in the following illustration.

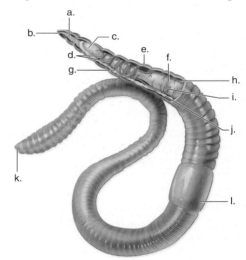

24. Which of the following statements about earthworms is false?
    a. Earthworms are hermaphroditic.
    b. An earthworm has no obvious head.
    c. An earthworm has no heart.
    d. An earthworm has a segmented body.

25. Which of the following is not a feature that has led to the success of the arthropods?
    a. They have two types of eyes.
    b. They have jointed appendages.
    c. They produce an exoskeleton.
    d. They have ears that can hear a wide range of sounds.
    e. They have well-developed eyes.

26. An insect does not have
    a. compound eyes.
    d. an exoskeleton.
    b. eight legs.
    e. jointed legs.
    c. antennae.

27. Which animal undergoes metamorphosis?

    a. grasshopper   c. earthworm
    b. human   d. More than one of these is correct.

28. Which animal is not an arachnid?

    a. spider   d. scorpion
    b. tick   e. beetle
    c. mite

29. Which feature is not characteristic of animals?

    a. heterotrophic   d. have contracting fibers
    b. diploid life cycle   e. single cells or colonial
    c. usually practice sexual reproduction

30. The evolutionary tree of animals shows that

    a. cnidarians evolved directly from sponges.
    b. flatworms evolved directly from roundworms.
    c. both sponges and cnidarians evolved from protists.
    d. coelomates gave rise to the acoelomates.
    e. All of these are correct.

31. Which pair is mismatched?

    a. sponges—spicules   d. crayfish—compound eye
    b. tapeworms—proglottids   e. roundworms—cilia
    c. cnidarians—nematocysts

32. The presence of mesoderm

    a. is necessary to mesoglea formation.
    b. restricts the development of a coelom.
    c. is associated with the organ level of organization.
    d. is associated with the development of muscles.
    e. Both c and d are correct.

33. *Ascaris* is a parasitic

    a. roundworm.   d. sponge.
    b. flatworm.   e. mollusc.
    c. hydra.

34. The classification of annelids and arthropods is based on

    a. the first embryonic opening becoming the mouth.
    b. there being a complete digestive tract.
    c. there being a ventral solid nerve cord.
    d. there being segments.
    e. there being a heart.

35. Which of these pairs is mismatched?

    a. clam—gills
    b. lobster—gills
    c. grasshopper—book lungs
    d. polychaete—parapodia
    e. All of these are matched correctly.

36. A radula is a unique organ for feeding found in

    a. molluscs.   d. arthropods.
    b. roundworms.   e. All of these are correct.
    c. annelids.

Understanding the Terms

acoelomate 628	bivalve 637
annelid 638	cephalization 628
arachnid 645	cephalopod 637
arthropod 640	chitin 641
asymmetrical 628	cnidarian 630
bilateral symmetry 628	coelom 628

coral 631	mollusc 635
crustacean 641	molt 641
deuterostome 628	nematocyst 631
filter feeder 630	nephridium 638
flatworm 632	protostome 628
gastropod 636	pseudocoelom 628
gastrovascular cavity 632	radial symmetry 628
hemocoel 642	roundworm 634
hermaphroditic 630	sac plan 628
hydrostatic skeleton 634	segmentation 629
insect 642	sessile 628
invertebrate 626	sponge 629
leech 640	tracheae 644
Malpighian tubule 643	tube-within-a-tube plan 628
metamorphosis 644	vertebrate 626

Match the terms to these definitions:

a._____ Air tubes in insects located between the spiracles and the tracheoles.

b._____ Body cavity lying between the digestive tract and the body wall that is completely lined by mesoderm.

c._____ Body plan having two corresponding or complementary halves.

d._____ To periodically shed the exoskeleton, as in arthropods.

e._____ Segmentally arranged, paired excretory tubules of many invertebrates, as in the earthworm.

Thinking Critically

1. Roundworms are tubular and have a complete digestive tract. What advantages can you associate with these anatomical features that might explain why roundworms are more plentiful than flatworms?

2. Cnidarians are radially symmetrical, while flatworms, roundworms, molluscs, annelids, and arthropods are bilaterally symmetrical. How does the type of symmetry relate to the lifestyle of each type of animal? How does their body plan complement their lifestyle?

3. What features of arthropods make them the most diverse animal phylum on Earth?

4. Compare and contrast locomotion in an annelid and an arthropod. What body features are important in their different styles of locomotion?

5. Why do you think coelomates far outnumber acoelomates or pseudocoelomates in terms of species diversity?

ARIS, the *Inquiry into Life* Website

ARIS, the website for *Inquiry into Life,* provides a wealth of information organized and integrated by chapter. You will find practice quizzes, interactive activities, labeling exercises, flashcards, and much more that will complement your learning and understanding of general biology.

www.aris.mhhe.com

chapter 31

Animals: Part II

Thought to be extinct for the past 60 years, *the ivory-billed woodpecker was rediscovered in 2005 in the Cache River National Wildlife Refuge in Arkansas. The ivory-billed is the largest woodpecker in the United States and the second largest in the world. After reports that biologists had heard the bird's call, Dr. John Fitzpatrick, Director of the Cornell Lab of Ornithology, and Scott Simon, Arkansas State Director of The Nature Conservancy, spent a year leading a team that searched for the bird. Its existence has been confirmed by video footage and seven eyewitness sightings. The Nature Conservancy and the U.S. Fish and Wildlife Service are cooperating to construct a "Corridor of Hope," an area about 120 miles long and up to 20 miles wide in the Big Woods of Arkansas to protect the bird. The Departments of Interior and Agriculture have committed $10 million to protecting the woodpecker, and private donations have added another $10 million.*

In Chapter 30, we learned about invertebrates, of which new species are being discovered all the time. We are also discovering and rediscovering vertebrates such as the ivory-billed woodpecker, which, like so many other animals in today's world, is threatened with extinction. Thanks to a large-scale cooperative conservation effort, the ivory-billed woodpecker now has a second chance. However, approximately 28% of all mammals, 25% of all fish, 12% of birds, and 40% of amphibians are threatened with extinction. The future of many vertebrates may, in large part, depend on our ability to protect them.

Chapter Concepts

31.1 Echinoderms

In Chapter 30, we introduced the invertebrates, the large group of animals characterized by their lack of an endoskeleton of bone or cartilage. This chapter will focus mainly on the echinoderms and the chordates, which include the vertebrates.

Among animals, chordates are most closely related to the **echinoderms** (phylum Echinodermata) as witnessed by their similar embryological development. Both echinoderms and chordates are deuterostomes, in which the second embryonic opening becomes the mouth and the coelom forms by outpocketing the primitive gut.

Characteristics of Echinoderms

Echinoderms are a diverse group of marine animals; there are no terrestrial echinoderms. They have an endoskeleton (internal skeleton) consisting of spine-bearing, calcium-rich plates. The spines, which stick out through their delicate skin, account for their name.

It may seem surprising that echinoderms, although related to chordates, lack those features we associate with vertebrates such as human beings. For example, echinoderms are often radially, not bilaterally, symmetrical. Their larva is a free-swimming filter feeder with bilateral symmetry, but it typically metamorphoses into a radially symmetrical adult. Recall that with radial symmetry, it is possible to obtain two mirror images, no matter how the animal is sliced longitudinally, whereas in bilaterally symmetrical animals, one longitudinal cut gives two mirror images.

Echinoderm Diversity

Echinoderms are quite diverse (Fig. 31.1). They include sea lilies, motile feather stars, brittle stars, and sea cucumbers. Sea cucumbers actually look like a cucumber, except they have feeding tentacles surrounding their mouth. You may be more familiar with sea urchins and sand dollars in the class Echinoidea, which have spines for locomotion, defense, and burrowing. Sea urchins are food for sea otters off the coast of California. We will use the sea star as an example of echinoderm characteristics.

a. Brittle star

b. Sea cucumber

c. Sea urchins

d. Feather star

Figure 31.1 Echinoderm diversity.
a. Brittle stars have slender, branched arms, which make them the most mobile of echinoderms. **b.** Sea cucumbers look like a cucumber; they lack arms but have tentacle-like tube feet with suckers around the mouth. **c.** Sea urchins have large, colored, external spines for protection. **d.** A feather star extends its arms to filter suspended food particles from the sea.

Sea Stars

Sea stars (starfish) are commonly found along rocky coasts, where they feed on clams, oysters, and other bivalve molluscs. The five-rayed body has an oral, or mouth, side (the underside) and an aboral, or anus, side (the upper side) (Fig. 31.2). Various structures project through the body wall: (1) spines from the

a.

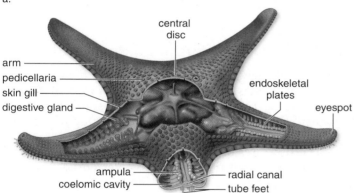

b. aboral side showing ray cross-section

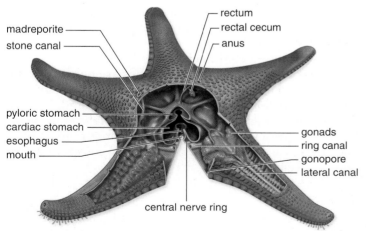

c. aboral side showing internal cross-section

Figure 31.2 Sea star anatomy and behavior.

a. A sea star uses the suction of its tube feet to open a bivalve mollusc, its primary source of food. **b.** Each arm of a sea star contains digestive glands, gonads, and portions of the water vascular system. This system (colored orange) terminates in tube feet. **c.** A different cross-section shows the orientation of mouth, digestive system, and anus.

endoskeletal plates offer some protection; (2) pincerlike structures called pedicellariae keep the surface free of small particles; and (3) skin gills, tiny fingerlike extensions of the skin, are used for gas exchange. On the oral surface, each arm has a groove lined by small tube feet.

To feed, a sea star positions itself over a bivalve such as a clam and attaches some of its tube feet to each side of the shell (Fig. 31.2a). By working its tube feet in alternation, it pulls the shell open. A very small crack is enough for the sea star to evert its cardiac stomach and push it through the crack so that it contacts the soft parts of the bivalve. The stomach secretes enzymes, and digestion begins even while the bivalve is attempting to close its shell. Later, partly digested food is taken into the sea star's body, where digestion continues in the pyloric stomach, using enzymes from digestive glands found in each arm. A short intestine opens at the anus on the aboral side.

In each arm, the well-developed coelomic cavity contains not only a pair of digestive glands, but also gonads (either male or female), which open on the aboral surface by very small pores. The nervous system consists of a central nerve ring from which radial nerves extend into each arm. A light-sensitive eyespot is at the tip of each arm. Sea stars are capable of coordinated but slow responses and body movements.

Locomotion depends on their **water vascular system.** Water enters this system through a structure on the aboral side called the madreporite, or sieve plate. From there it passes down a stone canal into a ring canal, which surrounds the mouth. The ring canal gives off a radial canal in each arm. From the radial canals, water enters the ampullae. Contraction of ampullae forces water into the tube foot, expanding it. When the foot touches a surface, the center is withdrawn, giving it suction so that it can adhere to the surface. By alternating the expansion and contraction of the tube feet, a sea star moves slowly along.

Echinoderms don't have a respiratory, excretory, or circulatory system. Fluids within the coelomic cavity and the water vascular system carry out many of these functions. For example, gas exchange occurs across the skin gills and the tube feet. Nitrogenous wastes diffuse through the coelomic fluid and the body wall. Cilia on the peritoneum lining the coelom keep the coelomic fluid moving.

Sea stars reproduce both asexually and sexually. If the body is fragmented, each fragment can regenerate a whole animal as long as the fragment contains part of the central disk. Fishermen who try to get rid of sea stars by cutting them up and tossing them overboard are merely propagating more sea stars! Sea stars also spawn, releasing either eggs or sperm. The larva is bilaterally symmetrical and metamorphoses to become the radially symmetrical adult.

Echinoderms have a well-developed coelom and internal organs despite typically being radially symmetrical. Spines project from their endoskeleton, and they have a unique water vascular system.

31.2 Chordates

To be classified as a **chordate** (phylum Chordata, about 45,000 species), an animal must have, at some time during its life history, the characteristics depicted in Figure 31.3 and listed here:

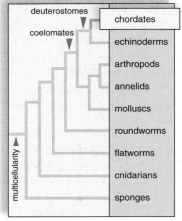

1. A **dorsal supporting rod** called a **notochord.** The notochord is located just below the nerve cord toward the back (i.e., dorsal). Vertebrates have an embryonic notochord that is replaced by the vertebral column during development.

2. A **dorsal tubular nerve cord.** Tubular means that the cord contains a canal filled with fluid. In vertebrates, the nerve cord is protected by the vertebrae. Therefore, it is called the spinal cord because the vertebrae form the spine.

3. **Pharyngeal pouches.** Most vertebrates have pharyngeal pouches only during embryonic development. In the nonvertebrate chordates, the fishes, and some amphibian larvae, the pharyngeal pouches become functioning gills. Water passing into the mouth and the pharynx goes through the gill slits, which are supported by gill arches. In terrestrial vertebrates that breathe by lungs, the pouches are modified for various purposes. In humans, the first pair of pouches becomes the auditory tubes. The second pair becomes the tonsils, while the third and fourth pairs become the thymus gland and the parathyroids.

4. **Postanal tail.** The tail extends beyond the anus, hence the term postanal.

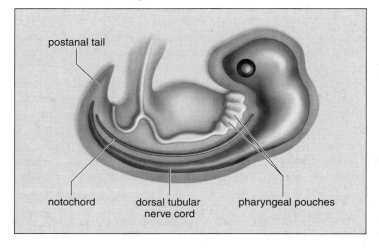

Figure 31.3 Chordate characteristics.

TABLE 31.1   Animal Phyla, Part II

Phylum Echinodermata: echinoderms

Radially symmetrical as adults; endoskeleton of calcium-rich plates; spiny skin; unique water vascular system with tube feet; able to regenerate lost body parts.

Phylum Chordata: chordates

Segmented, with a notochord; dorsal tubular nerve cord; pharyngeal pouches; postanal tail at some time during the life cycle.

Subphylum Urochordata

Sessile adult; encased in a tunic; lacks notochord and nerve cord. Example: tunicates

Subphylum Cephalochordata

Retains the four chordate characteristics as adults; obvious segmentation throughout body. Example: lancelets

Subphylum Vertebrata: vertebrates

The notochord is replaced by the vertebral column. Examples: fishes, amphibians, reptiles, birds, mammals

Evolutionary Trends Among the Chordates

Figure 31.4 depicts the phylogenetic tree of the chordates, and Table 31.1 previews the animal groups we will discuss in this chapter. The figure also lists at least one main evolutionary trend that distinguishes each group of animals from the preceding one.

The tunicates and lancelets are nonvertebrate chordates, meaning that they don't have vertebrae. The vertebrates are the fishes, amphibians, reptiles, birds, and mammals. The cartilaginous fishes were the first to have jaws, and some early bony fishes had lungs; however, not all bony fishes have lungs. Amphibians are the first group to clearly have jointed appendages and to invade land. However, the fleshy appendages of lobe-finned fishes from the Devonian era contained bones homologous to those of terrestrial vertebrates. These fishes are believed to be ancestral to the amphibians. Reptiles, birds, and mammals have a means of reproduction suitable for land. During development, an amnion and other extraembryonic membranes are present. These membranes support an embryo and prevent it from drying out as it develops into a particular species' offspring.

Chordates have four features (a notochord, a dorsal tubular nerve cord, pharyngeal pouches, and a tail) that distinguish them from other phyla. Vertebrates, in which the notochord is replaced by a vertebral column, evolved by way of the key innovations shown in Figure 31.4.

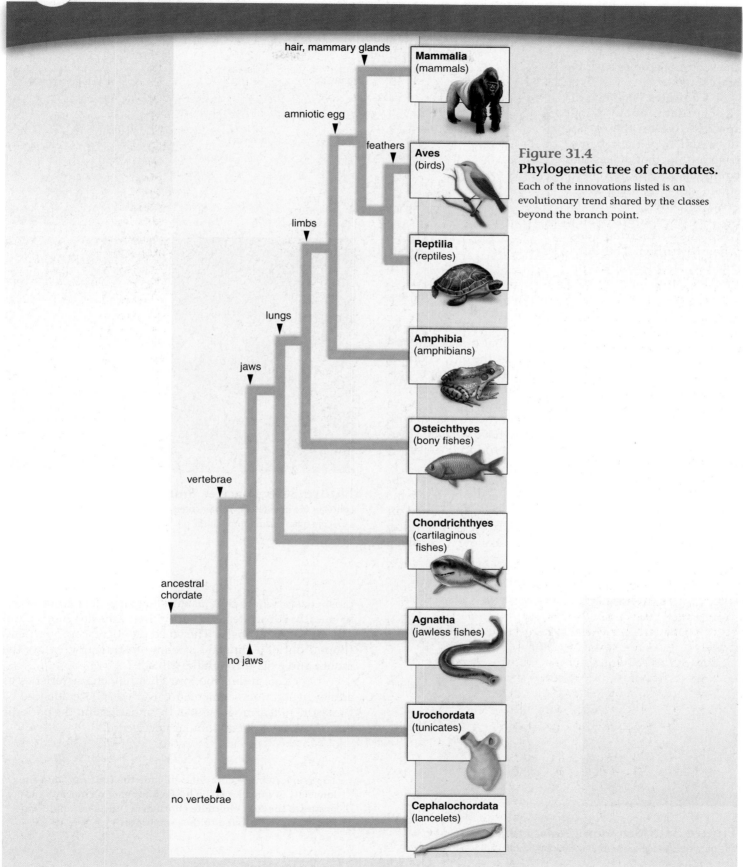

hair, mammary glands

amniotic egg

feathers

limbs

lungs

jaws

vertebrae

ancestral chordate

no jaws

no vertebrae

Mammalia (mammals)

Aves (birds)

Reptilia (reptiles)

Amphibia (amphibians)

Osteichthyes (bony fishes)

Chondrichthyes (cartilaginous fishes)

Agnatha (jawless fishes)

Urochordata (tunicates)

Cephalochordata (lancelets)

Figure 31.4
Phylogenetic tree of chordates.
Each of the innovations listed is an evolutionary trend shared by the classes beyond the branch point.

653

Nonvertebrate Chordates

There are a few nonvertebrate chordates—that is, the notochord never becomes a vertebral column.

Tunicates (subphylum Urochordata, about 1,250 species) live on the ocean floor and take their name from a tunic that makes the adults look like thick-walled, squat sacs (Fig. 31.5). They are also called sea squirts because they squirt water from one of their siphons when disturbed. The tunicate larva is bilaterally symmetrical and has the four chordate characteristics. Metamorphosis produces the sessile adult, which has incurrent and excurrent siphons.

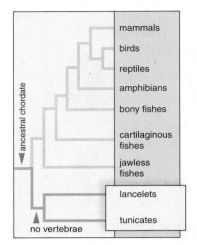

The pharynx is lined by numerous cilia. Beating of these cilia creates a current of water that moves into the pharynx and out the numerous gill slits, the only chordate characteristic that remains in the adult. Microscopic particles adhere to a mucous secretion and are digested.

Some biologists have suggested that tunicates are directly related to vertebrates. They believe that a larva with the four chordate characteristics may have become sexually mature, and evolution thereafter produced a fishlike vertebrate.

Lancelets, formerly referred to as amphioxus (subphylum Cephalochordata, about 23 species), are marine chordates only a few centimeters long. They are named for their resemblance to a lancet—a small, two-edged surgical knife (Fig. 31.6). Lancelets are found in shallow water along

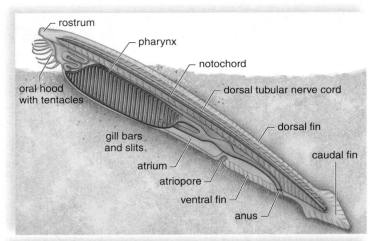

Figure 31.6 Lancelet, *Branchiostoma*.
Lancelets are filter feeders. Water enters the mouth and exits at the atriopore after passing through the gill slits.

most coasts, where they usually lie partly buried in sandy or muddy substrates with only their anterior mouth and gill apparatus exposed. They feed on microscopic particles filtered out of a constant stream of water that enters the mouth and exits through the gill slits.

Lancelets retain the four chordate characteristics as adults. In addition, segmentation is present, as evidenced by the segmental arrangement of the muscles and the periodic branches of the dorsal tubular nerve cord.

The nonvertebrate chordates include the tunicates and the lancelets. A lancelet is the best example of a chordate that possesses the four chordate characteristics as an adult and is most closely related to the vertebrates.

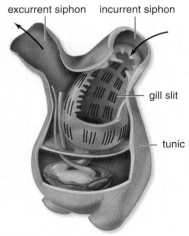

Figure 31.5 Sea squirt, *Halocynthia*.
The only chordate characteristic remaining in the adult is gill slits.

31.3 Vertebrates

At some time in their life history, **vertebrates** (subphylum Vertebrata, about 43,700 species) have all four chordate characteristics. The embryonic notochord, however, is generally replaced by a vertebral column composed of individual vertebrae. The vertebral column, which is a part of the flexible but strong jointed endoskeleton, provides evidence that vertebrates are seg-

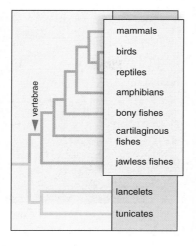

mented. The skeleton protects the internal organs and serves as a place of attachment for the muscles. Together, the skeleton and muscles form a system that permits rapid and efficient movement. Two pairs of appendages are characteristic. The pectoral and pelvic fins of fishes evolved into the jointed appendages that allowed vertebrates to move onto land.

The main axis of the endoskeleton consists of not only the vertebral column, but also a skull that encloses and protects the brain. The high degree of cephalization is accompanied by complex sense organs. The eyes develop as outgrowths of the brain. The ears are primarily equilibrium devices in aquatic vertebrates, but they also function as sound-wave receivers in land vertebrates.

The evolution of jaws in vertebrates has allowed some to take up a predatory way of life. Vertebrates have a complete digestive tract and a large coelom. Their circulatory system is closed, meaning that the blood is contained entirely within blood vessels. Vertebrates have an efficient means of obtaining oxygen from water or air, as appropriate. The kidneys are important excretory and water-regulating organs that conserve or rid the body of water as necessary. The sexes are generally separate, and reproduction is usually sexual. The evolution of the amnion allowed reproduction to take place on land. Reptiles, birds, and some mammals lay a shelled egg. In placental mammals, development takes place within the uterus of the female.

Figure 31.7 shows major milestones in the history of vertebrates: the evolution of jaws, limbs, and the amnion, an extraembryonic membrane that is first seen in the shelled **amniotic egg** of reptiles.

Vertebrates are distinguished in particular by
- strong, jointed endoskeleton
- vertebral column composed of vertebrae
- closed circulatory system
- efficient respiration and excretion
- high degree of cephalization

In short, vertebrates are adapted to an active lifestyle.

a. Great white shark, *Carcharodon*

b. Spotted salamander, *Ambystoma*

c. Veiled chameleons, *Chamaeleo*

Figure 31.7 Milestones in vertebrate evolution.

a. The evolution of jaws in fishes allows animals to be predators and to feed off other animals. **b.** The evolution of limbs in most amphibians is adaptive for locomotion on land. **c.** The evolution of an amnion and a shelled egg in reptiles is adaptive for reproduction on land.

Fishes: First Jaws, Then Lungs

The first vertebrates were jawless fishes that wiggled through the water and sucked up food from the ocean floor. Today, there are three living groups of fishes: jawless fishes, cartilaginous fishes, and bony fishes. The two latter groups have *jaws*, tooth-bearing bones of the head. Jaws are believed to have evolved from the first pair of gill arches, structures that ordinarily support gills. The second pair of gill arches became support structures for the jaws, instead of gills. The presence of jaws permits a predatory way of life in many species of fishes.

Fishes are adapted to life in the water. Usually, they shed their sperm and eggs into the water, where fertilization occurs. The zygote develops into a swimming larva, which must fend for itself until it develops into the adult form.

Jawless Fishes

Living representatives of the **jawless fishes** (superclass Agnatha, about 63 species) are cylindrical and up to a meter long. They have smooth, scaleless skin and no jaws or paired fins. The two groups of living jawless fishes are *hagfishes* and *lampreys*. The hagfishes are scavengers, feeding mainly on dead fishes, while some lampreys are parasitic. When parasitic, the round mouth of the lamprey serves as a sucker. The lamprey attaches itself to another fish and taps into its circulatory system. Unlike other fishes, the lamprey cannot take in water through its mouth. Instead, water moves in and out through the gill openings.

Cartilaginous Fishes

Cartilaginous fishes (class Chondrichthyes, about 850 species) are the sharks (see Fig. 31.7*a*), the rays (Fig. 31.8*a*), and the skates. This group of fishes is so named because they have skeletons of cartilage instead of bone. The small dogfish shark is often dissected in biology laboratories. One of the most dangerous sharks inhabiting both tropical and temperate waters is the hammerhead shark, although people are rarely attacked by sharks. Each year more people are injured or killed by dog attacks than shark attacks. The largest sharks, the whale sharks, feed on small fishes and marine invertebrates and do not attack humans. Skates and rays are rather flat fishes that live partly buried in the sand and feed on mussels and clams.

Three well-developed senses enable sharks and rays to detect their prey: (1) They have the ability to sense electric currents in water—even those generated by the muscle movements of animals. (2) They have a lateral line system, a series of pressure-sensitive cells along both sides of the body that can sense

a. Blue-spotted stingray, *Taeniura lymma*

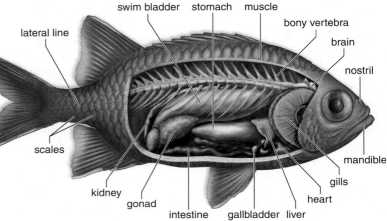

b. Soldierfish, *Myripristis jacobus*

Figure 31.8 Jawed fishes.

a. Rays, such as this blue-spotted stingray, are cartilaginous fishes. **b.** A soldierfish has the typical appearance and anatomy of a ray-finned fish, the most common type of bony fish.

pressure caused by nearby movement in the water. (3) They have a keen sense of smell; the part of the brain associated with this sense is enlarged relative to the other parts. Sharks can detect about one drop of blood in 25 gallons of water.

Bony Fishes

Bony fishes (class Osteichthyes, about 20,000 species) are by far the most numerous and diverse of all the vertebrates.

Most of the bony fishes we eat, such as cod, trout, salmon, and haddock, are **ray-finned fishes.** Their paired fins, which they use to balance and propel the body, are thin and supported by bony rays. Ray-finned fishes have various ways of life. Some, such as herring, are filter feeders; others, such as trout, are opportunists; and still others are predaceous carnivores, such as piranhas and barracudas.

Ray-finned fishes have a swim bladder, which usually regulates buoyancy (Fig. 31.8b). By secreting gases into or absorbing gases from the bladder, these fishes can change their density and thus go up or down in the water, respectively. The streamlined shape, fins, and muscle action of ray-finned fishes are all suited for locomotion in the water. Their skin is covered by bony scales that protect the body but do not prevent water loss. When ray-finned fishes respire, their gills are kept continuously moist by the passage of water through the mouth and out the gill slits. As the water passes over the gills, oxygen is absorbed by the blood, and carbon dioxide is given off. Ray-finned fishes have a single-circuit circulatory system. The heart is a simple pump, and the blood flows through the chambers, including a nondivided atrium and ventricle, to the gills. Oxygenated blood leaves the gills and is circulated throughout the body by the heart.

Another type of bony fish, called the **lobe-finned fishes,** evolved into the amphibians. These fishes not only had fleshy appendages that could be adapted to land locomotion, but most also had a **lung** that was used for respiration. One type of lobe-finned fish, called a coelacanth, was thought to have gone extinct about 20,000 years ago. However, some coelacanths were recently discovered off the coast of Madagascar, making them the only "living fossil" among these fishes.

Most fishes today are ray-finned fishes. They have the following characteristics:

- bony skeleton and scales
- swim bladder
- two-chambered heart
- paired fins
- jaws
- gills

Amphibians: Jointed Appendages

Amphibians (class Amphibia, about 5,700 species), whose class name means living on both land and in the water, are represented today by frogs, toads, newts, salamanders, and caecilians, which are strange, wormlike amphibians that spend most of their life underground. Aside from *jointed appendages,* amphibians have other features not seen in bony fishes: they usually have four limbs, eyelids for keeping their eyes moist, ears (a tympanum) for picking up sound waves, and a voice-producing larynx. The brain is larger than that of a fish. Adult amphibians usually have small lungs, but some species respire entirely through their skin. Air enters the mouth by way of nostrils, and when the floor of the mouth is raised, air is forced into the relatively small lungs. Respiration is supplemented by gas exchange through the smooth, moist, and glandular skin. The amphibian heart has a divided atrium but a single ventricle. The right atrium receives de-oxygenated blood from the body, and the left atrium receives oxygenated blood from the lungs. These two types of blood are partially mixed in the single ventricle (see Fig. 31.12c). Mixed blood is then sent to all parts of the body; some is sent to the skin, where it is further oxygenated.

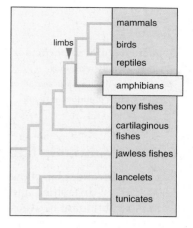

Most members of this group lead an amphibious life—that is, the larval stage lives in the water, and the adult stage lives on land. However, the adult usually returns to the water to reproduce. Figure 31.9 illustrates how a frog tadpole undergoes metamorphosis into a tetrapod (four-limbed) adult before taking up life on land. Currently, there is widespread concern about the future of amphibians because over 40% of all amphibian species are threatened with extinction. Because of their permeable skin and a life cycle that often depends on both water and land, they are thought to be "indicator species" of environmental quality. That is, they are the first to respond to environmental degradation, and disappearances

1. Tadpoles hatch. 2. Tadpole respires with external gills. 3. Front and hind legs are present. 4. Frog respires with lungs.

Figure 31.9 Frog metamorphosis.
During metamorphosis, the animal changes from an aquatic to a terrestrial organism.

of populations and species thus generate concern for other species—even humans.

These features in particular distinguish amphibians:
- usually tetrapods
- usually undergo metamorphosis
- three-chambered heart
- usually lungs in adults
- smooth, moist skin

Reptiles: Amniotic Egg

Reptiles (class Reptilia, about 6,000 species) diversified and were most abundant between 245 and 65 million years ago. These reptiles included the mammal-like reptiles, the ancestors of today's living reptiles, and the dinosaurs, which became extinct. Some dinosaurs are remembered for their great size. *Brachiosaurus,* a herbivore, was about 23 meters (75 feet) long and about 17 meters (56 feet) tall. *Tyrannosaurus rex,* a carnivore, was 5 meters (16 feet) tall when standing on its hind legs. The bipedal stance of some reptiles preceded the evolution of wings in birds. In fact, some say birds are actually living dinosaurs; new molecular, morphological, and behavioral studies support this hypothesis.

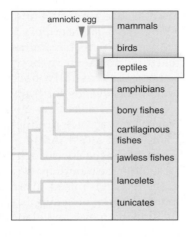

amniotic egg

mammals
birds
reptiles
amphibians
bony fishes
cartilaginous fishes
jawless fishes
lancelets
tunicates

The reptiles living today are mainly turtles, alligators, snakes, and lizards. The body is covered with hard, keratinized scales, which protect the animal from desiccation and from predators. Another adaptation for a land existence is the manner in which snakes typically use their tongue as a sense organ (Fig. 31.10). Reptiles have well-developed lungs enclosed by a protective rib cage. When the rib cage expands, a partial vacuum establishes a negative pressure, which causes air to rush into the lungs. The atrium of the heart is always separated into right and left chambers, but division of the ventricle varies. An interventricular septum

Figure 31.10
The tongue as a sense organ.
Snakes wave a forked tongue in the air to collect chemical molecules, which are then brought back into the mouth and delivered to an organ in the roof of the mouth. Analyzed chemicals help the snake trail a prey animal, recognize a predator, or find a mate.

is incomplete in certain species; therefore, some oxygenated and deoxygenated blood is mixed between the ventricles.

Perhaps the most outstanding evolutionary innovation of reptiles is that their means of reproduction is suitable to a land existence. The penis of the male passes sperm directly to the female. Fertilization is internal, and the female typically lays leathery, flexible, shelled eggs. This *amniotic egg* made development on land possible and eliminated the need for a swimming-larval stage during development. This egg provides the developing embryo with atmospheric oxygen, food, and water; it removes nitrogenous wastes; and it protects the embryo from drying out and from mechanical injury. This is accomplished by the presence of extraembryonic membranes such as the chorion. (Fig. 31.11).

Fishes, amphibians, and reptiles are **ectothermic,** meaning that their body temperature matches the temperature of the external environment. If it is cold externally, they are cold internally; if it is hot externally, they are hot internally. Reptiles try to regulate their body temperatures by basking in the sun if they need warmth or by hiding in the shadows if they need cooling off.

These features in particular distinguish reptiles:
- lungs with expandable rib cage
- shelled amniotic egg
- keratinized scaly skin

shell

jaws

a. American crocodile, *Crocodylus acutus*

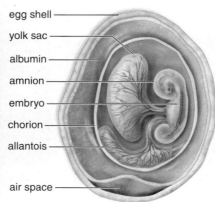

egg shell
yolk sac
albumin
amnion
embryo
chorion
allantois

air space

b.

Figure 31.11
The reptilian egg allows reproduction on land.
a. Baby American crocodile, *Crocodylus,* hatching out of its shell. Note that the shell is leathery and flexible, not brittle like birds' eggs. **b.** Inside the egg, the embryo is surrounded by extraembryonic membranes. The chorion aids gas exchange, the yolk sac provides nutrients, the allantois stores waste, and the amnion encloses a fluid that prevents drying out and provides protection.

Science Focus

Animals and Human Medicine

Hundreds of pharmaceutical products come from animals, and even animals that produce poisons and toxins give us beneficial medicines. The Thailand cobra paralyzes its victim's nerves and muscles with a potent venom that eventually leads to respiratory arrest. However, that venom is also the source of the drug Immunokine, which has been used for over ten years to treat multiple sclerosis patients. Immunokine, which is almost without side effects, actually protects patients' nerve cells from destruction by their immune system.

A compound known as ABT-594, derived from the skin of the poison-dart frog (Fig. 31A*a*), is approximately 50 times more powerful than morphine in relieving chronic and acute pain without the addictive properties. Recently, a study showed that certain natural proteins contained in frog skin (called antimicrobial peptides) are very effective in killing the HIV virus and preventing it from leaving dendritic cells, where the virus first "hides out" in the early stages of infection. The copperhead snake, the fer-de-lance viper, and the giant Israeli scorpion are also some unlikely animals that directly provide, or serve as a model for the development of pharmaceuticals such as pain killers, antibiotics, and anticancer drugs.

Other animals produce proteins similar enough to human proteins to be used for medical treatment. Until 1978, when recombinant DNA human insulin was produced, diabetics injected insulin purified from pigs. Currently, the flu vaccine is produced in fertilized chicken eggs. However, the production of these drugs is often time-consuming, labor intensive, and expensive.

In 2003, pharmaceutical companies used 90 million chicken eggs and took nine months to produce the flu vaccine. In 2004, several million doses of the flu vaccine were contaminated, creating a shortage.

Powerful applications of genetic engineering can be found in the development of treatments for human diseases. In fact, this new biotechnology has actually led to a new industry: animal pharming. Animal pharming uses genetically altered animals, such as mice, sheep, goats, cows, pigs, and chickens, to produce medically useful pharmaceutical products. A human gene for a particular useful product is inserted into the embryo of the animal. That embryo is implanted into a foster mother, which gives birth to a transgenic animal, so called because it contains genes from two sources. An adult transgenic animal produces large quantities of the pharmed product in its blood, eggs, or milk, which is then harvested and purified. The first such product, alpha-1-antitrypsin for the treatment of emphysema and cystic fibrosis, is now undergoing clinical trials.

Some species of vertebrates have amazing abilities to regenerate body parts. For example, salamanders can regenerate limbs, upper and lower jaws, and the retina of the eye, while zebrafish can regenerate fins, the spinal cord, and part of the heart. Scientists believe that because regeneration exists throughout the animal kingdom, humans may already have the genes to regenerate body parts, but these genes are turned off; recent research suggests that a gene associated with regeneration in zebrafish also exists in humans. Perhaps drugs could be

developed to turn these genes on. Or, it is possible that these genes or their products could be utilized for human medicine by animal pharming or other recombinant DNA methods.

Because there is an alarming shortage of human donor organs, xenotransplantation, the transplantation of animal tissues and organs into human beings, is another benefit of genetically altered animals. For example, in the late 1990s, two patients were kept alive using a pig liver outside of their body to filter their blood until a human organ was available for transplantation. Although baboons are genetically closer to humans than pigs, pigs are generally healthier, produce more offspring in a shorter time, and are already farmed for food. Currently, pig heart valves and skin are routinely used to treat humans. Miniature pigs, whose heart size is similar to that of humans, are being genetically engineered to make their tissues less foreign to the human body, in order to avoid rejection (Fig. 31A*b*).

Discussion Questions

1. Is it ethical to change the genetic makeup of animals to use them as drug or organ factories?
2. Given the potential health effects and concern of transfer of diseases from animals to humans, what guidelines should there be with regard to xenotransplantation?
3. Using the information you learned from this reading, develop an argument for the conservation of biodiversity.

a. Poison-dart frogs, source of a medicine

b. Pigs, source of organs

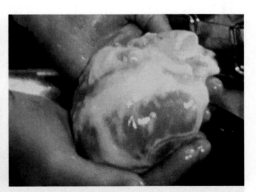

c. Heart for transplantation

Figure 31A Medical uses of animals
a. The poison-dart frog is the source of a pain medication. **b.** Pigs are now being genetically altered to provide a supply of hearts for **(c)** heart transplant operations.

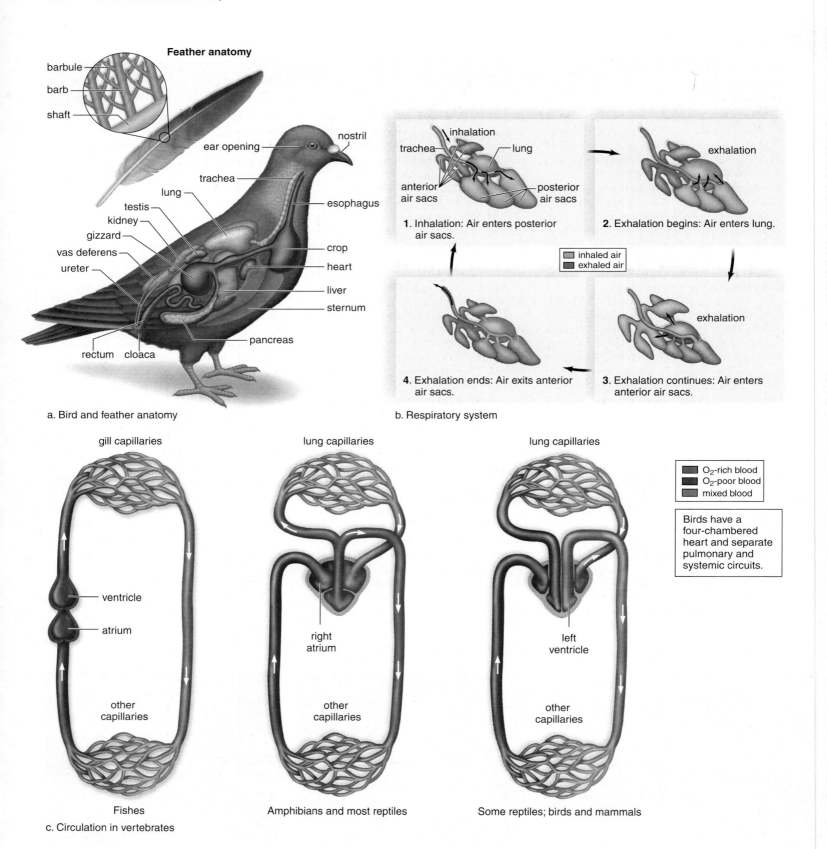

a. Bird and feather anatomy

b. Respiratory system

1. Inhalation: Air enters posterior air sacs.

2. Exhalation begins: Air enters lung.

3. Exhalation continues: Air enters anterior air sacs.

4. Exhalation ends: Air exits anterior air sacs.

inhaled air
exhaled air

c. Circulation in vertebrates

Fishes

Amphibians and most reptiles

Some reptiles; birds and mammals

O_2-rich blood
O_2-poor blood
mixed blood

Birds have a four-chambered heart and separate pulmonary and systemic circuits.

Figure 31.12 Bird anatomy and physiology.

a. The anatomy of a pigeon is representative of bird anatomy. **b.** In birds, air passes one way through the lungs. **c.** In fishes, a single-loop system utilizes a two-chambered heart. In amphibians and most reptiles the heart has three chambers, and some mixing of O_2-rich and O_2-poor blood takes place. In birds, mammals, and some reptiles, the heart has four chambers and sends only O_2-poor blood to the lungs and O_2-rich blood to the body.

Birds: Feathers

To many people, the **birds** (class Aves) are the most conspicuous, melodic, beautiful, and fascinating group of vertebrates. Over 9,000 species of birds have been described. Birds range in size from the tiny "bee" hummingbird at 1.8 g to the ostrich at a maximum weight of 160 kg and a height of 2.7 m. The fossil record of birds is unraveling evolutionary mysteries

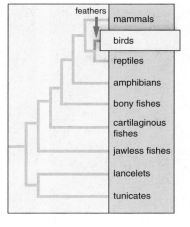

because *Archaeopteryx* and newly described fossils are providing evidence that birds evolved from bipedal dinosaurs. Birds may be classified as dinosaurs in the future; their legs have scales and their **feathers** (Fig. 31.12*a*) are modified reptilian scales. However, birds lay a hard-shelled amniotic egg rather than the leathery egg of reptiles.

Diversity of Birds

The majority of birds, including eagles, geese, and mockingbirds, have the ability to fly. However, some birds, such as emus, penguins, and ostriches, are flightless. Traditionally, birds have been classified based on beak and foot type (Fig. 31.13) and, to some extent, on their habitat and behavior. The various orders include birds of prey with notched beaks and sharp talons; shorebirds with long, slender, probing beaks and long, stiltlike legs; woodpeckers with sharp, chisel-like beaks and grasping feet; waterfowl with broad beaks and webbed toes; penguins with wings modified as paddles; and songbirds with perching feet.

Anatomy and Physiology of Birds

Nearly every anatomical feature of a bird can be related to its ability to fly (Fig. 31.12a). The forelimbs are modified

as wings. The hollow, very light bones are laced with air cavities. A horny beak has replaced jaws equipped with teeth, and a slender neck connects the head to a rounded, compact torso. The sternum is enlarged and has a keel, to which strong muscles are attached for flying. Respiration is efficient since the lobular lungs form anterior and posterior air sacs. The presence of these sacs means that the air circulates one way through the lungs and that gases are continuously exchanged across respiratory tissues (Fig. 31.12*b*). Another benefit of air sacs is that they lighten the body and thereby aid flying.

Birds have a four-chambered heart that completely separates O_2-rich blood from O_2-poor blood. The left ventricle pumps O_2-rich blood under pressure to the muscles. Birds are **endothermic.** Like mammals, their internal temperature is constant because they generate and maintain metabolic heat. This may be associated with their efficient nervous, respiratory, and circulatory systems. Also, their feathers provide insulation. Birds have no bladder and excrete uric acid in a semi-dry state.

Flight requires acute sense organs and an intricate nervous system. Birds have particularly good vision and well-developed brains. Their muscle reflexes are excellent. An enlarged portion of the brain seems to be the area responsible for instinctive behavior. A ritualized courtship often precedes mating. Many newly hatched birds require parental care before they are able to fly away and seek food for themselves. A remarkable aspect of bird behavior is the seasonal migration of many species over very long distances. Birds navigate by day and night, whether it's sunny or cloudy, by using the sun and stars and even the Earth's magnetic field to guide them.

These features in particular distinguish birds:

- feathers
- hard-shelled amniotic egg
- four-chambered heart
- usually wings for flying
- air sacs
- endothermic

a. Cardinal, *Cardinalis cardinalis*

b. Bald eagle, *Haliaetus leucocephalus*

c. Flamingo, *Phoenicopterus ruber*

Figure 31.13 Bird beaks.

a. A cardinal's beak allows it to crack tough seeds. **b.** A bald eagle's beak allows it to tear prey apart. **c.** A flamingo's beak strains food from the water with bristles that fringe the mandibles.

a. Duckbill platypus, *Ornithorhynchus anatinus*   b. Virginia opossum, *Didelphis virginianas*   c. Koala, *Phascolarctos cinereus*

Figure 31.14   Monotremes and marsupials.
a. The duckbill platypus is a monotreme that inhabits Australian streams. **b.** The opossum is the only marsupial in the Americas. The Virginia opossum is found in a variety of habitats. **c.** The koala is an Australian marsupial that lives in trees.

Mammals: Hair and Mammary Glands

Mammals (class Mammalia, about 4,500 species) also evolved from the reptiles. The first mammals were small, about the size of mice. During all the time the dinosaurs flourished (165–65 million years ago), mammals were typically small in size and changed little evolutionarily. Some of the earliest mammalian groups are still represented today by the monotremes and marsupials, but they are not abundant.

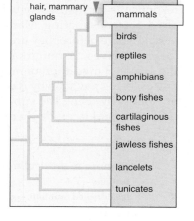

The placental mammals that evolved later went on to live in many habitats, including air, land, and sea.

The chief characteristics of mammals are body hair and milk-producing mammary glands. Almost all mammals are endothermic and maintain a constant internal temperature. Many of the adaptations of mammals are related to temperature control. *Hair,* for example, provides insulation against heat loss and allows mammals to be active, even in cold weather. Like birds, mammals have efficient respiratory and circulatory systems, which ensure a ready supply of oxygen to muscles whose contraction produces body heat. Also, like birds, mammals have double-loop circulation and a four-chambered heart (see Fig. 31.12*c*).

Mammary glands enable females to feed (nurse) their young without leaving them to find food. Nursing also creates a bond between mother and offspring that helps ensure parental care while the young are helpless. In all mammals (except monotremes), the young are born alive after a period of development in the uterus, a part of the female reproductive tract. Internal development shelters the young and allows the female to move actively about while the young are maturing. Mammals are classified as monotremes, marsupials, or placental mammals, according to how they reproduce.

Monotremes

Monotremes are mammals that, like birds, have a *cloaca,* a terminal region of the digestive tract that serves as a common chamber for feces, excretory wastes, and sex cells. They also lay hard-shelled amniotic eggs. They are represented by the spiny anteater and the duckbill platypus, both of which are found in Australia (Fig. 31.14*a*). The female duckbill platypus lays her eggs in a burrow in the ground. She incubates the eggs, and after hatching, the young lick up milk that seeps from modified sweat glands on the abdomens of both males and females. The spiny anteater has a pouch on the belly side formed by swollen mammary glands and longitudinal muscle. The egg moves from the cloaca to this pouch, where hatching takes place and the young remain for about 53 days. Then they stay in a burrow, where the mother periodically visits and nurses them.

Marsupials

The young of **marsupials** begin their development inside the female's body, but they are born in a very immature condition. Newborns crawl up into a pouch on their mother's abdomen. Inside the pouch, they attach to the nipples of mammary glands and continue to develop. Frequently, more are born than can be accommodated by the number of nipples, and it's "first come, first served."

The Virginia opossum (*Didelphis virginiana*) is the only marsupial that occurs north of Mexico (Fig. 31.14*b*). In Australia, however, marsupials underwent adaptive radiation for several million years without competition from pla-

a. White-tailed deer, *Odocoileus virginianus*

b. African lioness, *Panthera leo*

c. Squirrel monkey, *Saimiri sciureus*

d. Killer whale, *Orcinus orca*

Figure 31.15
Placental mammals.
Placental mammals have adapted to various ways of life. **a.** Deer are herbivores that live in forests. **b.** Lions are carnivores on the African plain. **c.** Monkeys inhabit tropical forests. **d.** Whales are sea-dwelling placental mammals.

cental mammals, which arrived there only recently. Thus, marsupial mammals are now found mainly in Australia, with some in Central and South America as well. Among the herbivorous marsupials, koalas are tree-climbing browsers (Fig. 31.14*c*), and kangaroos are grazers. The Tasmanian wolf or tiger, thought to be extinct, was a carnivorous marsupial about the size of a collie dog.

Placental Mammals

The vast majority of living mammals are **placental mammals** (Fig. 31.15). In these mammals, the extraembryonic membranes of the reptilian egg (see Fig. 31.11) have been modified for internal development within the uterus of the female. The chorion contributes to the fetal portion of the placenta, while a part of the uterine wall contributes to the maternal portion. Here, nutrients, oxygen, and waste are exchanged between fetal and maternal blood.

Mammals, with the exception of marine mammals, are adapted to life on land and have limbs that allow them to move rapidly. In fact, an evaluation of mammalian features leads us to the obvious conclusion that they lead active lives. The lungs are expanded not only by the action of the rib cage, but also by the contraction of the diaphragm, a horizontal muscle that divides the thoracic cavity from the abdominal cavity. The heart has four chambers. The internal temperature is constant, and hair, when abundant, helps insulate the body.

The mammalian brain is well-developed and enlarged due to the expansion of the cerebral hemispheres, which control the rest of the brain. The brain is not fully developed until after birth, and the young learn to take care of themselves during a period of dependency on their parents.

Mammals have differentiated teeth. Typically, in the front, the incisors and canine teeth have cutting edges for capturing and killing prey. On the sides, the premolars and molars chew food. The specific shape and size of the teeth may be associated with whether the mammal is an herbivore (eats vegetation), a carnivore (eats meat), or an omnivore (eats both meat and vegetation). For example, mice (order Rodentia) have continuously growing incisors; horses have large, grinding molars; and dogs (order Carnivora) have long canine teeth. Placental mammals are classified based on their methods of obtaining food and their mode of locomotion. For example, bats (order Chiroptera) have membranous wings supported by digits; horses (order Perissodactyla) have long, hoofed legs; and whales (order Cetacea) have paddlelike forelimbs.

These features in particular distinguish placental mammals:
- mammary glands
- body hair
- differentiated teeth
- infant dependency
- endothermic
- well-developed brain
- internal development

Primates

Primates are members of the order Primates. In contrast to the other orders of placental mammals, most primates are adapted to an arboreal life—that is, for living in trees. Primate limbs are mobile, and the hands and feet both have five digits each. Many primates have a big toe and a thumb that are both opposable, meaning the big toe or thumb can touch each of the other toes or fingers. (Humans don't have an opposable big toe, but the thumb is opposable, and this results in a powerful and precise grip.) The opposable thumb allows a primate to easily reach out and bring food such as fruit to the mouth. When locomoting, primates grasp and release tree limbs freely because they have nails instead of the claws of their evolutionary ancestors.

In primates, the snout is shortened considerably, allowing the eyes to move to the front of the head. The stereoscopic vision (or depth perception) that results permits primates to make accurate judgments about the distance and position of adjoining tree limbs. Some primates that are active during the day, such as humans, have color vision and strong visual acuity because the retina contains cone cells in addition to rod cells. Cone cells require bright light, but the image is sharp and in color. The lens of the eye focuses light directly on the fovea, a region of the retina where cone cells are concentrated.

The evolutionary trend among primates is toward a larger and more complex brain; the brain size is smallest in prosimians and largest in modern humans. The cerebral cortex, which has many association areas, expands so much that it becomes extensively folded in humans as we learn or experience new things. The portion of the brain devoted to smell gets smaller, and the portions devoted to sight increase in size and complexity during primate evolution. Also, more and more of the brain is involved in controlling and processing information received from the hands and the thumb. The result is good hand-eye coordination in humans.

It is difficult to care for several offspring while moving from limb to limb, and so one birth at a time is the norm in primates. The juvenile period of dependency on parental care is extended, and there is an emphasis on learned behavior and complex social interactions.

The order Primates has two suborders: the **prosimians** (lemurs, tarsiers, and lorises) and the **anthropoids** (monkeys, apes, and humans). Thus, humans are more closely related to the monkeys and apes than they are to the prosimians.

These characteristics especially distinguish primates from other mammals:

- Opposable thumb (and in some cases, big toe)
- Well-developed brain
- Nails (not claws)
- Usually single offspring per mating
- Extended period of parental care
- Emphasis on learned behavior

31.4 Human Evolution

The evolutionary tree in Figure 31.16 shows that all primates share one common ancestor and that the other types of primates diverged from the human line of descent (called a **lineage**) over time. Notice that prosimians, represented by lemurs and tarsiers, were the first types of primates to diverge from the human line of descent, and African apes were the last group to diverge from our line of descent. Also note that the tree suggests that humans are most closely related to African apes. One of the most unfortunate misconceptions concerning human evolution is the belief that Darwin and others suggested that humans evolved from apes. On the contrary, humans and apes shared a common apelike ancestor. Today's apes are our distant cousins, and we couldn't have evolved from our cousins because we are contemporaries—living on Earth at the same time.

Molecular data have been used to determine the date of the split between the human lineage and that of apes. When two lines of descent first diverge from a common ancestor, the genes of the two lineages are nearly identical. But as time goes by, each lineage accumulates genetic changes. Many genetic changes are neutral (not tied to adaptation) and accumulate at a fairly constant rate; such changes can be used as a kind of **molecular clock** to indicate the relatedness of the two groups and the point when they diverged from one another. Molecular data suggest that the split between the ape and human lineages occurred about 6 million years ago (MYA). Currently, humans and chimpanzees are probably the most closely related, sharing about 99% of our DNA.

Evolution of Hominids

The classification of humans is given in Table 31.2. To be a **hominid**, a fossil must have an anatomy suitable for standing erect and walking on two feet (called **bipedalism**).

Until recently, many scientists thought that hominids evolved because of a dramatic change in climate that caused forests to be replaced by grassland about 4 MYA. It is thought that hominids began to stand upright to keep the sun off their backs and to have better vision for detecting predators or

TABLE 31.2	Classification of Humans
Order: Primates (prosimians, monkeys, apes, hominids)	
Family: Hominidae (hominids)	
(Ardipithecines, Australopithecines, humans)	
Genus: *Homo* (humans)	
Early *Homo**:	
Homo habilis, Homo ergaster, Homo erectus	
More recent *Homo**:	
Homo heidelbergensis, Homo neandertalensis, Homo floresiensis	
Modern humans*:	
Homo sapiens sapiens	

* Category not in the classification of organisms but added here for clarity.

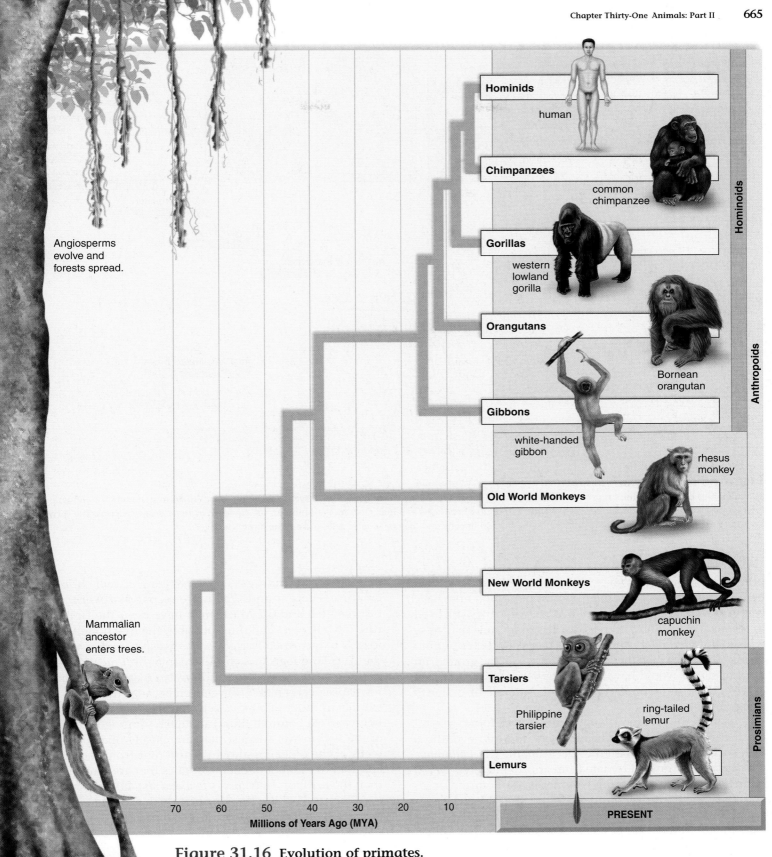

Angiosperms
evolve and
forests spread.

Mammalian
ancestor
enters trees.

Hominids

human

Chimpanzees

common
chimpanzee

Gorillas

western
lowland
gorilla

Orangutans

Bornean
orangutan

Gibbons

white-handed
gibbon

rhesus
monkey

Old World Monkeys

New World Monkeys

capuchin
monkey

Tarsiers

ring-tailed
lemur

Philippine
tarsier

Lemurs

Hominoids

Anthropoids

Prosimians

70   60   50   40   30   20   10

Millions of Years Ago (MYA)

PRESENT

Figure 31.16 Evolution of primates.

Primates are descended from an ancestor that may have resembled a tree shrew. The descendants of this ancestor adapted to
the new way of life and developed traits such as a shortened snout and nails instead of claws. The time when each type of
primate diverged from the main line of descent is known from the fossil record. A common ancestor was living at each point of
divergence: for example, there was a common ancestor for monkeys, apes, and hominids about 45 MYA: one for all apes and
hominids about 15 MYA: and one for just African apes and hominids about 7 MYA. The evolutionary split between the ape and
human lineage may have occurred about this time.

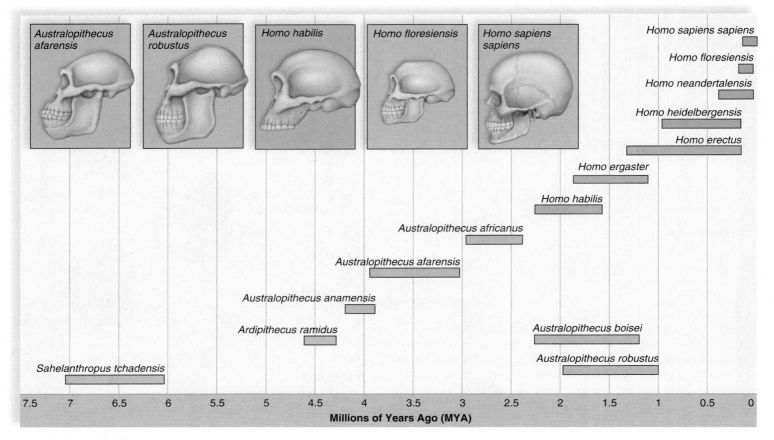

Figure 31.17 Human evolution.
Several groups of extinct hominids preceded the evolution of modern humans. These groups have been divided into the possible early hominids (orange), later hominids (green), early *Homo* species (lavender), and later *Homo* species (blue). Only modern humans are classified as *Homo sapiens sapiens*. The early hominid fossils are dated closest to the time when the human lineage and the ape lineage split and are the most primitive.

prey in grasslands. However, this is currently under debate because recent fossil findings suggest hominids may have begun to walk upright as early as 6 MYA, while they still lived in forests. While living in trees, the first hominids may have walked upright on large branches as they collected fruit from overhead. Then, when they began to forage on the ground, an upright stance would have made it easier for them to travel more easily from woodland to woodland and/or easier to forage among bushes.

Within the past ten years, scientists have found several fossils that may provide clues to early hominid evolution. One such skull fossil, with the species name *Sahelanthropus tchadensis*, is estimated to date from approximately 7 MYA, but researchers are debating whether it is a hominid or an ape. Only the braincase has been found, and it is very ape-like. However, a point at the back of the skull where the neck muscles attach suggests bipedalism. Also, the canines are smaller and the tooth enamel thicker than in the teeth of an ape. Another series of leg fossils (genus name *Orrorin*), which are about 6 million years old, may suggest upright walking because the femur appears strong and able to support more weight than that of the apes. An area called Middle Awash, Ethiopia, seems to be a "hotbed" of recent

primate fossil findings. There, remains of two chimplike creatures were found. The more recent discovery, dated between 5.8 and 5.2 MYA, may have been the first creature to walk erect and thus, possibly, the first hominid. The fossil's name, *Ardipithecus ramidus kadabba*, is derived from the Afar language spoken in Ethiopia. *Ardi* means ground or floor; *ramid* means root; and *kadabba* means family ancestor. This species name was chosen because the bones were similar to those of *Ardipithecus ramidus ramidus*, a 4.4-million-year-old hominid found about ten years ago. The bones of the two fossils suggest they are closely related, but the teeth of the older fossil, while different from those of an ape, have a number of characteristics that are more apelike than the teeth of *Ardipithecus ramidus ramidus*.

Figure 31.17 is a timeline of human evolution in which these potential early hominids are represented by orange-colored bars.

The evolution of bipedalism is believed to be the distinctive feature that separates hominids from apes. Scientists are still debating when and why bipedalism evolved.

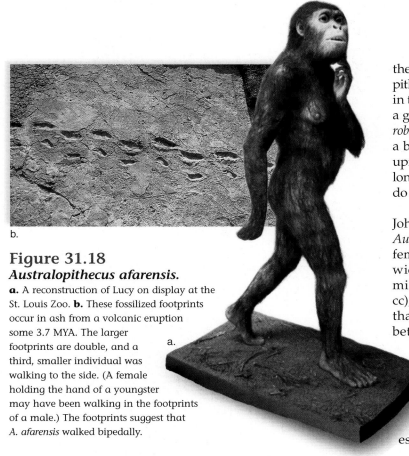

Figure 31.18
Australopithecus afarensis.
a. A reconstruction of Lucy on display at the
St. Louis Zoo. **b.** These fossilized footprints
occur in ash from a volcanic eruption
some 3.7 MYA. The larger
footprints are double, and a
third, smaller individual was
walking to the side. (A female
holding the hand of a youngster
may have been walking in the footprints
of a male.) The footprints suggest that
A. afarensis walked bipedally.

Australopithecines

Most recently, scientists have discovered a fossil, which they have named *Australopithecus anamensis*, in Middle Awash, where they have found eight hominid-like species so far. They believe this fossil may be the missing evolutionary link between *Ardipithecus* and the **australopithecines** (represented by green-colored bars in Figure 31.17), the possible direct hominid ancestors for humans (genus *Homo*). It is thought that the australopithecines evolved and diversified in Africa 4 MYA and went extinct about 1 MYA. Australopithecines had a small brain (an apelike characteristic) and walked erect (a humanlike characteristic). Thus, it seems that human-like characteristics did not all evolve at the same time. Australopithecines give evidence of **mosaic evolution,** meaning that different body parts changed at different rates and therefore at different times.

Australopithecines stood about 100–115 cm high and had a brain slightly larger than that of a chimpanzee (about 370–515 cc). The forehead was low, and the face projected forward. There is no evidence that they used tools. Some are thought to have been slight of frame, or "gracile" (slender), while others were robust (powerful), with strong upper bodies and massive jaws powered by chewing muscles attached to a prominent bony crest on top of the skull. The gracile types probably fed on soft fruits and leaves, while the robust types likely had a more fibrous diet that may have included hard nuts.

Some fossil remains of australopithecines have been found in southern Africa, while others have been found in eastern Africa. The exact evolutionary relationship between

these two groups is currently not known. The first australopithecine was discovered in southern Africa by Raymond Dart in the 1920s. This hominid, named *Australopithicus africanus,* is a gracile type. A second southern African fossil specimen, *A. robustus,* is a robust type. Both *A. africanus* and *A. robustus* had a brain size of about 500 cc. These hominids clearly walked upright, but the proportions of their limbs are apelike (i.e., longer forelimbs than hindlimbs). Therefore, most scientists do not believe that *A. africanus* is ancestral to early *Homo.*

More than 20 years ago, a team led by Donald Johanson unearthed nearly 250 fossils of a hominid called *Australopithecus afarensis* in eastern Africa. A now-famous female skeleton discovered in this group is known worldwide by its field name, "Lucy." This skeleton is about 3.2 million years old. Although her brain was quite small (400 cc), the shapes and relative proportions of her limbs indicate that she stood up and walked bipedally (Fig. 31.18*a*). Even better evidence of bipedal locomotion comes from a trail of footprints in Laetoli from about 3.7 MYA (Fig. 31.18*b*). For these reasons, *A. afarensis* is usually favored as being more directly related to *Homo* than *A. africanus. A. afarensis,* a gracile type, is believed to be ancestral to the robust types found in eastern Africa: such as *A. boisei,* which had a powerful upper body and the largest molars of any hominid.

> Australopithecines, which arose in Africa, were definitively the first hominids. Their remains show that bipedalism was the first humanlike feature to evolve. It is unknown at this time which australopithecine is ancestral to early *Homo.*

Evolution of Early *Homo*

Fossils designated as early *Homo* species are represented in Figure 31.17 by a lavender colored bar. These fossils appear in the fossil record approximately 2 MYA. They all have a brain size that is 600 cc or greater, their jaw and teeth resemble those of humans, and tool use is in evidence.

In general, *Homo habilis* has a more primitive anatomy than *Homo ergaster* and *Homo erectus,* which are all grouped together as early *Homo* species. The latter two species have a higher forehead, a flatter face, and a larger brain size than the first two species.

Homo habilis

Although the height of *Homo habilis* did not exceed those of the australopithecines, some of this species' fossils have a brain size as large as 800 cc, considerably larger than that of *A. afarensis.* The cheek teeth of these hominids tend to be smaller than even those of the gracile australopithecines. Therefore, it is likely that they were omnivorous and ate meat in addition to plant material. Bones at the campsites of *H. habilis* bear cut marks, indicating that these hominids used tools to strip off the meat.

The stone tools made by *H. habilis* are rather crude but, sharp enough to scrape away hide, cut tendons, and easily remove meat from bones.

The skulls of these two species suggest that the portions of the brain associated with speech were enlarged. The ability to speak may have led to hunting cooperatively. Some members of the group may have remained plant gatherers, and if so, both hunters and gatherers most likely ate together and shared their food. In this way, society and culture could have begun. **Culture,** which encompasses human behavior and products (such as technology and the arts), is dependent on the capacity to speak and transmit knowledge. We can further speculate that the advantages of culture to these hominids may have hastened the extinction of the australopithecines.

> *H. habilis* and *H. rudolfensis* warrant classification as *Homo* because of brain size, dentition, and tool use. These early *Homo* species may have had the rudiments of a culture.

Homo ergaster *and* Homo erectus

Homo ergaster evolved in Africa, perhaps from *H. habilis.* Similar fossils found in Asia are different enough to be classified as *Homo erectus.* These fossils span the dates between 1.9 and 0.3 MYA. A Dutch anatomist named Eugene Dubois was the first to unearth *H. erectus* bones in Java in 1891, and since that time many other fossils belonging to both species have been found in Africa and Asia.

Compared to *H. habilis, H. ergaster* had a larger brain (about 1,000 cc) and a flatter face, with a nose that projected. This type of nose is adaptive for a hot, dry climate because it permits water to be absorbed before air leaves the body. The recovery of an almost complete skeleton of a 10-year-old boy indicates that *H. ergaster* was much taller than the hominids discussed thus far (Fig. 31.19). Males were 1.8 m tall, and females were 1.55 m. Indeed, these hominids stood erect and most likely had a *striding gait* similar to modern humans. The robust and most likely heavily muscled skeleton still retained some australopithecine features. Even so, the size of the birth canal indicates that infants were born in an immature state that required an extended period of parental care.

H. ergaster first appeared in Africa but then migrated into Europe and Asia. At one time, the migration was thought to have occurred about 1 MYA, but recently *H. erectus* fossils in Java and the Republic of Georgia (formerly part of the U.S.S.R.) have been dated at 1.9 and 1.6 MYA, respectively. These remains push the evolution of *H. ergaster* in Africa to an earlier date than has yet been determined because most likely *H. erectus* evolved from *H. ergaster* after *H. ergaster* arrived in Asia. In any case, such an extensive population movement is a first in the history of humankind and a tribute to the intellectual and physical skills of the species.

These hominids were the first to use fire, and both types of hominids fashioned more advanced tools than did the other early *Homo* species. They used heavy, teardrop-shaped axes

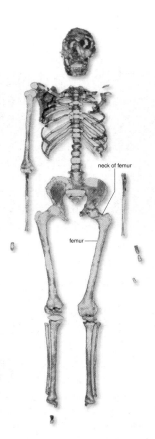

Figure 31.19
Homo ergaster.
This skeleton of a 10-year-old boy who lived 1.6 MYA in eastern Africa shows femurs that are angled because the neck of the femur is quite long.

neck of femur

femur

and cleavers as well as flakes, which were probably used for cutting and scraping. Some investigators believe that these hominids were systematic hunters and brought kills to the same site repeatedly. In one location, researchers found over 40,000 bones and 2,647 stones. These sites could have been "home bases" where social interaction occurred and a prolonged childhood allowed time for learning. Perhaps a language evolved and a culture developed.

Homo floresiensis   In 2004, scientists announced the discovery of the fossil remains of *Homo floresiensis.* The 18,000-year-old fossil of a 1 m tall, 25 kg adult female was discovered on the island of Flores in the South Pacific. This important finding suggests that a species of *Homo* coexisted with modern *Homo sapiens* much more recently than Neandertals, which went extinct about 28,000 years ago. The specimen was the size of a three-year-old *Homo sapiens sapiens* but possessed a braincase only one-third the size of that of a modern human. Researchers suspect that this diminutive hominid and her peers evolved from normal-sized, island-hopping *Homo erectus* populations that reached Flores about 840,000 years ago. Apparently *H. floresiensis* used tools and fire. Some scientists think that its small size is due to island dwarfing, the reduction in size of large animals when their gene pool is limited to a small environment. This phenomenon is seen in other island populations.

> *Homo ergaster* could have evolved from *H. rudolfensis* in Africa. This hominid had a striding gait, made well-fashioned tools (perhaps for hunting), and migrated out of Africa between 2 and 1 MYA. A related form, *H. erectus,* is also found in Asia.

Evolution of Modern Humans

The later species of *Homo* are represented in Figure 31.17 by blue-colored bars. Most researchers believe that modern humans *Homo sapiens* evolved from *H. ergaster,* but they differ as to the details. Many disparate early *Homo* species in Europe are now classified as *Homo heidelbergensis.* Just as *H. erectus* is believed to have evolved from *H. ergaster* in Asia, so *H. heidelbergensis* is believed to have evolved from *H. ergaster* in Europe. Further, for the sake of discussion, *H. ergaster* in Africa, *H. erectus* in Asia, and *H. heidelbergensis* in Europe can be grouped together as archaic humans who lived as long as a million years ago. The presence of archaic humans at these different locations suggests to some that modern humans evolved from archaic humans separately in all three places (Fig. 31.20*a*). This hypothesis, called the **multiregional continuity hypothesis,** proposes that modern humans arose from archaic humans in essentially the same manner. Even so, the humans of each region should show a continuity of unique anatomical characteristics from about the time of the archaic species.

Opponents argue that it is unlikely that evolution would have produced essentially the same result in these different places. They suggest, instead, the **out-of-Africa hypothesis,** which proposes that modern humans evolved from *H. ergaster* only in Africa, and then *H. sapiens* migrated to Europe and Asia, where it replaced the archaic species about 100,000 years BP (before the present) (Fig. 31.20*b*). If so, fossils dated 200,000 BP and 100,000 BP are expected to be markedly different from each other.

According to which hypothesis would modern humans be most genetically alike? The multiregional hypothesis states that human populations have been evolving separately for a long time; therefore, genetic differences are expected between groups. According to the out-of-Africa hypothesis, we are all descended from a few individuals from about 100,000 years BP. Therefore, the out-of-Africa hypothesis suggests that we are more genetically similar.

Several years ago, a study attempted to show that all the people of Europe (and the world for that matter) have essentially the same mitochondrial DNA. Called the "mitochondrial Eve" hypothesis by the press (note that this is a misnomer because no single ancestor is proposed), the raw data—which indicate a close genetic relationship among all Europeans—support the out-of-Africa hypothesis.

These opposing hypotheses have sparked many other innovative studies to test them. The final conclusions are still being debated, but evidence is leaning toward the out-of-Africa hypothesis.

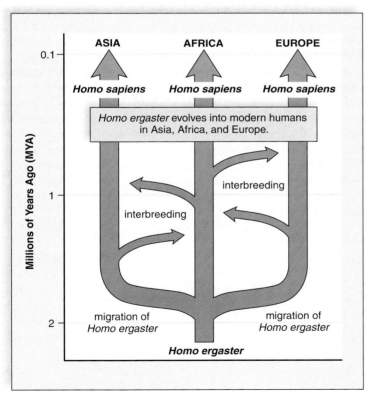

a. Multiregional continuity

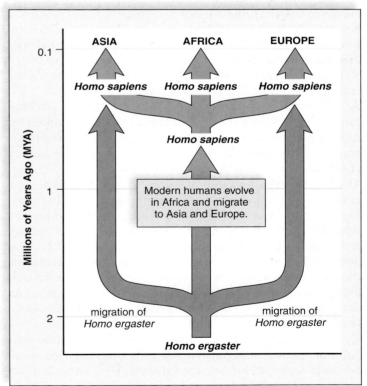

b. Out of Africa

Figure 31.20 Evolution of modern humans.

a. The multiregional continuity hypothesis proposes that *Homo sapiens* evolved separately in at least three different places: Asia, Africa, and Europe. Therefore, continuity of genotypes and phenotypes is expected in each region but not between regions. **b.** The out-of-Africa hypothesis proposes that *Homo sapiens* evolved only in Africa and then migrated out of Africa, as *H. ergaster* did many thousands of years before. *H. sapiens* would have supplanted populations of ancestral *Homo* in Asia and Europe about 100,000 years ago.

Figure 31.21 Neandertals.
This drawing shows that the nose and the mouth of the Neandertals protruded from their faces, and their muscles were massive. They made stone tools and were most likely excellent hunters.

Figure 31.22 Cro-Magnons.
Cro-Magnon people are the first to be designated *Homo sapiens sapiens.* Their tool-making ability and other cultural attributes, such as their artistic talents, are legendary.

> Investigators are currently testing two hypotheses: (1) modern humans evolved separately in Asia, Africa, and Europe, or (2) modern humans evolved in Africa and then migrated to Asia and Europe.

Neandertals

The **Neandertals,** sometimes classified as *Homo neandertalensis,* are an intriguing species of archaic humans that lived between 200,000 and 28,000 years ago. Neandertal fossils have been found from the Middle East throughout Europe. Neandertals take their name from Germany's Neander Valley, where one of the first Neandertal skeletons, estimated to be about 200,000 years old, was discovered.

According to the out-of-Africa hypothesis, archaic human species, including Neandertals, were supplanted by modern humans. Surprisingly, however, the Neandertal brain was, on the average, slightly larger than that of *Homo sapiens* (1,400 cc, compared with 1,360 cc in most modern humans). The Neandertals had massive brow ridges and wide, flat noses. They also had a forward-sloping forehead and a receding lower jaw. Their nose, jaws, and teeth protruded far forward. Physically, the Neandertals were powerful and heavily muscled, especially in the shoulders and neck (Fig. 31.21). The bones of Neandertals were shorter and thicker than those of modern humans. New fossils show that the pubic bone was long compared to that of modern humans. The Neandertals lived in Europe and Asia during the last Ice Age, and their sturdy build could have helped conserve heat.

Archaeological evidence suggests that Neandertals were culturally advanced. Some Neandertals lived in caves; however, others probably constructed shelters. They manufactured a variety of stone tools, including spear points, which they could have used for hunting, and scrapers and knives, which would have helped in food preparation. They most likely successfully hunted bears, woolly mammoths, rhinoceroses, reindeer, and other contemporary animals. They used and could control fire, which probably helped in cooking frozen meat and in keeping warm. They even buried their dead with flowers and tools and may have had a religion.

Cro-Magnons

Cro-Magnons are the oldest fossils to be designated *Homo sapiens sapiens.* In keeping with the out-of-Africa hypothesis, the Cro-Magnons, who are named after a fossil location in France, were the modern humans who entered Asia and Europe from Africa 100,000 years ago or even earlier. They probably reached western Europe about 40,000 years ago. Cro-Magnons had a thoroughly modern appearance (Fig. 31.22). They had lighter bones, flat high foreheads, domed skulls housing brains of 1,350 cc, small teeth, and a distinct chin. They made advanced stone tools, including compound tools, such as by fitting stone flakes to a wooden handle. They may have been the first to make knifelike blades and to throw spears, enabling them to kill animals from a distance. They were such accomplished hunters that some researchers believe they may have been responsible for the extinction of many larger mammals, such as the giant sloth, the mammoth, the saber-toothed tiger, and the giant ox, during the late Pleistocene epoch. This event is known as the Pleistocene overkill.

Cro-Magnons hunted cooperatively, and perhaps they were the first to have language. They are believed to have lived in small groups, with the men hunting by day while the women remained at home with the children. It's quite possible that this hunting way of life among prehistoric people influences our behavior even today. The Cro-Magnon culture included art. They sculpted small figurines out of reindeer bones and antlers. They also painted beautiful drawings of animals on cave walls in Spain and France (Fig. 31.22).

> Neandertals and Homo floresiensis likely coexisted with Cro-Magnons, the direct ancestor to modern humans. Today, there is only one *Homo* species in existence—namely, *Homo sapiens sapiens.*

Some people believe that animals should be protected in every way and should not be used in laboratory research. There is a growing recognition of what is generally referred to as animal rights. Psychologists with Ph.D.s earned in the 1990s are half as likely to express strong support for animal research as those who earned their Ph.D.s before 1970.

Those who approve of laboratory research involving animals point out that even today it would be difficult to develop new vaccines and medicines against infectious diseases, new surgical techniques for saving human lives, or new treatments for spinal cord injuries without using animals. Even so, most scientists today are in favor of what are now called the "three Rs": replacement of animals by in vitro, or test-tube, methods whenever possible; reduction of the numbers of animals used in experiments; and refinement of experiments to cause less suffering to animals. In the Netherlands, all scientists starting research that involves animals are well trained in the three Rs. After designing an experiment that uses animals, they are asked to find ways to answer the same questions without using animals. In the United States, the Animal Welfare Act regulates all federally funded institutions to minimize pain and suffering in vertebrate animal research.

F. Barbara Orlans of the Kennedy Institute of Ethics at Georgetown University says, "It is possible to be both pro research and pro reform." She feels animal activists need to accept that sometimes animal research is beneficial to humans, and all scientists need to consider the ethical dilemmas that arise when animals are used for laboratory research.

Form Your Own Opinion

1. Are you opposed to the use of animals in laboratory experiments? Always or under certain circumstances?
2. Are we re-defining the relationship between animals and humans to the detriment of both? Why or why not?
3. Do you feel that it would be possible for animal activists and scientists to find a compromise they can both accept? Discuss.

Summarizing the Concepts

31.1 Echinoderms

Both echinoderms and chordates are deuterostomes. In deuterostomes, the second embryonic opening becomes the mouth, the coelom forms by outpocketing of the primitive gut (they are enterocoelomates), and the dipleurula larva is found among some. Echinoderms (e.g., sea stars, sea urchins, sea cucumbers, sea lilies) have radial symmetry as adults (not as larvae) and spines from endoskeletal plates. Typical of echinoderms, sea stars have tiny skin gills, a central nerve ring with branches, and a water vascular system for locomotion. Each arm of a sea star contains branches from the nervous, digestive, and reproductive systems.

31.2 Chordates

Chordates (tunicates, lancelets, and vertebrates) have a notochord, a dorsal tubular nerve cord, pharyngeal pouches, and a postanal tail at some time in their life history. Tunicates and lancelets are the nonvertebrate chordates. Adult tunicates lack chordate characteristics except gill slits, but adult lancelets have the four chordate characteristics.

31.3 Vertebrates

Vertebrates have the four chordate characteristics as embryos, but the notochord is replaced by the vertebral column, a part of their endoskeleton. The internal organs are well-developed, and cephalization is apparent. The vertebrate classes trace their evolutionary history. The first vertebrates, represented by hagfishes and lampreys, lacked jaws and fins. Bony fishes have jaws and two pairs of fins; the bony fishes include those that are ray-finned and a few that are lobe-finned. Some of the lobe-finned fishes have lungs.

Amphibians evolved from the lobe-finned fishes and have two pairs of limbs. They are represented today by frogs and salamanders. Frogs usually return to the water to reproduce; frog tadpoles metamorphose into terrestrial adults.

Reptiles (today's snakes, lizards, turtles, and crocodiles) lay a shelled egg, which contains extraembryonic membranes, including an amnion that allows them to reproduce on land.

Birds are feathered, which helps them maintain a constant body temperature. They are adapted for flight: their bones are hollow; lungs have air sacs that penetrate the bones and allow one-way ventilation; their breastbone is keeled; and they have well-developed sense organs.

Mammals have hair and mammary glands. The former helps them maintain a constant body temperature, and the latter allow them to nurse their young. Monotremes lay eggs; marsupials have a pouch in which the newborn matures; and placental mammals, which are far more varied and numerous, retain offspring inside the uterus until birth.

31.4 Human Evolution

Primates are mammals adapted to living in trees. During the evolution of primates, various groups diverged in a particular sequence from the human line of descent. Prosimians (tarsiers and lemurs) diverged first, followed by the monkeys and then the apes. Molecular biologists say that humans are most closely related to the African apes, whose ancestry split from ours about 6–7 MYA.

Human evolution continued in eastern Africa with the evolution of the ardipithecines and the australopithecines. The most famous australopithecine is Lucy 3.2 MYA, whose brain was small but who walked bipedally. *Homo habilis,* present about 2 MYA is certain to have made tools. *Homo erectus,* with a brain capacity of 1,000 cc and a striding gait, was the first to migrate out of Africa.

Two contradicting hypotheses have been suggested about the origin of modern humans. The multiregional continuity hypothesis says that modern humans originated separately in Asia, Europe, and Africa. The out-of-Africa hypothesis says that modern humans originated in Africa and, after migrating into Europe and Asia, replaced the other *Homo* species (including the Neandertals) found there. Most recently, a fossil called *Homo floresiensis* suggests that a *Homo* species coexisted with modern humans more recently than the Neandertals.

Testing Yourself

Choose the best answer for each question.

1. Which of the following statements about echinoderms is false?
   a. They have an internal skeleton.
   b. They are found on land and in water.
   c. They are spiny.
   d. They are typically radially symmetrical.

2. Sea stars
   a. have no respiratory, excretory, or circulatory system.
   b. usually reproduce asexually.
   c. can evert their stomach to digest food.
   d. Both a and c are correct.

3. Sea stars move by
   a. expansion and contraction of their tube feet.
   b. producing jets of water.
   c. using their feet as legs and walking.
   d. taking advantage of water currents to carry them.

For questions 4–9, match the feature with the animal group in the key. Each answer may be used more than once. Each question may have more than one answer.

Key:
   a. echinoderms
   b. nonvertebrate chordates
   c. vertebrate chordates

4. Closed circulatory system

5. Notochord

6. Sea urchins and sand dollars

7. Vertebral column

8. Lancelets and tunicates

9. Paired appendages

10. Sharks and rays detect prey using their ability to sense
    a. smell.
    b. electric currents.
    c. pressure.
    d. All of these are correct.

11. Ray-finned fishes can change their buoyancy using
    a. their fins.
    b. a swim bladder.
    c. their gills.
    d. their kidneys.

12. Unlike bony fishes, amphibians have
    a. a type of ear.
    b. jaws.
    c. a circulatory system.
    d. a heart.

13. Which of the following associations is not correctly matched?
    a. birds and mammals—always a four-chambered heart
    b. fish—always a two-chambered heart
    c. amphibians—always a three-chambered heart
    d. All of these are correctly matched.

For questions 14–18, match the feature with the animal group in the key. Each answer may be used more than once. Each question may have more than one answer.

Key:
   a. fishes
   b. amphibians
   c. reptiles
   d. birds

14. Internal fertilization

15. Jointed appendages

16. Internal skeleton

17. Shelled eggs

18. Ectothermic

19. Label the parts of a bony fish in the following diagram.

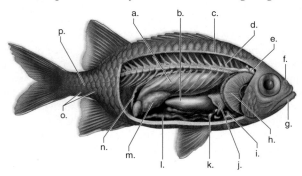

20. Feathers are modified
    a. hairs.
    b. spines.
    c. scales.

21. Which of the following is not an adaptation for flight in birds?
    a. air sacs
    b. modified forelimbs
    c. bones with air cavities
    d. enlarged sternum
    e. well-developed bladder

22. The classification of birds is based on features of
    a. beaks and bones.
    b. beaks and feet.
    c. bones and feathers.
    d. beaks and feathers.
    e. feet and feathers.

23. Unlike any other animal, mammals have
    a. hair and efficient respiratory systems.
    b. hair and efficient circulatory systems.
    c. milk-producing glands and hair.
    d. a four-chambered heart and hair.
    e. a four-chambered heart and milk-producing glands.

24. Examples of monotremes include the
    a. spiny anteater and duckbill platypus.
    b. opossum and koala.
    c. badger and skunk.
    d. porcupine and armadillo.

25. There are more marsupials in Australia than in North America because
    a. they did not need to compete with placental mammals until recently.
    b. their environment is more favorable for their survival in Australia.
    c. volcanoes led to the mass extinction of marsupials in North America.
    d. food sources are more diverse in Australia.

26. Placental mammals are classified based on
    a. size and hair type.
    b. mode of locomotion and method of obtaining food.
    c. number of limbs and method of caring for young.
    d. number of mammary glands and number of offspring.

27. Stereoscopic vision is possible in primates due to
    a. the presence of cone cells.
    b. an enlarged brain.
    c. a shortened snout.
    d. None of these are correct.

28. The first humanlike feature to evolve in the hominids was
    a. a large brain.
    b. massive jaws.
    c. a slender body.
    d. bipedal locomotion.

29. In *H. habilis*, enlargement of the portions of the brain associated with speech probably led to
    a. cooperative hunting.
    b. the sharing of food.
    c. the development of culture.
    d. All of these are correct.

30. Compared to *H. habilis*, *H. erectus*
    a. had a smaller brain.
    b. was shorter.
    c. had a flatter nose.
    d. had a flatter face.

31. Mitochondrial DNA data support which hypothesis for the evolution of humans?
    a. multiregional continuity   b. out-of-Africa

32. Neandertals had a larger brain than *H. sapiens sapiens*, probably in order to
    a. learn to use tools.
    b. control large muscles.
    c. develop speech.
    d. allow the senses to become more acute.

33. The first *Homo* species to use art appears to be
    a. *H. neandertalensis.*
    b. *H. erectus.*
    c. *H. habilis.*
    d. *H. sapiens.*

34. Which of these is not a chordate characteristic?
    a. dorsal supporting rod, the notochord
    b. dorsal tubular nerve cord
    c. pharyngeal pouches
    d. postanal tail
    e. vertebral column

35. Which of these pairs is incorrectly matched?
    a. *H. erectus*—made tools
    b. Neandertal—good hunter
    c. *H. habilis*—controlled fire
    d. Cro-Magnon—good artist
    e. *A. afarensis*—walked erect

36. What indicates that birds are related to reptiles?
    a. Birds have scales and feathers, which are modified scales.
    b. Birds are ectothermic, as are reptiles.
    c. Birds lay leathery, shelled eggs, as do reptiles.
    d. Birds have an open circulatory system, as do all reptiles.

37. Amphibians arose from
    a. tunicates and lancelets.
    b. cartilaginous fishes.
    c. jawless fishes.
    d. ray-finned fishes.
    e. bony fishes with lungs.

Understanding the Terms

amniotic egg 655
amphibian 657
anthropoid 664
australopithecine 667
bipedalism 664
bird 661
bony fish 657
cartilaginous fish 656
chordate 652
Cro-Magnon 670
culture 668
echinoderm 650
ectothermic 658
endothermic 661
feathers 661
hominid 664
jawless fish 656
lancelet 654
lineage 664
lobe-finned fish 657
lung 657
mammal 662
marsupial 662
molecular clock 664
monotreme 662
mosaic evolution 667
multiregional continuity hypothesis 669
Neandertal 670
notochord 652
out-of-Africa hypothesis 669
placental mammal 663
primate 664
prosimian 664
ray-finned fish 657
reptile 658
tunicate 654
vertebrate 655
water vascular system 651

Match the terms to these definitions:

a._____ Series of canals that take water to the tube feet of an echinoderm, allowing them to expand.
b._____ Group of primates that includes lemurs and tarsiers and may resemble the first primates to have evolved.
c._____ Dorsal supporting rod that exists in all chordates sometime in their life history; replaced by the vertebral column in vertebrates.
d._____ Maintenance of a constant body temperature independent of the environmental temperature.
e._____ Member of a family that contains humans and their direct ancestors, which are known only from the fossil record.

Thinking Critically

1. Compare locomotion in a starfish with that in vertebrates. What anatomical structures are particularly important for these differences?

2. Compare the hearts of birds and mammals, amphibians, and fish. Which heart is most efficient and why?

3. Researchers are now sequencing the genome of several vertebrate species, including the zebrafish and the African clawed frog. What information about the human genome might these studies give us that could be applied to the development of new medical treatments for humans?

4. Bipedalism has many selective advantages, including the increased ability to spot predators and prey. However, bipedalism has one particular disadvantage—upright posture leads to a smaller pelvic opening, which makes giving birth to an offspring with a large head very difficult. This situation results in a higher percentage of deaths (of both mother and child) during birth in humans compared to other primates. How can you explain the selection for a trait, such as bipedalism, that has both positive and negative consequences for fitness?

5. In studying recent fossils of the genus *Homo*, such as Cro-Magnon, biologists have determined that modern humans have not undergone much biological evolution in the past 50,000 years. Rather, cultural anthropologists argue that cultural evolution has been far more important than biological evolution in the recent history of modern humans. What do they mean by this? Support your argument with some examples.

ARIS, the *Inquiry into Life* Website

ARIS, the website for *Inquiry into Life,* provides a wealth of information organized and integrated by chapter. You will find practice quizzes, interactive activities, labeling exercises, flashcards, and much more that will complement your learning and understanding of general biology.

www.aris.mhhe.com

Behavior and Ecology

part **VII**

Emperor penguins tend to their young. *These penguins live in Antarctica and breed on glaciers, where temperatures can reach 100° below zero! As you can imagine, they have developed many adaptive behaviors in order to survive. The recent popular movie "March of the Penguins" revealed their secrets to survival.*

Emperor penguins breed in winter, which begins in March in Antarctica. After an extended courtship with a male, a female emperor lays a single egg. She then passes this egg—very carefully—to the male so that she can return to the water to feed; she has lost much of her body weight in the 50-mile migration to the breeding grounds. The male must then keep the egg on top of his feet and under a pouch of skin in order to keep it warm—for about two months, during which time he does not feed.

The female returns and feeds the newly hatched young; the male, who has not eaten in over four months, then returns 50 miles to the ocean and feeds. Meanwhile, the female keeps the young in her brood pouch for another two months. The non-feeding penguins huddle together in large groups in order to survive the cold winter temperatures. This social behavior is necessary for survival, and penguins cooperate, taking turns between the outside and the middle of the huddle where it's warm.

In this chapter you will learn about animal behavior in general and gain a better understanding of the complexity of the emperor penguin lifestyle. First, we will discuss the basis of behavior—that is, whether behaviors have a genetic or environmental basis. Then we will examine mating behaviors, social behaviors, and communication, necessary components of all animals' struggle to survive, including humans.

chapter 32

Animal Behavior

Chapter Concepts

32.1 Nature versus Nurture: Genetic Influences

The "nature versus nurture" question asks to what degree our genes (nature) and environmental influences (nurture) affect behavior. **Behavior** encompasses any action that can be observed and described. Because the anatomy and physiology of animals, including humans, determine what types of behavior are possible for them, we immediately know that the genes, to a degree, control behavior. Many experiments have been done to discover the degree to which genetics controls behavior.

Experiments with Lovebirds, Snakes, and Snails

Lovebirds are small, green and pink African parrots that nest in tree hollows. There are several closely related species of lovebirds in the genus *Agapornis*, whose behavior differs by the way they build nests. Fischer lovebirds, *Agapornis fischeri*, pick up a large leaf (or in the laboratory, a piece of paper) with their bills, perforate it with a series of bites along its length, and then cut out long strips. They carry the strips in their bills to the nest and weave them with others to make a deep cup (Fig. 32.1*a*). Peach-faced lovebirds, *Agapornis roseicollis*, cut somewhat shorter strips in a similar manner, but then carry them to the nest in a very unusual way. They pick up the strips in their bills and insert them deep into the rump feathers (Fig. 32.1*b*). In this way, they can carry several short strips with each trip to the nest, while Fischer lovebirds can carry only one of the longer strips at a time.

Researchers hypothesized that if the behavior for obtaining and carrying nesting material is inherited, then hybrids might show intermediate behavior. When the two species of birds were mated, the hybrid birds had difficulty carrying nesting materials. They cut strips of intermediate length and then attempted to tuck the strips into their rump feathers. They did not push the strips far enough into the feathers, however, and when they walked or flew, the strips always came out. Hybrid birds eventually (about three years in this study) learned to carry the cut strips in their bills, but still briefly turned their heads toward their rumps before flying off. These studies support the hypothesis that behavior has a genetic basis.

Several experiments have been conducted using the garter snake, *Thamnophis elegans*, to determine if food preference has a genetic basis. In California, inland garter snake populations are aquatic and commonly feed underwater on frogs and fish, white coastal populations are terrestrial and feed mainly on slugs. In the laboratory, inland adult snakes refused to eat slugs, while coastal snakes readily did so. The experimental results of matings between snakes from the two populations (inland and coastal) show their newborns have an overall intermediate incidence of slug acceptance, suggesting a genetic basis for feeding preference.

a. Fischer lovebird with nesting material in its beak.

b. Peach-faced lovebird with nesting material in its rump feathers.

Figure 32.1 Nest building behavior in lovebirds.
a. Fischer lovebirds carry strips of nesting material in their bills, as do most other birds. **b.** Peach-faced lovebirds tuck strips of nesting material into their rump feathers before flying back to the nest.

Differences between slug acceptors and slug rejecters appear to be inherited, but what physiological difference is there between the two populations? A clever experiment answered this question. Snakes use tongue flicks to recognize their prey, and their tongues carry chemi-

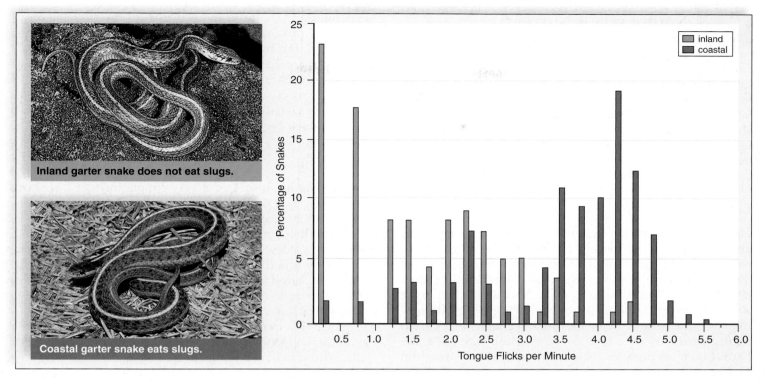

Figure 32.2 Feeding behavior in garter snakes.
The number of tongue flicks by inland and coastal garter snakes is measured in terms of their response to slug extract on cotton swabs. Coastal snakes tongue-flicked more than inland snakes, indicating that coastal snakes are more sensitive to the smell of slugs than inland snakes.

cals to an odor receptor in their mouth. Even newborns will flick their tongues at cotton swabs dipped in fluids of their prey. Swabs were dipped in slug extract, and the number of tongue flicks were counted for newborn inland and coastal snakes. Coastal snakes had a higher number of tongue flicks (indicating more interest in the smell) than inland snakes (Fig. 32.2). Apparently, inland snakes do not eat slugs because they are not sensitive to their smell. A genetic difference between the two populations of snakes has resulted in a physiological difference in their nervous systems. Although hybrids showed a great deal of variation in the number of tongue flicks, they were generally intermediate, as predicted by the genetic hypothesis.

The nervous and endocrine systems are both responsible for the coordination of body systems. Is the endocrine system also involved in behavior? Research studies support the involvement of the endocrine system. For example, the egg-laying behavior in the marine snail *Aplysia* involves a set sequence of movements. Following copulation, the animal extrudes long strings of more than a million egg cases. It takes the egg case string in its mouth, covers it with mucus, waves its head back and forth to wind the string into an irregular mass, and attaches the mass to a solid object, such as a rock. Several years ago, scientists isolated and analyzed an egg-laying hormone (ELH) that causes the snail to lay eggs even if it has not mated. ELH was found to be a small protein of 36 amino acids that diffuses into the circulatory system and excites the smooth muscle cells of the reproductive duct, causing them to contract and expel the egg string. Using recombinant DNA techniques, the investigators isolated the ELH gene. The gene's product turned out to be a protein with 271 amino acids. The protein can be cleaved into as many as 11 possible products, and ELH is one of these. ELH alone, or in conjunction with these other products, is thought to control all the components of egg-laying behavior in *Aplysia*.

Experiments with Humans

Human twins, on occasion, have been separated at birth and raised under different environmental conditions. Studies of separated twins show that they have similar food preferences and activity patterns, and even select mates with similar characteristics. These twin studies lend support to the hypothesis that at least certain types of behavior are primarily influenced by nature (i.e. genes).

The results of many studies support the hypothesis that many types of behavior have a genetic basis. Apparently, genes influence the development of neural and hormonal mechanisms that control behavior.

32.2 Nature versus Nurture: Environmental Influences

Even though genetic inheritance serves as a basis for behavior, environmental influences (nurture) also affect behavior. For example, behaviorists originally believed that some behaviors were **fixed action patterns (FAP)** elicited by a sign stimulus. But then they found that many behaviors formerly thought to be FAPs improve with practice—for example, by learning. In this context, **learning** is defined as a durable change in behavior brought about by experience.

Learning in Birds

Laughing gull chicks' begging behavior appears to be an FAP, because it is always performed the same way in response to the parent's red bill (the sign stimulus). A chick directs a pecking motion toward the parent's bill, grasps it, and strokes it downward (Fig. 32.3a). Parents can bring about the begging behavior by swinging their bill gently from side to side. After the chick responds, the parent regurgitates food onto the floor of the nest. If need be, the parent then encourages the chick to eat. This interaction between the chicks and their parents suggests that the begging behavior involves learning. To test this hypothesis, diagrammatic pictures of gull heads were painted on small cards, and then eggs were collected in the field. The eggs were hatched in a dark incubator to eliminate visual stimuli before the test. On the day of hatching, each chick was allowed to make about a dozen pecks at the model. The chicks were returned to the nest, and then each

was retested. The tests showed that on the average, only one-third of the pecks by a newly hatched chick strike the model. But one day after hatching, more than half of the pecks are accurate, and two days after hatching, the accuracy reaches a level of more than 75% (Fig. 32.3b, c). Investigators concluded that improvement in motor skills, as well as visual experience, strongly affect the development of chick begging behavior.

Imprinting

Imprinting is considered a very simple form of learning. Imprinting was first observed in birds when chicks, ducklings, and goslings followed the first moving object they saw after hatching. This object is ordinarily their mother, but investigators found that birds can imprint on any object, as long as it is the first moving object they see during a sensitive period of two to three days after hatching. The term *sensitive period* means that the behavior develops only during this time.

A chick imprinted on a red ball follows it around and chirps whenever the ball is moved out of sight. Social interactions between parent and offspring during the sensitive period seem key to normal imprinting. For example, female mallards cluck during the entire time imprinting is occurring, and it could be that vocalization before and after hatching is necessary to normal imprinting.

Song Learning

White-crowned sparrows sing a species-specific song, but males of a particular region have their own dialect. Birds

a.

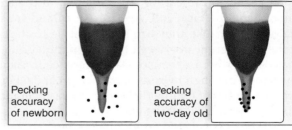

b.

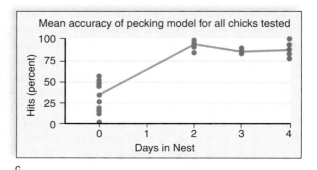

c.

Figure 32.3 Begging behavior in laughing gull chicks.
a. At about three days of age, a laughing gull chick grasps the red bill of a parent stroking it downward, and the parent then regurgitates food. **b.** The accuracy of a chick when striking a test probe, painted red. **c.** Chick-pecking accuracy illustrated graphically. Note from these diagrams that a chick markedly improves its ability (within only two days) to peck a bill, a behavior that normally causes a parent to regurgitate food.

were caged in order to test the hypothesis that young white-crowned sparrows learn how to sing from older members of their species.

Three groups of birds were tested. Birds in the first group *heard no songs at all.* When grown, these birds sang a song, but it was not fully developed. Birds in the second group *heard tapes of white-crowns singing.* When grown, they sang in that dialect, as long as the tapes had been played during a sensitive period from about age 10 to 50 days. White-crowned sparrows' dialects (or other species' songs) played before or after this sensitive period had no effect on their song. Birds in a third group did not hear tapes and instead were *given an adult tutor.* These birds sang the song of a still different species—no matter when the tutoring began—showing that social interactions apparently assist learning in birds.

Associative Learning

A change in behavior that involves an association between two events is termed **associative learning.** For example, birds that get sick after eating a monarch butterfly no longer prey on monarch butterflies, even though they may be readily available. Or the smell of fresh-baked bread may entice you to eat even though you may have just eaten. Because you really enjoyed bread in the past, you associate the smell with those past experiences, which makes you hungry. Both classical conditioning and operant conditioning are examples of associative learning.

Classical Conditioning

In **classical conditioning,** the presentation of two different types of stimuli at the same time causes an animal to form an association between them. The best-known laboratory example of classical conditioning is an experiment done by the Russian psychologist Ivan Pavlov. First, Pavlov observed that dogs salivate when presented with food. Then he rang a bell whenever the dogs were fed. Eventually, the dogs would salivate whenever the bell was rung, regardless of whether food was present (Fig. 32.4).

Classical conditioning suggests that an organism can be trained—that is, conditioned—to associate any response to any stimulus. Unconditioned responses are those that occur naturally, as when salivation follows the presentation of food. Conditioned responses are those that are learned, as when a dog learns to salivate when it hears a bell. Advertisements attempt to use classical conditioning. For example, commercials pair attractive people with a particular product in the hope that viewers will associate attractiveness with that product. This pleasant association may cause them to buy the product.

In a similar way, it's been suggested that holding children on your lap when reading to them facilitates an interest in reading. Why? Because they will associate a pleasant feeling with reading.

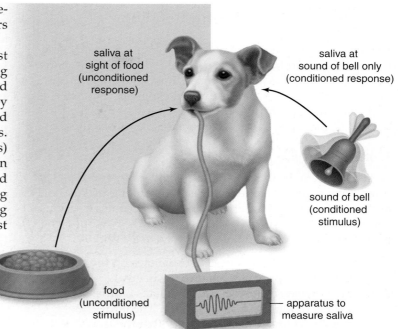

saliva at sight of food (unconditioned response)

saliva at sound of bell only (conditioned response)

sound of bell (conditioned stimulus)

food (unconditioned stimulus)

apparatus to measure saliva

Figure 32.4 Classical conditioning.
Ivan Pavlov discovered classical conditioning by performing this experiment with dogs. A bell is rung when a dog is fed. Salivation is noted. Eventually, the dog salivates when the bell is rung even though no food is presented. Food is an unconditioned stimulus, and the sound of the bell is a conditioned stimulus that brings about the response—that is, salivation.

Operant Conditioning

During **operant conditioning,** a stimulus-response connection is strengthened. Most people know that it is helpful to give an animal a reward, such as food or affection, when teaching them a trick. When we go to an animal show, it is quite obvious that trainers use operant conditioning. They present a stimulus, such as a hoop, and then give a reward (food) for the proper response (jumping through the hoop).

B. F. Skinner is well known for studying this type of learning in the laboratory. In Skinner's simplest type of experiment, a caged rat happens to press a lever and is rewarded with sugar pellets, which it avidly consumes. Thereafter, the rat regularly presses the lever whenever it is hungry. In more sophisticated experiments, Skinner even taught pigeons to play ping-pong by rewarding desired responses to stimuli.

In applying operant conditioning to child rearing, it's been suggested that parents who give a positive reinforcement for good behavior will be more successful than parents who punish behaviors they believe are undesirable.

Genetic influences (nature) and environmental influences (nurture), such as learning and conditioning, are both important in affecting behavior.

Sexual selection refers to adaptive changes in males and females that lead to an increased ability to secure a mate. Sexual selection in males may make them better able to compete with other males for a mate, while females may select a male with the best **fitness** (number of fertile offspring produced). In that way, the female increases her own fitness.

Female Choice

In general, female animals produce few eggs, and females often have a large energetic investment in egg production and parental care, so the choice of a mate becomes a serious consideration. In a study of satin bowerbirds, two opposing hypotheses regarding female choice were tested:

Good genes hypothesis: Females choose mates on the basis of traits that improve the chance of survival.

Run-away hypothesis: Females choose mates on the basis of traits that improve appearance. The term "run-away" pertains to the possibility that the trait will be exaggerated in the male until its mating benefit is checked by the trait's unfavorable survival cost.

As investigators observed the behavior of satin bowerbirds, they discovered that females usually chose aggressive males as mates. It could be that inherited aggressiveness

Figure 32.5
Raggiana bird of paradise.
Raggiana males have brilliantly colored plumage brought about by sexual selection. The drab females tend to choose brightly colored males as mates.

improves the chance of survival—or that aggressive males are good at stealing blue feathers from other males, since females prefer blue feathers as bower decorations. Therefore, the data did not clearly support either hypothesis.

The Raggiana bird of paradise is remarkably *dimorphic*, meaning that males and females differ in size and other traits. The males are larger than the females and have beautiful orange flank plumes, while the females are drab (Fig. 32.5). Female choice can explain why male birds are more ornate than females. Consistent with the two hypotheses, it is possible that the remarkable plumes of the male signify health and vigor to the female. Some investigators have hypothesized that extravagant male features indicate that they are relatively parasite-free. In barn swallows, females also chose those with the longest tails, and investigators have shown that males that are relatively free of parasites have longer tails than otherwise. Or, it's possible that females choose the most flamboyant males because of stronger sensory stimulation; their sons will then also display highly exaggerated traits. Eventually, the males' extreme coloration or plumage may become a hazzard—for example, making them more easily spotted or captured by predators.

Male Competition

Males can father many offspring because they continuously produce sperm in great quantity with a low energetic cost. In addition, parental care by males is generally less frequent than by females. We expect males to compete in order to inseminate as many females as possible. **Cost-benefit analyses** have been done to determine if the *benefit* of access to mating is worth the *cost* of competition among males.

Baboons, a type of Old World monkey, live together in a troop. Males and females have separate **dominance hierarchies** in which a higher-ranking animal has greater access to resources, such as females, than a lower-ranking animal. Dominance is decided by confrontations that result in one animal giving way to the other.

Baboons are dimorphic; the males are larger than the females, and they can threaten other members of the troop with their long, sharp canines (Fig. 32.6). One or more males become dominant by frightening the other males. However, the male baboon pays a cost for his dominant position. Being larger means that he needs more food, and being willing and able to fight predators means that he may get hurt, and so forth. Is there a reproductive benefit to his behavior? Yes, in that dominant males do indeed monopolize females when they are most fertile. Nevertheless, there may be other ways to attract mates and father offspring. A male may act as a helper to a female and her offspring; then, the next time she is in estrus, she may mate preferentially with him instead of with a dominant male. Or, subordinate males may form a friendship group that opposes a dominant male, making him give up a receptive female.

A **territory** is an area that is defended against competitors. Scientists are able to track an animal in the wild in order to determine the size of its territory. **Territoriality** includes the type of defensive behavior needed to defend a territory. Baboons travel within a territory, foraging for food each day

Figure 32.6 A male olive baboon displaying full threat. In olive baboons, males are larger than females and have enlarged canines. Competition between males establishes a dominance hierarchy for the distribution of resources.

and sleeping in trees at night. Dominant males decide where and when the troop will move. If the troop is threatened, dominant males protect the troop as it retreats and attack intruders when necessary. Vocalization and displays, rather than outright fighting, may be sufficient to defend a territory. In songbirds, for example, males use singing to announce their willingness to defend a territory. Other males of the species then become reluctant to make use of the same area.

Red deer stags (males) on the Scottish island of Rhum compete to be the harem master of a group of hinds (females) that mate only with them. The reproductive group occupies a territory that the harem master defends against other stags. Harem masters first attempt to repel challengers by roaring. If the challenger remains, the two lock antlers and push against one another (Fig. 32.7). If the challenger then withdraws, the master pursues him for a short distance, roaring the whole time. If the challenger wins, he becomes the harem master.

A harem master can father two dozen offspring at most, because he is at the peak of his fighting ability for only a short time. And there is a cost to being a harem master. Stags must be large and powerful in order to fight; therefore, they grow faster and have less body fat. During bad times, they are more likely to die of starvation, and in general, they have shorter lives. Harem master behavior will persist in the population only if its cost (reduction in the potential number of offspring because of a shorter life) is less than its benefit (increased number of offspring due to harem access).

Evolution by sexual selection occurs when females have the opportunity to select among potential mates and/or when males compete among themselves for access to reproductive females.

a.

Figure 32.7 Competition between male red deer. Male red deer compete for a harem within a particular territory. **a.** Roaring alone may frighten off a challenger, but **(b)** outright fighting may be necessary, and the victor is most likely the stronger of the two animals.

b.

Mating in Humans

A study of human mating behavior shows that the concepts of female choice and male competition apply to humans as well as to the animals we have been discussing. Again, increased fitness seems to influence behavior, this time mate choice behavior in men and women, in most human cultures. Of course, applying the principles of evolutionary biology to human behavior is not without controversy because we can't be reduced to being "pre-programmed" and still have the ability to make conscious choices. That said, understanding the evolutionary basis of animal behavior can provide some interesting insights into why people behave the way they do.

Human Males Compete

Consider that women, by nature, must invest more in having a child than men. After all, it takes nine months to have a child, and pregnancy is followed by lactation when a woman may nurse her infant. Men, on the other hand, need only contribute sperm during a sex act that may require only a few minutes. The result is that men are generally more available for mating than are women. Because more men are available, they necessarily have to compete with others for the privilege of mating.

Like many other animals, humans are dimorphic. Males tend to be larger and more aggressive than females, perhaps as a result of past sexual selection by females. As in other animals, males may pay a price due to high levels of testosterone, the energetic costs of male-male competition, and increased stress associated with finding mates. Male humans live on average six to seven years shorter than females do.

Females Choose

David Buss, an evolutionary psychologist at the University of Texas, conducted studies of female preference of a male mate across cultures in over 20 countries. His research, although somewhat controversial, suggests that the number-one trait females prefer in a male mate is his ability to obtain (financial) resources. Financial success means that men are more likely to provide females with the resources they need to raise their children, and thus increase the fitness (reproductive success) of the female.

Other recent studies have shown that symmetry in facial and body features is also important in female mate choice, possibly related to the "good genes" hypothesis discussed earlier. Symmetry in other animals is often a sign of good health and a strong immune system (e.g., animals with high parasite loads are often asymmetrical). Females may thus choose symmetrical males to pass these traits onto their offspring.

Men Also Have a Choice

Just as women choose men who can provide resources, men prefer youthfulness and attractiveness in females, signs that their partner can provide them with children. In one controversial study, men ages 8 to 80 across four continents

Figure 32.8 King Hussein and family.
The tendency of men to mate with fertile younger women is exemplified by King Hussein of Jordan, who was about 16 years older than his wife, Queen Noor. One of King Hussein's children is now King Abdullah of Jordan.

preferred women with a waist-to-hip ratio (WHR) of 0.7, regardless of weight. Research showed that a WHR of 0.7 is optimal for conception; that is, for each increase of 10% in WHR, the odds of conception during each ovulation event decrease by 30%. Men responding to questionnaires list attributes that biologists associate with a strong immune system and good health (e.g., symmetry), high estrogen levels (indicating fertility), and especially youthfulness. On average, men marry women 2.5 years younger than they are, but as men age, they tend to prefer women who are many years younger. Men are capable of reproduction for many more years than women. Therefore, by choosing younger women, older men can increase their fitness (Fig. 32.8). Other factors are also involved in human mate choice, as explained in the Science Focus, "Mate Choice and Smelly T-shirts."

Male and female mate choice can be influenced by characteristics that influence their fitness. Studies show that females tend to prefer males with resources, while men prefer women who are youthful and have physical features that indicate high fertility.

ScienceFocus

Mate Choice and Smelly T-Shirts

Mate choice has been studied extensively in humans, largely because of our curiosity about how evolutionary biology and behavior in other animals may give us insight into human behavior. And, of course, mate choice is one of the more fascinating human behaviors. In one unusual and recent study, scientists had several men wear T-shirts for a few nights while they slept, infusing the T-shirts with their unique smell. Then they asked women to rate the attractiveness of the male without seeing the person—that is, based only on the smell of the T-shirt. It turns out that women tended to choose men whose major histocompatability complex (MHC) alleles were different from their own. Recall from Chapter 13 that the MHC is associated with recognition of foreign antigens (e.g., viruses or bacteria) and enhances the immune system's ability to fight off infections. In choosing men with different (complementary) MHC alleles, females are maximizing the chance that their offspring have highly heterozygous MHC and potentially a more robust immune system. A more heterozygous MHC is associated with a better innate ability to recognize a more diverse group of

antigens. In fact, females in this study said unattractive shirt smells reminded them of their fathers, who, of course, have very similar MHC alleles. Where did scientists get the idea for such a strange experiment? It has already been well documented that mice preferentially choose mates with complementary MHC alleles based on smell.

Of course, while evolutionary and behavioral biology can provide insights into human mate choice, the process is clearly complex. Although males prefer young faithful females with physical attributes that indicate fertility, they also look for symmetry, intelligence, and a sense of humor. And while females prefer males with financial resources, they also list such qualities as dependability and emotional stability (indicators of good parenting), physical attractiveness (indicated by symmetry and testosterone markers, such as above-average strength and height, possibly to defend resources), sense of humor, and intelligence. The way a person smells also could play a factor, as indicated by the T-shirt study. Then there are unknowns—the context under which people meet, their professions, their likes/dislikes, etc.

Discussion Questions

1. Scientists have found a biological basis for infanticide. In lions, for example, males who kill the offspring of other males when they take over a pride then induce females to come into estrus, so they can mate with them. This increases male fitness. Should a biological basis for infanticide be used as a defense in human court cases where, say, a stepfather kills a stepchild?

2. As discussed in section 32.3, some studies suggest that men prefer women with a waist-to-hip ratio of 0.7. Women of different weights can have this ratio because it is based on bone structure, rather than weight. Nonetheless, women may misinterpret this information, which can contribute to the epidemic of eating disorders, such as bulimia and anorexia. How can we better educate people about the scientific information on mate choice and the dangers of eating disorders?

3. Studies of human mate choice are generally complex, and often controversial. Should federal agencies fund these studies? Why or why not?

32.4 Sociobiology and Animal Behavior

Sociobiology applies the principles of evolutionary biology to the study of social behavior in animals. Sociobiologists hypothesize that living in a society has a greater reproductive benefit than reproductive cost. Then they perform a cost-benefit analysis to see if this hypothesis is supported.

Group living does have its benefits. It can help an animal avoid predators, rear offspring, and find food. A group of impalas is more likely than a solitary one to hear an approaching predator. Schooling fish moving rapidly in different directions might distract a would-be predator. Pair bonding of trumpet manucodes helps the birds raise their young, since due to their particular food source, the female cannot rear as many offspring alone as she can with the male's help. Weaver birds form giant colonies that help protect them from predators, and the birds may also share information about food sources. Primate members of the same troop signal to one another when they have found an especially bountiful fruit tree. Lions working together are able to capture large prey, such as zebra and buffalo.

Group living also has its disadvantages. When animals are crowded together into a small area, there is competition for resources, such as the best feeding places and sleeping sites.

Dominance hierarchies are one way to apportion resources, but this puts subordinates at a disadvantage. Among red deer, sons are preferable because, as a harem master, a son will result in a greater number of grandchildren. However, sons, being larger than daughters, need to be nursed more frequently and for a longer period of time and are thus more energetically costly to the mothers. Subordinate females do not have access to enough food resources to adequately nurse sons, and therefore, they tend to rear daughters, not sons. Still, like the subordinate males in a baboon troop, subordinate females in a red deer harem may be better off in terms of fitness if they stay with a group, despite the cost involved.

Living in close quarters exposes individuals to illness and parasites, which can more easily pass from one animal to another. Social behavior helps offset some of the proximity disadvantages. For example, baboons and other types of social primates invest much time in grooming one another, and this likely helps them remain healthy. Humans utilize extensive medical care to help offset the health problems that arise from living in the densely populated cities of the world.

Sociobiology and Human Culture

Humans today rely on living in organized societies. Clearly, the benefits of social living must outweigh the costs because we have organized governments, with laws that tend to increase the potential for human survival. The **culture** of a human society involves a wide spectrum of customs, ranging from how to dress to forms of entertainment, marriage rituals, and types of food. Language and the use of tools are essential to human culture. Language is used to socialize children, teach them how to use tools, educate them, and train them in skills that may help them maintain a household or begin a career. Technological skills, such as knowing how to use a computer or how to fix plumbing, are taught from an early age. Certainly cultural evolution has surpassed biological evolution in the past few centuries. For example, medical advances, which are passed on through language, have increased the human life span from 40–45 years a century ago to nearly 80 years today.

Why did human societies originate? Perhaps the earliest organized societies were composed of "hunter-gatherers." Which of our ancestors first became hunters is debated among paleontologists (biologists who study fossils) and sociologists. Nonetheless, scientists speculate that a predatory lifestyle may have encouraged the evolution of intelligence and the development of language. That is, cooperation, communication, and tools (such as weapons) are necessary for hunting large animals. In hunter-gatherer societies, men tended to hunt, while women utilized wild and cultivated plants as a source of food. Sometimes, when animal food was scarce, hunter-gatherer societies relied on plant food cultivated by women. In this sense, cooperation ensured that people were fed.

> Social living has both advantages and disadvantages. Only if the benefits, in terms of individual reproductive success, outweigh the costs will societies come into existence.

Altruism versus Self-Interest

Altruism can be generally described as a self-sacrificing behavior for the good of another member of the society. In an evolutionary sense, altruism may compromise the fitness (i.e., lifetime reproductive success) of the altruist, while benefiting the fitness of the recipient. Are animals truly altruists, given that natural selection should eventually rid populations of altruists due to their decreased reproductive success? In humans, we can clearly think of examples—a volunteer firefighter dying while trying to save a stranger, a soldier losing his life for his country, a woman diving in front of a car to save an elderly person from being hit. But what about other animals?

In general, altruistic behavior in animals is explained by the concept of **kin selection.** In other words, because close relatives share many of your genes, it may make sense to self-sacrifice to save them. For example, your siblings share 50% of your genes, so sacrificing your life for two brothers or sisters makes sense evolutionarily. **Inclusive fitness** refers to an individual's personal reproductive success, as well as that of his or her relatives, and thus to an individual's total genetic contribution to the next generation. The concepts of kin selec-

tion and inclusive fitness are well supported by research in complex animal societies. For example, in a colony of army ants, the queen is inseminated only during her nuptial flight. Thereafter, she spends the rest of her life reproducing constantly, laying up to 30,000 eggs per day! The eggs hatch into three different sizes of sterile female workers whose jobs contribute to the colony, which can include over one million individuals. The smallest workers (3 mm), called nurses, take care of the queen and the larvae by feeding and cleaning them. The intermediate-sized workers (3–12 mm), which constitute most of the population, go out on raids to collect food. The soldiers (14 mm), which have huge heads and powerful jaws, surround raiding parties to protect them and the colony from attack by intruders. Why do the worker ants engage in this apparently altruistic behavior?

The queen ant is diploid (2n), but her mate is haploid. Thus, assuming the queen has had only one mate, her sister workers are more closely related to each other than normal siblings (75% versus 50%). Therefore, a worker can achieve higher inclusive fitness by helping her mother (the queen) produce additional sisters than by producing her own offspring, which would only share 50% of her genes. Thus, this behavior is not really altruistic; rather, it is adaptive because it is likely due to the increase in fitness of the helpers relative to those that hypothetically might defect and breed on their own. Similar social patterns are observed in certain species of bees and wasps.

Helpers at the Nest

In some bird species, offspring from a previous clutch of eggs may stay at the nest to help the parents rear the next batch of offspring. In a study of Florida scrub jays, the number of fledglings produced by an adult pair doubled when they had helpers. Some mammalian offspring, such as the meerkat in Figure 32.9, also help their parents. Among jackals in Africa, solitary pairs managed to rear

**Figure 32.9
Inclusive fitness.**
A meerkat is acting as a babysitter for its young sisters and brothers while their mother is away. Researchers point out that the helpful behavior of the older meerkat can lead to increased inclusive fitness.

an average of 1.4 pups, whereas pairs with helpers reared 3.6 pups. What are the benefits of staying behind to help? First, a helper is contributing to the survival of its own kin. Therefore, the helper actually gains a fitness benefit. Second, a helper is more likely than a nonhelper to inherit a parental territory—including other helpers. Helping, then, involves making a minimal, short-term reproductive sacrifice in order to maximize future reproductive potential. Once again, an apparently altruistic behavior turns out to be adaptive.

Inclusive fitness is measured by the genes an individual contributes to the next generation, either directly through their own offspring or indirectly by way of relatives. Many behaviors once thought to be altruistic turn out, on closer examination, to be examples of kin selection and are thus adaptive.

32.5 Animal Communication

Animals exhibit a wide diversity of social behaviors. Some animals are largely solitary and join with a member of the opposite sex only for the purpose of reproduction. Others pair, bond, and cooperate in raising offspring. Still others form a **society** in which members of species are organized in a cooperative manner, extending beyond sexual and parental behavior. We have already mentioned the social groups of baboons and red deer. Social behavior in these and other animals requires that they communicate with one another.

Communicative Behavior

Communication is a signal by a sender that influences the behavior of a receiver. The communication can be purposeful, but does not have to be. Bats send out a series of sound pulses and listen for the corresponding echoes in order to find their way through dark caves and locate food at night. Some moths have an ability to hear these sound pulses, and they begin evasive tactics when they sense that a bat is near. Are the bats purposefully communicating with the moths? No, bat sounds are simply a cue to the moths that danger is near.

The types of communication signals used by animals include chemical, auditory, visual, and tactile.

Chemical Communication

Chemical signals have the advantage of being effective both night and day. The term **pheromone** designates chemical signals in low concentration that are passed between members of the same species. For example, female moths secrete chemicals from special abdominal glands, which are detected downwind by receptors on male antennae. The antennae are especially sensitive, and this ensures that only male moths of the correct species (not predators) will be able to detect them.

Cheetahs and other cats mark their territories by depositing urine, feces, and anal gland secretions at the boundaries (Fig. 32.10). Klipspringers (small antelope) use secretions from a gland below the eye to mark twigs and grasses of their territory.

Auditory Communication

Auditory (sound) communication has some advantages over other kinds of communication. It is faster than chemical communication, and it too is effective both night and day. Further, auditory communication can be modified not only by loudness but also by pattern, duration, and repetition. In an experiment with rats, a researcher discovered that an intruder can avoid attack by increasing the frequency with which it makes an appeasement sound.

Male birds have songs for a number of different occasions. Such as one song for distress, another for courting, and still another for marking territories. Sailors have long heard the songs of humpback whales transmitted through the hull of a ship. But only recently has it been shown that the song has six basic themes, each with its own phrases, that can vary in length, be interspersed with sundry cries and chirps, and be heard for many miles. Interestingly, humpbacks songs change from year to year, but all the humpbacks in an area learn and converge on singing the same song. The purpose of the song is probably sexual, serving to advertise the availability of the singer.

Language is the ultimate auditory communication, but only humans have the biological ability to produce a large number of different sounds and to put them together in many different ways. Nonhuman primates have at most only 40 different vocalizations, each having a definite meaning, such as the one meaning "baby on the ground," which is uttered by a baboon when a baby baboon falls out of a tree. Although chimpanzees can be taught to use a visual language they never progress beyond the capability level of a two-year-old

Figure 32.10 Use of a pheromone.
This male cheetah is spraying urine onto a tree to mark its territory.

Figure 32.11 A chimpanzee with a researcher.
Chimpanzees are unable to speak but can learn to use a visual language consisting of symbols. Some believe chimps only mimic their teachers and never understand the cognitive use of a language. Here the researcher holds a sweater with the animal's name on it and it has learned to sign the word "me."

child (Fig. 32.11). It has also been difficult to prove that chimps understand the concept of grammar or can use their language to reason. Thus, human language suggests we possess a communication ability unparalleled in other animals.

Visual Communication

Visual signals are most often used by species that are active during the day. Contests between males make use of threat postures and possibly prevent outright fighting, a behavior that might result in reduced fitness. A male baboon displaying full threat is an awesome sight that establishes his dominance and keeps peace within the baboon troop (see Fig. 32.6). Hippopotamuses perform territorial displays that include mouth opening.

The plumage of a male Raggiana bird of paradise allows him to put on a spectacular courtship dance to attract a female and to give her a basis on which to select a mate (see Fig. 32.5). Defense and courtship displays are exaggerated and always performed in the same way so that their meaning is clear.

Visual communication allows animals to signal others of their intentions without providing any auditory or chemical messages. The body language of students during a lecture provides an example. Some students lean forward in their seats and make eye contact with the instructor. They want the instructor to know they are interested and find the material of value. Others lean back in their chairs and text-message their friends or doodle, indicating disinterest. Teachers can use students' body language to determine if they are effectively presenting the material and make changes accordingly.

Other human behaviors also send visual clues. The hairstyle and dress of a person or the way he or she walks and talks are ways to send messages to others. Psychologists have long tried to understand how visual clues can be used to better understand human emotions and behavior. Some studies have suggested that women are apt to dress in an appealing manner and be sexually inviting when they are ovulating. People who dress in black, move slowly, fail to make eye contact, and sit alone may be telling others that they are unhappy. Similarly, body language in animals is being used to suggest that they too have emotions, as discussed in the Science Focus, "Do Animals Have Emotions?"

Tactile Communication

Tactile communication occurs when one animal touches another. For example, laughing gull chicks peck at the parent's beak in order to induce the parent to feed them (see Fig. 32.3). A male leopard nuzzles the female's neck to calm her and stimulate her willingness to mate. In primates, grooming—one animal cleaning the coat and skin of another—helps cement social bonds within a group.

Honeybees use a combination of communication methods, including tactile ones, to convey information about their environment. When a foraging bee returns to the hive, it performs a waggle dance that communicates the distance and direction to the food source (Fig. 32.12). Inside the hive, other foragers crowd around her and touch the forager with their antennae, and as the bee moves between the two loops of a figure 8, it buzzes noisily and shakes its entire body in so-called waggles. The angle of the straight

Figure 32.12
Communication among bees.
a. Honeybees do a waggle dance to indicate the direction of food. **b.** If the dance is done outside the hive on a horizontal surface, the straight run of the dance will point to the food source. If the dance is done inside the hive on a vertical surface, the angle of the straightaway to that of the direction of gravity is the same as the angle of the food source to the sun.

a.

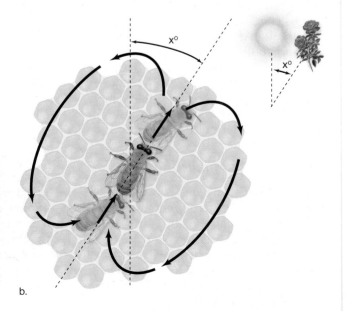
b.

Science Focus

Do Animals Have Emotions?

In recent years, investigators have become interested in determining if animals have emotions. The body language of animals suggests that animals do have feelings (Fig. 32A). When wolves reunite, they wag their tails to and fro, whine, and jump up and down; elephants vocalize—emit their "greeting rumble"—flap their ears, and spin about. Many young animals play with one another, or even by themselves, as when dogs chase their own tails. On the other hand, upon the death of a friend or parent, chimps are apt to sulk, stop eating, and even die. It seems reasonable to hypothesize that animals are "happy" when they reunite, "enjoy" themselves when they play, and are "depressed" over the loss of a close friend or relative. Even people who rarely observe animals usually agree about what the animal must be feeling.

In the past, scientists found it expedient to collect data only about observable behavior and to ignore the possible mental state of the animal. Why? Because emotions are personal and no one can ever know exactly how another animal is feeling. B. F. Skinner, whose research method is described in section 32.2, regarded animals as robots that become conditioned to respond automatically to a particular stimulus. He and others never considered that animals might have feelings. But now, some scientists believe sufficient data suggest that at least other vertebrates and/or mammals do have feelings, including fear, joy, embarrassment, jealousy, anger, love, sadness, and grief. And they believe that those who hypothesize otherwise should have to present the opposing data.

Perhaps it would be reasonable to consider the suggestion of Charles Darwin, the father of evolutionary biology, who said that animals are different in degree rather than in kind. This means that all animals can, say, feel love, but perhaps not to the degree that humans can. B. Würsig watched the courtship of two right whales. They touched, caressed, rolled side-by-side, and eventually swam off together. He wondered if their behavior indicated they felt love for one another. When you

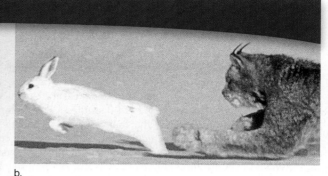

b.

Figure 32A Emotions in animals.
a. Do chimpanzees feel motherly love, as this photograph seems to suggest? **b.** Does the hare feel fear as the lynx closes in?

think about it, it is unlikely that emotions first appeared in humans with no evolutionary homologies in animals.

Neurobiological data support the hypothesis that animals other than humans are capable of enjoying themselves when they perform an activity, such as playing. Researchers have found a high level of dopamine in the brain when rats play, and the dopamine level increases even when rats anticipate the opportunity to play. Certainly even the staunchest critic is aware that many different species of animals have limbic systems and are capable of fight-or-flight responses in dangerous situations. Can we go further and suggest that animals feel fear even when no physiological response has yet occurred?

Laboratory animals may be too stressed to provide convincing data on emotions. This makes field research more useful. Then, too, we have to consider that emotions evolved under an animal's normal environmental conditions. It is possible to fit animals with devices that transmit information on heart rate, body temperature, and eye movements as they go about their daily routine. Such information will help researchers learn how animal emotions might correlate with their behavior. In humans, emotions influence behavior, and the same may be true of other animals. One possible definition of emotion is a psychological phenomenon that helps animals direct and manage their behavior.

M. Bekoff, who is prominent in studying the basis of animal behavior, encourages us to be open to the possibility that animals have emotions. He states:

> **By remaining open to the idea that many animals have rich emotional lives, even if we are wrong in some cases, little truly is lost. By closing the door on the possibility that many animals have rich emotional lives, even if they are very different from our own or from those of animals with whom we are most familiar, we will lose great opportunities to learn about the lives of animals with whom we share this wondrous planet.[1]**

Discussion Questions

1. Do you believe animals have emotions? Why or why not? If not, what experiment(s) would convince you that animals do have emotions?
2. Pet psychology, an emerging field, is based on the premise that pets have feelings and emotions. Would you spend $50–100 per hour to take your pet to a psychologist?
3. Does knowing of existing evidence that animals may have emotions change your opinion about how animal research should be conducted, as was discussed in Chapter 31?

[1]Bekoff, M. Animal emotions: Exploring passionate natures. October 2000. Bioscience 50:10, page 869.

run of the figure 8 to that of the direction of gravity is the same as that of the angle between the sun and the food source. In other words, a 40° angle to the left of vertical means that food is 40° to the left of the sun. Bees can use the sun as a compass because they have a biological clock that allows them to compensate for the movement of the sun in the sky. (A biological clock is an internal means of telling time—for example, darkness outside stimulates many animals to sleep, including humans.) Outside the hive, the

dance is done on a horizontal surface, and the straight-run part of the figure 8 indicates the direction of the food.

Communication is a signal by a sender that affects the behavior of a receiver. Animals use a number of different ways to communicate, thus facilitating cooperation.

Summarizing the Concepts

32.1 Nature versus Nurture: Genetic Influences
Investigators have long been interested in the degree to which nature (genetics) or nurture (environment) influence behavior. Experiments to test this question have involved birds, snakes, and snails among many other animals. The result of hybrid studies with lovebirds are consistent with the hypothesis that behavior has a genetic basis. Garter snake experiments indicate that the nervous system controls behavior. *Aplysia* DNA studies show that the endocrine system is also involved in controlling behavior. Twin studies in humans indicate that certain types of behavior are apparently inherited.

32.2 Nature versus Nurture: Environmental Influences
Even behaviors formerly thought to be fixed action patterns (FAPs), or otherwise inflexible, can sometimes be modified by learning. A red spot on the bill of laughing gulls initiates chick begging behavior. However, with experience, chick begging behavior improves and chicks demonstrate an increased ability to recognize their parents.

Other studies suggest that learning is involved in behaviors. Imprinting in birds during a sensitive period causes them to follow the first moving object they see. Song learning in birds involves various elements—including the existence of a sensitive period, during which an animal is primed to learn—and the effect of social interactions.

Associative learning includes classical conditioning and operant conditioning. In classical conditioning, the pairing of two different types of stimuli causes an animal to form an association between them. In this way, dogs will salivate at the sound of a bell. In operant conditioning, animals learn behaviors because they are rewarded when they perform them.

32.3 Adaptive Mating Behavior
Traits that promote reproductive success are expected to be advantageous overall, despite any possible disadvantage. Males, who produce many sperm, are expected to compete to inseminate females. Females, who produce few eggs, are expected to be selective about their mates. Studies of satin bowerbirds and birds of paradise have tested hypotheses regarding female choice. A cost-benefit analysis can be applied to competition between males for mates, in reference to a dominance hierarchy (e.g., baboons) or territoriality (e.g., red deer).

It is quite possible that evolutionary forces have shaped mating behavior and mate choice in humans as well. Biological differences between males and females may promote different mate preferences because they influence fitness differently in the two sexes.

32.4 Sociobiology and Animal Behavior
Living in a social group can have its advantages (e.g., ability to avoid predators, raise young, and find food). It also has disadvantages (e.g., competition between members, spread of illness and parasites, and reduced reproductive potential). When animals live in groups, the benefits must outweigh the costs or the behavior would not exist.

In most instances, the individuals of a society act to increase their own fitness (number of fertile offspring produced). It is speculated that human societies may have originally evolved because hunting required group effort. Today, much of human behavior is influenced by cultural learning in societies. Sometimes, animals perform altruistic acts, as when individuals help their parents rear siblings. In this context, it is necessary to consider inclusive fitness, which includes personal reproductive success and the reproductive success of relatives also. Social insects help their mother reproduce, but this behavior seems reasonable when we consider that siblings share 75% of their genes. Among mammals and birds, a parental helper may be likely to inherit the parent's territory.

These examples show that altruistic behavior most likely results in increased inclusive fitness.

32.5 Animal Communication
Animals that form social groups communicate with one another. In this context, communication is an action by a sender that affects the behavior of a receiver. Chemical, auditory, visual, and tactile signals are forms of communication that foster cooperation benefiting both the sender and the receiver. Pheromones are chemical signals that are passed between members of the same species. Auditory communication includes language, which occurs between other types of animals, not just humans. Visual communication allows animals to signal others without needing auditory or chemical messages. Tactile communication is especially associated with social behavior.

Testing Yourself

Choose the best answer for each question.

1. Behavior is
   a. any action that is learned.
   b. all responses to the environment.
   c. any action that can be observed and described.
   d. all activity that is controlled by hormones.
   e. unique to birds and humans.

2. A behaviorist would most likely study which one of the following items?
   a. the flow of energy through an ecosystem
   b. the way a bird digests seeds
   c. the number of times a fiddler crab flashes its claw to attract a mate
   d. the structure of a horse's leg
   e. All of these are correct.

3. Which one of the following is considered, by behaviorists, to control (in part or in whole) animal behavior?
   a. circulatory and respiratory systems
   b. respiratory and digestive systems
   c. digestive and nervous systems
   d. nervous and endocrine systems
   e. All systems of the body control behavior.

4. In lovebirds, if carrying strips in the bill is controlled by a single dominant gene, then hybrids between peach-faced lovebirds and Fischer lovebirds would
   a. carry strips in their bills.
   b. carry strips in their rump feathers.
   c. not carry strips.
   d. carry strips in both their bills and rump feathers.
   e. have decreased fitness.

5. Which of the following is not an example of a genetically based behavior?
   a. Inland garter snakes do not eat slugs, while coastal populations do.

b. One species of lovebird carries nesting strips one at a time, while another carries several.

c. One species of warbler migrates, while another one does not.

d. Snails lay eggs in response to egg-laying hormone.

e. Wild foxes raised in captivity are not capable of hunting for food.

6. How would the following graph differ if begging behavior in laughing gulls was a fixed action pattern?

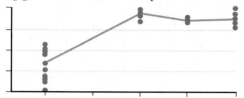

a. It would be a diagonal line with an upward incline.

b. It would be a diagonal line with a downward incline.

c. It would be a horizontal line.

d. It would be a vertical line.

e. None of these are correct.

7. Egg-laying hormone causes snails to lay eggs, implicating which system in this behavior?

a. digestive
b. endocrine
c. nervous
d. lymphatic
e. respiratory

8. Using treats to train a dog to do a trick is an example of

a. imprinting.
b. tutoring.
c. vocalization.
d. operant conditioning.

9. The benefits of imprinting (following and obeying the mother) generally outweigh the costs (following the wrong object instead of the mother) because

a. an animal that has imprinted on the wrong object can re-imprint on the mother.

b. imprinting behavior never lasts more than a few months.

c. animals in the wild rarely imprint on anything other than their mother.

d. animals that imprint on the wrong object generally die before they pass their genes on.

10. In white-crowned sparrows, social experience exhibits a very strong influence over the development of singing patterns. What observation led to this conclusion?

a. Birds only learned to sing when they were trained by other birds.

b. The window in which birds learn from other birds is wider than that in which birds learn from tape recordings.

c. Birds could learn different dialects only from other birds.

d. Birds that learned to sing from a tape recorder could, change their song when they listened to another bird.

11. Which reaction best provides physiological evidence that animals have emotions?

a. Adrenaline levels increase when a deer senses danger.

b. Dopamine levels in the brains of rats increase when they play.

c. Chimpanzees experience decreases in acetylcholine levels when they are content.

d. The blood pressure of wolves drops when they return to the pack.

12. Which of the following best describes classical conditioning?

a. the gradual strengthening of stimulus-response connections that are seemingly unrelated

b. a type of associative learning in which there is no contingency between response and reinforcer

c. the learning behavior in which an organism follows the first moving object it encounters

d. the learning behavior in which an organism exhibits a fixed action pattern from the time of birth

13. The observation that the most fit male bowerbirds are the ones that can keep their nests intact supports which hypothesis?

a. good genes hypothesis—females choose mates based on their improved chances of survival

b. run-away hypothesis—females choose mates based on their appearance

c. Either hypothesis could be supported.

d. Neither hypothesis is supported.

14. In some bird species, the female bird chooses a mate that is most similar to her in size. This supports the

a. good genes hypothesis.
b. run-away hypothesis.
c. Either hypothesis could be supported.
d. Neither hypothesis is supported.

15. Which of the following are costs that a dominant male baboon must pay in order to gain a reproductive benefit?

a. He requires more food and must travel longer distances.

b. He requires more food and must care for his young.

c. He is more prone to injury and requires more food.

d. He is more prone to injury and must care for his young.

e. He must care for his young and travel longer distances.

16. A red deer harem master typically dies earlier than other males because he is

a. likely to get expelled from the herd and cannot survive alone.

b. more prone to disease because he interacts with so many animals.

c. likely to starve to death.

d. apt to place himself between a predator and the herd to protect the herd.

For problems 17–21, match the type of communication in the key with its description. Answers may be used more than once.

Key:

a. chemical communication
b. auditory communication
c. visual communication
d. tactile communication

17. Aphids (insects) release an alarm pheromone when they sense they are in danger.

18. Male peacocks exhibit an elaborate display of feathers to attract females.

19. Ground squirrels give an alarm call to warn others of the approach of a predator.

20. Male silkmoths are attracted to females by a sex attractant released by the female moth.

21. Sage grouses perform an elaborate courtship dance.

Understanding the Terms

altruism 684

associative learning 679

behavior 676

classical conditioning 679

communication 685

cost-benefit analysis 680

culture 684

dominance hierarchy 680

fitness 680

fixed action pattern
    (FAP) 678

imprinting 678

inclusive fitness 684

kin selection 684

learning 678

operant conditioning 679

pheromone 685

sexual selection 680

society 685

sociobiology 683

territoriality 680

territory 680

Match the terms to these definitions:

a._____ Behavior related to defending a particular area used for feeding, mating, and caring for young.

b._____ Social interaction that has the potential to decrease the lifetime reproductive success of the member exhibiting the behavior.

c._____ Signal by a sender that influences the behavior of a receiver.

d._____ Chemical substance secreted into the environment by one organism that influences the behavior of another.

Thinking Critically

1. You are studying two populations of rats—one in New York and the other in Florida. Adult New York rats seem to prefer swiss cheese, while Florida rats seem to prefer cheddar. Design an experiment to determine the extent to which this behavior may be genetically controlled versus environmentally influenced.

2. Operant conditioning is often used to train animals, particularly pets. Describe an example of operant conditioning and why you think it works.

3. Until about 15 years ago, it was long thought that most bird species were monogamous—that is, at least within a breeding season, males and females paired and mated only with each other. However, with advances in DNA technology, researchers began to find out that offspring in a nest were often sired by males in neighboring territories. When females mate with neighboring males, it is called "extra pair copulation" or EPC. Why do you think EPC behavior might have evolved in many bird species?

4. If altruistic behavior has usually evolved to improve the altruist's inclusive fitness by helping relatives, then why do you think humans perform altruistic acts for completely unrelated individuals (e.g., jumping into a pool to save a drowning person)?

5. Various types of communication behavior have evolved so that males can attract females for mating purposes. These may consist of loud calls, bright or flamboyant mating displays, or the construction of elaborate structures. These communication behaviors could also attract predators, which could result in the death of the male, the female, or both. So, how could such behaviors evolve?

ARIS, the *Inquiry into Life* Website

ARIS, the website for *Inquiry into Life,* provides a wealth of information organized and integrated by chapter. You will find practice quizzes, interactive activities, labeling exercises, flashcards, and much more that will complement your learning and understanding of general biology.

www.aris.mhhe.com

Overfishing threatens the world's fish populations, such as that of the swordfish. *Global fisheries production has increased steadily over the past 50 years, from about 18 million tons per year in 1950 to over 133 million tons in 2003. By the year 2010, it is anticipated that the fish catch will need to increase by 15.5 million tons per year to keep up with the demand imposed by humans.*

Overfishing occurs when the catch of a particular fish species exceeds reproduction, resulting in declining numbers. In other words, the death rate exceeds the birthrate. Normally, fish populations exist at the carrying capacity of the environment (the numbers the environment can support). However, if humans reduce fish populations well below the carrying capacity, a lag in population growth occurs. If fishing continues at a high rate during this lag phase, it can drive fishery populations to extinction.

Population ecologists study the distribution and abundance of organisms, such as fish species that are commercially exploited. Our knowledge of population growth and regulation has led to controls on fishing, including size limits and catch limits, to try to sustain some of the world's fishery populations. Thus, the study of population ecology is necessary for the conservation of species and the maintenance of viable populations for commercial use.

chapter 33

Population Growth and Regulation

Chapter Concepts

33.1   The Scope of Ecology

Ecology is the study of the interactions of organisms with each other and with the physical environment. Ecology, like so many biological disciplines, is wide-ranging and involves several levels of study (Fig. 33.1).

At the simplest level, ecologists study how the individual organism is adapted to its environment. For example, they might study why a particular species of fish in a coral reef lives only in warm tropical waters and how it feeds. Most organisms do not exist singly; rather, they are part of a **population,** which is defined as all the organisms of the same species interacting with the environment at a particular locale. At the population level of study, ecologists describe the size of populations over time. For example, in a coral reef, an ecologist might study the relative sizes of parrotfishes in a population over time. A **community** consists of all the various populations at a particular locale. A coral reef community contains numerous populations of fish species, crustaceans, corals, and so forth. At the community level, ecologists study how various extrinsic factors (e.g., weather) and intrinsic factors. (e.g., species competition for resources) affect the size of these populations. An **ecosystem** encompasses a community of populations, as well as the nonliving environment. For example, energy flow and chemical cycling in a coral reef can affect the success of the organisms that inhabit it. Finally, the **biosphere** is that portion of the entire Earth's surface—air, water, land—where living things exist. Knowing the composi-

TABLE 33.1	Ecological Terms
Term	**Definition**
Ecology	Study of the interactions of organisms with each other and with the physical environment
Population	All the members of the same species that inhabit a particular area
Community	All the populations found in a particular area
Ecosystem	A community and its physical environment, including both nonliving (abiotic) and living (biotic) components
Biosphere	All the communities on Earth whose members exist in air and water and on land

tion and diversity of the coral reef ecosystem is important to the dynamics of the biosphere. Table 33.1 defines and summarizes the levels commonly studied in ecology.

Modern ecology is not just descriptive; it also focuses on developing testable hypotheses. A central goal of modern ecology is to develop models that explain and predict the distribution and abundance of populations. Ultimately, ecology considers not one particular area, but the distribution and abundance of populations and species in the entire biosphere. In this chapter, we particularly concentrate on the patterns of population growth, and how population growth is regulated by interactions with other populations. Then, at the community level, we examine how populations change through time during the process of ecological succession.

Figure 33.1 Levels of ecological organization.

The study of ecology encompasses levels of organization that proceed from the individual organism to the population, to the community, and to an ecosystem.

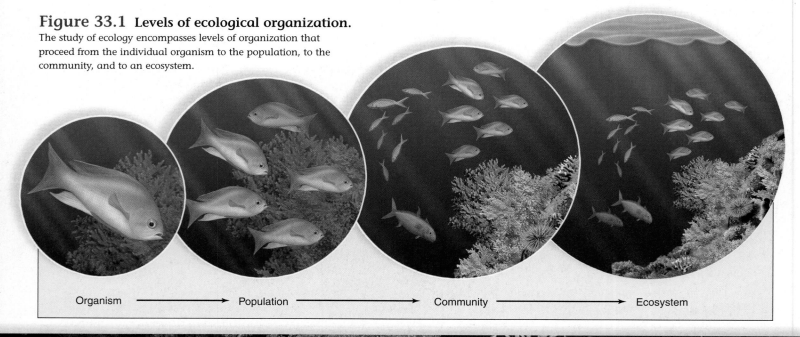

Organism ⟶ Population ⟶ Community ⟶ Ecosystem

Coral reef community

a.

b.

Figure 33.2 Biotic potential.

A population's maximum growth rate under ideal conditions—that is, its biotic potential—is greatly influenced by the number of offspring produced in each reproductive event. **a.** Pigs, which produce many offspring that quickly mature to produce more offspring, have a much higher biotic potential than (**b**) the rhinoceros, which produces only one or two offspring during infrequent reproductive events.

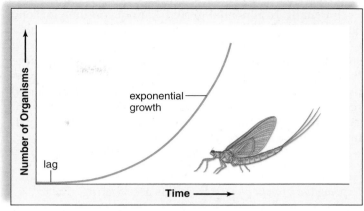

a.

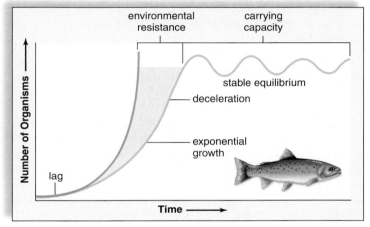

b.

Figure 33.3 Patterns of population growth.

a. Exponential growth results in a J-shaped growth curve because the growth rate is positive, and increasing, as in insects. **b.** Logistic growth results in an S-shaped growth curve because environmental resistance causes the population size to level off and be in a steady state at the carrying capacity of the environment, as in fish.

33.2 Patterns of Population Growth

Each population has particular patterns of growth. The population size can stay the same, increase, or decrease, according to a *per capita rate of increase* or growth rate. Suppose, for example, a human population presently has a size of 1,000 individuals, the birthrate is 30 per year, and the death rate is 10 per year. The growth rate per year will be:

$$30 - 10 / 1,000 = 0.02 = 2.0\% \text{ per year}$$

(Notice that this per capita rate of increase disregards both immigration and emigration, which for the purpose of our discussion can be assumed to be equal and thus to cancel each other out.) The highest possible per capita rate of increase for a population is called its **biotic potential** (Fig. 33.2). Whether the biotic potential is high or low depends on such factors as the following:

1. Average number of offspring per reproduction
2. Chances of survival until age of reproduction
3. Age at first reproduction
4. How often each individual reproduces

Suppose we are studying the growth of a population of insects that are capable of infesting and taking over an area. Under these circumstances, **exponential growth** is expected. An exponential pattern of population growth results in a J-shaped curve (Fig. 33.3*a*). This pattern of population growth can be likened to compound interest at the bank: the more your money increases, the more interest you will get. If the insect population has 2,000 individuals and the per capita rate of increase is 20% per month, there will be 2,400 insects after one month, 2,880 after two months, 3,456 after three months, and so forth.

Notice that a J-shaped curve has these phases:

Lag phase. Growth is slow because the population is small.
Exponential growth phase. Growth is accelerating, and the population is exhibiting its biotic potential.

Usually, exponential growth cannot continue for long because of environmental resistance. **Environmental resistance** is all those environmental conditions—such as limited food supply, accumulation of waste products, increased competition, or predation—that prevent populations from achieving their biotic potential. The intensity of environmental resistance increases as the population grows larger, until population growth levels off, resulting in an S-shaped pattern called **logistic growth** (Fig. 33.3b).

Notice that an S-shaped curve has these phases:

Lag phase. Growth is slow because the population is small.
Exponential growth phase. Growth is accelerating, due to biotic potential.
Deceleration phase. The rate of population growth slows down.
Stable equilibrium phase. Little if any growth takes place because births and deaths are about equal.

The stable equilibrium phase is said to occur at the **carrying capacity** of the environment. The carrying capacity is the number of individuals of a species that a particular environment can support.

Our knowledge of logistic growth has practical implications. The model predicts that exponential growth occurs only when population size is much lower than the carrying capacity. So, for example, if humans are using a fish population as a continuous food source, it would be best to maintain that population size in the exponential phase of growth when the birthrate is the highest. If we overfish, the fish population will sink into the lag phase, and it will be years before exponential growth recurs. On the other hand, if we are trying to limit the growth of a pest, it is best to reduce the carrying capacity. Reducing the population size only encourages exponential growth to begin again. Farmers can reduce the carrying capacity for a pest by alternating rows of different crops rather than growing one type of crop throughout the entire field.

> Exponential growth produces a J-shaped curve because population growth accelerates over time. Logistic growth produces an S-shaped curve because the population size stabilizes when the carrying capacity of the environment has been reached.

Survivorship

Population growth patterns assume that populations are made up of identically aged individuals. In real populations, however, individuals are in different stages of their life spans. Let us consider how many members of an original group of individuals born at the same time, called a **cohort,** are still alive after certain intervals of time. The result is a survivorship curve.

For the sake of discussion, three types of idealized survivorship curves are recognized (Fig. 33.4). The type I curve is characteristic of a population of humans, in which most individuals survive well past the midpoint, and death comes near the end of the maximum life span. On the other hand, the type

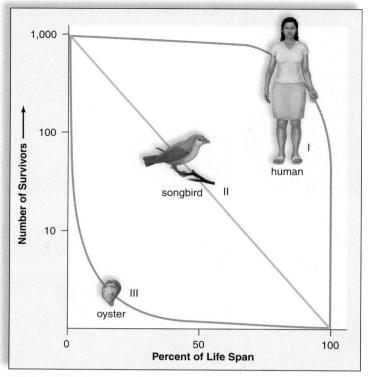

Figure 33.4 Survivorship curves.
Human beings have a type I survivorship curve: the individual usually lives a normal life span, and then death is increasingly expected. Songbirds have a type II curve: the chances of surviving are the same for any particular age. Oysters have a type III curve: most deaths occur during the free-swimming larva stage, but oysters that survive to adulthood usually live a normal life span.

III curve is typical for a population of oysters, in which most individuals die very young. In the type II curve, survivorship decreases at a constant rate throughout the life span. This has been found typical of a population of songbirds. Some species, however, do not fit any of these curves exactly.

Much can be learned about the life history of a species by studying its survivorship curve. Would you predict that most or only a few members of a population with a type III survivorship curve are contributing offspring to the next generation? Obviously, since death comes early for most members, only a few live long enough to reproduce.

> Populations have a pattern of survivorship that becomes apparent from studying the survivorship curve of a cohort.

Human Population Growth

Figure 33.5a illustrates human population growth. Growth in *less-developed countries* is still in the exponential phase and is not expected to level off until about 2040, while growth in *more-developed countries* has leveled off. The equivalent of a medium-sized city (225,000) is added to the world's population every day, and 82 million (the equivalent of the

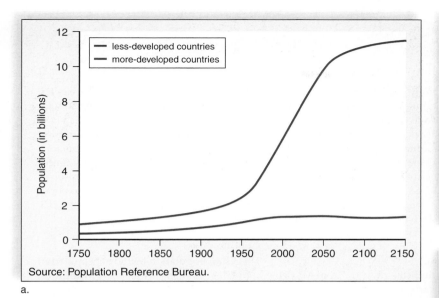

b.

c.

Figure 33.5 World population growth.
a. The graph shows world population size in the past, with estimates to 2150. People in (**b**) more-developed countries have a high standard of living and will contribute the least to world population growth, while people in (**c**) less-developed countries have a low standard of living and will contribute the most to world population growth.

combined populations of the United Kingdom, Norway, Ireland, Iceland, Finland, the Netherlands, and Denmark) are added every year.

The rapid growth of the human population can be appreciated by considering the **doubling time,** the length of time it takes for the population size to double. Currently, the doubling time is estimated to be 56 years. Such an increase in population size will put extreme demands on our ability to produce and distribute resources. In 56 years, the world will need to double the amount of food, jobs, water, energy, and so on just to maintain the present standard of living.

Many people are gravely concerned that the amount of time needed to add each additional billion persons to the world population has become shorter. The first billion was reached in 1800; the second billion in 1930; the third billion in 1960; and today there are nearly 6.5 billion. By contrast, **zero population growth,** in which the birthrate equals the death rate and population size remains steady, can only be achieved if the per capita rate of increase declines. The world's population may level off at 8, 10.5, or 14.2 billion, depending on the speed with which the per capita rate of increase declines.

More-Developed versus Less-Developed Countries
The **more-developed countries (MDCs),** typified by countries in North America and Europe, are those in which population growth is low and people enjoy a good standard of living (Fig. 33.5*b*). The **less-developed countries (LDCs),** such as countries in Latin America, Africa, and Asia, are those in which population growth is expanding rapidly and the majority of people live in poverty (Fig. 33.5*c*).

The MDCs doubled their populations between 1850 and 1950 due to a decline in the death rate, the development of modern medicine, and improved socioeconomic conditions. The decline in the death rate led to an upward trend in population size. However, the transition from an agricultural society to an urban society reduced the incentive for people to have large families (e.g., more people to help on the farm) because maintaining large families in urban areas is more costly than in rural areas, causing birthrates to decline. These factors, combined with the decreased death rate, caused populations in MDCs to experience only modest growth between 1950 and 1975. This sequence of events (i.e., decreased death rate followed by decreased birthrate) is termed a **demographic transition.**

Yearly growth of the MDCs as a whole has now stabilized at about 0.1%. The populations of several of the MDCs, including Germany, Greece, Italy, Hungary, and Sweden, are not growing or are actually decreasing in size. In contrast, there is no leveling off in U.S. population growth. Although yearly growth of the United States is only 0.6%, many people immigrate to the United States each year. In addition, an unusually large number of babies were born between 1947 and 1964 (called a baby boom). Therefore, a large number of women are still of reproductive age. The

The U.S. Constitution requires that a national census be taken every ten years. The latest census concluded that the U.S. population numbered 281,421,906 persons in April 2000. This is a gain of 32.7 million since the last census in 1990 and the largest increase ever between censuses. It is clear that the United States, unlike other more-developed countries, is still undergoing exponential growth. Because the U.S. population is very large to begin with, that adds up to a lot of people in a very short time. Also, many persons immigrate to the United States each year. In 2006, the U.S. population reached 300 million.

Minorities

Most people in the United States are Caucasian, but their proportion of the total population is decreasing because the major minorities—the Latino, African American, and Asian populations—are increasing at a faster rate than the Caucasian population (Fig. 33A). The Latino population grew the most over the past ten years, followed by the Asian and then the African American populations.

The Native American population increased only slightly, from 0.6% to 0.7% of the population. For the first time, the census allowed Americans to identify with more

Figure 33A Percent changes in U.S. population.

Over the past ten years, the Caucasian population dropped in percent of total population, while the Latino, Asian, and African American populations increased in percent of total population.

than one ethnic group, and about 2% did so. A significant number of these chose Native American as one of the ethnic groups, and then they could not be counted as belonging to this single minority.

California is now a "minority majority" state, meaning that those Californians who considered themselves Caucasian make up less than one-half of the population. Other statistics regarding minorities are also of interest.

Latinos

Latinos who indicate they are from Mexico account for nearly 60% of this particular seg-

Ethnicity: Caucasian
2000 69.1%
1990 75.6%
−6.5%

ment of the total population. Whereas ten years ago Latinos made up 9% of the total population, now they are 12.5%. This is the largest percent increase of all the minorities.

Altogether, the percentage of persons identified as Latino is now slightly larger than that of those who consider themselves African American. By 2010, the Latino population may be as high as 20% of the total U.S. population.

Asians

The Asian population includes people from countries such as China, the Philippines, Japan, Korea, India, and Pakistan. The

Ethnicity: Hispanic
2000 12.5%
1990 9.0%
+3.5%

Science Focus provides details about the current trends in U.S. population growth.

Although the death rate began to decline steeply in the LDCs following World War II with the importation of modern medicine from the MDCs, the birthrate remained high. The yearly growth of the LDCs peaked at 2.5% between 1960 and 1965. Since that time, a demographic transition has occurred: the decline in the death rate slowed, and the birthrate fell. The yearly growth is now 1.9%. Still, because of exponential growth, the population of LDCs may explode from 4.9 billion today to 11 billion in 2100. Most of this growth will occur in Africa, Asia, and Latin America. Suggestions for greatly reducing the expected increase include the following:

1. Establish and/or strengthen family planning programs. A decline in growth is seen in countries with good family planning programs supported by community leaders. Currently, 25% of women in sub-Saharan Africa say they would like to delay or stop childbearing, and yet they do not have access to birth control. Likewise, 15% of women in Asia and Latin America have an unmet need for birth control.

2. Use social progress to reduce the desire for large families. Many couples in LDCs presently desire as many as four to six children. But providing education, raising the status of women, reducing child mortality, and improving economic stability

Asian population increased by two-thirds, but even so they account for only 3.7% of the year 2000 population. About 60% of Asian Americans are foreign born; therefore, immigration accounts for most of the increase in the Asian population.

African Americans

The African American population stayed at just about 12% over the past ten years. Unlike the other minority populations, shown in Figure 33A, there is little immigration from other countries, and this accounts for why the African American population did not increase as fast as those of other minorities.

African Americans remain the predominant minority group in the South, however, where they make up about 19% of the population. During the 1990s, many African Americans left the North and returned to the South to take advantage of the region's booming economy.

What's Ahead?

The 2000 census found that nearly 40% of the population under age 18 (i.e., children) belong to a minority group. This means that minorities are sure to increase as a percentage of the total population during the next decade. More than one-half of the

children in California, Arizona, Hawaii, New Mexico, and Texas are members of a minority. By the next census, all these states will likely have minority majorities as California has now.

The percentage of people identifying themselves as multiracial will likely increase over the next ten years as well. It is this racial, cultural, and ethnic diversity that makes the U.S. population unique and special.

Discussion Questions

1. Although the U.S. population growth rate is currently estimated at 0.6%, which is far lower than that of many LDCs, people in the United States consume many more resources (5–100 times) than people in LDCs. What might happen in terms of resources if the U.S. population continues to grow?
2. Much of the current U.S. population growth is due to immigration. Current laws have been proposed to limit immigration into the United States. Should we do this? Should we grant illegal immigrants legal status, including access to health care?
3. Diversity of races, ethnicities, and cultures is part of the reason the United States was originally founded. Yet, racial tension still exists. If U.S. minorities continue to increase as predicted, how might we increase racial and ethnic tolerance in the United States?

Ethnicity: Asian
2000 3.7%
1990 2.8%
    +0.9%

Ethnicity: Black
2000 12.1%
1990 11.7%
    +0.4%

are desirable social improvements that could help decrease population growth.
3. Delay the onset of childbearing. A delay in the onset of childbearing and wider spacing of births could cause a temporary decline in the birthrate and reduce the present reproductive rate.

Age Distributions

An **age-structure diagram** divides the population into three age groups: preproductive, reproductive, and postreproductive (Fig. 33.6). The LDCs are experiencing a population momentum because they have more women entering the reproductive years than older women leaving them.

It is a misconception that if each couple has two children, referred to as **replacement reproduction,** that zero population growth will take place immediately. Actually, the population will continue to grow as long as there are more young women entering their reproductive years than there are older women leaving them. This is true of most LDCs, where a large proportion of the population is younger than 15, which will cause LDC populations to expand greatly even after replacement reproduction is attained. This type of population growth results in *unstable age structure* (Figure 33.6a). In LDCs, the more quickly replacement reproduction is achieved, the sooner zero population growth will occur. The Bioethical Focus at the end of this chapter discusses population growth in LDCs.

**Figure 33.6
Age structure
diagrams (2002).**
The diagrams illustrate that
(**a**) the LCDs will expand
rapidly due to their age
distributions, whereas (**b**)
the MDCs are approaching
stabilization.

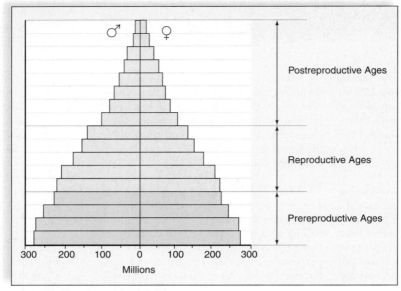

a. Less-developed countries (LDCs)

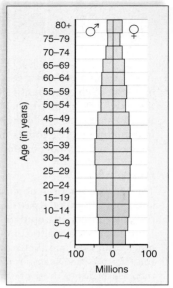

b. More-developed countries (MDCs)

In MDCs, on the other hand, it is more typical that the number of people in the pre-reproductive class roughly equals the number in the reproductive age class. This causes a *stable age structure,* and population numbers will remain about the same for the foreseeable future (Fig. 33.6*b*).

> **Currently, the less-developed countries are expanding dramatically because of exponential growth due to an unstable age structure.**

33.3 Regulation of Population Growth

Ecologists want to determine the factors that regulate population growth. They have observed two types of life history patterns characterized by how long it takes to reach reproductive maturity and how great is the reproductive output: an *opportunistic pattern* and an *equilibrium pattern* (Fig. 33.7). Members of opportunistic populations are small in size, mature early, and have a short life span. They tend to produce many relatively small offspring and to forego parental care in favor of a greater number of offspring. The more offspring, the more likely it is that some of them will survive a population crash. Classic examples of opportunistic species are many insects and weeds.

In contrast, the size of equilibrium populations remains pretty much at the carrying capacity. Resources such as food and shelter are relatively scarce for these individuals, and those who are best able to compete will have the largest number of offspring. These organisms allocate more energy than opportunists to their own growth and

survival and to the growth and survival of their offspring. Therefore, they are fairly large, are slow to mature, and have a fairly long life span. They are specialists rather than colonizers and tend to become extinct when their normal way of life is destroyed. The best examples of equilibrium species are found among birds and mammals. For example, the Florida panther is the largest animal in the Florida Everglades, requires a very large range, and produces few offspring, which must be cared for. Currently, the Florida panther is unable to compensate for a reduction in its range due to human destruction of its habitat and is therefore on the verge of extinction.

For some time, ecologists have recognized that the environment contains both abiotic and biotic components. Abiotic factors, such as weather and natural disasters, are **density-independent,** meaning that the effects of the factor are the same for all sizes of populations. For example, fires don't necessarily kill a larger percentage of individuals as the population increases in size. On the other hand, biotic factors, such as competition, predation, and parasitism, are called **density-dependent.** The effects of a density-dependent factor depend on the size of the population. The more dense a population is, the faster a disease might spread, for example. Populations that have the opportunistic life history pattern tend to be regulated by density-independent factors, and those that have the equilibrium life history pattern tend to be regulated by density-dependent factors.

> **Density-independent and density-dependent factors can often explain the dynamics of natural populations.**

Opportunistic Pattern

- Small individuals
- Short life span
- Fast to mature
- Many offspring
- Little or no care of offspring
- Many offspring die
  before reproducing
- Early reproductive age

Equilibrium Pattern

- Large individuals
- Long life span
- Slow to mature
- Few and large offspring
- Much care of offspring
- Most young survive to
  reproductive age
- Adapted to stable
  environment

Figure 33.7 **Life history patterns.**
Dandelions are an opportunistic species with the characteristics noted, and bears are an equilibrium species with the characteristics noted. Often the distinctions between these two possible life history patterns are not as clear-cut as they may seem.

Competition

Competition occurs when members of different species try to utilize a resource (such as light, space, or nutrients) that is in limited supply. Understanding the effects of competition first requires a grasp of the concept of ecological niche. The **ecological niche** is the role an organism plays in the community, including the **habitat** it requires (where the organism lives) and its interactions with other organisms. The niche is essentially a species' total way of life, and thus includes the resources needed to meet its energy, nutrient, survival, and reproductive demands.

According to the **competitive exclusion principle,** no two species can occupy the same ecological niche at the same time if resources are limiting. While it may seem as if several populations living in the same area are occupying the same niche, it is usually possible to find slight differences. When two species of paramecia are grown separately, each survives. But when they are grown in one test tube, resources are limiting, and only one species survives (Fig. 33.8). What does it take to have different ecological niches so that no species becomes extinct. In another laboratory experiment, two species of paramecia continue to occupy the same tube when one species feeds on bacteria at the bottom of the tube and the other feeds on bacteria suspended in solution. The division of feeding niches, called **resource partitioning,** decreases competition between the two species. As another example, consider that swallows, swifts, and martins all eat flying insects and parachuting spiders. These birds even frequently fly in mixed flocks. But each species of bird has different nesting sites and migrates at a slightly different time of year. Therefore, they are not competing for the same food source when they are feeding their young.

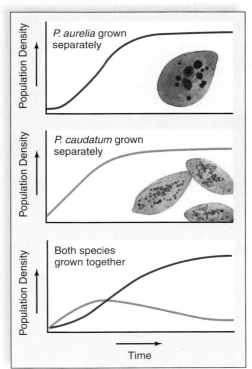

Figure 33.8 Competition between two laboratory populations of *Paramecium*. When grown alone in pure culture (top two graphs), *Paramecium caudatum* and *Paramecium aurelia* exhibit logistic growth. When the two species are grown together in mixed culture (bottom graph), *P. aurelia* is the better competitor, and *P. caudatum* dies out. This experiment illustrates the competitive exclusion principle.

On the Scottish coast, a small barnacle (*Chthamalus stellatus*) lives on the high part of the intertidal zone, and a large barnacle (*Balanus balanoides*) lives on the lower part (Fig. 33.9). Free-swimming larvae of both species attach themselves to rocks at any point in the intertidal zone, where they develop into the sessile adult forms. In the lower zone, the large *Balanus* barnacles seem to either force the smaller *Chthamalus* individuals off the rocks or grow over them. Competition is therefore restricting the range of *Chthamalus* on the rocks. *Chthamalus* is more resistant to drying out than is *Balanus*; therefore, it has an advantage that permits it to grow in the upper intertidal zone.

Competition may result in resource partitioning. When similar species seem to be occupying the same ecological niche, it is usually possible to find differences that indicate resource partitioning has occurred.

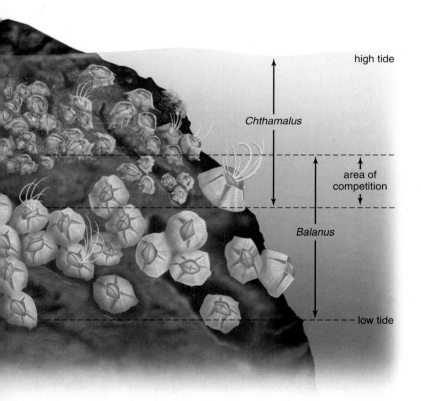

high tide

Chthamalus

area of competition

Balanus

low tide

Figure 33.9 Competition between two species of barnacles.
Competition prevents two species of barnacles from occupying as much of the intertidal zone as possible. Both exist in the area of competition between *Chthamalus* and *Balanus*. Above this area, only *Chthamalus* survives, and below it only *Balanus* survives.

Predation

Predation occurs when one organism, called the **predator,** feeds on another, called the **prey.** In the broadest sense, predators include not only animals such as lions that kill zebras, but also filter-feeding blue whales that strain krill from ocean waters, parasitic ticks that suck blood from their victims, and even herbivorous deer that browse on trees and bushes.

Predator-Prey Population Dynamics

Predators reduce the population density of prey, as shown by a laboratory study in which the protozoans *Paramecium caudatum* (prey) and *Didinium nasutum* (predator) were grown together in a culture medium. *Didinium* ate all the *Paramecium* and then died of starvation. In nature, we can find a similar example. When a gardener brought prickly-pear cactus to Australia from South America, the cactus spread out of control until millions of acres were covered with nothing but cacti. The cacti were brought under control when a moth from South America, whose caterpillar feeds only on the cactus, was introduced. Now, both cactus and moth are found at greatly reduced densities in Australia.

Mathematical formulas predict that predator and prey populations may sometimes cycle instead of maintaining a steady state. Cycling could occur if (1) the predator population overkills the prey and then the predator population also declines in number, and (2) the prey population overshoots the carrying capacity and suffers a crash, and then the predator population follows suit because of a lack of food. In either case, the result would be a series of peaks and valleys in population density, with the predator population lagging slightly behind the prey.

A famous case of predator/prey cycles occurs between the snowshoe hare and the Canadian lynx, a type of small cat (Fig. 33.10). The snowshoe hare is a common herbivore in the coniferous forests of North America, where it feeds on terminal twigs of various shrubs and small trees. The Canadian lynx feeds on snowshoe hares but also on ruffed grouse and spruce grouse, two types of birds. Investigators at first assumed that the lynx had brought about the decline of the hare population. But others noted that the decline in snowshoe hare abundance was accompanied by low growth and reproductive rates that could be signs of a food shortage. It appears that both explanations apply to the data. In other words, both a predator-hare cycle and a hare-food cycle have combined to produce an overall effect, which is observed in Figure 33.10. Although not shown on the graph, it is additionally interesting to note that densities of the grouse populations also cycle, perhaps because the lynx switches to this food source when the hare population declines. Predators and prey do not normally exist as simple, two-species systems, and therefore, abundance patterns should be viewed with the complete community in mind.

> **Interactions between predator and prey and between prey and its food source can produce complex cycles.**

Antipredator Defenses

While predators have evolved strategies to secure the maximum amount of food with minimal expenditure of energy, prey organisms have evolved strategies to escape predation. **Coevolution** is present when each of two species adapts in response to selective pressure imposed by the other. This phenomenon applies to predation, as well as to symbiotic interactions (discussed next).

In plants, the sharp spines of the cactus, the pointed leaves of holly, and the tough, leathery leaves of the oak tree

Figure 33.10 **Predator-prey cycling of a lynx and a snowshoe hare.**
The number of pelts received yearly by the Hudson Bay Company for almost 100 years shows a pattern of ten-year cycles in population densities. The snowshoe hare (prey) population reaches a peak before that of the lynx (predator) by a year or more.

Data from D.A. MacLulich, Fluctuations in the Numbers of the Varying Hare (Lepus americanus), University of Toronto Press, Toronto, 1937; reprinted 1974, p. 543.

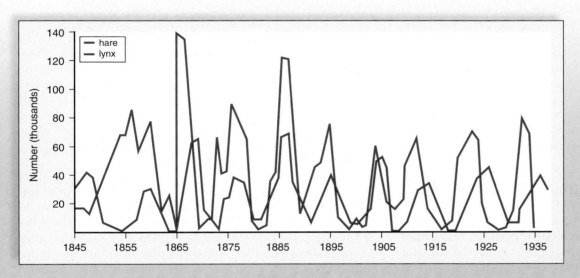

a.

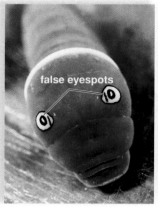
b.

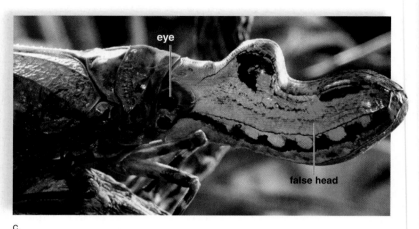
c.

eye

false eyespots

false head

Figure 33.11 Antipredator defenses.

a. The skin secretions of poison-dart frogs are so poisonous that they were used by natives to make their arrows instant lethal weapons. The bright coloration of these frogs is called "warning coloration" because brightly colored species are generally toxic. **b.** The caterpillar of the eastern tiger swallowtail butterfly has false eyespots used to confuse a predator. **c.** The South American lantern fly has a large false head that resembles that of an alligator. This may frighten a predator into thinking it is facing a dangerous animal.

all discourage predation by insects. Plants even produce poisonous chemical. Animals have varied antipredator defenses. Some effective defenses are poisonous secretions, concealment, fright, flocking together, and warning coloration (Fig. 33.11), as well as mimicry. A good example of concealment occurs with caddisfly larvae, which build protective cases out of sticks and sand, making themselves look like the stream bottom where they reside.

Mimicry   **Mimicry** occurs when one species resembles another species that has evolved to defend against predators or resembles an object in the environment to deceive prey. Mimicry can help a predator capture food or a prey avoid capture. For example, angler fishes have lures that resemble worms for the purpose of bringing fish within reach. To avoid capture, some inchworms resemble twigs, and some caterpillars can transform themselves into shapes resembling snakes.

Batesian mimicry (named for Henry Bates, who discovered it) occurs when a prey that is not harmful mimics another species that has a successful antipredator defense. Many examples of Batesian mimicry involve warning coloration. For example, stinging insects all tend to have black and yellow bands, like those of the wasp in Figure 33.12*a*. Once a predator has experienced the defense of the wasp, it will remember the coloration and avoid all animals that look similar, such as the nonstinging flower fly and longhorn beetle in Figure 33.12*b, c*, which can be considered Batesian mimics.

Another type of mimicry occurs when species that resemble each other all have successful defenses. For example, many coral snake species have brilliant red, black, and yellow body rings and are venomous. Mimics that share the same protective defense are called Müllerian mimics, after Fritz Müller, who discovered the phenomenon. The bumblebee in Figure 33.12*d* is a Müllerian mimic of the wasp in Figure 33.12*a*—that is, both insects have the same appearance and mode of defense.

Just as with other antipredator defenses, behavior plays a role in mimicry. Mimicry works better if the mimic acts like the model (the species it is mimicking). For example, beetles

a.

b.

c.

d.

Figure 33.12 Mimicry among insects.

a. A yellow jacket wasp uses stinging as its defense. **b.** A flower fly and **(c)** a longhorn beetle are Batesian mimics because they are incapable of stinging another animal, and yet they have the same appearance as the yellow jacket wasp. **d.** A bumblebee and the yellow jacket wasp are Müllerian mimics because they have a similar appearance, and both use stinging as a defense.

that phenotypically resemble a wasp actively fly from place to place and spend most of their time in the same habitat as the wasp model. Their behavior makes them resemble a wasp to an even greater degree.

> **Prey escape predation by utilizing concealment, fright, flocking together, warning coloration, and mimicry.**

Symbiosis

Symbiosis refers to close interactions between members of two species. Three types of symbiotic relationships have traditionally been defined—parasitism, commensalism, and mutualism. Table 33.2 lists these three categories in terms of their benefits to one species or the other. Of the three types, only mutualism may increase the population size of both species. However, some ecologists now consider it wasted effort to try to classify symbiotic relationships into these categories, since the amount of harm or good the individuals of two species do to one another is dependent on what the investigator chooses to measure. Although the following discussion describes the traditional classification system, bear in mind that symbiotic relationships do not always fall neatly into these three categories.

Parasitism

Parasitism is a symbiotic relationship in which the *parasite* derives nourishment from another organism, called the *host*. Therefore, the parasite benefits and the host is harmed. Parasites occur in all kingdoms of life. Bacteria (e.g., strep infection), protists (e.g., malaria), fungi (e.g., rusts and smuts), plants (e.g., mistletoe), and animals (e.g., leeches) all include parasitic species. The effects of parasites on the health of the host can range from slightly weakening them to actually killing them over time.

In addition to providing nourishment, some host organisms also provide their parasites with a place to live and reproduce, as well as a mechanism for dispersing offspring to new hosts. Many parasites have both a primary and secondary host. The secondary host may be a vector that transmits the parasite to the next primary host. As an example, consider the deer ticks *Ixodes dammini* and *I. ricinus* in the eastern and western United States, respectively. Deer ticks are arthropods that go through a number of stages (egg, larva, nymph, adult). They are so named because adults feed and mate on white-tailed deer in the fall. The female lays her eggs on the ground, and when the eggs hatch in the spring, they become larvae that feed primarily on white-footed mice. If a mouse is infected with the bacterium *Borrelia burgdorferi*, the larvae become infected also. The fed larvae overwinter and molt the next spring to become nymphs that can, by chance, take a blood meal from a human. At this time, the tick may pass the bacterium on to a human, who subsequently comes down with Lyme disease, characterized by arthritic-like symptoms and often a "bulls-eye" rash around the site of the tick bite. The fed nymphs develop into adults, and the cycle begins again.

Commensalism

Commensalism is a symbiotic relationship between two species in which one species is benefited and the other is neither benefited nor harmed. Often one species provides a home and/or transportation for the other species, as when barnacles attach themselves to the backs of whales.

Clownfishes live within the waving mass of tentacles of sea anemones. Because most fishes avoid the poisonous tentacles of the anemones, clownfishes are protected from predators. If clownfishes attract other fishes on which the anemone can feed, this relationship borders on mutualism. Other examples of commensalism may also be mutualistic. For example, birds called cattle egrets benefit from feeding near cattle because the cattle flush insects and other animals from the vegetation as they graze (Fig. 33.13).

Figure 33.13 Egret symbiosis.
Cattle egrets eat insects off and around various animals, such as this African cape buffalo.

TABLE 33.2	Symbiotic Relationships	
	Species 1	**Species 2**
Parasitism*	Benefited	Harmed
Commensalism	Benefited	No effect
Mutualism	Benefited	Benefited

*Can be considered a type of predation.

Mutualism

Mutualism is a symbiotic relationship in which both members of the association benefit, though not necessarily equally. An example is the relationship between plants and their animal pollinators, which began when herbivores, such as insects, feasted on pollen. The provision of nectar may have spared the pollen and at the same time allowed the animal to become an instrument of pollination.

Ants form mutualistic relationships with both plants and insects. In tropical America, the bullhorn acacia tree is adapted to provide a home for ants of the species *Pseudomyrmex ferruginea.* Unlike other acacias, this species has swollen thorns with a hollow interior, where ant larvae can grow and develop. In addition to housing the ants, the acacias provide them with food. The ants constantly protect the plant from the caterpillars of moths and butterflies by swarming and stinging them, and from other plants that might shade the plant. When the ants on experimental trees were poisoned, the trees died.

Cleaning symbiosis is a relationship in which the individual being cleaned is often a vertebrate. Crustaceans, fish, and birds act as cleaners and are associated with a variety of vertebrate clients. Large fish in coral reefs line up at cleaning stations and wait their turn to be cleaned by small fish that even enter the mouths of the large fish (Fig. 33.14). It's been suggested that cleaners may be exploiting the relationship by feeding on host tissues as well as on ectoparasites. On the other hand, cleaning could ultimately lead to net gains in client fitness.

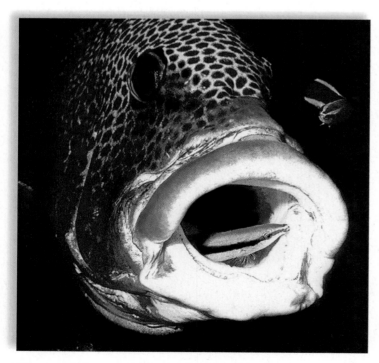

Figure 33.14 Cleaning symbiosis.
A cleaner wrasse, *Labroides dimidiatus,* in the mouth of a spotted sweetlip, *Plectorhincus chaetodontoides,* is feeding off parasites. Does this association improve the health of the sweetlip, or is the sweetlip being exploited? Investigation is under way.

> Symbiotic relationships occur between species, but it may be too simplistic to divide them into parasitic, commensalistic, and mutualistic relationships.

Ecological Succession

Like populations, communities, which are composed of all of the interacting populations in an area, can also be regulated by biotic interactions and ecological disturbances. **Ecological succession** is a change in a community's composition that is directional and follows a continuous pattern of extinction and colonization by new species. On land, *primary succession* is the establishment of a plant community (usually first by early colonizers such as lichens) in a newly formed area where there is no soil formation. Primary succession is typically instigated by a major disturbance such as a volcanic eruption or a glacial retreat. *Secondary succession* is the return of a community to its natural vegetation following a disturbance, as when a cultivated cornfield returns to its natural state.

The first species to begin the process of secondary succession are plants that are colonizers of disturbed habitats, called *pioneer species.* Then succession progresses through a series of stages, as shown in Figure 33.15. Note that, in this case, the process began with grasses followed by shrubs and then to a mixture of shrubs and trees, until finally there were only trees.

Succession can also occur in aquatic communities, as when lakes and ponds sometimes undergo stages whereby they become filled in and eventually disappear.

Models of Succession

Ecologists have developed various hypotheses to explain succession and predict future events. The *climax-pattern model* of succession says that particular areas will always lead to the same type of community in a particular area, called a **climax community,** that usually exists for a long time relative to its predecessors. This model is based on the observation that climate, in particular, determines whether a desert, a grassland, or a particular type of forest results. Therefore, for example, coniferous forests occur in northern latitudes, deciduous forests in temperate zones, and tropical rain forests in the tropics. The climax-pattern model of succession now recognizes that the exact composition of a community need not always be the same. That is, while we might expect to see a coniferous forest as opposed to a tropical rain forest in northern latitudes, the exact mix of plants and animals after each stage of succession can vary.

Does each stage of succession facilitate (help) or inhibit the next stage? To support a *facilitation model,* it can be observed that shrubs can't grow on sand dunes until dune grass has caused enough soil to develop. Similarly, in the example shown in Figure 33.15, shrubs can't arrive until grasses have made the

Bioethical Focus

Population experts have discovered that when women in the less-developed countries have rights equal to those of men they tend to have fewer children. Also, family planning leads to healthier women, and healthier women have healthier children, and the cycle continues. Under these circumstances, women do not have to have many babies for a few to survive. More education is also helpful because better-educated people are more interested in postponing childbearing and in promoting the status of women.

"There isn't any place where women who have the choice haven't chosen to have fewer children," says Beverly Winikoff at the Population Council in New York City. Bangladesh is a case in point. Bangladesh is one of the densest and poorest countries in the world. In 1990, the birthrate was 4.9 children per woman, and now it is 3.3. This achievement was due in part to the Dhaka-based Grameen Bank, which lends small amounts of money, mostly to destitute women, to start a business. The bank discovered that when women start making decisions about their lives, they also start making decisions about the size of their families. Family planning within Grameen families is twice as common as the national average; in fact, those women who get a loan promise to keep their families small! Also helpful has been the network of village clinics that counsel women who want to use contraceptives. The expression "contraceptives are the best contraceptives" refers to the fact that it is not necessary to wait for social changes to get people to use contraceptives—the two can occur simultaneously and reinforce each other.

Form Your Own Opinion

1. Do you think less-developed countries should simply make contraceptives available, or should more persuasive methods be employed? Explain.
2. Do you think that more-developed countries should be concerned about population growth in the less-developed countries? Why or why not?
3. Are you in favor of providing foreign financial aid to help countries develop family planning programs? Why or why not?

Figure 33.15 **Secondary succession in a forest.**
Secondary succession is a process that begins in areas that already have soil (not on bare rock as primary succession does). This example of secondary succession occurred in a large conifer plantation in central New York state. Notice that certain species are common to the particular stages. However, the process of regrowth shows approximately the same series of stages as secondary succession would in a former cornfield. (Arrows indicate the passage of times)

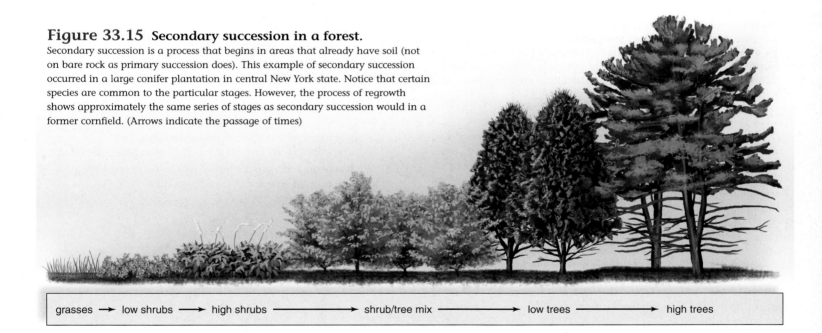

grasses → low shrubs → high shrubs → shrub/tree mix → low trees → high trees

soil suitable for them. Thus, it's possible that each successive community prepares the way for the next, and this is why grass-shrub-forest development occurs in a sequential way.

On the other hand, the *inhibition model* says that colonists hold on to their space and inhibit the growth of other plants until the colonists die or are damaged. Still another possible model, the *tolerance model*, predicts that different types of plants can colonize an area at the same time. Sheer chance determines which seeds arrive first, and successional stages may simply reflect the length of time it takes individuals to mature. This alone could account for the grass-shrub-forest development seen in Figure 33.15. In reality, the models we have mentioned are not mutually exclusive, and succession is recognized as a complex process.

Ecological succession, which occurs after a disturbance, often results in a progression of biotic communities through time. However, the process is complex, and the resulting community composition cannot always be predicted.

Summarizing the Concepts

33.1 The Scope of Ecology
Ecology is the study of the interactions of organisms with other organisms and with the physical environment. Ecology encompasses several levels of study: organism, population, community, ecosystem, and biosphere.

33.2 Patterns of Population Growth
The per capita rate of population increase is dependent on biotic potential and environmental resistance. The per capita rate of increase is calculated by subtracting the number of deaths from the number of births and dividing by the number of individuals in the population. Every population has a biotic potential, which results in the greatest possible per capita rate of increase under ideal circumstances.

Two patterns of population growth are possible. Exponential growth results in a J-shaped curve because, as the population increases in size, so does the number of new members. Exponential growth cannot continue indefinitely because of environmental resistance. Then logistic growth occurs, and an S-shaped growth curve results. When the population reaches carrying capacity, the population stops growing because environmental resistance opposes biotic potential.

Populations tend to have one of three types of survivorship curves, depending on whether most individuals live out the normal life span, die at a constant rate regardless of age, or die early.

The human population is expanding exponentially. Most of the expected increase will occur in certain LDCs (less-developed countries) of Africa, Asia, and Latin America. Support for family planning, human development, and delayed childbearing could help lessen the expected increase.

33.3 Regulation of Population Growth
The patterns of population growth have been used to suggest that life history patterns vary, from species that are opportunists to those that are in equilibrium with the carrying capacity of the environment. Opportunistic species produce many young within a short period of time and rely on rapid dispersal to new, unoccupied environments. Their population size is regulated by density-independent factors. Equilibrium species produce a limited number of young, which they nurture for a long time, and their population size is regulated by density-dependent factors, such as competition and predation.

According to the competitive exclusion principle, no two species can occupy the same ecological niche at the same time when resources are limiting. When such resources as food and living space are divided between two or more species, resource partitioning has occurred.

Predator-prey interactions between two species are influenced by environmental factors. Sometimes predation can cause prey populations to decline and remain at relatively low densities, or a cycling of population densities may occur. Prey defenses take many forms: concealment, use of fright, and warning coloration are three possible mechanisms. Batesian mimicry usually occurs when one species has the warning coloration but lacks the defense of another species. Müllerian mimicry occurs when two species with the same warning coloration have the same defense.

In a parasitic relationship, the parasite benefits and the host is harmed. In a commensalistic relationship, neither party is harmed. And in a mutualistic relationship, both partners benefit. Parasites often utilize more than one host. In a commensalistic relationship, one species often provides a home and/or transportation for another species. Examples of mutualistic relationships include flowers and their pollinators, ants and acacia trees, and cleaning symbiosis.

A change in community composition over time is called ecological succession. A climax community is associated with a particular geographic area.

Testing Yourself
Choose the best answer for each question.

1. Place the following levels of organization in order, from lowest to highest.
   a. population, organism, community, ecosystem
   b. community, ecosystem, population, organism
   c. organism, community, population, ecosystem
   d. population, ecosystem, organism, community
   e. organism, population, community, ecosystem

2. Assume that a deer population contains 500 animals. The birthrate is 105 animals per year, and the death rate is 100 animals per year. The per capita rate of increase per year is
   a. 5%.   d. 2%.
   b. 4%.   e. 1%.
   c. 3%.

3. Which of the following statements about a plant species is not relevant for determining its biotic potential?
   a. It produces 10 kg of mass per year.
   b. It produces its first flowers at five years of age.
   c. It reproduces annually.
   d. 50% of seedlings grow into mature plants.
   e. On average, 100 seedlings are produced by each plant every year.

4. If a farmer sprays a pesticide onto a field and kills half of the insect pests, he has caused a reduction in
   a. field capacity.   c. population size.
   b. carrying capacity.   d. More than one of these are correct.

5. What type of survivorship curve would you expect for a plant species in which most seedlings die?
   a. type I   c. type III
   b. type II

6. Label this S-shaped growth curve:

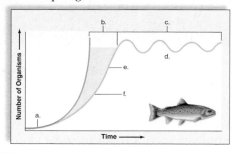

7. The U.S. human population is
   a. growing.   c. steady.
   b. declining.

8. The current doubling time for the world's population is 53 years. Therefore, if the population is 6 billion people, what will be the size of the population in 159 years if the doubling rate does not change?
   a. 18 billion   c. 36 billion
   b. 48 billion   d. 24 billion

9. The human population in less-developed countries is expected to explode in the near future chiefly because

    a. annual growth rates are very high.
    b. death rates are near zero.
    c. families are getting larger.
    d. the population is in an exponential growth stage.

10. If an age structure diagram indicates that there are more people in the prereproductive stage of life than in the reproductive stage, then replacement reproduction, over the long run, will result in

    a. zero population growth.
    b. an increase in population size.
    c. a decrease in population size.

11. Which of the following is not a feature of a organism with an opportunistic life history pattern?

    a. small size
    b. regulated by density-independent factors
    c. small number of offspring
    d. mature early
    e. offspring small in size

12. An ecological niche includes an organism's

    a. interactions with other organisms.
    b. habitat.
    c. resources.
    d. Both a and b are correct.
    e. All of these are correct.

13. During the cycling of predator-prey populations, the population peaks of the

    a. predator lag behind those of the prey.
    b. prey lag behind those of the predator.
    c. prey are identical to those of the predator.

14. A colorful, nonpoisonous frog that mimics a poison-dart frog is exhibiting

    a. Batesian mimicry.
    b. Müllerian mimicry.

For problems 15–19, indicate the type of symbiosis illustrated by each example. Some answers may be used more than once.

Key:

    a. parasitism
    b. commensalism
    c. mutualism

15. Unicellular algae live in the tissues of coral animals. The algae provide food for the coral, while the coral provides a stable home for the algae.

16. Cowbirds live near cattle, feeding on insects that are disturbed during grazing.

17. Flowering plants reward visiting insects with nectar. The insects, in turn, carry pollen to other flowers.

18. Small wasps lay eggs on other insects. The eggs hatch into larvae that feed on the insects and kill them.

19. Rust fungi grow on wheat plants, destroying tissues as they use them for a food source.

20. Which of these levels of ecological study involves an interaction between abiotic and biotic components?

    a. organisms
    b. populations
    c. communities
    d. ecosystem
    e. All of these are correct.

21. A J-shaped growth curve should be associated with

    a. exponential growth.
    b. biotic potential.
    c. no environmental resistance.
    d. high per capita rate of increase.
    e. All of these are correct.

22. An S-shaped growth curve

    a. occurs when there is no environmental resistance.
    b. includes an exponential growth phase.
    c. occurs if survivorship is short-lived.
    d. occurs in natural populations but not in laboratory ones.
    e. All of these are correct.

23. If a population has a type I survivorship curve (most of its members live the entire life span), which of these would you also expect?

    a. a single reproductive event per adult
    b. most individuals reproduce
    c. sporadic reproductive events
    d. reproduction occurring near the end of the life span
    e. None of these are correct.

24. A pyramid-shaped age distribution means that

    a. the prereproductive group is the largest group.
    b. the population will grow for some time in the future.
    c. there are more young women entering the reproductive years than older women leaving them.
    d. the country is more likely an LDC than an MDC.
    e. All of these are correct.

25. Which of these is a density-independent regulating factor?

    a. competition
    b. predation
    c. size of population
    d. weather
    e. resource availability

26. An equilibrium life history pattern does not include

    a. large individuals.
    b. long life span.
    c. individuals slow to mature.
    d. few offspring.
    e. little or no care of offspring.

27. The human population

    a. is undergoing exponential growth.
    b. is not subject to environmental resistance.
    c. fluctuates from year to year.
    d. grows only if emigration occurs.
    e. All of these are correct.

28. Six species of monkeys are found in a tropical forest. Most likely, they

    a. occupy the same ecological niche.
    b. eat different foods and occupy different ranges.
    c. spend much time fighting each other.
    d. are from different stages of succession.
    e. All of these are correct.

29. Leaf cutter ants keep fungal gardens. The ants provide food for the fungus but also feed on the fungus. This is an example of

    a. competition.
    b. predation.
    c. commensalism.
    d. parasitism.
    e. mutualism.

30. Clownfishes live among sea anemone tentacles, where they are protected. If the clownfish provides no service to the anemone, this is an example of

    a. competition.
    b. predation.
    c. parasitism.
    d. commensalism.
    e. mutualism.

31. Two species of barnacles vie for space in the intertidal zone. The one that remains is

    a. the better competitor.
    b. better adapted to the area.
    c. the better predator on the other.
    d. the better parasite on the other.
    e. Both a and b are correct.

32. A bullhorn acacia provides a home and nutrients for ants. If the relationship is mutualistic,
    a. the plant is under the control of pheromones produced by the ants.
    b. the ants protect the plant.
    c. the plants and the ants compete with one another.
    d. the plants and the ants have coevolved to occupy different ecological niches.
    e. All of these are correct.

33. The frilled lizard of Australia suddenly opened its mouth wide and unfurled folds of skin around its neck. Most likely, this was a way to
    a. conceal itself.   d. scare its prey.
    b. warn that it was noxious to eat.   e. All of these are correct.
    c. scare a predator.

34. Mosses growing on bare rock will eventually help to create soil. These mosses are involved in _____ succession.
    a. primary   c. tertiary
    b. secondary

35. Assume that a farm field is allowed to return to its natural state. By chance, the field is first colonized by native grasses, which begin the succession process. This is an example of which model of succession?
    a. climax pattern   c. facilitation
    b. tolerance   d. inhibition

Understanding the Terms

age-structure diagram 697
biosphere 692
biotic potential 693
carrying capacity 694
climax community 704
coevolution 701
cohort 694
commensalism 703
community 692
competition 699
competitive exclusion
    principle 699
demographic transition 695
density-dependent 698
density-independent 698
doubling time 695
ecological niche 699
ecological succession 704
ecology 692
ecosystem 692

environmental
    resistance 694
exponential growth 693
habitat 699
less-developed country
    (LDC) 695
logistic growth 694
mimicry 702
more-developed country
    (MDC) 695
mutualism 704
parasitism 703
population 692
predation 700
predator 700
prey 700
replacement
    reproduction 697
resource partitioning 699
symbiosis 703
zero population growth 695

Match the terms to these definitions:
a._____ Group of organisms of the same species occupying a certain area.
b._____ Maximum population growth rate under ideal conditions.
c._____ Growth, particularly of a population, in which the increase occurs in the same manner as compound interest.
d._____ Due to industrialization, a decline in the birthrate following a reduction in the death rate so that the population growth rate is lowered.
e._____ Community that results when succession has come to an end.

Thinking Critically

1. You are a farmer fighting off an insect crop pest. You know from evolutionary biology and natural selection that using the same pesticide causes resistance to evolve in the insect population. Knowing what you know about population growth and regulation, what are some strategies you might use to control the insect population?

2. Human population growth is perhaps our biggest environmental concern. The more people there are, the more of the world's resources we must use to feed, shelter, and provide energy for ourselves. Population growth has slowed in MDCs due to a demographic transition. However, more population growth is predicted to occur in LDCs. Describe strategies to help stimulate a demographic transition in LDCs.

3. You find two species of insects, both using bright coloration and similar color patterns as an antipredator defense. Design an experiment to tell whether this is an example of Müllerian or Batesian mimicry.

4. If predators tend to reduce the population densities of prey, what prevents predators from reducing prey populations to such low levels that they drive themselves extinct?

ARIS, the *Inquiry into Life* Website

ARIS, the website for *Inquiry into Life,* provides a wealth of information organized and integrated by chapter. You will find practice quizzes, interactive activities, labeling exercises, flashcards, and much more that will complement your learning and understanding of general biology.

www.aris.mhhe.com

An ecosystem is characterized by biotic (living) and abiotic (nonliving) interactions, as well as by energy flow and nutrient cycling. This cycling begins when algae and green plants use a small percentage of the sun's energy to transform inorganic chemicals into organic compounds. These compounds serve as nutrients and energy for plants as well as for other species in the ecosystem. Eventually, when decomposers break down organic matter, inorganic chemicals are liberated, but much of the energy is dissipated as heat.

Ecosystems are not self-contained; they interact, through inputs and outputs, with other ecosystems on Earth. Therefore, many ecologists think of the Earth as a global ecosystem, with humans an important part of it. Thus, human activities, such as large-scale forest and woodland removal, affect the global ecosystem. Approximately 120,000 km² of forests are cleared annually, equal to about 20 football fields per minute! Forest destruction affects other ecosystems by increasing soil erosion, adding silt to waterways, removing species' habitat, decreasing atmospheric oxygen, changing local weather patterns, and disrupting natural chemical cycles. Thus, destruction of forests and other ecosystems impacts the biosphere and threatens the existence of all living things, including humans.

chapter 34

Nature of Ecosystems

34.1   The Biotic Components of Ecosystems

An **ecosystem** possesses both abiotic and biotic components. The abiotic components include resources, such as sunlight and inorganic nutrients, and conditions, such as type of soil, water availability, prevailing temperature, and amount of wind. The biotic components of an ecosystem are the various populations of organisms that form a community.

Populations of an Ecosystem

The populations of an ecosystem are categorized according to their food source. **Autotrophs** (i.e., "self-feeders") require only inorganic nutrients and an outside energy source to produce organic nutrients for their own use and for all the other members of a community. Because they produce food, they are called **producers** (Fig. 34.1a). Photosynthetic organisms such as algae and plants produce most of the organic nutrients for the biosphere.

Some bacteria are chemoautotrophs. They reduce carbon dioxide to an organic compound by using energetic electrons derived from inorganic compounds, such as hydrogen sulfides, hydrogen gas, or ammonium ions. Chemoautotrophs have been found to support communities at hydrothermal vents along deep-sea oceanic ridges.

Heterotrophs (i.e., "other feeders") need a preformed source of organic nutrients. Because they consume food, they are called **consumers. Herbivores** are animals that graze directly on plants or algae (Fig. 34.1b). **Carnivores** feed on other animals; birds that feed on insects are carnivores, and so are hawks that feed on birds (Fig. 34.1c). In this example, herbivorous insects are primary consumers, birds are secondary consumers, and hawks are tertiary consumers. Sometimes tertiary consumers are called top predators. **Omnivores** are animals that feed on both plants and animals. As you likely know, most humans are omnivores.

Decomposers are heterotrophic bacteria and fungi, such as molds and mushrooms, that break down nonliving organic matter, including animal wastes (Fig. 34.1d). They perform a very valuable service because they release inorganic nutrients that are taken up by plants once more. **Detritus** is partially decomposed matter in the water or soil. Detritivores are found in all ecosystems. Earthworms and some beetles, termites, and maggots are soil detritivores.

Energy Flow and Chemical Cycling

Every ecosystem is characterized by two fundamental phenomena: energy flow and chemical cycling. Energy flow begins when producers absorb solar energy, and chemical cycling begins when producers take in inorganic nutrients from the physical environment (Fig. 34.2). Thereafter, producers make organic nutrients (food) directly for themselves and indirectly for the other populations of the ecosystem. Most

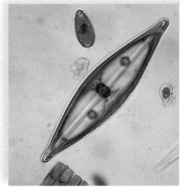

a. Producers

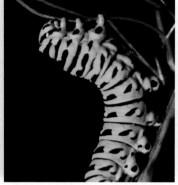

b. Herbivores

c. Carnivores

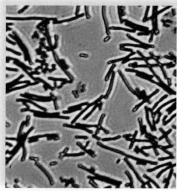

d. Decomposers

Figure 34.1 Biotic components.
a. Diatoms, a type of algae, and green plants are producers. **b.** Caterpillars and rabbits are herbivores. **c.** Spiders and ospreys are carnivores. **d.** Bacteria and mushrooms are decomposers.

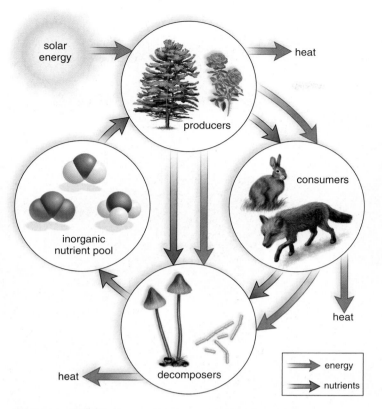

Figure 34.2 Energy flow and nutrient cycling.
Nutrients cycle, but energy flows through an ecosystem. As energy transformations repeatedly occur, all the energy derived from the sun eventually dissipates as heat.

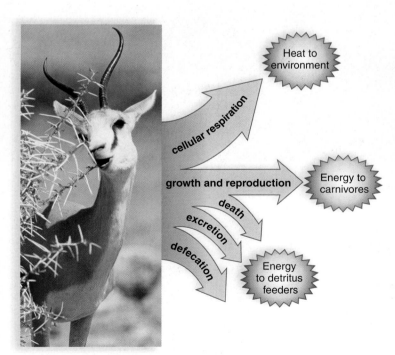

Figure 34.3 Energy balances.
Only about 10% of the food energy taken in by a herbivore is passed on to carnivores and much of the rest is lost as heat to the environment. A large portion goes to detritivores via defecation, excretion, and death, and another large portion is used for cellular respiration.

ecosystems cannot exist without a continual supply of solar energy. Chemicals cycle when inorganic nutrients are returned to the producers from the atmosphere or soil, as appropriate.

Only a portion of the organic nutrients made by autotrophs is passed on to heterotrophs because plants use organic molecules to fuel their own cellular respiration. Similarly, only a small percentage of nutrients taken in by heterotrophs is available to higher-level consumers. Figure 34.3 shows why. A certain amount of the food eaten by a herbivore is never digested and is eliminated as feces, which decomposers recycle. Metabolic wastes are excreted as urine. Of the assimilated energy, a large portion is utilized during cellular respiration to provide energy to the organism and thereafter becomes heat. Only the remaining food, which is converted into increased body weight (or additional off-spring), becomes available to carnivores.

In Chapter 6, you learned that energy flow is described by the laws of thermodynamics: (1) Energy cannot be created or destroyed, and (2) In every energy transformation, some energy is lost as heat. These principles explain why ecosystems are dependent on a continual outside source of energy, such as sunlight, and why only a part of the original energy from producers is available to consumers.

Energy flows through the populations of an ecosystem, while chemicals cycle within and among ecosystems.

34.2 Energy Flow

The principles we have been discussing can now be applied to an actual ecosystem—a forest, for example. The interconnecting paths of energy flow are represented by diagramming a **food web.** Figure 34.4*a* is a **grazing food web** because it begins with a producer, specifically the oak trees. Caterpillars feed on leaves in the trees, and mice, rabbits, and deer feed on leaves at or near the ground. Birds, chipmunks, and mice eat fruits and nuts, but are omnivores because they also feed on caterpillars. These herbivores and omnivores all provide food for a number of different carnivores.

Figure 34.4*b* is a **detrital food web,** which begins with detritus. Detritus is food for soil organisms such as earthworms and beetles. These animals are, in turn, fed on by salamanders and shrews. Because the members of the detrital food web may become food for aboveground carnivores, the detrital and grazing food webs are joined.

In this particular forest, the organic matter lying on the forest floor and mixed into the soil contains over twice as much energy as the leaves of living trees. Therefore, more energy in a forest may be funneling through the detrital food web than through the grazing food web.

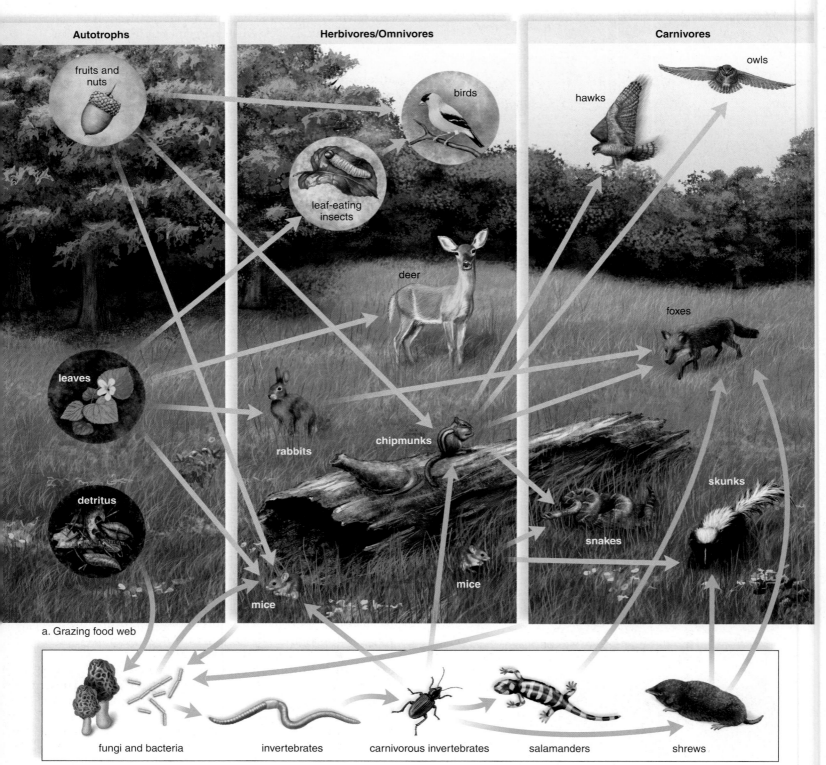

Autotrophs

fruits and nuts

leaves

detritus

Herbivores/Omnivores

birds

leaf-eating insects

deer

rabbits

chipmunks

mice

mice

Carnivores

owls

hawks

foxes

skunks

snakes

a. Grazing food web

fungi and bacteria   invertebrates   carnivorous invertebrates   salamanders   shrews

b. Detrital food web

Figure 34.4 Grazing and detrital food webs.

Food webs are descriptions of who eats whom. **a.** Tan arrows illustrate possible grazing food webs. For example, birds, which feed on nuts, may be eaten by a hawk. Autotrophs such as the tree are producers (first trophic level); the first series of animals are primary consumers (second trophic level); and the next group of animals are secondary consumers (third trophic level). **b.** Green arrows illustrate possible detrital food webs, which begin with detritus, the waste products and remains of dead organisms. A large portion of these remains are from the grazing food web illustrated in (**a**). The organisms in the detrital food web are sometimes consumed by animals in the grazing food web, as when robins feed on earthworms. Thus, the grazing food web and the detrital food web are connected.

Trophic Levels

You can see that Figure 34.4*a* would allow us to link organisms one to another in a straight line, according to who eats whom. Diagrams that show a single path of energy flow such as this are called **food chains.** For example, in the grazing food web, we could find this **grazing food chain:**

leaves ⟶ caterpillars ⟶ tree birds ⟶ hawks

And in the detrital food web (Fig. 34.4*b*), we could find this **detrital food chain:**

detritus ⟶ earthworms ⟶ shrews

A **trophic level** is composed of all the organisms that feed at a particular link in a food chain. In the grazing food web in Figure 34.4*a*, going from left to right, the trees are primary producers (first trophic level). The first series of animals, the herbivores, are primary consumers (second trophic level). The group of animals at the far right, the carnivores, are secondary consumers (third trophic level).

Ecological Pyramids

The shortness of food chains can be attributed to the loss of energy between trophic levels. In general, only about 10% of the energy of one trophic level is available to the next

Figure 34.5 Ecological pyramid.
An ecological pyramid reflects the loss of energy from one trophic level to the next. Energy is lost as top carnivores feed on carnivores, as carnivores feed on herbivores, and as herbivores feed on producers. This loss of energy results in decreasing biomass and numbers of organisms at higher trophic levels relative to lower trophic levels (levels not to scale).

trophic level. Therefore, if a herbivore population consumes 1,000 kg of plant material, only about 100 kg is converted to herbivore tissue, 10 kg to first-level carnivores, and 1 kg to second-level carnivores. The so-called 10% rule of thumb explains why few carnivores can be supported in a food web. The large energy losses that occur between successive trophic levels are sometimes depicted as an **ecological pyramid** (Fig. 34.5).

Biomass is the number of organisms multiplied by their weight. Thinking in terms of biomass rather than simply the number of organisms can be useful, since in terms of only numbers, one oak tree may have hundreds of caterpillars; therefore, it might appear that there are more herbivores than autotrophs. But we would certainly expect the biomass of the producers to be greater than the biomass of the herbivores and that of the herbivores to be greater than that of the carnivores.

In aquatic ecosystems, such as lakes and open seas where algae are the main producers, the herbivores may have a greater biomass than the producers when you take their measurements. The reason is that, over time, the algae reproduce rapidly, but they are also consumed at a high rate. Any pyramids like this one, which have more herbivores than producers, are called inverted pyramids:

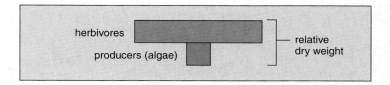

Ecologists are hesitant about using pyramids to describe ecological relationships for all situations. Detrital food chains, which are rarely included in pyramids, may represent a large portion of energy in many ecosystems, so conclusions based on aboveground pyramids may be inaccurate.

The flow of energy through the populations explains in large part the organization of an ecosystem depicted in food webs, food chains, and ecological pyramids.

34.3 Global Biogeochemical Cycles

Since the pathways by which chemicals circulate through ecosystems involve both biotic and geological components, they are known as **biogeochemical cycles.** For each element, chemical cycling may involve (1) a reservoir—a source normally unavailable to producers, such as fossilized remains, rocks, and deep-sea sediments; (2) an exchange pool—a source from which organisms generally take chemicals, such as the atmosphere or soil; and (3) the biotic community—through which chemicals move along food chains, perhaps never entering a pool (Fig. 34.6).

With the exception of water, which exists in gas, liquid, and solid forms, there are two types of biogeochemical cycles. In a gaseous cycle, exemplified by the carbon and nitrogen cycles, the element is withdrawn from and returns to the atmosphere as a gas. In a sedimentary cycle, exemplified by the phosphorus cycle, the element is absorbed from soil by plant roots, passed to heterotrophs, and eventually returned to the soil by decomposers, usually nearby.

The diagrams on the next few pages show how nutrients flow between terrestrial and aquatic ecosystems. In the nitrogen and phosphorus cycles, these nutrients run off from a terrestrial to an aquatic ecosystem and in that way enrich the aquatic ecosystem. However, too much of these nutrients can be harmful—for example, by causing plant overgrowth. Decaying organic matter in aquatic ecosystems can be a source of nutrients for intertidal inhabitants such as fiddler crabs. Seabirds feed on fish but deposit guano (droppings) on land, and in that way phosphorus from the water is deposited on land. Anything put into the environment in one ecosystem can find its way to another ecosystem. As proof, scientists find the soot from urban areas and pesticides from agricultural fields in the snow and animals of the Arctic.

The Water Cycle

The **water cycle** (or hydrologic cycle) is described in Figure 34.7. During the water cycle, fresh water is distilled from salt water. First, evaporation occurs. During **evaporation,** a liquid (in this case, water) changes from a liquid state to a gaseous state. The sun's rays cause fresh water to evaporate from seawater, and the salts are left behind. Next, condensation occurs. During **condensation,** a gas is changed to a liquid. For example, vaporized fresh water rises into the atmosphere, is stored in clouds, cools, and falls as rain over the oceans and the land.

Water also evaporates from land, from plants (transpiration), and from bodies of fresh water. Because land lies above sea level, gravity eventually returns all fresh water to the sea. In the meantime, water is contained within standing waters (lakes and ponds), flowing water (streams and rivers), and groundwater.

Some of the water from **precipitation** (e.g., rain, snow, sleet, hail, and fog) sinks, or percolates, into the ground and saturates the earth to a certain level. The top of the saturation zone is called the groundwater table, or simply, the water table. Because water infiltrates through the soil and rock layers, sometimes groundwater is also located in **aquifers,** rock layers that contain water and release it in appreciable quantities to wells or springs. Aquifers are recharged when rainfall and melted snow percolate into the soil.

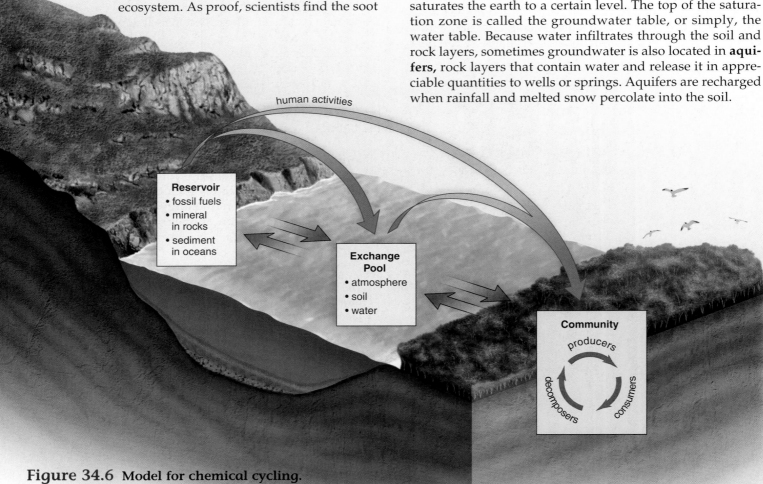

Figure 34.6 Model for chemical cycling.
Chemical nutrients cycle between these components of ecosystems. Reservoir which include fossi fuels minerals in rocks, and sediments in oceans, are normally relatively unavailable sources. However, exchange pools, such as those in the atmosphere, soil, and water are available sources of chemicals for the biotic community. When human activities (purple avows) remove chemicals from reservoirs and pools and make them available to the biotic community, pollution can result.

Human Activities

In some parts of the United States, especially the arid West and southern Florida, withdrawals from aquifers exceed any possibility of recharge. This is called "groundwater mining." In these locations, the groundwater level is dropping, and residents may run out of groundwater, at least for irrigation purposes, within a few years.

Fresh water, which makes up only about 3% of the world's supply of water, is called a renewable resource because a new supply is always being produced as a result of the water cycle. But it is possible to run out of fresh water when the available supply is not adequate or has become polluted so that it is not usable. Notice that the price of bottled water has gone up steadily over the past several years. This topic is discussed further in Chapter 36.

> In the water cycle, fresh water evaporates from bodies of water. Precipitation on land enters surface waters, the ground, or aquifers. Water ultimately returns to the ocean.

The Phosphorus Cycle

In the phosphorus cycle, phosphorus moves from rocks on land to the oceans, where it gets trapped in sediments; then phosphorus moves back onto land following a geological upheaval. You can verify this by following the appropriate arrows in Figure 34.8. However, on land, the very slow weathering of rocks makes phosphate ions (PO_4^{3-} and HPO_4^{2-}) available to plants, which take up phosphate from the soil.

Producers use phosphate to form a variety of molecules, including phospholipids, ATP, and the nucleotides that become a part of DNA and RNA. Animals incorporate some of the phosphate into teeth, bones, and shells that take many years to decompose. Eventually, however, phosphate ions become available to producers once again. Because the available amount of phosphate is already being utilized within food chains, phosphate is usually a limiting inorganic nutrient for plants—that is, their growth is limited by the amount of available phosphorus.

Some phosphate runs off into aquatic ecosystems, where algae acquire phosphate before it becomes trapped in sediments. Phosphate in marine sediments does not become available to producers on land again until a geological upheaval exposes sedimentary rocks to weathering once more. Phosphorus does not enter the atmosphere therefore, the phosphorus cycle is called a sedimentary cycle.

Human Activities

Human beings boost the amount of available phosphate by mining phosphate ores and using them to make fertilizers, animal feed supplements, and detergents. Fertilizers usually contain three basic ingredients: nitrogen, phosphorus, and potassium. Most laundry detergents contain approximately

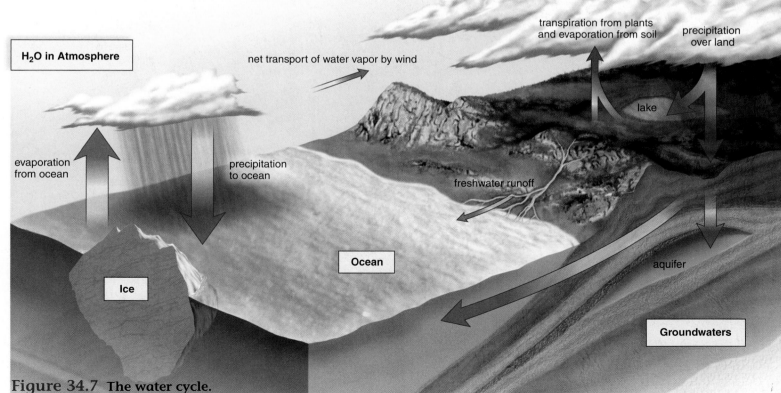

Figure 34.7 The water cycle.
Evaporation from the ocean exceeds precipitation so there is a net movement of water vapor onto land where precipitation results in the eventual flow of surface water and groundwater back to the sea. On land, transpiration by plants contributes to evaporation.

35% to 75% sodium triphosphate. The amount of phosphate in animal feed varies.

Animal wastes from livestock feedlots, fertilizers from lawns and cropland, and untreated and treated sewage discharged from cities all add excess phosphate to nearby waters. The end result is **cultural eutrophication** (over-enrichment), which can lead to an algal bloom, as indicated by green scum floating on the water or excessive mats of filamentous algae. When the algae die off, decomposers use up all available oxygen for cellular respiration. The result can be a massive fish kill.

In the phosphorus cycle, weathering makes phosphate available to producers from the reservoir (rocks) at a very slow rate. Therefore, phosphate is a limiting nutrient in communities.

The Nitrogen Cycle

Nitrogen gas (N_2) makes up about 78% of the atmosphere, but plants cannot use nitrogen gas. Therefore, nitrogen is also a limiting inorganic nutrient for plants.

Nitrogen Fixation

In the nitrogen cycle, **nitrogen fixation** occurs when nitrogen gas (N_2) is converted to ammonium ions (NH_4^+), a form plants can use (Fig. 34.9). Some cyanobacteria in aquatic ecosystems and some free-living bacteria in soil are able to fix atmospheric nitrogen in this way. Other nitrogen-fixing bacteria live in nodules on the roots of legumes (see Chapter 9). They make organic compounds containing nitrogen available to the host plants so that the plant can form proteins and nucleic acids.

Nitrification

Plants can also use nitrates as a source of nitrogen. The production of nitrates during the nitrogen cycle is called **nitrification**. Nitrification can occur in two ways: (1) Nitrogen gas is converted to nitrate (NO_3^-) in the atmosphere when cosmic radiation, meteor trails, and lightning provide the high energy needed for nitrogen to react with oxygen.

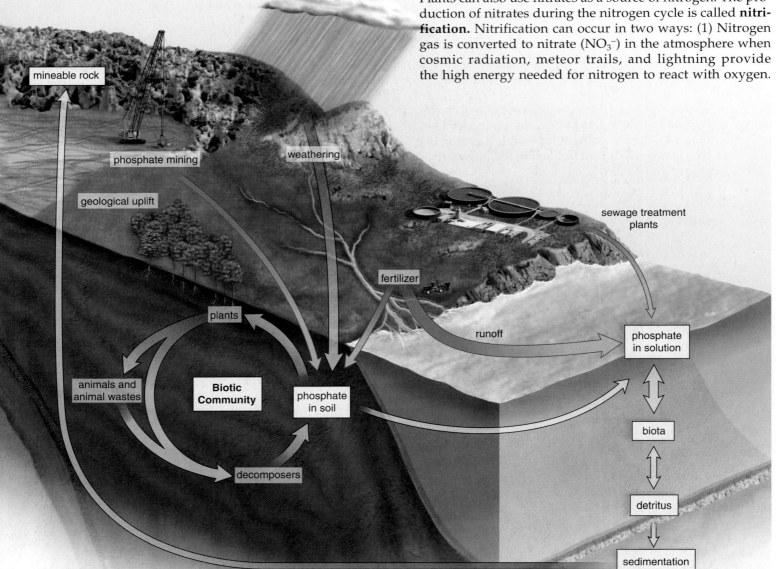

Figure 34.8 The phosphorus cycle.
Purple arrows represent human activities; gray arrows represent natural events.

(2) Ammonium ions in the soil from decomposition are converted to nitrate by soil bacteria in a two-step process. First, nitrite-producing bacteria convert ammonium to nitrite (NO_2^-), and then nitrate-producing bacteria convert nitrite to nitrate. These two groups of bacteria, called the nitrifying bacteria, are chemoautotrophs.

Denitrification

Denitrification is the conversion of nitrate back to nitrogen gas, which enters the atmosphere. Denitrifying bacteria living in the anaerobic mud of lakes, bogs, and estuaries carry out this process as a part of their own metabolism. In the nitrogen cycle, denitrification would counterbalance nitrogen fixation except for human activities.

Human Activities

Human activities significantly alter the transfer rates in the nitrogen cycle by producing fertilizers from N_2—in fact, they nearly double the nitrogen fixation rate. Fertilizer, which also contains phosphate, runs off into lakes and rivers and results in algal overgrowth and fish kill.

Fertilizer use also results in the release of nitrous oxide (N_2O), a greenhouse gas, component of acid rain, and contributor to ozone shield depletion (see Ecology Focus).

In the nitrogen cycle, nitrogen-fixing bacteria (in root nodules and in the soil) convert nitrogen gas to ammonium ions, which plants can use; nitrifying bacteria convert ammonium to nitrate; denitrifying bacteria convert nitrate back to nitrogen gas.

Figure 34.9 The nitrogen cycle.
Purple arrows represent human activities; gray arrows represent natural events.

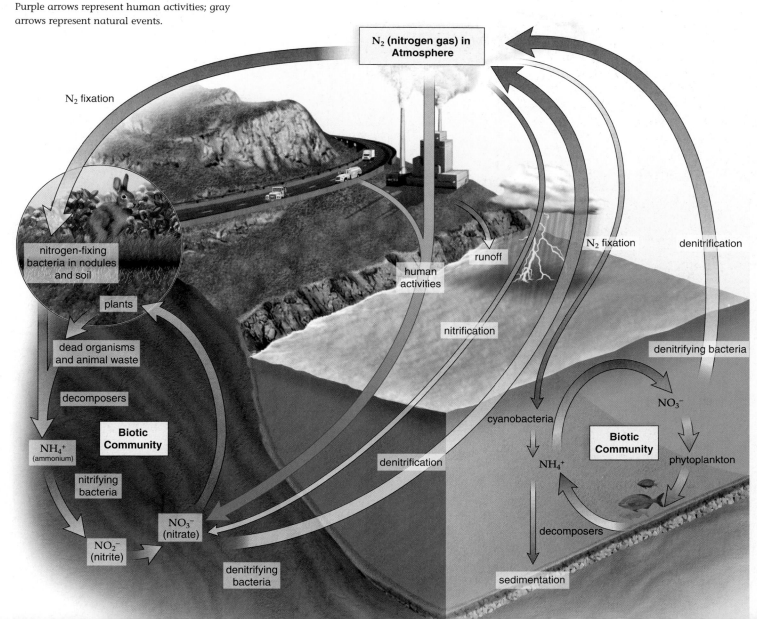

The Carbon Cycle

In the carbon cycle, plants in both terrestrial and aquatic ecosystems take up carbon dioxide (CO_2) from the air, and through photosynthesis, they incorporate carbon into food that is used by autotrophs and heterotrophs alike (Fig. 34.10). When all organisms, including plants, respire, a portion of this carbon is returned to the atmosphere as carbon dioxide.

In aquatic ecosystems, the exchange of carbon dioxide with the atmosphere is primarily indirect. However, there is a small quantity of free carbon dioxide in the water. Carbon dioxide from the air combines with water to produce bicarbonate ions (HCO_3^-), a source of carbon for algae that produce food for themselves and for heterotrophs. Similarly, when aquatic organisms respire, the carbon dioxide they give off becomes bicarbonate ion. The amount of bicarbonate in the water is in equilibrium with the amount of carbon dioxide in the air.

Reservoirs Hold Carbon

Living and dead organisms contain organic carbon and serve as one of the reservoirs for the carbon cycle. The world's biotic components, particularly trees, contain 800 billion tons of organic carbon, and an additional 1,000–3,000 billion metric tons are estimated to be held in the remains of plants and animals in the soil. Before decomposition can occur, some of these remains are subjected to physical processes that transform them into coal, oil, and natural gas. We call these materials the **fossil fuels.** Most of the fossil fuels were formed during the Carboniferous period, 286–360 million years ago, when an exceptionally large amount of organic matter was buried before decomposing. Another reservoir is the inorganic carbonate that accumulates in limestone and in calcium carbonate shells of many marine organisms.

Human Activities

The transfer rates of carbon dioxide due to photosynthesis and cellular respiration, which includes the work of decomposers, are just about even. However, due to human activities, more carbon dioxide is being released into the atmosphere than is being removed. In 1850, atmospheric

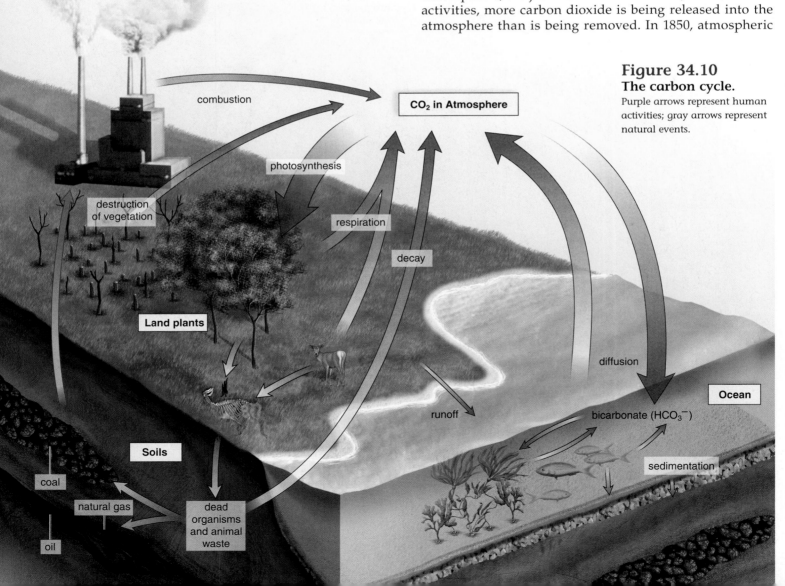

Figure 34.10
The carbon cycle.
Purple arrows represent human activities; gray arrows represent natural events.

EcologyFocus

Ozone Shield Depletion

Ozone is a gas that forms in Earth's atmosphere when ultraviolet radiation from the sun splits oxygen molecules (O_2), and then the single oxygen atoms (O) combine with other oxygen molecules to form ozone (O_3). When ozone is present in the troposphere, the layer of the atmosphere closest to the ground, it is considered a pollutant because it adversely affects plant growth and our ability to breathe oxygen. But in the stratosphere, some 50 kilometers above Earth, ozone forms the **ozone shield,** a layer that absorbs solar ultraviolet radiation.

The absorption of UV radiation by the ozone shield is critical for living things. In humans, UV radiation causes mutations that can lead to skin cancer and can make the lens of the eye develop cataracts. In addition, it adversely affects the immune system and our ability to resist infectious diseases. UV radiation also impairs crop and tree growth and kills off algae (phytoplankton) and tiny shrimplike animals (krill) that sustain oceanic life. Without an adequate ozone shield, therefore, our health and food sources are threatened.

It became apparent in the 1980s that ozone levels had become depleted worldwide and that the depletion was most severe above the Antarctic every spring. This so-called **ozone hole** is equivalent in area to two and a half times the size of Europe (Fig. 34A). It exposed not only Antarctica but also the southern tip of South America and vast areas of the Pacific and Atlantic oceans to harmful UV rays. Of even greater concern, an ozone hole has now appeared above the Arctic as well, and ozone holes have also been detected within northern and southern latitudes where many people live. A United Nations Environmental Program report predicts a 26% rise in cataracts and nonmelanoma skin cancers for every 10% drop in the ozone level. A 26% increase translates into 1.75 million additional cases of cataracts and 300,000 more skin cancer cases every year, worldwide.

Scientists found that the cause of ozone depletion is chlorine atoms (Cl), which can destroy up to 100,000 molecules of ozone before settling to Earth's surface as chloride years later. These chlorine atoms come primarily from the breakdown of **chlorofluorocarbons (CFCs)**. The best-known CFC is Freon, a coolant found in refrigerators and air conditioners. CFCs are also used as a foaming agent during the production of Styrofoam coffee cups, egg cartons, insulation, and padding. Formerly, CFCs were used as propellants in spray cans, but this application is now banned in the United States and several European countries.

Most countries in the world have stopped using CFCs, and the United States halted production in 1995. Since that time, satellite measurements indicate that the amount of harmful chlorine pollution in the stratosphere has started to decline. It is clear, however, that recovery of the ozone shield may take many more years and involve other pollution-fighting approaches, aside from lowering chlorine pollution. For example, as greenhouse gases increase global warming (see Bioethical Focus), the stratosphere becomes colder because less heat is reflected from the Earth's surface. Ozone depletion increases when the stratosphere is very cold.

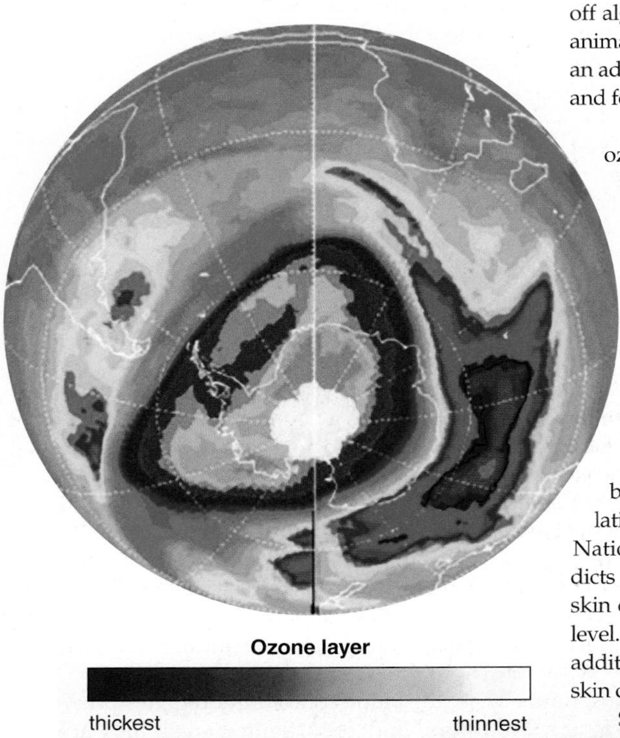

Discussion Questions

1. Should we care about ozone depletion, especially when the most serious ozone depletion is over the Antarctic, where almost no humans live?
2. Ozone depletion and global warming are connected in that warming of the Earth's surface causes less heat to radiate in the atmosphere. Should we decrease our burning of fossil fuels, which reduces the amount of greenhouse gases released into the atmosphere—even though this action might decrease profits for U.S. corporations such as oil companies and automobile manufacturers?
3. Some chemicals, such as industrial solvents, still release harmful chlorine into the atmosphere. As the owner of a factory, would you support using alternative "environmentally friendly" solvents even if they would cost more? Why or why not?

Ozone layer

thickest .... thinnest

Figure 34A
Ozone shield depletion.

Map of ozone levels in the atmosphere of the Southern Hemisphere, September 2000. The ozone depletion there, often called an ozone hole, is much larger than the size of Europe.

CO_2 was at about 280 parts per million; today, it is over 350 ppm. This increase is largely due to the burning of fossil fuels and the destruction of forests to make way for farmland and pasture. Today, the amount of carbon dioxide released into the atmosphere is about twice the amount that remains in the atmosphere. It's believed that most of this dissolves in the ocean.

The increased amount of carbon dioxide (and other gases) in the atmosphere is causing a rise in temperature called **global warming.** These gases allow the sun's rays to pass through, but they absorb and radiate heat back to Earth, a phenomenon called the **greenhouse effect.** Global warming is expected to cause dire consequences, as discussed in Chapter 36. The Bioethical Focus discusses worldwide efforts to cutback on greenhouse gases.

> In the carbon cycle, photosynthesis removes carbon dioxide from the atmosphere, but respiration returns it. Forests and dead organisms are carbon reservoirs, as is the ocean.

Global warming is predicted to upset normal weather cycles, which will most likely lead to outbreaks of such diseases as malaria, dengue and yellow fevers, filariasis, encephalitis, schistosomiasis, and cholera. That is, increased temperatures will lead to the spread of disease-carrying insects, such as mosquitoes that carry malaria, from tropical to temperate areas. Clearly, use of greenhouse gases should be curtailed.

In December 1997, 159 countries met in Kyoto, Japan, to work out a protocol that would reduce greenhouse gas emissions worldwide. Greenhouse gases are those that are produced in large part through the burning of fossil fuels. Carbon dioxide and methane are two of the most common ones. Such gases allow the sun's rays to pass through but then trap the heat from escaping. It is believed that the emission of greenhouse gases, especially from power plants, will cause Earth's temperature to rise 1.5–4.5° on average by 2060. Today, over 140 countries have ratified the Kyoto protocol, with the United States a notable exception. The U.S. Senate did not want to ratify the agreement because it did not include a binding emissions commitment from the less-developed countries. While the United States presently emits a large proportion of the world's greenhouse gases, China is expected to pass the United States in about 2020 to become the biggest source of greenhouse emissions.

Negotiations with the less-developed countries are still going on, and some creative ideas have been put forward. Why not have a trading program that allows companies to buy and sell emission credits across international boundaries? Greenhouse reduction techniques would also be for sale. Presumably, if it became monetarily worth their while, companies in less-developed countries would have an incentive to use such techniques to reduce greenhouse emissions.

Form Your Own Opinion

1. Should all countries be expected to reduce their greenhouse emissions? Why or why not?
2. Do you approve of giving companies monetary incentives to reduce greenhouse emissions? Why or why not?
3. Should the United States agree to the Kyoto protocol? Why or why not?

Summarizing the Concepts

34.1 The Biotic Components of Ecosystems

An ecosystem is composed of populations of organisms plus the physical environment. Some species are producers and some are consumers. Producers are autotrophs that produce their own organic food. Consumers are heterotrophs that take in organic food. Consumers may be herbivores, carnivores, omnivores, or decomposers.

Ecosystems are characterized by energy flow and chemical cycling. Energy is lost from the biosphere, but inorganic nutrients are not. They recycle within and among ecosystems. Decomposers return some proportion of inorganic nutrients to autotrophs, and other portions are imported or exported among ecosystems in global cycles.

Producers transform solar energy into food for themselves and all consumers. As herbivores feed on plants (or algae) and carnivores feed on herbivores, energy is converted to heat. Feces, urine, and dead bodies become food for decomposers. Eventually, all the solar energy that enters an ecosystem is converted to heat, and thus ecosystems require a continual supply of solar energy.

34.2 Energy Flow

Ecosystems contain food webs in which the various organisms are connected by eating relationships. In a grazing food web, food chains begin with a producer. In a detrital food web, food chains begin with detritus. The two food webs are joined when the same consumer is a link in both a grazing and a detrital food chain. A trophic level is composed of all the organisms that feed at a particular link in a food chain. Ecological pyramids show trophic levels stacked one on top of the other like building blocks. Generally, they illustrate that energy content, and therefore numbers of organisms and biomass, decreases from one trophic level to the next.

34.3 Global Biogeochemical Cycles

Biogeochemical cycles consist of reservoirs, exchange pools, and biotic communities. A reservoir pool is a component of an ecosystem, such as fossil fuels, sediments, and rocks, that contains elements available on a limited basis to living things. An exchange pool is a component of an ecosystem, such as the atmosphere, soil, and water, that is a ready source of nutrients for living things.

In the water cycle, evaporation over the ocean is not compensated for by rainfall. Evaporation from terrestrial ecosystems includes transpiration from plants. Rainfall over land results in bodies of fresh water plus groundwater, including aquifers. Eventually, all water returns to the oceans.

In the phosphorus cycle, the biotic community recycles phosphorus back to the producers, and only limited quantities are made available by the weathering of rocks. Phosphates are mined for fertilizer production; when phosphates and nitrates enter lakes and ponds, overenrichment occurs. Many kinds of wastes enter the rivers and then flow to the oceans, which become degraded from added pollutants.

In the nitrogen cycle, certain bacteria in water, soil, and nitrogen-fixing bacteria on root nodules can fix atmospheric nitrogen. Other bacteria return nitrogen to the atmosphere. Human activities convert atmospheric nitrogen to fertilizer, which is broken down by soil bacteria; humans also burn fossil fuels. In this way, nitrogen oxides can contribute to the formation of smog and acid rain, both of which are detrimental to animal and plant life.

In the carbon cycle, organisms add as much carbon dioxide to the atmosphere as they remove. Shells in ocean sediments, organic compounds in living and dead organisms, and fossil fuels are reservoirs for carbon. Human activities, such as burning fossil fuels and trees, are adding carbon dioxide to the atmosphere. Like the panes of a greenhouse, carbon dioxide and other gases allow the sun's rays to pass through but impede the release of infrared wavelengths. A buildup of these "greenhouse gases" is leading to global warming; a rise in sea level and a change in climate patterns could follow.

Testing Yourself

Choose the best answer for each question.

1. Which of the following would be a primary consumer in a vegetable garden?

   a. aphid sucking sap from cucumber leaves
   b. lady beetle eating aphids
   c. songbird eating lady beetles
   d. fox eating songbirds
   e. All of these are correct.

2. Detritus always contains

   a. bacteria and fungi.
   b. leaf litter.
   c. decaying logs.
   d. animal carcasses.

3. Producers are

   a. autotrophs.
   b. herbivores.
   c. omnivores.
   d. carnivores.
   e. Both a and b are correct.

4. Label the populations in this diagram of an ecosystem.

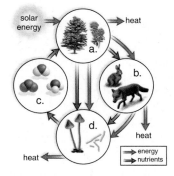

5. During chemical cycling, inorganic nutrients are typically returned to the soil by

   a. autotrophs.
   b. detritivores
   c. decomposers.
   d. tertiary consumers.

6. When a heterotroph takes in food, only a small percentage of the energy in that food is used for growth. The remainder is

   a. not digested and eliminated as feces.
   b. excreted as urine.
   c. given off as heat.
   d. All of these are correct.
   e. None of these are correct.

7. All ecosystems need a continual outside source of energy. This requirement is expected based on the

   a. first law of thermodynamics.
   b. second law of thermodynamics.
   c. first and second laws of thermodynamics.
   d. None of these are correct.

8. Because of the _____ law of thermodynamics, a fraction of the energy absorbed by plants is available to ecosystems.

   a. first
   b. second
   c. third

9. The first trophic level on a food web is composed of the

   a. producers.
   b. primary consumers.
   c. secondary consumers.
   d. tertiary consumers.

10. Which of the following is a grazing food chain?

    a. leaves → detritivores → deer → owls
    b. birds → mice → snakes
    c. nuts → leaf-eating insects → chipmunks → hawks
    d. leaves → leaf-eating insects → mice → snakes

11. In this diagram, label the trophic levels (blanks a–d), using two of these terms for each level: producers, top carnivores, secondary consumers, autotrophs, primary consumers, tertia consumers, carnivores, herbivores.

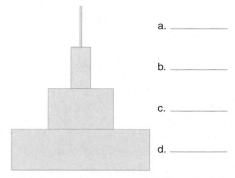

a. _____

b. _____

c. _____

d. _____

12. In a grazing food web, carnivores that eat herbivores are

    a. producers.
    b. primary consumers.
    c. secondary consumers.
    d. tertiary consumers.

13. Few carnivores can be supported in a food web because

    a. mineral nutrients do not cycle quickly enough.
    b. they produce only a few young per generation.
    c. their generation time is very long.
    d. a large amount of energy is lost at each trophic level.

14. According to the 10% rule of thumb for trophic level energy exchange, a rabbit that eats 5 kg of grass converts how much of that mass to rabbit tissue?

    a. 10 kg   b. 5 kg   c. 50 kg   d. 0.5 kg

15. Terrestrial ecological pyramids are typically based on

    a. numbers of organisms.
    b. mass of organisms.
    c. mass of organisms multiplied by numbers of organisms.
    d. mass of organisms divided by numbers of organisms.

16. Which of the following is a sedimentary biogeochemical cycle?

    a. carbon
    b. nitrogen
    c. phosphorus
    d. water

17. Underground oil is an example of a carbon

    a. cycle.
    b. pathway.
    c. reservoir.
    d. exchange pool.

18. Which of the following is not a component of the water cycle?

    a. evaporation
    b. filling of aquifers
    c. runoff
    d. photosynthesis
    e. condensation

19. Groundwater mining is

    a. removing phosphorus from groundwater.
    b. extracting water from a well.
    c. removing water from an aquifer faster than it can recharge.
    d. removing nitrogen from the groundwater.

20. The most abundant gas in the atmosphere is

    a. oxygen.
    b. nitrogen.
    c. carbon dioxide.
    d. hydrogen.

For questions 21–23, match each type of nitrogen conversion with a process in the key.

Key:

a. nitrogen fixation   b. nitrification   c. denitrification

21. Nitrate to nitrogen gas

22. Nitrogen gas to nitrate

23. Nitrogen gas to ammonium

24. Choose the statement that is true concerning this food chain:
grass → rabbits → snakes → hawks
    a. Each predator population has a greater biomass than its prey population.
    b. Each prey population has a greater biomass than its predator population.
    c. Each population is omnivorous.
    d. Each population returns inorganic nutrients and energy to the producer.
    e. Both a and c are correct.

25. In what ways are decomposers like producers?
    a. Either may be the first member of a grazing or a detrital food chain.
    b. Both produce oxygen for other forms of life.
    c. Both require a source of nutrient molecules and energy.
    d. Both are present only in the lithosphere.
    e. Both produce organic nutrients for other members of ecosystems.

26. How do nitrogen-fixing bacteria contribute to the nitrogen cycle?
    a. They return nitrogen (N_2) to the atmosphere.
    b. They change ammonium to nitrate.
    c. They change N_2 to ammonium.
    d. They withdraw nitrate from the soil.
    e. They decompose and return nitrogen to autotrophs.

27. How do plants contribute to the carbon cycle?
    a. When plants respire, they release CO_2 into the atmosphere.
    b. When plants photosynthesize, they consume CO_2 from the atmosphere.
    c. When plants photosynthesize, they provide oxygen to heterotrophs.
    d. When plants emigrate, they transport carbon molecules between ecosystems.
    e. Both a and b are correct.

Understanding the Terms

aquifer 714
autotroph 710
biogeochemical cycle 713
biomass 713
carnivore 710
chlorofluorocarbons (CFCs) 719
condensation 714
consumer 710
cultural eutrophication 716
decomposer 710
denitrification 717
detrital food chain 713
detrital food web 711
detritus 710
ecological pyramid 713
ecosystem 710
evaporation 714
food chain 713
food web 711
fossil fuel 718
global warming 719
grazing food chain 713
grazing food web 711
greenhouse effect 719
herbivore 710
heterotroph 710
nitrification 716
nitrogen fixation 716
omnivore 710
ozone hole 719
ozone shield 719
precipitation 714
producer 710
trophic level 713
water (hydrologic) cycle 714

Match the terms to these definitions:

a._____ Decomposers and partially decomposed remains of plants and animals found in soil and on the beds of bodies of water.
b._____ Formed from oxygen in the upper atmosphere, it protects Earth from ultraviolet radiation.
c._____ Remains of once-living organisms that are burned to release energy, such as coal, oil, and natural gas.
d._____ Process by which atmospheric nitrogen gas is changed to forms plants can use.
e._____ Complex pattern of interlocking and crisscrossing food chains.

Thinking Critically

1. A large forest has been removed by timber harvest (clear-cutting) and the land has not been replanted. After several years, humidity and rainfall in the area seem to have decreased. How can your knowledge of the water cycle be used to explain these observations?

2. Explain how human use of excess nitrogenous fertilizers and laundry detergents containing phosphorus can disrupt aquatic ecosystems.

3. In mountainous regions of the western United States, large predators (e.g., wolves in Yellowstone National Park), which were previously driven out of the area, are being reintroduced. People living in these areas are concerned that livestock may be lost if there is not enough wild food for the predators. What type of data are needed concerning food webs and ecological pyramids to ensure the successful reintroduction of these predators and minimal impact on livestock?

4. It is the year 2100 and world population has surpassed 13 billion people. Advances in technology have made it possible to create an artificial atmosphere on the moon that traps appropriate levels of oxygen and allows ample sunlight for life to persist. You are a member of a team of ecologists assigned the task of making the moon inhabitable. You realize that much more than a suitable atmosphere is necessary for survival. Construct an ecosystem that will support human and other life as we move people to the moon; justify your choices of organisms used in this ecosystem.

5. Most food webs in any particular ecosystem have four or five trophic levels at the most. Why?

ARIS, the *Inquiry into Life* Website

ARIS, the website for *Inquiry into Life,* provides a wealth of information organized and integrated by chapter. You will find practice quizzes, interactive activities, labeling exercises, flashcards, and much more that will complement your learning and understanding of general biology.

www.aris.mhhe.com

From space, the Earth looks like a pristine aqua globe, hovering in a vast backdrop of darkness.

As you move closer, Earth's churning atmosphere and immense water systems come into view. Not until the planet's surface is visible, as from an airplane, can you make out human developments, such as farms, towns, and cities, which have reduced the Earth's natural ecosystems to a mere fraction of their original size. Human activities have impacted most of the world's ecosystems so much that they no longer function as they once did. Pollution does not stay localized, and pollutants abound in the biosphere, even in undeveloped areas. As a result of the major environmental changes brought about by people, significant numbers of plants and animals have become extinct within the past few hundred years.

In this chapter, we see how climate determines the characteristics of major biomes, such as forests, deserts, and grasslands, as well as the distribution of ecosystems within them. Each biome has its own mix of species that are adapted to living under particular environmental conditions characteristic of that biome. It is critical that we learn to value the services of ecosystems and work toward conserving the major portions of the original biomes. In this way, we can help preserve species—even our own.

The Biosphere

Chapter Concepts

35.1   Climate and the Biosphere

Climate refers to the prevailing weather conditions in a particular region. Climate is dictated by temperature and rainfall, which are in turn influenced by the following factors: (1) variations in solar radiation distribution due to the tilt of the spherical Earth as it orbits about the sun; and (2) other effects, such as topography and whether a body of water is nearby.

Effect of Solar Radiation

Because the Earth is spherically shaped, the sun's rays are more direct at the equator and more spread out progressing toward the poles. Therefore, the tropical regions nearest the equator are warmer than the temperate regions farther away from the equator. In addition, the Earth does not face the sun directly. Rather, it is on a slight tilt (about 23°) away from the sun. Throughout the year, as the Earth orbits around the sun, different parts of the planet are tilted toward or away from the sun. For example, during winter, the Northern Hemisphere is tilted away from the sun (Fig. 35.1). At the same time, the Southern Hemisphere is tilted toward the sun. This is why it is summer in the Southern Hemisphere when it is winter in the Northern Hemisphere.

Because the Earth rotates on its axis daily and its surface consists of continents and oceans, the flows of warm and cold air form three large circulation patterns in each hemisphere. The direction in which the air rises and cools determines the direction of the wind (Fig. 35.2). At the equator, the equatorial doldrums (areas of low wind) are formed as the sun heats the air and evaporates water. The warm, moist air rises and loses most of its moisture as rain, with the greatest amounts of rainfall on Earth nearest the equator. The rising air flows toward the poles, but at about 30° north and south latitude,

it sinks toward Earth's surface and reheats. This forms the northeast trade winds, which blow from the northeast toward the southwest, and the southeast trade winds, which blow from the southeast toward the northwest. Trade winds are so called because early sailors depended on them to fill the sails and thus power the movement of trading ships. As this air descends and warms in these latitudes, it becomes very dry, creating zones of low rainfall. As a result, the great deserts of Africa, Australia, and the Americas occur at these latitudes.

At about 60° north and south latitude, the air rises and cools as it does near the equator, producing additional zones of high rainfall. Between 30° and 60° latitude, the strong wind patterns known as the westerlies occur in both the Northern and Southern Hemispheres. The westerlies move from west to east. The west coasts of the continents at these latitudes are wet, as in the Pacific Northwest of the United States, where a massive temperate (evergreen) rain forest is located. Weaker winds, called the polar easterlies, blow from east to west latitudes higher than 60° in both hemispheres.

It is the spinning of the Earth about its axis that affects wind patterns. In the Northern Hemisphere, large-scale winds generally move clockwise, while in the Southern Hemisphere, they move counterclockwise. This explains, for example, why the westerlies move from southwest toward the northeast in the Northern Hemisphere and from northwest toward the southeast in the Southern Hemisphere (Fig. 35.2).

The distribution of solar energy caused by the spherical Earth and the rotation of Earth around the sun affects how the winds blow and the amount of rainfall various regions receive.

Figure 35.1
Distribution of solar energy.
a. Since Earth is a sphere, beams of solar energy striking the planet near one of the poles are spread over a wider area than similar beams striking Earth at the equator.
b. The seasons of the Northern and Southern Hemispheres are due to the tilt of Earth on its axis as it rotates around the sun.

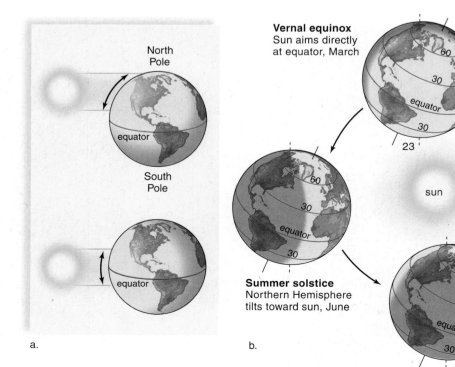

North Pole

equator

South Pole

equator

equator

a.

Vernal equinox
Sun aims directly at equator, March

60
30
equator
30
23

sun

Winter solstice
Northern Hemisphere tilts away from sun, December

60
30
equator
30

60
30
equator
30

Summer solstice
Northern Hemisphere tilts toward sun, June

60
30
equator
30

Autumnal equinox
Sun aims directly at equator, September

b.

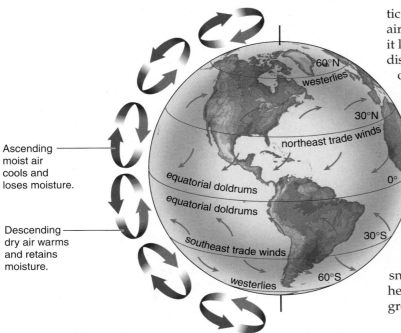

Figure 35.2 Global wind circulation.
Air ascends and descends as shown because the Earth rotates on its axis. Also, the trade winds blow from the northeast to the west in the Northern Hemisphere, and blow the southeast to the west in the Southern Hemisphere. The westerlies blow coward the east.

tic circulation of air: warm air rises over the land, and cooler air comes in off the ocean to replace it. As the warm air rises, it loses its moisture, and the monsoon season begins. As just discussed, rainfall is particularly heavy on the windward side of hills. Cherrapunji in northern India receives an annual average of 1,090 cm of rain a year because of its high altitude. The weather pattern has reversed by November, when the land has become cooler than the ocean; therefore, dry winds blow from the Asian continent across the Indian Ocean. In the winter, the air over the land is dry, the skies cloudless, and temperatures pleasant. The chief crop of India is rice, which starts to grow when the monsoon rains begin.

In the United States, people often speak of the "lake effect," meaning that in the winter, arctic winds blowing over the Great Lakes become warm and moisture-laden. When these winds rise and lose their moisture, snow begins to fall. Places such as Buffalo, New York, get heavy snowfalls due to the lake effect, and snow is on the ground there for an average of 90 to 140 days every year.

> Atmospheric circulations between the ocean and the landmasses influence regional climate conditions.

Other Effects

The term topography refers to the physical features, or "the lay," of the land. One physical feature that affects climate is the presence of mountains. As air blows up and over a mountain range, it rises and cools. One side of the mountain, called the windward side, receives more rainfall than the other side, called the leeward side. On the leeward side, the air descends, picks up moisture, and produces clear weather (Fig. 35.3). The difference between the windward side and the leeward side can be quite dramatic. In the Hawaiian Islands, for example, the windward side of the mountains typically receives more than 750 cm of rain a year while the leeward side, which is in a **rain shadow,** gets an average of only 50 cm of rain and is generally sunny. In the United States, the western side of the Sierra Nevada mountain range is lush, while the eastern side is a semidesert.

The oceans are slower to change temperature than are landmasses. This causes coasts to have a unique weather pattern that is not seen inland. During the day, the land warms more quickly than the ocean, and the air above the land rises. Then a cool sea breeze blows in from the ocean. At night, the reverse happens; the breeze blows from the land to the sea.

India and some other countries in southern Asia have a **monsoon** climate, in which wet ocean winds blow onshore for almost half the year. The land heats more rapidly than the waters of the Indian Ocean during spring. The difference in temperature between the land and the ocean causes a gigan-

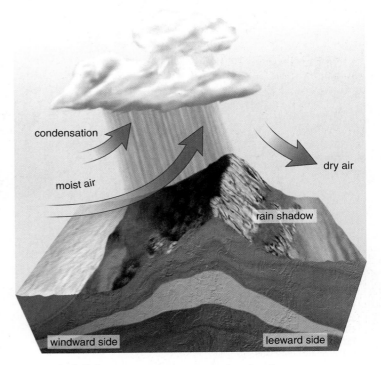

Figure 35.3 Formation of a rain shadow.
When winds from the sea cross a coastal mountain range, they rise and release their moisture as they cool the windward side of the mountain. The leeward side of a mountain is warmer and receives relatively little rain; therefore, it is said to lie in a "rain shadow."

Figure 35.4 Pattern of biome distribution.

a. Pattern of world biomes in relation to temperature and moisture. The dashed line encloses a wide range of environments in which either grasses or woody plants can dominate the area, depending on the soil type.
b. The same type of biome can occur in different regions of the world, as shown on this global map.

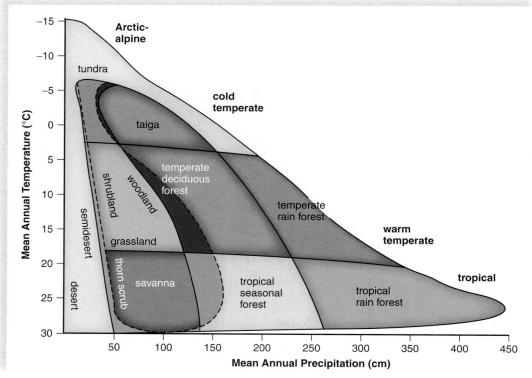

a.

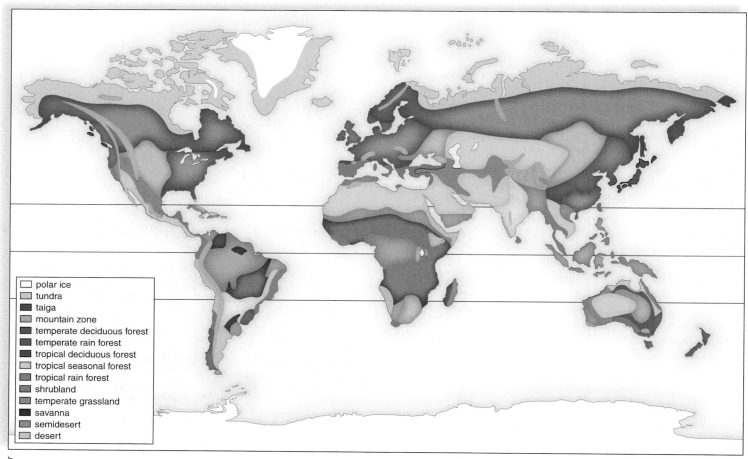

b.

35.2 Terrestrial Ecosystems

A major type of terrestrial ecosystem is called a **biome.** A biome has a particular mix of plants and animals that are adapted to living under certain environmental conditions, of which climate is an overriding influence. For example, when biomes are plotted according to their mean annual temperature and mean annual precipitation, a particular pattern results (Fig. 35.4*a*). The distribution of biomes is shown in Figure 35.4*b*. Even though Figure 35.4 shows definite demarcations, keep in mind that the biomes gradually change from one type to the other. To simplify our discussion in this chapter, we will group the biomes into the six general categories listed in Table 35.1. Also, although we will be discussing each type of biome separately, we should remember that each biome has inputs from and outputs to all the other terrestrial and aquatic ecosystems of the biosphere.

The distribution of the biomes and their corresponding populations of organisms are determined principally by differences in climate due to the distribution of solar radiation, water, and defining topographical features. Both latitude and altitude are responsible for temperature gradients. When traveling from the equator to the North Pole, it would be possible to observe first a tropical rain forest, followed by a temperate deciduous forest, a coniferous forest, and tundra, in that order, and this sequence is also seen when ascending a mountain (Fig. 35.5). The coniferous forest of a mountain is called a *montane coniferous forest,* and the tundra near

the peak of a mountain is called an *alpine tundra.* However, when going from the equator to the South Pole, it would not be possible to reach a region corresponding to the coniferous forest and tundra of the Northern Hemisphere because the distribution of the landmasses is shifted toward the north.

The distribution of biomes is determined by physical factors such as climate (principally temperature and rainfall), which varies according to latitude and altitude.

TABLE 35.1	Selected Biomes
Name	**Characteristics**
Tundra (arctic tundra, alpine tundra)	Around North Pole; average annual temperature is −12°C to −6°C, low annual precipitation (less than 25 cm); permafrost (permanent ice) year-round within a meter of surface
Taiga (coniferous forests, montane forests)	Large northern biome that circles just below the Arctic Circle; temperature is below freezing for half the year; moderate annual precipitation (30–85 cm); long nights in winter and long days in summer
Temperate deciduous forests	Eastern half of United States, Canada, Europe, and parts of Russia; four seasons of the year with hot summers and cold winters; relatively high annual precipitation (75–150 cm)
Tropical forests (tropical rain, tropical deciduous)	Located near the equator in Latin America, Southeast Asia, and West Africa; warm (20–25°C) and wet (190 cm/year); has wet/dry season
Shrublands	In the western U.S., western South America, and central Asia. Contains shrubs that are adapted to arid conditions with small, thick leaves. Low to moderate annual precipitation (20–100 cm/year).
Grasslands (savanna prairie, steppe)	Called prairies in North America, savannas in Africa, pampas in South America, steppes in Europe; hot in summer and cold in winter (United States); moderate annual precipitation; good soil for agriculture
Deserts	Northern and Southern Hemispheres at 30° latitude; hot (38°C) days and cold (7°C) nights; low annual precipitation (less than 25 cm)

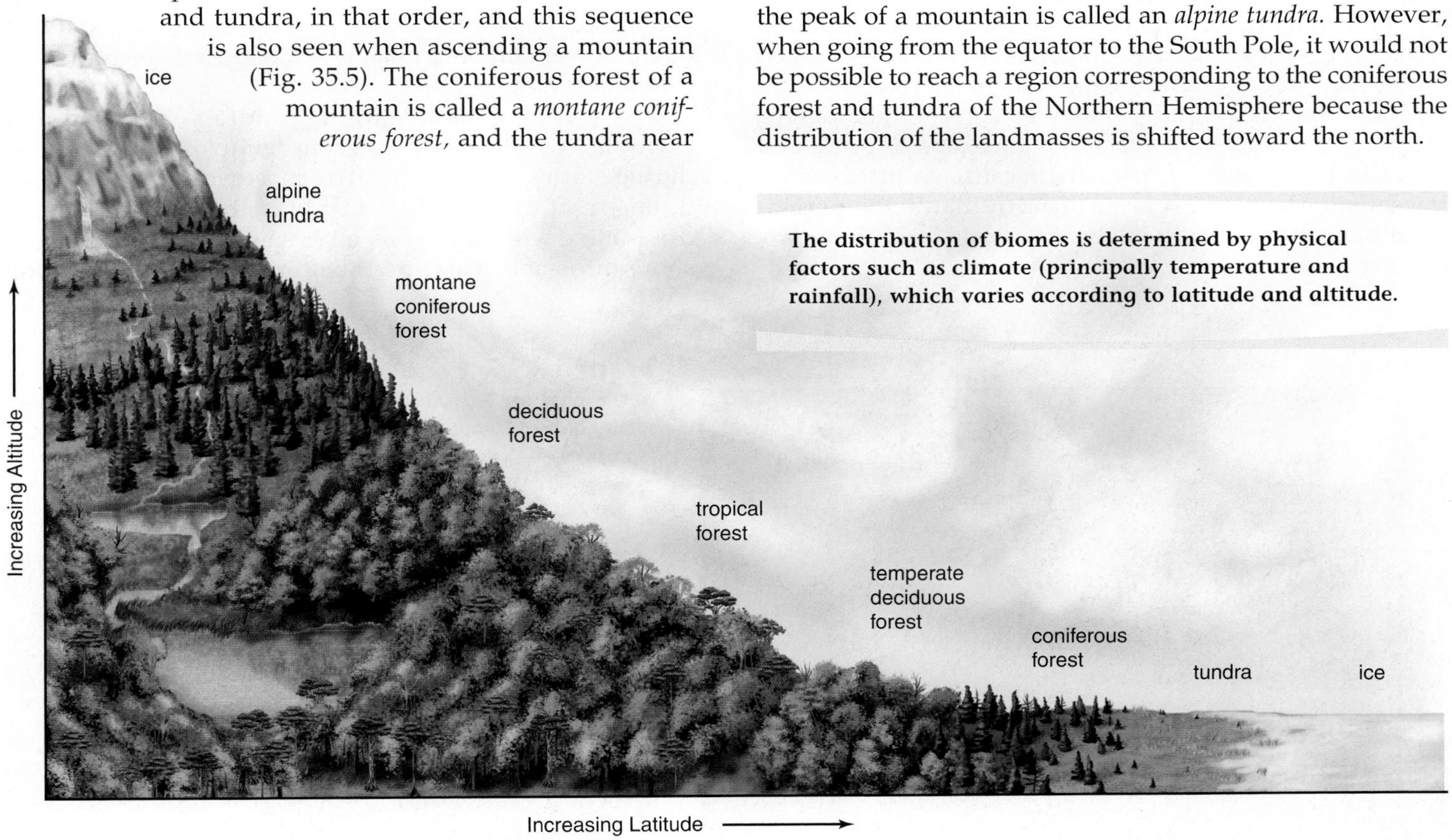

Figure 35.5 Climate and terrestrial biomes.
Biomes change with altitude just as they do with latitude because vegetation is partly determined by temperature. Precipitation also plays a significant role, which is one reason grasslands, instead of tropical or deciduous forests, are sometimes found at the base of mountains.

a. Tundra

b. Bull caribou

Figure 35.6 Tundra.
a. Vegetation consists principally of lichens, mosses, grasses, and low-growing shrubs. **b.** Caribou feed on the vegetation in the summer.

Tundra

The **tundra** biome, which encircles the arctic region just south of the ice-covered polar seas in the Northern Hemisphere, covers about 20% of Earth's land surface. As previously mentioned, a similar ecosystem, called the alpine tundra, occurs above the timberline on mountain ranges (Fig. 35.6). The *arctic tundra* is cold and dark much of the year. Because rainfall amounts to only about 20 cm a year, the tundra could possibly be considered a desert. However, melting snow creates a landscape of pools and mires in the summer, especially because so little evaporates. Only the topmost layer of earth thaws; the **permafrost** beneath this layer is always frozen, and therefore, drainage is minimal.

Trees are not found in the tundra because the growing season is too short, their roots cannot penetrate the permafrost, and they cannot become anchored in the boggy soil of summer. In the summer, the ground is covered with short grasses and sedges, as well as numerous patches of lichens and mosses (Fig. 35.6a). Dwarf woody shrubs, such as dwarf birch, flower and seed quickly while there is plentiful sun for photosynthesis.

A few animals live in the tundra year-round. For example, the ptarmigan, a grouse, burrows in the snow during storms, and the musk ox conserves heat because of its thick coat and short, squat body. In the summer, the tundra is alive with numerous insects and birds, particularly shore-birds and waterfowl that migrate inland. Caribou

a. Taiga

Figure 35.7 Coniferous forest.
a. The taiga, which means swampland, is a coniferous forest that spans northern Europe, Asia, and North America. **b.** The appellation "spruce-moose" refers to the dominant presence of spruce trees and moose, which frequent the ponds.

b. Bull moose

(Fig. 35.6*b*) and reindeer also migrate to and from the tundra, as do the wolves that prey upon them. Polar bears are common near coasts.

Coniferous Forests

Coniferous forests are found in three locations: in the **taiga**, which extends around the world in the northern part of North America and Eurasia; near mountaintops (where it is called a montane coniferous forest); and along the Pacific coast of North America, as far south as northern California.

The taiga typifies the coniferous forest with its cone-bearing trees, such as spruce, fir, and pine (Fig. 35.7). These trees are well adapted to the cold because both the leaves (reduced to needles) and bark have thick coverings. Also, the needlelike leaves can withstand the weight of heavy snow. There is a limited understory of plants, but the floor is covered by low-lying mosses and lichens beneath a layer of needles. Birds harvest the seeds of the conifers, and bears, deer, moose, beaver, and muskrat live around the cool lakes and along the streams. Wolves prey on these larger mammals. A montane coniferous forest also harbors the wolverine and the mountain lion.

The coniferous forest that runs along the west coast of Canada and the United States is sometimes called a *temperate rain forest.* The plentiful rainfall and rich soil have produced some of the tallest conifer trees ever in existence, including the coastal redwoods. This forest is also called an old-growth forest because some trees are as old as 800 years. It truly is an evergreen forest because mosses, ferns, and other plants grow on all the tree trunks. Whether the limited portion of old-growth forest that remains should be preserved from logging has been controversial.

Temperate Deciduous Forests

Temperate deciduous forests are found south of the taiga in eastern North America, eastern Asia, and much of Europe. The climate in these areas is moderate, with relatively high rainfall (75–150 cm per year). The seasons are well defined, and the growing season ranges between 140 and 300 days. The trees, such as oak, beech, and maple, have broad leaves and are called deciduous trees because they lose their leaves in fall and grow them again in spring.

The tallest trees form a canopy, an upper layer of leaves that are the first to receive sunlight (Fig. 35.8).

Figure 35.8 Temperate deciduous forest.
A temperate deciduous forest is home to many varied plants and animals. Millipedes can be found among leaf litter, chipmunks feed on acorns, and bobcats prey on these and other small mammals.

millipede

eastern chipmunk

bobcat

Even so, enough sunlight penetrates to provide energy for another layer of trees, called understory trees. Beneath these trees are shrubs and herbaceous plants that may flower in the spring before the trees have put forth their leaves, which create shade. Still another layer of plant growth—mosses, lichens, and ferns—resides beneath the shrub layer. This stratification provides a variety of habitats for insects and birds. Ground life is also plentiful. Squirrels, cottontail rabbits, shrews, skunks, woodchucks, and chipmunks are small herbivores. These and ground birds such as turkeys, pheasants, and grouse are preyed on by red foxes. White-tail deer and black bears have increased in number of late. In contrast to the taiga, a greater diversity of amphibians and reptiles occur in this biome because the winters are not as cold. Frogs and turtles generally prefer an aquatic existence, as do the beaver and muskrat, which are mammals.

Autumn fruits, nuts, and berries provide a supply of food for the winter, and the leaves, after turning brilliant colors and falling to the ground, contribute to the rich layer of humus. The minerals within the rich soil are washed far into the ground by spring rains, but the deep tree roots capture these and bring them back up into the forest system again.

Tropical Forests

The most common type of **tropical forest** is the *tropical rain forest,* which is found in areas of South America, Africa, and the Indo-Malayan region near the equator. The weather in a tropical rain forest is always warm (between 20° and 25°C), and rainfall is plentiful (a minimum of 190 cm per year). This may be the richest biome in terms of both number of species and their abundance. The diversity of species is enormous—a 10 km² area of tropical rain forest may contain 750 species of trees and 1,500 species of flowering plants.

A tropical rain forest has a complex structure, with many levels of life. Some of the broadleaf evergreen trees grow to heights of 15 to 50 meters or more. These tall trees often have trunks buttressed at ground level to prevent them from toppling over. Lianas, or woody vines, which often encircle they trees as grow, also help strengthen the trunk.

Although some animals live on the ground (e.g., pacas, agoutis, peccaries, and armadillos), most live in the trees (Fig. 35.9). Insect life is so abundant that the majority of species have not yet been identified. Termites play a vital role in the decomposition of woody plant material, and ants

Figure 35.9 Representative animals of the tropical rain forests of the world.

poison-dart frog

cone-headed katydid

ocelot

blue and gold macaw

brush-footed butterfly

lemur

arboreal lizard

a. California chaparral

b. Chemise

Figure 35.10 Shrublands.
Shrublands, such as the chaparral in California (**a**), are subject to raging fires, but the shrubs, such as chemise in (**b**), are adapted to quickly regrow.

are found everywhere, particularly in the trees. The various birds, such as hummingbirds, parakeets, parrots, and toucans, are often beautifully colored. Amphibians and reptiles are well-represented by many types of frogs, snakes, and lizards. Lemurs, sloths, and monkeys are well-known primates that feed on the fruits of the trees. The largest carnivores are the big cats—the jaguars in South America and the leopards in Africa and Asia.

Many animals spend their entire life in the canopy, as do some plants. **Epiphytes** are plants that grow on other plants but usually have roots of their own that absorb moisture and minerals leached from the canopy; others catch rain and debris in hollows produced by overlapping leaf bases. The most common epiphytes are related to pineapples, orchids, and ferns.

Whereas the soil of a temperate deciduous forest biome is rich enough for agricultural purposes, the soil of a tropical rain forest biome is not. Nutrients are cycled directly from the litter to the plants again. Productivity is high because of warm and consistent temperatures, a yearlong growing season, and the rapid recycling of nutrients from the litter. To make up for the soil's low fertility, trees are sometimes cut and burned, so that the resulting ashes can provide enough nutrients for several harvests. This practice, called Swidden agriculture or slash-and-burn agriculture, has been successful, but also destructive, in the tropics. Once the crops are harvested, the nutrients become depleted, consequently removing nutrients

from the system. Furthermore, soil erodes due to lack of tree roots, and this washes away nutrients as well.

While we usually think of tropical forests as nonseasonal rain forests, *tropical deciduous forests* that have wet and dry seasons are found in India, Southeast Asia, West Africa, South and Central America, the West Indies, and northern Australia. They have deciduous trees, with many layers of growth beneath them. In addition to the animals just mentioned, some of these forests also contain elephants, tigers, and hippopotamuses.

Shrublands

Shrublands contain shrubs that have small but thick evergreen leaves that are often coated with a waxy material that prevents loss of moisture. Shrubs are adapted to withstand arid conditions and can also quickly re-grow after a fire. The seeds of many species require the heat and scarring action of fire to induce germination. Other shrubs sprout from the roots after a fire. Shrublands tend to occur along coasts that have dry summers and receive most of their rainfall in winter. Typical shrubland species include coyotes, jackrabbits, gophers and other rodents, as well as fire-adapted plant species such as manzanita. The dense shrubland that occurs in California is known as **chaparral** (Fig. 35.10). This type of shrubland, called Mediterranean, lacks an understory and ground litter and is highly flammable.

Grasslands

Grasslands occur where rainfall is greater than 25 cm but generally insufficient to support trees. For example, in temperate areas, where rainfall is between 10 and 30 inches, it is too dry for forests and too wet for deserts to form. Natural grasslands once covered more than 40% of Earth's land surface, but most areas that were once grasslands are now used to grow crops such as wheat and corn.

Grasses are well adapted to a changing environment and can tolerate some grazing, as well as flooding, drought, and sometimes fire. Where rainfall is high, large tall grasses that reach more than 2 m in height (e.g., pampas grass) can flourish. In drier areas, shorter grasses between 5 and 10 cm are dominant (e.g., grama grass). Grasses also generally grow in different seasons; some grassland animals migrate, and ground squirrels hibernate, when there is little grass for them to eat.

The temperate grasslands include the Russian *steppes*, the South American *pampas*, and the North American *prairies* (Fig. 35.11). Large herds of bison—estimated at hundreds of thousands—once roamed the prairies, as did herds of pronghorn antelope. Now, small mammals, such as mice, prairie dogs, and rabbits, typically live below ground but usually feed above ground. Hawks, snakes, badgers, coyotes, and foxes feed on these mammals.

Savannas

Savannas, which are grasslands that contain some trees, occur in regions where a relatively cool dry season is followed by a hot, rainy one. One tree that can survive the severe dry season is the flat-topped acacia, which sheds its leaves during a drought. The African savanna supports the greatest variety and number of large herbivores of all

cheetah

giraffe

wildebeest

zebra (in foreground)

Tall-grass prairie

American bison

Figure 35.11 The prairie.
Tall-grass prairies are seas of grasses dotted by pines and junipers. Bison, once abundant, are now being reintroduced into certain areas.

Figure 35.12 The savanna.
The African savanna varies from grassland to widely spaced shrubs and trees. This biome supports a large assemblage of herbivores, including zebras, wildebeests, giraffes, and gazelles. Carnivores, such as the cheetah, prey on these.

the biomes (Fig. 35.12). Elephants and giraffes are browsers that feed on tree vegetation. Antelopes, zebras, wildebeests, water buffalo, and rhinoceroses are grazers that feed on grasses. Any plant litter that is not consumed by grazers is attacked by a variety of small organisms, among them termites. Termites build towering nests in which they tend fungal gardens, their source of food. The herbivores support a large population of carnivores. Lions and hyenas sometimes hunt in packs, cheetahs hunt singly by day, and leopards hunt singly by night.

Deserts

As discussed in section 35.1, **deserts** are usually found at latitudes of about 30°, in both the Northern and Southern Hemispheres. The winds that descend in these regions lack moisture. Therefore, the annual rainfall is less than 25 cm. Days are hot because a lack of cloud cover allows the sun's rays to penetrate easily, but nights are cold because heat escapes easily into the atmosphere.

The Sahara, which stretches all the way from the Atlantic coast of Africa to the Arabian Peninsula, and a few other deserts have little or no vegetation. But most deserts have a variety of plants (Fig. 35.13). The best-known desert perennials in North America are the succulent, spiny-leafed cacti, which have stems that store water and carry on photosynthesis. Also common are nonsucculent shrubs, such as the many-branched sagebrush with silvery gray leaves and the spiny-branched ocotillo that produces leaves during wet periods and sheds them during dry periods.

Some animals are adapted to the desert environment. For example, many desert animals are nocturnal; they are active at night when it is cooler. Reptiles and insects have waterproof outer coverings that conserve water. A desert has numerous insects, which pass through the stages of development from pupa to pupa again when there is rain. Reptiles, especially lizards and snakes, are perhaps the most characteristic group of vertebrates found in deserts, but running birds (e.g., the roadrunner) and rodents (e.g., the kangaroo rat) are also well known (Fig. 35.13). Larger mammals, such as the coyote, prey on the rodents, as do the hawks. Depending on where they are found in terms of latitude and longitude and the amount of precipitation they receive, terrestrial ecosystems differ in their species composition.

Figure 35.13 Deserts.

Plants and animals that live in a desert are adapted to arid conditions. The plants are either succulents, which retain moisture, or shrubs with woody stems and small leaves, which lose little moisture. The kangaroo rat feeds on seeds and other vegetation; the roadrunner preys on insects, lizards, and snakes. The kit fox is a desert carnivore.

bannertail kangaroo rat

greater roadrunner

kit fox

35.3 | Aquatic Ecosystems

Aquatic ecosystems are classified as two types: freshwater (inland) or saltwater (usually marine). Brackish water, however, is a mixture of fresh and salt water. In this section, we describe lakes as examples of freshwater ecosystems and oceans as examples of saltwater ecosystems. Coastal ecosystems will represent areas of brackish water.

Lakes

Lakes are bodies of fresh water often classified by their nutrient abundance. Oligotrophic (nutrient-poor) lakes are characterized by a small amount of organic matter and low productivity (Fig. 35.14*a*). Eutrophic (nutrient-rich) lakes are characterized by plentiful organic matter and high productivity (Fig. 35.14*b*). Such lakes are usually situated in naturally nutrient-rich regions or are enriched by agricultural or urban and suburban runoff. Oligotrophic lakes can become eutrophic through large inputs of nutrients, a process called **eutrophication.** Excess nutrient inputs, such as nitrogen fertilizer, can cause eutrophication and fish kills. The Bioethical Focus in this chapter discusses freshwater pollution legislation.

In the temperate zone, deep lakes are stratified in the summer and winter. In summer, lakes in the temperate zone have three layers of water that differ in temperature (Fig. 35.15). The surface layer, the epilimnion, is warm due to solar radiation; the middle thermocline layer experiences an abrupt

Figure 35.14 Types of lakes.

Lakes can be classified according to whether they are (**a**) oligotrophic (nutrient-poor) or (**b**) eutrophic (nutrient-rich). Eutrophic lakes tend to have large populations of algae and rooted plants, resulting in a large population of decomposers that use up much of the oxygen and leave little oxygen for fishes.

a.

b.

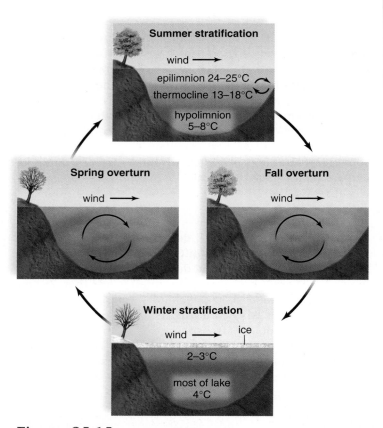

Figure 35.15 Lake stratification in a temperate region.

Temperature profiles of a large oligotrophic lake in a temperate region vary with the season. During the spring and fall overturns, the deep waters receive oxygen from surface waters, and surface waters receive inorganic nutrients from deep waters.

drop in temperature; and the lowest layer, the hypolimnion, is cold. These differences in temperature prevent mixing. The warmer, less dense water of the epilimnion "floats" on top of the colder, more dense water of the hypolimnion.

As the season progresses, the epilimnion becomes nutrient-poor, while the hypolimnion begins to be depleted of oxygen. Phytoplankton (see Chapter 28) found in the sunlit epilimnion use up nutrients as they photosynthesize. Photosynthesis releases oxygen, giving this layer a ready supply. Detritus naturally falls by gravity to the bottom of the lake, and here oxygen is used up as decomposition occurs. Decomposition releases nutrients, however.

In the fall, as the epilimnion cools, and in the spring, as it warms, an overturn occurs. In the fall, the upper epilimnion waters become cooler than the hypolimnion waters. This causes the surface water to sink and the deep water to rise. The **fall overturn** continues until the temperature is uniform throughout the lake. At this point, wind helps circulate the water so that mixing occurs.

As winter approaches, the water cools further. Ice formation begins at the top, and the ice floats because ice is less dense than cold water. Ice has an insulating effect, preventing further cooling of the water below. This permits aquatic organisms to live through the winter in the water beneath the surface of the ice.

In the spring, as the ice melts, the cooler water on top sinks below the warmer water below it. The **spring overturn** continues until the temperature is uniform through the lake. At this point, wind aids in the circulation of water as before. When the surface waters absorb solar radiation, thermal stratification occurs once more.

This vertical stratification and seasonal change of temperatures in a lake basin influence the seasonal distribution of fish and other aquatic life. For example, coldwater fish move to the deeper water in summer and inhabit the upper water in winter. In the fall and spring just after mixing occurs, phytoplankton growth at the surface is most abundant.

Life Zones

In both fresh and salt water, free-drifting microscopic **plankton,** which includes phytoplankton and also zooplankton (see Chapter 28), are important components of the ecosystem. Lakes and ponds can be divided into several life zones, as shown in Figure 35.16. Aquatic plants are rooted in the shallow littoral zone of a lake, and various microscopic organisms cling to these plants and to rocks. Some organisms, such as the water strider,

live at the water-air interface and can literally walk on water. In the limnetic zone, small fishes, such as minnows and killifish, feed on plankton and also serve as food for large fishes. In the profundal zone, zooplankton and fishes, such as whitefish, feed on debris that falls from above. Pike species are an example of "lurking predators." They wait among vegetation around the margins of lakes and surge out to catch passing prey.

A few insect larvae are in the limnetic zone, but they are far more abundant in both the littoral and profundal zones. Midge larvae and ghost worms are common members of the benthos, animals that live on the bottom in the benthic zone. In a lake, the benthos includes crayfishes, snails, clams, and various types of worms and insect larvae.

Freshwater aquatic ecosystems, such as lakes, have life zones at different depths determined by the amount of sunlight and the temperature of the water. These characteristics determine the composition of plants and animals in each zone.

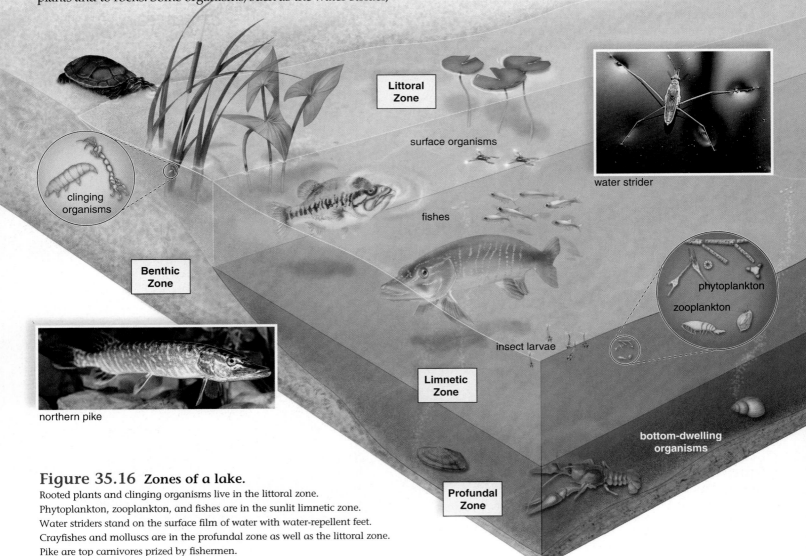

Littoral Zone

surface organisms

water strider

clinging organisms

Benthic Zone

fishes

phytoplankton

zooplankton

northern pike

insect larvae

Limnetic Zone

bottom-dwelling organisms

Profundal Zone

Figure 35.16 Zones of a lake.
Rooted plants and clinging organisms live in the littoral zone.
Phytoplankton, zooplankton, and fishes are in the sunlit limnetic zone.
Water striders stand on the surface film of water with water-repellent feet.
Crayfishes and molluscs are in the profundal zone as well as the littoral zone.
Pike are top carnivores prized by fishermen.

Coastal Ecosystems

Near the mouth of a river, a salt marsh in the temperate zone or a mangrove swamp in the subtropical and tropical zones is likely to develop. Also, the silt carried by a river may form mudflats.

Estuaries

An **estuary** is a partially enclosed body of water where fresh water and salt water meet and mix (Fig. 35.17). Coastal bays, tidal marshes, fjords (an inlet of water between high cliffs), some deltas (a triangular-shaped area of land at the mouth of a river), and lagoons (a body of water separated from the sea by a narrow strip of land) are all examples of estuaries.

Organisms living in an estuary must be able to withstand constant mixing of waters and rapid changes in salinity. Not many organisms are suited to this environment, but those that are suited find an abundance of nutrients. An estuary acts as a nutrient trap because the sea prevents the rapid escape of nutrients brought by a river. It has been estimated that well over half of all marine fishes develop in the protective environment of an estuary, which explains why estuaries are called the nurseries of the sea. Estuaries are also feeding grounds for many birds, fish, and shellfish because they offer a ready supply of food.

Salt marshes, dominated by salt marsh cordgrass, are often associated with estuaries. So are mudflats and mangrove swamps, where sediment and nutrients from the land collect (Fig. 35.18).

Seashores

Seashores, which may be rocky or sandy, are constantly bombarded by the sea as the tides roll in and out (Fig. 35.19). The **littoral zone** lies between the high- and low-tide marks. The littoral zone of a rocky beach is divided into subzones. In the upper portion of the littoral zone, barnacles are glued so tightly to stone by their own secretions that their calcareous outer plates remain in place even after the enclosed shrimplike animal dies. In the midportion of the littoral zone, brown algae known as rockweed may overlie the barnacles. In the lower portions of the littoral zone, oysters and mussels attach themselves to the rocks by filaments called *byssal threads*. Also present are snails called limpets and periwinkles. Periwinkles have a coiled shell and secure themselves by hiding in crevices or under seaweeds, while limpets press their single, flattened cone tightly to a rock. Below the littoral zone, macroscopic seaweeds, which are the main photosynthesizers, anchor themselves to the rocks by holdfasts.

Organisms cannot attach themselves to shifting, unstable sands on a sandy beach; therefore, nearly all the permanent residents dwell underground. They either burrow during the day and surface to feed at night, or they remain permanently within their burrows and tubes. Ghost crabs and sandhoppers (amphipods) burrow themselves above the high-tide mark and feed at night when the tide is out. Sandworms and sand (ghost) shrimp remain within their

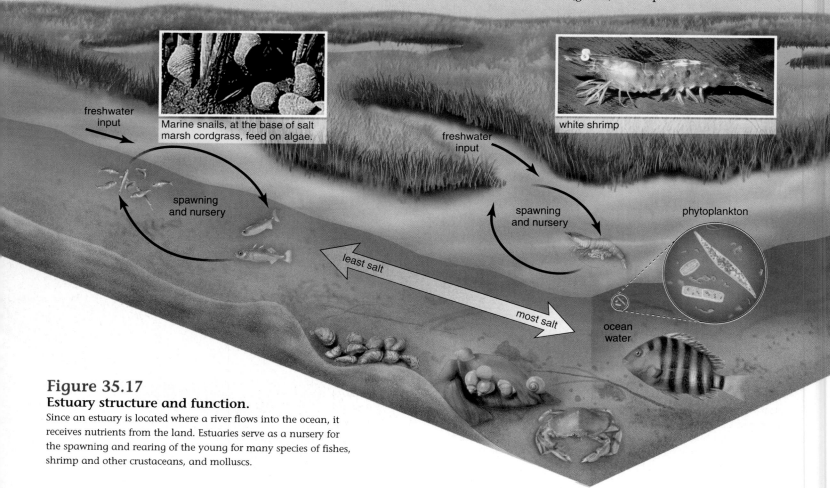

freshwater input

Marine snails, at the base of salt marsh cordgrass, feed on algae.

spawning and nursery

freshwater input

white shrimp

spawning and nursery

phytoplankton

least salt

most salt

ocean water

Figure 35.17
Estuary structure and function.
Since an estuary is located where a river flows into the ocean, it receives nutrients from the land. Estuaries serve as a nursery for the spawning and rearing of the young for many species of fishes, shrimp and other crustaceans, and molluscs.

a.

b.

Figure 35.18
Types of estuaries.
Some of the many types of regions that qualify as estuaries include (**a**) mudflats, which are frequented by migrant birds, and (**b**) mangrove swamps, which skirt the coastlines of many tropical and subtropical lands. The tangled roots of mangrove trees trap sediments and nutrients that sustain many immature forms of sea life. The salt marsh depicted in Figure 35.17 is yet another type of estuary.

burrows in the littoral zone and feed on detritus whenever possible. Still lower in the sand, clams, cockles, and sand dollars are found.

> Coastal ecosystems, such as estuaries and seashores, are strongly influenced by current patterns and can be quite diverse and productive.

Oceans

Climate is driven by the sun, but the oceans play a major role in redistributing heat in the biosphere. Water tends to be warm at the equator and much cooler at the poles because of the distribution of the sun's rays, as we have discussed before (see Fig. 35.1a). Air takes on the temperature of the water below, and warm air moves from the equator to the poles. In other words, the oceans make the winds blow. (The landmasses also play a role, but the oceans hold heat longer and remain cool longer during periods of changing temperature than do solid continents.)

When the wind blows strongly and steadily across a great expanse of ocean for a long time, friction from the mov-ing air begins to drag the water along with it. Once the water has been set in motion, its momentum, aided by the wind, keeps it moving in a steady flow we call a current. Because the ocean currents eventually strike land, they move in a circular path—clockwise in the Northern Hemisphere and counter-clockwise in the Southern Hemisphere. As the currents flow, they take warm water from the equator to the poles. One such current, called the Gulf Stream, brings tropical Caribbean water to the east coast of North America and the higher latitudes of western Europe. Without the Gulf Stream, Great Britain, which has a relatively warm temperature, would be as cold as Greenland. In the Southern Hemisphere, another major ocean current warms the eastern coast of South America.

Also, in the Southern Hemisphere, the Humboldt Current flows toward the equator. The Humboldt Current carries phosphorus-rich cold water northward along the west coast of South America. During a process called **upwelling,** cold offshore winds cause cold nutrient-rich waters to rise and take the place of warm nutrient-poor waters. In South America, the enriched waters nourish an abundance of marine life that supports the fisheries of Peru and northern Chile. When the Humboldt Current is not as cool as usual, upwelling does not occur, stagnation results, the fisheries decline, and climate patterns change globally. This phenomenon is discussed in the Ecology Focus, "El Niño–Southern Oscillation."

a. Sea stars at low tide

b. Shorebirds

Figure 35.19 Seashores.
a. Some organisms of a rocky coast live in tidal pools. **b.** A sandy shore looks devoid of life except for the birds that feed there.

Pelagic Division

Oceans cover approximately three-quarters of our planet. It is customary to place the organisms of the oceans into either the pelagic division (open waters) or the benthic division (ocean floor). The geographic areas and zones of an ocean are shown in Figure 35.20, and the animals that inhabit the different zones are shown in Figure 35.21. The **pelagic division** includes the neritic province and the oceanic province.

Neritic Province   The neritic province hosts a greater concentration of organisms than the oceanic province because the neritic province is sunlit and has a supply of inorganic

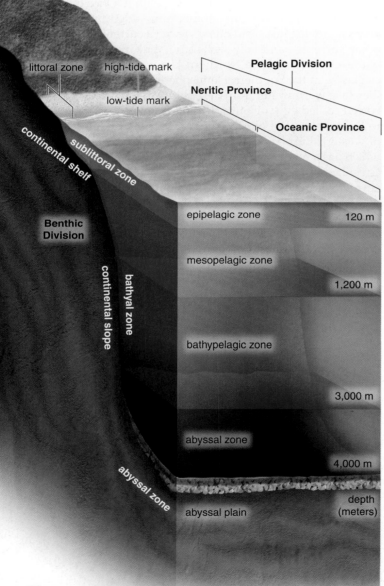

littoral zone   high-tide mark

low-tide mark

continental shelf

sublittoral zone

Pelagic Division

Neritic Province

Oceanic Province

Benthic Division

continental slope

bathyal zone

epipelagic zone   120 m

mesopelagic zone

1,200 m

bathypelagic zone

3,000 m

abyssal zone

4,000 m

abyssal zone

depth (meters)

abyssal plain

Figure 35.20 Marine environment.

Organisms reside in the pelagic division (blue) where waters are divided as indicated. Organisms also reside in the benthic division (brown) in the surfaces and zones indicated.

nutrients for photosynthesizers. Phytoplankton, consisting of suspended algae, is food not only for zooplankton but also for small fishes. These small fishes, in turn, are food for commercially valuable fishes.

Coral reefs are areas of high biological abundance and productivity found in shallow, warm, tropical waters just below the surface of the water. Their chief constituents are stony corals, animals that have a calcium carbonate (limestone) exoskeleton, and calcareous red and green algae. Corals do not usually occur individually; rather, they form colonies derived from individual corals that have reproduced by means of budding. Corals provide a home for a microscopic alga called zooxanthella. The corals, which feed at night, and the algae, which photosynthesize during the day, are mutualistic and share materials and nutrients. The close relationship of corals and zooxanthellae may be the reason coral reefs form only in shallow sunlit water—the algae need sunlight for photosynthesis.

A reef is densely populated with life. The large number of crevices and caves provide shelter for filter feeders (sponges, sea squirts, and fan worms) and for scavengers (crabs and sea urchins). There are many types of small, beautifully colored fishes. Parrot fishes feed directly on corals, and others feed on plankton or detritus. Small fishes become food for larger fishes, including snappers and other fish that are caught for human consumption. The barracuda, moray eel, and shark are top predators in coral reefs.

Oceanic Province   The oceanic province lacks the inorganic nutrients of the neritic province and, therefore, does not have as high a concentration of phytoplankton, even though the epipelagic zone is sunlit. Still, the photosynthesizers are food for a large assembly of zooplankton, which then become food for herrings and bluefishes. These, in turn, are eaten by larger mackerels, tunas, and sharks. Flying fishes, which glide above the surface, are preyed upon by dolphins (not to be confused with porpoises, which are also present). Whales are other mammals found in the epipelagic zone (Fig. 35.21). Baleen whales strain krill (small crustaceans) from the water, and toothed sperm whales feed primarily on the common squid.

Animals in the mesopelagic zone are carnivores, are adapted to the absence of light, and tend to be translucent, red in color, or even luminescent. There are luminescent shrimps, squids, and fishes, such as lantern and hatchet fishes.

The bathypelagic zone is in complete darkness except for an occasional flash of bioluminescent light. Carnivores and scavengers are found in this zone. Strange-looking fishes with distensible mouths and abdomens and small, tubular eyes feed on infrequent prey.

Benthic Division

The **benthic division** includes organisms that live on or in the soil of the continental shelf (sublittoral zone), the continental slope (bathyal zone), and the abyssal plain (abyssal zone) (see Fig. 35.20).

Figure 35.21 Ocean inhabitants.

Different organisms are characteristic of the epipelagic, mesopelagic, and bathypelagic zones of the pelagic division compared to the abyssal zone of the benthic division.

Seaweed grows in the sublittoral zone, and it can be found in batches on outcroppings as the water gets deeper. More diversity of life exists in the sublittoral and bathyal zones than in the abyssal zone. In the first two zones, clams, worms, and sea urchins are preyed upon by starfishes, lobsters, crabs, and brittle stars. Photosynthesizing algae occur in the sunlit sublittoral zone, but benthic organisms in the bathyal zone are dependent on the slow rain of detritus from the waters above.

The abyssal zone is inhabited by animals that live at the soil-water interface of the abyssal plain (see Fig. 35.21). It once was thought that few animals exist in this zone because of the intense pressure and the extreme cold. Yet many invertebrates live there by feeding on debris floating down from the mesopelagic zone. Sea lilies rise above the seafloor; sea cucumbers and sea urchins crawl around on the sea bottom; and tube worms burrow in the mud.

The flat abyssal plain is interrupted by enormous underwater mountain chains called oceanic ridges. Along the axes of the ridges, crustal plates spread apart, and molten magma rises to fill the gap. At **hydrothermal vents,** seawater percolates through cracks and is heated to about 350°C, causing sulfate to react with water and form hydrogen sulfide (H_2S). Free-living or mutualistic chemoautotrophic bacteria use energetic electrons derived from hydrogen sulfide to reduce bicarbonate to organic compounds. These compounds support an ecosystem that includes huge tube worms and clams. It was a surprise to find communities of organisms living so deep in the ocean, where light never penetrates. Unlike photosynthesizers, chemoautotrophs are producers that do not require solar energy.

The neritic province and the epipelagic zone of the oceanic province receive sunlight and contain the organisms with which we are most familiar. The organisms of the benthic division are dependent upon debris that floats down from above.

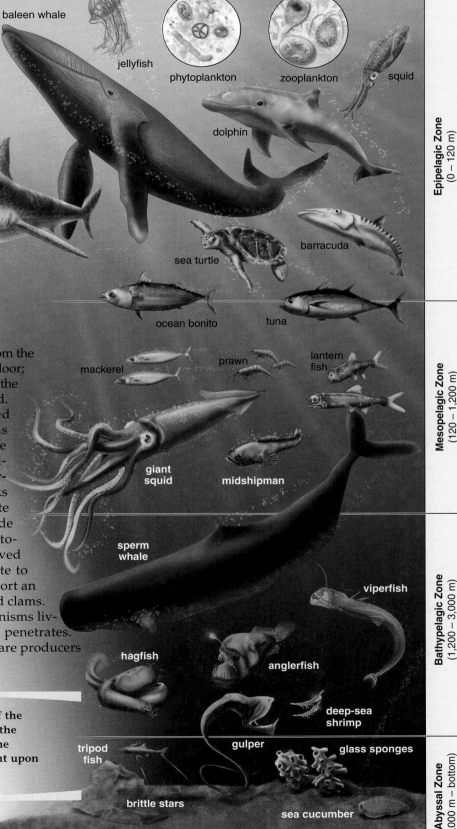

baleen whale
jellyfish
phytoplankton
zooplankton
squid
dolphin
shark
sea turtle
barracuda
ocean bonito
tuna
mackerel
prawn
lantern fish
giant squid
midshipman
sperm whale
viperfish
hagfish
anglerfish
deep-sea shrimp
tripod fish
gulper
glass sponges
brittle stars
sea cucumber

Epipelagic Zone (0 – 120 m)

Mesopelagic Zone (120 – 1,200 m)

Bathypelagic Zone (1,200 – 3,000 m)

Abyssal Zone (3,000 m – bottom)

El Niño–Southern Oscillation

Climate largely determines the distribution of life on Earth. Short-term variations in climate, which we call weather, also have a pronounced effect on living things. There is no better example than an El Niño. Originally, El Niño referred to a warming of the seas off the coast of Peru at Christmastime—hence, the name El Niño, "the boy child," for the Christ child Jesus.

Now scientists prefer the term **El Niño–Southern Oscillation (ENSO)** to denote a severe weather change brought on by an interaction between the atmosphere and ocean currents. Ordinarily, the southeast trade winds move along the coast of South America and turn west because of Earth's daily rotation on its axis. As the winds drag warm ocean waters from east to west, there is an upwelling of nutrient rich cold water from the ocean's depths, resulting in a bountiful Peruvian harvest of anchovies. When the warm ocean waters reach their western destination, the monsoons bring rain to India and Indonesia. Scientists have noted that these events correlate with a difference in the barometric pressure over the Indian Ocean and the southeastern Pacific—that is, the barometric pressure is low over the Indian Ocean and high over the southeastern Pacific. But when a "southern oscillation" occurs and the barometric pressures switch, an El Niño begins.

During an El Niño, both the northeast and the southeast trade winds slacken. Upwelling no longer occurs, and the anchovy catch off the coast of Peru plummets. During a severe El Niño, waters from the east never reach the west, and the winds lose their moisture in the middle of the Pacific instead of over the Indian Ocean. The monsoons fail, and drought occurs in India, Indonesia, Africa, and Australia. Harvests decline, cattle must be slaughtered, and famine is likely in highly populated India and Africa, where funds to import replacement supplies of food are limited.

A backward movement of winds and ocean currents may even occur so that the waters warm to more than 14° above normal along the west coast of the Americas. This is a sign that a severe El Niño has occurred, and the weather changes are dramatic in the Americas also. Southern California is hit by storms and even hurricanes, and the deserts of Peru and Chile receive so much rain that flooding occurs. A jet stream (strong wind currents) can carry moisture into Texas, Louisiana, and Florida, with flooding a near certainty. Or the winds can turn northward and deposit snow in the mountains along the west coast so that flooding occurs in the spring. Some parts of

the United States, however, benefit from an El Niño. The Northeast is warmer than usual, few if any hurricanes hit the east coast, and there is a lull in tornadoes throughout the Midwest. Altogether, a severe El Niño affects the weather over three-quarters of the globe.

Eventually, an El Niño dies out, and normal conditions return. The normal cold-water state off the coast of Peru is known as La Niña (the girl). Figure 35A contrasts the weather conditions of a La Niña with those of an El Niño. Since 1991, the sea surface has been almost continuously warm, and two record-breaking El Niños have occurred. What could be causing more of the El Niño state than the La Niña state? Some scientists are seeking data to relate this environmental change to global warming, a rise in environmental temperature due to greenhouse gases in the atmosphere. Like the glass of a greenhouse, the gases allow the sun's rays to pass through, but they trap the heat. Greenhouse gases, including carbon dioxide, are pollutants that human beings have been pumping into the atmosphere since the Industrial Revolution.

Discussion Questions

1. Recent research has suggested that global warming is contributing to the increased frequency and severity of El Niño events. This can cause severe drought in areas such as India, Indonesia, Africa, and Australia. In addition, it can greatly reduce fish catch in parts of the Southern Hemisphere. Based on this information and what you learned in Chapter 34 about global warming, do you support stricter laws to decrease greenhouse gas emissions?

2. Aside from the recent increases in gasoline prices, would you consider buying a more fuel-efficient car to do your part in decreasing the emission of greenhouse gases such as carbon dioxide?

3. Should the responsibility for decreasing greenhouse gas emissions be greater for more-developed countries (MDCs), since they produce a greater amount of greenhouse gases per capita than less-developed countries (LDCs), even though El Niños have more severe consequences for LDCs?

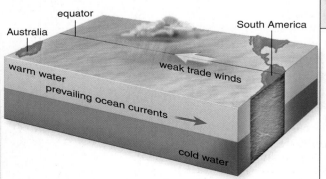
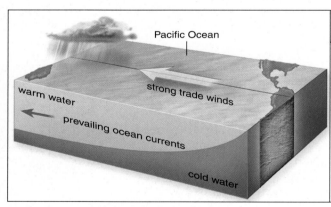

La Niña

- Upwelling off the west coast of South America brings cold waters to the surface.
- Barometric pressure is high over the southeastern Pacific.
- Monsoons associated with the Indian Ocean occur.
- Hurricanes occur off the east coast of the United States.

El Niño

- Great ocean warming occurs off the west coast of the Americas.
- Barometric pressure is low over the southeastern Pacific.
- Monsoons associated with the Indian Ocean fail.
- Hurricanes occur off the west coast of the United States.

Figure 35A La Niña and El Niño.

Pollution Legislation

Agricultural fertilizers are the chief cause of nitrate contamination of drinking-water wells. Excessive nitrates in a baby's bloodstream can lead to a slow suffocation known as blue-baby syndrome. Agricultural herbicides are suspected carcinogens in the tap water of scattered communities coast to coast. What can be done?

Some farmers are already using irrigation methods that deliver water directly to plant roots, no-till agriculture that reduces the loss of topsoil and cuts back on herbicide use, and integrated pest management that relies heavily on good bugs to kill bad bugs. Perhaps more should do so. Encouraged—in some cases, compelled—by state and federal agents, dairy farmers have built sheds, concrete containments, and underground liquid storage tanks to hold the wastes from rainy days. Then,

the manure can be trucked to fields and spread as fertilizer.

Like golf courses and ski resorts, home owners also contribute to the problem. Manicuring lawns, using motor vehicles, and constructing and using roads and buildings all add contaminants to streams, lakes, and aquifers. Citizens around Grand Traverse Bay on the eastern shore of Lake Michigan have gotten the message, especially because they want to keep on enjoying water-dependent activities such as boating, bathing, and fishing. James Haverman, a concerned member of the Traverse Bay Watershed Initiative, says, "If we can't change the way people live their everyday lives, we are not going to be able to make a difference." Builders in Traverse County are already required to control soil erosion with filter fences,

steer rainwater away from exposed soil, build sediment basins, and plant protective buffers. Home owners must have a 25-foot setback from wetlands and a 50-foot setback from lakes and creeks. They are also encouraged to pump out their septic systems every two years.

Form Your Own Opinion

1. Do you approve of legislation that requires farmers, businesses, and home owners to protect freshwater supplies? Why or why not?
2. Should landscapers and gardeners be required to follow certain restrictions? Why or why not?
3. Do you approve of legislation that restricts the number of homes in a particular area? Explain your reasons.

Summarizing the Concepts

35.1 Climate and the Biosphere
Because the Earth is a sphere, the sun's rays at the poles are spread out over a larger area than the vertical rays at the equator. The temperature at the Earth's surface, therefore, decreases from the equator to each pole. The planet is tilted on its axis, and the seasons change as Earth rotates annually around the sun.

Warm air rises near the equator, loses its moisture, and then descends at about 30° north and south latitude, and so forth, to the poles. When the air descends, it retains moisture, and therefore, the great deserts of the world are formed at 30° latitudes. Because Earth rotates on its axis daily, the winds blow in opposite directions above and below the equator. Air rising over coastal ranges loses its moisture on the windward side, making the leeward side arid.

35.2 Terrestrial Ecosystems
A biome is a major type of terrestrial ecosystem. Biomes tend to be distributed according to climate—that is, temperature and rainfall influence the pattern of biomes around the world. The effect of temperature causes the same sequence of biomes when traveling to northern latitudes as when traveling up a mountain.

The tundra is the northernmost biome and consists largely of short grasses and sedges and dwarf woody plants. Because of cold winters and short summers, most of the water in the soil is frozen year-round. This is called the permafrost.

The taiga, a coniferous forest, has less rainfall than other types of forests. The temperate deciduous forest has trees that gain and lose their leaves because of the alternating seasons of summer and winter. Tropical forests are the most complex and productive of all biomes.

Shrublands usually occur along coasts that have dry summers and receive most of their rainfall in the winter. Among grasslands, the savanna, a tropical grassland, supports the greatest number of different types of large herbivores. Prairies, such as those in the mid United States, have a limited variety of vegetation and animal life.

Deserts are characterized by a lack of water—they are usually found in places with rainfall less than 25 cm a year. Some plants, such as cacti, are succulents, and others are shrubs with thick leaves that they often lose during dry periods.

35.3 Aquatic Ecosystems
In deep lakes of the temperate zone, the temperature and the concentrations of nutrients and gases in the water vary with depth. The entire body of water is cycled twice a year, during spring and fall turnover, distributing nutrients from the bottom layers. Lakes and ponds have three life zones. Rooted plants and clinging organisms live in the littoral zone, plankton and fishes live in the sunlit limnetic zone, and bottom-dwelling organisms, such as crayfishes and molluscs, live in the profundal zone.

Marine ecosystems are divided into coastal ecosystems and the oceans. The coastal ecosystems, especially estuaries, are more productive than the oceans. Estuaries (e.g., salt marshes, mudflats, and mangrove forests) are near the mouth of a river. Estuaries are the nurseries of the sea.

An ocean is divided into the pelagic division and the benthic division. The oceanic province of the pelagic division (open waters) has three zones. The epipelagic zone receives adequate sunlight and supports the most life. The mesopelagic and bathypelagic zones contain organisms adapted to minimum light and no light, respectively. The benthic division (ocean floor) includes organisms living on the continental shelf in the sublittoral zone, the continental slope in the bathyal zone, and the abyssal plain in the abyssal zone.

Testing Yourself

Choose the best answer for each question.

1. The Northern Hemisphere is tilted toward the sun during the
    a. winter solstice.
    b. autumnal equinox.
    c. summer solstice.
    d. vernal equinox.

2. Which side of a mountain range lies in a rain shadow?
    a. windward
    b. leeward

3. A monsoon climate is produced
    a. when a landmass heats more rapidly than the ocean.
    b. when the ocean heats more rapidly than a landmass.
    c. in regions of the Earth that are tilting toward the sun.
    d. in regions of the Earth that are tilting away from the sun.

4. Place the following biomes in their correct positions in the following graph: desert, grassland, savanna, semidesert, shrubland, taiga, temperate deciduous forest, temperate rain forest, thorn scrub, tropical rain forest, tropical seasonal forest, tundra, woodland.

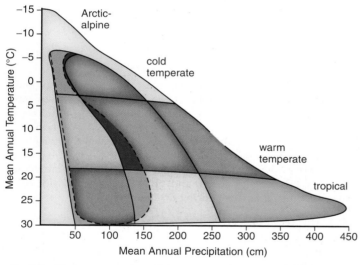

5. Whether you are ascending a mountain or traveling from the equator to the North Pole, you would observe terrestrial biomes in which order?
    a. coniferous forest, tropical rain forest, temperate deciduous forest, tundra
    b. tropical rain forest, coniferous forest, temperate deciduous forest, tundra
    c. tropical rain forest, temperate deciduous forest, coniferous forest, tundra
    d. tropical rain forest, coniferous forest, tundra, temperate deciduous forest
    e. tundra, temperate deciduous forest, tropical rain forest, coniferous forest

6. Which of the following is not true of the arctic tundra?
    a. Soil below the uppermost layer stays frozen year-round.
    b. Annual rainfall is low.
    c. Trees cannot grow because the growing season is too short.
    d. Large mammals are absent.

For questions 7–13, match the description to the biome in the key.

Key:
    a. tundra
    b. taiga
    c. temperate deciduous forests
    d. tropical forests
    e. shrublands
    f. grasslands
    g. deserts

7. Dominated by coniferous trees

8. Chaparral in California

9. Contains permafrost

10. Typically found at latitudes of about 30°

11. Has a complex structure, with many levels of life, from the canopy to the floor

12. Moderate climate, with relatively high rainfall and well-defined seasons

13. Include savannas

14. DNA analysis has been used to
    a. show that loggerhead turtles living in the Mediterranean Sea were hatched in the United States.
    b. prove that there are at least five different types of brown bears.
    c. verify that sushi in Japanese markets contains dolphin meat.
    d. prove that a trophy deer was shot illegally on a preserve.
    e. All of these are correct.

15. Epiphytes are most likely to be found in
    a. deserts.
    b. tundra.
    c. grasslands.
    d. tropical forests.
    e. shrublands.

16. The major environmental factor that determines whether a prairie has tall grass or short grass is
    a. annual rainfall.
    b. mean annual high temperature.
    c. mean annual low temperature.
    d. soil nutrient levels.
    e. types of grazers present.

17. Compared to the hypolimnion layer, the epilimnion layer of a lake in late summer is
    a. nutrient-rich and oxygen-rich.
    b. nutrient-rich and oxygen-poor.
    c. nutrient-poor and oxygen-rich.
    d. nutrient-poor and oxygen-poor.

18. An oligotrophic lake
    a. is nutrient-rich.
    b. is cold.
    c. is likely to be found in an agricultural or urban area.
    d. has poor productivity.

For questions 19–22, indicate the life zone that each lake organism is most likely to inhabit. A question may have more than one answer.

Key:

  a. littoral
  b. limnetic
  c. profundal
  d. benthic

19. Predatory fish

20. Water lily

21. Clams

22. Unicellular algae

23. An estuary

  a. does not contain a mixture of fresh and salt water.
  b. is not nutrient-rich.
  c. does not contain many large fish.
  d. is not a good feeding ground for birds.

24. In an ocean, the highest concentrations of phytoplankton would be found in the

  a. littoral zone.
  b. neritic province.
  c. oceanic province.
  d. All areas would have roughly equal concentrations.

25. Energy for the food chain near hydrothermal vents comes from

  a. dead organisms that fall down from above.
  b. highly efficient photosynthetic phytoplankton.
  c. chemoautotrophic bacteria.
  d. heat given off by the vents.

26. The mild climate of Great Britain is best explained by

  a. the winds called westerlies.
  b. the spinning of Earth on its axis.
  c. Great Britain being a mountainous country.
  d. the flow of ocean currents.

27. The location of deserts at 30° is best explained by which statement?

  a. Warm air rises and loses its moisture; cool air descends and retains moisture.
  b. Ocean currents give up heat as they flow from the equator to the poles.
  c. Ocean breezes cool the coast during the day.
  d. All of these are correct.

28. All of the following phrases describe a tropical rain forest except

  a. nutrient-rich soil.
  b. many arboreal plants and animals.
  c. canopy composed of many layers.
  d. broad-leaved evergreen trees.

29. Why are lush evergreen forests present in the Pacific Northwest region of the United States?

  a. Earth rotates.
  b. Winds blow toward the land from the ocean, bringing moisture.
  c. These forests are located on the leeward side of a mountain range.
  d. Both b and c are correct.

30. A wetland

  a. purifies water.
  b. helps absorb overflow and prevents flooding.
  c. provides a home for organisms that are the links in various food chains.
  d. All of these are correct.

31. An estuary acts as a nutrient trap because of the

  a. action of rivers and tides.
  b. depth at which photosynthesis can occur.
  c. amount of rainfall received.
  d. height of the water table.

32. Which area of the seashore is exposed only during low tide?

  a. upper portion of the littoral zone
  b. midportion of the littoral zone
  c. lower portion of the littoral zone
  d. None of these are correct.

Understanding the Terms

benthic division 738
biome 727
chaparral 731
climate 724
coniferous forests 729
coral reef 738
deserts 733
El Niño–Southern Oscillation (ENSO) 740
epiphyte 731
estuary 736
eutrophication 734
fall overturn 734
grasslands 732
hydrothermal vent 739
lake 734
littoral zone 736
monsoon 725
pelagic division 738
permafrost 728
plankton 735
rain shadow 725
shrubland 731
spring overturn 735
taiga 729
temperate deciduous forests 729
tropical forests 730
tundra 728
upwelling 737

Match the terms to these definitions:

  a._____ End of a river where fresh water and salt water mix as they meet.
  b._____ Major ecosystem characterized by certain climatic conditions and dominated by particular types of plants.
  c._____ Division of the ocean that includes the ocean floor and supports a unique set of organisms.
  d._____ A grassland in Africa characterized by few trees and a severe dry season.
  e._____ Shore zone between the high-tide mark and low-tide mark; also, the shallow water of a lake where light penetrates to the bottom.

Thinking Critically

1. Global warming is predicted to have the least severe consequences in terms of temperature change nearest the equator and more severe consequences progressing into the more temperate zones. Why do you think this is?

2. Scientists have documented a phenomenon called "biodiversity gradients," which shows that the greatest amounts of species diversity exist nearest the equator and that diversity tends to diminish moving through increasing latitudes. Based on your knowledge of climate, topography, and energy transport among trophic levels in ecological pyramids, why do you think biodiversity gradients exist?

3. Based on what you know about tropical forests versus temperate forests, why do you think temperate forest habitats are more suitable for long-term agriculture, despite the fact that they have a shorter growing season than tropical forests?

4. Human nutrient inputs, such as nitrogen and phosphorus, can convert oligotrophic lakes into eutrophic lakes through the process of eutrophication. Discuss how this process affects the diversity of fish and spring and fall overturn.

5. Explain why coral reefs are some of the most diverse ecosystems on Earth. They are also some of the most sensitive and are undergoing declines worldwide. Hypothesize why nearby logging that causes deposits of silt (dirt runoff) in coastal areas is thought to be a major cause of coral reef decline.

ARIS, the *Inquiry into Life* Website

ARIS, the website for *Inquiry into Life,* provides a wealth of information organized and integrated by chapter. You will find practice quizzes, interactive activities, labeling exercises, flashcards, and much more that will complement your learning and understanding of general biology.

www.aris.mhhe.com

The demands made on the Earth's natural resources by the growth of the human population have created a biodiversity crisis *through environmental degradation, habitat destruction, introduction of exotic species, pollution, overexploitation, and spread of disease.*

Biodiversity has both direct and indirect value to humans. Billions of dollars worth of medicines have come from plant and animal products. Plants and animals also provide valuable resources, such as timber and food, as well as helping to control pests that threaten our crops. Indirect values of biodiversity, often referred to as "ecosystem services," are estimated to be more valuable than the total gross national product of the entire world! For example, preserving biogeochemical cycles, such as the water cycle, provides clean drinking water, and conserving topsoil provides billions of dollars worth of indirect value for crop growth and erosion control.

Conservation biology is directed at protecting biodiversity for the good of all living things, including humans. As an illustration, consider the past plight of the bald eagle. In the twentieth century, the pesticide DDT entered food chains after being used worldwide to control mosquitoes and other insect pests. The concentration of DDT became magnified up the food chain. As a result, bald eagles, which are tertiary consumers, were driven to near extinction; DDT in their food caused their eggs to be extremely thin and easily crushed by incubating parents. The Environmental Protection Agency banned the use of DDT in the United States in 1972, and since then, bald eagle populations have recovered to the point that they are candidates for removal from the endangered species list.

chapter **36**

Environmental Concerns

Chapter Concepts

36.1   Human Use of Resources

Human beings have certain basic needs, and a **resource** is anything from the biotic or abiotic environment that helps meet those needs. Land, water, food, energy, and minerals are the maximally used resources to be discussed in this chapter (Fig. 36.1).

Some resources are nonrenewable, and some are renewable. **Nonrenewable resources** are limited in supply. For example, the amount of land, fossil fuels, and minerals is finite and can be exhausted. Better extraction methods can make more fossil fuels and minerals available. Efficient use, recycling, or substitution can make the supply last longer, but eventually these resources will run out.

Renewable resources are not limited in supply. We can use certain forms of energy (e.g., solar energy) or harvest plants and animals for food, and more supply will always be forthcoming. Still, we have to be careful not to squander renewable resources. Consider, for example, that most species have population thresholds below which they cannot recover, as when the huge herds of buffalo that once roamed the Western U.S. disappeared after being over-hunted.

Unfortunately, a side effect of resource consumption can be pollution. **Pollution** is any alteration of the environment in an undesirable way. Pollution is often caused by human activities. The impact of humans on the environment is proportional to the size of the global human population. As the population grows, so do the need for resources and the amount of pollution caused by using these resources. Consider that six people adding waste to the ocean may not be alarming, but six and a half billion people doing so are certainly affecting its cleanliness. Actually, in modern times, the consumption of mineral and energy resources has grown faster than population size, mostly due to increased development in less-developed countries (LDCs) and increased use in more-developed countries (MDCs).

Land Use Change

People need a place to live. Worldwide, there are currently more than 32 persons for each square kilometer (83 persons per square mile) of all available land, including Antarctica, mountain ranges, jungles, and deserts. Naturally, land is also needed for a variety of uses aside from homes, such as agriculture, electric power plants, manufacturing plants, highways, hospitals, schools, and so on.

Human population

Figure 36.1 Resources.
Human beings use land, water, food, energy, and minerals to meet their basic needs, such as a place to live, food to eat, and products that make their lives easier.

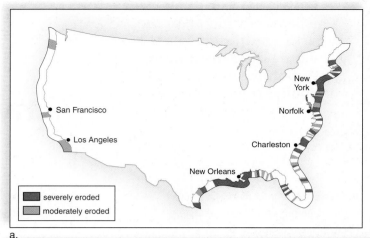

a.

b.

Figure 36.2 Beach erosion.
a. Most of the U.S. coastline is subject to beach erosion. **b.** People who choose to live near the coast may eventually lose their homes.

Beaches and Human Habitation

At least 40% of the world population lives within 100 km (60 mi) of a coastline, and this number is expected to increase. In the United States today, over one-half of the population lives within 80 km (50 mi) of the coasts (including the Great Lakes). Living right on the coast is an unfortunate choice because it leads to beach erosion and loss of habitat for marine organisms.

Beach Erosion   An estimated 70% of the world's beaches are eroding; Figure 36.2 shows how extensive the problem is in the United States. The seas have been rising for the past 12,000 years, ever since the climate turned warmer after the last Ice Age. Authorities are concerned that global warming, discussed later in this chapter, is also contributing to the melting of ice caps and glaciers and, therefore, to an increase in sea level.

Humans carry on other activities that divert more water to the oceans, and this contributes to rising of the seas and erosion of the beaches. For example, humans have filled in coastal wetlands, such as mangrove swamps in the southern United States and salt marshes in the northern United States. Greater than 80% of U.S. wetlands have been destroyed, with destruction as high as 90% in states such as Iowa and California. With growing recognition of the services provided by wetlands, the rate of wetland loss has been stemmed in the United States during the past 40 years, but such loss is just starting in South

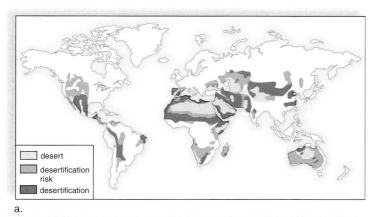

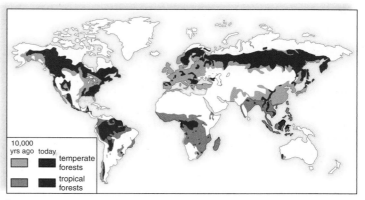

b.

Figure 36.3 Desertification.
a. Desertification is a worldwide occurrence that **(b)** reduces the amount of land suitable for human habitation.

b.

Figure 36.4 Deforestation.
a. Nearly half of the world's forest lands have been cleared for farming, logging, and urbanization. **b.** The soil of tropical rain forests is not suitable for long-term farming. Note the deforested mountainous area.

America, where a project to straighten the Paraná River will drain the world's largest wetland. One reason to protect coastal wetlands is that they are spawning areas for fish and other forms of marine life. They are also habitats for certain terrestrial species, including many types of birds (see Fig. 35.18a).

Humans often try to stabilize beaches by building groins (structures that extend from the beach into the water) and seawalls. Groins trap sand on one side, but erosion is worse on the other side. Seawalls, in the end, also increase erosion because ocean waves remove sand from in front of and to the side of the seawalls. Importing sand is a better solution, but it is very costly and can disturb plant and animal populations. It's estimated that today, the U.S. shoreline loses 40% more sediment than it receives, especially because the building of dams on rivers prevents sediment from reaching the coast.

Coastal Pollution The coast is particularly subject to pollution because toxic substances placed in freshwater lakes, rivers, and streams often combine and eventually find their way to the coast. Oil spills at sea cause localized harmful effects also.

Semiarid Lands and Human Habitation

Forty percent of the Earth's lands are already deserts, and land adjacent to a desert (called semiarid land) is in danger

of becoming unable to support human life if it is improperly managed by humans. **Desertification** is the conversion of semiarid land to desertlike conditions (Fig. 36.3).

Quite often, desertification begins when humans allow animals to overgraze the land. The soil can no longer hold rainwater, and it runs off instead of keeping the remaining plants alive or recharging wells. Humans then remove whatever vegetation they can find to use as fuel or fodder for their animals. The end result is a lifeless desert. Some estimate that nearly three-quarters of all rangelands worldwide are in danger of desertification. The recent famine in Ethiopia was due, at least in part, to degradation of the land to the point that it could no longer support human beings and their livestock.

Tropical Rain Forests and Human Habitation

Deforestation, the removal of trees, has long allowed humans to live in areas where forests once covered the land. The concern of late has been that people are settling in tropical rain forests, such as the Amazon, following the building of roads (Fig. 36.4). This land, too, is subject to desertification. Soil in the tropics is often thin and nutrient-poor because all the nutrients are tied up in the trees and other vegetation. When the trees are felled and the land is used for agriculture or grazing, it quickly loses its fertility and becomes subject to desertification.

Loss of Biodiversity   Logging in the Congo Republic has contributed to the growth of a city called Pokola, whose residents routinely dine on wild bush animals. From this and other examples, we know that development in tropical rain forests leads to loss of biodiversity.

> **People are more apt to have a detrimental effect on the environment when they live on the coast, in semiarid lands, or in tropical rain forests.**

Water

In some areas of the world, people do not have ready access to drinking water, and if they do, the water may be impure (Fig. 36.5). It's considered a human right for people to have clean drinking water, but actually, most fresh water is utilized by industry and agriculture. Worldwide, 70% of all fresh water is used to irrigate crops! Much of a recent surge in demand for water stems from increased industrial activity and irrigation-intensive agriculture, the type of agriculture that now supplies about 40% of the world's food crops. Domestically in MDCs, more water is usually used for bathing, flushing toilets, and watering lawns than for drinking and cooking.

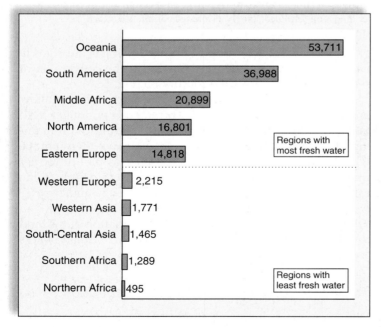

Figure 36.5 Freshwater resources.

Countries and regions within continents differ in the amount of fresh water available. Amounts given are average annual cubic meters per capita.

Source: Data from World Resources 2000–2001.

Increasing Water Supplies

Although the needs of the human population overall do not exceed the renewable supply of water, this is not the case in certain regions of the United States and the world. As illustrated in Figure 36.3, about 40% of the world's land is desert, and deserts are bordered by semiarid land. When needed, humans increase the supply of fresh water by damming rivers and withdrawing water from aquifers.

Dams   The world's 45,000 large dams catch 14% of all precipitation runoff, provide water for up to 40% of irrigated land, and give some 65 countries more than half their electricity. Damming of certain rivers has been so extensive that they are almost completely drained. The Yellow River in China fails to reach the sea most years; the Colorado River barely makes it to the Gulf of California; and even the Rio Grande dries up before it can merge with the Gulf of Mexico. The Nile in Egypt and the Ganges in India are also so overexploited that at some times of the year, they hardly reach the ocean.

Dams have other drawbacks: (1) Reservoirs lose water due to evaporation and seepage into underlying rock beds. The amount of water lost sometimes equals the amount dams make available! (2) The salt left behind by evaporation and agricultural runoff increases salinity and can make a river's water unusable farther downstream. (3) Dams hold back less water with time because of sediment buildup. Sometimes a reservoir becomes so full of silt that it is no longer useful for storing water.

Aquifers   To meet their freshwater needs, people are pumping vast amounts of water from **aquifers,** which are reservoirs just below or as much as 1 km below the ground's surface. Aquifers hold about 1,000 times the amount of water that falls on land as precipitation each year. This water accumulates from rain that fell in far-off regions even hundreds of thousands of years ago. In the past 50 years, groundwater depletion has become a problem in many areas of the world. In substantial portions of the High Plains Aquifer, which stretches from South Dakota to the Texas Panhandle, more than half of the water has been pumped out. In the 1950s, India had 100,000 motorized pumps in operation; today, India has 20 million pumps, a huge increase in groundwater pumping.

Environmental Consequences   Removal of water is causing land **subsidence,** settling of the soil as it dries out. In California's San Joaquin valley, an area of more than 13,000 square km has subsided at least 30 cm due to groundwater depletion, and in the worst spot, the surface of the ground has dropped more than 9 m! In some parts of Gujarat, India, the water table has dropped as much as 7 m. Subsidence damages canals, buildings, and underground pipes. Withdrawal of groundwater can also cause **sinkholes,** in which an underground cavern collapses when water no longer holds up its roof.

Saltwater intrusion is another consequence of aquifer depletion. The flow of fresh water into streams and aquifers usually keeps them fairly free of seawater. But as water is withdrawn, the water table can lower to the point that seawater backs up into streams and aquifers. Saltwater intrusion reduces the supply of fresh water along the coast.

Conservation of Water

By 2025, two-thirds of the world's population may be facing serious water shortages. Some solutions for expanding water supplies have been suggested. Planting drought- and salt-tolerant crops would help a lot. Using drip irrigation delivers more water to crops and saves about 50% over traditional methods while increasing crop yields as well. Although the first drip systems were developed in 1960, they are currently used on less than 1% of irrigated land. Most governments subsidize irrigation so heavily that farmers have little incentive to invest in drip systems or other water-saving methods. Reusing water and adopting conservation measures could help the world's industries cut their water demands by more than half.

Food

Generally speaking, food comes from three activities: growing crops, raising animals, and fishing. In 1950, the human population numbered 2.5 billion, and there was only enough food to provide less than 2,000 calories per person per day; now, with 6.5 billion people on Earth, the world food supply provides more than 2,500 calories per person per day. The increase in the food supply has largely been possible because of modern farming methods, which unfortunately include some harmful practices:

1. *Planting of a few genetic varieties.* The majority of farmers practice monoculture—that is, wheat farmers plant the same type of wheat, and corn farmers plant the same type of corn. Unfortunately, monoculture means that a single type of parasite can cause much devastation.
2. *Heavy use of fertilizers, pesticides, and herbicides.* Fertilizer production is energy intensive, and fertilizer runoff contributes to water pollution. Pesticides reduce soil fertility because they kill off beneficial soil organisms as well as pests, and some pesticides and herbicides are linked to the development of cancer. In addition, pesticide resistance evolves in insects, requiring the use of alternate or stronger pesticides and increasing costs.
3. *Generous irrigation.* As already discussed, water is sometimes taken from aquifers, and in the future their water content may become so reduced that it could become too expensive to pump out any more.
4. *Excessive fuel consumption.* Irrigation pumps require energy to remove water from aquifers, and large farming machines are used to spread fertilizers, pesticides, and herbicides, as well as to sow and harvest the crops. In effect, modern farming methods transform fossil fuel energy into food energy.

Figure 36.6 shows some ways to minimize the harmful effects of modern farming practices.

Soil Loss and Degradation

Land suitable for farming and grazing animals is being degraded worldwide. Topsoil, the topmost portion of the soil, is the richest in organic matter and the most capable of supporting grass and crops. When bare soil is acted on by water and wind, soil erosion occurs and topsoil is lost. As a result, marginal rangeland becomes desert, and farmland loses its productivity.

a. Polyculture

b. Contour farming

c. Biological pest control

Figure 36.6 Conservation methods.
a. Polyculture reduces the ability of one parasite to wipe out an entire crop and lessens the need to use a herbicide to kill weeds. This farmer has planted alfalfa between strips of corn, which also replenishes the nitrogen content of the soil (instead of adding fertilizers). Alfalfa, a legume, has root nodules that contain nitrogen-fixing bacteria. **b.** Contour farming conserves topsoil in gently sloping areas because water has less tendency to run off. **c.** Instead of pesticides, it is sometimes possible to use a natural predator. Here, ladybugs are feeding on cottony-cushion scale insects on citrus trees.

The custom of planting the same crop in straight rows that facilitate the use of large farming machines has caused the United States and Canada to have one of the highest rates of soil erosion in the world. Conserving the nutrients now being lost could save farmers $20 billion annually in fertilizer costs. Much of the eroded sediment ends up in lakes and streams, where it reduces the ability of aquatic species to survive.

Between 25% and 35% of the irrigated Western U.S. croplands are thought to have undergone **salinization,** an accumulation of mineral salts due to the evaporation of excess irrigation water. Salinization makes the land unsuitable for growing crops because there is too much salt in the soil.

Green Revolution

About 50 years ago, research scientists began to breed tropical wheat and rice varieties especially for farmers in LDCs. The dramatic increase in yield due to the introduction of these new varieties around the world was called "the green revolution." These plants helped the world food supply keep pace with the rapid increase in world population. Most green revolution plants are called "high responders" because they need high levels of fertilizer, water, and pesticides in order to produce a high yield. In other words, they require the same subsidies and create the same ecological problems as do modern farming methods.

Genetic Engineering   As we discussed in Chapter 24, genetic engineering can produce transgenic plants with new and different traits, including resistance to both insects and herbicides. When herbicide-resistant crops are planted, weeds are easily controlled, less tillage is needed, and soil erosion is minimized. Researchers also want to produce crops that tolerate salt, drought, and cold. Some progress has also been made in increasing the food quality of crops so that they will supply more of the proteins, vitamins, and minerals people need. Genetically engineered crops could result in still another green revolution.

Nevertheless, some citizens are opposed to the use of genetically engineered crops, fearing that they will damage the environment and may lead to health problems in humans. The Bioethical Focus in Chapter 24 discusses these problems.

Domestic Livestock

A low-protein, high-carbohydrate diet consisting only of grains such as wheat, rice, or corn can lead to malnutrition. In LDCs, kwashiorkor, caused by a severe protein deficiency, is seen in infants and children ages 1–3, usually after a new baby arrives in the family and the older children are no longer breastfed. Such children are lethargic and irritable, and have bloated abdomens. Mental retardation is expected.

In MDCs, many people tend to have more than enough protein in their diet. Almost two-thirds of U.S. cropland is devoted to producing livestock feed. This means that a large percentage of the fossil fuel, fertilizer, water, herbicides, and pesticides we use are actually for the purpose of raising livestock. Typically, cattle are range-fed for about four months, and then they are brought to crowded feedlots where they receive growth hormone and antibiotics while feeding on grain or corn. Most pigs and chickens spend their entire lives cooped up in crowded pens and cages.

If livestock eat a large proportion of the crops in the United States, then raising livestock accounts for much of the pollution associated with farming. Consider, also, that fossil fuel energy is presently needed not just to produce herbicides and pesticides and to grow food, but also to make the food available to the livestock. Raising livestock is extremely energy-intensive in MDCs. In addition, water is used to wash livestock wastes into nearby bodies of water, where they add significantly to water pollution. Whereas human wastes are sent to sewage treatment plants, raw animal wastes are not.

For these reasons, it is prudent to recall the ecological energy pyramid (see Fig. 34.5), which shows that as you move up the food chain, energy is lost. As a rule of thumb, for every 10 calories of energy from a plant, only 1 calorie is available for the production of animal tissue in a herbivore. In other words, it is extremely wasteful for the human diet to contain more protein than is needed to maintain good health. It is possible to feed ten times more people on protein-rich grain than on meat. Thus, judicious consumption of meat or a vegetarian diet can help conserve natural, nonrenewable resources.

Fishing

Worldwide, between 1970 and 1990, the number of large boats devoted to fishing doubled to 1.2 million. The U.S. fishing fleet participated in this growth due to the availability of federal loans for building fishing boats. The new boats have sonar and depth recorders, and their computers remember the sites of previous catches so that the boats can return. Helicopters, planes, and even satellite data are used to help find fish. The result of the increased number and efficiency of fishing boats is a severe reduction in fish catch (Fig. 36.7). For example, the number of North Atlantic swordfish caught in the United States declined 70% from 1980 to 1990, and the average weight per fish fell from 115 to 60 pounds. Many believe the Atlantic bluefin tuna is so overfished that it will never recover and will instead become extinct.

Modern fishing practices negatively impact biodiversity because a large number of marine animals, including endangered species, such as sea turtles, are caught by chance in the huge nets that some fishing boats use. These animals are discarded. The world's shrimp fishery has an annual catch of 1.8 million tons, but the other animals caught and discarded in the process amount to 9.5 million tons. This topic is discussed further in section 36.2.

Environmental problems can result from the manner in which farming, raising livestock, and fishing are carried out.

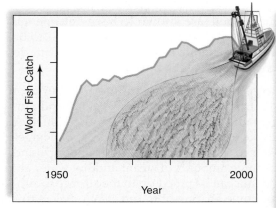

a.

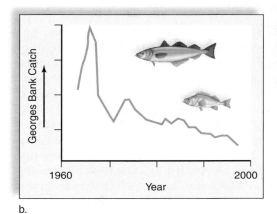

b.

c.

Figure 36.7 Fisheries.

a. The world catch increased from 1950 to the mid-1990s with better harvesting methods. It is now declining because of overexploitation. **b.** This figure depicts fish catches in the Georges Bank area off the Gulf of Maine. This historic area has recently become overfished. **c.** Modern fishing methods are overexploiting fisheries. *Data from U.S. Marine Fisheries Service.*

Energy

Modern society runs on various sources of energy. Some of these energy sources are nonrenewable, and others are renewable.

Nonrenewable Sources

Presently, about 6% of the world's energy supply comes from nuclear power, and 75% comes from fossil fuels; both of these are finite, nonrenewable sources. Although it was once predicted that the nuclear power industry would fill a significant portion of the world's energy needs, this has not happened for two reasons: (1) People are very concerned about nuclear power dangers, such as the meltdown that occurred in 1986 at the Chernobyl nuclear power plant in Russia. (2) Radioactive wastes from nuclear power plants remain a threat to the environment for thousands of years, and we still have not decided how best to safely store them.

Fossil fuels (oil, natural gas, and coal) are so named because they are derived from the compressed remains of plants and animals that died millions of years ago. Although the U.S. population makes up less than one-fifth of the world's population, it uses more than one-half of the fossil fuel energy supply. Comparatively speaking, each person in an MDC uses approximately as much energy in one day as a person in an LDC uses in one year.

Among the fossil fuels, oil burns more cleanly than coal, which may contain a considerable amount of sulfur.

So, despite the fact that the United States is well supplied with coal, imported oil is our preferred fossil fuel today. Even so, the burning of any fossil fuel causes environmental problems because as it burns, pollutants and greenhouse gases are emitted into the air.

Fossil Fuels and Global Climate Change In 1850, the level of carbon dioxide in the atmosphere was about 280 parts per million (ppm), and today it is about 350 ppm. This increase is largely due to the burning of fossil fuels and the burning and clearing of forests to make way for farmland and pasture. Human activities are causing the emission of other gases as well. For example, the amount of methane given off by oil and gas wells, rice paddies, and all sorts of organisms, including domesticated cows, is increasing by about 1% per year. These gases are known as **greenhouse gases** because, just like the panes of a greenhouse, they allow solar radiation to pass through but hinder the escape of infrared heat back into space.

Today, data collected around the world show a steady rise in the concentration of the various greenhouse gases. These data are used to generate computer models that predict the Earth may warm to temperatures never before experienced by living things. The global climate has already warmed by about 0.6°C since the Industrial Revolution. Computer models are unable to consider all possible variables, but Earth's temperature may rise 1.5–4.5°C by 2060 if greenhouse emissions continue at the current rates.

Because of global warming, scientists predict that as the oceans warm, temperatures in the polar regions will rise to a greater degree than in other regions. If so, glaciers will melt, and sea levels will rise, not only due to this melting but also because water expands as it warms. Water evaporation will increase, and most likely precipitation will increase along the coasts, while conditions inland become dryer. The occurrence of droughts will reduce agricultural yields, and also cause trees to die off. Coastal agricultural lands, such as the deltas of Bangladesh, India, and China, will be flooded, and billions will have to be spent to keep coastal cities, such as New York, Boston, Miami, and Galveston in the United States, from disappearing into the sea.

Renewable Energy Sources

Renewable types of energy include hydropower, geothermal energy, wind power, and solar energy.

Hydropower   Hydroelectric plants convert the energy of falling water into electricity (Fig. 36.8). Hydropower accounts for about 10% of the electric power generated in the United States and almost 98% of the total renewable energy used. Brazil, New Zealand, and Switzerland produce at least 75% of their electricity with hydropower, but Canada is the world's leading hydropower producer, accounting for 97% of its renewable energy generation. Worldwide, hydropower presently generates 20% of all electricity utilized, but this percentage is expected to rise because of increased use in certain countries. For example, Iceland has an ambitious hydropower project under way because presently it uses only 10% of its potential capacity.

Much of the hydropower development in recent years has been due to the construction of enormous dams that,

as previously mentioned, are known to have detrimental environmental effects. The better choice is believed to be small-scale dams that generate less power per dam but do not have the same environmental impact.

Geothermal Energy   The Earth has an internal source of heat. Elements such as uranium, thorium, radium, and plutonium undergo radioactive decay below Earth's surface and then heat the surrounding rocks to hundreds of degrees Celsius. When the rocks are in contact with underground streams or lakes, huge amounts of steam and hot water are produced. This steam can be piped up to the surface to supply hot water for home heating or to run steam-driven turbogenerators. The California's Geysers project is the world's largest geothermal electricity generating complex.

Wind Power   Wind power is expected to account for a significant percentage of our energy needs in the future (Fig. 36.9a). Despite the common belief that a huge amount of land is required for the "wind farms" that produce commercial electricity, the actual amount of space for a wind farm compares favorably to the amount of land required by a coal-fired power plant or a solar thermal energy system.

A community that generates its own electricity by using wind power can solve the problem of uneven energy production by selling electricity to a local public utility when an excess is available and buying electricity from the same facility when wind power is in short supply.

Figure 36.8 Hydropower.
Hydropower dams provide a clean form of energy but can be ecologically disastrous in other ways.

a.   c.

Figure 36.9 Other renewable energy sources.
a. Wind power requires land on which to place enough windmills to generate energy. **b.** Photovoltaic cells on rooftops and **(c)** the use of fuel-cell hybrid vehicles such as this prototype, will reduce air pollution and dependence on fossil fuels.

Energy and the Solar-Hydrogen Revolution   Solar energy is diffuse energy that must be (1) collected, (2) converted to another form, and (3) stored if it is to compete with other available forms of energy. Passive solar heating of a house is successful when the windows of the house face the sun, the building is well insulated, and heat can be stored in water tanks, rocks, bricks, or some other suitable material.

In a **photovoltaic (solar) cell,** a wafer of an electron-emitting metal is in contact with another metal that collects the electrons and passes them along into wires in a steady stream. Spurred by the oil shocks of the 1970s, the U.S. government has been supporting the development of photovoltaic cells ever since. As a result, the price of buying one has dropped from about $100 per watt to around $4. The photovoltaic cells placed on roofs, for example, generate electricity that can be used inside a building and/or sold back to a power company (Fig. 36.9*b*).

Several types of solar power plants are now operating in California. In one type, huge reflectors focus sunlight on a pipe containing oil. The heated pipes boil water, generating steam that drives a conventional turbogenerator. In another type, 1,800 sun-tracking mirrors focus sunlight onto a molten salt receiver mounted on a tower. The hot salt generates steam that drives a turbogenerator.

Now that we have better ways to capture solar energy, scientists are investigating the possibility of using solar energy to extract hydrogen from water via electrolysis. The hydrogen can then be used as a clean-burning fuel; when it burns, water is produced. Traditional cars have internal combustion engines that run on gasoline, but recently, hybrid cars have been developed that run on both gasoline and electricity, increasing mileage per gallon. In the future, vehicles are expected to be powered by fuel cells that use hydrogen to produce electricity. The electricity runs a motor that propels the vehicle. Fuel cells are now powering buses in Vancouver and Chicago, and fuel cell–powered autos are planned (Fig. 36.9*c*).

Hydrogen fuel can be produced locally or in central locations, using energy from photovoltaic cells. If in central locations, hydrogen can be piped to filling stations using the natural gas pipes already plentiful in the United States. The advantages of a solar-hydrogen revolution would be at least twofold: (1) The world would no longer be dependent on the the limited oil reserves, and (2) environmental problems, such as global warming, acid rain, and smog, would begin to lessen.

Minerals

Minerals are nonrenewable raw materials in Earth's crust that can be mined (extracted) and used by humans. Nonrenewable minerals include fossil fuels; nonmetallic raw materials, such as sand, gravel, and phosphate; and metals, such as aluminum, copper, iron, lead, and gold.

Nonrenewable resources are subject to *depletion*—that is, the supply that is mineable will eventually run out. How quickly this occurs is dependent on how fast the resource is used, whether new reserves can be found, and whether recycling and reuse are possible. We can extend our supply of fossil fuels if we conserve our use, increase our reliance on renewable energy sources, and find new reserves. In addition to these possibilities, metals can be recycled.

Most of the metals mined each year are consumed by the United States, Japan, and Europe, despite the fact that they are primarily mined in South America, South Africa, and the former Soviet Union. One of the greatest threats to the maintenance of ecosystems and biodiversity is surface mining, called *strip mining*. In the United States, huge machines can go as far as removing mountaintops in order to reach a particular mineral (Fig. 36.10). The land devoid of vegetation is destroyed completely, and rain washes toxic waste deposits into nearby streams and rivers. Legislation now requires that strip miners restore the land to its original condition, a process that can take many years.

The most dangerous metals to human health are the heavy metals: lead, mercury, arsenic, cadmium, tin, chromium, zinc, and copper. They are used to produce batteries, electronics, pesticides, medicines, paints, inks, and dyes. In the ionic form, they enter the body, where they inhibit vital enzymes and can lead to mental deficiencies, such as retardation. That's why these items should be discarded carefully and taken to hazardous waste sites.

Synthetic Organic Compounds

Thus far, we have concentrated on our use of metals, but synthetic organic compounds are another area of considerable ecological concern because of their detrimental effects on the health of living things, including humans. Synthetic organic compounds play a role in the production of plastics, pesticides, herbicides, cosmetics, coatings, solvents, wood preservatives, and hundreds of other products.

Figure 36.10 Modern mining capabilities.
Giant mining machines—some as tall as a 20-story building—can remove an enormous amount of the Earth's crust in one scoop in order to mine for coal or a metal.

Synthetic organic compounds include halogenated hydrocarbons, in which halogens (chlorine, bromine, fluorine) have replaced certain hydrogens. One such molecule comprises the **CFCs (chlorofluorocarbons),** a type of halogenated hydrocarbon in which both chlorine and fluorine atoms replace some of the hydrogen atoms. CFCs have brought about a thinning of the Earth's **ozone shield,** which protects terrestrial life from the dangerous effects of ultraviolet radiation (see the Ecology Focus in Chapter 34). Most MDCs have passed legislation to halt production of any more CFCs. Hydrofluorocarbons, which contain no chlorine, now take their place in coolants and other products. The ozone shield is predicted to recover by 2050; in the meantime, many more cases of skin cancer are expected.

Other synthetic organic chemicals pose a direct and serious threat to the health of living things, including humans. Rachel Carson's book *Silent Spring,* published in 1962, made the public aware of the deleterious effects of pesticides.

Wastes

Every year, the countries of the world discard billions of tons of solid wastes, some on land and some in fresh and marine waters.

Wastes are generated during the mining and manufacturing process. Clean water and clean air legislation in the early 1970s prevented companies from venting untreated wastes into the atmosphere and flushing them into waterways. Industry turned to land disposal, which was unregulated at the time. Utilization of deep-well injection, pits with plastic liners, and landfills led to much water pollution and human illness, including cancer.

An estimated 5 billion metric tons of highly toxic chemicals were improperly discarded in the United States between 1950 and 1975. The public's concern was so great that the Environmental Protection Agency (EPA) came into existence. Using an allocation of monies called the superfund, part of the EPA's job is to oversee the cleanup of hazardous waste disposal sites in the United States.

The ten most commonly found contaminants are heavy metals (lead, arsenic, cadmium, chromium) and organic compounds (trichloroethylene, benzene, polychlorinated biphenyls [PCBs], chloroform, and toluene). Some of these are endocrine-disrupting contaminants.

Endocrine-Disrupting Contaminants

Recently, we have become aware of the subtle hormonal effects of pollutants, as discussed in the Ecology Focus in Chapter 21. At first, the altered reproductive traits in wildlife seemed to indicate only the estrogenlike effects of certain contaminants, but we now know that pollutants can have all sorts of hormonal effects. Therefore, *endocrine-disrupting contaminant (EDC)* is the preferred term for chemicals used in pesticides, herbicides, plastics, food additives, and personal hygiene products that can affect the endocrine system. These pollutants can occur at a level 1,000 times greater in the environment than the natural hormone levels in human blood.

Sewage

Raw sewage causes oxygen depletion in lakes and rivers. Also, human feces can contain pathogenic microorganisms that cause cholera, typhoid fever, and dysentery. In regions of LDCs where sewage treatment is practically nonexistent, many children die each year from these diseases.

Typically, sewage treatment plants use bacteria to break down organic matter to inorganic nutrients, such as nitrates and phosphates, which then enter surface waters. The end result can be cultural eutrophication. First an algal bloom occurs. Then, when the algae die off, decomposition robs the water of oxygen, which can result in a massive fish kill and greatly reduce the diversity of life.

Industrial Wastes

Industrial wastes can include heavy metals and chlorinated hydrocarbons, such as those in some pesticides. Sometimes, they accumulate in the mud of deltas and estuaries of highly polluted rivers and cause environmental problems if disturbed. Industrial pollution is being addressed in many MDCs but usually has low priority in LDCs. These wastes enter bodies of water and are subject to **biological magnification.** Decomposers are unable to break down these wastes. They enter and remain in the bodies of organisms because they are not excreted. Therefore, they become more concentrated as they pass along a food chain. Biological magnification is most apt to occur in aquatic food chains, which have more links than terrestrial food chains. Humans are often the final consumers in both types of food chains, and in some areas, human milk contains detectable amounts of DDT, discussed at the beginning of this chapter, and PCBs, which are polychlorinated hydrocarbons.

At every step, from the mining of minerals to the manufacture and use of a product, pollution occurs. Pollution prevention measures are believed to be the best way to address this ecological concern.

36.2 Impact on Biodiversity

Biodiversity can be defined as the variety of life on Earth, described in terms of the number of different species. We are presently in a biodiversity crisis—the number of extinctions (loss of species) expected to occur in the near future is unparalleled in Earth's history. Researchers have found that habitat loss was involved in 85% of the recently documented cases (Fig. 36.11*a*). Exotic species had a hand in nearly 50%, pollution was a factor in 24%, overexploitation in 17%, and disease in 3%. The percentages add up to more than 100% because most of these species are imperiled for more than one reason. For example, not only

Figure 36.11
Loss of biodiversity.
a. Habitat loss, exotic species, pollution, overexploitation, and disease have been identified as causes of the extinction of organisms. **b.** Macaws that reside in South American tropical rain forests are an example of a species that is endangered for the reasons listed in the graph.

b.

have macaws seen their habitat reduced by encroaching timber and mining companies, but they are also hunted for food and collected for the pet trade (Fig. 36.11*b*). DNA technology may now help enforce laws to protect wildlife, as described in the Ecology Focus.

Habitat Loss

Human occupation of the coastline, semiarid lands, tropical rain forests, and other areas are the most important cause of the loss of biodiversity. Scientists are especially concerned about the tropical rain forests and coral reefs because they are particularly rich in species. Already, tropical rain forests have been reduced from their original 14% of total land-mass to less than 6%. Also, 60% of coral reefs have been destroyed or are on the verge of destruction; it is possible that all coral reefs may disappear during the next 40 years.

Exotic Species

Exotic species are nonnative members of an ecosystem. Although few exotics become invasive to the point that they reduce native biodiversity, exotics are the second most important reason for biodiversity loss. Humans have introduced exotic species into new ecosystems chiefly by the following means:

Colonization. Europeans, in particular, brought various familiar species with them when they colonized new places. For example, the pilgrims brought the dandelion to the United States as a familiar salad green.

Horticulture and agriculture. Some exotics now taking over vast tracts of land have escaped from cultivated areas. Kudzu is a vine from Japan that the U.S. Department of Agriculture thought would help prevent soil erosion. The plant now covers much of the landscape in the South.

Accidental transport. Global trade and travel accidentally bring many new species from one country to another. The zebra mussel from the Caspian Sea was accidentally introduced into the Great Lakes in 1988. It now forms dense beds that reduce biodiversity by decreasing the amount of food available for higher levels in the food chain because it is such an effective filter feeder. Zebra mussels cause millions in damage per year by clogging sewage and other pipes.

Exotics on Islands

Islands are particularly susceptible to environmental discord caused by the introduction of exotic species. Islands have unique assemblages of native species that are closely adapted to one another and cannot compete well against exotics. Myrtle trees, *Myrica faya*, introduced into the Hawaiian Islands from the Canary Islands, are symbiotic with a type of bacterium that is capable of nitrogen fixation. This feature allows the species to establish itself on nutrient-poor volcanic soil such as that found in Hawaii. Once established, myrtle trees halt normal ecological succession by outcompeting native plants on volcanic soil.

Figure 36.12 Exotic species.
Mongooses were introduced into Hawaii to control rats, but they also prey on native birds.

The brown tree snake has been introduced onto a number of islands in the Pacific Ocean. The snake eats eggs, nestlings, and adult birds. On Guam, it has reduced ten native bird species to the point of extinction. Mongooses introduced into the Hawaiian Islands to control rats have increased dramatically in abundance and also prey on native birds, causing some population declines. (Fig. 36.12).

Pollution

Pollution brings about environmental changes that adversely affect the lives and health of living things. Biodiversity is particularly threatened by the following types of environmental pollution:

Acid deposition. The Ecology Focus in Chapter 2 explains that acid deposition decimates forests because it causes trees to weaken and increases their susceptibility to disease and insects. Many lakes in northern states are now lifeless because of the effects of acid deposition.

Eutrophication. Lakes are also under stress due to overenrichment, leading to an abundance of algae and decomposers, decreased oxygen, and often fish kills.

Global warming. Global warming, an increase in the Earth's temperature due to the presence of greenhouse gases in the atmosphere, is expected to have many detrimental effects, including the destruction of coastal wetlands due to a rise in sea levels, a shift in suitable temperatures to locations where various species cannot live, and the death of coral reefs (Fig. 36.13).

Ozone depletion. The Ecology Focus in Chapter 34 tells how the ozone shield protects Earth from harmful ultraviolet (UV) radiation. The release of chlorofluorocarbons (CFCs), in particular, into the atmosphere causes the shield to break down, leading to impaired growth of crops and trees and the death of plankton that sustain oceanic life.

Organic chemicals. As discussed previously, many of the organic chemicals released into the environment are endocrine-disrupting contaminants that can possibly affect the endocrine system and reproductive potential of humans and animals.

Overexploitation

Overexploitation occurs when the number of individuals taken from a wild population is so great that the population becomes severely reduced in numbers. A positive feedback cycle explains overexploitation: the smaller the population, the more commercially valuable its members, and the greater the incentive to capture the few remaining organisms.

Markets for decorative plants and exotic pets support both legal and illegal trade in wild species. Rustlers dig up rare cacti, such as the single-crested saguaro, and sell them to gardeners. Parakeets and macaws are among the birds taken from the wild for sale to pet owners. For every bird delivered alive, many more have died in the smuggling process. The same holds true for tropical fish, which often come from the coral reefs of Indonesia and the Philippines. Divers dynamite reefs or use cyanide to stun the fish; in the process, many fish die.

Declining species of mammals are still hunted for their hides, tusks, horns, or bones. Because of its rarity, a single Siberian tiger is now worth more than $500,000—its bones are pulverized and used as a medicinal powder. The horns of rhinoceroses become ornate carved daggers, and their bones are ground up to sell as a medicine. The ivory of elephants' tusks is used to make art objects, jewelry, or piano keys. The fur of a Bengal tiger sells for as much as $100,000 in Tokyo.

Fish are a renewable resource if harvesting does not exceed the ability of the fish to reproduce. Our society uses larger and more efficient fishing fleets to decimate fishing stocks. Pelagic species, such as tuna, are captured by purse-seine fishing, in which a very large net surrounds a school of fish and then is closed like a draw-string purse. Dolphins that swim above schools of tuna are often captured and then incidentally killed in this type of net. Other fishing boats drag huge trawling nets, large enough to accommodate

a.   b.

Figure 36.13 Global warming and coral reefs.
A temperature rise of only a few degrees causes **(a)** healthy coral reefs to **(b)** bleach and become lifeless.

a.

b.

Figure 36.14 Trawling.
a. These Alaskan pollock were caught by dragging a net along the seafloor. **b.** Appearance of the seafloor before *(top)* and after *(bottom)* the net went by.

12 jumbo jets, along the seafloor to capture bottom-dwelling fish (Fig. 36.14*a*). Only large fish are kept; undesirable small fish and sea turtles are discarded, dying, back into the ocean. Trawling has been called the marine equivalent of clear-cutting trees because after the net goes by, the sea bottom is devastated (Fig. 36.14*b*). Today's fishing practices often don't allow fisheries to recover. Cod and haddock, once the most abundant fish along the northeast coast, are now often outnumbered by dogfish and skate.

A marine ecosystem can be disrupted by overfishing, as exemplified on the U.S. west coast. When sea otters began to decline in numbers, investigators found that they were being eaten by orcas (killer whales). Usually, orcas prefer seals and sea lions to sea otters, but they began eating sea otters when few seals and sea lions could be found. What caused a decline in seals and sea lions? Their preferred food sources—perch and herring—were no longer plentiful due to overfishing. Ordinarily, sea otters keep the population of sea urchins, which feed on kelp, under control. But with fewer sea otters around, the sea urchin population exploded and decimated the kelp beds. Thus, overfishing set in motion a chain of events that detrimentally altered the food web of an ecosystem.

Disease

Wildlife is subject to emerging diseases just as humans are. Due to the encroachment of humans on their habitat, wildlife have been exposed to new pathogens from domestic animals living nearby, and due to human intervention, they have been exposed to animals they would ordinarily not come in contact with. For example, canine distemper was spread from domesticated dogs to lions in the African Serengeti, causing population declines. Avian influenza likely emerged from domesticated fowl (e.g., chicken) populations and will likely lead to the deaths of millions of wild birds.

The significant effect of diseases on biodiversity is underscored by National Wildlife Health Center findings that almost half of sea otter deaths along the coast of California are due to infectious diseases. Scientists tell us the number of pathogens that cause disease is on the rise, and just as human health is threatened, so is that of wildlife. Extinctions due simply to disease may occur, and are currently implicated in the worldwide decline of amphibians (Fig. 36.15).

The five main causes of extinction are habitat loss, introduction of exotic species, pollution, overexploitation, and disease.

Figure 36.15 Amphibian at risk.
The harlequin toad is near extinct due to infections by a fungal pathogen. Diseases are implicated in global population declines and extinctions of amphibians.

36.3   Value of Biodiversity

Conservation biology strives to reverse the trend toward the possible extinction of thousands of plants and animals. To bring this about, it is necessary to make all people aware that biodiversity is a resource of immense value.

Direct Value

Various individual species perform services for human beings and contribute greatly to the value we should place on biodiversity. Figure 36.16 summarizes the direct value of wildlife.

Medicinal Value

Most of the prescription drugs used in the United States were originally derived from living organisms. The rosy periwinkle from Madagascar is an excellent example of a tropical plant that has provided us with useful medicines. Potent chemicals from this plant are now used to treat two forms of cancer: leukemia and Hodgkin disease. Because of these drugs, the survival rate for childhood leukemia has improved from 10% to 90%, and Hodgkin disease is usually curable. Although the value of saving a life cannot be calculated, it is still sometimes easier for us to appreciate the worth of a resource if it is explained in monetary terms. Thus, researchers estimate that, judging from the success rate in the past, an additional 328 types of drugs are yet to be found in tropical rain forests, and the value of this resource to society is probably $147 billion.

You may already know that the antibiotic penicillin is derived from a fungus and that certain species of bacteria produce the antibiotics tetracycline and streptomycin. These drugs have proven to be indispensable in the treatment of diseases, including sexually transmitted diseases such as chlamydia, the most common STD in the United States.

Leprosy is among those diseases for which there is, as yet, no cure. The bacterium that causes leprosy will not grow in the laboratory, but scientists discovered that it grows naturally in the nine-banded armadillo. Having a source of the bacterium may make it possible to find a cure for leprosy. The blood of horseshoe crabs contains a substance called limulus amoebocyte lysate, which is used to ensure that medical devices, such as pacemakers, surgical implants, and prosthetic devices, are free of bacteria. Blood is taken from 250,000 crabs a year, which are then returned to the sea unharmed.

Figure 36.16 Direct value of wildlife.

The direct services of wild species benefit human beings immensely. It is sometimes possible to calculate the monetary value, which is always surprisingly large.

Wild species, like the rosy periwinkle, are sources of many medicines.

Wild species, like many marine species, provide us with food.

Wild species, like the nine-banded armadillo, play a role in medical research.

Agricultural Value

Crops such as wheat, corn, and rice are derived from wild plants that have been modified to be high producers. The same high-yield, genetically similar strains tend to be grown worldwide. When rice crops in Africa were being devastated by a virus, researchers grew wild rice plants from thousands of seed samples until they found one that contained a gene for resistance to the virus. These wild plants were then used in a breeding program to transfer the gene into high-yield rice plants. If this variety of wild rice had become extinct before it could be discovered, rice cultivation in Africa might have collapsed.

Biological pest controls—natural predators and parasites—are often preferable to using chemical pesticides. When a rice pest called the brown planthopper became resistant to pesticides, farmers began to use natural brown planthopper enemies instead. The economic savings were calculated at well over $1 billion. Similarly, cotton growers in Cañete Valley, Peru, found that pesticides were no longer working against the cotton aphid because of resistance. Research identified natural predators that are now being used to an ever greater degree by cotton farmers. Again, savings have been enormous.

Most flowering plants are pollinated by animals such as bees, wasps, butterflies, beetles, birds, and bats. The honeybee, *Apis mellifera,* has been domesticated, and it pollinates almost $10 billion worth of food crops annually in the United States. The danger of this dependency on a single species is exemplified by the fact that mites have now wiped out more than 20% of the commercial honeybee population in the United States. Where can we get resistant bees? From the wild, of course. The value of wild pollinators to the U.S. agricultural economy has been calculated at $4.1 to $6.7 billion a year.

Consumptive Use Value

We have experienced much success in cultivating crops, keeping domesticated animals, growing trees in plantations, and so forth. But so far, aquaculture, the growing of fish and shellfish for human consumption, has contributed only minimally to human welfare. Instead, most freshwater and marine harvests depend on wild animals, such as fishes (e.g., trout, cod, tuna, and flounder) and crustaceans (e.g., lobsters, shrimps, and crabs). Obviously, these aquatic organisms are an invaluable biodiversity resource.

Wild species, like the lesser long-nosed bat, are pollinators of agricultural and other plants.

Wild species, like rubber trees, can provide a product indefinitely if the forest is not destroyed.

Wild species, like ladybugs, play a role in biological control of agricultural pests.

Wildlife Conservation and DNA

After DNA analysis, scientists were amazed to find that some 60% of loggerhead turtles drowning in the nets and hooks of fisheries in the Mediterranean Sea were from beaches in the southeastern United States. Since the unlucky creatures were a good representative sample of the turtles in the area, that meant more than half of the young turtles living in the Mediterranean Sea had hatched from nests on beaches in Florida, Georgia, and South Carolina (Fig. 36A*a*). Some 20,000–50,000 loggerheads die each year due to the Mediterranean fisheries, which may partly explain the decline in loggerheads nesting on southeastern U.S. beaches for the last 25 years.

The sequencing of DNA from Alaskan brown bears allowed Sandra Talbot (a graduate student at the University of Alaska's Institute of Arctic Biology) and wildlife geneticist Gerald Shields to conclude that there are two types of brown bears in Alaska. One type resides only on southeastern Alaska's Admiralty, Baranof, and Chichagof islands, known as the ABC Islands. The other brown bear in Alaska is found throughout the rest of the state, as well as in Siberia and western Asia (Fig. 36A*b*).

A third distinct type of brown bear, known as the Montana grizzly, resides in other parts of North America. These three types comprise all of the known brown bears in the New World.

The ABC bears' uniqueness may be bad news for the timber industry, which has expressed interest in logging parts of the ABC Islands. Says Shields, "Studies show that when roads are built and the habitat is fragmented, the population of brown bears declines. Our genetic observations suggest they are truly unique, and we should consider their heritage. They could never be replaced by transplants."

In what will become a classic example of how DNA analysis might be used to protect endangered species from future ruin, scientists from the United States and New Zealand carried out discreet experiments in a Japanese hotel room on whale sushi bought in local markets. Sushi, a staple of the Japanese diet, is rice and meat wrapped in seaweed. Armed with a miniature DNA sampling machine, the scientists found that, of the 16 pieces of whale sushi they examined, many were from whales that are endangered or protected under an international moratorium on whaling. "Their findings demon-

strated the true power of DNA studies," says David Woodruff, a conservation biologist at the University of California, San Diego.

One sample was from an endangered humpback, four were from fin whales, one was from a northern minke, and another from a beaked whale. Stephen Palumbi of Harvard University says the technique could be used for monitoring and verifying catches. Until then, he says, "no species of whale can be considered safe."

Meanwhile, Ken Goddard, director of the unique U.S. Fish and Wildlife Service Forensics Laboratory in Ashland, Oregon, is already on the watch for wildlife crimes in the United States and 122 other countries that send samples to him for analysis. "DNA is one of the most powerful tools we've got," says Goddard, a former California police crime-lab director.

The lab has blood samples, for example, for all of the wolves being released into Yellowstone Park—"for the obvious reason that we can match those samples to a crime scene," says Goddard. The lab has many cases currently pending in court that he cannot discuss. But he likes to tell the story of the lab's first DNA-matching case. Shortly after the lab opened in 1989, California wildlife authorities contacted Goddard. They had seized the carcass of a trophy-sized deer from a hunter. They believed the deer had been shot illegally on a 3,000-acre preserve owned by actor Clint Eastwood. The agents found a gut pile on the property but had no way to match it to the carcass. The hunter had two witnesses to deny the deer had been shot on the preserve.

Goddard's lab analysis made a perfect match between tissue from the gut pile and tissue from the carcass. Says Goddard: "We now have a cardboard cutout of Clint Eastwood at the lab saying 'Go ahead: Make my DNA.'"

Discussion Questions

1. Recently, distinctiveness in terms of genetic DNA sequence has been used to set policy. That is, if researchers find a population of a species, such as ABC brown bears, to be genetically distinct from other populations of the same species, then that population may be afforded special protection, such as threatened or endangered status under the U.S. Endangered Species Act. This

policy is then used to make major environmental decisions, such as halting logging in that area. Should DNA distinctness be used to make such major policy decisions? Why or why not?

2. Forensic DNA evidence has recently been used to prosecute hunters who illegally hunt animals, such as deer, in areas where they do not have a proper hunting license. Occasionally, such hunters face harsh sentences that may even include jail time. Should people who hunt illegally be put in jail or simply fined?

3. DNA studies, such as those conducted in Japan, opened our eyes to the fact that illegal whale hunting still occurs. Some people have recommended we screen the DNA of much of the imported fish meat coming from different countries. However, this can be quite expensive. Do you support genetic testing of imported fish meat, even if it increases the cost of the meat to the consumer?

a. Loggerhead turtle

b. Brown bears

Figure 36A DNA Studies.

a. Many loggerhead turtles found in the Mediterranean Sea are from the southeastern United States. **b.** These two brown bears appear similar, but one type, known as an ABC bear, resides only on southeastern Alaska's Admiralty, Baranof, and Chichagof islands.

The environment provides all sorts of other products that are sold in the marketplace worldwide, including wild fruits and vegetables, skins, fibers, beeswax, and seaweed. Also, some people obtain their meat directly from the wild. In one study, researchers calculated that the economic value of wild pig in the diet of native hunters in Sarawak, East Malaysia, was approximately $40 million per year.

Similarly, many trees are still felled in the natural environment for their wood. Researchers have calculated that a species-rich forest in the Peruvian Amazon region is worth far more if the forest is used for fruit and rubber production than for timber production. Fruit and the latex needed to produce rubber can be brought to market for an unlimited number of years, whereas once the trees are gone, no more timber or other resulting products can be harvested.

Indirect Value

The wild species we have been discussing live in ecosystems. If we want to preserve them, many scientists argue it is more economical to save the ecosystems than the individual species. Ecosystems perform many services for modern humans, who increasingly live in cities. These services are said to be indirect because they are pervasive and not easily discernible. Even so, our survival depends on the functions that ecosystems perform for us. Recent studies suggest the indirect value of ecosystem services surpasses that of the global gross national product—about $33 trillion per year!

Biogeochemical Cycles

You'll recall from Chapter 34 that ecosystems are characterized by energy flow and chemical cycling. The biodiversity within ecosystems contributes to the workings of the water, carbon, nitrogen, phosphorus, and other biogeochemical cycles. We are dependent on these cycles for fresh water, removal of carbon dioxide from the atmosphere, uptake of excess soil nitrogen, and provision of phosphate. When human activities upset the usual workings of biogeochemical cycles, the dire environmental consequences include the release of excess pollutants that are harmful to us. Technology is unable to artificially contribute to or recreate any of the biogeochemical cycles.

Waste Disposal

Decomposers break down dead organic matter and other types of wastes to inorganic nutrients that are used by the producers within ecosystems. This function aids humans immensely because we dump millions of tons of waste material into natural ecosystems each year. If it were not for decomposition, waste would soon cover the entire surface of our planet. We can build sewage treatment plants, but they are expensive. In addition, few of them break down solid wastes completely to inorganic nutrients. It is less expensive and more efficient to water plants and trees with partially treated wastewater and let soil bacteria cleanse it completely.

Biological communities are also capable of breaking down and immobilizing pollutants, such as heavy metals and pesticides, that humans release into the environment. A review of wetland functions in Canada assigned a value of $50,000 per hectare (100 acres or 10,000 square meters) per year to the ability of natural areas to purify water and take up pollutants.

Provision of Fresh Water

Few terrestrial organisms are adapted to living in a salty environment—they need fresh water. The water cycle continually supplies fresh water to terrestrial ecosystems. Humans use fresh water in innumerable ways, including for drinking and for irrigating their crops. Freshwater ecosystems, such as rivers and lakes, also provide us with fish and other types of organisms for food.

Unlike other commodities, there is no substitute for fresh water. We can remove salt from seawater to obtain fresh water, but the cost of desalinization is about four to eight times the average cost of fresh water acquired via the water cycle.

Forests and other natural ecosystems exert a "sponge effect." They soak up water and then release it at a regular rate. When rain falls in a natural area, plant foliage and dead leaves lessen its impact, and the soil slowly absorbs it, especially if the soil has been aerated by organisms. The water-holding capacity of forests reduces the possibility of flooding. The value of a marshland outside Boston, Massachusetts, has been estimated at $72,000 per hectare per year based solely on its ability to reduce floods. Flooding in New Orleans by Hurricane Katrina in 2005 would likely have been much less severe if the natural wetlands and marshlands in the Gulf of Mexico had still been intact.

Prevention of Soil Erosion

Intact ecosystems naturally retain soil and prevent soil erosion. The importance of this ecosystem attribute is especially observed following deforestation. In Pakistan, the world's largest dam, the Tarbela Dam, is losing its storage capacity of 12 billion cubic meters many years sooner than expected because silt is building up behind the dam due to deforestation. At one time, the Philippines exported $100 million worth of oysters, mussels, clams, and cockles each year. Now, silt carried down rivers following deforestation is smothering the mangrove ecosystem that serves as a nursery for the sea. Most coastal ecosystems are not as bountiful as they once were because of deforestation and a myriad of other assaults.

Regulation of Climate

At the local level, trees provide shade, reduce the need for fans and air conditioners during the summer, and provide buffers from noise. In tropical rain forests, trees maintain regional rainfall, and without them the forests become arid.

Globally, forests ameliorate the climate because they take up carbon dioxide and release oxygen. The leaves of trees use carbon dioxide when they photosynthesize, and the bodies of the trees store carbon. When trees are cut and burned, carbon dioxide is released into the atmosphere. Carbon dioxide makes a significant contribution to global warming, which is expected to be stressful for many plants and animals.

Ecotourism

In the United States, nearly 100 million people enjoy vacationing in a natural setting. To do so, they spend $4 billion each year on fees, travel, lodging, and food. Many tourists want to go sport fishing, whale watching, boat riding, hiking, birdwatching, and the like. Some merely want to immerse themselves in the beauty of a natural environment.

36.4 Working Toward a Sustainable Society

A **sustainable** society, like a sustainable ecosystem, would be able to provide the same goods and services for future generations of human beings as it does now. At the same time, biodiversity would be conserved.

Today's Society

Evidence indicates that human society today is most likely not sustainable (Fig. 36.17a). The following characteristics make the present human society unsustainable:

- A considerable proportion of land, and therefore of natural ecosystems, is being altered for human purposes (homes, agriculture, factories, etc.).
- Agriculture requires large inputs of nonrenewable fossil fuel energy, fertilizer, and pesticides, which create much pollution. More fresh water is used for agriculture than in homes.
- At least half of the agricultural yield in the United States goes toward feeding animals. According to the ten to one rule of thumb, it takes 10 pounds of grain to grow one pound of meat. Therefore, it is wasteful for citizens in MDCs to eat as much meat as they do. Also, animal sewage pollutes water.
- Even though fresh water is a renewable resource, we are running out of the available supply within a given time frame.
- Our society primarily utilizes nonrenewable fossil fuel energy, which leads to global warming, acid precipitation, and smog.
- Minerals are nonrenewable, and the mining, manufacture, and use of mineral products are responsible for much environmental pollution.

Characteristics of a Sustainable Society

A natural ecosystem can offer clues as to what a sustainable human society would be like. A natural ecosystem uses only renewable solar energy, and its materials cycle through the vari-

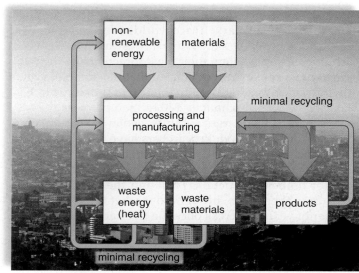

a. Human society at present

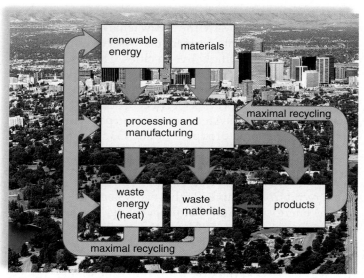

b. Sustainable society

Figure 36.17 Current human society versus a sustainable society.
a. At present, our "throw-away" society is characterized by high input of energy and raw materials, large output of waste materials and energy in the form of heat (*red arrows*), and minimal recycling (*purple arrows*).
b. A sustainable society would be characterized by the use of only renewable energy sources, reuse of heat and waste materials, and maximal recycling of products (*purple arrows*).

ous populations back to the producer once again. It is clear that if we want to develop a sustainable society, we too should use renewable energy sources and recycle materials (Fig. 36.17b).

While we are sometimes quick to realize that the growing populations of LDCs are putting a strain on the environment, we should realize that the disproportionate resource consumption of MDCs also greatly stresses the environment. Sustainability is, more than likely, incompatible with the kinds of consumption and waste patterns currently practiced in MDCs.

The term "conservation" was first adopted by President Theodore Roosevelt around 1908 to describe America's new environmental policies, including the establishment of many new forest reserves. Roosevelt was a farmer and an avid hunter, and by adding millions of acres to America's forest reserves, he hoped to conserve lands threatened with shrinking numbers of wild game or grazing lands that could no longer support cattle due to overgrazing. This kind of environmental protection is historically a movement to protect the environment for human use and to sustain it for future human generations. In the eyes of most conservationists, in any conflict between nature and people, people often come first.

In the past few decades, an environmentalism movement has emerged that tries to view the world through the eyes of the environment itself, giving only secondary consideration to human concerns. Such an approach sometimes limits people's freedoms in the name of "mother nature." Under its policies, people often find themselves in opposition to laws that put animals, plants, or rare habitats first. Logging companies can't log their own lands because of a nesting population of rare spotted owls, or mountain bikers are denied access to trails because of increasing soil erosion. In some cases, concerns for natural resources and their proper distribution are even closely linked to laws that limit families' reproductive choices. In China, laws limiting couples to only one child certainly put human interests second to large-scale environmental concerns. Is the benefit to the environment worth limiting such personal freedoms?

Environmentalists believe that these limits on freedom are only expanded cases of traditional conservation, which look further into the future, or perhaps take a more cautious approach to environmental problems. They believe that environmental regulations are not too strict, and that, in fact, much more needs to be done. Still, many industry lobbyists hope to limit environmental regulations by arguing for greater individual freedom and property rights, and they support a more traditional type of conservation, which often puts immediate human prosperity above long-term environmental prosperity.

Form Your Own Opinion

1. When considering environmental protection, should we put human concerns first, or take a larger view that puts the environment first?
2. Are any activities that you partake in sometimes limited by environmental concerns? Do you feel this is justified, or should you be allowed more personal freedoms?
3. Do you agree with the view of traditional conservationism or with environmentalism?

Assessing Economic Well-Being and Quality of Life

The gross national product (GNP) is a measure of the flow of money from consumers to businesses in the form of goods and services purchased. It can also be considered the total costs of all manufacturing, production, and services in the form of salaries and wages, mortgage and rent, interest and loans, taxes, and profit within and outside a country. In other words, GNP pertains solely to economic activities and does not take into consideration any activities that are socially or environmentally harmful. For example, the balance sheet would account for costs of drugs and hospital stays for lung cancer patients, but not for their pain and distress. Nor do GNP measures account for the damage done to the environment during clear-cutting of forests, strip mining, or land development.

Measuring Nonmonetary Values

Measures that include noneconomic indicators are probably better than the GNP at revealing our quality of life. The Index of Sustainable Economic Welfare (ISW) includes real per capita income, distributional equity, natural resource depletion, environmental damage, and the value of unpaid labor. Another such index is the Genuine Progress Indicator (GPI). The ISW and GPI both suggest that the global quality of life has gone down, despite the economic improvements indicated by the GNP. While some improvements have occurred over the past several years, such as the improved status of women in many LDCs, the overall quality of life has not improved.

Ecological economists are searching for a way to measure still other values, including the following:

Use value. The actual price we pay to use or consume a resource.

Option value. Preserving options for the future, such as saving a wetland or a forest.

Existence value. Saving things we might not realize exist yet, such as flora and fauna in a tropical rain forest that one day could be the sources of new drugs.

Aesthetic value. Appreciating an area or creature for its beauty and/or its contribution to biodiversity.

Cultural value. Factors such as language, mythology, and history that are important for cultural identity.

Scientific and educational value. Valuing the knowledge of naturalists, or even an experience of nature, as types of rational facts.

Today's society is not sustainable. Tomorrow's society might be sustainable if it becomes more like a natural ecosystem and begins to assign a monetary value to all kinds of environmental benefits.

Summarizing the Concepts

36.1 Human Use of Resources

People need a place to live, and sometimes they make poor choices. Beach erosion is common along coasts; the seas are rising, and people often fill in wetlands. They also build dams that retard sediment so that it never reaches the coast. Desertification is a possibility when people overuse semiarid lands and also after they remove trees from tropical rain forests, where the soil is thin and nutrient-poor.

Fresh water is available as surface water in rivers and lakes and also in underground sources called aquifers. To increase the supply of fresh water, people build dams, which may not be useful for long because of sediment buildup. Many rivers around the world now run dry and carry a heavy burden of salt. People also remove water from aquifers at a rate that cannot be sustained. Subsidence, sinkholes, and saltwater intrusion are environmental consequences to withdrawing too much water from aquifers.

Food production has kept pace with population growth. Modern farming is characterized by the use of few genetic varieties; heavy use of fertilizers, pesticides, and herbicides; generous irrigation; and excessive fuel consumption. Soil erosion and salinization are side effects of modern farming methods. The green revolution gave the LDCs hybrid plants called "high responders" that produce well as long as they have the subsidies already mentioned. Genetically engineered crops might usher in a second green revolution. Much of the grain grown in the MDCs is used to feed domesticated animals, which produce a great deal of sewage that gets washed into water-ways. Also, it takes 10 grain calories to produce one meat calorie; thus, it would be possible to feed ten times as many people on grain as on meat. Modern fishing methods have decimated fisheries, and the catches of the past may never be seen again.

Some types of energy are nonrenewable (e.g., fossil fuels and nuclear power), and some are renewable (e.g., hydropower, geothermal energy, wind power, and solar energy). The burning of fossil fuels has environmental consequences. Like the panes of a greenhouse, carbon dioxide and other gases allow the sun's rays to pass through but impede the release of infrared wavelengths. Scientists predict that a buildup of these "greenhouse gases" is already leading to global warming. The effects of global warming could include a rise in sea level and a change in climate patterns. An effect on agriculture could follow.

With regard to renewable energy supplies, only hydropower (dams) and geothermal energy are routinely utilized. However, wind and solar energy are expected to be used more in the future. Some predict a solar-hydrogen revolution, in which solar energy replaces fossil fuel energy and cars run on hydrogen fuel instead of gasoline. This would eventually do away with the environmental problems associated with using fossil fuel energy.

Minerals are a nonrenewable resource that we are using up at a rapid rate and the mining-process creates waste and pollution. Some synthetic organic chemicals are involved in the production of plastics, pesticides, and herbicides as well as all sorts of other products. Ozone shield destruction is particularly associated with CFCs.

The countries of the world discard billions of tons of solid wastes, some on land and some in fresh and marine waters. The sources of surface water pollution are many and varied, including human sewage and agricultural and industrial wastes.

36.2 Impact on Biodiversity

Researchers have identified the major causes of extinction. Habitat loss is the most frequent cause, followed by introduction of exotic species, pollution, overexploitation, and disease. Habitat loss has occurred in all parts of the biosphere, but concern has now centered on tropical rain forests and coral reefs, where biodiversity is especially high. Exotic species have been introduced into foreign ecosystems because of colonization, horticulture or agriculture, and accidental transport. Among the various causes of pollution (acid rain, eutrophication, ozone depletion, and organic chemicals), global warming is expected to cause the most instances of extinction. Overexploitation is exemplified by commercial fishing, which is so efficient that fisheries of the world are collapsing. Wildlife is subject to an increasing number of disease-causing pathogens.

36.3 Value of Biodiversity

The direct value of biodiversity is evidenced by the observable services of individual wild species. Wild species are our best source of new medicines to treat human ills, and they meet other medical needs as well. For example, the bacterium that causes leprosy grows naturally in armadillos, and horseshoe crab blood contains a bacteria-fighting substance.

Wild species have agricultural value. Domesticated plants and animals are derived from wild species, which also serve as a source of genes for the improvement of their phenotypes. Instead of pesticides, wild species can be used as biological controls, and most flowering plants make use of animal pollinators. Much of our food, particularly fish and shellfish, is still caught in the wild. Hardwood trees from natural forests supply us with lumber for various purposes, such as making furniture.

The indirect services provided by ecosystems are largely unseen but absolutely necessary to our well-being. These services include the workings of biogeochemical cycles, waste disposal, provision of fresh water, prevention of soil erosion, and regulation of climate. Many people also enjoy spending time in natural settings.

36.4 Working Toward a Sustainable Society

Our present-day society is not sustainable. Calculation of the GNP does not take into account overuse of environmental resources as it should. If we were to pattern our society after natural ecosystems, solar energy would supply our energy needs, and materials would be recycled. A new index such as the GPI would take into account nonmonetary ecological values.

Testing Yourself

Choose the best answer for each question.

1. Which type of resource is renewable and readily available?
   - a. oil
   - b. coal
   - c. solar energy
   - d. deep aquifers
   - e. phosphorus

2. Which activity could not be a cause of beach erosion?
   - a. filling in wetlands
   - b. building groins
   - c. building seawalls
   - d. warming the atmosphere
   - e. importing sand

3. Desertification is often caused by
   - a. overuse of aquifers.
   - b. urban sprawl.
   - c. air pollution.
   - d. overgrazing.

4. Soil in the tropics is often nutrient-poor because
   - a. nutrients are tied up in plants.
   - b. it is mostly sand.
   - c. it has a high pH.
   - d. rainfall leaches out minerals.

5. Most fresh water is used for
   - a. domestic purposes, such as bathing, flushing toilets, and watering lawns.
   - b. domestic purposes, such as cooking and drinking.
   - c. agriculture.
   - d. industry.

6. The damming of rivers does not cause a(n)
   - a. increase in salinity.
   - b. loss of water through evaporation.
   - c. change in the level of the water table.
   - d. buildup of sediment.

7. Removal of groundwater from aquifers may cause
   - a. pollution.
   - b. subsidence.
   - c. mineral depletion.
   - d. soil erosion.
   - e. All of these are correct.

8. Modern agriculture is not characterized by
   - a. dependency on chemical inputs.
   - b. frequent irrigation.
   - c. high fuel consumption.
   - d. high diversity of crop varieties planted.

9. Farmland becomes desertified as a result of
   - a. high fertilizer input.
   - b. frequent irrigation.
   - c. soil erosion.
   - d. deep tilling.
   - e. Both b and d are correct.

10. Green revolution crop varieties are
    - a. genetically diverse.
    - b. high responders.
    - c. drought tolerant.
    - d. deep-rooted plants.

11. The raising of domestic livestock
    - a. consumes large amounts of fossil fuels.
    - b. leads to water pollution.
    - c. is energetically wasteful.
    - d. All of these are correct.

12. In the world's shrimp fishery, the number of pounds of other animals caught and discarded
    - a. equals the number of pounds of shrimp caught.
    - b. is less than the number of pounds of shrimp caught.
    - c. is greater than the number of pounds of shrimp caught.

For questions 13–17, match the description to the type of fuel in the key. Each answer can be used more than once. Each question may have more than one answer.

Key:
- a. nuclear power
- b. fossil fuels
- c. hydropower
- d. solar power
- e. wind power
- f. geothermal power

13. Renewable source of power

14. Waste products are harmful

15. Detrimental environmental effects

16. More available in certain geographic locations

17. Contribute to global warming

18. Heavy metals are dangerous to humans because they
    - a. inhibit important enzymes.
    - b. cause apoptosis.
    - c. reduce the body's ability to carry oxygen.
    - d. break down mitochondria.

19. Earth's ozone shield has been damaged by
    - a. heavy metals.
    - b. strip mining.
    - c. chlorofluorocarbons.
    - d. fossil fuels.

20. Endocrine-disrupting contaminants most often affect
    - a. growth rate.
    - b. adrenaline response.
    - c. reproduction.
    - d. brain function.
    - e. heart tissue.

21. The most significant cause of the loss of biodiversity is
    - a. habitat loss.
    - b. pollution.
    - c. exotic species.
    - d. disease.
    - e. overexploitation.

22. Agricultural runoff of nutrients can result in a fish kill due to
    - a. toxic effects of the high nutrient levels on the fish.
    - b. toxic effects of the high nutrient levels of the food sources of the fish.
    - c. low oxygen levels in the water as decomposers break down algae.
    - d. the cooling of the water as algae shade the bottom.

23. A transition to hydrogen fuel technology will
   a. be long in coming and probably not of major significance.
   b. lessen many current environmental problems.
   c. not be likely since it will always be expensive and as polluting as natural gas.
   d. be of major consequence, but resource limitations for obtaining hydrogen will hinder its progress.

24. GNP measures all but
   a. total costs of all manufacturing and production.
   b. costs of services.
   c. cost of environmental degradation through waste costs.
   d. profit.
   e. All of these are correct.

25. Biological magnification is pronounced in
   a. aquatic food chains.
   b. terrestrial food chains.
   c. long food chains.
   d. energy pyramids.
   e. Both a and c are correct.

26. Global warming does not bring about
   a. the inability of species to migrate to cooler climates as environmental temperatures rise.
   b. the bleaching and drowning of coral reefs.
   c. a rise in sea levels and loss of wetlands.
   d. preservation of species because cold weather causes hardships.
   e. All of these results are expected.

Understanding the Terms

aquifer 748
biodiversity 754
biological magnification 754
CFCs (chlorofluorocarbons) 754
deforestation 747
desertification 747
exotic species 755
fossil fuel 751
greenhouse gases 751
mineral 753

nonrenewable resource 746
ozone shield 754
photovoltaic (solar) cell 753
pollution 746
renewable resource 746
resource 746
salinization 750
saltwater intrusion 748
sinkhole 748
subsidence 748
sustainable 762

Match the terms to these definitions:

a._____ Any environmental change that adversely affects the lives and health of living things.
b._____ Occurs when a portion of Earth's surface gradually settles downward.
c._____ Accumulation of O_3, formed from oxygen in the upper atmosphere; a filtering layer that protects Earth from ultraviolet radiation.
d._____ Denuding and degrading a once-fertile land, initiating a desert-producing cycle that feeds on itself and causes long-term changes in the soil, climate, and biota of an area.
e._____ Rock layers that contain water and will release it in appreciable quantities to wells or springs.

Thinking Critically

1. Why should we care about conserving biodiversity, even if doing so may pose a threat to individual people's livelihoods?

2. What are some ways we can reduce our reliance on nonrenewable resources and move toward increased reliance on renewable resources?

3. Discuss both the direct and indirect results of deforestation.

4. Why should people be generally more concerned about pollutants in our meat than in our vegetables?

5. List some ways that you, personally, can contribute to moving humans toward a more sustainable society.

ARIS, the *Inquiry into Life* Website

ARIS, the website for *Inquiry into Life,* provides a wealth of information organized and integrated by chapter. You will find practice quizzes, interactive activities, labeling exercises, flashcards, and much more that will complement your learning and understanding of general biology.

www.aris.mhhe.com

Appendix A
Answer Key

This appendix contains the answers to the Testing Yourself questions and the Understanding the Terms problems, which appear at the end of each chapter, and the Practice Problems and Additional Genetics Problems, which appear in Chapters 23 and 26. Answers to the Thinking Critically questions that appear at the end of each chapter can be found on the book's website at www.mhhe.com/maderinquiry12.

Chapter 1
Testing Yourself
1. c; 2. c; 3. d; 4. b; 5. d; 6. a; 7. b; 8. b; 9. c; 10. b; 11. d; 12. a; 13. d; 14. a; 15. d; 16. c; 17. b; 18. d; 19. c; 20. a; 21. e; 22. d; 23. b; 24. **a.** energy flow, **b.** chemical cycling; 25. c; 26. b; 27. c; 28. e

Understanding the Terms
a. principle or law; **b.** hypothesis; **c.** energy; **d.** adaptation; **e.** homeostasis

Chapter 2
Testing Yourself
1. d; 2. c; 3. d; 4. d; 5. a; 6. d; 7. c; 8. b; 9. d; 10. d; 11. d; 12. a; 13. c; 14. a; 15. d; 16. d; 17. a; 18. c; 19. c; 20. false; 21. false; 22. true; 23. false; 24. hydrogen; 25. a; 26. d; 27. a; 28. d; 29. b

Understanding the Terms
a. protein; **b.** enzyme; **c.** cellulose; **d.** neutron; **e.** acid

Chapter 3
Testing Yourself
1. a; 2. d; 3. d; 4. c; 5. d; 6. d; 7. a; 8. c; 9. b; 10. d; 11. c; 12. b; 13. d; 14. c; 15. d; 16. c; 17. a; 18. b; 19. e; 20. d; 21. d; 22. b; 23. d; 24. a; 25. **a.** rougher, **b.** nucleus, **c.** nucleolus, **d.** Golgi; 26. c; 27. c; 28. a; 29. c; 30. a; 31. d

Understanding the Terms
a. matrix; **b.** nucleolus; **c.** cytoskeleton; **d.** Golgi apparatus; **e.** endoplasmic reticulum

Chapter 4
Testing Yourself
1. d; 2. c; 3. d; 4. d; 5. c; 6. b; 7. true; 8. a; 9. d; 10. a; 11. b; 12. b; 13. c; 14. b; 15. c; 16. b; 17. a; 18. e; 19. d; 20. b; 21. d; 22. a; 23. c; 24. b; 25. b; 26. a; 27. e; 28. c; 29. b; 30. b; 31. b; 32. a; 33. a; 34. e; 35. d

Understanding the Terms
a. diffusion; **b.** turgor pressure; **c.** isotonic; **d.** facilitated transport; **e.** exocytosis

Chapter 5
Testing Yourself
1. a; 2. b; 3. a; 4. b; 5. S; 6. S; 7. G_2; 8. M; 9. c; 10. c; 11. e; 12. a; 13. d; 14. b; 15. c; 16. h, i; 17. a, c, d, f; 18. b, e; 19. j; 20. c; 21. d; 22. d; 23. d; 24. a; 25. b; 26. e; 27. d; 28. a; 29. c; 30. c; 31. b; 32. c; 33. c; 34. c; 35. c; 36. e

Understanding the Terms
a. spermatogenesis; **b.** polar body; **c.** secondary oocyte; **d.** synapsis; **e.** cytokinesis

Chapter 6
Testing Yourself
1. d; 2. a; 3. a; 4. c; 5. b; 6. c; 7. a; 8. d; 9. b; 10. d; 11. c; 12. a; 13. d; 14. d; 15. a; 16. a; 17. a; 18. b; 19. a; 20. d; 21. c; 22. c; 23. a, c; 24. b; 25. b, d; 26. e; 27. d; 28. d; 29. e; 30. d; 31. c; 32. b; 33. b; 34. d; 35. b

Understanding the Terms
a. metabolism; **b.** cofactor; **c.** kinetic energy; **d.** vitamin; **e.** denatured

Chapter 7
Testing Yourself
1. b; 2. d; 3. a; 4. c; 5. e; 6. b; 7. a; 8. b; 9. a; 10. substrate; 11. substrate; 12. product; 13. product; 14. c; 15. b; 16. c; 17. b; 18. c; 19. d; 20. b; 21. c; 22. d; 23. c; 24. b; 25. a; 26. e; 27. b; 28. b; 29. c; 30. d; 31. d; 32. c; 33. c; 34. a; 35. b; 36. c

Understanding the Terms
a. citric acid cycle; **b.** aerobic; **c.** fermentation; **d.** pyruvate; **e.** electron transport chain

Chapter 8
Testing Yourself
1. d; 2. b; 3. c; 4. d; 5. b; 6. d; 7. c; 8. a; 9. c; 10. d; 11. d; 12. a; 13. c; 14. c; 15. a; 16. d; 17. b; 18. c; 19. **a.** electron acceptor; **b.** electron transport chain; **c.** ATP; 20. e; 21. **a.** thylakoid; **b.** O_2; **c.** stroma; **d.** Calvin cycle; **e.** grana; 22. e; 23. a; 24. e; 25. a; 26. e; 27. c; 28. e; 29. e; 30. e; 31. d

Understanding the Terms
a. light reactions; **b.** thylakoid; **c.** C_3 plant; **d.** chlorophyll; **e.** photosystem

Chapter 9
Testing Yourself
1. c; 2. c; 3. b; 4. b; 5. b; 6. c; 7. a; 8. **a.** epidermis, **b.** cortex, **c.** endodermis, **d.** phloem, **e.** xylem. See also Figure 9.19; 9. **a.** upper epidermis, **b.** palisade mesophyll, **c.** leaf vein, **d.** spongy mesophyll, **e.** lower epidermis. See also Figure 9.8; 10. d; 11. c; 12. d; 13. c; 14. e; 15. d; 16. d; 17. c; 18. d; 19. b; 20. a; 21. c; 22. **a.** blade, **b.** leaflet, **c.** petiole, **d.** axillary bud; 23. c; 24. b; 25. d; 26. b

Understanding the Terms

a. mesophyll; b. cork; c. root hair; d. phloem; e. transpiration

Chapter 10

Testing Yourself

1. b; 2. c; 3. d; 4. d; 5. a; 6. b; 7. c; 8. e; 9. a; 10. d; 11. c; 12. d; 13. d; 14. b; 15. c; 16. d; 17. d; 18. c; 19. d; 20. a. stigma, b. style, c. carpel, d. sepal, e. anther; 21. a. sporophyte, b. meiosis, c. microspore, d. megaspore, e. male gametophyte (pollen grain), f. female gametophyte (embryo sac), g. egg, h. fertilization, i. mitosis; 22. d; 23. b; 24. d; 25. a; 26. e; 27. c; 28. c

Understanding the Terms

a. tropism; b. gametophyte; c. stigma; d. microspore; e. phytochrome

Chapter 11

Testing Yourself

1. c; 2. d; 3. d; 4. b; 5. d; 6. c; 7. d; 8. d; 9. b; 10. b; 11. d; 12. a; 13. d; 14. c; 15. a; 16. b; 17. c; 18. c; 19. a. hair shaft, b. epidermis, c. dermis, d. subcutaneous layer, e. adipose tissue, f. sweat gland, g. hair follicle, h. arrector pili muscle, i. oil gland, j. sensory receptor, k. melanocytes, l. sweat pore; 20. a. ventral, b. thoracic, c. abdominal, d. pelvic, e. dorsal, f. cranial, g. vertebral; 21. d; 22. e

Understanding the Terms

a. ligament; b. epidermis; c. striated; d. negative feedback; e. spongy bone

Chapter 12

Testing Yourself

1. b; 2. a; 3. b; 4. d; 5. a; 6. b; 7. a; 8. b; 9. c; 10. d 11. b; 12. d; 13. c; 14. a; 15. e; 16. b; 17. a. P wave, b. R wave, c. T wave; 18. c; 19. e; 20. a. internal jugular vein, b. common carotid artery, c. inferior vena cava, d. common iliac vein, e. aorta, f. femoral vein; 21. b; 22. b; 23. e

Understanding the Terms

a. diastole; b. vena cava; c. fibrinogen; d. hemoglobin; e. systemic circuit

Chapter 13

Testing Yourself

1. b; 2. d; 3. c; 4. a; 5. c; 6. d; 7. e; 8. a; 9. b; 10. d; 11. c; 12. d; 13. d; 14. b; 15. a; 16. a. thymus, primary, b. lymph node, secondary, c. bone marrow, primary, d. spleen, secondary; 17. a. antigen-binding sites, b. light chain, c. heavy chain, V = variable, C = constant; 18. e; 19. a; 20. a; 21. a; 22. e; 23. b; 24. c

Understanding the Terms

a. vaccine; b. lymph; c. antigen; d. apoptosis; e. T cell

Chapter 14

Testing Yourself

1. d; 2. d; 3. c; 4. d; 5. c; 6. d; 7. d; 8. b; 9. e; 10. d; 11. b; 12. d; 13. a; 14. e; 15. c; 16. d; 17. a; 18. b; 19. a. large intestine, b. small intestine, c. cecum, d. vermiform appendix; 20. b; 21. a. bile canals, b. branch of hepatic artery, c. branch of hepatic portal vein, d. bile duct, e. central vein; 22. a; 23. b; 24. d; 25. d; 26. a; 27. c; 28. a; 29. d

Understanding the Terms

a. duodenum; b. sphincter; c. defecation; d. lipase; e. gallbladder

Chapter 15

Testing Yourself

1. a. nasal cavity, b. nostril, c. pharynx, d. epiglottis, e. glottis, f. larynx, g. trachea, h. bronchus, i. bronchiole; 2. d; 3. d; 4. a; 5. b; 6. d; 7. d; 8. d; 9. c; 10. a; 11. a; 12. b; 13. d; 14. c; 15. e; 16. a; 17. b; 18. d; 19. c; 20. a. sinus, b. hard palate, c. epiglottis, d. larynx, e. uvula, f. glottis; 21. c; 22. e; 23. a. inspiratory reserve volume, b. tidal volume, c. vital capacity, d. residual volume, e. total lung volume; 24. b; 25. b; 26. d; 27. c; 28. e; 29. e; 30. b; 31. a

Understanding the Terms

a. pharynx; b. residual volume; c. bicarbonate ion; d. expiration; e. alveoli

Chapter 16

Testing Yourself

1. b; 2. d; 3. b; 4. a; 5. c; 6. b; 7. b; 8. c; 9. c; 10. a; 11. c; 12. b; 13. d; 14. a; 15. d; 16. d; 17. a; 18. c; 19. d; 20. d; 21. c; 22. a; 23. d; 24. c; 25. a; 26. b; 27. b; 28. c; 29. c; 30. a. renal artery, b. renal vein, c. aorta, d. inferior vena cava, e. kidneys, f. ureters, g. urinary bladder, h. urethra; 31. a. renal cortex, b. renal vein, c. ureter, d. renal pelvis; 32. b

Understanding the Terms

a. loop of the nephron; b. tubular reabsorption; c. aldosterone; d. renal cortex; e. peritubular capillary network

Chapter 17

Testing Yourself

1. b; 2. b; 3. a; 4. c; 5. d; 6. a; 7. d; 8. d; 9. d; 10. d; 11. d; 12. d; 13. b; 14. b; 15. c; 16. a; 17. a; 18. c; 19. d; 20. b; 21. d; 22. d; 23. a. skull, b. meninges, c. corpus callosum, d. pituitary gland, e. midbrain, f. pons, g. medulla oblongata, h. thalamus, i. hypothalamus, j. pineal gland, k. spinal cord; 24. a. depolarization, b. repolarization, c. action potential, d. resting potential

Understanding the Terms

a. nerve impulse; b. cerebral hemisphere; c. integration; d. motor neuron; e. hippocampus

Chapter 18

Testing Yourself

1. e; 2. d; 3. e; 4. d; 5. b; 6. c; 7. e; 8. a; 9. d; 10. b; 11. a; 12. c; 13. a; 14. e; 15. a. thalamic nucleus, b. optic chiasma, c. optic nerve; 16. a. stapes (stirrup), b. incus (anvil), c. malleus (hammer), d. tympanic membrane, e. semicircular canals, f. cochlea, g. auditory tube; 17. e; 18 a. retina, b. choroid, c. sclera, d. optic nerve, e. fovea centralis, f. ciliary body, g. lens, h. iris, i. pupil, j. cornea; 19. c; 20. d; 21. c; 22. b; 23. e; 24. d; 25. a

Understanding the Terms

a. cochlea; b. chemoreceptor; c. rhodopsin; d. sensation; e. mechanoreceptor

Chapter 19

Testing Yourself

1. b; 2. c; 3. d; 4. c; 5. c; 6. b; 7. e; 8. a; 9. b; 10. b; 11. a; 12. e; 13. c; 14. d; 15. e; 16. a; 17. c; 18. a; 19. a. frontal bone, b. maxilla, c. mandible, d. occipital bone, e. temporal bone, f. parietal bone; 20. d ; 21. b; 22. e; 23. a; 24. d; 25. c; 26. b; 27. f; 28. c; 29. e; 30. T tubule; 31. sarcoplasmic reticulum; 32. myofibril; 33. Z line; 34. sarcomere; 35. sarcolemma; 36. b; 37. d

Understanding the Terms

a. neuromuscular junction; b. ligament; c. menisci; d. myosin; e. red bone marrow

Chapter 20

Testing Yourself

1. d; 2. b; 3. c; 4. c; 5. e; 6. d; 7. d; 8. b; 9. a; 10. d; 11. e; 12. c; 13. b; 14. e; 15. b; 16. d; 17. a. pineal gland, b. hypothalamus, c. pituitary gland, d. thymus, e. pancreas, f. adrenal glands, g. thyroid gland; 18. d; 19. b; 20. a. inhibits, b. inhibits, c. releasing hormone, d. stimulating hormone, e. target gland hormone; 21. f; 22. b; 23. c; 24. a; 25. e; 26. a. ANH, b. renin, c. aldosterone; 27. e; 28. d; 29. e

Understanding the Terms

a. thyroid gland; b. diabetes mellitus; c. adrenocorticotropic hormone (ACTH); d. glucagon; e. oxytocin

Chapter 21

Testing Yourself

1. d; 2. c; 3. b; 4. e; 5. c; 6. b; 7. d; 8. a; 9. e; 10. c; 11. b; 12. c; 13. b; 14. c; 15. e; 16. a. oviduct, b. ovary, c. uterus, d. urethra, e. clitoris, f. anus, g. vagina, h. cervix; 17 a. seminal vesicle, b. ejaculatory duct, c. prostate gland, d. bulbourethral gland, e. anus, f. vas deferens, g. epididymis, h. testis, i. scrotum, j. foreskin, k. glans penis, l. penis, m. urethra, n. vas deferens, o. urinary bladder; 18. c; 19. c; 20. d; 21. d; 22. d; 23. e; 24. b; 25. f; 26. a; 27. g; 28. c; 29. d; 30. e; 31. c; 32. d; 33. c

Understanding the Terms

a. vagina; b. endometrium; c. prostate gland; d. follicle-stimulating hormone; e. vulva

Chapter 22

Testing Yourself

1. d; 2. a; 3. a; 4. b; 5. d; 6. a; 7. c; 8. c; 9. c; 10. c; 11. b; 12. c; 13. b; 14. a. zygote, b. morula, c. blastula, d. early gastrula, e. late gastrula; 15. d; 16. d; 17. a. neural tube, b. somite, c. notochord, d. gut, e. coelom; 18. c; 19. a; 20. d; 21. a; 22. b; 23. b; 24. b

Understanding the Terms

a. induction; b. germ layer; c. morphogenesis; d. placenta; e. cleavage

Chapter 23

Practice Problems 1

1. a. W; b. WS, Ws; c. T, t; d. Tg, tg; e. AB, Ab, aB, ab 2. a. gamete; b. genotype; c. gamete; d. genotype

Practice Problems 2

1. 75% or 3/4; 2. mother and father: Ee and Ee; 3. father: DD; mother: dd; children: Dd

Practice Problems 3

1. Dihybrid; 2. 1/16; 3. child: ddff; father: DdFf; mother: ddff

Practice Problems 4

1. No; 2. child: ii, mother: $I^A i$, father: ii, $I^A i$, $I^B i$; 3. baby 1 = Doe; baby 2 = Jones

Practice Problems 5

1. His mother; mother $X^B X^b$, father $X^B Y$, son $X^b Y$; 2. 100%; none; 100%; 3. color blind; The husband is not the father; 4. AABBCC; aabbcc

Practice Problems 6

1. 9:3:3:1; gene linkage; 2. crossing-over

Testing Yourself

1. e; 2. c; 3. a; 4. c; 5. d; 6. a; 7. c; 8. c; 9. a; 10. c; 11. see Fig. 23.6; 12. a; 13. e; 14. d; 15. d; 16. see Fig. 23.7; 17. c; 18. e; 19. e; 20. a; 21. e; 22. b; 23. b; 24. a; 25. c; 26. c

Additional Genetic Problems

1. c; 2. c, because the genotypes are not known; 3. e; 4. e; 5. b

Understanding the Terms

a. Punnett square; b. allele; c. dominant allele; d. testcross; e. genotype

Chapter 24

Testing Yourself

1. b; 2. c; 3. a; 4. c; 5. b; 6. c; 7. b; 8. a; 9. c; 10. see Fig. 24.2; 11. c; 12. b; 13. a; 14. b; 15. c; 16. b; 17. b; 18. d; 19. c; 20. a; 21. c; 22. e; 23. a; 24. c; 25. c; 26. b; 27. d

Understanding the Terms

a. mutagen; b. RNA polymerase; c. transcription; d. translation; e. anticodon

Chapter 25

Testing Yourself

1. c; 2. e; 3. a; 4. a; 5. b; 6. a. active X chromosome, b. inactive X, c. Barr bodies; 7. e; 8. c ; 9. a. transcription activator, b. transcription factors, c. RNA polymerase, d. enhancer, e. promoter; 10. e; 11. b; 12. c; 13. e; 14. b; 15. c; 16. e; 17. a; 18. d; 19. b; 20. d; 21. e

Understanding the Terms

a. apoptosis; b. growth factor; c. metastasis; d. transcription activator; e. oncogene

Chapter 26

Pedigree Questions

Figure 26.9, page 534 The individual does not have the disorder but has a child with the disorder. Therefore, the individual has a recessive allele; Figure 26.10, page 535 The individual has the disorder and has a child without the disorder. Therefore, the individual has a recessive allele.

Testing Yourself

1. e; 2. d; 3. a; 4. b; 5. d; 6. c; 7. a; 8. e; 9. d; 10. c; 11. a; 12. b; 13. $X^B X^b$; 14. e; 15. b; 16. c; 17. b; 18. d; 19. d; 20. d; 21. b

Additional Genetic Problems

1. a. 100%; b. None, but they will be carriers. 2. a. $X^A X^a$ or $X^a X^a$; b. None, but she will be a carrier. 3. a. AA or Aa because he shows signs of the disease. b. Aa or AA or aa; 4. a. and b. Yes,

if he is heterozygous and the mother is homozygous normal or heterozygous. The son or daughter must receive a recessive allele from both parents. **5. a.** Aa × aa, or aa × Aa; **b.** He has one normal allele, and therefore, could live long enough to father a child. **6.** Depends on the result of the testing. If one or both are homozygous dominant, there is no chance of a child with cystic fibrosis. If they are both carriers, each child has a 25% chance of cystic fibrosis.

Understanding the Terms

a. chromosomal mutation; **b.** trisomy; **c.** Barr body; **d.** inversion; **e.** duplication

Chapter 27
Testing Yourself

1. c; **2.** a; **3.** c; **4.** a; **5.** c; **6.** b; **7.** d; **8.** a; **9.** d; **10.** a; **11.** d; **12.** d; **13.** a; **14.** b; **15.** d; **16.** c; **17.** f; **18.** b; **19.** e; **20.** b; **21.** b; **22.** b; **23.** a; **24.** b

Understanding the Terms

a. natural selection; **b.** directional selection; **c.** punctuated equilibrium; **d.** homologous structure; **e.** gene flow

Chapter 28
Testing Yourself

1. c; **2.** a; **3.** e; **4.** b; **5.** d; **6.** a; **7.** b; **8.** c; **9.** b; **10.** d; **11.** b; **12.** d; **13.** b; **14.** c; **15.** e; **16.** e; **17.** e; **18.** a; **19.** d; **20.** e; **21.** c; **22.** b; **23.** c; **24.** d; **25.** b; **26.** e; **27.** a; **28.** e; **29.** b; **30. 1.** attachment, **2.** entry, **3.** replication, **4.** biosynthesis, **5.** assembly, **6.** budding; **31.** e; **32.** b; **33.** c; **34.** b; **35.** a; **36.** a; **37.** c; **38.** c

Understanding the Terms

a. viroid; **b.** Gram stain; **c.** mycoses; **d.** contractile vacuole; **e.** lichen

Chapter 29
Testing Yourself

1. b; **2.** a; **3.** b; **4.** see Figure 29.2; **5.** c; **6.** a; **7.** b; **8.** b,c,d; **9.** c,d; **10.** a,b; **11.** b; **12.** c; **13.** d; **14.** a; **15.** b; **16.** c; **17.** a; **18.** see Figure 29.16; **19.** c; **20.** a; **21.** c

Understanding the Terms

a. archegonium; **b.** gymnosperm; **c.** sporophyte; **d.** seed; **e.** carpel

Chapter 30
Testing Yourself

1. a; **2.** b; **3.** c; **4.** b; **5.** a; **6.** c; **7.** b; **8.** b; **9.** c; **10.** a; **11.** c; **12.** a; **13.** b, c; **14.** a; **15.** a,b,c; **16.** a,b; **17.** e; **18.** d; **19.** a; **20.** d; **21.** d; **22.** a; **23.** see Figure 30.14; **24.** c; **25.** d; **26.** b; **27.** a; **28.** e; **29.** e; **30.** c; **31.** e; **32.** e; **33.** a; **34.** a; **35.** c; **36.** a

Understanding the Terms

a. trachea; **b.** coelom; **c.** bilateral symmetry; **d.** molting; **e.** nephridia

Chapter 31
Testing Yourself

1. b; **2.** d; **3.** a; **4.** c; **5.** b, c; **6.** a; **7.** c; **8.** b; **9.** c; **10.** d; **11.** b; **12.** a; **13.** d; **14.** c, d; **15.** b, c, d; **16.** a, b, c, d; **17.** c, d; **18.** a, b, c; **19.** see Fig. 31.8; **20.** c; **21.** e; **22.** b; **23.** c; **24.** a; **25.** a; **26.** b; **27.** c; **28.** d; **29.** d; **30.** d; **31.** b; **32.** b; **33.** d; **34.** e; **35.** c; **36. a**; **37.** e

Understanding the Terms

a. water vascular system; **b.** prosimian; **c.** notochord; **d.** endothermic; **e.** hominid

Chapter 32
Testing Yourself

1. c; **2.** c; **3.** d; **4.** a; **5.** e; **6.** c; **7.** b; **8.** d; **9.** c; **10.** b; **11.** b; **12.** a; **13.** c; **14.** c; **15.** c; **16.** c; **17.** a; **18.** c; **19.** b; **20.** a; **21.** c

Understanding the Terms

a. territoriality; **b.** altruism; **c.** communication; **d.** pheromone

Chapter 33
Testing Yourself

1. e; **2.** e; **3.** a; **4.** c; **5.** c; **6. a.** lag, **b.** environmental resistance, **c.** carrying capacity, **d.** stable equilibrium, **e.** deceleration, **f.** exponential growth; **7.** a; **8.** b; **9.** a; **10.** b; **11.** c; **12.** e; **13.** a; **14.** a; **15.** c; **16.** b; **17.** c; **18.** a; **19.** a; **20.** d; **21.** e; **22.** b; **23.** b; **24.** e; **25.** d; **26.** e; **27.** a; **28.** b; **29.** e; **30.** d; **31.** e; **32.** b; **33.** c; **34.** a; **35.** b

Understanding the Terms

a. population; **b.** biotic potential; **c.** exponential growth; **d.** demographic transition; **e.** climax community

Chapter 34
Testing Yourself

1. a; **2.** a; **3.** a; **4. a.** producers, **b.** consumers, **c.** nutrients, **d.** decomposers; **5.** c; **6.** d; **7.** c; **8.** b; **9.** a; **10.** d; **11. a.** tertiary consumers, **b.** secondary consumers, **c.** primary consumers, **d.** producers; **12.** c; **13.** d; **14.** d; **15.** b; **16.** c; **17.** c; **18.** d; **19.** c; **20.** b; **21.** c; **22.** b; **23.** a; **24.** b; **25.** c; **26.** c; **27.** e

Understanding the Terms

a. detritus; **b.** ozone shield; **c.** fossil fuel; **d.** nitrogen fixation; **e.** food web

Chapter 35
Testing Yourself

1. c; **2.** b; **3.** a; **4.** see Figure 35.4; **5.** c; **6.** d; **7.** b; **8.** e; **9.** a; **10.** g; **11.** d; **12.** c; **13.** f; **14.** d; **15.** d; **16.** a; **17.** c; **18.** d; **19.** b,c; **20.** a; **21.** c,d; **22.** b; **23.** c; **24.** b; **25.** c; **26.** d; **27.** a; **28.** a; **29.** d; **30.** d; **31.** a; **32.** c

Understanding the Terms

a. estuary; **b.** biome; **c.** benthic division; **d.** savanna; **e.** littoral zone

Chapter 36
Testing Yourself

1. c; **2.** e; **3.** d; **4.** a; **5.** c; **6.** c; **7.** b; **8.** d; **9.** c; **10.** b; **11.** d; **12.** c; **13.** c,d,e,f; **14.** a,b; **15.** a,b,c; **16.** c,d,e,f; **17.** b; **18.** a; **19.** c; **20.** c; **21.** a; **22.** c; **23.** b; **24.** c; **25.** e; **26.** d

Understanding the Terms

a. pollution; **b.** subsidence; **c.** ozone shield; **d.** desertification; **e.** aquifer

Appendix B

Classification of Organisms

Domain Bacteria

Prokaryotic, unicellular organisms that lack a membrane-bounded nucleus and reproduce asexually. Metabolically diverse being heterotrophic by absorption; autotrophic by chemosynthesis or by photosynthesis. Motile forms move by flagella consisting of a single filament. (577)

Domain Archaea

Prokaryotic, unicellular organisms that lack a membrane-bounded nucleus and reproduce asexually. Many are autotrophic by chemosynthesis; some are heterotrophic by absorption. Most live in extreme or anaerobic environments. Archaea are distinguishable from bacteria by their unique rRNA base sequence and their distinctive plasma membrane and cell wall chemistry. (583)

Domain Eukarya

Eukaryotic, unicellular to multicellular organisms that have a membrane-bounded nucleus containing several chromosomes. Sexual reproduction is common. Phenotypes and nutrition are diverse; each kingdom has specializations that distinguish it from the other kingdoms. Flagella, if present, have a 9 + 2 organization. (584)

Kingdom Protista

Eukaryotic, unicellular organisms and their immediate multicellular descendants. Asexual reproduction is common, but sexual reproduction as a part of various life cycles does occur. Phenotypically and nutritionally diverse, being either photosynthetic or heterotrophic by various means. Locomotion, if present, utilizes flagella, cilia, or pseudopods. (584)

Algae*

> Phylum Chlorophyta: green algae (584)
>
> Phylum Chrysophyta: diatoms, golden-brown algae (586)
>
> Phylum Pyrrophyta: dinoflagellates (586)
>
> Phylum Rhodophyta: red algae (586)
>
> Phylum Phaeophyta: brown algae (587)
>
> Phylum Euglenophyta: euglenoids (587)

Protozoans*

> Phylum Ciliophora: ciliates (588)
>
> Phylum Rhizopoda: amoeboids (589)
>
> Phylum Zoomastigophora: zooflagellates (589)
>
> Phylum Apicomplexa: sporozoans (590)

Water Molds*

> Phylum Oomycota: water molds (591)

Slime Molds*

> Phylum Myxomycota: plasmodial slime molds (591)
>
> Phylum Acrasiomycota: cellular slime molds (592)

Kingdom Fungi

Multicellular eukaryotes that form nonmotile spores during both asexual and sexual reproduction as a part of the haplontic life cycle. The only multicellular forms of life to be heterotrophic by absorption. They lack flagella in all life-cycle stages. (582)

> Phylum Zygomycota: zygospore fungi (593)
>
> Phylum Ascomycota: sac fungi (593)
>
> Phylum Basidiomycota: club fungi (594)
>
> Phylum Deuteromycota: imperfect fungi (means of sexual reproduction is not known) (594)

*Not a classification category, but added for clarity

Kingdom Plantae

Multicellular, primarily terrestrial eukaryotes with well-developed tissues. Plants have an alternation of generations life cycle and are usually photosynthetic. Like green algae, they contain chlorophylls *a* and *b*, carotenoids; store starch in chloroplasts; and have a cell wall that contains cellulose. (608)

Nonvascular Plants*

Phylum Anthocerophyta: hornworts (610)

Phylum Hepatophyta: liverworts (610)

Phylum Bryophyta: mosses (610)

Seedless Vascular Plants*

Phylum Psilotophyta: whisk ferns (613)

Phylum Lycophyta: club mosses, spike mosses, quillworts (613)

Phylum Sphenophyta: horsetails (613)

Phylum Pterophyta: ferns (614)

Gymnosperms*

Phylum Coniferophyta: conifers, such as pines, firs, yews, redwoods, spruces (615)

Phylum Cycadophyta: cycads (617)

Phylum Ginkgophyta: maidenhair tree (618)

Phylum Gnetophyta: gnetophytes (618)

Angiosperms*

Phylum Anthophyta: flowering plants (618)

Class Monocotyledones: monocots (618)

Class Eudicotyledones: eudicots (618)

Kingdom Animalia

Multicellular organisms with well-developed tissues that have the diplontic life cycle. Animals tend to be mobile and are heterotrophic by ingestion, generally in a digestive cavity. Complexity varies; the more complex forms have well-developed organ systems. More than a million species have been described. (626)

Invertebrates*

Phylum Porifera: sponges (629)

Phylum Cnidaria: hydras, jellyfishes (630)

Class Anthozoa: sea anemones, corals (631)

Class Hydrozoa: *Hydra, Obelia* (631)

Class Scyphozoa: jellyfishes (631)

Phylum Platyhelminthes: flatworms (632)

Class Turbellaria: planarians (632)

Class Cestoda: tapeworms (633)

Class Trematoda: flukes (633)

Phylum Nematoda: roundworms (634)

Phylum Mollusca: molluscs (635)

Class Polyplacophora: chitons (635)

Class Gastropoda: snails, slugs, nudibranchs (637)

Class Cephalopoda: squid, chambered nautilus, octopus (637)

Class Bivalvia: clams, scallops, oysters, mussels (637)

Phylum Annelida: annelids (638)

Class Polychaeta: clam worms, tube worms (638)

Class Oligochaeta: earthworms (639)

Class Hirudinea: leeches (639)

Phylum Arthropoda: arthropods (640)

Subphylum Crustacea: crustaceans (shrimps, crabs, lobsters, barnacles) (641)

Subphylum Uniramia: insects, millipedes, centipedes (642)

Subphylum Chelicerata: spiders, scorpions, horseshoe crabs (645)

Phylum Echinodermata: echinoderms (650)

Class Crinoidea: sea lilies, feather stars (650)

Class Holothuroidea: sea cucumbers (650)

Class Ophiuroidea: brittle stars (650)

Class Echinoidea: sea urchins, sand dollars (650)

Class Asteroidea: sea stars (651)

Phylum Chordata: chordates (652)

Subphylum Urochordata: tunicates (654)

Subphylum Cephalochordata: lancelets (654)

Vertebrates*

Subphylum Vertebrata: Vertebrates (655)

Superclass Agnatha: jawless fishes (hagfishes, lampreys) (656)

Superclass Gnathostomata: jawed fishes, tetrapods (656, 657)

Class Chondrichthyes: cartilaginous fishes (sharks, skates) (656)

Class Osteichthyes: bony fishes (lobe-finned fishes and ray-finned fishes) (656)

Class Amphibia: amphibians (frogs, toads) (657)

Class Reptilia: reptiles (turtles, snakes, lizards, crocodiles) (658)

Class Aves: birds (owls, woodpeckers, kingfishers, cuckoos) (661)

Class Mammalia: mammals (monotremes—duckbill platypus, spiny anteater; marsupials—opposums, kangaroos, koalas; placental mammals—shrews, whales, rats, rabbits, dogs, cats) (662)

Order Primates: primates (prosimians, monkeys, apes) (664)

Family hominids (664)

Genus *Homo*: humans (667)

*Not a classification category, but added for clarity

Appendix C

Metric System

Unit and Abbreviation	Metric Equivalent	Approximate English-to-Metric Equivalents
Length		
nanometer (nm)	$= 10^{-9}\,m\ (10^{-3}\,\mu m)$	
micrometer (μm)	$= 10^{-6}\,m\ (10^{-3}\,mm)$	
millimeter (mm)	$= 0.001\ (10^{-3})\,m$	
centimeter (cm)	$= 0.01\ (10^{-2})\,m$	1 inch = 2.54 cm 1 foot = 30.5 cm
meter (m)	$= 100\ (10^{2})\,cm$ $= 1{,}000\,mm$	1 foot = 0.30 m 1 yard = 0.91 m
kilometer (km)	$= 1{,}000\ (10^{3})\,m$	1 mi = 1.6 km
Weight (mass)		
nanogram (ng)	$= 10^{-9}\,g$	
microgram (μg)	$= 10^{-6}\,g$	
milligram (mg)	$= 10^{-3}\,g$	
gram (g)	$= 1{,}000\,mg$	1 ounce = 28.3 g 1 pound = 454 g
kilogram (kg)	$= 1{,}000\ (10^{3})\,g$	= 0.45 kg
metric ton (t)	$= 1{,}000\,kg$	1 ton = 0.91 t
Volume		
microliter (μl)	$= 10^{-6}\,l\ (10^{-3}\,ml)$	
milliliter (ml)	$= 10^{-3}\,liter$ $= 1\,cm^{3}\ (cc)$ $= 1{,}000\,mm^{3}$	1 tsp = 5 ml 1 fl oz = 30 ml
liter (l)	$= 1{,}000\,ml$	1 pint = 0.47 liter 1 quart = 0.95 liter 1 gallon = 3.79 liter
kiloliter (kl)	$= 1{,}000\,liter$	

Units of Temperature

°F scale values: 230, 220, 212°–210, 200, 190, 180, 170, 160°–160, 150, 140, 134°, 131°–130, 120, 110, 105.8°, 98.6°–100, 90, 80, 70, 60, 56.66°, 50, 40, 32°–30, 20, 10, 0, -10, -20, -30, -40

°C scale values: 110, 100–100°, 90, 80, 70–71°, 60–57°, 50, 40–41°, 37°, 30, 20, 13.7°, 10, 0–0°, -10, -20, -30, -40

To convert temperature scales:

$$^{\circ}C = \frac{(^{\circ}F - 32)}{1.8}$$

$$^{\circ}F = 1.8\,(^{\circ}C) + 32$$

°C	°F	
100	212	Water boils at standard temperature and pressure.
71	160	Flash pasteurization of milk
57	134	Highest recorded temperature in the United States, Death Valley, July 10, 1913
41	105.8	Average body temperature of a marathon runner in hot weather
37	98.6	Human body temperature
13.7	56.66	Human survival is still possible at this temperature.
0	32.0	Water freezes at standard temperature and pressure.

Appendix D

Periodic Table of the Elements

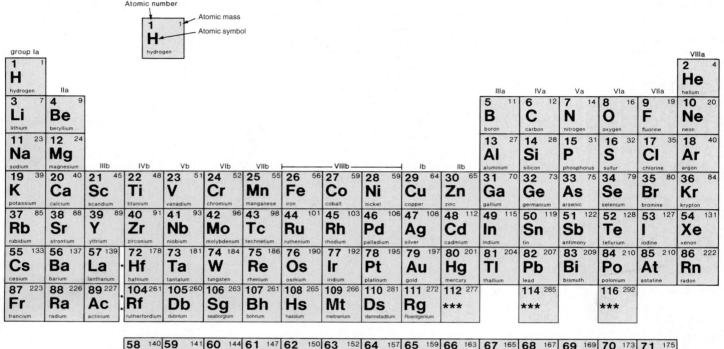

atrophy Wasting away or decrease in size of a tissue or body part, usually due to disuse or injury. 388

auditory canal Tube in the outer ear that leads to the tympanic membrane. 357

auditory tube Extension from the middle ear to the nasopharynx that equalizes air pressure on the eardrum; sometimes called the eustachian tube. 296, 358

australopithecine One of several species of Australopithecus, a genus that contains the first generally recognized hominids. 667

autoimmune disease Disease that results when the immune system mistakenly attacks the body's own tissues. 255

autonomic system Branch of the peripheral nervous system that has control over the internal organs; consists of the sympathetic and parasympathetic divisions. 337

autosome Any chromosome other than the sex chromosomes. 481

autotroph Organism that can capture energy and synthesize organic molecules from inorganic nutrients. 132, 552, 710

auxin Plant hormone regulating growth, particularly cell elongation; most often called indoleacetic acid (IAA). 188

AV (atrioventricular) node Small region of neuromuscular tissue that transmits impulses received from the SA node to the ventricular walls. 222

axial skeleton Portion of the skeleton that supports and protects the organs of the head, the neck, and the trunk. 373

axillary bud Bud located in the axil of a leaf. 149

axon Fiber of a neuron that conducts nerve impulses away from the cell body. 320

axon terminal Small swelling at the tip of one of many endings of the axon. 324

B

Bacteria One of the three domains of life; contains prokaryotic cells called bacteria that differ from archaea because they have their own unique genetic, biochemical, and physiological characteristics. 7, 570

bacterial vaginosis Inflammation of the vagina caused by a bacterial infection. 436

ball-and-socket joint The most freely movable type of joint; e.g., the shoulder or hip joint. 380

bark External part of a tree, containing cork, cork cambium, and phloem. 158

Barr body Dark-staining body (discovered by M. Barr) in the nuclei of female mammals that contains a condensed, inactive X chromosome. 528

basal nuclei Subcortical nuclei deep within the white matter that serve as relay stations for motor impulses and produce dopamine to help control skeletal muscle activities. 330

base Molecules tending to lower the hydrogen ion concentration in a solution and raise the pH numerically. 29

basement membrane Layer of nonliving material that anchors epithelial tissue to underlying connective tissue. 198

basophil White blood cell with a granular cytoplasm; able to be stained with a basic dye. 229

B cell Lymphocyte that matures in the bone marrow and, when stimulated by the presence of a specific antigen, gives rise to antibody-producing plasma cells. 245

B-cell receptor (BCR) A molecule on the surface of a B cell that binds to a specific antigen. 245

behavior Observable, coordinated responses to environmental stimuli. 4, 676

benign prostatic hyperplasia (BPH) Enlargement of the prostate gland. 437

benthic division Ocean or lake floor, extending from the high-tide mark to the deepest depths, which supports a unique set of organisms. 738

bicarbonate ion Ion that participates in buffering the blood, and the form in which carbon dioxide is transported in the bloodstream. 293

bilateral symmetry Body plan having two corresponding or complementary halves. 628

bile Secretion of the liver that is temporarily stored and concentrated in the gallbladder before being released into the small intestine, where it emulsifies fat. 264

binary fission Bacterial reproduction into two daughter cells without the utilization of a mitotic spindle. 578

biodiversity Total number of species, the variability of their genes, and the communities in which they live. 9, 754

biogeochemical cycle Circulating pathway of elements such as carbon and nitrogen involving exchange pools, storage areas, and biotic communities. 713

biogeography Study of the geographical distribution of organisms. 556

bioinformatics Computer technologies used to study the genome. 543

biological magnification Process by which nonexcreted substances such as DDT become more concentrated in organisms in the higher trophic levels of a food chain. 754

biology Scientific study of life. 10

biomass The number of organisms multiplied by their weight. 713

biome One of the biosphere's major communities, characterized in particular by certain climatic conditions and particular types of plants. 727

biosphere That portion of Earth's surface (air, water, and land) where living things exist. 7, 692

biotechnology Term that encompasses genetic engineering and other techniques that make use of natural biological systems to create a product or achieve a particular result desired by humans. 504

biotic potential Maximum population growth rate under ideal conditions. 693

bipedalism Walking erect on two feet. 664

bird Endothermic vertebrate that has feathers and wings, is often adapted for flight, and lays hard-shelled eggs. 661

birth control pill Oral contraception containing estrogen and progesterone. 428

bivalve Type of mollusc with a shell composed of two valves; includes clams, oysters, and scallops. 637

blade Broad, expanded portion of a plant leaf. 149

blastocoel Fluid-filled cavity of a blastula. 447

blastocyst Early stage of human embryonic development that consists of a hollow, fluid-filled ball of cells. 454

blastopore Opening into the primitive gut formed at gastrulation. 448

blastula Hollow, fluid-filled ball of cells occurring during animal development prior to gastrula formation. 447

blind spot Region of the retina, lacking rods or cones, where the optic nerve leaves the eye. 356

blood Type of connective tissue in which cells are separated by a liquid called plasma. 202

blood pressure Force of blood pushing against the inside wall of blood vessels. 225

bone Connective tissue having protein fibers and a hard matrix of inorganic salts, notably calcium salts. 201

bony fish Member of a class of vertebrates (class Osteichthyes) containing numerous diverse fishes that have a bony rather than a cartilaginous skeleton. 657

bottleneck effect Type of genetic drift; occurs when a majority of genotypes are prevented from participating in the production of the next generation as a result of a natural disaster or human interference. 563

brain stem Portion of the brain consisting of the medulla oblongata, pons, and midbrain. 330

Broca's area Region of the frontal lobe that coordinates complex muscular actions of the mouth, tongue, and larynx, making speech possible. 329

bronchioles Smaller air passages in the lungs that begin at the bronchi and terminate in alveoli. 288

bronchus (pl., bronchi) One of two major divisions of the trachea leading to the lungs. 288

buffer Substance or group of substances that tend to resist pH changes in a solution, thus stabilizing its relative acidity and basicity. 29, 311

bulbourethral gland Either of two small structures located below the prostate gland in males; adds secretions to semen. 418

bulimia nervosa Eating disorder characterized by binge eating followed by purging via self-induced vomiting or use of a laxative. 278

bursa Saclike, fluid-filled structure, lined with synovial membrane, that occurs near a synovial joint. 380

bursitis Inflammation of any of the friction-easing sacs called bursae, which are found near joints. 380

C

C3 plant Plant that fixes carbon dioxide via the Calvin cycle; the first stable product of C3 photosynthesis is a 3-carbon compound. 141

C4 plant Plant that fixes carbon dioxide to produce a C4 molecule that releases carbon dioxide to the Calvin cycle. 141

calcitonin Hormone secreted by the thyroid gland that increases the blood calcium level. 402

calorie Amount of heat energy required to raise the temperature of water 1°C. 27

Calvin cycle reaction Pathway of photosynthesis that converts carbon dioxide to carbohydrate. 134

calyx The sepals collectively; the outermost flower whorl. 619

capillary Microscopic vessel connecting arterioles to venules; exchange of substances between blood and tissue fluid occurs across their thin walls. 198, 218

capsid Protein-containing layer that surrounds and protects the genetic material of a virus. 596

capsule Gelatinous layer surrounding the cells of blue-green algae and certain bacteria. 48

carbaminohemoglobin Hemoglobin-carrying carbon dioxide. 295

carbohydrate Class of organic compounds that includes monosaccharides, disaccharides, and polysaccharides. 32

carbon dioxide (CO_2) fixation Photosynthetic reaction in which carbon dioxide is attached to an organic compound. 139

carbonic anhydrase Enzyme in red blood cells that speeds the formation of carbonic acid from water and carbon dioxide. 293

carcinogen Environmental agent that causes mutations leading to the development of cancer. 519

carcinoma Cancer arising in epithelial tissue. 198

cardiac cycle One complete cycle of systole and diastole for all heart chambers. 222

cardiac muscle Striated, involuntary muscle tissue found only in the heart. 203

cardiovascular system Organ system in which blood vessels distribute blood under the pumping action of the heart. 206

carnivore Secondary or higher consumer in a food chain that therefore eats other animals. 710

carotenoid Yellow or orange pigment that serves as an accessory to chlorophyll in photosynthesis. 132

carotid body Structure located at the branching of the carotid arteries that contains chemoreceptors sensitive to the hydrogen ion concentration as well as to the levels of carbon dioxide and oxygen in blood. 293

carpel Ovule-bearing unit that is a part of a pistil. 176, 619

carrier Heterozygous individual who has no apparent abnormality but can pass on an allele for a recessively inherited genetic disorder. 534

carrier protein Protein that combines with and transports a molecule or ion across the plasma membrane. 69

carrying capacity The limit of the number of individuals of a particular species that the environment can support in a particular area. 694

cartilage Connective tissue in which the cells lie within lacunae embedded in a flexible, proteinaceous matrix. 200

cartilaginous fish Member of a class of vertebrates (class Chondrichthyes) with a cartilaginous rather than a bony skeleton; includes sharks, rays, and skates. 656

Casparian strip Layer of impermeable lignin and suberin bordering four sides of root endodermal cells; prevents water and solute transport between adjacent cells. 163

catabolism Metabolic process that breaks down large molecules into smaller ones; catabolic metabolism. 116

cecum Small pouch. In humans, a cecum lies below the entrance of the small intestine and is the blind end of the large intestine. 266

cell Structural and functional unit of an organism; the smallest structure capable of performing all the functions necessary for life. 3

cell body Portion of a neuron that contains a nucleus and from which dendrites and an axon extend. 320

cell cycle Repeating sequence of events in eukaryotes that involves cell growth and nuclear division; consists of the stages G1, S, G2, and M. 82

cell-mediated immunity Specific mechanism of defense in which T cells destroy antigen-bearing cells. 247

cell plate Structure that precedes the formation of the cell wall as a part of cytokinesis in plant life. 88

cell recognition protein Glycoprotein that helps the body defend itself against pathogens. 69

cell theory One of the major theories of biology; states that all organisms are made up of cells and that cells are capable of self-reproduction and come only from preexisting cells. 46

cellular differentiation Process and developmental stages by which a cell becomes specialized for a particular function. 450

cellular respiration Metabolic reactions that use the energy from carbohydrate, fatty acid, or amino acid breakdown to produce ATP molecules. 117

cellular slime mold Free-living protist that exists as individual amoeboid cells; they are common in soil, where they feed on bacteria and yeasts. 592

cellulose Polysaccharide that is the major complex carbohydrate in plant cell walls. 33

cell wall Structure that surrounds a plant, protistan, fungal, or bacterial cell and maintains the cell's shape and rigidity. 47

central canal Tube within the spinal cord that is continuous with the ventricle of the brain and contains cerebrospinal fluid. 326

central nervous system (CNS) Portion of the nervous system consisting of the brain and spinal cord. 320

centriole Cell organelle, existing in pairs, that occurs in the centrosome and may help organize a mitotic spindle for chromosome movement during animal cell division. 61, 86

centromere Constricted region of a chromosome where sister chromatids are attached to one another and where the chromosome attaches to a spindle fiber. 85

centrosome Central microtubule organizing center of cells. In animal cells, it contains two centrioles. 60

cephalization Having a well-recognized anterior head with a brain and sensory receptors. 628

cephalopod Type of mollusc in which a modified foot develops into the head region; includes squids, cuttlefish, octopuses, and nautiluses. 637

cerebellum Part of the brain located posterior to the medulla oblongata and pons that coordinates skeletal muscles to produce smooth, graceful motions. 330

cerebral cortex Outer layer of the cerebral hemispheres; receives sensory information and controls motor activities. 329

cerebral hemisphere One of the large, paired structures that together constitute the cerebrum of the brain. 328

cerebrospinal fluid (CSF) Fluid found in the ventricles of the brain, in the central canal of the spinal cord, and in association with the meninges. 326

cerebrum Main part of the brain consisting of two large masses, or cerebral hemispheres; the largest part of the brain in mammals. 327

cervix Narrow end of the uterus, which leads into the vagina. 422

channel protein Forms a channel to allow a particular molecule or ion to cross the plasma membrane. 69

chaparral Biome characterized by broad-leafed evergreen shrubs forming dense thickets. 731

chemical energy Energy associated with the interaction of atoms in a molecule. 102

chemical signal Molecule(s) released by cells, or even by organisms, that affect the metabolism of target cells or even the behavior of other individuals. 398

chemical synapse Junction between neurons consisting of the presynaptic (axon) membrane, the synaptic cleft, and the postsynaptic (usually dendrite) membrane. 324

chemiosmosis Ability of certain membranes to use a hydrogen ion gradient to drive ATP formation. 124

chemoautotroph Organism able to synthesize organic molecules by using carbon dioxide as the carbon source and the oxidation of an inorganic substance (such as hydrogen sulfide) as the energy source. 579

chemoreceptor Sensory receptor that is sensitive to chemical stimulation; e.g., receptors for taste and smell. 348

chemotherapy The administration of drugs to a patient to selectively kill cancer cells. 520

chitin Strong but flexible nitrogenous polysaccharide found in fungal cell walls and the exoskeleton of arthropods. 641

chlamydia Sexually transmitted disease caused by the bacterium *Chlamydia trachomatis*; often causes painful urination and swelling of the testes in men; is usually symptomless in women but can cause inflammation of the cervix or uterine tubes. 435

chlorofluorocarbons (CFCs) Organic compounds containing carbon, chlorine, and fluorine atoms. CFCs such as Freon can deplete the ozone shield by releasing chlorine atoms in the upper atmosphere. 719, 754

chlorophyll Green pigment that captures solar energy during photosynthesis. 133

chloroplast Membranous organelle that contains chlorophyll and is the site of photosynthesis. 57, 133

chordae tendineae Tough bands of connective tissue that attach the papillary muscles to the atrioventricular valves within the heart. 220

chordate Member of the phylum Chordata, which includes lancelets, tunicates, fishes, amphibians, reptiles, birds, and mammals; characterized by a notochord, dorsal tubular nerve cord, pharyngeal gill pouches, and postanal tail at some point in the life cycle. 652

chorion Extraembryonic membrane functioning for respiratory exchange in birds and reptiles; contributes to placenta formation in mammals. 454

chorionic villi Treelike extensions of the chorion, an extraembryonic membrane, projecting into the maternal tissues at the placenta. 457

chorionic villi sampling (CVS) Prenatal test in which a sample of chorionic villi cells is removed for diagnostic purposes. 526

choroid Vascular, pigmented middle layer of the eyeball. 353

chromatin Network of fibrils consisting of DNA and associated proteins observed within a nucleus that is not dividing. 52, 85

chromosomal mutation Variation in regard to the normal number of chromosomes inherited or in regard to the normal sequence of alleles on a chromosome; the sequence can be inverted, translocated from a nonhomologous chromosome, deleted, or duplicated. 531

chromosome Rodlike structure in the nucleus seen during cell division; contains the hereditary units, or genes. 52, 85

chronic bronchitis Obstructive pulmonary disorder that tends to recur; marked by inflamed airways filled with mucus and degenerative changes in the bronchi, including loss of cilia. 296

chronic disease Disease that persists over a long duration. 212

chyme Thick, semiliquid food material that passes from the stomach to the small intestine. 264

ciliary body In the eye, the structure that contains the ciliary muscle, which controls the shape of the lens. 354

ciliary muscle Muscle that controls the curvature of the eye. 354

ciliate Complex unicellular protist that moves by means of cilia and digests food in food vacuoles. 588

cilium (pl., cilia) Motile, short, hairlike extensions on the exposed surfaces of cells. 61, 198

circadian rhythm Regular physiological or behavioral activity that occurs on an approximately 24-hour cycle. 406

cirrhosis Chronic, irreversible injury to liver tissue; commonly caused by frequent alcohol consumption. 281

citric acid cycle Cyclic metabolic pathway found in the matrix of mitochondria that participates in cellular respiration; breaks down acetyl groups to carbon dioxide and hydrogen. 119

clade Taxon or group of taxa consisting of an ancestral species and all of its descendants, forming a distinct branch on a phylogenetic tree. 570

cladogram A branching diagram that shows the relationship among species in regard to their shared, derived characters. 570

class One of the categories, or taxa, used by taxonomists to group species; class is the taxon above the order level. 569

classical conditioning Type of learning whereby an unconditioned stimulus that elicits a specific response is paired with a neutral stimulus so that the response becomes conditioned. 679

cleavage Cell division without cytoplasmic addition or enlargement; occurs during the first stage of animal development. 88, 447

cleavage furrow Indentation that begins the process of cleavage, by which animal cells undergo cytokinesis. 88

climate Prevailing weather conditions (e.g., temperature, precipitation) in a region. 724

climax community In ecology, the community that results when succession has come to an end. 704

clonal selection theory The concept that an antigen selects which lymphocyte will undergo clonal expansion and produce more lymphocytes bearing the same type of antigen receptor. 245

cloning Production of identical copies. In organisms, the production of organisms with the same genes; in genetic engineering, the production of many identical copies of a gene. 501

clotting Process of blood coagulation, usually after an injury. 229

club fungi Members of the phylum Basidiomycota. 593

club moss Type of seedless vascular plants that are also called ground pines because they appear to be miniature pine trees. 613

cnidarian Invertebrate in the phylum Cnidaria existing as either a polyp or a medusa with two tissue layers and radial symmetry. 630

cochlea Portion of the inner ear that resembles a snail's shell and contains the spiral organ, the sense organ for hearing. 358

cochlear canal Canal within the cochlea that bears the spiral organ. 358

codominance Pattern of inheritance in which both alleles of a gene are equally expressed. 480

codon Three-base sequence of messenger RNA that causes the insertion of a particular amino acid into a protein or termination of translation. 496

coelom Embryonic body cavity lying between the digestive tract and the body wall that is completely lined by mesoderm; in humans, the embryonic coelom becomes the thoracic and abdominal cavities. 205, 628

coenzyme Nonprotein organic molecule that aids the action of the enzyme to which it is loosely bound. 109

coevolution Interaction of two species such that each influences the evolution of the other species. 701

cofactor Nonprotein adjunct required by an enzyme in order to function; many cofactors are metal ions, while others are coenzymes. 109

cohesion-tension model Explanation for upward transport of water in xylem based upon transpiration-created tension and the cohesive properties of water molecules. 166

cohort A group of individuals in a population in the same age group (e.g., prereproductive). 694

collagen fiber White fiber in the matrix of connective tissue, giving flexibility and strength. 200

collecting duct Duct within the kidney that receives fluid from several nephrons; the reabsorption of water occurs here. 307

collenchyma Plant tissue composed of cells with unevenly thickened walls; supports growth of stems and petioles. 152

colon The major portion of the large intestine, consisting of the ascending colon, the transverse colon, and the descending colon. 266

color blindness Deficiency in one or more of the three kinds of cones responsible for color vision. 538

color vision Perception of and ability to distinguish colors. 355

colostrum Thin, milky fluid rich in proteins, including antibodies, that is secreted by the mammary glands a few days prior to or after delivery before true milk is secreted. 464

columnar epithelium Type of epithelial tissue with cylindrical cells. 498

commensalism Symbiotic relationship in which one species is benefited, and the other is neither harmed nor benefited. 703

common ancestor In the field of systematics, species or lineages are grouped together by having a common ancestor, which is usually an extinct species that is hypothesized or reconstructed based on the evolutionary relationships of current species. 570

communication Signal by a sender that influences the behavior of a receiver. 685

community Assemblage of populations interacting with one another within the same environment. 7, 692

compact bone Type of bone that contains osteons consisting of concentric layers of matrix and osteocytes in lacunae. 201, 370

companion cell Cell associated with sieve-tube members in phloem of vascular plants. 153

competition Interaction between two organisms in which both require the same limited resource, which results in harm to both. 699

competitive exclusion principle The conclusion that no two species can occupy the same niche at the same time. 699

complementary base pairing Hydrogen bonding between particular bases. In DNA, thymine (T) pairs with adenine (A), and guanine (G) pairs with cytosine (C); in RNA, uracil (U) pairs with A, and G pairs with C. 492

complement system Group of plasma proteins that form a nonspecific defense mechanism, often by puncturing microbes; it complements the activities of antibodies. 244

compound Substance having two or more different elements united chemically in a fixed ratio. 24

concentration gradient Gradual change in chemical concentration from one point to another. 70

conclusion Statement made following an experiment as to whether the results support the hypothesis. 11

condensation The process by which water vapor turns into liquid. 714

cone cell Photoreceptor in the retina of eye that responds to bright light; detects color and provides visual acuity. 355

conidium (pl., conidia) Structure that asexually produces fungal spores. 593

conifer Cone-bearing gymnosperm plants that include pine, cedar, and spruce trees. 617

coniferous forest Forested area generally found in areas with temperate climate and characterized by cone-bearing trees that are mostly evergreens with needle-shaped or scale-like leaves. The taiga is a coniferous forest extending in a broad belt across northern Eurasia and North America. A montane coniferous forest is an area having similar characteristics that is found on a mountain. 729

conjugation Transfer of genetic material from one cell to another. 578

connective tissue Type of tissue characterized by cells separated by a matrix that often contains fibers. 200

consumer Organism that feeds on another organism in a food chain; primary consumers eat plants, and secondary (or higher) consumers eat animals. 710

continental drift Movement of continents with respect to one another over Earth's surfaces. 556

contraceptive Medication or device used to reduce the chance of pregnancy. 428

contraceptive implant Birth control method utilizing synthetic progesterone; prevents ovulation by disrupting the ovarian cycle. 429

contraceptive injection Birth control method utilizing progesterone or estrogen and progesterone; prevents ovulation by disrupting the ovarian cycle. 429

contraceptive vaccine Birth control method that immunizes against the hormone hCG, crucial to maintaining implantation of the embryo. 429

contractile vacuole Organelle that pumps water out of the cell. 588

control Sample that goes through all the steps of an experiment but does not contain the variable being tested; a standard against which the results of an experiment are checked. 10

coral Group of cnidarians having a calcium carbonate skeleton that participate in the formation of coral reefs. 631

coral reef Structure found in tropical waters that is formed by the buildup of coral skeletons and is home to many and various types of organisms. 738

cork Outer covering of the bark of trees; made of dead cells that may be sloughed off. 151

cork cambium Lateral meristem that produces cork. 158

cornea Transparent, anterior portion of the outer layer of the eyeball. 353

corolla The petals, collectively; usually the conspicuously colored flower whorl. 619

coronary artery Artery that supplies blood to the wall of the heart. 225

corpus callosum Mass of white matter within the brain, composed of nerve fibers connecting the right and left cerebral hemispheres. 328

corpus luteum Yellow body that forms in the ovary from a follicle that has discharged its secondary oocyte; it secretes progesterone and some estrogen. 425

cortex In plants, ground tissue bounded by the epidermis and vascular tissue in stems and roots; in animals, outer layer of an organ, such as the cortex of the kidney or adrenal gland. 163

cortisol Glucocorticoid secreted by the adrenal cortex that responds to stress on a long-term basis; reduces inflammation and promotes protein and fat metabolism. 403

cost-benefit analysis A weighing-out of the costs and benefits (in terms of contributions to reproductive success) of a particular strategy or behavior. 680

cotyledon Seed leaf for the embryo of a flowering plant; provides nutrient molecules for the developing plant before photosynthesis begins. 150, 180

coupled reactions Reactions that occur simultaneously; one is an exergonic reaction that releases energy, and the other is an endergonic reaction that requires an input of energy in order to occur. 105

covalent bond Chemical bond in which atoms share one pair of electrons. 25

cranial nerve Nerve that arises from the brain. 334

creatine phosphate Compound unique to muscles that contains a high-energy phosphate bond. 386

creatinine Nitrogenous waste; the end product of creatine phosphate metabolism. 304

crenation The shriveling of an animal cell in a hypertonic solution due to osmosis. 73

cretinism Condition resulting from improper development of the thyroid in an infant; characterized by stunted growth and mental retardation. 410

cristae (sing., crista) Short, fingerlike projections formed by the folding of the inner membrane of mitochondria. 58, 122

Cro-Magnon Common name for the first fossils to be designated *Homo sapiens sapiens*. 670

crossing-over Exchange of corresponding segments of genetic material between nonsister chromatids of homologous chromosomes during synapsis of meiosis. 90

crustacean Member of a group of marine arthropods that contains, among others, shrimps, crabs, crayfish, and lobsters. 641

cuboidal epithelium Type of epithelial tissue characterized by cube-shaped cells. 198

cultural eutrophication Overenrichment caused by human activities leading to excessive bacterial growth and oxygen depletion of a body of water. 716

culture Total pattern of human behavior; includes technology and the arts, and is dependent upon the capacity to speak and transmit knowledge. 668, 684

Cushing syndrome Condition resulting from hypersecretion of glucocorticoids; often characterized by thin arms and legs and a "moon face," and accompanied by high blood glucose and sodium levels. 409

cutaneous receptor Sensory receptors for pressure and touch found in the dermis of the skin. 351

cuticle Waxy layer covering the epidermis of plants that protects the plant against water loss and disease-causing organisms. 151

cyanobacterium Photosynthetic bacterium that contains chlorophyll and releases oxygen; formerly called a blue-green alga. 579

cycad Type of gymnosperm with palmate leaves and massive cones; cycads are most often found in the tropics and subtropics. 617

cyclic electron pathway Portion of the light reaction that involves only photosystem I and generates ATP. 137

cyclin Protein that cycles in quantity as the cell cycle progresses; combines with and activates the kinases that function to promote the events of the cycle. 83

cystic fibrosis (CF) A generalized, autosomal recessive disorder of infants and children characterized by widespread dysfunction of the exocrine glands and accumulation of thick mucus in the lungs. 297, 536

cystitis Inflammation of the urinary bladder. 314

cytochrome Any of several iron-containing protein molecules that serve as electron carriers in photosynthesis and cellular respiration. 123

cytokine Chemical messenger secreted by cells of the immune system that stimulates other cells to perform their various functions. 243

cytokinesis Division of the cytoplasm following mitosis and meiosis. 82

cytokinin Plant hormone that promotes cell division; often works in combination with auxin during organ development in plant embryos. 189

cytolysis Disruption or bursting of a cell; can be in response to osmosis in a hypotonic solution. 73

cytoplasm Contents of a cell between the nucleus and the plasma membrane that contains the organelles. 47

cytoplasmic segregation In development, the parceling of signals present in the egg cytoplasm to the cells that result from cleavage. 451

cytosine (C) One of four nitrogen-containing bases in the nucleotides composing the structure of DNA and RNA; pairs with guanine. 39

cytoskeleton Internal framework of the cell, consisting of microtubules, actin filaments, and intermediate filaments. 59

cytotoxic T cell Type of T cell that attacks and kills antigen-bearing cells. 247

D

data Facts or pieces of information collected through observation and/or experimentation. 11

daughter chromosome Separate chromatids become daughter chromosomes during anaphase of mitosis and anaphase II of meiosis. 85

day-neutral plant Plant whose flowering is not dependent on day length; e.g., tomato and cucumber. 191

deamination Removal of an amino group (–NH2) from an amino acid or other organic compound. 116

deciduous tree Tree that sheds its leaves annually. 149

decomposer Organism, usually a bacterium or fungus, that breaks down organic matter into inorganic nutrients that can be recycled in the environment. 583, 710

deductive reasoning Process of logic and reasoning using "if . . . then" statements. 10

defecation Discharge of feces from the rectum through the anus. 266

deforestation Removal of trees from a forest in a way that ever reduces the size of the forest. 747

dehydration reaction Chemical reaction resulting in a covalent bond with the accompanying loss of a water molecule. 31

delayed allergic response Allergic response initiated at the site of the allergen by sensitized T cells, involving macrophages and regulated by cytokines. 252

deletion Change in chromosome structure in which the end of a chromosome breaks off, or two simultaneous breaks lead to the loss of an internal segment; often causes abnormalities—e.g., cri du chat syndrome. 532

demographic transition Decline in the birthrate following reduction in the death rate, usually due to urbanization, such that the overall population growth rate is lowered. 695

denaturation Loss of normal shape by an enzyme so that it no longer functions; caused by a less than optimal pH and temperature. 39

dendrite Part of a neuron that sends signals toward the cell body. 320

dendritic cell Type of white blood cell that presents fragments of antigens to T cells. 229, 244

denitrification Conversion of nitrate or nitrite to nitrogen gas by bacteria in soil. 717

dense fibrous connective tissue Type of connective tissue containing many collagen fibers packed together; found in tendons and ligaments, for example. 200

density-dependent Biotic factor, such as disease or competition, that affects population size according to the population's density. 698

density-independent Abiotic factor, such as fire or flood, that affects population size independent of the population's density. 698

dental caries Tooth decay that occurs when bacteria within the mouth metabolize sugar and give off acids that erode teeth; a cavity. 261

dermis Region of the skin that lies beneath the epidermis. 208

desert Ecological biome characterized by a limited amount of rainfall; deserts have hot days and cool nights. 733

desertification Denuding and degrading a once-fertile land, initiating a desert-producing cycle that feeds on itself and causes long-term changes in the soil, climate, and biota of an area. 747

detection Conversion of environmental signals, such as light or sound, to nerve impulses by the various sensory systems. 348

detrital food chain Straight-line linking of organisms according to who eats whom, beginning with detritus. 713

detrital food web Complex pattern of interlocking and crisscrossing food chains, beginning with detritus. 711

detritus Partially decomposed organic matter derived from tissues and animal wastes. 710

deuterostome Group of coelomate animals in which the second embryonic opening is associated with the mouth; the first embryonic opening, the blastopore, is associated with the anus. 628

development Series of stages by which a zygote becomes an organism or by which an organism changes during its life span; includes puberty and aging, for example. 5

diabetes insipidus (DI) A disorder caused by insufficient secretion of antidiuretic hormone (ADH) by the posterior pituitary gland, causing production of large amounts of dilute urine and excessive thirst. 408

diabetes mellitus Condition characterized by a high blood glucose level and the appearance of glucose in the urine due to a deficiency of insulin production or inability of cells to respond normally to insulin. 411

dialysate Material that passes through the membrane in dialysis. 313

diaphragm Dome-shaped horizontal sheet of muscle and connective tissue that divides the thoracic cavity from the abdominal cavity. Also, a birth control device consisting of a soft rubber or latex cup that fits over the cervix. 429

diastole Relaxation period of a heart chamber during the cardiac cycle. 222

diastolic pressure Arterial blood pressure during the diastolic phase of the cardiac cycle. 225

diatom Golden-brown alga with a cell wall in two parts, or valves; significant part of phytoplankton. 585

diencephalon Portion of the brain in the region of the third ventricle that includes the thalamus and hypothalamus. 330

dietary supplement Product added to a person's diet that is not considered food; usually contains vitamins, minerals, amino acids, or a concentrated energy source. 275

differentially permeable Ability of plasma membranes to regulate the passage of substances into and out of the cell, allowing some to pass through and preventing the passage of others; sometimes called selectively permeable. 70

diffusion Movement of molecules or ions from a region of higher concentration to one of lower concentration; it requires no energy and stops when the distribution is equal. 71

digestive system Organ system that includes the mouth, esophagus, stomach, small intestine, and large intestine (colon), which receives food and digests it into nutrient molecules. Also has associated organs: teeth, tongue, salivary glands, liver, gallbladder, and pancreas. 206

digit Referring to fingers or toes. 378

dihybrid Individual that is heterozygous for two traits; shows the phenotype governed by the dominant alleles but carries the recessive alleles. 478

dinoflagellate Photosynthetic unicellular protist with two flagella, one a whiplash and the other located within a groove between protective cellulose plates; significant part of phytoplankton. 585

diploid (2n) number Twice the number of chromosomes found in the gametes. 85

directional selection Natural selection in which an extreme phenotype is favored, usually in a changing environment. 564

disaccharide Sugar that contains two units of a monosaccharide; e.g., maltose. 32

disease Illness; state of homeostatic imbalance in which part or all of the body does not function properly. 212

disruptive selection Natural selection in which extreme phenotypes are favored over the average phenotype, leading to more than one distinct form. 564

distal convoluted tubule (DCT) Final portion of a nephron that joins with a collecting duct; associated with tubular secretion. 307

diuretic Drug used to counteract hypertension by causing the excretion of water. 311

DNA (deoxyribonucleic acid) Nucleic acid found in cells; the genetic material that specifies protein synthesis in cells. 39, 490

DNA fingerprinting The use of DNA fragment lengths resulting from restriction enzyme cleavage to identify particular individuals. 503

DNA ligase Enzyme that links DNA fragments; used during production of recombinant DNA to join foreign DNA to vector DNA. 502

DNA probe Piece of single-stranded DNA that will bind to a complementary piece of DNA. 539

DNA profiling In forensic genetics, DNA profiling is a way of matching a genetic sample found at a crime scene, for example, to that of a potential suspect. It is similar to DNA fingerprinting in that each person's genome is unique and thus can be thought of as a genetic fingerprint. 539

DNA replication Synthesis of a new DNA double helix prior to mitosis and meiosis in eukaryotic cells and during prokaryotic fission in prokaryotic cells. 492

domain Largest of the categories, or taxa, used by taxonomists to group species; the three domains are Archaea, Bacteria, and Eukarya. 6, 569

dominance hierarchy Organization of animals in a group that determines the order in which the animals have access to resources. 680

dominant allele Allele that exerts its phenotypic effect in the heterozygote; it masks the expression of the recessive allele. 473

dormancy In plants, a cessation of growth under conditions that seem appropriate for growth. 182

dorsal root ganglion Mass of sensory neuron cell bodies located in the dorsal root of a spinal nerve. 335

double fertilization In flowering plants, one sperm fuses with the egg to produce a zygote and a second sperm fuses with the polar nuclei within the embryo sac to produce a 3n endosperm nucleus. 179

double helix Double spiral; describes the three-dimensional shape of DNA. 40, 492

doubling time Number of years it takes for a population to double in size. 695

Duchenne muscular dystrophy Chronic progressive disease affecting the shoulder and pelvic girdles, commencing in early childhood. Characterized by increasing weakness of the muscles, followed by atrophy and a peculiar swaying gait with the legs kept wide apart. The condition is transmitted as an X-linked trait, and affected individuals, predominantly males, rarely survive to maturity. Death is usually due to respiratory weakness or heart failure. 538

duodenum First part of the small intestine, where chyme enters from the stomach. 264

duplicated chromosome Chromosome that consists of two chromatids held together at a centromere. 85

duplication Change in chromosome structure in which a particular segment is present more than once. 532

E

echinoderm Phylum of marine animals that includes sea stars, sea urchins, and sand dollars; characterized by radial symmetry and a water vascular system. 650

ecological niche Role an organism plays in its community, including its habitat and its interactions with other organisms. 699

ecological pyramid Pictorial graph based on the biomass, number of organisms, or energy content of various trophic levels in a food web—from the producer to the final consumer populations. 713

ecological succession Gradual replacement of communities in an area following a disturbance (secondary succession) or the creation of new soil (primary succession). 704

ecology Study of the interactions of organisms with other organisms and the physical environment. 692

ecosystem Biological community together with the associated abiotic environment; characterized by energy flow and chemical cycling. 8, 692, 710

ectoderm Outermost germ layer of the embryonic gastrula; it gives rise to the nervous system and skin. 448

ectothermic Having a body temperature that varies according to the environmental temperature. 658

edema Swelling due to tissue fluid accumulation in the intercellular spaces. 241, 312

elastic cartilage Type of cartilage composed of elastic fibers, allowing greater flexibility. 201

elastic fiber Yellow fiber in the matrix of connective tissue, providing flexibility. 200

electrocardiogram (ECG) Recording of the electrical activity associated with the heartbeat. 223

electroencephalogram (EEG) Graphic recording of the brain's electrical activity. 330

electron Subatomic particle that has almost no weight and carries a negative charge; orbits the nucleus of an atom in a shell. 20

electronegativity The ability of an atom to attract electrons toward itself in a chemical bond. 20

electron transport chain Chain of electron carriers in the cristae of mitochondria and the thylakoid membrane of chloroplasts. As the electrons pass from one carrier to the next, energy is released and used to establish a hydrogen ion gradient. This gradient is associated with the production of ATP molecules. 119

element Substance that cannot be broken down into substances with different properties; composed of only one type of atom. 20

El Niño–Southern Oscillation (ENSO) Warming of water in the Eastern Pacific equatorial region such that the Humboldt Current is displaced; possible negative results include the reduction in marine life. 740

elongation During DNA replication or the formation of mRNA during transcription, elongation is the step whereby the bases are added to the new or "daughter" strands of DNA or the mRNA transcript strand is lengthened, respectively. 498

embolus Moving blood clot that is carried through the bloodstream. 232

embryo Stage of a multicellular organism that develops from a zygote before it becomes free-living; in seed plants, the embryo is part of the seed. 179

embryonic disk During human development, flattened area during gastrulation from which the embryo arises. 456

embryonic period Time from approximately the second to the eighth week of human development during which the major organ systems are organized. 454

embryo sac Female gametophyte of flowering plants that produces an egg cell. 179

emphysema Degenerative disorder in which the bursting of alveolar walls reduces the total surface area for gas exchange. 297

emulsification Breaking up of fat globules into smaller droplets by the action of bile salts or any other emulsifier. 34

endergonic reaction Chemical reaction that requires an input of energy; opposite of exergonic reaction. 104

endocrine gland Ductless organ that secretes hormone(s) into the bloodstream. 396

endocrine system Organ system involved in the coordination of body activities; uses hormones as chemical signals secreted into the bloodstream. 207

endocytosis Process by which substances are moved into the cell from the environment by phagocytosis (cellular eating) or pinocytosis (cellular drinking); includes receptor-mediated endocytosis. 76

endoderm Inner germ layer of the embryonic gastrula that becomes the lining of the digestive and respiratory tracts and associated organs. 448

endodermis Plant root tissue that forms a boundary between the cortex and the vascular cylinder. 163

endometriosis The presence of endometrial tissue in places other than the uterus, usually within the abdominal cavity, that often results in severe pain and infertility. 438

endometrium Lining of the uterus; becomes thickened and vascular during the uterine cycle. 423

endoplasmic reticulum (ER) Membranous system of tubules, vesicles, and sacs in cells, sometimes having attached ribosomes. Rough ER has ribosomes; smooth ER does not. 54

endosperm In angiosperms, the 3n tissue that nourishes the embryo and seedling and is formed as a result of a sperm joining with two polar nuclei. 179

endospore Thick-walled, resilient structure formed within certain bacteria. 578

endosymbiotic theory Possible explanation for the evolution of eukaryotic organelles by phagocytosis of prokaryotes. 63

endothelium A single layer of simple squamous epithelium that forms the innermost lining of blood vessels, lymphatic vessels, and the heart. 218

endothermic Maintenance of a constant body temperature independent of the environmental temperature. 661

energy Capacity to do work and bring about change; occurs in a variety of forms. 3, 102

energy of activation (E_a) Energy that must be added to cause molecules to react with one another. 106

enhancer Element that regulates transcription from nearby genes; functions by acting as a binding site for transcription factors. 516

entropy Measure of disorder or randomness. 103

envelope Lipid layer around some viruses. 597

environmental resistance Total of factors in the environment that limit the numerical increase of a population in a particular region. 693

enzymatic protein Protein that catalyzes a specific reaction. 69

enzyme Organic catalyst, usually a protein, that speeds up a reaction in cells due to its particular shape. 37, 106

enzyme inhibition Means by which cells regulate enzyme activity; may be competitive or noncompetitive. 109

eosinophil White blood cell containing cytoplasmic granules that stain with acidic dye. 229

epidermal tissue Exterior tissue, usually one cell thick, of leaves, young stems, roots, and other parts of plants. 151

epidermis In plants, tissue that covers roots, leaves, and stems of a nonwoody organism; in mammals, the outer protective region of the skin. 151, 163, 208

epididymis (pl., epididymides) Coiled tubule next to the testes where sperm mature and may be stored for a short time. 418

epiglottis Structure that covers the glottis, the opening to the air tract, during the process of swallowing. 262, 288

epinephrine Hormone secreted by the adrenal medulla in times of stress; also called adrenaline. 403

epiphyte Plant that takes its nourishment from the air because its placement in other plants gives it an aerial position. 169, 731

episiotomy Surgical procedure performed during childbirth in which the opening of the vagina is enlarged to avoid tearing. 464

episodic memory Capacity of the brain to store and retrieve information about persons and events. 331

epithelial tissue Type of tissue that lines hollow organs and covers surfaces; also called epithelium. 198

erectile dysfunction (ED) Failure of the penis to achieve erection. 436

erythropoietin Hormone, produced by the kidneys, that speeds red blood cell formation. 228, 304

esophagus Muscular tube for moving swallowed food from the pharynx to the stomach. 262

essential amino acids Amino acids required in the human diet because the body cannot make them. 271

essential fatty acid Fatty acid that is necessary in the diet because the body is unable to manufacture it. 273

estrogen Female sex hormone that helps maintain sexual organs and secondary sex characteristics. 406, 426

estuary Portion of the ocean located where a river enters and fresh water mixes with salt water. 736

ethylene Plant hormone that causes ripening of fruit and is also involved in abscission. 190

eudicot Abbreviation of eudicotyledon, a member of the class of flowering plants called Eudicotyledones. These plants have several common characteristics, including two embryonic leaves (cotyledons) in the seed, net-veined leaves, cylindrical arrangement of vascular bundles, and flower parts in fours or fives or multiples of four or five. 150, 618

euglenoid Flagellated and flexible freshwater unicellular protist that usually contains chloroplasts and has a semirigid cell wall. 586

Eukarya One of the three domains of life, consisting of eukaryotes, which are organisms in the kingdoms Protista, Fungi, Plantae, and Animalia. 7, 570

eukaryotic cell Type of cell that has a membrane-bounded nucleus and organelles; found in organisms within the domain Eukarya. 49

eutrophication Enrichment of water by inorganic nutrients used by phytoplankton; often overenrichment caused by human activities, leading to excessive bacterial growth and oxygen depletion. 734

evaporation The process by which liquid water turns into gaseous form (water vapor). 714

evolution Changes that occur in the members of a species with the passage of time, often resulting in increased adaptation of organisms to the environment. 5, 552

excretion Removal of metabolic wastes from the body. 304

exergonic reaction Chemical reaction that releases energy; opposite of endergonic reaction. 104

exocytosis Process in which an intracellular vesicle fuses with the plasma membrane so that the vesicle's contents are released outside the cell. 76

exon A segment of a gene that codes for a protein. 495

exophthalmia An abnormal protrusion of the eyes, often caused by hyperactivity of the thyroid gland. 410

exotic species Nonnative species that migrate or are introduced by humans into a new ecosystem. 755

experimental design A test that controls the conditions under which experimental observations are made. The experimental design should control all factors except the one that researchers are interested in. 10

experimental variable A treatment that is varied in an experiment to test a hypothesis. For example, an experimental variable may be four different nitrogen levels added to a plant species to determine how each affects plant growth. 11

expiration Act of expelling air from the lungs; exhalation. 286

expiratory reserve volume Volume of air that can be forcibly exhaled after normal exhalation. 291

exponential growth Growth, particularly of a population, in which the increase occurs in the same manner as compound interest. 693

external respiration The exchange of gases between air in the lung alveoli and blood in the pulmonary capillaries. 293

exteroceptor Sensory receptor that detects stimuli from outside the body; e.g., taste, smell, vision, hearing, and equilibrium. 348

extraembryonic membrane Membrane that is not a part of the embryo but is necessary to the continued existence and health of the embryo. 454

F

facilitated transport Use of a plasma membrane carrier to move a substance into or out of a cell from a higher to a lower concentration; no energy required. 74

FAD (flavin adenine dinucleotide) Coenzyme of oxidation-reduction that becomes $FADH_2$ as oxidation of substrates occurs and then delivers electrons to the electron transport system in mitochondria during cellular respiration. 119

fall overturn Mixing process that occurs in fall in stratified lakes whereby oxygen-rich top waters mix with nutrient-rich bottom waters. 734

familial hypercholesterolemia (FH) Inability to remove cholesterol from the bloodstream; predisposes individual to heart attack. 537

family One of the categories, or taxa, used by taxonomists to group species; the taxon above the genus level. 6, 569

farsighted Abnormal vision due to a shortened eyeball from front to back; light rays focus in back of the retina when viewing close objects. 362

fat Organic molecule that contains glycerol and fatty acids and is found in adipose tissue. 34

fate map Diagram that traces the differentiation of cells during development, from their origin to their final structure and function. 452

fatty acid Molecule that contains a hydrocarbon chain and ends with an acid group. 34

female condom Birth control method that blocks the entrance of sperm to the uterus; also prevents STDs. 429

female gametophyte In seed plants, the gametophyte that produces an egg; in flowering plants, an embryo sac. Sometimes called a megagametophyte. 178

fermentation Anaerobic breakdown of glucose that results in a gain of two ATP and end products such as alcohol and lactate. 119

fern Member of a group of plants that have large fronds; in the sexual life cycle, the independent gametophyte produces flagellated sperm, and the vascular sporophyte produces windblown spores. 614

fertilization Union of a sperm nucleus and an oocyte nucleus to create a zygote with the diploid number of chromosomes. 90, 446

fiber Structure resembling a thread; also, plant material that is nondigestible. 266

fibrin Insoluble protein threads formed from fibrinogen during blood clotting. 229

fibrinogen Plasma protein that is converted into fibrin threads during blood clotting. 229

fibroblast Cell in connective tissues that produces fibers and other substances. 200

fibrocartilage Cartilage with a matrix of strong collagenous fibers. 201

fibrous root system In most monocots, a mass of similarly sized roots that cling to the soil. 164

filament End-to-end chains of cells that form as cell division occurs in only one plane; in plants, the elongated stalk of a stamen. 176, 619

filter feeder Method of obtaining nourishment by certain animals that strain minute organic particles from the water in a way that deposits them in the digestive tract. 630

fimbria (pl., fimbriae) Fingerlike extension from the oviduct near the ovary. 48, 422

first messenger An extracellular chemical, such as a hormone, that binds to a cellular receptor and initiates intracellular activity. 398

fitness Lifetime reproductive success; the relative ability of an individual to produce fertile offspring as measured against other individuals of the same species in the same environment. 564, 680

fixed action pattern (FAP) Innate behavior pattern that is stereotyped, spontaneous, genetically encoded, and independent of individual learning. 678

flagellum (pl., flagella) Long, slender extension used for locomotion by some bacteria, protozoans, and sperm. 48, 578

flatworm Unsegmented worm lacking a body cavity; phylum Platyhelminthes. 632

flower Reproductive organ of plants that contains the structures for the production of pollen grains and covered seeds. 176, 608

fluid-mosaic model Model of the plasma membrane based on the changing location and pattern of protein molecules in a fluid phospholipid bilayer. 68

focus Bending of light rays by the cornea, lens, and humors so that they converge and create an image on the retina. 354

follicle Structure in the ovary that produces the oocyte and, in particular, the female sex hormones estrogen and progesterone. 425

follicle-stimulating hormone (FSH) Hormone secreted by the anterior pituitary gland that stimulates the development of an ovarian follicle in a female or the production of sperm in a male. 421

fontanel Membranous region located between certain cranial bones in the skull of a fetus or infant. 374, 458

food chain The order in which one population feeds on another in an ecosystem, from detritus (detrital food chain) or producer (grazing food chain) to final consumer. 713

food web Complex pattern of energy flow in an ecosystem represented by interlocking and crisscrossing food chains. 711

foramen magnum Opening in the occipital bone of the vertebrate skull through which the spinal cord passes. 374

formed element Constituent of blood that is cellular (red blood cells and white blood cells) or at least cellular in origin (platelets). 226

fossil Any past evidence of an organism that has been preserved in Earth's crust. 553

fossil fuel Remains of once-living organisms that are burned to release energy, such as coal, oil, and natural gas. 718, 751

founder effect Type of genetic drift in which only a fraction of the total genetic diversity of the original gene pool is represented as a result of a few individuals founding a colony. 563

fovea centralis Region of the retina, consisting of densely packed cones, that is responsible for the greatest visual acuity. 354

free energy Energy in a system that is capable of performing work. 104

fruit In a flowering plant, the structure that forms from an ovary and associated tissues and encloses seeds. 176, 621

functional group Specific cluster of atoms attached to the carbon skeleton of organic molecules that enters into reactions and behaves in a predictable way. 31

fungus (pl., fungi) Saprotrophic decomposer; the body is made up of filaments called hyphae that form a mass called a mycelium; e.g., mushrooms and molds. 7, 592

G

G3P (glyceraldehyde 3-phosphate) Significant metabolite in both the glycolytic pathway and the Calvin cycle; G3P is oxidized during glycolysis and reduced during the Calvin cycle. 120

gallbladder Organ attached to the liver that stores and concentrates bile. 268

gamete Haploid sex cell; an oocyte or a sperm that joins during fertilization to form a zygote. 90, 441

gametophyte Haploid generation of the alternation of generations life cycle of a plant; produces gametes that unite to form a diploid zygote. 176, 609

ganglion Collection or bundle of neuron cell bodies usually outside the central nervous system. 334

gap junction Region between cells formed by the joining of two adjacent plasma membranes; it lends strength and allows ions, sugars, and small molecules to pass between cells. 200

gastric gland Gland within the stomach wall that secretes gastric juice. 264

gastropod Mollusc with a broad flat foot for crawling; e.g., snails and slugs. 636

gastrovascular cavity Blind digestive cavity that also serves a circulatory (transport) function in animals lacking a circulatory system. 632

gastrula In animal development, the embryonic stage that follows formation of the germ layers. 448

gastrulation Process during animal development by which the germ layers form. 448

gene Unit of heredity existing as alleles on the chromosomes; in diploid organisms, typically two alleles are inherited—one from each parent. 4, 493

gene cloning Production of one or more copies of the same gene. 501

gene flow Sharing of genes between two populations through interbreeding. 563

gene mutation Alteration in a gene due to a change in DNA composition. 500

gene pool Total of all the genes of all the individuals in a population. 559

gene therapy Correction of a detrimental mutation by the addition of new DNA and its insertion in a genome. 544

genetically modified plant A plant that carries the genes of another organism as a result of DNA technology; also, a transgenic plant. 186

genetic counseling Prospective parents consult a counselor who determines the genotype of each and whether an unborn child will have a genetic disorder. 526

genetic drift Mechanism of evolution due to random changes in the allelic frequencies of a population; more likely to occur in small populations or when only a few individuals of a large population reproduce. 562

genetic engineering Alteration of genomes for medical or industrial purposes. 501

genetic marker Abnormality in the sequence of a base at a particular location on a chromosome, signifying a disorder. 539

genetic profiling An individual's complete genotype, including any possible mutations. 539

genetic recombination Due to crossing-over and independent assortment during meiosis, offspring do not have the same genetic makeup as either parent. 90

genetic test Laboratory test done to determine whether or not a patient carries a genetic disease. 519

genital herpes Sexually transmitted disease caused by herpes simplex virus and sometimes accompanied by painful ulcers on the genitals. 434

genital warts Warts on or near the genitals that are caused by human papillomaviruses. 434

genome Full set of genetic information of a species or a virus. 542

genomics The study of all the nucleotide sequences, including structural genes, regulatory sequences, and noncoding DNA segments, in the chromosomes of an organism. 542

genotype Genes of an individual for a particular trait or traits; often designated by letters, such as BB or Aa. 473

genus One of the categories, or taxa, used by taxonomists to group species; contains those species that are most closely related through evolution. 6, 569

germinate Beginning of the growth of a seed, spore, or zygote, especially after a period of dormancy. 182

germ layer Developmental layer of the body—that is, ectoderm, mesoderm, or endoderm. 448

gerontology Study of aging. 465

gibberellin Plant hormone promoting increased stem growth; also involved in flowering and seed germination. 188

ginkgo One of four phyla of gymnosperms. 618

girdling Removal of a strip of bark from around a tree. 170

gland Epithelial cell or group of epithelial cells that are specialized to secrete a substance. 198

glaucoma Increasing loss of the field of vision caused by blockage of the ducts that drain the aqueous humor, creating pressure buildup and nerve damage. 354

global warming Predicted increase in the Earth's temperature, due to the greenhouse effect, which will lead to the melting of polar ice and a rise in sea levels. 719

glomerular capsule Double-walled cup that surrounds the glomerulus at the beginning of the nephron. 307

glomerular filtrate Filtered portion of blood contained within the glomerular capsule. 309

glomerular filtration Movement of small molecules from the glomerulus into the glomerular capsule due to blood pressure. 309

glomerulus Capillary network within the glomerular capsule of a nephron, where glomerular filtration takes place. 306

glottis Opening for airflow in the larynx. 262, 288

glucagon Hormone secreted by the pancreas that causes the liver to break down glycogen and raises the blood glucose level. 405

glucocorticoid Type of hormone secreted by the adrenal cortex that influences carbohydrate, fat, and protein metabolism; see cortisol. 403

glucose Six-carbon sugar that organisms degrade as a source of energy during cellular respiration. 32

glycemic index (GI) The blood glucose response of a given food compared to, for example, white bread. 271

glycogen Storage polysaccharide, found in animals, that is composed of glucose molecules joined in a linear fashion but having numerous branches. 32

glycolipid Lipid in plasma membranes that bears a carbohydrate chain attached to a hydrophobic tail. 68

glycolysis Anaerobic metabolic pathway found in the cytoplasm that participates in cellular respiration and fermentation; it converts glucose to two molecules of pyruvate. 119

glycoprotein Protein in plasma membranes that bears a carbohydrate chain. 68

gnetophyte Division of plants composed of three related woody gymnosperms. 618

Golgi apparatus Organelle, consisting of flattened saccules and also vesicles, that processes, packages, and distributes molecules within or from the cell. 55

gonad Organ that produces gametes; the ovary produces oocytes, and the testis produces sperm. 405

gonadotropic hormone Substance secreted by the anterior pituitary that regulates the activity of the ovaries and testes; principally, follicle-stimulating hormone (FSH) and luteinizing hormone (LH). 400

gonadotropin-releasing hormone (GnRH) Hormone secreted by the hypothalamus that stimulates the anterior pituitary to secrete follicle-stimulating hormone and luteinizing hormone. 421

gonorrhea Sexually transmitted disease caused by the bacterium Neisseria gonorrhoeae that causes painful urination and swollen testes in men; it is usually symptomless in women but can cause inflammation of the cervix and uterine tubes. 435

gout Joint inflammation caused by accumulation of uric acid. 304

Gram stain Common laboratory test performed to help identify bacteria and aid in determining their susceptibility to antibiotics. 578

granular leukocyte White blood cell with prominent granules in the cytoplasm. 229

granum (pl., grana) Stack of chlorophyll-containing thylakoids in a chloroplast. 58, 133

grassland A biome characterized by grasses and small plants, but no large trees. Precipitation, ranging from 25 cm to 76 cm per year, limits the vegetation types in grasslands. Savanna is a grassland in Africa characterized by few trees and a severe dry season. Other types of grasslands are prairie, steppe, and pampas. 732

gravitropism Growth response of roots and stems of plants to Earth's gravity. Roots experience positive gravitropism, while stems experience negative. 188

gray crescent Gray area that appears in an amphibian egg after being fertilized by the sperm; thought to contain chemical signals that turn on the genes that control development. 450

gray matter Nonmyelinated nerve fibers and cell bodies in the central nervous system. 321

grazing food chain A flow of energy to a straight-line linking of organisms according to who eats whom. 713

grazing food web Complex pattern of interlocking and crisscrossing food chains that begins with populations of autotrophs serving as a producers. 711

green alga Member of a diverse group of photosynthetic protists that contains chlorophylls a and b and has other biochemical characteristics like those of plants. 585

greenhouse effect Reradiation of solar heat toward Earth caused by gases in the atmosphere. 719

greenhouse gases Atmospheric gases, such as carbon dioxide, methane, nitrous oxide, water vapor, ozone, and nitrous oxide, that are involved in the greenhouse effect. 751

ground tissue Tissue that constitutes most of the body of a plant; consists of parenchyma, collenchyma, and sclerenchyma cells that function in storage, basic metabolism, and support. 151

growth factor Chemical signal that regulates mitosis and differentiation of cells that have receptors for it; important in such processes as fetal development, tissue maintenance and repair, and hematopoiesis; sometimes a contributing cofactor in cancer. 408, 518

growth hormone (GH) Substance secreted by the anterior pituitary; controls size of individual by promoting cell division, protein synthesis, and bone growth. 400

growth plate The area of a long bone between the epiphysis and the diaphysis where growth in length occurs. 372

guanine (G) One of four nitrogen-containing bases in nucleotides composing the structure of DNA and RNA; pairs with cytosine. 39

guard cell One of two cells that surround a leaf stoma; changes in the turgor pressure of these cells cause the stoma to open or close. 151

gymnosperm Type of woody seed plant in which the seeds are not enclosed by fruit and are usually borne in cones, such as those of the conifers. 617

H

habitat Place where an organism lives and is able to survive and reproduce. 699

hair cell Cell with stereocilia (long microvilli) that is sensitive to mechanical stimulation; mechanoreceptor for hearing and equilibrium in the inner ear. 357

hair follicle Tubelike depression in the skin in which a hair develops. 209

haploid (n) number Half the diploid number; the number characteristic of gametes that contain only one set of chromosomes. 85

hard palate Bony, anterior portion of the roof of the mouth. 260

hay fever Seasonal variety of allergic reaction to a specific allergen. Characterized by sudden attacks of sneezing, swelling of nasal mucosa, and often asthmatic symptoms. 252

heart Muscular organ located in the thoracic cavity whose rhythmic contractions maintain blood circulation. 220

heart attack Damage to the myocardium due to blocked circulation in the coronary arteries; also called a myocardial infarction (MI). 232

heartburn Burning pain in the chest that occurs when part of the stomach contents escape into the esophagus. 263

helper T cell Type of T cell that releases cytokines and stimulates certain other immune cells to perform their respective functions. 247

hemocoel Residual coelom found in arthropods, which is filled with hemolymph. 642

hemodialysis Cleansing of blood by using an artificial membrane that causes substances to diffuse from blood into a dialysis fluid. 313

hemoglobin Iron-containing pigment in red blood cells that combines with and transports oxygen. 228

hemophilia X-linked, recessive genetic disease in which one or more clotting factors are missing. 538

hepatic portal system Portal system that begins at the villi of the small intestine and ends at the liver. 225

hepatic portal vein A vein formed by capillaries in the digestive tract that carries nutrients to the liver. 225

hepatic vein Vein that runs between the liver and the inferior vena cava. 225

hepatitis Inflammation of the liver. Viral hepatitis occurs in several forms. 281

herbaceous stem Nonwoody stem. 157

herbivore Primary consumer in a grazing food chain; a plant eater. 710

hermaphroditic Animals having both male and female sex organs. 630

heterotroph Organism that cannot synthesize organic molecules from inorganic nutrients and therefore must take in organic nutrients (food). 132, 552, 710

heterozygous Having two different alleles (e.g., Aa) for a given trait. 473

hexose Six-carbon sugar. 32

hinge joint Type of joint that allows movement as a hinge does; e.g., the movement of the knee. 380

hippocampus Part of the cerebral cortex where memories form. 331

histamine Substance, produced by basophils and mast cells in connective tissue, that causes capillaries to dilate. 243

homeobox Nucleotide sequence located in all homeotic genes and serving to identify portions of the genome in many different types of organisms that are active in pattern formation. 452

homeostasis Maintenance of normal internal conditions in a cell or an organism by means of self-regulating mechanisms. 4, 209

homeotic gene Gene that controls the overall body plan by controlling the fate of groups of cells during development. 452

hominid Member of the family Hominidae, which contains australopithecines and humans. 664

homologous chromosomes Similarly constructed chromosomes that have the same shape and contain genes for the same traits; also called homologues. 89

homologous structure Structure that is similar in two or more species because of common ancestry. 556

homologue Member of a homologous pair of chromosomes. 468

homozygous Having identical alleles (e.g., AA or aa) for a given trait; pure breeding. 473

hormone Chemical signal produced in one part of the body that controls the activity of other parts. 265, 396

horsetail Member of a seedless vascular plant phylum that has only one genus (Equisetum) in existence today; characterized by rhizomes, scale-like leaves, strobili, and tough, rigid stems. 613

human chorionic gonadotropin (hCG) Hormone produced by the chorion that functions to maintain the uterine lining. 427, 455

Human Genome Project Initiative to determine the complete sequence of the human genome and to analyze this information. 542

Huntington disease Genetic disease marked by progressive deterioration of the nervous system due to deficiency of a neurotransmitter. 537

hyaline cartilage Cartilage whose cells lie in lacunae separated by a white translucent matrix containing very fine collagen fibers. 201

hybridization Crossing of different species. 186

hydrogen bond Weak bond that arises between a slightly positive hydrogen atom of one molecule and a slightly negative atom of another molecule or between parts of the same molecule. 26

hydrogen ion Hydrogen atom that has lost its electron and therefore bears a positive charge (H^+). 29

hydrolysis reaction Splitting of a compound by the addition of water, with the H^+ being incorporated in one fragment and the OH^- in the other. 31

hydrolytic enzyme Enzyme that catalyzes a reaction in which the substrate is broken down by the addition of water. 268

hydrophilic Type of molecule that interacts with water by dissolving in water and/or by forming hydrogen bonds with water molecules. 28

hydrophobic Type of molecule that does not interact with water because it is nonpolar. 28

hydrostatic skeleton Fluid-filled body compartment that provides support for muscle contraction resulting in movement; seen in cnidarians, flatworms, roundworms, and segmented worms. 634

hydrothermal vent Hot springs in the seafloor along ocean ridges where heated seawater and sulfate react to produce hydrogen sulfide; here, chemosynthetic bacteria support a community of varied organisms. 739

hydroxide ion One of two ions that results when a water molecule dissociates; it has gained an electron and therefore bears a negative charge ($OH-$). 29

hyperthyroidism A disorder in which the thyroid gland produces excess thyroid hormone(s). 410

hypertension Elevated blood pressure, particularly the diastolic pressure. 235

hypertonic solution Higher solute concentration (less water) than the cell; causes cell to lose water by osmosis. 73

hypertrophy Increase in size of a tissue or body part. 388

hypha Filament of the vegetative body of a fungus. 591

hypothalamus Part of the brain located below the thalamus that helps regulate the internal environment of the body and produces releasing factors that control the anterior pituitary. 330, 399

hypothesis Supposition formulated after making an observation; it can be tested by obtaining more data, often by experimentation. 10

hypothyroidism A disorder in which the thyroid gland produces insufficient thyroid hormone(s). 410

hypotonic solution Lower solute (more water) concentration than the cytosol of a cell; causes cell to gain water by osmosis. 72

I

immune system All the cells and molecules in the body that protect the body against foreign organisms and substances, and also against cancerous cells. 206

immunity Ability of the body to protect itself from foreign substances and cells, including disease-causing agents. 242

immunization Use of a vaccine to protect the body against specific disease-causing agents. 248

immunoglobulin (Ig) Globular plasma protein that functions as an antibody. 245

immunosuppressive Any condition that inhibits the activity of the immune system. Sometimes done intentionally, for example, to treat allergies or to prevent organ rejection, usually via a drug. 255

implant Attachment and penetration of the embryo into the lining of the uterus (endometrium). 422

imprinting Learning to make a particular response to only one type of animal or object. 678

inclusive fitness Fitness that results from personal reproduction and from helping nondescendant relatives reproduce. 684

incomplete dominance Occurs when the heterozygote is intermediate in phenotype between the two homozygotes. 480

incontinence Inability to control excretory functions, especially urination. 305

incus The middle of the three ossicles of the ear that conduct vibrations from the tympanic membrane to the oval window of the inner ear; the other two ossicles are the malleus and the stapes. 358

independent assortment Alleles of unlinked genes segregate independently of each other during meiosis so that the gametes contain all possible combinations of alleles. 91

induced fit model Change in the shape of an enzyme's active site that enhances the fit between the active site and its substrate(s). 107

induction Ability of a chemical or a tissue to influence the development of another tissue. 451

inductive reasoning Using specific observations and the process of logic and reasoning to arrive at a hypothesis. 10

inferior vena cava Large vein that enters the right atrium from below and carries blood from the trunk and lower extremities. 224

infertility Inability to have as many children as desired. 440

inflammatory reaction Tissue response to injury, characterized by redness, swelling, pain, and heat. 243

initiation During DNA replication or transcription, initiation is the step whereby DNA replication begins or transcription begins; this step is catalyzed by specific enzymes such that the DNA "unravels" and forms a bubble. 497

innate immunity Protective responses of the body that are fully functional without previous exposure to a disease agent. 242

inner ear Portion of the ear consisting of a vestibule, semicircular canals, and the cochlea, where equilibrium is maintained and sound is transmitted. 358

inorganic molecule Type of molecule that is not derived from a living organism. 31

insect Member of a group of arthropods having a head with antennae, compound eyes, and simple eyes; a thorax with three pairs of legs and often wings; and an abdomen containing internal organs. 642

insertion End of a muscle that is attached to a movable bone. 380

inspiration Taking air into the lungs; inhalation. 286

inspiratory reserve volume Volume of air that can be forcibly inhaled after normal inhalation. 291

insulin Hormone secreted by the pancreas that lowers the blood glucose level by promoting the uptake of glucose by cells and the conversion of glucose to glycogen by the liver and skeletal muscles. 405

integumentary system Organ system consisting of skin and various organs, such as hair, that are found in skin. 206

intercalated disk Region that holds adjacent cardiac muscle cells together and appears as dense bands at right angles to the muscle striations. 203

interferon Protein formed by a cell infected with a virus that blocks the infection of another cell. 244

interkinesis Period of time between meiosis I and meiosis II during which no DNA replication takes place. 92

interleukin Cytokine produced by immune cells, such as macrophages and T cells, that functions as a regulator of the immune response. 251

intermembrane space The space between the inner and outer membrane of the mitochondrial membrane; important for the establishment of gradients necessary for chemiosmosis. 124

internal respiration Exchange of oxygen and carbon dioxide between blood and tissue fluid. 294

interneuron Neuron located within the central nervous system that conveys messages between parts of the central nervous system. 321

internode In vascular plants, the region of a stem between two successive nodes. 149

interoceptor Sensory receptor that detects stimuli from inside the body; e.g., pressoreceptors, osmoreceptors, chemoreceptors. 348

interphase Stages of the cell cycle (G1, S, G2) during which growth and DNA synthesis occur when the nucleus is not actively dividing. 82

interstitial cell Hormone-secreting cell located between the seminiferous tubules of the testes. 421

interstitial cell-stimulating hormone (ICSH) Name sometimes given to luteinizing hormone in males; controls the production of testosterone by interstitial cells. 421

intervertebral disk Layer of cartilage located between adjacent vertebrae. 326, 376

intrauterine device (IUD) Birth control device consisting of a small piece of molded plastic inserted into the uterus; believed to alter the uterine environment so that fertilization does not occur. 429

intron Segment of a gene that does not code for a protein. 495

inversion Change in chromosome structure in which a segment of a chromosome is turned around 180°; this reversed sequence of genes can lead to altered gene activity and abnormalities. 532

invertebrate Referring to an animal without a serial arrangement of vertebrae. 626

ion Particle that carries a negative or positive charge. 24

ionic bond Chemical bond in which ions are attracted to one another by opposite charges. 24

iris Muscular ring that surrounds the pupil and regulates the passage of light through its opening. 353

isotonic solution Solution that is equal in solute concentration to that of the cytoplasm of a cell; causes cell to neither lose nor gain water by osmosis. 72

isotope Atom of the same element having the same atomic number but a different mass number due to the number of neutrons. 22

J

jaundice Yellowish tint to the skin caused by an abnormal amount of bilirubin (bile pigment) in the blood, indicating liver malfunction. 281

jawless fish Type of fish that has no jaws; includes today's hagfishes and lampreys. 656

joint Union of two or more bones; an articulation. 372

juxtaglomerular apparatus Structure located in the walls of arterioles near the glomerulus that regulates renal blood flow. 310

K

karyotype Chromosomes arranged by pairs according to their size, shape, and general appearance in mitotic metaphase. 526

kidney One of a pair of organs in the urinary system that produce and excrete urine. 304

kinetic energy Energy associated with motion. 102

kingdom The taxonomic grouping above phylum. There are six kingdoms: Animalia, Plantae, Fungi, Protista, Archaea, and Bacteria. 6, 569

L

lacteal Lymphatic vessel in an intestinal villus; aids in the absorption of lipids. 264

lactose intolerance Inability to digest lactose because of an enzyme deficiency. 269

lacuna Small pit or hollow cavity, as in bone or cartilage, where a cell or cells are located. 200

lake Body of fresh water, often classified by nutrient status, such as oligotrophic (nutrient-poor) or eutrophic (nutrient-rich). 734

lancelet Invertebrate chordate with a body that resembles a lancet and has the four chordate characteristics as an adult. 654

lanugo Short, fine hair that is present during the later portion of fetal development. 458

large intestine Last major portion of the digestive tract, extending from the small intestine to the anus and consisting of the cecum, the colon, the rectum, and the anus. 266

laryngitis Inflammation of the larynx with accompanying hoarseness. 295

larynx Cartilaginous organ located between the pharynx and the trachea that contains the vocal cords; also called the voice box. 288

law Theory that is generally accepted by an overwhelming number of scientists. 11

law of independent assortment Alleles of unlinked genes assort independently of each other during meiosis so that the gametes contain all possible combinations of alleles. 478

law of segregation Separation of alleles from each other during meiosis so that the gametes contain one from each pair. Each resulting gamete has an equal chance of receiving either allele. 472

leaf Lateral appendage of a stem, highly variable in structure, often containing cells that carry out photosynthesis. 149

leaf vein Vascular tissue within a leaf. 153

learning Relatively permanent change in an animal's behavior that results from practice and experience. 331, 678

leech Blood-sucking annelid, usually found in fresh water, with a sucker at each end of a segmented body. 640

lens Clear, membranelike structure found in the vertebrate eye behind the iris; brings objects into focus. 354

lenticel Frond of usually numerous, lightly raised, somewhat spongy groups of cells in the bark of woody plants. Permits gas exchange between the interior of a plant and the external atmosphere. 151

leptin Hormone produced by adipose tissue that acts on the hypothalamus to signal satiety. 407

less-developed country (LDC) Country in which population growth is expanding rapidly and the majority of people live in poverty. 695

leukemia Cancer of the blood-forming tissues leading to the overproduction of abnormal white blood cells. 518

lichen Fungi and algae coexisting in a symbiotic relationship. 594

ligament Tough connective tissue that usually connects bone to bone. 200, 372

light reactions Portion of photosynthesis that captures solar energy and takes place in thylakoid membranes. 134

lignin Chemical that hardens the cell walls of plants. 152

limbic system Association of various brain centers, including the amygdala and hippocampus; governs learning and memory and various emotions such as pleasure, fear, and happiness. 331

lineage An evolutionary group held together by common ancestry. 664

linkage group Alleles of different genes that are located on the same chromosome and tend to be inherited together. 480

lipase Fat-digesting enzyme secreted by the pancreas. 267, 484

lipid Organic compound that is insoluble in water; notably fats, oils, and steroids. 34

lipopolysaccharide (LPS) Molecule containing both lipid and polysaccharide, which is important in the outer membrane of the gram-negative cell wall. 578

liposome A microscopic artificial sac composed of fatty substances and used in experimental research on the cell. Liposomes may resemble the sac that separated genetic material from the environment in the early evolution of the cell. 552

littoral zone Shore zone between high-tide mark and low-tide mark; also, shallow water of a lake where light penetrates to the bottom. 736

liver Large, dark-red internal organ that produces urea and bile, detoxifies the blood, stores glycogen, and produces the plasma proteins, among other functions. 267

liverwort Type of bryophyte. 610

lobe-finned fish Type of fish with limblike fins. 657

localized Condition that affects only a limited area of the body. 212

locus Particular site where a gene is found on a chromosome. Homologous chromosomes have corresponding gene loci. 472

logistic growth An "S-shaped" population growth curve that goes through an exponential phase; then, when population size approaches carrying capacity, the growth rate starts to flatten, thus flattening the curve. 694

long-day plant Plant that flowers when day length is longer than a critical length; e.g., wheat, barley, clover, and spinach. 191

long-term memory Retention of information that lasts longer than a few minutes. 331

loop of the nephron Portion of a nephron between the proximal and distal convoluted tubules; functions in water reabsorption. 307

loose fibrous connective tissue Tissue composed mainly of fibroblasts widely separated by a matrix containing collagen and elastic fibers. 200

lung cancer Malignant tumors in the lungs that are highly associated with smoking. 299

lungs Paired, cone-shaped organs within the thoracic cavity; function in internal respiration and contain moist surfaces for gas exchange. 288, 657

luteinizing hormone (LH) Hormone produced by the anterior pituitary gland that stimulates the development of the corpus luteum in females and the production of testosterone in males. 421

lymph Fluid, derived from tissue fluid, that is carried in lymphatic vessels. 232, 240

lymphatic capillaries Small vessels that form a blind pouch at one end and collect lymphatic fluid in various organs and tissues. 232

lymphatic system Organ system consisting of lymphatic vessels and lymphoid organs; transports lymph and lipids, and aids the immune system. 206, 240

lymphatic vessel Vessel that carries lymph. 240

lymph node Mass of lymphoid tissue located along the course of a lymphatic vessel. 242

lymphocyte Specialized white blood cell that functions in acquired immunity; occurs in two forms—T lymphocytes and B lymphocytes; also called T cells and B cells. 229

lymphoid (lymphatic) organ Collections of lymphocytes and other immune cells; divided into primary lymphoid organs where lymphocytes develop and secondary lymphoid organs where most lymphocyte responses occur. 241

lymphoma Cancer of the lymphocytes (white blood cells). 518

lysosome Membrane-bounded vesicle that contains hydrolytic enzymes for digesting macromolecules. 56

M

macrophage Large phagocytic cell, derived from a monocyte, that ingests microbes and debris. 229, 243

malaria Serious infectious illness caused by the parasitic protozoan Plasmodium. Malaria is characterized by bouts of high chills and fever that occur at regular intervals. 589

male condom Sheath used to cover the penis during sexual intercourse; used as a contraceptive and, if latex, to minimize the risk of transmitting infection. 429

male gametophyte In seed plants, the gametophyte that produces sperm; a pollen grain. Sometimes called a microgametophyte. 178

malleus The first of the three ossicles of the ear that conduct vibrations from the tympanic membrane to the oval window of the inner ear; the other two ossicles are the incus and the stapes. 358

Malpighian tubule Blind, threadlike excretory tubule attached to the gut of an insect. 643

maltase Enzyme produced in the small intestine that breaks down maltose to two glucose molecules. 269

mammal Homeothermic vertebrate characterized especially by the presence of hair and mammary glands. 662

Marfan syndrome Congenital disorder of connective tissue characterized by abnormal length of the limbs. 537

marsupial Member of a group of mammals bearing immature young nursed in a marsupium, or pouch; e.g., kangaroo and opossum. 662

mast cell Cell to which IgE antibodies, which form in response to allergens, attach, causing the cell to release allergy mediators that cause symptoms. 243

matrix Unstructured semifluid substance that fills the space between cells in connective tissues or inside organelles. 58, 122, 200

matter Anything that takes up space and has mass. 20

mechanical energy A type of kinetic energy, such as walking or running. 102

mechanoreceptor Sensory receptor that is sensitive to mechanical stimulation, such as pressure, sound waves, or gravity. 348

medulla oblongata Part of the brain stem that is continuous with the spinal cord; controls heartbeat, blood pressure, breathing, and other vital functions. 330

megakaryocyte Large bone marrow cell that gives rise to blood platelets. 229

megaspore One of the two types of spores produced by seed plants; develops into a female gametophyte (embryo sac). 176

megasporocyte Megaspore mother cell; produces megaspores by meiosis. Only one megaspore persists. 178

meiosis Type of nuclear division that occurs as part of sexual reproduction in which the daughter cells receive the haploid number of chromosomes in varied combinations. 89

melanocyte Specialized cell in the epidermis that produces melanin, the pigment responsible for skin color. 208

melanocyte-stimulating hormone (MSH) Substance that causes melanocytes to secrete melanin in lower vertebrates. 400

melatonin Hormone, secreted by the pineal gland, that is involved in biorhythms. 406

memory The brain function of recalling something that has been learned. 331

meninges Protective membranous coverings of the central nervous system. 205, 326

meningitis Inflammation of the meninges, the membranes that surround the brain and spinal cord. 341

menisci (sing., meniscus) Cartilaginous wedges that separate the surfaces of bones in synovial joints. 379

menopause Termination of the ovarian and uterine cycles in older women. 427

menstruation Loss of blood and tissue from the uterus at the end of a uterine cycle. 426

meristem Undifferentiated embryonic tissue in the active growth regions of plants. 151

mesoderm Middle germ layer of embryonic gastrula; gives rise to muscles, the connective tissue, and the circulatory system. 448

mesophyll Inner, thickest layer of a leaf consisting of palisade and spongy mesophyll; the site of most photosynthesis. 154

messenger RNA (mRNA) Ribonucleic acid whose sequence of codons specifies the sequence of amino acids during protein synthesis. 494

metabolic pathway Series of linked reactions, beginning with a particular reactant and terminating with an end product. 106

metabolism All of the chemical changes that occur within a cell. 4, 104

metamorphosis Change in shape and form that some animals, such as amphibians and insects, undergo during development. 644

metaphase Mitotic phase during which chromosomes are aligned at the metaphase plate. 86

metaphase plate A disk formed during metaphase in which all of a cell's chromosomes lie in a single plane at right angles to the spindle fibers. 86

methanogen Any of the various species of archaea that produce methane gas as a metabolic by-product. 584

microevolution Change in gene frequencies between populations of a species over time. 559

microsatellite Tandem repeat of DNA nucleotides (e.g., CACACA) that varies in length among individuals; used in forensic investigations. 504

microspore One of two types of spores produced by seed plants; develops into a male gametophyte (pollen grain). 176

microsporocyte Microspore mother cell; produces microspores by meiosis. 178

microtubule Small, cylindrical structure that contains 13 rows of the protein tubulin around an empty central core; present in the cytoplasm, centrioles, cilia, and flagella. 59

microvillus Cylindrical process that extends from an epithelial cell of a villus of the intestinal wall and serves to increase the surface area of the cell. 198

micturition Emptying of the bladder; urination. 306

midbrain Part of the brain located below the thalamus and above the pons; contains reflex centers and tracts. 330

middle ear Portion of the ear consisting of the tympanic membrane, the oval and round windows, and the ossicles, where sound is amplified. 357

mimicry Superficial resemblance of two or more species; mechanism by which an organism avoids predation by appearing to be noxious. 702

mineral Naturally occurring inorganic substance containing two or more elements; certain minerals are needed in the diet. 276, 753

mineralocorticoids Hormones secreted by the adrenal cortex that regulate salt and water balance, leading to increases in blood volume and blood pressure. 403

mitochondrion Membrane-bounded organelle in which ATP molecules are produced during the process of cellular respiration. 57

mitosis Type of cell division in which daughter cells receive the exact chromosomal and genetic makeup of the parental cell; occurs during growth and repair. 82

model In an experiment, an organism that serves as a representative for the organism of interest. For example, mice are a model for much human disease or pharmaceutical research. 10

molecular clock Idea that the rate at which mutational changes accumulate in certain genes is constant over time and is not involved in adaptation to the environment. 664

molecule Union of two or more atoms of the same element; also, the smallest part of a compound that retains the properties of the compound. 24

mollusc Member of the phylum Mollusca, which includes squid, clams, snails, and chitons; characterized by a visceral mass, a mantle, and a foot. 635

molt To periodically shed the exoskeleton (in arthropods). 641

monoclonal antibody One of many antibodies produced by a clone of hybridoma cells that all bind to the same antigen. 251

monocot Abbreviation of monocotyledon, a member of the class of flowering plants called Monocotyledones. Monocots have several common characteristics, including one cotyledon in the seed, scattered vascular bundles in the stem, and flower parts in threes or multiples of three. 150, 618

monocyte Type of agranular leukocyte that functions as a phagocyte, particularly after it becomes a macrophage, which is also an antigen-presenting cell. 229, 243

monohybrid Individual that is heterozygous for one trait; shows the phenotype of the dominant allele but carries the recessive allele. 475

monomer Small molecule that is a subunit of a polymer; e.g., glucose is a monomer of starch. 31

monosaccharide Simple sugar; a carbohydrate that cannot be decomposed by hydrolysis. 32

monosomy One less chromosome than usual. 528

monotreme Egg-laying mammal; e.g., duckbill platypus and spiny anteater. 662

monsoon Climate in India and southern Asia caused by wet ocean winds that blow onshore for almost half the year. 725

more-developed country (MDC) Country in which population growth is low and the people enjoy a good standard of living. 695

morphogenesis Emergence of shape in tissues, organs, or an entire embryo during development. 450

morphogen gene Unit of inheritance that controls a gradient that influences morphogenesis. 452

morula Spherical mass of cells resulting from cleavage during animal development prior to the blastula stage. 447

mosaic evolution Concept that human characteristics did not evolve at the same rate; e.g., some body parts are more humanlike than others in early hominids. 667

moss Type of bryophyte. 610

motor molecule Proteins that attach to cytoskeletal elements and allow for cell/organelle movement. 60

motor neuron Nerve cell that conducts nerve impulses away from the central nervous system and innervates effectors (muscles and glands). 321

motor unit Motor neuron and all the muscle fibers it innervates. 387

mucous membrane Membrane that lines a cavity or tube that opens to the outside of the body; also called mucosa. 205

multiple allele Inheritance pattern in which there are more than two alleles for a particular trait; each individual has only two of all possible alleles. 480

multiple sclerosis (MS) Disease in which the outer, myelin layer of nerve fiber insulation becomes damaged, interfering with normal conduction of nerve impulses. 255, 341

multiregional continuity hypothesis The idea that modern humans left Africa about 1.8 million years ago and evolved several times through interbreeding between *Homo erectus* and other hominid species. 669

muscle tone Contraction of some fibers in skeletal muscle at any given time. 388

muscle twitch Contraction of a whole muscle in response to a single stimulus. 387

muscular system System of muscles that produces both movement within the body and movement of its limbs; principal components are skeletal muscle, smooth muscle, and cardiac muscle. 207

muscular tissue Type of tissue composed of fibers that can shorten and thicken. 202

mutagen An environmental agent, such as ultraviolet radiation, that causes mutations in DNA. 500

mutation Alteration in chromosome structure or number; also, an alteration in a gene due to a change in DNA composition. 562

mutualism Symbiotic relationship in which both species benefit. 704

myasthenia gravis Chronic disease characterized by muscles that are weak and easily fatigued. It results from the immune system's attack on neuromuscular junctions so that stimuli are not transmitted from motor neurons to muscle fibers. 255

mycelium Mass of hyphae that makes up the body of a fungus. 592

mycorrhiza Mutually beneficial symbiotic relationship between a fungus and the roots of vascular plants. 164

mycorrhizal fungi Fungi that grow on the roots of plants. 595

mycosis Any disease caused by a fungus. 595

myelin sheath White, fatty material—derived from the membrane of Schwann cells—that forms a covering for nerve fibers. 320

myocardium Cardiac muscle in the wall of the heart. 220

myofibril Specific muscle cell organelle containing a linear arrangement of sarcomeres, which shorten to produce muscle contraction. 382

myogram Recording of a muscular contraction. 387

myosin Muscle protein making up the thick filaments in a sarcomere; it pulls actin to shorten the sarcomere, yielding muscle contraction. 384

myxedema Condition resulting from a deficiency of thyroid hormone in an adult. 410

N

NAD+ (nicotinamide adenine dinucleotide) Coenzyme of oxidation-reduction that accepts electrons and hydrogen ions to become NADH + H+ as oxidation of substrates occurs. During cellular respiration, NADH carries electrons to the electron transport system in mitochondria. 118

NADP+ (nicotinamide adenine dinucleotide phosphate) Coenzyme of oxidation-reduction that accepts electrons and hydrogen ions to become NADPH + H+. During photosynthesis, NADPH participates in the reduction of carbon dioxide to glucose. 134

nasal cavity One of two canals in the nose, separated by a septum. 287

nasopharynx Region of the pharynx associated with the nasal cavity. 262

natural killer (NK) cell Lymphocyte-like cell that causes certain virus-infected or cancerous cells to undergo apoptosis. 244

natural selection Mechanism of evolution caused by environmental selection of organisms most fit to reproduce; results in adaptation to the environment. 5, 564

Neandertal Hominid with a sturdy build that lived during the last Ice Age in Europe and the Middle East; hunted large game and left evidence of being culturally advanced. 670

nearsighted Vision abnormality due to an elongated eyeball from front to back; light rays focus in front of the retina when viewing distant objects. 362

negative feedback Mechanism of homeostatic response by which the output of a system suppresses or inhibits the activity of the system. 209, 397

nematocyst In cnidarians, a capsule that contains a threadlike fiber whose release aids in the capture of prey. 631

nephridium Segmentally arranged, paired excretory tubules of many invertebrates, as in the earthworm. 638

nephron Anatomical and functional unit of the kidney; a kidney tubule. 306

nerve Bundle of nerve fibers outside the central nervous system. 204, 334

nerve impulse Action potential (electrochemical change) traveling along a neuron. 321

nervous system Organ system consisting of the brain, spinal cord, and associated nerves that coordinates the other organ systems of the body. 207

nervous tissue Tissue that contains nerve cells (neurons), which conduct impulses, and neuroglial cells, which support, protect, and provide nutrients to neurons. 204

neural plate Region of the dorsal surface of the chordate embryo that marks the future location of the neural tube. 449

neural tube Tube formed by closure of the neural groove during development. In vertebrates, the neural tube develops into the spinal cord and brain. 449

neuroglia Nonconducting nerve cells that are intimately associated with neurons and function in a supportive capacity. 204, 320

neuromuscular junction Region where an axon bulb approaches a muscle fiber; contains a presynaptic membrane, a synaptic cleft, and a postsynaptic membrane. 385

neuron Nerve cell that characteristically has three parts: a cell body, dendrites, and an axon. 204, 320

neurotransmitter Chemical stored at the ends of axons that is responsible for transmission across a synapse. 324

neurula The early embryo during the development of the neural tube from the neural plate, marking the first appearance of the nervous system; the next stage after the gastrula. 449

neutron Subatomic particle that has the weight of one atomic mass unit, carries no charge, and is found in the nucleus of an atom. 20

neutrophil Granular leukocyte that is the most abundant of the white blood cells; first to respond to most bacterial infections. 229, 243

nitrification Process by which nitrogen in ammonia and organic compounds is oxidized to nitrites and nitrates by soil bacteria. 716

nitrogen fixation Process whereby free atmospheric nitrogen is converted into compounds, such as ammonium and nitrates, usually by bacteria. 716

node In plants, the place where one or more leaves attach to a stem. 149

node of Ranvier Gap in the myelin sheath around a nerve fiber; increases the speed of nerve impulse conduction. 320

noncyclic electron pathway Portion of the light-dependent reaction of photosynthesis that involves both photosystem I and photosystem II. It generates both ATP and NADPH. 136

nondisjunction Failure of homologous chromosomes or daughter chromosomes to separate during meiosis I and meiosis II, respectively. 93, 528

nonpolar covalent bond Bond in which the sharing of electrons between atoms is fairly equal. 25

nonrandom mating Mating among individuals on the basis of their phenotypic similarities or differences, rather than mating on a random basis. 563

nonrenewable resource Minerals, fossil fuels, and other materials present in essentially fixed amounts (within the human timescale) in our environment. 746

nonvascular plant Bryophytes such as mosses and liverworts that have no vascular tissue and either occur in moist locations or have special adaptations for living in dry locations. 610

norepinephrine (NE) Neurotransmitter of the postganglionic fibers in the sympathetic division of the autonomic system; also, a hormone produced by the adrenal medulla. 325, 403

normal microflora Microbes, mostly bacteria, that live on and in the body and usually have beneficial effects. 577

nostril One of the two openings in the nose that allow air to enter the nasal cavities. 287

notochord Dorsal supporting rod that exists in all chordates at some time in their life history; replaced by the vertebral column in vertebrates. 449, 652

nuclear envelope Double membrane that surrounds the nucleus and is continuous with the endoplasmic reticulum. 52

nuclear pore Opening in the nuclear envelope that permits the passage of proteins into the nucleus and ribosomal subunits out of the nucleus. 52

nucleoid Region of prokaryotic cells where DNA is located; it is not bounded by a nuclear envelope. 48, 578

nucleolus (pl., nucleoli) Dark-staining, spherical body in the nucleus that produces ribosomal subunits. 52

nucleoplasm Semifluid medium of the nucleus containing chromatin. 52

nucleotide Monomer of DNA and RNA consisting of a 5-carbon sugar bonded to a nitrogenous base and a phosphate group. 39

nucleus Membrane-bounded organelle within a eukaryotic cell that contains chromosomes and controls the structure and function of the cell. 49

nutrients Chemical substances in foods that are essential to the diet and contribute to good health. 270

O

obesity Excess adipose tissue; exceeding ideal weight by more than 20%. 278

observation Step in the scientific method by which data are collected before a conclusion is drawn. 10

oil Triglyceride, usually of plant origin, that is composed of glycerol and three fatty acids and is liquid in consistency due to many unsaturated bonds in the hydrocarbon chains of the fatty acids. 34

olfactory cell Modified neuron that is a sensory receptor for the sense of smell. 352

oligodendroglial cells Cells that form the myelin sheaths surrounding axons in the central nervous system. 321

omnivore Organism in a food chain that feeds on both plants and animals. 710

oncogene Cancer-causing gene. In its normal state, involved in regulation of the growth of cells; when mutated, allows cells to grow out of control. 518

oocyte The female gamete, or egg, prior to fertilization. 422

oogenesis Production of oocytes in females by the process of meiosis and maturation. 95, 422

operant conditioning Learning that results from rewarding or reinforcing a particular behavior. 679

operon In bacteria, a group of structural and regulatory genes that acts as a unit. 512

opportunistic infection Disease that arises in the presence of an impaired immune system. 432

optic chiasma X-shaped structure on the underside of the brain formed by partial crossing-over of optic nerve fibers. 356

optic nerve Cranial nerve that carries nerve impulses from the retina of the eye to the brain, thereby contributing to the sense of sight. 354

order One of the categories, or taxa, used by taxonomists to group species; the taxon above the family level. 6, 569

organ Combination of two or more different tissues performing a common function. 3, 148

organelle Small, often membranous structure in the cytoplasm having a specific structure and function. 46

organic molecule Molecule that always contains carbon and hydrogen, and often contains oxygen as well; organic molecules are associated with living things. 31

organism Individual living thing. 3

organ system Group of related organs working together. 3

origin End of a muscle that is attached to a relatively immovable bone. 380

osmosis Diffusion of water through a differentially permeable membrane. 72

osmotic pressure Measure of the tendency of water to move across a differentially permeable membrane; visible as an increase in liquid on the side of the membrane with higher solute concentration. 72

ossicle One of the small bones of the vertebrate middle ear—malleus, incus, or stapes. 358

osteoarthritis (OA) Disintegration of the cartilage between bones at a synovial joint. 390

osteoblast Bone-forming cell. 372

osteoclast Cell that causes erosion of bone. 372

osteocyte Mature bone cell located within the lacunae of bone. 370

osteon Cylindrical-shaped unit containing bone cells that surround an osteonic canal; also called a Haversian system. 370

osteoporosis Condition in which bones break easily because calcium is removed from them faster than it is replaced. 277, 390

otitis media Inflammation of the middle ear, often caused by a bacterial infection, characterized by pain and possibly by a sense of fullness, hearing loss, vertigo, and fever. 296

otolith Calcium carbonate granule associated with ciliated cells in the utricle and saccule. 360

outer ear Portion of the ear consisting of the pinna and the auditory canal. 357

out-of-Africa hypothesis Proposal that modern humans originated only in Africa; then they migrated and supplanted populations of Homo in Asia and Europe about 100,000 years ago. 669

oval opening In the fetus, a shunt for the flow of blood from the right atrium to the left atrium, thus bypassing the lungs. 458

oval window The oval-shaped opening between the middle ear and the vestibule to which the stapes bone attaches. 357

ovarian cycle Monthly follicle changes occurring in the ovary that control the level of sex hormones in the blood and the uterine cycle. 425

ovary In animals, the female gonad, the organ that produces oocytes, estrogen, and progesterone; in flowering plants, the base of the pistil that protects ovules and, along with associated tissues, becomes a fruit. 176, 405, 422, 619

oviduct Tube that transports oocytes to the uterus; also called a uterine tube. 422

ovulation Release of a secondary oocyte from the ovary. 422

ovule In seed plants, a structure where the female megaspore becomes an egg-producing female gametophyte that develops into a seed following fertilization. 177, 617

oxidation Loss of one or more electrons from an atom or molecule; in biological systems, generally the loss of hydrogen atoms. 110

oxygen debt Amount of oxygen needed to metabolize lactate, a compound that accumulates during vigorous exercise. 126, 386

oxyhemoglobin Compound formed when oxygen combines with hemoglobin. 293

oxytocin Hormone released by the posterior pituitary that causes contraction of the uterus and milk letdown. 400

ozone hole Seasonal thinning of the ozone shield in the lower stratosphere at the North and South Poles. 719

ozone shield Accumulation of O_3, formed from oxygen in the upper atmosphere; a filtering layer that protects Earth from ultraviolet radiation. 719, 754

P

palisade mesophyll In a plant leaf, the layer of mesophyll containing elongated cells with many chloroplasts. 154

pancreas Large gland that lies behind the stomach in humans; has both digestive and endocrine functions. 267, 405

pancreatic amylase Enzyme in the pancreas that digests starch to maltose. 267

pancreatic islet (of Langerhans) Distinctive group of cells within the pancreas that secretes insulin and glucagon. 405

pancreatitis Inflammation of the pancreas. 280

Pap smear Sample of cells removed from the tip of the cervix and then stained and examined microscopically. 434

parasitism Symbiotic relationship in which one species (the parasite) benefits in terms of growth and reproduction to the detriment of the other species (the host). 703

parasympathetic division That part of the autonomic system that is active under normal conditions; uses acetylcholine as a neurotransmitter. 337

parathyroid gland One of four glands embedded in the posterior surface of the thyroid gland; produces parathyroid hormone. 402

parathyroid hormone (PTH) Hormone secreted by the four parathyroid glands that increases the blood calcium level and decreases the phosphate level. 402

parenchyma Thin-walled, minimally differentiated cell that photosynthesizes or stores the products of photosynthesis. 152

Parkinson disease (PD) Progressive deterioration of the central nervous system due to a deficiency in the neurotransmitter dopamine. 341

parturition Processes that lead to and include birth and the expulsion of the afterbirth. 461

passive immunity Protection against infection acquired by transfer of antibodies to a susceptible individual. 248

pathogen Disease-causing agent such as a virus, parasitic bacterium, fungus, or animal. 242

pattern formation During development, how tissues and organs are arranged in the body. 450

pectoral girdle Portion of the skeleton that provides support and attachment for the upper limbs. 377

pedigree Chart showing the relationships of relatives and their particular traits. 479

pelagic division Open portion of the sea. 738

pelvic girdle Portion of the skeleton to which the lower limbs are attached. 378

pelvic inflammatory disease (PID) Latent infection of gonorrhea or chlamydia in the vasa deferentia or uterine tubes. 435

penis External organ in males through which the urethra passes; also serves as the organ of sexual intercourse. 418

pentose Five-carbon sugar. Deoxyribose is the pentose sugar found in DNA; ribose is the pentose sugar found in RNA. 32

pepsin Enzyme secreted by gastric glands that digests proteins to peptides. 264

peptidase Intestinal enzyme that breaks down short chains of amino acids to individual amino acids that are absorbed across the intestinal wall. 269

peptide bond Covalent bond that joins two amino acids. 37

peptide hormone Type of hormone that is a protein, a peptide, or derived from an amino acid. 398

perception Mental awareness of sensory stimulation. 348

perennial Flowering plant that lives more than one growing season because the underground parts regrow each season. 148

pericardium Protective serous membrane that surrounds the heart. 220

pericycle Layer of cells surrounding the vascular tissue of roots; produces branch roots. 163

periderm Protective tissue that replaces the epidermis of a stem and includes the cork and cork cambium. 151

periosteum Fibrous connective tissue covering the surface of bone. 370

peripheral nervous system (PNS) Nerves and ganglia that lie outside the central nervous system. 320

peristalsis Wavelike contractions that propel substances along a tubular structure such as the esophagus. 262

peritonitis Generalized infection of the lining of the abdominal cavity. 266

peritubular capillary network Capillary network that surrounds a nephron and functions in reabsorption during urine formation. 306

permafrost Permanently frozen ground, usually in the tundra, a biome of arctic regions. 728

peroxisome Enzyme-filled vesicle in which fatty acids and amino acids are metabolized to hydrogen peroxide that is then broken down to harmless products. 56

petal A flower part just inside the sepals; often conspicuously colored to attract pollinators. 176, 619

petiole The part of a plant leaf that connects the blade to the stem. 149

phagocytosis Process by which amoeboid cells engulf large substances, forming an intracellular vacuole. 76

pharyngitis Inflammation of the pharynx. 295

pharynx Portion of the digestive tract between the mouth and the esophagus that serves as a passageway for food on its way to the stomach and air on its way to the trachea. 262, 287

phenomenon Something that a scientist observes in nature, such as the law of gravity. If this phenomenon is unexplained, a scientist may wish to test it by formulating a hypothesis to explain it and then conducting an experiment to test that hypothesis. 10

phenotype Outward appearance of an organism caused by the genotype and environmental influences. 473

phenylketonuria (PKU) Condition caused by the accumulation of phenylalanine; characterized by mental retardation, light pigmentation, eczema, and neurologic manifestations unless treated with a diet low in phenylalanine. 536

pheromone Chemical signal that works at a distance and alters the behavior of another member of the same species. 408, 685

phloem Vascular tissue that conducts organic solutes in plants; contains sieve-tube members and companion cells. 149, 612

phospholipid Molecule that forms the bilayer of the cell's membranes; has a polar, hydrophilic head bonded to two nonpolar, hydrophobic tails. 35

photoperiod Relative lengths of daylight and darkness that affect the physiology and behavior of an organism. 191

photoreceptor Sensory receptor in the retina that responds to light stimuli. 348

photosynthesis Process by which plants and algae make their own food using the energy of the sun. 132

photosystem Cluster of light-absorbing pigment molecules within thylakoid membranes. 136

phototropism Growth response of plant stems to light; stems demonstrate positive phototropism. 188

photovoltaic (solar) cell An energy-conversion device that captures solar energy and directly converts it to electrical current. 753

pH scale Measurement scale for hydrogen ion concentration; ranges from 0 (acid) to 14 (basic), with 7 neutral. 29

phyletic gradualism Evolutionary model that proposes that evolutionary change resulting in a new species can occur gradually in an unbranched lineage. 567

phylogenetics The study of how organisms are evolutionarily related; used to build evolutionary trees to describe these relationships. 570

phylogeny Evolutionary history of a group of organisms. 569

phylum One of the categories, or taxa, used by taxonomists to group species; the taxon above the class level. 6, 569

phytochrome Photoreversible plant pigment involved in photoperiodism and other responses of plants such as etiolation. 192

phytoplankton Part of plankton containing organisms that photosynthesize, releasing oxygen into the atmosphere and serving as food producers in aquatic ecosystems. 585

phytoremediation The use of plants to clean up pollution. 165

pineal gland Endocrine gland located in the third ventricle of the brain that produces melatonin. 406

pinna Outer, funnel-like structure of the ear that picks up sound waves. 357

pinocytosis Process by which vesicle formation brings macromolecules into the cell. 76

pit Any depression or opening; usually the small openings in the cell walls of xylem cells that provide a continuum between adjacent xylem cells. 153

pith Parenchyma tissue in the center of some stems and roots. 163

pituitary dwarfism Condition characterized by normal proportions but small stature; caused by inadequate growth hormone. 409

pituitary gland Endocrine gland that lies just inferior to the hypothalamus; consists of the anterior pituitary and the posterior pituitary. 399

pivot joint Type of joint, such as that between the radius and ulna, that only allows rotation. 380

placenta Structure that forms from the chorion and the uterine wall and allows the embryo, and then the fetus, to acquire nutrients and rid itself of wastes. 427, 460

placental mammal A group of species that rely on internal development whereby the fetus exchanges nutrients and wastes with its mother via a placenta. 663

plankton Freshwater and marine organisms suspended on or near the surface of the water; includes phytoplankton and zooplankton. 735

plant Multicellular, usually photosynthetic organism belonging to the plant kingdom. 7

plant hormone Chemical signal that is produced by various plant tissues and coordinates the activities of plant cells. 188

plaque Accumulation of soft masses of fatty material, particularly cholesterol, beneath the inner linings of the arteries. 232, 273

plasma Liquid portion of blood. 202, 226

plasma membrane Membrane surrounding the cytoplasm that consists of a phospholipid bilayer with embedded proteins; regulates the entrance and exit of molecules from the cell. 47

plasmid Self-duplicating ring of accessory DNA in the cytoplasm of bacteria. 48, 502

plasmodial (acellular) slime mold Free-living mass of cytoplasm that moves by pseudopods on a forest floor or in a field, feeding on decaying plant material by phagocytosis; reproduces by spore formation. 591

plasmolysis Contraction of the cell contents due to the loss of water. 73

platelet Cell fragment that is necessary to blood clotting; also called a thrombocyte. 202, 229

pleura Serous membrane that encloses the lungs. 289

plumule In flowering plants, the embryonic plant shoot that bears young leaves. 183

pneumonectomy Surgical removal of all or part of a lung. 299

pneumonia Infection of the lungs that causes alveoli to fill with mucus and pus. 296

polar body Nonfunctioning daughter cell, formed during oogenesis, that has little cytoplasm. 96

polar covalent bond Bond in which the sharing of electrons between atoms is unequal; results in one atom with a partial positive charge and the other with a partial negative charge. 25

pollen cone One of two types of cones produced by gymnosperms; contains windblown pollen (male gametophyte). 617

pollen grain In seed plants, the sperm-producing male gametophyte. 178, 617

pollen tube In seed plants, a tube that forms when a pollen grain lands on the stigma and germinates. The tube grows, passing between the cells of the stigma and the style to reach the egg inside an ovule, where fertilization occurs. 617

pollination In seed plants, the delivery of pollen to the vicinity of the egg-producing female gametophyte. 179, 617

pollution Any environmental change that adversely affects the lives and health of living things. 746

polygenic inheritance Inheritance pattern in which a trait is controlled by several allelic pairs; each dominant allele contributes to the phenotype in an additive and like manner. 482

polymer Macromolecule consisting of covalently bonded monomers; for example, a polypeptide is a polymer of monomers called amino acids. 31

polymerase chain reaction (PCR) Technique that uses the enzyme DNA polymerase to produce copies of a particular piece of DNA within a test tube. 502

polyp Small, abnormal growth that arises from the epithelial lining. 280

polypeptide Polymer of many amino acids linked by peptide bonds. 37

polyribosome String of ribosomes simultaneously translating regions of the same mRNA strand during protein synthesis. 497

polysaccharide Polymer made from sugar monomers; the polysaccharides starch and glycogen are polymers of glucose monomers. 32

pons Portion of the brain stem above the medulla oblongata and below the midbrain; assists the medulla oblongata in regulating the breathing rate. 330

population Group of organisms of the same species occupying a certain area and sharing a common gene pool. 7, 559, 692

positive feedback Mechanism of homeostatic response in which the output intensifies and increases the likelihood of response instead of countering and canceling it. 211, 400

posterior pituitary Portion of the pituitary gland that stores and secretes oxytocin and antidiuretic hormone, which are produced by the hypothalamus. 400

postzygotic isolating mechanism A mechanism that prevents an offspring from developing or becoming sexually mature after fertilization has taken place. 566

potential energy Stored energy as a result of location or spatial arrangement. 102

precipitation The process by which water leaves the atmosphere and falls to the ground as water, ice, or snow. 714

predation Interaction in which one organism uses another, called the prey, as a food source. 700

predator Any nonherbivorous species that is not a producer. 700

prediction A prediction results from a hypothesis and is usually stated in an "if . . . then" way. For example, if the hypothesis is "My car won't start because the battery is dead," then the prediction should be "If I charge the battery, then my car will start." 10

prefrontal area Association area in the frontal lobe that receives information from other association areas and uses it to reason and plan actions. 329

preparatory (prep) reaction Reaction that converts pyruvate from glycolysis into acetyl CoA for entrance into the citric acid cycle. 119

pressure-flow model Explanation for phloem transport; osmotic pressure following active transport of sugar into phloem brings a flow of sap from a source to a sink. 170

prey Organism that provides nourishment for a predator. 700

prezygotic isolating mechanism A mechanism that prevents fertilization. 566

primary motor area Area in the frontal lobe where voluntary commands begin; each section controls a part of the body. 329

primary mRNA After transcription, primary mRNA is formed. This mRNA is edited so that introns can be removed before translation can occur. 494

primary root Original root that grows straight down and remains the dominant root of the plant; contrast with fibrous root system. 164

primate Member of the order Primate; includes prosimians, monkeys, apes, and hominids, all of whom have adaptations for living in trees. 664

principle Theory that is generally accepted by an overwhelming number of scientists; also called a law. 11

prion Infectious particle, consisting of protein only and no nucleic acid, that is believed to be linked to several diseases of the central nervous system; stands for proteinlike infectious agent. 602

producer Photosynthetic organism at the start of a grazing food chain that makes its own food; e.g., green plants on land and algae in water. 710

product Substance that forms as a result of a reaction. 104

progesterone Female hormone that helps maintain sexual organs and secondary sex characteristics. 406, 426

prokaryote Lacking a membrane-bounded nucleus and organelles; a prokaryotic cell is the cell type within the domains Bacteria and Archaea. 7

prolactin (PRL) Hormone secreted by the anterior pituitary that stimulates the production of milk from the mammary glands. 400

promoter In an operon, a sequence of DNA where RNA polymerase binds prior to transcription. 494

prophase Mitotic phase during which chromatin condenses so that chromosomes appear; chromosomes are scattered. 86

proprioceptor Sensory receptor that assists the brain in knowing the position of the limbs. 349

prosimian Group of primates that includes lemurs and tarsiers, and may resemble the first primates to have evolved. 664

prostaglandin (PG) Hormone that has various and powerful local effects. 408

prostate gland Gland located around the male urethra below the urinary bladder; adds secretions to semen. 418

protein Organic macromolecule composed of one or several polypeptides. 37

protein-first hypothesis In chemical evolution, the proposal that protein originated before other macromolecules and made possible the formation of protocells. 551

proteinoid Abiotically polymerized amino acids that, when exposed to water, become microspheres, which have cellular characteristics. 551

proteomics The study of all the proteins in an organism. 543

prothrombin Plasma protein converted to thrombin during the steps of blood clotting. 229

prothrombin activator Enzyme that catalyzes the transformation of the precursor prothrombin to the active enzyme thrombin. 229

protist Member of the kingdom Protista, one of the six kingdoms in the classification of organisms. 7, 584

protocell In biological evolution, a possible cell forerunner that became a cell once it could reproduce. 550

proton Positive subatomic particle, located in the nucleus and having a weight of approximately one atomic mass unit. 20

protoplast Plant cell from which the cell wall has been removed. 185, 505

protostome Group of coelomate animals in which the first embryonic opening (the blastopore) is associated with the mouth. 628

protozoan Heterotrophic unicellular protist that moves by flagella, cilia, or pseudopodia, or is immobile. 587

provirus Latent form of a retrovirus in which the viral DNA is incorporated into the chromosome of the host. 599

proximal convoluted tubule (PCT) Highly coiled region of a nephron near the glomerular capsule, where tubular reabsorption takes place. 307

pseudocoelom Body cavity lying between the digestive tract and the body wall that is incompletely lined by mesoderm. 628

psilotophyte Type of seedless vascular plant represented by the whisk ferns, which have no roots or leaves; their branches carry on photosynthesis. 613

pulmonary artery Blood vessel that takes blood away from the heart to the lungs. 221

pulmonary circuit Circulatory pathway that consists of the pulmonary trunk, the pulmonary arteries, and the pulmonary veins; takes O_2-poor blood from the heart to the lungs and O_2-rich blood from the lungs to the heart. 224

pulmonary fibrosis Accumulation of fibrous connective tissue in the lungs; usually caused by inhaling irritating particles, such as silica, coal dust, or asbestos. 297

pulmonary tuberculosis Tuberculosis of the lungs, caused by the bacterium Mycobacterium tuberculosis. 296

pulmonary vein Blood vessel that takes blood to the heart from the lungs. 221

pulse Rhythmic expansion and recoil of arteries, due to contraction of the heart, that can be felt in certain areas at the body surface. 221

punctuated equilibrium Evolutionary model that proposes that periods of rapid change dependent on speciation are followed by long periods of stasis. 569

Punnett square Grid created to calculate the expected results of simple genetic crosses. 475

pupil The opening in the iris of the eye through which light passes. 353

purine Type of nitrogen-containing base, such as adenine or guanine, having a double-ring structure. 492

Purkinje fibers Specialized muscle fibers that conduct the cardiac impulse from the AV bundle into the ventricles. 222

pyelonephritis Inflammation of the renal pelvis and kidney tissue. 312

pyrenoid Spherical protein body embedded in an algal chloroplast; surrounded by starch granules and containing RuBP carboxylase. 584

pyrimidine Type of nitrogen-containing base, such as cytosine, thymine, or uracil, having a single-ring structure. 492

pyruvate End product of glycolysis; its further fate, involving fermentation or entry into a mitochondrion, depends on oxygen availability. 119

R

radial symmetry Body plan in which similar parts are arranged around a central axis, like the spokes of a wheel. 628

radioactive isotope Unstable form of an atom that spontaneously emits radiation in the form of radioactive particles or radiant energy. 22

rain shadow A drier climate area, usually found on the leeward (nonwindy) side of a mountain range. 725

ray-finned fish Group of bony fishes having fins supported by parallel bony rays connected by webs of thin tissue. 657

reactant Substance that participates in a reaction. 104

receptacle Area where a flower attaches to a floral stalk. 619

receptor-mediated endocytosis Selective uptake of molecules into a cell by vacuole formation after they bind to specific receptor proteins in the plasma membrane. 76

receptor protein Protein in the plasma membrane or within the cell that binds to a substance that alters some metabolic aspect of the cell. 69

recessive allele Hereditary factor that expresses itself in the phenotype only when the genotype is homozygous. 473

recombinant DNA (rDNA) DNA that contains genes from more than one source. 502

rectum Terminal end of the digestive tube between the sigmoid colon and the anus. 266

red algae Marine photosynthetic protist with a notable abundance of phycobilin pigments; includes coralline algae of coral reefs. 586

red blood cell Formed element that contains hemoglobin and carries oxygen from the lungs to the tissues; often called an erythrocyte. 202, 228

red bone marrow Vascularized modified connective tissue sometimes found in the cavities of spongy bone; site of blood cell formation. 242, 370

redox reaction Oxidation-reduction reaction; one molecule loses electrons (oxidation), while another molecule simultaneously gains electrons (reduction). 110

red tide Occurs frequently in coastal areas and is often associated with population blooms of dinoflagellates. Dinoflagellate pigments are responsible for the red color of the water. Under these conditions, the dinoflagellates often produce saxitoxin, which can lead to paralytic shellfish poisoning. 586

reduced hemoglobin Hemoglobin that is carrying hydrogen ions. 295

reduction Chemical reaction that results in the addition of one or more electrons to an atom, ion, or compound. The reduction of one substance occurs simultaneously with the oxidation of another. 110

referred pain Pain perceived as having come from a site other than that of its actual origin. 351

reflex Automatic, involuntary response of an organism to a stimulus. 335

reflex action An action performed automatically, without conscious thought; e.g., swallowing. 262

refractory period Time following an action potential when a neuron is unable to conduct another nerve impulse. 323

regulator gene In an operon, a gene coding for a protein that regulates the expression of other genes. 514

renal artery Vessel that originates from the aorta and delivers blood to the kidney. 304

renal cortex Outer portion of the kidney that appears granular. 306

renal medulla Inner portion of the kidney that consists of renal pyramids. 306

renal pelvis Hollow chamber in the kidney that receives freshly prepared urine from the collecting ducts. 306

renal vein Vessel that takes blood from the kidney to the interior vena cava. 304

renewable resource Resource normally replaced or replenished by natural processes and not depleted by moderate use. Examples include: solar energy; biological resources, such as forests and fisheries; biological organisms; and some biogeochemical cycles. 746

renin Enzyme released by the kidneys that leads to the secretion of aldosterone and a rise in blood pressure. 310, 404

replacement reproduction The number of offspring necessary to keep the population at a constant size through time. 697

repressor In an operon, protein molecule that binds to an operator, preventing transcription of structural genes. 514

reproductive cloning Genetically identical to the original individual. 513

reproductive system Organ system that contains male or female organs and specializes in the production of offspring. 207

reptile Member of a class of terrestrial vertebrates with internal fertilization, scaly skin, and an egg with a leathery shell; includes snakes, lizards, turtles, and crocodiles. 658

residual volume Amount of air remaining in the lungs after a forceful expiration. 291

resource In economic terms, anything with the potential to create wealth or give satisfaction. 746

resource partitioning Mechanism that increases the number of niches by apportioning the supply of a resource, such as food or living space, between species. 699

respiratory center Group of nerve cells in the medulla oblongata that send out rhythmic nerve impulses, resulting in involuntary inspiration on an ongoing basis. 293

respiratory system Organ system consisting of the lungs and tubes that bring oxygen into the lungs and take carbon dioxide out. 206

response variable A variable that is measured in an experiment due to a treatment (such as plant growth under natural versus enhanced nitrogen levels). 11

resting potential Polarity across the plasma membrane of a resting neuron due to an unequal distribution of ions. 322

restriction enzyme Bacterial enzyme that stops viral reproduction by cleaving viral DNA; used to cut DNA at specific points during the production of recombinant DNA. 502

reticular connective tissue Composed of reticular cells and reticular fibers; found in bone marrow and lymphoid organs, and in lesser amounts elsewhere. 200

reticular fiber Very thin collagen fibers in the matrix of connective tissue, highly branched and forming delicate supporting networks. 200

retina Innermost layer of the vertebrate eyeball; contains the rod cells and the cone cells. 354

retinal Light-absorbing molecule that is a derivative of vitamin A and a component of rhodopsin. 355

retrovirus RNA virus containing the enzyme reverse transcriptase that synthesizes DNA from RNA. 599

rheumatoid arthritis (RA) Persistent inflammation of synovial joints due to an autoimmune reaction, often causing cartilage destruction, bone erosion, and joint deformities. 255, 390

rhizome Rootlike underground stem. 160

rhodopsin Visual pigment found in the rods whose activation by light energy leads to vision. 355

ribosomal RNA (rRNA) Type of RNA found in ribosomes where protein synthesis occurs. 494

ribosome RNA and protein in two subunits; site of protein synthesis in the cytoplasm. 48, 497

ribozyme A strand of RNA that attaches to other strands of RNA at specific sites so that it can break the strands apart. 495

RNA (ribonucleic acid) Nucleic acid produced by covalent bonding of nucleotide monomers that contain the sugar ribose; occurs in three forms: messenger RNA, ribosomal RNA, and transfer RNA. 39, 494

RNA-first hypothesis In chemical evolution, the proposal that RNA originated before other macromolecules and allowed the formation of the first cell(s). 551

RNA polymerase During transcription, an enzyme that joins nucleotides complementary to a portion of DNA. 494

rod cell Photoreceptor in the retina of the eye that responds to dim light. 355

root cap Protective covering of the root tip, whose cells are constantly replaced as they are ground off when the root pushes through rough soil particles. 162

root hair Extension of a root epidermal cell that increases the surface area for the absorption of water and minerals. 151

root nodule Structure on plant root that contains nitrogen-fixing bacteria. 164

root system Includes the main root and any and all of its lateral (side) branches. 148

round window Membrane-covered opening between the inner ear and the middle ear. 357

roundworm Member of the phylum Nematoda, having a cylindrical body, a complete digestive tract, and a pseudocoelom. Some forms are free-living in water and soil; many are parasitic. 634

RuBP carboxylase Enzyme that catalyzes the reaction of ribulose bisphosphate and carbon dioxide during the Calvin cycle. 139

rugae Deep folds, as in the wall of the stomach. 264

S

sac fungi Members of the phylum Ascomycota. 592

sac plan Body with a digestive cavity that has only one opening, as in cnidarians and flatworms. 628

saccule Saclike cavity in the vestibule of the inner ear; contains sensory receptors for gravitational equilibrium. 360

salinization Process by which mineral salts accumulate in the soil, killing plants; occurs when soils in dry climates are irrigated profusely. 750

salivary amylase Secreted by the salivary glands; the first enzyme to act on starch. 261

saltwater intrusion Movement of salt water into freshwater aquifers in coastal areas where groundwater is withdrawn faster than it is replenished. 748

SA (sinoatrial) node Small region of neuromuscular tissue that initiates the heartbeat; also called the pacemaker. 222

sarcolemma The plasma membrane enclosing a muscle fiber. 382

sarcomere One of many units, arranged linearly within a myofibril, whose contraction produces muscle contraction. 383

sarcoplasmic reticulum Smooth endoplasmic reticulum of skeletal muscle cells; surrounds the myofibrils and stores calcium ions. 382

saturated fatty acid Molecule that lacks double bonds between the carbons of its hydrocarbon chain. The chain bears the maximum number of hydrogens. 34

Schwann cell Cell that surrounds a peripheral nerve fiber and forms the myelin sheath. 320

scientific theory Concept supported by a broad range of observations, experiments, and conclusions. 11

sclera White, fibrous, outer layer of the eyeball. 353

sclerenchyma Plant tissue composed of cells with heavily lignified cell walls; functions in support. 152

scoliosis Abnormal lateral (side-to-side) curvature of the vertebral column. 375

scrotum Pouch of skin that encloses the testes. 418

secondary oocyte In oogenesis, the functional product of meiosis I. 96

secondary spermatocyte In spermatogenesis, the functional product of meiosis I; becomes sperm. 95

second messenger Chemical produced within a cell in response to the binding of a messenger to a membrane receptor; triggers a metabolic reaction in the cell. 398

secretion Release of a substance by exocytosis from a cell that may be a gland or part of a gland. 55

seed Mature ovule that contains a sporophyte embryo with stored food enclosed by a protective coat. 176, 608

seed cone One of two types of cones produced by gymnosperms; contains windblown seeds (female gametophyte). 617

segmentation Repetition of body parts as segments along the length of the body; seen in annelids, arthropods, and chordates. 629

semantic memory Capacity of the brain to store and retrieve information with regard to words or numbers. 331

semen Thick, whitish fluid consisting of sperm and secretions from several glands of the male reproductive tract. 418

semicircular canal One of three tubular structures within the inner ear that contain the sensory receptors responsible for the sense of rotational equilibrium. 358

semilunar valve Valve resembling a half moon located between the ventricles and their attached vessels. 220

seminal vesicle Convoluted, saclike structure attached to the vas deferens near the base of the urinary bladder in males; adds secretions to semen. 418

seminiferous tubule Highly coiled duct within the male testes that produces and transports sperm. 421

senescence Sum of processes involving the aging, decline, and eventual death of a plant or plant part. 189

sensation Conscious awareness of a stimulus due to a nerve impulse sent to the brain from a sensory receptor by way of sensory neurons. 348

sensory adaptation Phenomenon in which a sensation becomes less noticeable once it has been recognized by constant repeated stimulation. 349

sensory neuron Nerve cell that transmits nerve impulses to the central nervous system after a sensory receptor has been stimulated. 320

sensory receptor Structure that receives either external or internal environmental stimuli and is a part of a sensory neuron or transmits signals to a sensory neuron. 348

sepal Outermost, sterile, leaflike covering of the flower; usually green in color. 176, 619

serous membrane Membrane that covers internal organs and lines cavities without an opening to the outside of the body; also called the serosa. 205

serum Light yellow liquid left after clotting of blood. 230

sessile Animal that tends to stay in one place. 628

severe combined immunodeficiency disease (SCID) Congenital illness in which both antibody- and cell-mediated immunity are lacking. 255

sex chromosome Chromosome that determines the sex of an individual; in humans, females have two X chromosomes, and males have an X and a Y chromosome. 481

sex-linked Allele that occurs on the sex chromosomes but may control a trait having nothing to do with the sexual characteristics of an individual. 481

sexual selection Changes in males and females, often due to male competition and female selectivity, leading to increased fitness. 680

shoot apical meristem Group of actively dividing embryonic cells at the tips of plant shoots. 156

shoot system Aboveground portion of a plant consisting of the stem, leaves, and flowers. 148

short-day plant Plant that flowers when day length is shorter than a critical length; e.g., cocklebur, poinsettia, and chrysanthemum. 191

short-term memory Retention of information for only a few minutes, such as remembering a telephone number long enough to dial it. 331

shrubland Arid terrestrial biome characterized by shrubs and tending to occur along coasts that have dry summers and receive most of their rainfall in the winter. 731

sickle cell disease Hereditary disease due to a mutation in the hemoglobin gene, in which red blood cells are narrow and curved so that they are unable to pass through capillaries and are destroyed, leading to chronic anemia. 536

sieve-tube member Member that joins with others in the phloem tissue of plants as a means of transport for nutrient sap. 153

signal transduction pathway Activation and inhibition of intracellular targets after binding of chemical messengers to a cell surface receptor. 516

simple goiter Condition in which an enlarged thyroid produces low levels of thyroxine. 410

sink In the pressure-flow model of phloem transport, the location (roots) from which sugar is constantly being removed. Sugar will flow to the roots from the source. 171

sinkhole Large surface crater caused by the collapse of an underground channel or cavern; often triggered by groundwater withdrawal. 748

sinus Cavity or hollow space in an organ such as the skull. 287, 374

sinusitis Inflammation of the sinuses, usually caused by an infection, which leads to blockage of the openings to the sinuses, and is characterized by postnasal discharge and facial pain. 295

sister chromatid One of two genetically identical chromosomal units that are the result of DNA replication and are attached to each other at the centromere. 85

skeletal muscle Striated, voluntary muscle tissue that comprises skeletal muscles; also called striated muscle. 203

skeletal system System of bones, cartilage, and ligaments that works with the muscular system to protect the body and provide support for locomotion and movement. 207

skill memory Capacity of the brain to store and retrieve information necessary to perform motor activities, such as riding a bike. 331

skin Outer covering of the body; part of the integumentary system, which also includes organs such as the sense organs. 208

skull Bony framework of the head, composed of cranial bones and the bones of the face. 374

sliding filament theory An explanation for muscle contraction based on the movement of actin filaments in relation to myosin filaments. 384

slime layer Gelatinous sheath surrounding the cell walls of certain bacteria. 48

small intestine Long, tubelike chamber of the digestive tract between the stomach and the large intestine. 264

smooth muscle Nonstriated, involuntary muscle tissue found in the walls of internal organs; also called visceral muscle. 203

society Organization of the members of species in a cooperative manner, extending beyond sexual and parental behavior. 685

sociobiology Application of evolutionary principles to the study of the social behavior of animals, including humans. 683

sodium-potassium pump Carrier protein in the plasma membrane that moves sodium ions out of and potassium ions into cells; important in nerve and muscle cells. 75, 322

soft palate Entirely muscular posterior portion of the roof of the mouth. 260

solute Substance that is dissolved in a solvent, forming a solution. 28

solvent Fluid, such as water, that dissolves solutes. 71

somatic cell A body cell; excludes cells that undergo meiosis and become a sperm or an oocyte. 82

somatic embryo Plant cell embryo that is asexually produced through tissue culture techniques. 185

somatic system The portion of the peripheral nervous system containing motor neurons that control skeletal muscles. 335

source In the pressure-flow model of phloem transport, the location (leaves) of sugar production. Sugar will flow from the leaves to the sink. 171

species Group of similarly constructed organisms capable of interbreeding and producing fertile offspring; organisms that share a common gene pool. 5, 6, 566, 569

sperm Male sex cell with three distinct parts at maturity; head, middle piece, and tail. 95, 421

spermatid In spermatogenesis, the functional product of meiosis II; becomes sperm. 95

spermatogenesis Production of sperm in males by the process of meiosis and maturation. 95, 421

sphincter Muscle that surrounds a tube and closes or opens the tube by contracting and relaxing. 263

spinal cord Part of the central nervous system; the nerve cord that is continuous with the base of the brain and housed within the vertebral column. 326

spinal nerve Nerve that arises from the spinal cord. 335

spindle Microtubule structure that brings about chromosomal movement during nuclear division. 86

spiral organ Organ in the cochlear duct of the inner ear that is responsible for hearing; also called the organ of Corti. 358

spleen Large organ located in the upper left region of the abdomen; stores and purifies blood. 242

sponge Invertebrate animal of the phylum Porifera; pore-bearing filter feeder whose inner body wall is lined by collar cells. 629

spongy bone Type of bone that has an irregular, meshlike arrangement of thin plates of bone; often contains red bone marrow. 201, 370

spongy mesophyll Layer of tissue in a plant leaf containing loosely packed cells, increasing the amount of surface area for gas exchange. 154

sporangium Structure that produces spores. 592

spore Haploid reproductive cell, sometimes resistant to unfavorable environmental conditions, which is capable of producing a new individual that is also haploid. 176, 592, 609

sporophyte Diploid generation of the alternation of generations life cycle of a plant; produces haploid spores that develop into the haploid generation. 176, 609

sporozoan Spore-forming protist that has no means of locomotion and is typically a parasite with a complex life cycle having both sexual and asexual phases. 590

spring overturn Mixing process that occurs in spring in stratified lakes whereby the oxygen-rich top waters mix with the nutrient-rich bottom waters. 735

squamous epithelium Type of epithelial tissue that contains flat cells. 198

stabilizing selection Outcome of natural selection in which extreme phenotypes are eliminated and the average phenotype is conserved. 564

stamen In flowering plants, the portion of the flower that consists of a filament and an anther containing pollen sacs where pollen is produced. 176, 619

stapes The last of the three ossicles of the ear that conduct vibrations from the tympanic membrane to the oval window membrane of the inner ear; the other two ossicles are the malleus and the incus. 358

starch Storage polysaccharide found in plants; composed of glucose molecules joined in a linear fashion with few side chains. 32

stem Usually the upright, vertical portion of a plant that transports substances to and from the leaves. 148

stem cell Any cell that can divide and differentiate into more functionally specific cell types; embryonic cells or adult cells, such as those in red bone marrow that give rise to blood cells. 230

stent Metal or plastic tube inserted into a blood vessel to help keep it open; often used in the treatment of vascular disease due to atherosclerosis. 233

steroid Type of lipid molecule having a complex of four carbon rings; examples are cholesterol, progesterone, and testosterone. 35

steroid hormone Type of hormone that has a complex of four carbon rings but different side chains from those of other steroid hormones. 398

stigma In flowering plants, portion of the pistil where pollen grains adhere and germinate before fertilization can occur. 176, 619

stimulus Change in the internal or external environment that a sensory receptor can detect, leading to nerve impulses in sensory neurons. 348

stolon Stem that grows horizontally along the ground and may give rise to new plants where it contacts the soil; e.g., the runners of a strawberry plant. 160

stomach Muscular sac that mixes food with gastric juices to form chyme, which enters the small intestine. 264

stoma (pl., stomata) Microscopic opening bordered by guard cells in the leaves of plants through which gas exchange takes place. 133

striated Having bands; in cardiac and skeletal muscle, alternating light and dark crossbands produced by the distribution of contractile proteins. 203

stroke Condition resulting when an arteriole in the brain bursts or becomes blocked by an embolism; also called a cerebrovascular accident. 232, 341

stroma Fluid within a chloroplast that contains enzymes involved in the synthesis of carbohydrates during photosynthesis. 58, 133

style Elongated, central portion of the pistil between the ovary and the stigma. 176, 619

subcutaneous layer A sheet that lies just beneath the skin and consists of loose connective and adipose tissue. 208

subsidence Occurs when a portion of Earth's surface gradually settles downward. 748

substrate Reactant in a reaction controlled by an enzyme. 106

substrate-level ATP synthesis Process in which ATP is formed by transferring a phosphate from a metabolic substrate to ADP. 120

superior vena cava Large vein that enters the right atrium from above and carries blood from the head, thorax, and upper limbs to the heart. 224

surfactant Agent that reduces the surface tension of water; in the lungs, a surfactant prevents the alveoli from collapsing. 289

sustainable Ability of a society or ecosystem to maintain itself while also providing services to human beings. 762

suture Type of immovable joint articulation found between the bones of the skull. 379

sweat gland Skin gland that secretes a fluid substance for evaporative cooling; also called a sudoriferous gland. 209

symbiosis Relationship that occurs when two different species live together in a unique way; it may be beneficial, neutral, or detrimental to one and/or the other species. 703

sympathetic division The part of the autonomic system that usually promotes activities associated with emergency (fight-or-flight) situations; uses norepinephrine as a neurotransmitter. 337

sympatric speciation Origin of new species in populations that overlap geographically. 567

synapsis Pairing of homologous chromosomes during prophase I of meiosis. 89

synaptic cleft Small gap between the presynaptic and postsynaptic membranes of a synapse. 324

synaptic integration Summing up of excitatory and inhibitory signals by a neuron or by some part of the brain. 325

syndrome Group of symptoms that appear together and tend to indicate the presence of a particular disorder. 526

synovial joint Freely moving joint in which two bones are separated by a cavity. 379

synovial membrane Membrane that forms the inner lining of the capsule of a freely movable joint. 205, 379

syphilis Sexually transmitted disease caused by the bacterium *Treponema pallidum*; characterized by a painless chancre on the penis or cervix; if untreated, can lead to cardiac and central nervous system disorders. 436

systematics The discipline of identifying and classifying organisms according to their evolutionary relationships. 6

systemic Occurring throughout the body; invading many compartments and organs via circulation. 212

systemic circuit That part of the circulatory system that serves body parts other than the gas-exchanging surfaces in the lungs. 224

systemic lupus erythematosis (SLE) Autoimmune disease that causes a wide variety of symptoms and usually damages the kidneys. 255

systole Contraction period of the heart during the cardiac cycle. 222

systolic pressure Arterial blood pressure during the systolic phase of the cardiac cycle. 225

T

taiga A coniferous forest extending in a broad belt across northern Eurasia and North America. 729

taproot Main axis of a root that penetrates deeply and is used by certain plants, such as carrots, for food storage. 164

taste bud Sense organ containing the receptors associated with the sense of taste. 351

Tay-Sachs disease Lethal genetic disease in which the newborn has a faulty lysosomal digestive enzyme. 536

T cell Lymphocyte that matures in the thymus; cytotoxic T cells kill antigen-bearing cells outright; helper T cells release cytokines that stimulate other immune system cells. 246

T-cell receptor (TCR) A molecule on the surface of a T cell that can bind to a specific antigen fragment in combination with an MHC molecule. 246

technology The science or study of the practical or industrial arts. 14

tectorial membrane Membrane that lies above and makes contact with the hair cells in the spiral organ. 358

telomere The end of a chromosome; shortens with every round of replication and therefore helps determine the number of times a cell can divide. 518

telophase Mitotic phase during which daughter cells are located at each pole. 86

temperate deciduous forest Forest found south of the taiga, characterized by the following: deciduous trees such as oak, beech, and maple; moderate climate; relatively high rainfall; stratified plant growth; and plentiful ground life. 729

temperate rain forest Coniferous forest—e.g., that running along the west coast of Canada and the United States; characterized by plentiful rainfall and rich soil. 729

tendon Strap of fibrous connective tissue that connects skeletal muscle to bone. 200, 372

terminal bud Bud that develops at the apex of a shoot. 156

territoriality Behavior related to the act of marking or defending a particular area against invasion by another species member; area often used for feeding, mating, and caring for young. 680

territory Area that an animal or group of animals occupies and defends. 680

testcross Cross between an individual with the dominant phenotype and an individual with the recessive phenotype. The resulting phenotypic ratio indicates whether the dominant phenotype is homozygous or heterozygous. 476

testes Male gonads that produce sperm and the male sex hormones. 405, 418

testosterone Male sex hormone that helps maintain sexual organs and secondary sex characteristics. 405, 421

tetanus Sustained muscle contraction without relaxation. 387

tetany Severe twitching caused by involuntary contraction of the skeletal muscles due to a calcium imbalance. 410

thalamus Part of the brain located in the lateral walls of the third ventricle that serves as the integrating center for sensory input; plays a role in arousing the cerebral cortex. 330

therapeutic cloning Used to create mature cells of various cell types. Also used to learn about specialization of cells and provide cells and tissue to treat human illnesses. 513

thermoreceptor Sensory receptor that is sensitive to changes in temperature. 348

threshold Electrical potential level (voltage) at which an action potential or nerve impulse is produced. 323

thrombin Enzyme that converts fibrinogen to fibrin threads during blood clotting. 229

thromboembolism Obstruction of a blood vessel by a thrombus that has dislodged from the site of its formation. 232

thrombus Blood clot that remains in the blood vessel where it formed. 232

thylakoid Flattened sac within a granum whose membrane contains chlorophyll and where the light-dependent reactions of photosynthesis occur. 48, 133

thymine (T) One of four nitrogen-containing bases in nucleotides composing the structure of DNA; pairs with adenine. 39

thymus gland Lymphoid organ, located along the trachea behind the sternum, involved in the maturation of T cells in the thymus gland. 242, 406

thyroid gland Endocrine gland in the neck that produces several important hormones, including thyroxine, triiodothyronine, and calcitonin. 400

thyroid-stimulating hormone (TSH) Substance produced by the anterior pituitary that causes the thyroid to secrete thyroxine and triiodothyronine. 400

thyroxine (T₄) Hormone secreted by the thyroid gland that promotes growth and development; in general, it increases the metabolic rate of cells. 402

tidal volume Amount of air normally moved in the human body during an inspiration or an expiration. 291

tight junction Region between cells where adjacent plasma membrane proteins join to form an impermeable barrier. 198

tinea Name for many different kinds of superficial fungal infections of the skin, nails, and hair. 596

tissue Group of similar cells that perform a common function. 3, 198

tissue culture Process of growing tissue artificially, usually in a liquid medium in laboratory glassware. 185

tissue fluid Fluid that surrounds the body's cells; consists of dissolved substances that leave the blood capillaries by filtration and diffusion. 232

tonsillectomy Surgical removal of the tonsils. 295

tonsillitis Infection of the tonsils that causes inflammation and can spread to the middle ears. 295

tonsils Partially encapsulated lymph nodules located in the pharynx. 295

total artificial heart (TAH) A mechanical replacement for the heart as opposed to replacement with a natural human heart. 235

totipotent Cell that has the full genetic potential of the organism and the potential to develop into a complete organism. 185

trachea Windpipe from the larynx to the bronchi; in insects, tracheae are tubes that transport air to the tissues. 288, 644

tracheid In flowering plants, type of cell in xylem that has tapered ends and pits through which water and minerals flow. 153

tracheostomy Creation of an artificial airway by incision of the trachea and insertion of a tube. 296

tract Bundle of myelinated axons in the central nervous system. 326

transcription Process resulting in the production of a strand of mRNA that is complementary to a segment of DNA. 494

transcription activator Protein that speeds transcription. 516

transcription factor Protein that initiates transcription by RNA polymerase and thereby starts the process that results in gene expression. 515

transduction The transfer of genetic material between bacteria by viruses. 578

transfer rate Amount of a substance that moves from one component of the environment to another within a specified period of time. 714

transfer RNA (tRNA) Type of RNA that transfers a particular amino acid to a ribosome during protein synthesis; at one end, it binds to the amino acid, and at the other end, it has an anticodon that binds to an mRNA codon. 494

transformation The taking up of extraneous genetic material from the environment by bacteria. 578

transgenic organism Free-living organism in the environment that has had a foreign gene in its cells. 504

transgenic plant A plant that carries the genes of another organism as a result of DNA technology; also, a genetically modified plant. 186

transitional link A form, usually extinct, that bears a resemblance to two groups that in the present day are classified separately. 553

translation Process by which the sequence of codons in mRNA dictates the sequence of amino acids in a polypeptide. 494

translocation Movement of a chromosomal segment from one chromosome to another, nonhomologous chromosome, leading to abnormalities; e.g., Down syndrome. 533

transpiration Plant's loss of water to the atmosphere, mainly through evaporation at leaf stomata. 167

transposon DNA sequence capable of randomly moving from one site to another in the genome. 500

trichome In plants, a specialized outgrowth of the epidermis, such as root hairs. 151

trichomoniasis Sexually transmitted disease caused by the parasitic protozoan *Trichomonas vaginalis*. 436

triglyceride Neutral fat composed of glycerol and three fatty acids. 34

triplet code A three-base-pair segment of DNA that, after transcription, results in the formation of an mRNA codon. Each triplet code of DNA then eventually codes for a particular amino acid. 495

trisomy One more chromosome than usual. 528

trophic level Feeding level of one or more populations in a food web. 713

trophoblast Outer membrane surrounding the embryo in mammals; when thickened by a layer of mesoderm, it becomes the chorion, an extraembryonic membrane. 454

tropical forest Can be wet or dry. Tropical rain forests and tropical deciduous forests are characterized by high annual precipitation (125–660 cm/year), lush vegetation, an annual mean temperature of 20–25°C, high humidity (>80%) and characterized by both a wet and dry season. The difference between the two forest types is in the length of the dry season. Tropical deciduous forests have a longer dry season and consequently, trees lose their leaves. Tropical rain forests are noted for some of the highest biodiversity in the world. 730

tropism In plants, a growth response toward or away from a directional stimulus. 191

trypsin Protein-digesting enzyme secreted by the pancreas. 267

T (transverse) tubules Extensions of the sarcolemma that protrude into the muscle cell. 382

tubercle Small, rounded nodular lesion produced by *Mycobacterium tuberculosis*. 581

tube-within-a-tube plan Body having a digestive tract that has both a mouth and an anus. 628

tubular reabsorption Movement of primarily nutrient molecules and water from the contents of the nephron into the blood at the proximal convoluted tubule. 309

tubular secretion Movement of certain molecules from the blood into the distal convoluted tubule of a nephron so that they are added to urine. 309

tumor marker test Blood test for tumor antigens/antibodies. 519

tumor suppressor gene Gene coding for a protein that ordinarily suppresses cell division; inactivity can lead to a tumor. 518

tundra Biome characterized by permanently frozen subsoil found between the ice cap and the tree line of regions of the Arctic, just south of the ice-covered polar seas in the Northern Hemisphere; also known as the arctic tundra. The area is treeless and usually characterized by vegetation such as lichens, mosses, and small shrubs. Alpine tundra, having similar characteristics, is found near the peak of a mountain. 728

tunicate Type of primitive invertebrate chordate. 654

turgor pressure In plant cells, pressure of the cell contents against the cell wall when the central vacuole is full. 73

tympanic membrane Membranous region that receives air vibrations in an auditory organ; in humans, the eardrum. 357

U

ulcer Open sore in the lining of the stomach; frequently caused by bacterial infection. 279

umbilical artery Fetal blood vessel that travels to and from the placenta. 458

umbilical cord Cord connecting the fetus to the placenta through which blood vessels pass. 457

umbilical vein Fetal blood vessel that travels to and from the placenta. 458

unsaturated fatty acid Fatty acid molecule that has one or more double bonds between the atoms of its carbon chain. 35

upwelling Upward movement of deep, nutrient-rich water along coasts; it replaces surface waters that move away from shore when the direction of prevailing wind shifts. 737

uracil (U) One of the four nitrogen-containing bases in nucleotides composing the structure of RNA; pairs with adenine. 39

urea Primary nitrogenous waste of humans derived from amino acid breakdown. 304

ureter One of two tubes that take urine from the kidneys to the urinary bladder. 304

urethra Tubular structure that receives urine from the bladder and carries it to the outside of the body. 305, 418

urethritis Inflammation of the urethra. 314

uric acid Waste product of nucleotide metabolism. 304

urinary bladder Organ where urine is stored before being discharged by way of the urethra. 305

urinary system Organ system consisting of the kidneys and urinary bladder; rids the body of nitrogenous wastes and helps regulate the water-salt balance of the blood. 207

uterine cycle Monthly-occurring changes in the characteristics of the uterine lining (endometrium). 426

uterus Organ located in the female pelvis where the fetus develops; also called the womb. 422

utricle Saclike cavity in the vestibule of the inner ear that contains sensory receptors for gravitational equilibrium. 360

V

vaccine Antigens prepared in such a way that they can promote active immunity without causing disease. 248

vacuole Membrane-bounded sac that holds fluid and a variety of other substances. 56

vagina Organ that leads from the uterus to the vestibule and serves as the birth canal and organ of sexual intercourse in females. 423

valve Membranous extension of a vessel of the heart wall that opens and closes, ensuring one-way flow. 220

varicose veins Irregular dilation of superficial veins, seen particularly in the lower legs, due to weakened valves within the veins. 226

vascular bundle In plants, primary phloem and primary xylem enclosed by a bundle sheath. 153

vascular cambium In plants, lateral meristem that produces secondary phloem and secondary xylem. 156

vascular cylinder In dicot roots, a core of tissues bounded by the endodermis, consisting of vascular tissues and pericycle. 153

vascular plant Plant that has vascular tissue (xylem and phloem); includes seedless vascular plants (e.g., ferns) and seed plants (gymnosperms and angiosperms). 612

vascular tissue Transport tissue in plants, consisting of xylem and phloem. 151, 163, 608

vas deferens (pl., vasa deferentia) Tube that leads from the epididymis to the urethra in males. 418

vector In genetic engineering, a means of transferring foreign genetic material into a cell; e.g., a plasmid. 502

vein Vessel usually having nonelastic walls that takes blood to the heart from venules; in plants, veins consist of vascular bundles that branch into the leaves. 218

vena cava One of two large veins that convey oxygen-poor blood to the right atrium of the heart. 221

venous duct Fetal connection between the umbilical vein and the inferior vena cava; also called the ductus venosus. 458

ventilation Process of moving air into and out of the lungs; breathing. 286

ventricle Cavity in an organ, such as a lower chamber of the heart or the ventricles of the brain. 220, 326

venule Vessel that takes blood from capillaries to a vein. 219

vermiform appendix Small, tubular appendage that extends outward from the cecum of the large intestine. 266

vernix caseosa Cheeselike substance covering the skin of the fetus. 458

vertebral column Portion of the vertebrate endoskeleton that houses the spinal cord; consists of many vertebrae separated by intervertebral disks. 375

vertebrate Chordate in which the notochord is replaced by a vertebral column. 626, 655

vertigo Symptom associated with disorders of the inner ear in which the patient experiences a sense of movement or dizziness. 364

vesicle Small, membrane-bounded sac that stores substances within a cell. 54

vessel element Cell that joins with others to form a major conducting tube found in xylem. 153

vestibule Space or cavity at the entrance to a canal, such as the cavity that lies between the semicircular canals and the cochlea. 358

vestigial structure Underdeveloped structure that was functional in some ancestor but is no longer functional in a particular organism. 557

villus Fingerlike projection from the wall of the small intestine that functions in absorption. 264

viroid Infectious strand of RNA devoid of a capsid and much smaller than a virus. 601

virus Noncellular obligate parasite of living cells consisting of an inner core of nucleic acid surrounded by a protein capsid. Some viruses have an additional lipid layer called an envelope. 597

visual accommodation Ability of the eye to focus at different distances by changing the curvature of the lens. 354

vital capacity Maximum amount of air moved in or out of the human body with each breathing cycle. 291

vitamin Essential organic requirement in the diet, needed in small amounts. Vitamins are often part of coenzymes. 109, 273

vitreous humor Clear, gelatinous material between the lens of the eye and the retina. 354

vocal cord Fold of tissue within the larynx; creates vocal sounds when it vibrates. 288

vulva External genitals of the female that surround the opening of the vagina. 423

W

water (hydrologic) cycle Interdependent and continuous circulation of water from the ocean to the atmosphere, to the land, and back to the ocean. 714

water mold Filamentous organism having cell walls made of cellulose; typically a decomposer of dead freshwater organisms, but some are parasites of aquatic or terrestrial organisms. 590

water vascular system Series of canals that takes water to the tube feet of an echinoderm, allowing them to expand. 651

Wernicke's area Brain area involved in language comprehension. 329

white blood cell Leukocyte, of which there are several types, each having a specific function in protecting the body from invasion by foreign substances and organisms. 202, 229

white matter Myelinated axons in the central nervous system. 321

wood Secondary xylem that builds up year after year in woody plants and becomes the annual rings. 158

X

X-linked Allele that is located on an X chromosome but may control a trait that has nothing to do with the sex characteristics of an individual. 481

xylem Vascular tissue that transports water and mineral solutes upward through the plant body; contains vessel elements and tracheids. 149, 612

Y

yeast Unicellular fungus that has a single nucleus and reproduces asexually by budding or fission, or sexually through spore formation. 594

yolk sac Extraembryonic membrane that encloses the yolk of birds; in humans, it is the first site of blood cell formation. 454

Z

zero population growth No growth in population size. 695

zooflagellate Nonphotosynthetic protist that moves by flagella; zooflagellates typically enter into symbiotic relationships, and some are parasitic. 588

zooplankton Part of plankton containing protozoans and other types of microscopic animals. 587

zygospore fungi Members of the phylum Zygomycota. 592

zygote Diploid cell formed by the union of sperm and oocyte; the product of fertilization. 95, 422

Credits

Inside Front Cover

Screen capture of "Vanishing White Matter" video courtesy of ScienCentral, Inc. ScienCentral news stories are supported in part by the National Science Foundation (http://www.nsf.gov), Division of Informal Science Education under Grants No. ESI0206184, ESI021155, and ESI9904457. Any opinions, findings, and conclusions or recommendations expressed in this material are those of the author(s) and do not necessarily reflect the views of the National Science Foundation.

History of Biology

Leeuwenhoek, Darwin, Pasteur, Koch, Pavlov, Lorenz, Pauling: © Corbis-Bettmann; McClintock: © AP/Wide World Photos; Franklin: © Photo Researchers, Inc.

Chapter 1

Opener: © Brand X Pictures/PunchStock; **1.1(giraffes):** © Dr. Sylvia S. Mader; **1.1(Monarch):** © Dave Cavagnaro/Peter Arnold, Inc.; **1.1(Sequoia):** © Getty R.F.; **1.1(Earth):** © NASA/TSADO/Tom Stack & Associates; **1.1(mushroom):** © IT Stock/Age Fotostock; **1.1(city street):** © David Grossman/Photo Researchers, Inc.; **1.1(peacock):** © Brand X Pictures/PunchStock; **1.3:** © John Cancalosi/Peter Arnold, Inc.; **1.4(mature tree):** © PhotoDisc Website; **1.4(seedling):** © Herman Eisenbeiss/Photo Researchers, Inc.; **1.4(sapling):** Courtesy Paul Wray, Iowa State University; **1.5a:** © Ralph Robinson/Visuals Unlimited; **1.5b:** © A.B. Dowsett/SPL/Photo Researchers, Inc.; **1.8:** Courtesy Leica Microsystems, Inc.; **1.9:** © Dr. Jeremy Burgess/Photo Researchers, Inc.; **1.10(all):** Courtesy Jim Bidlack.

Chapter 2

PHOTOS: Opener: © Vstock.LLC; **2.1:** © Gunter Ziesler/Peter Arnold, Inc.; **2.4a:** © Biomed Comm./Custom Medical Stock Photo; **2.4b(left):** © Mazzlota et al./Photo Researchers, Inc.; **2.4b(right):** © Hank Morgan/Rainbow; **2.5a(both):** © Tony Freeman/PhotoEdit; **2.5b:** © Geoff Tompkinson/SPL/Photo Researchers, Inc.; **2.7b(crystals):** © Charles M. Falco/Photo Researchers, Inc.; **2.7b(shaker):** © The McGraw-Hill Companies, Inc./ Evelyn Jo Hebert, photographer; **2.10b:** © Grant Taylor/Stone/Getty Images; **2Aa:** © Frederica Georgia/Photo Researchers, Inc.; **2Ab:** © The McGraw-Hill Companies, Inc./Louis Rosenstock, photographer; **2.13:** © The McGraw Hill Companies, Inc./John Thoeming, photographer; **2.17:** © Jeremy Burgess/SPL/Photo Researchers, Inc.; **2.18:** © Don W. Fawcett/Photo Researchers, Inc.; **2.19:** © Science Source/J.D. Litvay/Visuals Unlimited; **2.26a:** Courtesy Ealing Corporation. **TEXT: Box 2B:** U.S. Department of Agriculture.

Chapter 3

Opener: © Jack Bostrack/Visuals Unlimited; **3.3:** © Ralph A. Slepecky/Visuals Unlimited; **3.4:** © Alfred Pasieka/Photo Researchers, Inc.; **3.5:** © Newcomb/Wergin/Biological Photo Service; **3.6(top):** Courtesy Ron Milligan/Scripps Research Institute; **3.6(bottom):** Courtesy E.G. Pollock; **3Aa:** © Robert Brons/Biological Photo Service; **3Ab:** © Dr. Don W. Fawcett/Visuals Unlimited; **3Ac:** © Biophoto Assoc./Photo Researchers, Inc.; **3.7:** © R. Bolender & D. Fawcett/Visuals Unlimited; **3.9:** © S.E. Frederick & E.H. Newcomb/Biological Photo Service; **3.10a:** Courtesy Herbert W. Israel, Cornell University; **3.11a:** Courtesy Dr. Keith Porter; **3.12a(immuno-fluorescence):** © M. Schliwa/Visuals Unlimited; **3.12a(Chara):** McGraw-Hill Companies/photo by Dennis Strete and Darrell Vodopich; **3.12b(immunofluorescence):** © K.G. Murti/Visuals Unlimited; **3.12b(girls):** James Morgan/Getty Images; **3.12c (immunofluorescence):** © K.G. Murti/Visuals Unlimited; **3.12c(chameleon):** © Photodisc/Vol 6/Getty Images; **3.13(top):** Courtesy Kent McDonald, University of Colorado Boulder; **3.13(bottom):** From Manley McGill, D.P. Highfield, T.M. Monahan, and B.R. Brinkley, Journal of Structural Biology Vol. 57, 1976 pp. 43-53 (pg. 48, fig. 6.) Academic Press; **3.14(sperm):** Sperm: © David M. Phillips/Photo Researchers, Inc.; **3.14(cilia):** Cilia: © Dr. G. Moscoso/Photo Researchers, Inc.; **3.14(flagellum):** © William L. Dentler/Biological Photo Service.

Chapter 4

Opener: © Comstock/Superstock; **4.7(all RBCs):** © David M. Phillips/Photo Researchers, Inc.; **4.7(normal and hypotonic plant cells):** © Dwight Kuhn; **4.7(hypertonic plant cells):** © Ed Reschke/Peter Arnold, Inc.; **4.11a:** © Eric Grave/Phototake Inc.; **4.11b:** © Don W. Fawcett/Photo Researchers, Inc.; **4.11c(both):** Courtesy of Mark Bretscher.

Chapter 5

Opener: © Junebug Clark/Photo Researchers, Inc.; **5.2:** Courtesy Douglas R. Green/LaJolla Institute for Allergy and Immunology; **5A:** © Dr. Dennis Kunkel/Visuals Unlimited; **5.4 (early prophase, prophase, anaphase, metaphase, telophase):** © Ed Reschke; **5.4(late prophase):** © Michael Abbey/Photo Researchers, Inc.; **5.5(Prophase, Metaphase, Anaphase):** © R. Calentine/Visuals Unlimited; **5.5 (Telophase):** © Jack M. Bostrack/Visuals Unlimited; **5.6(both):** © R.G. Kessel and C.Y. Shih, "Scanning Electron Microscopy in Biology: A Student's Atlas on Biological Organization," 1974 Springer-Verlag, New York; **5.7:** © B.A. Palevitz & E.H. Newcomb/BPS/Tom Stack & Associates.

Chapter 6

Opener: © Royalty-Free/Corbis; **6.3b:** © Darwin Dale/Photo Researchers, Inc.; **6.8b:** © James Watt/Visuals Unlimited; **6.8c:** © Digital Vision/PunchStock; **6A:** © Brand X Pictures/PunchStock; **6.11(left):** © Comstock/PunchStock; **6.11(right):** © PhotoDisc/Getty Images.

Chapter 7

Opener: © Royalty-Free/Corbis; **7.1:** © C Squared Studios/Getty Images; **7.3:** © Getty Images/SW Productions.

Chapter 8

6.5 © Chuck Davis/Stone/Getty; **8.1c:** © Thomas Wiewandt/ChromoSohm Media Inc./Photo Researchers, Inc.; **8.2:** © Dr. George Chapman/Visuals Unlimited; **8.10:** © Herman Eisenbeiss/Photo Researchers, Inc.; **8.11a:** © The McGraw-Hill Companies, Inc./ Evelyn Jo Hebert, photographer; **8.11b:** © Royalty-Free/Corbis; **8.11c:** © S. Alden/PhotoLink/Getty Images; **8A(fruit):** © Comstock/PunchStock; **8A(fabric):** © The McGraw-Hill Companies, Inc./ Evelyn Jo Hebert, photographer; **8A(garden):** © Alan & Linda Detrick/Photo Researchers, Inc.; **8.11(tulips):** © The McGraw-Hill Companies, Inc./ Evelyn Jo Hebert, photographer; **8.11(corncob):** © Royalty-Free/Corbis. **8.11(pineapple):** © S. Alden/PhotoLink/Getty Images.

Chapter 9

Opener: The McGraw-Hill Companies Inc./Ken Cavanagh Photographer; **9.2a, b, c:** © Dwight Kuhn; **9.4a:** © Runk/Schoenberger/Grant Heilman Photography; **9.4b:** © J.R. Waaland/Biological Photo Service; **9.4c:** © Kingsley Stern; **9.5a:** © Dr. Ken Wagner/ Visuals Unlimited; **9.5b:** © George Wilder/ Visuals Unlimited; **9.5c:** © Biophoto Assoc./Photo Researchers, Inc.; **9.6a:** © J. Robert Waaland/Biological Photo Service; **9.7a:** © George Wilder/Visuals Unlimited; **9.8:** © Jeremy Burgess/SPL/Photo Researchers, Inc.; **9.10a:** © Royalty-free/CORBIS; **9.10b:** © Gerald & Buff Corsi/Visuals Unlimited; **9.10c:** © P. Goetgheluck/Peter Arnold, Inc.; **9.13(top):** © Ed Reschke; **9.13(bottom):** Courtesy Ray F. Evert/University of Wisconsin Madison; **9.14(top):** © CABISCO/Phototake; **9.14(bottom):** © Kingsley Stern; **9.17a:** © Ardea London Limited; **9.18a:** © Stanley Schoenberger/Grant Heilman Photography; **9.18b:** © Wally Eberhart/Visuals Unlimited; **9.18c, d:** © The McGraw Hill Companies, Inc./Carlyn Iverson, photographer; **9A:** © James Schnepf Photography, Inc.; **9.19a:** Courtesy Ray F. Evert/University of Wisconsin Madison; **9.19b:** © CABISCO/Phototake; **9.20:** © Dwight Kuhn; **9.21a:** © John D. Cunningham/Visuals Unlimited; **9.21b:** Courtesy of George Ellmore, Tufts University; **9.22a:** © Dr. Robert Calentine/Visuals Unlimited; **9.22b:** © Ed Degginger/Color Pic; **9.22c:** © David Newman/Visuals Unlimited; **9.22d:** © Terry Whittaker/Photo Researchers, Inc.; **9.22e(left):** © Pat Pendarvis; **9.22e(right):** © Biophot; **9B:** Courtesy Gary Banuelos/Agriculture Research Service/USDA; **9.24a, b:** Courtesy Wilfred A. Cote, from H.A. Core, W.A. Cote, and A.C. Day, Wood: Structure and Identification 2/e; **9.24c:** Courtesy W.A. Cote, Jr., N.C. Brown Center for Ultrastructure Studies, SUNY-ESF; **9.25a, b:** © Jeremy Burgess/SPL/Photo Researchers, Inc.; **9.26a:** © Dwight Kuhn; **9.26b:** © E.H. Newcomb & S.R. Tardon/Biological Photo Service; **9.27(left):** © Science VU/R. Roncadori/Visuals Unlimited; **9.27(right):** © Dana Richter/Visuals Unlimited; **9.28a:** Reprinted with permission from M.H. Zimmerman "Movement of Organic Substances in Trees" in SCIENCE 133 (13) January 1961 page 667, Fig. 3 page 73, © 1961 AAAS; **9.28b:** © Bruce Iverson/SPL/Photo Researchers, Inc.

Chapter 10

Opener: © Mitch Hrdlicka/Getty Images; **10.3a:** © Farley Bridges; **10.3b:** © Pat Pendarvis; **10.4a:** © Arthur C. Smith, III/Grant Heilman Photography, Inc.; **10.4b:** © Larry Lefever/Grant Heilman Photography, Inc.; **10.5(top):** Courtesy Graham Kent; **10.5(bottom):** © Ed Reschke; **10.6a:** © George Bernard/Animals Animals/Earth Scenes; **10.6b:** © Simko/Visuals Unlimited; **10.6c:** © Dwight Kuhn; **10.8a:** © James Mauseth; **10.8b:** © Edward S. Ross; **10.8c:** © Runk/Schoenberger/Grant Heilman Photography, Inc.; **10.8d:** © BJ Miller/Biological Photo Service; **10.9b:** © BJ Miller/Biological Photo Service; **10.10b(left):** © James Mauseth; **10.10b(right):** © Barry L. Runk/Grant Heilman, Inc.; **10Aa:** © Jupiterimages; **10Ab:** © H. Eisenbeiss/Photo Researchers, Inc.; **10.11(all):** Courtesy Prof. Dr. Hans-Ulrich Koop, from _Plant Cell Reports_, 17:601-604; **10.12(all):** Courtesy Monsanto; **10.13b:** Courtesy Eduardo Blumwald; **10.15(both):** Courtesy Prof. Malcolm B. Wilkins; **10.16:** © Sylvan Wittwer/Visuals Unlimited; **10.18a, b:** © Kingsley Stern; **10.18c:** © Kent Knudson/PhotoLink/Getty Images; **10.19a:** © Runk/Schoenberger/Grant Heilman Photography; **10.19b:** © Kingsley Stern; **10.21a, b:** © Grant Heilman Photography.

Chapter 11

Opener: © Tim de Waele/Corbis; **11.1(all):** © Ed Reschke; **11.2a:** © David M. Phillips/Visuals Unlimited; **11.2b:** From Douglas E. Kelly, J. Cell Biol. 28 (1966): 51. Reproduced by copyright permission of The Rockefeller University Press; **11.2c:** Courtesy Camillo Peracchia, M.D.; **11.3a, c, d, e):** © Ed Reschke; **11.3b:** © McGraw-Hill Higher Education, Dennis Strete, Photographer; **11.5a, c:** © McGraw-Hill Higher Education, Dennis Strete, Photographer; **11.6b:** © Ed Reschke; **11Aa:** © Ken Greer/Visuals Unlimited; **11Ab:** © Dr. P. Marazzi/SPL/Photo Researchers, Inc.; **11Ac:** © James Stevenson/SPL/Photo Researchers, Inc.

Chapter 12

Opener: © Susan Van Etten/Photo Edit; **12.1d:** © Ed Reschke; **12.6b, c:** © Ed Reschke; **12.11c:** © Ed Reschke; **12.12a:** © Ed Reschke/Peter Arnold, Inc.; **12.12b:** © Andrew Syred/Photo Researchers, Inc.; **12.13:** © Boehringer Ingelheim International GmbH, photo by Lennart Nilsson; **12.14b:** © Eye of Science/Photo Researchers, Inc.; **12A:** © Biophoto Associates/Photo Researchers, Inc.; **12B:** © Astrid & Hanns-Frieder Michler/Photo Researchers, Inc.; **12.20:** ABIOMED, Inc.

Chapter 13

PHOTOS: Opener: © AJPhoto/Photo Researchers, Inc.; **13.2(thymus, spleen):** © Ed Reschke/Peter Arnold, Inc.; **13.2(lymph node):** © Fred E. Hossler/Visuals Unlimited; **13.2(bone marrow):** © R. Valentine/Visuals Unlimited; **13.6b:** Courtesy Dr. Arthur J. Olson, Scripps Institute; **13.8b:** © Bohringer Ingelheim International, photo by Lennart Nilsson: 13.9: © Michael Newman/Photo Edit; **13.10b:** © Digital Vision/Getty Images; **13.10c:** © Aaron Haupt/Photo Researchers, Inc.; **13B(pollen):** © David Scharf/SPL/Photo Researchers; **13B(girl):** © Damien Lovegrove/

Index

that asshole predicted the
future!

History of Biology

Antonie van Leeuwenhoek

Charles Darwin

Louis Pasteur

Robert Koch

Ivan Pavlov

Year	Name	Country	Contribution
1628	William Harvey	Britain	Demonstrates that the blood circulates and the heart is a pump.
1665	Robert Hooke	Britain	Uses the word *cell* to describe compartments he sees in cork under the microscope.
1668	Francesco Redi	Italy	Shows that decaying meat protected from flies does not spontaneously produce maggots.
1673	Antonie van Leeuwenhoek	Holland	Uses microscope to view living microorganisms.
1735	Carolus Linnaeus	Sweden	Initiates the binomial system of naming organisms.
1809	Jean B. Lamarck	France	Supports the idea of evolution but thinks there is inheritance of acquired characteristics.
1825	Georges Cuvier	France	Founds the science of paleontology and shows that fossils are related to living forms.
1828	Karl E. von Baer	Germany	Establishes the germ layer theory of development.
1838	Matthias Schleiden	Germany	States that plants are multicellular organisms.
1839	Theodor Schwann	Germany	States that animals are multicellular organisms.
1851	Claude Bernard	France	Concludes that a relatively constant internal environment allows organisms to survive under varying conditions.
1858	Rudolf Virchow	Germany	States that cells come only from preexisting cells.
1858	Charles Darwin	Britain	Presents evidence that natural selection guides the evolutionary process.
1858	Alfred R. Wallace	Britain	Independently comes to same conclusions as Darwin.
1865	Louis Pasteur	France	Disproves the theory of spontaneous generation for bacteria; shows that infections are caused by bacteria, and develops vaccines against rabies and anthrax.
1866	Gregor Mendel	Austria	Proposes basic laws of genetics based on his experiments with garden peas.
1882	Robert Koch	Germany	Establishes the germ theory of disease and develops many techniques used in bacteriology.
1900	Walter Reed	United States	Discovers that the yellow fever virus is transmitted by a mosquito.
1902	Walter S. Sutton Theodor Boveri	United States Germany	Suggest that genes are on the chromosomes, after noting the similar behavior of genes and chromosomes.
1903	Karl Landsteiner	Austria	Discovers ABO blood types.
1904	Ivan Pavlov	Russia	Shows that conditioned reflexes affect behavior, based on experiments with dogs.
1910	Thomas H. Morgan	United States	States that each gene has a locus on a particular chromosome, based on experiments with *Drosophila*.
1922	Sir Frederick Banting Charles Best	Canada	Isolate insulin from the pancreas.
1924	Hans Spemann Hilde Mangold	Germany	Show that induction occurs during development, based on experiments with frog embryos.
1927	Hermann J. Muller	United States	Proves that X rays cause mutations.
1929	Sir Alexander Fleming	Britain	Discovers the toxic effect of a mold product he called penicillin on certain bacteria.